HANDBOOKS IN OPERATIONS RESEARCH
AND MANAGEMENT SCIENCE
VOLUME 7

Handbooks ir
Operations Researcl
and Management Scienc

Edito

G.L. Nemhaus
Georgia Institute of Technolo

A.H.G. Rinnooy Ka
Erasmus University Rotterda

Volume

ELSEVI
Amsterdam – Lausanne – New York – Oxford – Shannon – Tok

Network Models

Edited by

M.O. Ball
University of Maryland

T.L. Magnanti
Massachusetts Institute of Technology

C.L. Monma
Bellcore

G.L. Nemhauser
Georgia Institute of Technology

1995
ELSEVIER
Amsterdam – Lausanne – New York – Oxford – Shannon – Tokyo

ELSEVIER SCIENCE B.V.
Sara Burgerhartstraat 25
P.O. Box 211, 1000 AE Amsterdam, The Netherlands

ISBN: 0 444 89292 3

This book is printed on acid-free paper.

Printed in The Netherlands

Preface

Networks pervade everyday life in a modern technological society. When we travel to work or to a place to shop, we do so over a transportation network. When we make a telephone call or watch television, we receive electronic signals delivered to us through a telecommunications network. When we try to advance our careers, we must deal with the vagaries of social networks. This handbook considers the scientific analysis of network models. It covers methodology, algorithms, and computer implementations, and a variety of network models used extensively by business and government to design and manage the networks they encounter each day.

Network models have played a key role in the development of operations research and management science since the initial development of these disciplines. Network flow theory developed alongside the theory of linear programming. Network models have been the basis for many of the fundamental developments in integer programming. Early on, researchers recognized that network flow problems define a class of linear programs that always have integer extreme point optimal solutions. Attempts to understand and generalize this finding led to many new results, culminating in an entire branch of optimization known as polyhedral combinatorics. Work on the matching problem was fundamental to the development of both combinatorial optimization and complexity theory. The traveling salesman problem has served as the prototypical problem for nearly all developments in integer programming and combinatorial optimization. The development of fast network flow codes spurred the development of strong interactions between operations research and computer science and the application of optimization models in a wide range of industries.

The set of papers in this Handbook reflect both the rich theory and wide range of applications of network models. Two of the most vibrant applications areas of network models are telecommunications and transportation. Several chapters explicitly model issues arising in these problem domains. Research on network models has been closely aligned with the field of computer science both in developing data structures for efficiently implementing network algorithms and in analyzing the complexity of network problems and algorithms. The basic structure underlying all network problems is a graph. Thus, there has historically been strong ties between network models and graph theory. The papers contained in this volume reflect these various relationships.

The first four chapters treat core network models and applications. Chapter 1 by Ahuja, Magnanti, Orlin and Reddy describes a variety of network applications. The diversity of the problems discussed in this chapter shows why practitioners and researchers have applied network models so extensively in practice. The field of network optimization is most commonly associated with the minimum cost flow problem and with several of its classical specializations: the shortest path problem the maximum flow problem and the transportation problem. The first volume in this series, the Handbook on Optimization, covers the fundamentals of network flows. Chapter 2 of this volume, by Helgason and Kennington, analyzes alternate approaches for implementing network flow algorithms and direct generalizations of the network flow problem. Analysts have used these techniques to develop highly efficient codes for network flows, generalized network flows and related problems. The matching problem and the traveling salesman problem are two network optimization problems that in many ways set the stage for all combinatorial optimization problems. Chapter 3 treats the matching problem. The polynomial time solution algorithm for this problem exploits both the structure of the underlying polyhedron as well as the problem's special graphical properties. The traveling salesman problem has served as the development ground and testbed for the entire range of techniques for difficult (i.e., NP-hard) combinatorial optimization problems. Chapter 4 by Junger, Reinelt and Rinaldi reviews a variety of approaches, while concentrating on those that have been shown to be effective at solving problems of practical size.

The second group of papers present recent fundamental advances in network algorithms. Significant advances in hardware architectures for computationally intensive operations are likely to increasingly involve parallel computing. Chapter 5 by Bertsekas, Castanon, Eckstein and Zenios treats the design and analysis of parallel algorithms for network optimization. A second important general trend in computer science is probabilistic algorithms and probabilistic analysis of algorithms. Chapter 6 by Steele and Snyder treats these topics. During the past few years, and in many problem settings, researchers have recognized the possibility of designing more efficient algorithms for certain problems by directly modeling and exploiting the underlying geometric problem structure. Chapter 7 by Mitchell and Suri covers this topic. One of the most significant developments in combinatorics in the past ten years is the Graph Minor Project due to Robertson, Seymour and Thomas. In Chapter 8, Bienstock and Langston discuss the extensive implications of this body of work.

The next two chapters cover methodology for constructing networks with certain properties. Much of this work is motivated by telecommunications network design problems. Chapter 9, by Magnanti and Wolsey, addresses the problem of designing tree networks. This class of problems includes a problem fundamental to both network optimization and matroid optimization, the minimum spanning tree problem and several of its variants and extensions. In several applications, networks must be able to withstand the failure/deletion of a single arc. This requirement leads to survivable network design problems which Grötschel, Monma and Stoer treat in Chapter 10. Once a network is constructed, we often wish to compute a measure of its reliability given the reliability (probability of operation) of its components. Chapter 11 by Ball, Colbourn and Provan covers these reliability analysis problems.

A companion volume in the Handbook series, entitled Network Routing, examines problems related to the movement of commodities over a network. The problems treated arise in several application areas including logistics, telecommunications, facility location, VLSI design, and economics.

The broad set of material covered in both these volumes attests to the richness of networks as both a significant scientific field of inquiry and as an important pragmatic modeling and problem solving tool. In this sense, networks is a problem domain that reflects the best tradition of operations research and management science and allied fields such as applied mathematics and computer science. We hope that the set of papers assembled in this volume will serve as a valuable summary and synthesis of the extensive network literature and that this material might inspire its readers to develop additional theory and practice that builds upon this fine tradition.

Michael Ball
Tom Magnanti
Clyde Monma
George Nemhauser

Contents

M.O. Ball et al., Eds., *Handbooks in OR & MS, Vol. 7*

Chapter 1

Applications of Network Optimization

Ravindra K. Ahuja
Department of Industrial and Management Engineering, I.I.T., Kanpur – 208 016, India

Thomas L. Magnanti, James B. Orlin
Sloan School of Management, M.I.T., Cambridge, MA 02139, U.S.A.

M.R. Reddy
Department of Industrial and Management Engineering, I.I.T., Kanpur – 208 016, India

1. Introduction

Highways, telephone lines, electric power systems, computer chips, water delivery systems, and rail lines: these physical networks, and many others, are familiar to all of us. In each of these problem settings, we often wish to send some good(s) (vehicles, messages, electricity, or water) from one point to another, typically as efficiently as possible – that is, along a shortest route or via some minimum cost flow pattern. Although these problems trace their roots to the work of Gustav Kirchhoff and other great scientists of the last century, the topic of network optimization as we know it today has its origins in the 1940's with the development of linear programming and, more broadly, optimization as an independent field of scientific inquiry, and with the parallel development of digital computers capable of performing massive computations. Since then, the field of network optimization has grown at an almost dizzying pace with literally thousands of scientific papers and multitudes of applications modeling a remarkably wide range of practical situations.

Network optimization has always been a core problem domain in operations research, as well as in computer science, applied mathematics, and many fields of engineering and management. The varied applications in these fields not only occur 'naturally' on some transparent physical network, but also in situations that apparently are quite unrelated to networks. Moreover, because network optimization problems arise in so many diverse problem contexts, applications are scattered throughout the literature in several fields. Consequently, it is sometimes difficult for the research and practitioner community to fully appreciate the richness and variety of network applications.

This chapter is intended to introduce many applications and, in doing so, to highlight the pervasiveness of network optimization in practice. Our coverage is not intended to be encyclopedic, but rather attempts to demonstrate a range of applications, chosen because they are (i) 'core' models (e.g., a basic production planning model), (ii) depict a range of applications including such fields as medicine and the molecular biology that might not be familiar to many readers, and (iii) cover many basic model types of network optimization: (1) shortest paths; (2) maximum flows; (3) minimum cost flows; (4) assignment problems; (5) matchings; (6) minimum spanning trees; (7) convex cost flows; (8) generalized flows; (9) multicommodity flows; (10) the traveling salesman problem; and (11) network design. We present five applications for each of the core shortest paths, maximum flows, and minimum cost flow problems, four applications for each of the matching, minimum spanning tree, and traveling salesman problems, and three applications for each of the remaining problems.

The chapter describes the following 42 applications, drawn from the fields of operations research, computer science, the physical sciences, medicine, engineering, and applied mathematics:

1. System of difference constraints;
2. Telephone operator scheduling;
3. Production planning problems;
4. Approximating piecewise linear functions;
5. DNA sequence alignment;
6. Matrix rounding problem;
7. Baseball elimination problem;
8. Distributed computing on a two-processor computer;
9. Scheduling on uniform parallel machines;
10. Tanker scheduling;
11. Leveling mountainous terrain;
12. Reconstructing the left ventricle from X-ray projections;
13. Optimal loading of a hopping airplane;
14. Directed Chinese postman problem;
15. Racial balancing of schools;
16. Locating objects in space;
17. Matching moving objects;
18. Rewiring of typewriters;
19. Pairing stereo speakers;
20. Determining chemical bonds;
21. Dual completion of oil wells;
22. Parallel saving heuristics;
23. Measuring homogeneity of bimetallic objects;
24. Reducing data storage;
25. Cluster analysis;
26. System reliability bounds;
27. Urban traffic flows;
28. Matrix balancing;
29. Stick percolation problem;
30. Determining an optimal energy policy;

31. Machine loading;
32. Managing warehousing goods and funds flow;
33. Routing of multiple commodities;
34. Racial balancing of schools;
35. Multivehicle tanker scheduling;
36. Manufacturing of printed circuit boards;
37. Identifying time periods for archeological finds;
38. Assembling physical mapping in genetics;
39. Optimal vane placement in turbine engines;
40. Designing fixed cost communication and transportation systems;
41. Local access telephone network capacity expansion;
42. Multi-item production planning.

In addition to these 42 applications, we provide references for 140 additional applications.

2. Preliminaries

In this section, we introduce some basic notation and definitions from graph theory as well as a mathematical programming formulation of the minimum cost flow problem, which is the core network flow problem that lies at the heart of network optimization. We also present some fundamental transformations that we frequently use while modeling applications as network problems.

Let $G = (N, A)$ be a directed network defined by a set N of n *nodes*, and a set A of m *directed arcs*. Each arc $(i, j) \in A$ has an associated *cost* c_{ij} per unit flow on that arc. We assume that the flow cost varies linearly with the amount of flow. Each arc $(i, j) \in A$ also has a *capacity* u_{ij} denoting the maximum amount that can flow on the arc, and a *lower bound* l_{ij} denoting the minimum amount that must flow on the arc. We associate with each node $i \in N$ an integer $b(i)$ representing its supply/demand. If $b(i) > 0$, then node i is a *supply node*; if $b(i) < 0$, then node i is a *demand node*; and if $b(i) = 0$, then node i is a *transshipment node*.

The minimum cost flow problem is easy to state: we wish to determine a least cost shipment of a commodity through a network that will satisfy the flow demands at certain nodes from available supplies at other nodes. The decision variables in the minimum cost flow problem are arc flows; we represent the flow on an arc $(i, j) \in A$ by x_{ij}. The minimum cost flow problem is an optimization model formulated as follows:

$$\text{minimize} \quad \sum_{(i,j)\in A} c_{ij} x_{ij} \tag{1a}$$

subject to

$$\sum_{\{j:(i,j)\in A\}} x_{ij} - \sum_{\{j:(j,i)\in A\}} x_{ji} = b(i), \qquad \text{for all } i \in N, \tag{1b}$$

$$l_{ij} \le x_{ij} \le u_{ij}, \qquad \text{for all } (i, j) \in A. \tag{1c}$$

The data for this model satisfies the feasibility condition $\sum_{i=1}^{n} b(i) = 0$ (that is, total supply must equal the total demand). We refer to the constraints in (1b) as *mass balance constraints*. The mass balance constraints state that the net flow out of each node (outflow minus inflow) must equal the supply/demand of the node. The flow must also satisfy the lower bound and capacity constraints (1c) which we refer to as the *flow bound constraints*. The flow bounds typically model physical capacities or restrictions imposed upon the flows' operating ranges. In most applications, the lower bounds on arc flows are zero; therefore, if we do not state lower bounds explicitly, we assume that they have value zero.

We now collect together several basic definitions and describe some notation. A *walk* in $G = (N, A)$ is a sequence of nodes and arcs i_1, (i_1, i_2), i_2, (i_2, i_3), i_3, $\ldots$, (i_{r-1}, i_r), i_r satisfying the property that either $(i_k, i_{k+1}) \in A$ or $(i_{k+1}, i_k) \in A$ for each $k = 1, \ldots, r-1$. A walk might revisit nodes. A *path* is a walk whose nodes (and, hence, arcs) are all distinct. For simplicity, we often refer to a path as a sequence of nodes i_1–i_2–$\ldots$–i_k when its arcs are apparent from the problem context. A *directed path* is defined similarly except that for any two consecutive nodes i_k and i_{k+1} on the path, the path must contain the arc (i_k, i_{k+1}). A *directed cycle* is a directed path together with the arc (i_r, i_1), and a *cycle* is a path together with the arc (i_r, i_1) or (i_1, i_r).

A graph $G' = (N', A')$ is a *subgraph* of $G = (N, A)$ if $N' \subseteq N$ and $A' \subseteq A$. A graph $G' = (N', A')$ is a spanning subgraph of $G = (N, A)$ if $N' = N$ and $A' \subseteq A$. Two nodes i and j are said to be *connected* if the graph contains at least one undirected path between these nodes. A graph is said to be *connected* if every pair of its nodes are connected; otherwise, it is *disconnected*. The connected subgraphs of a graph are called its *components*. A tree is a connected graph that contains no cycle. A subgraph T is a *spanning tree* of G if T is a tree of G containing all its nodes. A *cut* of G is any set $Q \subseteq A$ satisfying the property that the graph $G' = (N, A - Q)$ is disconnected, and no subset of Q has this property. A cut partitions the graph into two sets of nodes, X and N–X. We shall sometimes represent the cut Q as the node partition $[X, N\text{–}X]$. A cut $[X, N\text{–}X]$ is an s–t cut for two specially designated nodes s and t if $s \in X$ and $t \in N$–X.

Transformations

Frequently, we require network transformations to model an application context, to simplify a network problem, or to show equivalencies between different network problems. We now briefly describe some of these transformations.

Removing undirected networks. Sometimes minimum cost flow problems contain undirected arcs. Whereas a directed arc (i, j) permits flow only from node i to node j, an undirected arc $\{i, j\}$ permits flow from node i to node j as well as flow from node j to node i. To transform the undirected case into the directed case, we replace each undirected arc $\{i, j\}$ of cost c_{ij} and capacity u_{ij}, by two directed arcs (i, j) and (j, i), both of cost c_{ij} and capacity u_{ij}. It is easy to see that if x_{ij} and x_{ji} are the flows on arcs (i, j) and (j, i) in the directed network respectively, then $x_{ij} - x_{ji}$ or $x_{ji} - x_{ij}$, whichever is nonnegative, is the associated flow on the undirected arc $\{i, j\}$.

Removing nonzero lower bounds. Suppose arc (i, j) has a nonzero lower bound l_{ij} on its flow x_{ij}. We can eliminate this lower bound by sending l_{ij} units of flow on arc (i, j), which decreases $b(i)$ by l_{ij} units and increases $b(j)$ by l_{ij} units, and then we measure (by the variable x'_{ij}) the incremental flow on the arc beyond the flow value l_{ij} (therefore, we reduce the capacity of the arc (i, j) to $u_{ij} - l_{ij}$).

Node splitting. Node splitting transforms each node i into two nodes i' and i'' corresponding to the node's output and input functions. This transformation replaces each original arc (i, j) by an arc (i', j'') of the same cost and capacity. It also adds an arc (i'', i') of zero cost and with infinite capacity for each i. The input side of node i (i.e., node i'') receives all the node's inflow, the output side (i.e., node i') sends all the node's outflow, and the additional arc (i'', i') carries flow from the input side to the output side. We define the supplies/demands of nodes in the transformed network in accordance with the following two cases: (i) if $b(i) \geq 0$, then $b(i'') = b(i)$ and $b(i') = 0$; and (ii) if $b(i) < 0$, then $b(i'') = 0$ and $b(i') = b(i)$. It is easy to show a one-to-one correspondence between a flow in the original network and the corresponding flow in the transformed network.

We can use the node splitting transformation to handle situations in which nodes as well as arcs have associated capacities and costs. In these situations, each unit of flow passing through node i incurs a cost c_i and the maximum flow that can pass through the node is u_i. We can reduce this problem to the standard 'arc flow' form of the network flow problem by performing the node splitting transformation and letting c_i and u_i be the cost and capacity of arc (i'', i').

3. Shortest paths

The shortest path problem is among the simplest network flow problems. For this problem, we wish to find a path of minimum cost (length) from a specified *source node* s to another specified *sink node* t in either a directed or undirected network assuming that each arc $(i, j) \in A$ has an associated cost (or length) c_{ij}. In the formulation of the minimum cost flow problem given in (1) for a directed network, if we set $b(s) = 1$, $b(t) = -1$, and $b(i) = 0$ for all other nodes, and set each $l_{ij} = 0$ and each $u_{ij} \geq 1$, then the solution to this problem will send one unit of flow from node s to node t along the shortest directed path. If we want to determine shortest paths from the source node s to every other node in the network, then in the minimum cost flow problem, we set $b(s) = (n - 1)$ and $b(i) = -1$ for all other nodes. We can set each arc capacity u_{ij} to $n - 1$. The minimum cost flow solution would then send one unit of flow from node s to every other node i along a shortest path. We will consider several applications defined on directed networks. We can model problems defined on undirected networks as special cases of the minimum cost network flow problem by using the transformations described in Section 2.

Shortest path problems are alluring to both researchers and to practitioners for several reasons: (i) they arise frequently in practice since in a wide variety of application settings we wish to send some material (for example, a computer data packet, a telephone call, or a vehicle) between two specified points in a network as quickly, as cheaply, or as reliably as possible; (ii) they are easy to solve efficiently; (iii) as the simplest network models, they capture many of the most salient core ingredients of network flows and so they provide both a benchmark and a point of departure for studying more complex network models; and (iv) they arise frequently as subproblems when solving many combinatorial and network optimization problems. In this section, we describe a few applications of the shortest path problem that are indicative of its range of applications. The applications arise in applied mathematics, biology, computer science, production planning, and work force scheduling. We conclude the section with a set of references for many additional applications in a wide variety of fields.

Application 1. System of difference constraints [Bellman, 1958]

In some linear programming applications (see Application 2) with constraints of the form $\mathbf{A}x \leq b$, the $n \times m$ constraint matrix $\mathbf{A}$ contains one $+1$ and one -1 in each row; all the other entries are zero. Suppose that the kth row has a $+1$ entry in column j_k and a -1 entry in column i_k; the entries in the vector b have arbitrary signs. This linear program defines the following set of m *difference constraints* in the n variables $x = (x(1), x(2), \ldots, x(n))$:

$$x(j_k) - x(i_k) \leq b(k), \qquad \text{for each } k = 1, \ldots, m. \tag{2}$$

We wish to determine whether the system of difference constraints given by (2) has a feasible solution, and if so, we want to identify one. This model arises in a variety of applications; Application 2 describes the use of this model in the telephone operator scheduling; additional applications arise in the scaling of data [Orlin & Rothblum, 1985] and just-in-time scheduling [Elmaghraby, 1978; Levner & Nemirovsky, 1991].

Each system of difference constraints has an associated graph G, which we call a *constraint graph*. The constraint graph has n nodes corresponding to the n variables and m arcs corresponding to the m difference constraints. We associate an arc (i_k, j_k) of length $b(k)$ in G with the constraint $x(j_k) - x(i_k) \leq b(k)$. As an example, consider the following system of constraints:

$$x(3) - x(4) \leq 5, \tag{3a}$$

$$x(4) - x(1) \leq -10, \tag{3b}$$

$$x(1) - x(3) \leq 8, \tag{3c}$$

$$x(2) - x(1) \leq -11, \tag{3d}$$

$$x(3) - x(2) \leq 2. \tag{3e}$$

To model the system of difference constraints as a shortest path problem, we use the two following well-known results about shortest paths (see, e.g., Cormen, Leiserson & Rivest [1990], and Ahuja, Magnanti & Orlin [1993]):

Observation 1. The shortest path distance $d(i)$ from source node s to node i for every node $i \in N$ satisfy the following optimality conditions: $d(j) - d(i) \leq c_{ij}$ for every arc $(i, j) \in A$.

Observation 2. The shortest path distances in a network G exist if and only if G contains no negative cycle (i.e., a cycle whose total cost, summed over all its arcs, is negative).

First, notice that the structure of the shortest path optimality conditions given in Observation 1 is similar to the inequalities of the system of difference constraints (3). In fact, Figure 1a gives the network for which (3) become the shortest path optimality conditions. The second observation implies that the system of difference constraints has a feasible solution if and only if the corresponding network contains no negative cycle. For instance, the network shown in Figure 1a contains a negative cycle 1–2–3–1 of length -1, and the corresponding constraints (i.e., $x(2) - x(1) \leq -11$, $x(3) - x(2) \leq 2$, and $x(1) - x(3) \leq 8$) are inconsistent because summing these constraints yields $0 \leq -1$. We can thus conclude that the system of difference constraints given by (3) has no feasible solution.

We can detect the presence of a negative cycle in a network by using a label correcting algorithm. Label correcting algorithms require that all the nodes in the network are reachable by a directed path from some node, which we use as the source node for the shortest path problem. To satisfy this requirement, we introduce a new node s and join it to all the nodes in the network with arcs of zero cost. For our example, Figure 1b shows the modified network. Since all the arcs incident to node s are directed out of this node, node s is not contained in any

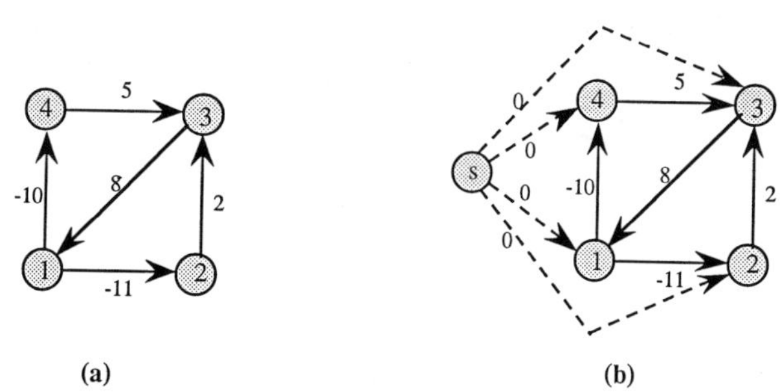

Fig. 1. Graph corresponding to a system of difference constraints.

directed cycle and so the modification does not create any new directed cycles, and so does not introduce any cycles with negative costs. Label correcting algorithms either detect the presence of a negative cycle or provide the shortest path distances. In the former case, the system of difference constraints has no solution, and in the latter case, the shortest path distances constitute a solution of (2).

Application 2. Telephone operator scheduling [Bartholdi, Orlin & Ratliff, 1980]

The following telephone operator scheduling problem is an application of the system of difference constraints. A telephone company needs to schedule operators around the clock. Let $b(i)$ for $i = 0, 1, 2, \ldots, 23$, denote the minimum number of operators needed for the ith hour of the day ($b(0)$ denotes number of operators required between midnight and 1 AM). Each telephone operator works in a shift of 8 consecutive hours and a shift can begin at any hour of the day. The telephone company wants to determine a 'cyclic schedule' that repeats daily, i.e., the number of operators assigned to the shift starting at 6 AM and ending at 2 PM is the same for each day. The optimization problem requires that we identify the fewest operators needed to satisfy the minimum operator requirement for each hour of the day. If we let y_i denote the number of workers whose shift begins at the ith hour, then we can state the telephone operator scheduling problem as the following optimization model:

$$\text{minimize} \quad \sum_{i=0}^{23} y_i \tag{4a}$$

subject to

$$y_{i-7} + y_{i-6} + \ldots + y_i \geq b(i), \qquad \text{for all } i = 8 \text{ to } 23, \tag{4b}$$

$$y_{17+i} + \ldots + y_{23} + y_0 + \ldots + y_i \geq b(i), \qquad \text{for all } i = 0 \text{ to } 7, \tag{4c}$$

$$y_i \geq 0 \qquad \text{for all } i = 0 \text{ to } 23. \tag{4d}$$

Notice that this linear program has a very special structure because the associated constraint matrix contains only 0's and 1's and the 1's or 0's in each row appear consecutively. In this application, we study a restricted version of the telephone operator scheduling problem: we wish to determine whether some feasible schedule uses p or fewer operators. We convert this restricted problem into a system of difference constraints by redefining the variables. Let $x(0) = y_0$, $x(1) = y_0 + y_1$, $x(2) = y_0 + y_1 + y_2, \ldots$, and $x(23) = y_0 + y_2 + \ldots + y_{23} = p$. In terms of these new variables, we can rewrite each constraint in (4b) as

$$x(i) - x(i-8) \geq b(i), \qquad \text{for all } i = 8 \text{ to } 23, \tag{5a}$$

and each constraints in (4c) as

$$x(23) - x(16+i) + x(i) = p - x(16-i) + x(i) \geq b(i), \qquad \text{for all } i = 0 \text{ to } 7. \tag{5b}$$

Finally, the nonnegativity constraints (4d) become

$$x(i) - x(i-1) \geq 0. \tag{5c}$$

By virtue of this transformation, we have reduced the restricted version of the telephone operator scheduling problem into a problem of finding a feasible solution of the system of difference constraints. Application 1 shows that we can further reduce this problem into a shortest path problem.

Application 3. Production planning problems [Veinott & Wagner, 1962; Zangwill, 1969; Evans, 1977]

Many optimization problems in production and inventory planning are network optimization models. All of these models address a basic *economic order quantity* issue: when we plan a production run of any particular product, how much should we produce? Producing in large quantities reduces the time and cost required to set up equipment for the individual production runs; on the other hand, producing in large quantities also means that we will carry many items in inventory awaiting purchase by customers. The *economic order quantity* strikes a balance between the set up and inventory costs to find the production plan that achieves the lowest overall costs. The models that we consider in this section all attempt to balance the production and inventory carrying costs while meeting known demands that vary throughout a given planning horizon. We present one of the simplest models: a single product, single stage model with concave costs and backordering, and transform it to a shortest path problem. This is an example not naturally as a shortest path problem, but that becomes a shortest path problem because of an underlying 'spanning tree property' of the optimal solution.

In this model, we assume that the production cost in each period is a concave function of the level of production. In practice, the production x_j in the jth period frequently incurs a fixed cost F_j (independent of the level of production) and a per unit production cost c_j. Therefore, for each period j, the production cost is 0 for $x_j = 0$, and $F_j + c_j x_j$ if $x_j > 0$, which is a concave function of the production level x_j. The production cost might also be concave due to other economies of scale in production. In this model, we also permit backordering, which implies that we might not fully satisfy the demand of any period from the production in that period or from current inventory, but could fulfill the demand from production in future periods. We assume that we do not lose any customer whose demand is not satisfied on time and who must wait until his or her order materializes. Instead, we incur a penalty cost for backordering any item. We assume that the inventory carrying and backordering costs are linear, and that we have no capacity imposed upon production, inventory, or backordering volumes.

In this model, we wish to meet a prescribed demand d_j for each of K periods $j = 1, 2, \ldots, K$ by either producing an amount x_j in period j, by drawing upon the inventory I_{j-1} carried from the previous period, and/or by backordering the item from the next period. Figure 2a shows the network for modeling this problem. The network has $K + 1$ nodes: the jth node, for $j = 1, 2, \ldots, K$, represents the

jth planning period; node 0 represents the 'source' of all production. The flow on the *production arc* $(0, j)$ prescribes the production level x_j in period j, the flow on *inventory carrying arc* $(j, j+1)$ prescribes the inventory level I_j to be carried from period j to period $j+1$, and the flow B_j on the backordering arc $(j, j-1)$ represents the amount backordered from the next period.

The network flow problem in Figure 2a is a concave cost flow problem, because the cost of flow on every production arc is a concave function. The following well-known result about concave cost flow problems helps us to solve the problem (see, for example, Ahuja, Magnanti & Orlin [1993]):

Spanning tree property. A concave cost network flow minimization problem whose objective function is bounded from below over the set of feasible solutions always has an *optimal spanning tree solution*, that is, an optimal solution in which only the arcs in a spanning tree have positive flow (all other arcs, possibly including some arcs in the spanning tree, have zero flow).

Figure 2b shows an example of a spanning tree solution. This result implies the following property, known as the *production property*: In the optimal solution, each time we produce, we produce enough to meet the demand for an integral number of contiguous periods. Moreover, in no period do we both produce and carry inventory from the previous period or into next period.

The production property permits us to solve the production planning problem very efficiently as a shortest path problem on an auxiliary network G' shown in Figure 2c, which is defined as follows: The network G' consists of nodes 1 through $K+1$ and contains an arc (i, j) for every pair of nodes i and j with $i < j$. We set the cost of arc (i, j) equal to the sum of the production, inventory carrying and backorder carrying costs incurred in satisfying the demands of periods $i, i+1, \ldots, j-1$ by producing in some period k between i and $j-1$; we select this period k that gives the least possible cost. In other words, we vary k from i to $j-1$, and for each k, we compute the cost incurred in satisfying the demands of periods i through $j-1$ by the production in period k; the minimum of these values defines the cost of arc (i, j) in the auxiliary network G'. Observe that for every production schedule satisfying the production property, G' contains a directed path from node 1 to node $K+1$ with the same objective function value, and vice-versa. Therefore, we can obtain the optimal production schedule by solving a shortest path problem.

Several variants of the production planning problem arise in practice. If we impose capacities on the production, inventory, or backordering arcs, then the production property does not hold and we cannot formulate this problem as a shortest path problem. In this case, however, if the production cost in each period is linear, the problem becomes a minimum cost flow model. The minimum cost flow problem also models multistage situations in which a product passes through a sequence of operations. To model this situation, we would treat each production operation as a separate stage and require that the product pass through each of the stages before completing its production. In a further multi-item generalization, common manufacturing facilities are used to manufacture multiple products in

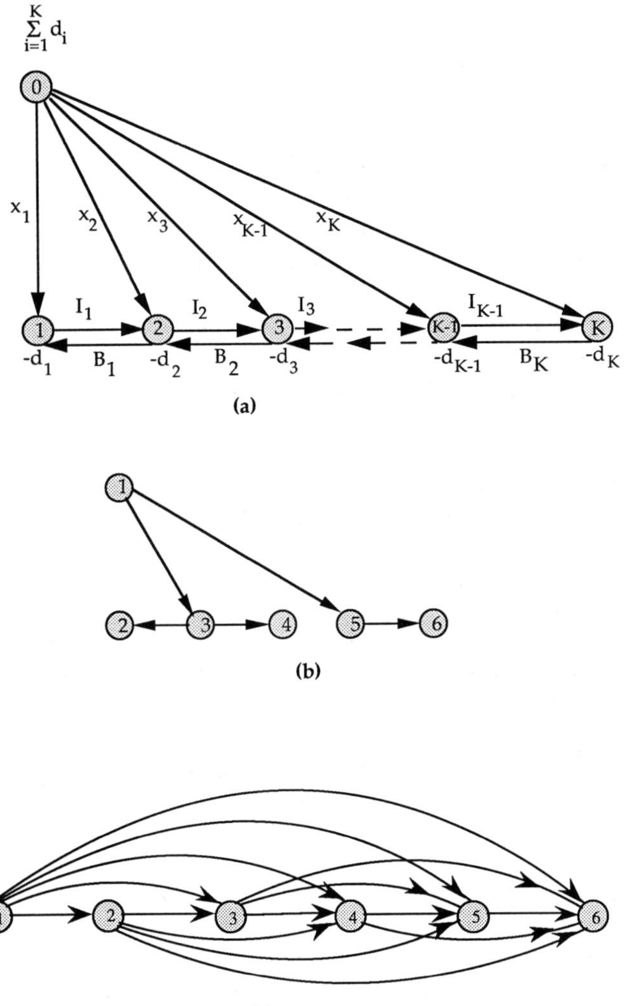

Fig. 2. Production planning problem. (a) Underlying network. (b) Graphical structure of a spanning tree solution. (c) The resulting shortest path problem.

multiple stages; this problem is a multicommodity flow problem. The references cited for this application describe these various generalizations.

Application 4. Approximating piecewise linear functions [Imai & Iri, 1986]

Numerous applications encountered within many different scientific fields use piecewise linear functions. On several occasions, because these functions contain a large number of breakpoints, they are expensive to store and to manipulate (for example, even to evaluate). In these situations, it might be advantageous to replace the piecewise linear function by another approximating function that uses fewer breakpoints. By approximating the function, we will generally be able to save on storage space and on the cost of using the function; we will, however, incur a cost because the approximating function introduces inaccuracies in representing the function. In making the approximation, we would like to make the best possible tradeoff between these conflicting costs and benefits.

Let $f_1(x)$ be a piecewise linear function of a scalar x. We represent the function in the two-dimensional plane: it passes through n points $a_1 = (x_1, y_1)$, $a_2 = (x_2, y_2), \ldots, a_n = (x_n, y_n)$. Suppose that we have ordered the points so that $x_1 \leq x_2 \leq \cdots \leq x_n$. We assume that the function varies linearly between every two consecutive points x_i and x_{i+1}. We consider situations in which n is very large and for practical reasons we wish to approximate the function $f_1(x)$ by another function $f_2(x)$ that passes through only a subset of the points $a_1, a_2, \ldots, a_n$ (but including a_1 and a_n). As an example, consider Figure 3a: in this figure, we have approximated a function $f_1(x)$ passing through 10 points by a function $f_2(x)$ (drawn with dotted lines) passing through only 5 of the points.

This approximation results in a savings in storage space and in the use (evaluation) of the function. Suppose that the cost of storing one data point is a. Therefore, by storing fewer data points, we incur less storage costs. But the

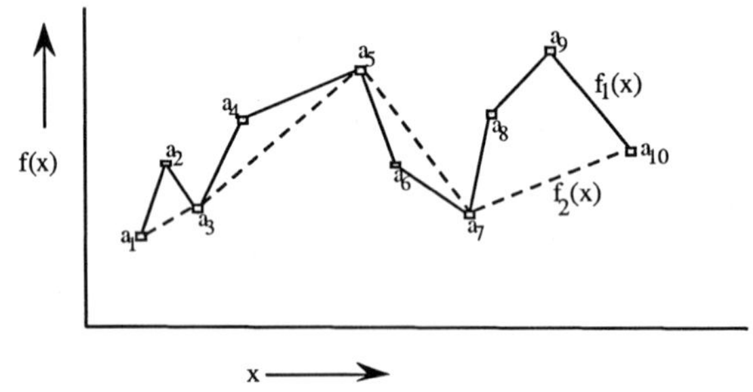

(a)

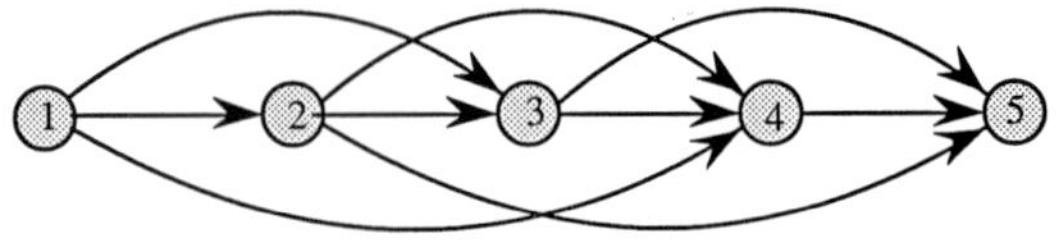

(b)

Fig. 3. Approximating precise linear functions. (a) Approximating a function $f_1(x)$ passing through 10 points by a function $f_2(x)$ passing through only 5 points. (b) Corresponding shortest path problem.

approximation introduces errors with an associated penalty. We assume that the error of an approximation is proportional to the sum of the squared errors between the actual data points and the estimated points. In other words, if we approximate the function $f_1(x)$ by $f_2(x)$, then the penalty is

$$\beta \sum_{k=1}^{p} [f_1(x_k) - f_2(x_k)]^2 \tag{6}$$

for some constant β. Our decision problem is to identify the subset of points to be used to define the approximation function $f_2(x)$ so we incur the minimum total cost as measured by the sum of the cost of storing and the cost of the errors incurred by the approximation.

We formulate this problem as a shortest path problem on a network G with n nodes, numbered 1 through n, as follows. The network contains an arc (i, j) for each pair of nodes i and j. Figure 3b gives an example of the network with $n = 5$ nodes. The arc (i, j) in this network signifies that we approximate the linear segments of the function $f_1(x)$ between the points $a_i, a_{i+1}, \ldots, a_j$ by one linear segment joining the points a_i and a_j. Each directed path from node 1 to node n in G corresponds to a function $f_2(x)$, and the cost of this path equals the total cost for storing this function and for using it to approximate the original function. For example, the path 1–3–5 corresponds to the function $f_2(x)$ passing through the points a_1, a_3 and a_5. The cost c_{ij} of the arc (i, j) has two components: the storage cost a and the penalty associated with approximating all the points between a_i and a_j by the line joining a_i and a_j. Observe that the coordinates of a_i and a_j are given by $[x_i, y_i]$ and $[x_j, y_j]$. The function $f_2(x)$ in the interval $[x_i, x_j]$ is given by the line $f_2(x) = (x - x_i)\{[f_1(x_j) - f_1(x_i)]/(x_j - x_i)\}$. This interval contains the data points with x-coordinates as $x_i, x_{i+1}, \ldots, x_j$, and so we must associate the corresponding terms of (6) with the cost of the arc (i, j). Consequently, we define the cost c_{ij} of an arc (i, j) as

$$c_{ij} = \alpha + \beta \sum_{k=i}^{j} [f_1(x_k) - f_2(x_k)]^2.$$

As a consequence of the preceding observations, we see that the shortest path from node 1 to node n specifies the optimal set of points needed to define the approximating function $f_2(x)$.

Application 5. DNA sequence alignment [Waterman, 1988]

Scientists model strands of DNA as a sequence of letters drawn from the alphabet {A,C,G,T}. Given two sequences of letters, say $B = b_1 b_2 \ldots b_p$ and $D = d_1 d_2 \ldots d_q$ of possibly different lengths, molecular biologists are interested in determining how similar or dissimilar these sequences are to each other. (These sequences are subsequences of the nucleic acids of DNA in a chromosome typically containing several thousand letters.) A natural way of measuring the dissimilarity between the two sequences B and D is to determine the minimum

$$
\begin{array}{lllllll}
B' & = & \oplus & \oplus & A & G & T \\
D & = & C & T & A & G & C
\end{array}
\quad \Longrightarrow \quad
\begin{array}{lllll}
C & T & A & G & C \\
C & T & A & G & C
\end{array}
$$

Fig. 4. Transforming the sequence B into the sequence D.

'cost' required to transform sequence B into sequence D. To transform B into D, we can perform the following operations: (i) insert an element into B (at any place in the sequence) at a 'cost' of a units; (ii) delete an element from B (at any place in the sequence) at a 'cost' of b units; and (iii) mutate an element b_i into an element d_j at a 'cost' of $g(b_i, d_j)$ units. Needless to say, it is possible to transform the sequence B into the sequence D in many ways and so identifying a minimum cost transformation is a nontrivial task. We show how we can solve this problem using dynamic programming, which we can also view as solving a shortest path problem on an appropriately defined network.

Suppose that we conceive of the process of transforming the sequence B into the sequence D as follows. First, add or delete elements from the sequence B so that the modified sequence, say B', has the same number of elements as D. Next 'align' the sequences B' and D to create a one-to-one alignment between their elements. Finally, mutate the elements in the sequence B' so that this sequence becomes identical with the sequence D. As an example, suppose that we wish to transform the sequence B = AGTT into the sequence D = CTAGC. One possible transformation is to delete one of the elements T from B and add two new elements at the beginning, giving the sequence $B' = \oplus \oplus$ AGT (we denote any new element by a placeholder $\oplus$ and later assign a letter to this placeholder). We then align B' with D, as shown in Figure 4, and mutate the element T into C so that the sequences become identical. Notice that because we are free to assign values to the newly added elements, they do not incur any mutation cost. The cost of this transformation is $b + 2a + g(\text{T,C})$.

We now describe a dynamic programming formulation of this problem. Let $f(i, j)$ denote the minimum cost of transforming the subsequence $b_1 b_2 \ldots b_i$ into the subsequence $d_1 d_2 \ldots d_j$. We are interested in the value $f(p, q)$, which is the minimum cost of transforming B into D. To determine $f(p, q)$, we will determine $f(i, j)$ for all $i = 0, 1, \ldots, p$, and for all $j = 0, 1, \ldots, q$. We can determine these intermediate quantities $f(i, j)$ using the following recursive relationships:

$$f(i, 0) = \beta i \quad \text{for all } i; \tag{7a}$$

$$f(0, j) = \alpha j \quad \text{for all } j; \text{ and} \tag{7b}$$

$$f(i, j) = \min\{f(i-1, j-1) + \\ + g(b_i, d_j),\ f(i, j-1) + \alpha,\ f(i-1, j) + \beta\}. \tag{7c}$$

We now justify this recursion. The cost of transforming a sequence of i elements into a null sequence is the cost of deleting i elements. The cost of transforming a null sequence into a sequence of j elements is the cost of adding j elements.

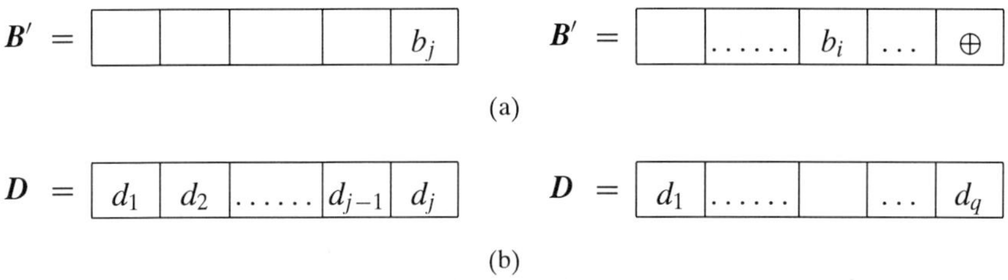

Fig. 5. Explaining the dynamic programming recursion.

Next consider $f(i, j)$. Let B' denote the optimal aligned sequence of B (i.e., the sequence just before we create the mutation of B' to transform it into D). At this point, B' satisfies exactly one of the following three cases:

Case 1. *B' contains the letter b_i which is aligned with the letter d_j of D* (as shown in Figure 5a). In this case, $f(i, j)$ equals the optimal cost of transforming the subsequence $b_1b_2 \ldots b_{i-1}$ into $d_1d_2 \ldots d_{j-1}$ and the cost of transforming the element b_i into d_j. Therefore, $f(i, j) = f(i-1, j-1) + g(b_i, d_j)$.

Case 2. *B' contains the letter b_i which is not aligned with the letter d_j* (as shown in Figure 5b). In this case, b_i is to the left of d_j and so a newly added element must be aligned with b_j. As a result, $f(i, j)$ equals the optimal cost of transforming the subsequence $b_1b_2 \ldots b_i$ into $d_1d_2 \ldots d_{j-1}$ plus the cost of adding a new element to B. Therefore, $f(i, j) = f(i, j-1) + \alpha$.

Case 3. *B' does not contains the letter b_i*. In this case, we must have deleted b_i from B and so the optimal cost of the transformation equals the cost of deleting this element and transforming the remaining sequence into D. Therefore, $f(i, j) = f(i-1, j) + \beta$.

The preceding discussion justifies the recursive relationships specified in (7). We can use these relationships to compute $f(i, j)$ for increasing values of i and, for a fixed value of i, for increasing values of j. This method allows us to compute $f(p, q)$ in $O(pq)$ time, that is, time proportional to the product of the number of elements in the two sequences.

We can alternatively formulate the DNA sequence alignment problem as a shortest path problem. Figure 6 shows the shortest path network for this formulation for a situation with $p = 3$ and $q = 3$. For simplicity, in this network we let g_{ij} denote $g(b_i, d_j)$. We can establish the correctness of this formulation by applying an induction argument based upon the induction hypothesis that the shortest path length from node 0^0 to node i^j equals $f(i, j)$. The shortest path from node 0^0 to node i^j must contain one of the following arcs as the last arc in the path: (i) arc $(i-1^{j-1}, i^j)$; (ii) arc (i^{j-1}, i^j), or (iii) arc $(i-1^j, i^j)$. In these three cases, the lengths of these paths will be $f(i-1, j-1) + g(b_i, d_j)$, $f(i, j-1) + \alpha$, and $f(i-1, j) + \beta$. Clearly, the shortest path length $f(i, j)$ will

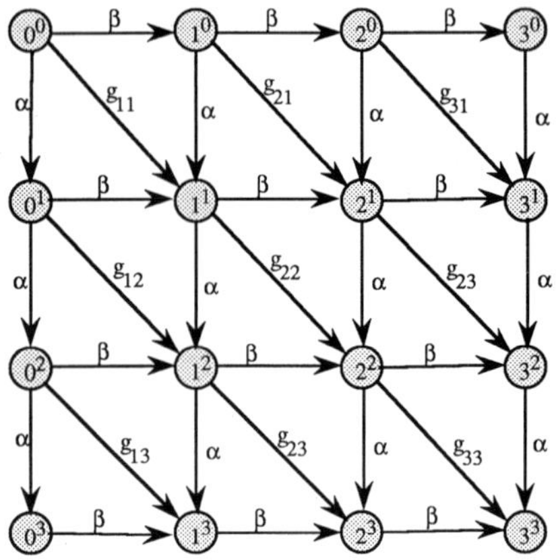

Fig. 6. Sequence alignment problem as a shortest path problem.

equal the minimum of these three numbers, which is consistent with the dynamic programming relationships stated in (4).

This application shows a relationship between shortest paths and dynamic programming. We have seen how to solve the DNA sequence alignment problem through the dynamic programming recursion or by formulating and solving it as a shortest path problem on an acyclic network. The recursion we use to solve the dynamic programming problem is just a special implementation of one of the standard algorithms for solving shortest path problems on acyclic networks. This observation provides us with a concrete illustration of the meta-statement that '(deterministic) dynamic programming is a special case of the shortest path problem'. Accordingly, shortest path problems model the enormous range of applications in many disciplines that are solvable by dynamic programming.

Additional applications

Some additional applications of the shortest path problem include: (1) *knapsack problem* [Fulkerson, 1966]; (2) *tramp steamer problem* [Lawler, 1966]; (3) *allocating inspection effort on a production line* [White, 1969]; (4) *reallocation of housing* [Wright, 1975]; (5) *assortment of steel beams* [Frank, 1965]; (6) *compact book storage in libraries* [Ravindran, 1971]; (7) *concentrator location on a line* [Balakrishnan, Magnanti, Shulman & Wong, 1991]; (8) *manpower planning problem* [Clark & Hasting, 1977]; (9) *equipment replacement* [Veinott & Wagner, 1962]; (10) *determining minimum project duration* [Elmaghraby, 1978]; (11) *assembly line balancing* [Gutjahr & Nemhauser, 1964]; (12) *optimal improvement of transportation networks* [Goldman & Nemhauser, 1967]; (13) *machining process optimization* [Szadkowski, 1970]; (14) *capacity expansion* [Luss, 1979]; (15) *routing in computer communication networks* [Schwartz & Stern, 1980]; (16) *scaling of matrices*

[Golitschek & Schneider, 1984]; (17) *city traffic congestion* [Zawack & Thompson, 1987]; (18) *molecular confirmation* [Dress & Havel, 1988]; (19) *order picking in an isle* [Goetschalckx & Ratliff, 1988]; and (20) *robot design* [Haymond, Thornton & Warner, 1988]. Shortest path problems often arise as important subroutines within algorithms for solving many different types of network optimization problems. These applications are too numerous to mention.

4. Maximum flows

In the maximum flow problem we wish to send the maximum amount of flow from a specified source node s to another specified sink node t in a network with arc capacities u_{ij}'s. If we interpret u_{ij} as the maximum flow rate of arc (i, j), then the maximum flow problem identifies the maximum steady-state flow that the network can send from node s to node t per unit time. We can formulate this problem as a minimum cost flow problem in the following manner. We set $b(i) = 0$ for all $i \in N$, $l_{ij} = c_{ij} = 0$ for all $(i, j) \in A$, and introduce an additional arc (t, s) with cost $c_{ts} = -1$ and flow bounds $l_{ts} = 0$ and capacity $u_{ts} = \infty$. Then the minimum cost flow solution maximizes the flow on arc (t, s); but since any flow on arc (t, s) must travel from node s to node t through the arcs in A (since each $b(i) = 0$), the solution to the minimum cost flow problem will maximize the flow from node s to node t in the original network.

Some applications impose nonzero lower bounds l_{ij} as well as capacities u_{ij} on the arc flows. We refer to this problem as the *maximum flow problem with lower bounds*.

The maximum flow problem is very closely related to another fundamental network optimization problem, known as the *minimum cut problem*. Recall from Section 2 that an s–t cut is a set of arcs whose deletion from the network creates a disconnected network with two components, one component S contains the source node s and the other component $\overline{S}$ contains the sink node t. The capacity of an s–t cut $[S, \overline{S}]$ is the sum of the capacities of all arcs (i, j) with $i \in S$ and $j \in \overline{S}$. The minimum cut problem seeks an s–t cut of minimum capacity.

The *max-flow min-cut theorem*, a celebrated theorem in network optimization, establishes a relationship between the maximum flow problem and the minimum cut problem, namely, the value of the maximum flow from the source node s to the sink node t equals the capacity of a minimum s–t cut. This theorem allows us to determine in $O(m)$ time a minimum cut from the optimal solution of the maximum flow problem.

The maximum flow problem arises in a wide variety of situations and in several forms. Examples of the maximum flow problem include determining the maximum steady state flow of (i) petroleum products in a pipeline network, (ii) cars in a road network; (iii) messages in a telecommunication network; and (iv) electricity in an electrical network. Sometimes the maximum flow problem occurs as a subproblem in the solution of more difficult network problems such as the minimum cost flow problem or the generalized flow problem. The maximum flow problem also arises in a number of combinatorial applications that on the surface might not appear to be maximum flow problems at all. In this section, we describe a few such applications.

Application 6. Matrix rounding problem [Bacharach, 1966]

This application is concerned with consistent rounding of the elements, row sums, and column sums of a matrix. We are given a $p \times q$ matrix of *real* numbers $D = \{d_{ij}\}$, with row sums α_i and column sums β_j. At our discretion, we can round any real number α to the next smaller integer $\lfloor \alpha \rfloor$ or to the next larger integer $\lceil \alpha \rceil$. The matrix rounding problem requires that we round the matrix elements, and the row and column sums of the matrix so that the sum of the rounded elements in each row equals the rounded row sum, and the sum of the rounded elements in each column equals the rounded column sum. We refer to such a rounding as a *consistent rounding*.

This matrix rounding problem arises is several application contexts. For example, the U.S. Census Bureau uses census information to construct millions of tables for a wide variety of purposes. By law, the bureau has an obligation to protect the source of its information and not disclose statistics that could be attributed to any particular individual. We might disguise the information in a table as follows. We round off each entry in the table, including the row and column sums, either up or down to a multiple of a constant k (for some suitable value of k), so that the entries in the table continue to add to the (rounded) row and column sums, and the overall sum of the entries in the new table adds to a rounded version of the overall sums in the original table. This Census Bureau problem is the same as the matrix rounding problem discussed earlier except that we need to round each element to a multiple of $k \geq 1$ instead of rounding it to a multiple of 1. For the moment, let us suppose that $k = 1$ so that this application is a matrix rounding problem in which we round any element to a nearest integer.

We shall formulate this problem and some of the subsequent applications as a problem known as the *feasible flow problem*. In the feasible flow problem, we wish to determine a flow x in a network $G = (N, A)$ satisfying the following constraints:

$$\sum_{\{j:(i,j)\in A\}} x_{ij} - \sum_{\{j:(j,i)\in A\}} x_{ji} = b(i), \quad \text{for } i \in N, \tag{8a}$$

$$0 \leq x_{ij} \leq u_{ij}, \quad \text{for all } (i, j) \in A. \tag{8b}$$

As we noted in Section 2, this model includes feasible flow applications with nonzero lower bounds on arc flows since we can transform any such problem into an equivalent problem with zero lower bounds on the arc flows.

We assume that $\sum_{i\in N} b(i) = 0$. We can solve the feasible flow problem by solving a maximum flow problem defined on an augmented network as follows. We introduce two new nodes, a source node s and a sink node t. For each node i with $b(i) > 0$, we add an arc (s, i) with capacity $b(i)$, and for each node i with $b(i) < 0$, we add an arc (i, t) with capacity $-b(i)$. We refer to the new network as the *transformed network*. Then we solve a maximum flow problem from node s to node t in the transformed network. It is easy to show that the problem (8) has a feasible solution if and only if the maximum flow saturates all the arcs emanating from the source node.

We show how we can discover such a rounding scheme by solving a feasible flow problem for a network with nonnegative lower bounds on arc flows. Figure 7b shows the maximum flow network for the matrix rounding data shown in Figure 7a. This network contains a node i corresponding to each row i and a node j' corresponding to each column j. Observe that this network contains an arc (i, j') for each matrix element d_{ij}, an arc (s, i) for each row sum, and an arc (j', t) for each column sum. The lower and upper bounds of arc (k, l) corresponding to the matrix element, row sum, or column sum of value a are $\lfloor \alpha \rfloor$ and $\lceil \alpha \rceil$, respectively. It is easy to establish a one-to-one correspondence between the consistent roundings of the matrix and feasible integral flows in the associated

				row sum
	3.1	6.8	7.3	17.2
	9.6	2.4	0.7	12.7
	3.6	1.2	6.5	11.3
column sum	16.3	10.4	14.5	

(a)

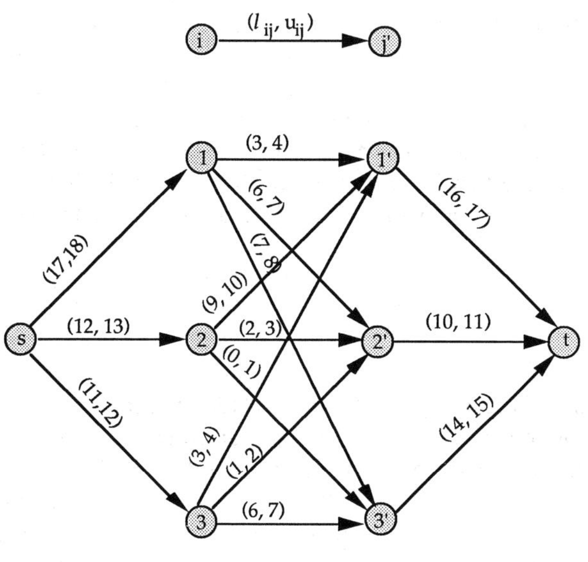

(b)

Fig. 7. (a) Matrix rounding problem. (b) Associated network.

network. We know that there is a feasible integral flow since the original matrix elements induce a feasible fractional flow, and maximum flow algorithms produce all integer flows. Consequently, we can find a consistent rounding by solving a maximum flow problem with lower bounds.

The solution of a matrix rounding problem in which we round every element to a multiple of some positive integer k, as in the Census application we mentioned previously, is similar. In this case, we define the associated network as before, but now defining the lower and upper bounds for any arc with an associated real number a as the greatest multiple of k less than or equal to a and the smallest multiple of k greater than or equal to a.

Application 7. Baseball elimination problem [Schwartz, 1966]

At a particular point in the baseball season, each of $n+1$ teams in the American League, which we number as 0, 1, ..., n, has played several games. Suppose that team i has won w_i of the games it has already played and that g_{ij} is the number of games that teams i and j have yet to play with each other. No game ends in a tie. An avid and optimistic fan of one of the teams, the Boston Red Sox, wishes to know if his team still has a chance to win the league title. We say that we can *eliminate* a specific team 0, the Red Sox, if for every possible outcome of the unplayed games, at least one team will have more wins than the Red Sox. Let $w_{\max}$ denote w_0 (i.e., the number of wins of team 0) plus the total number of games team 0 has yet to play, which, in the best of all possible worlds, is the number of victories the Red Sox can achieve. Then, we cannot eliminate team 0 if in some outcome of the remaining games to be played throughout the league, $w_{\max}$ is at least as large as the possible victories of every other team. We want to determine whether we can or cannot eliminate team 0.

We can transform this baseball elimination problem into a feasible flow problem on a bipartite network shown in Figure 8, whose node set is $N_1 \cup N_2$. The node set for this network contains (i) a set N_1 of n *team nodes* indexed 1 through n, (ii) $n(n-1)/2$ *game nodes* of the type i–j for each $1 \leq i \leq j \leq n$, and (iii) a *source node* s. Each game node i–j has a demand of g_{ij} units and has two incoming arcs $(i, i\text{–}j)$ and $(j, i\text{–}j)$ The flows on these two arcs represent the number of victories for team i and team j, respectively, among the additional g_{ij} games that these two teams have yet to play against each other (which is the required flow into the game node i–j). The flow x_{si} on the source arc (s, i) represents the total number of additional games that team i wins. We cannot eliminate team 0 if this network contains a feasible flow x satisfying the conditions

$$w_{\max} \geq w_i + x_{si}, \quad \text{for all } i = 1, \ldots, n,$$

which we can rewrite as

$$x_{si} \leq w_{\max} - w_i, \quad \text{for all } i = 1, \ldots, n.$$

This observation explains the capacities of arcs shown in the figure. We have thus shown that if the feasible flow problem shown in Figure 8 admits a feasible

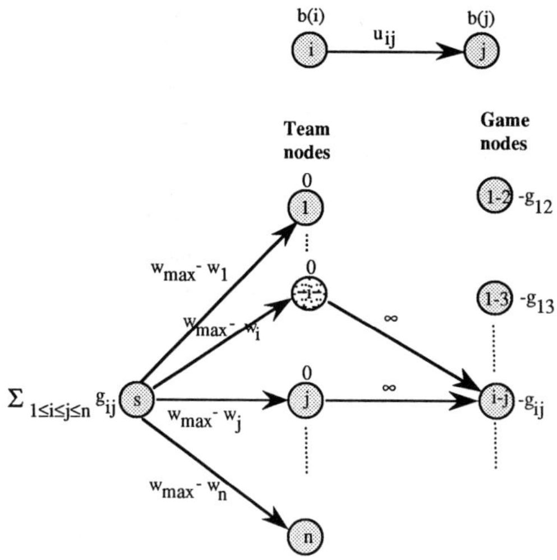

Fig. 8. Network formulation of the baseball elimination problem.

flow, then we cannot eliminate team 0; otherwise, we can eliminate this team and our avid fan can turn his attention to other matters.

Application 8. Distributed computing on a two-processor computer [Stone, 1977]

This application concerns assigning different modules (subroutines) of a program to two processors of a computer system in a way that minimizes the collective costs of interprocessor communication and computation. The two processors need not be identical. We wish to execute a large program consisting of several modules that interact with each other during the program's execution. The cost of executing each module on the two processors is known in advance and might vary from one processor to the other because of differences in the processors' memory, control, speed, and arithmetic capabilities. Let a_i and b_i denote the cost of computation of module i on processors 1 and 2, respectively. Assigning different modules to different processors incurs relatively high overhead costs due to interprocessor communication. Let c_{ij} denote the interprocessor communication cost if modules i and j are assigned to different processors; we do not incur this cost if we assign modules i and j to the same processor. The cost structure might suggest that we allocate two modules to different processors – we need to balance this cost against the communication costs that we incur by allocating the jobs to different processors. Therefore, we wish to allocate modules of the program on the two processors so that we minimize the total processing and interprocessor communication cost.

To formulate this problem as a minimum cut problem on an undirected network, we define a source node s representing processor 1, a sink node t representing processor 2, and a node for every module of the program. For every node i, other than the source and sink nodes, we include an arc (s, i) of capacity β_i and an

i	1	2	3	4
α_i	6	5	10	4
β_i	4	10	3	8

(a)

$\{c_{ij}\} =$

	1	2	3	4
1	0	5	0	0
2	5	0	6	2
3	0	6	0	1
4	0	2	1	0

(b)

Fig. 9. Data for the distributed computing model.

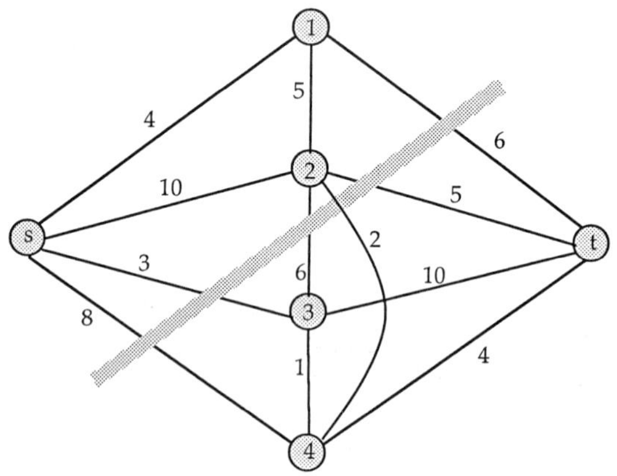

Fig. 10. Network for the distributed computing model.

arc (i, t) of capacity α_i. Finally, if module i interacts with module j during the program's execution, we include arc (i, j) with a capacity equal to c_{ij}. Figures 9 and 10 give an example of this construction. Figure 9 gives the data of this problem and Figure 10 gives the corresponding network.

We now observe a one-to-one correspondence between s–t cuts in the network and assignments of modules to the two processors; moreover, the capacity of a cut equals the cost of the corresponding assignment. To establish this result, let A_1 and A_2 be an assignment of modules to processors 1 and 2, respectively. The cost of this assignment is $\sum_{i \in A_1} a_i + \sum_{i \in A_2} b_i + \sum_{(i,j) \in A_1 \times A_2} c_{ij}$. The s–t cut corresponding to this assignment is $(\{s\} \cup A_1, \{t\} \cup A_2)$. The approach we used to construct the network implies that this cut contains an arc (i, t) of capacity α_i for every $i \in A_1$, an arc (s, i) of capacity β_i for every $i \in A_2$, and all arcs (i, j) with $i \in A_1$ and $j \in A_2$ with capacity c_{ij}. The cost of the assignment A_1 and A_2 equals the capacity of the cut $(\{s\} \cup A_1, \{t\} \cup A_2)$. (The reader might wish to verify this conclusion on the example given in Figure 10 with $A_1 = \{1, 2\}$ and $A_2 = \{3, 4\}$.) Consequently, the minimum s–t cut in the network gives the minimum cost assignment of the modules to the two processors.

Application 9. Scheduling on uniform parallel machines [Federgruen & Groenevelt, 1986]

In this application, we consider the problem of scheduling a set J of jobs on M uniform parallel machines. Each job $j \in J$ has a processing requirement p_j (denoting the number of machine days required to complete the job); a release date r_j (representing the beginning of the day when job j becomes available for processing); and a due date $d_j \geq r_j + p_j$ (representing the beginning of the day by which the job must be completed). We assume that a machine can work on only one job at a time and that each job can be processed by at most one machine at a time. However, we allow *preemptions*, i.e., we can interrupt a job and process it on different machines on different days. The scheduling problem is to determine a feasible schedule that completes all jobs before their due dates or to show that no such schedule exists.

This type of preemptive scheduling problem arises in batch processing systems when each batch consists of a large number of units. The feasible scheduling problem, described in the preceding paragraph, is a fundamental problem in this situation and can be used as a subroutine for more general scheduling problems, such as the maximum lateness problem, thc (weighted) minimum completion time problem, and the (weighted) maximum utilization problem.

To illustrate the formulation of the feasible scheduling problem as a maximum flow problem, we use the scheduling data described in Figure 11.

First, we rank all the release and due dates, r_j and d_j for all j, in ascending order and determine $P \leq 2|J| - 1$ mutually disjoint intervals of dates between consecutive milestones. Let $T_{k,l}$ denote the interval that starts at the beginning of date k and ends at the beginning of date $l + 1$. For our example, this order of release and due dates is 1, 3, 4, 5, 7, 9. We have five intervals represented by $T_{1,2}$, $T_{3,3}$, $T_{4,4}$, $T_{5,6}$ and $T_{7,8}$. Notice that within each interval $T_{k,l}$, the set of available jobs (that is, those released, but not yet due) does not change: we can process all jobs j with $r_j \leq k$ and $d_j \geq l + 1$ in the interval.

We formulate the scheduling problem as a maximum flow problem on a bipartite network G as follows. We introduce a source node s, a sink node t, a node corresponding to each job j, and a node corresponding to each interval $T_{k,l}$, as shown in Figure 12. We connect the source node to every job node j with an arc with capacity p_j, indicating that we need to assign a minimum of p_j machine

Job (j)	1	2	3	4
Processing time (p_j)	1.5	1.25	2.1	3.6
Release time (r_j)	3	1	3	5
Due date (d_j)	5	4	7	9

Fig. 11. A scheduling problem.

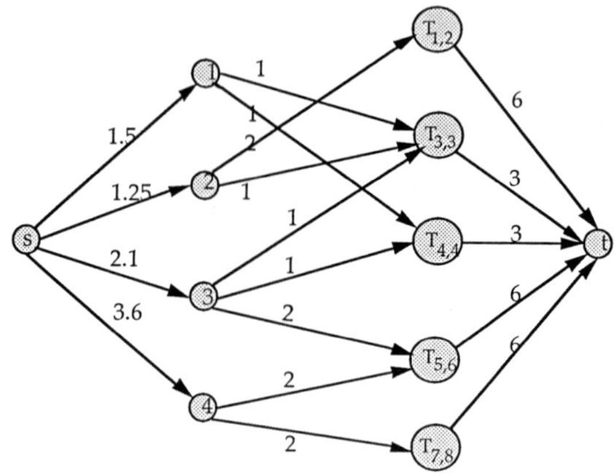

Fig. 12. Network for scheduling uniform parallel machines.

days to job j. We connect each interval node $T_{k,l}$ to the sink node t by an arc with capacity $(l-k+1)M$, representing the total number of machine days available on the days from k to l. Finally, we connect job node j to every interval node T_l if $r_j \leq k$ and $d_j \geq l+1$ by an arc with capacity $(l-k+1)$ which represents the maximum number of machines that we can allot to job j on the days from k to l. We next solve a maximum flow problem on this network: the scheduling problem has a feasible schedule if and only if the maximum flow value equals $\sum_{j \in J} p_j$ (alternatively, for every node j, the flow on arc (s, j) is p_j). The validity of this formulation is easy to establish by showing a one-to-one correspondence between feasible schedules and flows of value equal to $\sum_{j \in J} p_j$ from the source to the sink.

Application 10. Tanker scheduling [Dantzig & Fulkerson, 1954]

A steamship company has contracted to deliver perishable goods between several different origin–destination pairs. Since the cargo is perishable, the customers have specified precise dates (i.e., delivery dates) when the shipments must reach their destinations. (The cargoes may not arrive early or late.) The steamship company wants to determine the minimum number of ships needed to meet the delivery dates of the shiploads.

To illustrate a modeling approach for this problem, we consider an example with four shipments; each shipment is a full shipload with the characteristics shown in Figure 13a. For example, as specified by the first row in this figure, the company must deliver one shipload available at port A and destined for port C on day 3. Figures 13b and 13c show the transit times for the shipments (including allowances for loading and unloading the ships) and the return times (without a cargo) between the ports.

We solve this problem by constructing a network shown in Figure 14a. This network contains a node for each shipment and an arc from node i to node j if it

Shipment	Origin	Destination	Delivery date
1	Port A	Port C	3
2	Port A	Port C	8
3	Port B	Port D	3
4	Port B	Port C	6

(a)

	C	D
A	3	2
B	2	3

(b)

	A	B
C	2	1
D	1	2

(c)

Fig. 13. Data for the tanker scheduling problem.

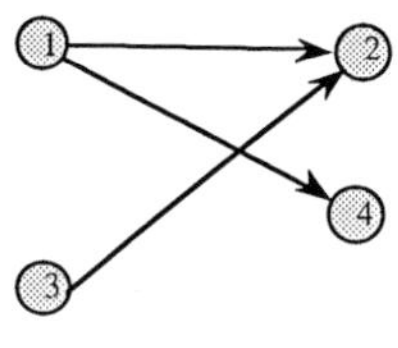

(a)

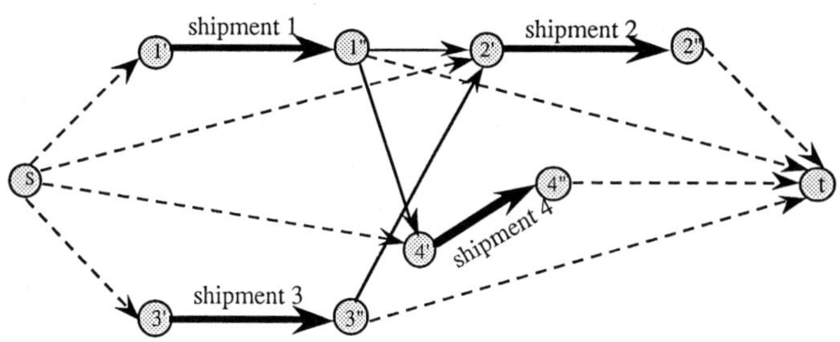

(b)

Fig. 14. Network formulation of the tanker scheduling problem. (a) Network of feasible sequences of two consecutive shipments. (b) Maximum flow model.

is possible to deliver shipment j after completing shipment i; that is, the start time of shipment j is no earlier than the delivery time of shipment i plus the travel time from the destination of shipment i to the origin of shipment j. A directed path in this network corresponds to a feasible sequence of shipment pickups and deliveries. The tanker scheduling problem requires that we identify the minimum number of directed paths that will contain each node in the network on exactly one path.

We can transform this problem into the framework of the maximum flow problem as follows. We split each node i into two nodes i' and i'' and add the arc (i', i''). We set the lower bound on each arc (i', i''), called the *shipment arc*, equal to one so that at least unit flow passes through this arc. We also add a source node s and connect it to the origin of each shipment (to represent putting a ship into service), and we add a sink node t and connect each destination node to it (to represent taking a ship out of service). We set the capacity of each arc in the network to value one. Figure 14b shows the resulting network for our example. In this network, each directed path from the source node s to the sink node t corresponds to a feasible schedule for a single ship. As a result, a feasible flow of value v in this network decomposes into schedules of v ships, and our problem reduces to identifying a feasible flow of minimum value. We note that the zero flow is not feasible because shipment arcs have unit lower bounds. We can solve this problem, which is known as the *minimum value problem*, in the following manner. We first remove lower bounds on the arcs using the transformation described in Section 2. We then establish a feasible flow in the network by solving a maximum flow problem, as described in Application 6. Although the feasible flow x satisfies all of the constraints, the amount of flow sent from node s to node t might exceed the minimum. In order to find a minimum flow, we need to return the maximum amount of flow from t to s. To do so, we find a maximum flow from node t to node s in the residual network, which is defined as follows: For each arc (i, j) in the original network, the residual network contains an arc (i, j) with capacity $u_{ij} - x_{ij}$, and another arc (j, i) with capacity $x_{ij} - l_{ij}$. A maximum flow between nodes t to s in the residual network corresponds to returning a maximum amount of flow from node t to node s, and provides optimal solution of the minimum value problem.

As exemplified by this application, finding a minimum (or maximum) flow in the presence of lower bounds on arc flows typically requires solving two maximum flow problems. The solution to the first maximum flow problem is a feasible flow. The solution to the second maximum flow problem establishes a minimum (or maximum) flow.

Several other applications that bear a close resemblance to the tanker scheduling problem are solvable using the same technique. We next briefly introduce some of these applications.

Optimal coverage of sporting events. A group of reporters wants to cover a set of sporting events in an Olympiad. The sports events are held in several stadiums throughout a city. We know the starting time of each event, its duration, and the stadium where it is held. We are also given the travel times between different stadiums. We want to determine the least number of reporters required to cover the sporting events.

Bus scheduling problem. A mass transit company has p routes that it wishes to service using the fewest possible number of buses. To do so, it must determine the most efficient way to combine these routes into bus schedules. The starting time for route i is a_i and the finishing time is b_i. The bus requires r_{ij} time to travel from the point of destination of route i to the point of origin of route j.

Machine setup problem. A job shop needs to perform eight tasks on a particular day. We know the start and end times of each task. The workers must perform these tasks according to this schedule and so that exactly one worker performs each task. A worker cannot work on two jobs at the same time. We also know the setup time (in minutes) required for a worker to go from one task to another. We wish to find the minimum number of workers to perform the tasks.

Additional applications

The maximum flow problem arises in many other applications, including: (1) *problem of representatives* [Hall, 1956]; (2) *open pit mining* [Johnson, 1968]; (3) *the tournament problem* [Ford & Johnson, 1959]; (4) *police patrol problem* [Khan, 1979]; (5) *nurse staff scheduling* [Khan & Lewis, 1987]; (6) *solving a system of equations* [Lin, 1986]; (7) *statistical security of data* [Gusfield, 1988; Kelly, Golden & Assad, 1992]; (8) *minimax transportation problem* [Ahuja, 1986]; (9) *network reliability testing* [Van Slyke & Frank, 1972]; (10) *maximum dynamic flows* [Ford & Fulkerson, 1958]; (11) *preemptive scheduling on machines with different speeds* [Martel, 1982]; (12) *multifacility rectilinear distance location problem* [Picard & Ratliff, 1978]; (13) *selecting freight handling terminals* [Rhys, 1970]; (14) *optimal destruction of military targets* [Orlin, 1987]; and (15) *fly away kit problem* [Mamer & Smith, 1982]. The following papers describe additional applications of the maximum flow problem or provide additional references: Berge & Ghouila-Houri [1962]; McGinnis & Nuttle [1978], Picard & Queyranne [1982], Abdallaoui [1987], Gusfield, Martel & Fernandez-Baca [1987], Gusfield & Martel [1992], and Gallo, Grigoriadis & Tarjan [1989].

5. Minimum cost flows

The minimum cost flow model is the most fundamental of all network flow problems. In this problem, as described in Section 1, we wish to determine a least cost shipment of a commodity through a network that will satisfy demands at certain nodes from available supplies at other nodes. This model has a number of familiar applications: the distribution of a product from manufacturing plants to warehouses, or from warehouses to retailers; the flow of raw material and intermediate goods through the various machining stations in a production line; the routing of automobiles through an urban street network; and the routing of calls through the telephone system. Minimum cost flow problems arise in almost all industries, including agriculture, communications, defense, education, energy, health care, manufacturing, medicine, retailing, and transportation. Indeed, minimum cost flow problems are pervasive in practice. In this section, by considering a few selected applications, we illustrate some of these possible uses of minimum cost flow problems.

Application 11. Leveling mountainous terrain [Farley, 1980]

This application was inspired by a common problem facing civil engineers when they are building road networks through hilly or mountainous terrain. The

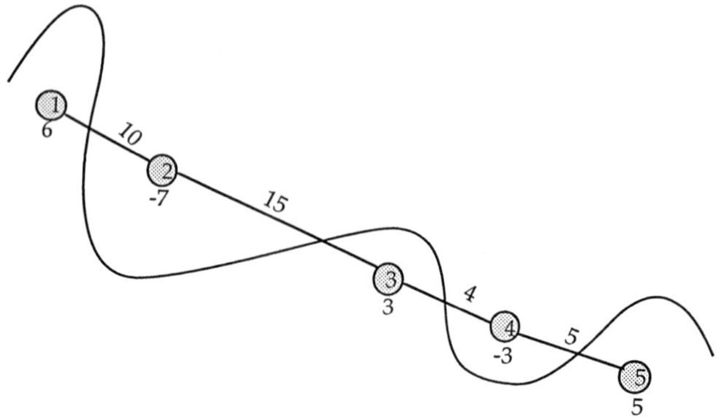

Fig. 15. A portion of the terrain graph.

problem concerns the distribution of earth from high points to low points of the terrain in order to produce a leveled road bed. The engineer must determine a plan for leveling the route by specifying the number of truck loads of earth to move between various locations along the proposed road network.

We first construct a *terrain graph* which is an undirected graph whose nodes represent locations with a demand for earth (low points) or locations with a supply of earth (high points). An arc of this graph represents an available route for distributing the earth and the cost of this arc represents the cost of per truck load of moving earth between the two points. (A *truckload* is the basic unit for redistributing the earth.) Figure 15 shows a portion of the terrain graph.

A leveling plan for a terrain graph is a flow (set of truckloads) that meets the demands at nodes (levels the low points) by the available supplies (by earth obtained from high points) at the minimum cost (for the truck movements). This model is clearly a minimum cost flow problem in the terrain graph.

Application 12. Reconstructing the left ventricle from X-ray projections [Slump & Gerbrands, 1982]

This application describes a minimum cost flow model for reconstructing the three-dimensional shape of the left ventricle from biplane angiocardiograms that the medical profession uses to diagnose heart diseases. To conduct this analysis, we first reduce the three-dimensional reconstruction problem into several two-dimensional problems by dividing the ventricle into a stack of parallel cross sections. Each two-dimensional cross section consists of one connected region of the left ventricle. During a cardiac catheterization, doctors inject a dye known as Roentgen contrast agent into the ventricle; by taking X-rays of the dye, they would like to determine what portion of the left ventricle is functioning properly (that is, permitting the flow of blood). Conventional biplane X-ray installations do not permit doctors to obtain a complete picture of the left ventricle; rather, these

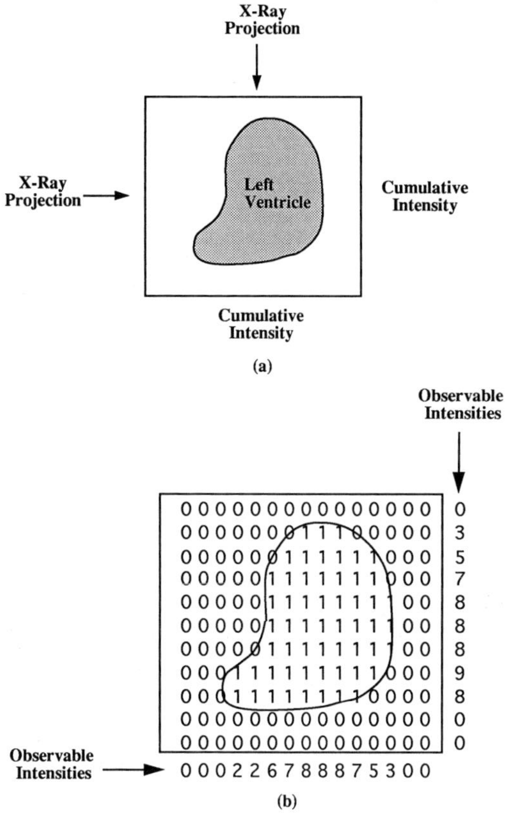

Fig. 16. Using X-Ray projections to measure a left ventricle.

X-rays provide one-dimensional projections that record the total intensity of the dye along two axes (see Figure 16). The problem is to determine the distribution of the cloud of dye within the left ventricle, and thus the shape of the functioning portion of the ventricle, assuming that the dye mixes completely with the blood and fills the portions that are functioning properly.

We can conceive of a cross section of the ventricle as a $p \times r$ binary matrix: a 1 in a position indicates that the corresponding segment of the ventricle allows blood to flow and a 0 indicates that it doesn't permit blood to flow. The angiocardiograms gives the cumulative intensity of the contrast agent in two planes which we can translate into row and column sums of the binary matrix. The problem is then to construct the binary matrix given its row and column sums. This problem is a special case of the feasible flow problem (discussed in Application 6) on a network with a node i for each row i of the matrix with the supply equal to the cumulative intensity of the row, a node j' for each column j of the matrix with demand equal to the cumulative intensity of the column, and a unit capacity arc from each row node i to each column node j'.

Typically, the number of feasible solutions are quite large; and, these solutions might differ substantially. To constrain the feasible solutions, we might use certain facts from our experience that indicate that a solution is more likely to contain certain segments rather than others. Alternatively, we can use *a priori* information: for example, after some small time interval, the cross sections might resemble cross sections determined in a previous examination. Consequently, we might attach a probability p_{ij} that a solution will contain an element (i, j) of the binary matrix and might want to find a feasible solution with the largest possible cumulative probability. This problem is equivalent to a minimum cost flow problem on the network we have already described.

Application 13. Optimal loading of a hopping airplane [Gupta, 1985; Lawania, 1990]

A small commuter airline uses a plane, with a capacity to carry at most p passengers, on a 'hopping flight' as shown in Figure 17a. The hopping flight visits the cities 1, 2, 3, ..., n, in a fixed sequence. The plane can pick up passengers at any node and drop them off at any other node. Let b_{ij} denote the number of passengers available at node i who want to go to node j, and let f_{ij} denote the fare per passenger from node i to node j. The airline would like to determine the number of passengers that the plane should carry between the various origins to destinations in order to maximize the total fare per trip while never exceeding the plane capacity.

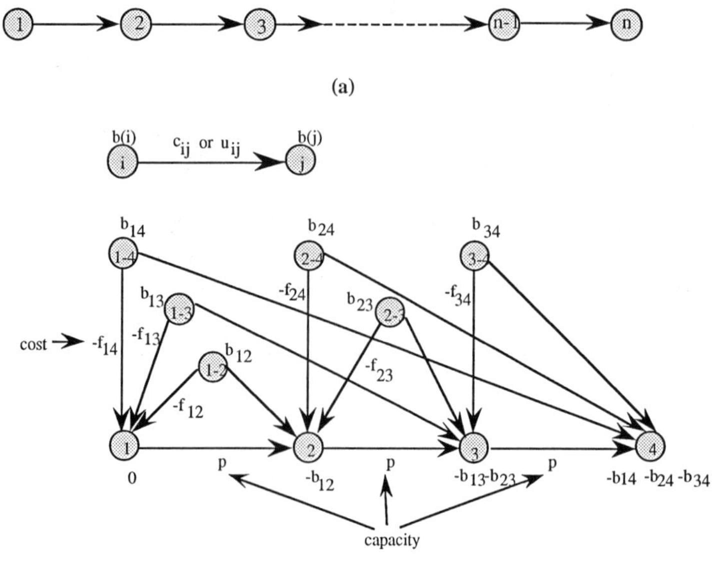

Fig. 17. Formulating the hopping plane flight problem as a minimum cost flow problem.

Figure 17b shows a minimum cost flow formulation of this hopping plane flight problem. The network contains data for only those arcs with nonzero costs and with finite capacities: any arc without an associated cost has a zero cost; any arc without an associated capacity has an infinite capacity. Consider, for example, node 1. Three types of passengers are available at node 1, those whose destination is node 2, node 3, or node 4. We represent these three types of passengers by the nodes 1-2, 1-3, and 1-4 with supplies b_{12}, b_{13}, and b_{14}. A passenger available at any such node, say 1-3, either boards the plane at its origin node by flowing through the arc (1-3, 1), and thus incurring a cost of $-f_{13}$ units, or never boards the plane which we represent by the flow through the arc (1-3, 3). Therefore, each loading of the plane has a corresponding feasible flow of the same cost in the network. The converse result is true (and is easy to establish): each feasible flow in the network has a corresponding feasible airplane loading. Thus, this formulation correctly models the hopping plane application.

Application 14. Directed Chinese postman problem [Edmonds & Johnson, 1973]

Leaving from his home post office, a postman needs to visit the households on each block in his route, delivering and collecting letters, and then returning to the post office. He would like to cover this route by travelling the minimum possible distance. Mathematically, this problem has the following form: Given a network $G = (N, A)$ whose arcs (i, j) have an associated nonnegative length c_{ij}, we wish to identify a walk of minimum length that starts at some node (the post office), visits each arc of the network at least once, and returns to the starting node. This problem has become known as the *Chinese postman problem* because it was first discussed by a Chinese mathematician, K. Mei-Ko. The Chinese postman problem arises in other application settings as well; for instance, in patrolling streets by a police officer, routing of street sweepers and household refuse collection vehicles, fuel oil delivery to households, and the spraying of roads with sand during snow storms. In this application, we discuss the Chinese postman problem on directed networks.

In the Chinese postman problem on directed networks, we are interested in a closed (directed) walk that traverses each arc of the network at least once. The network need not contain any such walk! Figure 18 shows an example. The network contains the desired walk if and only if every node in the network is reachable from every other node; that is, it is *strongly connected*. The strong connectivity

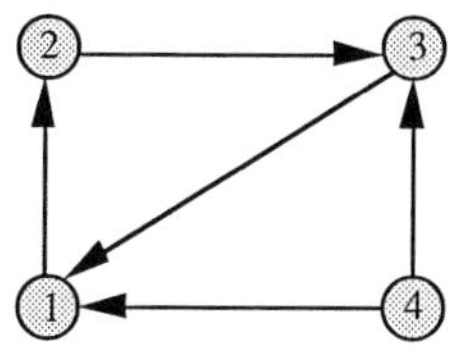

Fig. 18. Network containing no feasible solution for the Chinese postman problem.

of a network can be easily determined in $O(m)$ time [see, e.g., Ahuja, Magnanti & Orlin, 1993]. We shall henceforth assume that the network is strongly connected.

In an optimal walk, a postman might traverse arcs more than once. The minimum length walk minimizes the sum of lengths of the arcs that the walk repeats. Let x_{ij} denote the number of times the postman traverses an arc (i, j) in a walk. Any walk of the postman must satisfy the following conditions:

$$\sum_{\{j:(i,j)\in A\}} x_{ij} - \sum_{\{j:(j,i)\in A\}} x_{ji} = 0, \qquad \text{for all } i \in N, \tag{9a}$$

$$x_{ij} \geq 1, \qquad \text{for all } (i, j) \in A. \tag{9b}$$

The constraints (9a) state that the postman enters a node the same number of times that he/she leaves it. The constraints (9b) state that the postman must visit each arc at least once. Any solution x satisfying (9a) and (9b) defines a postman's walk. We can construct such a walk in the following manner. We replace each arc (i, j) with flow x_{ij} with x_{ij} copies of the arc, each carrying a unit flow. Let A' denote the resulting arc set. Since the outflow equals inflow for each node in the flow x, once we have transformed the network, the outdegree of each node will equal its indegree. This implies that we can decompose the arc set A' into a set of at most m directed cycles. We can connect these cycles together to form a closed walk as follows. The postman starts at some node in one of the cycles, say W_1, and visits the nodes (and arcs) of W_1 in order until he either returns to the node he started from, or encounters a node that also lies in a directed cycle not yet visited, say W_2. In the former case, the walk is complete; and in the latter case, the postman visits cycle W_2 first before resuming his visit of the nodes in W_1. While visiting nodes in W_2, the postman follows the same policy, i.e., if he encounters a node lying on another directed cycle W_3 not yet visited, then he visits W_3 first before visiting the remaining nodes in W_2, and so on. We illustrate this method on a numerical example. Let A' be as indicated in Figure 19a. This solution decomposes into three directed cycles W_1, W_2 and W_3. As shown in Figure 19b, the postman starts at node a and visits the nodes in the following order: a–b–d–g–h–c–d–e–b–c–f–a.

This discussion shows that the solution x defined by a feasible walk for the postman satisfies (9), and, conversely, every feasible solution of (9) defines a walk of the postman. The length of a walk equals $\sum_{(i,j)\in A} c_{ij}x_{ij}$. Therefore, the Chinese postman problem seeks a solution x that minimizes $\sum_{(i,j)\in A} c_{ij}x_{ij}$, subject to the set of constraints (9). This problem is clearly an instance of the minimum cost flow problem.

Application 15. Racial balancing of schools [Belford & Ratliff, 1972]

Suppose a school district has S schools. For the purpose of this formulation, we divide the school district into L district locations and let b_i and w_i denote the number of black and white students at location i. These locations might, for example, be census tracts, bus stops, or city blocks. The only restrictions on the

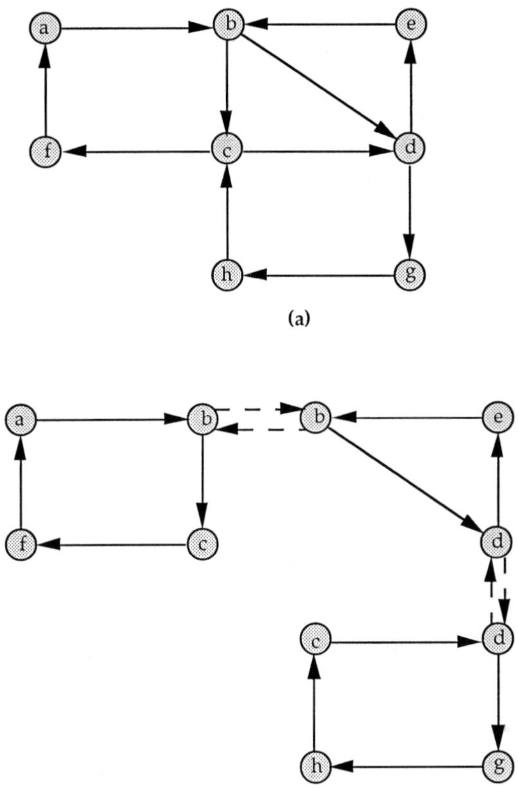

(b)

Fig. 19. Constructing a closed walk for the postman.

locations is that they be finite in number and that there be a single distance measure d_{ij} that reasonably approximates the distance any student at location i must travel if he or she is assigned to school j. We make the reasonable assumption that we can compute the distances d_{ij} before assigning students to schools. School j can enroll u_j students. Finally, let $\underline{p}$ denote a lower bound and $\overline{p}$ denote an upper bound on the percentage of blacks assigned to each school (we choose these numbers so that school j has same percentage of blacks as does the school district). The objective is to assign students to schools in a manner that maintains the stated racial balance and minimizes the total distance travelled by the students.

We model this problem as a minimum cost flow problem. Figure 20 shows the minimum cost flow network for a three-location, two-school problem. Rather than describe the general model formally, we will merely describe the model ingredients for this figure. In this formulation, we model each location i as two nodes l_i' and l_i'' and each school j as two nodes s_j' and s_j''. The decision variables for this problem

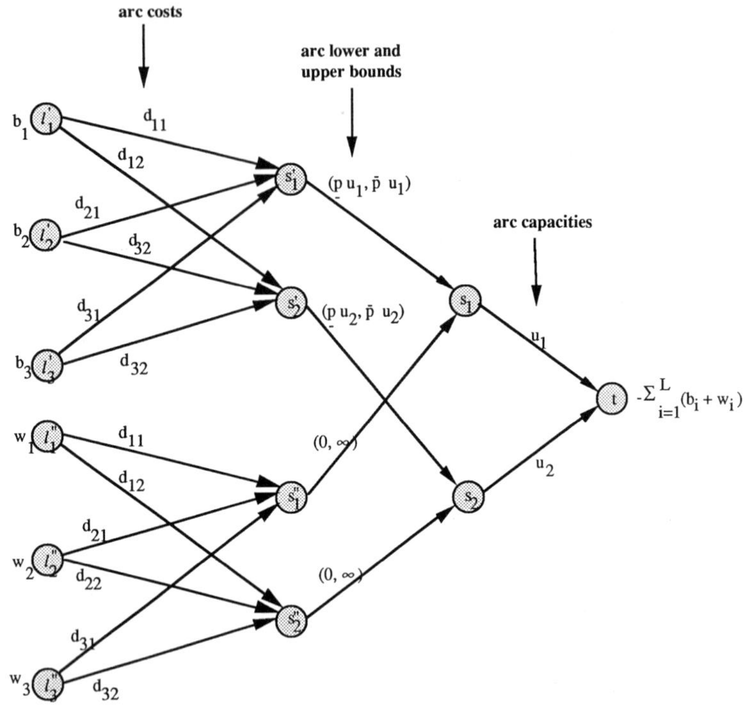

Fig. 20. Network for the racial balancing of schools.

are the number of black students assigned from location i to school j (which we represent by an arc from node l_i' to node s_j') and the number of white students assigned from location i to school j (which we represent by an arc from node l_i'' to node s_j''). These arcs are uncapacitated and we set their per unit flow cost equal to d_{ij}. For each j, we connect the nodes s_j' and s_j'' to the school node s_j. The flow on the arcs (s_j', s_j) and (s_j'', s_j) denotes the total number of black and white students assigned to school j. Since each school must satisfy lower and upper bounds on the number of black students it enrolls, we set the lower and upper bounds of the arc (s_j', s_j) equal to $(\underline{p}u_j, \overline{p}u_j)$. Finally, we must satisfy the constraint that school j enrolls at most u_j students. We incorporate this constraint in the model by introducing a sink node t and joining each school node j to node t by an arc of capacity u_j. As is easy to verify, this minimum cost flow problem correctly models the racial balancing application.

Additional applications

A complete list of additional applications of the minimum cost flow problem is too vast to mention here. A partial list of additional references is: (1) *distribution problems* [Glover & Klingman, 1976]; (2) *building evacuation models* [Chalmet,

Francis & Saunders, 1982]; (3) *scheduling with consecutive ones in columns* [Veinott & Wagner, 1962]; (4) *linear programs with consecutive circular ones in rows* [Bartholdi, Ratliff & Orlin, 1980]; (5) *the entrepreneur's problem* [Prager, 1957]; (6) *optimal storage policy for libraries* [Evans, 1984]; (7) *zoned warehousing* [Evans, 1984]; (8) *allocation of contractors to public works* (Cheshire, McKinnon & Williams, 1984]; (9) *phasing out capital equipment* [Daniel, 1973]; (10) *the terminal assignment problem* [Esau & Williams, 1966]; (11) *capacitated maximum spanning trees* [Garey & Johnson, 1979]; (12) *caterer problem* [Jacobs, 1954]; (13) *allocating receivers to transmitters* [Dantzig, 1962]; (14) *faculty-course assignment* [Mulvey, 1979]; (15) *automatic karyotyping of chromosomes* [Tso, Kleinschmidt, Mitterreiter & Graham, 1991]; (16) *just-in-time scheduling* [Elmaghraby, 1978; Levner & Nemirovsky, 1991]; (17) *time-cost tradeoff in project management* [Fulkerson, 1961; Kelley, 1961]; (18) *warehouse layout* [Francis & White, 1976]; (19) *rectilinear distance facility location* [Cabot, Francis & Stary, 1970]; (20) *dynamic lot-sizing* [Zangwill, 1969]; (21) *multistage production-inventory planning* [Evans, 1977]; (22) *mold allocation* [Love & Vemuganti, 1978]; (23) *a parking model* [Dirickx & Jennergren, 1975]; (24) *the network interdiction problem* [Fulkerson & Harding, 1977]; (25) *truck scheduling* [Gavish & Schweitzer, 1974]; and (26) *optimal deployment of firefighting companies* [Denardo, Rothblum & Swersey, 1988]; (27) *warehousing and distribution of a seasonal product* [Jewell, 1957]; (28) *economic distribution of coal supplies in the gas industry* [Berrisford, 1960]; (29) *upsets in round robin tournaments* [Fulkerson, 1965]; (30) *optimal container inventory and routing* [Horn, 1971]; (31) *distribution of empty rail containers* [White, 1972]; (32) *optimal defense of a network* [Picard & Ratliff, 1973]; (33) *telephone operator scheduling* [Segal, 1974]; (34) *multifacility minimax location problem with rectilinear distances* [Dearing & Francis, 1974]; (35) *cash management problems* [Srinivasan, 1974]; (36) *multiproduct multifacility production-inventory planning* [Dorsey, Hodgson & Ratliff, 1975]; (37) *'hub' and 'wheel' scheduling problems* [Arisawa & Elmaghraby, 1977]; (38) *warehouse leasing problem* [Lowe, Francis & Reinhardt, 1979]; (39) *multi-attribute marketing models* [Srinivasan, 1979]; (40) *material handling systems* [Maxwell & Wilson, 1981]; (41) *microdata file merging* [Barr & Turner, 1981]; (42) *determining service districts* [Larson & Odoni, 1981]; (43) *control of forest fires* [Kourtz, 1984]; (44) *allocating blood to hospitals from a central blood bank* [Sapountzis, 1984]; (45) *market equilibrium problems* [Dafermos & Nagurney, 1984]; (46) *automatic chromosome classifications* [Tso, 1986]; (47) *city traffic congestion problem* [Zawack & Thompson, 1987]; (48) *satellite scheduling* [Servi, 1989]; (49) *determining k disjoint cuts in a network* [Wagner, 1990]; (50) *controlled rounding of matrices* [Cox & Ernst, 1982]; (51) *scheduling with deferral costs* [Lawler, 1964].

6. The assignment problem

The assignment problem is a special case of the minimum cost flow problem and can be defined as follows: Given a weighted bipartite network $G = (N_1 \cup N_2, A)$

with $|N_1| = |N_2|$ and arc weights c_{ij}, find a one-to-one assignment of nodes in N_1 with nodes in N_2, so that the sum of the costs of arcs in the assignment is minimum. It is easy to notice that if we set $b(i) = 1$ for each node $i \in N_1$ and $b(i) = -1$ for each node $i \in N_2$, then the optimal solution of the minimum cost flow problem in G will be an assignment. For the assignment problem, we allow the network G to be directed or undirected. If the network is directed, then we require that each arc $(i, j) \in A$ has $i \in N_1$ and $j \in N_2$. If the network is undirected, then we make it directed by designating all arcs as pointing from nodes in N_1 to nodes in N_2. We shall, therefore, henceforth assume that G is a directed graph.

Examples of the assignment problem include assigning people to projects, jobs to machines, tenants to apartments, swimmers to events in a swimming meet, and medical school graduates to available internships. We now describe three more clever applications of the assignment problem.

Application 16. Locating objects in space [Brogan, 1989]

This application concerns locating objects in space. To identify an object in (three-dimensional) space, we could use two infrared sensors, located at geographically different sites. Each sensor provides an angle of sight of the object and, hence, the line on which the object must lie. The unique intersection of the two lines provided by the two sensors (provided that the two sensors and the object are not co-linear) determines the unique location of the object in space.

Consider now the situation in which we wish to determine the locations of p objects using two sensors. The first sensor would provide us with a set of lines L_1, $L_2, \ldots, L_p$ for the p objects and the second sensor would provide us a different set of lines $L'_1, L'_2, \ldots, L'_p$. To identify the location of the objects – using the fact that if two lines correspond to the same object, then the lines intersect one-another – we need to match the lines from the first sensor to the lines from the second sensor. In practice, two difficulties limit the use of this approach. First, a line from a sensor might intersect more than one line from the other sensor, so the matching is not unique. Second, two lines corresponding to the same object might not intersect because the sensors make measurement errors in determining the angle of sight. We can overcome this difficulty in most situations by formulating this problem as an assignment problem.

In the assignment problem, we wish to match the p lines from the first sensor with the p lines from the second sensor. We define the cost c_{ij} of the assignment (i, j) as the minimum Euclidean distance between the lines L_i and L_j. We can determine c_{ij} using standard calculations from geometry. If the lines L_i & L_j correspond to the same object, then c_{ij} would be close to zero. An optimal solution of the assignment problem would provide an excellent matching of the lines. Simulation studies have found that in most circumstances, the matching produced by the assignment problem defines the correct location of the objects.

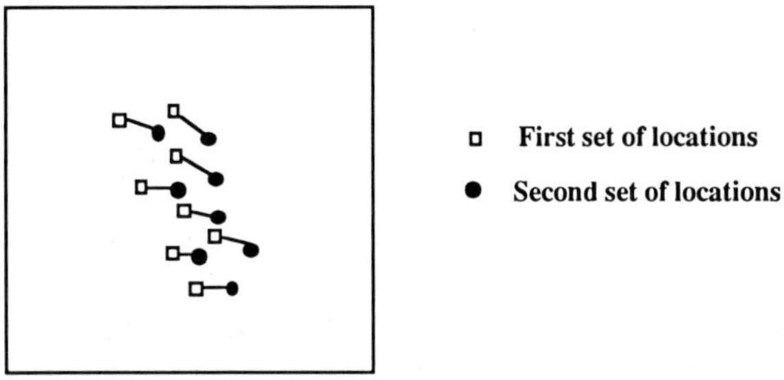

Fig. 21. Two snapshots of a set of 8 objects.

Application 17. Matching moving objects [Brogan, 1989; Kolitz, 1991]

In several different application contexts, we might wish to estimate the speeds and the directions of movement of a set of p objects (e.g., enemy fighter planes, missiles) that are moving in space. Using the method described in the preceding application, we can determine the location of the objects at any point in time. One plausible way to estimate the movement directions of the objects at which the objects are moving is to take two snapshots of the objects at two distinct times and then to match one set of points with the other set of points. If we correctly match the points, then we can assess the speed and direction of movement of the objects. As an example, consider Figure 21 which denotes the objects at time 1 by squares and the objects at time 2 by circles.

Let (x_i, y_i, z_i) denote the coordinates of object i at time 1 and (x_i', y_i', z_i') denote the coordinates of the same object at time 2. We could match one set of points with the other set of points in many ways. Minimizing the sum of the squared Euclidean distances between the matched points is reasonable in this scenario because it attaches a higher penalty to larger distances. If we take the snapshots of the objects at two times that are sufficiently close to each other, then the optimal assignment will often match the points correctly. In this application of the assignment problem, we let $N_1 = \{1, 2, \ldots, p\}$ denote the set of objects at time 1, let $N_2 = \{1', 2', \ldots, p'\}$ denote the set of objects at time 2, and we define the cost of an arc (i, j') as $[(x_i - x_i')^2 + (y_i - y_i')^2 + (z_i - z_i')^2]$. The optimal assignment in this graph will specify the desired matching of the points. From this matching, we obtain an estimate of the movement directions and speeds of the individual objects.

Application 18. Rewiring of typewriters [Machol, 1961]

For several years, a company had been using special electric typewriters to prepare punched paper tapes to enter data into a digital computer. Because the typewriter is used to punch a six-hole paper tape, it can prepare $2^6 = 64$ binary hole–no-hole patterns. The typewriters have 46 characters and each punches one of the 64 patterns. The company acquired a new digital computer that uses different coding hole–no-hole patterns to represent characters. For example, using 1 to represent a hole and 0 to represent a no-hole, the letter A is 111100 in the code for the old computer and 011010 in the code for the new computer. The typewriter presently punches the former and must be modified to punch the latter.

Each key in the typewriter is connected to a steel code bar, and so changing the code of that key requires mechanical changes in the steel bar system. The extent of the changes depends upon how close the new and old characters are to each other. For the letter A, the second, third and sixth bits are identical in the old and new codes and no changes need be made for these bits; however, the first, fourth and fifth bits are different and so we would need to make three changes in the steel code bar connected to the A-key. Each change involves removing metal at one place and adding metal at another place. When a key is pressed, its steel code bar activates six cross-bars (which are used by all the keys) that are connected electrically to six hole punches. If we interchange the fourth and fifth wires of the cross-bars to the hole punches (which is essentially equivalent to interchanging the fourth and fifth bits of all characters in the old code), then we would reduce the number of mechanical changes needed for the A-key from three to one. However, this change of wires might increase the number of changes for some of the other 45 keys. The problem, then, is how to optimally connect the wires from the six cross-bars to the six punches so that we can minimize the number of mechanical changes on the steel code bars.

We formulate this problem as an assignment problem as follows. Define a network $G = (N_1 \cup N_2, A)$ with node sets $N_1 = \{1, 2, \ldots, 6\}$ and $N_2 = \{1', 2', \ldots, 6'\}$, and an arc set $A = N_1 x N_2$; the cost of the arc $(i, j') \in A$ is the number of keys (out of 46) for which the ith bit in the old code differs from the jth bit in the new code. Thus if we assign cross-bar i to the punch j, then the number of mechanical changes needed to correctly print the ith bit of each symbol is c_{ij}. Consequently, the minimum cost assignment will minimize the number of mechanical changes.

Additional applications

Additional applications of the assignment problem include: (1) *bipartite personnel assignment* [Machol, 1970; Ewashko & Dudding, 1971]; (2) *optimal depletion of inventory* [Derman & Klein, 1959]; (3) *scheduling of parallel machines* [Horn, 1973]; (4) *solving shortest path problems in directed networks* [Hoffman & Markowitz, 1963]; (5) *discrete location problems* [Francis & White, 1976]; and (6) *vehicle and crew scheduling* [Carraresi & Gallo, 1984].

7. Matchings

A *matching* in an undirected graph $G = (N, A)$ is a set of arcs with the property that every node is incident to at most one arc in the set; thus a matching induces a pairing of (some of) the nodes in the graph using the arcs in A. In a matching, each node is matched with at most one other node, and some nodes might not be matched with any other node. In some applications, we want to match all the nodes (that is, each node must be matched to some other node); we refer to any such matching as a *perfect matching*. Suppose that each arc (i, j) in the network has an associated cost c_{ij}. The *(perfect) matching problem* seeks a matching that minimizes the total cost of the arcs in the (perfect) matching. Since any perfect matching problem can be formulated as a matching problem if we add the same large negative cost to each arc, in the following discussion, we refer only to matching problems.

Matching problems on bipartite graphs (i.e., on a graph $G = (N_1 \cup N_2, A)$, where $N = N_1 \cup N_2$ and each arc $(i, j) \in A$ has $i \in N_1$ and $j \in N_2$), are called *bipartite matching problems* and those on general graphs that need not be bipartite are called *nonbipartite matching problems*. There are two further ways of categorizing matching problems: *cardinality matching problems,* that maximize the number of pairs of nodes matched, and *weighted matching problems* that minimize the weight of the matching. The weighted matching problem on a bipartite graph is the same as the *assignment problem*, whose applications we described in the previous section. In this section, we describe applications of matching problems on nonbipartite graphs.

The matching problem arises in many different problem settings since we often wish to find the best way to pair objects or people together to achieve some desired goal. Some direct applications of nonbipartite matching problems include matching airplane pilots to planes, and assigning roommates to rooms in a hostel. We describe now three indirect applications of matching problems.

Application 19. Pairing stereo speakers [Mason & Philpott, 1988]

As a part of its manufacturing process, a manufacturer of stereo speakers must pair individual speakers before it can sell them as a set. The performance of the two speakers depends upon their frequency response. In order to measure the quality of the pairs, the company generates matching coefficients for each possible pair. It calculates these coefficients by summing the absolute differences between the responses of the two speakers at twenty discrete frequencies, thus giving a matching coefficient value between 0 and 30,000. Bad matches yield a large coefficient, and a good pairing produces a low coefficient.

The manufacturer typically uses two different objectives in pairing the speakers: (i) finding as many pairs as possible whose matching coefficients do not exceed a specification limit; or (ii) pairing speakers within specification limits in order to minimize the total sum of the matching coefficients. The first objective minimizes the number of pairs outside of specification, and so the number of speakers that the firm must sell at a reduced price. This model is an application of the nonbipartite cardinality matching problem on an undirected graph: the nodes of this graph represent speakers and arcs join two nodes if the matching coefficients of the corresponding speakers are within the specification limit. The second model is an application of the nonbipartite weighted matching problem.

Application 20. Determining chemical bonds [Dewar & Longuet-Higgins, 1952]

Matching problems arise in the field of chemistry as chemists attempt to determine the possible atomic structures of various molecules. Figure 22a specifies the partial chemical structure of a molecule of some hydrocarbon compound. The molecule contains carbon atoms (denoted by nodes with the letter 'C' next to them) and hydrogen atoms (denoted by nodes with the letter 'H' next to them). Arcs denote bonds between atoms. The bonds between the atoms, which can be either single or double bonds, must satisfy the 'valency requirements' of all the nodes. (The valency of an atom is the sum of its bonds.) Carbon atoms must have a valency of 4 and hydrogen atoms a valency of 1.

In the partial structure shown in Figure 22a, each arc depicts a single bond and, consequently, each hydrogen atom has a valency of 1, but each carbon atom has a valency of only 3. We would like to determine which pairs of carbon atoms to connect by a double bond so that each carbon atom has valency 4. We can formulate this problem of determining some feasible structure of double bonds to determining whether or not the network obtained by deleting the hydrogen atoms contains a maximum cardinality matching with all nodes are matched. Figure 22b gives one feasible bonding structure of the compound; the bold lines in this network denote double bonds between the atoms.

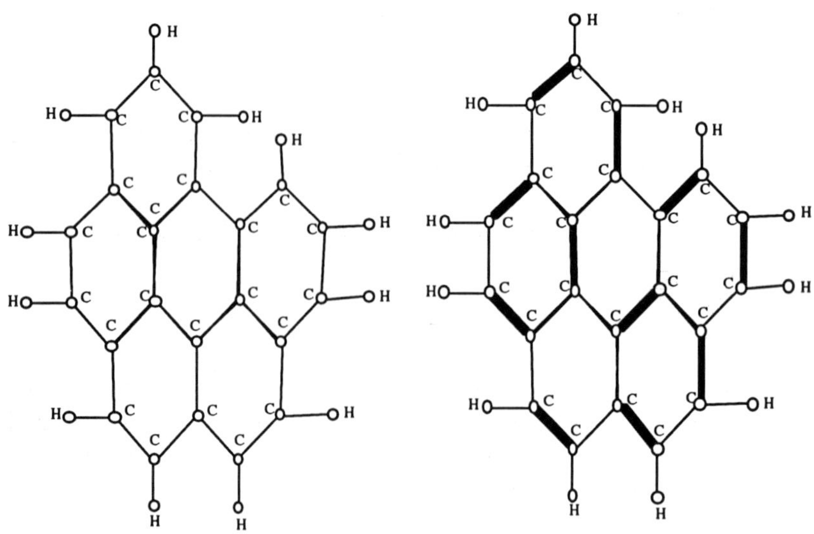

(a) (b)

Fig. 22. Determining the chemical structure of a hydrocarbon.

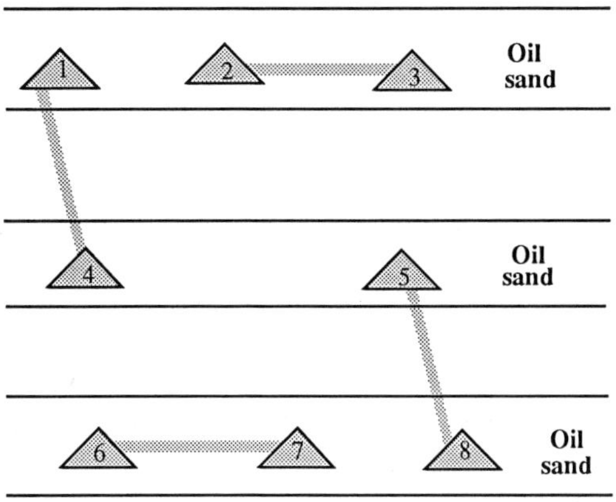

Fig. 23. Targets and matchings for the dual completion problem.

Application 21. Dual completion of oil wells [Devine, 1973]

An oil company has identified several individual oil traps, called *targets*, in an offshore oil field and wishes to drill wells to extract oil from these traps. Figure 23 illustrates a situation with eight targets. The company can extract any target separately (so called *single completion*) or extract oil from any two targets together by drilling a single hole (so called *dual completion*). It can estimate the cost of drilling and completing any target as a single completion or any pair of targets as a dual completion. This cost will depend on the three-dimensional spatial relationships of targets to the drilling platform and to each other. The decision problem is to determine which targets (if any) to drill as single completions and which pairs to drill together as duals, so as to minimize the total drilling and completion costs. If we restrict the solution to use only dual completions, then the decision problem is a nonbipartite weighted matching problem.

Application 22. Parallel savings heuristics [Ball, Bodin & Dial, 1983; Altinkemer & Gavish, 1991]

A number of combinatorial optimization problems have an underlying clustering structure. For example, in many facility location applications we wish to cluster customers into service regions, each served by a single facility (warehouses in distribution planning, concentrators in telecommunication systems design, hospitals within a city). In vehicle routing, we wish to form customer clusters, each representing the customer deliveries allocated to a particular vehicle. The cost for any particular cluster might be simple, for example, the total cost of traveling from each customer to its service center, or complex: in vehicle routing, the cost of a solution is the sum of the costs required to service each cluster (which usually is computed by solving a traveling salesman problem – see Section 12) plus a cost that depends on the number of clusters. Airline or mass transit crew scheduling provide another example. In this context, we need to assign crews to flight legs or transit trips. A cluster is a set of flight legs or trips that forms the workload for one crew.

In these examples, as well as in other applications, the generic problem has an input set T together with a cost function $c(S)$ defined on subsets S of T. The problem is to partition T into subsets $S_1, S_2, \ldots, S_k$ in a way that minimizes the sum $c(S_1) + c(S_2) + \ldots + c(S_k)$.

In these settings, matching can be used as a tool for designing 'smart' heuristic solution procedures. A standard 'greedy' savings heuristic, for example, starts with an initial solution consisting of all subsets of size 1. Suppose we are given an intermediate solution that partitions the elements of T into disjoint subsets S_1, $S_2, \ldots, S_k$. For any pair of subsets, S_i and S_j, we define the savings obtained by combining S_i and S_j as:

$$c(S_i) + c(S_j) - c(S_i \cup S_j).$$

The general step of the greedy algorithm computes the savings obtained by combining all pairs of subsets and combines the two that provide the largest savings. A parallel savings algorithm considers the sets $S_1, S_2, \ldots, S_k$ simultaneously rather than just two at a time. At each iteration, it simultaneously combines the set of pairs – e.g., S_1 with S_6, S_2 with S_{11}, S_3 with S_4, etc. – that yields the greatest savings. We can find this set of combinations by solving a matching problem. The matching graph contains a node for each subset S_j. The graph contains the arc (i, j) whenever combining the two end nodes (i.e., the sets S_i and S_j) is feasible and yields positive savings. We allow nodes to remain unmatched. A maximum weight matching in this graph yields the set of combinations producing the greatest savings.

As stated, the savings algorithm always combines sets to form larger ones. Using a similar approach, we can construct a parallel savings improvement algorithm for this same class of problems. We start with any set of subsets that constitute a 'good' feasible solution (if we had obtained this partition by first using the parallel savings algorithm, combining no pair of subsets would induce a positive savings). We construct a matching graph as before, except that the savings associated with two subsets S_i and S_j becomes:

$$c(S_i) + c(S_j) - [c(S_i^*) + c(S_j^*)].$$

In this expression, S_i^* and S_j^* are the minimum cost partitions of $S_i \cup S_j$. We then replace S_i and S_j with S_i^* and S_j^*. We then iterate on this process until the matching graph contains no positive cost arcs.

In the most general setting, the minimum cost partitions could involve more than 2 sets. If finding the minimum cost partitions of $S_i \cup S_j$ is too expensive, we might instead use some heuristic method to find a 'good' partition of these sets (the parallel saving heuristic is a special case in which we always choose the partition as the single set $S_i \cup S_j$).

We then iterate on this process until the matching graph contains no positive cost edges.

Analysts have devised parallel savings algorithms and shown them to be effective for problems arising in vehicle routing, crew scheduling, and telecommunications.

Additional applications

Additional applications of the matching problems include: (1) *solving shortest path problems in undirected networks* [Edmonds, 1967]; (2) *the undirected Chinese postman problem* [Edmonds & Johnson, 1973]; (3) *two-processor scheduling* [Fujii, Kasami & Ninomiya, 1969]; (4) *vehicle and crew scheduling* [Carraresi & Gallo, 1984]; (5) *determining the rank of a matrix* [Anderson, 1975]; and (6) *making matrices optimally sparse* [Hoffman & McCormick, 1984].

8. Minimum spanning trees

As we noted previously, a spanning tree is a tree (i.e., a connected acyclic graph) that spans (touches) all the nodes of an undirected network. The cost of a spanning tree is the sum of the costs (or lengths) of its arcs. In the minimum spanning tree problem, we wish to identify a spanning tree of minimum cost (or length). Minimum spanning tree problems generally arise in one of two ways, directly or indirectly. In *direct* applications, we typically wish to connect a set of points using the least cost or least length collection of arcs. Frequently, the points represent physical entities that need to be connected to each other. In *indirect* applications, we either (i) wish to connect some set of points using a measure of performance that on the surface bears little resemblance to the minimum spanning tree objective (sum of arc costs), or (ii) the problem itself bears little resemblance to an 'optimal tree' problem – these instances often require creativity in modeling so that they become a minimum spanning tree problem. In this section, we consider several indirect applications. Section 12 on the network design problem describes several direct applications of the minimum spanning tree problem.

Application 23. Measuring homogeneity of bimetallic objects [Shier, 1982; Filliben, Kafadar & Shier, 1983]

This application shows how a minimum spanning tree problem can be used to determine the degree to which a bimetallic object is homogeneous in composition. To use this approach, we measure the composition of the bimetallic object at a set of sample points. We then construct a network with nodes corresponding to the sample points and with an arc connecting physically adjacent sample points. We assign a cost with each arc (i, j) equal to the product of the physical (Euclidean) distance between the sample points i and j and a homogeneity factor between 0 and 1. This homogeneity factor is 0 if the composition of the corresponding samples is exactly alike, and is 1 if the composition is very different; otherwise, it is a number between 0 and 1. Note that this measure gives greater weight to two points if they are different and are far apart. The cost of the minimum spanning tree is a measure of the homogeneity of the bimetallic object. The cost of the tree is 0 if all the sample points are exactly alike, and high cost values imply that the material is quite nonhomogeneous.

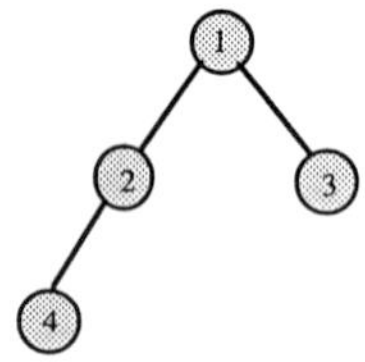

Fig. 24. Compact storage of a matrix.

Application 24. Reducing data storage (Kang, Lee, Chang & Chang, 1977]

In several different application contexts, we wish to store data specified in the form of a two-dimensional array more efficiently than storing all the elements of the array (that is, to save memory space). We assume that the rows of the array have many similar entries and differ only at a few places. Since the entities in the rows are similar, one approach for saving memory is to store one row, called the *reference row*, completely, and to store only the differences between some of the rows so that we can derive each row from these differences and the reference row. Let c_{ij} denote the number of different entries in rows i and j; that is, if we are given row i, then by making c_{ij} changes to the entries in this row we can obtain row j, and vice-versa. Suppose that the array contains four rows, represented by R_1, R_2, R_3 and R_4, and we decide to treat R_1 as a reference row. Then one plausible solution is to store the differences between R_1 and R_2, R_2 and R_4, and R_1 and R_3. Clearly, from this solution, we can obtain rows R_2 and R_3 by making c_{12} and c_{13} changes to the elements in row R_1. Having obtained row R_2, we can make c_{24} changes to the elements of this row to obtain R_4. We can depict this storage scheme in the form of a spanning tree shown in Figure 24.

It is easy to see that it is sufficient to store differences between those rows that correspond to arcs of a spanning tree. These differences permit us to obtain each row from the reference row. The total storage requirement for a particular storage scheme will be the length of the reference row (which we can take as the row with the least amount of data) plus the sum of the differences between the rows. Therefore, a minimum spanning tree would provide the least cost storage scheme.

Application 25. Cluster analysis (Gower & Ross, 1969; Zahn, 1971]

In this application, we describe the use of spanning tree problems to solve a class of problems that arises in the context of cluster analysis. The essential issue in cluster analysis is to partition a set of data into 'natural groups'; the data points within a particular group of data, or a cluster, should be more 'closely related' to each other than the data points not in that cluster. Cluster analysis is important in a variety of disciplines that rely upon empirical investigations. Consider, for example, an instance of a cluster analysis arising in medicine. Suppose we have data on a set of 350 patients, measured with respect to 18 symptoms. Suppose,

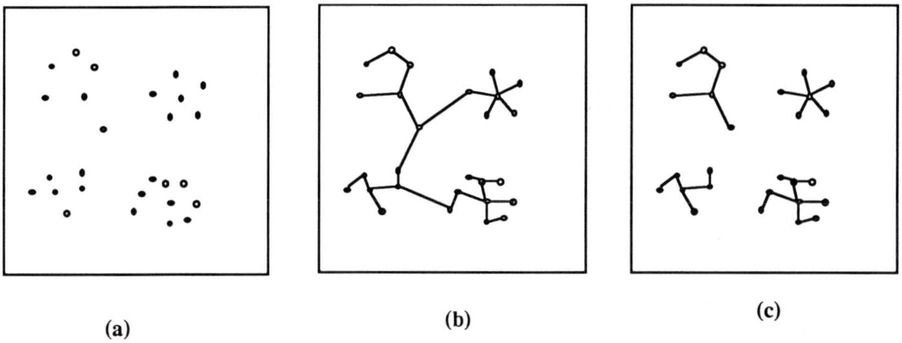

Fig. 25. Identifying clusters by finding a minimum spanning tree.

further, that a doctor has diagnosed all of these patients as having the same disease which is not well understood. The doctor would like to know if he or she can develop a better understanding of this disease by categorizing the symptoms into smaller groupings using cluster analysis. Doing so might permit the doctor to find more natural disease categories to replace or subdivide the original disease.

Suppose we are interested in finding a partition of a set of n points in two-dimensional Euclidean space into clusters. A popular method for solving this problem is by using Kruskal's algorithm for solving the minimum spanning tree problem [see, e.g., Ahuja, Magnanti & Orlin, 1993]. Kruskal's algorithm maintains a forest (i.e., a collection of node-disjoint trees) and adds arcs in a nondecreasing order of their costs. We can regard the components of the forest at intermediate steps as different clusters. These clusters are often excellent solutions for the clustering problem and, moreover, we can obtain them very efficiently. Kruskal's algorithm can be thought of providing n partitions: the first partition contains n clusters, each cluster containing a single point, and the last partition contains just one cluster containing all the points. Alternatively, we can obtain n partitions by starting with a minimum spanning tree and deleting tree arcs one by one in nonincreasing order of their lengths. We illustrate the later approach using an example. Consider a set of 27 points shown in Figure 25a. Suppose that Figure 25b shows a minimum spanning tree for these points. Deleting the three largest length arcs from the minimum spanning tree gives a partition with four clusters shown in Figure 25c.

Analysts can use the information obtained from the preceding analysis in several ways. The procedure we have described yields n partitions. Out of these, we might select the 'best' partition by simple visualization or by defining an appropriate objective function value. A good choice of the objective function depends upon the underlying features of the particular clustering application. We might note that this analysis is not limited to points in the two-dimensional space; we can easily extend it to multi-dimensional space if we define inter-point distances appropriately.

Application 26. System reliability bounds [Hunter, 1976; Worsley, 1982]

All systems/products are subject to failure. Typically, the reliability of a system (as measured by the probability that it will operate) depends upon the reliability of the system's individual components as well as the manner in which these components interact; that is, the reliability of the system depends upon the both the component reliabilities and the system structure. To model these situations, let us first consider a stylized problem setting with a very simple system structure. After analyzing this situation, we will comment on how this same analysis applies to more general, and often more realistic, systems.

In our simple setting, several components $k = 1, 2, \ldots, K$ of a system can perform the same function so a system fails only if all its components fail. Suppose we model this situation as follows. Let E_k denote the event that the kth component is operable and let E_k^c denote the complementary event that the kth component fails. Then, since the systems operates if and only if at least one of the components $1, 2, \ldots, K$ operates (or, equivalently, all its components don't fail),

$$\text{Prob(system operates)} = \text{Prob}\left(\bigcup_{k=1}^{K} E_k\right) = 1 - \text{Prob}\left(\bigcap_{k=1}^{K} E_k^c\right).$$

If component failures are independent events, then $\text{Prob}(\cap_{k=1}^{K} E_k^c) = \prod_{k=1}^{K} \text{Prob}(E_k^c)$, and so we can determine the system's operating probability if we know the failure probabilities of each component. In more complex situations, however, the component failure probabilities will be dependent (for example, because all the components wear together as we use the system or because the components are subject to similar environmental conditions (e.g., dust)).

In these situations, in theory we can compute the system operating probability using the principle of inclusion/exclusion (also known as the Boole or Poincaré formula), which requires knowledge of $\text{Prob}(\cap_{i \in S} E_i)$ for all subsets S of the components $k = 1, 2, \ldots, K$. In practice, the use of this formula is limited because of the difficulty of assessing the probability of joint events of the form $E_1 \cap E_2 \cap \ldots \cap E_q$, particularly for systems that have many components. As a result, analysts frequently attempt to find bounds on the system operating probability using less information than all the joint probabilities. One approach is to use more limited information, for example, the probability of single event of the form E_i and the joint probability of only two events at a time (i.e., events of the form $E_{ij} \equiv E_i \cap E_j$). Using this information, we have the following bound on the system's operating probability.

$$\text{Prob(system operates)} = \text{Prob}\left(\bigcup_{k=1}^{K} E_k\right) \leq \sum_{k=1}^{K} \text{Prob}(E_k).$$

Can we obtain better bounds using only the probabilities $\text{Prob}(E_{ij})$ of joint events?

Figure 26a shows a situation with three events. In this familiar Venn diagram, the three squares represent the three events, the intersection of two squares the joint events E_{ij}, and the intersection of all three squares the joint event $E_1 \cap E_2 \cap$

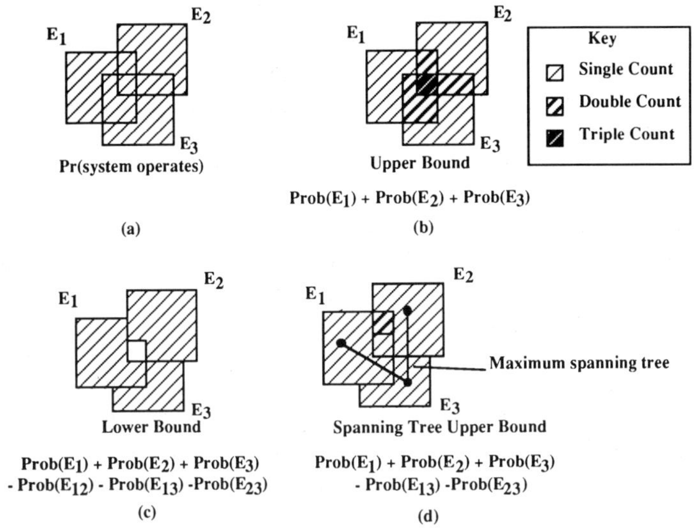

Fig. 26. Computing bounds on systems reliability.

E_3. Suppose, for convenience, that the area of any event (or joint event) equals the probability of that event. Summing the three events $\text{Prob}(E_1)$, $\text{Prob}(E_2)$, $\text{Prob}(E_3)$ gives an upper bound on the system's operating probability since the sum double counts the areas (probabilities) of the events $E_{ij} - E_1 \cap E_2 \cap E_3$ and triple counts the areas of the event $E_1 \cap E_2 \cap E_3$ (see the shading in Figure 26b). Note that the probability $\sum_{k=1}^{K} \text{Prob}(E_k) - \sum_{i<j} \text{Prob}(E_i \cap E_j)$ gives a lower bound on the system's operating probability since it by (generally) overcompensating for the double and triple counting: it eliminates the double counting of areas of all the events $E_{ij} - E_1 \cap E_2 \cap E_3$, but also eliminates all three counts of the event $E_1 \cap E_2 \cap E_3$ and so doesn't account for this event at all (see Figure 26c). Note, however, that in this case if instead of subtracting all three events of the form $\text{Prob}(E_{ij})$ from $\sum_{k=1}^{K} \text{Prob}(E_k)$, we subtract any two of them, then as shown by the shading in Figure 26d, we obtain an upper bound on the system operating probability.

We can generalize this observation for situations with more than three events E_k as follows: suppose we construct an undirected network with a node j for each event E_j. For any two events E_i and E_j, the network contains an arc (i, j) with a cost $c_{ij} = \text{Prob}(E_{ij})$. If T is any spanning tree on this network, we then obtain the *spanning tree upper bound*

$$\text{Prob}\left(\bigcup_{k=1}^{K} E_k\right) \leq \sum_{k=1}^{K} \text{Prob}(E_k) - \sum_{(i,j)\in T} \text{Prob}(E_{ij}).$$

To see that this bound is valid, let us decompose the events $E_1, E_2, \ldots, E_K$ into disjoint events $F_1, F_2, \ldots, F_Q$ satisfying the following property: no subset F_i is partially contained within a subset E_j for any i and j, that is either $F_i \subseteq E_j$ or else

$F_i \cap E_j = \emptyset$. For example, in Figure 26, the subset $E_1 \cap E_2$ is partially contained within E_3, but it can be decomposed into two disjoint subsets $F_1 = E_1 \cap E_2 \cap E_3$ and $F_2 = E_1 \cap E_2 - E_3$. Neither F_1 nor F_2 is partially contained within a subset E_j. In general, it is easy to decompose the events $E_1, E_2, \ldots, E_K$ in this manner.

Since $F_i \cap F_j = \emptyset$ whenever $i \neq j$,

$$\text{Prob(system operates)} = \\ = \text{Prob}\left(\cup_{k=1}^{K} E_k\right) = \text{Prob}\left(\cup_{q=1}^{Q} F_q\right) = \sum_{q=1}^{Q} \text{Prob}(F_q),$$

and so we wish to show that

$$\sum_{q=1}^{Q} \text{Prob}(F_q) \leq \sum_{k=1}^{K} \text{Prob}(E_k) - \sum_{(i,j)\in T} \text{Prob}(E_{ij}).$$

To see that this inequality is valid, note that each region F_i contributes $\text{Prob}(F_i)$ to the left-hand side of the inequalities. Now let us consider what it contributes to the right-hand side. Without loss of generality, renumber the events so that F_i is contained in events $E_1, \ldots, E_r$ for some $r \geq 1$. Then the region F_i contributes *exactly* $r\text{Prob}(F_i)$ to the term $\sum_{k=1}^{K} \text{Prob}(E_k)$. F_i also contributes $\text{Prob}(F_i)$ to each term $\text{Prob}(E_{jk})$ on the right-hand side when $(j,k) \in T$ with $1 \leq j, k \leq r$ since in these circumstances $F_i \subseteq E_{jk}$. However, because the set of arcs $(j,k) \in A$ with $1 \leq j, k \leq r$ forms a subgraph with only r nodes, it contains at most $r - 1$ arcs (j,k) from the tree T. We conclude that F_i contributes at most $(r-1)\text{Prob}(F_i)$ to $\sum_{(i,j)\in T} \text{Prob}(E_{ij})$, and so the contribution of F_i to the right-hand side of the inequality is at least $\text{Prob}(F_i)$.

To obtain the best spanning tree upper bound, we choose T as the maximum spanning tree with respect to the costs c_{ij}. Figure 26d shows the network for the three event example. In this case, the figure provides the upper bound for the maximum spanning tree with the arcs (1,3) and (2,3). We would obtain inferior, but valid upper bounds by choosing any other spanning tree, for example, the tree with the arcs (1,2) and (2,3). Using this approach, by assessing the joint probabilities of two events at a time, and by performing a maximum spanning tree computation, we obtain an improved upper bound on the system's operating probability.

In many contexts, the system structure is more complicated than the simple structure we have considered in which the system operates only if one of its components operates. 'Coherent binary systems' provide one such generic set of applications. One way to represent the system structure in this setting is to produce a list of 'pathsets': a pathset is a component subset S satisfying the property that if all elements in S operate then so does the system. (In the simple setting we have analyzed, each set S is a single component.) If we assume that all component states are independent, then the probability that an individual pathset operates is the product of the reliabilities of the components contained in that pathset. In more general problem settings, the components in any pathset might not be independent and so obtaining the probability that any pathset operates might be more difficult. In any event, if E_k represents the event that the kth pathset operates, then we can use our prior analysis to obtain the system reliability from the pathset reliabilities. Further development of this approach (particularly, how to define pathsets) would take us beyond the scope of our coverage here. We simple note that the analysis we have considered applies to much more general problems settings, including problem contexts broader than system reliability.

Additional applications

Some additional applications of the minimum spanning tree problem arise in the following situations: (1) *optimal message passing* [Prim, 1957]; (2) *all-pairs minimax path problem* [Hu, 1961]; (3) *solving a special case of the traveling salesman problem* [Gilmore & Gomory, 1964]; (4) *chemical physics* [Stillinger, 1967]; (5) *Lagrangian relaxation techniques* [Held & Karp, 1970]; (6) *network reliability analysis* [Van Slyke & Frank, 1972]; (7) *pattern classification* [Dude & Hart, 1973]; (8) *picture processing* [Osteen & Lin, 1974]; and (9) *network design* [Magnanti & Wong, 1984]. The survey paper of Graham & Hell [1985] provides references for additional applications of the minimum spanning tree problem.

9. Convex cost flows

In the minimum cost flow problem, we assume that the cost of the flow on any arc varies linearly with the amount of flow. Convex cost flow problems have a more general cost structure: the cost is a convex function of the flow. Flow costs vary in a convex manner in numerous problem settings including (i) power losses in a electrical network due to resistance; (ii) congestion costs in a city transportation network; and (iii) expansion costs of a communication network.

Many of the linear network flow models that we have considered in our previous discussions arise in more general forms with nonlinear costs. System congestion and queueing effects are one source of these nonlinearities (since queueing delays vary nonlinear with flows). In finance, we often are interested in not only the returns on various investments, but also in their risks, which analysts often measure by quadratic functions. In some other applications, cost functions assume different forms over different operating ranges, and so the resulting cost function is piecewise linear. For example, in production applications, the cost of satisfying customers demand is different if we meet the demand from current inventory or by backordering items. To give a flavor of the applications of convex cost network flow models, in this section, we describe three applications.

Application 27. Urban traffic flows (Magnanti, 1984; Sheffi, 1985; Florian & Hearn, 1992]

In road networks, as more vehicles use any road segment, the road become increasingly congested and so the delay on that road increases. For example, the delay on a particular road segment, as a function of the flow x on that road, might be $ax/(u-x)$. In this expression, u denotes a theoretical capacity of the road and a is another constant: as the flow increases, so does the delay; moreover, as the flow x approaches the theoretical capacity of that road segment, the delay on the link becomes arbitrarily large. In many instances, as in this example, the delay function on each road segment is a convex function of the road segment's flow and so finding the flow plan that achieves the minimum overall delay, summed over all road segments, is a convex cost network flow model.

The most general model of this form has a multicommodity flow structure (see Section 11) with a commodity, and hence a separate set of mass balance equations (1b), defined for each pair of nodes that serve as each origins and destinations for travelers in the transportation system. In this case, the congestion effects (e.g., the function of the form $ax/(u-x)$) applies to the total flow x on any arc, which is the sum of flows by all commodities. The problem is a single commodity model whenever all the trips either originate or terminate at a single node; the flow x on any arc then is the total number of trips from the common origin (or destination) that use that arc.

Another model of urban traffic flow rests upon the behavioral assumption that users of the system will travel, with respect to any prevailing system flow, from their origin to their destination using a path with the minimum delay. So if $C_{ij}(x_{ij})$ denotes the delay on arc (i, j) as a function of the arc's flow x_{ij}, each user of the system will travel along a shortest delay path with respect to the total delay cost $C_{ij}(x_{ij})$ on the arcs of that path. This problem is a complex equilibrium model because the delay that one user incurs depends upon the flow of other users, and all of the users are simultaneously choosing their shortest paths. In this problem setting, we can find the equilibrium flow by solving a convex network flow model with the objective function

$$\sum_{(i,j)\in A} \int_0^{x_{ij}} C_{ij}(y)\,\mathrm{d}y.$$

If the delay function is nondecreasing for each variable x_{ij}, then the integral within the summation is convex and, since the sum of convex functions is convex, the overall objective function is convex. Moreover, if we solve the network optimization problem defined by this objective function and the network flow constraints, the optimality conditions are exactly the shortest path conditions for the users. (See the references for the details of these claims.)

This example is a special case of a more general result, known as *variational principle*, that arises in many settings in the physical and social sciences. The variational principle says that to find an equilibrium of a system, we can solve an associated optimization problem: the optimality conditions for the problem are equivalent with the equilibrium conditions.

Application 28. Matrix balancing [Schneider & Zenios, 1990]

Statisticians, social scientists, agricultural specialists and many other practitioners frequently use experimental data to try to draw inferences about the effects of certain control parameters at their disposal (for example, the effects of using different types of fertilizers). These practitioner often use contingency tables to classify items according to several criteria, with each criteria partitioned into a finite number of categories. As an example, consider classifying r individuals in a population according to the criteria of marital status and age, assuming that we have divided these two criteria into p and q categories, respectively. As categories for marital status, we might choose single, married, separated, or divorced; and as categories for age, we might choose below 20, 21 – 25, 26 – 30, and so forth. This categorization gives a table of pq cells. Dividing the entry in each cell by the total number of items r in the population gives the probability

In many applications, it is important to estimate the current cell probabilities which continually change with time. We can estimate these cell probabilities very accurately using a census, or approximately by using a statistical sampling of the population; however, even this sampling procedure is expensive. Typically, we would calculate the cell probabilities by sampling only occasionally (for some applications, only once in several years), and at other times revise the most recent cell probabilities based on partial observations. Suppose we let a_{ij} denote the most recent cell probabilities. Suppose, further, that we know some aggregate data in each category with high precision; in particular, suppose we know the row sums and column sums. Let u_i denote the number of individuals in the ith marital category and let v_j denote the number of individuals in the jth age category. Let $r = \sum_{i=1}^{p} u_i$. We want to obtain an estimate x_{ij}'s of the current cell probabilities so that the cumulative sum of the cell probabilities for ith row equals u_i/r, the cumulative sum of the cell probabilities for jth column equals v_j/r, and the matrix x is, in a certain sense, *nearest* to the most recent cell probability matrix α. One popular measure of defining the *nearness* is to minimize the weighted cumulative squared deviation of the individual cell probabilities. With this objective, our problem reduces to the following convex cost flow problem:

$$\text{minimize} \quad \sum_{i=1}^{p} \sum_{j=1}^{q} w_{ij}(x_{ij} - \alpha_{ij})^2$$

subject to

$$\sum_{j=1}^{q} x_{ij} = \frac{u_i}{r}, \quad \text{for all } i = 1, \ldots, p,$$

$$\sum_{i=1}^{p} x_{ij} = -\frac{v_j}{r}, \quad \text{for all } j = 1, \ldots, q,$$

$$x_{ij} \geq 0, \quad \text{for all } i = 1, \ldots, p, \text{ and for all } j = 1, \ldots, q.$$

This type of *matrix balancing problem* arises in many other application settings. The inter-regional migration of people provides another important application. In the United States, using the general census of the population, taken once every ten years, the federal government produces flow matrices with detailed migration characteristics. It uses this data for a wide variety of purposes, including the allocation of federal funds to the states. Between the ten year census, net migration estimates for every region become available as by-products of annual population estimates. Using this information, the federal government updates the migration matrix so that it can reconcile the out-of-date detailed migration patterns with the more recent net figures.

Application 29. Stick percolation problem [Ahlfeld, Dembo, Mulvey & Zenios, 1987]

One method for improving the structural properties of (electrically) insulating materials is to embed sticks of high strength in the material. The current approach used in practice is to add, at random, sticks of uniform length to the insulating material. Because the sticks are generally conductive, if they form a connected path from one side of the material to the other, then the configuration will destroy the material's desired insulating properties. Material scientists call this phenomenon *percolation*. Although longer sticks offer better structural properties, they are more likely to cause percolation. Using simulation, analysts would like to know the effect of stick length on the possibility that a set of sticks causes percolation, and when percolation does occur what is the resulting heat loss due to (electrical) conduction.

Analysts use simulation in the following manner: the computer randomly places p sticks of a given length L in a square region; see Figure 27a for an example. We experience heat loss because of the flow of current through the intersection of two sticks. We assume that each such intersection has unit resistance. We can identify whether percolation occurs and determine the associated power dissipation by creating an equivalent resistive network as follows. We assign a resistor of one unit to every intersection of the sticks. We also associate a current source of one unit with one of the boundaries of the insulant and a unit sink with the opposite boundary. The problem then becomes of determining the minimum power dissipation of the resistive network.

Figure 27b depicts the transformation of the stick percolation problem into a network model. In this network model, each node represents a resistance and contributes to the power dissipation. The node splitting transformation described in Section 2 permits us to model the node resistances as arc resistances. Recall from Ohm's law that a current flow of x amperes across a resistor of r Ω creates a power dissipation of rx^2 W. Moreover, the current flows in an electrical network in a way that minimizes the rate of power dissipation (i.e., follows the path of least resistance). Consequently, we can state the stick percolator problem as the following convex cost flow problem.

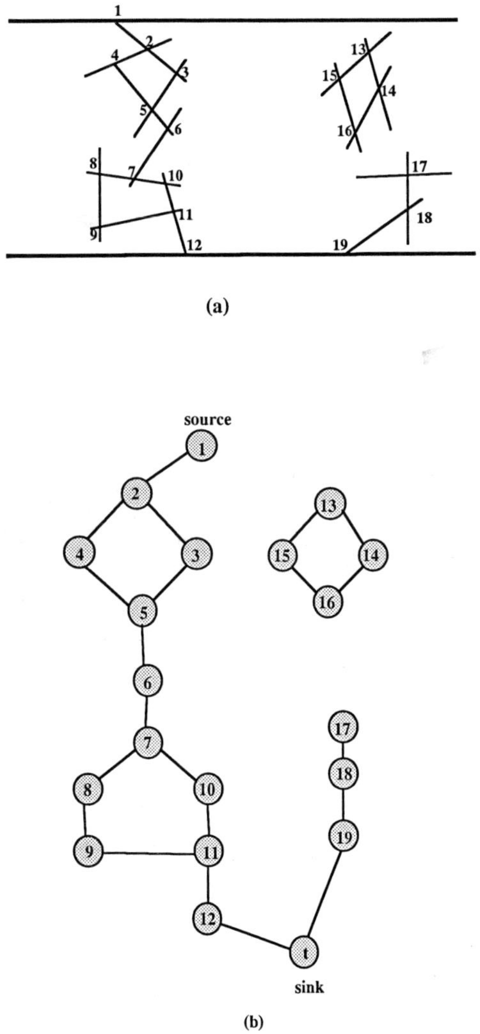

Fig. 27. Formulating the stick percolator problem. (a) Placement of sticks. (b) Corresponding network.

$$\text{minimize} \quad \sum_{(i,j)\in A} r_{ij} x_{ij}^k$$

subject to

$$\sum_{\{j:(i,j)\in A\}} x_{ij} - \sum_{\{j:(j,i)\in A\}} x_{ji} = \begin{cases} 1, & \text{for } i = s \\ 0, & \text{for all } i \in N - \{s \text{ and } t\} \\ -1, & \text{for } i = t \end{cases}$$

$$x_{ij} \geq 0, \qquad \text{for all } (i, j) \in A.$$

In this model, x_{ij} is the current flow on arc (i, j). The solution of this convex cost flow model indicates whether percolation occurs (that is, if the problem has a feasible solution), and if so, then the solution specifies the value of the associated power loss.

Additional applications

Some additional applications of the convex cost flow problem are (1) *flows in electrical networks* [Hu, 1966]; (2) *area transfers in communication networks* [Monma & Segal, 1982]; (3) *the target-assignment problem* [Manne, 1958]; (4) *solution of Laplace's equation* [Hu, 1967]; (5) *production scheduling problems* [Ratliff, 1978]; (6) *pipeline network analysis problem* [Collins, Cooper, Helgason, Kennington & Leblanc, 1978]; (7) *microdata file merging* [Barr & Turner, 1981]; and (8) *market equilibrium problems* [Barros & Weintraub, 1986]. Papers by Ali, Helgason & Kennington [1978], Dembo, Mulvey & Zenios [1989], and Schneider & Zenios [1990] provide additional references on applications of the convex cost flow problem.

10. Generalized flows

In the minimum cost flow problem, arcs conserve flows, i.e., the flow entering an arc equals the flow leaving the arc. In generalized flow problems, arcs might 'consume' or 'generate' flow. If x_{ij} units of flow enter an arc (i, j), then $m_{ij}x_{ij}$ units arrive at node j; m_{ij} is a positive *multiplier* associated with the arc. If $0 < m_{ij} < 1$, then the arc is *lossy*, and if $1 < m_{ij} < \infty$, then the arc is *gainy*. Generalized network flow problems arise in several application contexts: for example (i) power transmission through electric lines where power is lost with distance; (ii) flow of water though pipelines or canals that lose water due to seepage or evaporation; (iii) transportation of a perishable commodity; and (iv) cash management scenarios in which arcs represent investment opportunities and multipliers represent appreciation or depreciation of an investment's value.

Generalized networks can successfully model many application settings that cannot adequately be represented as minimum cost flow problems. Two common interpretations of the arc multipliers underlie many uses of generalized flows. In the first interpretation, we view the arc multipliers as modifying the amount of flow of some particular item. Using this interpretation, generalized networks model situations involving physical transformations such as evaporation, seepage, deterioration, and purification processes with various efficiencies, as well as administrative transformations such as monetary growth due to interest rates. In the second interpretation, we view the multiplication process as transforming one type of item into another. This interpretation allows us to model processes such as manufacturing, currency exchanges, and the translation of human resources into job requirements. The following applications of generalized network flows use one or both of these interpretations of the arc multipliers.

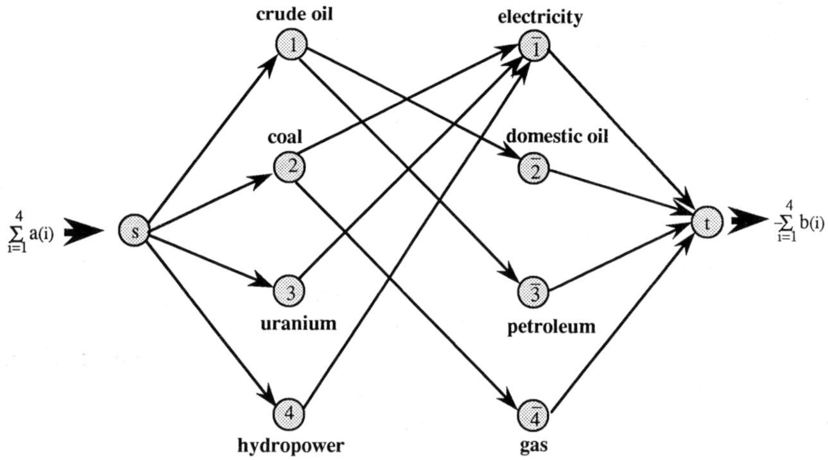

Fig. 28. Energy problem as a generalized network flow problem.

Application 30. Determining an optimal energy policy [Gondran & Minoux, 1984]

As part of their national planning effort, most countries need to decide upon an energy policy, i.e., how to utilize the available raw materials to satisfy their energy needs. Assume, for simplicity, that a particular country has four basic raw materials: crude oil, coal, uranium, and hydropower; and it has four basic energy needs: electricity, domestic oil, petroleum, and gas. The country has the technological base and infrastructure to convert each raw material into one or more energy forms. For example, it can convert crude oil into domestic oil or petrol, coal into electricity, and so forth. The available technology base specifies the efficiency and the cost of each conversion. The objective is to satisfy, at the least possible cost of energy conversion, certain annual consumption levels of various energy needs from a given annual production of raw materials.

Figure 28 shows the formulation of this problem as a generalized network flow problem. The network has three types of arcs: (i) *source arcs* (s, i) emanating from the source node s; (ii) *sink arcs* $(\bar{j}, t)$ entering the sink node t; and *conversion arcs* $(i, \bar{j})$. The source arc (s, i) has a capacity equal to the availability $a(i)$ of the raw material i and a flow multiplier of value one. The sink arc $(\bar{j}, t)$ has capacity equal to the demand $b(j)$ of type j energy need and flow multiplier of value one. Each conversion arc $(i, \bar{j})$ represents the conversion of raw material i into the energy form j; the multiplier of this arc is the efficiency of the conversion, i.e., units of energy j obtained from one unit of raw material i; the cost of the arc $(i, \bar{j})$ is the cost of this conversion.

Application 31. Machine loading [Dantzig, 1962]

Machine loading problems arise in a variety of application domains. In one of the most popular contexts, we would like to schedule the production of r products

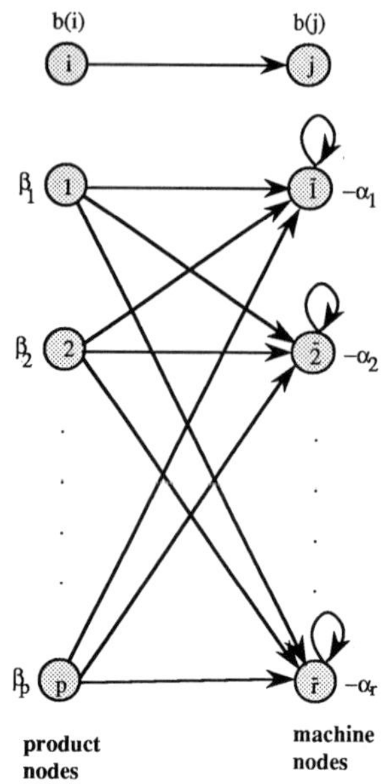

Fig. 29. Formulating a machine loading problem as a generalized network flow problem.

on p machines. Suppose that machine i is available for a_i hours and that any of the p machines can produce each product. Producing one unit of product i on machine j consumes a_{ij} hours of the machine's time and costs c_{ij} dollars. To meet the demands of the products, we must produce β_j units of product j. In the machine loading problem, we wish to determine how we should produce, at the least possible production cost, r products on p machines.

In this problem setting, products compete with each other for the use of the more efficient, faster machines; the limited availability of these machines forces us to use the less economical and slower machines to process some of the products. To achieve the optimal allocation of products to the machines, we can formulate the problem as a generalized network flow problem as shown by Figure 29. The network has p product nodes, 1, 2, ..., p, and r machine nodes, $\bar{1}$, $\bar{2}$, ..., $\bar{r}$. Product node i is connected to every machine node $\bar{j}$. The multiplier a_{ij} on arc $(i, \bar{j})$ indicates the hours of machine capacity needed to produce one unit of product i on machine j. The cost of the arc $(i, \bar{j})$ is c_{ij}. The network also has arcs $(\bar{j}, \bar{j})$ for each machine node $\bar{j}$; the multiplier of each of these arcs in 2 and the cost is zero. The purpose of these arcs is to account for the unfulfilled capacities of the machines: we can send additional flow along these arcs to generate enough flow (or items) to exactly consume the available machine capacity.

Application 32. Managing warehousing goods and funds flows [Cahn, 1948]

An entrepreneur owns a warehouse of fixed capacity H that she uses to store a price-volatile product. Knowing the price of this product over the next K time

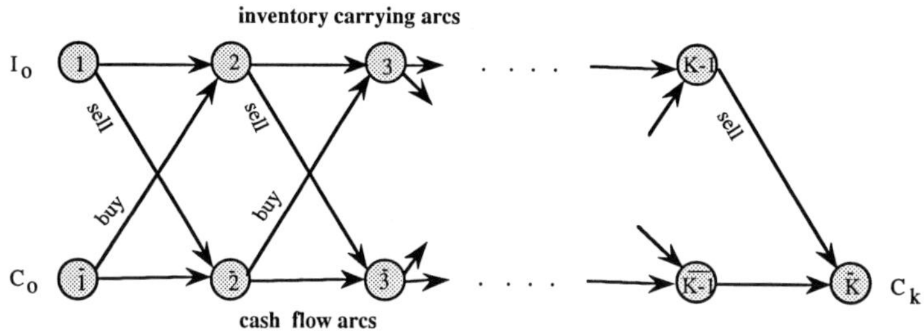

Fig. 30. Formulating the warehouse funds flow model.

periods, she needs to manage her purchases, sales, and storage patterns. Suppose that she holds I_0 units of the good and C_0 dollars as her initial assets. In each period, she can either buy more goods, or sell the goods in the warehouse to generate additional cash. The price of the product varies from period to period and ultimately all goods must be sold. The problem is to identify a buy–sell strategy that maximizes the amount of cash C_K available at the end of the Kth period.

Figure 30 gives a generalized network flow formulation of the warehousing goods and funds flow problem. This formulation has the following three types of arcs:

(i) *Inventory carrying arcs.* The cost of this type of arc is the inventory carrying cost per unit; its capacity is H. The multiplier on this arc is less than one or equal to one, depending upon whether carrying inventory incurs any loss.

(ii) *Cash flow arcs.* These arcs are uncapacitated and have zero cost. The multiplier of this type of arc is the bank interest rate for that period.

(iii) *Buy and sell arcs.* These arcs also are uncapacitated and have zero cost. If p_i is the purchase cost of the product in period i, then a buy arc has a multiplier of value $1/p_i$ and a sell arc has a multiplier of value p_i.

It is easy to establish a one-to-one correspondence between buy–sell strategies of the entrepreneur and flows in the underlying network. To enrich this model, we could easily incorporate various additional features such as policy limits on the maximum and minimum amount of cash held or goods flow in each period or introduce alternative investments such as certificates of deposits that extend beyond a single period.

Additional applications

Additional applications of the generalized network flow problem arise in (1) *resort development* [Glover & Rogozinski, 1982]; (2) *airline seat allocation problems* [Glover, Hultz, Klingman & Stutz, 1978; Dror, Trudeau & Ladany, 1988]; (3) *personnel planning* [Gorham, 1963]; (4) *a consensus ranking model* [Barzilai, Cook & Kress, 1986]; (5) *cash flow management in an insurance company* [Crum & Nye, 1981]; and (6) *land management* [Glover, Glover & Martinson, 1984]. The survey papers of Glover, Hultz, Klingman & Stutz [1978], and Glover,

Klingman & Phillips [1990] contain additional references concerning applications of generalized network flow problems.

11. Multicommodity flows

The minimum cost flow problem models the flow of a single commodity over a network. Multicommodity flow problems arise when several commodities use the same underlying network. The commodities might be differentiated either by their physical characteristics or simply by their origin–destination pairs. Different commodities have different origins and destinations, and commodities have separate mass balance constraints at each node. However, the sharing of common arc capacities binds the commodities together. In fact, the essential issue addressed by the multicommodity flow problem is the allocation of the capacity of each arc to the individual commodities in a way that minimizes overall flow costs. Multicommodity flow problems arise in many practical situations including (i) the transportation of passengers from different origins to different destinations within a city; (ii) the routing of nonhomogeneous tankers (nonhomogeneous in terms of speed, carrying capability and operating costs); (iii) the worldwide shipment of different varieties of grains (such as corn, wheat, rice, and soybeans) from countries that produce grain to those that consume it; and (iv) the transmission of messages in a communication network between different origin–destination pairs.

Multicommodity flow problems arise in a wide variety of application contexts. In this section, we consider several instances of one very general type of application as well as a racial balancing example and a vehicle fleet planning example.

Application 33. Routing of multiple commodities [Golden, 1975; Crainic, Ferland & Rousseau, 1984]

In many applications of the multicommodity flow problem, we distinguish commodities because they are different physical goods, and/or because they have different points of origin and destination; that is, either (i) several physically distinct commodities, for example different manufactured goods, share a common network, or (ii) a single physical good (e.g., messages or products) flows on a network, but the good has multiple points of origin and destination defined by different pairs of nodes in the network that need to send the good to each other. This second type of application arises frequently in problem contexts such as communication systems or distribution/transportation systems. In this section, we briefly introduce several application domains of both types.

In some of these applications, the costs sometimes are convex, as in the case of the urban traffic equilibrium (see Application 27), due to congestion effects (e.g., delays) on arc flows.

Communication networks. In a communication network, nodes represent origin and destination stations for messages, and arcs represent transmission lines.

Messages between different pairs of nodes define distinct commodities; the supply and demand for each commodity is the number of messages to be sent between the origin and destination nodes of that commodity. Each transmission line has a fixed capacity (in some applications the capacity of each arc is fixed, in others, we might be able to increase the capacity at a certain cost per unit). In this network, the problem of determining the minimum cost routing of messages is a multicommodity flow problem.

Computer networks. In a computer communication network, the nodes represent storage devices, terminals, or computer systems. The supplies and demands correspond to the data transmission rates between the computer, terminals, and storage devices, and the transmission line capacities define the bundle constraints.

Railroad transportation networks. In a rail network, nodes represent yard and junction points, and arcs represent track sections between the yards. The demand is measured by the number of cars (or, any other equivalent measure of tonnage) to be loaded on any train. Since the system incurs different costs for different goods, we divide traffic demand into different classes. Each commodity in this network corresponds to a particular class of demand between a particular origin–destination pair. The bundle capacity of each arc is the number of cars that we can load on the trains that are scheduled to be dispatched on that arc (over some period of time). The decision problem in this network is to meet the demands of cars at the minimum possible operating cost.

Distribution networks. In distribution systems planning, we wish to distribute multiple (nonhomogeneous) products from plants to retailers using a fleet of trucks or railcars, and using a variety of railheads and warehouses. The products define the commodities of the multicommodity flow problem, and the joint capacities of the plants, warehouses, railyards, and the shipping lanes define the bundle constraints. Note that this application has important bundle constraints imposed upon the nodes (plants, warehouses) as well as the arcs.

Foodgrain export–import network. The nodes in this network correspond to geographically dispersed locations in different countries, and the arcs correspond to shipments by rail, truck, and ocean freighter. Between these locations, the commodities are various foodgrains such as corn, wheat, rice and soybean. The capacities at the ports define the bundle constraints.

Application 34. Racial balancing of schools [Clarke & Surkis, 1968]

In 1968, the Supreme Court of the United States ruled that all school systems in the country should begin admitting students to schools on a nondiscriminatory basis, and should employ faster techniques to promote desegregated schools across the nation. This decision made it necessary for many school systems to develop radically different procedures for assigning students to schools. Since the

Supreme Court did not specify what constitutes an acceptable racial balance, the individual school boards used their own best judgments to arrive at acceptable criteria upon which to base their desegregation plans. This application describes a multicommodity flow model for determining an optimal assignment of students to schools that minimizes the total distance travelled by the students, given a specification of lower and upper limits on the required racial balance in each school.

Suppose that a school district has S schools and the capacity of the jth school is u_j. For the purpose of this formulation, we divide the school district into L population centers. These locations might, for example, be census tracts, bus stops, or city blocks. The only restriction on the population centers is that they be finite in number, and that a single distance measure reasonably approximates the distance any student at center i must travel if he or she is assigned to school j. Let S_{ik} denote the available number of students of the kth ethnic group at the ith population center. The objective is to assign students to schools in a way that achieves the desired ethnic composition for each school and minimizes the total distance traveled by the students. Each school j has the ethnic requirement that it must enroll at least l_{jk} and no more than u_{jk} students from the kth ethnic group.

We can model this problem as a multicommodity flow problem on an appropriately defined network. Figure 31 shows this network representation for a problem with three population centers and three schools. This network has one node for each population center and for each school as well as a 'source' and a 'sink' node for each ethnic group. The flow commodities represent the students of different ethnic groups. The students of the kth ethnic group flow from the source a_k to the sink e_k via population center and school nodes. We set the upper bound on arc (a_k, b_i) connecting the kth ethnic group source node and the ith population center equal to s_{ik} and the cost of the arc (b_i, c_j) connecting the ith population

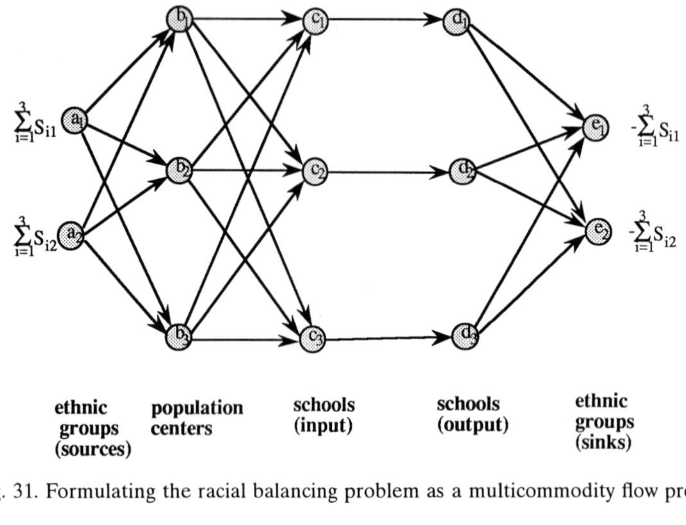

Fig. 31. Formulating the racial balancing problem as a multicommodity flow problem.

center and jth school equal to f_{ij}, the distance between that population center and that school. By setting the capacity of the arc (c_j, d_j) equal to u_j, we ensure that the total number of students (of all ethnic groups) allocated to the jth school do not exceed the maximum student population for this school. The students of all ethnic groups must share the capacity of each school. Finally, we incorporate constraints on the ethnic compositions of the schools by setting the lower and upper bounds on the arc (d_j, e_k) equal to l_{jk} and u_{jk}. It is fairly easy to verify that the multicommodity flow problem models the racial balancing problem and so a minimum multicommodity flow will specify an optimal assignment of students to the schools.

Application 35. Multivehicle tanker scheduling [Bellmore, Bennington & Lubore, 1971]

Suppose we wish to determine the optimal routing of fuel oil tankers required to achieve a prescribed schedule of deliveries: each delivery is a shipment of some commodity from a point of supply to a point of demand with a given delivery date. In the simplest form, this problem considers a single product (e.g., aviation gasoline or crude oil) to be delivered by a single type of tanker. As shown by Application 10, it is possible to determine the minimum tanker fleet to meet the delivery schedule for this simple version of the problem by solving a maximum flow problem. The multivehicle tanker scheduling problem, studied in this application, considers the scheduling and routing of a fixed fleet of nonhomogeneous tankers to meet a prespecified set of shipments of multiple products. The tankers differ in their speeds, carrying capabilities, and operating costs.

To formulate the multivehicle tanker scheduling problem as a multicommodity flow problem, we let the different commodities correspond to different tanker types. The network corresponding to the multivehicle tanker scheduling problem is similar to that of the single vehicle type, shown in Figure 32, except that each distinct type of tanker originates at a unique source node s^k. This network has four types of arcs (see Figure 32 for a partial example with two tankers types): in-service arcs, out-of-service arcs, delivery arcs, and return arcs. An in-service arc

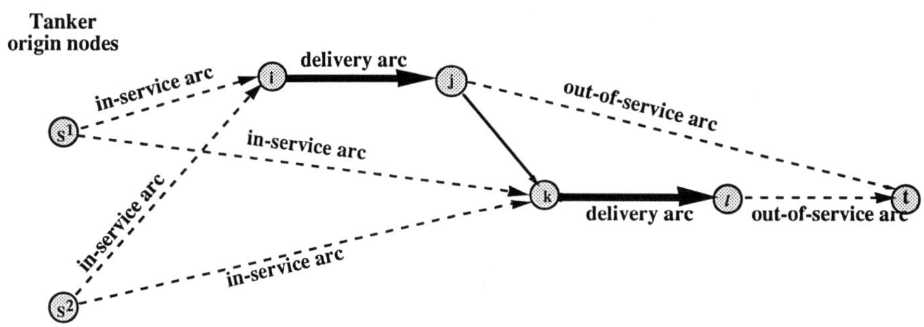

Fig. 32. Multivehicle tanker scheduling problem as a multicommodity flow problem.

corresponds to the initial employment of a tanker type; the cost of this arc is the cost of deploying the tanker at the origin of the shipment. Similarly, an out-of-service arc corresponds to the removal of the tanker from service. A delivery arc (i, j) represents a shipment from origin i to destination j; the cost c_{ij}^k of this arc is the operating cost of carrying the shipment by a tanker of type k. A return arc (j, k) denotes the movement ('backhaul') of an empty tanker, with an appropriate cost, between two consecutive shipments (i, j) and (k, l).

Each arc in the network has a capacity of one. The shipment arcs have a bundle capacity ensuring that at most one tanker type services that arc. Each shipment arc also has a lower flow bound of one unit, which ensures that the chosen schedule does indeed deliver the shipment. Some arcs might also have commodity-based capacities u_{ij}^k. For instance, if tanker type 2 is not capable of handling the shipment on arc (i, j), then we set $u_{ij}^2 = 0$. Moreover, if tanker type 2 can use the return arc (j, k), but the tanker type 1 cannot (because it is too slow to make the connection between shipments), then we set $u_{jk}^1 = 0$.

Airline scheduling is another important application domain for this type of model. In this problem context, the vehicles are different type of airplanes in an airline's fleet (for example, Boeing 727s or 747s or McDonald Douglas DC 10s). The delivery arcs in this problem context are the flight legs that the airline wishes to cover.

We might note that in this formulation of the multivehicle tanker scheduling problem, we are interested in integer solutions of the multicommodity flow problem. The solutions obtained by the multicommodity flow algorithms [see Kennington & Helgason, 1980; Ahuja, Magnanti & Orlin, 1993] need not be integral. Nevertheless, the fractional solution might be useful in several ways. For example, we might be able to convert the non-integral solution into a (possibly suboptimal) integral solution by minor tinkering or we might use the non-integral solution as a bound in solving the integer-valued problem by a branch and bound enumeration procedure.

Additional applications

Additional applications of the multicommodity flow problem include: (1) *warehousing of seasonal products* [Jewell, 1957]; (2) *optimal deployment of resources* [Kaplan, 1973]; (3) *multiproduct multistage production-inventory planning* [Evans, 1977]; (4) *multicommodity distribution system design* [Geoffrion & Graves, 1974]; (5) *rail freight planning* [Bodin, Golden, Assad & Ball, 1983; and Assad, 1980]; and (6) *VLSI chip design* [Korte, 1988].

12. The traveling salesman problem

The traveling salesman problem (TSP) is perhaps the most well known problem in the field of network optimization, or in fact in all of operations research. The problem is deceptively easy to state: We are given n cities denoted $1, 2, 3, \ldots, n$, as well as an inter city distance c_{ij} between each pair i, j of cities. The problem is to determine a tour (cycle) that passes through each city exactly once and minimizes the total distance traveled. In general this problem is known to be difficult to solve. In the parlance of computational complexity theory, it is NP-hard.

The simplicity of the problem belies the difficulty of solving it to optimality. The problem has attracted an immense amount of attention from researchers in combinatorial optimization, who have written hundreds of papers on this particular topic. In fact, the classic book on this problem, *The Traveling Salesman Problem: A Guided Tour of Combinatorial Optimization*, lists over 400 references.

There are a number of ways of mathematically describing the traveling salesman problems. We describe them in terms of permutations.

Let $N = \{1, 2, \ldots, n\}$. We represent a tour by a function t, letting $t(i)$ denotes the ith city visited on the tour. Let T denote the set of all tours. Using this notation, we can describe the traveling salesman problem as follows:

$$\begin{aligned} &\text{minimize} \quad c_{t(n)t(1)} + \sum_{i=1}^{n-1} c_{t(i)t(i+1)} \\ &\text{subject to} \quad t \in T. \end{aligned}$$

In some problem instances, the tour need not return to its starting city. In this case, the tour is a path, often referred to as a *Hamiltonian path*, and we refer to the problem as the *minimum cost Hamiltonian path problem*, or more briefly as the *Hamiltonian path problem*. To formulate the Hamiltonian path problem mathematically, we can use of slight variant of the previous TSP formulation:

$$\begin{aligned} &\text{minimize} \quad \sum_{i=1}^{n-1} c_{t(i)t(i+1)} \\ &\text{subject to} \quad t \in T. \end{aligned}$$

We might note that the Hamiltonian path problem is easily transformed into a traveling salesman problem. We simply add a dummy node whose distance from all other nodes is some constant, say 1.

Matrix representations of traveling salesman problems

To describe two of our applications, we use a particularly useful representation of the traveling salesman problem that we next describe.

Suppose that u and v are each vectors with m 0 or 1 components. The *hamming distance* from u to v, which we denote as $d(u, v)$, is the number of components in which u differs from v. For example, if $u = 1\,0\,0\,0\,1\,1\,0$ and $v = 1\,0\,0\,1\,1\,0\,0$, then $d(u, v) = 2$.

We say that a traveling salesman problem (or the Hamiltonian path problem) is matrix representable if each city i has an associated 0–1 vector A_i and $c_{ij} = d(A_i, A_j)$. Any traveling salesman problem with integral distances can be matrix represented if we permit ourselves to add a constant M to each c_{ij} prior to representing it (we will not establish this fact). Suppose that a traveling salesman problem is matrix represented. Can we interpret the optimal tour in terms of the representing vectors? To illustrate how to do so, consider the following representing vectors.

$$\begin{aligned} A_1 &= 1\,1\,1\,0 \\ A_2 &= 0\,1\,0\,0 \\ A_3 &= 1\,1\,0\,0 \\ A_4 &= 0\,0\,0\,1 \\ A_5 &= 0\,0\,0\,1 \\ A_6 &= 1\,0\,1\,1 \\ A_7 &= 1\,1\,1\,1 \\ A_8 &= 0\,0\,0\,0 \end{aligned}$$

In this case, $c_{12} = c_{21} = d(A_1, A_2) = 2$; $c_{13} = c_{31} = d(A_1, A_3) = 1$; and so forth.

In this context, the rows of the matrix correspond to the cities and a tour is a permutation of the rows. That is, $t(i)$ denote the ith row 'visited' by the tour. For example, we can represent the tour 1–2–3–7–6–5–4–8–1 by the following permuted version of the matrix:

$$\begin{aligned} A_1 &= 1\,1\,1\,0 \\ A_2 &= 0\,1\,0\,0 \\ A_3 &= 1\,1\,0\,0 \\ A_7 &= 1\,1\,1\,1 \\ A_6 &= 1\,0\,1\,1 \\ A_5 &= 0\,0\,0\,1 \\ A_4 &= 0\,0\,0\,1 \\ A_8 &= 0\,0\,0\,0 \end{aligned}$$

This matrix has a row of all 0's, which we may assume without loss of generality to be the last city of the tour. Note that column 1 contributes a distance of 4 to the Hamiltonian path since its has different elements in rows 1 and 2, rows 2 and 3, rows 5 and 6, and rows components 8 and 1. In general, if the last row of the matrix consists of all 0's, each block of consecutive ones in a column contributes 2 to the tour length. Therefore, the traveling salesman distance is twice the number of consecutive blocks of ones in the columns.

Put another way, in this case the traveling salesman problem is equivalent to the problem of determining how to permute the rows of a matrix so that the columns of the resulting matrix has the fewest possible blocks of consecutive ones.

Application 36. Manufacturing of printed circuit boards [Magirou, 1986]

In the manufacturing of printed circuit boards (PCB's), lasers need to drill hundreds of small holes at pin locations. Subsequently, a manufacturer will solder electronic components to the PCB's at these points. To expedite the production of each chip, the laser must move efficiently from one drilling site to the next, ultimately visiting each drilling site along its 'tour.' This manufacturing problem can be solved as a traveling salesman problem. More precisely, we can model the problem as a Hamiltonian path problem: the drill must visit each node at least once, but the first and last nodes need not be the same.

In practice, we would not list the between site transit times for each pair of drilling sites, but would instead list the coordinates of each drilling site. Typically, the distance between sites is the maximum of the distance from site i to site j in the x coordinate and the distance from site i to site j in the y coordinate. This distance models two independent motors, one working in the x direction and one working in the y direction.

This problem also arises in the manufacturing of VLSI circuits. The primary difference computationally is one of size. Bentley [1990] writes 'VLSI-designs in the near future could employ the solution of 10^7-city TSPs.'

Application 37. Identifying time periods for archeological finds [Shuchat, 1984]

One part of an archeologist's work is to assign dates to archeological finds or deposits. Consider a collection of deposits from m different excavation sites containing artifacts (e.g., pottery, jewelry), sorted into $n - 1$ different types. We assume that each artifact type is associated with an (unknown) time period during which it was in use, and that each excavation site contains artifacts from consecutive (or nearly consecutive) time periods.

Associated with type i is a vector A_i, whose jth component a_{ij} is one if excavation site j contains an artifact of type i and is zero otherwise. If the artifact types were in their true order, then the ones of each column of the corresponding matrix would be in consecutive (or nearly consecutive) time positions. As in the matrix-represented TSP discussion, we introduce an artificial type n with an associated vector A_n of all 0's. Without loss of generality, we assume that n will be the last node of the tour. If t denotes a traveling salesman tour, the interpretation is that the oldest artifact is $t(1)$, the second oldest is $t(2)$, and so forth. A gap between two blocks of consecutive ones corresponds to a time period omitted from an excavation site between two time periods that are represented at the site.

By solving the traveling salesman problem, we are assigning periods to types and thus implicitly permuting the rows of A so as to minimize the number of blocks of consecutive ones in the resulting matrix. Equivalently, we are minimizing the number of time period gaps at the excavation sites.

Application 38. Assembling physical mapping in genetics [Alizadeh, Karp, Newberg & Weisser, 1992; Arratia, Lander, Tivaré & Waterman, 1991]

The human genome consists of 23 pairs of chromosomes. Each chromosome contains a single very long molecule of DNA, which is constructed as a linear arrangement of 4 distinct nucleic acids, each represented by a letter A, C, T or G. Thus a chromosome can be thought of as a long sequence of letters such as ACTGACCTGGATTCG... Biologists estimate that in total all the human chromosomes contain 3 billion letters. As a pictorial image, if printed in a single line of type, a line containing all of the letters would be over 6,000 miles long.

DNA has often been described as the 'language of genetics' and the chromosomes might be thought of as a 'blueprint' for the human body. In order to decode the information stored in the chromosomes, the United States government has funded The Human Genome Project, whose goal is to determine the sequence of genes in each chromosome and ultimately to determine the sequence of nucleic acids that makes up the DNA of each chromosome. An interim goal of the Human Genome Project is to create rough maps of the chromosomes known as 'physical maps.' Here we show how to construct a reasonably accurate physical map by solving a related traveling salesman problem.

A clone is a fragment of DNA, stored in such a way that copies are easy to make. A common method of storing extremely large clones (large in terms of the amount of DNA of the clone) is to store them inside a yeast cell. When the yeast cell duplicates, the clone also duplicates. By controlling the replication of the yeast, biologist can create as many copies of a clone as they want.

Biologists often refer to a collection of clones from the same DNA source as a library. Suppose for example that we have a library of chromosome number 1 from humans. Such a library is a very useful source of genetic material for biologists studying chromosome 1 (and perhaps looking for some genetic disease); unfortunately, in creating a library from a genetic source such as a human chromosome, we lose all positional information. The biologist has no simple way of determining where each clone lies on the chromosome. Rather, through experimental means, the scientists infer where clones lie; a physical map is a plausible (and hopefully accurate) layout of the clones on the chromosome.

A *probe* may be thought of as a unique point on the genome. Biologists can select probes randomly and they can experimentally determine whether a clone contains a probe. Let $a_{ij} = 1$ if clone j contains probe i, $a_{ij} = 0$ otherwise. As before, we insert a dummy probe n with $a_{nj} = 0$ for all j, that is, this 'artificial' probe does not occur in any clone. Assuming that probes occur uniquely within the genome, and assuming that experimental data is error free, then any two clones that have a common probe must overlap in the genome. Moreover, if the probes are arranged in the correct linear order on the chromosome, then the collection of probes contained by a given clone will occur in consecutive positions. Determining the order of the probes on the genome will simultaneously order the clones as well. Such an ordering of either the clones or the probes is referred to as a physical map. Figure 33 gives an example of some probe-clone data and a physical map.

Reordering the probes so that the probes within each clone are consecutive is equivalent to ordering the rows of A so that the 1's in each column are consecutive. This matrix ordering problem is known as 'determining whether A has the consecutive 1's property in columns' and is efficiently solvable.

Unfortunately, in practice, the data is not so perfect. Many clones contain not one but two fragments of DNA that were accidentally 'glued' together in the

Probes \ **Clones**	1	2	3	4	5	6
A	+					
B		+			+	+
C	+		+	+		
D		+	+	+		
E		+	+			+
F	+			+		
G					+	+

(a)

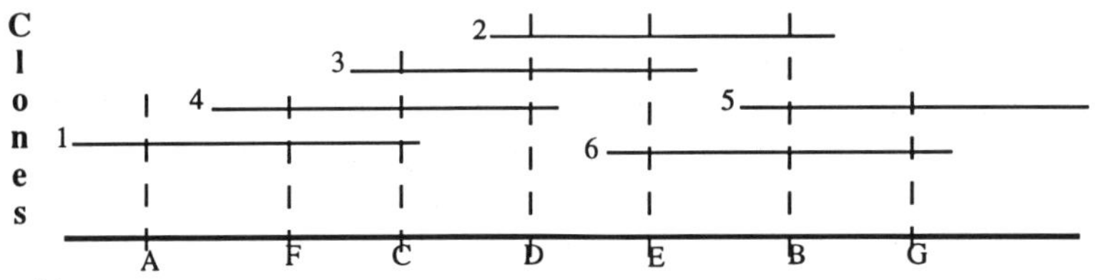

(b)

Fig. 33. (a) A '+' in entry (i, j) indicates that clone j contains probe i. (b) A physical map based on the data from (a).

cloning process. Two-fragment clones are referred to as chimeras. In addition, the data is subject to experimental error. Quite frequently an experiment falsely reports that a probe is not contained within a clone. (Error rates of 10% or higher are common.) Occasionally, an experiment also falsely reports that a probe is contained within a clone. To account for the problems with cloning and with experimental errors, an alternative objective is to order the rows of A in a way that minimizes the number of blocks of consecutive 1's, summed over all the columns. This is the matrix-represented traveling salesman problem in which the vector associated with the ith probe is A_i. In the case of perfect data, this problem reduces to the previous problem of identifying consecutive ones since each column will have exactly one block of consecutive ones.

Application 39. Optimal vane placement in turbine engines [Plante, Lowe & Chandresekaran, 1987]

A nozzle guide vane is a critical component of a gas turbine engine used in both military and commercial aircraft. The nozzle guide vane accelerates, deflects,

Fig. 34. Optimal vane placement in turbine engines. (a) Nozzle guide vane. (b) Nozzle guide assembly.

and distributes the gases that drive the turbine rotor. The nozzle assembly (see Figure 34) consists of a number of vanes (generally 46 to 100) placed around the circumference of the nozzle guide vane. Vane i has two associated parameters A_i and B_i that can be effectively measured. If vane j is placed immediately clockwise in relation to vane i, then the area between these two vanes is estimated to be $A_i + B_j$. Ideally, all of these areas should be approximately the same value, say d. A good choice in practice for ordering the vanes is a tour t that minimizes

$$(d - A_{t(n)} - B_{t(1)})^2 + \sum_{i=1}^{n-1}(d - A_{t(i)} - B_{t(i+1)})^2.$$

This order places the vanes so as to minimize the squared deviations from the target value d.

This problem is a special case of the traveling salesman problem, with $c_{ij} = (d - A_i - B_j)^2$. Observe that minimizing the displayed quantities is equivalent to minimizing $A_{t(n)}B_{t(1)} + \sum A_{t(i)}B_{t(i+1)}$ since the other terms in the expansion, $nd^2 - nd\sum A_{t(i)} - n\sum B_{t(i)} + \sum(A_{t(i)})^2 + \sum(B_{t(i)})^2$, are independent of the permutation. In this instance, the traveling salesman problem has the additional property that $c_{ij} = A_iB_j$. This special case of the traveling salesman problem is known as the product form, and it too is NP-hard.

Additional applications

The traveling salesman problem has also been applied to *cutting wallpaper* [Garfinkel, 1977], *job sequencing* [Gilmore & Gomory, 1964], *computer wiring* [Lenstra & Rinnooy Kan, 1975], *reading records from a rotating storage device* [Fuller, 1972], *clustering and aggregation* [McCormick, Schweitzer & White, 1972; Lenstra & Rinnooy Kan, 1975], *order picking in a warehouse* [Ratliff & Rosenthal, 1983].

13. Network design

In traditional network flow models (shortest paths, minimum cost and maximum flows, generalized flows, and multicommodity flows), we are given a network $G = (N, A)$ and we wish to route flow on this network in order to fulfill some underlying flow requirements. In network design, we are given a node set N and a set $\overline{A}$ of candidate arcs, and we wish to simultaneously define the underlying network $G = (N, A)$ with $A \subseteq \overline{A}$, and determine an optimal flow on it. In one standard version of this problem, we incur a fixed cost F_{ij} for including any arc (i, j) from A in the network.

The following model is an optimization formulation of the network design problem:

$$\text{minimize} \quad \sum_{k=1}^{K} \sum_{(i,j)\in\overline{A}} c_{ij}^k f_{ij}^k + \sum_{(i,j)\in\overline{A}} F_{ij} y_{ij} \tag{10a}$$

subject to

$$\sum_{j\in N} f_{ij}^k - \sum_{j\in N} f_{ji}^k = \begin{cases} 1, & \text{if } i = O(k) \\ -1, & \text{if } i = D(k) \\ 0, & \text{otherwise} \end{cases} \quad \text{for all } 1 \le k \le K \tag{10b}$$

$$\sum_{k=1}^{K} r_k f_{ij}^k \le u_{ij} y_{ij} \quad \text{for all } (i, j) \in \overline{A} \tag{10c}$$

$$f_{ij}^k \le y_{ij} \quad \text{for all } (i, j) \in \overline{A} \text{ and all } 1 \le k \le K \tag{10d}$$

$$f_{ij}^k \ge 0 \quad \text{for all } (i, j) \in \overline{A} \text{ and all } 1 \le k \le K \tag{10e}$$

$$y \in Y \tag{10f}$$

$$y_{ij} \ge 0 \quad \text{and integer for all arcs } (i, j). \tag{10g}$$

In this multicommodity flow version of the problem, each commodity $k = 1, 2, \ldots, K$, has an origin node $O(k)$, a destination node $D(k)$, and a flow requirement r_k. f_{ij}^k is the fraction of commodity k's flow requirement that flows on arc (i, j), and y_{ij} is an integer variable indicating how many copies of arc (i, j) we install on the network, each copy providing u_{ij} units of capacity. Another popular model, a 0–1 model, imposes the restriction $y_{ij} \le 1$ stating that we can add each arc at most once. Constraints (10b) are mass balance constraints. Constraints (10c) state that the total flow on arc (i, j) cannot exceed its installed capacity. Constraint (10d) is a forcing constraint stating that we can't flow on arc (i, j) if we don't install it, e.g., if $y_{ij} = 0$; this constraint is redundant in the integer programming version of this model, but not in its linear programming relaxation and so it is useful to include it in the model. Constraint (10f) permits us to impose restrictions on the design variables y_{ij}, for example, degree constraints on the nodes. F_{ij} is the fixed cost for installing any copy of arc (i, j) and c_{ij}^k is the cost for routing all of the flow requirement of commodity k on arc (i, j).

This model in quite general. If we choose the data appropriately, it includes the following problems as special cases:

(1) *The minimum spanning tree problem*: If d_{ij} are the minimum spanning tree costs, then we set each $c_{ij}^k = 0$ and each $F_{ij} = F_{ji} = d_{ij}$ (note that in the network design model, the arcs are directed, so we replace each arc in the minimum spanning tree graph with two directed arcs, each with the same fixed cost); $u_{ij} > \sum_{k=1}^{K} r_k$, each $O(k)$ is the same node (say, node 1), Y states that the design has $|N| - 1$ arcs, and $r_k = 1$ for each node $k \neq 1$. Note that if each $d_{ij} > 0$, then we can eliminate constraints (10f) entirely and every optimal solution to the network design problem will be a spanning tree.

(2) *The traveling salesman problem*: If d_{ij} are the TSP costs, then we set each $c_{ij}^k = 0$ and each $F_{ij} = d_{ij}$; Y states that the design has $|N|$ arcs, each pair of nodes defines a commodity and each $r_k = 1$. Alternatively, if $F_{ij} = d_{ij} + M$ (M is a large constant), then we can eliminate the constraints Y and every optimal solution will be an optimal TSP tour.

(3) *Facility location models*: In this case, we use the familiar node splitting device of network flows to convert the decision as to whether we place a facility on a node into a decision about placing a facility on an arc.

We next describe several network design applications.

Application 40. Designing fixed cost communication and transportation systems

The primary purpose of a communication (transportation) system, such as telephone or computer networks, is to provide communication (transportation) links between the system's users. In many of these application contexts, routing costs are zero or negligible. Therefore, once we have constructed the system, we can send as many messages (trucks, passengers) as we like (within the system's operating capacity) at no cost. There are many variants on this general theme: for example, models with or without capacity restrictions, models with or without reliability considerations, and models that impose special topological restrictions (e.g., node degree constraints) on the system's design. The following generic models illustrate a few of these features. In each case, the only costs in the model are the fixed costs imposed upon the arcs.

Minimum spanning tree. If we need to connect all the nodes of the network, then we need to find a minimum spanning tree on the given network. As we have previously indicated, this model is a special case of the network design problem.

Steiner tree [Maculan, 1987; Winter, 1987]. If we need only a subset of the nodes of the network (customer nodes), then we need to find a minimum cost spanning tree that includes these nodes as well perhaps as some or all of the other nodes (called *Steiner nodes*). This generalization of the minimum spanning tree is known as the *Steiner tree problem*. The formulation of this model as a special case of the network design model is similar to that of the minimum spanning tree except that now we assume node 1 is one of the customer nodes and we define commodities only for the customer nodes $k \neq 1$; for each such node k, we set $O(k) = 1$, $D(k) = k$ and $r_k = 1$.

Private network leasing [Magnanti, Mirchandani & Vachani, 1991]. In this instance, one or more types of facilities (communication lines) are available to carry flow on each arc: for example, DS0 and DS1 lines (digital service type 0 and 1 lines). DS1 lines cost more than DS0 lines, but have 24 times the flow carrying capacity. Given demands between the users of the system, we need to find a minimum cost network configuration that has sufficient capacity to carry the flow. Therefore, we need to specify how many units of each type of transmission line to load on each arc of the network and how to route the flow through the resulting network. This model is a variant of the network design model, with zero routing costs for all arcs and with two different types of parallel arcs joining each arc in A, a DSO arc and a DS1 arc. If we load x_{ij} units of DS0 lines and y_{ij} units of DS1 lines between nodes i and j, then one of these parallel arcs has a capacity x_{ij} and the other a capacity of $24y_{ij}$. (Therefore, the capacities on the righthand side of the constraints (10c) will be $u_{ij} = 1$ and $u_{ij} = 24$.)

In a freight transportation system, the arcs correspond to different types of trucks, for example 24 and 36 footers, each with a different capacity and cost, that we can dispatch on the arc.

Network survivability [Monma & Shallcross, 1989; Groetschel, Monma & Stoer, 1992]. Almost all networks are subject to failure, in the sense that periodically the nodes and/or arcs fail and therefore become unavailable to carry flow. With the advent of new high capacity transmission lines, for example, fiber optic lines, the failure of a single line can lead to enormous service disruptions (major 'brown outs'). To deal with these contingencies, network planners sometimes need to build redundancy into a network's deign. That is, rather than designing a spanning tree with a single path between each node, we might impose a redundancy or survivability requirement by specifying the need to include at least r_{ij} arc-disjoint paths (if the arcs are subject to failure) between nodes i and j. For this application, each pair of nodes i and j with $r_{ij} > 0$ defines a commodity. In this version of the 0–1 network design model, the routing costs are again zero, the capacity u_{ij} for each arc (i, j) is one, and each variable is a 0–1 variable.

Application 41. Local access telephone network capacity expansion [Balakrishnan, Magnanti & Wong, 1991; Balakrishnan, Magnanti, Shulman & Wong, 1991]

As the demand for services (voice, data, video) increases, telephone companies must determine the optimal investment policy for adding capacity to their networks. The local access component of the overall telephone system is the part of the service network that connects end subscribers to local switching centers and to the long distance, or backbone, network. Typically, the local access network uses

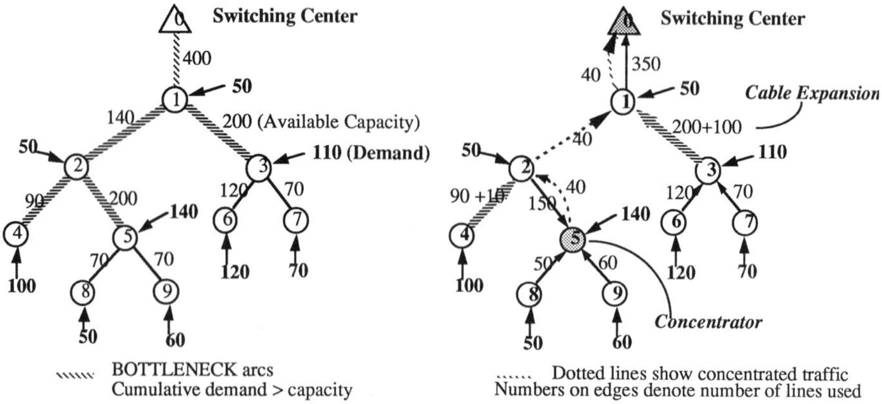

Fig. 35. Capacity expansion example. (a) Problem data; (b) expansion plan.

copper cables to carry messages and network has a tree structure as shown in Figure 35a. In this example, the current capacity of links (0,1), (1,2), (1,3), (2,4) and (2,5) are insufficient to meet future demand.

Telephone companies have several options for adding capacity. They can (i) add more copper cable to any link of the network, (ii) add fiber optics links to the network, and (iii) add remote switches and message concentrators to the nodes of the network. Figure 35b shows a possible expansion plan for this problem (without the fiber optic cable option). This expansion loads the network with additional capacity on links (1,3) and (2,4) and adds a concentrator (with a compression ratio of 10) at node 5 which handles all the traffic from nodes 2, 5, 8, and 9. The concentrator makes it possible to compress messages so that they use less capacity on all arcs that lie 'upstream' towards the local switch.

In one popular practical version of this problem, (i) all the traffic from any subscriber node i must home to (i.e., be routed to) the same node j which must contain a concentrator or the local switch, and (ii) the traffic pattern must satisfy a *contiguity property* stating that if node i homes to node j and node k lies on the tree path P_{ij} between nodes i and j, then node k must also home to node j. Therefore, any feasible solution is a 'packing' of subtrees in the given tree T; that is, is a collection of subtrees $T_1, T_2, \ldots, T_q$ of T that are node-disjoint and that collectively contain all the nodes of T. In Figure 35b, the tree T_1 contains the nodes 2, 4, 5, 8, and 9 and the tree T_2 contains the nodes 1, 3, 6, and 7. We assume that we either connect any concentrator directly to the switching center or that the compression ratio of any concentrator is so high that we can ignore the capacity used by any compressed traffic. Therefore, the cost of any feasible solution is just the sum of costs of the subtrees $T_1, T_2, \ldots, T_q$. Then the cost of the subtree is (i) the fixed cost F_j of placing a concentrator at some node j in the subtree, (ii) the cost a_{ij} for assigning any node i in the tree to node j (this is the variable cost of adding more capacity to a concentrator), and (iii) the cost, if any, of expanding any arc of the network beyond its current capacity

B_{ij} so that it has sufficient capacity to carry the required flow in the subtree to node j.

Letting node 0 be the location of the switching center in the given local access tree, we can view this problem is a variant of the network design model (10) if our underlying network has three type of arcs: (i) an arc for each arc (i, j) in the tree with $u_{ij} = B_{ij}$ and with no associated fixed or routing costs, (ii) a parallel arc to (i, j) with the fixed cost of cable expansion on that arc and a routing cost equal to the variable cable expansion cost, and (iii) an arc $(j, 0)$ for each node j in T with fixed and routing costs equal to the fixed and variable costs for installing a concentrator at that node.

In the special instance of this problem without any existing capacity on any arc, the problem is easy to solve by dynamic programming [see Aghezzaf, Magnati & Wolsey, 1992; Barany, Edmonds & Wolsey, 1986]. In this case, we can include the arc expansion cost for assigning node i to node j (that is, the cost of expanding all the arcs on the path P_{ij}) in the assignment cost a_{ij}, and so the problem has only the costs (i) and (ii). In the general case, the model will contain added complexity (this version of the problem is NP-hard), but can be solved fairly efficiently by a Lagrangian relaxation decomposition method that solves the version of the problem without any existing arc capacities as a subproblem. This modeling and solution approach extends to more general situations as well – for example, a version of the problem with alternate types of concentrators, each with its own fixed and variable costs.

Application 42. Multi-item production planning [Leung, Magnati & Vachani, 1989; Pochet & Wolsey, 1991]

In Application 3, we showed how to convert certain single-item production planning problems into shortest path and minimum cost flow problems. In more general problem settings with production capacities or with multiple items that share scarce resources (e.g., machine capacity), the problems often become network design models. Even the special case that we examined in Application 3 is a special network design problem with the following interpretation (see Figure 2): (i) the production arcs $(0, t)$ for $t = 1, 2, \ldots, T$ incur a fixed cost F_t; (ii) we can define a commodity t for each time period t with $O(t) = 1$, $D(t) = t$, and $r_t = d_t$; (iii) the routing cost of the arcs $(0, t)$ equals the variable production cost c_t; (iv) the arcs $(t, t + 1)$ and $(t + 1, t)$ have zero fixed costs and have a routing cost for commodity k equal to the one period inventory holding cost and backorder cost for the entire demand d_k of that item. For the uncapacitated version of the problem that we considered in Application 3, $u_{ij} > \sum_{t=1}^{T} d_t$ and so we can eliminate the constraints (10c). We can also eliminate the constraints (10f).

When we impose production capacities, we include constraint (10c) for each arc $(0, t)$ with u_{0t} equal to the production capacity in period t (often $u_{0t} = U$, a constant capacity throughout the planning horizon).

For multi-item version of the problem with Q items, we can define a 'multi-layered' version of the network in Figure 3, with a demand node qt for each item

$q = 1, 2, \ldots, Q$, and time period t. We also insert a new node t in each period with an arc $(0, t)$ that carries the total production in period t, and arcs (t, qt) that carries the production in period t of item $q = 1, 2, \ldots, Q$. The flow on arc $(0, t)$ equals the sum of the flows on the arcs (t, qt) for $q = 1, 2, \ldots, Q$. For each commodity $O(q, t) = 1$, $D(q, t) = t$, and $r_{qt} = d_{qt}$, the demand for item q in period t. Arc $(0, t)$ imposes a total production capacity for all items in period t and the arcs (t, qt) impose a production capacity, if any, for item q in period t.

Other versions of this problem can also be viewed as network design problems, for example, in some applications we can produce only one item in any period and we incur a changeover cost whenever we switch between the production of any two items (for example, to clean dies or to reconfigure machine settings). Magnanti & Vachani [1990] and Van Hoesel, Wagelmans & Wolsey [1991] provide details about the network design formulation of these problems.

Additional applications

Magnanti & Wong [1984] and Minoux [1989] provide many additional references on network design, including several additional applications, and Magnanti & Wolsey [1993] describe applications, models and algorithms for several tree versions of network design problems. Bertsekas & Gallagar [1992], Graves, Rinnooy Kan & Zipkin [1993], and Mirchandani & Francis [1990] contain extensive discussions of communication networks, production planning, and facility location. Some additional applications of network design include (1) *airline route planning* [Jaillet & Yu, 1993]; (2) *centralized network design* [Gavish, 1984]; (3) *computer teleprocessing networks* [Chandy & Russell, 1972; Esau & Williams, 1966; Kershenbaum & Boorstyn, 1975]; (4) *electricity distribution planning* [Bousba & Wolsey, 1990; Gascon, Benchakroun & Ferland, 1991]; (5) *less-than truckload (LTL) consolidation* [Leung, Magnanti & Singhal, 1990; Braklow, Graham, Hassler, Peck & Powell, 1992]; (6) *regional water resources systems* [Jarvis, Rardin, Unger, Moore & Schimpeler, 1978]; and (7) *VLSI design* [Korte, Promel & Steger, 1990].

14. Summary

In this paper, we have described several applications of the following network optimization problems: shortest paths, maximum flows, minimum cost flows, assignment and matchings, minimum spanning trees, convex cost flows, generalized flows, multicommodity flows, the traveling salesman problem, and network design. We have adapted about 75% of these applications from Ahuja, Magnanti & Orlin [1993] who describe many other applications as well. The survey papers of Bennington [1974], Glover & Klingman [1976], Bodin, Golden, Assad & Ball [1983], Aronson [1989], and Glover, Klingman & Phillips [1990] provide many additional application references. The books by Gondran & Minoux [1984], Evans & Minieka [1992], and Glover, Klingman & Phillips [1992] also describe a variety of applications of network optimization.

Some of the models included in this discussion have direct relevance to practitioners in such fields as communications, manufacturing and transportation; other models are generic mathematical problems of some interest to applied mathematicians (e.g., finding a solution to certain types of inequality systems) and that themselves have applications in varied practical settings (for example, in personnel scheduling). Some models are formulated naturally as network optimization problems; others don't appear naturally as network models, but are transformable into the form of network optimization problems (for example, by taking linear programming duals of 'natural' formulations). Taken as a whole, the models we have considered, and those we have cited, establish network optimization as an unusually powerful and rich modeling environment, adding evidence to justify the claim that applied mathematics, computer science, and operations research do indeed have much to offer to the world of practice.

Acknowledgments

The applications described in this paper are drawn from a number of operations research, computer science, and applied mathematics journals. The integer programming bibliographies by Kastning [1976], Hausman [1978] and Von Randow [1982, 1985] have been particularly valuable in helping us to identify these applications. Many of the applications we cover are excerpted from the book, *Network Flows: Theory, Algorithms, and Applications*, written by Ahuja, Magnanti & Orlin [1993]; this book describes over 150 applications of network optimization problems. We are indebted to Prentice-Hall for its permission to collect together and summarize these applications from our book in this chapter.

Figures 34a and 34b are reprinted by permission of the Operations Research Society of America and the authors of the paper 'The product matrix traveling salesman problem: An application and solution heuristic', *Operations Research* **35**, 772–783.

We are grateful to Professor Michael Ball for a number of valuable comments on an earlier version of this chapter, and, in particular, for pointing out two of the applications we have used.

The first and third authors have been supported by grants from the United Parcel Service, Prime Computers, and the Air Force Office of Scientific Research (Grant Number AFORS-88-0088).

References

Abdallaoui, G. (1987). Maintainability of a grade structure as a transportation problem. *J. Oper. Res. Soc.* 38, 367–369.

Aghezzaf, E.H., T.L. Magnanti and L.A. Wolsey (1992). Optimizing constrained subtrees of trees, CORE Discussion Paper No. 9250, Université Catholique de Louvain.

Ahlfeld, D.P., R.S. Dembo, J.M. Mulvey and S.A. Zenios (1987). Nonlinear programming on generalized networks. *ACM Trans. Math. Software* 13, 350–367.

Ahuja, R.K. (1986). Algorithms for the minimax transportation problem. *Nav. Res. Logistics Q.* 33, 725–740.

Ahuja, R.K., T.L. Magnanti and J.B. Orlin (1993). *Network Flows: Theory, Algorithms and Applications*. Prentice-Hall, Inc., Englewood Cliffs, N.J.

Ali, A.I., D. Barnett, K. Farhangian, J.L. Kennington, B. Patty, B. Shetty, B. McCarl and P. Wong (1984). Multicommodity network problems: Applications and computations. *AIIE Trans.* 16, 127–134.

Ali, A.I., R.V. Helgason and J.L. Kennington (1978). The convex cost network flow problem: A state-of-the-art survey. Technical Report OREM 78001, Southern Methodist University, Texas.

Alizadeh, F., R.M. Karp, L.A. Newberg and D.K. Weisser (1992). Physical mapping of chromosomes: A combinatorial problem in molecular biology. Unpublished manuscript.

Altinkemer, K., and B. Gavish (1991). Parallel savings based heuristics for the delivery problem, *Oper. Res.* 39, 456–469.

Anderson, W.N. (1975). Maximum matching and the rank of a matrix. *SIAM J. Appl. Math.* 28, 114–123.

Arisawa, S., and S.E. Elmaghraby (1977). The 'hub' and 'wheel' scheduling problems. *Transp. Sci.* 11, 124–146.

Aronson, J.E. (1989). A survey of dynamic network flows. *Ann. Oper. Res.* 20, 1–66.

Arratia, R., E.S. Lander, S. Tivaré and M.S. Waterman (1991). Genomic mapping by anchoring random clones: A mathematical analysis. *Genomics* 11, 806–827.

Assad, A.A. (1980). Models for rail transportation. *Transp. Res.* 14A, 205–220.

Bacharach, M. (1966). Matrix rounding problems. *Manage. Sci.* 9, 732–742.

Balakrishnan, A., T.L. Magnanti and R. Wong (1991). A decomposition algorithm for local access telecommunications network expansion planning. Working Paper, Operations Research Center, MIT.

Balakrishnan, A., T.L. Magnanti, A. Shulman and R.T. Wong (1991). Models for planning capacity expansion in local access telecommunication networks. *Ann. Oper. Res.* 33, 239–284.

Ball, M.O., L. Bodin and R. Dial (1983). A matching based heuristic for scheduling mass transit crews and vehicles. *Transp. Sci.* 17, 4–31.

Barany, I., J. Edmonds and L.A. Wolsey (1986). Packing and covering a tree by subtrees. *Combinatorica* 6, 245–257.

Barr, R.S., and J.S. Turner (1981). Microdata file merging through large scale network technology. *Math. Program. Study B* 15, 1–22.

Barros, O., and A. Weintraub (1986). Spatial market equilibrium problems as network models. *Discr. Appl. Math.* 13, 109–130.

Bartholdi, J.J., J.B. Orlin and H.D. Ratliff (1980). Cyclic scheduling via integer programs with circular ones. *Oper. Res.* 28, 1074–1085.

Barzilai, J., W.D. Cook and M. Kress (1986). A generalized network formulation of the pairwise comparison consensus ranking model. *Manage. Sci.* 32, 1007–1014.

Belford, P.C., and H.D. Ratliff (1972). A network-flow model for racially balancing schools. *Oper. Res.* 20, 619–628.

Bellman, R. (1958). On a routing problem. *Q. Appl. Math.* 16, 87–90.

Bellmore, M., G. Bennington and S. Lubore (1971). A multivehicle tanker scheduling problem. *Transp. Sci. B* 5, 36–47.

Bennington, G.E. (1974). Applying network analysis. *Ind. Eng.* 6, 17–25.

Bentley, J.L. (1990). Experiments on traveling salesman heuristics. *Procedings of the First Annual ACM-SIAM Symposium on Discrete Algorithms*, San Francisco, Calif., pp. 91–99.

Berge, C., and A. Ghouila-Houri (1962). *Programming, Games and Transportation Networks*. John Wiley and Sons, Chichester.

Berrisford, H.G. (1960). The economic distribution of coal supplies in the gas industry: An application of the linear programming transport theory. *Oper. Res. Q.* 11, 139–150.

Bertsekas, D., and R. Gallagar (1992). *Data Networks* (2nd ed.), Prentice-Hall, Englewood Cliffs, N.J.

Bousba, C., and L.A. Wolsey (1990). Finding minimum cost directed trees with demands and capacities. Discussion Paper No. 8913, Center for Oper. Res. and Econometrics, University of Louvain, Belgium.
Bodin, L.D., B.L. Golden, A.A. Assad and M.O. Ball (1983). Routing and scheduling of vehicles and crews: The state of the art. *Comput. Oper. Res.* 10, 69–211.
Braklow, J.W., W.W. Graham, S. Hassler, K.E. Peck and W.B. Powell (1992). Interactive optimization improves service and systems performance for yellow freight. *Interfaces* 22, 147–172.
Brogan, W.L. (1989). Algorithm for ranked assignments with application to multiobject tracking. *J. Guidance*, 357–364.
Cabot, A.V., R.L. Francis and M.A. Stary (1970). A network flow solution to a rectilinear distance facility location problem. *AIIE Trans. B* 2, 132–141.
Cahn, A.S. (1948). The warehouse problem (Abstract). *Bull. Am. Math. Soc.* 54, 1073.
Carraresi, P., and G. Gallo (1984). Network models for vehicle and crew scheduling. *Eur. J. Oper. Res.* 16, 139–151.
Chalmet, L.G., R.L. Francis and P.B. Saunders (1982). Network models for building evacuation. *Manage. Sci.* 28, 86–105.
Chandy, K.M., and R.A. Russell (1972). The design of multipoint linkages in a teleprocessing tree network. *IEEE Trans. Comput.* C-21, 1062–1066.
Cheshire, M., K.I.M. McKinnon and H.P. Williams (1984). The efficient allocation of private contractors to public works. *J. Oper. Res. Q.* 35, 705–709.
Clark, J.A., and N.A.J. Hastings (1977). Decision networks. *Oper. Res. Q.* 20, 51–68.
Clarke, S., and J. Surkis (1968). An operations research approach to racial desegregation of school systems. *Socio-Economic Planning Sci.* 1, 259–272.
Collins, M., L. Cooper, R. Helgason, J. Kennington and L. Leblanc (1978). Solving the pipe network analysis problem using optimization techniques. *Manage. Sci.* 24, 747–760.
Cox, L.H., and L.R. Ernst (1982). Controlled rounding. *INFOR* 20, 423–432.
Cormen, T.H., C.L. Leiserson, and R.L. Rivest (1990). *Introduction to Algorithms* MIT Press and McGraw-Hill, New York.
Crainic, T., J.A. Ferland and J.M. Rousseau (1984). A tactical planning model for rail freight transportation. *Transp. Sci.* 18, 165–184.
Crum, R.L., and D.J. Nye (1981). A network model of insurance company cash flow management. *Math. Program.* 15, 86–101.
Dafermos, S., and A. Nagurney (1984). A network formulation of market equilibrium problems and variational inequalities. *Oper. Res. Lett.* 5, 247–250.
Daniel, R.C. (1973). Phasing out capital equipment. *Oper. Res. Q.* 24, 113–116.
Dantzig, G.B. (1962). *Linear Programming and Extensions*. Princeton University Press, Princeton, NJ.
Dantzig, G.B., and D.R. Fulkerson (1954). Minimizing the number of tankers to meet a fixed schedule. *Nav. Res. Logistics Q.* 1, 217–222.
Dearing, P.M., and R.L. Francis (1974). A network flow solution to a multifacility minimax location problem involving rectilinear distances. *Transp. Sci.* 8, 126–141.
Dembo, R.S., J.M. Mulvey and S.A. Zenios (1989). Large-scale nonlinear network models and their applications. *Oper. Res. B* 37, 353–372.
Denardo, E.V., U.G. Rothblum and A.J. Swersey (1988). A transportation problem in which costs depend on the order of arrival. *Manage. Sci.* 34, 774–783.
Derman, C., and M. Klein (1959). A note on the optimal depletion of inventory. *Manage. Sci.* 5, 210–214.
Devine, M.V. (1973). A model for minimizing the cost of drilling dual completion oil wells. *Manage. Sci.* 20, 532–535.
Dewar, M.S.J., and H.C. Longuet-Higgins (1952). The correspondence between the resonance and molecular orbital theories. *Proc. R. Soc. London B* A214, 482–493.
Dirickx, Y.M.I., and L.P. Jennergren (1975). An analysis of the parking situation in the downtown area of West Berlin. *Transp. Res.* 9, 1–11.

Dorsey, R.C., T.J. Hodgson and H.D. Ratliff (1975). A network approach to a multifacility multi-product production scheduling problem without backordering. *Manage. Sci.* 21, 813–822

Dress, A.W.M., and T.F. Havel (1988). Shortest path problems and molecular conformation. *Discr. Appl. Math.* 19, 129–144.

Dror, M., P. Trudeau and S.P. Ladany (1988). Network models for seat allocation on flights. *Transp. Res.* 22B, 239–250.

Dude, R.O., and P.E. Hart (1973). *Pattern Classification and Science Analysis*. Wiley-Interscience, New York, N.Y.

Edmonds, J. (1967). An introduction to matching. Mimeographed notes, Engineering Summer Conference, The University of Michigan, Ann Arbor.

Edmonds, J., and E.L. Johnson (1973). Matching, Euler tours and the Chinese postman. *Math. Program.* 5, 88–124.

Elmaghraby, S.E. (1978). *Activity Networks: Project Planning and Control by Network Models*. Wiley-Interscience, New York, N.Y.

Esau, L.R., and K.C. Williams (1966). On teleprocessing system design II. *IBM Systems J., B* 5, 142–147.

Evans, J.R. (1977). Some network flow models and heuristics for multiproduct production and inventory planning. *AIIE Trans.* 9, 75–81.

Evans, J.R. (1984). The factored transportation problem. *Manage. Sci.* 30, 1021–1024.

Evans, J.R., and E. Minieka (1992). *Optimization Algorithms for Networks and Graphs*. Marcel Dekker, Inc., New York, N.Y.

Ewashko, T.A., and R.C. Dudding (1971). Application of Kuhn's Hungarian assignment algorithm to posting servicemen. *Oper. Res. B* 19, 991.

Farley, A.R. (1980). Levelling terrain trees: A transshipment problem. *Inf. Process. Lett. B* 10, 189–192.

Federgruen, A., and H. Groenevelt (1986). Preemptive scheduling on uniform machines by network flow techniques. *Manage. Sci.* 32, 341–349.

Filliben, J.J., K. Kafadar and D.R. Shier (1983). Testing for homogeneity of two-dimensional surfaces. *Math. Modelling* 4, 167–189.

Florian, M., and D.W. Hearn (1992). Network equilibrium: Models and algorithms, to appear.

Ford, L.R., and D.R. Fulkerson (1958). Constructing maximal dynamic flows from static flows. *Oper. Res.* 6, 419–433.

Ford, L.R., and S.M. Johnson (1959). A tournament problem. *Am. Math. Mon.* 66, 387–389.

Francis, R.L., and J.A. White (1976). *Facility Layout and Location*. Prentice-Hall, Englewood Cliffs, N.J.

Frank, C.R. (1965). A note on the assortment problem. *Manage. Sci.* 11, 724–726.

Fujii, M., T. Kasami and K. Ninomiya (1969). Optimal sequencing of two equivalent processors. *SIAM J. Appl. Math.* 17, 784–789. Erratum, same journal 18, 141.

Fulkerson, D.R. (1961). A network flow computation for project cost curve. *Manage. Sci. B* 7, 167–178.

Fulkerson, D.R. (1965). Upsets in a round robin tournaments. *Can. J. Math.* 17, 957–969.

Fulkerson, D.R. (1966). Flow networks and combinatorial operations research. *Am. Math. Mon.* 73, 115–138.

Fulkerson, D.R., and G.C. Harding (1977). Maximizing the minimum source–sink path subject to a budget constraint. *Math. Program.* 13, 116–118.

Fuller, S.H. (1972). An optimal drum scheduling algorithm. *IEEE Trans. Comput.* C-21, 1153–1165.

Gallo, G., M.D. Grigoriadis and R.E. Tarjan (1989). A fast parametric maximum flow algorithm and applications. *SIAM J. Comput.* 18, 30–55.

Garey, M.S., and D.S. Johnson (1979). *Computers and Intractability: A Guide to the Theory of NP-Completeness*. W.H. Freeman.

Garfinkel, R.S. (1977). Minimizing wallpaper waste part I: A class of traveling salesman problems. *Oper. Res.* 25, 741–751.

Gascon, V., A. Benchakroun and J.A. Ferland (1991). Electricity distribution planning model: A

network design approach solving the master problem of the Benders decomposition method. Working Paper, University on Montreal.
Gavish, B. (1984). Augmented Lagrangean based algorithms for centralized network design. *IEEE Trans. Commun.* 33, 1247–1275.
Gavish, B., and P. Schweitzer (1974). An algorithm for combining truck trips. *Transp. Sci.* 8, 13–23.
Geoffrion, A.M., and G.W. Graves (1974). Multicommodity distribution system design by Benders decomposition. *Manage. Sci. B* 20 , 822–844.
Gilmore, P.C., and R.E. Gomory (1964). Sequencing a one state-variable machine: A solvable case of the travelling salesman problem. *Oper. Res.* 12, 655–679.
Glover, F., and D. Klingman (1976). Network applications in industry and government. *AIIE Trans. B* 9, 363–376.
Glover, F., and J. Rogozinski (1982). Resort development: A network-related model for optimizing sites and visits. *J. Leisure Res.*, 235–247.
Glover, F., D. Klingman and N. Phillips (1990). Netform modeling and applications. *Interfaces* 20, 7–27.
Glover, F., D. Klingman and N. Phillips (1992). *Network Models in Optimization and Their Applications*. John Wiley and Sons, Chichester.
Glover, F., J. Hultz, D. Klingman and J. Stutz (1978). Generalized networks: A fundamental computer based planning tool. *Manage. Sci.* 24, 1209–1220.
Glover, F., R. Glover and F.K. Martinson (1984). A netform system for resource planning in the U.S. Bureau of land management. *J. Oper. Res. Soc.* 35, 605–616.
Goetschalckx, M., and H.D. Ratliff (1988). Order picking in an aisle. *IIE Trans. B* 20, 53–62.
Golden, B.L. (1975). A minimum cost multicommodity network flow problem concerning imports and exports. *Networks* 5, 331–356.
Goldman, A.J., and G.L. Nemhauser (1967). A transport improvement problem transformable to a best-path problem. *Transp. Sci.* 1, 295–307.
Golitschek, M.V., and H. Schneider (1984). Applications of shortest path algorithms to matrix scalings. *Numer. Math. B* 44, 111–126.
Gondran, M., and M. Minoux (1984). *Graphs and Algorithms*. Wiley-Interscience, New York, N.Y.
Gorham, W. (1963). An application of a network flow model to personnel planning. *IEEE Trans. Eng. Manage.* 10, 113–123.
Gower, J.C., and G.J.S. Ross (1969). Minimum spanning trees and single linkage cluster analysis. *Appl. Stat.* 18, 54–64.
Groetschel, M., C.L. Monma and M. Stoer (1992). Design of surviavable networks, to appear.
Graham, R.L., and P. Hell (1985). On the history of minimum spanning tree problem. *Ann. Hist. Comput.* 7, 43–57.
Graves, S., A.H.G. Rinnooy Kan and P. Zipkin (eds.) (1993). In *Handbooks of Oper. Res. and Manage. Sci.*, Volume: *Logistics of Production and Inventory*, North-Holland.
Gupta, S.K. (1985). *Linear Programming and Network Models*. Affiliated East–West Press Private Limited, New Delhi.
Gusfield, D. (1988). A graph theoretic approach to statistical data security. *SIAM J. Comput* 17, 552–571.
Gusfield, D., and C. Martel (1992). A fast algorithm for the generalized parametric minimum cut problem and applications. *Algorithmica* 7, 499–519.
Gusfield, D., C. Martel and D. Fernandez-Baca (1987). Fast algorithms for bipartite network flow. *SIAM J. Comput.* 16, 237–251.
Gutjahr, A.L., and G.L. Nemhauser (1964). An algorithm for the line balancing problem. *Manage. Sci.* 11, 308–315.
Hall, M. (1956). An algorithm for distinct representatives. *Am. Math. Mon.* 63, 716–717.
Hausman, H. (1978). *Integer Programming and Related Areas: A Classified Bibliography*. Lecture Notes in Economics and Mathematical Systems, Vol. 160, Springer-Verlag, Berlin.
Haymond, R.E., J.R. Thornton and D.D. Warner (1988). A shortest path algorithm in robotics and its implementation on the FPS T-20 hypercube. *Ann. Oper. Res. B* 14, 305–320.

Held, M., and R. Karp (1970). The traveling salesman problem and minimum spanning trees. *Oper. Res.* 18, 1138–1162.

Hoffman, A.J., and H.M. Markowitz (1963). A note on shortest path, assignment and transportation problems. *Nav. Res. Logistics Q. B* 10, 375–379.

Hoffman, A.J., and S.T. McCormick (1984). A fast algorithm that makes matrices optimally sparse, in: *Progress in Combinatorial Optimization*, Academic Press.

Horn, W.A. (1971). Determining optimal container inventory and routing. *Transp. Sci.* 5, 225–231.

Horn, W.A. (1973). Minimizing average flow time with parallel machines. *Oper. Res.* 21, 846–847.

Hu, T.C. (1961). The maximum capacity route problem. *Oper. Res. B* 9, 898–900.

Hu, T.C. (1966). Minimum cost flows in convex cost networks. *Nav. Res. Logistics Q.* 13, 1–9.

Hu, T.C. (1967). Laplace's equation and network flows. *Oper. Res.* 15, 348–354.

Hunter, D. (1976). An upper bound for the probability of a union. *J. Appl. Probab. B* 13, 597–603.

Imai, H., and M. Iri (1986). Computational-geometric methods for polygonal approximations of a curve. *Comput. Vision, Graphics Image Process.* 36, 31–41.

Jaillet, P., and G. Yu (1993). An integer programming model for the airline network design problem, *Bulletin, National Meeting of the Oper. Res. Soc. Am. Inst. Manage. Sci.*, Phoenix, Arizona.

Jarvis, J.J., R.L. Rardin, V.E. Unger, R.W. Moore and C.C. Schimpeler (1978). Optimal design of regional wastewater systems: a fixed-charge network flow model, *Oper. Res.* 26, 538–550.

Jacobs, W.W. (1954). The caterer problem. *Nav. Res. Logistics Q. B* 1, 154–165.

Jewell, W.S. (1957). Warehousing and distribution of a seasonal product. *Nav. Res. Logistics Q. B* 4, 29–34.

Johnson, T.B. (1968). Optimum pit mine production scheduling. Technical Report, University of California, Berkeley.

Kang, A.N.C., R.C.T. Lee, C.L. Chang and S.K. Chang (1977). Storage reduction through minimal spanning trees and spanning forests. *IEEE Trans. Comput.* C-26, 425-434.

Kaplan, S. (1973). Readiness and the optimal redeployment of resources. *Nav. Res. Logistics Q.* 20, 625–638.

Kastning, C. (1976). *Integer Programming and Related Areas: A Classified Bibliography*. Lecture Notes in Economics and Mathematical Systems, Vol. 128, Springer-Verlag, Berlin.

Kelley, J.R., (1961). Critical path planning and scheduling: Mathematical basis. *Oper. Res. B* 9, 296–320.

Kelly, J.P., B.L. Golden and A.A. Assad (1992). Cell suppression: Disclosure protection for sensitive tabular data. *Networks* 22, 397–417.

Kennington, J.L., and R.V. Helgason (1980). *Algorithms for Network Programming*. Wiley-Interscience, New York, N.Y.

Kershenbaum, A., and R.B. Boorstyn (1975). Centralized teleprocessing network design, *Proc. National Telecommunications Conference*, 27.11–27.14.

Khan, M.R. (1979). A capacitated network formulation for manpower scheduling. *Ind. Manage. B* 21, 24–28.

Khan, M.R., and D.A. Lewis (1987). A network model for nursing staff scheduling. *Z. Oper. Res.* 31, B161–171.

Kolitz, S. (1991). Personal communication.

Korte, B. (1988). Applications of combinatorial optimization. Technical Report No. 88541-OR. Institute für Okonometrie und Operations Research, Nassestrasse 2, D-5300, Bonn.

Korte, B., H.J. Promel and A. Steger (1990). Steiner trees in VLSI layout, in: B. Korte, L. Lovasz, H.J. Promel, A. Schijver (eds.), *Paths, Flows and VLSI Layout*, Springer-Verlag, Berlin.

Kourtz, P. (1984). A network approach to least cost daily transfers of forest fire control resources. *INFOR B* 22, 283–290.

Larson, R.C., and A.R. Odoni (1981). *Urban Operations Research*. Prentice-Hall, Englewood Cliffs, N.J.

Lawania (1990). Personal communication.

Lawler, E.L. (1964). On scheduling problems with deferral costs. *Management Science* 11, 280–287.

Lawler, E.L. (1966). Optimal cycles in doubly weighted linear graphs. *Theory of Graphs: International Symposium,* Dunod, Paris and Gordon and Breach, New York, N.Y., pp. 209–213.
Lawler, E.L. (1976). *Combinatorial Optimization: Networks and Matroids.* Holt, Rinehart and Winston.
Lawler, E.L., J.K. Lenstra, A.H.G. Rinnooy Kan and D.B. Shmoys, eds. (1985).. *The traveling salesman problem: A guided tour of combinatorial optimization.* John Wiley and Sons, Chichester.
Lenstra, J.K., and A.H.G. Rinnooy Kan (1975). Some simple applications of the travelling salesman problem. *Oper. Res. Q.* 26, 717–733.
Leung, J., T.L. Magnanti and V. Singhal (1990). Routing in point to point delivery systems. *Transp. Sci.* 24, 245–260.
Leung, J., T.L. Magnanti and R. Vachani (1989). Facets and algorithms for capacitated lot sizing. *Math. Program.* 45, 331–360.
Levner, E.V., and A.S. Nemirovsky (1991). A network flow algorithm for just-in-time project scheduling. Memorandum COSOR 91-21, Dept. of Mathematics and Computing Science, Eindhoven Univesity of Technology, Eindhoven.
Lin, T.F. (1986). A system of linear equations related to the transportation problem with application to probability theory. *Discr. Appl. Math.* 14, 47–56.
Love, R.R., and R.R. Vemuganti (1978). The single-plant mold allocation problem with capacity and changeover restriction. *Oper. Res. B* 26, 159–165.
Lowe, T.J., R.L. Francis and E.W. Reinhardt (1979). A greedy network flow algorithm for a warehouse leasing problem. *AIIE Trans.* 11, 170–182.
Luss, H. (1979). A capacity expansion model for two facilities. *Nav. Res. Logistics Q.* 26, 291–303.
Machol, R.E. (1961). An application of the assignment Problem. *Oper. Res. B* 9, 585–586.
Machol, R.E. (1970). An application of the assignment Problem. *Oper. Res. B* 18, 745–746.
Maculan, N. (1987). The steiner problem in graphs. *Ann. Discr. Math.* 31, 185–212.
Magirou, V.F. (1986). The efficient drilling of printed circuit boards. *Interfaces* 16, 13–23.
Magnanti, T.L. (1984). Models and algorithms for predicting urban traffic equilibria, in: M. Florian (ed.), *Transportation Planning Models*, North-Holland, Amsterdam, pp. 153–186.
Magnanti, T.L., P. Mirchandani and R. Vachani (1991). Modeling and solving the capacitated network loading problem. Working Paper, Oper. Res. Center, MIT.
Magnanti, T.L., and R. Vachani (1990). A strong cutting plane algorithm for production scheduling with changeover costs. *Oper. Res.* 38, 456–473.
Magnanti, T.L., and R. Wong (1984). Network design and transportation planning: Models and algorithms. *Transp. Sci.* 18, 1–55.
Magnanti, T.L., and L.A. Wolsey (1993). Optimal Trees, to appear.
Mamer, J.W., and S.A. Smith (1982). Optimizing field repair kits based on job completion rate. *Manage. Sci. B* 28, 1328–1334.
Manne, A.S. (1958). A target-assignment problem. *Oper. Res. B* 6, 346–351.
Martel, C. (1982). Preemptive scheduling with release times, deadlines and due times. *J. ACM* 29, 812–829.
Mason, A.J., and A.B. Philpott (1988). Pairing stereo speakers using matching algorithms. *Asia-Pacific J. Oper. Res.* 5, 101–116.
Maxwell, W.L., and R.C. Wilson (1981). Dynamic network flow modelling of fixed path material handling systems. *AIIE Trans.* 13, 12–21.
McCormick, W.T. Jr., P.J. Schweitzer and T.W. White (1972). Problem decomposition and data reorganization by a clustering technique. *Oper. Res.* 20, 993–1009.
McGinnis, L.F., and H.L.W. Nuttle (1978). The project coordinators problem. *OMEGA* 6, 325–330.
Minoux, M. (1989). Network synthesis and optimum network design problems: Models, solution methods and applications. *Networks* 19, 313–360.
Mirchandani P.B., and R.L. Francis (1990). *Discrete Location Theory.* John Wiley and Sons, Chichester.
Monma, C.L., and D.F. Shallcross (1989). Methods for designing communication networks with certain two-connected survivability constraints. *Oper. Res. B* 37, 531–541.

Monma, C.L., and M. Segal (1982). A primal algorithm for finding minimum-cost flows in capacitated networks with applications. *Bell Systems Tech. J.* 61, 949–468.

Mulvey, J.M. (1979). Strategies in modeling: A personal scheduling example. *Interfaces* 9, 66–76.

Orlin, D. (1987). Optimal weapons allocation against layered defenses. *Nav. Res. Logistics B* 34, 605–617.

Orlin, J.B., and U.G. Rothblum (1985). Computing optimal scalings by parametric network algorithms. *Math. Program.* 32, 1–10.

Osteen, R.E., and P.P. Lin (1974). Picture skeletons based on eccentricities of points of minimum spanning trees. *SIAM J. Comput.* 3, 23–40.

Picard, J.C., and H.D. Ratliff (1973). Minimum cost cut equivalent networks. *Manage. Sci.* 19, 1087–1092.

Picard, J.C., and H.D. Ratliff (1978). A cut approach to the rectilinear distance facility location problem. *Oper. Res.* 26, 422–433.

Picard, J.C., and M. Queyranne (1982). Selected applications of minimum cuts in networks. *INFOR B* 20, 394–422.

Plante, R.D., T.J. Lowe and R. Chandresekaran (1987). The product matrix traveling salesman problem: An application and solution heuristic. *Oper. Res.* 35, 772–783.

Pochet, Y., and L.A. Wolsey (1991). Solving multi-item lot-sizing problems with strong cutting planes. *Manage. Sci.* 37, 53–67.

Prager, W. (1957). On warehousing problems. *Oper. Res.* 5, 504–512.

Prim, R.C. (1957). Shortest connection networks and some generalizations. *Bell Systems Tech. J.* 36, 1389–1401.

Ratliff, H.D. (1978). Network models for production scheduling problems with convex cost and batch processing. *AIIE Trans.* 10, 104–108.

Ratliff, H.D., and A.S. Rosenthal (1983). Order picking in a rectangular warehouse: A solvable case of the traveling salesman problem. *Oper. Res.* 31, 507–521.

Ravindran, A. (1971). On compact book storage in libraries. *Opsearch* 8, 245–252.

Rhys, J.M.W. (1970). A selection problem of shared fixed costs and network flows. *Manage. Sci.* 17, 200–207.

Sapountzis, C. (1984). Allocating blood to hospitals from a central blood bank. *Eur. J. Oper. Res.* 16, 157–162.

Schneider, M.H., and S.A. Zenios (1990). A comparative study of algorithms for matrix balancing. *Oper. Res.* 38, 439–455.

Schwartz, B.L. (1966). Possible winners in partially completed tournaments. *SIAM Rev.* 8, 302–308.

Schwartz, M., and T.E. Stern (1980). Routing techniques used in computer communication networks. *IEEE Trans. Commun.* COM-28, 539–552.

Segal, M. (1974). The operator-scheduling problem: A network flow approach. *Oper. Res.* 22, 808–824.

Sheffi, Y. (1985). *Urban Transportation Networks: Equilibrium Analysis with Mathematical Programming Methods.* Prentice-Hall, Englewood Cliffs, N.J.

Servi, L.D. (1989). A network flow approach to a satellite scheduling problem. Research Report, GTE Laboratories, Inc., Waltham, MA.

Shier, D.R. (1982). Testing for homogeneity using minimum spanning trees. *The UMAP J.* 3, 273–283.

Slump, C.H., and J.J. Gerbrands (1982). A network flow approach to reconstruction of the left ventricle from two projections. *Comput Graphics Image Process.* 18, 18–36.

Srinivasan, V. (1974). A transshipment model for cash management decisions. *Manage. Sci.* 20, 1350–1363.

Srinivasan, V. (1979). Network models for estimating brand-specific effects in multiattribute marketing models. *Manage. Sci.* 25, 11–21.

Stillinger, F.H. (1967). Physical clusters, surface tension and critical phenomenon. *J. Chem. Phys. B* 47, 2513–2533.

Stone, H.S. (1977). Multiprocessor scheduling with the aid of network flow algorithms. *IEEE Trans. Software Eng.* 3, 85–93.

Shuchat, A. (1984). Matrix and network models in archeology. *Math. Mag.* 57, 3–14.

Szadkowski. (1970). An approach to machining process optimization. *Int. J. Prod. Res. B* 9, 371–376.

Tso, M. (1986). Network flow models in image processing. *J. Oper. Res. Society* 37, 31–34.

Tso, M., P. Kleinschmidt, I. Mitterreiter and J. Graham (1991). An efficient transportation algorithm for automatic chromosome karyotyping. *Pattern Recognition Lett.* 12, 117–126.

Van Hoesel, C.P.M., A.P.M. Wagelmans and L.A. Wolsey (1991). Economic lot-sizing with start-up costs: The convex hull. Core Discussion Paper 9109, Universite Catholique de Louvain (to appear in *SIAM J. Discr. Math.*).

Van Slyke, R., and H. Frank (1972). Network reliability analysis: Part I. *Networks* 1, 279–290.

Veinott, A.F., and H.M. Wagner (1962). Optimal capacity scheduling — Part I and II. *Oper. Res.* 10, 518–547.

Von Randow, R. (1982). *Integer Programming and Related Areas: A Classified Bibliography 1978–1981*. Lecture Notes in Economics and Mathematical Systems, Vol. 197, Springer-Verlag, Berlin.

Von Randow, R. (1985). *Integer Programming and Related Areas: A Classified Bibliography 1981–1984*. Lecture Notes in Economics and Mathematical Systems, Vol. 243, Springer-Verlag, Berlin.

Wagner, D.K. (1990). Disjoint (s, t)-cuts in a network. *Networks* 20, 361–371.

Waterman, M.S. (1988). *Mathematical Methods for DNA Sequences*. CRC Press.

White, L.S. (1969). Shortest route models for the allocation of inspection effort on a production line. *Manage. Sci.* 15, 249–259.

White, W.W. (1972). Dynamic transshipment networks: An algorithm and its application to the distribution of empty containers. *Networks B* 2, 211–230.

Winter, P. (1987). Steiner problem in networks: A survey. *Networks* 17, 129–167.

Worsley, R.J. (1982). An improved Bonferroni inequality. *Biometrika* 69, 297–302.

Wright, J.W. (1975). Reallocation of housing by use of network analysis. *Oper. Res. Q.* 26, 253–258.

Zahn, C.T. (1971). Graph-theoretical methods for detecting and describing gestalt clusters. *IEEE Trans. Comput.* C20, 68–86.

Zangwill, W.I. (1969). A backlogging model and a multi-echelon model of a dynamic economic lot size production system--A network approach. *Manage. Sci.* 15, 506–527.

Zawack, D.J., and G.L. Thompson (1987). A dynamic space–time network flow model for city traffic congestion. *Transp. Sci.* 21, 153–162.

M.O. Ball et al., Eds., *Handbooks in OR & MS, Vol. 7*

Chapter 2

Primal Simplex Algorithms for Minimum Cost Network Flows

Richard V. Helgason and Jeffery L. Kennington
Department of Computer Science and Engineering, Southern Methodist University, Dallas, Texas 75275, U.S.A.

1. Introduction

Due to the special structure of bases for the linear network flow model, specialized simplex-based software can solve these problems in from one to two orders of magnitude faster than general linear programming software. The objectives of this chapter are to (1) summarize the ideas fundamental to efficient software implementation of the primal simplex algorithm for minimum cost network flow problems and (2) indicate how these ideas have been extended for generalized networks, multicommodity networks, and networks with arbitrary side constraints.

1.1. Set notation

For the most part we adopt standard set notation conventions. Sets will usually be denoted by upper case Roman letters such as X. The empty set will be denoted by Φ. For a finite set X we let $\#X$ denote the number of elements in X. We let $\imath n = \{1, \ldots, n\}$ and $\jmath n = \{0, ..., n\}$. Given set X, we define the *equality relation on* X to be $X \doteq X \equiv \{(x, x) : x \in X\}$. We will also use multisets in which a repetition factor is allowed for set elements. For a finite multiset Y then, $\#Y$ will also incorporate multiplicities.

1.2. Matrix and vector notation

Matrices will usually be denoted by upper case Roman letters such as A. Row vectors will usually be denoted by lower case Greek letters such as π. Column vectors will usually be denoted by lower case Roman letters such as x. The element in the ith row and jth column of matrix A will be denoted by A_{ij}. The row (column) vector whose entries are from the ith row (jth column) of A will be denoted by $A_{i.}$ ($A_{.j}$). The ith element of a vector such as x will be denoted by x_i. We will allow extensive subscripting and superscripting of matrices and vectors for identification purposes. Inasmuch as this may interfere with the subscripting convention for element identification above, we also adopt the functional notation $(\cdot)_i$ and $(\cdot)_{ij}$ for vector and matrix element identification, respectively, so that $(X_{ij})_{pq}$ is the pqth element of matrix X_{ij}. We will use e_i (e^i) for the column (row) vector whose ith element is a 1 and whose other elements are all zeros. We will use e_{ij} to denote the column vector whose ith element is a 1, whose jth element is a -1, and whose other elements are all zeros, so that e_{ij} is $e_i - e_j$. We will use $\hat{0}$ and $\hat{1}$ as row or column vectors with orientation and dimension given by context, having as uniform elements 0 or 1, respectively. We abuse notation by allowing $\imath n = \{1, \ldots, n\}$ to also be used as a row vector. The *diagonal* of matrix A is the set of elements $\{A_{ii}\}$. The matrix A is said to be *upper* (*lower*) *triangular* if $A_{ij} = 0$ when $j > i$ ($i > j$) and, more simply, *triangular* in either case. The matrix A is said to be *diagonal* if it is both upper and lower triangular. A triangular matrix will be nonsingular when its diagonal elements are all nonzero. The matrix A is said to be *triangularizable* if it can be brought to nonsingular triangular form by a sequence of row and column interchanges.

1.3. Graph notation

We define a set of *nodes* or *vertices* V to be any set of consecutive integers which we typically take to be $\imath n$ or $\jmath n$. Given a set of nodes V, we define an *arc* or *edge* for V to be any ordered pair (i, j) with $i \in V$, $j \in V$, and $i \neq j$. The arc (i, j) is said to be *incident on* (touch) both i and j, to *connect* i and j (or j and i), and to be *directed from* i *to* j. Formally, a *network* or *directed graph* is defined to be $G = \langle V, E\rangle$ where V is a set of nodes and E is a set of arcs for V. Apparently then $E \subseteq (V \times V) \setminus (V \dot{=} V)$. When $V = \Phi$ then also $E = \Phi$ and in this case G is called the *trivial* graph. We shall also allow E to be a multiset when it is desirable to have more than one arc connect two nodes. In this case one could more properly refer to G as a *multigraph*. For $\#E = m$, we will find it convenient to label the arcs with elements from $\imath m$.

1.4. Visual representation

The nodes of a network may be viewed as locations or terminals that a given commodity can be moved from, to, or through and the arcs of a network may be viewed as unidirectional means of commodity transport connecting or serving those nodes. Hence arcs may represent streets and highways in an urban transportation network, pipes in a water distribution network, or telephone lines in a communication network. The structure of the network can be displayed by means of a labeled drawing in which nodes are represented by circles or boxes and arcs are represented by line segments incident on two nodes. Each line segment will have an arrowhead placed somewhere on it to indicate the direction of the associated commodity transport. Typically the arrowhead will be incident on the node to which the commodity is being transported. An example network illustration is given in Figure 1.

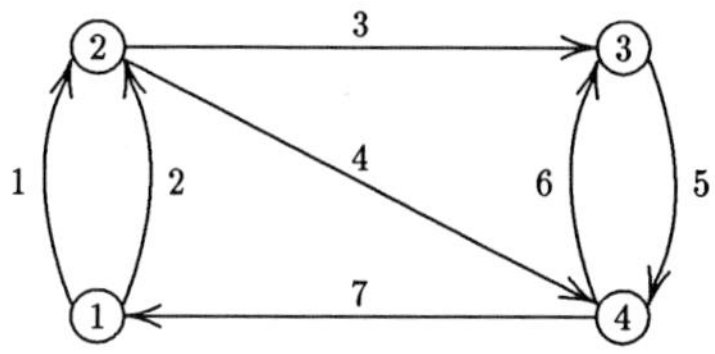

Fig. 1. Example network.

1.5. Node–arc matrix representation

The structure of a network may also be described using a *node–arc incidence matrix* A given by

$$A_{ik} = \begin{cases} +1 & \text{if arc } k \text{ is directed away from node } i, \\ -1 & \text{if arc } k \text{ is directed toward node } i, \\ 0 & \text{otherwise.} \end{cases}$$

Apparently then $A_{.k} = e_{ij}$ for some i and j, and we shall allow ourselves to abuse notation by saying that in this case the kth arc *is* e_{ij}. An example node–arc incidence matrix corresponding to Figure 1 is given below.

$$\begin{array}{cc} & \text{arcs} \\ & \begin{array}{ccccccc} 1 & 2 & 3 & 4 & 5 & 6 & 7 \end{array} \\ \text{nodes} \begin{array}{c} 1 \\ 2 \\ 3 \\ 4 \end{array} & \begin{bmatrix} 1 & 1 & 0 & 0 & 0 & 0 & -1 \\ -1 & -1 & 1 & 1 & 0 & 0 & 0 \\ 0 & 0 & -1 & 0 & 1 & -1 & 0 \\ 0 & 0 & 0 & -1 & -1 & 1 & 1 \end{bmatrix} \end{array}$$

Since each column of A contains only a $+1$, a -1, and zeros, summing all rows of A produces the vector $\hat{0}$. Hence A in not of full rank.

1.6. Subgraphs

A graph $G' = \langle V', E' \rangle$ is said to be a *subgraph* of $G = \langle V, E \rangle$ if $V' \subseteq V$ and $E' \subseteq E$. Note that G' is required to be a graph itself, so that V' and E' cannot simply be arbitrary subsets of V and E, respectively. Further, G' is said to *span* G or G' is said to be a *spanning subgraph* for G when $V' = V$. Given a node subset $V' \subseteq V$, we define the *subgraph generated by* V' to be $G(V') \equiv \{(i, j) \in G : i \in V'$ and $j \in V'\}$. Example subgraphs corresponding to Figure 1 are given in Figure 2.

1.7. Paths and cycles

Given a graph $G = \langle V, E \rangle$, a finite odd length sequence

$$P = \{v_1, e_{i_1 j_1}, v_2, e_{i_2 j_2}, \ldots, v_q, e_{i_q j_q}, v_{q+1}\}$$

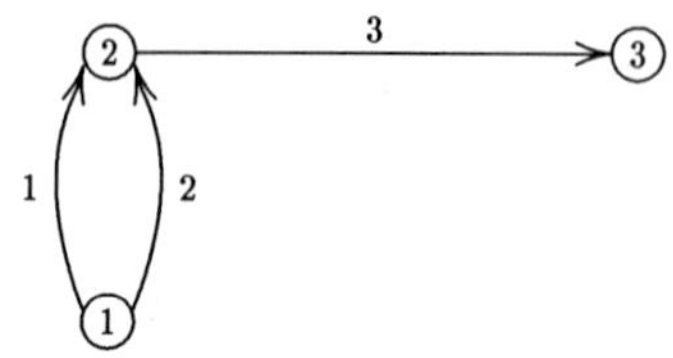

(a) Generated subgraph $G(\{1, 2, 3\})$

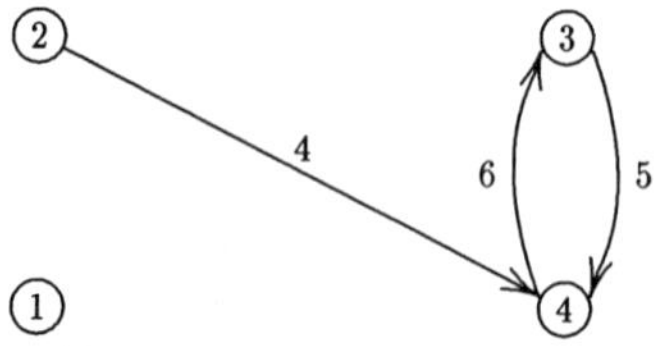

(b) A spanning subgraph for G

Fig. 2. Example network subgraphs.

whose odd elements are nodes of V and whose even elements are arcs of E is defined to be a *walk* in G of *length* $q \geq 0$ in G if: (1) P has at least one node, and (2) for $0 < r \leq q$, arc $e_{i_r j_r}$ connects v_r and v_{r+1}. Apparently then from (2), $e_{i_r j_r}$ could be either (v_r, v_{r+1}) or (v_{r+1}, v_r). The sequence formed by reversing the elements of P is also a walk and will be denoted by $rev(P)$. If we envision moving from v_1 *to* v_{q+1}, utilizing the sequential elements of P in order, we can assign an (implied) orientation to the arcs in the walk by defining the *orientation function*

$$O(e_{i_r j_r}) = \begin{cases} +1 & \text{if } e_{i_r j_r} = (v_r, v_{r+1}), \\ -1 & \text{if } e_{i_r j_r} = (v_{r+1}, v_r). \end{cases}$$

If the sequence of nodes $\{v_1, \ldots, v_{q+1}\}$ from P is composed of distinct nodes, the walk P is said to be a (*simple*) *path* which *links* v_1 *to* v_{q+1}. It follows that the arcs of a path are distinct. Apparently then $rev(P)$ is also a path which links v_{q+1} *to* v_1. It also follows that any walk P of length 0 is a path which links v_1 to itself. If the walk P (1) is of length at least two, (2) $\{v_1, e_{i_1 j_1}, \ldots, v_q\}$ is a path, (3) $\{v_2, e_{i_2 j_2}, \ldots, v_{q+1}\}$ is a path, and (4) $v_1 = v_{q+1}$, the walk P is said to be a *cycle*. Example walks in the graph of Figure 1 are given in Figure 3.

If we form a linear combination of the columns of A corresponding to arcs of a cycle using $O(e_{i_r j_r})$ as the combining coefficient for $e_{i_r j_r}$ for each r the vector $\hat{0}$ is produced. Hence the columns of A corresponding to a cycle are linearly dependent.

Given a cycle P it is possible to form other cycles using the sequential elements of P in wrap-around order, i.e., starting at v_m we can define a cycle

$$P' = \{v_m, e_{i_m j_m}, v_{m+1}, \ldots, v_q, e_{i_q j_q}, v_1, e_{i_1 j_1}, v_2, \ldots, v_m\}$$

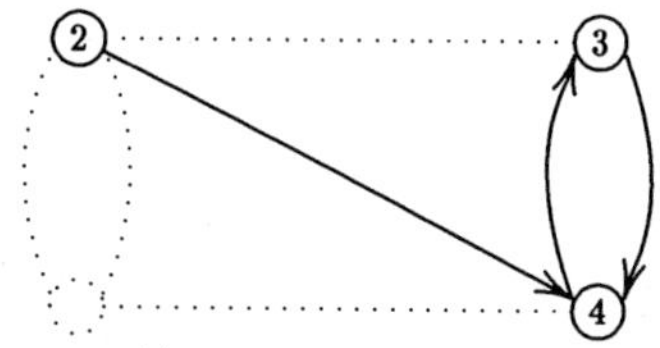

$P^{(a)} = \{4, e_{43}, 3, e_{34}, 4, e_{24}, 2\}$

(a) Nonpath walk from node 4 to node 2

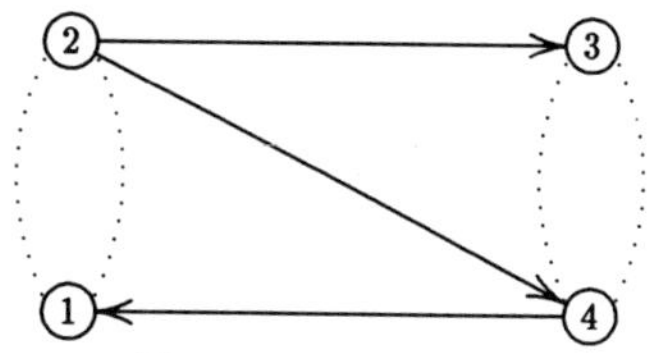

$P^{(b)} = \{1, e_{41}, 4, e_{24}, 2, e_{23}, 3\}$

(b) Path from node 1 to node 3

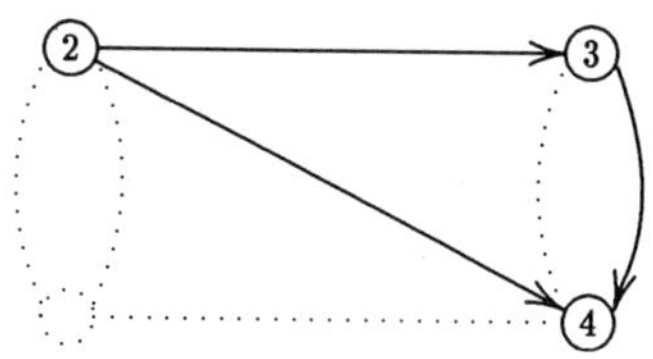

$P^{(c)} = \{2, e_{23}, 3, e_{34}, 4, e_{24}, 2\}$

(c) A cycle on nodes $\{2, 3, 4\}$

Fig. 3. Walks in the example network.

which retains the essential arc and node orders and arc orientations of P when we envision moving from v_m *to* v_m on P'. Thus we will consider cycles such as P and P' to be the same cycle and also refer to any of this set of equivalent representations as *a cycle on nodes* $\{v_1, \ldots, v_q\}$. The arcs of a cycle are generally distinct, except for two special cases which can arise when considering cycles of length two. Cycles of the form

$$\{v_1, (v_1, v_2), v_2, (v_1, v_2), v_1\}$$

and

$$\{v_1, (v_2, v_1), v_2, (v_2, v_1), v_1\}$$

do have arcs which are not distinct and will be called *inadmissible* cycles. All other cycles (which have distinct arcs) will be called *admissible*. Apparently then if P is an admissible cycle on nodes $\{v_1, \ldots, v_q\}$ then $rev(P)$ is a distinct cycle on nodes $\{v_1, \ldots, v_q\}$. A graph G in which no admissible cycle can be formed is said to be *acyclic* and is also said to *contain no cycles*.

1.8. Connectedness and components

A graph $G = \langle V, E \rangle$ is said to be connected if any two vertices u and v can be linked by a path in G. The maximal connected subgraphs of G serve to partition G and are called *components* of G. If $G' = \langle \{v\}, \Phi \rangle$ is a component of G, v is said to be an *isolated* node of G. Summing all rows of A which correspond to a particular component of G produces the vector $\hat{0}$. Hence A cannot be of rank greater than $\#E$ less the number of components of G.

1.9. Trees

A nontrivial connected acyclic graph is called a *tree*. A graph which consists of an isolated node only will be called a *trivial tree*. A graph whose components are all trees is called a *forest*. An *endnode* of a tree is a node which has only one arc of the tree incident on it. A *leaf* of a tree is an arc of the tree incident on an endnode. A component of G must necessarily contain a spanning tree.

A tree $G = \langle V, E \rangle$ has several important properties:

(1) E has one less arc than V has nodes, i.e. $\#E = \#V - 1$,

(2) if an endnode and a leaf incident on it are removed from G, the resulting subgraph is also a tree,

(3) if G has an arc (i, j) incident on two endnodes, then $V = \{i, j\}$ and $E = \{(i, j)\}$ or $E = \{(j, i)\}$.

(4) if $\#E = 1$, G has exactly one leaf,

(5) if $\#E > 1$, G has at least two leaves,

(6) for every distinct pair of nodes $u, v \in E$, there is a unique path in G linking u to v.

An example tree is given in Figure 4.

A *root* is a node of a tree which we wish to distinguish from the others, usually

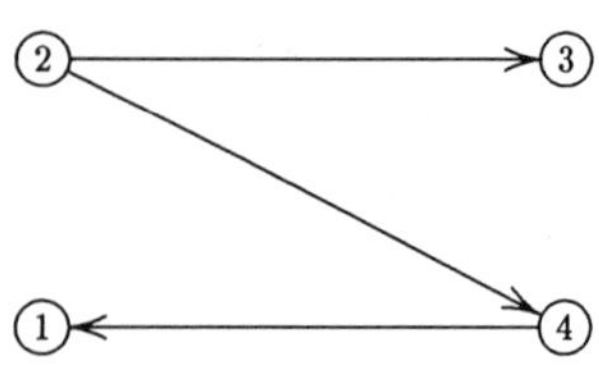

Fig. 4. Example tree.

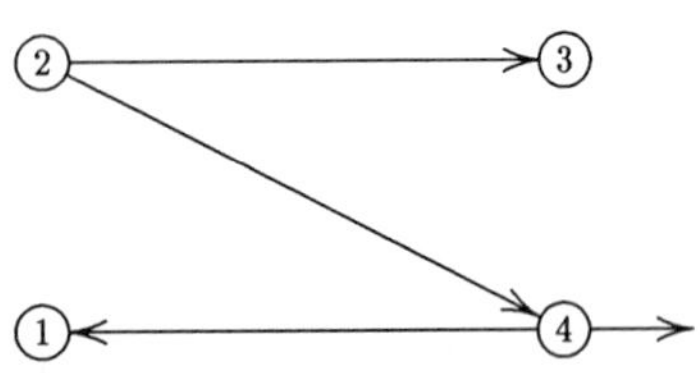

Fig. 5. Example rooted tree.

for some algorithmic purpose. Occasionally this may be made explicit by drawing a tree oriented with the root at the top of the diagram. Alternatively, this may be made explicit in a diagram by drawing a short line segment (with or without an arrowhead) incident on only the root node. An example rooted tree is given in Figure 5.

1.10. Tree solution algebra

Consider the solution of the system

$$Ax = b, \tag{1}$$

where A is the $n \times (n-1)$ node–arc incidence matrix for a tree $T = \langle V, E \rangle$ with n nodes and $n-1$ arcs and b is a given n-vector. A procedure which can be used to reduce the system to a permuted diagonal form is given below:

procedure DIAGONAL REDUCTION

inputs:	$T = \langle V, E \rangle$	–	nontrivial tree with n nodes
	ρ	–	root node for T
	A	–	node–arc incidence matrix for T
	b	–	n-vector
output:	$\bar{A}x = \bar{b}$	–	permuted diagonal system equivalent to $Ax = b$

begin
  [Initialize]
  $\bar{V} \Leftarrow V, \bar{E} \Leftarrow E, \bar{A} \Leftarrow A, \bar{b} \Leftarrow b$;
  [Iterate until the tree becomes trivial]
  while $\bar{E} \neq \Phi$ **do**
    [Pick a leaf]
    select an endnode p not the root ρ in the tree $\bar{T} = \langle \bar{V}, \bar{E} \rangle$.
    let r be the other node of the leaf incident on p.
    let (i, j) be the leaf incident on p and r.
    let c be the column of $\bar{A}$ corresponding to the leaf (i, j).
    [Pivot the system on the selected endnode]
    $\bar{b}_r \Leftarrow \bar{b}_r + \bar{b}_p, \bar{A}_{r.} \Leftarrow \bar{A}_{r.} + \bar{A}_{p.}$;
    $\bar{b}_p \Leftarrow \bar{A}_{pc}(\bar{b}_p), \bar{A}_{p.} \Leftarrow \bar{A}_{pc}(\bar{A}_{p.})$;
    [Update the tree by removing the leaf]
    $\bar{V} \Leftarrow \bar{V} \setminus \{p\}, \bar{E} \Leftarrow \bar{E} \setminus \{(i, j)\}$;
  endwhile
end

Note that at each pivot step in the above procedure, a partial sum of components of b is produced in the component $\bar{b}_r$, and in the last pivot, $\bar{b}_\rho = \sum_{k=1}^{n} b_k$ is produced. Also, after each pivot and tree update, the subset of the rows and columns of $\bar{A}$ corresponding to the nodes of $\bar{V}$ and the arcs of $\bar{E}$, respectively, is the node–arc incidence matrix of the updated tree $\bar{T}$.

In a node–arc incidence matrix A for a tree, let c be the column corresponding to a leaf with an endnode p not the root node ρ and let r be the other leaf node. Row p contains only one nonzero in column c, and row r contains the only other nonzero in column c. Thus when a pivot is made at the matrix position A_{pc}, A_{rc} will be zeroed and A_{pc} will become 1 if it was -1, but no other entries in A will be altered.

Now $\bar{A}$ initially has $2n-2$ nonzeros and $n-1$ pivots are made overall, so that the final pivot produces a $\bar{A}$ with $n-1$ nonzeros, all in the pivot rows. Thus row ρ of $\bar{A}$, the only row in which no pivot occurred, must contain all zeros. It follows that the system (1) has a solution if and only if $\bar{b}_\rho = 0$, hence no solution is possible unless $\sum_{k=1}^{n} b_k = 0$. Furthermore, since $n-1$ pivots have been made, the matrix A has rank $n-1$ so that when $\sum_{k=1}^{n} b_k = 0$, the solution produced by the algorithm is the unique solution to system (1).

To illustrate the use of the algorithm consider the tree in Figure 4.

The original system corresponding to (1) is

$$\begin{bmatrix} -1 & 0 & 0 \\ 0 & 1 & 1 \\ 0 & -1 & 0 \\ 1 & 0 & -1 \end{bmatrix} \begin{bmatrix} x_1 \\ x_2 \\ x_3 \end{bmatrix} = \begin{bmatrix} b_1 \\ b_2 \\ b_3 \\ b_4 \end{bmatrix}$$

Selecting node 4 as the root and using the sequence of selected endnodes 3, 2, 1 produces the following sequence of systems.

$$\begin{bmatrix} -1 & 0 & 0 \\ 0 & 0 & 1 \\ 0 & 1 & 0 \\ 1 & 0 & -1 \end{bmatrix} \begin{bmatrix} x_1 \\ x_2 \\ x_3 \end{bmatrix} = \begin{bmatrix} b_1 \\ b_2 + b_3 \\ -b_3 \\ b_4 \end{bmatrix}$$

$$\begin{bmatrix} -1 & 0 & 0 \\ 0 & 0 & 1 \\ 0 & 1 & 0 \\ 1 & 0 & 0 \end{bmatrix} \begin{bmatrix} x_1 \\ x_2 \\ x_3 \end{bmatrix} = \begin{bmatrix} b_1 \\ b_2 + b_3 \\ -b_3 \\ b_2 + b_3 + b_4 \end{bmatrix}$$

$$\begin{bmatrix} 1 & 0 & 0 \\ 0 & 0 & 1 \\ 0 & 1 & 0 \\ 0 & 0 & 0 \end{bmatrix} \begin{bmatrix} x_1 \\ x_2 \\ x_3 \end{bmatrix} = \begin{bmatrix} -b_1 \\ b_2 + b_3 \\ -b_3 \\ b_1 + b_2 + b_3 + b_4 \end{bmatrix}$$

Now let us consider adjoining an additional column e_k, where $1 \le k \le n$, to the right of A and lengthening x to accommodate the additional column. The expanded system is then

$$[A \mid e_k]x = b. \tag{2}$$

Suppose that we agree to choose k as the root node ($\rho = k$) and apply the above procedure to the expanded matrix and original tree, i.e. in the initialization step we set $\bar{A} \Leftarrow [A \mid e_\rho]$ instead of $\bar{A} \Leftarrow A$.

The same vector $\bar{b}$ and matrix $\bar{A}$ is produced in the first $n-1$ (original) columns and $\bar{A}_{.n} = e_k = e_\rho$. The system (2) is also permuted diagonal but is now of rank n and in its solution has $x_n = \sum_{k=1}^{n} b_k$. Since $[A \mid e_k]$ is square, the solution produced by the procedure must be unique. Furthermore, the solution to (1) produced by the above procedure when $\sum_{k=1}^{n} b_k = 0$ must have the same first $n-1$ variable values as those produced for the enlarged system (2).

In the above example, the original system corresponding to (2) with root node 4 is

$$\begin{bmatrix} -1 & 0 & 0 & 0 \\ 0 & 1 & 1 & 0 \\ 0 & -1 & 0 & 0 \\ 1 & 0 & -1 & 1 \end{bmatrix} \begin{bmatrix} x_1 \\ x_2 \\ x_3 \\ x_4 \end{bmatrix} = \begin{bmatrix} b_1 \\ b_2 \\ b_3 \\ b_4 \end{bmatrix}$$

And after the same sequence of pivots, the final equivalent system produced is

$$\begin{bmatrix} 1 & 0 & 0 & 0 \\ 0 & 0 & 1 & 0 \\ 0 & 1 & 0 & 0 \\ 0 & 0 & 0 & 1 \end{bmatrix} \begin{bmatrix} x_1 \\ x_2 \\ x_3 \\ x_4 \end{bmatrix} = \begin{bmatrix} -b_1 \\ b_2 + b_3 \\ -b_3 \\ b_1 + b_2 + b_3 + b_4 \end{bmatrix}$$

We remark that this usage of an extra e_k column in conjunction with the solution of system (1) provides strong impetus to extend the node–arc matrix representation to include a representation of a root k by a column e_k, when the underlying graph is a tree. We will find it useful to do so even when the underlying graph is a tree with additional arcs.

2. Primal simplex algorithm

All the network models presented in this Chapter may be viewed as special cases of the linear program and many of the algorithms to be presented are specializations of the primal simplex algorithm. Familiarity with the simplex method is assumed and this section simply presents a concise statement of the primal simplex method which will be specialized for several of the network models.

Let A be an $m \times n$ matrix with full row rank, c and u be n-component vectors, and b be an m-component vector. Let $X = \{x : Ax = b,\ \hat{0} \leq x \leq u\}$ and assume that $X \neq \Phi$. The *linear program* is to find an n-component vector $\bar{x}$ such that $c\bar{x} = \min\{cx : x \in X\}$. We adopt the convention that the symbol x refers to the vector of unknown decision variables and the symbol $\bar{x}$ refers to some known point in X.

Since A has full row rank, by a sequence of column interchanges it may be displayed in the partitioned form $[B \mid N]$, where B is nonsingular. With a corresponding partitioning of both x and u, we may write

$$X = \left\{(x^B \mid x^N) : Bx^B + Nx^N = b,\ \hat{0} \le x^B \le u^B, \hat{0} \le x^N \le u^N\right\}$$

A point $(\bar{x}^B \mid \bar{x}^N) \in X$ is called a *basic feasible solution* if $\bar{x}_j^N \in \{0, u_j^N\}$ for all $j \in \iota\,(n-m)$. The variables in x^B (x^N) are called the *basic* (*nonbasic*) variables. It is well known from linear programming theory that for every linear program with $X \ne \Phi$, there exists at least one basic feasible solution which is optimal.

We say that two basic feasible solutions are *adjacent* if x^{B_1} and x^{B_2} differ by exactly one variable, so that B_1 may differ from B_2 in exactly one column. The primal simplex algorithm is an iterative procedure which begins with some basic feasible solution and moves through a sequence of adjacent basic feasible solutions until a stopping criterion is met which guarantees that the final basic feasible solution is optimal.

A mathematical description of the primal simplex algorithm follows:

procedure PRIMAL SIMPLEX

inputs:	A	–	$m \times n$ constraint matrix
	c	–	m-component vector of unit costs
	u	–	n-component vector of upper bounds
	b	–	m-component right-hand-side vector

output: $(\bar{x}^B \mid \bar{x}^N)$, an optimal basic feasible solution for $\min\{cx : x \in X\}$

assumptions:
1. $rank(A) = m$
2. $X = \{x : Ax = b,\ \hat{0} \le x \le u\} \ne \Phi$

begin
[Initialization]
optimal $\Leftarrow$ 'no';
let $(\bar{x}^B \mid \bar{x}^N) \in X$ be any basic feasible solution with A, c, and u partitioned in corresponding fashion as $[B \mid N]$, $(c^B \mid c^N)$, and $(u^B \mid u^N)$;
while optimal = 'no' **do**
[Dual Calculation]
$\pi \Leftarrow c^B B^{-1}$
[Pricing]
$L \Leftarrow \{j : \pi N_{\cdot j} - c_j^N > 0 \text{ and } \bar{x}_j^N = 0\}$;
$M \Leftarrow \{j : \pi N_{\cdot j} - c_j^N < 0 \text{ and } \bar{x}_j^N = u_j^N\}$;
if $L \cup M = \Phi$
then
optimal $\Leftarrow$ 'yes'
else
select $k \in L \cup M$;
[Column Update]
$y \Leftarrow B^{-1} N_{\cdot k}$;

[Ratio Test]
$R \Leftarrow \{i : y_i > 0\}$ and $S \Leftarrow \{i : y_i < 0\}$;
if $k \in L$
then
$\Delta \Leftarrow \min\{u_k^N, \min_{i \in R}(\bar{x}_i^B / y_i), \min_{i \in S}((u_i^B - \bar{x}_i^B)/ - y_i)\}$;
$T \Leftarrow \{i \in R : (\bar{x}_i^B / y_i) = \Delta\} \cup \{i \in S : ((u_i^B - \bar{x}_i^B)/ - y_i) = \Delta\}$;
else
$\Delta \Leftarrow \min\{u_k^N, \min_{i \in R}((u_i^B - \bar{x}_i^B)/y_i), \min_{i \in S}(\bar{x}_i^B / - y_i)\}$;
$T \Leftarrow \{i \in R : ((u_i^B - \bar{x}_i^B)/y_i) = \Delta\} \cup \{i \in S : (\bar{x}_i^B / - y_i) = \Delta\}$;
endif
[Value Update]
if $k \in L$
then
$\bar{x}_k^N \Leftarrow \Delta$ and $\bar{x}^B \Leftarrow \bar{x}^B - \Delta y$;
else
$\bar{x}_k^N \Leftarrow u_k^N - \Delta$ and $\bar{x}^B \Leftarrow \bar{x}^B + \Delta y$;
endif
if $\Delta \neq u_k^N$
then
[Basis Exchange Update]
select $j \in T$;
interchange x_j^B with x_k^N to form a new partitioning $[B \mid N]$;
endif
endif
endwhile
end

The only sticky issue concerning the above algorithm is that it may be possible to move through a sequence of basis interchanges, all of which correspond to the same actual point, only changing the representations to correspond to the varying bases. That is, there may exist a sequence of basis interchanges all having $\Delta = 0$ which could result in the above procedure being stuck in a nonterminating loop. This phenomenon is known as *cycling*. Much discussion of cycling can be found in the literature along with anticycling rules [1] which have been developed to ensure convergence.

3. Linear network models

Let $\langle V, E\rangle$ be a network through which some commodity will be flowing. Associated with each node $v \in V$ we define a requirement r_v. A node having a supply of the commodity is assigned a positive requirement equal to the supply.

[1] We have never observed cycling in any of our software implementations of this algorithm and have not incorporated any of the anticycling rules in our software.

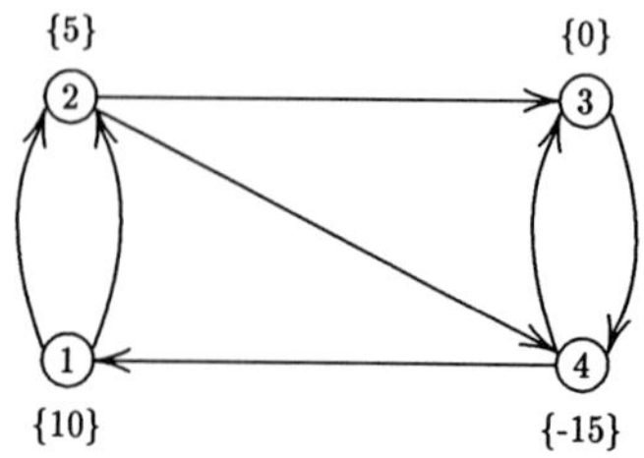

Fig. 6. Example network with requirements.

Conversely, negative requirements correspond to demands for the commodity at the specified nodes.

Suppose that for the example network illustrated in Figure 1, that nodes 1 and 3 are supply nodes with supply of 10 and 5, respectively; and that node 4 is a demand point with a demand of 15. This network and the corresponding requirements are illustrated in Figure 6.

For linear network problems, one is seeking a set of flows on the arcs which satisfy the supply and demand restrictions at each node. For the example network, the total flow into node 2 plus the five supply units which originate at node 2 must equal the total flow departing node 2. We say that a *feasible flow* satisfies *flow conservation* which implies that it is an element of $\{x : Ax = r\}$, where A is the corresponding node–arc incidence matrix for $\langle V, E\rangle$.

For the example network, the flow conservation equations are

$$\begin{bmatrix} 1 & 1 & 0 & 0 & 0 & 0 & -1 \\ -1 & -1 & 1 & 1 & 0 & 0 & 0 \\ 0 & 0 & -1 & 0 & 1 & -1 & 0 \\ 0 & 0 & 0 & -1 & -1 & 1 & 1 \end{bmatrix} \begin{bmatrix} x_1 \\ x_2 \\ x_3 \\ x_4 \\ x_5 \\ x_6 \\ x_7 \end{bmatrix} = \begin{bmatrix} 10 \\ 5 \\ 0 \\ -15 \end{bmatrix}$$

Associated with each of the arcs in E, we define a vector of unit flow costs c and a vector of flow capacities or bounds u. Thus, the cost for each unit of flow in arc k is given by c_k and the flow on arc k is restricted to the interval $[0, u_k]$. Mathematically, the linear network model on $\langle V, E\rangle$ with node–arc incidence matrix A is given by

$$\min_x \{cx : Ax = r, 0 \le x \le u\}$$

A sample model for the network illustrated in Figure 6 is

$$\text{minimize} \quad x_1 + 3x_2 + 5x_3 - 7x_4 + 7x_5 - x_6 + 9x_7$$

subject to

$$\begin{bmatrix} 1 & 1 & 0 & 0 & 0 & 0 & -1 \\ -1 & -1 & 1 & 1 & 0 & 0 & 0 \\ 0 & 0 & -1 & 0 & 1 & -1 & 0 \\ 0 & 0 & 0 & -1 & -1 & 1 & 1 \end{bmatrix} \begin{bmatrix} x_1 \\ x_2 \\ x_3 \\ x_4 \\ x_5 \\ x_6 \\ x_7 \end{bmatrix} = \begin{bmatrix} 10 \\ 5 \\ 0 \\ -15 \end{bmatrix}$$

$$\begin{array}{ccccc} 0 & \le & x_1 & \le & 6 \\ 0 & \le & x_2 & \le & 8 \\ 0 & \le & x_3 & \le & 10 \\ 0 & \le & x_4 & \le & 10 \\ 0 & \le & x_5 & \le & 8 \\ 0 & \le & x_6 & \le & 8 \\ 0 & \le & x_7 & \le & 8 \end{array}$$

3.1. Basis characterization

Recall that one of the assumptions for the primal simplex algorithm is that the constraint matrix has full row rank, i.e. the $m \times n$ constraint matrix A has $rank(A) = m$. It is clear from Sections 1.5 and 1.8 that if A is a node–arc incidence matrix, $rank(A) < m$. We will henceforth assume that $\langle V, E \rangle$ is a connected graph, so that the rank of the node–arc incidence matrix is one less than the number of nodes. In addition, a *root arc* is appended to the problem. Let

$$X^E = \left\{ (x \mid a) : [A \mid e_\rho] \left[\frac{x}{a} \right] = r, \quad 0 \le x \le u \right\}$$

The revised model is then simply

$$\min_x \left\{ cx : (x \mid a) \in X^E \right\},$$

The connected graph must have a spanning tree and the results in Section 1.8 imply that

$$[A \mid e_\rho] \left[\frac{x}{a} \right] = e_k$$

may now be solved for any k. Thus the enlarged constraint matrix $[A \mid e_\rho]$ has full row rank, so that the primal simplex algorithm can be applied directly to this model. In Section 1.10 it was shown that the root arc will carry no flow ($a = 0$) when $\sum_i r_i = 0$ and we are only solving the linear system. This will also be true when any nonbasic arcs are set to upper bound, since setting a nonbasic flow $x_k = u_k$, where x_k is the flow on arc (i, j) is equivalent to adding u_k to r_j and subtracting u_k from r_i, thus preserving the condition $\sum_i r_i = 0$.

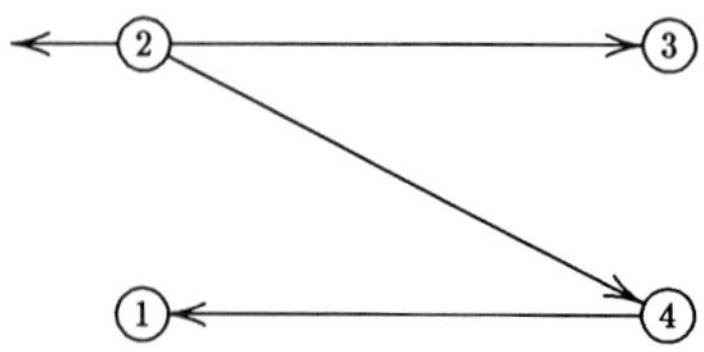

(a) Basis tree 1

(b) Basis tree 2

Fig. 7. Basis trees for sample network.

Let B be a basis for $[A \mid e_\rho]$ so that the entire matrix is partitioned as $[B \mid N]$. B may be further partitioned into $[S \mid e_\rho]$. Since without e_ρ the constraint matrix is not of full rank, e_ρ must be part of the basis. In Section 1.7 we noted that the columns of A corresponding to a cycle are linearly dependent. Thus the other $m-1$ columns of B must be acyclic and will form a spanning tree for $\langle V, E\rangle$. We may write the corresponding basic solution as $(x^S \mid a \mid x^N)$.

By row and column interchanges B may be displayed in lower triangular form. The trees corresponding to the bases

$$\begin{array}{cccc} e_{41} & e_{24} & e_{23} & e_2 \end{array} \qquad\qquad \begin{array}{cccc} e_{12} & e_{41} & e_{43} & e_3 \end{array}$$
$$\begin{bmatrix} -1 & 0 & 0 & 0 \\ 0 & 1 & 1 & 1 \\ 0 & 0 & -1 & 0 \\ 1 & -1 & 0 & 0 \end{bmatrix} \quad \text{and} \quad \begin{bmatrix} 1 & -1 & 0 & 0 \\ -1 & 0 & 0 & 0 \\ 0 & 0 & -1 & 1 \\ 0 & 1 & 1 & 0 \end{bmatrix}$$

are illustrated in Figure 7.

The trees can be used to determine the row and column interchanges required to display the bases in lower triangular form. The root node and root arc are always placed in the last row and column. The first $m-1$ rows and columns are determined recursively by a process in which we first select an endnode with its corresponding leaf arc and then remove them from the tree after appropriately reordering the matrix. Note that this ordering may not be unique since multiple endnodes are always present.

A procedure which can be used to display a linear network basis in lower triangular form is given below:

procedure DISPLAY LOWER TRIANGULAR

input:	$[S \mid e_\rho]$	–	$m \times m$ basis for a linear network problem, where S corresponds to a spanning tree
outputs:	*nodeorder*$[i]$	–	the node in row position i
	arcorder$[j]$	–	the arc in column position j

begin
  [Initialization]
  let $\langle V, T\rangle$ be a network corresponding to S
  nodeorder$[m] \Leftarrow \rho$, *arcorder*$[m] \Leftarrow e_\rho$;
  $i \Leftarrow 1$;
  while $i < m$ **do**
    [Tree Reduction]
    let $v \in V$ be an endnode of $\langle V, T\rangle$;
    let e_{jk} be the leaf arc incident to v;
    nodeorder$[i] \Leftarrow v$, *arcorder*$[i] \Leftarrow e_{jk}$, $V \Leftarrow V \setminus \{v\}$, $T \Leftarrow T \setminus \{e_{jk}\}$, $i \Leftarrow i+1$;
  endwhile
end

An application of procedure DISPLAY LOWER TRIANGULAR to the tree in Figure 7a with certain choices in the tree reduction yields *nodeorder* $= [3, 1, 4, 2]$ and *arcorder* $= [e_{23}, e_{41}, e_{24}, e_2]$. The corresponding matrix is:

$$\begin{array}{c} \text{(nodes)} \\ 3 \\ 1 \\ 4 \\ 2 \end{array} \begin{array}{c} \begin{array}{cccc} e_{23} & e_{41} & e_{24} & e_2 \end{array} \\ \begin{bmatrix} -1 & 0 & 0 & 0 \\ 0 & -1 & 0 & 0 \\ 0 & 1 & -1 & 0 \\ 1 & 0 & 1 & 1 \end{bmatrix} \end{array}$$

Another application of the procedure with different choices made in the tree reduction would yield *nodeorder* $= [1, 3, 4, 2]$ and *arcorder* $= [e_{41}, e_{23}, e_{24}, e_2]$, with corresponding matrix is:

$$\begin{array}{c} \text{(nodes)} \\ 1 \\ 3 \\ 4 \\ 2 \end{array} \begin{array}{c} \begin{array}{cccc} e_{41} & e_{23} & e_{24} & e_2 \end{array} \\ \begin{bmatrix} -1 & 0 & 0 & 0 \\ 0 & -1 & 0 & 0 \\ 1 & 0 & -1 & 0 \\ 0 & 1 & 1 & 1 \end{bmatrix} \end{array}$$

3.2. Dual calculation

Based on the results in Section 3.1, it is known that every basis for a linear network problem takes the form $[S \mid e_\rho]$, where the columns of S correspond to a spanning tree $\langle V, T\rangle$ and S is triangularizable. Therefore the dual calculation

$$\pi[S \mid e_\rho] = [c^S \mid 0]$$

can be specialized to exploit the underlying network structure. Since S consists of columns from a node–arc incidence matrix, every column of S is a vector e_{ij} for some i and j. Hence $S = [e_{i_1 j_1}, e_{i_2 j_2}, \ldots, e_{i_{m-1} j_{m-1}}]$ and the dual calculation reduces to the system

$$\left\{\begin{array}{ccc} \pi_{i_1} - \pi_{j_1} & = & c_1^S \\ \pi_{i_2} - \pi_{j_2} & = & c_2^S \\ \vdots & \vdots & \vdots \\ \pi_{i_{m-1}} - \pi_{j_{m-1}} & = & c_{m-1}^S \\ \pi_\rho & = & 0 \end{array}\right\}$$

Since this system is upper triangular it can be solved by back substitution. Beginning with $\pi_\rho = 0$, the duals for all basic arcs incident on node ρ can be determined. Once these duals are known, all duals for basic arcs incident on those arcs can be determined. Continuing in this manner, eventually all duals will be determined. The following procedure formally states how this can be accomplished without actually triangularizing a matrix.

procedure DUAL CALCULATION

inputs: $[S \mid e_\rho]$ – $m \times m$ basis for a linear network problem
c^S – $m-1$ vector of basic costs, where c_{ij}^S is the unit cost for basic arc e_{ij}

output: π – $\pi = [c^S \mid 0][S \mid e_\rho]^{-1}$

begin
  [Initialization]
  let $\langle V, T\rangle$ be the spanning tree corresponding to S;
  for $i = 1, \ldots, m$
    $label[i]$ = 'no';
  endfor
  $\pi_\rho \Leftarrow 0$, $label[\rho] \Leftarrow$ 'yes', $k \Leftarrow 1$;
  while $k < m$ **do**
    let $e_{ij} \in T$ such that $label[i] \neq label[j]$;
    if $label[i]$ = 'yes'
      then
        $\pi_j = \pi_i - c_{ij}^S$, $label[j]$ = 'yes';
      else
        $\pi_i = \pi_j + c_{ij}^S$, $label[i]$ = 'yes';
    endif
    $k \Leftarrow k + 1$;
  endwhile
end

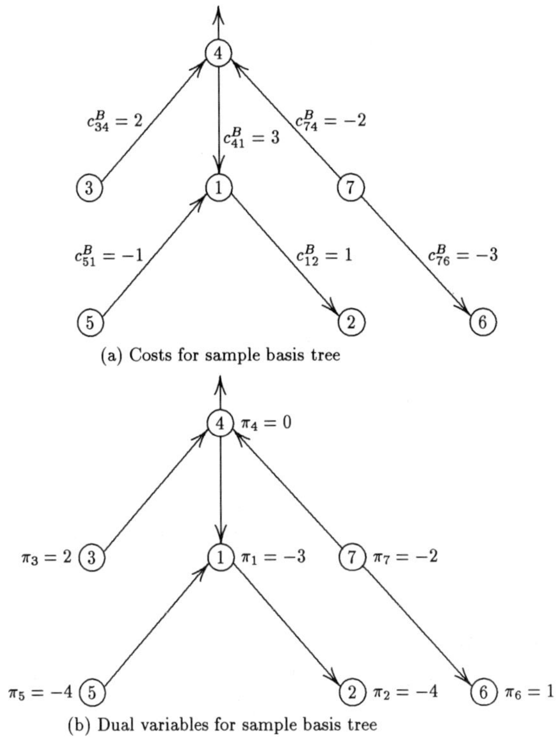

Fig. 8. Dual variable calculation example.

Consider the basis tree in Figure 8 with

$$[c_{12}^S, c_{34}^S, c_{41}^S, c_{51}^S, c_{74}^S, c_{76}^S] = [1, 2, 3, -1, -2, -3]$$

Since node 4 is the root node, $\pi_4 = 0$. The equations

$$\left\{ \begin{array}{rcr} \pi_3 - \pi_4 & = & 2 \\ \pi_4 - \pi_1 & = & 3 \\ \pi_7 - \pi_4 & = & -2 \end{array} \right\}$$

yield $\pi_3 = 2$, $\pi_1 = -3$, and $\pi_7 = -2$. Then the equations

$$\left\{ \begin{array}{rcr} \pi_5 - \pi_1 & = & -1 \\ \pi_1 - \pi_2 & = & 1 \\ \pi_7 - \pi_6 & = & -3 \end{array} \right\}$$

yield $\pi_5 = -4$, $\pi_2 = -4$, and $\pi_6 = 1$. Also note that the dual calculation only involves addition and subtraction. Hence, if the cost coefficients are all integer, then the dual variables will also be integral valued.

3.3. Column update

Suppose e_{st} denotes the nonbasic arc which prices favorably and is selected for a potential flow change. There is no guarantee that the flow will actually change since the ratio test could yield $\Delta = 0$. The column update step of the primal simplex algorithm requires that $y = [S \mid e_\rho]^{-1} e_{st}$ be determined. Since the arcs of S correspond to a spanning tree and $[S \mid e_\rho]$ is lower triangular, the calculation of y can be simplified.

The updated column y is a vector which solves the lower triangular system $[S \mid e_\rho]y = e_{st}$ or, in component form,

$$S_{.1}y_1 + S_{.2}y_2 + \cdots + S_{.m-1}y_{m-1} + e_\rho y_m = e_{st}$$

That is, a set of scalars $y_1, \ldots, y_m$ are sought which when multiplied by the columns of S and the vector e_ρ and added together form the vector e_{st}. Let $\langle V, T\rangle$ denote the tree corresponding to S and let

$$P = \{v_1, e_{i_1 j_1}, v_2, e_{i_2 j_2}, \ldots, v_q, e_{i_q j_q}, v_{q+1}\}$$

denote the simple path in $\langle V, T\rangle$ linking s to t. Such a path is illustrated in Figure 9. Since some arcs in the path may be directed from some v_i toward v_{i+1} and others from some v_{j+1} toward v_j, arrow heads have been omitted from the illustration.

By reordering the rows and the arcs in S, the system of equations corresponding to the arcs in the path may be illustrated as shown below.

$$\pm \begin{bmatrix} 1 \\ -1 \\ 0 \\ 0 \\ \vdots \\ 0 \\ 0 \\ 0 \end{bmatrix} y_1 \pm \begin{bmatrix} 0 \\ 1 \\ -1 \\ 0 \\ \vdots \\ 0 \\ 0 \\ 0 \end{bmatrix} y_2 \pm \begin{bmatrix} 0 \\ 0 \\ 1 \\ -1 \\ \vdots \\ 0 \\ 0 \\ 0 \end{bmatrix} y_3 \pm \cdots \pm \begin{bmatrix} 0 \\ 0 \\ 0 \\ 0 \\ \vdots \\ 0 \\ 1 \\ -1 \end{bmatrix} y_q = \begin{bmatrix} 1 \\ 0 \\ 0 \\ 0 \\ \vdots \\ 0 \\ 0 \\ -1 \end{bmatrix}$$

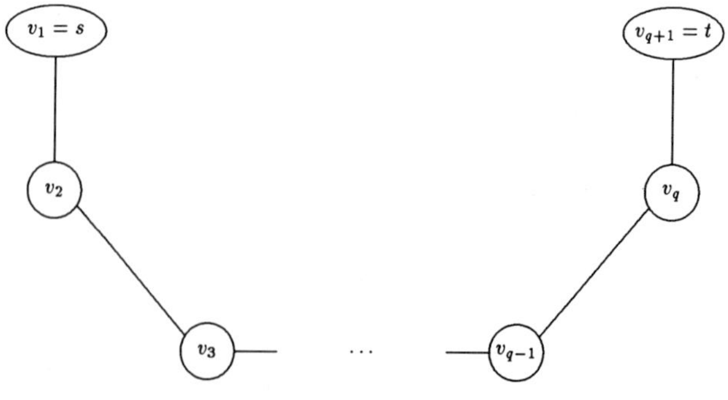

Fig. 9. A simple path linking s to t.

Since the first row of the above system has a single nonzero, y_1 is uniquely determined and must be either 1 or -1. Once y_1 is known, y_2 is then uniquely determined and must be either 1 or -1. Similarly, $y_3, \ldots, y_q$ can be determined successively. Therefore, a solution to $[S \mid e_\rho]y = e_{st}$ can be constructed by setting the components of y corresponding to the path from s to t in $\langle V, T\rangle$ to $\pm 1s$ as described above and setting all other components to zero. This is generally called a *cycle trace* and is formally presented in the following procedure.

procedure CYCLE TRACE

inputs: $[S \mid e_\rho]$ – $m \times m$ basis for a linear network problem
e_{st} – m vector corresponding to a selected nonbasic arc

output: y – $y = [S \mid e_\rho]^{-1} e_{st}$

begin
  [Initialization]
  let $\langle V, T\rangle$ be the spanning tree corresponding to S;
  let $P = \{v_1, e_{i_1 j_1}, v_2, e_{i_2 j_2}, \ldots, v_q, e_{i_q j_q}, v_{q+1}\}$
  be the simple path in $\langle V, T\rangle$ linking node s to node t;
  for $i = 1, \ldots, m$
    $y_i = 0$;
  endfor
  $k \Leftarrow 1$;
  while $k \leq q$ **do**
    let c be the column index of S corresponding to arc $e_{i_k j_k}$;
    if $e_{i_k j_k} = e_{v_k} - e_{v_{k+1}}$
      then
        $y_c = 1$;
      else
        $y_c = -1$;
    endif
    $k \Leftarrow k + 1$;
  endwhile
end

Consider the basis tree in Figure 8 and suppose $s = 5$ and $t = 6$. The simple path and corresponding values of the updated column are illustrated in Figure 10. Note that the nonzero components of the updated column y are identical to the orientation function (see Section 1.7) on the simple path from s to t. Furthermore, the components of y are from $\{1, -1\}$.

3.4. Basis update

As seen in Sections 3.2 and 3.3, the key operations for the primal simplex algorithm can be performed directly on the spanning tree $\langle V, T\rangle$. In this section a

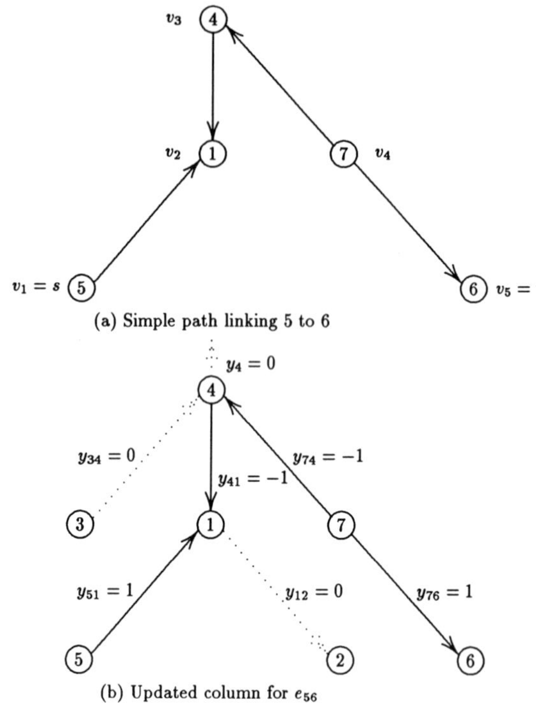

Fig. 10. Updated column example.

data structure used to store the spanning tree in computer memory is presented along with an algorithm which will perform the basis exchange update using this data structure.

Suppose the rooted tree is drawn in the plane placing the root node at the top with the branches extending downward as illustrated in Figure 11a. One may imagine tracing a line around the contours of the tree as illustrated in Figure 11b. Traversing a tree in this way has become known as a depth-first search. For the example, the nodes in this search could be ordered as 4,3,4,1,5,1,2,1,4,7,6,7,4. By eliminating all duplicate occurrences an ordering known as *preorder* is obtained. For this example, the corresponding preorder is 4,3,1,5,2,7,6. The label which gives the next node in the preorder for node v is known as the *thread*, denoted by $t(v)$. The thread for the example is illustrated in Figure 11c.

The other two labels are related to the path from a given node v to the root ρ in the basis tree $\langle V, T \rangle$. Let

$$P = \{v = v_1, e_{i_1 j_1}, v_2, e_{i_2 j_2}, \ldots, v_q, e_{i_q j_q}, v_{q+1} = \rho\}$$

denote the simple path in $\langle V, T \rangle$ which links v to ρ. The predecessor label $p(v)$ is v_2 (the second node on the path) and the distance label $d(v)$ is q (the number of arcs on the path). Both the predecessor and the distance labels of the root ρ are

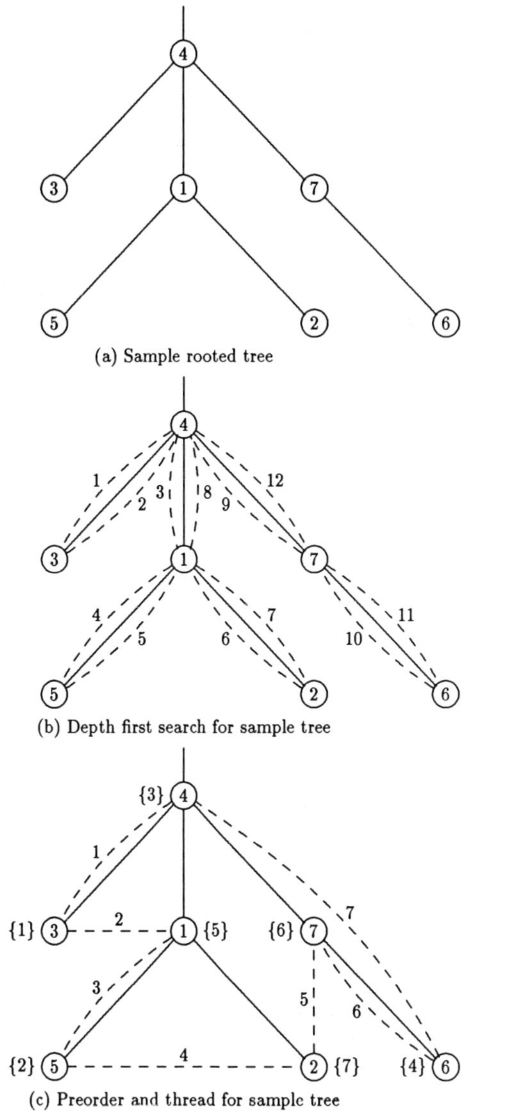

Fig. 11. Illustration of thread labels for sample rooted tree.

defined to be zero. The labels for the rooted tree shown in Figure 11 are given in Table 1.

Both the dual calculation (Section 3.2) and the column update (Section 3.3) can be easily implemented in software using the triple labels to represent the basis tree $\langle V, T\rangle$. The only tricky part is the technique needed for a basis exchange update. For example, suppose the arc (2, 7) is exchanged with the arc (1, 4) in the rooted tree illustrated in Figure 11.

Table 1
Labels for the rooted tree in Figure 11a

Node v	Predecessor $p(v)$	Thread $t(v)$	Distance $d(v)$
1	4	5	1
2	1	7	2
3	4	1	1
4	0	3	0
5	1	2	2
6	7	4	2
7	4	6	1

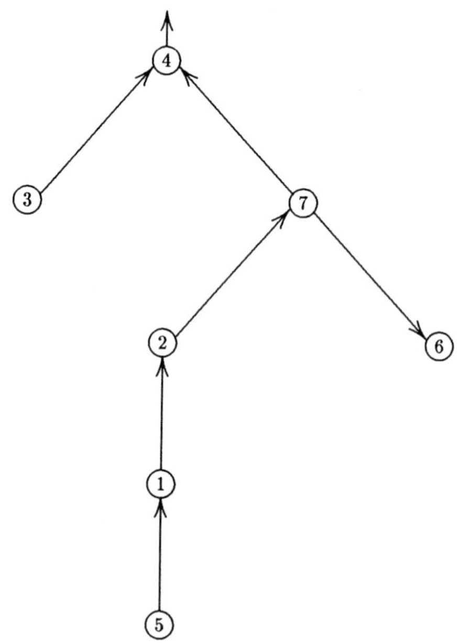

Fig. 12. New rooted tree after basis exchange update.

Table 2
Labels after the basis exchange update

Node v	Predecessor $p(v)$	Thread $t(v)$	Distance $d(v)$
1	2	5	3
2	7	1	2
3	4	7	1
4	0	3	0
5	1	6	4
6	7	4	2
7	4	2	1

The new tree is illustrated in Figure 12 and the corresponding triple labels are given in Table 2. An algorithm which will perform this exchange is given below.

procedure BASIS EXCHANGE UPDATE

inputs:	$\langle V, T\rangle$	–	the current basis tree
	$p[i]$	–	the predecessor label of node i
	$t[i]$	–	the thread label of node i
	$d[i]$	–	the distance label of node i
	(u, v)	–	the arc selected to become part of the new basis
	k	–	node such that k and $p[k]$ are the end nodes of the arc which is to be removed from the current basis, with k and $p[k]$ on the path from u to the root node of the current basis tree
outputs:	$p[i]$	–	updated predecessor label of node i
	$t[i]$	–	updated thread label of node i
	$d[i]$	–	updated distance label of node i

begin
  [Initialization]
  $q \Leftarrow u,\ q' \Leftarrow v,\ i \Leftarrow q,\ j \Leftarrow p[q],\ k' \Leftarrow p[k],\ l \Leftarrow d[q'] + 1$, end $\Leftarrow$ 'no';
  while end = 'no' **do**
    $l' \Leftarrow d[i],\ m \Leftarrow l - d[i],\ d[i] \Leftarrow l,\ z \Leftarrow i,\ x \Leftarrow t[i]$;
    while $d[x] > l'$ **do**
      $d[x] \Leftarrow d[x] + m,\ z \Leftarrow x,\ x \Leftarrow t[x]$;
    endwhile
    $r \Leftarrow j$;
    while $t[r] \neq i$ **do**
      $r \Leftarrow t[r]$;
    endwhile
    $t[r] \Leftarrow x$;
    if $i = k$
      then
        $t[z] \Leftarrow t[q'],\ t[q'] \Leftarrow q,\ p[q] \Leftarrow q'$, end $\Leftarrow$ 'yes';
      else
        $t[z] \Leftarrow j,\ r \Leftarrow i,\ i \Leftarrow j,\ j \Leftarrow p[j],\ p[i] \Leftarrow r,\ l \Leftarrow l + 1$;
    endif
  endwhile
end

A slightly more advanced data structure which uses four node-length arrays has been incorporated into some software implementations of the network simplex algorithm. Node i will be called a *successor* of node j, if j is on the simple path in the basis tree $\langle V, T\rangle$ which links i to the root ρ. For $j \in V$ with $j \neq \rho$, let $C_j = \{i : j$ is on the simple path in $\langle V, T\rangle$ linking i to $\rho\}$. That is, C_j is the set of

Table 3
Enhanced labels for the rooted tree in Figure 11a

Node v	Predecessor $p(v)$	Thread $t(v)$	Cardinality $c(v)$	Last successor $l(v)$
1	4	5	3	2
2	1	7	1	2
3	4	1	1	3
4	0	3	7	6
5	1	2	1	5
6	7	4	1	6
7	4	6	2	6

successors of j. We also define $C_\rho = V$. Note that for all $j \in V$, $j \in C_j$. The *cardinality* of node j, denoted by $c(j)$ is simply $|C_j|$. The *last successor* of node $j \neq \rho$ is defined to be the unique $v \in C_j$ such that the thread of v is not an element of C_j, i.e. $t(v) \notin C_j$. Also, define the last successor of ρ as the node k such that $t(k) = \rho$. The labels for the rooted tree shown in Figure 11 are given in Table 3.

This data structure eliminates the search for the last successor required in the previous procedure at the expense of maintaining this label. For each basis tree exchange, the number of nodes whose cardinality changes is usually fewer than the number of nodes whose distance changes. Also, knowing the cardinality of a subtree rooted at a given node can be used to speedup the dual variable update.

In our software implementation of the primal simplex algorithm for linear network problems, the basis exchange update is also integrated with the value update. In addition, since the set of nodes whose dual values change and the set of nodes whose distance labels change is the same set, the dual update calculation is also integrated with the basis exchange update. All of these specializations can result in software which can execute one hundred times faster than general purpose linear programming software. Most of the modern commercial linear programming systems now contain a specialized network solution module.

4. Generalized networks

Let G be an $m \times n$ matrix with full row rank such that every column of G has at most two nonzero entries. Let c and u be n-component vectors and b be an m-component vector. Let $Y = \{x : Gx = b,\ \hat{0} \leq x \leq u\}$ and assume that $Y \neq \emptyset$. The *generalized network problem* is to find an n-component vector $\bar{x}$ such that

$$c\bar{x} = \min_x \{cx : x \in Y\}$$

Note that the linear network model is a special case of the generalized network problem. The revised model developed in Section 3.1 is simply

$$\min_x \{cx : (x, a) \in X^E\}$$

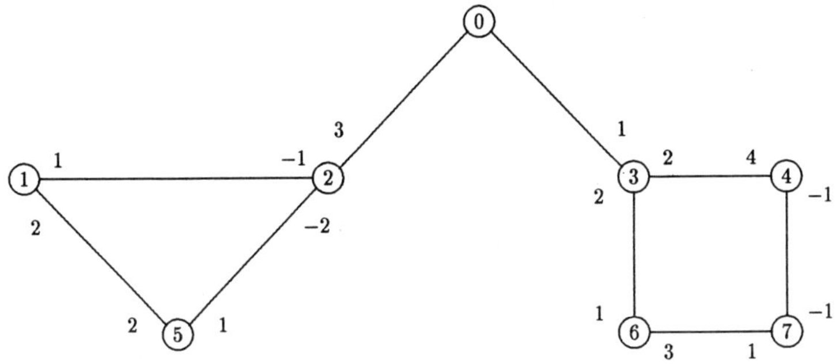

Fig. 13. Example generalized network.

where

$$X^E = \left\{(x \mid a) : Ax + e_\rho a = r, \hat{0} \le x \le u\right\}$$

The corresponding generalized model has $G = [A \mid e_\rho]$.

Let $V = \iota m = \{1, \ldots, m\}$. Then n arcs on the node set V can be constructed using the n columns of G. If $G_{.j}$ has two nonzero entries in rows i and k, we let the corresponding arc be (i, k). If $G_{.j}$ had only a single nonzero entry in row i, we let the corresponding arc be a root (i).

Consider the following matrix.

$$\begin{bmatrix} 1 & 2 & 0 & 0 & 0 & 0 & 0 & 0 & 0 \\ -1 & 0 & 3 & -2 & 0 & 0 & 0 & 0 & 0 \\ 0 & 0 & 0 & 0 & 1 & 2 & 2 & 0 & 0 \\ 0 & 0 & 0 & 0 & 0 & 4 & 0 & -1 & 0 \\ 0 & 2 & 0 & 1 & 0 & 0 & 0 & 0 & 0 \\ 0 & 0 & 0 & 0 & 0 & 0 & 1 & 0 & 3 \\ 0 & 0 & 0 & 0 & 0 & 0 & 0 & -1 & 1 \end{bmatrix}$$

Using the rule presented above, the following network can be constructed:

$$\langle\{1, 2, 3, 4, 5, 6, 7\}, \{(1, 2), (1, 5), (2), (2, 5), (3), (3, 4), (3, 6), (4, 7), (6, 7)\}\rangle$$

A visual representation of this network is given in Figure 13. The numbers associated with the ends of the arcs correspond to the nonzero entries in the columns of G.

4.1. Basis characterization

Let G be an $m \times n$ matrix with full row rank such that every column of G has at most two nonzero entries. Let B be a basis for G and let $\bar{B}$ be a reordering of rows and columns of B such that $\bar{B}$ is displayed in block diagonal form. That is,

$$\bar{B} = \begin{bmatrix} \bar{B}^1 & & & \\ & \bar{B}^2 & & \\ & & \ddots & \\ & & & \bar{B}^p \end{bmatrix}$$

For example:

$$B = \begin{bmatrix} 1 & 2 & 0 & 0 & 0 & 0 & 0 \\ -1 & 0 & -2 & 0 & 0 & 0 & 0 \\ 0 & 0 & 0 & 1 & 2 & 2 & 0 \\ 0 & 0 & 0 & 0 & 4 & 0 & -1 \\ 0 & 2 & 1 & 0 & 0 & 0 & 0 \\ 0 & 0 & 0 & 0 & 0 & 1 & 0 \\ 0 & 0 & 0 & 0 & 0 & 0 & -1 \end{bmatrix}$$

and

$$\bar{B} = \left[\begin{array}{ccc|cccc} 1 & 0 & 2 & & & & \\ -1 & -2 & 0 & & 0 & & \\ 0 & 1 & 2 & & & & \\ \hline & & & 1 & 0 & 0 & 0 \\ & 0 & & 0 & -1 & 0 & 0 \\ & & & 0 & -1 & 4 & 0 \\ & & & 2 & 0 & 2 & 1 \end{array}\right] \quad \begin{array}{c} \text{(row)} \\ 1 \\ 2 \\ 5 \\ 6 \\ 7 \\ 4 \\ 3 \end{array}$$

with $p = 2$. The number of blocks, p, can be as many as m. Let the networks associated with each of the blocks be denoted by $\langle V^1, E^1 \rangle, \ldots, \langle V^p, E^p \rangle$. For the above example, $\langle V^1, E^1 \rangle = \langle \{1, 2, 5\}, \{(1, 2), (2, 5), (1, 5)\} \rangle$ and $\langle V^2, E^2 \rangle = \langle \{3, 4, 6, 7\}, \{(3), (3, 4), (3, 6), (4, 7)\} \rangle$.

The corresponding networks are illustrated in Figure 14.

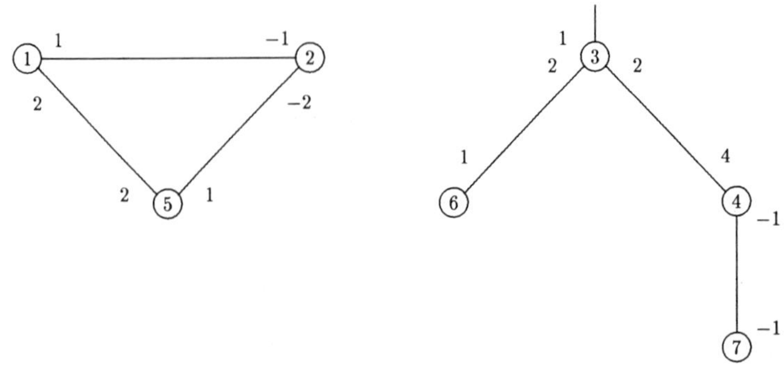

Fig. 14. Basis for a generalized network.

A connected network having exactly one cycle is traditionally called a *one-tree* and a connected network having exactly one root arc is traditionally called a *rooted tree*. It is well known that the connected components from a basis of a generalized network are either one-trees or rooted trees (see e.g. Section 8.1 of Murty [1992]). That is, $\langle V^1, E^1\rangle, \ldots, \langle V^p, E^p\rangle$ are either one-trees or rooted trees. For the basis illustrated in Figure 14, one component is a rooted tree and the other is a one-tree.

4.2. *Dual calculation*

The special structure of the basis $\bar{B}$ can be exploited in the calculation of the dual variables from $\pi\bar{B} = \bar{c}$. Since $\bar{B}$ is block diagonal we obtain

$$\left[\ \pi^1 \mid \pi^2 \mid \cdots \mid \pi^p\ \right]\begin{bmatrix} \bar{B}^1 & & & \\ & \bar{B}^2 & & \\ & & \ddots & \\ & & & \bar{B}^p \end{bmatrix} = \left[\ \bar{c}^1 \mid \bar{c}^2 \mid \cdots \mid \bar{c}^p\ \right]$$

Hence, the p systems $\pi^q\bar{B}^q = \bar{c}^q$ can be solved independently.

Two cases must be considered. Suppose $\bar{B}^q$ corresponds to a rooted tree. Then by row and column interchanges, $\bar{B}^q$ can be displayed in lower triangular form and $\pi^q\bar{B}^q = \bar{c}^q$ can be solved by back-substitution.

Consider the example illustrated in Figure 15 with

$$\left[\ \pi_6 \mid \pi_7 \mid \pi_4 \mid \pi_3\ \right]\left[\begin{array}{c|c|c|c} 1 & 0 & 0 & 0 \\ 0 & -1 & 0 & 0 \\ 0 & -1 & 4 & 0 \\ 2 & 0 & 2 & 1 \end{array}\right] = \left[\ -1 \mid 2 \mid 4 \mid -2\ \right]$$

From the fourth equation, $\pi_3 = -2$. Then from the third equation, $\pi_4 = 2$. The first two equations yield $\pi_6 = 3$ and $\pi_7 = -4$.

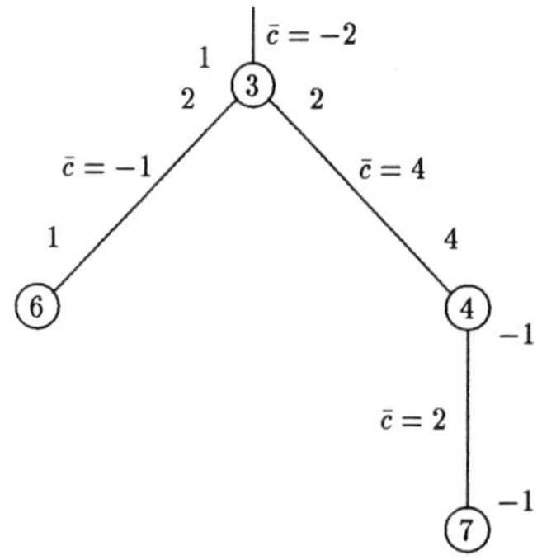

Fig. 15. Rooted tree component of a basis for a generalized network.

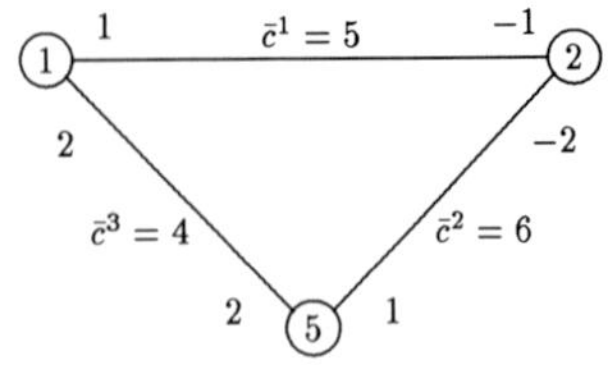

Fig. 16. One-tree component of a basis for a generalized network.

For the rooted tree case, a procedure similar to the one given in Section 3.2 can be developed.

Consider the example illustrated in Figure 16 with

$$[\ \pi_1 \mid \pi_2 \mid \pi_5\ ] \left[\begin{array}{r|r|r} 1 & 0 & 2 \\ -1 & -2 & 0 \\ 0 & 1 & 2 \end{array} \right] = [\ 5 \mid 6 \mid 4\ ]$$

This system of equations for a cycle is almost lower triangular (only the last column has a nonzero entry above the diagonal) and takes the special form:

$$\pi \left[\begin{array}{c|c|c|c|c|c} a_1 & & & & & b_n \\ b_1 & a_2 & & & & \\ & b_2 & a_3 & & & \\ & & b_3 & \ddots & & \\ & & & \ddots & a_{n-1} & \\ & & & & b_{n-1} & a_n \end{array} \right] = c$$

which corresponds to the cycle illustrated in Figure 17. Under the assumption that the above system has full row rank, then the unique solution is given by:

$$\pi_1 = \frac{c_1/a_1 + w_1 c_2/a_2 + w_1 w_2 c_3/a_3 + \cdots + w_1 w_2 \cdots w_{n-1} c_n/a_n}{1 - w_1 w_2 \cdots w_n} \tag{3}$$

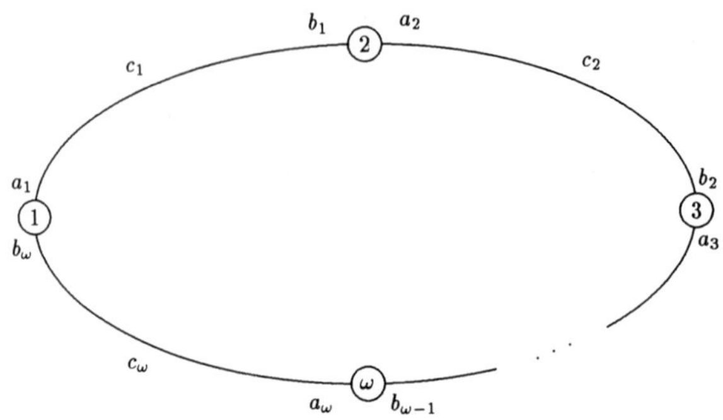

Fig. 17. Cycle for a generalized network.

and

$$\pi_{i+1} = \frac{c_i - a_i \pi_i}{b_i} \text{ for } i = 1, \ldots, n-1, \tag{4}$$

where $w_i = -b_i/a_i$ for $i = 1, \ldots, n$.
For the example illustrated in Figure 16, $w_1 = 1, w_2 = 1/2, w_3 = -1, \pi_1 = 2, \pi_2 = -3$, and $\pi_5 = 0$.

An extension of the procedure DUAL CALCULATION given in Section 3.2 can be developed for the case of generalized networks. The basic idea is that the components corresponding to rooted trees are triangularizable and the components corresponding to one-trees are nearly triangularizable. For the components corresponding to one-trees, (3) and (4) are used for calculations involving the cycle and all other calculations involve a lower triangular system of equations. Hence, once the duals on the cycle are known, all other duals can be determined by back-substitution.

4.3. Column update

Suppose that the nonbasic arc corresponding to $G_{.j}$ prices favorably and is selected for a potential flow change. The column update step of the primal simplex algorithm requires that y be determined where $y = \bar{B}^{-1}G_{.j}$ or $\bar{B}y = G_{.j}$. Since $\bar{B}$ is block diagonal, this may be written in the form:

$$\begin{bmatrix} \bar{B}^1 & & & \\ & \bar{B}^2 & & \\ & & \ddots & \\ & & & \bar{B}^p \end{bmatrix} \begin{bmatrix} y^1 \\ y^2 \\ \vdots \\ y^p \end{bmatrix} = \begin{bmatrix} g^1 \\ g^2 \\ \vdots \\ g^p \end{bmatrix}$$

Therefore, one must solve p systems of the form $\bar{B}^q y^q = g^q$ where the corresponding network $\langle V^q, E^q \rangle$ is either a rooted tree or a one-tree. Furthermore, g^q will have at most two nonzero entries. If $g^q = \hat{0}$, then $y^q = \hat{0}$.

Suppose g^q has two nonzero entries a and b in positions i and j, respectively. Then $g^q = ae_i + be_j$. Let y_a^q and y_b^q solve $\bar{B}^q y_a^q = ae_i$ and $\bar{B}^q y_b^q = be_j$, respectively. Then $y^q = y_a^q + y_b^q$. Hence, a specialized algorithm for the column update can be developed from an algorithm to efficiently solve the system $\bar{B}^q y_c^q = ce_k$ where $\langle V^q, E^q \rangle$ is either a rooted tree or a one-tree.

If $\langle V^q, E^q \rangle$ is a rooted tree, then $\bar{B}^q$ is triangularizable and y_c^q can be found by forward substitution. Consider the example illustrated in Figure 18 with

$$\left[\begin{array}{c|c|c|c} 1 & 0 & 0 & 0 \\ 0 & -1 & 0 & 0 \\ 0 & -1 & 4 & 0 \\ 2 & 0 & 2 & 1 \end{array}\right] \begin{bmatrix} y_{36} \\ y_{47} \\ y_{34} \\ y_{30} \end{bmatrix} = \begin{bmatrix} 0 \\ 5 \\ 0 \\ 0 \end{bmatrix}$$

From the first equation, $y_{36} = 0$. From the second equation, $y_{47} = -5$. From the third equation, $y_{34} = -5/4$ and from the fourth equation, $y_{30} = 5/2$.

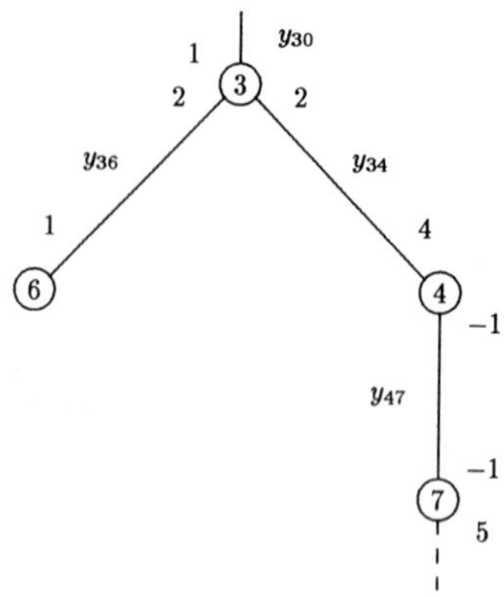

Fig. 18. Column update for a rooted tree component of a basis for a generalized network.

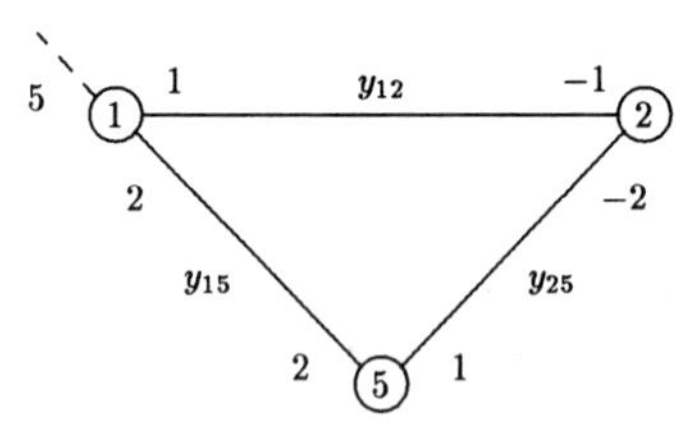

Fig. 19. Column update for a one-tree component of a basis for a generalized network.

Consider the one-tree illustrated in Figure 19 with

$$\left[\begin{array}{r|r|r} 1 & 0 & 2 \\ -1 & -2 & 0 \\ 0 & 1 & 2 \end{array}\right]\begin{bmatrix} y_{12} \\ y_{25} \\ y_{15} \end{bmatrix} = \begin{bmatrix} 5 \\ 0 \\ 0 \end{bmatrix}$$

The system of equations for a cycle is almost lower triangular (only the last column has a nonzero entry above the diagonal) and takes the special form:

$$\left[\begin{array}{c|c|c|c|c|c} a_1 & & & & & b_n \\ b_1 & a_2 & & & & \\ & b_2 & a_3 & & & \\ & & b_3 & \ddots & & \\ & & & \ddots & a_{n-1} & \\ & & & & b_{n-1} & a_n \end{array}\right] y = \alpha e_1$$

which corresponds to the cycle illustrated in Figure 17. Under the assumption that the above system has full row rank, the unique solution is given by:

$$y_1 = \frac{\alpha/a_1}{1 - w_1 w_2 \cdots w_n} \tag{5}$$

and

$$y_{i+1} = \frac{-b_i y_i}{a_{i+1}} \text{ for } i = 1, \ldots, n-1, \tag{6}$$

where $w_i = -b_i/a_i$ for $i = 1, \ldots, n$.

For the example illustrated in Figure 19, $w_1 = 1$, $w_2 = 1/2$, $w_3 = -1$, $y_{12} = 10/3$, $y_{25} = -5/3$, and $y_{15} = 5/6$.

An extension of the procedure cycle trace given in Section 3.3 can be developed for the case of generalized networks. The basic idea is that the rooted trees are triangularizable and the one-trees are nearly triangularizable. For the components corresponding to one-trees, (5) and (6) are used for calculations involving a lower triangular system of equations. If the nonbasic column of interest, $G_{.j}$, has two nonzero entries, then the procedure is applied twice and the results are added.

4.4. A data structure

The data structure presented in Section 3.4 has been extended to accommodate the one-tree components of a generalized network basis. With respect to the distance label, the nodes in the cycle of a one-tree and the root node in a rooted tree are treated in the same way with all nodes in a cycle having distance label of zero. The predecessor label of nodes in a cycle point to the next node in the cycle. Beginning with any node in a cycle, say v, the sequence v, $p(v)$, $p(p(v))$, ... will identify all nodes in the cycle and will eventually return to node v. For root nodes, v we adopt the convention $p(v) = v$.

The thread has also been extended in an obvious way with the convention that traversal around a cycle using the thread is in the opposite direction to that using the predecessor. The data structure corresponding to the generalized basis illustrated in Figure 20 is given in Table 4.

Efficient procedures have been developed for updating this data structure after a basis exchange. As with the linear network case, the flow or value update and the dual variable update is usually integrated with the basis exchange update. All of these specialized techniques can result in software from one to two order of magnitude faster than general linear programming software. All calculations involving the flows, dual variables, and updated columns must be performed using real arithmetic as opposed to the linear network case which only requires addition and subtraction of integers when the data is integral.

Some additional computational efficiencies can be achieved by scaling columns of G so that every column has at least one $+1$ entry. The disadvantage of using this scaling is that the resulting software cannot be easily incorporated within a branch-and-bound algorithm which can accommodate integrality restrictions on some or all of the flow variables.

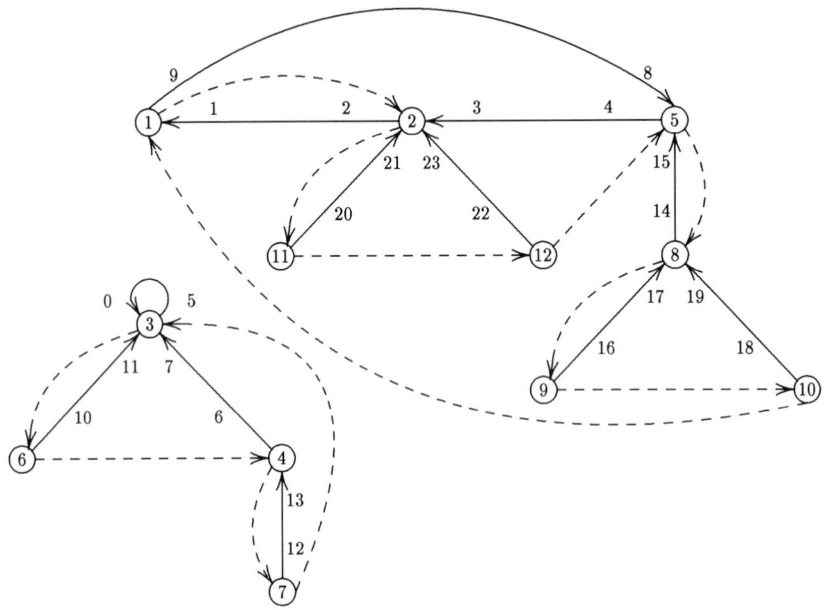

Fig. 20. Basis for a generalized network with the predecessor and thread labels illustrated.

Table 4
Data structure for the basis illustrated in Figure 20

Node v	Predecessor $p(v)$	Thread $t(v)$	Distance $d(v)$	Multipliers a	Multipliers b
1	5	2	0	9	8
2	1	11	0	2	1
3	3	6	0	5	0
4	3	7	1	6	7
5	2	8	0	4	3
6	3	4	1	10	11
7	4	3	2	12	13
8	5	9	1	14	15
9	8	10	2	16	17
10	8	1	2	18	19
11	2	12	1	20	21
12	2	5	1	22	23

5. Multicommodity networks

Multicommodity networks arise in practice when more than one type of commodity must share the capacity of the arcs in a network. Typical examples include Air Force cargo routing models in which cargo with different origin-destination pairs must share the capacity of the various aircraft which represent arcs in the model. For this model, a commodity represents all cargo with a given base of origin. The well-known LOGAIR Model is a model of this type.

Another well-known Air Force model is the family of Patient Distribution System (PDS) Models. The PDS generator was developed to assist in evaluating the airlift capacity of moving patients from a European theatre to U.S. hospitals. One of the input parameters for the PDS generator is the number of days in the model. A 20 day model has over 100,000 columns in the corresponding linear program.

Let $G = \langle V, E \rangle$ be the underlying network through which K commodities will be flowing and let the $m \times n$ matrix A denote the corresponding node–arc incidence matrix. Let the m-component vector r^k denote the requirements vector for commodity k and the n-component vector x^k denote the flows for commodity k with corresponding unit costs and upper bounds of c^k and u^k, respectively. Let $Y^k = \{x^k : Ax^k = r^k,\ \hat{0} \le x^k \le u^k\}$ and assume that $Y^k \neq \Phi$. Suppose the shared (also called mutual) capacity for arc i is b_i, an entry of the n-component vector b. The multicommodity network flow problem is to find K n-component vectors $\bar{x}^1, \ldots, \bar{x}^K$ such that

$$\sum_k c^k \bar{x}^k = \min \left\{ \sum_k c^k x^k : \sum_k x^k \le b,\ x^k \in Y^k \right\}$$

It is possible to generalize the multicommodity network model to allow for both commodity dependent networks ($A^k x^k = r^k$) and for multipliers on the mutual capacity constraints ($\sum_k D^k x^k \le b$, where each D^k is an $n \times n$ diagonal matrix with nonnegative entries). It is also common that the mutual capacity constraints do not involve all n of the arcs. These generalizations present no mathematical difficulties in the algorithms to follow but greatly complicate the notation.

Consider the sample two commodity network illustrated in Figure 21. The matrices and vectors corresponding to this model are as follows:

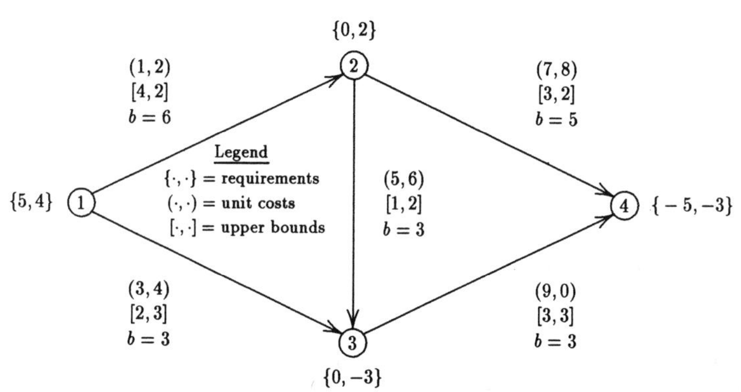

Fig. 21. Example multicommodity (two commodity) network model.

$$A = \begin{bmatrix} 1 & 1 & 0 & 0 & 0 \\ -1 & 0 & 1 & 1 & 0 \\ 0 & -1 & -1 & 0 & 1 \\ 0 & 0 & 0 & -1 & -1 \end{bmatrix}$$

$$\begin{aligned} r^1 &= [\ 5 \quad 0 \quad 0 \quad -5\ ] \\ r^2 &= [\ 4 \quad 2 \quad -3 \quad -3\ ] \\ c^1 &= [\ 1 \quad 3 \quad 5 \quad 7 \quad 9\ ] \\ c^2 &= [\ 2 \quad 4 \quad 6 \quad 8 \quad 0\ ] \\ u^1 &= [\ 4 \quad 2 \quad 1 \quad 3 \quad 3\ ] \\ u^2 &= [\ 2 \quad 3 \quad 2 \quad 2 \quad 3\ ] \\ b &= [\ 6 \quad 3 \quad 3 \quad 5 \quad 3\ ] \end{aligned}$$

After adding slack variables to the mutual capacity constraints the system of equations corresponding to this model is

$$M \begin{bmatrix} x^1 \\ x^2 \\ s \end{bmatrix} = \begin{bmatrix} r^1 \\ r^2 \\ b \end{bmatrix}$$

where M is the matrix

$$\left[\begin{array}{ccccc|ccccc|ccccc} 1 & 1 & 0 & 0 & 0 & 0 & 0 & 0 & 0 & 0 & 0 & 0 & 0 & 0 & 0 \\ -1 & 0 & 1 & 1 & 0 & 0 & 0 & 0 & 0 & 0 & 0 & 0 & 0 & 0 & 0 \\ 0 & -1 & -1 & 0 & 1 & 0 & 0 & 0 & 0 & 0 & 0 & 0 & 0 & 0 & 0 \\ 0 & 0 & 0 & -1 & -1 & 0 & 0 & 0 & 0 & 0 & 0 & 0 & 0 & 0 & 0 \\ \hline 0 & 0 & 0 & 0 & 0 & 1 & 1 & 0 & 0 & 0 & 0 & 0 & 0 & 0 & 0 \\ 0 & 0 & 0 & 0 & 0 & -1 & 0 & 1 & 1 & 0 & 0 & 0 & 0 & 0 & 0 \\ 0 & 0 & 0 & 0 & 0 & 0 & -1 & -1 & 0 & 1 & 0 & 0 & 0 & 0 & 0 \\ 0 & 0 & 0 & 0 & 0 & 0 & 0 & 0 & -1 & -1 & 0 & 0 & 0 & 0 & 0 \\ \hline 1 & 0 & 0 & 0 & 0 & 1 & 0 & 0 & 0 & 0 & 1 & 0 & 0 & 0 & 0 \\ 0 & 1 & 0 & 0 & 0 & 0 & 1 & 0 & 0 & 0 & 0 & 1 & 0 & 0 & 0 \\ 0 & 0 & 1 & 0 & 0 & 0 & 0 & 1 & 0 & 0 & 0 & 0 & 1 & 0 & 0 \\ 0 & 0 & 0 & 1 & 0 & 0 & 0 & 0 & 1 & 0 & 0 & 0 & 0 & 1 & 0 \\ 0 & 0 & 0 & 0 & 1 & 0 & 0 & 0 & 0 & 1 & 0 & 0 & 0 & 0 & 1 \end{array}\right]$$

In general, the linear programming constraint matrix corresponding to a multicommodity network model takes the form:

$$\begin{pmatrix} A & & & & \\ & A & & & \\ & & \ddots & & \\ & & & A & \\ I & I & \cdots & I & I \end{pmatrix} \tag{7}$$

Hence, the columns either take the form e_i for some i or have exactly three nonzero entries (two +1s and one −1).

There are three types of specialized algorithms which have been used to exploit the underlying structure of the multicommodity model, *primal partitioning*, *price-directive decomposition*, and *resource-directive decomposition*. Primal partitioning is a specialization of the primal simplex method which exploits the network structure of the basis. Price-directive decomposition is a specialization of Dantzig-Wolfe decomposition which exploits the network structure of the subproblems. Resource-directive decomposition allocates the mutual capacity among the K commodities and solves K linear network problems per allocation. The trick is to develop a reallocation scheme which will guarantee that an optimum can be found.

5.1. Primal partitioning

By appending a root arc to the network for each of the K commodities, we convert the matrix (7) into a matrix having full row rank. That is, the full rank linear programming constraint matrix takes the form:

$$\begin{array}{c} \\ m \\ m \\ \vdots \\ m \\ n \end{array}\begin{array}{c} \begin{array}{ccccc} n+1 & n+1 & & n+1 & n \end{array} \\ \left[\begin{array}{ccccc} A \mid e_\rho & & & & \\ & A \mid e_\rho & & & \\ & & \ddots & & \\ & & & A \mid e_\rho & \\ I \mid \hat{0} & I \mid \hat{0} & \cdots & I \mid \hat{0} & I \end{array}\right] \end{array} \tag{8}$$

where A is a node–arc incidence matrix. It is well known that every basis for (8) may be partitioned in the following form:

$$\begin{array}{c} m \\ \vdots \\ m \\ p \\ n-q \end{array}\left(\begin{array}{ccccccc} B^1 & & & R^1 & & & \\ & \ddots & & & \ddots & & \\ & & B^K & & & R^K & \\ P^1 & \cdots & P^K & T^1 & \cdots & T^K & \\ S^1 & \cdots & S^K & U^1 & \cdots & U^K & I \end{array}\right)$$

where $B^1, \ldots, B^K$ correspond to rooted trees. This basis takes the general form:

$$\begin{array}{c} \\ mK \\ q \\ p \end{array}\begin{array}{c} \begin{array}{ccc} mK & n-p & p \end{array} \\ \left[\begin{array}{ccc} L_1 & R_1 & \\ L_2 & R_2 & \\ L_3 & R_3 & I \end{array}\right] \end{array} \tag{9}$$

where

$$L_1 = \begin{pmatrix} B^1 & & \\ & \ddots & \\ & & B^K \end{pmatrix}$$

$$R_1 = \begin{pmatrix} R^1 & & \\ & \ddots & \\ & & R^K \end{pmatrix}$$

$$L_2 = [\ P^1 \ \cdots \ P^K \ ]$$
$$R_2 = [\ T^1 \ \cdots \ T^K \ ]$$
$$L_3 = [\ S^1 \ \cdots \ S^K \ ]$$

and

$$R_3 = [\ U^1 \ \cdots \ U^K \ ]$$

By partitioning the dual variables to be compatible with (9), the dual calculation requires the solution to the following system:

$$[\ \pi_1 \mid \pi_2 \mid \pi_3\ ] \left[\begin{array}{c|c|c} L_1 & R_1 & \\ \hline L_2 & R_2 & \\ \hline L_3 & R_3 & I \end{array}\right] = [\ c_1 \mid c_2 \mid 0\ ]$$

Let $Q = [R_2 - L_2 L_1^{-1} R_1]$. Then $\pi_3 = \hat{0}$, $\pi_2 = (c_2 - c_1 L_1^{-1} R_1) Q^{-1}$, and $\pi_1 = (c_1 - \pi_2 L_2) L_1^{-1}$. Note that L_1 corresponds to K rooted trees so that all calculations involving L_1^{-1} can be executed directly on the rooted trees and do not require the use of explicit matrices. The only matrix operations involve the $q \times q$ matrix Q which is called the *working basis*.

Partitioning the updated column in a similar way, the column update calculation requires the solution of the following system:

$$\left[\begin{array}{c|c|c} L_1 & R_1 & \\ L_2 & R_2 & \\ L_3 & R_3 & I \end{array}\right] \begin{bmatrix} x \\ y \\ z \end{bmatrix} = \begin{bmatrix} u \\ t \\ w \end{bmatrix}$$

The solution is given by $y = Q^{-1}(r - L_2 L_1^{-1} u)$, $x = L_1^{-1} u - L_1^{-1} R_1 y$, and $z = w - L_3 L_1^{-1} u - (R_3 - L_3 L_1^{-1} R_1) y$. As before, the only matrix operations involve the working basis Q.

Each basis exchange can alter the dimension of the working basis by at most one. Efficient techniques have been developed for maintaining Q^{-1} as the basis is updated. If q is small, then most of the work associated with a simplex pivot can be performed efficiently using the K rooted trees. If q is large, then the pivots can be more expensive than a regular simplex pivot which does not exploit the special network structure.

5.2. Price-directive decomposition

For each $k \in \iota K$, suppose that $Y^k = \{x^k : Ax^k = r^k,\ \hat{0} \le x^k \le u^k\} \ne \Phi$ and is bounded, and let $\bar{w}_1^k, \ldots, \bar{w}_{n_k}^k$ denote the extreme points of Y^k. Let the $n \times n_k$ matrix W^k be given by $[\ \bar{w}_1^k \mid \bar{w}_2^k \mid \cdots \mid \bar{w}_{n_k}^k\ ]$. Then any $\bar{x}^k \in Y^k$ can be represented as a convex combination of those extreme points (the columns of W^k). That is, $\bar{x}^k = W^k\lambda^k$ where $\hat{1}\lambda^k = 1$ and $\lambda^k \ge \hat{0}$. Let

$$W = \Big\{ (\ \lambda^1 \mid \cdots \mid \lambda^k \mid s\ ) : \sum_k W^k\lambda^k + s = b,\ s \ge \hat{0},\ \hat{1}\lambda^k = 1,\ \lambda^k \ge \hat{0} \Big\}$$

Then an equivalent statement of the multicommodity network problem is to find $K+1$ vectors $(\ \bar{\lambda}^1 \mid \cdots \mid \bar{\lambda}^k \mid \bar{s}\ )$ such that

$$\sum_k c^k W^k \bar{\lambda}^k = \min \Big\{ \sum_k c^k W^k \lambda^k : (\ \lambda^1 \mid \cdots \mid \lambda^k \mid s\ ) \in W \Big\}$$

The disadvantage of this statement of the problem is that it is quite difficult to find all the extreme points for Y^k (which form the columns of W^k). However, the price-directive algorithm provides a mechanism for applying the primal simplex algorithm to this formulation by providing columns of W^k only when they are needed. In fact, the complete matrix W^k is never actually generated.

Suppose we have a nonnegative basic feasible solution to the system of equations:

$$\begin{aligned} &\sum_k W^k\lambda^k + s = b &&(\pi) \\ &\hat{1}\lambda^1 \qquad\qquad = 1 &&(\gamma_1) \\ &\qquad\qquad\quad \vdots && \\ &\hat{1}\lambda^K \qquad\qquad = 1 &&(\gamma_K) \end{aligned}$$

with the corresponding duals $(\ \pi \mid \gamma_1 \mid \cdots \mid \gamma_k\ )$. From Section 2, we see that the ith nonbasic slack variable prices favorably if $\pi_i > 0$. Also, λ_j^k prices favorably if $\bar{w}_j^k\pi + \gamma_k - c^k\bar{w}_j^k > 0$. Note that the extreme point from Y^k which prices most favorably can be found by solving the linear network problem

$$\min \big\{ (c^k - \pi)w^k : w^k \in Y^k \big\}$$

using an extreme point algorithm. If the extreme point produced from solving the linear network problem prices unfavorably, then none of the columns of W^k will price favorably. Hence, columns of W^k are generated only as they are needed.

The linear program

$$\min \Big\{ \sum_k c^k W^k \lambda^k : (\ \lambda^1 \mid \cdots \mid \lambda^k \mid s\ ) \in W \Big\}$$

is known as the *master problem* and the linear network problems $\min\{c^k x^k : x^k \in Y^k\}$ for $k \in \iota K$ are called the *slave subproblems*. Columns of the master problem (extreme points of the Y^k) are found by solving linear network problems.

A mathematical description of the price-directive algorithm follows:

procedure PRICE-DIRECTIVE DECOMPOSITION

inputs:	A	–	$m \times n$ node–arc incidence matrix
	K	–	number of commodities
	$c^1, \ldots, c^K$	–	n-component vectors of unit costs
	$u^1, \ldots, u^K$	–	n-component vectors of upper bounds
	$r^1, \ldots, r^K$	–	m-component vectors of requirements
	b	–	n-component vector of mutual capacities
outputs:	$\bar{y}^1, \ldots, \bar{y}^K$	–	n-component vectors of optimal flows

assumption: $Y^k = \{x^k : Ax^k = r^k, \hat{0} \le x^k \le u^k\} \ne \Phi$ and is bounded for all $k \in \iota K$.

begin
  [Initialize Master Problem]
  obtain a basic feasible solution $[\bar{y}^1 \mid \ldots \mid \bar{y}^K]$ for W with corresponding dual variables $[\pi \mid \gamma]$ (if a basic feasible solution is not readily available, then artificial variables and a two-phase procedure may be used);
  optimal ⇐ 'no';
  while optimal = 'no' **do**
    [Price Slacks]
    $i \Leftarrow 1$;
    while $i \le n$ **do**
      if $\pi_i \le 0$
        then
          perform one simplex iteration in the master problem using $[e_i \mid \hat{0}]$ as the nonbasic column which prices favorably;
          update $[\bar{y}^1 \mid \ldots \mid \bar{y}^K]$ and $[\pi \mid \gamma]$;
          $i \Leftarrow 1$;
        else
          $i \Leftarrow i + 1$;
      endif
    endwhile
    [Price Extreme Points]
    $k \Leftarrow 1$, favorable ⇐ 'no';
    while $k \le K$ **and** favorable = 'no' **do**
      obtain an optimal extreme point w^k for $\min\{(c^k - \pi)x^k : Ax^k = r^k, \bar{0} \le x^k \le u^k\}$;

if $(c^k - \pi)x^k \leq \gamma_k$
then
favorable $\Leftarrow$ 'yes';
else
$k \Leftarrow k + 1$;
endif
endwhile
if favorable = 'yes'
then
perform one primal simplex iteration in the master problem using $[\bar{w}^k \mid e_k]$ as the nonbasic column which prices favorably;
update $[\bar{y}^1 \mid \ldots \mid \bar{y}^K]$ and $[\pi \mid \gamma]$;
else
optimal = 'yes';
endif
endwhile
end

A pictorial description of the price-directive decomposition algorithm is given in Figure 22. The master program produces dual variables used to construct the objective function coefficients for each of the slave subproblems. Each slave subproblem, in turn, produces an extreme point which may or may not provide an objective-improving column relative to the current basis for the master problem. A favorable column will be used for a simplex iteration in the master problem, producing new dual variables for the slave subproblems. Note that successive calls to a routine which solves a given subproblem involve only a change in the objective function and do not involve any changes to the constraints. Hence each such call can make use of the optimal basis tree produced by the previous call. We also recommend that a wrap-around list be maintained to organize the calls to the subproblems, a minor variation on the above procedure. That is, if commodity k produced a favorably priced column at iteration i, then begin the new iteration by pricing the columns associated with commodity $(k + 1) \bmod K$.

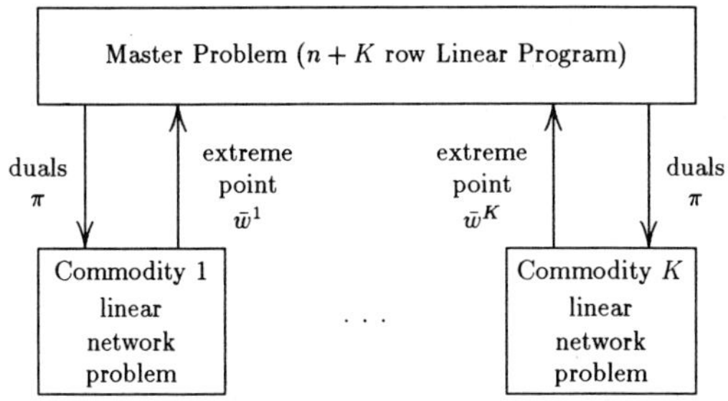

Fig. 22. The price-directive decomposition algorithm.

5.3. Resource-directive decomposition

It is the mutual capacity constraints $\sum_k x^k \leq b$ that make the multicommodity problem difficult. If these constraints could be eliminated or ignored, then the problem would decompose into K linear network problems. Let $[\ \bar{y}^1 \mid \cdots \mid \bar{y}^K\ ]$ denote an allocation of b among the K commodities such that $\hat{0} \leq \bar{y}^k \leq u^k$ for each k and $\sum_k \bar{y}^k \leq b$. Let

$$z^k(\bar{y}^k) = \min\left\{c^k x^k : Ax^k = r^k,\ \hat{0} \leq x^k \leq \bar{y}^k\right\}$$

An equivalent statement of the multicommodity network problem is to find K vectors $[\ \bar{y}^1 \mid \cdots \mid \bar{y}^K\ ]$ such that

$$\sum_k z^k(\bar{y}^k) = \min\left\{\sum_k c^k x^k : Ax^k = r^k,\ \hat{0} \leq x^k \leq \bar{y}^k\right\}$$

The objective function $g([\ y^1 \mid \cdots \mid y^K\ ]) = \sum_k z^k(y^k)$ is piece-wise linear and convex. Hence, the classic subgradient algorithm can be applied to this problem.

Consider the nonlinear program $\min\{g(y) : y \in Y\}$, where $g(\cdot)$ is piece-wise linear and convex and $Y \neq \Phi$ is formed by the intersection of a finite number of closed half-spaces. To apply the subgradient algorithm, one must be able to solve the problem

$$\min\left\{\left[\sum_i (y_i - \bar{p}_i)^2\right]^{1/2} : y \in Y\right\}$$

for any point $\bar{p}$. The solution of this problem is called the *projection* of $\bar{p}$ onto Y. This is traditionally expressed as $\bar{y} \longleftarrow P[\bar{p}]$.

Mathematically the subgradient algorithm may be stated as follows:

procedure SUBGRADIENT

inputs:	$g(\cdot)$	–	the objective function
	Y	–	the feasible region
	$s_1, s_2, \ldots$	–	the sequence of step size values
output:	$\bar{y}$	–	an optimum for $\min\{g(y) : y \in Y\}$

assumptions: $g(\cdot)$ is piece-wise linear and convex
$Y \neq \Phi$ is formed by the intersection of a finite number of closed half-spaces

begin
[Initialization]
obtain a point $\bar{y} \in Y$, optimal $\Leftarrow$ 'no', $i \Leftarrow 1$;
while optimal = 'no' **do**
obtain a subgradient $\bar{v}$ of $g(\cdot)$ at $\bar{y}$;

```
      if v̄=0̂
        then
          optimal ⇐ 'yes';
        else
          ȳ ⇐ P[ȳ − s_i v̄], i ⇐ i + 1;
      endif
    endwhile
end
```

Several convergence results for this algorithm have appeared in the literature. These results all provide restrictions and guidance on the selection of the sequence of step sizes.

The computationally expensive part of the above algorithm when applied to the multicommodity network flow problem is the calculation of the subgradient. That is, we need a subgradient of the function $g([\ y^1 \mid \cdots \mid y^K\ ])$ at the point $[\ \bar{y}^1 \mid \cdots \mid \bar{y}^K\ ]$, where $g([\ \bar{y}^1 \mid \cdots \mid \bar{y}^K\ ]) = \sum_k z^k(\bar{y}^k)$. By duality theory,

$$
\begin{aligned}
z^k(\bar{y}^k) &= \min\left\{c^k x^k : Ax^k = r^k,\ \hat{0} \le x^k \le \bar{y}^k\right\} \\
&= \max\left\{r^k \mu^k - \bar{y}^k \nu^k : \mu^k A - \nu^k \le c^k,\ \nu^k \ge \hat{0}\right\}
\end{aligned}
$$

Let $[\bar{\mu}^k \mid \bar{\nu}^k]$ for all $k \in \iota K$ denote optimal dual variables for $g([\ \bar{y}^1 \mid \cdots \mid \bar{y}^K\ ])$. It can easily be shown that $[\ -\bar{\nu}^1 \mid \cdots \mid -\bar{\nu}^K\ ]$ is a subgradient for $g(\cdot)$ at the point $([\ \bar{y}^1 \mid \cdots \mid \bar{y}^K\ ])$. Efficient projection algorithms for the special case of the multicommodity network problem are also available in the literature.

The major attraction of the subgradient algorithm is that the multicommodity network problem can be solved by solving a sequence of single commodity network problems. The major disadvantage is that it is quite difficult to select a set of step sizes so that the software implementation is robust over a wide range of problems. Our experience has indicated that much skill is required in the selection of the step sizes. Many skilled mathematical programming software developers have discovered that the step size selection is more of an art than a science.

6. Networks with side constraints

Let $G = \langle V, E\rangle$ be a network and let the $m \times n$ matrix A denote the corresponding node–arc incidence matrix. Let the m-component vector r denote the requirements vector, and the n-component vectors x, c, and u denote the flows, unit costs, and upper bounds, respectively. Let the $p \times n$ matrix S, $p \times k$ matrix P, the p-component vector b, and the k-component vector v be arbitrary. Mathematically, the *network with side constraint model* is given by

$$\min_{x,z} \{cx + \mathrm{d}z : Ax = r,\quad Sx + Pz = b,\quad 0 \le x \le u,\quad 0 \le z \le v\}$$

Note that the multicommodity network flow model is a special case of the network with side constraint model.

Typical side constraints impose capacity restrictions on the sum of flows in several arcs (as in the multicommodity network flow model), impose requirements that given pairs of arcs must have identical flow, and/or impose requirements that flows in certain arcs must maintain a given proportionality ratio. Examples of side constraints are presented for the network illustrated in Figure 6.

Recall that the feasible region for this sample network is defined by the following system of equations and inequalities:

$$\begin{bmatrix} 1 & 1 & 0 & 0 & 0 & 0 & -1 \\ -1 & -1 & 1 & 1 & 0 & 0 & 0 \\ 0 & 0 & -1 & 0 & 1 & -1 & 0 \\ 0 & 0 & 0 & -1 & -1 & 1 & 1 \end{bmatrix} \begin{bmatrix} x_1 \\ x_2 \\ x_3 \\ x_4 \\ x_5 \\ x_6 \\ x_7 \end{bmatrix} = \begin{bmatrix} 10 \\ 5 \\ 0 \\ -15 \end{bmatrix}$$

$$\begin{array}{ccccc} 0 & \le & x_1 & \le & 6 \\ 0 & \le & x_2 & \le & 8 \\ 0 & \le & x_3 & \le & 10 \\ 0 & \le & x_4 & \le & 10 \\ 0 & \le & x_5 & \le & 8 \\ 0 & \le & x_6 & \le & 8 \\ 0 & \le & x_7 & \le & 8 \end{array}$$

Suppose the total flow on arc $(3, 4)$ together with the flow on each of the (duplicate) arcs $(1, 2)$ is limited to 12 units. This can be modeled with the two side constraints $x_1 + x_5 \le 12$ and $x_2 + x_5 \le 12$. Suppose the flows on all three of those arcs must be identical. This can be imposed by the two side constraints $x_1 - x_2 = 0$ and $x_2 - x_5 = 0$. Suppose the flows leaving node 2 via arcs $(2, 3)$ and $(2, 4)$ must have a ratio of 1 to 2. This can be modeled with the side constraint $2x_3 - x_4 = 0$. Flows which satisfy this network with side constraint model are illustrated in Figure 23.

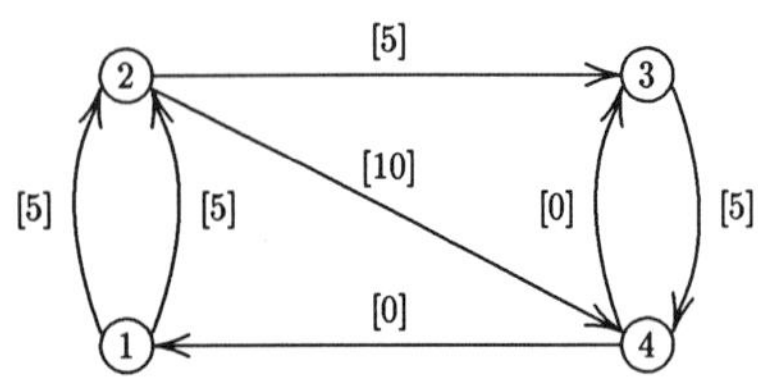

Fig. 23. Example flows for side constraints.

6.1. Basis characterization

We assume that the matrix $\left[\ S \mid P\ \right]$ has full rank. If this is not the case, then artificial variables can be appended to the system so that this assumption is satisfied. As in Section 3.1 a root arc is appended to $Ax = r$ so that the first m rows of the constraint matrix will have full row rank. Let $Z = \{(\ x \mid z \mid a\ ) : Ax + e_j a = r,\ \ Sx + Pz = b,\ \ 0 \le x \le u,\ \ 0 \le z \le v\}$. The revised model is simply

$$\min_{x,z} \{cx + \mathrm{d}z : (\ x \mid z \mid a\ ) \in Z\}$$

and the constraint matrix is:

$$\bar{A} = \begin{matrix} m \\ p \end{matrix} \begin{pmatrix} A & & e_j \\ S & P & \end{pmatrix}$$

Every basis for $\bar{A}$ takes the form

$$\bar{B} = \begin{matrix} \\ m \\ p \end{matrix} \begin{matrix} \begin{matrix} m & p \end{matrix} \\ \begin{bmatrix} B & C \\ D & F \end{bmatrix} \end{matrix}$$

where the nonsingular matrix B is a submatrix of $\left[\ A \mid e_j\ \right]$ and e_j is one of the columns of B. That is, the columns of B correspond to a spanning tree with root at node j. It also follows that

$$\bar{B}^{-1} = \left[\begin{array}{c|c} B^{-1} + B^{-1}CQ^{-1}DB^{-1} & -B^{-1}CQ^{-1} \\ \hline -\ Q^{-1}DB^{-1} & Q^{-1} \end{array} \right]$$

where $Q = F - DB^{-1}C$.

Note that $\bar{B}^{-1}$ can be constructed from C and D (original data) and the two matrices B^{-1} and Q^{-1}. Since B corresponds to the rooted spanning tree $\langle V, T\rangle$, calculations involving B^{-1} can be performed directly on the graph. Hence, the only matrix operations are those involving the inverse of the $p \times p$ working basis Q. For the specialization of the simplex algorithm, it is assumed that $\langle V, T\rangle$ and Q^{-1} are maintained corresponding to the current basis $\bar{B}$.

6.2. Dual calculation

We seek an efficient algorithm which exploits the underlying network structure to obtain π, where $\pi\bar{B} = c^B$. Let $\pi = \left[\ \pi^1 \mid \pi^2\ \right]$ and $c^B = \left[\ c^1 \mid c^2\ \right]$ be partitioned to be compatible with the partitioning of $\bar{B}$. Then

$$\pi^2 = (c^2 - c^1B^{-1}C)Q^{-1}$$

$$\pi^1 = (c^1 - \pi^2 D)B^{-1}$$

Consider the following algorithm which performs the above calculations:

procedure DUAL CALCULATION WITH SIDE CONSTRAINTS

inputs: $\left[\begin{array}{c|c} B & C \\ \hline D & F \end{array}\right]$ – current basis $\bar{B}$ in partitioned form

Q^{-1} – inverse of current working basis

$[c^1 \mid c^2]$ – basic costs in partitioned form

outputs: $[\pi^1 \mid \pi^2]$ – dual variables

begin
[Use procedure DUAL CALCULATION (Section 3.2)]
$\gamma^1 \Leftarrow c^1 B^{-1}$;
$\pi^2 \Leftarrow (c^2 - \gamma^1 C) Q^{-1}$;
$\gamma^2 \Leftarrow c^1 - \pi^2 D$;
[Use procedure DUAL CALCULATION (Section 3.2)]
$\gamma^2 \Leftarrow \gamma^2 B^{-1}$;
end

The first and last steps can be executed directly on the tree $\langle V, T \rangle$ and only require the operations of addition and subtraction. Further, if c^1 is an integer vector, then γ^1 will be an integer vector. If B is much larger than Q, i.e. $m \gg p$, then performing those calculations using the partitioned basis may yield a substantial savings in computational time.

6.3. Column update

We seek an efficient algorithm to obtain y, where $\bar{B}y = \bar{d}$ and $\bar{d}$ is some column of $\bar{A}$. Let

$$y = \left[\begin{array}{c} y^1 \\ \hline y^2 \end{array}\right]$$

and

$$\bar{d} = \left[\begin{array}{c} d^1 \\ \hline d^2 \end{array}\right]$$

be partitioned so as to be compatible with the partitioning of $\bar{B}$. Then

$$y^2 = Q^{-1}(d^2 - DB^{-1}d^1)$$
$$y^1 = B^{-1}(d^1 - Cy^2)$$

When $m \gg p$ these calculations can be performed efficiently by the following algorithm:

procedure COLUMN UPDATE WITH SIDE CONSTRAINTS

inputs: $\left[\begin{array}{c|c} B & C \\ \hline D & F \end{array}\right]$ – current basis $\bar{B}$ in partitioned form

Q^{-1} – inverse of current working basis

$\left[\begin{array}{c} d^1 \\ \hline d^2 \end{array}\right]$ – column of $\bar{A}$ in partitioned form

outputs: $\left[\begin{array}{c} y^1 \\ \hline y^2 \end{array}\right]$ – updated column

begin

[Use procedure CYCLE TRACE (Section 3.3)]

$\gamma^1 \Leftarrow B^{-1}d^1$;

$y^2 \Leftarrow Q^{-1}(d^2 - D\gamma^1)$;

$\gamma^2 \Leftarrow d^1 - Cy^2$;

$y^1 \Leftarrow B^{-1}\gamma^2$;

end

The first step is performed on the spanning tree $\langle V, T\rangle$ and requires no matrix operations. Since B and B^{-1} are triangularizable, the last step can also be performed directly on the tree $\langle V, T\rangle$.

6.4. Basis update

In this section we use subscripts to denote the iteration count. Let Q_i denote the working basis (at iteration i) corresponding to the partitioned basis $\bar{B}_i$. The first m columns of $\bar{B}_i$ are called the *key columns* while the others are called *nonkey*. That is,

$$\bar{B}_i = \begin{array}{c} \begin{array}{cc} \text{key} & \text{nonkey} \end{array} \\ \begin{bmatrix} B_i & C_i \\ D_i & F_i \end{bmatrix} \end{array}$$

To maintain the proper partitioning it may be necessary to interchange a nonkey column with a key column before the simplex pivot is actually performed. Even though $\bar{B}_i$ and $\bar{B}_{i+1}$ differ only in that two columns have been interchanged, Q_i^{-1} and Q_{i+1}^{-1} may be quite different. For either an interchange or a regular pivot it is well known that $\bar{B}_{i+1}^{-1} = E\bar{B}_i^{-1}$, where E is either an elementary column matrix (i.e., a matrix that differs from the identity matrix in only one column) or a permutation matrix. By partitioning E to be compatible with $\bar{B}_i$, we obtain

$$\left[\begin{array}{c|c} & \\ \hline & Q_{i+1}^{-1} \end{array}\right] = \left[\begin{array}{c|c} E_1 & E_2 \\ \hline E_3 & E_4 \end{array}\right] \left[\begin{array}{c|c} & -B_iC_iQ_i^{-1} \\ \hline & Q_{i+1}^{-1} \end{array}\right]$$

which implies $Q_{i+1}^{-1} = (E_4 - E_3B_iC_i)Q_i^{-1}$.

Table 5
Basis update for side constraint model

Leaving column	Switch with nonkey possible?	Modifications required	
		$\langle V, T\rangle$	Q^{-1}
key	yes	1	2
key	no	1	0
nonkey	no	0	1

If the column selected to leave the basis is nonkey, then E_3 is a matrix of zeroes and $Q_{i+1}^{-1} = E_4 Q_i^{-1}$. If the column selected to leave the basis is key, then we attempt to execute a permutation which makes it nonkey. After a successful switch, the leaving column is nonkey and the usual pivot is executed. If no such permutation is possible, then both the entering and leaving arcs are key and $Q_{i+1}^{-1} = Q_i^{-1}$. All modifications to B_i and its spanning tree $\langle V_i, T_i\rangle$ are made using the techniques described in Section 3.4.

In summary, there are three types of pivots as shown in Table 5. One of the updates is rather expensive while the other two are rather straightforward.

6.5. *Applicability*

We advocate the use of the partitioned simplex algorithm only when the network portion of the structural constraints is very large compared to the nonnetwork portion. Our rule-of-thumb is that the number of nodes should be at least an order of magnitude larger than the number of side constraints. We also advocate using the optimal solution to the pure network relaxed subproblem (without the side constraints) as an advanced start for the solution to the network with side constraint model.

It is also tempting to dualize the side constraints and use Lagrangean duality in an attempt to solve the network with side constraint model. While we have used this approach successfully for special cases, this strategy is fraught with pitfalls. Finding good procedures for updating the Lagrange multipliers can be a challenging task.

7. Reference notes

The specialized algorithms presented in this chapter all rely on the graphical interpretation of the simplex steps when applied to a linear program possessing an underlying network structure, in whole or in part. Graphical characterizations for bases for both the linear network problem and the generalized network problem can be found in Dantzig [1963]. Additional elaboration on the interpretation of the algebraic operations on a graphical structure were provided by Johnson [1965]. The first software implementations which empirically demonstrated the merit of those specialized algorithms were developed by Srinivasan & Thompson [1973] and by Glover and Klingman and their colleagues at the University of Texas (see Glover, Karney & Klingman [1974], Glover, Karney, Klingman & Napier [1974], and Glover, Klingman & Stutz [1974]). Since these seminal papers, hundreds of papers with various extensions and specializations have appeared in the literature. Bradley, Brown & Graves [1977] and Barr, Glover & Klingman [1979] are two of our favorites. Empirical results from parallel versions of this algorithm may be found in Peters [1990] and in Barr & Hickman [1993]

The first specialized primal simplex code which exploited the graphical nature of the basis of a generalized network was developed by Glover, Klingman & Stutz [1973]. Since then, many excellent software implementations have been developed. A table of codes and references may be found in Clark, Kennington, Meyer & Ramamurti [1992]. One of our favorite generalized network codes is GENNET which was developed by Brown & McBride [1984].

The primal partitioning method for multicommodity problems is due primarily to Hartman & Lasdon [1970, 1972]. Additional results may be found in Helgason & Kennington [1977]. The price-directive decomposition procedure was first developed by Ford & Fulkerson [1958]. It is purported that the well-known Dantzig–Wolfe decomposition algorithm (see Dantzig & Wolfe [1960]) was inspired by the Ford & Fulkerson paper on multicommodity problems. The subgradient algorithm was first applied to the multicommodity network problem by Held, Wolfe & Crowder [1974]. A brief description of the LOGAIR problem may be found in Ali et al. [1984], and a brief description of the PDS family of multicommodity models may be found in Carolan, Hill, Kennington, Niemi & Wichmann [1990].

The update for the inverse of the working basis in the network with side constraint section is due to Hartman & Lasdon [1970]. An empirical analysis with various updating algorithms may be found in Barr, Farhangian & Kennington [1986]. A specialized algorithm having equal flow side constraints may be found in Ali, Kennington & Shetty [1988].

In this chapter, the algorithms presented were generally based on a specialization of the simplex method. There are three other basic approaches which are now competing quite successfully in empirical analyses. The oldest is the relaxation method which was developed by Bertsekas and his colleagues at MIT (see Bertsekas [1991]). The second is the algorithm of Goldberg which has been implemented in software by Anderson & Setubal [1992]. The third algorithm which could have an impact on the field is the network interior point method of Resende & Veiga [1992]. All of these approaches have their advocates and we expect improvements for each of these relatively new algorithms to appear in the literature in the near future.

There are other well-known network problems which can be modeled as linear programs. Included among these are the one-to-one shortest path problem, the maximal flow problem, the assignment problem, the semi-assignment problem, and the transportation problem. Our work and that of others has concluded that simplex-based methods are not best for many of these models. The best one-to-one shortest path algorithm may be found in Helgason, Kennington & Stewart

[1993] and the best semi-assignment algorithm is described in Kennington & Wang [1992]. Bertsekas and his colleagues at MIT have developed a family of algorithms which are also very effective for many of these special models (see Bertsekas [1981, 1991]).

Many books exist which contain excellent presentations of the various algorithms for a variety of network problems. Our favorites include Papadimitriou & Steiglitz [1982], Chvátal [1983], and Ahuja, Manganti & Orlin [1989]. (For some excellent texts which have recently appeared see Ahuja, Magnanti & Orlin [1993], Bertsekas [1991], Evans & Minieka [1992], and Murty [1992].) This chapter would not be complete without mentioning our own book (Kennington & Helgason [1980]) which contains a wealth of technical details not found in other publications.

References

Ahuja, R., T. Magnanti and J. Orlin (1989). Network flows, in: G. L. Nemhauser, A. H. G. Rinnooy Kan and M. J. Todd (eds.), *Optimization*, Handbooks in Operations Research and Management Science, Vol. 1, North-Holland Publishing Company, Amsterdam, Chapter IV.

Ahuja, R., T. Magnanti and J. Orlin (1993). *Network Flows: Theory, Algorithms, and Applications*, Prentice-Hall, Inc., Englewood Cliffs, N.J.

Ali, I., D. Barnett, K. Farhangian, J. Kennington, B. McCarl, B. Patty, B. Shetty and P. Wong (1984). Multicommodity network problems: Applications and computations, *IIE Trans.* 16, 127–134.

Ali, A., J. Kennington and B. Shetty (1988). The equal flow problem. *Eur. J. Oper. Res.* 36, 107–115.

Anderson, R.J., and J.C. Setubal (1992). Goldberg's algorithm for maximal flow in perspective: A computational study. *Operations Research*, 42, 65–80.

Barr, R., F. Glover and D. Klingman (1979). Enhancement of spanning tree labeling procedures for network optimization. *INFOR* 17, 16–34.

Barr, R., K. Farhangian and J. Kennington (1986). Networks with side constraints: An LU factorization update. *Ann. Soc. Logist. Eng.* 1, 66–85.

Barr, R., and B. Hickman (1993). Parallel simplex for large pure network problems: Computational testing and sources of speedup, *Oper. Res.* 42, 65–80.

Bertsekas, D. P. (1981). A new algorithm for the assignment problem. *Math. Program.* 21, 152–171.

Bertsekas, D. P. (1991). *Network Optimization: Algorithms and Codes*, The MIT Press, Cambridge, MA.

Bradley, G., G. Brown and G. Graves (1977). Design and implementation of large scale primal transshipment algorithms. *Manage. Sci.* 24, 1–38.

Brown, G., and R. McBride (1984). Solving generalized networks. *Manage. Sci.* 30, 1497–1523.

Carolan, W., J. Hill, J. Kennington, S. Niemi and S. Wichmann (1990). An empirical evaluation of the Korbx algorithms for military airlift applications. *Oper. Res.*, 38, 240–248.

Chvátal, V. (1983). *Linear Programming*, W. H. Freeman and Company, New York, NY.

Clark, R., J. Kennington, R. Meyer and M. Ramamurti (1992). Generalized networks: Parallel algorithms and an empirical analysis. *ORSA J. Comput.* 4, 132–145.

Dantzig, G.B. (1963). *Linear Programming and Extensions*, Princeton University Press, Princeton, NJ.

Dantzig, G.B., and P. Wolfe (1960). Decomposition principle for linear programs. *Oper. Res.* 8, 101–111.

Evans, J.R., and E. Minieka (1992). *Optimization Algorithms for Networks and Graphs*, 2nd Ed., Marcel Dekker, Inc., New York, NY.

Ford, L.R., and D.R. Fulkerson (1958). A suggested computation for maximal multicommodity network flow. *Manage. Sci.* 5, 97–101.

Glover, F., D, Karney and D. Klingman (1974). Implementation and computational comparisons of primal, dual, and primal-dual computer codes for minimum cost network flow problems. *Networks* 4, 191–212.

Glover, F., D, Karney, D. Klingman and A. Napier (1974). A computational study on start procedures, basis change criteria, and solution algorithms for transportation problems. *Manage. Sci.* 20, 793–813.

Glover, F., D. Klingman and N. Phillips (1992). *Network Models in Optimization and Their Applications in Practice*, John Wiley and Sons, Inc., New York, NY.

Glover, F., D. Klingman and J. Stutz (1973). Extension of the augmented predecessor index method to generalized network problems. *Transp. Sci.* 7, 377–384.

Glover, F., D. Klingman and J. Stutz (1974). Augmented threaded index method for network optimization. *INFOR* 12, 293–298.

Hartman, J.K., and L.S. Lasdon (1970). A generalized upper bounding method for doubly coupled linear programs. *Nav. Res. Logist. Q.* 17, 4, 411–429.

Hartman, J.K., and L.S. Lasdon (1972). A Generalized upper bounding algorithm for multicommodity network flow problems. *Networks* 1, 333–354.

Held, M., P. Wolfe and H. Crowder (1974). Validation of subgradient optimization. *Math. Program.* 6, 62–88.

Helgason, R., and J. Kennington (1977). A product form representation of the inverse of a multicommodity cycle matrix. *Networks* 7, 297–322.

Helgason, R., J. Kennington and B. Stewart (1993). Computational comparison of sequential and parallel algorithms for the one-to-one shortest parth problem, *Computational Optimization and Applications* 2, 47–75.

Johnson, E. (1965). Programming in networks and graphs, Technical Report ORC 65-1, Operations Research Center, University of California–Berkeley, Berkeley, CA.

Kennington, J., and R. Helgason (1980). *Algorithms for Network Programming*, John Wiley and Sons, Inc., New York, NY.

Kennington, J., and Z. Wang (1992). A shortest augmenting path algorithm for the semi-assignment problem. *Oper. Res.* 40, 178–187.

Murty, K. (1992). *Network Programming*, Prentice-Hall, Inc., Englewood Cliffs, NJ.

Papadimitriou, C., and K. Steiglitz (1982). *Combinatorial Optimization: Algorithms and Complexity*, Prentice-Hall, Inc., Englewood Cliffs, NJ.

Peters, J. (1990). The network simplex method on a multiprocessor. *Networks* 20, 845–859.

Resende, M., and G. Veiga (1992). An efficient implementation of a network interior point method, Technical Report, AT&T Bell Laboratories, Murray Hill, NJ 07974.

Srinivasan, V., and G. Thompson (1973). Benefit–cost analysis of coding techniques for the primal transportation algorithm. *J. Assoc. Comput. Mach.* 20, 194–213.

M.O. Ball et al., Eds., *Handbooks in OR & MS, Vol. 7*

Chapter 3

Matching

A.M.H. Gerards
CWI, P.O. Box 94079, 1090 GB Amsterdam, The Netherlands

1. Introduction

Matching is pairing: dividing a collection of objects into pairs. Typically the objective is to maximize total profit (or minimize cost), where the profit of each possible pair is known in advance.

For a more formal definition, let G be an undirected graph with node set V and edge set E. A subset M of E such that no two edges in M are incident to a common node is a *matching*. If M has exactly one edge incident to each node $v \in V$, we call M a *perfect matching*. The *maximum weight matching problem* with respect to the *weights* w on the edges is:

Find a matching M with total weight $\sum_{e \in M} w_e$ as large as possible. (1)

The *minimum weight perfect matching problem* is:

Find a perfect matching M with total weight $\sum_{e \in M} w_e$ as small as possible. (2)

A matching problem is defined by two parameters: the graph G and the weights w. We classify matching problems by these different parameters:

The weights – cardinality or weighted matching. Finding a maximum cardinality matching, i.e. considering all edges to have weight one, is easier than dealing with more general weights. Moreover, an algorithm for finding a maximum cardinality matching can be, and in our presentation is, used as a subroutine for solving the general weighted problem. Therefore, we discuss the cardinality case first, in Section 2, and study general weights later, in Section 6.

The graph – bipartite or non-bipartite matching. Matching problems are significantly easier in bipartite graphs. So, we discuss several topics in bipartite matching before venturing into the complications of the non-bipartite case.

Matching theory is one of the cornerstones of mathematical programming. Yet, matchings are not as ubiquitous in practice as network flow problems (for applications of network flows, see Ahuja, Magnanti & Orlin [1989]) and, when they do arise in practice, it is most often in bipartite graphs where they can be modeled as network flow problems anyway (see Section 3.1). So, to what does 'Matching' owe its prominence?

First, there is the position of matching problems between the 'easier' problems like network flows, and the hard (NP-hard) problems like general integer linear programming. This position has been pointed out by Edmonds & Johnson [1970]. It is probably best expressed by the qualification: "*Optimum matching problems ... constitute the only class of genuine integer programs for which a good solution is known*" [Cunningham & Marsh, 1978]. In a footnote, Cunningham and Marsh clarify this statement with: "*... Every other class of well-solved combinatorial problems is not 'genuine' because either: (a) No explicit formulation as an integer program using a reasonable amount of data is known (example: minimum spanning tree problems); or (b) When such a formulation is known, the resulting linear program already has integer-valued optimal solutions (example: network flow problems).*"

Second, and perhaps more intrinsic to the importance of matching, there is the intricate structure of matching theory. Just a glimpse in the excellent book *Matching Theory* by Lovász & Plummer [1986] should convince the reader of this. For instance, there are the structural descriptions of the class of all maximum cardinality matchings in a graph. In this chapter we see how one of these, the Edmonds–Gallai structure (see Section 4.1), helps in understanding an algorithm for finding a minimum weight perfect matching (see Section 6.2). Third, there is the 'self-refining' property of matching theory: it contains a wide class of its generalizations as special cases (see Sections 3 and 7).

1.1. Examples of matching problems

We begin with four examples of matching problems. A classic example is:

The assignment problem. Suppose n tasks are to be carried out and each must be assigned to a single person. We have a staff of n people available and each person can be assigned only one task. Moreover, we know for each person p and task t a number $w_{p,t}$ quantifying the productivity of p when carrying out t. Now, the question is: How do we assign the tasks to the people, i.e. make task-person pairs, so that the total productivity is as large as possible?

Clearly, this is a bipartite matching problem. An example of a non-bipartite matching problem is:

The oil well drilling problem [Devine, 1973]. Suppose we are given the locations of oil deposits. We want to exploit all the deposits at minimum drilling cost. It is technically feasible to access two deposits with one well: drill a hole to the first deposit, and then continue drilling from that same hole, possibly at a different angle, to the second deposit. The oil is then brought up from both deposits via concentric pipes. We know the savings possible from combined drilling operations for each pair of deposits. The question is: How do we combine the drilling operations so that the savings are as large as possible?

In both of these cases, the matching character of the problem is immediately obvious from the description. A more disguised matching problem is:

Plotting street maps [Iri & Taguchi, 1980; Iri, Murota & Matsui, 1983]. A pen-plotter has to draw a street map of a city. Since the total length of the lines to be drawn is fixed, this amounts to minimizing the total distance of the 'non-drawing' moves, or *shifts*, the pen makes to change position. A natural assumption is that the pen starts from some prescribed point and returns to it when the drawing is finished. For simplicity, we assume that this common starting and ending point is a point of the map to be drawn. Moreover, we assume that the graph of the map (edges are streets, nodes are intersections) is connected, i.e., that you can travel between any two intersections in the city along streets. The question is: In what order should the edges be drawn to minimize the total drawing time?

An old and famous theorem of Euler's [1736] asserts that if all the nodes in a connected graph have even degree, one can draw the graph without making any shifts. However, when there are nodes with odd degree, shifts are unavoidable. In this case, the problem amounts to adding edges to the graph, which will become the shifts, so that every node in the resulting graph has even degree. Distances in the city are Euclidean. So we may assume, by the triangle inequality, that no two shifts have a common endpoint. Hence, finding the fastest way to draw the map amounts to pairing up the nodes with odd degree so that the total length of the line segments connecting the two nodes in each pair is as short as possible. This is a minimum weight perfect matching problem.

The following application is even less obvious. Again, we solve this problem as a matching problem, but now the resulting matching must be modified in a non-trivial way to obtain a solution to the original problem.

Scheduling [Fujii, Kasami & Ninomiya, 1969]. Suppose, you and your partner want to restore this lovely old house you just bought. You have divided the whole project into a number of one-week jobs that must be carried out in accordance with certain precedence relations. For example, it is difficult to paper a new wall before putting it up. As the project puts already enough pressure on the relationship, you have agreed not to work together on any job. You both have about the same skills, however, so either of you can do each job equally well. The question is: How do you allocate the jobs between you so you can finish the project as early as possible?

We have a list of one-week jobs $J_1, \ldots, J_k$, together with a partial order $\prec$ on them. The relation $J_i \prec J_j$ means that J_i should be carried out before J_j and two jobs are *incomparable* if neither must precede the other. Minimizing the project duration amounts to scheduling the jobs, in accordance with the precedence constraints, so that in as many weeks as possible you and your partner are both assigned jobs. Clearly, if you and your partner are both working in a given week, your jobs must be incomparable. Therefore, as a first attempt at finding an optimal schedule, we look for a largest set of disjoint, incomparable pairs: a maximum cardinality matching problem.

If a largest matching $\mathcal{P}$ consists of ℓ disjoint, incomparable pairs, then we know that the whole project will take at least $k - \ell$ weeks. In fact, we can complete the project within that period. The idea is to schedule the jobs from each pair in $\mathcal{P}$ in a single week. Each of the remaining jobs, called the *singleton jobs*, is assigned to a week by itself. We might, however, have to change $\mathcal{P}$ to make this possible.

The proof that we can complete the project in $k - \ell$ weeks is by induction on k, the number of jobs. Let $\mathcal{E}$ denote the jobs that are minimal with respect to $\prec$. If there is a pair, say $(J_1, J_2) \in \mathcal{P}$ with both J_1 and J_2 in $\mathcal{E}$, we schedule these two jobs in the first week. $\mathcal{P} \setminus (J_1, J_2)$ is still a largest collection of incomparable pairs among the remaining jobs. Hence the induction hypothesis assures us we can schedule the jobs in $k - \ell$ weeks. A similar argument applies if $\mathcal{E}$ contains a singleton job.

Now suppose $\mathcal{E}$ contains neither a singleton job nor a pair in $\mathcal{P}$. Among all pairs $(J_1, J_2) \in \mathcal{P}$ with $J_1 \in \mathcal{E}$ choose one with J_2 minimal with respect to $\prec$. As $J_2 \notin \mathcal{E}$ there exists a $J_3 \in \mathcal{E}$ with $J_3 \prec J_2$. Moreover, as J_3 is not a singleton job, there exists a job J_4 with $(J_3, J_4) \in \mathcal{P}$. As $J_3 \prec J_2$, J_2 does not precede J_4. Moreover, by the choice of J_2, J_4 does not precede J_2. So jobs J_2 and J_4 are incomparable. Hence, $(\mathcal{P} \setminus \{(J_1, J_2), (J_3, J_4)\}) \cup \{(J_1, J_3), (J_2, J_4)\}$ is also a largest collection of disjoint, incomparable pairs. As this collection does have pair in $\mathcal{E}$, we can proceed as before, schedule that pair in the first week, and by the induction hypothesis the other jobs in the remaining $k - \ell - 1$ weeks.

For other applications of matching see Ball, Bodin & Dial [1983] and Murota [1993].

1.2. Generalizations of the matching problem

There are two obvious directions in which the matching problem can be generalized.

Degree-constrained optimization. Matchings are subgraphs in which each edge appears at most once and each node has degree at most one. This suggests extensions allowing more general degree constraints and allowing an edge to appear more than once in a feasible solution. These generalizations lead to so-called 'b-factors' and 'b-matchings', and – if we limit the number of times an edge can appear – to 'capacitated' b-matchings. We can also impose lower bounds on the degrees of nodes and on the number of times edges appear. Thus we can extend the matching problem to a quite general degree-constrained (multi-)graph optimization problem. Surprisingly, many results for ordinary matchings extend to these generalizations. This phenomenon, which we explain in Section 7, is part of the self-refining nature of matching theory previously mentioned.

This self-refining nature persists even when we impose parity conditions on the degrees of nodes, e.g. demanding that feasible solutions have an odd number of edges incident to particular nodes (see Section 7.4).

Set-packing problems. As matching is finding disjoint pairs, a natural generalization of the matching problem is trying to find, in some sense optimal, collections of disjoint triples, quadruples or otherwise structured sets. This gives rise to the very general *set-packing* problem, which includes many combinatorial optimization problems as special cases. Unfortunately the set-packing problem is too general, it is NP-hard (even if we only consider packing triples [Karp, 1972]). Except for a few lines on stable sets in perfect graphs (in Section 3.2), we do not study them here but confine ourselves to matchings.

1.3. Other sources on matching

The number of publications concerning matching problems is enormous, the references in this chapter constitute only a very limited part of them. There are many good books on matching. We mentioned already *Matching Theory* by Lovász & Plummer [1986], which really is the most complete source on matchings available at this moment. Other highly recommendable books that deal (partly) with matchings are: *Graphs* by Berge [1985], *Combinatorial Optimization: Networks and Matroids* by Lawler [1976], *Graphs and Algorithms* by Gondran & Minoux [1984] and *Programming in Networks and Graphs* by Derigs [1988a]. Sources on general integer programming (including matching) are: *Theory of Linear and Integer Programming* by Schrijver [1986] and *Integer and Combinatorial Optimization* by Nemhauser & Wolsey [1988].

Excellent introductions to matchings are: Schrijver's [1983a] survey paper on min–max relations in combinatorial optimization, with special emphasis on the self-refining properties of matching and the paper by Pulleyblank [1995], with more emphasis on structural results then one will find in this chapter. A very nice historical overview of matching theory – from its birth to today's state of the art – is the paper by Plummer [1992]. Recently, the same author gave an overview on how 'hard' or 'easy' various matching and vertex-packing problems are [Plummer, 1993].

As mentioned, bipartite matchings are really network flows. Network flows are also discussed in most of the just mentioned publications. For extensive treatments of network flows we refer to the surveys by Ahuja, Magnanti & Orlin [1989], by Goldberg, Tardos & Tarjan [1990] and by Helgason & Kennington [1995, this volume].

1.4. Outline

The first part of this chapter considers algorithms for finding maximum cardinality as well as maximum weight matchings (Sections 2 and 6). We start with the Hungarian method for finding a maximum cardinality matching in a bipartite graph (Section 2.1). We then extend the method in two directions: to Edmonds' blossom algorithm for finding a maximum cardinality matching in a general graph (Section 2.2) and to the Hungarian method for finding a maximum weight matching in a bipartite graph (Section 6.1). Finally, the ideas of these two methods are combined in Edmonds' algorithm for the weighted matching problem in general graphs (Section 6.2). From the insight provided by the Hungarian method for maximum cardinality matching in bipartite graphs the classical theorems of Fröbenius and König on bipartite matchings easily follow (Section 3). Some of the self-refining properties of matchings can be found in that section. Edmonds' blossom algorithm for cardinality matching yields the theorems of Tutte and Berge on matchings in general, non-bipartite, graphs and the Edmonds–Gallai structure theorem (Section 4). This structure theorem facilitates the description and analysis of Edmonds' blossom algorithm for weighted matching. Because algorithms for the weighted matching problem are closely related to the formulation of this problem as a linear program, we discuss the matching polytope in Section 5. In that section we also briefly mention stable matchings (Section 5.2). Together these sections contain the basic algorithmic and structural aspects of matchings.

The second part of this chapter consists of four sections. In Section 7 we consider general degree constraints and discuss some of the self-refining aspects of matching. In Section 8 we present other algorithms for matching problems, including randomized algorithms for maximum matching and for counting matchings. In Section 9, we discuss applications of matchings to other combinatorial optimization problems like the traveling salesman problem. This chapter concludes with a (short) section on the computer implementation of matching algorithms and on heuristics for matching problems.

We conclude this section with some notation and conventions.

1.5. Notation and conventions

1.5.1. Graphs

Typically we denote the edge set of an undirected graph G by E and its node set by V. When ambiguity might arise, we write $E(G)$ and $V(G)$. We write $uv \in E$ to mean that uv is an edge with endpoints u and v. In case of parallel edges this might seem a bit ambiguous, but generally it is not. Parallel edges are not particularly relevant for matching problems and we can always assume that there are none. Yet, parallel edges hardly complicate the problem. In fact, in solving matching problems we construct graphs with parallel edges. So we do not explicitly forbid them.

We assume the reader is familiar with the basic notions of graph theory [see Bondy & Murty, 1976] and only discuss those most important for this chapter. For each subset $U \subseteq V(G)$, we define: $\delta(U) := \{uv \in E(G) \mid u \in U, v \notin U\}$ and $\langle U \rangle := \{uv \in E(G) \mid u \in U, v \in U\}$. We let $G|U$ denote the subgraph of G *induced* by U, i.e., $V(G|U) = U$, $E(G|U) = \langle U \rangle$, and we let $G \setminus U := G|(V(G) \setminus U)$. For each node $u \in V(G)$, $G \setminus u := G \setminus \{u\}$, $\delta(u) := \delta(\{u\})$ and $\deg(u) := |\delta(u)|$, the *degree* of u. For each edge $e \in E(G)$, $G \setminus e$ is the graph with node set $V(G)$ and edge set $E(G) \setminus \{e\}$. We use $G \cup e$ to denote the graph obtained by adding the new edge $e \notin E(G)$ to G, i.e., $V(G \cup e) = V(G)$ and $E(G \cup e) = E(G) \cup \{e\}$. The notions of connected graphs, components, paths, trees and forests are so standard that we omit their definitions here. Circuits and Eulerian graphs are standard notions too, but there is a rather wide-spread babel as far as the terminology is concerned. We refer to a *circuit* (of length k) as a graph C with k distinct nodes $V(C) := \{v_1, \ldots, v_k\}$ and the k distinct edges $E(C) := \{v_1v_2, v_2v_3, \ldots, v_{k-1}v_k, v_kv_1\}$. A *cycle* is a graph with all degrees even. So a cycle is an edge-disjoint union of circuits (in other writings one might find 'cycle' where we use 'circuit'). A connected cycle is called an *Eulerian* graph. We often identify a circuit C with its edge set $E(C)$, and so write $e \in C$, meaning $e \in E(C)$. Similarly, we identify other subgraphs like paths or trees with their edge sets.

We denote a bipartite graph by $G = (V_1 \cup V_2, E)$ where V_1 and V_2 are the color classes (so each edge has one endpoint in V_1 and one in V_2).

Given a directed graph D, $V(D)$ denotes the node set, $A(D)$ denotes the arc set, and $\overrightarrow{uv}$ denotes an arc from u to v. For each subset $U \subseteq V$, $\delta^-(U) := \{\overrightarrow{uv} \in A(G) \mid u \notin U, v \in U\}$ and $\delta^+(U) := \{\overrightarrow{uv} \in A(G) \mid u \in U, v \notin U\}$. Again, for each node $u \in V(D)$, we abbreviate $\delta^-(\{u\})$ as $\delta^-(u)$ and $\delta^+(\{u\})$ as $\delta^+(u)$.

1.5.2. Numbers, vectors, polytopes and polyhedra

For $\alpha \in \mathbb{R}$, $\lfloor \alpha \rfloor$ denotes the largest integer not greater than α. Similarly $\lceil \alpha \rceil$ denotes the smallest integer not smaller than α.

Given a set R and a finite set S, R^S denotes the collection of vectors indexed over S with components in R. So, for example, $\mathbb{R}^S$ is the collection of real vectors and $\mathbb{Z}^S$ is the collection of integral vectors with components indexed by S. We use $\mathbb{R}_+$ to denote the set of non-negative reals and $\mathbb{Z}_+$ to denote the set of non-negative integers. For each subset $T \subseteq S$, $\chi^T \in \{0,1\}^S$ denotes the *characteristic vector* of T as a subset of S, i.e., $(\chi^T)_t = 1$ if $t \in T$ and $(\chi^T)_t = 0$ if $t \notin T$. Given $x \in \mathbb{R}^S$ and $T \subseteq S$, we frequently use $x(T)$ to denote $\sum_{t \in T} x_t$.

The *node-edge incidence matrix* $N = (N_{u,e})$ of an undirected graph G is the $V(G) \times E(G)$ matrix with $N_{u,e} = 1$ if u is an endpoint of edge e and $N_{u,e} = 0$ otherwise.

Let $x^1, \ldots, x^k \in \mathbb{R}^S$. Any vector of the form $\sum_{i=1}^k \lambda_i x^i$ with $\sum_{i=1}^k \lambda_i = 1$ and $\lambda_i \geq 0$ for each $i \in \{1, \ldots, k\}$, is called a *convex combination* of $x^1, \ldots, x^k$. The *convex hull* of a set X is the collection of all convex combinations of finite subsets of X. A *polytope* is the convex hull of a finite set. A *polyhedron* is the solution set of a finite system of linear inequalities, i.e., a set of the form $\{x \in \mathbb{R}^n \mid Ax \leq b\}$ for some matrix A and vector b.

1.5.3. Algorithms

One measure of the running time of an algorithm is the number of arithmetic steps it requires. Here, an arithmetic step is the addition, multiplication, division or comparison of two numbers. We report the running time of an algorithm by giving its asymptotic behavior as a function of the size of the input. So, we say for instance that the running time is $O(n)$ meaning that there exists a constant γ so that given input of size n the algorithm takes no more than γn steps before it produces the output. An algorithm is *polynomial* if its running time is $O(n^p)$ for some $p \in \mathbb{Z}_+$.

The input for matching problems is in the form of graphs and rational numbers. We assume that the graph G is represented by its *adjacency lists*, i.e., for each node v we have a list of the nodes adjacent to v. Thus, the size of the input of a graph is proportional to $|V(G)| + |E(G)|$. (Note that every edge is represented twice in this way.)

Because we measure the running time of algorithms with respect to the number of arithmetic steps, we can consider the input size of a rational number to be 1. However, when implementing the algorithms on a computer, rationals take more space to encode. A rational number p/q, where p and q are integers, can be represented with $\log(|p|+1) + \log(|q|+1)$ binary bits. On the other hand, an arithmetic operation on two rationals can be carried out in a number of binary operations that is polynomial in the number of binary bits required to encode the two rationals. So, if the number of arithmetic operations an algorithm requires is polynomial in the number of input rationals, the number of binary operations it requires will be polynomial in the number of binary digits needed to encode those rationals; that is, provided the numbers that are calculated in the process do not become too big! Though not true in general, in this chapter the numbers computed do not become too large and so, it is enough to argue that the number of arithmetic operations an algorithm requires is polynomial in the size of the graph considered.

2. Finding a matching of maximum cardinality

The maximum cardinality of a matching in an undirected graph $G = (V, E)$ is denoted by $\nu(G)$. The main question of this section is: How do we find – in polynomial time – a *maximum matching*, that is, a matching of maximum cardinality? We consider this problem separately from the more general weighted problem because it is easier and it contains the main ideas and notions of the weighted problem: 'alternating paths' and 'shrinking'.

2.1. Alternating paths and forests

A path P in a graph $G = (V, E)$ is said to be *alternating* with respect to a matching M, or *M-alternating*, if the edges of P are alternately in M and not in

Fig. 1. The bold edges are in the matchings M, respectively M'

M (see Figure 1a; for instance, the paths $\{e_4, e_7, e_6\}$ and $\{e_1, e_2, e_3, e_4, e_5, e_6\}$ are M-alternating). So, every node in an M-alternating path P, except possibly its end nodes, is incident to an edge in $M \cap E(P)$.

If an edge uv is in a matching M, we say that the node u is *matched by* M and that the two nodes u and v are *matched*. We also write u_M to denote v. Nodes not matched by M are called *exposed* and we denote the set of exposed nodes by $\exp(M)$. We define the *deficiency* of G as $\operatorname{def}(G) := |V(G)| - 2\nu(G)$. So the deficiency is minimum cardinality of $\exp(M)$. An alternating path P is *augmenting*, or more precisely *M-augmenting*, if both its end nodes are exposed (see Figure 1a; the dotted line indicates an M-augmenting path). Augmenting paths obviously yield larger matchings (we refer to the operation described in (3) and illustrated in Figure 1, as AUGMENT):

If P is an augmenting path with respect to a matching M, then the symmetric difference $M' := P \triangle M$ is a matching too. Moreover, $|M'| = |M| + 1$. (3)

In fact, the converse is also true: if there is no augmenting path, there is no larger matching.

Theorem 1 [Berge, 1957; Norman & Rabin, 1959]. *A matching M in a graph G is a maximum matching if and only if there is no M-augmenting path in G.*

Proof. Let M' be a matching in G with $|M'| > |M|$. The graph consisting of the edges in $M' \triangle M$ has maximum degree 2. Hence, each of its components is either a path or a circuit in which the edges are alternately in M' and in M. Clearly one of these components must contain more edges from M' than from M and that component must be an augmenting path with respect to M. The converse is (3). □

Mulder [1992] pointed out that this result was in fact already known by Julius Petersen [1891], probably the first to study matchings in graphs.

So, searching for maximum matchings amounts to searching for augmenting paths. We search for augmenting paths by 'growing alternating forests'.

Let M be a matching in a graph $G = (V, E)$. A tree T in G is called *alternating* if the following hold (see Figure 2):

– T contains exactly one exposed node, denoted by r_T,

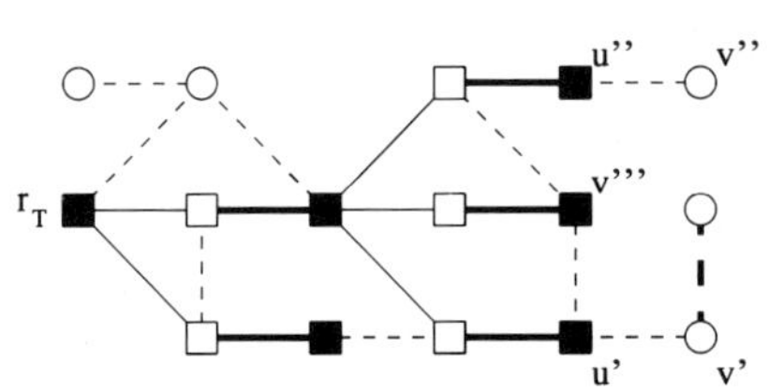

Fig. 2. The solid edges form an alternating tree. The bold edges, dashed or not, are in the matching. Open square nodes are odd; filled square nodes are even.

– for each node $v \in V(T)$, the path from r_T to v in T is alternating, and

– for each node v of degree one in T, other than r_T, the matching-edge vv_M is in T.

An *alternating forest* is a node-disjoint union of alternating trees such that each exposed node is in one of the trees. So, the forest consisting only of the exposed nodes without any edges is alternating. For each node v in an alternating forest F, F_v denotes the alternating tree in F containing v and $r_{v,F}$ denotes the unique exposed node in F_v. We call a node v in F *even* (*odd*) if the unique path in F from $r_{v,F}$ to v contains an even (odd) number of edges. We denote the set of even nodes of an alternating forest F by even(F) and the set of odd nodes by odd(F).

The following procedure uses alternating forests to search for augmenting paths.

GROW: Either (1) or (2) below applies:

(1) If there is a node $u \in$ even(F) adjacent to a node $v \notin$ odd(F), then exactly one of the following occurs:

$v \notin$ even(F). In this case, we extend F to a larger alternating forest by adding the edges uv and vv_M (v is matched). We refer to this as 'GROWING the alternating forest'. (See Figure 2; $u = u'$, $v = v'$.)

$v \in$ even(F) **and** $F_u \neq F_v$. In this case, we have FOUND an AUGMENTING PATH, namely the union of: the path in F_u from $r_{u,F}$ to u, the edge uv, and the path in F_v from v to $r_{v,F}$. (See Figure 2; $u = u''$, $v = v''$.)

$v \in$ even(F) **and** $F_u = F_v$. In this case, the procedure HALTS. (See Figure 2; $u = u'$, $v = v'''$.)

(2) If no node $u \in$ even(F) is adjacent to a node $v \notin$ odd(F), the procedure TERMINATES. In this case, the forest F is called *Hungarian*.

Lemma 2. *Let M be a matching in a graph $G = (V, E)$ and let F be an alternating forest with respect to M. If F is Hungarian, then M is a maximum matching.*

Proof. Since F is Hungarian, nodes in even(F) are adjacent only to nodes in odd(F). So, each matching in G has at least $|\text{even}(F)| - |\text{odd}(F)|$ exposed nodes. On the other hand, from the definition of alternating forest it follows that $|\exp(M)| = |\text{even}(F)| - |\text{odd}(F)|$. So M is a maximum matching. □

Lemma 2 implies that GROW either finds an augmenting path, terminates with a maximum matching, or HALTS. When G is bipartite, each even node v is in the same color class of G as $r_{v,F}$. So, no two adjacent even nodes can be in the same component of F and GROW cannot HALT. Thus, we can find a maximum matching in a bipartite graph by iteratively applying GROW and AUGMENT. This algorithm has been introduced by Kuhn [1955], in the context of matchings in bipartite graphs, and Hall [1956], in the context of 'systems of distinct representatives' (see Section 3). Kuhn called it the *Hungarian method* in recognition of König and Egerváry's contributions to the theory of matchings. The Hungarian method is easy to implement.

Theorem 3. *The Hungarian method finds a matching of maximum cardinality in a bipartite graph in $O(|E|\min(|V_1|, |V_2|))$ time.*

Proof. It takes $O(|E|)$ time for GROW to find an augmenting path or construct a Hungarian alternating forest. Each augmentation takes $O(|E|)$ time as well. Since we AUGMENT $\nu(G) \leq \min(|V_1|, |V_2|)$ times, the theorem follows. (Note that we apply GROW $\nu(G) + 1$ times.) □

Hopcroft & Karp [1971, 1973] improved on this running time by searching for a collection of disjoint, shortest augmenting paths and then augmenting along all the paths simultaneously.

Given a matching M, we define $\ell(M)$ to be the number of edges in a shortest M-augmenting path. A collection of node-disjoint, shortest augmenting paths $P_1, \ldots, P_t$ is called *maximal* if there is no shortest augmenting path in G node-disjoint from each of the paths $P_1, \ldots, P_t$. Hopcroft and Karp's algorithm is based on the following two observations.

We can find a maximal collection of node-disjoint shortest augmenting paths in $O(|E|)$ time. (4)

Indeed, breadth-first search starting from $\exp(M) \cap V_1$ accomplishes this.

If $P_1, \ldots, P_t$ is a maximal collection of shortest M-augmenting paths, then $\ell(M \triangle P_1 \triangle \cdots \triangle P_t) > \ell(M)$. (5)

To see this, let Q be an augmenting path with respect to $M' := M \triangle P_1 \triangle \cdots \triangle P_t$. Now, observe that in proving Theorem 1 we actually proved:

Let M_1 and M_2 be two matchings with $k := |M_2| - |M_1| > 0$. Then there exists a collection of k mutually node-disjoint M_1-augmenting paths in $M_1 \triangle M_2$. (6)

Applying (6) to M and $M' \triangle Q$, we get $t + 1$ node-disjoint M-augmenting paths $Q_1, \ldots, Q_{t+1}$ in $M \triangle M' \triangle Q$. Now one easily verifies:

$$\begin{aligned}\ell(M)(t+1) &\leq |Q_1| + \cdots + |Q_{t+1}| \leq |M \triangle M' \triangle Q| = \\ &= |P_1 \triangle \cdots \triangle P_t \triangle Q| \\ &= |(P_1 \cup \cdots \cup P_t) \triangle Q| = |P_1 \cup \cdots \cup P_t| + |Q| - \\ &\quad - 2|(P_1 \cup \cdots \cup P_t) \cap Q| \\ &= \ell(M)t + |Q| - 2|(P_1 \cup \cdots \cup P_t) \cap Q|. \end{aligned} \tag{7}$$

Hence

$$|Q| \geq \ell(M) + 2|(P_1 \cup \cdots \cup P_t) \cap Q|. \tag{8}$$

So $|Q| \geq \ell(M)$. Suppose $|Q| = \ell(M)$. Then Q has no edge in common with any of $P_1, \ldots, P_t$ Hence, since Q is augmenting with respect to $M \triangle P_1 \triangle \cdots \triangle P_t$, it must also have no node in common with any of $P_1, \ldots, P_t$. This contradicts the maximality of the collection $P_1, \ldots, P_t$. So we conclude that $|Q| > \ell(M)$, which proves (5).

This is the algorithm of Hopcroft and Karp. In each *phase* we are given an (initially empty) matching M. We find a maximal collection of node-disjoint shortest M-augmenting paths $P_1, \ldots, P_t$ and augment along all of them to obtain the larger matching $M' := M \triangle P_1 \triangle \cdots \triangle P_t$. We iteratively repeat this procedure using as input to each successive phase the matching M' constructed in the previous phase.

The algorithm of Hopcroft and Karp finds a maximum cardinality matching in a bipartite graph in $O(|E|\sqrt{|V|})$ time. (9)

Since each phase can be carried out in $O(|E|)$ time, it suffices to prove that we need only $O(\sqrt{|V|})$ phases. Let M_1 be the matching found after $\sqrt{|V|}$ phases. By (5), $\ell(M_1) \geq \sqrt{|V|}$. So there can be at most $|V|/\sqrt{|V|} = \sqrt{|V|}$ mutually edge-disjoint M_1-augmenting paths. Applying (6) to M_1 and some maximum matching M_2, we see that $|M_1| \geq \nu(G) - \sqrt{|V|}$. Hence, after at most $\sqrt{|V|}$ further phases we obtain a maximum cardinality matching.

Improvements on (9) are the $O(|V|^{1.5}\sqrt{|E|/\log|V|})$ algorithm by Alt, Blum, Mehlhorn & Paul [1991], which is faster for dense graphs, and the $O(|E|\sqrt{|V|}\log_{|V|}(|V|^2/E))$ algorithm by Feder & Motwani [1991].

In Section 3.1 we show how to find a maximum cardinality matching by solving a max-flow problem. In fact, Even & Tarjan [1975] observed that we can interpret Hopcroft and Karp's algorithm as Dinic's max-flow algorithm [Dinic, 1970] applied to matchings in bipartite graphs. Recently, Balinski & Gonzalez [1991] developed an $O(|E||V|)$ algorithm for finding maximum matchings in bipartite graphs that is not based on augmenting path methods.

2.2. Non-bipartite graphs – shrinking blossoms

In non-bipartite graphs the procedure GROW may HALT even when there are augmenting paths. Indeed, consider the example in Figure 3. Nodes u and v are even, adjacent and belong to the same alternating tree. Clearly, we cannot grow the tree any further. On the other hand, there is an augmenting path (in this case it is unique) and it contains edge uv. So, we must modify our procedure for finding augmenting paths.

2.2.1. Alternating circuits and blossoms

A circuit C is said to be *alternating* with respect to a matching M if $M \cap E(C)$ is a maximum matching in C. So, when C is an alternating odd circuit with respect to a matching M, exactly one node in C is exposed with respect to $M \cap E(C)$. We call this node the *tip* of the alternating odd circuit C. If the tip t of an alternating odd circuit C is connected to an exposed node by an even alternating path P with $V(P) \cap V(C) = \{t\}$, then C is called a *blossom* and P is called a *stem* of C.

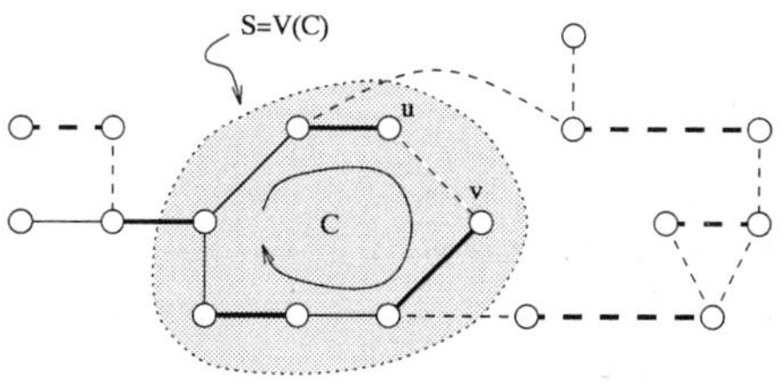

Fig. 3. Solid edges are in the alternating forest; bold edges, dashed or not, are in the matching. The shaded region indicates a blossom.

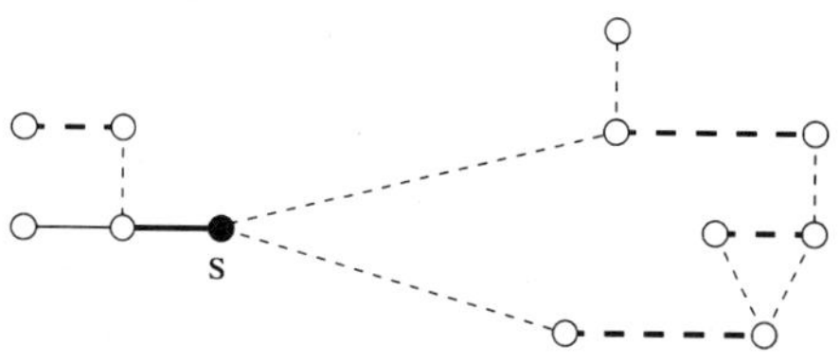

Fig. 4.

When the procedure GROW HALTS, *G contains a blossom.* (10)

Indeed, suppose we have two adjacent even nodes u and v in an alternating forest F, both belonging to the same alternating tree T of F (see Figure 3). Consider the paths P_u from r_T to u and P_v from r_T to v in F. Then $E(P_u) \triangle E(P_v)$ together with uv forms a blossom and the intersection of P_u and P_v is one of its stems.

Having detected a blossom C, we 'shrink' it. That is, we apply the procedure SHRINK to $V(C)$. Figure 4 illustrates the effect of shrinking $V(C)$ in Figure 3.

SHRINK: The graph $G \times S$ obtained from G by *shrinking* $S \subseteq V$ is constructed as follows. Remove S from V and add the new node s, called a *pseudo-node*. Remove $\langle S \rangle$ from E and replace each edge uv with one endpoint, v, in S with an edge us. We denote by $M \times S$ the edges of M in $G \times S$, i.e., $M \times S = (M \setminus \langle S \rangle) \cup \{us \mid uv \in M \cap \delta(S) \text{ and } v \in S\}$. Similarly, $F \times S$ denotes the edges of F in $G \times S$. If no confusion is likely, we write M and F in place of the more cumbersome $M \times S$ and $F \times S$.

When we apply SHRINK to a blossom C with node set S, $M \times S$ is a matching and $F \times S$ is an $M \times S$-alternating forest in $G \times S$. In fact, we can continue our search for an augmenting path in the shrunken graph.

Each augmenting path Q in $G \times S$ can be extended to an augmenting path Q' in G. (11)

Indeed, if $s \notin V(Q)$ then take $Q' = Q$. Otherwise, there is a unique even path P in C with the tip as one of the endpoints, such that adding P to Q yields a path Q' in G. It is easy to see that Q' is an augmenting path in G.

So, finding an augmenting path in $G \times S$, amounts to finding one in G. We can EXPAND the blossom to extend the alternating path Q in $G \times S$ to an alternating path Q' in G, and augment the matching in G. Therefor, when GROW HALTS we SHRINK.

The next theorem shows that alternately applying GROW and SHRINK finds an augmenting path, if one exists.

Theorem 4. *Let S be a blossom with respect to the matching M in the graph G. Then M is a maximum cardinality matching in G if and only if $M \times S$ is a maximum cardinality matching in $G \times S$.*

Proof. By (11), M is a maximum cardinality matching only if $M \times S$ is. To prove the converse, we assume that M is not maximum in G.

We may assume that the tip t of S and the pseudo-node s corresponding to S are exposed nodes. Indeed, if this is not the case, we simply take the stem P of S and replace M and $M \times S$ with the matchings $M \triangle P$ and $(M \times S) \triangle P$ having the same cardinalities.

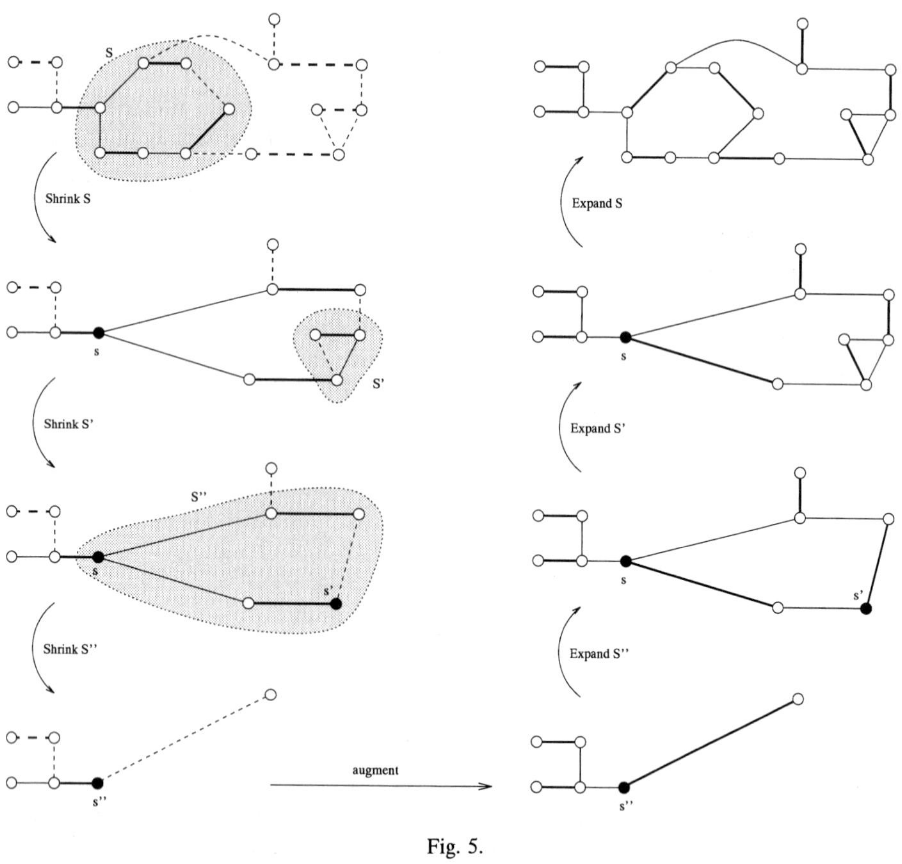

Fig. 5.

Let Q be an augmenting path in G with endpoints u and v, where $u \neq t$. If Q is disjoint from S, it is augmenting in $G \times S$. Otherwise, $Q \times S$ contains a unique us-path P. Clearly, P is augmenting in $G \times S$, so $M \times S$ is not maximum in $G \times S$. □

Thus we have an algorithm for finding a largest matching in a non-bipartite graph. This algorithm, called the *blossom algorithm* has been developed by Edmonds [1965c] (Figure 5 continues the examples in Figures 3 and 4). Witzgall & Zahn [1965] developed an $O(|V|^3)$ algorithm for general non-bipartite matching that does not rely on shrinking.

Clearly, the blossom algorithm can be implemented to run in polynomial time. Edmonds' implementation runs in $O(|V|^4)$. Balinski [1969], Gabow [1973, 1976], and Lawler [1976] developed $O(|V|^3)$ versions (but only the latter two authors explicitly state the running time). Developing an $O(|V|^3)$ version requires careful implementation of SHRINK.

2.2.2. *Implementation of the blossom algorithm*

Assume that during the search for an augmenting path, the blossom algorithm successively generates the graphs $G = G_0, G_1, \ldots, G_k$, i.e., GROW identifies the blossom S_i in G_i and SHRINK produces the graph $G_{i+1} = G_i \times S_i$ by shrinking S_i to the pseudo-node s_i $(i = 0, \ldots, k-1)$. We identify each node v in G_i, not in S_i, with the corresponding node in G_{i+1}. Pseudo-nodes are considered to be new elements. (So, $V(G_{i+1}) = (V(G_i) \setminus S_i) \cup \{s_i\}$ for $i = 0, \ldots, k-1$.)

We use the following notation to represent the relations between the pseudo-node s_i and the set S_i of nodes and pseudo-nodes 'contained' in it. For each node $s \in \bigcup_{i=0}^{k} V(G_i)$ $(= V(G) \cup \{s_0, \ldots, s_{k-1}\})$ we define:

$$\text{SHALLOW}[s] := \begin{cases} \emptyset & \text{if } s \in V(G) \\ [t_0, \ldots, t_\ell] & \text{if } s = s_i \ (i = 0, \ldots, k-1),\ t_0t_1, t_1t_2, \ldots, t_\ell t_0 \\ & \text{is the alternating odd circuit defining the} \\ & \text{blossom } S_i, \text{ and } t_0 \text{ is the tip of } S_i, \end{cases}$$

(so SHALLOW$[s]$ is an ordered list) and

$$\text{BLOSSOM}[s] := \begin{cases} t & \text{if } s \in \text{SHALLOW}[t],\ t \in \{s_0, \ldots, s_{k-1}\} \\ \emptyset & \text{otherwise.} \end{cases}$$

We define, recursively, the relations:

$$\text{DEEP}[s] := \begin{cases} s & \text{if } s \in V(G) \\ \bigcup_{t \in \text{SHALLOW}[s]} \text{DEEP}[t] & \text{otherwise} \end{cases}$$

and

$$\text{OUTER}_i[s] := t \text{ if } s \in \text{DEEP}[t],\ t \in V(G_i).$$

During the execution of the algorithm, we are mainly interested in the most recently constructed graph G_k and so denote OUTER$_k$ by OUTER. Together with the adjacency lists representing G, OUTER represents the current shrunken graph G_k.

Implementation 1 – explicitly updating OUTER: We maintain the functions SHALLOW, DEEP and OUTER as data structures. Each time we detect a blossom S, we introduce a new pseudo-node s. We can determine SHALLOW$[s]$ and the nodes in S in $O(|S|)$ time and we can determine DEEP$[s]$ and update the array OUTER in $O(\sum_{t\in \text{SHALLOW}[s]} |\text{DEEP}[t]|) = O(|V(G)|)$ time. So, since we shrink at most $\frac{1}{2}|V(G)|$ times between successive augmentations, we spend $O(|V(G)|^2)$ time updating SHALLOW, DEEP and OUTER between successive augmentations.

It is easy to implement GROW with these data structures. We grow an alternating forest F' in the current shrunken graph G' represented by OUTER by scanning the edges uv in G such that OUTER$[u] \in \text{even}(F')$. Note that we can determine OUTER$[u]$ and OUTER$[v]$ in constant time. We keep track of the forest by creating a label FOREST[OUTER$[v]$]= u for the node OUTER$[v]$ each time we decide to add the edge uv to F'. When we shrink a blossom S with tip t into a pseudo-node s, we easily create $F' \times S$ in the new graph $G' \times S$ by setting FOREST$[s]$:= FOREST$[t]$. Implemented in this way, GROW takes $O(|E(G)|)$ time (disregarding the time spent on updating SHALLOW, DEEP and OUTER), just as in the Hungarian method.

When we detect an augmenting path in G', we can construct it using FOREST and expand it to an augmenting path in G using OUTER and the ordered list SHALLOW. This, and carrying out the augmentation, takes $O(|E(G)|)$ time plus $O(|V(G)|^2)$ time for updating OUTER (which is necessary each time we expand a blossom).

As there are at most $|V(G)|$ augmentations, we obtain the following result:

> *The blossom algorithm can be implemented to run in* $O(|V(G)|^3)$ *time.* (12)

In sparse graphs, when $|E(G)|$ is significantly smaller than $\binom{|V(G)|}{2}$, the running time is dominated by the time required to update DEEP and OUTER. The other operations take $O(|V(G)||E(G)|)$ time. So, to improve the running time one could economize on the implementation of the blossoms. This can be done by employing a more implicit method of updating OUTER.

Implementation 2 – implicitly updating OUTER: To represent the blossoms, we maintain a single (dynamic) function IN, with the property that IN$[u]$ = OUTER$_i[u]$ for some (unspecified) i. Initially, IN$[u] = u$. When we shrink a blossom S to a pseudo-node s we update IN by resetting IN$[u] := s$ for $u \in S$ and setting IN$[s] := s$. This takes $O(|S|)$ time. So, between two successive augmentations we spend, overall, $O(|E(G)|)$ time updating IN. To determine OUTER$[u_0]$, we iterate $u_i :=$ IN$[u_{i-1}]$ until, at some point $u_k =$ IN$[u_k]$, then OUTER$[u_0] = u_k$. This can take as many as $\frac{1}{2}|V(G)|$ iterations, but – in the hope that the next time we need OUTER$[u_0]$, we get it almost for free – we reset IN$[u_i] :=$ OUTER$[u_0](= u_k)$ for each $i = 0, \ldots, k$.

The disadvantage of this approach is that we no longer have the data structure SHALLOW to help expand augmenting paths. We can, however, overcome this by extending the labels in FOREST to include labels on the nodes in a blossom indicating how to trace an augmenting path through the blossom. It is quite straightforward to find such labeling, but some care is needed to handle the tips of the blossoms properly. The labeling can be implemented so that finding the labels and using them to find an augmenting path can be carried out in $O(|E(G)|)$ time per augmentation [see Lawler, 1976; Lovász & Plummer, 1986; or Tarjan, 1983].

Implementing the blossoms with IN, *the blossom algorithm uses 'almost a constant' times* $|E(G)||V(G)|$ *steps.* (13)

Gabow [1973, 1976] demonstrated this result. To make the statement (13) precise, 'almost a constant' refers to a function $\alpha(|E(G)|, |V(G)|)$ (the inverse of the Ackerman function) that grows very slowly. The procedure is in fact a standard implementation of the 'set union' problem: blossoms are sets, which are united into new sets every time we shrink. For a precise definition of α and a proof of (13), see Aho, Hopcroft & Ullman [1974] or Tarjan [1983]. Gabow & Tarjan [1983] developed a linear time algorithm for set union problems with special structure. As blossoms have that structure [cf. Gabow & Tarjan, 1983], this implies:

The blossom algorithm can be implemented in $O(|E(G)||V(G)|)$ *time.* (14)

The same time bound has been achieved by Kameda & Munro [1974], but in a different manner. Instead of fine tuning the implementation of the set union problem, they obtain the $O(|V||E|)$ time bound by growing the alternating forest in a depth-first manner.

The result (14) and the way it is achieved reflect the following perspective on Edmond's blossom algorithm: The algorithm searches for augmenting paths as though the graph were bipartite and, as soon as it encounters an odd circuit, it 'shrinks the trouble away'. So, we might hope that by implementing shrinking efficiently, we could achieve the same time bound for non-bipartite matching as for the bipartite matching algorithm used as a subroutine. We have just seen that this is indeed the case when the bipartite matching subroutine is the Hungarian method. Generally, it appears that the hope is idle. For instance, we cannot simply apply this idea to Hopcroft and Karp's $O(\sqrt{|V|}|E|)$ algorithm for bipartite matching because shrinking changes the length of augmenting paths. On the other hand, the $O(\sqrt{|V|}|E|)$ bound can be achieved for non-bipartite graphs. Indeed, Hopcroft and Karp's algorithm can be generalized to achieve this bound, but the generalization is far from trivial. Even and Kariv developed an $O(|V|^{5/2})$ algorithm (as did Bartnik [1978]) and an $O(\sqrt{|V|}|E|\log|V|)$ algorithm [Even & Kariv, 1975; Kariv, 1976]. The gap was finally bridged by Micali & Vazirani [1980], who developed an $O(\sqrt{|V|}|E|)$ algorithm for general cardinality matching (see Vazirani [1994] for a proof of correctness). Blum [1990a, b] and Gabow & Tarjan [1991] achieved the same time bound in different manners.

3. Bipartite matching duality

A subset $N \subseteq V(G)$ is called a *node cover* if each edge has at least one endpoint in N. $\tau(G)$ denotes the minimum cardinality of a node cover in G. Because a node cover is always at least as large as a matching we have that, for any graph G, bipartite or not:

$$\nu(G) \leq \tau(G). \tag{15}$$

In the complete graph K_3 on 3 nodes: $\nu(K_3) = 1 \neq 2 = \tau(K_3)$. So, ν and τ need not be equal. But, when G is bipartite we have equality in (15):

Theorem 5 [König, 1931, 1933]. *For each bipartite graph G, $\nu(G) = \tau(G)$.*

Proof. Let F be a Hungarian forest with respect to a maximum matching M. Then $N := (V_1 \setminus \text{even}(F)) \cup (V_2 \cap \text{odd}(F))$ is a node cover with $|N| = |M|$. □

Many equivalent versions of *König's theorem* (Theorem 5) appeared in the first half of this century. The oldest of these is probably the following result due to Frobenius [1917]. (For a short 'linear algebra'-proof see Edmonds [1967].)

> *The determinant of a square matrix A viewed as a polynomial in its non-zero coefficients is identically zero, i.e., is zero for all values of its non-zero coefficients, if and only if there exists, for some p with $0 < p \leq n$, a $p \times (n - p + 1)$ submatrix of A having only zero coefficients.* (16)

On the other hand, the best known version addresses the existence of a system of distinct representatives. A *system of distinct representatives* for a finite collection of finite sets $S_1, \ldots, S_n$ is a collection of distinct elements $s_1, \ldots, s_n$ with $s_i \in S_i$ for each $i \in \{1, 2, \ldots, n\}$.

> *There exists a system of distinct representatives for $S_1, \ldots, S_n$ if and only if $|\bigcup_{i \in I} S_i| \geq |I|$ for each $I \subseteq \{1, 2, \ldots, n\}$.* (17)

Formulated in terms of matchings in bipartite graphs, (16) and (17) become:

Theorem 6 [Frobenius, 1917; Hall, 1935]. *In each bipartite graph $G = (V_1 \cup V_2, E)$, exactly one of the following holds:*

- *$\nu(G) = |V_1|$, i.e. there exists a matching 'of V_1 into V_2';*
- *there exists a set $U \subseteq V_1$, such that $|\Gamma(U)| < |U|$.*

Here, $\Gamma(U) := \{v \in V_2 \mid \text{there is a node } u \in U \text{ with } uv \in E\}$. (See Ore [1955] for a version of König's theorem in terms of $\Gamma(U)$.) In fact, (16) is the special case of Theorem 6 in which $|V_1| = |V_2|$. Clearly, Theorem 6 follows from König's theorem. If $\nu(G) < |V_1|$, there is a node cover N with $|N| = \nu(G) < |V_1|$. Since N is a node cover, $U := V_1 \backslash N$ satisfies $\Gamma(U) \subseteq V_2 \cap N$, i.e., $|\Gamma(U)| \leq |V_2 \cap N| < |V_1 \backslash N| = |U|$. On the other hand, even though Theorem 6 seems a rather special case of König's theorem, the latter can easily be proved from the former. (Add $|V_1| - \tau(G)$ new nodes to V_2, each adjacent to every node in V_1.) This is an example of the self-refining nature of matching theory.

Frobenius used (16) to simplify the proof of one of his earlier results [Frobenius, 1912] describing when the determinant of a matrix viewed as a polynomial in its non-zero coefficients can be factored into polynomials of lower degree. König [1915] also gave a simpler proof of Frobenius' factoring result in which he pointed out the relation to matchings in graphs. Frobenius did not appreciate this relation: *'Die Theorie der Graphen, mittels deren Hr. König den obigen Satz abgeleitet hat, is nach meiner Ansicht ein wenig geeignetes Hilfsmittel für die Entwicklung der Determinantentheorie. In diesem Falle führt sie zu einem ganz speziellen Satz von geringem Werte* [essentially statement (18) below]. *Was von seinem Inhalt Wert hat, ist in dem Satze II* [statement (16)] *ausgesprochen'* [Frobenius, 1917]. Moreover he did not acknowledge König's proof, though König had sent it to him. Apparently, this did not please König [see König, 1933, 1936]. Like so many fields, matching theory also has its controversies. (Schneider [1977], in trying to reconstruct the issue, speculates that Frobenius had not penned this criticism. He hypothesized that because Frobenius was already very ill at the time – he died in August 1917 – someone else finished the paper and wrote the criticism. See Schneider [1977] for Mirsky's refutation of this hypothesis.)

König also proved the following consequence of Theorem 6 for *regular* graphs, i.e., graphs in which all nodes have the same degree.

> *Each regular bipartite graph admits a perfect matching* [König, 1916a, b]. (18)

An immediate consequence of (18) deals with edge colorings in bipartite graphs. An *edge coloring* is an assignment of colors to the edges so that if two edges share an end node they get different colors. The minimum number of colors needed to color the edges of G in this way is denoted by $\chi_e(G)$. Clearly, $\chi_e(G)$ is at least the maximum degree of a node in G, denoted $\Delta(G)$. In finding an edge coloring we may assume that the graph is regular (just add edges and nodes to G until this is the case; of course without changing the maximum degree). Since an edge coloring is just a partition of the edge set into matchings, (18) implies the following.

> *For each bipartite graph G, $\chi_e(G) = \Delta(G)$* [König, 1916a, b]. (19)

We digress briefly from our discussion of bipartite graphs to point out that although $\chi_e(G)$ is either $\Delta(G)$ or $\Delta(G) + 1$ for each simple (non-bipartite) graph G [Vizing, 1964, 1965], determining whether or not $\chi_e(G) = \Delta(G)$ is $\mathcal{NP}$-hard, even when G is regular of degree 3 [Holyer, 1981].

Another form of edge coloring asks for a coloring of the edges of G so that there is at least one edge of each color incident to each node. A coloring of this form corresponds to a partition of $E(G)$ into edge covers. An *edge cover* in a graph G is a collection of edges F such that each node in V is an endpoint of at least one of the edges in F. Clearly, when the minimum degree of a node in G is $\delta(G)$, a coloring of this form cannot use more than $\delta(G)$ colors. König [1916a, b] showed that when G is bipartite this upper bound is achievable, i.e., there is a coloring of the edges using $\delta(G)$ colors such that each node is incident to each color.

König's theorem establishes a strong relationship between matchings and node covers in bipartite graphs. There is also a strong relationship in bipartite graphs between edge covers and matchings and between node covers and stable sets. A *stable set* in G is a collection of mutually non-adjacent nodes in G. We use $\alpha(G)$ to denote the maximum cardinality of a stable set in G and $\rho(G)$ to denote the minimum cardinality of an edge cover in G. Clearly $\alpha(G) \leq \rho(G)$. The following relationship among these problems is true for all graphs, bipartite or not.

Theorem 7 [Gallai, 1959]. *For each graph $G = (V, E)$ without isolated nodes, $\alpha(G) + \tau(G) = |V| = \nu(G) + \rho(G)$.*

Proof. The first equality is trivial: stable sets are exactly the complements of node covers. To prove the second equality we use edge/node covers instead of edge covers. An *edge/node cover* of G is a covering of the nodes of G by edges and nodes. It is easy to see that the minimum cardinality of an edge/node cover is also $\rho(G)$: each edge cover is an edge/node cover, and conversely, each edge/node cover can be turned into an edge cover of the same cardinality by replacing each node by an incident edge (G has no isolated nodes). Now observe that the edges a minimum cardinality edge/node cover may assumed to form a matching – a maximum matching in fact. So: $\rho(G) = \mathrm{def}(G) + \nu(G) = |V| - \nu(G)$. □

An immediate consequence of this result is:

For each bipartite graph $G = (V_1 \cup V_2, E)$ without isolated nodes,
$\alpha(G) = \rho(G)$ [König, 1931]. (20)

Bipartite graphs are not the only graphs with $\nu(G) = \tau(G)$ and, equivalently, $\alpha(G) = \rho(G)$. The class of graphs with this property, called the *König property*, has been characterized by Deming [1979] and Sterboul [1979] (see also Bourjolly & Pulleyblank [1989], Lovász [1983] and Lovász & Plummer [1986]).

Intermezzo: min–max relations and good characterizations

Both (20) and Theorem 7 are characterizations for the maximum size of a stable set. Even though (20) applies only to bipartite graphs we consider it to be a 'better' characterization than Theorem 7. The reason is that (20) assures that whatever the answer to the question: Given a bipartite graph G, is $\alpha(G) \leq k$? is, we can always give a polynomial length certificate of it. Namely, either a stable set with k nodes or an edge-cover with less than k edges. This is not the case with Theorem 7. If $\alpha(G) < k$ it only guarantees that **all** node-covers of G are larger than $|V| - k$, and it is not clear how to verify that.

A problem is called *well-characterized* if whatever the answer is, there exists a polynomial length certificate for the correctness of that answer. Note that the existence of the certificate might not guarantee us that we can find it in polynomial time. A theorem asserting that a problem is well-characterized is called a *good characterization* for the problem. (In NP-language: a decision problem is well-characterized if it belongs to NP $\cap$ co-NP.)

So '$\alpha(G) \leq k$?' is well-characterized for bipartite graphs and (20) is a good characterization for it. Theorem 7 is not a good characterization for '$\alpha(G) \leq k$?'. For non-bipartite graphs, '$\alpha(G) \leq k$?' is not known to be well-characterized. (If it would be, then NP = co-NP, which is generally believed not to be true.) It was Edmonds who first explicitly made the distinction between characterizations that are good and those that are not.

Good characterizations for optimization problems often come in the form of a min–max relation, like Theorem 5 or (20). However, they can have other forms as well. For instance a polynomial time algorithm is a good-characterization (implying that P $\subseteq$ NP $\cap$ co-NP). On the other hand, many polynomial time algorithms for optimization problems use a good characterization, mostly a min–max relation, as stopping criterion – also in this chapter. The theorems in this section (except for Theorem 7) are good characterizations, and there are more to come in this chapter.

3.1. Network flows

There is a strong relation between bipartite matching problems and flow problems; essentially they are identical.

Given a directed graph $D = (V(D), A(D))$ and two nodes s and t. An s,t-*flow* is a function f from $A(D)$ to $\mathbb{R}_+$ such that for each $v \in V(D) \setminus \{s,t\}$: $f(\delta^-(v)) = f(\delta^+(v))$. The *value* of a flow f is defined by $f(\delta^+(s)) - f(\delta^-(s))$ $(= f(\delta^-(t)) - f(\delta^+(t)))$. The *max-flow problem* is to find a flow f of maximum value subject to the *capacity constraints*: $f(a) \leq c(a)$ for each $a \in A$. Here, c is a given *capacity function* from $A(D)$ to $\mathbb{R}_+ \cup \{\infty\}$. For each $U \subseteq V(D)$ with $s \in U, t \notin V$, the *capacity* of the s,t-*cut* $\delta^+(U)$ is defined to be $c(\delta^+(U))$. Flow values and cut capacities satisfy the following min–max relation known as the *Max-flow min-cut theorem*.

Theorem 8 [Ford & Fulkerson, 1956; Elias, Feinstein & Shannon, 1956]. *The maximum value of an s,t-flow with respect to a given capacity function is equal to the minimum capacity of an s,t-cut. Moreover, if the capacity function is integral, then there is an integral maximum flow.*

The following construction demonstrates the relation between flow problems and bipartite matching problems. Given a bipartite graph $G = (V_1 \cup V_2, E)$; construct a directed graph $D = (V(D), A(D))$ as follows. Add a node s to V and directed edges from s to each node of V_1. Orient all edges in G from V_1 to V_2. Add a node t and a directed edge from each node of V_2 to t. Assign the

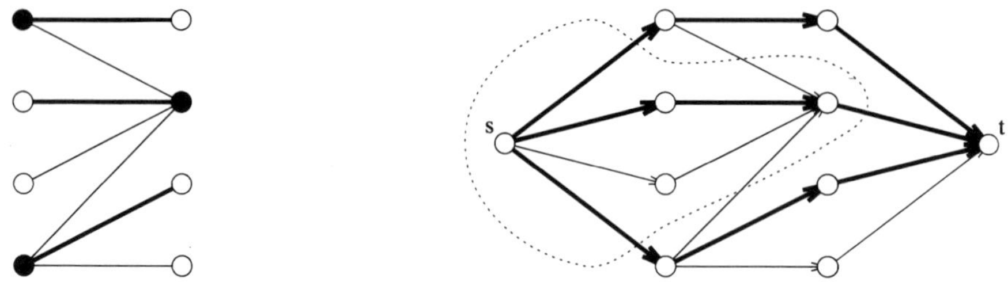

Fig. 6.

following capacities to the directed edges. Each arc corresponding to an edge in G has infinite capacity. The arcs out of s or into t each have capacity 1. There is a one-to-one correspondence between matchings in G and integral flows in D and between node covers in G and cuts with finite capacity in D. Figure 6 illustrates these relations: bold edges form a matching and bold arcs indicate a corresponding flow (bold arcs carry flow 1; the other arcs carry flow 0). The black nodes indicate a node cover and the dotted balloon identifies the set U such that $\delta^+(U)$ is the corresponding cut. Thus, König's theorem follows from the Max-flow min-cut theorem. Similarly, the algorithm for finding a maximum matching in a bipartite graph is essentially Ford and Fulkerson's maximum flow algorithm.

Conversely, it is possible to derive the Max-flow min-cut theorem from König's theorem. Instead, we use König's theorem to prove another closely related result on directed graphs, namely *Menger's theorem* (Theorem 9).

Let D be a directed graph and let $S, T \subseteq V(D)$. An *S, T-path* in D is a directed path from a node in S to a node in T. A set $U \subseteq V(D)$ is called an *S, T-separator* if it intersects each S, T-path.

Theorem 9 [Menger, 1927]. *Let D be a directed graph and let $S, T \subseteq V(D)$. Then the maximum number of mutually node-disjoint S, T-paths is equal to the minimum cardinality of an S, T-separator.*

Proof. Clearly, we may assume that S and T are disjoint. Let $W := V(D)\setminus(S\cup T)$ and construct a bipartite graph G as follows. For each node $u \in W$ we have two nodes u^+ and u^- in G. For each node $u \in S$ we have only one node u^+ in G. Similarly, for each node $u \in T$ we have one node u^- in G. (We refer to the set $\{u^+ \mid u \in S\}$ of nodes in G as S and to the set $\{u^- \mid u \in T\}$ of nodes in G as T.) There are two types of edges in G: for each arc $\overrightarrow{uv}$ in D, u^+v^- is an edge in G and for each $u \in W$, u^+u^- is an edge in G.

Let M be a maximum matching in G and assume, without loss of generality, that $\exp(M) \subseteq S \cup T$. Let $k := |S \setminus \exp(M)| = |T \setminus \exp(M)|$. It is easy to see that the collection $\{\overrightarrow{uv} \in A(D) \mid u^+v^- \in M\}$ forms the node-disjoint union of k directed S, T-paths and a collection of directed circuits in D.

To prove the theorem it suffices to show the existence of an S, T- separator with cardinality k. Let N be a minimum node cover. By König's theorem, $|N| = \nu(G) =$

$\frac{1}{2}(|V(G)|-|\exp(M)|) = \frac{1}{2}(2|W|+|S|+|T|-|\exp(M)|) = \frac{1}{2}(2|W|+2k) = |W|+k$. For each node u in W, to cover the edge u^+u^- in G either u^+ or u^- must be in N. Hence the set $U := (S \cap N) \cup (T \cap N) \cup \{u \in W \mid u^+ \text{ and } u^- \in N\}$ has cardinality k. It remains to prove that U is an S, T-separator. Let $u_0, \ldots, u_t$ be a directed S, T-path. Since $|N \cap \{u_0^+, u_1^-, u_1^+, \ldots, u_t^-\}| \geq t$, either u_0^+ is in N in which case u_0 is in U, u_t^- is in N in which case u_t is in U, or $\{u_i^-, u_i^+\} \subseteq N$ for some $0 < i < t$ in which case u_i is in U. □

There are many equivalent versions of Menger's theorem. One can, for instance, consider internally node-disjoint directed s, t-paths (shrink S and T to the single nodes s and t). Further, similar min–max relations hold for arc-disjoint s, t-paths as well as node- or edge-disjoint s, t-paths in undirected graphs. All these results are equivalent in the sense that we can easily derive one from another. In fact, all these results can be seen as versions of König's theorem (or conversely).

Menger's theorem for the number of arc-disjoint s, t-paths is a special case of the Max-flow min-cut theorem in which all capacities are one. And conversely, one can derive the Max-flow min-cut theorem from this version of Menger's theorem. So there is a close relationship between matchings in bipartite graphs and flows. It extends to almost every problem on bipartite graphs discussed in this chapter. For instance, the minimum weight matching problem in bipartite graphs corresponds to the *min-cost flow problem* in which we are given the unit costs of flow through each arc and are asked to find a maximum flow of minimum total cost.

For further consequences of König's theorem, e.g., Dilworth's theorems on chains and anti-chains in partially ordered sets [Dilworth, 1950], see Lovász & Plummer [1986] or Mirsky [1971]. For a tour along all the above mentioned equivalences, see Reichmeider [1984].

3.2. Intermezzo: perfect graphs and matroids

We conclude this section with a short discussion on extensions of the results on matchings in bipartite graphs described in this section. This brings us outside the context of bipartite graphs. We go here in two directions: first, node coloring and stable sets in general graphs, and second, matroids.

3.2.1. Node coloring and stable sets

The *chromatic number* $\chi(G)$ of G is the minimum number of colors needed to color the nodes of G such that adjacent nodes receive different colors. $\chi(G)$ is at least the maximum size $\omega(G)$ of a collection of mutually adjacent nodes in G. A graph is called *perfect* if and only if $\chi(H) = \omega(H)$ for each induced subgraph H in G. Odd circuits are not perfect. On the other hand, bipartite graphs are, trivially, perfect. König's results stated in the beginning of this section yield three other, less trivial, classes of perfect graphs.

We need a few definitions. The *complement* $\overline{G}$ of a graph G has the same nodes as G; nodes are adjacent in $\overline{G}$ when they are non-adjacent in G. The *line graph* of G has the edges of G as nodes; two edges in G are adjacent in the line graph when they share an end node in G. Clearly, $\chi_e(G)$ is the chromatic number of the line graph of G. Similarly, matchings in G are stable sets in the line graph of G.

With these definitions we get the following results. By (20), the complement of bipartite graph is perfect. By (19), the line graph of a bipartite graph is perfect. And, by Theorem 5, the complement of the line graph of a bipartite graph is perfect.

So perfect graphs not only generalize bipartite graphs but also their line graphs. Moreover, also the complements of all these graphs are perfect. The latter is not so much of a coincidence: *The complement of a perfect graph is perfect*. This is the famous *Perfect graph theorem* proved by Lóvasz [1972b]. Although, many classes of perfect graphs have been discovered over the last decennia, the main conjecture on perfect graphs is still open: *If a graph is not perfect, then it or its complement contains an odd circuit with 5 or more edges as an induced subgraph* [Berge, 1962].

The *stable set problem* is: Given a graph G and a weight function on $V(G)$, find a stable set of maximum weight. In general this problem is NP-hard [Karp, 1972]. On the other hand, by the Perfect graph theorem, we have a min–max relation for the maximum cardinality stable set in a perfect graph: $\alpha(G) = \chi(\overline{G})$. In fact, the stable set problem can be solved in polynomial time when G is perfect [Grötschel, Lovász & Schrijver, 1981, 1984, 1988].

There is another class of graphs for which the stable set problem is polynomially solvable. In fact, it is very strongly related to matchings. A graph is *claw-free* if it has no node with three mutually non-adjacent neighbors. Line graphs are claw-free. Sbihi [1980] and Minty [1980] proved that one can find a maximum cardinality stable set in a claw-free graph in polynomial time. Minty [1980] did this by reducing the problem to a matching problem [see also Lovász & Plummer, 1986]. His algorithm also solves the weighted case.

3.2.2. *Matroid intersection*

In this section we will see an extension of König's theorem.

A matroid $\mathcal{M} = (E, \mathcal{I})$ consists of a finite set E, the *ground set*, and a collection $\mathcal{I}$ of subsets of E satisfying the following three axioms: $\emptyset \in \mathcal{I}$; $I \in \mathcal{I}$ and $J \subseteq I$ implies that $J \in \mathcal{I}$; and, finally, $I, J \in \mathcal{I}$ and $|I| < |J|$ implies that there exist an $e \in J \setminus I$ such that $I \cup \{e\} \in \mathcal{I}$. The members of $\mathcal{I}$ are called *the independent sets of* $\mathcal{M}$. The *rank-function* $r_{\mathcal{M}}$ of a matroid $\mathcal{M}$ is defined by $r_{\mathcal{M}}(F) := \max\{|I| \mid I \in \mathcal{I}; I \subseteq F\}$. Examples of matroids are: the edge sets of forests in a graph and the linearly independent collections of columns of a matrix.

There is a vast theory on matroids [see Welsh, 1976; Recski, 1989; Truemper, 1992; Oxley, 1992] and many of the results there are inspired by results on matchings [see Lovász & Plummer, 1986]. Here we just mention one of these results.

The *matroid intersection* problem is: given two matroids on the same ground set find the largest set that is independent in both matroids. The maximum matching problem in a bipartite graph is a matroid intersection problem: given $G = (V_1 \cup V_2, E)$; define $\mathcal{M}_1$ and $\mathcal{M}_2$ with ground set $E(G)$, and with $\mathcal{I}_i := \{I \subseteq E \| I \cap \delta(v) \leq 1$ for each $v \in V_i\}$ for $i = 1, 2$. It is easy to check that these are matroids, and that a collection of edges is independent in both $\mathcal{M}_1$ and $\mathcal{M}_2$ if and only if it is a matching.

Edmonds [1970] derived the following min–max relation for the matroid intersection problem.

> *Let $\mathcal{M}_1 := (E, \mathcal{I}_1)$ and $\mathcal{M}_2 := (E, \mathcal{I}_2)$ be two matroids on the same ground set E. Then* $\max\{|I| \mid I \in \mathcal{I}_1 \cup \mathcal{I}_2\} = \min\{r_{\mathcal{M}_1}(F) + r_{\mathcal{M}_2}(E \setminus F) | F \subseteq E\}$. (21)

Königs theorem is a special case of (21). Indeed, let $G = (V_1 \cup V_2, E)$ be a bipartite graph and let $\mathcal{M}_1$ and $\mathcal{M}_2$ be the two matroids defined above. If F is a set of edges in G then $r_{\mathcal{M}_i}(F)$ is the cardinality of set of nodes in V_i incident to at least one of the edges in F. From that the relation between sizes of node covers and $r_{\mathcal{M}_1}(F) + r_{\mathcal{M}_2}(E \setminus F)$ is easy.

A similar extension of non-bipartite matching to a problem on matroids is the 'matroid matching' problem [see Lovász & Plummer, 1986].

4. Non-bipartite matching duality

We begin our discussion of non-bipartite matching by developing a min–max relation for the size of a maximum matching. Although $\tau(G)$ does not, as we have seen, provide a good characterization for the size of a maximum matching in a non-bipartite graph G, there is a min–max relation for $\nu(G)$. An *odd component* of a graph G is a connected component of G with an odd number of nodes. We let $c_o(G)$ denote the number of odd components of $G = (V, E)$.

> *For each matching M and subset* $B \subseteq V$, $|\exp(M)| \geq c_o(G \setminus B) - |B|$. (22)

To see this, let M_1 be the set of the edges in M with both endpoints in $G \setminus B$. As M_1 leaves at least one node exposed in each odd component of $G \setminus B$, $|\exp(M_1)| \geq c_o(G \setminus B) + |B|$. Since, $|\exp(M)| = |\exp(M_1)| - 2|M \setminus M_1| \geq |\exp(M_1)| - 2|B|$, (22) follows.

Theorem 10 [Berge, 1958]. *For each graph $G = (V, E)$:*

$$\operatorname{def}(G) = \max_{B \subseteq V} c_o(G \setminus B) - |B|,$$

$$\nu(G) = \min_{B \subseteq V} \tfrac{1}{2}\left(|V| - c_o(G \setminus B) + |B|\right).$$

Proof. Clearly, it suffices to prove the formula for def(G). Let M be a maximum matching in G. Apply the procedures GROW and SHRINK to G and M until we get a shrunken graph G' with a Hungarian forest F'. Each odd node in F' is a node of the original graph G and so is not contained in a pseudo-node. Each odd component of $G \setminus \mathrm{odd}(F')$ has been shrunk into a different even node of F' (or is equal to an even node). Moreover each even node arises in this way. Hence, $c_o(G \setminus \mathrm{odd}(F')) = |\mathrm{even}(F')|$. So, $c_o(G \setminus \mathrm{odd}(F')) - |\mathrm{odd}(F')| = |\mathrm{even}(F')| - |\mathrm{odd}(F')| = \mathrm{def}(G') = \mathrm{def}(G)$. Combining this with (22), the theorem follows. □

Theorem 10 generalizes Tutte's characterization of those graphs that contain a perfect matching.

Theorem 11 [Tutte, 1947, 1954]. *The graph $G = (V, E)$ has a perfect matching if and only if*

$$c_o(G \setminus B) \leq |B| \textit{ for each } B \subseteq V.$$

Tutte's used matrix-techniques ('Pfaffians') to prove this. The first proof of his theorem that uses alternating paths has been given by Gallai [1950].

Edmonds' *Odd set cover theorem* is a version of Berge's theorem that more explicitly extends König's theorem to non-bipartite graphs. An *odd set cover* is a collection B of nodes together with a collection $\{S_1, \ldots, S_k\}$ of subsets of V, each with odd cardinality, such that for each edge uv either $\{u, v\} \cap B \neq \emptyset$, or $\{u, v\} \subseteq S_i$ for some $i = 1, \ldots, k$.

Theorem 12 [Edmonds, 1965c]. *For each graph G,*

$$\nu(G) = \min\Big\{|B| + \sum_{i=1}^{k} \tfrac{1}{2}(|S_i| - 1) \mid B \textit{ and } S_1, \ldots, S_k \textit{ form an odd set cover of } G\Big\}.$$

Proof. Among all $B \subseteq V$ with $\frac{1}{2}(|V| - c_o(G \setminus B) + |B|) = \nu(G)$, choose one with $|B|$ maximum, and let $S_1, \ldots, S_k$ denote the components of $G \setminus B$. Then all S_i are odd (if $|S_i|$ would be even, then $B \cup \{v\}$ with $v \in S_i$ would contradict the choice of B). Hence B and $S_1, \ldots, S_k$ form an odd set cover. It satisfies $|B| + \sum_{i=1}^{k} \frac{1}{2}(|S_i| - 1) = |B| + \frac{1}{2}|V \setminus B| - \frac{1}{2}k = \frac{1}{2}(|V| - c_o(G \setminus B) + |B|) = \nu(G)$. As obviously the minimum is at least $\nu(G)$ the theorem follows. □

The following special case of Tutte's theorem is over a hundred years old. It clearly demonstrates the long and careful attention paid to perfect matchings. A *cubic graph* is one in which each node has degree three. An *isthmus* is an edge that, when deleted from G, disconnects the graph.

Theorem 13 [Petersen, 1891]. *Every connected, cubic graph with at most two isthmi contains a perfect matching.*

Proof. Let $B \subseteq V$ and let $S_1, \ldots, S_k$ denote the odd components of $G \setminus B$. As G is cubic, $|\delta(S_i)|$ is odd for all $i = 1, \ldots, k$. Hence, $3|B| \geq |\delta(B)| \geq \sum_{i=1}^{k} |\delta(S_i)| \geq 3(k-2)+2 = 3c_o(G \setminus B) - 4$. So, $|B| \geq c_o(G \setminus B) - \frac{4}{3}$. Which implies that $|B| \geq c_o(G \setminus B)$ (as $|B| - c_o(G \setminus B)$ is even). So, by Theorem 11, we may conclude that G has a perfect matching. □

4.1. The Edmonds–Gallai structure theorem

A graph G can have many different maximum matchings and applying GROW and SHRINK to one of them can lead to many different Hungarian forests. The ultimate shrunken graph, however, is independent of the choice of matching or the order in which we apply GROW and SHRINK. This observation is one aspect of the Edmonds–Gallai structure theorem [Gallai, 1963, 1964; Edmonds, 1965c]. In this section we discuss the main features of the Edmonds–Gallai structure, which plays a role in the development of algorithms for finding maximum weight matchings. In fact, every polynomial time maximum matching algorithm that calculates the Edmonds–Gallai structure can be embedded in a 'primal-dual framework' for solving the weighted matching problem in polynomial time (see Section 6.2).

Suppose we have applied GROW and SHRINK to a graph G and a maximum matching M, yielding a Hungarian forest F in a shrunken graph G^*. We use the notation OUTER$[u]$ (for nodes u in $V(G)$) and DEEP$[u]$ (for nodes u in $V(G^*)$) introduced in Section 2.2. The *Edmonds–Gallai structure* of a graph G is the partition of $V(G)$ into three sets, $D(G)$, $A(G)$ and $C(G)$, defined by $D(G) := \{u \in V \mid \nu(G \setminus u) = \nu(G)\}$, $A(G) := \{u \in V(G) \setminus D(G) \mid u \text{ is adjacent to a node in } D(G)\}$ and $C(G) := V(G) \setminus (D(G) \cup A(G))$.

> *The set $D(G)$ is the union of the sets* DEEP$[u]$ *with* $u \in \text{even}(F)$. *In fact, the components of $G|D(G)$ are exactly the sets* DEEP$[u]$ *with* $u \in \text{even}(F)$. *Moreover,* $A(G) = \text{odd}(F)$. (23)

So, G^* is obtained from G by shrinking the components of $G|D(G)$. We call the shrunken graph G^* the *Edmonds–Gallai* graph of G. The result (23) follows from the definitions of GROW and SHRINK. All statements of (23) follow easily from the first one: if F is a Hungarian forest in G^*, $D(G)$ is the union of the sets DEEP$[u]$ with $u \in \text{even}(F)$. To see this, consider a node u in G. Construct a new graph H by adding a new node v to G adjacent to u and construct a new graph H^* by adding a new node v^* to G^* adjacent to OUTER$[u]$. Now, we have the following equivalences:

$$u \in D(G) \Longleftrightarrow \nu(H) = \nu(G)+1 \Longleftrightarrow \nu(H^*) = \nu(G^*)+1 \qquad (24)$$

$$\text{OUTER}[u] \in \text{even}(F) \Longleftrightarrow u \in \text{DEEP}[w] \textit{ for some } w \textit{ in } \text{even}(F). \qquad (25)$$

So we only need to establish the equivalence of the third statement in (24) with the first one in (25). If OUTER$[u] \in \text{even}(F)$, consider the alternating forest F^* in H^* obtained by adding v^* as a component to F. The nodes v^* and OUTER$[u]$ are both in even(F^*) and in different components of F^*. So, in that case GROW will find an augmenting path using the edge connecting these two nodes, implying that $\nu(H^*) = \nu(G^*) + 1$. On the other hand, if OUTER$[u] \notin$ even(F), then

$$\begin{aligned} \mathrm{def}(H^*) &\geq c_o(H^* \setminus \mathrm{odd}(F)) - |\mathrm{odd}(F)| = \\ &= |\mathrm{even}(F)| + 1 - |\mathrm{odd}(F)| = \mathrm{def}(G^*) + 1. \end{aligned} \tag{26}$$

So in that case, $\nu(H^*) = \nu(G^*)$. Thus (23) follows.

The relation between the Edmonds–Gallai structure and the Hungarian forest in the Edmonds–Gallai graph provides insight into the structure of all maximum cardinality matchings in G. We need a few definitions. A graph G is *factor-critical* if $\nu(G \setminus v) = \nu(G) = \frac{1}{2}(|V(G)| - 1)$ for each $v \in V(G)$. A matching is *near-perfect* if it has exactly one exposed node. We let $\kappa(G)$ denote the number of components of $G|D(G)$.

> *Each component of $G|D(G)$ is factor-critical and* $\mathrm{def}(G) = \kappa(G) - |A(G)|$. *Moreover, a matching M in G is maximum if and only if it consists of:*
> – *a perfect matching in $C(G)$,*
> – *a near-perfect matching in each component of $G|D(G)$,*
> – *a matching of each node $u \in A(G)$ to a node in a distinct component of $G|D(G)$.* (27)

This is the *Edmonds–Gallai structure theorem*. Note that it implies all the results on non-bipartite matching duality stated earlier in this section. Every statement in (27) follows easily from the first one: each component of $G|D(G)$ is factor-critical. This follows inductively from (23) together with the fact that each graph spanned by an odd circuit – like a blossom – is factor-critical, and the following:

> *Let S be a subset of nodes in a graph G such that $G|S$ is factor-critical. If $G \times S$ is factor-critical then so is G.* (28)

And, (28) is in turn immediate from:

> *Let S be a subset of nodes in a graph G such that $G|S$ is factor-critical. Let s be the pseudo-node in $G \times S$ obtained by shrinking S. Then, for each matching M in $G \times S$, there exists a near-perfect matching M_S in S such that $M \cup M_S$ is a matching in G with*
> – $\exp(M \cup M_S) = \exp(M)$ *if* $s \notin \exp(M)$,
> – $\exp(M \cup M_S) = (\exp(M) \setminus \{s\}) \cup \{s_1\}$ *for some* $s_1 \in S$, *if* $s \in \exp(M)$. (29)

Dulmage & Mendelsohn [1958, 1959, 1967] derived the Edmonds–Gallai structure for bipartite graphs. The components of $G|D(G)$ in a bipartite graph G each consist of a single node (because, GROW can never HALT in a bipartite graph, or equivalently, because the only factor-critical bipartite graph consists of a single node and no edges).

The Edmonds–Gallai structure describes the maximum matchings of a graph as unions of (near-)perfect matchings in certain subgraphs and a matching in a bipartite subgraph (between $A(G)$ and $D(G)$). So, more detailed information on the structure of maximum matchings requires structural knowledge about perfect and near-perfect matchings. Indeed, such structural results exist. In addition to the 'Dulmage–Mendelsohn decomposition' for bipartite graphs [Dulmage & Mendelsohn, 1958], we have the 'ear-decomposition' of bipartite graphs developed by Brualdi & Gibson [1977]. Hetyei [1964], Lovász & Plummer [1975], Lovász [1983] extended the 'ear-decomposition' to, not necessarily bipartite, graphs with perfect matchings and Lovász [1972c] developed the 'ear-decomposition' of factor-critical graphs. Finally, Kotzig [1959a, b, 1960], Lovász [1972d] and Lovász & Plummer [1975] developed the 'brick decomposition' of graphs with perfect matchings. See Lovász [1987] for an overview of some of these results and Chapters 3, 4, and 5 of the book by Lovász & Plummer [1986] for an exhaustive treatment of the subject.

We conclude this section with some observations on how the Edmonds–Gallai structure behaves under certain changes of the graph. These observations facilitate our analysis of an algorithm for weighted matching in Section 6.2.

For each edge $e = uv \in E(G)$, where $u \in A(G)$ and $v \in A(G) \cup C(G)$, G and $G \setminus e$ have the same Edmonds–Gallai structure. (30)

It is easy to see that this is true. A bit more complicated is:

For each pair of nodes $u \in D(G)$ and $v \in C(G) \cup D(G)$ not both in the same component of $G|D(G)$, $\mathrm{def}(G \cup e) \leq \mathrm{def}(G)$, where $e = uv$. Moreover, if $\mathrm{def}(G \cup e) = \mathrm{def}(G)$, then $D(G \cup e) \supsetneq D(G)$. (31)

To see this, first observe that $\mathrm{def}(G \cup e) \leq \mathrm{def}(G)$. Further, if $\mathrm{def}(G \cup e) = \mathrm{def}(G)$, then $D(G \cup e) \supseteq D(G)$. Now, assume that $\mathrm{def}(G \cup e) = \mathrm{def}(G)$ and $D(G \cup e) = D(G)$. Obviously, this implies that $\kappa(G \cup e) \leq \kappa(G)$ and $|A(G \cup e)| \geq |A(G)|$. By (27), this yields $\kappa(G \cup e) = \kappa(G)$ and $|A(G \cup e)| = |A(G)|$ (otherwise $\mathrm{def}(G \cup e) < \mathrm{def}(G)$). But, this contradicts the definition of edge e.

Let $S \subseteq V(G)$ such that $G|S$ is factor-critical and such that the pseudo-node s in $G \times S$, obtained by shrinking S, is contained in $A(G \times S)$. Then $\mathrm{def}(G) \leq \mathrm{def}(G \times S)$. Moreover, if $\mathrm{def}(G) = \mathrm{def}(G \times S)$ then $D(G) \supseteq D(G \times S)$. Finally, if $\mathrm{def}(G) = \mathrm{def}(G \times S)$ and $D(G) = D(G \times S)$, then $C(G) \supsetneq C(G \times S)$. (32)

The proof of (31) is similar to the proof of (31). By (29), $\mathrm{def}(G) \leq \mathrm{def}(G \times S)$. Further, if $\mathrm{def}(G) = \mathrm{def}(G \times S)$, then $D(G) \supseteq D(G \times S)$. Now assume that $\mathrm{def}(G) = \mathrm{def}(G \times S)$ and $D(G) = D(G \times S)$. Then $C(G) \supseteq C(G \times S)$ and $\kappa(G \times S) = \kappa(G)$. By (27), $|A(G)| = |A(G \times S)|$. Combining all this with $|V(G)| > |V(G \times S)|$ yields $|C(G)| > |C(G \times S)|$.

5. Matching and integer and linear programming

In the next section we discuss the problem of finding a maximum weight matching. In this section we show that the problem is a linear programming problem. We first observe that, like many combinatorial optimization problems, the maximum weight matching problem is an integer linear programming problem:

$$\begin{array}{lllll} \text{maximize} & \sum_{e\in E} w_e x_e & & & \\ \text{subject to} & x(\delta(v)) & \leq & 1 & (v\in V) \\ & x_e & \geq & 0 & (e\in E) \\ & x_e & \in & \mathbb{Z} & (e\in E). \end{array} \tag{33}$$

In general, integer linear programming problems are hard to solve; they are NP-hard [Cook, 1971]. On the other hand, linear programming problems are easy to solve. There are not only practically efficient (but non-polynomial) procedures like the simplex method [Dantzig, 1951], but also polynomial time algorithms [Khachiyan, 1979; Karmarkar, 1984] for solving linear programs. Moreover, we have a min–max relation for linear programming, namely the famous *LP-duality theorem* [Von Neumann, 1947; Gale, Kuhn & Tucker, 1951]:

$$\max\{w^\top x \mid Ax \leq b\} = \min\{y^\top b \mid y^\top A = w^\top;\ y^\top \geq 0\}. \tag{34}$$

This min–max relation provides a good characterization for linear programming. In this chapter one of problems in such a pair of linear programming problems will typically be a maximum or minimum weight matching problem. In that case we will refer to the other problem as the *dual problem*. Its feasible (optimal) solutions will be called the *dual feasible (optimal) solutions*.

One consequence of the LP-duality theorem is that a pair of solutions, a feasible solution x to the maximization problem and a feasible solution y to the minimization problem, are both optimal if and only if they satisfy the *complementary slackness conditions*:

$$y^\top(b - Ax) = 0. \tag{35}$$

The complementary slackness conditions, more than the linear programming algorithms, guide the algorithms for maximum weight matching in Sections 6 and 8.1. An obvious first attempt at a linear programming formulation of the weighted matching problem is the *LP-relaxation*:

$$\begin{array}{lllll} \text{maximize} & \sum_{e\in E} w_e x_e & & & \\ \text{subject to} & x(\delta(v)) & \leq & 1 & (v\in V) \\ & x_e & \geq & 0 & (e\in E). \end{array} \tag{36}$$

If (36) admits an integral optimum solution, that solution also solves (33). Hence the question arises: When does (36) admit an integral optimum solution for every weight function w? This question is equivalent to: When is the polyhedron

$$\mathrm{Fract}(G) := \{x \in \mathbb{R}_+^E \mid x(\delta(v)) \le 1 \quad (v \in V)\} \tag{37}$$

equal to the *matching polytope*:

$$\begin{aligned}\mathrm{Match}(G) &:= \text{convex hull}\,\{x \in \mathbb{Z}_+^E | x(\delta(v)) \le 1 \quad (v \in V)\} \\ &= \text{convex hull}\,\{\chi^M | M \text{ is a matching in } G\}\ ?\end{aligned} \tag{38}$$

If $\mathrm{Fract}(G) \ne \mathrm{Match}(G)$, can we find another system of inequalities $Ax \le b$ such that $\mathrm{Match}(G) := \{x \in \mathbb{R}^{E(G)} | Ax \le b\}$ (and thus, write (33) as $\max\{w^\top x | Ax \le b\}$)? In this section we answer these questions.

5.1. Bipartite graphs – the assignment polytope

Theorem 14. *Let G be an undirected graph. Then* $\mathrm{Match}(G) = \mathrm{Fract}(G)$ *if and only if G is bipartite.*

Proof. First, consider a bipartite graph G and a vector $x \in \mathrm{Fract}(G)$. Let $F := \{e \in E | 0 < x_e < 1\} \ne \emptyset$ and select $K \subseteq F$ as follows. If F is a forest, let K be a path in F between two nodes of degree 1 in F. If F is not a forest, let K be a circuit in F, which – since G is bipartite – is even. In either case K is the disjoint union of two matchings, M_1 and M_2. It is easy to see that for some sufficiently small, positive ϵ both $x + \epsilon(\chi^{M_1} - \chi^{M_2})$ and $x - \epsilon(\chi^{M_1} - \chi^{M_2})$ are in $\mathrm{Fract}(G)$. Moreover x is a convex combination of these two vectors. Thus, each extreme point of $\mathrm{Fract}(G)$ is an integral vector. (A vector x in a set P is an *extreme point* of P if it cannot be expressed as a convex combination of other vectors in P.) In other words, each extreme point of $\mathrm{Fract}(G)$ is the characteristic vector of a matching and so $\mathrm{Fract}(G) \subseteq \mathrm{Match}(G)$. The reverse inclusion holds trivially.

Next, consider a non-bipartite graph G. Since G is not bipartite, it contains an odd circuit C. Clearly, $x := \frac{1}{2}\chi^C$ is in $\mathrm{Fract}(G)$. If x is also in $\mathrm{Match}(G)$, then there are matchings M_i $(i = 1, \ldots, k)$ and non-negative numbers λ_i $(i = 1, \ldots, k)$, such that: $x = \sum_{i=1}^k \lambda_i \chi^{M_i}$ and $\sum_{i=1}^k \lambda_i = 1$. This implies that: $\frac{1}{2}|C| = x(C) = \sum_{i=1}^k \lambda_i \chi^{M_i}(C) \le \sum_{i=1}^k \lambda_i \frac{1}{2}(|C| - 1) = \frac{1}{2}(|C| - 1)$; a contradiction. So, $x \notin \mathrm{Match}(G)$ and $\mathrm{Fract}(G) \ne \mathrm{Match}(G)$. □

So, when G is bipartite, (36) has an integral optimum solution (the characteristic vector of a maximum weight matching).

Egerváry proved the following strengthening of Theorem 14.

Theorem 15 [Egerváry, 1931]. *Let $G = (V_1 \cup V_2, E)$ be a bipartite graph and $w \in \mathbb{Z}^E$. Then both of the following linear programming problems have integral optimum solutions.*

$$\begin{array}{llll}\text{maximum} & \displaystyle\sum_{e \in E} w_e x_e & = \text{minimum} & \pi(V_1 \cup V_2) \\ \text{subject to} & x(\delta(v)) \le 1 \ (v \in V_1 \cup V_2) & \text{subject to} & \pi_u + \pi_v \ge w_{uv} \ (uv \in E) \\ & x_e \quad \ge 0 \ (e \in E) & & \pi_v \quad \ge 0 \quad (v \in V_1 \cup V_2).\end{array}$$

Proof. That the maximization problem admits an integral optimum solution is Theorem 14. So, we consider only the dual problem (i.e. the minimization problem).

Let π' be a, not necessarily integral, dual optimal solution. We show that $\widetilde{\pi} \in \mathbb{Z}_+^{V_1 \cup V_2}$ defined by:

$$\widetilde{\pi}_v := \begin{cases} \lfloor \pi'_v \rfloor & \text{if} \quad v \in V_1 \\ \lceil \pi'_v \rceil & \text{if} \quad v \in V_2 \end{cases} . \tag{39}$$

is an integral dual optimal solution.

To see that $\widetilde{\pi}$ is feasible, observe that for each $uv \in E$

$$\widetilde{\pi}_u + \widetilde{\pi}_v = \lfloor \pi'_u \rfloor + \lceil \pi'_v \rceil = \lfloor \pi'_u + \lceil \pi'_v \rceil \rfloor \geq \lfloor \pi'_u + \pi'_v \rfloor \geq \lfloor w_{uv} \rfloor = w_{uv}. \tag{40}$$

Define for $\alpha \in \mathbb{R}$: $V_1^\alpha := \{u \in V_1 | \pi'_u - \lfloor \pi'_u \rfloor = \alpha\}$; $V_2^\alpha := \{u \in V_2 | \lceil \pi'_u \rceil - \pi'_u = \alpha\}$ and $V^\alpha := V_1^\alpha \cup V_1^\alpha$. For each $\alpha > 0$, $|V_2^\alpha| - |V_1^\alpha| \leq 0$. Indeed, for some sufficiently small $\epsilon > 0$, $\pi^\epsilon := \pi' - \epsilon(\chi^{V_2^\alpha} - \chi^{V_1^\alpha})$ is a dual feasible solution. So $\pi'(V_1 \cup V_2) \leq \pi^\epsilon(V_1 \cup V_2) = \pi'(V_1 \cup V_2) - \epsilon(|V_2^\alpha| - |V_1^\alpha|)$. And thus we get:

$$\widetilde{\pi}(V_1 \cup V_2) = \pi'(V_1 \cup V_2) + \sum_{\alpha > 0, V^\alpha \neq \emptyset} \alpha(|V_2^\alpha| - |V_1^\alpha|) \leq \pi'(V_1 \cup V_2). \tag{41}$$

So $\widetilde{\pi}$ is an integral dual optimal solution. □

So, when the weight function w is integral, the linear programming dual of (36) has an integral optimum solution.

A system of linear inequalities $Ax \leq b$ with the property that – like (36) – $\min\{y^\top b | y^\top A = w; y \geq 0\}$ is attained by an integral y for each integral objective function w for which the minimum exists, is called *totally dual integral*. Edmonds & Giles [1977] (and Hoffman [1974] for the special case of *pointed polyhedra*, i.e., polyhedra with extreme points) proved that a totally dual integral system $Ax \leq b$, with A and b integral, describes an integral polyhedron. (A polyhedron in $\mathbb{R}^n$is *integral* if it is the convex hull of vectors in $\mathbb{Z}^n$.) Thus, when G is bipartite, the fact that the minimization problem in Theorem 15 admits an integral optimal solution implies that Fract(G) = Match(G).

The *perfect matching polytope* of a graph G is the convex hull Perfect(G) of the characteristic vectors of perfect matchings in G. In the next section, when studying the weighted matching problem, we concentrate on perfect matchings. In the context of bipartite graphs the perfect matching polytope is often referred to as the *assignment polytope*. The following characterization of the assignment polytope follows easily from Theorem 14.

Corollary 16. *Let G be a bipartite graph. Then*

$$\text{Perfect}(G) = \{x \in \mathbb{R}_+^E \mid x(\delta(v)) = 1 \ (v \in V)\}. \tag{42}$$

Note that, unlike Theorem 14, there exist non-bipartite graphs G for which (42) holds.

Theorem 16 is probably best known in terms of doubly stochastic matrices and by the names of its re-inventors. A matrix $A = (a_{ij})$ is *doubly stochastic* if all its entries are non-negative and all its row and column sums are equal to one, i.e., $\sum_j a_{ij} = 1$ for each row i and $\sum_i a_{ij} = 1$ for each column j.

Theorem 17 [Birkhoff, 1946; Von Neumann, 1953]. *Each doubly stochastic matrix is a convex combination of permutation matrices.*

5.2. Intermezzo: stable matchings

Shapley & Shubik [1972] give the following economic interpretation of Theorem 15. Consider the vertices in V_1 and V_2 as disjoint sets of players and each edge uv in G as a possible coalition between u and v. If u and v form a coalition they may share out the *worth* w_{uv} of uv as payoffs π_u and π_v among themselves. Suppose that M is a matching of G (i.e. a collection of coalitions) and that $\pi_u (u \in V)$ is a corresponding collection of payoffs, i.e. $\pi_u + \pi_v = w_{uv}$ if $uv \in M$. If there exists an edge $uv \notin M$ such that $\pi_u + \pi_v < w_{uv}$, then the payoffs are *not stable for* M: u and v could increase their profits by breaking their respective coalitions and joining together. A matching is *payoff-stable* if there exists a collection of payoffs for M without such instability. By Theorem 15 and the complementary slackness conditions (35) payoff-stable matchings exist, they are exactly the maximum weight matchings. The optimal dual solutions compress all the possible payoffs without any instability.

Gale & Shapley [1962] introduced another notion of stability for matchings, the *stable marriage problem*. Suppose we have a marriage-market with n man and n women. Each person u has linear order $\prec_u$ on the persons of opposite sex, where $v \prec_u w$ means that u prefers to be married with w rather than with v. Modeling this on a bipartite graph, a collection of monogamous marriages is a matching. A perfect matching M between the men and women is *stable*, if for each $uv \notin M$, with $uu', vv' \in M$, either $v \prec_u u'$ or $u \prec_v v'$. In other words, a perfect matching is stable if no unmarried couple would prefer to get divorced and marry each other.

Gale & Shapley [1962] showed that a stable matching always exists. To see this, consider the following procedure in which each man starts proposing to his most preferred woman. If she rejects, he proposes to the next woman on his preference list and so on. Each woman keeps her best received proposal under consideration, rejecting all the other ones. When all women hold a proposal, the procedure stops, a stable matching has been obtained. Indeed, suppose there is unmarried couple u and v. If u did not propose to v he prefers his partner above v. If he did proposed to v, she rejected him, which means that she got a better proposal.

Instead of letting the above 'propose and reject' game decide on how the couples are made, there also could be a match-maker that arranges the marriages. His arrangement has to be stable, but additionally to the preferences he has a weight function on the possible pairs and he wishes to arrange a stable matching with maximum weight. The following polyhedral characterization of stable matchings, due to Vande Vate [1989], shows that this maximal weight stable matching problem is a linear programming problem.

The convex hull of the characteristic vectors of stable matchings in a complete bipartite graph $G = (V_1 \cup V_2, E)$ is:

$$\Big\{x \in \text{Perfect}(G) \mid \sum_{v \prec_u w} x_{uw} + \sum_{u \prec_v w} x_{wv} + x_{uv} \geq 1 (u \in V_1, v \in V_2)\Big\}. \tag{43}$$

In fact, it turned out that many of the properties of stable matchings can be derived form this polyhedral result [see Roth, Rothblum and Vande Vate, 1993].

The stable matching problem can also be formulated for non-bipartite graphs. However, in that case, no stable matching might exist; a 4-node example is easily constructed. On the other hand, Irving [1985] derived a polynomial time algorithm that finds a stable matching if it exists. For further reading on stable matchings we refer to the book of Gusfield and Irving [1989].

5.3. Non-bipartite graphs – Edmonds' matching polytope

So, when G is not bipartite, $\text{Fract}(G) \neq \text{Match}(G)$. In trying to formulate the weighted matching problem in a non-bipartite graph G as a linear programming problem, we begin with the inequalities defining $\text{Fract}(G)$. Then, we search for inequalities that 'cut off' the fractional extreme points of $\text{Fract}(G)$. The following lemma characterizes the fractional extreme points of $\text{Fract}(G)$. Its proof is similar to that of Theorem 14.

Lemma 18 [Balinski, 1965]. *A vector $x \in \mathbb{R}^E$ is an extreme point of* $\text{Fract}(G)$ *if and only if there exists a matching M and a collection of odd circuits $C_1, \ldots, C_k$ in the graph G, such that the matching and the odd circuits are pairwise node-disjoint and*

$$x = \chi^M + \tfrac{1}{2}(\chi^{C_1} + \cdots + \chi^{C_k}). \tag{44}$$

Let $U \subseteq V(G)$, with $|U| \geq 3$ and odd. Add up all the inequalities $\sum_{e \in \delta(v)} x_e \leq 1$ with $v \in U$, and all inequalities $-x_e \leq 0$ with $e \in \delta(U)$. Dividing the resulting inequality by 2 yields

$$x(\langle U \rangle) = \tfrac{1}{2}\Big(\sum_{v \in U} x(\delta(v)) - \sum_{e \in \delta(U)} x_e\Big) \leq \tfrac{1}{2}|U|. \tag{45}$$

Obviously, each $x \in \text{Fract}(G)$ satisfies (45). Rounding down the right hand side we get the following *blossom constraint*

$$x(\langle U \rangle) \leq \tfrac{1}{2}(|U| - 1). \tag{46}$$

The characteristic vectors of matchings in G satisfy all the blossom constraints. However, the fractional extreme points of $\text{Fract}(G)$ do not. Indeed, the fractional extreme point x of $\text{Fract}(G)$ violates the blossom constraint obtained when U is chosen to be the node set of one of the odd circuits defining x. So, if we add all the blossom constraints to the constraints defining Fract(G), we get a polyhedron Blossom(G), which contains Match(G), but is, for non-bipartite graphs, properly contained in Fract(G). In particular, the blossom constraints 'cut off' all the fractional extreme points of Fract(G). In the process, however, we might have introduced new fractional extreme points. Edmonds showed that adding the blossom constraints does not introduce any new fractional extreme points.

Theorem 19 [Edmonds, 1965b]. *For each graph G,* Match(G) = Blossom(G).

Proof. Edmonds [1965b] originally proved this result via his weighted matching algorithm (cf. Section 6.2). Since then, others including Balinski [1972], Hoffman & Oppenheim [1978], Lovász [1979a], Seymour [1979], Aráoz, Cunningham, Edmonds & Green-Krótki [1983] and Schrijver [1983b] have offered different proofs. We essentially follow the proof by Aráoz and coworkers and Schrijver.

Suppose that for some graph G, Match(G) $\neq$ Blossom(G). Among all such graphs, suppose $G = (V, E)$ has $|V| + |E|$ as small as possible. So G is connected and non-bipartite. Consider a fractional extreme point x of Blossom(G). Since no fractional extreme point of Fract(G) is in Blossom(G), x is not an extreme point of Fract(G). Hence, there exists a subset S of V with $|S| \geq 3$ and odd, such that

$$x(\langle S\rangle) = \tfrac{1}{2}(|S| - 1). \tag{47}$$

Among all such subsets, choose S so that $|S|$ is as small as possible.

Claim 1. $|S| < |V|$.

Proof of Claim 1. If not, $x(E) = \frac{1}{2}(|V| - 1)$ must be the only blossom constraint x satisfies with equality. Further, since $|V| + |E|$ is as small as possible, $x_e > 0$ for each $e \in E$. Otherwise, if $x_e = 0$ for some $e \in E$, $G \setminus e$ would be a smaller counterexample. Finally, since x is an extreme point of Blossom(G), x satisfies with equality at least $|E|$ constraints from the defining system. Hence there are at least $|E| - 1$ nodes u in $V(G)$ with $x(\delta(u)) = 1$. On the other hand, since $x(\delta(u)) > 0$ for each node u and

$$\sum_{v\in V}(1 - x(\delta(v))) = |V| - 2x(E) = 1, \tag{48}$$

there are at least two nodes u such that $x(\delta(u)) < 1$. Hence, $|V| - 1 \geq |E|$. But connected non-bipartite graphs have at least as many edges as nodes – contradiction! *End of proof of Claim 1.*

Partition x into $x = [x^1, x^2]$, where x^1 is the restriction of x to edges in $\langle S\rangle$. Consider the graph $G \times S$ obtained by shrinking S to the pseudo-node s.

Claim 2. $x^1 \in$ Blossom($G|S$) *and* $x^2 \in$ Blossom($G \times S$).

Proof of Claim 2. Since x is in Blossom(G), x^1 satisfies $x^1(\delta(v)) \leq 1$ for each $v \in S$ and the blossom constraints for $G|S$. So, $x^1 \in$ Blossom($G|S$). Likewise,

x^2 satisfies $x^2(\delta(v)) \leq 1$ for each $v \in V(G) \setminus S$ and the blossom constraint $x^2(\langle U \rangle) \leq \frac{1}{2}(|U|-1)$ for each subset $U \subseteq V \setminus S$ with $|U| \geq 3$ and odd. Further, since $|S| \geq \sum_{v \in S} x(\delta(v)) = 2x(\langle S \rangle) + x(\delta(S))$ and $2x(\langle S \rangle) = |S|-1$, $x^2(\delta(s)) = x(\delta(S)) \leq 1$. Finally, for each $U \subseteq V(G \times S)$ containing s,

$$\begin{aligned} x^2(\langle U \rangle) &= x(\langle (U \setminus \{s\}) \cup S \rangle) - x(\langle S \rangle) \leq \\ &\leq \tfrac{1}{2}(|(U \setminus \{s\}) \cup S| - 1) - \tfrac{1}{2}(|S|-1) = \tfrac{1}{2}(|U|-1) \end{aligned}$$

and so $x^2 \in \text{Blossom}(G \times S)$. *End of proof of Claim 2.*

Since G is a smallest counterexample,

$$\begin{aligned} &\text{Match}(G|S) = \text{Blossom}(G|S) \\ &\quad \text{and} \quad \text{Match}(G \times S) = \text{Blossom}(G \times S). \end{aligned} \tag{49}$$

Hence, x^1 can be expressed as a convex combination of the characteristic vectors of matchings in $G|S$ and x^2 can be expressed as a convex combination of the characteristic vectors of matchings in $G \times S$. This implies that there is a non-negative integer k, matchings $M_1, \ldots, M_k$ in $G|S$, and matchings $N_1, \ldots, N_k$ in $G \times S$, such that

$$x_1 = \frac{1}{k}(\chi^{M_1} + \cdots + \chi^{M_k}) \quad \text{and} \quad x_2 = \frac{1}{k}(\chi^{N_1} + \cdots + \chi^{N_k}). \tag{50}$$

Note that the matchings M_i, and similarly the matchings N_i, need not all be distinct.

Claim 3. *We can renumber the matchings N_i $(i = 1, \ldots, k)$, so that $M_i \cup N_i$ is a matching in G for each $(i = 1, \ldots, k)$.*

Proof of Claim 3. By (47) each M_i has exactly one exposed node in $G|S$. Thus, we need only prove that for each $u \in S$: $|\{i \mid |M_i \cap \delta(u)| = 0\}| \geq |\{i \mid |N_i \cap \delta(u)| = 1\}|$. To see this, observe that:

$$\begin{aligned} &|\{i \mid |M_i \cap \delta(u)| = 0\}| - |\{i \mid |N_i \cap \delta(u)| = 1\}| \\ &\quad = k - |\{i \mid |M_i \cap \delta(u)| = 1\}| - |\{i \mid |N_i \cap \delta(u)| = 1\}| \\ &\quad = k - \Big(\sum_{i=1}^{k} |M_i \cap \delta(u) \cap \langle S \rangle| + \sum_{i=1}^{k} |N_i \cap \delta(u) \cap \delta(S)|\Big) \\ &\quad = k - k\left(x^1(\delta(u) \cap \langle S \rangle) + x^2(\delta(u) \cap \delta(S))\right) \\ &\quad = k - kx(\delta(u)) \geq 0. \end{aligned}$$

End of proof of Claim 3.

So, as $x = 1/k(\chi^{M_1 \cup N_1} + \cdots + \chi^{M_k \cup N_k})$, it is the convex combination of characteristic vectors of matchings; contradicting the assumption that it is a fractional extreme point of Blossom(G). □

We can sharpen this result in the sense that we can specify which of the non-negativity, degree and blossom constraints are necessary to describe the matching polytope of a given graph. Indeed, although we need all the non-negativity constraints, we do not need the *degree constraint* $x(\delta(v)) \leq 1$ for a node $v \in V(G)$ if there is another node u with $\delta(v) \subsetneq \delta(u)$ or there is an edge uw with $\delta(v) \subseteq \delta(u) \cup \delta(w)$. Moreover, we only need the blossom constraint $x(\langle S \rangle) \leq \frac{1}{2}(|S| - 1)$ for $S \subseteq V(G)$ such that $|S| \geq 3$ and odd, $G|S$ is factor-critical and $G|S$ has no cut node. (A node u is called a *cut node* of a connected graph G if $G \setminus u$ is not connected.) Pulleyblank & Edmonds [1974] showed that these are exactly the constraints we need in order to have a minimum system of linear inequalities defining the matching polytope. In geometric terms these constraints correspond to the 'facets' of the matching polytope [see Pulleyblank, 1989].

The fact that the dual problem in Theorem 15 has integral optimum solutions extends to non-bipartite graphs: the non-negativity, degree and blossom constraints form a totally dual integral system [Cunningham & Marsh, 1978; Hoffman & Oppenheim, 1978; Schrijver & Seymour, 1977; Schrijver, 1983a, b]. In fact, this remains true if we restrict ourselves to Pulleyblank and Edmonds' minimal description of the matching polytope [Cunningham & Marsh, 1978].

The following characterization of the perfect matching polytope follows easily from Theorem 19.

Corollary 20. *For each graph* $G = (V, E)$, Perfect(G) *is the solution set of the system:*

$$\begin{array}{lcll} x(\delta(v)) & = & 1 & (v \in V) \\ x(\langle U \rangle) & \leq & \frac{1}{2}(|U| - 1) & (U \subseteq V,\ |U| \text{ odd and at least } 3) \\ x_e & \geq & 0 & (e \in E), \end{array} \tag{51}$$

which is equivalent to the system

$$\begin{array}{lcll} x(\delta(v)) & = & 1 & (v \in V) \\ x(\delta(U)) & \geq & 1 & (U \subseteq V,\ |U| \text{ odd and at least } 3) \\ x_e & \geq & 0 & (e \in E). \end{array} \tag{52}$$

Proof. That (51) describes the perfect matching polytope is trivial. We need only prove that (51) and (52) are equivalent.

Consider $U \subseteq V$, with $|U|$ odd, and let $x \in \mathbb{R}^E$ be such that $x(\delta(v)) = 1$ for each $v \in V$. Then

$$\begin{aligned} x(\langle U \rangle) \leq \tfrac{1}{2}(|U| - 1) &\iff x(\langle U \rangle) - \tfrac{1}{2} \sum_{v \in U} x(\delta(v)) \leq \tfrac{1}{2}(|U| - 1) - \tfrac{1}{2}|U| \\ &\iff -\tfrac{1}{2} x(\delta(U)) \leq -\tfrac{1}{2} \\ &\iff x(\delta(U)) \geq 1. \end{aligned}$$

□

So, we have two descriptions of the perfect matching polytope. We call (51) the *blossom description* of the perfect matching polytope and (52) the *odd cut description*. The inequalities $x(\delta(U)) \geq 1$ in (52) are called the *odd cut constraints*.

Like the blossom description of the matching polytope, the blossom description of the perfect matching polytope is totally dual integral (the 'perfect matching case' follows directly from the 'matching case'). This is not the case for the odd cut description – the complete graph on 4 nodes, K_4, provides a counterexample. However, from the proof of Corollary 20 and the fact that (51) is totally dual integral, it can be shown that when the weight function is integral, the odd cut description admits half-integral dual optimal solutions.

The (perfect) matching polytope is a geometric object: namely the convex hull of points in a Euclidean space. There are many other interesting geometric objects related to matchings, e.g., the cone generated by the characteristic vectors of perfect matchings, the linear hull of the characteristic vectors of matchings, etc. Edmonds, Lovász & Pulleyblank [1982], Naddef [1982], Naddef & Pulleyblank [1982] have obtained results in this vein. Structural results on matchings like those mentioned in Section 4.1 ('ear-decomposition' and 'brick-decomposition') often play a crucial role in characterizing these geometric objects. An especially noteworthy example is Lovász's beautiful characterization of the matching lattice, i.e., the set $\{\sum_{M \in \mathcal{M}} \lambda_M \chi^M \mid \lambda_M \in \mathbb{Z} \, (M \in \mathcal{M})\}$ where $\mathcal{M}$ denotes the set of perfect matchings [Lovász, 1987; cf. Murty, 1994].

Karzanov [1992] derived a polynomial time algorithm for calculating the Euclidean distance from the origin to the perfect matching polytope of a bipartite graph.

In this section we have formulated matching problems as linear programming problems. Making combinatorial problems accessible to linear programming techniques in this way is one of the main goals of 'polyhedral combinatorics'. In general, this area could be described as the study of methods for solving combinatorial problems using the theory of linear inequalities. Over the years, the results in this section have been among the principal paradigms of this polyhedral approach. (However, even for matchings not all polyhedral questions have been resolved, see Cunningham & Green-Krotki [1986].) For surveys on polyhedral combinatorics, see Pulleyblank [1983, 1989] and Schrijver [1983a, 1995]. The standard reference for the theory of integer and linear programming – the toolbox for polyhedral combinatorics – is Schrijver's book [1986].

6. Finding maximum and minimum weight matchings

In this section we give polynomial time algorithms for finding a (perfect) matching of maximum or minimum weight. Actually we consider only the problem of finding a minimum weight perfect matching, but this is not really a restriction. Indeed, suppose we are given a non-negative weight function $w \in \mathbb{R}^E$ and want to find a maximum weight matching in a graph G. By adding nodes and edges with zero weight, we can transform the problem into one of finding a maximum weight matching in a complete graph with an even number of vertices, or if G is bipartite, in a complete bipartite graph in which the two color classes have the same number of nodes. Since each matching in these graphs is contained in a perfect matching, we may find a maximum weight matching in the original graph by finding a maximum weight perfect matching in the complete (bipartite) graph. Replacing each weight w_e by $-w_e$, we turn the problem into a minimization problem and, if we prefer non-negative weights, we may add a suitable constant to each weight.

A weighted matching problem is a linear programming problem, so the duality theorem of linear programming (34) provides a stopping criterion. In fact, using the complementary slackness conditions, we reduce the weighted matching problem to a series of cardinality matching problems.

6.1. *Bipartite graphs*

Throughout this section $G = (V_1 \cup V_2, E)$ is a bipartite graph and $w \in \mathbb{R}_+^E$. For convenience, we assume that G contains a perfect matching. From Corollary 16 and linear programming duality (34) the minimum weight of a perfect matching is equal to the maximum in

$$\begin{array}{lll} \text{maximize} & \pi(V_1 \cup V_2) & \\ \text{subject to} & \pi_u + \pi_v \quad \leq \quad w_{uv} & (uv \in E). \end{array} \tag{53}$$

We refer to (53) as the *dual* problem and to each feasible solution π to (53) as *dual feasible*. For each dual feasible solution π we define the *reduced cost function* $w^\pi \in \mathbb{R}^E$ by: $w^\pi_{uv} := w_{uv} - \pi_u - \pi_v$ for each $uv \in E$, and the graph G_π by $V(G_\pi) := V(G)$ and $E(G_\pi) := \{uv \in E(G) \mid w^\pi_{uv} = 0\}$. Thus, G_π is the subgraph on the nodes of G that includes only those edges, called *admissible*, with zero reduced cost under π. The edges not in $E(G_\pi)$ are called *inadmissible*.

The complementary slackness conditions (35) imply that:

A dual feasible solution π is optimal if and only if G_π admits a perfect matching. Moreover, if G_π admits a perfect matching, then the perfect matchings in G_π are exactly the minimum weight perfect matchings in G. (54)

So, given a dual feasible π, we check whether G_π has a perfect matching M. If it does, M is a minimum weight perfect matching in G and we are done. If it does not, we change π as follows.

DUAL CHANGE (in bipartite graphs): Let M be a maximum matching and F a Hungarian forest in G_π. Define π' by

$$\pi'_u := \begin{cases} \pi_u + \epsilon & \text{if } u \in V_1 \cap \text{even}(F) \\ \pi_u - \epsilon & \text{if } u \in V_2 \cap \text{odd}(F) \\ \pi_u & \text{otherwise,} \end{cases} \tag{55}$$

where

$$\epsilon := \min\{w^\pi_{uv} | u \in \text{even}(F) \cap V_1, v \in V_2 \setminus \text{odd}(F), uv \in E\}. \tag{56}$$

By (56) and because F is Hungarian, π' is dual feasible. Moreover,

M and F are contained in $G_{\pi'}$. (57)

So, we can apply GROW, starting with the matching M and the alternating forest F, to search for a maximum matching in $G_{\pi'}$.

Note that, since $uv \in E(G_{\pi'})$ for each $u \in V_1 \cap \text{even}(F)$ and $v \in V_2 \setminus \text{odd}(F)$ with $w^{\pi}_{uv} = \epsilon$,

$$F \textit{ is not Hungarian in } G_{\pi'}. \tag{58}$$

Thus, we have the following algorithm for finding a minimum weight perfect matching in a bipartite graph. We begin with the dual feasible solution π with $\pi_v = 0$ for each $v \in V_1 \cup V_2$.

We apply GROW and AUGMENT until we find either a perfect matching M or a Hungarian alternating forest F in G_{π}. If we find a perfect matching M in G_{π}, we are done: M is a minimum weight perfect matching in G. Otherwise, we apply DUAL CHANGE. Clearly this algorithm, called the *weighted Hungarian method*, runs in polynomial time. It was originally introduced by Kuhn [1955, 1956]. For variants of his method, see Flood [1956], Ford & Fulkerson [1957], Motzkin [1956] and Munkres [1957]. Bertsekas [1979] proposed a so-called auction method [cf. Bertsekas, 1990].

Kuhn's algorithm is a 'dual' algorithm in the sense that at any stage it keeps a dual feasible solution and a primal infeasible solution (namely a non-perfect matching) satisfying complementary slackness. Primal feasibility, i.e. the matching being perfect, acts as the stopping criterion. Another possible approach is the 'primal' algorithm of Balinski & Gomory [1964]. It keeps at any stage a perfect matching which is changed until it becomes optimal. The stopping criterion in their algorithm is dual feasibility.

Note that although the weighted Hungarian method was motivated by Theorem 14, its correctness does not rely on that result. In fact, the algorithm provides a separate proof of Theorem 14 as well as of Theorem 15 (the initial dual feasible solution is integral and, when the weights are integral, each dual change maintains integrality).

6.1.1. Implementation of the weighted Hungarian method

After $|V_1|$ augmentations, the weighted Hungarian method finds a perfect matching M in G_{π} and, by the complementary slackness conditions (35), M is a minimum weight perfect matching in G. Thus, the running time of the algorithm depends on the computations required between consecutive augmentations, called a *phase*.

Note that if an admissible edge becomes inadmissible after a dual change, it cannot be come admissible again until after the next augmentation. This means that if we ignore the effort required to make the dual changes, each phase of the algorithm is essentially an application of GROW. One could say that the 'dual changer' confounds the 'grower': any time the alternating forest becomes Hungarian, the 'dual changer' adjusts the graph so that the forest is no longer Hungarian. Consequently, if we disregard the effort spent on dual changes, each phase can be carried out in $O(|E|)$ time. We sketch two implementations of the dual changes; one for dense graphs and one for sparse graphs.

Dual changes in dense graphs: In dense graphs, where $|V|^2 = O(|E|)$, we maintain an array CLOSE so that for each $v \in V_2$, $\text{CLOSE}[v] = u$, where $u \in V_1 \cap \text{even}(F)$ and $w^{\pi}_{uv} = \min\{w^{\pi}_{u'v} \mid u' \in V_1 \cap \text{even}(F)\}$. Using CLOSE, we can make each dual change in $O(|V|)$ time and, each time we add a node to $V_1 \cap \text{even}(F)$, we can update CLOSE in $O(|V|)$ time. During a phase we make at most $|V_1|$ dual changes and add at most $|V_1|$ nodes to $V_1 \cap \text{even}(F)$. So, with this implementation we can carry out the dual changes for each phase in $O(|V|^2)$ time.

The weighted Hungarian method can be implemented to run in $O(|V|^3)$ time. (59)

Dual changes in sparse graphs: In sparse graphs, where $|E|$ is significantly smaller than $|V_1|^2$, we can improve the running time by implementing the dual changes more efficiently. First, we concentrate on finding the value of ϵ. For each node v in $V_2 \setminus \text{odd}(F)$, we maintain the value $\text{SLACK}[v] := w^{\pi}_{\text{CLOSE}[v]v}$. Each time we add a node u to $V_1 \cap \text{even}(F)$, we scan each of its $\deg(u)$ neighbors v to update CLOSE and SLACK. So, during an entire phase we require only $O(|E|)$ time to maintain these two arrays.

Standard data structures such as 'd-heaps' or 'priority queues' [see Tarjan, 1983, or Aho, Hopcroft & Ullman, 1974] for storing $V_2 \setminus \text{odd}(F)$ according to the entries in SLACK facilitate quick determination of ϵ. Using such a data structure we can not only find ϵ, but also update the data structure itself in $O(\log |V|)$ time whenever SLACK changes or a node leaves $V_2 \setminus \text{odd}(F)$. So, it is possible to determine the values of ϵ in $O(|E| \log |V|)$ time per phase.

Instead of making dual changes explicitly, which can take up to $|V|$ steps each, we make them implicitly. We maintain a variable ϵ_{total} and, for each node $u \in V$, we keep two variables: π^{old}_u and π^{cor}_u. Together these represent the dual variable π_u according to the following rule:

$$\pi_u := \begin{cases} \pi^{old}_u + (\epsilon_{total} - \pi^{cor}_u) & \text{if } u \in V_1 \cap \text{even}(F) \\ \pi^{old}_u - (\epsilon_{total} - \pi^{cor}_u) & \text{if } u \in V_2 \cap \text{odd}(F) \\ \pi^{old}_u & \text{otherwise} \end{cases} \tag{60}$$

To make a dual change implicitly, we replace ϵ_{total} by $\epsilon_{total} + \epsilon$. When a node u enters the alternating forest, we set $\pi^{cor}_u := \epsilon_{total}$; and when an augmentation leads us to remove the node u from the alternating forest we set

$$\pi^{old}_u := \begin{cases} \pi^{old}_u + (\epsilon_{total} - \pi^{cor}_u) & \text{if } u \in \text{even}(F) \\ \pi^{old}_u - (\epsilon_{total} - \pi^{cor}_u) & \text{if } u \in \text{odd}(F). \end{cases}$$

Clearly, this 'delayed' revision can be carried out within GROW and AUGMENT.

Thus, we have the following result:

The weighted Hungarian method can be implemented to run in $O(|E||V| \log |V|)$ time. (61)

Fredman & Tarjan [1987] improved this time bound to $O(|V|(|E| + |V| \log |V|))$ using 'Fibonacci-heaps'. Brezovec, Cornuéjols, & Glover [1988] obtained the same time bound based on algorithm for a special case of matroid intersection.

6.2. Non-bipartite graphs

In this section we consider the problem of finding a minimum weight perfect matching in a non-bipartite graph G. For convenience, we assume that G admits a perfect matching and that the weights $w \in \mathbb{R}^E$ are non-negative.

By Corollary 20, a minimum weight perfect matching solves the linear programming problem

$$\begin{array}{lrcll} \text{maximize} & \sum_{e\in E} w_e x_e & & & \\ \text{subject to} & x(\delta(v)) & = & 1 & (v \in V) \\ & x(\delta(S)) & \geq & 1 & (S \in \Omega(G);\ |S| \neq 1) \\ & x_e & \geq & 0 & (e \in E), \end{array} \tag{62}$$

where $\Omega(G) := \{S \subseteq V(G) \mid |S| \text{ is odd}\}$. Thus, by linear programming duality (34), the minimum weight of a perfect matching is equal to the maximum in

$$\begin{array}{lrcll} \text{maximize} & \sum_{S\in\Omega(G)} \pi_S & & & \\ \text{subject to} & \sum_{S\in\Omega(G);\delta(S)\ni e} \pi_S & \leq & w_e & (e \in E) \\ & \pi_S & \geq & 0 & (S \in \Omega(G);\ |S| \neq 1). \end{array} \tag{63}$$

We refer to (63) as the *dual* problem and to each feasible solution π to (63) as *dual feasible*. For each dual feasible solution π the *reduced cost function* $w^\pi \in \mathbb{R}^E$ is defined by: $w_e^\pi := w_e - \sum_{S\in\Omega(G);\delta(S)\ni e} \pi_S$ for each $e \in E$, and the graph G_π on the node set $V(G_\pi) := V(G)$ has edge set defined by $E(G_\pi) := \{uv \in E(G) \mid w_{uv}^\pi = 0\}$. So, again, G_π is the subgraph on the nodes of G that includes only those edges, called *admissible*, with zero reduced cost under π. Finally, we define $\Omega_\pi := \Omega_\pi(G) := \{S \in \Omega(G) \mid \pi_S > 0 \text{ and } |S| \neq 1\}$.

The complementary slackness conditions (35) imply the following characterization of minimum weight perfect matchings:

> *A dual feasible solution π is optimal if and only if G_π admits a perfect matching M such that $M \cap \delta(S) = 1$ for each $S \in \Omega_\pi(G)$. If π is optimal, the collection of all such perfect matchings in G_π is exactly the collection of minimum weight perfect matchings in G.* (64)

So minimum weight perfect matchings are perfect matchings in G_π, for some optimal π, that satisfy additional conditions. These additional conditions prevent us from finding a minimum weight perfect matching in G by simply searching for a maximum cardinality matching in G_π as we did in the case of bipartite graphs. To overcome this, we restrict attention to those dual feasible solutions π, called *structured*, that satisfy the following two conditions:

$$\Omega_\pi \text{ is } \textit{nested}\text{, i.e., if } S, T \in \Omega_\pi, \text{ then } S \subseteq T,\ T \subseteq S \text{ or } S \cap T = \emptyset. \tag{65}$$

> If $S \in \Omega_\pi$, and $S_1, \ldots, S_k$ are the inclusion-wise maximal members of Ω_π properly contained in S, then $(G_\pi \times S_1 \times \cdots \times S_k)|S$ is factor-critical. (66)

Note that by (28), (66) implies that $G_\pi|S$ and $G|S$ are also factor-critical.

For each structured dual feasible π we define $\widetilde{G_\pi}$ to be the graph obtained from G_π by shrinking the members of Ω_π. The motivation for considering only structured dual solutions is apparent from the following consequence of (64) and (29):

> *A structured dual feasible solution π is optimal if and only if $\widetilde{G_\pi}$ admits a perfect matching.* (67)

This suggests the following algorithm, developed by Edmonds [1965b], for finding a minimum weight perfect matching in a non-bipartite graph G.

EDMONDS' ALGORITHM: Given a structured dual feasible π – initially identical to 0 – construct $\widetilde{G_\pi}$. If $\widetilde{G_\pi}$ admits a perfect matching M, then the dual feasible solution π is optimal; extend M to a minimum weight perfect matching in G_π. Otherwise, determine the Edmonds–Gallai structure of $\widetilde{G_\pi}$ and revise the dual solution according to (68) below.

We rely on notation similar to that in Section 2.2 to describe the relation between G and $\widetilde{G_\pi}$. If $S \in \Omega_\pi(G)$ is shrunk into pseudo-node $s \in \widetilde{G_\pi}$, we define $\text{DEEP}_\pi[s] = S$ and $\text{OUTER}_\pi[u] = s$ for each $u \in S$. For each node u in G that is also a node in $\widetilde{G_\pi}$, we define $\text{DEEP}_\pi[u] = \{u\}$ and $\text{OUTER}_\pi[u] = u$. For each $T \subseteq V(\widetilde{G_\pi})$, $\text{DEEP}_\pi[T] := \cup_{s \in T} \text{DEEP}_\pi[s]$.

DUAL CHANGE (in non-bipartite graphs): Given a structured dual feasible solution π, define

$$\pi'_S := \begin{cases} \pi_S + \epsilon & \text{if } S = \text{DEEP}_\pi[D] \text{ for some component } D \text{ of } D(\widetilde{G_\pi}) \\ \pi_S - \epsilon & \text{if } S = \text{DEEP}_\pi[s] \text{ for some } s \in A(\widetilde{G_\pi}) \\ \pi_S & \text{otherwise,} \end{cases} \tag{68}$$

where

$$\begin{aligned} \epsilon &:= \min\{\epsilon_1, \tfrac{1}{2}\epsilon_2, \epsilon_3\}, \quad \text{and} \\ \epsilon_1 &:= \min\{w^\pi_{uv} \mid \text{OUTER}_\pi[u] \in D(\widetilde{G_\pi}); \text{OUTER}_\pi[v] \in C(\widetilde{G_\pi})\}; \\ \epsilon_2 &:= \min\{w^\pi_{uv} \mid \text{OUTER}_\pi[u] \text{ and } \text{OUTER}_\pi[v] \\ &\qquad \text{in different components of } D(\widetilde{G_\pi})\}; \\ \epsilon_3 &:= \min\{\pi_S \mid S = \text{DEEP}_\pi[s] \text{ for some } s \in A(\widetilde{G_\pi}), \\ &\qquad |\text{DEEP}_\pi[s]| \neq 1\}. \end{aligned} \tag{69}$$

Determining whether or not $\widetilde{G_\pi}$ has a perfect matching, computing the Edmonds–Gallai structure of $\widetilde{G_\pi}$, and extending a perfect matching in $\widetilde{G_\pi}$ to a perfect matching in G_π can all be accomplished via any maximum cardinality algorithm that determines the Edmonds–Gallai structure, like the blossom algorithm or the algorithm in Section 8.4. A perfect matching in G_π obtained by extending a perfect matching in $\widetilde{G_\pi}$ satisfies the complementary slackness conditions (64) and hence is a minimum weight perfect matching in G. If $\widetilde{G_\pi}$ does not admit a perfect matching, DUAL CHANGE increases the dual objective function value by $\epsilon\,(\kappa(\widetilde{G_\pi}) - |A(\widetilde{G_\pi})|) = \epsilon\,\mathrm{def}(\widetilde{G_\pi}) > 0$. So Edmonds' algorithm only stops, when a minimum weight perfect matching has been obtained. That the algorithm does stop follows from the following lemma.

Lemma 21. *Given a structured dual feasible solution π,* DUAL CHANGE *yields a structured dual feasible solution π', such that:*

- $\mathrm{def}(\widetilde{G_{\pi'}}) \le \mathrm{def}(\widetilde{G_\pi})$;
- *if* $\mathrm{def}(\widetilde{G_{\pi'}}) = \mathrm{def}(\widetilde{G_\pi})$, *then* $\mathrm{DEEP}_{\pi'}(D(\widetilde{G_{\pi'}})) \supseteq \mathrm{DEEP}_\pi(D(\widetilde{G_\pi}))$;
- *if* $\mathrm{def}(\widetilde{G_{\pi'}}) = \mathrm{def}(\widetilde{G_\pi})$ *and* $\mathrm{DEEP}_{\pi'}(D(\widetilde{G_{\pi'}})) = \mathrm{DEEP}_\pi(D(\widetilde{G_\pi}))$, *then* $\mathrm{DEEP}_{\pi'}(C(\widetilde{G_{\pi'}})) \supsetneq \mathrm{DEEP}_\pi(C(\widetilde{G_\pi}))$.

Proof. For each component D of $D(\widetilde{G_\pi})$, the sets in Ω_π are either disjoint from $\mathrm{DEEP}_\pi[D]$ or contained in $\mathrm{DEEP}_\pi[D]$. So $\Omega_{\pi'}$ is nested. Moreover, by (27) $\widetilde{G_\pi}|D$ is factor-critical, so π' satisfies (66) and hence is structured.

To prove the remainder of the lemma, observe that $\widetilde{G_{\pi'}}$ can be obtained from $\widetilde{G_\pi}$ in two steps. First, shrink the node-sets S that are not in Ω_π but are in $\Omega_{\pi'}$, this yields $\widetilde{G_\pi}^*$ (i.e. the Edmonds–Gallai graph of $\widetilde{G_\pi}$). The nodes in $D(\widetilde{G_\pi}^*)$ and in $C(\widetilde{G_\pi}^*)$ are contained in $V(\widetilde{G_{\pi'}})$, Hence:

$$\begin{aligned} &- \mathrm{def}(\widetilde{G_\pi}^*) = \mathrm{def}(\widetilde{G_\pi});\\ &- \mathrm{DEEP}_{\pi'}(D(\widetilde{G_\pi}^*)) = \mathrm{DEEP}_\pi(D(\widetilde{G_\pi}));\\ &- \mathrm{DEEP}_{\pi'}(C(\widetilde{G_\pi}^*)) = \mathrm{DEEP}_\pi(C(\widetilde{G_\pi})) \end{aligned} \tag{70}$$

$\widetilde{G_{\pi'}}$ can be obtained from $\widetilde{G_{\pi'}}^*$ by applying the operation (31) if $\epsilon = \epsilon_1$ or $\frac{1}{2}\epsilon_2$, the operation in (32) if $\epsilon = \epsilon_3$ and the operation (30). So, by (30), (31) and (32):

$$\begin{aligned} &- \mathrm{def}(\widetilde{G_{\pi'}}) \le \mathrm{def}(\widetilde{G_\pi}^*);\\ &- \textit{if } \mathrm{def}(\widetilde{G_{\pi'}}) = \mathrm{def}(\widetilde{G_\pi}^*) \textit{ then } D(\widetilde{G_{\pi'}}) \supseteq D(\widetilde{G_\pi}^*);\\ &- \textit{if both } \mathrm{def}(\widetilde{G_{\pi'}}) = \mathrm{def}(\widetilde{G_\pi}^*) \textit{ and } D(\widetilde{G_{\pi'}}) = D(\widetilde{G_\pi}^*),\\ &\quad \textit{then } C(\widetilde{G_{\pi'}}) \supsetneq C(\widetilde{G_\pi}^*). \end{aligned} \tag{71}$$

From (71) and (70), the lemma follows. □

As a consequence, there are at most $O(|V(G)|^3)$ dual changes. Since we can find a maximum cardinality matching and the Edmonds–Gallai structure of $\widetilde{G_\pi}$ in polynomial time,

Edmonds' algorithm finds a minimum weight perfect matching in polynomial time. (72)

Note that the algorithm provides a constructive proof of Corollary 20.

6.2.1. Implementing Edmonds' algorithm

In implementing Edmonds' algorithm for finding a minimum weight perfect matching in a non-bipartite graph, we can exploit the efficient algorithms discussed in Section 2.2 for finding a maximum cardinality matching. Note, however, that unlike the cardinality problem, in solving the weighted problem we must be able to expand a pseudo-node without expanding the pseudo-nodes contained in it. We can similarly exploit the efficient methods discussed in Section 6.1 for revising the dual solution but the continual shrinking and expanding of blossoms gives rise to certain complications [see Galil, Micali & Gabow, 1986].

Lawler [1976] developed an $O(|V|^3)$ implementation of Edmonds' algorithm. Galil, Micali & Gabow [1986] derived an $O(|E||V|\log|V|)$ algorithm. Gabow, Galil & Spencer [1989] derived an implementation that runs in $O(|V|(|E|\log_2\log_2\log_{\max\{|E|/|V|,2\}}|V| + |V|\log|V|))$ time. This, in turn, has been improved by Gabow's $O(|V|(|E| + |V|\log|V|))$ bound [Gabow, 1990]. Nice reviews of these implementations are Ball & Derigs [1983] and Galil [1986a].

Gabow & Tarjan [1991] and Gabow [1985] have developed algorithms whose running times depend on the edge weights. These algorithms require $O(\sqrt{|V|\alpha(|V|,|E|)\log|V|}|E|\ \log(|V|N))$ and $O(|V|^{3/4}|E|\log N)$ time, respectively, where N is an upper bound on the edge weights.

These running times can be further improved when we confine ourselves to restricted classes of weighted matching problems. Lipton & Tarjan [1980] derived an $O(|V|^{3/2}\log|V|)$ algorithm for matching in planar graphs. This algorithm is based on their *Separator theorem* for planar graphs: If $G = (V, E)$ is planar we can partition V into three sets A, B and C with $|A|, |B| \le \frac{2}{3}|V|$ and $|C| \le 2\sqrt{2|V|}$ such that no edge connects A with B [Lipton & Tarjan, 1979]. The *separator* C can be found in linear time and can be used to recursively decompose a matching problem in a planar graph into matching problems in smaller planar graphs.

Vaidya [1989] showed that *Euclidean matching problems*, in which the nodes are given as points in the plane and the weight of an edge between the two points u and v is the distance between the two points in the plane, can be solved in $O(|V|^{5/2}(\log|V|)^4)$ time. When the points lie on a convex polygon, a minimum weight matching can be found in $O(|V|\log|V|)$ time [Marcotte & Suri, 1991].

7. General degree constraints

Matching can be viewed as a 'degree-constrained subgraph problem': find a maximum cardinality, or maximum weight, subgraph in which each node has degree at most one. In this section we consider more general degree constraints.

Let $G = (V, E)$ be an undirected graph, possibly with loops. The collection of loops incident to node v is denoted by $\lambda(v)$. The *general matching problem* is: Given edge weights $w \in \mathbb{R}^E$, edge capacities $c \in (\mathbb{R} \cup \{\infty\})^E$ and degree bounds $a, b \in (\mathbb{R} \cup \{\infty\})^V$ find a minimum or maximum weight integral vector x

satisfying:

$$\begin{array}{lclcll} a_v & \leq & x(\delta(v)) + 2x(\lambda(v)) & \leq & b_v & (v \in V) \\ 0 & \leq & x_e & \leq & c_e & (e \in E). \end{array} \tag{73}$$

We call an integral vector x satisfying (73) a *general matching*. We call a the *degree lower bounds*, b the *degree upper bounds* and c the *capacities*. The corresponding constraints are called the *lower* and *upper degree constraints* and the *capacity constraints*. We did not impose more general lower bounds on the values of x_e since this does not yield a more general problem. (Given a lower bound $d \neq 0$ on the edges, replace each degree lower bound a_v by $a_v - d(\delta(v)) - 2d(\lambda(v))$, each degree upper bound b_v by $b_v - d(\delta(v)) - 2d(\lambda(v))$, each capacity c_e by $c_e - d_e$, and each edge variable x_e by $x_e - d_e$.)

In addition to matching and perfect matching, the general matching problem includes: the *simple b-matching problem* in which $a = 0$, b is arbitrary and $c = 1$; the *b-matching problem* in which $a = 0$, b is arbitrary and $c = \infty$; the *capacitated b-matching problem* in which $a = 0$ and b and c are arbitrary; and the *edge cover problem* in which $a = 1$, $b = \infty$, and $c = 1$. The general matching problem also includes the *perfect* versions of the (capacitated) b-matching problems in which $a = b$ and the simple perfect b-matching problem, also called the *b-factor problem*.

We can use loops to express the degree constraints as parity conditions. For instance, to force the degree of a node v to be an odd number between 3 and 11 we add a loop ℓ to v and impose the constraints: $x(\delta(v)) + 2x_\ell = 11$; $0 \leq x_\ell \leq 4$. We discuss parity conditions in Section 7.4.

An even more general degree-constrained subgraph problem is the *D-matching problem*: Given a set $D_v \subseteq \mathbb{Z}_+^E$ for each $v \in V$, find a subgraph G' of G with $\deg_{G'}(v) \in D_v$ for each $v \in V$. Lovász [1972a] proved that finding a D-matching is NP-complete, even when D_v is restricted to be either $\{1\}$ or $\{0, 3\}$. When for each $v \in V$, $\mathbb{Z}_+ \setminus D_v$ contains no consecutive integers, the D-matching problem is polynomially solvable [Lovász, 1973; Cornuéjols, 1988; and Sebö, 1993].

Related to the degree-constrained subgraph problem is the question: For which $d \in \mathbb{Z}_+^{V(G)}$ has (73) an integral solution x with $x(\delta(v)) = d_v$ for all $v \in V(G)$? A polyhedral answer to this question has been given by Cunningham & Green-Krótki [1991], generalizing results by Balas & Pulleyblank [1983, 1989], who solved cases $a = 0$, $b = 1$ and $c = 1$, and by Koren [1973]. Koren considered the special instances of the question that G is the complete graph, $a = 0$, $b = \infty$ and $c = 1$, in other words he derived a system of inequalities for the convex hull of all the degree-sequences of simple graphs on $V(G)$ (see also Peled & Srinivasan [1989] and Cunningham & Zhang [1992]). The inequalities in this system are exactly the well-known necessary and sufficient conditions derived by Erdös & Gallai [1960] for a sequence of integers to be the sequence of degrees of a simple graph. For a separation algorithm (cf. Section 8.3) for the polyhedron given by Balas & Pulleyblank [1989] see Cunningham & Green-Krótki [1994].

For other generalizations of matchings see: Cornuéjols & Hartvigsen [1986], Cornuéjols, Hartvigsen & Pulleyblank [1982], Giles [1982a, b, c], and Lovász [1970b]. In this chapter we restrict attention to general matchings as defined in (73).

7.1. *Reducing the general matching problem*

One aspect of the self-refining nature of matching theory is that the general matching problem not only includes matching as a special case, but can also be reduced to the matching problem. There are two main steps in the reduction of general matching to matching. First, one reduces the general matching problem to a perfect b-matching problem in a new graph with no loops and no capacity constraints. Second, one further reduces the perfect b-matching problem to a perfect matching problem. The reductions are due to Tutte [1954].

In view of these reductions it is reasonable to expect that results analogous to those discussed earlier in this chapter extend to general matching. Indeed, this is the case. Tutte [1952, 1974] generalized his perfect matching theorem (Theorem 11) to give necessary and sufficient conditions for the existence of an f-factor (see also Tutte [1981] and, for an algorithmic proof, Anstee [1985]). Lovász [1970a] further generalized this result to *(f, g)-factors* or general matchings with $a_v = \deg(v) - g_v$ $(v \in V)$, $b = f$ and $c = 1$. Lovász also generalized the Edmonds–Gallai structure theorem to a structure theorem for (f, g)-factors [Lovász, 1970a, c, 1972a] and f-factors [Lovász, 1972e]. For a discussion of these latter results, see Lovász & Plummer [1986].

The polyhedral results and polynomial-time solvability of matching also extend to the general matching problem. After explaining the reductions from general matching problems to the (perfect) matching problem, we first consider polyhedral results for general matchings and next deal with algorithmic issues. It is possible to derive general matching results via the reductions [see Aráoz, Cunningham, Edmonds & Green-Krótki, 1983], but direct proofs are typically simpler.

In explaining the reductions we only show how the new graphs should be constructed and what the new bounds on the degrees should be. We leave it to the reader to determine appropriate weights on the edges and to prove that indeed the original problem can be solved by solving the newly constructed problem.

7.1.1. *Reduction to perfect b-matching*

We first show how to transform the general matching problem in $G = (V, E)$ to a perfect b-matching problem in a graph with no loops and no capacity constraints.

Reduction to $c_{uv} < \infty$ for each edge uv and $b_v < \infty$ for each node v: First, replace for each edge uv the capacity c_{uv} with $\min\{c_{uv}, b_u, b_v\}$ or, if this minimum is infinite, with $\max\{a_u, a_v\}$. With these new, finite, capacities, replace for each node v the degree upper bound b_v with $\min\{b_v, c(\delta(v)) + 2c(\lambda(v))\}$.

Reduction to $a_v = b_v$ for each node v: Next, construct a new graph G' as follows. Make two copies G_1 and G_2 of G. For each node v in G add an edge v_1v_2 with capacity $c_{v_1v_2} := b_v - a_v$ between the copies v_1 in G_1 and v_2 in G_2 of v. A copy (in G_1 or G_2) of an edge e in G gets the same capacity as e. For each node v in G the degree bounds of its two copies v_1 and v_2 in G' are: $a_{v_1} := b_{v_1} := a_{v_2} := b_{v_2} := b_v$.

Reduction to a loopless graph with $c \equiv \infty$: Finally, replace each edge $e = uv$ in G' with two new nodes, u_e and v_e, and three new edges, uu_e, $u_e v_e$, and $v_e v$. The capacities of these new edges are infinite and the degree bounds of the new nodes are: $a_{u_e} := a_{v_e} := b_{u_e} := b_{v_e} := c_{uv}$.

7.1.2. Reduction to perfect matching

Now, we further reduce the perfect b-matching problem in the loopless graph G' to a perfect matching problem in a graph G''. Replace each node v in G' with b_v copies $v_1, v_2, \ldots, v_{b_v}$. Replace each edge uv in G' by a collection of edges in G'', namely one between each copy u_i of u and each copy v_j of v. The b-matching problem in G' is now a perfect matching problem in the new graph G''.

7.2. General matching polyhedra

As polyhedral results for bipartite graphs are easier, we consider them first.

7.2.1. Bipartite graphs

Theorem 22. *The polyhedron described by (73) is integral for all integral vectors a, b and c if and only if G is bipartite. (Note that bipartiteness excludes loops.)*

Proof. Rather than derive this result by combining Corollary 16 with the above reductions, we present a proof based on the well-known result of Hoffman & Kruskal [1956] on totally unimodular matrices. An $m \times n$-matrix A is called *totally unimodular* if each square submatrix of A has determinant equal to 0 or ± 1. The following is easy to prove:

The node-edge incidence matrix of a graph G is totally unimodular if and only if G is bipartite. (74)

Hence, the theorem follows from:

Given an $m \times n$-matrix A, the polyhedron $\{x \in \mathbb{R}^n \mid a \le Ax \le b;\ 0 \le x \le c\}$ is integral for each $a, b \in \mathbb{Z}^m$, $c \in \mathbb{Z}^n$ if and only if A is totally unimodular [Hoffman & Kruskal, 1956]. (75)

□

7.2.2. Network flows and bidirected graphs

Given a directed graph $D = (V(D), A(D))$, $a, b \in \mathbb{Z}^{V(D)}$ and $c \in \mathbb{Z}^{A(D)}$ a *general network flow* is a vector x satisfying:

$$\begin{array}{lllll} a_v & \le & x(\delta^+(v)) - x(\delta^-(v)) & \le & b_v \quad (v \in V(D)) \\ 0 & \le & x_a & \le & c_a \quad (a \in A(D)). \end{array} \tag{76}$$

That (76) defines an integral polyhedron follows from network flow theory as well as from Hoffman and Kruskal's theorem (75). By a construction similar to that used in Section 3.1 to reduce bipartite matching problems to max-flow problems, this in turn implies Theorem 22.

The general network flow problem and the general matching problem are similar. Both are constraint by a system of the form $a \le Ax \le b; 0 \le x \le c$, where each column of A has at most two non-zero coefficients. In the matching case both non-zero coefficients are 1, whereas in the network flow case one is 1 and the other is -1. The other difference is that in the matching case we also allow columns with a single coefficient of 2 as the only non-zero entry. Edmonds & Johnson [1970] proposed a common generalization of these two models: *the general matching problem for bidirected graphs*. A *bidirected graph* is a matrix A in which each column either contains two non-zero entries both ± 1 or contains a single non-zero entry equal to ± 1 or ± 2. A general matching in a bidirected graph A is an integral vector x satisfying: $a \le Ax \le b; 0 \le x \le c$. The results in this section also hold for these more general objects [Edmonds & Johnson, 1970].

7.2.3. Non-bipartite graphs

We derived the matching polytope (45) and (46) of a non-bipartite graph from the degree constraints by adding constraints obtained in the following manner. Add up degree and non-negativity constraints so that the coefficients in the left hand side of the resulting inequality are even, divide the resulting inequality by 2 and round the right hand side down to the nearest integer. Applying the same construction to (73) yields the inequalities:

$$x(\langle V_1\rangle)-x(\langle V_2\rangle)+x(F_1)-x(F_2^*) \le \left\lfloor \tfrac{1}{2}(b(V_1)-a(V_2)+c(F_1 \cup F_2)) \right\rfloor$$
for each pair V_1, V_2 of disjoint subsets of V, each $F_1 \subseteq \delta(V_1) \setminus \delta(V_2)$, and each partition F_2, F_2^* of $\delta(V_2) \setminus \delta(V_1)$. (77)

In fact these inequalities, also called *blossom constraints*, describe the convex hull of general matchings. This can be derived from Theorem 19 or Corollary 20 via the above reductions [see Cook, 1983b; Schrijver, 1983a].

Theorem 23 [Edmonds, 1965b; Edmonds & Johnson, 1970]. *For each graph $G = (V, E)$, $a, b \in \mathbb{Z}^V$ and $c \in \mathbb{Z}^E$, the convex hull of all integral solutions to (73) is the solution set of the system of inequalities defined by (73) and (77). Moreover, this system is totally dual integral.*

Note that many of the inequalities (77) are redundant, e.g., when $b(V_1) - a(V_2) + c(F_1 \cup F_2)$ is even (though this is not the only case!). Although restricting the formulation to the inequalities (77) with $b(V_1) - a(V_2) + c(F_1 \cup F_2)$ odd gives a description of the convex hull of general matchings, we can no longer be assured that the system is totally dual integral. So, unlike ordinary matchings, the systems that are non-redundant and those that are minimally totally dual integral are distinct [see Cook & Pulleyblank, 1987; Cook, 1983a, b; Pulleyblank, 1980, 1981].

Below we list consequences of Theorem 23 for some of the more prominent special cases of general matching.

The b-matching polytope: The convex hull of all b-matchings is given by:

$$\begin{array}{lll} x(\delta(v)) & \leq & b_v & (v \in V) \\ x(\langle U \rangle) & \leq & \lfloor \frac{1}{2} b(U) \rfloor & (U \subseteq V) \\ x_e & \geq & 0 & (e \in E) \end{array} \tag{78}$$

(Edmonds [1965b], see Hoffman & Oppenheim [1978] and Schrijver & Seymour [1977] for alternative proofs).

If we replace the constraints $x(\delta(v)) \leq b_v$ with the constraints $x(\delta(v)) = b_v$, we get the convex hull of perfect b-matchings. In the case of perfect b-matchings, as with perfect matchings, we can replace the blossom constraints with the odd cut constraints to get the following description of the convex hull of perfect b-matchings:

$$\begin{array}{llll} x(\delta(v)) & = & b_v & (v \in V) \\ x(\delta(U)) & \geq & 1 & (U \subseteq V \text{ with } b(U) \text{ odd}) \\ x_e & \geq & 0 & (e \in E). \end{array} \tag{79}$$

When all components of b are even, the b-matching polytope is described by the degree and the non-negativity constraints, regardless of whether the graph is bipartite or not. In fact, when b has only even components, we can reduce the b-matching problem in a non-bipartite graph to one on a bipartite graph, or equivalently, to a general network flow problem. Consider the perfect b-matching problem on a graph G where all the components of b are even. Construct a directed graph $D := (V(D), A(D))$ as follows: For each node v in G there are two nodes, v^- and v^+, in $V(D)$, and for each edge uv in G there are two directed edges, one from u^- to v^+ and one from v^- to u^+, in D. Now, solving the perfect b-matching problem in G is equivalent to solving a general network flow problem in D, subject to the following constraints (note that, $\delta^-(v^+) = \delta^+(v^-) = \emptyset$ for each $v \in V$):

$$\begin{array}{llll} x(\delta^+(v^+)) - x(\delta^-(v^+)) & = & \frac{1}{2} b_v & (v \in V) \\ x(\delta^+(v^-)) - x(\delta^-(v^-)) & = & -\frac{1}{2} b_v & (v \in V) \\ x_a & \geq & 0 & (a \in A(D)). \end{array} \tag{80}$$

Since the right-hand-side in the general network flow problem is integral, it admits an integral optimum solution.

The 2-factor polytope: A 2-factor, or simple perfect 2-matching, in $G = (V, E)$ is a collection of node-disjoint circuits covering V. Note the difference with perfect 2-matchings, in which we may use edges twice. The perfect 2-matching problem is a special case of the perfect b-matching problem with b even, the 2-factor problem is not. Theorem 23 implies that the convex hull of 2-factors is described by:

$$\begin{array}{lll} x(\delta(v)) & = 2 & (v \in V) \\ 0 \leq x_e & \leq 1 & (e \in E) \\ x(\langle U \rangle) + x(F) & \leq |U| + \frac{1}{2}(|F| - 1) & (U \subseteq V,\ F \subseteq \delta(U),\ |F| \text{ odd}). \end{array} \tag{81}$$

Again, we may replace the blossom constraints with the odd cut constraints:

$$x(\delta(U) \setminus F) - x(F) \leq 1 - |F| \; (U \subseteq V, F \subseteq \delta(U), \; |F| \text{ odd}). \tag{82}$$

Using (81), and (82), one can derive a characterization of those graphs with simple perfect 2-matchings [see Belck, 1950; Tutte, 1952].

The edge-cover polyhedron: The convex hull of edge covers is described by:

$$\begin{array}{rlcll} & x(\delta(v)) & \geq & 1 & (v \in V) \\ 0 \leq & x_e & \leq & 1 & (e \in E) \\ & x(\delta(U) \cup \langle U \rangle) & \geq & \frac{1}{2}(|U|+1) & (U \subseteq V, \; |U| \text{ odd}). \end{array} \tag{83}$$

7.3. Algorithms for general matchings

In this section we consider the polynomiality of the algorithms for general matching problems.

The reduction of a general matching problem to a perfect b-matching problem requires $O(|V| + |E|)$ time and results in a graph with $O(|V| + |E|)$ nodes and edges. So, given a polynomial time algorithm for the perfect b-matching problem, we may solve the general matching problem in polynomial time. The reduction from the perfect b-matching problem to the perfect matching problem, on the other hand, requires $O(b(V))$ steps and results in a perfect matching problem in a graph with $O(b(V))$ nodes. Hence we get:

There exists an algorithm for the b-matching problem with running time bounded by a polynomial in $|E(G)|$ and $b(V)$ (84)

[Edmonds, 1965b, see also Edmonds, Johnson & Lockhart, 1969, and Pulleyblank, 1973]. This time bound, however, is not very good. It grows polynomially with the values of b_v's and hence exponentially with the space required to encode them. (Recall that the integer b_v can be encoded in $\log(|b_v| + 1) + 1$ binary digits.) So, the given time bound is exponential in the size of the input of the problem. (An algorithm like this, whose running time is polynomial in the values of the numbers involved in the input, is called *pseudo-polynomial.*) If we restrict attention to those instances of the general matching problem in which the degree bounds and capacities are bounded by some fixed constant (or by a polynomial in $|V(G)|$), (84) yields polynomial algorithms:

The (simple) (perfect) 2-matching problem and the edge cover problem can be solved in time bounded by a polynomial in $|V(G)|$. (85)

A different approach is required to solve the general b-matching problem in polynomial time.

7.3.1. A strongly polynomial algorithm for perfect b-matching

We describe an algorithm, due to Edmonds, that solves a b-matching problem by first solving a single general network flow problem and then a single perfect matching problem. It is based on the following 'sensitivity' result.

Theorem 24. *Let $G = (V, E)$ be an undirected graph and $b, b' \in \mathbb{Z}_+^V$. If x' is a minimum weight perfect b'-matching with respect to a given weight function $w \in \mathbb{Z}_+^E$, then there exists a minimum weight perfect b-matching x (with respect to w) such that*

$$|x_e - x'_e| \leq \sum_{v \in V} |b_v - b'_v| \quad \textit{for each } e \in E.$$

Proof. Let $d := b - b'$. Clearly, it suffices to prove the theorem for the special case that $\sum_{v \in V} |d_v| = 2$. In fact, we will additionally assume that $b \in \{0, 1\}^V$. (In applying this theorem, we only need that case anyway. Moreover, the other cases are proved similarly.) So there exist $u_1, u_2 \in V$ such that $d_{u_1} = d_{u_2} = 1$ and $d_u = 0$ if $u \notin \{u_1, u_2\}$.

Let x' be a minimum weight perfect b'-matching, and x'' be a minimum weight perfect b-matching. Let B be the collection of all $y \in \mathbb{Z}^E$ such that:

$$\begin{array}{rcccll} & & y(\delta(v)) & = & d_v & (v \in V) \\ 0 & \leq & y_e & \leq & x''_e - x'_e & (e \in E \text{ and } x''_e \geq x'_e) \\ x''_e - x'_e & \leq & y_e & \leq & 0 & (e \in E \text{ and } x''_e \leq x'_e). \end{array} \tag{86}$$

Note that if $y \in B$, $x'' - y$ is a perfect b'-matching and $x' + y$ is a perfect b-matching. Hence $w^\top(x'' - y) \geq w^\top x'$, and thus $w^\top(x' + y) \leq w^\top x''$. Which implies that for each $y \in B$, $x' + y$ is a minimum weight perfect b-matching. So it suffices to prove that B contains a vector y with $|y_e| \leq 2$ for each $e \in E$.

Take a sequence $v_0, e_1, v_1, e_2, v_2, \ldots, e_k, v_k$ of edges and nodes such that the following conditions are satisfied: $v_0 = u_1$ and $v_k = u_2$; $e_i = v_{i-1}v_i$ for $i = 1, \ldots, k$; if i is odd then $x''_{e_i} > x'_{e_i}$; if i is even then $x''_{e_i} < x'_{e_i}$; and, for each edge e at most $|x''_e - x'_e|$ edges e_i are equal to e. It is not difficult to see that, since $x'' - x' \in B$, such a sequence exists. Assume that the sequence is as short as possible. This implies that we do not use an edge more than twice in the sequence. Let $y \in \mathbb{Z}^E$ be defined by $y_e := \sum_{i=1, e_i = e}^{k} (-1)^{i+1}$. Then $y \in B$ and $|y_e| \leq 2$, so the theorem follows. □

We can apply this theorem in solving perfect b-matching problems as follows: Let x' be a minimum weight perfect b'-matching, where $b'_v := 2\lfloor \frac{1}{2} b_v \rfloor$ for each $v \in V$. Next define $d := b - b'$ ($\in \{0, 1\}^V$) and search for a minimum weight general matching $\overline{x}$ subject to the constraints:

$$\begin{array}{rcll} x(\delta(v)) & = & d_v & (v \in V) \\ x_e & \geq & \max\{-|V|, -x'_e\} & (e \in E). \end{array} \tag{87}$$

Then, by Theorem 24, $x' + \overline{x}$ is a minimum weight perfect b-matching.

By the remarks following (73) and the reductions in Section 7.1 we can transform the general matching problem subject to (87) into perfect matching problem on a graph whose size is a polynomial in the size of G. As b' has only even components the perfect b'-matching problem is a general network flow problem. So, we have:

A b-matching problem in a graph G can be solved by solving **one** *polynomially sized general network flow problem and* **one** *polynomially sized perfect matching problem (Edmonds).* (88)

The general network flow problem with constraints (80) is essentially equivalent to the min-cost flow (or circulation) problem. The first polynomial algorithm for the min-cost circulation problem was developed by Edmonds & Karp [1970, 1972] and has running time polynomial in $\sum_{v \in V(D)} \log(|b_v| + 1)$. This algorithm combines the pseudo-polynomial 'out-of-kilter' method [Yakovleva, 1959; Minty, 1960; and Fulkerson, 1961] with a scaling technique. Cunningham and Marsh [see Marsh, 1979] and Gabow [1983] found algorithms for b-matching that are polynomial in $\sum_{v \in V(D)} \log(|b_v| + 1)$, also using a scaling technique. The disadvantage of these algorithms is that the number of arithmetic steps grows with $\sum_{v \in V(D)} \log(|b_v| + 1)$. So, larger numbers in the input not only involve more work for each arithmetic operation, but also require more arithmetic operations. This raised the question of whether there is an algorithm such that the number of arithmetic operations it requires is bounded by a polynomial in $|V(D)|$ and the size of the numbers calculated during its execution is bounded by a polynomial in $\sum_{v \in V(D)} \log(|b_v| + 1)$ (this guarantees that no single arithmetic operation requires exponential time). For a long time this issue remained unsettled, until Tardos [1985] showed that, indeed, there exists such a, *strongly polynomial*, algorithm for the min-cost circulation problem [see also Goldberg & Tarjan, 1989]. Combining this with (88) we get:

Theorem 25. *There exists a strongly polynomial algorithm for the general matching problem.*

For a similar strongly polynomial algorithm for b-matching, see Anstee [1987].

7.4. Parity constraints

7.4.1. The Chinese postman problem [Kwan Mei-Ko, 1962; Edmonds, 1965a]

Given a connected graph $G = (V, E)$ and a length function $\ell \in \mathbb{Z}_+^E$: find a closed walk $e_1, \ldots, e_k$ in the graph using each edge at least once – we call this a *Chinese postman tour* – such that its length $\ell(e_1) + \cdots + \ell(e_k)$ is as small as possible.

If G is *Eulerian*, i.e., the degree of each node is even, then there exists an *Eulerian walk*, that is a closed walk using each edge exactly once. This is Euler's [1736] famous resolution of the Königsberger bridge problem. So, for Eulerian graphs the Chinese postman problem is trivial (actually finding the Eulerian walk takes $O(|E|)$ time). On the other hand, if G has nodes of odd degree, every Chinese postman tour must use some edges more than once. We call a vector $x \in \mathbb{Z}^E$ such that $x_e \geq 1$ for each edge e and $\sum_{e \in \delta(v)} x_e$ is even for each node v, *Eulerian*. By Euler's theorem it is clear that for each Eulerian vector x there is a Chinese postman tour that uses each edge e exactly x_e times. Thus, solving the Chinese postman problem amounts to finding an Eulerian vector x of minimum length $\ell^\top x$. Clearly, a minimum length Eulerian vector can be assumed to be $\{1, 2\}$-valued. Hence, searching for a shortest Eulerian vector x amounts to searching for a set $F := \{e \in E \mid x_e = 2\}$ with $\ell(F)$ minimum such that duplicating the edges of F leads to an Eulerian graph. Duplicating the edges in F leads to an Eulerian graph exactly when each node v is incident to an odd number of edges in F if and only if the degree of v in G is odd. So, the Chinese postman problem is a 'T-join problem' discussed below.

There are other versions of the Chinese postman problem. In a directed graph, the problem is a general network flow problem. Other versions, including: the rural postman problem in which we need only visit a subset of the edges; the mixed Chinese postman problem in which some edges are directed and others are not; and the windy postman problem in which the cost of traversing an edge depends on the direction, are NP-hard.

7.4.2. The T-join problem

Given a graph $G = (V, E)$ and an even subset T of the node set V, a subset F of edges such that $|\delta_F(v)|$ is odd for each node v in T and even for each v not in T is called a *T-join*. The *T-join problem* is: Given a length function $\ell \in \mathbb{Z}^E$ find a T-join F of minimum length $\ell(F)$.

The T-join problem is the special case of the general matching problem with no upper bound constraints on the edges and no degree constraints other than the parity constraints.

7.4.3. Algorithms for T-joins

We describe two algorithms for finding a shortest T-join with respect to a length function $\ell \in \mathbb{Z}_+^{E(G)}$. The two algorithms rely on matchings in different ways.

The first algorithm is due to Edmonds & Johnson [1973]. Let H be the complete graph with $V(H) = T$. Define the weight function $w \in \mathbb{Z}_+^{E(H)}$ as follows. For each edge $uv \in E(H)$, w_{uv} is the length, with respect to ℓ, of a shortest uv-path P_{uv} in G. Find a minimum weight perfect matching, $u_1u_2, u_3u_4, \ldots, u_{k-1}u_k$ say, in H. The symmetric difference of the edge sets of the shortest paths $P_{u_1u_2}, P_{u_3u_4}, \ldots, P_{u_{k-1}u_k}$ is a shortest T-join.

Since the shortest paths and a minimum weight perfect matching can be found in polynomial time, the algorithm runs in polynomial time. In fact, we can find shortest paths in polynomial time when some of the edges have negative length, as long as G has no negative length circuit (see Section 9.2). But, when we allow negative length circuits, the shortest path problems become NP-hard. Nevertheless, the T-join problem with a general length function can be solved in polynomial time (which implies that we also can find a T-join of maximum length).

In the second algorithm we construct a graph H as follows. For each node u in G and each edge e incident to u we have a node u_e. For each node u in T with even degree or not in T with odd degree, we have the node $\widetilde{u}$ and the edges $\widetilde{u}u_e$ for each edge e incident to u. For each node u in G and each pair e, f of edges in $\delta(u)$, we have an edge $u_e u_f$. Finally, for each edge $e = uv$ in G we have an edge $u_e v_e$ in H; we call these the *G-edges* of H. Each collection of G-edges is a matching in H and it corresponds to a T-join in G if and only if it is contained in a perfect matching of H. So, if we give each G-edge $u_e v_e$ weight ℓ_e and all other the edges in H weight 0, we have transformed the minimum length T-join problem in G into a minimum weight perfect matching problem in H.

Clearly, this algorithm allows edges with negative weights. Another way to solve a T-join problem with negative weights is by the following transformation to a T'-join problem with all weights non-negative. Let $N := \{e \in E \mid w_e < 0\}$ and $T_N := \{v \in V \mid \deg_N(v) \text{ is odd}\}$. Moreover, define $w^+ \in \mathbb{R}^E_+$ be defined by $w^+_e := |w_e|$ for each $e \in E$ and $T' := T \triangle T_N$. Then $\min\{w(F) \mid F \text{ is a } T\text{-join}\} = w(N) + \min\{w^+(F) \mid F \text{ is a } T'\text{-join}\}$. F is a minimum weight T-join with respect to w if and only if then $F \triangle N$ is a minimum weight T'-join with respect to w^+.

Edmonds & Johnson [1973] derived a direct algorithm for the T-join problem, which, like Edmonds' weighted matching algorithm, maintains a dual feasible solution and terminates when there is a feasible primal solution that satisfies the complementary slackness conditions. Barahona [1980] and Barahona, Maynard, Rammal, & Uhry [1982] derived a 'primal' algorithm using dual feasibility as a stopping criterion (similar to the primal matching algorithm of Cunningham and Marsh (see Section 8.1)). Like the matching algorithm, these algorithms can be implemented to run in $O(|V|^3)$ and $O(|E||V|\log|V|)$ time, respectively. In planar graphs the T-join problem can be solved in $O(|V|^{3/2}\log|V|)$ time [Matsumoto, Nishizeki & Saito, 1986; Gabow, 1985; Barahona, 1990].

7.4.4. Min–max relations for T-joins – the T-join polyhedron

For each $U \subseteq V(G)$ with $U \cap T$ odd, we call $\delta(U)$ a *T-cut*. Clearly, the maximum number $\nu(G,T)$ of pairwise edge-disjoint T-cuts cannot exceed the smallest number $\tau(G,T)$ of edges in a T-join. Equality need not hold. For example, $\nu(K_4, V(K_4)) = 1 < 2 = \tau(K_4, V(K_4))$. Seymour proved [Seymour, 1981]:

> *In a bipartite graph G, $\nu(G,T) = \tau(G,T)$ for each even subset T of nodes.* (89)

Frank, Sebö & Tardos [1984] and Sebö [1987] derived short proofs of this result. In a bipartite graph, a maximum collection of pairwise edge-disjoint T-cuts can be found in polynomial time. (Korach [1982] gives an $O(|E||V|^4)$ procedure and Barahona [1990] showed that the above mentioned $O(|V|^3)$ and $O(|E||V|\log|V|)$

T-join algorithms can be modified to produce a maximum collection of disjoint T-cuts when the graph is bipartite.)

When the length function ℓ is non-negative and integral, we have the following min–max relation for shortest T-joins in arbitrary graphs [Lovász, 1975]:

The minimum length of a T-join with respect to a length function $\ell \in \mathbb{Z}_+^{E(G)}$ is equal to half the maximum number of T-cuts such that each edge e is in at most $2\ell(e)$ of them. (90)

This can be proved from (89) or from the algorithm of Edmonds & Johnson [1993]. Let H be the bipartite graph obtained from G by replacing each edge e by a path of length $2\ell(e)$. If $\ell(e)$ is 0, contract e. A minimum length T-join in G corresponds to a minimum cardinality T-join in H. Applying (89) to H yields (90).

As a consequence, we obtain a linear inequality description of the *T-join polyhedron*, i.e., the set of vectors $x \in \mathbb{R}^E$ such that there exists a convex combination y of characteristic vectors of T-joins with $x \geq y$.

Corollary 26 [Edmonds & Johnson, 1973]. *Let T be an even subset of the node set of a graph $G = (V, E)$. Then the T-join polyhedron is the solution set of:*

$$\begin{array}{lcll} x(\delta(U)) & \geq & 1 & (U \subseteq V;\ |U \cap T| \text{ is odd}) \\ x_e & \geq & 0 & (e \in E). \end{array} \tag{91}$$

Note that this result immediately yields Corollary 20. Conversely, Corollary 26 follows from Corollary 20 via the reduction to perfect matchings used in Schrijver's T-join algorithm. Alternatively, we can prove Corollary 26 in a manner analogous to our proof of Theorem 19. For a generalization of Corollary 26, see Burlet & Karzanov [1993].

The system (91) is not totally dual integral. The complete graph on four nodes, K_4, with $T = V(K_4)$ again provides a counterexample. In a sense, this is the only counterexample. One consequence of Seymour's characterization of 'binary clutters with the max-flow min-cut property' [Seymour, 1977] is:

If G is connected and T is even, then (91) *is totally dual integral if and only if $V(G)$ cannot be partitioned into four sets $V_1, \ldots, V_4$ such that $V_i \cap T$ is odd and $G|V_i$ is connected for each $i = 1, \ldots, 4$ and for each pair V_i and V_j among $V_1, \ldots, V_4$, there is an edge uv with $u \in V_i$ and $v \in V_j$.* (92)

An immediate consequence of (92) is that, like bipartite graphs, series parallel graphs are *Seymour graphs*, meaning that $\nu(G, T) = \tau(G, T)$ for each even subset T of nodes. Other classes of Seymour graphs have been derived by Gerards [1992] and Szigeti [1993]. It is unknown whether recognizing Seymour graphs is in NP. Just recently, Ageev, Kostochka & Szigeti [1994] showed that this problem is in co-NP by proving a conjecture of Sebö.

Sebö [1988] derived a (minimal) totally dual integral system for the T-join polyhedron of a general graph. (For a short proof of this result and of (92) see

Frank & Szigeti [1994].) Sebö [1986, 1990] also developed a structure theory for T-joins analogous to the Edmonds–Gallai structure for matchings. The core of this structure theory concerns structural properties of shortest paths in undirected graphs with respect to length functions that may include negative length edges, but admit no negative length circuits. Frank [1993] derived a good characterization for finding a node set T in G that maximizes $\tau(G, T)$.

8. Other matching algorithms

In this section we discuss other algorithms for both cardinality and weighted matchings.

8.1. A primal algorithm

Edmonds' algorithm for finding a minimum weight perfect matching maintains a (structured) dual feasible solution and a non-perfect, and so infeasible, matching that together satisfy the complementary slackness conditions. At each iteration it revises the dual solution so that the matching can be augmented. When the matching becomes perfect it is optimal. An alternative approach is to maintain a perfect matching and a (structured) dual solution that satisfy the complementary slackness conditions. At each iteration, revise the matching so that the dual solution approaches feasibility. Cunningham & Marsh [1978] developed such a 'primal' algorithm. In outlining their algorithm we return to the notation of Section 6.2.

Let G be an undirected graph and suppose $w \in \mathbb{R}_+^{E(G)}$. Moreover, let $\pi \in \mathbb{R}^{\Omega(G)}$ be a structured dual solution, i.e., π satisfies (65) and (66). We also assume that $\pi_S \geq 0$ for each $S \in \Omega(G)$ with $|S| \neq 1$ and that M is a perfect matching in $\widetilde{G_\pi}$. If all the edges have non-negative reduced cost $w_e^\pi = w_e - \sum_{S\in\Omega(G);\delta(S)\ni e} \pi_S$, then π is dual feasible and, since M can be extended to a minimum weight perfect matching in G, π is optimal. Otherwise, we 'repair' π and M as follows:

REPAIR: Let $\widetilde{uv} = e \in E(G)$ with $w_e^\pi < 0$ and suppose there exists an alternating path P in $\widetilde{G_\pi}$ from OUTER$_\pi[u]$ to OUTER$_\pi[v]$ starting and ending with a matching edge. We call such a path a *repairing path*. Carry out the following repairs ($R :=$ DEEP$_\pi$[OUTER$_\pi[u]$]):

EXPANDING R: If $\pi_R \leq -w_e^\pi$ and $|R| \neq 1$, revise the dual solution by changing π_R to 0. This means that we must expand R in $\widetilde{G_\pi}$ and extend M accordingly. Moreover, since π satisfies (66), we can extend P to an alternating path from the new node OUTER$_\pi[u]$ to OUTER$_\pi[v]$, again starting and ending with a matching edge.

We repeat EXPANDING R until $\pi_R > -w_e^\pi$ or $|R| = 1$. Note that each EXPANSION of R causes a matching edge, namely the starting edge of P, to receive positive reduced cost. Once we have finished EXPANDING, we call REPAIRING e to find a new perfect matching and a revised dual solution that satisfy the complementary slackness conditions.

REPAIRING e: If $|R| = 1$, or $\pi_R > -w_e^\pi$, replace M by $M\Delta(P \cup \{e\})$ and change the dual solution by adding w_e^π to π_R.

So, all that remains is the question of how to find a repairing path. Assume u is a node incident to an edge with negative reduce cost and let $r := \text{OUTER}_\pi[u]$. We create an auxiliary graph H by adding a new node u^* to G and an edge from u^* to u. Similarly, we construct $\widetilde{H}_\pi$ by adding the edge u^*r to $\widetilde{G}_\pi$. Consider the Edmonds–Gallai structure of $\widetilde{H}_\pi$. There are two possibilities:

1. There is an edge between u and $\text{DEEP}_\pi[v]$ with negative reduced cost for some node $v \in D(\widetilde{H}_\pi)$. In this case, let Q be an M-alternating u^*v-path (Q exists because $v \in D(\widetilde{H}_\pi)$ and u^* is the only node in H exposed with respect to M). Clearly $Q \setminus \{u^*r\}$ is a repairing path.

2. If there is no such node v, we change the dual variables according to the definitions in (68) and (69), but with the understanding that in (69) we ignore those edges with negative reduced cost. We also ignore a dual change in u^* (note that u^* is a singleton component of $D(\widetilde{H}_\pi)$). We repeat this operation until 1. applies or until all the edges incident to u receive a non-negative reduced cost.

Needless to say, in implementing the algorithm we do not need to find the Edmonds–Gallai structure explicitly, but instead use GROW and SHRINK. The algorithm can be implemented in $O(|V(G)|^3)$ time.

8.2. Shortest alternating paths and negative alternating circuits

Given a matching M, a weight function $w \in \mathbb{R}^{E(G)}$ and a set of edges F, we define $w_M(F) := w(F \setminus M) - w(F \cap M)$. A *negative* circuit is an even alternating circuit C with $w_M(C) < 0$. A matching M is called *extreme* if it admits no negative circuit. In a manner similar to the proof of Theorem 1, one can prove that a perfect matching is extreme if and only if it is a minimum weight perfect matching. This suggests the following algorithm for finding a minimum weight perfect matching:

NEGATIVE CIRCUIT CANCELLING: Given a perfect matching M, look for a negative circuit. If none exists, M is extreme and hence optimal. If M admits a negative circuit C, replace M with $M\Delta C$ and repeat the procedure.

Given a matching M and an exposed node u, an augmenting path P starting at u is called a *shortest augmenting path from* u if it minimizes $w_M(P)$. It is easy to prove that if M is extreme and P is an augmenting path starting at an exposed node u, then $M\Delta P$ is extreme if and only if P is a shortest augmenting path from u. This also suggests an algorithm:

SHORTEST AUGMENTING PATHS: Given an extreme matching M, (initially $M = \emptyset$), look for a shortest augmenting path. If none exists, M is a minimum weight maximum cardinality matching. If M admits a shortest augmenting path P, replace M by $M\Delta P$ and repeat the procedure.

So the question arises: How to find negative circuits or shortest augmenting paths? The answer is not so obvious. We can hardly check all possible alternating circuits or augmenting paths. In fact, the observations above are weighted analogues of the theorem of Berge and Norman and Rabin (Theorem 1). However, Edmonds' algorithm for minimum weight perfect matching can be viewed as a shortest augmenting path algorithm and Cunningham and Marsh's primal algorithm is a negative circuit cancelling method. Derigs [1981] [see also Derigs, 1988b] developed versions of these algorithms in which shortest augmenting path occur more explicit. Not surprisingly, these algorithms also rely on alternating forests, shrinking and the use of dual variables.

8.3. Matching, separation and linear programming

In Section 5 we formulated the weighted matching problem as a linear programming problem. Can we solve it as a linear program? The main problem is the number of inequalities. There are, in general, an exponential number of blossom constraints (viz. odd cut constraints). A first approach to overcoming this is, in fact, the development of algorithms like Edmonds' algorithm and the primal algorithm by Cunningham and Marsh that can be viewed as special purpose versions of simplex methods in which only the constraints corresponding to non-zero dual variables are explicitly considered. A second approach is to use the ellipsoid method, the first polynomial time algorithm for linear programming [Khachiyan, 1979]. Grötschel, Lovász & Schrijver [1981], Karp & Papadimitriou [1982] and Padberg & Rao [1982] observed that the polynomial time performance of this method is relatively insensitive to the size of the system of linear constraints. The only information the ellipsoid method needs about the constraint system is a polynomial time separation algorithm for the set of feasible solutions. A *separation algorithm* for a polyhedron solves the following problem.

Separation problem: Given a polyhedron $P \subseteq \mathbb{R}^n$ and a vector $\tilde{x} \in \mathbb{R}^n$, decide whether $\tilde{x} \in P$ and, if it is not, give a *violated inequality*, that is an inequality $a^\top x \leq \alpha$ satisfied by each $x \in P$, but such that $a^\top \tilde{x} > \alpha$.

Padberg & Rao [1982] developed a separation algorithm for the perfect matching polytope. It is easy to check whether a given $x \in \mathbb{R}^{E(G)}$ satisfies the non-negativity and degree constraints. So, the separation problem for the perfect matching polyhedron is essentially: Given a non-negative vector $x \in \mathbb{R}^{E(G)}$, find an odd collection S of nodes such that $x(\delta(S)) < 1$ or decide that no such S exists. This problem can be solved by solving the following problem (with $T = V(G)$).

Minimum capacity T-cut problem: Given an even collection T of nodes and $x \in \mathbb{R}_+^{E(G)}$, find $S \subsetneq V(G)$ with $|S \cap T|$ odd and $x(\delta(S))$ as small as possible.

We call a set $\delta(S)$ with $S \subsetneq V(G)$ a *T-separator* if $S \cap T$ and $S \setminus T$ are not empty. A *minimum T-cut* is a T-cut $\delta(S)$ with $x(\delta(S))$ as small as possible. We define a *minimum T-separator* similarly.

Crucial to the solution of this problem is the following fact.

Let $\delta(W)$ be a minimum T-separator, then there exists a minimum T-cut $\delta(S)$ with $S \subseteq W$ or $S \subseteq V(G) \setminus W$ [Padberg & Rao, 1982]. (93)

To prove this, let $\delta(W)$ be a minimum T-separator and $\delta(Z)$ be a minimum T-cut. If $\delta(W)$ is a T-cut, $Z \subseteq W$ or $Z \subseteq V(G) \setminus W$, we are done. So, suppose none of these is the case. By interchanging W and $V(G) \setminus W$ or Z and $V(G) \setminus Z$ (or both) we may assume that $|Z \cap W \cap T|$ is odd and $V(G) \setminus (W \cup Z)$ contains a node of T. Hence $\delta(W \cap Z)$ is a T-cut and $\delta(W \cup Z)$ is a T-separator. So, $x(\delta(W)) \leq x(\delta(W \cup Z))$. Now, straightforward calculations show that: $x(\delta(Z)) - x(\delta(W \cap Z)) \geq x(\delta(Z)) - x(\delta(W \cap Z)) + x(\delta(W)) - x(\delta(W \cup Z)) = 2\sum_{u \in Z \setminus W} \sum_{u \in W \setminus Z} x_{uv} \geq 0$, which completes the proof of (93).

This suggests the following recursive algorithm. Determine a minimum T-separator $\delta(W)$. If $|W \cap T|$ is odd we are done, $\delta(W)$ is a minimum T-cut. Otherwise, we search for a minimum $(T \setminus W)$-cut in $G \times W$ and a minimum $(T \cap W)$-cut in $G \times (V(G) \setminus W)$. By (93), one of these two yields a minimum T-cut in G. It is easy to see that this recursive method requires at most $|T| - 1$ searches for a minimum T-separator. Each search for a T-separator can be carried out by solving $|T| - 1$ max-flow problems. (Indeed, fix $s \in T$ and use a max-flow algorithm to find a minimum s, t-cut for each $t \in T \setminus \{s\}$.) So the minimum odd cut problem and the separation problem for the perfect matching polytope can be solved in polynomial time by solving a series of $O(|T|^2)$ max-flow problems. Thus, the ellipsoid method provides a new polynomial time algorithm for the minimum weight perfect matching problem. (In fact, the minimum T-cut algorithm can be improved so that only $|T| - 1$ max-flow problems are required, by calculating a 'Gomory-Hu' tree [see Padberg & Rao, 1982].)

This method is not practical because the ellipsoid method performs poorly in practice. On the other hand, the separation algorithm can be used in a *cutting plane approach* for solving matching problems via linear programming. Start by solving a linear program consisting of only the non-negativity and degree constraints. If the optimal solution x^* to this problem is integral, it corresponds to a perfect matching and we are done. Otherwise, use Padberg and Rao's procedure to find an odd cut constraint violated by x^*. Add this to the list of constraints and resolve the linear programming problem. Grötschel & Holland [1985] built a matching code based on this idea. At that time, their code was competitive with existing combinatorial codes (based on Edmonds' or Cunningham and Marsh's algorithm). This contradicted the general belief that the more fully a method exploits problem structure, the faster it should be. This belief has been reconfirmed, at least for the matching problem, by new and faster combinatorial matching codes.

An entirely different approach to solving matching problems with linear programming is to construct a polynomial size system of linear inequalities $Ax + By \leq c$ such that $\{x \in \mathbb{R}^{E(G)} \mid Ax + By \leq c\}$ is the (perfect) matching polytope. We call such a linear system a *compact system* for the (perfect) matching polytope. Although, perfect matching polytopes of planar graphs [Barahona, 1993a] and, in fact, perfect matching polytopes of graphs embeddable on a fixed surface [Gerards, 1991] have compact systems, no compact system is known for the matching problem in general graphs. It should be noted that compact systems for matching polytopes that use no extra variables do not exist, not even for planar graphs [Gamble, 1989]. Yannakakis [1988] proved that there is no compact *symmetric* system for the matching polytope. (Here, 'symmetric' refers to an additional symmetry condition imposed on the systems.)

Barahona [1993b] proposes yet a different approach. Given a matching M, one can find a negative circuit with respect to M by searching for an even alternating circuit C with minimum average weight $w_M(C)/|C|$. When using these special negative circuits, $O(|E|^2 \log |V|)$ negative circuit cancellations suffice for finding a minimum weight perfect matching. An even alternating circuit of minimum average weight can be found by solving a polynomially sized linear programming problem. Hence, we can find a minimum weight perfect matching by solving $O(|E|^2 \log |V|)$ compact linear programming problems.

8.4. An algorithm based on the Edmonds–Gallai structure

Next we present an algorithm, due to Lovász & Plummer [1986], for finding a largest matching in a graph. Like the blossom algorithm, it searches for alternating paths, but in a quite different manner. For instance, it does not shrink blossoms. The algorithm is inspired by the Edmonds–Gallai structure theorem. The algorithm maintains a list $\mathcal{L}$ of matchings all of size k. Given the list $\mathcal{L}$, define: $D(\mathcal{L}) := \cup_{M \in \mathcal{L}} \exp(M)$, $A(\mathcal{L}) := \Gamma(D(\mathcal{L})) \setminus D(\mathcal{L})$, and $C(\mathcal{L}) := V(G) \setminus (D(\mathcal{L}) \cup A(\mathcal{L}))$. So, if k is $\nu(G)$ and $\mathcal{L}$ is the list all maximum matchings, then $D(\mathcal{L})$, $A(\mathcal{L})$, $C(\mathcal{L})$ is the Edmonds–Gallai structure of G. During the algorithm, however, k ranges from 0 to $\nu(G)$ and $\mathcal{L}$ never contains more than $|V(G)|$ matchings. The clue to the algorithm is the following fact, which will serve as the stopping criterion.

> *If $M \in \mathcal{L}$ is such that $M \cap \langle D(\mathcal{L}) \rangle$ has exactly one exposed node in each component of $G|D(\mathcal{L})$ and no node in $A(\mathcal{L})$ is matched to a node in $A(\mathcal{L}) \cup C(\mathcal{L})$, then M is a maximum matching in G.* (94)

Indeed, in this case each component of $G|D(\mathcal{L})$ is odd and each node in $A(\mathcal{L})$ is matched to a different component of $G|D(\mathcal{L})$. Hence, it is easy to see that $|\exp(M)| = c_0(A(\mathcal{L})) - |A(\mathcal{L})|$; proving that M is maximum (cf. (22)).

The following notions facilitate the exposition of the algorithm. For $u \in D(\mathcal{L})$ we define $\mathcal{L}_u := \{M \in \mathcal{L} \mid u \in \exp(M)\}$. For each $M \in \mathcal{L}_u$ and $M' \in \mathcal{L}$, we denote the maximal path in $M \Delta M'$ starting at u by $P(u; M, M')$ (if M' is also in $\mathcal{L}_u$, this path consists of u only). An M-alternating path from a node in $\exp(M)$ to a node in $A(\mathcal{L})$ with an even number of edges is called *M-shifting*. If P is M-shifting for $M \in \mathcal{L}$, then $M \Delta P$ is a matching of size k with an exposed node $v \notin D(\mathcal{L})$. So adding $M \Delta P$ to $\mathcal{L}$ adds the node v to $D(\mathcal{L})$. If Q is a path and u and v are nodes on Q then Q_{uv} denotes the uv-path contained in Q.

The algorithm works as follows:

Initially $\mathcal{L} := \{\emptyset\}$. Choose a matching M from $\mathcal{L}$. If it satisfies the conditions in (94) we are done: M is a maximum matching. Otherwise, apply the steps below to M to find an M'-augmenting or an M'-shifting path P with respect to some matching M'. If we find an M'-augmenting path P, we AUGMENT by setting $\mathcal{L}$ equal to $\{M' \Delta P\}$. If we find an M'-shifting path P, we SHIFT by adding $M' \Delta P$ to $\mathcal{L}$. The algorithm continues until we find a matching M satisfying (94).

Step 1: *If there is an edge $uv \in M$, with $u \in A(\mathcal{L})$ and $v \notin D(\mathcal{L})$,* choose $w \in \Gamma(u) \cap D(\mathcal{L})$ and $M_w \in \mathcal{L}_w$. If $P(w; M, M_w)$ has an odd number of edges it is M_w-augmenting and we AUGMENT. If $P(w; M, M_w)$ has an even number of edges and $uv \notin P(w; M, M_w)$ then $P(w; M, M_w) \cup \{wu, uv\}$ is M-shifting. Otherwise, either $P_{wu}(w; M, M_w)$ or $P_{wv}(w; M, M_w)$ does not contain uv and so is M_w-shifting. Select the appropriate path and SHIFT.

Step 2: *If there is a component S of $G|D(\mathcal{L})$ such that $M \cap \langle S \rangle$ is a perfect matching in $G|S$,* choose $w \in S$ and $M_w \in \mathcal{L}$. Since S is even, Step 3 below applies to M_w. Replace M by M_w and go to Step 3.

Step 3: *If there is a path Q in $G|D(\mathcal{L})$, such that $M \cap \langle D(\mathcal{L}) \rangle$ leaves the endpoints u and v of Q exposed,* then, if $uv \in E(G)$, go to Step 4. Otherwise, choose a node w in Q, different from u and v. If $w \in \exp(M)$, apply Step 3 with w in place of v and Q_{uw} in place of Q. If $w \notin \exp(M)$, choose $M_w \in \mathcal{L}_w$. If $P(w; M, M_w)$ is odd, it is M_w-augmenting, AUGMENT. If $P(w; M, M_w)$ is even, then $M' := M \Delta P(w; M, M_w)$ is a matching of size k that leaves w and (at least) one of u and v exposed. Assume $u \in \exp(M')$. Add M' to $\mathcal{L}$ and apply Step 3 with M' in place of M, w in place of v and Q_{uw} in place of Q.

(Note that each time we repeat Step 3, the path Q gets shorter.)

Step 4: *If there is an edge $uv \in G|D(\mathcal{L})$ such that $M \cap \langle D(\mathcal{L}) \rangle$ leaves u and v exposed*, consider the following two cases:

Step 4′: *If $u, v \notin \exp(M)$*, let $M_u \in \mathcal{L}_u$. If $P(u; M, M_u)$ is odd, it is M_u-augmenting. Otherwise, define $M' := M \Delta P(u; M, M_u)$. M' has size k and has $u \in \exp(M')$ and $v \in \exp(M' \cap \langle D(\mathcal{L}) \rangle)$. Add M' to $\mathcal{L}$ and go to Step 4" with M' in place of M.

Step 4″: *If $u \in \exp(M)$ or $v \in \exp(M)$*, we may assume that $u \in \exp(M)$. If $v \in \exp(M)$ too, then uv is M-augmenting. If $v \notin \exp(M)$ and $vw \in M$ then $\{uv, vw\}$ is M-shifting.

The correctness of the algorithm follows from its description. It runs in $O(|V(G)|^4)$ time.

8.5. Parallel and randomized algorithms – matrix methods

The invention of parallel computers raised the question which problems can be solved substantially quicker on a parallel machine than on a sequential one. For problems that are polynomially solvable on a sequential machine a measure could be that the parallel running time is 'better than polynomial'. To make this explicit, Pippenger [1979] invented the class NC of problems solvable by an NC-algorithm.

A parallel algorithm is called NC-*algorithm* if its running time is a polynomial in the logarithm of the input size and requires only a polynomial number of processors. For more precise definitions see Karp & Ramachandran [1990].

Many problems have been shown to be in NC [see Karp & Ramachandran, 1990; Bertsekas, Castañon, Eckstein & Zenion, 1995, this volume]. But for matching the issue is still open. Partial answers have been obtained: Goldberg, Plotkin, & Vaidya [1993] proved that bipartite matching can be solved in sub-linear time using a polynomial number of processors [see also Vaidya, 1990; Goldberg, Plotkin, Shmoys & Tardos, 1992; Grover, 1992]. NC-algorithms for matching problems for special classes of graphs (or weights) have been derived by Kozen, Vazirani & Vazirani [1985], Dahlhaus & Karpinski [1988], Grigoriev & Karpinski [1987], He [1991], and Miller & Naor [1989]. But, whether or not: Has G a perfect matching? is in NC remains open.

On the other hand, if we allow algorithms to take some random steps, and also allow some uncertainty in the output, we can say more: there exist randomized NC-algorithms for matching. They rely on matrix methods (for a survey, see Galil [1986b]).

In Section 3 (see (16)) we already saw a relation between matchings in bipartite graphs and matrices. Tutte [1947] extended this to non-bipartite graphs.

> Let $G = (V(G), E(G))$ be an undirected graph, and let $\vec{G} = (V(G), A(\vec{G}))$ be a directed graph obtained by orienting the edges in G. For each edge e in G we have a variable x_e. Then the *Tutte matrix* of G (with respect to $\vec{G}$) is the $V(G) \times V(G)$ matrix $\vec{G}(x)$ defined by:

$$\vec{G}(x)_{uv} := \begin{cases} x_{uv} & \text{if } \overrightarrow{uv} \in A(\vec{G}) \\ -x_{uv} & \text{if } \overrightarrow{vu} \in A(\vec{G}) \\ 0 & \text{if } uv \notin E(G) \end{cases} . \tag{95}$$

Note that the Tutte matrix essentially just depends on G: reversing the orientation of and edge e in G just amounts to substituting $-x_e$ for x_e in $\vec{G}(x)$.

> *G has a perfect matching if and only if the determinant of $\vec{G}(x)$ is a non-vanishing polynomial in the variables x_e ($e \in E(G)$)* [Tutte, 1947]. (96)

To see this, let $\mathcal{F} \subseteq \{0, 1, 2\}^{E(G)}$ denote the collection of perfect 2-matchings. Then $\det(\boldsymbol{G}(x)) = \sum_{f \in \mathcal{F}} a_f \prod_{e \in E(G)} x_e^{f_e}$. Moreover, it is not hard to show that $a_f = 0$ if and only if the 2-matching f contains an odd circuit. On the other hand, perfect 2-matchings without odd circuits contain a perfect matching.

By itself (95) is not that useful for deciding whether or not G has a perfect matching. Determinants can be calculated in polynomial time if the matrix contains specific numbers as entries, but evaluating a determinant of a matrix with variable entries takes exponential time (in fact the resulting polynomial may have an exponential number of terms). However, by the following lemma, we can still use the Tutte matrix computationally.

Lemma 27 [Schwartz, 1980]. *Let $p(x_1, \ldots, x_m)$ be a non-vanishing polynomial of degree d. If $\hat{x}_1, \ldots, \hat{x}_m$ are chosen independently and uniformly at random from $\{1, \ldots, n\}$ then the probability that $p(\hat{x}_1, \ldots, \hat{x}_m) = 0$ is at most d/n.*

Proof. Write $p(x_1, \ldots, x_m)$ as $\sum_{\ell=0}^{d_m} p_{d-\ell}(x_1, \ldots, x_{m-1})x_m^\ell$, where each p_k is a polynomial in $x_1, \ldots, x_{m-1}$ of degree at most k.

By induction to the number of variables, the probability that $p_{d-d_m}(\hat{x}_2, \ldots, \hat{x}_m) = 0$ is at most $(d - d_m)/n$. On the other hand, if $p_{d-d_m}(\hat{x}_2, \ldots, \hat{x}_m) \neq 0$ then $p(\hat{x}_1 \ldots, \hat{x}_{m-1}, x_m)$ is a non-vanishing polynomial in x_m of degree d_m, so has at most d_m roots. In other words, if $p_{d-d_m}(\hat{x}_1, \ldots, \hat{x}_{m-1}) \neq 0$, the probability that $p(\hat{x}_1, \ldots, \hat{x}_m) = 0$ is at most d_m/n. Hence the probability that $p(\hat{x}_1, \ldots, \hat{x}_m) = 0$ is at most $(d - d_m)/n + d_m/n = d/n$. □

If we apply Lemma 27 to $p(x) = \det \vec{G}(x)$, which has degree $|V(G)|$ if it is non-vanishing, and take $n = 2|V(G)|$, we get a randomized polynomial time algorithm with the property that if G has a perfect matching the algorithm discovers this with probability at least $\frac{1}{2}$ [Lovász, 1979b]. Although this randomized algorithm is slower that the fastest deterministic ones, it has the advantage that it can be parallelized. The reason is that calculating a determinant is in NC [Csansky, 1976]. So we have:

There exists a randomized NC-*algorithm that gives output '$\nu(G) = |V(G)|$' with probability at least $\frac{1}{2}$ if the input graph G has a perfect matching* [Lovász, 1979b; Csanski, 1976]. (97)

(Note that by running this algorithm several times, we can improve the probability of success as much as we want.)

More generally, we have a randomized NC-algorithm for deciding whether $\nu(G) \geq k$ (just add $|V(G)| - 2k$ mutually non-adjacent nodes to G, each of them adjacent to all nodes of G and than decide whether the new graph has a perfect matching). If we, combine this with a binary search on k, we get NC-algorithm that gives a number $\ell \leq \nu(G)$, that is equal to $\nu(G)$ with high probability.

These randomized algorithms have one big disadvantage: they are 'Monte Carlo' type algorithms. If G has no perfect matching the Lovász-Csanski algorithm does not discover this. The algorithm presented for $\nu(G)$ always gives an output $\ell \leq \nu(G)$, but never tells us that $\ell = \nu(G)$ (unless by change $\ell = \frac{1}{2}|V(G)|$). Karloff [1986] resolved this problem by deriving a randomized NC-algorithm that determines a set $B \subseteq V(G)$ such that with high probability $c_0(G \setminus B) - |B| = \text{def}(G)$ (cf. Theorem 10). Combining this with the previously described Monte Carlo algorithm for $\nu(G)$ we get a randomized NC-algorithm that provides an upper and a lower bound for $\nu(G)$, which are equal with high probability.

Knowing $\nu(G)$ does not provide us with a maximum matching. Of course, we can delete edges one by one from G, making G smaller and smaller, and keep track what happens with the maximum size of a matching. If we store the edges whose deletion decreased the maximum size of a matching of the current graph, we get a maximum matching of our original graph. Combining this with a randomized algorithm for the size of a maximum matching we get a randomized algorithm for actually finding a maximum matching. However, this algorithm is highly sequential. Moreover, it is not obvious at all how to parallelize it, how to make sure that the different processors are searching for the same matching. (See Rabin & Vazirani [1989] for another sequential randomized algorithm for finding a maximum matching.) The first randomized NC-algorithm that finds a perfect matching with high probability if it exists is due to Karp, Upfal & Widgerson [1986]. It runs in $O(\log^3(|V(G)|))$ time. Below we sketch a randomized NC-algorithm due to Mulmuley, Vazirani, & Vazirani [1987] that runs in $O(\log^2(|V(G)|))$ time.

The main trick of this algorithm, besides using the Tutte matrix, is that it first implicitly selects a canonical perfect matching; which than is explicitly found. Let $\mathcal{U}(G)$ denote the set of all $w \in \mathbb{Z}_+^{E(G)}$ such that the minimum perfect matching, denoted by M_w, is unique. Mulmuley, Vazirani, and Vazirani proved the following fact: (If $e = uv \in E(G)$, then $\hat{x}_e^w := 2^{w_e}$ and $\vec{G}_e(x)$ denotes the submatrix of $\vec{G}(x)$ obtained by removing the row indexed by u and the column indexed by v.)

If $w \in \mathcal{U}(G)$, then
(1) $2^{-2w(M_w)} \det \vec{G}(\hat{x}^w)$ *is an odd integer;*
(2) $uv \in M_w \iff 2^{2(w_e - w(M_w))} \det \vec{G}_e(\hat{x}^w)$ *is an odd integer.* (98)

So as soon as we have found a $w \in \mathcal{U}(G)$, we can find a perfect matching by calculating the determinants in (98), which can be done in parallel by Csanski's NC-algorithm. The following lemma yields a randomized algorithm for selecting a weight function in $\mathcal{U}(G)$.

Lemma 28 [Mulmuley, Vazirani, & Vazirani, 1987]. *Let $S = \{x_1, \ldots, x_n\}$ be a finite set and $\mathcal{F}$ a collection of subsets of S. Assume that $w_1, \ldots, w_n$ are chosen uniformly and independently at random from $\{1, \ldots, 2n\}$. Then the probability that there is a unique $F \in \mathcal{F}$ minimizing $w(F)$ is at least $\frac{1}{2}$.*

Proof. The probability p that the minimum weight set is not unique is at most n times the probability p_1 that there exists a minimum weight set in $\mathcal{F}$ containing x_1 and a minimum weight set in $\mathcal{F}$ not containing x_1. For each fixed $w_2, \ldots, w_n$ this probability is either 0 or $1/2n$. Hence p_1 is at most $1/2n$. So $p \leq np_1 \leq \frac{1}{2}$. □

Hence, there exists a randomized NC-algorithm for finding a perfect matching and thus also for finding a maximum matching. It requires $O(|E| \log |V|)$ random bits. Chari, Rohatgi & Srinivasan [1993] found a very nice generalization of Lemma 28 that enables the design of randomized NC-algorithms that require only $O(|V| \log(|E|/|V|))$ random bits.

8.5.1. Counting perfect matchings

So randomization can help us where determinism does not (seem to) work. The same feature comes up when considering another computational task related to matchings: Count the number of perfect matchings in G. Over the years this problem has received a lot of attention, leading to many beautiful results. For many of these, and many references, see Lovász & Plummer [1986] and Minc [1978]. As the topic lies beyond the scope of this chapter, we will only mention a few results relevant from a computational point of view.

Valiant [1979] proved that counting the perfect matchings in a graph is as hard as solving any problem in NP, even for bipartite graphs (it is '#P-complete'). So, assuming P $\neq$ NP, there exists no polynomial time algorithm for calculating the number $\phi(G)$ of perfect matchings in G.

Kasteleyn [1963, 1967], however, derived a polynomial algorithm for counting perfect matchings in a planar graph. The main idea behind this algorithm is as follows (for details see Lovász & Plummer [1986]). If $\vec{G}$ is an orientation of G we denote by $p(\vec{G})$ the determinant of the matrix obtained by substituting 1 for each variable x_e of the Tutte matrix $\vec{G}(x)$. It can be shown that $p(\vec{G}) \leq \phi(G)^2$. $\vec{G}$ is called a *Pfaffian orientation* of G if $p(\vec{G}) = \phi(G)^2$. Kasteleyn proved that a planar graph has a Pfaffian orientation which can be found in polynomial time. So counting perfect matchings in planar graphs reduces to calculating a determinant. Not all graphs have Pfaffian orientations. For instance, $K_{3,3}$ does not have one. Little [1974] extended Kasteleyn's result by proving that if a graph has no subdivision of $K_{3,3}$ as a subgraph it has Pfaffian orientation. Vazirani [1989] showed that counting perfect matchings in these graphs is in fact in NC (by deriving an NC-algorithm for finding the Pfaffian orientation for these graphs).

So far for deterministic algorithms. Based on an idea of Broder [1986], Jerrum & Sinclair [1989] derived a polynomial time algorithm to approximate $\phi(G)$ with high probability when G has minimum degree $\frac{1}{2}|V(G)|$. Their algorithm gives a number Y such that $|Y - \phi(G)| \leq \epsilon\phi(G)$ with probability at least $1 - \delta$. The algorithm is polynomial in $|V(G)|$, $\log 1/\epsilon$, $\log 1/\delta$. The existence of such an algorithm for general graphs is still open. The main idea of the algorithm of Jerrum and Sinclair is as follows. Let for each k, $\phi_k(G)$ denote the of matchings of size k. Jerrum and Sinclair approximate $\phi(G)$ by approximating the ratios $r_k := \phi_k/\phi_{k-1}$ and multiplying them. So the problem reduces to approximating r_k. We restrict ourselves to $r_{1/2(|V(G)|)}$ and approximate it by choosing uniformly and independently *almost perfect* matchings (i.e. matchings of size at least $\frac{1}{2}|V(G)| - 1$) at random and counting how many of these are perfect and how many not. Clearly, in this way we can get a good estimate of $r_{1/2(|V(G)|)}$. Remains the question how to select an almost perfect matching at random. The problem is that there are exponentially many of them. This problem is overcome by defining a random walk on the set of almost perfect matchings. The steps in this random walk are as follows. Given an almost perfect matching M, with probability $\frac{1}{2}$ choose $e = uv \in E$ uniformly and independently at random. If M is perfect and $e \in M$, we move from M to $M \setminus \{e\}$. If M is not perfect, $e \notin M$ and $|M \cap \delta(\{u, v\})| \leq 1$, we move from M to $(M \setminus \delta(\{u, v\})) \cup \{e\}$. In all other cases we stay at M. Thus we get a Markov chain. It has a uniform stationary distribution and each random walk in the Markov chain converges to this stationary distribution. So if we start the Markov process with some arbitrary matching and 'walk forever', the matching will become 'more and more random'. The point is that we do not have to walk forever. This Markov chain is 'rapidly mixing', meaning that the probability distribution after a polynomial number of steps is very close to uniform (irrespective the matching we start off with). As we only need to approximate $r_{1/2(|V(G)|)}$, it suffices to select the almost perfect matching 'almost uniformly at random'. Explaining all the technicalities in full detail would go to far here, but we can sketch the main ideas.

We need some definitions. Let $\mathcal{G} = (\mathcal{V}, \mathcal{E})$ be the undirected graph with the almost perfect matchings in G as its nodes and with $M'M \in \mathcal{E}$ if $M \triangle M'$ is a path with at most two edges. Let $\mathcal{A} \in \mathbb{Z}^{\mathcal{V}\times\mathcal{V}}$ be defined by $\mathcal{A}_{M'M} = -1$ if $M'M \in \mathcal{E}$, $\mathcal{A}_{M'M} = 0$ if $M'M \notin \mathcal{E}$, and $\mathcal{A}_{MM} = \deg_{\mathcal{G}}(M)$ if $M \in \mathcal{V}$. Then the transition matrix of the above defined Markov chain is $\mathcal{P} := I - 1/(2|E|)\mathcal{A}$: the probability to move to M' being in M is $\mathcal{P}_{M'M}$. Finally we define $q \in \mathbb{R}^{\mathcal{V}}$ by $q_M := 1/|\mathcal{V}|$ for each $M \in \mathcal{V}$, and $x^M \in \{0,1\}^{\mathcal{V}}$ by $x^M_{M'} = 1$ if and only if $M' = M$.

$\mathcal{P}$ is a symmetric doubly stochastic matrix with only positive eigenvalues. The largest these eigenvalue is 1 and has multiplicity 1 (as $\mathcal{G}$ is connected). The corresponding eigenvector is q, i.e. the uniform distribution on $\mathcal{V}$ and the stationary distribution of the Markov chain. If we start our Markov chain with an almost perfect matching M then the probability distribution after k steps is $\mathcal{P}^k x^M$ which tends to q if k goes to ∞. The rate of convergence can be expressed in the second largest eigenvalue λ_2 of $\mathcal{P}$: $|(\mathcal{P}^k x^M)_{M'} - 1/|\mathcal{V}|| \leq \lambda_2^k$ for each $M' \in \mathcal{V}$. So far everything is just standard matrix theory.

Sinclair & Jerrum [1989] derived the following bound on the second largest eigenvalue of $\mathcal{P}$:

$$\lambda_2 \leq 1 - \tfrac{1}{2}\Phi(\mathcal{G})^2. \tag{99}$$

where

$$\Phi(\mathcal{G}) := \frac{1}{2|E|} \min\left\{ \frac{|\delta(S)|}{|S|} \,\middle|\, S \subseteq \mathcal{V}, |S| \leq \frac{1}{2}|\mathcal{V}| \right\}, \tag{100}$$

is the *conductance* of $\mathcal{G}$. Jerrum & Sinclair [1989], in turn, derived the following bound on the conductance.

$$\Phi(\mathcal{G}) \geq |V|^{-6}. \tag{101}$$

Combining all this we get:

$$\textit{If } k \geq \tau(\epsilon) := 2|V|^{12}\left(\log|\mathcal{V}| + \log\frac{1}{\epsilon}\right), \textit{ then } \left|(\mathcal{P}^k x^M)_{M'} - \frac{1}{|\mathcal{V}|}\right| \leq \epsilon\frac{1}{|\mathcal{V}|}. \tag{102}$$

So making $\lceil\tau(\epsilon)\rceil = O(|V|^{12}(|V|\log|V| + \log 1/\epsilon))$ steps in the Markov chain results in a random selection of an almost perfect matching from a probability distribution which is close to uniform.

These are the main ideas of Jerrum and Sinclair's algorithm for counting perfect matchings in graphs with minimum degree $\frac{1}{2}|V|$. The relation between the rate of convergence of a Markov chain and its conductance extends, under mild conditions, to other Markov chains, not related to matchings in graphs. Over the last decennium rapidly mixing Markov chains have become more and more important in the design of randomized counting or optimization algorithms.

9. Applications of matchings

In this section we discuss applications of matchings to other combinatorial optimization problems. In particular, we discuss the traveling salesman problem, shortest path problems, a multi-commodity flow problem in planar graphs, and the max-cut problem in planar graphs.

9.1. *The traveling salesman problem*

A *traveling salesman tour*, or *Hamiltonian circuit* in a graph $G = (V, E)$ is the edge set of a circuit that spans all the nodes, i.e., a closed walk through G that visits every node exactly once. Given a distance function $d \in \mathbb{R}^E$, the traveling salesman problem is to find a traveling salesman tour F of minimum length $d(F)$. The problem has many applications in many environments: routing trucks for pick-up and delivery services, drilling holes in manufacturing printed circuit boards, scheduling machines, etc.. The traveling salesman problem is NP-hard. In fact, simply finding a traveling salesman tour is NP-hard [Karp, 1972]. The problem has served pre-eminently as an example of a hard problem. For example, Lawler, Lenstra, Rinnooy Kan & Schmoys [1985] chose it as the guide in their tour through combinatorial optimization. Their volume provides a wide overview of research on this problem. For an update of what has emerged since then, see Jünger, Reinelt & Rinaldi [1995, this volume].

In this section we discuss a heuristic for the traveling salesman problem that uses matchings. We also discuss the relation between matching and polyhedral approaches to the traveling salesman problem. We assume from now on that $G = (V, E)$ is complete.

9.1.1. *Christofides' heuristic*

The problem is NP-hard and therefore is unlikely to be solvable in polynomial time. It makes sense then to take a heuristic approach, i.e., to find a hopefully good, but probably not optimal solution quickly. The heuristic we present here is due to Christofides [1976] and is meant for the case in which the distance function d is non-negative and satisfies the *triangle inequality*: $d_{uv} + d_{vw} \geq d_{uw}$ for each three nodes u, v, and w in G.

Let F be a minimum length spanning tree of G and let T be the set of nodes v in G with $\deg_F(v)$ odd (so F is a T-join). Find a minimum weight perfect matching M in $G|T$ with weight function d. Consider the union of F and M in the sense that if an edge occurs in both sets it is to be taken twice as a pair of parallel edges. This union forms an Eulerian graph and an Eulerian walk in this graph visits each node of G at least once. The length of the walk is $d(F) + d(M)$. Since G is complete, we may transform the Eulerian walk into a traveling salesman tour by taking short cuts and, by the triangle inequality, the length of this tour is at most $d(F) + d(M)$.

The heuristic runs in polynomial time. There are many polynomial time algorithms for finding a minimum weight spanning tree, e.g., Borůvka's algorithm [Borůvka, 1926], Kruskal's algorithm [Kruskal, 1956], or Jarník's algorithm (Jarník [1930], better known by the names of its re-inventors Prim [1957] and Dijkstra [1959]). Kruskal's algorithm, for instance, runs in $O(|E| \log |V|)$ time. Edmonds' matching algorithm, described in Section 6, finds a minimum weight matching in polynomial time. Once the tree and the matching are known, an Eulerian walk and a traveling salesman tour can be found in linear time. Gabow & Tarjan [1991] showed that the heuristic can be implemented in $O(|V|^{2.5}(\log |V|)^{1.5})$. (Instead of a minimum weight matching, their version finds a matching with weight at most $1 + 1/|V|$ times the minimum weight.)

The following theorem shows that the heuristic produces a tour that is at most 50% longer than the shortest traveling salesman tour.

Theorem 29 [Christofides, 1976]. *Let $G = (V, E)$ be a complete graph and $d \in \mathbb{R}_+^E$ be a distance function satisfying the triangle inequality. Then $\lambda^* \leq \frac{3}{2}\lambda$, where λ is the length of a shortest traveling salesman tour, and λ^* is the length of the tour found by Christofides' heuristic.*

Proof. Let C be a shortest traveling salesman tour, and let F and M be the tree and matching found by the heuristic. Let T be the nodes $t_1, \ldots, t_k$, with odd degree in F, where the numbering corresponds to the order in which C visits these nodes. Let $\widetilde{C}$ be the circuit with edges $t_1t_2, t_2t_3, \ldots, t_kt_1$. By the triangle inequality, $\widetilde{C}$ is shorter than C. Let $\widetilde{M}$ be the shorter of the two perfect matchings on T contained in $\widetilde{C}$. Then $\widetilde{M}$ is a perfect matching of $G|T$. So, $\lambda = d(C) \geq d(\widetilde{C}) \geq 2d(\widetilde{M}) \geq 2d(M)$. On the other hand, C contains a spanning tree – just delete an edge – so $\lambda = d(C) \geq d(F)$. Combining these inequalities, we see that $\lambda^* \leq d(F) + d(M) \leq \frac{3}{2}\lambda$. □

9.1.2. A polyhedral approach to the traveling salesman problem

Traveling salesman tours are connected 2-factors. So, the characteristic vectors of traveling salesman tours satisfy the following system of inequalities (compare with (81) and (82)):

$$\begin{array}{llll} x_e & \geq 0 & (e \in E) & \\ x_e & \leq 1 & (e \in E) & \\ x(\delta(v)) & = 2 & (v \in V) & (103) \\ x(\delta(U) \setminus F) - x(F) & \geq 1 - |F| & (U \subseteq V,\ F \subseteq \delta(U)) & \\ x(\delta(U)) & \geq 2 & (U \subseteq V;\ \emptyset \neq U \neq V). & \end{array}$$

In fact, every integral solution to (103) is the characteristic vector of a traveling salesman tour. Thus, the cutting plane approach described in Section 8.3 for solving the matching problem can be applied to the system (103) to solve the traveling salesman problem. In this case, however, the polyhedron defined by (103) has fractional extreme points and so success is not guaranteed. (Note that without the last set of inequalities, the system describes an integral polyhedron, namely the convex hull of 2-factors. The last set of inequalities, called the *subtour elimination constraints*, are necessary to 'cut-off' each 2-factor that is not a traveling salesman tour. However, adding these constraints introduces new, fractional, extreme points.) One could try to overcome this by adding more constraints to the system [see Grötschel & Padberg, 1985; Jünger, Reinelt & Rinaldi, 1995], but no complete description of the traveling salesman polytope is known. In fact, unless NP = co-NP, no 'tractable' system describing the traveling salesman polytope exists [see Karp & Papadimitriou, 1982].

'Partial' descriptions like (103), however, can be useful for solving traveling salesman problems. Minimum cost solutions to such systems provide lower bounds for the length of a shortest traveling salesman tour. These lower bounds can be used, for instance, to speed up branch-and-bound procedures. In fact, over the last decennium much progress has been made in this direction [see Jünger, Reinelt & Rinaldi, 1995].

The cutting plane approach requires a separation algorithm, or at least good separation heuristics, for the partial descriptions. We have separation algorithms for (103). Determining whether a given solution x satisfies the non-negativity, capacity and degree constraints is trivial. We can use a max-flow algorithm to determine whether x satisfies the subtour elimination constraints. So, all that remains is to find a polynomial time algorithm for the problem:

> Given $x \in \mathbb{R}_+^{E(G)}$, find a subset $U \subseteq V(G)$ and an odd subset F of $\delta(U)$ such that $x(\delta(U) \setminus F) - x(F) < 1 - |F|$ or decide that no such subsets exist. (104)

These constraints are the 'odd cut constraints' for the 2-factor problem and, in view of the reductions of general matching to perfect matching, it should not be surprising that we can solve this problem in much the same way as we solved the separation problem for the odd cut constraints for perfect matching. Construct an auxiliary graph as follows. Replace each edge $e = uv$ in G with two edges in series: $e_1 := uw$ and $e_2 := wv$. Define $x^*_{e_1} := x_e$ and $x^*_{e_2} := 1 - x_e$. Let T be the set of nodes in the resulting graph G^* meeting an odd number of the edges e_2. Consider the problem:

> Find $U \subseteq V(G^*)$, such that $|U \cap T|$ is odd and $x(\delta(U)) < 1$, or show that no such U exists. (105)

It is not so hard to see that if U in (105) exists, then we may choose U so that for each edge e in G, at most one of e_1 and e_2 is contained in $\delta(U)$. Hence, (104) is equivalent to (105) and the separation problem (105) amounts to finding a minimum weight T-cut.

Above we considered the traveling salesman problem as a matching problem with side constraints, namely of finding shortest connected 2-factors. Other papers on matchings with side constraints are: Ball, Derigs, Hilbrand & Metz [1990], Cornuéjols & Pulleyblank [1980a, b, 1982, 1983], and Derigs & Metz [1992].

9.2. *Shortest path problems*

The *shortest path problem* is: Given two nodes s and t in G, find an s,t-path of shortest length $d(P)$ with respect to a length function $d \in \mathbb{R}^E$. In general, this problem is NP-hard, it includes the traveling salesman problem; but it is polynomially solvable when no circuit C in G has negative length $d(C)$. When all edge lengths are non-negative, the problem can be solved by the labeling methods of Bellman [1958] & Ford [1956], Dijkstra [1959], and Floyd [1962a, b] and Warshall [1962]. The algorithms also find shortest paths in directed graphs, even when negative length edges are allowed (though some adaptations are required) as long as no directed circuit has negative length. The presence of negative length edges makes the problem in undirected graphs more complicated. Simple labeling techniques no longer work. In fact, the problem becomes a matching, or more precisely, a T-join problem. Indeed, let $T := \{s, t\}$. Since no circuit has negative length, a shortest T-join is a shortest s, t-path (possibly joined by circuits of length 0). So we can find a shortest path in an undirected graph with negative length edges but no negative length circuits, by solving a T-join problem. Alternatively we can model the shortest path problem as a generalized matching problem. For each node $v \in V(G) \setminus \{s, t\}$, add a loop $\ell(v)$, then the shortest path problem is the generalized matching problem subject to the constraints:

$$\begin{array}{rlcll} 0 \leq & x_e & \leq & 1 & e \in E(G) \\ 0 \leq & x_{\ell(v)} & \leq & 1 & v \in V(G) \setminus \{s,t\} \\ & x(\delta(v)) + 2x_{\ell(v)} & = & 2 & v \in V(G) \setminus \{s,t\} \\ & x(\delta(v)) & = & 1 & v \in \{s,t\} \end{array} \tag{106}$$

The reductions described in Section 7.1 reduce this problem to a perfect matching problem in an auxiliary graph.

9.2.1. *Shortest odd and even paths*

The *shortest odd path problem* asks for a shortest path from s to t with an odd number of edges. Similarly, the *shortest even path problem* asks for a shortest s, t-path with an even number of edges. In general these problems are NP-hard. The special case in which no circuit has negative length is, to my knowledge, still unsettled: the problems are not known to be hard, but neither do we know of any polynomial algorithm for them. If all edge lengths are non-negative, the problems are solvable in polynomial time: they are matching problems. We show this by a reduction due to Edmonds [see Grötschel & Pulleyblank, 1981]. We only consider the case of odd paths. The shortest even path problem can be solved by an easy reduction to the shortest odd path problem, or alternatively, by a similar reduction to the matching problem.

To find a shortest odd path between s and t, construct an auxiliary graph H as follows. Add to G a copy G' of G with the nodes s and t deleted (denote the copy in G' of node u by u' and the copy of edge e by e'). For each $u \in V(G) \setminus \{s, t\}$ add an edge from u to its copy u' in G'. The weight function w on H is defined by $w_e := w_{e'} := d_e$ for each $e \in E(G)$ and $w_{u'u} := 0$ for each $u \in V(G) \setminus \{s, t\}$. Let M be a perfect matching in H and define $P_M := \{e \in E(G) \mid e \in M \cap E(G) \text{ or } e' \in M \cap E(G')\}$. It is easy to see that P_M is the node-disjoint union of an odd s, t-path and a collection of circuits. If M has minimum length with respect to w, each of the circuits has length 0 and so minimum weight perfect matchings in H correspond to shortest odd s, t-paths in G.

Recently, Schrijver & Seymour [1994] characterized the *odd s, t-path polyhedron*, i.e., the convex hull of the subsets of $E(G)$ containing an odd s, t-path, thus proving a conjecture of Cook and Sebö. The inequalities describing the polyhedron are: $0 \leq x_e \leq 1$ for all $e \in E(G)$, and

$$2x(\langle W \rangle \setminus F) + x(\delta(W)) \geq 2 \text{ for each subgraph } H = (W, F) \text{ of } G \text{ such that both } s \text{ and } t \text{ are in } W \text{ but no } s, t\text{-path in } H \text{ is odd.} \quad (107)$$

9.3. *Max-cut and disjoint paths in planar graphs*

We conclude this section with the application of T-joins and planar duality to the max-cut problem and a disjoint paths problem in planar graphs. A graph G is *planar* if it can be embedded in the plane so that its edges do not cross. The *planar dual* G^* of G with respect to an embedding is defined as follows. The graph G divides the plane into several connected regions each corresponding to a node in $V(G^*)$. Each edge e in G separates at most two regions of the plane in the sense that, if we removed e, these regions would combine into one. For each edge $e \in E(G)$ there is an edge e^* in G^* joining the nodes in G^* corresponding to the regions separated by e. If e does not separate two regions, then it lies entirely in a single region and e^* is a loop at the corresponding node of $V(G^*)$. We identify each edge e in G with the corresponding edge e^* in G^*.

The graph G^* is planar and its definition suggests a natural embedding. If G is connected and G^* is embedded in the natural way, then $(G^*)^*$ is again G. The most prominent property of planar duality is that $C \subseteq E(G)$ $(= E(G^*))$ is a cycle in G if and only if it is a cut in G^* (recall that a *cycle* is a graph in which the degree of each node is even). The same relation exists between cuts in G and cycles in G^*.

The *max-cut problem* is: Given a weight function $w \in \mathbb{R}^{E(G)}$, find a cut $\delta(U)$ in G with $w(\delta(U))$ maximum. The problem is NP-hard in general [Karp, 1972], but polynomially solvable when G is planar. To see this, consider a planar graph G and a planar dual G^*. Define $T := \{v \in V(G^*) \mid \deg_{G^*}(v) \text{ is odd}\}$. Clearly, $F \in E(G^*)$ is a T-join if and only if $E(G^*) \setminus F$ is a cycle in $E(G^*)$. So, T-joins in G^* correspond to complements of cuts in G. Hence the max-cut problem in G is a T-join problem in G^* [Hadlock, 1975].

Combining planar duality with Seymour's theorem (89) for T-joins and T-cuts in bipartite graphs we obtain the following:

Theorem 30 [Seymour, 1981]. *Let G be a graph and let H be a collection pairs $\{s_1, t_1\}, \ldots, \{s_k, t_k\}$ of nodes. If the graph $G + H$, obtained from G by adding as extra edges the pairs in H, is planar and Eulerian, then the following are equivalent:*

(i) There exist edge-disjoint paths $P_1, \ldots, P_k$ in G such that each P_i goes from s_i to t_i;

(ii) For each $U \subseteq V(G)$, $|\delta_G(U)| \geq |\delta_H(U)|$.

Proof. Clearly, (ii) is necessary for (i), we show that it is also sufficient. Assume that (ii) holds and let $(G + H)^*$ be the planar dual of $G + H$ with respect to some embedding. Since $G + H$ is Eulerian, $E(G + H)$ is a cycle in $G + H$. In other words, $E((G + H)^*)$ is a cut in $(G + H)^*$ and so $(G + H)^*$ is bipartite.

Let T be the set of nodes in $V((G + H)^*)$ that meet an odd number of edges in H. Then H is a T-join in $(G + H)^*$. In fact, H is a minimum cardinality T-join in $(G + H)^*$. To see this, observe that for any other T-join F the symmetric difference $F \Delta H$ is a cycle in $(G + H)^*$ and so a cut in G. By (ii), $F \Delta H$ contains at least as many edges from F as from H. So, $|H| \leq |F|$ and H is a minimum cardinality T-join in $(G + H)^*$.

Now, applying (89) to $(G + H)^*$ and T, we see that there must be $|H| =: k$ disjoint odd cuts $C_1 = \delta(U_1), \ldots, C_k = \delta(U_k)$ in $(G + H)^*$. Clearly, each of these cuts has at least one edge in common with H and so each edge in H must be in exactly one of them. Assume $(s_i t_i)^* \in C_i$ for $i = 1, \ldots, k$. Without loss of generality, we may assume that the cuts are inclusion-wise minimal and so circuits in $G + H$. Then, $P_1 := C_1 \setminus s_1 t_1, \ldots, P_k := C_k \setminus s_k t_k$ are the desired paths. □

Matsumoto, Nishizeki, & Saito [1986] showed that the paths can be found in $O(|V(G)|^{5/2} \log |V(G)|)$ time. When $G + H$ is not Eulerian the problem becomes NP-hard [Middendorf & Pfeiffer, 1990]. For a general overview of the theory of disjoint paths, see Frank [1990].

10. Computer implementations and heuristics

10.1. Computer implementations

Over the years several computer implementations for solving matching problems have been designed, e.g. Pulleyblank [1973], Cunningham & Marsh [1978], Burkhard & Derigs [1980], Derigs [1981, 1986a, b, 1988b], Derigs & Metz [1986, 1991], Lessard, Rousseau & Minoux [1989] and Applegate & Cook [1993]. Grötschel & Holland [1985] used a cutting plane approach and Crocker [1993] and Mattingly & Ritchey [1993] implemented Micali and Vazirani's $O(\sqrt{|V|}|E|)$ algorithm for finding a maximum cardinality matching.

Designing efficient matching codes, especially those intended for solving large problems, involves many issues. Strategic decisions must be made, e.g., what algorithm and data structures to use. Moreover, tactical decisions must be made, e.g., how to select the next edge in the alternating forest and when to shrink blossoms. Finally, of course, numerous programming details affect the efficiency of the code. We restrict our attention to a few key strategic issues.

In solving large problems two paradigms appear to be important. The first of these is 'Find a 'good' starting solution quickly (the 'jump-start')' and the second is 'Avoid dense graphs'. We discuss the second paradigm first.

One feature of Grötschel and Holland's code [1985] (see Section 8.3) that competed surprisingly well with the existing combinatorial codes (based on Edmonds algorithm for instance), was that it first solved a matching problem in a sparse subgraph and then tuned the solution to find a matching in the original graph. Incorporating this approach sped up existing combinatorial codes significantly [Derigs & Metz, 1991]. The idea is to solve a minimum weight perfect matching problem on a (dense) graph G by first selecting a sparse subgraph G_{sparse} of G. A matching code, e.g., Edmonds' algorithm, can find a minimum weight perfect matching M and an optimal (structured) solution π in G_{sparse} quickly. In G the matching may not be of minimum weight and the dual solution may not be feasible. The second phase of the procedure corrects this. A primal algorithm, e.g., Cunningham and Marsh's algorithm described in Section 8.1, is ideal for this phase. Weber [1981], Ball & Derigs [1983], and Applegate & Cook [1993] have developed alternative methods for this.

The typical choice of G_{sparse} is the *k-nearest neighbor graph* of G, which is constructed by taking for each node u the k shortest edges incident to u. Typical choices for k run from 5 to 15. To give an impression of how few edges G_{sparse} can have: Applegate & Cook [1993] used their code to solve an Euclidean problem on 101230 nodes (i.e., the nodes lie in the Euclidean plane and the weight of an edge is given by the L_∞ distance between its endpoints). So, G is complete and has $0.5 \cdot 10^{10}$ edges. When k is 10, G_{sparse} has 10^6 edges or less then 0.05% of the all the edges in G. In fact, Applegate and Cook solved this 101230 node problem – a world record. For more moderately sized problems (up to twenty thousand nodes) their code seems dramatically faster than previously existing matching codes.

Many matching codes incorporate a jump-start to find a good matching and a good dual solution quickly before executing the full matching algorithm. Originally these initial solutions were typically produced in a greedy manner. Derigs and Metz [1986] suggested a jump-start from the fractional matching problem (or equivalently the 2-matching problem). First, solve the 2-matching problem: $\max\{w^\top x \mid x \geq 0;\ x(\delta(v)) = 2\ (v \in V)\}$. Let x^* and π^* be primal and dual optimal solutions to this linear programming problem (which can, in fact, be solved as a bipartite matching problem or a network flow problem). The set $\{e \in E \mid x^*_e > 0\}$ is the node-disjoint union of a matching $M' := \{e \in E \mid x^*_e = 2\}$ and a collection of odd circuits. Jump-start with the matching M obtained from M' and a maximum matching in each of the odd circuits and the dual solution π^* (setting the dual variables corresponding to the blossoms equal to zero). Since x^* and π^* are primal and dual optimal solutions to the 2-matching problem, they satisfy the complementary slackness conditions. If G is dense, the 2-matching problem is first solved on a sparse subgraph. In fact, Applegate and Cook use different sparse graphs for finding the jump-start and for solving the actual problem (the latter is the k-nearest neighbor graph using the reduced costs with respect to the jump-start dual solution).

10.2. Heuristics

When solving large matching problems, searching for a good jump-start, or applying matchings in a heuristic for some other problem (e.g., Christofides' heuristic for the traveling salesman problem described in Section 9.1) it is often useful to use a heuristic to find a good matching quickly. A straightforward approach, called the greedy heuristic, attempts to construct a minimum weight perfect matching by starting with the empty matching and iteratively adding a minimum weight edge between two exposed nodes. The greedy heuristic runs in $O(|V|^2 \log |V|)$ time and finds a solution with weight at most $\frac{4}{3}|V|^{\log 3/2}$ times the minimum weight of a perfect matching [Reingold & Tarjan, 1981]. The version of the greedy heuristic designed to find a maximum weight matching, finds a solution with at least half the weight of a maximum weight matching. Results on greedy heuristics appear in Avis [1978, 1981], Avis, Davis & Steele [1988], Reingold & Tarjan [1981], Frieze, McDiarmid & Reed [1990] and Grigoriadis, Kalantari & Lai [1986].

Several heuristics have been developed for Euclidean matching problems where the set of points that have to be matched lie in the unit square. Many of these heuristics find the heuristic matching by dividing the unit square into subregions, finding a matching in each subregion and combining these matchings to a perfect matching between all the points. Other heuristics match the points in the order in which they lie on a space-filling curve. For detailed description and the analysis of such heuristics see: Bartholdi & Platzman [1983], Imai [1986], Imai, Sanae & Iri [1984]. Iri, Murota & Matsui [1981, 1982], Papadimitriou [1977], Reingold & Supowit [1983], Steele [1981], Supowit, Plaisted & Reingold [1980], Supowit & Reingold [1983], Supowit, Reingold & Plaisted [1983].

For a good overview on matching heuristics, see the survey of Avis [1983]. Here we mention some recent heuristics in more detail.

When the weight function w satisfies the triangle inequality, each minimum weight V-join is a perfect matching (or, when some edges have weight 0, can be transformed easily into a perfect matching with the same weight). So, when w satisfies the triangle inequality, we can use T-join heuristics as matching heuristics.

Plaisted [1984] developed a T-join heuristic that runs in $O(|V|^2 \log |V|)$ time and produces a T-join with weight at most $2\log_3(1.5|V|)$ times the weight of an optimal solution.

Given a graph $G = (V, E)$, an even subset T of V and $w \in \mathbb{R}^E$, construct a T-join J as follows. (Note that w need not satisfy the triangle inequality, it would not survive the recursion anyway.)

AUXILIARY GRAPH: If $T = \emptyset$, then set $J := \emptyset$. Otherwise, construct the weighted complete graph H on the node set T. The weight w'_{uv} of each edge uv in H is the length of a shortest uv-path P_{uv} in G (with respect to w).
SHRINK: For each $u \in T$, define $n_u := \min\{w'_{uv} \mid v \in T\}$. Construct a forest F in H as follows. Scan each node in order of increasing n_u. If the node u is not yet covered by F, add to F an edge uv with $w'_{uv} = n_u$. Let $F_1, \ldots, F_k$ denote the trees of F and let $G' := H \times V(F_1) \times \cdots \times V(F_k)$. (If parallel edges occur select one of minimum weight to be in G'.) The pseudo-node corresponding to $V(F_i)$ is in T' if and only if $|V(F_i)|$ is odd. Apply the procedure recursively to G', w' and T' (starting with AUXILIARY GRAPH) and let J' be the resulting T'- join.
EXPAND: Let J^* denote the set of edges in H corresponding to the edges of J' in G. Choose T^* so that J^* is a T^*-join. Then $T_i^* := (T \triangle T^*) \cap V(F_i)$ is even for each i=1, ..., k. Let J_i be the unique T_i^*-join in each tree F_i. Then $J_H := J' \cup J_1 \cup \cdots \cup J_k$ is a T-join in H.
T-JOIN. Let J be the symmetric difference of the shortest paths $\{P_{uv} : uv \in J_H\}$.

Note that each tree F_i contains at least 2 nodes. So, if $|V(F_i)|$ is odd it is at least three. Hence, the depth of the recursion is bounded by $\log_3 |T|$.

Goemans & Williamson [1992] proposed a heuristic that not only yields a T-join F but also a feasible solution π of

$$\begin{array}{llll} \text{maximize} & \sum_{S \in \Omega} \pi_S & & \\ \text{subject to} & \sum_{S \in \Omega; \delta(S) \ni e} \pi_S \leq & w_e & (e \in E) \\ & \pi_S \geq & 0 & (S \in \Omega), \end{array} \tag{108}$$

where $\Omega := \{S \subseteq V \mid |S \cap T| \text{ odd}\}$. (108) is the dual linear programming problem of the T-join problem (cf. (91)). The weight of the heuristic T-join will be at most $(2 - 2/|T|) \sum_{S \in \Omega; \delta(S) \ni e} \pi_S$, so at most $2 - 2/|T|$ times the minimum weight of a T-join.

During the procedure we keep a forest F' (initially $V(F') := V(G)$ and $E(F') := \emptyset$). For each $v \in V(G)$, F'_v denotes the component of F' containing v. We also keep a feasible solution π of (108) (initially, $\pi \equiv 0$).

The basic step of the heuristic is as follows: among all edges $e = uv$ in G with $F'_u \neq F'_v$ and $F'_u \in \Omega$, select one, e^* say, that minimizes the quantity:

$$\frac{1}{p(F'_u) + p(F'_v)} (w_{uv} - \sum_{S \in \Omega; \delta(S) \ni uv} \pi_S), \tag{109}$$

where $p(S) := 1$ if $S \in \Omega$ and $p(S) := 0$ if $S \notin \Omega$. Let ϵ be the value of (109) when $uv = e^*$. Add ϵ to π_S for each component S of F' that is in Ω and replace F' by $F' \cup e^*$. This basic step is repeated until no component of F' is in Ω. Then F' contains a unique T-join, which is the output of the heuristic.

The heuristic can be implemented $O(|V|^2 \log |V|)$ time. Note that when $|T| = 2$, so when the T-join problem is a shortest path problem, the heuristic T-join is in fact a shortest path. The heuristic also applies to other minimum weight forest problems with side constraints [see Goemans & Williamson, 1992].

Grigoriadis & Kalantari [1988] developed an $O(|V|^2)$ heuristic that constructs a matching with weight at most $2(|V|^{\log_3 7/3})$ times the optimum weight. Given a matching M, let G_M denote the 1-nearest neighbor graph of $G|(\exp(M))$. Begin with the empty matching M. In each component G_i of G_M choose a tour visiting each edge twice. Shortcut the tour to obtain a traveling salesman tour T_i of G_i. Greedily select a matching M_i of small weight from T_i (thus $|M_i| \geq \frac{1}{3}|T_i|$) and add it to M. Repeat the procedure until M is perfect.

The final matching heuristic we describe is due to Jünger & Pulleyblank [1991]. It runs in $O(|V|\log|V|)$ on Euclidean problems. Given a set of points in the plane, construct a complete graph G with a node for each point and let the length of each edge be the Euclidean distance between the corresponding points. So each node u has two coordinates u_1 and u_2 and each edge uv has weight (or length) $w_{uv} := \sqrt{(u_1 - v_1)^2 + (u_2 - v_2)^2}$. Construct a matching in G as follows.

Let F be a minimum weight spanning tree in G. (The maximum degree of a node in T is five [see Jünger & Pulleyblank, 1991].)

DECOMPOSE: If $|V| \leq 6$, find a minimum weight matching in G. Otherwise, T has a non-pendant edge (i.e., an edge not incident to a node of degree 1). Let uv be a maximum weight non-pendant edge in T, then $T \setminus \{uv\}$ consists of two trees: T_u containing u and T_v containing v. We consider two cases:

Both T_u and T_v contain an even number of nodes: Apply DECOMPOSE, recursively, to $G|V(T_u)$ and T_u, and to $G|V(T_v)$ and T_v. Note that T_u is a minimum spanning tree in $G|V(T_u)$ and T_v is a minimum spanning tree in $G|V(T_v)$. Return $M_u \cup M_v$, where M_u is the matching constructed in $G|V(T_u)$ and M_v is the matching constructed in $G|V(T_v)$.

Both T_u and T_v contain an odd number of nodes: Apply DECOMPOSE to $G|(V(T_u) \cup \{v\})$ and $T_u \cup \{uv\}$ (which is again a minimum spanning tree) to construct a matching M_u. Let x be the node matched to u in M_u and choose $y \in V(T_v)$ with w_{xy} minimum. Then $T_v \cup \{xy\}$ is a minimum spanning tree in $G|(V(T_v) \cup \{x\})$. Applying DECOMPOSE again yields a matching M_v in $G|(V(T_v) \cup \{x\})$. Return $(M_u \setminus \{ux\}) \cup M_v$.

Note that the heuristic computes only one minimum spanning tree and the minimum spanning trees for the decomposed problems are easily obtained from it. Jünger & Pulleyblank [1991] also give a heuristic for finding a dual feasible solution, again based on minimum spanning tree calculations.

We conclude with a result of Grigoriadis & Kalantari [1986]: The running time of a heuristic for the Euclidean matching problem that finds a matching of weight at most $f(|V|)$ times the minimum weight, can be bounded from below by a constant times $|V|\log|V|$. If the heuristic yields a matching of weight at most $f(|V|)$ times the minimum weight for all matching problems, its running time is at least a constant times $|V|^2$.

Acknowledgements

I would like to thank Michele Conforti, Jack Edmonds, Mike Plummer, Bill Pulleyblank, Lex Schrijver, Leen Stougie and John Vande Vate for many helpful comments. John Vande Vate made a tremendous, and highly appreciated, effort editing the paper; improving its English as well as its organization. Needless to say that all remaining failings are on my account.

References

Ageev, A.A., A.V. Kostochka and Z. Szigeti (1994). *A characterization of Seymour graphs*, preprint.

Aho, A.V., J.E. Hopcroft and J.D. Ullman (1974). *The Design and Analysis of Computer Algorithms*, Addison-Wesley, Reading, MA.

Ahuja, R.K., T.L. Magnanti and J.B. Orlin (1989). Network flows, in: G.L. Nemhauser, A.H.G. Rinnooy Kan and M.J. Todd (eds.), *Optimization*, Handbooks in Operations Research and Management Science, Vol. 1, North-Holland, Amsterdam, pp. 211–369.

Alt, H., N. Blum, K. Mehlhorn and M. Paul (1991). Computing a maximum cardinality matching in a bipatite graph in time $O(n^{1.5}\sqrt{m/\log n})$, *Inf. Process. Lett.* 37, 237–240.

Anstee, R.P. (1985). An algorithmic proof of Tutte's f-factor theorem. *J. Algorithms* 6, 112–131.

Anstee, R.P. (1987). A polynomial algorithm for b-matchings: an alternative approach. *Inf. Process. Lett.* 24, 153–157.

Applegate, D., and W. Cook (1993). Solving large-scale matching problems, in: D.S. Johnson and C.C. McGeoch (eds.), *Network Flows and Matchings: First DIMACS Implementation Challenge*, DIMACS Series in Discrete Mathematics and Theoretical Computer Science, Vol. 12, American Mathematical Society, Providence, RI, pp. 557–576.

Aráoz, J., W.H. Cunningham, J. Edmonds and J. Green-Krótki (1983). Reductions to 1-matching polyhedra. *Networks* 13, 455–473.

Avis, D. (1978). Two greedy heuristics for the weighted matching problem. *Congr. Numerantium* XXI, 65–76.

Avis, D. (1981). Worst case bounds for the Euclidean matching problem. *Comput. Math. Appl.* 7, 251–257.

Avis, D. (1983). A survey of heuristics for the weighted matching problem. *Networks* 13, 475–493.

Avis, D., B. Davis and J.M. Steele (1988). Probabilistic analysis for a greedy heuristic for Euclidean matching. *Probab. Eng. Inf. Sci.* 2, 143–156.

Balas, E., and W. Pulleyblank (1983). The perfectly matchable subgraph polytope of a bipartite graph. *Networks* 13, 495–516.

Balas, E., and W.R. Pulleyblank (1989). The perfectly matchable subgraph polytope of an arbitrary graph. *Combinatorica* 9, 321–337.

Balinski, M.L. (1965). Integer programming: methods, uses and computation. *Manage. Sci.* 12 (A), 253–313.

Balinski, M.L. (1969). Labeling to obtain a maximum matching (with discussion), in: R.C. Bose and T.A. Dowling (eds.), *Combinatorial Mathematics and its Applications*, The University of North California Monograph Series in Probability and Statistics, No. 4, University of North California Press, Chapel Hill, pp. 585–602.

Balinski, M.L. (1972). Establishing the matching polytope. *J. Comb. Theory, Ser. B* 13, 1–13.

Balinski, M.L., and R.E. Gomory (1964). A primal method for the assignment and transportation problems. *Manage. Sci.* 10, 578–593.

Balinski, M.L., and J. Gonzalez (1991). Maximum matchings in bipartite graphs via strong spanning trees. *Networks* 21, 165–179.

Ball, M.O., L.D. Bodin and R. Dial (1983). A matching based heuristic for scheduling mass transit crews and vehicles. *Transp. Sci.* 17, 4–31.

Ball, M.O., and U. Derigs (1983). An analysis of alternative strategies for implementing matching algorithms. *Networks* 13, 517–549.
Ball, M.O., U. Derigs, C. Hilbrand and A. Metz (1990). Matching problems with generalized upper bound side constraints. *Networks* 20, 703–721.
Barahona, F. (1980). *Application de l'Optimisation Combinatoire à Certains modèles de Verres de Spins: Complexité et Simulation*, Master's thesis, Université de Grenoble, France.
Barahona, F. (1990). Planar multicommodity flows, max cut and the Chinese postman problem, in: W. Cook and P.D. Seymour (eds), *Polyhedral Combinatorics*, DIMACS Series in Discrete Mathematics and Theoretical Computer Science, Vol. 1, American Mathematical Society, Providence, RI, pp. 189–202.
Barahona, F. (1993a). On cuts and matchings in planar graphs. *Math. Program.* 60, 53–68.
Barahona, F. (1993b). Reducing matching to polynomial size linear programming. *SIAM J. Opt.* 3, 688–695.
Barahona, F., R. Maynard, R. Rammal and J.P. Uhry (1982). Morphology of ground states of a two-dimensional frustration model. *J. Phys. A: Mathematical and General* 15, 673–699.
Bartholdi III, J.J., and L.K. Platzman (1983). A fast heuristic based on spacefilling curves for minimum-weight matching in the plane. *Inf. Process. Lett.* 17, 177–188.
Bartnik, G.W. (1978). Algorithmes de couplages dans les graphes, Thèse Doctorat 3e cycle, Unversité Paris VI.
Belck, H.-B. (1950). Reguläre Faktoren von Graphen. *J. Reine Angew. Math.* 188, 228–252.
Bellman, R. (1958). On a routing problem. *Q. Appl. Math.* 16, 87–90.
Berge, C. (1957). Two theorems in graph theory. *Proc. Nat. Acad. Sci. U.S.A.* 43, 842–844.
Berge, C. (1958). Sur le couplage maximum d'un graphe. *C.R. Acad. Sci., Sér. I (Mathematique)* 247, 258–259.
Berge, C. (1962). Sur une conjecture relative au problème des codes optimaux, Commun., 13éme Assemblée Générale de l'URSI, Tokyo.
Berge, C. (1985). *Graphs*, North-Holland, Amsterdam [revised edition of first part of: C. Berge, *Graphs and Hypergraphs*, North-Holland, Amsterdam, 1973].
Berstekas, D.P., D.A. Castañon, J. Eckstein and S.A. Zenion (1995). Parallel computing in network optimization, in: M.O. Ball, T.L. Magnanti, C. Monma and G.L. Nemhauser (eds.), *Network Models*, Handbooks in Operations Research and Management Science, Vol. 7, North-Holland, Amsterdam, Chapter 5, pp. 331–400, this volume.
Bertsekas, D.P. (1979). A distributed algorithm for the assignment problem, Working paper, Laboratory for Information and Decision Systems, M.I.T., Cambridge, MA.
Bertsekas, D.P. (1990). The auction algorithm for assignment and other network flow problems: a tutorial. *Interfaces* 20(4), 133–149.
Birkhoff, G. (1946). Tres observaciones sobre el algebra lineal. *Rev. Fac. Cie. Exactas Puras Apl. Univ. Nac. Tucuman, Ser. A (Matematicas y Fisica Teoretica)* 5, 147–151.
Blum, N. (1990a). A new approach to maximum matching in general graphs (extended abstract), in: M.S. Paterson (ed.), *Proc. 17th Int. Colloq. on Automata, Languages and Programming*, Lecture Notes in Computer Science, Vol. 443, Springer-Verlag, Berlin, pp. 586–597.
Blum, N. (1990b). *A New Approach to Maximum Matching in General Graphs*, Report No. 8546-CS, Institut für Informatik der Universität Bonn.
Bondy, J.A., and U.S.R. Murty (1976). *Graph theory with Applications*, MacMillan Press, London.
Borůvka, O. (1926). O jistém problému minimálním. *Práca Moravské Přírodovědecké Společnosti* 3, 37–48 (in Czech).
Bourjolly, J.-M., and W.R. Pulleyblank (1989). König-Egerváry graphs, 2-bicritical graphs and fractional matchings. *Discrete Appl. Math.* 24, 63–82.
Brezovec, C., G. Cornuéjols and F. Glover (1988). A matroid algorithm and its application to the efficient solution of two optimization problems on graphs. *Math. Program.* 42, 471–487.
Broder, A.Z. (1986). How hard is it to marry at random? (on the approximation of the permanent), in: *Proc. 18th Annual ACM Symp. on Theory of Computing*, Association for Computing Machinery, New York, NY, pp. 50–58 [Erratum in: *Proc. 20th ACM Symp. on Theory of Computing*,

1988, Association for Computing Machinery, New York, p. 551].

Brualdi, R.A., and P.M. Gibson (1977). Convex polyhedra of doubly stochastic matrices I. Applications of the permanent function. *J. Comb. Theory, Ser. A* 22, 194–230.

Burkard, R.E., and U. Derigs (1980). *Assignment and Matching Problems: Solution Methods with FORTRAN-Programs*, Lecture Notes in Economics and Mathematical Systems, Vol. 184, Springer-Verlag, Berlin, Heidelberg.

Burlet, M., and A.V. Karzanov (1993). *Minimum Weight* T, *d-Joins and Multi-Joins*, Rapport de Recherche RR929-M, Laboratoire ARTEMIS, Université Joseph Fourier, Grenoble.

Chari, S., P. Rohatgi and A. Srinivasan (1993). Randomness-optimal unique element isolation, with applications to perfect matching and related problems, preprint.

Christofides, N. (1976). *Worst-case Analysis of a New Heuristic for the Travelling Salesman Problem*, Technical report, GSIA Carnegie-Mellon University, Pittsburgh, Pennsylvania.

Cook, S.A. (1971). The complexity of theorem-proving procedures, in: *Proc. 3rd Annual ACM Symp. on Theory of Computing*, Association for Computing Machinery, New York, NY, pp. 151–158.

Cook, W. (1983a). A minimal totally dual integral defining system for the b-matching polyhedron. *SIAM J. Algebraic Discrete Methods* 4, 212–220.

Cook, W. (1983b). *On some Aspects of Totally Dual Integral Sytems*, PhD thesis, Department of Combinatorics and Optimization, University of Waterloo, Waterloo, Ontario.

Cook, W. and W.R. Pulleyblank (1987). Linear systems for constrained matching problems. *Math. Oper. Res.* 12, 97–120.

Cornuéjols, G. (1988). General factors of graphs. *J. Comb. Theory, Ser. B* 45, 185–198.

Cornuéjols, G., and D. Hartvigsen (1986). An extension of matching theory. *J. Comb. Theory, Ser. B* 40, 285–296.

Cornuéjols, G., D. Hartvigsen and W. Pulleyblank (1982). Packing subgraphs in a graph. *Oper. Res. Lett.* 1, 139–143.

Cornuéjols, G., and W. Pulleyblank (1980a). A matching problem with side conditions. *Discrete Math.* 29, 135–159.

Cornuéjols, G., and W.R. Pulleyblank (1980b). Perfect triangle-free 2-matchings. *Math. Program. Study* 13, 1–7.

Cornuéjols, G., and W. Pulleyblank (1982). The travelling salesman polytope and $\{0, 2\}$-matchings. *Ann. Discrete Math.* 16, 27–55.

Cornuéjols, G., and W.R. Pulleyblank (1983). Critical graphs, matchings and tours or a hierarchy of relaxations for the travelling salesman problem. *Combinatorica* 3, 35–52.

Crocker, S.T. (1993). An experimental comparison on two maximum cardinality matching programs, in: D.S. Johnson and C.C. McGeoch (eds), *Network Flows and Matchings: First DIMACS Implementation Challenge*, DIMACS Series in Discrete Mathematics and Theoretical Computer Science, Vol. 12, American Mathematical Society, Providence, RI, pp. 519–537.

Csansky, L. (1976). Fast parallel matrix inversion algorithms. *SIAM J. Comp.* 5, 618–623.

Cunningham, W.H., and J. Green-Krotki (1986). Dominants and submissives of matching polyhedra. *Math. Program.* 36, 228–237.

Cunningham, W.H., and J. Green-Krótki (1991). b-Matching degree-sequence polyhedra. *Combinatorica* 11, 219–230.

Cunningham, W.H., and J. Green-Krótki (1994). A separation algorithm for the matchable set polytope, *Math. Program.* 65, 139–190.

Cunningham, W.H., and A.B. Marsh III (1978). A primal algorithm for optimum matching, in: M.L. Balinski and A.J. Hoffman (eds.), *Polyhedral Combinatorics (dedicated to the memory of D.R. Fulkerson), Mathematical Programming Study* 8, North-Holland, Amsterdam, pp. 50–72.

Cunningham, W.H., and F. Zhang (1992). Subgraph degree-sequence polyhedra, in: E. Balas, G. Cornuéjols and R. Kannan (eds.), *Integer Programming and Combinatorial Optimization*, Proc. Conf. of the Mathematical Progamming Society, Carnegie-Mellon University, May 25–27, 1992, pp. 246–259.

Dahlhaus, E., and M. Karpinski (1988). Parallel construction of perfect matchings and Hamiltonian cycles on dense graphs. *Theor. Comput. Sci.* 61, 121–136.

Dantzig, G.B. (1951). Maximization of a linear function of variables subject to linear inequalities, in: Tj.C. Koopmans (ed.), *Activity Analysis of Production and Allocation*, John Wiley, New York, NY, pp. 339–347.

Deming, R.W. (1979). Independence numbers of graphs – an extension of the Koenig-Egervary theorem. *Discrete Math.* 27, 23–33.

Derigs, U. (1981). A shortest augmenting path method for solving minimal perfect matching problems. *Networks* 11, 379–390.

Derigs, U. (1986a). A short note on matching algorithms. *Math. Program. Study* 26, 200–204.

Derigs, U. (1986b). Solving large-scale matching problems efficiently: a new primal matching approach. *Networks* 16, 1–16.

Derigs, U. (1988a). *Programming in Networks and Graphs*, Lecture Notes in Economics and Mathematical Systems, Vol. 300, Springer-Verlag, Berlin.

Derigs, U. (1988b). Solving non-bipartite matching problems via shortest path techniques. *Ann. Oper. Res.* 13, 225–261.

Derigs, U., and A. Metz (1986). On the use of optimal fractional matchings for solving the (integer) matching problem. *Computing* 36, 263–270.

Derigs, U., and A. Metz (1991). Solving (large scale) matching problems combinatorially. *Math. Program.* 50, 113–121.

Derigs, U., and A. Metz (1992). A matching-based approach for solving a delivery/pick-up vehicle routing problem with time constraints. *Oper. Res. Spektrum* 14, 91–106.

Devine, M.D. (1973). A model for minimizing the cost of drilling dual completion oil wells. *Manage. Sci.* 20, 532–535.

Dijkstra, E.W. (1959). A note on two problems in connexion with graphs. *Numer. Math.* 1, 269–271.

Dilworth, R.P. (1950). A decomposition theorem for partially ordered sets. *Ann. Math.* (2) 51, 161–166.

Dinic, E.A. (1970). Algorithm for solution of a problem of maximum flow in a network with power estimation (in Russian). *Dokl. Akad. Nauk SSSR* 194, 745–757 [English translation: *Soviet Mathemathics Doklady*, 11, 1277–1280].

Dulmage, A.L., and N.S. Mendelsohn (1958). Coverings of bipartite graphs. *Can. J. Math.* 10, 517–534.

Dulmage, A.L., and N.S. Mendelsohn (1959). A structure theory of bipartite graphs of finite exterior dimension. *Trans. R. Soc. Can., Ser. III* 53, 1–13.

Dulmage, A.L., and N.S. Mendelsohn (1967). Graphs and matrices, in: F. Harary (ed.), *Graph Theory and Theoretical Physics*, Academic Press, New York, NY, pp. 167–277.

Edmonds, J. (1965a). The Chinese postman's problem. *Bull. Oper. Res. Soc.* 13, B–73.

Edmonds, J. (1965b). Maximum matching and a polyhedron with 0,1-vertices. *J. Res. Nat. Bur. Stand. – B. Math. Math. Phys.* 69B, 125–130.

Edmonds, J. (1965c). Paths, trees and flowers. *Can. J. Math.* 17, 449–467.

Edmonds, J. (1967). Systems of distinct representatives and linear algebra. *J. Res. Nat. Bur. Stand. – B. Math. Math. Phys.* 71B, 241–245.

Edmonds, J. (1970). Submodular functions, matroids, and certain polyhedra, in: R. Guy, H. Hanani, N. Sauer and J. Schönheim (eds.), *Combinatorial Structures and their Applications*, Gordon and Breach, New York, NY, pp. 69–87.

Edmonds, J., and R. Giles (1977). A min–max relation for submodular functions on graphs. *Ann. Discrete Math.* 1, 185–204.

Edmonds, J., and E. Johnson (1970). Matching: a well-solved class of integer linear programs, in: R. Guy, H. Hanani, N. Sauer and J. Schönheim (eds.), *Combinatorial Structures and their Applications*, Gordon and Breach, New York, NY, pp. 89–92.

Edmonds, J., E.L. Johnson and S.C. Lockhart (1969). *Blossom I, a Code for Matching*, unpublished report, IBM T.J. Watson Research Center, Yorktown Heights, NY.

Edmonds, J., and E. L. Johnson (1973). Matching, Euler tours and the Chinese postman. *Math. Program.* 5, 88–124.

Edmonds, J., and R.M. Karp (1970). Theoretical improvements in algorithmic efficiency for network flow problems, in: R. Guy, H. Hanani, N. Sauer and J. Schönheim (eds.), *Combinatorial Structures and their Applications*, Gordon and Breach, New York, NY, pp. 93–96.

Edmonds, J., and R.M. Karp (1972). Theoretical improvements in algorithmic efficiency for network flow problems. *J. Assoc. Comput. Mach.* 19, 248–264.

Edmonds, J., L. Lovász and W.R. Pulleyblank (1982). Brick decompositions and the matching rank of graphs. *Combinatorica* 2, 247–274.

Egerváry, E. (1931). Matrixok kombinatorius tulajdonságairol (in Hungarian). *Matematikai és Fizikai Lapok* 38, 16–28.

Elias, P., A. Feinstein and C.E. Shannon (1956). Note on the maximum flow through a network. *IRE Trans. Inf. Theory* IT 2, 117–119.

Erdös, P., and T. Gallai (1960). Gráfok elöírt fokú pontokkal (in Hungarian). *Mat. Lapok* 11, 264–274.

Euler, L. (1736). Solutio problematis ad geometriam situs pertinentis. *Comment. Acad. Sci. Imp. Petropolitanae* 8, 128–140.

Even, S., and O. Kariv (1975). An $O(n^{2.5})$ algorithm for maximum matching in general graphs, in: *Proc. 16th Annual Symp. on Foundations of Computer Science*, IEEE, New York, NY, pp. 100–112.

Even, S., and R.E. Tarjan (1975). Network flow and testing graph connectivity. *SIAM J. Comput.* 4, 507–518.

Feder, T., and R. Motwani (1991). Clique partitions, graph compression and speeding-up algorithms, in: *Proc. 23rd Annual ACM Symp. on Theory of Computing*, Association for Computing Machinery, New York, NY, pp. 123–133.

Flood, M.M. (1956). The traveling-salesman problem. *Oper. Res.* 4, 61–75.

Floyd, R.W. (1962a). Algorithm 96: ancestor. *Commun. Assoc. Comput. Mach.* 5, 344–345.

Floyd, R.W. (1962b). Algorithm 97: shortest path. *Commun. Assoc. Comput. Mach.* 5, 345.

Ford Jr., L.R. (1956). *Network Flow Theory*, Paper P-923, RAND Corporation, Santa Monica, CA.

Ford Jr., L.R., and D.R. Fulkerson (1956). Maximal flow through a network. *Can. J. Math.* 8, 399–404.

Ford Jr., L.R., and D.R. Fulkerson (1957). A simple algorithm for finding maximal network flows and an application to the Hitchcock problem. *Can. J. Math.* 9, 210–218.

Frank, A. (1990). Packing paths, circuits and cuts – a survey, in: B. Korte, L. Lovász, H.J. Prömel and A. Schrijver (eds.), *Paths, Flows and VLSI-Layout*, Springer-Verlag, Berlin, Heidelberg, pp. 47–100.

Frank, A. (1993). Conservative weightings and ear-decompositions of graphs. *Combinatorica* 13, 65–81.

Frank, A., A. Sebö and É. Tardos (1984). Covering directed and odd cuts. *Math. Program. Study* 22, 99–112.

Frank, A., and Z. Szigeti (1994). On packing T-cuts, *J. Comb. Theory, Ser. B* 61, 263–271.

Fredman, M.L., and R.E. Tarjan (1987). Fibonacci heaps and their uses in improved network optimization algorithms. *J. Assoc. Comput. Mach.* 34, 596–615.

Frieze, A., C. McDiarmid and B. Reed (1990). Greedy matching on the line. *SIAM J. Comput.* 19, 666–672.

Frobenius, G. (1912). Über Matrizen aus nicht negativen Elementen. *Sitszungsberichte der königlich preussischen Akademie der Wissenschaften zu Berlin,* 456–477.

Frobenius, G. (1917). Über zerlegbare Determinanten. *Sitszungsberichte der königlich preussischen Akademie der Wissenschaften zu Berlin,* 274–277.

Fujii, M., T. Kasami and N. Ninomiya (1969). Optimal sequencing of two equivalent processors. *SIAM J. Appl. Math.* 17, 784–789 [Erratum in: *SIAM J. Appl. Math.* 20 (1971), 141].

Fulkerson, D.R. (1961). An out-of-kilter method for minimal cost flow problems. *SIAM J. Appl. Math.* 9, 18–27.

Gabow, H.N. (1973). *Implementation of Algorithms for Maximum Matching on Non-bipartite Graphs*, PhD thesis, Stanford University, Department of Computer Science, 1973.

Gabow, H.N. (1976). An efficient implementation of Edmonds' algorithm for maximum matching on graphs. *J. Assoc. Comput. Mach.* 23, 221–234.

Gabow, H.N. (1983). An efficient reduction technique for degree-constraint subgraph and bidirected network flow problems, in: *Proc. 15th Annual ACM Symp. on Theory of Computing*, Association for Computing Machinery, New York, NY, pp. 448–456.

Gabow, H.N. (1985). A scaling algorithm for weighted matching on general graphs, in: *Proc. 26th Annual Symp. on Foundations of Computer Science*, IEEE, New York, NY, pp. 90–100.

Gabow, H.N. (1990). Data structures for weighted matching and nearest common ancestors with linking, in: *Proc. 1st Annual ACM–SIAM Symp. on Discrete Algorithms*, Association for Computing Machinery, New York, NY, pp. 434–443.

Gabow, H.N., Z. Galil and T.H. Spencer (1989). Efficient implementation of graph algorithms using contraction. *J. Assoc. Comput. Mach.* 36, 540–572.

Gabow H.N., and R.E. Tarjan (1983). A linear-time algorithm for a special case of disjoint set union, in: *Proc. 15th Annual ACM Symp. on Theory of Computing*, Association for Computing Machinery, New York, NY, pp. 246–251.

Gabow, H.N., and R.E. Tarjan (1991). Faster scaling algorithms for general graph-matching problems. *J. Assoc. Comput. Mach.* 38, 815–853.

Gale, D., H.W. Kuhn and A.W. Tucker (1951). Linear programming and the theory of games, in: Tj.C. Koopmans (ed.), *Activity Analysis of Production and Allocation*, New York, NY, pp. 317–329.

Gale, D., and L.S. Shapley (1962). College admissions and the stability of marriage. *Am. Math. Mon.* 69, 9–15.

Galil, Z. (1986a). Efficient algorithms for finding maximum matching in graphs. *ACM Comput. Surv.* 18, 23–38.

Galil, Z. (1986b). Sequential and parallel algorithms for finding maximum matchings in graphs. *Annu. Rev. Comput. Sci.* 1, 197–224.

Galil, Z., S. Micali and H. Gabow (1986). An $O(EV \log V)$ algorithm for finding a maximal weighted matching in general graphs. *SIAM J. Comput.* 15, 120–130.

Gallai, T. (1950). On factorisation of graphs. *Acta Math. Acad. Sci. Hung.* 1 133–153.

Gallai, T. (1959). Über extreme Punkt- und Kantenmengen. *Ann. Univ. Sci. Budap. Rolando Eötvös Nominatae, Sect. Math.* 2, 133–138.

Gallai, T. (1963). Kritische Graphen II. *Mag. Tud. Akad. Mat. Kut. Intéz. Közl.* 8, 373–395.

Gallai, T. (1964). Maximale Systeme unabhängiger Kanten. *Mag. Tud. Akad. Mat. Kut. Intéz. Közl.* 9, 401–413.

Gamble, A.R. (1989). *Polyhedral Extentions of Matching Theory*, PhD thesis, Department of Combinatorics and Optimization, University of Waterloo, Waterloo, Ontario.

Gerards, A.M.H. (1991). Compact systems for T-join and perfect matching polyhedra of graphs with bounded genus, *Oper. Res. Lett.* 10, 377–382.

Gerards, A.M.H. (1992). On shortest T-joins and packing T-cuts. *J. Comb. Theory, Ser. B* 55, 73–82.

Giles, R. (1982a). Optimum matching forests I: special weights. *Math. Program.* 22, 1–11.

Giles, R. (1982b). Optimum matching forests II: general weights. *Math. Program.* 22, 12–38.

Giles, R. (1982c). Optimum matching forests III: facets of matching forest polyhedra. *Math. Program.* 22, 39–51.

Goemans, M.X., and D.P. Williamson (1992). A general approximation technique for constrained forest problems, in: *Proc. 3rd Annual ACM–SIAM Symp. on Discrete Algorithms*, Association for Computing Machinery, New York, NY, pp. 307–316.

Goldberg, A.V., S.A. Plotkin, D.B. Shmoys and E. Tardos (1992). Using interior-point methods for fast parallel algorithms for bipartite matching and related problems. *SIAM J. Comput.* 21, 140–150.

Goldberg, A.V., S.A. Plotkin and P.M. Vaidya (1993). Sublinear-time parallel algorithms for matching and related problems. *J. Algorithms* 14, 180–213.

Goldberg, A.V., É. Tardos and R.E. Tarjan (1990). Network flow algorithms, in: B. Korte, L. Lovász, H.J. Prömel and A. Schrijver (eds.), *Paths, Flows and VLSI-Layout*, Springer-Verlag,

Berlin, Heidelberg, pp. 101–164.
Goldberg, A.V., and R.E. Tarjan (1989). Finding minimum-cost circulations by canceling negative cycles. *J. Assoc. Comput. Mach.* 36, 873–886.
Gondran, M., and M. Minoux (1984). *Graphs and Algorithms*, Wiley/Interscience, New York, NY.
Grigoriadis, M.D., and B. Kalantari (1986). A lower bound to the complexity of Euclidean and rectilinear matching algorithms. *Inf. Process. Lett.* 22, 73–76.
Grigoriadis, M.D., and B. Kalantari (1988). A new class of heuristic algorithms for weighted perfect matching. *J. Assoc. Comput. Mach.* 35, 769–776.
Grigoriadis, M.D., B. Kalantari and C.Y. Lai (1986). On the existence of weakly greedy matching heuristics. *Oper. Res. Lett.* 5, 201–205.
Grigoriev, D.Y., and M. Karpinski (1987). The matching problem for bipartite graphs with polynomially bounded permanents is in NC, in: *28th Annual Symposium on Foundations of Computer Science*, IEEE, New York, NY, pp. 166–172.
Grötschel, M., and O. Holland (1985). Solving matching problems with linear programming. *Math. Program.* 33, 243–259.
Grötschel, M., L. Lovász and A. Schrijver (1981). The ellipsoid method and its consequences in combinatorial optimization. *Combinatorica* 1, 169–197 [corrigendum in: *Combinatorica* 4 (1984), 291-295].
Grötschel, M., L. Lovász and A. Schrijver (1984). Polynomial algorithms for perfect graphs. *Ann. Discrete Math.* 21, 325–356.
Grötschel, M., L. Lovász and A. Schrijver (1988). *Geometric Algorithms and Combinatorial Optimization*, Springer-Verlag, Berlin.
Grötschel, M., and M.W. Padberg (1985). Polyhedral theory, in: E.L. Lawler, J.K. Lenstra, A.H.G. Rinnooy Kan, and D.B. Shmoys (eds.), *The Traveling Salesman Problem, A Guided tour of Combinatorial Optimization*, John Wiley and Sons, Chichester, pp. 251–305.
Grötschel, M., and W.R. Pulleyblank (1981). Weakly bipartite graphs and the max-cut problem. *Oper. Res. Lett.* 1, 23–27.
Grover, L.K. (1992). Fast parallel algorithms for bipartite matching, in: E. Balas, G. Cornuéjols, and R. Kannan (eds.). *Integer Programming and Combinatorial Optimization*, Proc. Conf. of the Mathematical Programming Society, Carnegie-Mellon University, May 25–27, 1992, pp. 367–384.
Gusfield, D., and R.W. Irving (1989). *The Stable Marriage Problem: Structure and Algorithms*, MIT Press, Cambridge, Massachusetts.
Hadlock, F. (1975). Finding a maximum cut of a planar graph in polynomial time. *SIAM J. Comput.* 4, 221–225.
Hall Jr., M. (1956). An algorithm for distinct representatives. *Am. Math. Mon.* 716–717.
Hall, P. (1935). On representatives of subsets. *J. Lond. Math. Soc.* 10, 26–30.
Helgason, R.V., and J.L. Kennington (1995). Primal simplex algorithms for minimum cost network flows, in: M.O. Ball, T.L. Magnanti, C. Monma and G.L. Nemhauser (eds.), *Network Models*, Handbooks in Operations Research and Management Science, Vol. 7, North-Holland, Amsterdam, Chapter 2, pp. 85–134, this volume.
He, X. (1991). An efficient parallel algorithm for finding minimum weight matching for points on a convex polygon, *Inf. Process. Lett.* 37, 111–116.
Hetyei, G. (1964). 2 × 1-es téglalapokkal lefedhetö idomokról (in Hungarian). *Pécsi Tanárképzö Föiskola Tud. Közl.* 8, 351–367
Hoffman, A.J. (1974). A generalization of max flow-min cut. *Math. Program.* 6, 352–359.
Hoffman, A.J., and J.B. Kruskal (1956). Integral boundary points of convex polyhedra, in: H.W. Kuhn and A.W. Tucker (eds.). *Linear Inequalities and Related Systems*, Annals of Mathematical Studies, Vol. 38, Princeton University Press, Princeton, NJ, pp. 223–246.
Hoffman, A.J., and R. Oppenheim (1978). Local unimodularity in the matching polytope. *Ann. Discrete Math.* 2, 201–209.
Holyer, I. (1981). The NP-completeness of edge-coloring. *SIAM J. Comput.* 10, 718–720.
Hopcroft, J.E., and R.M. Karp (1971). An $n^{5/2}$ algorithm for maximum matchings in bipartite graphs, in: *Conf. Record 1971 12th Annual Symp. on Switching and Automata Theory*, IEEE, New

York, NY, pp. 122–125.

Hopcroft, J.E., and R.M. Karp (1973). An $n^{5/2}$ algorithm for maximum matchings in bipartite graphs. *SIAM J. Comput.* 2, 225–231.

Imai, H. (1986). Worst-case analysis for planar matching and tour heuristics with bucketing techniques and spacefilling curves. *J. Oper. Res. Soc. Jap.* 29, 43–67.

Imai, H., H. Sanae and M. Iri (1984). A planar-matching heuristic by means of triangular buckets, in: *Proc. 1984 Fall Conf. of the Operations Research Society of Japan*, 2-D-4, pp. 157–158 (in Japanese).

Iri, M., K. Murota and S. Matsui (1981). Linear-time approximation algorithms for finding the minimum-weight perfect matching on a plane. *Inf. Process. Lett.* 12, 206–209.

Iri, M., K. Murota and S. Matsui (1982). An approximate solution for the problem of optimizing the plotter pen movement, in: R.F. Drenick and F. Kozin (eds.), *System Modeling and Optimization*, Proc. 10th IFIP Conf., New York, 1981, Lecture Notes in Control and Information Sciences, Vol. 38, Springer-Verlag, Berlin, pp. 572–580.

Iri, M., K. Murota and S. Matsui (1983). Heuristics for planar minimum-weight perfect matchings. *Networks* 13, 67–92.

Iri, M., and A. Taguchi (1980). The determination of the pen-movement of an XY-plotter and its computational complexity, in: *Proc. 1980 Spring Conf. of the Operations Research Society of Japan*, P-8, pp. 204–205 (in Japanese).

Irving, R.W. (1985). An efficient algorithm for the "stable roommates" problem. *J. Algorithms* 6, 577–595.

Jarník, V. (1930). O jistém problému minimálním (in Czech). *Práca Moravské Přírodovědecké Společnosti* 6, 57–63

Jerrum, M., and A. Sinclair (1989). Approximating the permanent. *SIAM J. Comput.* 18, 1149–1178.

Jünger, M., and W. Pulleyblank (1991). New primal and dual matching heuristics, Report No 91.105, Institut für Informatik, Universität zu Köln.

Jünger, M., G. Reinelt and G. Rinaldi (1995). The traveling salesman problem, in: M.O. Ball, T.L. Magnanti, C. Monma and G.L. Nemhauser (eds.), *Network Models*, Handbooks in Operations Research and Management Science, Vol. 7, North-Holland, Amsterdam, Chapter 4, pp. 225–330, this volume.

Kameda, T., and I. Munro (1974). An $O(|V| \cdot |E|)$ algorithm for maximum matching of graphs. *Computing* 12, 91–98.

Kariv, O. (1976). *An $O(n^{5/2})$ Algorithm for Maximum Matching in General Graphs*, PhD thesis, Weizman Institute of Science, Rehovot.

Karloff, H.J. (1986). A Las Vegas RNC algorithm for maximum matching. *Combinatorica* 6, 387–391.

Karmarkar, N. (1984). A new polynomial-time algorithm for linear programming. *Combinatorica* 4, 373–395.

Karp, R.M. (1972). Reducibility among combinatorial problems, in: R.E. Miller and J.W. Thatcher (eds.), *Complexity of Computer Computations*, Plenum Press, New York, NY, pp. 85–103.

Karp, R.M., and E. Upfal an A. Wigderson (1986). Constructing a perfect matching is in random NC. *Combinatorica* 6, 35–48.

Karp, R.M., and C. H. Papadimitriou (1982). On linear characterizations of combinatorial optimization problems. *SIAM J. Comput.* 11, 620–632.

Karp, R.M., and V. Ramachandran (1990). Parallel algorithms for shared-memory machines, in: J. van Leeuwen (ed.), *Handbook of Theoretical Computer Science, Vol. A: Algorithms and Complexity*, Elsevier, Amsterdam, pp. 869–941.

Karzanov, A. (1992). Determining the distance to the perfect matching polytope of a bipartite graph, preprint.

Kasteleyn, P.W. (1963). Dimer statistics and phase transitions. *J. Math. Phys.* 4, 287–293.

Kasteleyn, P.W. (1967). Graph theory and crystal physics, in: F. Harary (ed.), *Graph Theory and Theoretical Physics*, Academic Press, New York, NY, pp. 43–110.

Khachiyan, L.G. (1979). A polynomial algorithm in linear programming (in Russian). *Dokl. Akad. Nauk SSSR* 224, 1093–1096.

König, D. (1915). Vonalrendszerek és determinánsok (in Hungarian). *Mat. Természettudományi Értesitö* 33, 221–229.

König, D. (1916a). Graphok és alkalmazásuk a determinánsok és a halmazok elméletében (in Hungarian). *Mat. Természettudományi Értesitö* 34, 104–119.

König, D. (1916b). Graphen und ihre Anwendung auf Determinantentheorie und Mengenlehre. *Math. Ann.* 77, 453–465.

König, D. (1931). Graphok és matrixok (in Hungarian). *Mat. Fizikai Lapok* 38, 116–119.

König, D. (1933). Über trennende Knotenpunkte in Graphen (nebst Anwendungen auf Determinanten und Matrizen). *Acta Litt. Sci. Regiae Univ. Hung. Francisco-Josephinae (Szeged), Sectio Sci. Math.* 6, 155–179.

König, D. (1936). *Theorie der endlichen und unendlichen Graphen*, Akademischen Verlagsgesellschaft, Leipzig [reprinted: Chelsea, New York, 1950, and Teubner, Leipzig, 1986].

Korach, E. (1982). *On Dual Integrality, Min-Max Equalities and Algorithms in Combinatorial Programming*, PhD thesis, Department of Combinatorics and Optimization. University of Waterloo, Waterloo, Ontario.

Koren, M. (1973). Extreme degree sequences of simple graphs. *J. Comb. Theory, Ser. B* 15, 213–234.

Kotzig, A. (1959a). Z teórie konečných grafov s lineárnym faktorom I (in Slovak). *Mat.-Fyz. Časopis Slovenskej Akad. Vied* 9, 73–91.

Kotzig, A. (1959b). Z teórie konečných grafov s lineárnym faktorom II (in Slovak). *Mat.-Fyz. Časopis Slovenskej Akad. Vied* 9, 136–159.

Kotzig, A. (1960). Z teórie konečných grafov s lineárnym faktorom III (in Slovak). *Mat.-Fyz. Časopis Slovenskej Akad. Vied* 10, 205–215

Kozen, D., U.V. Vazirani and V.V. Vazirani (1985). NC algorithms for comparibility graphs, interval graphs, and testing for unique perfect matching, in: S.N. Maheshwari (ed.). *Foundations of Software Technology and Theoretical Computer Science, Fifth Conference, New Delhi, 1985*, Lecture Notes in Computer Science, Vol. 206, Springer-Verlag, Berlin, pp. 496–503.

Kruskal, J.B. (1956). On the shortest spanning subtree of a graph and the traveling salesman problem. *Proc. Amer. Math. Soc.* 7, 48–50.

Kuhn, H.W. (1955). The Hungarian method for the assignment problem. *Nav. Res. Logist. Q.* 2, 83–97.

Kuhn, H.W. (1956). Variants of the Hungarian method for assignment problems. *Nav. Res. Logist. Q.* 3, 253–258.

Kwan Mei-Ko (1962). Graphic programming using odd and even points. *Chin. Math.* 1, 273–277.

Lawler, E.L. (1976). *Combinatorial Optimization: Networks and Matroids*, Holt, Rinehart and Winston, New York, NY.

Lawler, E.L., J.K. Lenstra, A.H.G. Rinnooy Kan and D.B. Shmoys (1985). *The Traveling Salesman Problem, A Guided tour of Combinatorial Optimization*, John Wiley and Sons, Chichester.

Lessard, R., J.-M. Rousseau and M. Minoux (1989). A new algorithm for general matching problems using network flow subproblems. *Networks* 19, 459–479.

Lipton, R.J., and R.E. Tarjan (1979). A separator theorem for planar graphs. *SIAM J. Appl. Math.* 36, 177–189.

Lipton, R.J., and R.E. Tarjan (1980). Applications of a planar separator theorem. *SIAM J. Comput.* 9, 615–627.

Little, C.H.C. (1974). An extension of Kasteleyn's method for enumerating the 1-factors of planar graphs, in: D.A. Holton (ed.), *Combinatorial Mathematics*, Proc. 2nd Australian Conf., Lecture Notes in Mathematics, Vol. 403, Springer-Verlag, Berlin, pp. 63–72.

Lovász, L. (1970a). The factorization of graphs, in: R. Guy, H. Hanani, N. Sauer and J. Schönheim (eds.), *Combinatorial Structures and their Applications*, Gordon and Breach, New York, NY, pp. 243–246.

Lovász, L. (1970b). Generalized factors of graphs, in: P. Erdös, A. Rényi and V.T. Sós (eds.), *Combinatorial Theory and its Applications II*, Coloq. Math. Soc. János Bolyai, 4, North-Holland,

Amsterdam, pp. 773–781.
Lovász, L. (1970c). Subgraphs with prescribed valencies. *J. Comb. Theory* 8, 391–416.
Lovász, L. (1972a). The factorization of graphs II. *Acta Math. Acad. Sci. Hung.* 23, 223–246.
Lovász, L. (1972b). Normal hypergraphs and the perfect graph conjecture. *Discrete Math.* 2, 253–267.
Lovász, L. (1972c). A note on factor-critical graphs. *Stud. Sci. Math. Hung.* 7, 279–280.
Lovász, L. (1972d). On the structure of factorizable graphs. *Acta Math. Acad. Sci. Hung.* 23, 179–195.
Lovász, L. (1972e). On the structure of factorizable graphs, II. *Acta Math. Acad. Sci. Hung.* 23, 465–478.
Lovász, L. (1973). Antifactors of graphs. *Period. Math. Hung.* 4, 121–123.
Lovász, L. (1975). 2-matchings and 2-covers of hypergraphs. *Acta Math. Acad. Sci. Hung.* 26 (1975) 433–444.
Lovász, L. (1979a). Graph theory and integer programming. *Ann. Discrete Math.* 4, 141–158.
Lovász, L. (1979b). On determinants, matchings and random algorithms, in: L. Budach (ed.), *Fundamentals of Computation Theory*, FCT '79, Proc. Conf. on Algebraic, Arithmetic and Categorial Methods in Computation Theory, Akademie-Verlag, Berlin, pp. 565–574.
Lovász, L. (1983). Ear-decompositions of matching-covered graphs. *Combinatorica* 3, 105–117.
Lovász, L. (1987). Matching structure and the matching lattice. *J. Comb. Theory, Ser. B* 43, 187–222.
Lovász, L., and M.D. Plummer (1975). On bicritical graphs, in: A. Hajnal, R.Rado and V.T. Sós (eds.), *Infinite and Finite Sets, Vol. II*, North-Holland, Amsterdam, pp. 1051–1079.
Lovász, L., and M.D. Plummer (1986). *Matching Theory*, Akadémiai Kiadó, Budapest [also published as: *North-Holland Mathematics Studies* Vol. 121, North-Holland, Amsterdam, 1986].
Marcotte, O., and S. Suri (1991). Fast matching algorithms for points on a polygon. *SIAM J. Comput.* 20, 405–422.
Marsh III, A.B. (1979). *Matching Algorithms*, PhD thesis, The Johns Hopkins University, Baltimore.
Matsumoto, K., T. Nishizeki and N. Saito (1986). Planar multicommodity flows, maximum matchings and negative cycles. *SIAM J. Comput.* 15, 495–510.
Mattingly, R.B., and N.P. Ritchey (1993). Implementing an $O(\sqrt{N}M)$ cardinality matching algorithm, in: D.S. Johnson and C.C. McGeoch (eds.), *Network Flows and Matchings: First DIMACS Implementation Challenge*, DIMACS Series in Discrete Mathematics and Theoretical Computer Science, Vol. 12, American Mathematical Society, Providence, RI, pp. 539–556.
Menger, K. (1927). Zur allgemeinen Kurventheorie. *Fundam. Math.* 10, 96–115.
Micali, S., and V.V. Vazirani (1980). An $O(\sqrt{|v|}|E|)$ algorithm for finding maximum matching in general graphs, in: *Proc. 21th Annual Symp. on Foundations of Computer Science*, IEEE, New York, NY, pp. 17–27.
Miller, G.L., and J. Naor (1989). Flow in planar graphs with multiple sources and sinks, in: *Proc. 30th Annual Symp. on Foundations of Computer Science*, IEEE, New York, NY, pp. 112–117.
Minc, H. (1978). *Permanents*, Addison-Wesley, Reading.
Minty, G.J. (1960). Monotone networks. *Proc. R. Soc. Lond.* 257, 194–212.
Minty, G.J. (1980). On maximal independent sets of vertices in claw-free graphs. *J. Comb. Theory, Ser. B* 28, 284–304.
Mirsky, L. (1971). *Transversal Theory*, Academic Press, London.
Middendorf, M., and F. Pfeiffer (1990). On the complexity of the disjoint paths problem (extended abstract), in: W. Cook and P.D. Seymour, (eds.), *Polyhedral Combinatorics*, DIMACS Series in Discrete Mathematics and Theoretical Computer Science, Vol. 1, American Mathematical Society, Providence, RI, pp. 171–178.
Motzkin, T.S. (1956). The assignment problem, in: J.H. Curtiss (ed.), *Numerical Analysis*, Proc. Symp. in Applied Mathematics, Vol. IV, McGraw-Hill, New York, NY, pp. 109–125.
Mulder, H.M. (1992). Julius Petersen's theory of regular graphs. *Discrete Math.* 100, 157–175.
Mulmuley, K., U.V. Vazirani and V.V. Vazirani (1987). Matching is as easy as matrix inversion. *Combinatorica* 7, 105–113.

Munkres, J. (1957). Algorithms for the assignment and transportation problems. *J. Soc. Ind. Appl. Math.* 32–38.

Murota, K. (1993). *Combinatorial Relaxation Algorithm for the Maximum Degree of Subdeterminants: Computing Smith-McMillan Form at Infinity and Structural Indices in Kronecker Form*, RIMS-954, Research Institute for Mathematical Sciences, Kyoto University.

Murty, U.S.R. (1994). *The Matching Lattice and Related Topics*, preliminary report, University of Waterloo, Waterloo, Ontario.

Naddef, D. (1982). Rank of maximum matchings in a graph. *Math. Program.* 22, 52–70.

Naddef, D.J., and W.R. Pulleyblank (1982). Ear decompositions of elementary graphs and GF_2-rank of perfect matchings. *Ann. Discrete Math.* 16, 241–260.

Nemhauser, G.L., and L.A. Wolsey (1988). *Integer and Combinatorial Optimization*, John Wiley and Sons, New York, NY.

von Neumann, J. (1947). *Discussion of a maximum problem*, unpublished working paper, Institute for Advanced Studies, Princeton, NJ [Reprinted in: A.H. Taub (ed.), *John von Neumann, Collected works, Vol. VI,* Pergamon Press, Oxford, 1963, pp. 89–95].

von Neumann, J. (1953). A certain zero-sum two-person game equivalent to the optimal assignment problem, in: H.W. Kuhn and A.W. Tucker (eds.), *Contributions to the Theory of Games II*, Annals of Mathematical Studies, Vol. 28, Princeton University Press, Princeton, NJ, pp. 5–12.

Norman, R.Z., and M.O. Rabin (1959). An algorithm for a minimum cover of a graph. *Proc. Am. Math. Soc.* 10, 315–319.

Ore, O. (1955). Graphs and matching theorems. *Duke Math. J.* 22, 625–639.

Oxley, J.G. (1992). *Matroid Theory*, Oxford University Press, New York, NY.

Padberg, M.W. and M.R. Rao (1982). Odd minimum cut-sets and *b*-matchings. *Math. Oper. Res.* 7, 67–80.

Papadimitriou, C.H. (1977). The probabilistic analysis of matching heuristics, in: *Proc. 15th Annual Allerton Conf. on Communication, Control, and Computing*, pp. 368–378.

Peled, U.N., and M.N. Srinivasan (1989). The polytope of degree sequences, *Linear Algebra Appl.* 114/115, 349–373.

Petersen, J. (1891). Die Theorie der regulären graphs. *Acta Math.* 15, 193–220.

Pippinger, N. (1979). On simultaneous resource bounds, in: *Proc. 20th Annual Symp. on Foundations of Computer Science*, IEEE, New York, NY, pp. 307–311.

Plaisted, D.A. (1984). Heuristic matching for graphs satisfying the triangle inequality. *J. Algorithms* 5, 163–179.

Plummer, M.D. (1992). Matching theory – a sampler: from Dénes König to the present. *Discrete Math.* 100, 177–219.

Plummer, M.D. (1993). Matching and vertex packing: how "hard" are they? in: J. Gimbel, J.W. Kennedy and L.V. Quintas (eds.), *Quo Vadis, Graph Theory? A Source Book for Challenges and Directions, Ann. Discrete Math.* 55, 275–312.

Prim, R.C. (1957). Shortest connection networks and some generalizations. *Bell System Tech. J.* 36, 1389–1401.

Pulleyblank, W.R. (1973). *Faces of Matching Polyhedra*, PhD thesis, Department of Combinatorics and Optimization, University of Waterloo, Waterloo, Ontario.

Pulleyblank, W. (1980). Dual integrality in *b*-matching problems. *Math. Program. Study* 12, 176–196.

Pulleyblank, W.R. (1981). Total dual integrality and *b*-matchings. *Oper. Res. Lett.* 1, 28–30.

Pulleyblank, W.R. (1983). Polyhedral combinatorics, in: A. Bachem, M. Grötschel and B. Korte (eds.), *Mathematical Programming, the State of the Art: Bonn 1982*, Springer-Verlag, Berlin, pp. 312–345.

Pulleyblank, W.R. (1989). Polyhedral combinatorics, in: G.L. Nemhauser, A.H.G. Rinnooy Kan and M.J. Todd (eds.), *Optimization*, Handbooks in Operations Research and Management Science, Vol. 1, North-Holland, Amsterdam, pp. 371–446.

Pulleyblank, W.R. (1995). Matchings and stable sets, in: R. Graham, M. Grötschel, and L. Lovász (eds.), *Handbook of Combinatorics*, to appear.

Pulleyblank, W., and J. Edmonds (1974). Facets of 1-matching polyhedra, in: C. Berge and D. Ray-Chaudury (eds.), *Hypergraph Seminar*, Springer-Verlag, Berlin, pp. 214–242.
Rabin, M.O., and V.V. Vazirani (1989). Maximum matchings in general graphs through randomization. *J. Algorithms* 10, 557–567.
Recski, A. (1989). *Matroid Theory and its Applications in Electrical Networks and Statics*, Springer-Verlag, Heidelberg.
Reichmeider, P.F. (1984). *The Equivalence of some Combinatorial Matching Problems*, Polygonal Publishing House, Washington DC.
Reingold, E.W., and K.J. Supowit (1983). Probabilistic analysis of devide-and-conquer heuristics for minimum weighted Euclidean matching. *Networks* 13, 49–66.
Reingold, E.M., and R.E. Tarjan (1981). On a greedy heuristic for complete matching. *SIAM J. Comput.* 10, 676–681.
Roth, A.E., U.G. Rothblum and J.H. Vande Vate (1993). Stable matchings, optimal assignments and linear programming. *Math. Oper. Res.* 18, 803–828.
Sbihi, N. (1980). Algorithme de recherche d'un stable de cardinalité maximum dans une graphe sans étoile. *Discrete Math.* 29, 53–76.
Schneider, H. (1977). The concepts of irreducibility and full indecomposability of a matrix in the works of Frobenius, König and Markov. *Linear Algebra Appl.* 18, 139–162.
Schrijver, A. (1983a). Min–max results in combinatorial optimization, in: A. Bachem, M. Grötschel and B. Korte (eds.), *Mathematical Programming, the State of the Art: Bonn 1982*, Springer-Verlag, Berlin, pp. 439–500.
Schrijver, A. (1983b). Short proofs on the matching polyhedron. *J. Comb. Theory, Ser. B* 34, 104–108.
Schrijver, A. (1986). *Theory of Linear and Integer Programming*, John Wiley and Sons, Chichester.
Schrijver, A. (1995). Polyhedral combinatorics, in: R.Graham, M. Grötschel, and L. Lovász (eds.), *Handbook of Combinatorics*, to appear.
Schrijver, A., and P.D. Seymour (1977). A proof of total dual integrality of matching polyhedra, Mathematical Centre Report ZN 79/77, Mathematisch Centrum, Amsterdam.
Schrijver, A., and P.D. Seymour (1994). Packing odd paths, *J. Comb. Theory, Ser. B* 62, 280–288.
Schwartz, J.T. (1980). Fast probabilistic algorithms for verification of polynomial identities. *J. Assoc. Comput. Mach.* 27, 701–717.
Sebö, A. (1986). Finding the *t*-join structure of graphs. *Math. Program.* 36, 123–134.
Sebö, A. (1987). A quick proof of Seymour's theorem on *t*-joins. *Discrete Math.* 64, 101–103.
Sebö, A. (1988). The Schrijver system of odd join polyhedra. *Combinatorica* 8, 103–116.
Sebö, A. (1990). Undirected distances and the postman-structure of graphs. *J. Comb. Theory, Ser. B* 49, 10–39.
Sebö, A. (1993). General antifactors of graphs. *J. Comb. Theory, Ser. B* 58, 174–184.
Seymour, P.D. (1977). The matroids with the max-flow min-cut property. *J. Comb. Theory, Ser. B* 23, 189–222.
Seymour, P.D. (1979). On multi-colourings of cubic graphs, and conjectures of Fulkerson and Tutte. *Proc. Lond. Math. Soc., Third Ser.* 38, 423–460.
Seymour, P.D. (1981). On odd cuts and plane multicommodity flows. *Proc. Lond. Math. Soc., Third Ser.* 42, 178–192.
Shapley, L.S., and M. Shubik (1972). The assignment game I: the core. *Int. J. Game Theory* 1, 111–130.
Sinclair, A., and M. Jerrum (1989). Approximate counting, uniform generation and rapidly mixing Markov chains. *Inf. Comput.* 82, 93–133.
Steele, J.M. (1981). Subadditive Euclidean functionals and nonlinear growth in geometric probability. *Ann. Probab.* 9, 365–376.
Sterboul, F. (1979). A characterization of the graphs in which the transversal number equals the matching number. *J. Comb. Theory, Ser. B* 27, 228–229.
Supowit, K.J., D.A. Plaisted and E.M. Reingold (1980). Heuristics for weighted perfect matching, in: *Proc. 12th Annual ACM Symp. on Theory of Computing*, Association for Computing Machinery,

New York, NY, pp. 398–419.

Supowit, K.J., and E.M. Reingold (1983). Devide and conquer heuristics for minimum weighted Euclidean matching. *SIAM J. Comput.* 12, 118–143.

Supowit, K.J., E.M. Reingold and D.A. Plaisted (1983). The traveling salesman problem and minimum matching in the unit square. *SIAM J. Comput.* 12, 144–156.

Szigeti, Z. (1993). *On Seymour Graphs*, Technical report, Department of Computer Science, Eötvös Loránd University, Budapest.

Tardos, É. (1985). A strongly polynomial minimum cost circulation algorithm. *Combinatorica* 5, 247–255.

Tarjan, R. E. (1983). *Data Structures and Network Algorithms*, Society for Industrial and Applied Mathematics, Philadelphia, PA.

Truemper, K. (1992). *Matroid Decomposition*, Academic Press, San Diego.

Tutte, W.T. (1947). The factorization of linear graphs. *J. Lond. Math. Soc.* 22, 107–111.

Tutte, W.T. (1952). The factors of graphs. *Can. J. Math.* 4, 314–328.

Tutte, W.T. (1954). A short proof of the factor theorem for finite graphs. *Can. J. Math.* 6, 347–352.

Tutte, W.T. (1974). Spanning subgraphs with specified valancies. *Discrete Math.* 9, 97–108.

Tutte, W.T. (1981). Graph factors. *Combinatorica* 1, 79–97.

Vaidya, P.M. (1989). Geometry helps in matching. *SIAM J. Comput.* 18, 1201–1225.

Vaidya, P.M. (1990). Reducing the parallel complexity of certain linear programming problems, in: *Proc. 31th Annual Symp. on Foundations of Computer Science*, IEEE, New York, NY, pp. 583–589.

Valiant, L.G. (1979). The complexity of computing the permanent. *Theor. Comput. Sci.* 8, 189–201.

Vande Vate, J.H. (1989). Linear programming brings marital bliss. *Oper. Res. Lett.* 8, 147–153.

Vazirani, V.V. (1989). NC algorithms for computing the number of perfect matchings in $K_{3,3}$-free graphs and related problems. *Inf. Comput.* 80, 152–164.

Vazirani, V.V. (1994). A theory of alternating paths and blossoms for proving correctness of the $O(\sqrt{V}E)$ general graph maximum matching algorithm. *Combinatorica* 14, 71–109.

Vizing, V.G. (1964). On an estimate of the chromatic class of a p-graph (in Russian), *Diskretnyĭ Analiz* 3, 25–30

Vizing, V.G. (1965). The chromatic class of a multigraph (in Russian), *Kibernetika* 3, 29–39 [English translation: Cybernetics 1 (3) (1965) 32–41].

Warshall, S. (1962). A theorem on Boolean matrices. *J. Assoc. Comput. Mach.* 9, 11–12.

Weber, G.M. (1981). Sensitivity analysis of optimal matchings. *Networks* 11, 41–56.

Welsh, D.J.A. (1976). *Matroid Theory*, Academic Press, London.

Witzgall, C., and C.T. Zahn, Jr. (1965). Modification of Edmonds' maximum matching algorithm. *J. Res. Nat. Bur. Stand. – B. Math. Math. Phys.* 69B, 91–98.

Yakovleva, M.A. (1959). A problem on minimum transportation cost, in: V.S. Nemchinov (ed.), *Applications of Mathematics in Economic Research*, Izdat. Social'no-Ekon. Lit., Moscow, pp. 390–399.

Yannakakis, M. (1988). Expressing combinatorial optimization problems by linear programs, Working paper, AT&T Bell Laboratories [Extended abstract in: *Proc. 20th Annual ACM Symp. on Theory of Computing*, Association for Computing Machinery, New York, NY, pp. 223–228].

M.O. Ball et al., Eds., *Handbooks in OR & MS, Vol. 7*

Chapter 4

The Traveling Salesman Problem

Michael Jünger
Institut für Informatik der Universität zu Köln, Pohligstraße 1, D-50969 Köln, Germany

Gerhard Reinelt
Institut für Angewandte Mathematik, Universität Heidelberg, Im Neuenheimer Feld 294, D-69120 Heidelberg, Germany

Giovanni Rinaldi
Istituto di Analisi dei Sistemi ed Informatica, Viale Manzoni 30, I-00185 Roma, Italy

1. Introduction

A traveling salesman wants to visit each of a set of towns exactly once, starting from and returning to his home town. One of his problems is to find the shortest such trip.

The traveling salesman problem, TSP for short, has model character in many branches of Mathematics, Computer Science, and Operations Research. Heuristics, linear programming, and branch and bound, which are still the main components of todays most successful approaches to hard combinatorial optimization problems, were first formulated for the TSP and used to solve practical problem instances in 1954 by Dantzig, Fulkerson and Johnson.

When the theory of NP-completeness developed, the TSP was one of the first problems to be proven NP-hard by Karp in 1972. New algorithmic techniques have first been developed for or at least have been applied to the TSP to show their effectiveness. Examples are branch and bound, Lagrangean relaxation, Lin–Kernighan type methods, simulated annealing, and the field of polyhedral combinatorics for hard combinatorial optimization problems (polyhedral cutting plane methods and branch and cut).

This chapter presents a self-contained introduction into algorithmic and computational aspects of the traveling salesman problem along with their theoretical prerequisites as seen from the point of view of an operations researcher who wants to solve practical instances. Lawler, Lenstra, Rinnooy Kan & Shmoys [1985] motivated considerable research in this area, most of which became apparent at the specialized conference on the TSP which took place at Rice University in 1990. This chapter is intended to be a guideline for the reader confronted with the question of how to attack a TSP instance depending on its size, its structural properties (e.g., metric), the available computation time, and the desired quality of the solution (which may range from, say, a 50% guarantee to optimality). In contrast to previous surveys, here we are concerned with practical problem solving, i.e., theoretical results are presented in a form which make clear their importance in the design of algorithms for approximate but provably good, and optimal solutions of the TSP. For space reasons, we concentrate on the symmetric TSP and discuss related problems only in terms of their practical importance and the structural and algorithmic insights they provide for the symmetric TSP.

For the long history of the TSP we refer to Hoffman & Wolfe [1985]. The relevant algorithmic approaches, however, have all taken place in the last 40 years. The developments until 1985 are contained in Lawler, Lenstra, Rinnooy Kan & Shmoys [1985]. This chapter gives the most recent significant developments. Historical remarks are confined to achievements which appear relevant from our point of view.

Let $K_n = (V_n, E_n)$ be the complete undirected graph with $n = |V_n|$ nodes and $m = |E_n| = \binom{n}{2}$ edges. An edge e with endpoints i and j is also denoted by ij, or by (i, j). We denote by $\mathbb{R}^{E_n}$ the space of real vectors whose components are indexed by the elements of E_n. The component of any vector $z \in \mathbb{R}^{E_n}$ indexed by the edge $e = ij$ is denoted by z_e, z_{ij}, or $z(i, j)$.

Given an objective function $c \in \mathbb{R}^{E_n}$, that associates a 'length' c_e with every edge e of K_n, the *symmetric traveling salesman problem* consists of finding a *Hamiltonian cycle* (a cycle visiting every node exactly once) such that its c-length (the sum of the lengths of its edges) is as small (large) as possible. Without loss of generality, we only consider the minimization version of the problem. From now on we use the abbreviation *TSP* only for the symmetric traveling salesman problem.

Of special interest are the Euclidean instances of the traveling salesman problem. In these instances the nodes defining the problem correspond to points in the 2-dimensional plane and the distance between two nodes is the Euclidean distance between their corresponding points. More generally, instances that satisfy the triangle inequality, i.e., $c_{ij} + c_{jk} \geq c_{ik}$ for all three distinct i, j, and k, are of particular interest.

The reason for using a complete graph in the definition of the TSP is that for such a graph the existence of a feasible solution is always guaranteed, while for general graphs deciding the existence of a Hamiltonian cycle is an NP-complete problem. Actually, the number of Hamiltonian cycles in K_n, i.e., the size of the set of feasible solutions of the TSP, is $(n-1)!/2$. The TSP defined on general graphs is shortly described in Section 2 along with other combinatorial optimization problems whose relation to the TSP is close enough to make the algorithmic techniques covered in this chapter promising for the solution with various degrees of suitability. In Section 3 we discuss a selection of practical applications of the TSP or one of its close relatives. The algorithmic treatment of the TSP starts in Section 4 in which we cover approximation algorithms that cannot guarantee to find the optimum, but which are the only available techniques for finding good solutions to large problem instances. To assess the quality of a solution, one has to be able to compute a lower bound on the value of the shortest Hamiltonian cycle. Section 5 presents several relaxations on which lower bound computations can be based. Special emphasis is given to linear programming relaxations, which serve as a basis for finding optimal and provably good solutions within an enumerative environment to be discussed in Section 6. We do not address the algorithmic treatment of special cases of the TSP, where the special structure of the objective function can be exploited to find the optimal solution in polynomial time. Surveys on this subject are, e.g., Burkard [1990], Gilmore, Lawler & Shmoys [1985], van Dal [1992], van der Veen [1992], and Warren [1993]. Finally, in Section 7 we report on computational experiments for several TSP instances.

2. Related problems

We begin with some transformations showing that the TSP can be applied in a more general way than suggested by its definition (for some further examples see, e.g., Garfinkel [1985]).

We give transformations to some related problems or variants of the TSP. It is often convenient to assume that all edge lengths are positive. By adding a suitable constant to all edge lengths we can bring any TSP instance into this form. However we do have to keep in mind that there are algorithms whose performance may be sensitive to such a transformation.

Since we are concerned with practical computation, we can assume rational, and thus, integer data.

Traveling salesman problems in general graphs

There may be situations where we want to find shortest Hamiltonian cycles in arbitrary graphs $G = (V, E)$, in particular in graphs which are not complete. Depending on the requirements we can treat such cases in two ways. We discuss the first possibility here, the second one is given below in the discussion of the graphical TSP.

If it is required that each node is visited exactly once and that only edges of the given graph must be used then we do the following. Add all missing edges giving them a sufficiently large weight M (e.g., $M > \sum_{e \in E} c_e$) and apply an algorithm for the TSP in complete graphs. If this algorithm terminates with an optimal solution containing none of the edges with weight M then this solution is also optimal for the original problem. If an edge with weight M is contained in the optimal solution then the original graph does not contain a Hamiltonian cycle. Heuristics cannot guarantee to find a Hamiltonian cycle in G even if one exists, such a guarantee can only be provided by exact algorithms.

The second way to treat such problems is to allow that nodes may be visited more than once and edges be traversed more than once. If the given graph is connected we can always find a feasible round trip under this relaxation. This leads us to the so-called graphical traveling salesman problem.

The graphical traveling salesman problem

As in the case of the TSP we are given n cities, a set of connections between the cities represented in a graph $G = (V, E)$, and a 'length' c_e for each connection

$e \in E$. We assume that G is connected, otherwise no feasible solution exists. The *graphical traveling salesman problem* consists of finding a trip for the salesman to visit every city requiring the least possible total distance.

To define a feasible trip the salesman has to leave the home town (any node in the graph), visit any other town at least once, and go back to the home town. It is possible that a town is actually visited more than once and that an edge of G is 'traveled' more than once. Such a feasible trip is called a *tour*. To avoid unbounded situations every edge has nonnegative weight. Otherwise we could use an edge as often as we like in both directions to achieve an arbitrarily negative length of the solution.

This is sometimes a more practical definition of the TSP because we may have cases where the underlying graph of connections is not Hamiltonian.

We transform a graphical TSP to a TSP as follows. Consider the TSP on the complete graph $K_n = (V_n, E_n)$, where for each edge $ij \in E_n$ the objective function coefficient d_{ij} is given by the c-length of a shortest path from i to j in the graph G. Solving the TSP in K_n gives a Hamiltonian cycle $H \subseteq E_n$. The solution of the graphical TSP can be obtained by replacing each edge in H that is not in G with the edges of a shortest path that connects its endpoints in G.

Hamiltonian and semi-Hamiltonian graphs

A graph is called *Hamiltonian* if it contains a Hamiltonian cycle and it is called *semi-Hamiltonian* if it contains a Hamiltonian path, i.e., a path joining two nodes of the graph and visiting every node exactly once. Checking if a graph $G = (V, E)$ is Hamiltonian or semi-Hamiltonian can be done by solving a TSP in a complete graph where all edges of the original graph obtain weight 1 and all other edges obtain weight 2. If the length of an optimal Hamiltonian cycle in the complete graph is n, then G is Hamiltonian and therefore semi-Hamiltonian. If the length is $n + 1$, then G is semi-Hamiltonian, but not Hamiltonian. And, finally, if the length is $n + 2$ or more, G is not semi-Hamiltonian.

The asymmetric traveling salesman problem

In this case the cost of traveling from city i to city j is not necessarily the same as for traveling from city j to city i. This is reflected by formulating the *asymmetric traveling salesman problem* (*ATSP*) as finding a shortest directed Hamiltonian cycle in a weighted digraph. Let $D = (W, A)$, $W = \{1, 2, \ldots, n\}$, $A \subseteq W \times W$, be the digraph for which the ATSP has to be solved. Let d_{ij} be the distance from node i to node j, if there is an arc in A with tail i and head j. We define an undirected graph $G = (V, E)$ by

$$\begin{aligned} V &= W \cup \{n+1, n+2, \ldots, 2n\}, \\ E &= \{(i, n+i) \mid i = 1, 2, \ldots, n\} \cup \{(n+i, j) \mid (i, j) \in A\}. \end{aligned}$$

Edge weights are computed as follows

$$\begin{aligned} c_{i,n+i} &= -M && \text{for } i = 1, 2, \ldots, n, \\ c_{n+i,j} &= d_{ij} && \text{for } (i, j) \in A, \end{aligned}$$

where M is a sufficiently large number, e.g., $M = \sum_{(i,j)\in A} d_{ij}$. It is easy to see that for each directed Hamiltonian cycle in D with length d_D there is a Hamiltonian cycle in G with length $c_G = d_D - nM$. In addition, all edges with weight $-M$ are contained in an optimal Hamiltonian cycle in G. Therefore, this cycle induces a directed Hamiltonian cycle in D.

In our discussion on computational results in Section 7 we report on the solution of a hard asymmetric TSP instance that we attacked with symmetric TSP methods.

The multisalesmen problem

Instead of just one salesman we have m salesmen available who are all located in city $n+1$ and have to visit cities $1, 2, \ldots, n$. The cost of the solution is the total distance traveled by all salesmen together (all of them must travel). This is the basic situation when in vehicle routing m vehicles, located at a common depot, have to serve n customers.

We can transform this problem to the TSP by splitting city $n+1$ into m cities $n+1, n+2, \ldots, n+m$. The edges $(i, n+k)$, with $1 \le i \le n$ and $2 \le k \le m$, receive the weight $c(i, n+k) = c(i, n+1)$, and all edges connecting the nodes $n+1, n+2, \ldots, n+m$ receive a large weight M.

The rural postman problem

We are given a graph $G = (V, E)$ with edge weights $c(i, j)$ and a subset $F \subseteq E$. The *rural postman problem* consists of finding a shortest tour, containing all edges in F, in the subgraph of G induced by some subset of V. We call such a tour a *rural postman tour* of G. As for the graphical TSP we have to assume nonnegative edge weights to avoid unbounded situations.

If F induces a connected subgraph of G, then we have the special case of a *Chinese postman problem* which can be solved in polynomial time using matching techniques [Edmonds & Johnson, 1973]. In general the problem is NP-hard, since the TSP can easily be transformed to it. First add a sufficiently large number to all edge weights to guarantee that triangle inequality holds. Then split each node i into two nodes i and i'. For any edge (i, j) generate edges (i', j) and (i, j') with weights $c(i', j) = c(i, j') = c(i, j)$, and the edges connecting i to i' and j to j' receive zero weights. F consists of all the edges (i, i').

Conversely, we can transform the rural postman problem to the TSP as follows. Let $G_F = (V_F, F)$ be the subgraph of G induced by F. With every node $i \in V_F$ we associate a set $S_i = \{s_i^j \mid j \in N(i)\}$ where $N(i)$ is the set of neighbors of node i in G_F. Construct the weighted complete graph $G' = (V', E', c')$ on the set $V' = \bigcup_{i\in V_F} S_i$. The edge weights c' are defined as follows

$$c'(s_i^h, s_i^k) = 0 \quad \text{for } i \in V_F \text{ and } h, k \in N(i),\ h \neq k$$

$$c'(s_i^h, s_j^k) = \begin{cases} -M & \text{if } i = k \text{ and } j = h \\ d(i, j) & \text{otherwise} \end{cases} \quad \begin{array}{l} \text{for all } i, j \in V_F,\ i \neq j, \\ h \in N(i), \\ k \in N(j), \end{array}$$

where we denote by $d(i, j)$ the c-length of a shortest path between i and j in G.

It is trivial to transform an optimal Hamiltonian cycle in G' to an optimal rural postman tour in G. We can easily generalize this transformation for the case in which not only edges, but also some nodes are required to be in the tour. Such nodes are simply added to the resulting TSP instance, and all new edges receive as weights the corresponding shortest path lengths. In Section 7 we report on the solution of some instances that we obtained using this transformation.

The shortest Hamiltonian path problem

We are given a graph $G = (V, E)$ with edge weights c_{ij}. Two special nodes, say v_s and v_t, of V are also given. The task is to find a path from v_s to v_t visiting each node of V exactly once with minimum length, i.e., to find the shortest Hamiltonian path in G from v_s to v_t.

This problem can be solved as a standard TSP in two ways.

a) Choose M sufficiently large and assign weight $-M$ to the edge from v_s to v_t (which is created if it does not belong to E). Then compute the shortest Hamiltonian cycle in this graph. This cycle must contain edge $v_s v_t$ and thus solves the Hamiltonian path problem.

b) Add a new node 0 to V and edges from 0 to v_s and to v_t with weight 0. Each Hamiltonian cycle in this new graph corresponds to a Hamiltonian path from v_s to v_t in the original graph with the same length.

If only the starting point v_s of the Hamiltonian path is fixed we can solve the problem by introducing a new node 0 and adding edges from all nodes $v \in V \setminus \{v_s\}$ to 0 with zero length. Now we can solve the Hamiltonian path problem with starting point v_s and terminating point $v_t = 0$ which solves the original problem.

If also no starting point is specified, we just add node 0 and connect all other nodes to 0 with edges of length zero. In this new graph we solve the standard TSP.

The bottleneck traveling salesman problem

Instead of Hamiltonian cycles with minimum total length one searches in this problem for those whose longest edge is as short as possible. This *bottleneck traveling salesman problem* can be solved by a sequence of TSP instances. To see this, observe that the exact values of the distances are not of interest under this objective function, only their relative order matters. Hence we may assume that we have at most $\frac{1}{2}n(n-1)$ different integral distances and that the largest of them is not greater than $\frac{1}{2}n(n-1)$. We now solve problems of the following kind for some parameter b:

Is the graph consisting of all edges with weights at most b Hamiltonian?

This is exactly the problem discussed above. By performing a binary search on the parameter b (starting, e.g., with $b = \frac{1}{4}n(n-1)$) we can identify the smallest such b leading to a 'yes' answer by solving at most $O(\log n)$ TSP instances.

Computational results for the bottleneck TSP are reported in Carpaneto, Martello & Toth [1984].

We have seen that a variety of related problems can be transformed to the TSP. However, each such transformation has to be considered with some care, before actually trying to use it for practical problem solving. For example, the shortest path computations necessary to treat a graphical TSP as a TSP take time $O(n^3)$ which might not be acceptable in practice. Many transformations require the introduction of a large number M. This can lead to numerical problems or may even prevent the finding of feasible solutions at all using heuristics. In particular, for LP-based approaches, the usage of the 'big M' cannot be recommended. Here it is preferable use 'variable fixing techniques' (see Section 6) to force edges with cost $-M$ into the solution and prevent those with cost M in the solution. Moreover, in general, the transformations described above may produce TSP instances that are difficult to solve both for heuristic and exact algorithms.

3. Practical applications

Since we are aiming at the development of algorithms and heuristics for practical traveling salesman problem solving, we give a survey on some of the possible applications. The list is not complete but covers some important cases. We start with applications that can be modeled directly as one of the variants given in the previous section.

Drilling of printed circuit boards

A direct application of the TSP is the drilling problem whose solution plays an important rôle in economical manufacturing of printed circuit boards (PCBs). A computational study in an industry application of a large electronics company can be found in Grötschel, Jünger & Reinelt [1991].

To connect a conductor on one layer with a conductor on another layer, or to position (in a later stage of the PCB production) the pins of integrated circuits, holes have to be drilled through the board. The holes may be of different diameters. To drill two holes of different diameters consecutively, the head of the machine has to move to a tool box and change the drilling equipment. This is quite time consuming. Thus it is clear at the outset that one has to choose some diameter, drill all holes of the same diameter, change the drill, drill the holes of the next diameter, etc.

Thus, this drilling problem can be viewed as a sequence of TSP instances, one for each hole diameter, where the 'cities' are the initial position and the set of all holes that can be drilled with one and the same drill. The 'distance' between two cities is given by the time it takes to move the drilling head from one position to the other. The aim here is to minimize the travel time for the head of the machine.

X-Ray crystallography

An important application of the TSP occurs in the analysis of the structure of crystals [Bland & Shallcross, 1989; Dreissig & Uebach, 1990]. Here an X-ray diffractometer is used to obtain information about the structure of crystalline material. To this end a detector measures the intensity of X-ray reflections of the crystal from various positions. Whereas the measurement itself can be accomplished quite fast, there is a considerable overhead in positioning time since up to hundreds of thousands positions have to be realized for some experiments. In the two examples that we refer to, the positioning involves moving four motors. The time needed to move from one position to the other can be computed very accurately. The result of the experiment does not depend on the sequence in which the measurements at the various positions are taken. However, the total time needed for the experiment depends on the sequence. Therefore, the problem consists of finding a sequence that minimizes the total positioning time. This leads to a traveling salesman problem.

Overhauling gas turbine engines

This application was reported by Plante, Lowe & Chandrasekaran [1987] and occurs when gas turbine engines of aircraft have to be overhauled. To guarantee a uniform gas flow through the turbines there are so-called nozzle-guide vane assemblies located at each turbine stage. Such an assembly basically consists of a number of nozzle guide vanes affixed about its circumference. All these vanes have individual characteristics and the correct placement of the vanes can result in substantial benefits (reducing vibration, increasing uniformity of flow, reducing fuel consumption). The problem of placing the vanes in the best possible way can be modeled as a TSP with a special objective function.

The order-picking problem in warehouses

This problem is associated with material handling in a warehouse [Ratliff & Rosenthal, 1983]. Assume that at a warehouse an order arrives for a certain subset of the items stored in the warehouse. Some vehicle has to collect all items of this order to ship them to the customer. The relation to the TSP is immediately seen. The storage locations of the items correspond to the nodes of the graph. The distance between two nodes is given by the time needed to move the vehicle from one location to the other. The problem of finding a shortest route for the vehicle with minimum pickup time can now be solved as a TSP. In special cases this problem can be solved easily, see van Dal [1992] for an extensive discussion and for references.

Computer wiring

A special case of connecting components on a computer board is reported in Lenstra & Rinnooy Kan [1974]. Modules are located on a computer board and a given subset of pins has to be connected. In contrast to the usual case where a Steiner tree connection is desired, here the requirement is that no more than two wires are attached to each pin. Hence we have the problem of finding a shortest Hamiltonian path with unspecified starting and terminating points.

A similar situation occurs for the so-called testbus wiring. To test the manufactured board one has to realize a connection which enters the board at some specified point, runs through all the modules, and terminates at some specified point. For each module we also have a specified entering and leaving point for this test wiring. This problem also amounts to solving a Hamiltonian path problem with the difference that the distances are not symmetric and that starting and terminating point are specified.

Scheduling with sequence dependent process times

We are given n jobs that have to be performed on some machine. The time to process job j is t_{ij} if i is the job performed immediately before j (if j is the first job then its processing time is t_{0j}). The task is to find an execution sequence for the jobs such that the total processing time is as short as possible. Clearly, this problem can be modeled as a shortest (directed) Hamiltonian path problem.

Suppose the machine in question is an assembly line and that the jobs correspond to operations which have to be performed on some product at the workstations of the line. In such a case the primary interest would lie in balancing the line. Therefore, instead of the shortest possible time to perform all operations on a product, the longest individual processing time needed on a workstation is important. To model this requirement a bottleneck TSP is more appropriate.

Sometimes the TSP comes up as a subproblem in more complex combinatorial optimization processes that are devised to deal with production problems in industry. In such cases there is often no hope for algorithms with guaranteed performance, but hybrid approaches proved to be practical. We give three examples that cannot be transformed to the TSP, but share some characteristics of the TSP, or in which the TSP comes up as a subproblem.

Vehicle routing

Suppose that in a city n mail boxes have to be emptied every day within a certain period of time, say 1 hour. The problem is to find the minimum number of trucks to do this and the shortest time to do the collections using this number of trucks. As another example, suppose that n customers require certain amounts of some commodities and a supplier has to satisfy all demands with a fleet of trucks. The problem is to find an assignment of customers to the trucks and a delivery schedule for each truck so that the capacity of each truck is not exceeded and the total travel distance is minimized. Several variations of these two problems, where time and capacity constraints are combined, are common in many real-world applications.

This problem is solvable as a TSP if there are no time and capacity constraints and if the number of trucks is fixed (say m). In this case we obtain an m-salesmen problem. Nevertheless, one may apply methods for the TSP to find good feasible solutions for this problem (see Lenstra & Rinnooy Kan [1974]).

Mask plotting in PCB production

For the production of each layer of a printed circuit board, as well as for layers of integrated semiconductor devices, a photographic mask has to be produced. In our case for printed circuit boards this is done by a mechanical plotting device. The plotter moves a lens over a photosensitive coated glass plate. The shutter may be opened or closed to expose specific parts of the plate. There are different apertures available to be able to generate different structures on the board.

Two types of structures have to be considered. A line is exposed on the plate by moving the closed shutter to one endpoint of the line, then opening the shutter and moving it to the other endpoint of the line. Then the shutter is closed. A point type structure is generated by moving (with the appropriate aperture) to the position of that point then opening the shutter just to make a short flash, and then closing it again. Exact modeling of the plotter control problem leads to a problem more complicated than the TSP and also more complicated than the rural postman problem. A real-world application in the actual production environment is reported in Grötschel, Jünger & Reinelt [1991].

Control of robot motions

In order to manufacture some workpiece a robot has to perform a sequence of operations on it (drilling of holes of different diameters, cutting of slots, planishing, etc.). The task is to determine a sequence of the necessary operations that leads to the shortest overall processing time. A difficulty in this application arises because there are precedence constraints that have to be observed. So here we have the problem of finding the shortest Hamiltonian path (where distances correspond to times needed for positioning and possible tool changes) that satisfies certain precedence relations between the operations.

4. Approximation algorithms

When trying to solve practical TSP instances to optimality, one quickly encounters several difficulties. It may be possible that there is no algorithm at hand to solve an instance optimally and that time or knowledge do not permit the development and implementation of such an algorithm. The instances may be simply too large and therefore beyond the capabilities of even the best algorithms for attempting to find optimal solutions. On the other hand, it may also be possible that the time allowed for computation is not enough for an algorithm to reach the optimal solution. In all these cases there is a definite need for approximation algorithms (heuristics) which determine solutions of good quality and yield the best results achievable subject to the given side constraints.

It is the aim of this section to survey heuristics for the TSP and to give guidelines for their potential incorporation for the treatment of practical problems. We will first consider construction heuristics which build an initial Hamiltonian cycle. Procedures for improving a given cycle are discussed next. The third part is concerned with particular advantages one can exploit if the given problem instances are of geometric nature. A survey of other recently developed techniques concludes this section.

There is a huge number of papers dealing with finding near optimal solutions for the TSP. We therefore confine ourselves to the approaches that we think provide the most interesting ideas and that are important for attacking practical problems. This section is intended to give the practitioner enough detail to be able to design successful heuristics for large-scale TSP instances without studying additional literature. For further reading we recommend Golden & Stewart [1985], Bentley [1992], Johnson [1990] and Reinelt [1992, 1994].

An important point is the discussion of implementation issues. Although sometimes easily formulated, heuristics will often require extensive effort to obtain computer implementations that are applicable in practice. We will address these questions along with the presentation of the heuristics. We do not discuss techniques in detail. The reader should consult a good reference on algorithms and data structures (e.g., Cormen, Leiserson & Rivest [1989]) when doing own implementations.

Due to limited space we will not present many detailed computational results, rather we will give conclusions that we have drawn from computational testing. For our experiments we used problem instances from the public problem library TSPLIB [Reinelt, 1991a, b]. In this chapter we refer to a set of 30 Euclidean sample problems with sizes ranging from 105 to 2392 with known optimal solutions. The size of each problem instance appears in its name, e.g., *pcb442* is a TSP on 442 nodes. Since these problems come from real applications, our findings may be different from experiments on randomly generated problems. CPU times are given in seconds on a SUN SPARCstation 10/20. Some effort was put into the implementation of computer codes. However, it was not our intention to achieve ultimate performance, but to demonstrate the speedup that can be gained by careful implementation. Except for specially selected edges, distances were not stored but always computed by evaluating the Euclidean distance function. All CPU times include the time for distance computations.

Before starting to derive approximation algorithms, it is an interesting theoretical question, whether efficient heuristics can be designed that produces solutions with requested or at least guaranteed quality in polynomial time (polynomial in the problem size and in the desired accuracy). Whereas for other NP-hard problems such heuristics do exist, there are only negative results for the general TSP. For a problem instance let c_H denote the length of the Hamiltonian cycle produced by heuristic H and let c_{opt} denote the length of an optimal cycle. Sahni & Gonzales [1976] show that, unless P = NP, for any constant $r \geq 1$ there does not exist a polynomial time heuristic H such that $c_H \leq r \cdot c_{\text{opt}}$ for all problem instances.

A *fully polynomial approximation scheme* for a minimization problem is a heuristic H that for a given problem instance and any $\varepsilon > 0$ computes a feasible solution satisfying $c_H/c_{\text{opt}} \leq 1 + \varepsilon$ in time polynomial in the size of the instance and in $1/\varepsilon$. Such schemes are very unlikely to exist for the traveling salesman problem. Johnson & Papadimitriou [1985] show that, unless P = NP, there does not exist a fully polynomial approximation scheme for the Euclidean traveling salesman problem. This also holds in general for TSP instances satisfying the triangle inequality. The results tell us that for every heuristic there are problem instances where it fails badly. There are approximation results for problems satisfying the triangle inequality some of which will be addressed below.

It should be pointed out that running time and quality of an algorithm derived by theoretical (worst case or average case) analysis is usually insufficient to predict its behavior when applied to real-world problem instances.

In addition, the reader should be aware that polynomial time algorithms can still require a substantial amount of CPU time, if the polynomial is not of low degree. In certain applications algorithms having running time as low as $O(n^2)$ may not be acceptable. So, polynomiality by itself is not a sufficient criterion for efficiency in practice. It is our aim to show that, in the case of the traveling salesman problem, algorithms can be designed that are capable of finding good approximate solutions to even large sized real-world instances within rather moderate time limits. Thus, the NP-hardness of the TSP does not imply the nonexistence of reasonable algorithms for practical problem instances. Furthermore, we want to make clear that designing efficient heuristics is not a straightforward task. Although ideas often seem elementary, it requires substantial effort to design practically useful computer codes.

The performance of a heuristic is best assessed by comparing the value of the approximate solution it produces with the value of an optimal solution. We say that a heuristic solution value c_H has *quality* $p\%$ if $100 \cdot (c_H - c_{\text{opt}})/c_{\text{opt}} = p$. If no provably optimal solutions are known, then the quality can only be estimated from above by comparing the heuristic solution value with a lower bound for the optimal value. A frequently used such lower bound is the *subtour elimination lower bound* (see Section 5). This bound can be computed exactly using LP techniques or it can at least be approximated using iterative approaches to be discussed in Section 6.

4.1. *Construction heuristics*

For the beginning we shall consider pure *construction procedures*, i.e., heuristics that determine a Hamiltonian cycle according to some construction rule, but do not try to improve upon this cycle. In other words, a Hamiltonian cycle is successively built, and parts already built remain in a certain sense unchanged throughout the algorithm. We will confine ourselves to some of the most commonly used construction principles.

Nearest neighbor heuristics

One of the simplest heuristics for the TSP is the so-called *nearest neighbor heuristic* which attempts to construct Hamiltonian cycles based on connections to near neighbors. The standard version is stated as follows.

procedure NEAREST_NEIGHBOR

(1) Select an arbitrary node j, set $l = j$ and $W = \{1, 2, \ldots, n\} \setminus \{j\}$.
(2) As long as $W \neq \emptyset$ do the following.
  (2.1) Let $j \in W$ such that $c_{lj} = \min\{c_{li} \mid i \in W\}$.
  (2.2) Connect l to j and set $W = W \setminus \{j\}$ and $l = j$.
(3) Connect l to the node selected in Step (1) to form a Hamiltonian cycle.

A possible variation of the standard nearest neighbor heuristic is the *double-sided nearest neighbor heuristic* where the current path can be extended from both of its endnodes.

The standard procedure runs in time $O(n^2)$. No constant worst case performance guarantee can be given. In fact, Rosenkrantz, Stearns & Lewis [1977] show that for arbitrarily large n there exist TSP instances on n nodes such that the nearest neighbor solution is $\Theta(\log n)$ times as long as an optimal Hamiltonian cycle. This also holds if the triangle inequality is satisfied.

If one displays nearest neighbor solutions one realizes the reason for this poor performance. The procedure proceeds very well and produces connections with short edges in the beginning. But as can be seen from a graphics display, several nodes are 'forgotten' during the algorithm and have to be inserted at high cost in the end.

Although usually rather bad, nearest neighbor solutions have the advantage that they only contain a few severe mistakes. Therefore, they can serve as good starting solutions for subsequently performed improvement methods, and it is reasonable to put some effort in designing heuristics that are based on the nearest neighbor principle. For nearest neighbor solutions we obtained an average quality of 24.2% for our set of sample problems (i.e., on the average Hamiltonian cycles were 1.242 times longer than an optimal Hamiltonian cycle). In Johnson [1990] an average excess of 24% over an approximation of the subtour elimination lower bound is reported for randomly generated problems.

The standard procedure is easily implemented with a few lines of code. But, since running time is quadratic, this implementation can be too slow for large problems with 10,000 or 100,000 nodes, say. Therefore, even for this simple heuristic, it is worthwhile to think about speedup possibilities.

A basic idea, that we will apply for other heuristics as well, is the use of a *candidate subgraph*. A candidate subgraph is a subgraph of the complete graph on n nodes containing reasonable edges in the sense that they are 'likely' to be contained in a short Hamiltonian cycle. These edges are taken with priority in the various heuristics, thus avoiding the consideration of the majority of edges that are assumed to be of no importance. For the time being we do not address the question of how to choose such subgraphs and of how to compute them efficiently. This will be discussed in the subsection on geometric instances.

Because a major problem with nearest neighbor heuristics is that, in the end, nodes have to be connected at high cost, we modify it to avoid isolated nodes. To do this we first compute the 10 nearest neighbor subgraph, i.e., the subgraph containing for every node all edges to its 10 nearest neighbors. Whenever a node is connected to the current path we remove its incident edges in the subgraph. As soon as a node not contained in the path so far is connected to fewer than four nodes in the subgraph, we insert that node immediately into the path (eliminating all of its incident edges from the subgraph). To reduce the search for an insertion point, we only consider insertion after or before those nodes of the path that are among the 10 nearest neighbors of the node to be inserted. If all isolated nodes are added, the selection of the next node to be appended to the path is accomplished as follows. We first look for nearest neighbors of the node within its adjacent nodes in the candidate subgraph. If all nodes adjacent in the subgraph are already contained in the partial Hamiltonian cycle then we compute the nearest neighbor among all free nodes. The worst case time complexity is not affected by this modification.

Substantially less CPU time was needed to perform the modified heuristic compared to the standard implementation (even if the preprocessing time to compute the candidate subgraph is included). For example, whereas it took 15.3 seconds to perform the standard nearest neighbor heuristic for problem *pr2392*, the improved version required a CPU time of only 0.3 seconds. As expected, however, the variant still seems to have a quadratic component in its running time.

With respect to quality, insertion of forgotten nodes indeed improves the length of the nearest neighbor solutions. In contrast to the quality of 24.2% on average for the standard implementation, the modified version gave the average quality 18.6%. In our experiments we have chosen the starting node at random. The performance of nearest neighbor heuristics is very sensitive to the choice of the starting node. Choosing a different starting node can result in a solution whose quality differs by more than 10 percentage points.

Insertion heuristics

Another intuitively appealing approach is to start with cycles visiting only small subsets of the nodes and then extend these cycles by inserting the remaining nodes. Using this principle, a cycle is built containing more and more nodes of the problem until all nodes are inserted and a Hamiltonian cycle is found.

procedure INSERTION

(1) Select a starting cycle on k nodes $v_1, v_2, \ldots, v_k$ ($k \geq 1$) and set $W = V \setminus \{v_1, v_2, \ldots, v_k\}$.

(2) As long as $W \neq \emptyset$ do the following.

  (2.1) Select a node $j \in W$ according to some criterion.

  (2.2) Insert j at some position in the cycle and set $W = W \setminus \{j\}$.

Of course, there are several possibilities for implementing such an insertion scheme. The main difference is the choice of the selection rule in Step (2.1). The starting cycle can be just some cycle on three nodes or, in degenerate cases, a loop ($k = 1$) or an edge ($k = 2$). The selected node to be inserted is usually inserted into the cycle at the point causing shortest increase of the length of the cycle.

The following are some choices for extending the current cycle (further variants are possible). We say that a node is a *cycle node* if it is already contained in the partial Hamiltonian cycle. For $j \in W$ we define $d_{\min}(j) = \min\{c_{ij} \mid i \in V \setminus W\}$.

NEAREST INSERTION: Insert the node that has the shortest distance to a cycle node, i.e., select $j \in W$ with $d_{\min}(j) = \min\{d_{\min}(l) \mid l \in W\}$.

FARTHEST INSERTION: Insert the node whose minimum distance to a cycle node is maximum, i.e., select $j \in W$ with $d_{\min}(j) = \max\{d_{\min}(l) \mid l \in W\}$.

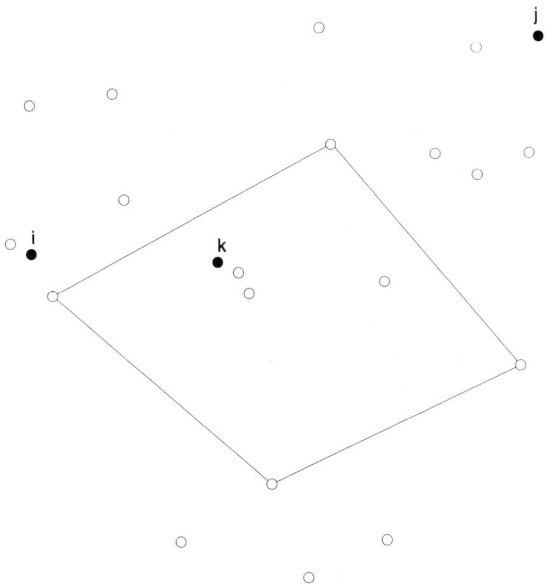

Fig. 1. Insertion heuristics.

CHEAPEST INSERTION: Insert the node that can be inserted at the lowest increase in cost.

RANDOM INSERTION: Select the node to be inserted at random and insert it at the best possible position.

Figure 1 visualizes the difference between the insertion schemes. Nearest insertion adds node i to the partial Hamiltonian cycle in the following step, farthest insertion chooses node j, and cheapest insertion chooses node k.

All heuristics except for cheapest insertion are easily implementable to run in time $O(n^2)$. Cheapest insertion can be executed in time $O(n^2 \log n)$ by storing for each external node a heap based on the insertion cost at the possible insertion points. Due to an $O(n^2)$ space requirement it cannot be used for large instances.

For some insertion type heuristics we have worst-case performance guarantees. For instances of the TSP obeying the triangle inequality, Hamiltonian cycles computed by the nearest insertion and cheapest insertion heuristic are less than twice as long as an optimal Hamiltonian cycle [Rosenkrantz, Stearns & Lewis, 1977]. The result is sharp in the sense that there exist instances for which these heuristics yield solutions that are $2 - 2/n$ times larger than an optimal solution. Hurkens [1991] gave examples where random and farthest insertion yield Hamiltonian cycles that are $13/2$ times longer than an optimal Hamiltonian cycle (although the triangle inequality is satisfied).

On our set of sample problems we obtained average qualities 20.0%, 9.9%, and 11.1% for nearest, farthest, and random insertion, respectively. An average excess over the subtour bound of 27% for the nearest insertion and of 13.5% for the farthest insertion procedure is reported in Johnson [1990] for random problem instances. Though appealing at first sight, the cheapest insertion heuristic only yields an average quality of 16.8% (with substantially longer running times). Performance of insertion heuristics does not depend as much on the starting configuration as in the nearest neighbor heuristic. One can expect deviations of about 6% for the random insertion variant and about 7-8% for the others.

There are also variants of the above ideas where the node selected is not inserted at cheapest insertion cost but in the neighborhood of the cycle node that is nearest to it. These variants are usually named 'addition' instead of insertion. Bentley [1992] reports that the results are slightly inferior.

Heuristics based on spanning trees

The heuristics considered so far construct Hamiltonian cycles 'from scratch' in the sense that they do not exploit any additional knowledge. The two heuristics to be described next use a minimum spanning tree as a basis for generating Hamiltonian cycles. They are particularly suited for TSP instances obeying the triangle inequality, but can, in principle, also be applied to general problems.

Before describing these heuristics we observe that, if the triangle inequality is satisfied, we can derive from any given tour a Hamiltonian cycle that is not longer than this tour. Let $v_{i_0}, v_{i_1}, \ldots, v_{i_k}$ be the sequence in which the nodes (including repetitions) are visited in the tour starting at v_{i_0} and returning to $v_{i_k} = v_{i_0}$. The following procedure obtains a Hamiltonian cycle.

procedure OBTAIN_CYCLE

(1) Set $T = \{v_{i_0}\}$, $v = v_{i_0}$, and $l = 1$.
(2) As long as $|T| < n$ perform the following steps.
  (2.1) If $v_{i_l} \notin T$ then set $T = T \cup \{v_{i_l}\}$, connect v to v_{i_l} and set $v = v_{i_l}$.
  (2.2) Set $l = l + 1$.
(3) Connect v to v_{i_0} to form a Hamiltonian cycle.

If the triangle inequality is satisfied, then every connection made in this procedure is either an edge of the tour or is a shortcut replacing a subpath of the tour by an edge connecting its two endnodes. Hence the resulting Hamiltonian cycle cannot be longer than the tour.

Both heuristics to be discussed next start with a minimum spanning tree and differ only in how a tour is generated from the tree.

procedure DOUBLETREE

(1) Compute a minimum spanning tree.
(2) Take all tree edges twice to obtain a tour.
(3) Call OBTAIN_CYCLE to get a Hamiltonian cycle.

The running time of the algorithm is dominated by the time needed to obtain a minimum spanning tree. Therefore we have time complexity $\Theta(n^2)$ for the general TSP and $\Theta(n \log n)$ for Euclidean problems (see Section 4.3).

If we compute the minimum spanning tree with Prim's algorithm [Prim, 1957], we could as well construct a Hamiltonian cycle along with the tree computation. We always keep a cycle on the nodes already in the tree (starting with the loop consisting of only one node) and insert the node into the current cycle which is added to the spanning tree. If this node is inserted at the best possible position this algorithm is identical to the nearest insertion heuristic. If it is inserted before or after its nearest neighbor among the cycle nodes, then we obtain the nearest addition heuristic.

In Christofides [1976] a more sophisticated method is suggested to make tours out of spanning trees. Namely, observe that it is sufficient to add a perfect matching on the odd-degree nodes of the tree. (A *perfect matching* of a node set W, $|W| = 2k$, is a set of k edges such that each node of W is incident to exactly one of these edges.) After addition of all matching edges all node degrees are even and hence the graph is a tour. The cheapest way (with respect to edge weights) to obtain a tour is to add a minimum weight perfect matching.

procedure CHRISTOFIDES

(1) Compute a minimum spanning tree.
(2) Compute a minimum weight perfect matching on the odd-degree nodes of the tree and add it to the tree to obtain a tour.
(3) Call OBTAIN_CYCLE to get a Hamiltonian cycle.

This procedure takes considerably more time than the previous one. Computation of a minimum weight perfect matching on k nodes can be performed in time $O(k^3)$ [Edmonds, 1965]. Since a spanning tree may have $O(n)$ leaves, Christofides' heuristic has cubic worst case time.

Figure 2 displays the principle of this heuristic. Solid lines correspond to the edges of a minimum spanning tree, broken lines correspond to the edges of a perfect matching on the odd-degree nodes of this tree. The union of the two edge sets gives a tour. The sequence of the edges in the tour is not unique. So one can try to find better solutions by determining different sequences.

A minimum spanning tree is not longer than a shortest Hamiltonian cycle and the matching computed in Step (2) of CHRISTOFIDES has weight at most half of the length of an optimal Hamiltonian cycle. Therefore, for every instance of the TSP obeying the triangle inequality, the double tree heuristic produces a solution which is at most twice as large as an optimal solution, and Christofides' heuristic produces a solution which is at most 1.5 times as large as an optimal solution. Cornuéjols & Nemhauser [1978] show that there are instances where Christofides' heuristic yields a Hamiltonian cycle that is $1.5 - 1/(2n)$ times longer than an optimal cycle.

Computing exact minimum weight perfect matchings in Step (2) is very time

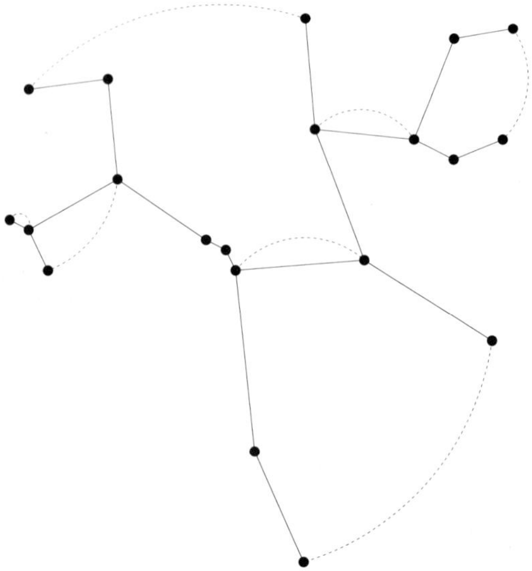

Fig. 2. Christofides' heuristic.

consuming. Therefore, the necessary matching is usually computed by a heuristic. We have used the following one. First we double all edges incident with the leaves of the spanning tree, and then we compute a farthest insertion cycle on the remaining (and newly introduced) odd-degree nodes. This cycle induces two perfect matchings and we add the shorter one to our subgraph. Time complexity is reduced to $O(n^2)$ this way. It was observed in many experiments that the Christofides heuristic does not perform as well as it might have been expected. Although it has the best known worst case bound of any TSP heuristic, the experiments produced solutions which rarely yield qualities better than 10%.

For our set of sample problems the average quality was 38.08% for the double tree and 19.5% for the modified Christofides heuristic which coincides with the findings in Johnson [1990]. Running times for *pr2392* were 0.2 seconds for the double tree and 0.7 seconds for the modified Christofides heuristic (not including the time for the minimum spanning tree computation). With their improved version of Christofides' heuristic, Johnson, Bentley, McGeoch & Rothberg [1994] achieve an average quality of about 11%. By using exact minimum matchings the quality can be further improved to about 10%.

Savings methods

The final type of heuristic to be discussed in this subsection was originally developed for vehicle routing problems [Clarke & Wright, 1964]. But it can also be usefully applied in our context, since the traveling salesman problem can be considered as a special vehicle routing problem involving only one vehicle with unlimited capacity. This heuristic successively merges subtours to eventually obtain a Hamiltonian cycle.

procedure SAVINGS

(1) Select a base node $z \in V$ and set up the $n-1$ subtours (z, v), $v \in V \setminus \{z\}$ consisting of two nodes each.
(2) As long as more than one subtour is left perform the following steps.
  (2.1) For every pair of subtours T_1 and T_2 compute the savings that is achieved if they are merged by deleting in each of them an edge to the base node and connecting the two open ends.
  (2.2) Merge the two subtours which provide the largest savings. (Note that this operation always produces a subtour which is a cycle.)

An iteration step of the savings heuristic is depicted in Figure 3. Two subtours are merged by deleting the edges from nodes i and j to the base node z and adding the edge ij.

In the implementation we have to maintain a list of possible mergings. The crucial point is the update of this list. We can consider the system of subtours as a system of paths (possibly having only one node) whose endnodes are thought of as being connected to the base node. A merge operation essentially consists of connecting two ends of different paths. For finding the best merge possibility we have to know for each endnode its best possible connection to an endnode of another path ('best' with respect to the cost of merging the corresponding subtours). Suppose that in Step (2.2) the two paths $[i_1, i_2]$ and $[j_1, j_2]$ are merged by connecting i_2 to j_1. The best merging now changes only for those endnodes whose former best merging was the connection to i_2 or to j_1, or for the endnode i_1 (j_2) if its former best merging was to j_1 (i_2). Because we do not know how many nodes are affected we can only bound the necessary update time by $O(n^2)$

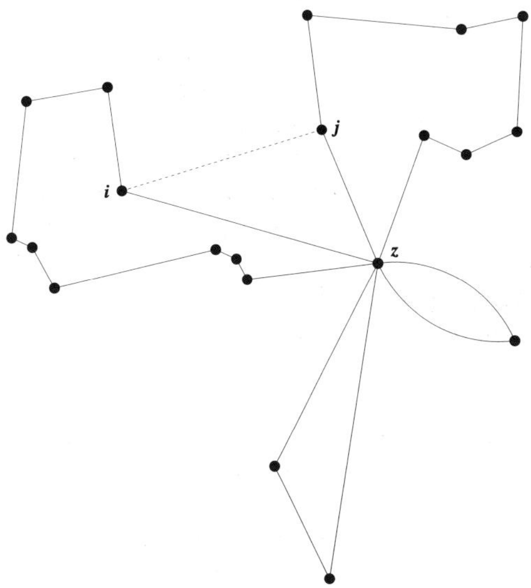

Fig. 3. A savings heuristic.

giving an overall heuristic with running time $O(n^3)$. For small problems we can achieve running time $O(n^2 \log n)$, but we have to store the matrix of all possible savings which requires $O(n^2)$ storage space. Further remarks on the Clarke/Wright algorithm can be found in Potvin & Rousseau [1990]. The average quality that we achieved for our set of problems was 9.8%.

We apply ideas similar to those above to speed up this heuristic. We again assume that we have a candidate subgraph of reasonable connections at hand. Now, merge operations are preferred that use a candidate edge for connecting two paths. The update is simplified in that for a node whose best merge possibility changes, only candidate edges incident to that node are considered for connections. If during the algorithm an endnode of a path becomes isolated, since none of its incident subgraph edges is feasible anymore, we compute its best merging by enumeration.

Surprisingly, the simplified heuristic yields solutions of similar average quality (9.6%) in much shorter time. For problem *pr2392* CPU time was 5.1 seconds with quality 12.4% compared to 194.6 seconds and quality 12.2% for the original version. We have also conducted experiments concerning the stability of this heuristic depending on the choice of the base node. It turns out that the savings heuristic gives much better results and is more stable than nearest neighbor or insertion heuristics.

Often, we will not apply the savings heuristic for constructing Hamiltonian cycles from scratch. Our purpose for employing this heuristic is to connect systems of paths in the following way. If we have a collection of paths we join all endnodes to a base node and proceed as in the usual heuristic. If we have long paths then the heuristic is started with few subtours and the necessary CPU time will be acceptable even without using more sophisticated speed up methods. If the paths are close to an optimal Hamiltonian cycle, we can obtain very good results.

This concludes our survey of construction heuristics suitable for general traveling salesman problem instances. In the special case of geometric instances there are further ideas that can be employed. Some of these are addressed in Section 4.3.

Table 1 contains for our sample problem set the qualities of the solutions (i.e., the deviations in percent from an optimal solution) found by the standard nearest neighbor heuristic started at node $\lfloor n/2 \rfloor$ (NN1), the variant of the nearest neighbor heuristic using candidate sets started at node $\lfloor n/2 \rfloor$ (NN2), the farthest insertion heuristic started with the loop $\lfloor n/2 \rfloor$ (FI), the modified Christofides heuristic (CH), and the savings heuristic with base node $\lfloor n/2 \rfloor$ (SA). All heuristics (except for the standard nearest neighbor and the farthest insertion heuristic) were executed in their fast versions using the 10 nearest neighbor candidate subgraph. Table 2 lists the corresponding CPU times (without times for computing the candidate sets).

From our computational experiments with these construction heuristics we have drawn the following conclusions. The clear winners are the savings heuristics, and because of the considerably lower running time we declare the fast implementation of the savings heuristic as the best construction heuristic. This is in conformity with other computational testings, for example Arthur & Frendeway [1985]. If one

Table 1
Results of construction heuristics (Quality)

Problem	NN1	NN2	FI	CH	SA
lin105	33.31	10.10	11.22	19.76	5.83
pr107	6.30	9.20	2.13	8.95	9.22
pr124	21.02	8.16	11.87	16.49	4.20
pr136	34.33	17.90	8.59	27.83	6.73
pr144	4.96	13.68	3.12	15.55	9.97
pr152	19.53	20.44	4.24	19.75	9.44
u159	15.62	30.43	10.34	20.95	12.05
rat195	17.86	16.23	12.87	24.41	5.42
d198	25.79	17.57	3.85	15.40	6.96
pr226	22.76	20.87	1.42	20.95	8.93
gil262	25.95	19.47	5.93	19.05	8.86
pr264	20.32	17.38	9.12	17.60	10.56
pr299	27.77	19.71	9.13	19.93	11.95
lin318	26.85	18.68	10.87	18.42	8.24
rd400	23.13	23.37	9.61	21.48	9.00
pr439	27.04	15.74	12.24	17.39	13.31
pcb442	21.36	16.09	13.83	18.59	10.20
d493	28.52	17.82	11.61	17.44	8.84
u574	29.60	19.20	11.39	20.02	12.36
rat575	24.82	18.81	10.20	21.87	9.07
p654	31.02	27.99	6.89	21.73	10.66
d657	31.26	16.66	11.87	17.50	10.20
u724	23.16	20.34	11.65	21.00	10.44
rat783	27.13	18.66	12.09	21.34	9.88
pr1002	24.35	24.28	10.85	20.67	10.24
pcb1173	28.18	19.00	14.22	18.77	10.53
rl1304	28.58	21.59	17.81	15.92	9.86
nrw1379	24.43	18.89	9.71	24.14	10.54
u1432	25.50	19.07	12.59	24.05	10.41
pr2392	24.96	20.27	14.32	18.70	12.40

has to employ a substantially faster heuristic then one should use the variant of the nearest neighbor heuristic where forgotten nodes are inserted. For geometric problems minimum spanning trees are readily available. In such a case the fast variant of Christofides' heuristic can be used instead of the nearest neighbor variant.

4.2. *Improvement heuristics*

The Hamiltonian cycles computed by the construction heuristics in the previous subsection were only of moderate quality. Although they can be useful for some applications, they are not satisfactory in general.

In this subsection we address the question of how to improve these cycles. In general, the heuristics we will discuss here are defined using a certain type of basic move to alter the current cycle. We will proceed from fairly simple moves to more

Table 2
CPU times for construction heuristics

Problem	NN1	NN2	FI	CH	SA
lin105	0.03	0.01	0.06	0.01	0.02
pr107	0.03	0.01	0.07	0.01	0.03
pr124	0.05	0.01	0.10	0.01	0.02
pr136	0.05	0.02	0.10	0.01	0.03
pr144	0.06	0.02	0.13	0.01	0.03
pr152	0.06	0.02	0.13	0.01	0.06
u159	0.07	0.01	0.16	0.01	0.03
rat195	0.11	0.02	0.23	0.01	0.04
d198	0.10	0.02	0.23	0.02	0.05
pr226	0.13	0.02	0.30	0.02	0.08
gil262	0.18	0.03	0.40	0.02	0.07
pr264	0.19	0.03	0.43	0.02	0.11
pr299	0.24	0.04	0.52	0.02	0.07
lin318	0.27	0.04	0.59	0.03	0.09
rd400	0.42	0.05	0.94	0.05	0.13
pr439	0.51	0.05	1.12	0.04	0.20
pcb442	0.51	0.05	1.14	0.04	0.15
d493	0.64	0.07	1.43	0.05	0.20
u574	0.95	0.07	1.93	0.07	0.28
rat575	0.93	0.08	1.94	0.06	0.24
p654	1.19	0.09	2.51	0.07	0.53
d657	1.14	0.09	2.56	0.06	0.34
u724	1.49	0.10	3.10	0.10	0.39
rat783	1.63	0.11	3.63	0.11	0.48
pr1002	2.63	0.14	6.02	0.17	0.86
pcb1173	3.65	0.16	8.39	0.17	1.12
rl1304	4.60	0.18	10.43	0.13	1.85
nrw1379	5.16	0.22	11.64	0.30	1.70
u1432	5.54	0.17	12.52	0.34	1.64
pr2392	15.27	0.33	35.42	0.72	5.07

complicated ones. Further types of moves can be found in Gendreau, Hertz & Laporte [1992].

Two-opt exchange

This improvement approach is motivated by the following observation for Euclidean problems. If a Hamiltonian cycle crosses itself it can be easily shortened. Namely, erase two edges that cross and reconnect the resulting two paths by edges that do not cross (this is always possible). The new cycle is shorter than the old one.

A *2-opt move* consists of eliminating two edges and reconnecting the two resulting paths in a different way to obtain a new cycle. The operation is depicted in Figure 4, where we obtain a better solution if edges ij and kl are replaced by edges ik and jl. Note that there is only one way to reconnect the paths, since adding edges il and jk would result in two subtours. The 2-opt improvement heuristic is then outlined as follows.

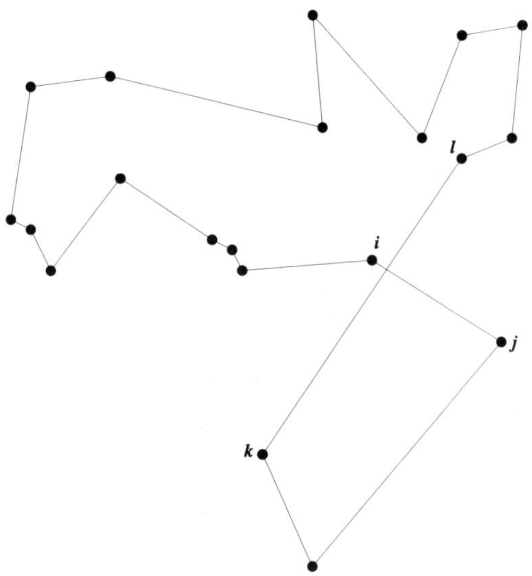

Fig. 4. A 2-opt move.

procedure 2-OPT

(1) Let T be the current Hamiltonian cycle.
(2) Perform the following until *failure* for every node i is obtained.
  (2.1) Select a node i.
  (2.2) Examine all 2-opt moves involving the edge between i and its successor in the cycle. If it is possible to decrease the cycle length this way, then choose the best such move, otherwise declare *failure* for node i.
(3) Return T.

Assuming integral data, the procedure runs in finite time. But, there are classes of instances where the running time cannot be bounded by a polynomial in the input size. Checking whether an improving 2-opt move exists takes time $O(n^2)$ because we have to consider all pairs of cycle edges.

The implementation of 2-OPT can be done in a straightforward way. But, observe that it is necessary to have an imposed direction on the cycle to be able to decide which two edges have to be added in order not to generate subtours. Having performed a move, the direction has to be reversed for one part of the cycle. CPU time can be saved if the update of the imposed direction is performed such that the direction on the longer path is maintained and only the shorter path is reversed. One can incorporate this shorter path update by using an additional array giving the rank of the nodes in the current cycle (an arbitrary node receives rank 1, its successor gets rank 2, etc.). Having initialized these ranks we can determine in constant time which of the two paths is shorter, and the ranks have to be updated only for the nodes in the shorter path. With such an implementation it still took 88.0 seconds to perform the 2-opt heuristic on a nearest neighbor solution for problem *pr2392*. The quality of the final solution was 8.3%.

Speedup possibilities are manifold. First of all, we can make use of a candidate subgraph. The number of possible 2-opt moves that are examined can then be reduced by requiring that in every 2-opt move at least one candidate edge is used to reconnect the paths.

Another modification addresses the order in which cycle edges are considered for participating in a 2-opt move. A straightforward strategy could use a fixed enumeration order, e.g., always scanning the nodes in Step (2.1) of the heuristic in the sequence $1, 2, \ldots, n$ and checking if a move containing the edge from node i to its successor in the current cycle can participate in an allowed move (taking restrictions based on the candidate set into account). But usually, one observes that, in the neighborhood of a successful 2-opt move, more improving moves can be found. The fixed enumeration order does not consider this. We have therefore implemented the following dynamic order. The nodes of the problem are stored in a list (initialized according to the sequence of the nodes in the cycle). In every iteration step the first node is taken from the list, scanned as described below, and reinserted at the end of the list. If i is the current node to be scanned, we examine if we can perform an improving 2-opt move which introduces a candidate edge having i as one endnode. If an improving move is found then all four nodes involved in that move are stored at the beginning of the node list (and therefore reconsidered with priority). The reduction in running time is considerable, because many fewer moves are examined. For example, when starting with a random Hamiltonian cycle for problem *rl5934*, with the dynamic enumeration only 85,762,731 moves were considered instead of 215,811,748 moves with the fixed enumeration. The reduction is less significant if one starts the 2-opt improvement with reasonable starting solutions. Since the 2-opt heuristic is very sensitive with respect to the sequence in which moves are performed, one can obtain quite different results for the two versions even for the same start. However, with respect to average quality both variants perform equally well.

Another point for speeding up computations further is to reduce the number of distance function evaluations, which accounts for a large portion of the running time. A thorough discussion of this issue can be found in Bentley [1992]. For example, one can inhibit evaluation of a 2-opt move that cannot be improving in the following way. When considering a candidate edge ij for taking part in a 2-opt move, we check if i and j have the same neighbors in the cycle as when ij was considered previously. If ij could not be used before in an improving move it can also not be used now. Furthermore, one can restrict attention to those moves where one edge ij is replaced by a shorter edge ik, since this must be true for one of the pairs.

Using an implementation of 2-opt based on the above ideas we can now perform the heuristic on a nearest neighbor solution for *pr2392* in 0.4 seconds achieving a Hamiltonian cycle of length 9.5% above the optimum. The average quality for our set of sample problems was 8.3%.

Performance of 2-opt can be improved by incorporating a simple additional move, namely *node insertion*. Such a move consists of removing one node from the current cycle and reinserting it at a different position. Since node insertion is not difficult to implement, we suggest to combine 2-opt and node insertion. On our set of sample problems we achieved an average quality of 6.5% using this combination. For problem *pr2392* we obtained a solution with quality 7.3% in 2.2 seconds. With his 2-opt implementation starting with a nearest neighbor solution Johnson [1990] achieved an excess of 6.5% over an approximation of the subtour bound. Bentley [1992] reports an excess of 8.7% for 2-opt and of 6.7% for a combination of 2-opt and node insertion. In both cases classes of random problems were used.

A further general observation for speeding up heuristics is the following. Usually, decrease in the objective function value is considerable in the first steps of the heuristic and then tails off. In particular, it takes a final complete round through all allowed moves to verify that no further improving move is possible. Therefore, if one stops the heuristics early (e.g., if only a very slow decrease is observed over some period) not too much quality is lost.

The 3-opt heuristic and variants

To have more flexibility for modifying the current Hamiltonian cycle we could break it into three parts instead of only two and combine the resulting paths in the best possible way. Such a modification is called *3-opt move*. The number of combinations to remove three edges of the cycle is $\binom{n}{3}$, and there are eight ways to connect three paths to form a cycle (if each of them contains at least one edge).

Note that node insertion and 2-opt exchange are special 3-opt moves. Node insertion is obtained if one path of the 3-opt move consists of just one node, a 2-opt move is a 3-opt move where one eliminated edge is used again for reconnecting the paths.

To examine all 3-opt moves takes time $O(n^3)$. Update after a 3-opt move is also more complicated than in the 2-opt case. The direction of the cycle may change on all but the longest of the three involved paths.

Therefore we decided to not consider full 3-opt (which takes 4661.2 seconds for problem *pcb442* when started with a nearest neighbor solution), but to limit in advance the number of 3-opt moves that are considered. The implemented procedure is the following.

procedure 3-OPT

(1) Let T be the current Hamiltonian cycle.
(2) For every node $i \in V$ define some set of nodes $N(i)$.
(3) Perform the following until *failure* is obtained for every node i.
  (3.1) Select a node i.
  (3.2) Examine all possibilities to perform a 3-opt move which eliminates three edges each having at least one endnode in $N(i)$. If it is possible to decrease the cycle length this way, then choose the best such move, otherwise declare *failure* for node i.
(4) Return T.

If we limit the cardinality of $N(i)$ by some fixed constant independent of n then checking in Step (3.2) if an improving 3-opt move exists at all takes time $O(n)$ (but with a rather large constant hidden by the O-notation).

We implemented the 3-opt routine using a dynamic enumeration order for node selection and maintaining the direction of the cycle on the longest path. Search in Step (3.2) is terminated as soon as an improving move is found. For a given candidate subgraph G_C we defined $N(i)$ as the set of all neighbors of i in G_C. In order to limit the CPU time (which is cubic in the cardinality of $N(i)$ for Step (3.2)) the number of nodes in each set $N(i)$ is bounded by 50 in our implementation.

With this restricted 3-opt version we achieved an average quality of 3.8% when started with a nearest neighbor solution and of 3.9% when started with a random solution (G_C was the 10 nearest neighbor subgraph augmented by the Delaunay graph to be defined in 4.3). CPU time is significantly reduced compared to the full version. Time for *pcb442* is now 18.2 seconds with the nearest neighbor start. Johnson, Bentley, McGeoch & Rothberg [1994] have a very much improved version of 3-opt that is only about four times slower than 2-opt.

One particular further variant of 3-opt is the so-called *Or-opt* procedure [Or, 1976]. Here it is required that one of the paths involved in the move has exactly l edges. Results obtained with this procedure lie between 2-opt and 3-opt (as can be expected) and it does not contribute significantly to the quality of the final solution if values larger than 3 are used for l.

Better performance than with the 3-opt heuristic can be obtained with general *k-opt* exchange moves, where k edges are removed from the cycle and the resulting paths are reconnected in the best possible way. A complete check of the existence of an improving k-opt move takes time $O(n^k)$ and is therefore only applicable for small problems. One can, of course, design restricted searches for higher values of k in the same way as we did for $k = 3$. For a discussion of update aspects see Margot [1992].

One might suspect that with increasing k the k-opt procedure should yield provably better approximations to the optimal solution. However, Rosenkrantz, Stearns & Lewis [1977] show that for every $n \geq 8$ and every $k \leq n/4$ there exists a TSP instance on n nodes and a k-optimal solution such that the optimal and k-optimal values differ by a factor of $2 - 2/n$. Nevertheless, this is only a worst case result. One observes that for practical applications it does pay to consider larger values of k and design efficient implementations of restricted k-opt procedures.

Lin–Kernighan type exchange

The final heuristic to be discussed in this subsection was originally described in Lin & Kernighan [1973]. The motivation for this heuristic is based on experience gained from practical computations. Namely, one observes that the more flexible and powerful the possible cycle modifications, the better are the obtained results. In fact, simple moves quickly run into local optima of only moderate quality. On the other hand, the natural consequence of applying k-opt for larger k requires

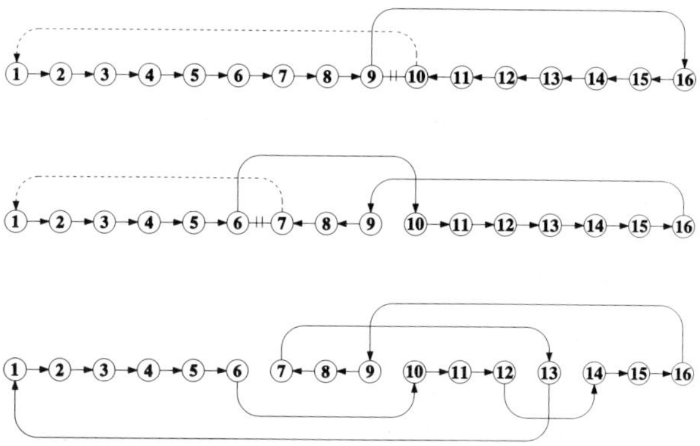

Fig. 5. The Lin–Kernighan heuristic.

a substantially increasing running time. Therefore, it seems more reasonable to follow an approach suggested by Lin and Kernighan.

Their idea is based on the observation that sometimes a modification slightly increasing the cycle length can open up new possibilities for achieving considerable improvement afterwards. The basic principle is to build complicated modifications that are composed of simpler moves where not all of these moves necessarily have to decrease the cycle length. To obtain reasonable running times, the effort to find the parts of the composed move has to be limited. Many variants of this principle are possible. We do not describe the original version of this algorithm which contains a 3-opt component, but discuss a somewhat simpler version where the basic components are 2-opt and node insertion moves. General complexity issues concerning the Lin–Kernighan heuristic are addressed in Papadimitriou [1990].

When building a move, in each substep we have some node from which a new edge is added to the cycle according to some criterion. We illustrate our procedure by the example of Figure 5. Suppose we start with the canonical Hamiltonian cycle $1, 2, \ldots, 16$ for a problem on 16 nodes and we decide to construct a modification starting from node 16.

In the first step it is decided to eliminate edge $(1, 16)$ and introduce the edge from node 16 to node 9. Adding this edge creates a subtour, and therefore edge $(9, 10)$ has to be deleted. To complete the cycle, node 10 is connected to node 1.

If we stop at this point we have simply performed a 2-opt move. The fundamental new idea is not to connect node 10 to node 1, but to search for another move starting from node 10. Suppose we now decide to add edge $(10, 6)$. Again, one edge, namely $(6, 7)$, has to be eliminated to break the subtour. The sequence of moves could be stopped here, if node 7 is joined to node 1.

As a final extension we perform a node insertion for node 13 instead, and place this node between 1 and 7. Thus we remove edges $(12, 13)$ and $(13, 14)$ and add edges $(12, 14)$, $(7, 13)$ and $(1, 13)$.

Note that the direction changes on some parts of the cycle while performing these moves and that these new directions have to be considered in order to be able to perform the next moves correctly. When building the final move we obtained three different solutions on the way. The best of these solutions (which is not necessarily the final one) can now be chosen as the new current Hamiltonian cycle.

Realization of this procedure is possible in various ways. We have chosen the following options.

– To speed up search for submoves a candidate subgraph is used. Edges to be added from the current node to the cycle are only taken from this set and are selected according to a local gain criterion. Let i be the current node. We define the local gain g_{ij} that is achieved by adding edge ij to the cycle as follows. If jk is the edge to be deleted if a 2-opt move is to be performed, then we set $g_{ij} = c_{jk} - c_{ij}$. If jk and jl are the edges to be deleted if a node insertion move is to be performed, then $g_{ij} = c_{jk} + c_{jl} - c_{lk} - c_{ij}$. The edge with the maximum local gain is chosen to enter the solution and the corresponding move is performed.

– The number of submoves in a move is limited in advance, and a dynamic enumeration order is used to determine the starting node for the next move.

– Examination of more than one candidate edge to enter the cycle is possible. The maximum number of candidates examined from the current node and the maximum number of submoves up to which alternative edges are taken into account are specified in advance. This option introduces an enumeration component for selecting the first few submoves.

The basic outline of the heuristic is then given as follows.

procedure LIN-KERNIGHAN

(1) Let T be the current Hamiltonian cycle.
(2) Perform the following computation until *failure* is obtained for every node i.
  (2.1) Select a node i to serve as a start for building a composed move.
  (2.2) Try to find an improving move starting from i according to the guidelines and the parameters discussed above. If no such move can be found, then declare *failure* for node i.
(3) Return T.

A central implementation issue concerns the management of the tentative moves. Since most of them do not lead to an improvement of the solution, it is reasonable to avoid an explicit cycle update for every such move, but only update as little information as possible.

We use an idea that was reported by Applegate, Chvátal & Cook [1990]. Consider for example a 2-opt move. Its effect on the current solution is completely characterized by storing how the two resulting paths are reconnected and if their direction has changed. To this end it suffices to know the endnodes of every path and the edges connecting them. For every other node its neighbors are unchanged, and, since we have ranks associated with the nodes, we can easily identify the path in which a node is contained. In general, the current Hamiltonian cycle is represented by a cycle of intervals of ranks where each interval represents a subpath of the starting Hamiltonian cycle. For an efficient identification of the interval to which a specific node belongs the intervals are kept in a balanced binary search tree. Therefore, the interval containing a given node can be identified in time $O(\log m)$ if we have m intervals. Note, that also in the interval representation we have to reorient paths of the sequence. But, as long as we have few intervals (i.e., few tentative submoves), this can be done fast. Of course, the number of intervals should not become too large because the savings in execution time decreases with the number of intervals that have to be managed. Therefore, if we have too many intervals, we clear the interval structure and generate correct successor and predecessor pointers to represent the current cycle. The longest path represented as an interval can remain unchanged, i.e., for its interior nodes successors, predecessors, and ranks do not have to be altered.

Possible choices for the parameters of this heuristic are so numerous that we cannot document all experiments here. We only discuss some basic insights.

The observations we gained from our experiments can be summarized as follows.

– At least 15 submoves should be allowed for every move in order to be able to generate reasonably complicated moves.

– It is better not to start out with a random solution, but with a locally good Hamiltonian cycle. But, this is of less importance when more elaborate versions of the Lin–Kernighan procedure are used.

– It is advisable to consider several alternative choices for the edge to be added from the first node.

– Exclusion of node insertion moves usually leads to inferior results.

We report about two variants of the Lin–Kernighan approach for our set of sample problems.

In the first variant, the candidate subgraph is the 6 nearest neighbor subgraph augmented by the Delaunay graph. Up to 15 submoves are allowed to constitute a move. Three alternative choices for the edge to be added to the cycle in the first submove are considered. Submoves are 2-opt and node insertion moves.

In the second variant, the candidate subgraph is the 8 nearest neighbor subgraph augmented by the Delaunay graph. Up to 15 submoves are allowed to constitute a move. Two alternative entering edges are considered for each of the first three submoves of a move (This gives a total of eight possibilities examined for the first three submoves of a move). Submoves are 2-opt and node insertion moves.

In contrast to simpler heuristics, the dependence on the starting solution is not very strong. Results and CPU times differ only slightly for various types of starting solutions. Even if one starts with a random Hamiltonian cycle not much quality is lost. Starting with a nearest neighbor solution we obtained an average quality of 1.9% for variant 1 and 1.5% for variant 2. Running time of the Lin–Kernighan exchange for problem *pr2392* was 61.7 and 122.3 seconds for the respective variants. Variant 2 is more expensive since more possibilities for moves are enumerated (larger candidate set and deeper enumeration level). In general, higher effort usually leads to better results. Similar results are given in Johnson [1990]. Another variant of the Lin–Kernighan heuristic is discussed in Mak & Morton [1993].

As a final experiment we ran an extension of the Lin–Kernighan heuristic first proposed by Johnson [1990]. The Lin–Kernighan heuristic, as every other heuristic, terminates in a local optimum which depends on the start and on the moves that are performed. To have a chance of finding good local optima one can start the procedure several times with different starting solutions. A more reasonable idea is not to restart with a completely new starting solution but only to perturb the current solution. This way one escapes the local optimum by making a move that increases the length but still has a solution that is close to an optimal one at least in some segments. Computational experiments show that this approach is superior. Johnson [1990] suggests that after termination of the Lin–Kernighan heuristic a random 4-opt move is performed and the heuristic is reapplied. Using this method several optimal solutions of some larger problems (e.g., *pr2392*) were found. In our experiment we used the second variant of the Lin–Kernighan heuristic described above, but this time allowing 40 submoves per move. In addition, we performed a restricted 3-opt after termination of each Lin–Kernighan heuristic. This approach was iterated 20 times. We now obtained an average quality of 0.6%.

Table 3 gives the quality of the solutions found by 2-opt (2-O), 3-opt (3-O), and the two versions of the Lin–Kernighan heuristic described above (LK1 and LK2). Column (ILK) displays the results obtained with the iterated Lin–Kernighan heuristic. The improvement heuristics were started with a nearest neighbor solution. Table 4 lists the corresponding CPU times.

From our computational tests we draw the following conclusions. If we want to achieve very good results, simple basic moves are not sufficient. If simple moves are employed, then it is advisable to apply them to reasonable starting solutions since they are not powerful enough for random starts. Nearest neighbor like solutions are best suited for simple improvement schemes since they consist of rather good pieces of a Hamiltonian cycle and contain few bad ones that can be easily repaired. For example, the 2-opt improvement heuristic applied to a farthest insertion solution would lead to much inferior results, although the farthest insertion heuristic delivers much better Hamiltonian cycles than those found by the nearest neighbor heuristic.

If one attempts to find solutions in the range of 1% above optimality one has to use the Lin–Kernighan heuristic since it can avoid bad local minima. However, applying it to large problems requires that considerable effort is spent in deriving an efficient implementation. A naïve implementation would consume an enormous amount of CPU time. If time permits, the iterated version of the Lin–Kernighan heuristic is the method of choice for finding good approximate solutions. For a more general discussion of local search procedures see Johnson, Papadimitriou & Yannakakis [1988].

It is generally observed that quality of heuristics degrades with increasing problem size, therefore more tries are necessary for larger problems. In Johnson [1990] and Bentley [1992] some interesting insights are reported for problems with up to a million nodes.

Table 3
Results of improvement heuristics

Problem	2-O	3-O	LK1	LK2	ILK
lin105	8.42	0.00	0.77	0.00	0.00
pr107	3.79	2.05	1.53	0.81	0.00
pr124	2.58	1.15	2.54	0.39	0.00
pr136	10.71	6.14	0.55	0.72	0.38
pr144	3.79	0.39	0.56	0.06	0.00
pr152	2.93	1.85	0.00	0.19	0.00
u159	14.00	11.49	2.20	1.59	0.00
rat195	6.46	3.01	1.55	1.55	0.47
d198	3.85	6.12	0.63	1.51	0.16
pr226	13.17	1.72	0.72	0.49	0.00
gil262	10.26	3.07	1.18	2.44	0.55
pr264	4.39	6.04	0.12	0.01	0.49
pr299	10.46	4.37	1.55	1.36	0.15
lin318	9.54	2.67	1.87	1.17	0.53
rd400	5.01	3.42	2.34	1.41	0.75
pr439	6.52	3.61	2.73	2.68	0.38
pcb442	8.74	3.01	1.41	1.94	0.90
d493	9.37	3.32	2.23	1.47	0.84
u574	7.85	4.61	2.05	0.98	0.60
rat575	7.93	4.46	2.48	1.68	1.03
p654	14.89	0.62	4.14	2.95	0.03
d657	7.57	3.52	3.10	1.65	0.74
u724	8.09	4.20	2.60	1.38	0.67
rat783	9.07	4.22	1.94	1.77	0.91
pr1002	8.46	3.80	2.92	2.72	1.51
pcb1173	10.72	5.26	2.18	3.22	1.46
rl1304	13.21	7.08	5.07	1.73	1.62
nrw1379	8.25	3.65	2.48	1.76	1.13
u1432	10.48	5.39	1.51	2.45	0.99
pr2392	9.48	5.26	2.95	2.90	1.75

4.3. Special purpose algorithms for geometric instances

TSP instances that arise in practice often are of geometric nature in that the points defining the problem instance correspond to locations in a space. The length of the edge connecting nodes i and j is the distance of the points corresponding to the nodes according to some metric, i.e., a function that satisfies the triangle inequality. Usually the points are in 2-dimensional space and the metric is the Euclidean (L_2), the Maximum (L_∞), or the Manhattan (L_1) metric.

In this subsection we discuss advantages that can be gained from geometric instances. Throughout this subsection we assume that the points defining a problem instance correspond to locations in the plane and that the distance of two points is their Euclidean distance.

Table 4
CPU times for improvement heuristics

Problem	2-O	3-O	LK1	LK2	ILK
lin105	0.02	4.10	1.15	4.09	168.40
pr107	0.01	2.73	0.71	2.25	121.34
pr124	0.02	3.82	1.03	3.08	219.52
pr136	0.03	3.76	1.05	2.98	221.96
pr144	0.02	6.37	1.20	3.85	304.52
pr152	0.02	3.44	1.03	2.85	260.19
u159	0.02	6.03	1.46	4.26	314.53
rat195	0.02	4.41	1.93	4.86	409.74
d198	0.03	7.22	5.27	6.04	520.23
pr226	0.02	12.85	2.64	7.16	488.87
gil262	0.04	7.84	2.84	8.37	575.52
pr264	0.03	9.83	3.53	8.29	455.71
pr299	0.04	10.27	3.47	10.97	750.62
lin318	0.04	12.56	5.30	11.98	825.53
rd400	0.05	13.19	4.57	13.33	1153.41
pr439	0.06	14.60	7.34	16.04	1086.08
pcb442	0.08	18.23	5.03	17.60	1079.45
d493	0.08	17.69	8.88	15.89	1465.07
u574	0.10	34.12	7.92	27.09	1677.75
rat575	0.07	17.38	7.13	29.37	1547.93
p654	0.08	41.65	9.09	17.17	1303.30
d657	0.10	22.19	12.51	26.00	1958.84
u724	0.10	30.27	9.37	26.55	1921.41
rat783	0.14	27.50	12.78	39.24	2407.84
pr1002	0.19	41.69	17.78	42.01	2976.47
pcb1173	0.17	55.41	17.07	55.10	3724.98
rl1304	0.21	112.02	22.88	54.73	4401.12
nrw1379	0.26	52.68	21.22	73.58	4503.37
u1432	0.19	61.85	16.91	66.21	3524.59
pr2392	0.40	148.63	61.72	122.33	8505.42

Geometric heuristics

Bartholdi & Platzman [1982] introduced the so-called *space filling curve heuristic* for problem instances in the Euclidean plane. It is particularly easy to implement and has some interesting theoretical properties. The heuristic is based on a bijective mapping $\psi : [0,1] \to [0,1] \times [0,1]$, a so-called *space filling curve*. The name comes from the fact that when varying the arguments of ψ from 0 to 1 the function values fill the unit square completely. Surprisingly, such functions exist and, what is interesting here, they can be computed efficiently and also for a given $y \in [0,1] \times [0,1]$ a point $x \in [0,1]$ such that $\psi(x) = y$ can be found in constant time.

The function used by Bartholdi and Platzman models the recursive subdivision of squares into four equally sized subsquares. The space filling curve is obtained by patching the four respective subcurves together. The heuristic is given as follows.

procedure SPACEFILL

(1) Scale the points to the unit square.
(2) For every point i with coordinates x_i and y_i compute z_i such that $\psi(z_i) = (x_i, y_i)$.
(3) Sort the numbers z_i in increasing order.
(4) Connect the points by a cycle respecting the sorted sequence of the z_i's (to complete the cycle connect the two points with smallest and largest z-value).

Since the values z_i can be computed in constant time the overall computation time is dominated by the time to sort these numbers in Step (3) and hence this heuristic runs in time $\Theta(n \log n)$.

It can be shown, that if the points are contained in a rectangle of area F then the Hamiltonian cycle is not longer than $2\sqrt{nF}$. Bartholdi and Platzman have also shown that the quotient of the length of the heuristic and the length of an optimal solution is bounded by $O(\log n)$.

At this point, we comment briefly on average case analysis for the Euclidean TSP. Suppose that the n points are drawn independently from a uniform distribution on the unit square and that c_{opt} denotes the length of an optimal solution. Beardwood, Halton & Hammersley [1959] show that there exists a constant C such that $\lim_{n\to\infty} c_{\text{opt}}/\sqrt{n} = C$ and they give the estimate $C \approx 0.765$. Such behavior can also be proved for the space filling curves heuristic with a different constant $\overline{C}$. Bartholdi & Platzman [1982] give the estimate $\overline{C} \approx 0.956$. Therefore, for this class of random problems the space filling curves heuristic can be expected to yield solutions that are approximately 25% larger than an optimal solution as n tends to infinity.

Since in space filling curves heuristic adding or deleting points does not change the relative order of the other points in the Hamiltonian cycle, this heuristic cannot be expected to perform too well. In fact, for our set of sample problems we achieved an average quality of 35.7%. As expected, running times are very low, e.g., 0.2 seconds for problem *pr2392*. Experiments show that space filling curves solutions are not suited as starts for improvement heuristics. They are useful if only extremely short computation times are allowed.

In the well-known *strip heuristic* the problem area is cut into $\sqrt{n}$ parallel vertical strips of equal width. Then Hamiltonian paths are constructed that collect the points of every strip sorted by the vertical coordinate, and finally these paths are combined to form a solution. The procedure runs in time $O(n \log n)$. Such a straightforward partition into strips is very useful for randomly generated problems, but can and will give poor results on real-world instances. The reason is that partition into parallel strips may not be adequate for the given point configuration.

To overcome this drawback, other approaches do not divide the problem area into strips but into segments, for example into squares or rectangles. In *Karp's partitioning heuristic* [Karp, 1977] the problem area is divided by horizontal and vertical cuts such that each segment contains no more than a certain number k of points. Then, a dynamic programming algorithm is used to compute an optimal Hamiltonian cycle on the points contained in each segment. In a final step all subtours are glued together according to some scheme to form a Hamiltonian cycle through all points. For fixed k the optimal solutions of the respective subproblems can be determined in linear time (however, depending on k, a large constant associated with the running time of the dynamic programming algorithm is hidden).

We give another idea to reduce the complexity of a large scale problem instance. Here the number of nodes of the problem is reduced in such a way that the remaining nodes still give a satisfactory representation of the geometry of the original points. Then a Hamiltonian cycle on this set of representative nodes is computed in order to serve as an approximation for the cycle through all nodes. In the final step the original nodes are inserted into this cycle (where the number of insertion points that will be checked can be specified) and the representative nodes (if not original nodes) are removed. More precisely, we use the following bucketing procedure.

procedure NODE_REDUCTION

(1) Compute an enclosing rectangle for the given points.
(2) Recursively subdivide each rectangle into four equally sized parts by a horizontal and a vertical line until each rectangle
  - contains no more than 1 point, or
  - is the result of at least m recursive subdivisions and contains no more than k points.
(3) Represent each (nonempty) rectangle by the center of gravity of the points contained in it.
(4) Compute a Hamiltonian cycle through the representative nodes.
(5) Insert the original points into this cycle. To this end at most $l/2$ insertion points are checked before and after the corresponding representative nodes in the current cycle. The best insertion point is then chosen.
(6) Remove all representative nodes that are not original nodes.

The parameters m, k, and l, and the heuristic needed in Step (4) can be chosen with respect to the available CPU time.

This heuristic is only suited for very large problems and we did not apply it to our sample set. One can expect qualities of 15% to 25% depending on the point configuration.

The heuristic is similar to a clustering algorithm given in Litke [1984], where clusters of points are represented by a single point. Having computed an optimal Hamiltonian cycle through the representatives, clusters are expanded one after another. A further partitioning heuristic based on geometry is discussed in Reinelt [1994]. Decomposition is also a topic of Hu [1967]. Since many geometric heuristics are fairly simple, they are amenable to probabilistic analysis. Some interesting results on average behavior can be found in Karp & Steele [1985].

Success of such simple approaches is limited, because global view is lost and parts of the final solution are computed independently from each other. In

Johnson [1990] a comparison of various geometric heuristics is given, concluding that the average excess over the subtour bound for randomly generated problems is 64.0%, 30.2%, and 23.2% for Karp's partitioning heuristic, strip heuristic, and Litke's clustering method, respectively. These results show that it is necessary to incorporate more sophisticated heuristics into simple partitioning schemes as we did in our procedure. Then, one can expect qualities of about or below 20%. In any case, these approaches are very fast and can virtually handle arbitrary problem sizes. If the given point configuration decomposes in a natural way, then much better results can be expected.

Convex hull starts

Let $v_1, v_2, \ldots, v_k$ be those points of the problem defining the boundary of the convex hull of all given points (in this order). Then in any optimal Hamiltonian cycle this sequence is respected, otherwise it would contain crossing edges and hence could not be optimal. Therefore it is reasonable to use the cycle $(v_1, v_2, \ldots, v_k)$ as start for the insertion heuristics.

Convex hulls can be computed very quickly (in time $\Theta(n \log n)$, see e.g., Graham [1972]). Therefore, only negligible additional CPU time is necessary to compute a good starting cycle for the insertion heuristics in the Euclidean case.

It turns out that the quality of solutions delivered by insertion heuristics is indeed improved if convex hull start is used. But, gain in quality is only moderate. In particular, also with this type of start, our negative assessment of insertion heuristics still applies.

Delaunay graphs

A very powerful tool for getting insight into the geometric structure of a Euclidean problem is the Voronoi diagram, or its dual: the Delaunay triangulation. Although known for a long time [Voronoï, 1908; Delaunay, 1934], these structures have only recently received significant attention in the literature on computation.

Let $S = \{P_1, P_2, \ldots, P_n\}$ be a finite subset of $\mathbb{R}^2$ and let $d : \mathbb{R}^2 \times \mathbb{R}^2 \longrightarrow \mathbb{R}$ denote the Euclidean metric. We define the *Voronoi region* $\mathrm{VR}(P_i)$ of a point P_i by $\mathrm{VR}(P_i) = \{P \in \mathbb{R}^2 \mid d(P, P_i) \leq d(P, P_j) \text{ for all } j = \{1, 2, \ldots, n\}\}$, i.e., $\mathrm{VR}(P_i)$ is the set of all points that are at least as close to P_i as to any other point of S. The set of all n Voronoi regions is called the *Voronoi diagram* $V(S)$ of S. Figure 6 shows the Voronoi diagram for a set of 15 points in the plane.

Given the Voronoi diagram of S, the *Delaunay triangulation* $G(S)$ is the undirected graph $G(S) = (S, D)$ where $D = \{\{P_i, P_j\} \mid \mathrm{VR}(P_i) \cap \mathrm{VR}(P_j) \neq \emptyset\}$. It is easy to see that $G(S)$ is indeed a triangulated graph.

In the following we use an alternative definition which excludes those edges (P_i, P_j) for which $|\mathrm{VR}(P_i) \cap \mathrm{VR}(P_j)| = 1$. In this case the name is misleading, because we do not necessarily have a triangulation anymore, and to avoid misinterpretation from now on we speak about the *Delaunay graph*. In contrast to the Delaunay triangulation defined above, the Delaunay graph is guaranteed to be a planar graph (implying $|D| = O(n)$). Moreover, as the Delaunay triangulation, it contains a minimum spanning tree of the complete graph on S with edge weights

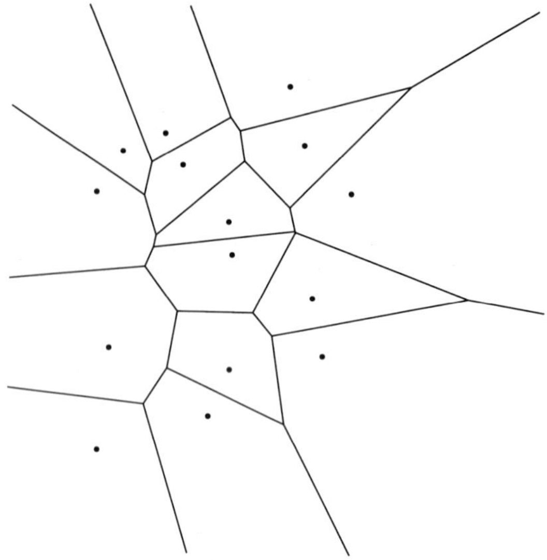

Fig. 6. A Voronoi diagram.

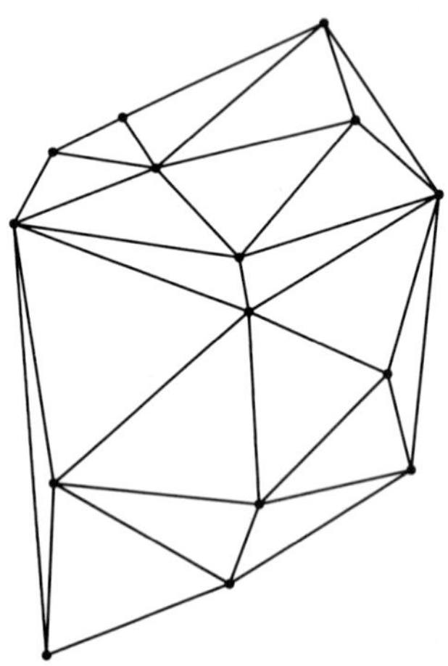

Fig. 7. A Delaunay graph.

$d(P_i, P_j)$ and contains for each node an edge to a nearest neighbor. Figure 7 shows the Delaunay triangulation corresponding to the Voronoi diagram displayed in Figure 6.

The Delaunay graph can be computed very efficiently. There are algorithms computing the Voronoi diagram (and hence the Delaunay triangulation) in time $O(n \log n)$ (see e.g., Shamos & Hoey [1975]). For practical purposes an algorithm given in Ohya, Iri & Murota [1984] seems to perform best. It has worst case running time $O(n^2)$, but linear running time can be observed for real problems. In the same paper some evidence is given that the linear expected running time for randomly generated problems can be mathematically proven. A rigorous proof, however, is still missing. We used an implementation of this algorithm in our experiments. CPU times are low, 4.5 seconds for computing the Delaunay graph for a set of 20,000 points. We note that computing Voronoi diagrams and Delaunay graphs in a numerically stable way is a nontrivial task. In Jünger, Reinelt & Zepf [1991] and Kaibel [1993] it is shown how round-off errors during computation can be avoided to obtain reliable computer codes.

There are heuristics which try to directly exploit information from the Voronoi diagram [Rujén, Evertsz & Lyklema, 1988; Cronin, 1990; Segal, Zhang & Tsai, 1991].

The Delaunay graph can be exploited to speed up computations, as can be seen from what follows.

Minimum spanning trees

For Euclidean problem instances, one can compute minimum spanning trees very fast because computation can be restricted to the edge set of the Delaunay graph. Now we can use Kruskal's algorithm [Kruskal, 1956] which runs (if properly implemented using fast union-find techniques) in time $O(n \log m)$ where m is the number of edges of the graph. In the Euclidean case we thus obtain a running time of $O(n \log n)$. For example, it takes time 1.3 seconds to compute a minimum spanning tree for problem *pr2392* from the Delaunay graph. Using more sophisticated data structures the theoretical worst-case running time can be improved further (see Tarjan [1983] and Cormen, Leiserson & Rivest [1989]), but this does not seem to be of practical importance.

Implementing the nearest neighbor heuristic efficiently

Using the Delaunay graph, we can improve the running time of the standard nearest neighbor heuristic for Euclidean problem instances. Namely, if we want to determine the k nearest neighbors of some node, then it is sufficient to consider only nodes which are at most k edges away in the Delaunay graph. Using breadth-first search starting at a node, say i, we compute for $k = 1, 2, \ldots$ the k-th nearest neighbor of i until a neighbor is found that is not yet contained in the current partial Hamiltonian cycle. Due to the properties of the Delaunay graph we should find the nearest neighbor of the current node by examining only a few edges. Since in the last steps of the algorithm we have to collect the forgotten nodes (which are far away from the current node) it makes no sense to use the Delaunay graph any further. We found that it is faster, if the final nodes are just added using the standard nearest neighbor approach.

The worst case time complexity of this modified nearest neighbor search is still $O(n^2)$ but, in general, reduction of running time is considerable. For *rl5934* we reduced the running time to 0.7 seconds (adding the final 200 nodes by the standard nearest neighbor routine) compared to 40.4 seconds for the standard implementation. Plotting CPU times versus problem sizes exhibits that we can expect an almost linear growth of running time.

Computing candidate sets efficiently

We have observed the importance of limiting search for improving moves (for example by using candidate sets). In this subsection we address the question of which candidate sets to use and of how to compute them efficiently for geometric problems. Three types of such sets were considered.

An obvious one is the *nearest neighbor candidate set*. Here, for every node the edges to its k nearest neighbors (where k is usually between 5 and 20) are determined. The candidate set consists of the collection of the corresponding edges. For example, optimal solutions for the problems *pcb442*, *rat783*, or *pr2392* are contained in their 8 nearest neighbor subgraphs. On the other hand, in problem *d198* the points form several clusters, so even the 20 nearest neighbor subgraph is still disconnected. Nevertheless, the edges to neighbors provide promising candidates to be examined. The idea of favoring near neighbors has already been used by Lin and Kernighan to speed up their algorithm. They chose $k = 5$ for their computational experiments.

A straightforward enumeration procedure computes the k nearest neighbors in time $O(n^2)$. As described above neighbor computations can be performed much faster if the Delaunay graph is available. For example, computation of the 10 nearest neighbor subgraph for a set of 20,000 points takes 8.3 seconds. In our practical experiments we observed a linear increase of the running time with the problem size.

Another candidate set is derived from the Delaunay graph itself, since it seems to give useful information about the geometric structure of a problem. It is known, however, that this graph does not have to contain a Hamiltonian cycle [Dillencourt, 1987a, b]. First experiments showed that it provides a candidate set too small for finding good Hamiltonian cycles. We therefore decided to use the *Delaunay candidate set*. This set is composed of the Delaunay graph as defined above and transitive edges of order 2, i.e., if node i is connected to node j, and node j is connected to node k in the Delaunay graph, then the edge from i to k is also taken into the candidate set. (This set may contain some very long edges that can be deleted in a heuristic way.)

Also this candidate subgraph can be computed very efficiently (e.g., in 14.5 seconds for 20,000 nodes). The average cardinality of the three subgraphs for our test of sample problems nodes is $2.75n$ for the Delaunay graph, $5.73n$ for the 10 nearest neighbor graph, and $9.82n$ for the Delaunay set.

Another efficient way of computing nearest neighbors is based on k-d-trees (see Bentley [1992] and Johnson [1990]). Using the Delaunay graph we can achieve running times competitive with this approach.

Experiments have shown that the nearest neighbor candidate set fails on clustered point configurations, whereas the Delaunay candidate set seems to have advantages for such configurations but contains too many edges. The *combined candidate set* attempts to combine the advantages of the two previous ones. For every node the edges to its k nearest neighbors (where k between about 5 and 20) are determined. The candidate set consists of the collection of these edges and those of the Delaunay graph.

We found that, in general and if applicable, the combined candidate set is preferable. It was therefore used in most of our practical computations. Of course, further candidate sets can be constructed. One possibility is to partition the plane into regions according to some scheme and then give priority to edges connecting points in adjacent regions.

Note that the very fast computation of these candidate sets strongly relies on the geometric nature of the problem instances. In general, one has to find other ways for deriving suitable candidate sets.

We have outlined some ideas which are useful for the handling of very large geometric traveling salesman problems. Though applied here only to Euclidean problems, the methods or variants of them are also suitable for other types of geometric problems. Delaunay triangulations for the Manhattan or maximum metric have the same properties as for the Euclidean metric and can be computed as efficiently. For treating geometric problems on several hundred thousand nodes it is necessary to use methods of the type discussed above. Exploitation of geometric information is an active research field, and we anticipate further interesting contributions in this area.

4.4. A survey of other recent approaches

The heuristics discussed so far have a chance to find optimal solutions. But, even if we apply the best heuristic of the previous subsections, namely the Lin–Kernighan heuristic, we will usually encounter solutions of quality only about 1%. This is explained by the fact that, due to limited modification capabilities, every improvement heuristic will only find a local minimum. The weaker the moves that can be performed, the larger is the difference between a locally optimal solution and a true optimal solution. One way to overcome this drawback is to start improvement heuristics many times with different (randomly generated) starts because this increases the chance of finding better local minima. Success is limited, though.

Most of the heuristics we consider in this subsection try to escape from local minima or avoid local minima in a more systematic way. A basic ingredient is the use of randomness or *stochastic search* in contrast to the purely deterministic heuristics we have discussed so far. The first random approach is the so-called *Monte-Carlo algorithm* [Metropolis, Rosenbluth, Rosenbluth, Teller & Teller, 1953]. In some cases the design of a particular method is influenced by the desire of imitating nature (which undoubtedly is able to find solutions to highly complex problems) in the framework of combinatorial optimization. We have not implemented the heuristics of this subsection, but give references to the literature.

Simulated annealing

The approach of *simulated annealing* is based on a correspondence between the process of searching for an optimal solution in a combinatorial optimization problem and phenomena occurring in physics [Kirkpatrick, Gelatt & Vecchi, 1983; Cerny, 1985].

To visualize this analogy consider the physical process of cooling a liquid to its freezing point with the goal of obtaining an ordered crystalline structure. Rapid cooling would not achieve this, one rather has to slowly cool (anneal) the liquid in order to allow improper structures to readjust and to have a perfect order (ground state) at the crystallization temperature. At each temperature step the system relaxes to its state of minimum energy.

Simulated annealing is based on the following analogy between such a physical process and an optimization method for a combinatorial minimization problem. Feasible solutions correspond to states of the system (an optimal solution corresponding to a ground state, i.e., a state of minimum energy). The objective function value resembles the energy in the physical system. Relaxation at a certain temperature is modeled by allowing random changes of the current feasible solution which are controlled by the level of the temperature. Depending on the temperature, alterations that increase the energy (objective function) are more or less likely to occur. At low temperatures it is very improbable that the energy of the system increases. System dynamics is imitated by local modifications of the current feasible solution.

Modifications that increase the length of a Hamiltonian cycle are possible, but only accepted with a certain probability. Pure improvement heuristics as we have discussed so far can be interpreted in this context as rapid quenching procedures that do not allow the system to relax.

The general outline of a simulated annealing procedure for the TSP is the following.

procedure SIMULATED_ANNEALING

(1) Compute an initial Hamiltonian cycle T and choose an initial temperature ϑ and a repetition factor r.
(2) As long as the stopping criterion is not satisfied perform the following steps.
  (2.1) Do the following r times.
    (2.1.1) Perform a random modification of the current cycle to obtain the cycle T' and let $\Delta = c(T') - c(T)$ (difference of lengths).
    (2.1.2) If $\Delta < 0$ then set $T = T'$. Otherwise compute a random number x, $0 \leq x \leq 1$ and set $T = T'$ if $x < e^{-\frac{\Delta}{\vartheta}}$
  (2.2) Update ϑ and r.
(3) Output the best solution found.

Simulated annealing follows the general principle, that improving moves are always accepted, whereas moves increasing the length of the current cycle are only accepted with a certain probability depending on the increase and current value of ϑ.

The formulation has several degrees of freedom and various realizations are possible. Usually 2-opt or 3-opt moves are employed as basic modification in Step (2.1.1). The temperature ϑ is decremented in Step (2.2) by setting $\vartheta = \gamma\vartheta$ where γ is a real number close to 1, and the repetition factor r is usually initialized with the number of cities and updated by $r = \alpha r$ where α is some factor between

1 and 2. Realization of Step (2.2) determines the so-called *annealing schedule* or *cooling scheme* (much more complicated schedules are possible). The scheme given above is named geometric cooling. The procedure is stopped if the length of the current Hamiltonian cycle was not altered during several temperature steps. Expositions of general issues for the development of simulated annealing procedures can be found in Aarts & Korst [1989] and Johnson, Aragon, McGeoch & Scheron [1991], a bibliography is given in Collins, Eglese & Golden [1988].

Computational experiments for the TSP are for example reported in Kirkpatrick [1984], van Laarhoven [1988], Johnson [1990]. It is generally observed that simulated annealing can find very good or even optimal solutions and beats Lin–Kernighan concerning quality. To be certain of this, however, one has to spend considerable CPU time because temperature has to be decreased very slowly and many repetitions at each temperature step are necessary.

We think that the most appealing property of simulated annealing is its fairly simple implementation. The principle can be used to approach very complicated problems if only a basic subroutine is available that turns a feasible solution into another feasible solution by some modification. Hajek [1985] proved convergence of the algorithm to an optimal solution with probability 1, if the basic move satisfies a certain property. Unfortunately, the theoretically required annealing scheme is not suited for practical use. The proper choice of an annealing scheme should not be underestimated. It is highly problem dependent and only numerous experiments can find the most suitable parameters.

A variant of simulated annealing enhanced by deterministic local improvement (3-opt) leads to so-called *large-step Markov chain methods* (see Martin, Otto & Felten [1992]). When such methods are properly implemented, near optimal solutions can be found faster than with pure simulated annealing. A related heuristic motivated by phenomena from physics is *simulated tunneling* described in Rujframework [1988].

A simplification of simulated annealing, called *threshold accept*, is proposed in Dueck & Scheuer [1990]. This heuristic removes the probability involved for the acceptance of a bad move in the original method. Rather, in each major iteration (Step (2.1)) an upper bound is given by which the length of the current cycle is allowed to be increased by the basic move. This threshold value is decreased according to some rule. The procedure is stopped if changes of the solution are not registered for several steps. Computational results are shown to display the same behavior as simulated annealing. A theoretical convergence result can also be obtained [Althöfer & Koschnick, 1991].

An even simpler variant is discussed in Dueck [1993] under the name of *great-deluge* heuristic. Here for each major iteration there is an upper limit on the length of Hamiltonian cycles that are accepted. Every random move yielding a cycle better than this length is accepted (note the difference from the threshold accept approach). The name of this approach comes from the interpretation that (for a maximization problem) the limit corresponds to a rising level of water and moves leading 'into the water' are not accepted. This method is reported to yield good results with fairly moderate computation times for practical traveling salesman problems arising from drilling printed-circuit boards.

Evolutionary strategies and genetic algorithms

The development of these two related approaches was motivated by the fact that many very good (or presumably optimal) solutions to highly complex problems can be found in nature itself.

The first approach is termed *evolutionary strategy* since it is based on analogues of 'mutation' and 'selection' to derive an optimization heuristic [Rechenberg, 1973]. Its basic principle is the following.

procedure EVOLUTION

(1) Compute an initial Hamiltonian cycle T.
(2) As long as the stopping criterion is not satisfied perform the following steps.
  (2.1) Generate a modification of T to obtain the cycle T'.
  (2.2) If $c(T') - c(T) < 0$ then set $T = T'$.
(3) Output the best solution found.

In contrast to previous methods of this subsection, moves increasing the length of the Hamiltonian cycle are not accepted. The term 'evolution' is used because the moves generated in Step (2.1) are biased by knowledge acquired so far, i.e., somehow moves that lead to a decrease of cycle length should influence the generation of the next move. This principle, however, is hardly followed in practice, moves taken into account are usually k-opt moves generated at random. Formulated this way the procedure cannot leave local minima and experiments show that it indeed gets stuck in poor local minima. Moreover, convergence is slow, justifying the name 'creeping random search' which is also used for this method. To leave local minima one has to incorporate the possibility of perturbations that increase the cycle length [Ablay, 1987]. Then this method resembles a mixture of pure evolutionary strategy, simulated annealing, threshold accept, and tabu search (see below).

More powerful in nature than mutation-selection is genetic recombination. Interpreted in terms of the TSP this means that new solutions should not be constructed from just one parent solution but rather be a suitable combination of two or more. Heuristics following this principle are termed *genetic algorithms*.

procedure GENETIC_ALGORITHM

(1) Compute an initial set $\mathcal{T}$ of Hamiltonian cycles.
(2) As long as the stopping criterion is not satisfied perform the following steps.
  (2.1) Recombine two or more cycles of $\mathcal{T}$ to obtain a new cycle T which is added to $\mathcal{T}$.
  (2.2) Reduce the set $\mathcal{T}$ according to some rule.
(3) Output the best solution found during the heuristic.

We see that Step (2.1) mimics reproduction in the population $\mathcal{T}$ and that Step (2.2) corresponds to a 'survival of the fittest' rule.

There are numerous possible realizations. Usually, subpaths of given cycles are connected to form new cycles and reduction is just keeping the set of k

best solutions of $\mathcal{T}$. One can also apply deterministic improvement methods to the newly generated cycle T before performing Step (2.2). Findings of optimal solutions are reported for some problem instances (the largest one being problem *att532*) with an enormous amount of CPU time. For further reading we refer to Mühlenbein, Gorges-Schleuter & Krämer [1988], Goldberg [1989], and Ulder, Pesch, van Laarhoven, Bandelt & Aarts [1990].

Tabu search

Some of the above heuristics allow length-increasing moves, so local minima can be left during computation. No precaution, however, is taken to prevent the heuristic to revisit a local minimum several times. This absence was the starting point for the development of *tabu search* where a built-in mechanism is used to forbid (tabu) returning to the same feasible solution. In principle the heuristic works as follows.

procedure TABU_SEARCH

(1) Compute an initial Hamiltonian cycle T and start with an empty tabu list $\mathcal{L}$.
(2) As long as the stopping criterion is not satisfied perform the following steps.
  (2.1) Perform the best move that is not forbidden by $\mathcal{L}$.
  (2.2) Update the tabu list $\mathcal{L}$.
(3) Output the best solution found.

Again, there are various possibilities to realize a heuristic based on the tabu search principle. Basic difficulties are the design of a reasonable tabu list, the efficient management of this list, and the selection of the most appropriate move in Step (2.1). A thorough discussion of these issues can be found in Glover [1990]. Computational results for the TSP are reported in Knox & Glover [1989], Malek, Guruswamy, Owens & Pandya [1989], and Malek, Heap, Kapur & Mourad [1989].

Neural networks

This approach tries to mimic the mode of operation of the human brain. Basically one models a set of neurons connected by a certain type of interconnection network. Based on the inputs that a neuron receives, a certain output is computed which is propagated to other neurons. A variety of models addresses activation status of neurons, determination of outputs and propagation of signals in the net with the basic goal of realizing some kind of learning mechanism. The result computed by a neural network either appears explicitly as output or is given by the state of the neurons.

In the case of the TSP there is, for example, the 'elastic band' approach [Durban & Willshaw, 1987] for Euclidean problem instances. Here a position in the plane is associated with each neuron. In the beginning, the neurons are ordered along a circle. During the computation, neurons are 'stimulated' and approach a cycle through the given set of points. Applications for the TSP can also be found in Fritzke & Wilke [1991]. For further reading on *neural networks* or *connectionism* see Hopfield & Tank [1985], Kemke [1988] and Rumelhart, Hinton & McClelland [1986]. Computational results are not yet convincing.

Summarizing, we would classify all heuristics presented in this subsection as *randomized improvement heuristics*. Also the iterated Lin–Kernighan heuristic falls into this class, since it performs a random move after each iteration. The analogues drawn from physics or biology are entertaining, but we think that they are a bit overstressed. The central feature is the systematic use of randomness which may avoid local minima and therefore yield a chance of finding optimal solutions (if CPU time is available). It is interesting from a theoretical point of view that convergence to an optimal solution with probability 1 can be shown for some variants, but the practical impact of these results is limited. The approaches have the great advantage, however, that they are generally applicable to combinatorial optimization problems and other types of problems. They can be implemented routinely with little knowledge about problem structure. If enough CPU and real time is available they can be applied (after spending some time for parameter tuning) to large problems with a good chance of finding solutions close to the optimum.

For many practical applications the heuristics presented in this subsection may be sufficient for treating the problems satisfactorily. But, if one is (or has to be) more ambitious and searches for proven optimal solutions or solutions meeting a quality guarantee, one has to go beyond these methods. The remainder of this chapter is concerned with solving TSP instances to optimality or computing near optimal solutions with quality guarantees.

5. Relaxations

A *relaxation* of an optimization problem P is another optimization problem R, whose set of feasible solutions $\mathcal{R}$ properly contains all feasible solutions $\mathcal{P}$ of P. The objective function of R is an arbitrary extension on $\mathcal{R}$ of the objective function of P. Consequently, the objective function value of an optimal solution to R is less than or equal to the objective function value of an optimal solution to P. If P is a hard combinatorial problem and R can be solved efficiently, the optimal value of R can be used as a *lower bound* in an enumeration scheme to solve P. The closer the optimal value of R to the optimal value of P, the more efficient is the enumeration algorithm.

Since the TSP is an NP-hard combinatorial optimization problem, the standard technique to solve it to optimality is based on an enumeration scheme, and so the study of effective relaxations is fundamental in the process of devising good exact algorithms. We consider here discrete and continuous relaxations, i.e., relaxations with discrete and continuous feasible sets.

Before we describe these relaxations we give some notation and recall some basic concepts.

For any edge set $F \subseteq E_n$ and any $x \in \mathbb{R}^{E_n}$, $x(F)$ denotes the sum $\sum_{e \in F} x_e$. For a node set $W \subset V_n$, $E_n(W) \subset E_n$ denotes $\{uv \in E_n \mid u, v \in W\}$ and $\delta_n(W) \subset E_n$

denotes $\{uv \in E_n \mid u \in W, v \in V_n \setminus W\}$. We call $\delta_n(W)$ a *cut* with *shores* W and $V_n \setminus W$.

The solution set of the TSP is the set $\mathcal{H}_n$ of all Hamiltonian cycles of K_n. A Hamiltonian cycle, as defined in Section 1, is a subgraph $H = (V_n, E)$ of K_n satisfying the following requirements:

(a) all nodes of H have degree 2; (5.1a)
(b) H is connected. (5.1b)

The edge set of a subgraph of K_n whose nodes have all degree 2 is a *perfect 2-matching*, i.e., a collection of simple disjoint cycles of at least three nodes and with no chords such that each node of K_n belongs to some of these cycles. Consequently, a Hamiltonian cycle can be defined as a connected perfect 2-matching.

It is easy to see that if a perfect 2-matching is connected, then it is also biconnected, i.e., it is necessary to remove at least two edges to disconnect it. Therefore, the requirements (5.1a) and (5.1b) can be replaced by

(a) all nodes of H have degree 2; (5.2a)
(b) H is biconnected. (5.2b)

With every $H \in \mathcal{H}_n$ we associate a unique *incidence vector* $\chi^H \in \mathbb{R}^{E_n}$ by setting

$$\chi_e^H = \begin{cases} 1 & \text{if } e \in H \\ 0 & \text{otherwise.} \end{cases}$$

The incidence vector of every Hamiltonian cycle satisfies the system of equations

$$A_n x = \mathbf{2}, \tag{5.3}$$

where A_n is the node-edge incidence matrix of K_n and $\mathbf{2}$ is an n-vector having all components equal to 2. The equations $A_n x = \mathbf{2}$ are called the *degree equations* and translate the requirement (5.2a) into algebraic terms. In addition, for any nonempty $S \subset V_n$ and for any Hamiltonian cycle H of K_n, the number of edges of H with an endpoint in S and the other in $V_n - S$ is at least 2 (and even). Therefore, the intersection of the edge set of H with the cut $\delta_n(S)$ has cardinality at least 2 (and even), and so χ^H must satisfy the following set of inequalities:

$$x(\delta_n(S)) \geq 2 \qquad \text{for all } \emptyset \neq S \subset V_n. \tag{5.4}$$

These inequalities are called *subtour elimination inequalities* because they are not satisfied by the incidence vector of nonconnected 2-matchings (i.e., the union of two or more subtours), and so they translate the requirement (5.2b) into algebraic terms.

Given an objective function $c \in \mathbb{R}^{E_n}$ that associates a 'length' c_e with every edge e of K_n, the TSP can be solved by finding a solution to the following integer linear program:

Problem 5.1.

$$\begin{aligned} &\text{minimize} && cx && \\ &\text{subject to} && A_n x = \mathbf{2}, && && (5.5)\\ &&& x(\delta_n(S)) \geq 2 && \text{for all } \emptyset \neq S \subset V_n, && (5.6)\\ &&& \mathbf{0} \leq x \leq \mathbf{1}, && && (5.7)\\ &&& x \text{ integer}. && && (5.8) \end{aligned}$$

This is the most commonly used integer linear programming formulation of the TSP.

5.1. Subtour relaxation

An LP relaxation of the TSP is the linear program obtained by relaxing the integrality condition (5.8) of any integer programming formulation of the problem. Consequently, an LP relaxation has a polyhedron in $\mathbb{R}^{E_n}$ as a feasible set and thus it is a *continuous relaxation*.

A simple LP relaxation is the one defined by the constraints (5.5)–(5.7). The polyhedron defined by these constraints is called the *subtour elimination polytope* and the corresponding relaxation is called the *subtour relaxation*.

The number of constraints defined in (5.5)–(5.7) is $n + 2^n - 2 + 2m$. Some of them are redundant, though, and it is not difficult to show that the system (5.5)–(5.7) can be reduced to one not containing redundant constraints and having n equations and $2^{n-1} - n - 1 + m$ inequalities, still a number of constraints given by an exponential function of n. Such a huge linear program obviously cannot be solved by a direct application of the simplex method or of any other efficient algorithm for linear programming. However, as we will describe in Section 5.5, an optimal solution of a linear objective function over the subtour elimination polytope can be found in polynomial time. Moreover, the optimization over this polytope can be carried out very efficiently in practical computation by a simple iterative procedure. We start, for example, by solving the linear program defined by the constraints (5.5) and (5.7) and by a small subset (possibly empty) of the constraints (5.6). Then we check if the optimal solution satisfies all the constraints (5.6). If so, we terminate since we have found the optimal solution over the subtour elimination polytope. Otherwise, we add some of the violated constraints to the linear program and we solve it again. The number of iterations that are necessary before obtaining the optimum over the subtour polytope is, in practice, a small fraction of n, and the constraints of the linear programs that are solved at each iteration form a small subset of (5.5)–(5.7).

We discuss this procedure in more detail in Section 5.5. However, the reader who is not familiar with polyhedral combinatorics should keep it in mind while reading Section 5.4, where other continuous relaxations, also involving constraints sets of exponential size, are considered.

It would be interesting to know how close the bound c_L, obtained by solving the subtour relaxation is to the length of an optimal Hamiltonian cycle c_{opt}.

Wolsey [1980] and Shmoys & Williamson [1990] show that for any cost function c satisfying the triangle inequality, the ratio c_L/c_{opt} is at least 2/3. The 2/3 bound is not shown to be tight and actually it is conjectured in Goemans [1993] that $c_L/c_{\mathrm{opt}} \geq 3/4$. To prove this conjecture a deeper knowledge of the structure of the subtour elimination polytope would probably be necessary. Some results in this direction are given by Boyd & Pulleyblank [1990] who characterize some of its vertices. But it is still an open problem how to completely characterize all of them.

Computational experiments show that for many instances the above ratio is very close to 1 (see the fourth column of Table 5). There are actually classes of instances for which the optimization over the subtour elimination polytope always yields the incidence vector of an optimal Hamiltonian cycle (thus the ratio is 1 for these instances). Padberg & Sung [1988] show that some instances of the TSP that are very hard for k-opt heuristic algorithms, fall into this category.

The excellent quality of the lower bound obtained from the subtour relaxation is most probably the main reason of the successful computation of the optimal solutions of several large TSP instances reported in the literature (see Section 7).

5.2. 1-tree relaxation

A discrete relaxation of the TSP can be obtained by defining a combinatorial problem whose feasible solutions are subgraphs of K_n satisfying a subset of the requirements (5.1a) and (5.1b) and whose optimal solutions can be found in polynomial time.

The requirements (5.1a) and (5.1b) imply that any subgraph of K_n that satisfies them has exactly n edges and spans K_n, i.e., has an edge incident with any node of K_n. Therefore, we only consider relaxations having as feasible solutions spanning subgraph of K_n with n edges.

As a first relaxation we can, for example, drop the requirement (5.1a) and consider as feasible solutions the connected spanning subgraphs of K_n with n edges. These graphs consist of a spanning tree plus an extra edge. With respect to the edge weights given by c, a minimal weight graph of this kind can be found by taking a minimal spanning tree and adding an edge not belonging to the tree with minimal weight. The minimal spanning tree can be found in polynomial time (see Section 4.3).

A better relaxation can be obtained by dropping the requirement (5.1a) for all nodes except one, say node 1. Denote by $K_n \setminus \{1\}$ the subgraph of K_n induced by $V_n \setminus \{1\}$. The subgraph of a Hamiltonian cycle induced by the nodes in $V_n \setminus \{1\}$ is a Hamiltonian path of $K_n \setminus \{1\}$. This is a subgraph H_p of $K_n \setminus \{1\}$ satisfying the following requirements:

(a) H_p spans $K_n \setminus \{1\}$;
(b) H_p has $n-2$ edges;
(c) H_p is connected;
(d) all nodes of H_p but two have degree 2. (5.9)

The union of a Hamiltonian path on $K_n \setminus \{1\}$ and of two edges incident with node 1 is definitely a relaxation of a Hamiltonian cycle, but not a useful one, since finding a minimum cost Hamiltonian path in $K_n \setminus \{1\}$ is as difficult as finding a minimum cost Hamiltonian cycle in K_n. However, instead of a Hamiltonian path we can consider its relaxation obtained by dropping the requirement (5.9). It can be shown that the resulting graph is a tree that spans $K_n \setminus \{1\}$.

Now we can produce a useful relaxation of the TSP by taking the union of a pair of edges incident with node 1 and a tree spanning $K_n \setminus \{1\}$. Such a graph is called a *1-tree*. The 1-tree that minimizes the objective function c can be easily obtained by finding the minimum spanning tree in $K_n \setminus \{1\}$ and adding to it the shortest two edges incident with node 1. Thus the complexity of finding a minimum 1-tree amounts to that of finding a minimum spanning tree.

Unfortunately, this relaxation is not very strong. To strengthen the lower bound obtained by finding a minimum cost 1-tree, we modify the objective function in the following way. If for all $u \in V_n$ we add a constant λ_u to the objective function coefficients of all edges of $\delta(u)$, the length of all Hamiltonian cycles of K_n increases by the same amount $\sum_{u \in V_n} 2\lambda_u$ (since any Hamiltonian cycle has exactly two edges in each edge set $\delta(u)$). For any edge (u, v) of K_n, the corresponding coefficient of the new objective function is $c'(u, v) = c(u, v) + \lambda_u + \lambda_v$ (since (u, v) belongs to both $\delta(u)$ and $\delta(v)$). Consequently, the optimal value of the problem

$$\min_{H \in \mathcal{H}_n} \left\{ \sum_{i<j} (c_{ij} + \lambda_i + \lambda_j) \chi_{ij}^H \right\} - 2 \sum_{i \in V} \lambda_i$$

does not depend on the vector $\lambda \in \mathbb{R}^{V_n}$.

The same does not hold for 1-trees, though, because in general not all nodes of a 1-tree satisfy (5.1a). Therefore, the length of an optimal 1-tree

$$L(\lambda) = \min_{\chi^{1T}} \left\{ \sum_{i<j} (c_{ij} + \lambda_i + \lambda_j) \chi_{ij}^{1T} \right\} - 2 \sum_{i \in V} \lambda_i,$$

where χ^{1T} is the incidence vector of a 1-tree in K_n, is a (nonconstant) function of $\lambda \in \mathbb{R}^{V_n}$. $L(\lambda)$ is a lower bound for the c-length of a Hamiltonian cycle and in general different vectors λ^1 and λ^2 yield different lower bounds $L(\lambda^1)$ and $L(\lambda^2)$. Thus the tightest bound is obtained by solving the maximization problem

$$\max_{\lambda} \{L(\lambda)\}, \tag{5.10}$$

called the Lagrangean dual problem. The bound obtained by solving (5.10) is proposed in Held & Karp [1970, 1971] and it is known as the Held–Karp bound.

It is interesting to note that the lower bound produced by the subtour relaxation is equivalent to the Held–Karp bound (see, e.g., Nemhauser & Wolsey [1988], pp. 469–475). Although these two bounds are identical they are quite different from the methodological viewpoint. Problem (5.10) is a piecewise linear nondifferentiable concave optimization problem and can be solved by an iterative procedure (see Section 6) whose convergence can be very slow. In practical computation the procedure is stopped prematurely, thus providing a lower bound that is worse than the theoretical one given in (5.10). On the contrary, the bound obtained from optimizing over the subtour elimination polytope can be computed by solving a (usually short) sequence of linear programs, the last of which provides the exact theoretical bound (5.10). For these reasons the latter method is preferable. However, the computation based on the subtour relaxation requires a linear program optimizer and a quite complex algorithm, while the computation of an approximation of the Held–Karp bound can be carried out with a very simple procedure that requires moderate amounts of computer memory (see Section 6). Therefore, when the instance is too large to be attacked by the available linear program optimizers or when one is willing to afford only moderate implementation costs, the Held–Karp approach seems more suitable.

Any LP relaxation tighter than the subtour relaxation produces a lower bound superior to the Held–Karp bound. This explains why the algorithms based on the LP relaxations perform much better than those based on the 1-tree relaxation. For this reason we will spend most of this section describing the LP relaxations in more detail.

5.3. 2-matching relaxation

A 2-matching relaxation for the TSP is another discrete relaxation obtained by relaxing the requirement (5.1b), i.e., by solving Problem 5.1 without the constraints (5.6). The resulting relaxation is a minimum cost 2-matching that can be found in polynomial time (see Edmonds & Johnson [1970], Cunningham & Marsh [1978], and Padberg & Rao [1982]). However, implementing an algorithm for the minimum cost 2-matching problem efficiently is not as simple as for the minimum cost spanning tree problem. The lower bound obtained by finding the minimum cost 2-matching is in general poor (see, e.g., Table 5) and, like for the 1-tree, it can be improved using techniques in the same spirit as for the Held–Karp bound. The improved bound is better than the one obtained by solving the problem (5.10) (see, e.g., Nemhauser & Wolsey [1988], pp. 469–475).

5.4. Strong LP relaxations

An LP relaxation of the TSP is not unique, since every linear program obtained from Problem 5.1 by adding to (5.5)–(5.7) any number of linear constraints which are *valid*, i.e., satisfied by the incidence vectors of all Hamiltonian cycles of K_n, is also an LP relaxation.

To produce a relaxation that is stronger than the subtour relaxation it is necessary to add inequalities to (5.5)–(5.7). They must satisfy two requirements. In order to produce a mathematically correct relaxation they must be valid. To produce a tight relaxation they must be 'strong', in the sense that they must define a polytope substantially smaller than the subtour polytope. For two valid inequalities $a_1x \geq b_1$ and $a_2x \geq b_2$ we say that the first is *stronger* than the second (or that the first *dominates* the second) if the polytope defined by $a_1x \geq b_1$ and by

(5.5)–(5.7) is properly contained in the polytope defined by $a_2 x \geq b_2$ and (5.5)–(5.7). To produce good relaxations it is quite natural to look for valid inequalities that are not dominated by any other. The derivation of these inequalities is intimately related to the study of the structure of a polytope associated with the TSP, called the symmetric traveling salesman polytope.

The *symmetric traveling salesman polytope* (STSP(n)) is the convex hull of the set of the incidence vectors of all Hamiltonian cycles of K_n, i.e.,

$$\mathrm{STSP}(n) = \mathrm{conv}\left\{\chi^H \mid H \in \mathcal{H}_n\right\}.$$

It is known that there exists a finite minimal set $\mathcal{B}^=$ of linear equations and a finite minimal set $\mathcal{B}^\leq$ of linear inequalities whose set of solutions is precisely STSP(n). The sets $\mathcal{B}^=$ and $\mathcal{B}^\leq$ are minimal in the sense that the removal of any of their elements results in a polytope that properly contains STSP(n). The equations of $\mathcal{B}^=$ are precisely the degree equations (5.3). Each of the inequalities in $\mathcal{B}^\leq$ defines a *facet* of STSP(n) (for the definition of facet and for other basic concepts of polyhedral theory we refer to Grötschel & Padberg [1985], Nemhauser & Wolsey [1988], and Pulleyblank [1983]). The equations of $\mathcal{B}^=$ and the inequalities of $\mathcal{B}^\leq$ are the constraints of the best LP relaxation of the TSP. In fact there is always an optimal extreme solution to this relaxation that is the incidence vector of an optimal Hamiltonian cycle, for any objective function c. Unfortunately, the set $\mathcal{B}^\leq$ contains an enormous number of inequalities and its size grows exponentially with n. Presently, a complete description of a minimal system of inequalities defining STSP(n) is known only for $n \leq 8$. For $n = 6$ the set $\mathcal{B}^\leq$ contains 100 inequalities and is described by Norman [1955]. For $n = 7$ the complete description of $\mathcal{B}^\leq$, which contains 3,437 inequalities, is given by Boyd & Cunningham [1991]. The set $\mathcal{B}^\leq$ for $n = 8$ (containing 194187 inequalities) is described by Christof, Jünger & Reinelt [1991]. It is very unlikely that a complete description of this system of inequalities can be found for all n. Nevertheless, good LP relaxations can be obtained using only subsets of $\mathcal{B}^\leq$. For this reason and because the description of STSP(n) with linear inequalities is a challenging mathematical task by itself, many researchers have been interested in studying this polytope. The work in this area has been focused mostly on characterizing large families of valid inequalities for STSP(n) and, in many cases, in showing that some of these families are subsets of $\mathcal{B}^\leq$.

The first systematic study of the polyhedral structure of STSP(n) was done by Grötschel and Padberg. The results of their work are published in the doctoral dissertation of Grötschel [1977] and in the papers Grötschel & Padberg [1974, 1977, 1978, 1979a, b]. Their main discovery was a large family of facet-defining inequalities, the *comb inequalities*, that are the major contributors to the LP relaxations used by the most successful algorithms for the solution of the TSP to optimality (see Sections 5.6 and 6). Another important piece of the current knowledge of the TSP polytope is provided by Cornuéjols, Fonlupt & Naddef [1985], who describe many new classes of inequalities that are facet-defining for GTSP, the polyhedron associated with the graphical traveling salesman problem. In fact, many results on GTSP can be extended to STSP, due to a strong relationship between the two polyhedra that is described and exploited in Naddef & Rinaldi [1993].

Many other important results on STSP and on related issues appeared in the literature. We cite only those that provide new members of the set $\mathcal{B}^{\leq}$ of facet-defining inequalities for this polytope. These results are reported in the next subsection. They may appear too technical to those who are mainly interested in the algorithmic issues. However, it may be worth to observe that the possibility to produce faster and more robust algorithms for finding optimal or provably good solutions of the TSP seems to depend right on the exploitation of these results.

5.5. *Known facets of STSP(n)*

The incidence vectors of all Hamiltonian cycles of K_n satisfy the degree equations (5.5), and so STSP(n) is not a full dimensional polytope (its dimension is $m - n$), i.e., it is contained in the intersection of the n hyperplanes defined by the degree equations. A consequence of this fact is that, unlike in the case of a full dimensional polyhedron, a facet-defining inequality for STSP(n) is not uniquely defined (up to a multiplication by a scalar). If $hx \geq h_0$ defines a facet of STSP(n), then the inequality $fx \geq f_0$, with $f = \lambda A_n + \pi h$, $f_0 = \lambda \mathbf{2} + \pi h_0$, $\pi > 0$, and $\lambda \in \mathbb{R}^{V_n}$, defines the same facet. The two inequalities are said to be *equivalent*.

As a consequence of this lack of uniqueness the facet-defining inequalities of STSP(n) are described in the literature in different forms. We describe the two most frequently used forms: the closed form and the tight triangular form. The direction of the inequalities is '$\leq$' for the first form and '$\geq$' for the second.

Let $\mathcal{S} = \{S_1, S_2, \ldots, S_t\}$ be a collection of subsets of V_n and let $r(\cdot)$ denote a suitable function of $\mathcal{S}$, which only depends on the number of its members but not on their size. An inequality is written in *closed form* if it is as follows:

$$\sum_{S \in \mathcal{S}} \alpha_S x(E_n(S)) \leq \sum_{S \in \mathcal{S}} \alpha_S |S| - r(\mathcal{S}), \tag{5.11}$$

where α_S is an integer associated with $S \in \mathcal{S}$. The closed form is nice for describing an inequality, but has two major drawbacks. Not all facet-defining inequalities of STSP(n) can be written in closed form. In addition, the closed form is not unique: for example, it can be shown that replacing any set S in (5.11) by its complement $V_n \setminus S$ produces a different inequality which is equivalent to (5.11).

A more interesting form for the facet-defining inequalities of STSP(n) is the tight triangular form. An inequality $fx \geq f_0$ defined on $\mathbb{R}^{E_n}$ is said to be in *tight triangular form* (or in *TT form*) if the following conditions are satisfied:

(a) the coefficients of f satisfy the triangle inequality, i.e., $f(u, v) \leq f(u, w) + f(w, v)$ for every triple u, v, w of distinct nodes in V_n;

(b) for all $u \in V_n$ there exists a pair of distinct nodes $v, w \in V_n \setminus \{u\}$ such that $f(v, w) = f(u, v) + f(u, w)$.

Let $hx \geq h_0$ be any inequality defined on $\mathbb{R}^{E_n}$. An inequality $fx \geq f_0$ in TT form that is equivalent to $hx \geq h_0$, with $f = \lambda A_n + \pi h$ and $f_0 = \lambda \mathbf{2} + \pi h_0$, can be obtained by setting π to any positive value and

$$\lambda_u = \frac{\pi}{2} \max \left\{ h(v, w) - h(u, v) - h(u, w) \mid v, w \in V_n \setminus \{u\},\ v \neq w \right\}$$

for all $u \in V_n$.

The properties of the TT form of the inequalities can be used to explain the tight relationship between STSP(n) and GTSP(n). In particular, being in TT form is a necessary and sufficient condition for a facet-defining inequality of STSP(n) to be facet-defining for GTSP(n). For the details see Naddef & Rinaldi [1993].

Although two equivalent inequalities define the same facet, using a form rather than another may not be irrelevant in computation. The inequalities that we consider are used as a constraint of some linear program and all current LP optimizers are very sensitive to the density (percentage of nonzero coefficients) of the constraints. The lower the density the faster is the LP optimizer. The inequalities in TT form are in general denser than those in closed form. However, when only a subset of the variables is explicitly represented in a linear program, which is often the case when solving large TSP instances (see Section 6), the inequalities in closed form tend to be denser.

We now describe the basic inequalities that define facets of STSP(n).

Trivial inequalities

The inequalities $x_e \geq 0$ for $e \in E_n$ are called the *trivial inequalities* (this is the only form used for these inequalities). A proof that they are facet-defining for STSP(n) (with $n \geq 5$) is given in Grötschel & Padberg [1979b].

Subtour elimination inequalities

The subtour elimination inequalities (5.4) define facets of STSP(n). In (5.4) they are written in TT form. The corresponding closed form, obtained by setting $\mathcal{S} = \{S\}$, $\alpha_S = 1$ and $r(\mathcal{S}) = 1$, is

$$x(E_n(S)) \leq |S| - 1.$$

These inequalities are introduced by Dantzig, Fulkerson & Johnson [1954] who do not address the issue of whether these inequalities are facet-defining for the TSP polytope. A proof that if $2 \leq |S| \leq n - 2$, they are facet-defining for STSP(n) (with $n \geq 4$) is given in Grötschel & Padberg [1979b].

Comb inequalities

A *comb inequality* is defined by setting $\mathcal{S} = \{W, T_1, \ldots, T_k\}$, $\alpha_S = 1$ for all $S \in \mathcal{S}$, and $r(\mathcal{S}) = (3k + 1)/2$. The set W is called the *handle* and the sets T_i are called the *teeth* of the comb. The inequality is facet-defining if the handle and the teeth satisfy the following conditions:

(i) $|T_i \cap W| \geq 1$ for $i = 1, \ldots, k$,
ii) $|T_i \setminus W| \geq 1$ for $i = 1, \ldots, k$,
(iii) $T_i \cap T_j = \emptyset$ for $1 \leq i < j \leq k$,
(iv) $k \geq 3$ and odd.

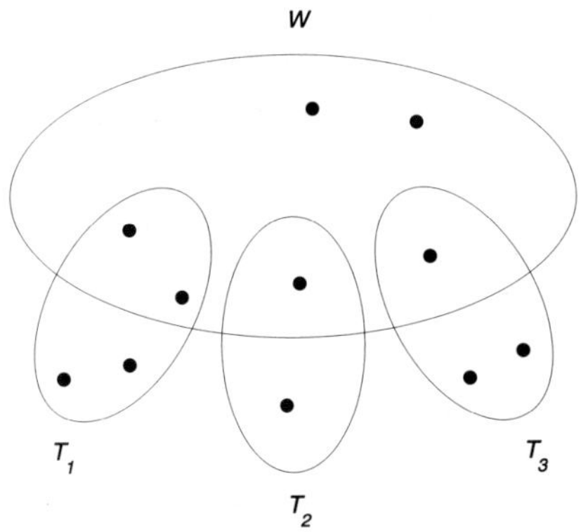

Fig. 8. Handle and teeth of a comb.

Special cases of comb inequalities have different names in the literature. If (i) is satisfied *with equality*, the inequality is called a *Chvátal comb*. If all teeth have cardinality 2 the comb inequality is also called *2-matching* inequality. Figure 8 shows an example of a comb inequality.

The 2-matching inequalities were discovered by Edmonds [1965], who used them to provide a complete description of the polytope associated with the 2-matching problem. Chvátal [1973] defined a class of valid inequalities as a generalization of the 2-matching inequalities and called them *comb inequalities*. Now we refer to the inequalities in this class as *Chvátal combs*. Grötschel & Padberg [1979a, b] generalized Chvátal's combs to a larger class that they called the *comb* inequalities and showed that the inequalities of this class are facet-defining for STSP(n) (with $n \geq 6$).

Clique-tree inequalities

A *clique-tree inequality* is defined by setting $\mathcal{S} = \{W_1, \dots, W_r, T_1, \dots, T_k\}$ (the sets W_i are called the *handles* and the sets T_i are called the *teeth* of the clique-tree), $\alpha_S = 1$ for all $S \in \mathcal{S}$, and $r(\mathcal{S}) = (2\sum_{i=1}^{k} h(T_i) + k + 1)/2$, where $h(T)$ is the number of handles that have nonempty intersection with T. A *clique-tree* is a subgraph of K_n whose cliques are the handles and the teeth. The inequality is facet-defining if the clique-tree is connected and the following conditions are satisfied:

(i) no two teeth intersect;
(ii) no two handles intersect;
(iii) each tooth contains at least two and at most $n - 2$ nodes;
(iv) each tooth contains at least one node not belonging to any handle;
(v) each handle intersects an odd number (≥ 3) of teeth;
(vi) if a tooth T and a handle W have a nonempty intersection, then $W \cap T$ is an articulation set of the clique-tree, i.e., the removal of the nodes in $W \cap T$ from K_n disconnects the clique-tree.

The clique-tree inequalities are a proper generalization of the comb inequalities.

A clique-tree inequality is displayed in Figure 9. These inequalities are introduced and proved to define facets of STSP(n) (with $n \geq 11$) by Grötschel & Pulleyblank [1986].

PWB inequalities

By *simple PWB inequalities* we denote three classes of inequalities, namely the path, the wheelbarrow, and the bicycle inequalities.

The simple path inequalities are defined by graphs that are called path configurations. For any odd $k \geq 3$ and any k-tuple of positive integers $(n_1, \ldots, n_k)$, with $n_i \geq 2$ for $i \in \{1, \ldots, k\}$, let $P(n_1, \ldots, n_k) = (V_P, E_P)$ be the graph with node set and edge set given by

$$V_P = \{Y, Z\} \cup \left\{u_j^i \mid j \in \{1, \ldots, n_i\},\ i \in \{1, \ldots, k\}\right\}$$

$$E_P = \left\{u_j^i u_{j+1}^i \mid j \in \{0, \ldots, n_i\},\ i \in \{1, \ldots, k\}\right\},$$

respectively, where for convenience we label $u_0^i = Y$, and $u_{n_{i+1}}^i = Z$.

We call $P(n_1, \ldots, n_k)$ a *k-path configuration*. A k-path configuration is the union of k disjoint paths connecting Y to Z. The length of a path i is n_i, i.e., the number of the internal nodes. The nodes Y and Z are called the *odd nodes* of the configuration, all the other nodes are called *even*. The edges of a path configuration are called *path-edges*, see Figure 10.

A *simple path inequality* associated with the k-path configuration $P(n_1, \ldots, n_k)$ is the inequality on $\mathbb{R}^{E_n}$, with $n = 2 + \sum_{i=1}^k n_i$, defined (in TT form) by

$$fx \geq f_0 = 1 + \sum_{i=1}^{k} \frac{n_i + 1}{n_i - 1}, \tag{5.12}$$

where

$$f_e = \begin{cases} \dfrac{|j-q|}{n_i - 1} & \text{for } e = u_j^i u_q^i,\ i \in \{1, \ldots, k\},\ j, q \in \{0, \ldots, n_i + 1\},\ j \neq q, \\[2ex] \dfrac{1}{n_i - 1} + \dfrac{1}{n_r - 1} + \left| \dfrac{j-1}{n_i - 1} - \dfrac{q-1}{n_r - 1} \right| & \text{for } e = u_j^i u_q^r,\ i \neq r \in \{1, \ldots, k\},\ j \in \{1, \ldots, n_i\},\ q \in \{1, \ldots, n_r\}. \\[2ex] 1 & \text{for } e = YZ, \end{cases}$$

A *simple wheelbarrow inequality* associated with the k-path configuration $P(n_1, \ldots, n_k)$ is the inequality on $\mathbb{R}^{E_n}$, with $n = 1 + \sum_{i=1}^k n_i$, defined (in TT form) by (5.12), where the coefficients and the right hand side are defined as above, but node Y and all edges incident with it are removed. If both Y and Z and all

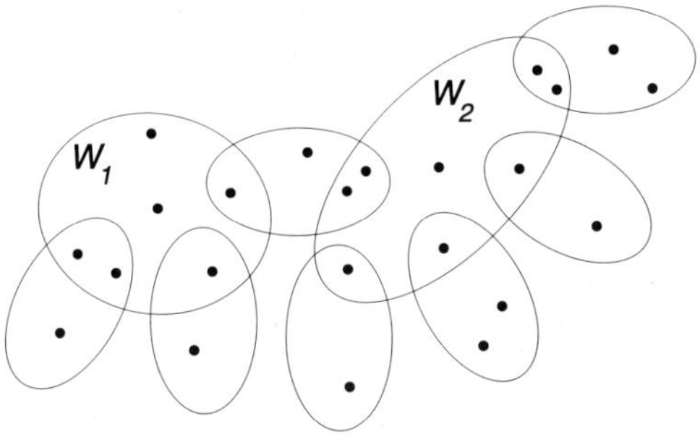

Fig. 9. Handles and teeth of a clique-tree.

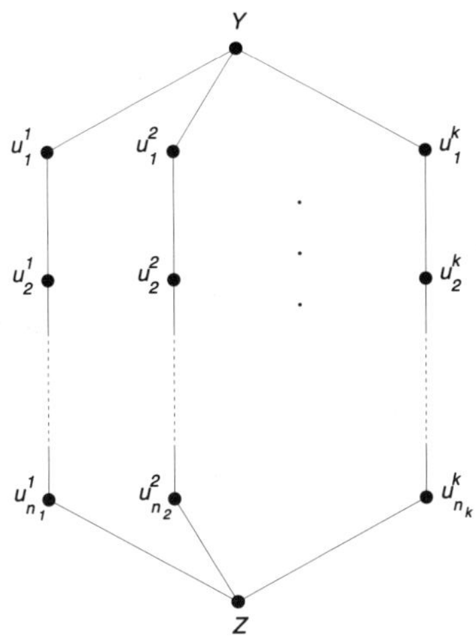

Fig. 10. A k-path configuration.

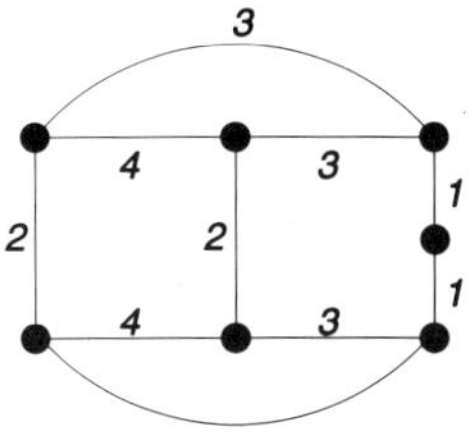

Fig. 11. A bicycle inequality with 7 nodes.

edges incident with them are removed, the inequality of $\mathbb{R}^{E_n}$, with $n = \sum_{i=1}^{k} n_i$, defined as above is called *simple bicycle inequality*. In Figure 11, a simple bicycle inequality with 7 nodes is illustrated. The coefficient of a missing edge is given by the f-length of a shortest path between its endnodes.

If $n_i = t$ for $t \geq 2$ and integer and for $i \in \{1, \ldots, k\}$, then the above inequalities are called *regular* (or *t-regular*).

The PWB inequalities are defined in Cornuéjols, Fonlupt & Naddef [1985] and proved to be facet-defining for GTSP, the polyhedron associated with the graphical traveling salesman problem. A proof that simple PWB inequalities define facets of STSP(n) (with $n \geq 6$) is given in Naddef & Rinaldi [1988].

Ladder inequalities

A *ladder inequality* is defined by a family $\mathcal{S}$ of sets

$$\mathcal{S} = \{W_1, W_2, P_1, P_2, T_1, \ldots, T_t, D_1, \ldots, D_m\},$$

with $t \geq 0$ and $m \geq 0$. The sets W_i, P_i, T_i, and D_i are called *handles*, *pendant teeth*, *regular teeth*, and *degenerate teeth*, respectively. A ladder inequality associated with $\mathcal{S}$ is defined as follows:

$$\sum_{S \in \mathcal{S}} \alpha_S x(E_n(S)) + x(E_n(P_1 \cap W_1 : P_2 \cap W_2)) \leq \sum_{S \in \mathcal{S}} \alpha_S |S| - 2t - 3m - 4,$$

where $\alpha_S = 2$ if $S \in \{D_1, \ldots, D_m\}$ and $\alpha_S = 1$ otherwise, and where we denote the set of edges of E_n with one endpoint in X and the other in Y by $E_n(X : Y)$. Observe that the inequality is not in closed form due to the last term of its left hand side. The ladder inequality is facet-defining for STSP(n), ($n \geq 8$) if the following conditions are satisfied:

(i) no two teeth intersect;

(ii) the two handles do not intersect;

(iii) P_1 intersects only W_1 and P_2 intersects only W_2;

(iv) each regular or degenerate tooth intersects both handles;

(v) each regular or pendant tooth contains at least one node not belonging to any handle;

(vi) each degenerate tooth does not contain nodes that do not belong to one of the two handles;

(vii) $t + m$ is even and at least 2.

A ladder inequality with 16 nodes is shown in Figure 12.

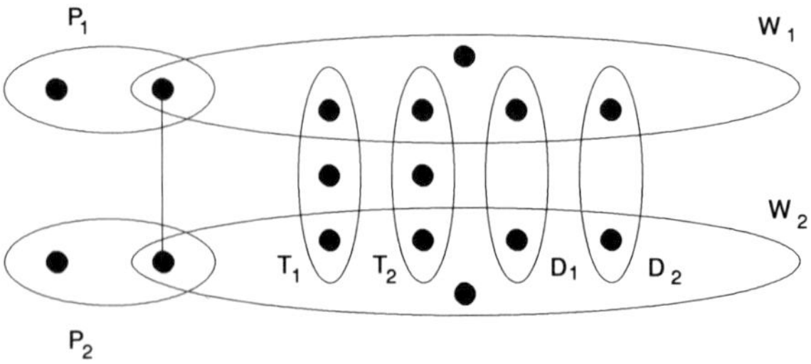

Fig. 12. Handles and teeth of a ladder inequality.

The first ladder inequality with 8 nodes is described in Boyd & Cunningham [1991]. The proof that the ladder inequalities are facet-defining for STSP(n), with $n \geq 8$ is given in Boyd, Cunningham, Queyranne & Wang [1993].

Crown inequalities

For any integer $k \geq 2$ let $C(k) = (V_C, E_C)$ be the graph with the following node and edge sets

$$V_C = \{u_i \mid i \in \{1, \ldots, 4k\}\},$$
$$E_C = \{u_i u_{[i+1]} \mid i \in \{1, \ldots, 4k\}\},$$

where $[i]$ stands for the expression $((i-1) \bmod 4k) + 1)$. We call $C(k)$ a *crown configuration.*

A *simple crown* inequality associated with $C(k)$ is the inequality $fx \geq f_0$ (in TT form), where $f_0 = 12k(k-1) - 2$, and, for $i \in \{1, \ldots 4k\}$,

$$f(u_i, u_{[i+j]}) = \begin{cases} 4k - 6 + |j| & \text{for } 1 \leq |j| \leq 2k-1, \\ 2(k-1) & \text{for } j = 2k. \end{cases}$$

The edges $u_i u_{[2k+i]}$ are called *diameters* of the simple crown inequality. A crown inequality with 8 nodes is shown in Figure 13.

Simple crown inequalities were discovered and proved to define facets of STSP(n) (with $n \geq 8$) by Naddef & Rinaldi [1992].

Extensions of facet-defining inequalities

Due to the very complex structure of STSP(n) it is very difficult to describe all inequalities known to define facets of this polytope. A technique used to simplify the description is to define some operations on the inequalities that allow the derivation of new inequalities from others that have already been characterized. Many inequalities can be described in such a constructive way, using the inequalities described above as building blocks. We describe here two

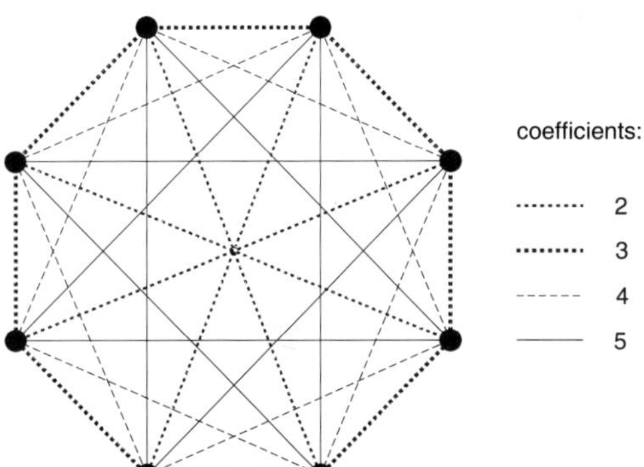

Fig. 13. A crown inequality with 8 nodes.

kinds of operations: the *zero node-lifting* of a node and the *edge cloning*. Below we show an application of another operation, the *2-sum composition* of inequalities. These operations are better described on a graph associated with an inequality. The graph $G_h = (V_n, E_n, h)$ associated with an inequality $hx \geq h_0$ of $\mathbb{R}^{E_n}$ is a weighted complete graph with n nodes, with a weight for each edge that is given by the corresponding inequality coefficient.

Let $hx \geq h_0$ be a facet-defining inequality for STSP(n) and let $G_h = (V_n, E_n, h)$ be its associated graph. Let u be a node of G_h and $G_{h^*} = (V_{n+k}, E_{n+k}, h^*)$ be the weighted complete graph obtained by adding k copies of node u and of its star to G_h. More precisely, G_{h^*} contains G_h as a subgraph and $h^*_{ij} = h_{ij}$ for all $e \in E_n$, $h^*_{ij} = h_{uj}$ for all $i \in V_{n+k} \setminus V_n$ and all $j \in V_n$, and $h^*_{ij} = 0$ for all i and j in $V_{n+k} \setminus V_n$. The inequality $h^*x^* \geq h_0$ defined on $\mathbb{R}^{E_{n^*}}$ and having G_{h^*} as associated graph is said to be obtained by *zero node-lifting* of node u.

An inequality in TT form with all the coefficients strictly positive is called *simple*. A facet-defining inequality in TT form that has a zero coefficient is always derivable from a simple inequality in TT form by a repeated application of the zero node-lifting. In Naddef & Rinaldi [1993] a simple sufficient condition is given for an inequality in TT form, obtained by zero node-lifting of a facet-defining inequality, to inherit the property of being facet-defining. This condition is verified by all the inequalities known to date that are facet-defining for STSP(n) and in particular, by the inequalities described above. For an inequality obtained by zero node-lifting of a simple inequality, we use the convention of keeping the same name of the simple inequality but dropping the word 'simple'. Consequently, the *PWB inequalities* and the *crown inequalities* are obtained by zero node-lifting of their corresponding simple archetypes and are all facet-defining for STSP(n).

Let uv be an edge of G_h and let $G_{h'} = (V_{n+2}, E_{n+2}, h')$ be the weighted complete graph with $V_{n+2} = V_n \cup \{u', v'\}$ defined as follows. The graph G_h is a subgraph of $G_{h'}$ and h' is defined as follows:

$$\begin{aligned} h'(u', j) &= h(u, j) && \text{for all } j \in V_n \setminus \{u\}, \\ h'(v', j) &= h(v, j) && \text{for all } j \in V_n \setminus \{v\}, \\ h'(u, u') &= h'(v, v') = 2h(u, v), \\ h'(u', v') &= h(u, v). \end{aligned}$$

The inequality $h'x' \geq h_0 + 2h(u, v)$ defined on $\mathbb{R}^{E_{n+2}}$ and having $G_{h'}$ as associated graph is said to be obtained from $hx \geq h_0$ by *cloning* the edge uv.

In Figure 14, we give an example of the cloning of an edge. The inequality represented by the graph on the right hand side is obtained by cloning the edge e. The cloning of an edge can be repeated any number of times. In Naddef & Rinaldi [1993] sufficient conditions are given for an edge of the graph associated with a facet-defining inequality in TT form to be clonable, i.e., to be cloned as described before, while producing a facet-defining inequality. A path-edge of a PWB inequality belonging to a path of length 2 is clonable. The inequalities obtained by cloning any set of these edges any number of times are called *extended*

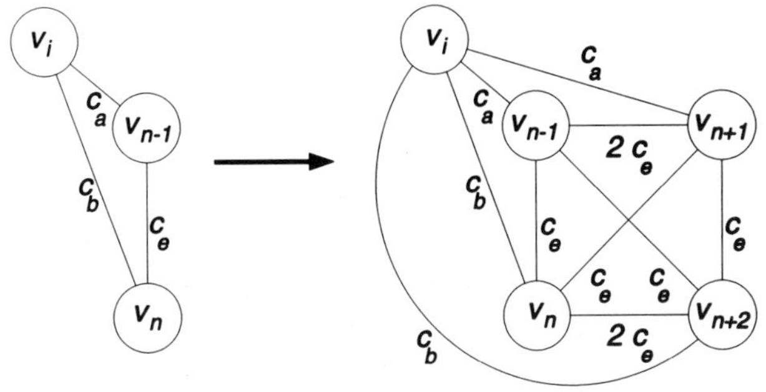

Fig. 14. Cloning of an edge.

PWB inequalities (see Naddef & Rinaldi [1988]). A diameter edge of a crown inequality is clonable. The inequalities obtained by cloning any set of diameters any number of times are called *extended crowns* [see Naddef & Rinaldi, 1992].

A generalization of the zero node-lifting is described in Naddef & Rinaldi [1993] and called 1-node lifting. Several sufficient conditions for an inequality, obtained by zero node-lifting of a facet-defining inequality, to be facet-defining for STSP(n) are given in Queyranne & Wang [1993]. Some of them are very simply to check and apply to basically all known inequalities facet-defining for STSP(n).

2-sum composition of path inequalities

The *2-sum composition* of inequalities is an operation that produces new facet-defining inequalities by merging two inequalities known to be facet-defining. Instead of describing the operation in general, we give an example of its application that produces a large class of facet-defining inequalities for STSP(n), called the *regular parity path-tree inequalities*. We define these inequalities recursively.

A simple regular PWB inequality is a *regular parity path-tree inequality*. Let $f^1x^1 \geq f_0^1$ and $f^2x^2 \geq f_0^2$ be two regular parity path-tree inequalities and let $G^1 = (V_{n_1}, E_{n_1}, f^1)$ and $G^2 = (V_{n_2}, E_{n_2}, f^2)$ be their corresponding associated graphs. Let u_1v_1 be a path-edge of the first inequality and u_2v_2 a path-edge of the second, satisfying the following conditions:

(i) the nodes u_1 and u_2 have the same parity (they are either both odd or both even) and so do the nodes v_1 and v_2;

(ii) $f^1(u_1, v_1) = f^2(u_2, v_2) = \varepsilon$.

The condition (ii) can always be satisfied by multiplying any of the two inequalities by a suitable positive real number. Let $G' = (V, E', f')$ be the weighted graph with $n = n_1 + n_2 - 2$ nodes obtained from G^1 and G^2 by identifying the nodes u_1 and u_2 and the nodes v_1 and v_2. We call the nodes that result from the identification of the two pairs u and v, respectively. Each of these two nodes is *odd* if it arises by the identification of two odd nodes, otherwise it is *even*. The edge uv qualifies as a *path-edge*. The node and the edge set of G' are given by $V' = (V_{n_1} \cup V_{n_2} \setminus \{u_1, v_1, u_2, v_2\}) \cup \{u, v\}$ and $E' = E_{n_1} \cup E_{n_2} \setminus \{u_1v_1, u_2v_2\} \cup \{uv\}$.

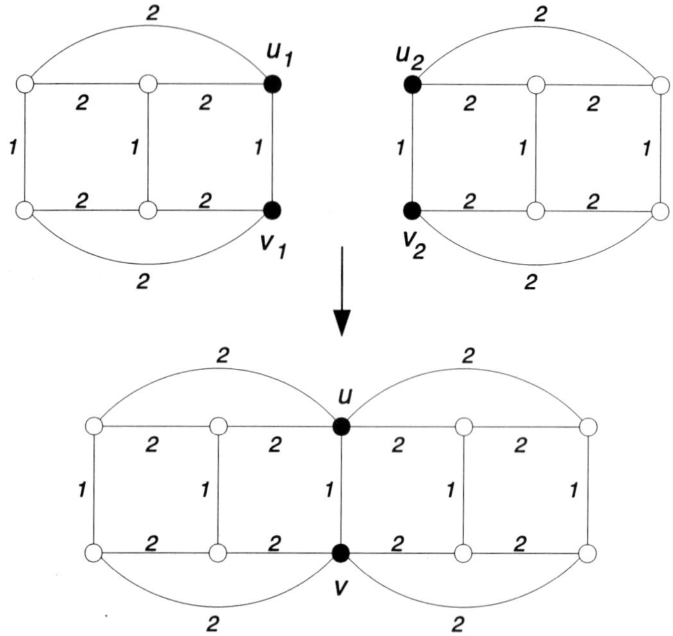

Fig. 15. The composition of two bicycle inequalities.

Let $G = (V, E, f)$ be the weighted graph obtained from G' by adding the edge ij for all $i \in V_{n_1} \setminus \{u_1, v_1\}$ and all $j \in V_{n_2} \setminus \{u_2, v_2\}$, with weight $f(i, j)$ given by the f-length of the shortest path from i to j in G'.

The inequality $fx \geq f_0 = f_0^1 + f_0^2 - 2\varepsilon$, having G as associated graph, is a *regular parity path-tree* inequality.

The PWB inequalities that are used in the composition of a regular parity path-tree are called *components* of the inequality. Figure 15 illustrates the composition of two bicycle inequalities by a 2-sum operation.

The s-sum composition of inequalities (of which the 2-sum is a special case) is introduced in Naddef & Rinaldi [1991] in the context of GTSP, as a tool to produce new facet-defining inequalities from known ones. The 2-sum composition for STSP is described in Naddef & Rinaldi [1993]. A proof that regular parity path-tree inequalities define facets of STSP(n), with $n \geq 10$, is given in Naddef & Rinaldi [1988]. Other composition operations for facet-defining inequalities of STSP are described in Queyranne & Wang [1990].

Relations between TT and other inequalities

The inequalities in TT form described above include most of the inequalities presently known that define facets of STSP(n). We conclude this subsection on the facets of STSP(n) by briefly showing how these inequalities are related to the other known facet-defining inequalities, described in closed form.

– The *2-matching inequalities* are 2-regular PWB inequalities derived from simple PWB inequalities by allowing zero node-lifting only on the nodes Y and Z.

– The *Chvatál comb inequalities* are 2-regular PWB inequalities derived from simple PWB inequalities by allowing zero node-lifting on all nodes but u_2^i for $i \in \{1, \ldots, k\}$.

– The *comb inequalities* are 2-regular PWB inequalities.

– The *chain inequalities* (see below) are 2-regular PWB inequalities where *only one* path-edge is cloned any number of times (consequently the chain inequalities are a special case of the extended PWB inequalities, and so they are facet-defining for STSP(n)).

– The *clique-tree inequalities* are regular parity path-tree inequalities obtained from 2-regular PWB inequalities with the condition that the nodes Z of all the component PWB inequalities are identified together in a single node.

Other facet-defining inequalities

To complete the list of all known facet-defining inequalities for STSP we mention a few more. Chvátal [1973] shows that an inequality defined by the Petersen graph is facet-defining for STSP(10). A generalization of this inequality, which is facet-defining for $n > 10$, is given in Maurras [1975]. Three inequalities facet-defining for STSP(8) are described in Christof, Jünger & Reinelt [1991]. The inequalities have to be added to trivial, subtour elimination, PWB, chain, ladder and crown inequalities to provide a complete description of STSP(8).

Other valid inequalities for STSP(n)

The fact that collections of node sets, satisfying some conditions, can be used to describe facet-defining inequalities for STSP(n) in closed form has motivated many researchers to proceed along these lines and consider collections of node sets satisfying more complex conditions.

A first generalization of comb inequalities obtained by replacing a tooth by a more complex structure leads to the *chain* inequalities, described in Padberg & Hong [1980], where only a proof of validity is given.

Another generalization of the comb inequalities is obtained by allowing not just a single handle but a nested family of handles. The inequalities obtained in this way are called *star* inequality and are described in Fleischmann [1988] where it is proved that they are valid for GTSP. The star inequalities properly contain the PWB inequalities but for those which are not PWB inequalities only a proof of validity for GTSP is currently known. Actually, some of them do not define facets of STSP(n) (see Naddef [1990]) and some others do (see Queyranne & Wang [1990]).

A generalization of the clique-tree inequalities is produced by relaxing the conditions (iii) and (vi) of the definition. The resulting inequalities are called *bipartition* inequalities [Boyd & Cunningham, 1991]. Further generalizations lead to the *hyperstar* inequalities [Fleischmann, 1987] and to the *binested* inequalities [Naddef, 1992].

For all these inequalities only a proof of validity is given in the cited papers. Therefore, these inequalities provide good candidates for members of the set $\mathcal{B}^{\leq}$ of all facet-defining inequalities of STSP(n), and can be used to provide stronger LP relaxations to the TSP.

For a complete survey on these classes of inequalities see Naddef [1990].

5.6. *The separation problem for STSP(n)*

In order to have a relaxation that produces a good lower bound, it is necessary that the LP relaxation contains at least the subtour elimination inequalities. The number of these constraints is exponential in n (it is precisely $2^{n-1} - n - 1$) and it becomes much larger if other inequalities, like 2-matching or comb inequalities, are added to the relaxation. Consequently, to find the optimal solution of an LP relaxation we cannot apply a linear programming algorithm directly to the matrix that explicitly represents all these constraints.

Let $\mathcal{L}$ be a system that contains a huge number of inequalities which are valid for STSP(n) and suppose that we want to solve the problem

Problem 5.2.

$$\begin{array}{lll} \text{minimize} & cx & \\ \text{subject to} & A_n x = \mathbf{2}, & \\ & lx \leq l_0 & \text{for all } (l, l_0) \in \mathcal{L}, \\ & \mathbf{0} \leq x \leq \mathbf{1}. & \end{array}$$

In principle Problem 5.2 can be solved by the following cutting-plane procedure:

procedure CUTTING_PLANE

Input: n, c, a family of 'known' inequalities $\mathcal{L}$.

(1) Set $\mathcal{L}' = \emptyset$

(2) Solve $\min\{cx \mid A_n x = \mathbf{2},\ lx \leq l_0 \text{ with } (l, l_0) \in \mathcal{L}',\ \mathbf{0} \leq x \leq \mathbf{1}\}$ and let z be its solution.

(3) Find one or more inequalities in $\mathcal{L}$ violated by z.

(4) If none is found, then stop. Otherwise add the violated inequalities to $\mathcal{L}'$ and go to (2).

Procedure CUTTING_PLANE stops after a finite number of steps, because $\mathcal{L}$ is finite. The *core* of the procedure is the problem solved in Step (3), which is called the *separation problem* and is formally stated as follows:

Problem 5.3. *Given a point $z \in \mathbb{R}^{E_n}$ and a family $\mathcal{L}$ of inequalities in $\mathbb{R}^{E_n}$,* identify *one (or more) inequalities in $\mathcal{L}$ violated by z or* prove *that no such inequality exists.*

An *exact separation procedure* for a family of inequalities $\mathcal{L}$ is one that solves Problem 5.3, a *heuristic separation procedure* is one that may find violated inequalities, but that in case it cannot find any, is unable to guarantee that no violated inequalities exist in $\mathcal{L}$.

The faster, the more efficient and the more 'prolific' the separation procedure used in Step (3) of the cutting plane procedure, the faster is the resolution of Problem 5.2. This statement is substantiated by a result of Grötschel, Lovász & Schrijver [1981,1988] and Padberg & Rao [1981] that can be stated as follows:

Proposition 5.4 *Problem 5.2 is solvable in polynomial time* if and only if *Problem 5.3 is solvable in polynomial time.*

Unfortunately an exact separation procedure that runs in polynomial time is presently known only for two classes of TSP inequalities. These classes are the subtour elimination and 2-matching inequalities. It follows that the lower bound produced by an LP relaxation that has all subtour and all 2-matching inequalities can be computed in polynomial time. Heuristic separation procedures have been developed for the comb and the clique-tree inequalities.

Let $z \in \mathbb{R}^{E_n}$ be a point that we want to separate from STSP(n) with a facet-defining inequality belonging to a given class. It is assumed, without loss of generality, that z satisfies (5.5) and (5.7). Usually, all separation algorithms operate on a graph associated with z, the support graph of z. The *support graph* $G_z = (V_n, E, z)$ of z is a weighted graph whose edge set E contains all edges of E_n corresponding to a positive component of z. The weight associated with $e \in E$ is z_e. The edges of G_z with weight 1 are called *1-edges*.

The separation for subtour elimination inequalities

The point z violates a subtour elimination inequality (5.6) if and only if the minimum weight cut in G_z has weight less than 2. Since the minimum weight cut in a graph with nonnegative edge weights can be found in polynomial time with the algorithm proposed by Gomory & Hu [1961], Problem 5.3 can be solved in polynomial time for the subtour elimination inequalities. Therefore, by Proposition 5.4, the subtour relaxation is solvable in polynomial time.

The Gomory–Hu algorithm is based on the computation of $n-1$ maximum flow problems on some weighted graphs derived from G_z. The complexity of a maximum flow algorithm is $O(|V||E|\log(|V|^2/|E|))$ (see Goldberg & Tarjan [1988]), and so the complexity of the algorithm is $O(|V|^2|E|\log(|V|^2/|E|))$. For large instances of the TSP such a complexity is expensive in terms of actual computation time, since Problem 5.3 has to be solved several times in a branch and cut algorithm (see Section 6). For this reason many heuristic procedures have been proposed to find violated subtour elimination inequalities quickly (see, e.g., Crowder & Padberg [1980] and Grötschel & Holland [1991]). Padberg & Rinaldi [1990a] describe an exact algorithm that finds the minimum weight cut in a graph with a drastic reduction in the number of maximum flow computations. Even though the algorithm has the same worst case time bound as the Gomory–Hu algorithm, it runs much faster in practice and it allows the execution of an exact separation algorithm at every iteration of a branch and cut algorithm. The idea of this algorithm is to exploit some simple sufficient conditions on G_z that guarantee that two nodes belong to the same shore of a minimum cut. If two nodes satisfy one of these conditions then they are *contracted*. The *contraction* of two nodes in G_z produces a new weighted graph where the two nodes are identified into a single node; loops are removed and any two parallel edges are replaced by a single edge with weight equal to the sum of their weights. The resulting graph has one node less and the shores of a minimum cut in it can be turned into the shores of a minimum cut in G_z, by replacing the node that results from the identification with the two original nodes. The contraction of a pair of nodes can be applied recursively until no more reductions apply. At this point the Gomory–Hu algorithm can be applied to the resulting reduced graph.

A different algorithm also based on the contraction of pairs of nodes is proposed by Nagamochi & Ibaraki [1992a, b]. After each major step of the algorithm a pair of nodes u and v is identified that have the following property. Either $\delta(\{u\})$ is a minimum cut or $\delta(\{v\})$ is a minimum cut or u and v belong to the same shore of a minimum cut. After recording the better of the two cuts $\delta(\{u\})$ and $\delta(\{v\})$, u and v are contracted. After $|V| - 1$ contractions the graph reduces to a single node and the best of the recorded cuts is the minimum cut. The algorithm does not require the computation of a maximum flow and runs in $O(|E||V| + |V|^2 \log |V|)$ time.

Another algorithm for the minimum cut is proposed in Hao & Orlin [1992]. It is a modified version of a maximum flow algorithm and it is able to compute the minimum cut in the same running time required by the computation of a single maximum flow ($O(|V||E| \log(|V|^2/|E|))$).

Finally we mention a randomized algorithm for computing a minimum cut with high probability. The algorithm runs in $O(|E||V|^2 \log^3 |V|)$ time [Karger, 1993] and an improved version in $O(|V|^2 \log^3 |V|)$ time [Karger & Stein, 1993].

All these algorithms can be conveniently utilized to solve the separation problem of the subtour elimination inequalities efficiently.

The separation for the 2-matching inequalities

Let G_z be defined as before and let us apply the following operations on it. First every edge e of G_z is split in two edges e' and e'' that have the same weight as e and are incident with a new node v_e. The resulting graph G'_z has $n + |E|$ nodes and $2|E|$ edges. All its nodes are labeled *even*. Then for each pair of edges $\{e', e''\}$ that comes from splitting, either e' or e'' is complemented, i.e., its weight z_e is replaced by $1 - z_e$ and the label of each endpoint becomes *odd* if it was *even* (and *even* if it was *odd*). Call G^*_z the new weighted and labeled graph. The number of odd nodes of G^*_z is even and at least $|E|$ (each of the nodes that is produced by splitting is odd). An *odd cut* of G^*_z is a cut whose shores have an odd number of odd nodes. Padberg & Rao [1982] propose an algorithm that finds the minimum weight odd cut of a labeled weighted graph in polynomial time. They also prove that the vector z satisfies all 2-matching inequalities if and only if the minimum odd cut in G^*_z has weight at least 1. Consequently, Problem 5.3 can be solved in polynomial time for the 2-matching inequalities.

Suppose that G_z^* has an odd cut with weight less than 1. Let us see how a 2-matching inequality violated by z can be generated. At most one of the two edges e' and e'' produced by splitting an edge e of G_z belongs to the cut. If such an edge has been complemented, then the endpoints of e are taken as a tooth of the inequality. All nodes of the original graph G_z that belong to one of the two shores of the cut are taken as the handle of the inequality. It may happen that after this construction two teeth of the 2-matching intersect in a node u, thus violating the condition (iii) of the definition of comb inequalities. In this case the two teeth are removed from the set of teeth. In addition, if u belongs to the handle, then it is removed from it; if it does not, then it is added to it.

The Padberg–Rao algorithm, which is based on the Gomory–Hu algorithm, requires as many max-flow calculations on G_z^* as the number of its odd nodes. As observed for the separation of the subtour elimination inequalities, this algorithm may be very time consuming. Therefore some reductions of the number of nodes, odd nodes, and edges of the graph to which the Padberg–Rao algorithm is applied have been proposed (see Padberg & Grötschel [1985], Grötschel & Holland [1987], and Padberg & Rinaldi [1990b]). A detailed description of an implementation of the Padberg–Rao algorithm is given in Grötschel & Holland [1987]. A simple implementation of the Gomory–Hu algorithm is described in Gusfield [1987].

Although the reductions applied to G_z^* may produce a sensible speed up in the solution of Problem 5.3, an exact solution of this problem may still be too expensive for large TSP instances. For this reason heuristic separation procedures are often used for 2-matching inequalities.

A first procedure, similar to the exact one sketched above, is proposed in Padberg & Rinaldi [1990b]. In this procedure all nodes of G_z are labeled *even* and all edges with weight greater than or equal to 0.5 are complemented. Then the reductions mentioned before are applied. Finally the Padberg–Rao algorithm is applied to the resulting graph, which is smaller (in terms of nodes, odd nodes and edges) than G_z^*.

Another heuristic, proposed first in Padberg & Hong [1980], follows a completely different approach. All 1-edges (or all edges with weight close to 1) are removed from G_z. The resulting graph is decomposed into its biconnected components. Each biconnected component with at least three nodes is considered as the handle of a possibly violated 2-matching inequality. The teeth of the inequality are the endpoints of edges of the original graph G_z with only one endpoint in the handle and with 'big' weight (usually greater than or equal to 0.5). The procedure is fast and quite successful. Implementations and variations of this heuristic separation algorithm are described in Padberg & Grötschel [1985], Grötschel & Holland [1987], and Padberg & Rinaldi [1990b]. In the last paper a variation is given that also produces violated Chvatál comb inequalities.

The separation for comb inequalities

As said above there is no known exact polynomial separation procedure for comb inequalities. The heuristic procedures proposed in the literature exploit the following two facts:

(a) comb inequalities expressed in TT form (i.e., 2-regular PWB inequalities) can be obtained by zero node lifting of 2-matching inequalities (i.e., simple 2-regular PWB inequalities);

(b) a separation procedure for 2-matching inequalities is available.

Let us assume that for a family $\mathcal{F}$ of simple inequalities in TT form, a separation procedure is available and that we want to exploit this procedure to find violated inequalities that are obtained by zero node-lifting of inequalities in $\mathcal{F}$. We can proceed as follows. Let $S \subset V_n$ be a set of nodes whose corresponding subtour elimination inequality is satisfied by z with equality, i.e.,

$$z(\delta(S)) = 2. \tag{5.13}$$

Let $\hat{G}_{\hat{z}} = (V_{\hat{n}}, \hat{E}, \hat{z})$, with $\hat{n} = n - |S| + 1$, be the weighted graph obtained by recursively contracting pairs of nodes in S until S is replaced by a single node s. It is easy to see that $\hat{z} \in \mathbb{R}^{E_{\hat{n}}}$ satisfies all inequalities (5.5), (5.6), and (5.7), and so it can be thought of as the solution of an LP relaxation of a TSP on a complete graph of $\hat{n}$ nodes. Suppose that we are able to separate $\hat{z}$ from STSP($\hat{n}$) with an inequality $\hat{h}\hat{x} \geq h_0$ in $\mathcal{F}$. It follows that $\hat{h}\hat{z} < h_0$. Let $hx \geq h_0$, with $h \in \mathbb{R}^{E_n}$, be the inequality obtained by zero lifting of node s. It follows that $hz = \hat{h}\hat{z}$, and so the inequality $hx \geq h_0$ is violated by z.

In conclusion, a separation procedure for inequalities obtained by zero node-lifting of inequalities in $\mathcal{F}$ can be devised by contracting any number of sets satisfying (5.13) by applying the separation procedure for the inequalities in the family $\mathcal{F}$ to the reduced graph, and finally by producing a violated inequality in $\mathbb{R}^{E_n}$ by zero lifting all nodes that are produced by contraction.

Unfortunately this procedure does not work in all cases. It may be the case that $z \notin$ STSP(n), i.e., there exists a valid inequality for STSP(n) that is violated by z, but the vector $\hat{z}$, produced by contracting a set S satisfying (5.13), belongs to STSP($\hat{n}$), and so all valid inequalities for STSP($\hat{n}$) are satisfied by $\hat{z}$. Put differently, the operation associated with the contraction of a set S may map a point outside STSP(n) to a point inside STSP($\hat{n}$), which is an undesirable situation. Figure 16 gives an example of this bad case. The graph in Figure 16a is the support graph of a point z (solid lines are edges with weight 1, while dotted lines are edges with value 0.5). The set $S = \{5, 6, 7, 8\}$ satisfies (5.13). If we contract this set we get the graph of Figure 16b. This is the support graph of a vector $\hat{z}$ which is the convex combination $z^1/2 + z^2/2$ of the incidence vectors z^1 and z^2 of the two Hamiltonian cycles $\{1, 2, s, 3, 4, 1\}$ and $\{1, 2, 3, 4, s, 1\}$, respectively. Consequently, $\hat{z}$ is a point of STSP($\hat{n}$). However if in the graph of Figure 16a the set $S = \{5, 8\}$ is contracted, in the resulting graph shown in Figure 16c there is a violated Chvátal comb with handle $\{1, s, 4\}$ and teeth $\{1, 2\}$, $\{4, 3\}$, and $\{s, 6, 7\}$. By zero-lifting node s we obtain a violated comb with handle $\{1, 4, 5, 8\}$ and teeth $\{1, 2\}$, $\{4, 3\}$, and $\{5, 6, 7, 8\}$.

In Padberg & Rinaldi [1990b] a set S satisfying (5.13) is called *shrinkable* if the graph obtained by contracting S is not the support graph of a point of STSP($\hat{n}$). Therefore, the shrinkable sets are those that can be safely contracted. In the same paper sufficient conditions are given for a set to be shrinkable and

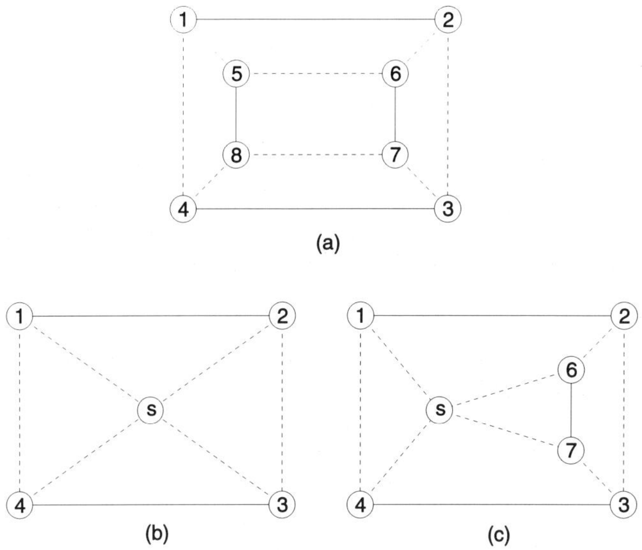

Fig. 16. An example of a nonshrinkable set satisfying (5.13).

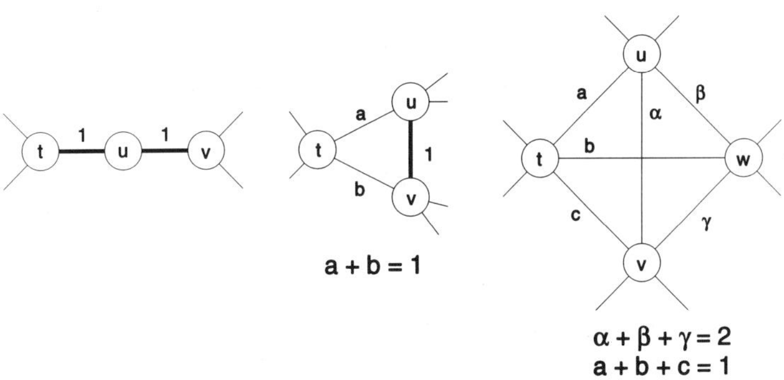

Fig. 17. Shrinkable set.

some cases of shrinkable sets are provided (see Figure 17). Finally a heuristic separation procedure for comb inequalities is described, which is based on the scheme outlined before and utilizes those special cases of shrinkable sets. The heuristic separation procedure for comb inequalities described in Grötschel & Holland [1991] is similar.

The separation for clique-tree inequalities

The separation problem for clique-tree inequalities has not been studied to the same extent as for comb inequalities. Two simple procedures are published in Padberg & Rinaldi [1990b]. One finds violated clique-tree inequalities with two handles. The other is an extension of the procedure for the 2-matching inequalities, based on the decomposition of a subgraph of G_z into biconnected components.

The improvement in the relaxation due to these procedures is not impressive.

The separation for PWB inequalities

A simple heuristic procedure for finding violated PWB inequalities is described in Clochard & Naddef [1993]. The procedure starts by finding a violated comb inequality (a 2-regular PWB) and then tries to 'extend' its paths (that all have length 2) to paths of bigger length. The procedure works well when the point z to be separated has many integer components. This is often the case when z is the optimal solution over the subtour elimination polytope. The support graph G_z of such a point often has long paths of consecutive 1-edges. These edges are good candidates to become path-edges of a violated PWB inequality.

The results obtained by applying this procedure, as reported in Clochard & Naddef [1993], are very promising.

5.7. Comparison of LP relaxations

To conclude this section, we show how different LP relaxations behave in terms of the value of the lower bound that they produce. To do so, we have computed the lower bound associated with each relaxation for each of the instances of our standard set of test problems. The results of this computation are reported in Table 5. For each test problem we have computed the *fractional 2-matching* relaxation, i.e., the relaxation obtained by solving Problem 5.1 with only the constraints (5.5) and (5.7). This relaxation is not properly an LP relaxation as defined in Section 5.1, because it may have an integer solution that is the incidence vector of a nonconnected 2-matching, since the constraints (5.6) are not imposed. The fractional 2-matching is often the first relaxation to be produced in any cutting-plane algorithm (see Section 6) since it requires only a small number (polynomial in n) of constraints and its polytope properly contains the polytope of any other LP relaxation considered in this section. In addition, we have computed the lower bound of the subtour relaxation, of the 2-matching relaxation and of an LP relaxation that includes subtour elimination, 2-matching, comb and clique-tree inequalities. The latter relaxation has been computed with the algorithm described in Padberg & Rinaldi [1991] that is based on heuristic separation procedures for comb and clique-tree inequalities. Consequently, the lower bound that we report is inferior to the one that would be obtained by optimizing over the polytope defined by subtour elimination, comb, and clique-tree inequalities. The lower bound has been computed from the relaxation constructed by the Padberg-Rinaldi algorithm just before going to the branching phase. Since the algorithm tries to minimize the overall computation time, it may resort to the branching phase even though more inequalities can be added to the relaxation. Therefore, this lower bound is an underestimate of the bound obtainable with the separation procedures described in Padberg & Rinaldi [1990b] and used in the algorithm. We denote this relaxation by 'all cuts' in the header of Table 5.

Table 5
Comparison of LP relaxations

Problem	r f2-m.	r 2-m.	r subtour	r all cuts	R 2-m.	R subtour	R all cuts
lin105	98.8	99.6	99.9	100.0	67.4	94.9	100.0
pr107	55.5	55.5	100.0	100.0	0.0	100.0	100.0
pr124	85.0	87.2	98.4	99.9	14.8	89.1	99.4
pr136	92.2	92.3	99.1	99.8	1.0	88.9	97.0
pr144	56.0	57.5	99.4	99.9	3.4	98.7	99.8
pr152	60.4	60.4	99.4	99.6	0.0	98.4	99.1
u159	96.7	97.8	99.6	100.0	34.3	88.9	100.0
rat195	97.8	98.9	99.0	99.8	48.5	53.0	90.6
d198	74.8	75.3	99.7	100.0	2.0	98.7	100.0
pr226	68.7	71.1	99.7	100.0	7.7	98.9	100.0
gil262	93.5	94.5	99.0	99.9	16.4	84.9	99.0
pr264	74.7	75.7	99.8	100.0	4.2	99.1	99.9
pr299	93.3	94.7	98.3	99.9	21.0	74.9	98.9
lin318	92.7	93.5	99.7	99.9	10.7	95.4	99.1
rd400	95.2	96.0	99.2	99.8	17.3	83.0	96.7
pr439	87.4	89.0	98.8	99.7	12.4	90.4	97.9
pcb442	98.7	99.2	99.5	99.9	40.5	58.7	95.8
d493	93.3	94.6	99.5	99.9	18.7	92.6	99.2
u574	92.8	93.3	99.5	100.0	6.4	92.8	100.0
rat575	98.1	98.7	99.3	99.9	29.9	61.0	93.6
p654	82.5	82.8	99.9	100.0	1.5	99.2	100.0
d657	95.4	96.4	99.1	99.8	21.2	79.6	96.7
u724	96.7	96.8	99.4	100.0	3.6	83.0	98.7
rat783	97.3	97.5	99.6	100.0	6.5	86.1	99.2
pr1002	93.0	93.3	99.1	99.9	4.7	87.5	98.9
pcb1173	97.7	98.6	99.0	99.9	40.1	58.1	96.9
rl1304	90.9	91.1	98.5	99.9	1.8	83.2	98.9
nrw1379	98.2	98.5	99.6	100.0	21.1	76.9	98.1
u1432	99.2	99.5	99.7	100.0	38.7	66.4	99.3
pr2392	95.1	96.5	98.8	100.0	27.4	75.3	100.0

In the first five columns of the table the names of the test problems are reported along with the value $r = 100 \times (LB/OPT)$ for each lower bound, where LB is the lower bound value and OPT is the objective function value of an optimal solution. As mentioned in Section 2, two TSP instances are equivalent if one is obtained by adding a constant C to each component of the objective function of the other. However, the value r of any of the four bounds is not the same for two equivalent instances and tends to 100 when C tends to infinity.

To overcome this problem, we have also computed a different index to compare the bounds. This index is the same for two equivalent instances and shows, better than the previous one, how the different relaxations contribute to covering the gap between the fractional 2-matching bound and the optimal solution value. The index is given by the ratio $R = 100 \times (LB - F2M)/(OPT - F2M)$, where $F2M$ is the value of the fractional 2-matching bound. A value of R equal to 100 shows

Table 6
Quality of GTSP facet-defining inequalities

Class of inequalities	strength
Comb	10/9
Clique tree	8/7
Path	4/3
Crown	11/10

that the relaxation is sufficient to provide an optimal solution to the problem, i.e., that no recourse to branch and cut is necessary.

In Section 5.5 we have seen many inequalities that define facets of STSP. Only for a few of them a satisfactory separation procedure is available at present. It would be interesting to know for which inequalities it would be profitable to invest in further research to solve the corresponding separation problem. In other words, given a class of valid inequalities $\mathcal{C}$, it would be interesting to know the improvement on the lower bound that we would get by adding them all to the constraints of the subtour relaxation. This improvement would measure the strength of the inequalities contained in the class $\mathcal{C}$. In Goemans [1993] this analysis is carried out for GTSP. For a class of inequalities $\mathcal{C}$ a theoretical computation is provided for the improvement that would result when all the inequalities are added to the subtour relaxation of GTSP. The subtour relaxation of GTSP has the nonnegativity constraints and the constraints (5.4). The results, which can serve as a useful indication also for STSP, are summarized in Table 6.

6. Finding optimal and provably good solutions

The discussion of the last two sections gives us a variety of tools for computing Hamiltonian cycles and lower bounds of the length of Hamiltonian cycles. The lower bounds enable us to make statements of the form: 'The solution found by the heuristic algorithm is at most $p\%$ longer than the shortest Hamiltonian cycle'. Probably most practitioners would be completely satisfied with such a quality guaranteed solution and consider the problem solved, if the deviation p is only small enough. In this section, we will consider algorithms which can achieve any desired quality (expressed in terms of p), including the special case $p = 0$, i.e., optimality.

6.1. Branch and bound

All published methods satisfying the above criterion are variants of the branch and bound principle. Branch and bound algorithms are well known so that we can omit a formal definition here. The flowchart of Figure 18 gives the basic control structure of a branch and bound algorithm for a combinatorial optimization problem whose objective function has to be minimized.

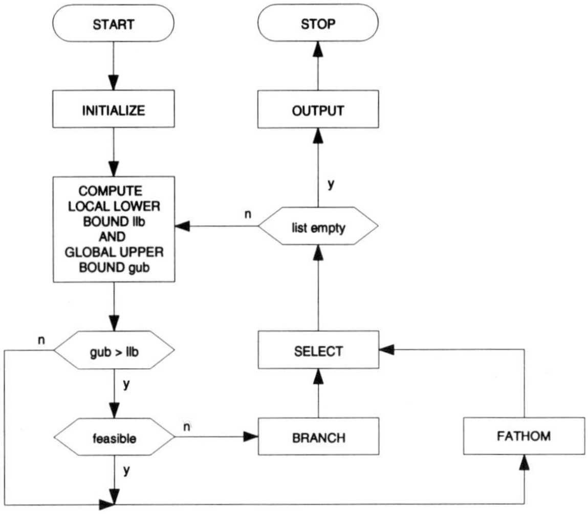

Fig. 18. Flowchart of a branch and bound algorithm.

A branch and bound algorithm maintains a list of subproblems of the original problem whose union of feasible solutions contains all feasible solutions of the original problem. This list is initialized with the original problem itself.

In each major iteration the algorithm selects a current subproblem from this list and tries to 'fathom' it in either of the following ways: a lower bound for the value of an optimal solution of the current subproblem is derived that is at least as high as the value of the best feasible solution found so far, or it is shown that the subproblem does not contain any feasible solution, or the current subproblem is solved to optimality. If the current subproblem cannot be fathomed according to one of these criteria, then it is split into new subproblems whose union of feasible solutions contains all feasible solutions of the current problem. These newly generated problems are added to the list of subproblems. This iteration process is performed until the list of subproblems is empty.

Whenever a feasible solution is found in the process, its value constitutes a global upper bound for the value of an optimal solution. Similarly, the minimum of the local lower bounds at any point of the computation is a global lower bound for the optimal value of the objective function. For the TSP, this means that during the execution of a branch and bound algorithm a sequence of feasible solutions of decreasing lengths and a sequence of lower bounds of increasing values is produced. The algorithm terminates with the optimal solution as soon as the lower bound coincides with the length of the shortest Hamiltonian cycle found. If we take the point of view that practical problem solving consists of producing a solution of a prescribed quality $p\%$, then a branch and bound algorithm can achieve this goal if it stops as soon as it has found a solution T of length $c(T)$ and lower bound l such that $(c(T) - l)/l \leq p/100$. A general survey of the branch and bound method applied to the TSP has been given in Balas & Toth [1985]. One of the crucial parts of such an algorithm is the lower bounding technique. Lower bounds are usually computed by solving an appropriate relaxation of the TSP.

Several relaxations have been considered for the TSP. Among them are *n-path relaxation*, *assignment relaxation*, or also so-called *additive bounding procedures*. For information on these approaches see Balas & Toth [1985] and Carpaneto, Fischetti & Toth [1989]. A branch and bound algorithm that uses the *2-matching relaxation* has been implemented by Miller, Pekny & Thompson [1991].

We shall discuss branch and bound algorithms based on the two relaxation methods discussed in Section 5, the *1-tree relaxation* and the *LP relaxations*.

6.2. 1-tree relaxation

First we discuss the method of determining lower bounds based on the approximation of the optimum value of the Lagrangean dual based on the 1-tree relaxation (see Section 5.1). This technique is due to Held & Karp [1970, 1971]. Recall that the corresponding Lagrangean problem is of the form $\max_\lambda\{L(\lambda)\}$. Determining $L(\lambda)$ for a given vector of node multipliers λ amounts to computing a 1-tree with respect to the modified edge weights $c_{ij} + \lambda_i + \lambda_j$ and subtracting $2\sum_{i=1}^{n} \lambda_i$.

The function L is piecewise linear, concave, and nondifferentiable. A method for maximizing L is the *subgradient method*. This method is a generalized gradient ascent method where in each iteration step the next iterate (i.e., the next vector of node multipliers) is determined by taking a step in the direction of a *subgradient*.

A minimum 1-tree readily supplies a subgradient as follows. Namely, let δ_i be the degree of node i in the minimum 1-tree with respect to node multipliers λ. Then the vector $(\delta_1 - 2, \delta_2 - 2, \ldots, \delta_n - 2)$ is a subgradient of L at λ.

If L is bounded from above (which is the case for the TSP) and if the step lengths α_k satisfy both $\lim_{k\to\infty} \alpha_k = 0$ and $\sum_{k=0}^{\infty} \alpha_k = \infty$, then the method converges to the maximum of L [Polyak, 1978]. However, it turned out in practice that such step lengths formulae lead to very slow convergence. There are update formulas for the α_k that do not satisfy the requirement $\sum_{k=0}^{\infty} \alpha_k = \infty$, but lead to better convergence in practice.

Based on the references Balas & Toth [1985], Volgenant & Jonker [1982], Christofides [1979] and on own experiments we use the following implementation.

procedure 1TREE_BOUND

(1) Let α_1 be the initial step length and γ a decrement factor for the step length.

(2) Set $\lambda_i^1 = 0$ for every node i, and $k = 1$.

(3) Perform the following steps for a specified number of iterations or until no significant increase in the lower bound can be observed for several iterations.

  (3.1) Compute a minimum spanning tree with respect to the edge weights $c_{ij} + \lambda_i + \lambda_j$.

  (3.2) Compute the best 1-tree obtainable from this spanning tree as follows. For each leaf v of the spanning tree we determine its second shortest incident edge. The length of the minimum spanning tree plus the length of this edge gives the length of the minimum 1-tree with v as fixed degree-2 node.

(3.3) Define the vector d^k by $d_i^k = \delta_i - 2$ where δ_i is the degree of node i in the 1-tree computed in Step (3.2).

(3.4) For every node i set $\lambda_i^{k+1} = \lambda_i^k + \alpha_k(0.7d_i^k + 0.3d_i^{k-1})$ (where $d_i^0 = 0$).

(3.5) Set $\alpha_{k+1} = \gamma\alpha_k$ and increment k by 1.

(4) Return the best bound computed.

Differences to straightforward realizations are that the direction of the subgradient step is a convex combination of the current and the preceding subgradient, that the direction vector is not normalized, and that the special node for the 1-tree computations is not fixed. In theory the same optimal value of the Lagrange dual is attained whatever node is fixed. But practical experiments have shown that better bounds are obtained if additional time is spent for computing various 1-trees. The chosen value of γ influences the running time of the method. The closer γ is to 1, the more iterations are performed and better bounds are obtained. Note that, since the step length is fixed a priori, there is no guarantee that each iteration step improves the bound.

Some authors propose to update the multipliers according to the formula

$$\lambda_i^{k+1} = \lambda_i^k + t^k(U - L(\lambda^k))\frac{d_i^k}{||d^k||}$$

where U is an estimate for the optimal solution value. For our set of sample problems, we found this formula inferior to the one above.

The running times are not very encouraging. Since edge weights are arbitrary we can compute the best tree in Step (3.1) only in time $O(n^2)$ and the best 1-tree in Step (3.2) in time $O(ln)$ where l is the number of leaves of the spanning tree. To obtain reasonable lower bounds more quickly, we use the following approach. The subgradient method is only performed on a sparse subgraph (e.g., for geometric instances a 10 nearest neighbor subgraph augmented by the Delaunay graph). This has the consequence that the computed bound may not be valid for the original problem. To get a valid lower bound, we compute a minimum 1-tree in the complete graph with respect to the node multipliers obtained with the subgradient method in the sparse graph.

Bounds obtained for problems *pcb442*, *rd783*, *pr1002*, and *pr2392* were 0.6%, 0.4%, 1.4%, and 1.3%, respectively, below the length of a shortest Hamiltonian cycle. The final iteration changed the bounds only slightly. In practical applications we can safely omit the final step and assume that the bound determined in the first phase is correct.

The subgradient method is not the only way for attacking nondifferentiable optimization problems occurring in the context of Lagrangean relaxation. A more elaborate approach is the so-called *bundle method* [Kiwiel, 1989]. It is also based on subgradients, but in every iteration the new direction is computed as a convex combination of several (10–20) previous subgradients. Moreover, line searches are performed to determine the step lengths. In this sense our approach is a rather simple version of the bundle method keeping only a 'bundle' of two subgradients (which are combined in a fixed way) and not performing line searches.

In Schramm [1989] an extension of this principle is discussed which combines the bundle approach with trust-region methods. From a theoretical point of view, it is interesting that finite convergence of this method to the optimal solution of the Lagrangean dual can be shown. Therefore, in the case of the TSP, this approach provides a finite procedure for computing the exact subtour elimination lower bound. But the running time is considerable, due to the many costly evaluations of $L(\lambda)$ in the line search.

6.3. LP relaxations

The first successful attempt to solve a 'large' instance of the traveling salesman problem is reported in the seminal paper by Dantzig, Fulkerson & Johnson [1954] who solved a 48-city instance. This paper is one of the cornerstones on which much of the methodology of using heuristics, linear programming and separation to attack combinatorial optimization problems is founded. It took a long time for the ideas of Dantzig, Fulkerson and Johnson to be pursued again, and this must probably be attributed to the fact that the systematic way of using cutting planes in integer programming, which had been put on a solid basis by the work of Gomory [1958, 1960, 1963], was not successful in practice. An important development is the systematic study of the traveling salesman polytope described in the previous section. Grötschel used the knowledge of the polytope to solve a 120 city instance to optimality, using IBM's linear programming package MPSX to optimize over relaxations of the traveling salesman polytope, visually inspecting the fractional solutions, adding violated facet defining inequalities, resolving etc., until the optimal solution was the incidence vector of a Hamiltonian cycle. He needed 13 iterations of this process (see Grötschel [1977, 1980]). Since the early eighties, more insight into the facial structure of the traveling salesman polytope and improved cutting plane based algorithms developed gradually.

On the computational side, the next steps were the papers Padberg & Hong [1980] and Crowder & Padberg [1980].

In the first paper, a primal cutting plane approach is used to obtain good bounds on the quality of solutions generated in the following way. An initial Hamiltonian cycle is determined by the Lin–Kernighan heuristic, and the first linear programming problem is given by (5.5) and (5.7) (i.e., by the fractional 2-matching relaxation). The initial basis corresponds to the initial Hamiltonian cycle. Then a pivoting variable is selected by the steepest edge criterion. If the adjacent basic solution after the pivot is the incidence vector of a Hamiltonian cycle, the pivot is carried out, and the algorithm proceeds with the new Hamiltonian cycle. Otherwise, one tries to identify a violated inequality which is satisfied with equality by the current solution but violated by the adjacent fractional solution. If such an inequality is found, it is appended to the current LP, a degenerate pivot is made on the selected pivoting variable, and the next pivoting variable is selected.

Otherwise, the current (final) linear program is solved to optimality in order to obtain a lower bound on the length of the shortest Hamiltonian cycle. Out of 74 sample problems ranging from 15 to 318 cities, 54 problems could be solved to optimality in this way. The whole algorithm was written by the authors including an implementation of the simplex algorithm in rational arithmetic.

In the second paper, IBM's MPSX LP-package is used instead, and IBM's MPSX-MIP integer programming package is used to find the incidence vector of an optimal solution as follows. MIP is applied to the final LP to find an optimal integral solution. If this solution is the incidence vector of a Hamiltonian cycle, this cycle is returned as the optimal solution. Otherwise the solution is necessarily a collection of subtours, and the corresponding subtour elimination inequalities are appended to the integer program and the process is iterated. Thus for the first time a fully automatic computer program involving no human interaction was available to solve traveling salesman problems by heuristics, linear programming, separation and enumeration in the spirit of Dantzig, Fulkerson and Johnson. Using their computer code, the authors were able to solve all 74 sample problems to optimality. The 318 city instance was solved in less than an hour of CPU time on an IBM 370/168 computer under the MVS operating system.

A similar, yet more sophisticated approach using MPSX/MIP is described in Grötschel & Holland [1991]. They use a (dual) cutting plane procedure to obtain a tight linear programming relaxation. Then they proceed as Crowder and Padberg. An additional important enhancement is the use of sparse graphs, a prerequisite for attacking larger problem instances. Furthermore, improved separation routines are used, partly based on new results by Padberg & Rao [1982] on the separation of 2-matching inequalities as described in the previous section. The code was used to solve geometric instances with up to 666 nodes and random instances with up to 1000 nodes. Depending on parameter settings, the former took between 9 and 16 hours of CPU time, and the latter between 23 and 36 minutes of CPU time on an IBM 3081D under the operating system VM/CMS. Random problems where the edge weights are drawn from a uniform distribution appear to be much easier than geometric instances. From a software engineering point of view, the codes by Padberg and Hong, Crowder and Padberg, and Grötschel and Holland had the advantage that any general purpose branch and bound software for integer programming could be used to find integer solutions. However, if such an integer solution contained subtours, the corresponding subtour elimination inequalities were added to the LP-relaxation and the branch and bound part was started from scratch, again using a fixed linear programming relaxation in each node of the branch and bound tree.

On the other hand, the iterated 'solving from scratch', whenever the addition of further subtour elimination inequalities was necessary, is a definite disadvantage. An even bigger drawback is the fact that the possibility of generating further globally valid cutting planes in non-root nodes of the branch and bound tree is not utilized. Furthermore, general purpose branch and bound software typically allows very little influence on the optimization process such as variable fixing based on structural properties of the problem. Such disadvantages are eliminated by the natural idea of applying the cutting plane algorithm with globally valid (preferably facet defining) inequalities in every node of the enumeration tree.

Such an approach was first published for the linear ordering problem by Grötschel, Jünger & Reinelt [1984]. In Padberg & Rinaldi [1987] a similar approach was outlined for the TSP and called 'branch and cut'. By reporting the solution to optimality of three large unsolved problems of 532, 1002, and 2392 cities, it was shown for the first time in this paper how the new approach could be successfully used to solve instances of the traveling salesman problem that probably could not be solved with other available techniques.

The first state-of-the-art branch and cut algorithm for the traveling salesman problem is the algorithm published in Padberg & Rinaldi [1991]. The major new features of the Padberg-Rinaldi algorithm are the branch and cut approach in conjunction with the use of column/row generation/deletion techniques, sophisticated separation procedures and an efficient use of the LP optimizer. The LPs are solved using the packages XMP of Marsten [1981] on the DIGITAL computers microVAX II, VAX 8700 and VAX 780, as well as on the Control Data computer CYBER 205, and the experimental version of the code OSL by John Forrest of IBM Research on an IBM 3090/600 supercomputer. With the latter version of the code, the 2392-node instance is solved to optimality in about 4.3 hours of CPU time.

The rest of this section is devoted to a more detailed outline of such a state-of-the-art branch and cut algorithm for the TSP. Our description is based on the original implementation of Padberg & Rinaldi [1991] and a new implementation of Jünger, Reinelt & Thienel [1994]. We use the terminology and the notation of the latter paper in which a new component is added to the Padberg-Rinaldi algorithm: a procedure that exploits the LP solution of the current relaxation to improve the current heuristic solution.

As in the original version of the algorithm, in the implementation described here, subtour elimination, 2-matching, comb and clique-tree inequalities are used as cutting planes. Since we use exact separation of subtour elimination inequalities, all integral LP solutions are incidence vectors of Hamiltonian cycles, as soon as no more subtour elimination inequalities are generated.

In our description, we proceed as follows. First we describe the enumerative part of the algorithm, i.e., we discuss in detail how branching and selection operations are implemented. Then we explain the work done in a subproblem of the enumeration. Finally we explain some important global data structures. There are two major ingredients of the processing of a subproblem, the computation of local lower and global upper bounds. The lower bounds are produced by performing an ordinary cutting plane algorithm for each subproblem. The upper bounds are obtained by exploiting fractional LP solutions in the construction of Hamiltonian cycles which are improved by heuristics.

The branch and cut algorithm for the TSP is outlined in the flowchart of Figure 19. Roughly speaking, the two leftmost columns describe the cutting plane phases within a single subproblem, the third column shows the preparation and execution of a branching operation, and in the rightmost column, the fathoming

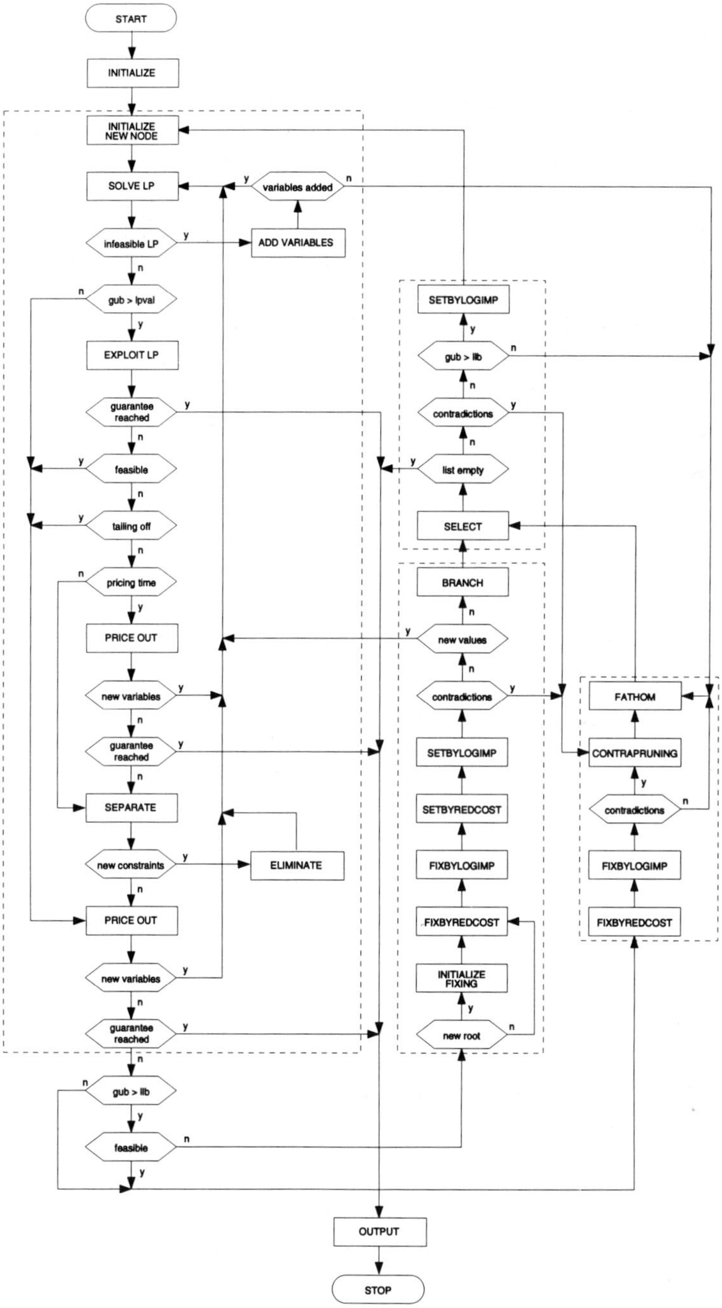

Fig. 19. Flowchart of the branch and cut algorithm.

of a subproblem is performed. We give informal explanations of all steps of the flowchart.

Before going into detail, we have to define some terminology. Since in a branching step two new subproblems are generated, the set of all subproblems can be represented by a binary tree, which we call the *branch and cut tree*. Hence we call a subproblem also a *branch and cut node*. We distinguish between three different types of branch and cut nodes. The node which is currently processed is called the *current branch and cut node*. The other unfathomed leaves of the branch and cut tree are called the *active nodes*. These are the nodes which still must be processed. Finally, there are the already processed *nonactive nodes*.

The terms *edge* of a graph and *variable* in the integer programming formulation are used interchangeably, as they are in a one to one correspondence. Each variable (edge) has one of the following status values during the computation: *atlowerbound*, *basic*, *atupperbound*, *settolowerbound*, *settoupperbound*, *fixedtolowerbound*, *fixedtoupperbound*. When we say that a variable is *fixed* to zero or one, it means that it is at this value for the rest of the computation. If it is *set* to zero or one, this value remains valid only for the current branch and cut node and all branch and cut nodes in the subtree rooted at the current one in the branch and cut tree. The meanings of the other status values are obvious: As soon as an LP has been solved, each variable which has not been fixed or set receives one of the values *atlowerbound*, *basic* or *atupperbound* by the revised simplex method with lower and upper bounds.

The global variable *lpval* always denotes the optimal value of the last LP that has been solved, the global variable *llb* (local lower bound) is a lower bound for the currently processed node, the global variable *gub* (global upper bound) gives the value of the currently best known solution. The minimal lower bound of all active branch and cut nodes and the current branch and cut node is the global lower bound *glb* for the whole problem, whereas the global variable *rootlb* is the lower bound found while processing the root node of the remaining branch and cut tree. As we will see later, *lpval* and *llb* may differ, because we use sparse graph techniques, i.e., the computation of the lower bounds is processed only on a small subset of the edges and only those edges are added which are necessary to guarantee the validity of the bounds on the complete graph.

By the *root of the remaining branch and cut tree* we denote the highest common ancestor in the branch and cut tree of all branch and cut nodes which still must be processed. The values of *gub* and *glb* can be used to terminate the computation as soon as the guarantee requirement is satisfied. As in branch and bound terminology we call a subproblem *fathomed*, if the local lower bound *llb* of this subproblem is greater than or equal to the global upper bound *gub* or the subproblem becomes infeasible (e.g., branching variables have been set in a way that the graph does not contain a Hamiltonian cycle). Following TSPLIB [Reinelt, 1991a, b] all distances are integers. So all terms of the computation which express a lower bound may be rounded up, e.g., one can fathom a node with global upper bound *gub* and local lower bound *llb*, if $\lceil llb \rceil \geq gub$. Since this is only correct for the distances defined in TSPLIB we neither outline this feature in the flowchart nor in the following explanations.

The algorithm consists of three different parts: The enumerative frame, the computation of upper bounds and the computation of lower bounds. It is easy to identify the boxes of the flowchart of Figure 18 with the dashed boxes of the flowchart of Figure 19. The upper bounding is done in EXPLOIT LP, the lower bounding in all other parts of the dashed bounding box. There are three possibilities to enter the bounding part and three to leave it. Normally we perform the bounding part after the startup phase in INITIALIZE or the selection of a new subproblem in SELECT. Furthermore it is advantageous, although not necessary for the correctness of the algorithm, to reenter the bounding part if variables are fixed or set to new values by FIXBYLOGIMP or SETBYLOGIMP, instead of creating two new subproblems in BRANCH. Normally, the bounding part is left if no variables are added by PRICE OUT. In this case we know that the bounds for the just processed subproblem are valid for the complete graph. Sometimes an infeasible subproblem can be detected in the bounding part. This is the second way to leave the bounding part after ADD VARIABLES. We also stop the computations of bounds and output the currently best known solution, if our guarantee requirement is satisfied (*guarantee reached*), but we ignore this, if we want to find the optimal solution.

6.4. Enumerative frame

In this paragraph we explain the implementation of the implicit enumeration. Nearly all parts of this enumerative frame are not TSP specific. Hence it is easy to adapt it to other combinatorial optimization problems.

INITIALIZE

The problem data is read. We distinguish between several problem types as defined in Reinelt [1991a, b] for the specifications of TSPLIB data. In the simplest case, all edge weights are given explicitly in the form of a triangular matrix. In this case very large problems are prohibitive because of the storage requirements for the problem data. But very large instances are usually generated by some algorithmic procedure, which we utilize. The most common case is the metric TSP instance, in which the nodes defining the problem correspond to points in d-dimensional space and the distance between two nodes is given by some metric distance between the respective points. Therefore, distances can be computed as needed in the algorithm and we make use of this fact in many cases.

In practical experiments it has been observed that most of the edges of an optimal Hamiltonian cycle connect near neighbors. Often, optimal solutions are contained in the 10-nearest neighbor subgraph of K_n. In any case, a very large fraction of the edges contained in an optimal Hamiltonian cycle are already contained in the 5-nearest neighbor subgraph of K_n. Depending on two parameters k_s and k_r we compute the k_s-nearest neighbor subgraph and augment it by the edges of a Hamiltonian cycle found by a simple heuristic so that the resulting *sparse graph* $G = (V, E)$ is Hamiltonian. Using this solution, we can also initialize the value of the global upper bound *gub*. We also compute a list of edges which have to be added to E to contain the k_r-nearest neighbor subgraph. These edges form the *reserve graph*, which is used in PRICE OUT and ADD VARIABLES. We will start working on G, adding and deleting edges (variables) dynamically during the optimization process. We refer to the edges in G as *active* edges and to the other edges as *nonactive* edges. All global variables are initialized. The set of active branch and cut nodes is initialized as the empty set. Afterwards the root node of the complete branch and cut tree is processed by the bounding part.

BOUNDING

The computation of the lower and upper bounds will be outlined in Section 6.5. We continue the explanation of the enumerative frame at the ordinary exit of the bounding part (at the end of the first column of the dashed bounding box). In this case it is guaranteed that the lower bound on the sparse graph *lpval* becomes a local lower bound *llb* for the subproblem on the complete graph.

Since we use exact separation of subtour elimination inequalities, all integral LP solutions are incidence vectors of Hamiltonian cycles, as soon as no more subtour elimination inequalities are generated.

We check if the current branch and cut node cannot contain a better solution than the currently best known one ($gub \leq llb$). If this is the case, the current branch and cut node can be fathomed (rightmost column of the flowchart), and if no further branch and cut nodes have to be considered, the currently best known solution must be optimal (*list empty* after SELECT). Otherwise we have to check if the current LP-solution is already a Hamiltonian cycle. If this is the case (*feasible*) we can fathom the node (possibly giving a new value to *gub*), otherwise we prepare a branching operation and the selection of another branch and cut node for further processing (third column of the flowchart).

INITIALIZE FIXING, FIXBYREDCOST

If we are preparing a branching operation, and the current branch and cut node is the root node of the currently remaining branch and cut tree, the reduced cost of the nonbasic active variables can be used to fix them forever at their current values. Namely, if for an edge e the variable x_e is nonbasic and the reduced cost is r_e, we can fix x_e to zero if $x_e = 0$ and $rootlb + r_e > gub$ and we can fix x_e to one if $x_e = 1$ and $rootlb - r_e > gub$.

During the computational process, the value of *gub* decreases, so that at some later point in the computation, one of these criteria can be satisfied, even though it is not satisfied at the current point of the computation. Therefore, each time when we get a new root of the remaining branch and cut tree, we make a *list of candidates for fixing* of all nonbasic active variables along with their values (0 or 1) and their reduced costs and update *rootlb*. Since storing these lists in every node, which might eventually become the root node of the remaining active nodes in the branch and cut tree, would use too much memory space, we process the complete bounding part a second time for the node, when it becomes the new root. If we could initialize the constraint system for the recomputation by those constraints, which were present in the last LP of the first processing of this node, we would need only a single call of the simplex algorithm. However, this would require too much memory. So we initialize the constraint system with the constraints of the last solved LP. As some facets are separated heuristically, it is not guaranteed that we can achieve the same local lower bound as in the previous bounding phase. Therefore we not only have to use the reduced costs and status values of the variables of this recomputation, but also the corresponding local lower bound as *rootlb* in the subsequent calls of the routine FIXBYREDCOST.

If we initialize the basis by the variables contained in the best known Hamiltonian cycle and call the primal simplex algorithm, we can avoid phase 1 of the simplex method. Of course this recomputation is not necessary for the root of the complete branch and cut tree, i.e., the first processed node. The list of candidates for fixing is checked by the routine FIXBYREDCOST whenever it has been freshly compiled or the value of the global upper bound *gub* has improved since the last call of FIXBYREDCOST.

FIXBYREDCOST may find that a variable can be fixed to a value opposite to the one it has been set to (*contradiction*). This means that earlier in the computation, somewhere on the path of the current branch and cut node to the root of the branch and cut tree, we have made an unfavorable decision which led to this setting either directly in a branching operation or indirectly via SETBYREDCOST or SETBYLOGIMP (to be discussed below). Contradictions are handled by CONTRAPRUNING, whenever FIXBYREDCOST has set *contradiction* to *true* using such a condition.

Before starting a branching operation and if no contradiction has occurred, some fractional (basic) variables may have been fixed to new values (0 or 1). In this case we solve the new LP rather than performing the branching operation.

FIXBYLOGIMP

After variables have been fixed by FIXBYREDCOST, we call FIXBYLOGIMP. This routine tries to fix more variables by logical implication as follows: If two edges incident to a node v have been fixed to 1, all other edges incident to v can be fixed to 0 (if not fixed already). As in FIXBYREDCOST, contradictions to previous variable settings may occur. Upon this condition the variable *contradiction* is set to *true*. If variables are fixed to new values, we proceed as explained in FIXBYREDCOST.

In principle also fixing or setting variables to zero could have logical implications. If all incident edges of a node but two are fixed or set to zero, these two edges can be fixed or set to one. However, as we work on sparse graphs, this occurs quite rarely so that we omit this check.

SETBYREDCOST

While fixings of variables are globally valid for the whole computation, variable settings are only valid for the current branch and cut node and all branch and cut nodes in the subtree rooted at the current branch and cut node. SETBYREDCOST sets variables by the same criteria as FIXBYREDCOST, but based on the local reduced cost and the local lower bound *llb* of the current subproblem rather than 'globally valid reduced cost' and the lower bound of the root node *rootlb*. Contradictions are possible if in the meantime the variable has been fixed to the opposite value. In this case we go to CONTRAPRUNING. The variable settings are associated with the current branch and cut node, so that they can be undone when necessary. All set variables are inserted together with the branch and cut node into the hash table of the set variables, which is explained in Section 6.6.

SETBYLOGIMP

This routine is called whenever SETBYREDCOST has successfully fixed variables, as well as after a SELECT operation. It tries to set more variables by logical implication as follows: If two edges incident to a node v have been set or fixed to 1, all other edges incident to v can be set to 0 (if not fixed already). As in SETBYREDCOST, all settings are associated with the current branch and cut node. If variables are set to new values, we proceed as explained in FIXBYREDCOST. As in SETBYREDCOST, the set variables are stored in a hash table, see Section 6.6.

After the selection of a new node in SELECT, we check if the branching variable of the father is set to 1 for the selected node. If this is the case, SETBYLOGIMP may also set additional variables.

BRANCH

Some fractional variable is chosen as the branching variable and, accordingly, two new branch and cut nodes, which are the two sons of the current branch and cut node, are created and added to the set of active branch and cut nodes. In the first son the branching variable is set to 1 in the second one to 0. These settings are also registered in the hash table.

SELECT

A branch and cut node is selected and removed from the set of active branch and cut nodes. Our strategy is to select the candidate with the minimal local lower bound, a variant of the 'best first search' strategy which compares favorably with commonly used strategies such as 'depth first search' or 'breadth first search'. If the list of active branch and cut nodes is empty, we can conclude optimality of the best known Hamiltonian cycle. Otherwise we start processing the selected node. After a successful selection, variable settings have to be adjusted according to the information stored in the branch and cut tree. If it turns out that some variable must be set to 0 or 1, yet has been fixed to the opposite value in the meantime, we have a contradiction similar as discussed above. In this case we prune the branch and cut tree accordingly by going to CONTRAPRUNING and fathom the node in FATHOM. If the local lower bound *llb* of the selected node is greater than or equal to the global upper bound *gub*, we fathom the node immediately and continue the selection process. A branch and cut node has pointers to its father and its two sons. So it is sufficient to store a set variable only once in any path from the root to a leaf in the branch and cut tree. If we select a new problem, i.e., proceed with the computation at some leaf of the tree, we only have to determine the highest common ancestor of the old node and the new leaf, reset the set variables on the path from the old node to the common ancestor and set the variables on the path from the common ancestor to the new leaf.

CONTRAPRUNING

Not only the current branch and cut node, where we have found the contradiction, can be deleted from further consideration, but all active nodes with the same 'wrong' setting can be fathomed. Let the variable with the contradiction be e. Via the hash table of the set variables we can efficiently determine all branch and cut nodes where e has been set. If in a branch and cut node b the variable e is set to the 'wrong' bound we remove all active nodes (unfathomed leaves) in the subtree below b from the set of active nodes.

FATHOM

If for a node the global upper bound *gub* does not exceed the local lower bound *llb*, or a contradiction occurred, or an infeasible branch and cut node has been generated, the current branch and cut node is deleted from further consideration. Even though a node is fathomed, the global upper bound *gub* may have changed during the last iteration, so that additional variables may be fixed by FIXBYREDCOST and FIXBYLOGIMP. The fathoming of nodes in FATHOM and CONTRAPRUNING may lead to a new root of the branch and cut tree for the remaining active nodes.

OUTPUT

The currently best known Hamiltonian cycle, which is either optimal or satisfies the desired guarantee requirement, is written to an output file.

6.5. Computation of lower and upper bounds

The computation of lower bounds consists of all elements of the dashed bounding box except EXPLOIT LP, where the upper bounds are computed.

During the whole computation, we keep a *pool* of active and nonactive facet defining inequalities of the traveling salesman polytope. The *active inequalities* are the ones in the current LP and are both stored in the pool and in the constraint matrix, whereas the *inactive inequalities* are only present in the pool. An inequality becomes inactive, if it is nonbinding in the last LP solution. When required, it is easily regenerated from the pool and made active again later in the computation. The pool is initially empty. If an inequality is generated by a separation algorithm, it is stored both in the pool and added to the constraint matrix.

INITIALIZE NEW NODE

Let A_G be the node-edge incidence matrix corresponding to the sparse graph G. If the node is the root node of the branch and cut tree the LP is initialized to

$$\begin{aligned} \text{minimize} \quad & cx \\ \text{subject to} \quad & A_G x = \mathbf{2} \\ & \mathbf{0} \le x \le \mathbf{1} \end{aligned}$$

and the feasible basis obtained from the initial Hamiltonian cycle is used as a starting basis. In subsequent subproblems, we initialize the constraint matrix by the equations induced by the node-edge incidence matrix of the sparse graph and by the inequalities which were active when the last LP of the father of the branch and cut node was solved. These inequalities can be regenerated from the pool. Since the final basis of the father is dual feasible for the initial LP of its sons, we start with this basis to avoid phase 1 of the simplex method. Every column of a nonbasic set or fixed variable is removed from the constraint matrix and if its status value is *settoupperbound* or *fixedtoupperbound*, the right hand side of every constraint containing it has to be adjusted. The corresponding coefficient of the objective function must be added to the optimal value returned by the simplex algorithm in order to get the correct value of the variable *lpval*. Set or fixed basic variables are not deleted, because this would lead to an infeasible basis and require phase 1 of the simplex method. We perform the adjustment of these variables by adapting their upper and lower bounds.

SOLVE LP

The LP is solved, either by the two phase primal simplex method, if the current basis is neither primal nor dual feasible, by the primal simplex method, if the basis is primal feasible (e.g., if variables have been added) or by the dual simplex method if the basis is dual feasible (e.g., if constraints have been added or more variables have been set). As LP solver we use CPLEX by R.E. Bixby (see CPLEX [1993]).

If the LP has no feasible solution we go to ADD VARIABLES, otherwise we proceed downward in the flowchart.

ADD VARIABLES

Variables have to be added to the sparse graph if indicated by the reduced costs (handled by PRICE OUT) or if the current LP is infeasible. The latter may be caused by three reasons. First, equations may not be satisfiable because the variables associated with all but at most one edge incident to a node v in the sparse graph may be fixed or set to 0. Such an infeasibility can either be removed by adding an additional edge incident to v, or, if all edges are present already, we can fathom the branch and cut node.

Second, suppose that all equations are satisfiable, yet some active inequality has a void left hand side, since all involved variables are fixed or set, but is violated. As is clear from our strategy for variable fixings and settings, this also means that the branch and cut node is fathomed, since all constraint coefficients are nonnegative in our implementation.

Finally, neither of the above conditions may apply, and the infeasibility is detected by the LP solver. In this case we perform a pricing step in order to find out if the dual feasible LP solution is dual feasible for the entire problem. We check for variables that are not in the current sparse graph (i.e., are assumed to be at their lower bound 0) and have negative reduced cost. Such variables are added to the current sparse graph. An efficient way of the computation of the reduced costs is outlined in PRICE OUT.

If variables have been added, we solve the new LP. Otherwise, we try to make the LP feasible by a more sophisticated method. The LP value *lpval*, which is the objective function value corresponding to the dual feasible basis where primal infeasibility is detected, is a lower bound for the objective function value obtainable in the current branch and cut node. So if $lpval \geq gub$, we can fathom the branch and cut node.

Otherwise, we try to add variables that may restore feasibility. First we mark all infeasible variables, including negative slack variables.

Let e be a nonactive variable and r_e be the reduced cost of e. We take e as a candidate only if $lpval + r_e \leq gub$. Let B be the basis matrix corresponding to the dual feasible LP solution, at which the primal infeasibility was detected. For each candidate e let a_e be the column of the constraint matrix corresponding to e and solve the system $B\overline{a}_e = a_e$. Let $\overline{a}_e(b)$ be the component of $\overline{a}_e$ corresponding to basic variable x_b. Increasing x_e reduces some infeasibility if one of the following holds.

- x_b is a structural variable (i.e., corresponding to an edge of G) and $x_b < 0$ and $\overline{a}_e(b) < 0$ or $x_b > 1$ and $\overline{a}_e(b) > 0$;
- x_b is a slack variable and $x_b < 0$ and $\overline{a}_e(b) < 0$.

In such a case we add e to the set of active variables and remove the marks from all infeasible variables whose infeasibility can be reduced by increasing x_e. We do this in the same hierarchical fashion as in the the procedure PRICE OUT that is described below.

If variables can be added, we regenerate the constraint structure and solve the new LP, otherwise we fathom the branch and cut node. Note that all systems of linear equations that have to be solved have the same matrix B, and only the right hand side a_e changes. We utilize this by computing a factorization of B only once, in fact, the factorization can be obtained from the LP solver for free. For further details on this algorithm, see Padberg & Rinaldi [1991].

EXPLOIT LP

We check if the current LP solution is the incidence vector of a Hamiltonian cycle. If this is the case, the variable *feasible* is set to *true*. Otherwise, the LP solution is exploited in the construction of a Hamiltonian cycle. To this end we use the following heuristic. Edges are sorted according to decreasing values in the current LP solution. This list is scanned and edges become part of the Hamiltonian cycle if they do not produce a subtour. Then the savings heuristic as described in Section 4 is used to combine the produced system of paths to form a Hamiltonian cycle. Then the Lin–Kernighan heuristic is applied. If the final solution has smaller cost than the currently best known one, it is made the incumbent solution, *upperbound* is updated and *improved* is set to *true*. For details of this step, see Jünger, Reinelt & Thienel [1994].

SEPARATE

This part implements separation for the TSP as described in the previous section. In a first phase, the pool is checked for inactive violated inequalities. If an inactive inequality is violated, it is added to the active set of constraints. While checking the pool, we remove, under certain conditions, all those inequalities from the pool which have been inactive for a long time. If violated inequalities have been added from the pool, we terminate the separation phase. Otherwise, we try to identify new violated constraints as outlined in the previous section, store them as active inequalities in the pool and add them to the LP. For details of the separation process, we refer to the original articles mentioned in Section 5.

ELIMINATE

Before the LP is solved after a successful cutting plane generation phase, all active inequalities which are nonbinding in the current LP solution are eliminated from the constraint structure and marked inactive in the pool. We can safely do this to keep the constraint structure as small as possible, because as soon as the inequality becomes violated in a later cutting plane generation phase, it can be generated anew from the pool (if it has not been removed in the meantime).

PRICE OUT

Pricing is necessary before a branch and cut node can be fathomed. Its purpose is to check if the LP solution computed on the sparse graph is valid for the complete graph, i.e., all nonactive variables 'price out' correctly. If this is not the case, nonactive variables with negative reduced cost are added to the sparse graph and the new LP is solved using the primal simplex method starting with the previous (now primal feasible) basis, otherwise we can update the local lower bound *llb* and possibly the global lower bound *glb*. If the global lower bound has changed, our guarantee requirement might be satisfied and we can stop the computation after the output of the currently best known Hamiltonian cycle.

Although the correctness of the algorithm does not require this, we perform additional pricing steps every k solved LPs (see Padberg & Rinaldi [1991]). The effect is that nonactive variables which are required in a good or optimal Hamiltonian cycle tend to be added to the sparse graph early in the computation.

In a first phase, only the variables in the reserve graph are considered. If the 'partial pricing' considering only the edges of the reserve graph has not added variables, we have to check the reduced costs of all nonactive variables which takes a lot of computational effort. But this second step of PRICE OUT can be processed more efficiently. If our current branch and cut node is the root of the remaining branch and cut tree, we can check if the reduced cost r_e of a nonactive variable e satisfies the relation $lpval + r_e > gub$. In this case we can discard this nonactive candidate edge forever. During the systematic enumeration of all edges of the complete graph, we can make an explicit list of those edges which remain possible candidates. In the early steps of the computation, too many such edges remain, so that we cannot store this list completely with reasonable memory consumption. Rather, we predetermine a reasonably sized buffer and mark the point where the systematic enumeration has to be resumed after considering the edges in the buffer. In later steps of the computation there is a good chance that the complete list fits into the buffer, so that later calls of the pricing routine become much faster than early ones.

To process PRICE OUT efficiently, for each node v a list of those constraints containing v is made. Whenever an edge $e = vw$ is considered, we initialize the reduced cost by c_e, then v's and w's constraint lists are compared, and the value of the dual variable y_f times the corresponding coefficient is subtracted from the reduced cost whenever the two lists agree in a constraint f. The format of the pool, which is explained in Section 6.6, provides us with an efficient way to compute the constraint lists and the coefficients.

6.6. Data structures

A suitable choice of data structures is essential for an efficient implementation of a branch and cut algorithm. This issue is discussed in detail in Jünger, Reinelt & Thienel [1994].

Sparse graphs

In INITIALIZE we select only a very small subset of the edges for our computations: the set of active edges, which remains small during the computations. For the representation of the resulting sparse graph we choose a data structure which saves memory and enables us to efficiently perform the operations scanning all incident edges of a node, scanning all adjacent nodes of a node, determining the endnodes of an edge and adding an edge to the sparse graph.

Branch and cut nodes

Although a subproblem is completely defined by the fixed variables and the variables that are set temporarily, it is necessary to store additional information at each node for an efficient implementation. Every branch and cut node has pointers to its father and sons. A branch and cut node contains the arrays *set* of its set variables and *setstat* with the corresponding status values (*settolowerbound, settoupperbound*). The first variable in this array is the branching variable of the father. There may be further entries to be made in case of successful calls of SETBYREDCOST and SETBYLOGIMP while the node is processed. The set variables of a branch and cut node are all the variables in the arrays *set* of all nodes in the path from the root to the node.

In a branch and cut node we store the local lower bound of the corresponding subproblem. After creation of a new leaf of the tree in BRANCH this is the bound of its father, but after processing the node we can in general improve the bound and update this value.

Of course it would be correct to initialize the constraint system of the first LP of a new selected node with the inequalities of the last processed node, since all generated constraints are facets of STSP. However, this would lead to tedious recomputations, and it is not guaranteed that we can regenerate all heuristically separated inequalities. So it is preferable to store in each branch and cut node pointers to those constraints in the pool, which are in the constraint matrix of the last solved LP of the node. We initialize with these constraints the first LP of each son of that node.

As we use an implementation of the simplex method to solve the linear programs, we store the basis of the last processed LP of each node, i.e., the status values of the variables and the constraints. Therefore we can avoid phase 1 of the simplex algorithm, if we carefully restore the LP of the father and solve this first LP with the dual simplex method. Since the last LP of the father and the first LP of the son differ only by the set branching variable, variables set by SETBYLOGIMP, and variables that have been fixed in the meantime, the basis of the father is dual feasible for the first LP of the son.

Active nodes

In SELECT a node is extracted from the set of active nodes for further processing. Every selection strategy defines an order on the active nodes. The minimal node is the next selected one. The representing data structure must allow efficient implementations of the operations *insert*, *extractmin* and *delete*. The operation *insert* is used after creation of two new branch and cut nodes in BRANCH, *extractmin* is necessary to select the next node in SELECT and *delete* is called if we remove an arbitrary node from the set of active nodes in CONTRAPRUNING. These operations are very well supported by a height balanced binary search tree. We have implemented a red-black tree [Bayer, 1972; Guibas & Sedgewick, 1978; see also Cormen, Leiserson & Rivest, 1989] which provides $O(\log m)$ running time for these operations, if the tree consists of m nodes. Each node of the red-black tree contains a pointer to the corresponding leaf of the branch and cut tree and vice versa.

Temporarily set variables

A variable is either set if it is the branching variable or it is set by SETBYREDCOST or SETBYLOGIMP. In CONTRAPRUNING it is essential to determine efficiently all nodes where a certain variable is set. To avoid scanning the complete branch and cut tree, we apply a hash function to a variable right after setting and store in the slot of the hash table the set variable and a pointer to the corresponding branch and cut node. So it is quick and easy to find all nodes with the same setting by applying an appropriate hashing technique. We have implemented a Fibonacci hash with chaining (see Knuth [1973]).

Constraint pool

The data structure for the pool is very critical concerning running time and memory requirements. It is not appropriate to store a constraint in the pool just as the corresponding row of the constraint matrix, because we also have to know the coefficients of variables which are not active. This is necessary in PRICE OUT, to avoid recomputation from scratch after addition of variables and in INITIALIZE NEW NODE. Such a format would require too much memory. We use a node oriented sparse format. The pool is represented by an array. Each component (constraint) of the pool is again an array, which is allocated dynamically with the required size. This last feature is important, because the required size for a constraint of STSP(n) can range from four entries for a subtour elimination constraint to about $2n$ entries for a comb or a clique-tree inequality.

A subtour elimination inequality is defined by the node set $W = \{w_1, \ldots, w_l\}$. It is sufficient to store the size of this node set and a list of the nodes.

2-matching inequalities, comb inequalities and clique-tree inequalities are defined by a set of handles $\mathcal{H} = \{H_1, \ldots, H_r\}$ and a set of teeth $\mathcal{T} = \{T_1, \ldots, T_k\}$, with the sets $H_i = \{h_{i_1}, \ldots, h_{i_{n_i}}\}$ and $T_j = \{t_{j1}, \ldots, t_{jm_j}\}$. In our pool format a clique-tree inequality with h handles and t teeth is stored as:

$$r, n_1, h_{1_1}, \ldots, h_{1_{n_1}}, \ldots, n_r, h_{r_1} \ldots, h_{r_{n_r}}, k, m_1, t_{1_1}, \ldots, t_{1_{m_1}}, \ldots, m_k, t_{k_1} \ldots, t_{k_{m_k}}$$

For each constraint in the pool, we also store its storage type (subtour or clique-tree).

This storage format of a pool constraint provides us with an easy method to compute the coefficient of every involved edge, even if it is not present in the sparse graph at generation time. In case of a subtour elimination inequality, the coefficient of an edge is 1 if both endnodes of the edge belong to W, otherwise it is zero. The computation of the coefficients of other constraints is straightforward. A coefficient of an edge of a 2-matching inequality is 1 if both endnodes of the edge belong to the handle or to the same tooth, 0 otherwise. Some more care is needed for comb inequalities and clique-tree inequalities. The coefficient of an edge is 2 if both endnodes belong to the same intersection of a handle and a tooth, 1 if both endnodes belong to the same handle or (exclusive) to the same tooth and 0 in all other cases.

Since the pool is the data structure using up the largest amount of memory, only those inactive constraints are kept in the pool, which have been active, when the last LP of the father of at least one active node has been solved. These inequalities are used to initialize the first LP of a new selected node. In the current implementation the maximal number of constraints in the pool is $50n$ for TSP(n). After each selection of a new node we try to eliminate those constraints from the pool which are neither active at the current branch and cut node nor necessary to initialize the first LP of an active node. If, nevertheless, more constraints are generated than free slots of the pool are available, we remove nonactive constraints from the pool. But now we cannot restore the complete LP of the father of an active node. In this case we proceed as in INITIALIZE FIXING to initialize the constraint matrix and to get a feasible basis.

7. Computation

Computational experience with the algorithmic techniques presented in the previous sections has been given along with the algorithms in Section 4 and parts of Sections 5 and 6. In this final section, we would like to report on computational results of linear programming based algorithms, in particular, the branch and cut algorithm, for various kinds of problem instances. We report both on optimal and provably good solutions.

7.1. Optimal solutions

For most instances of moderate size arising in practice, optimal solutions can indeed be found with the branch and cut technique. On the other hand, there are small instances that have not been solved yet.

Some Euclidean instances from TSPLIB

Computational results for a branch and cut algorithm for solving symmetric traveling salesman problems to optimality have been published by Padberg & Rinaldi [1991]. In order to have a common basis for comparison, the performance of their algorithm on a SUN SPARCstation 10/20 for our standard set of test problems defined in Section 4 is presented in Table 7. For each instance we show the number of nodes of the tree (not including the root node), the total number of distinct cuts generated by the separation algorithm, the maximum cardinality of the set of active constraints (including the degree equations), the maximum cardinality of the set of active edges, the number of times the LP solver is executed, the percentage of time spent in computing and improving a heuristic solution, the percentage of time spent by the LP solver, and the overall computation time in seconds.

All the problem instances have been solved with the same setting for the parameters that can be used to tune the algorithm. Tailoring parameters for each instance individually often gives better results. E.g., with a different setting the instance *pr2392* is solved without branching. The fact that all instances of Table 7 are Euclidean is not exploited in the implementation. For other computational results, in particular for non Euclidean instances, see Padberg & Rinaldi [1991].

Further computational results for branch and cut algorithms for solving the TSP to optimality have been reported in Clochard & Naddef [1993], Jünger, Reinelt & Thienel [1994] and Applegate, Bixby, Chvátal & Cook [1994]. In the latter the authors report on the optimal solution of several problem instances from TSPLIB obtained with their branch and cut implementation based on a new separation procedure for comb inequalities (not described in Section 5, since the reference was not available at the time of writing and was added in the proof). The instances were solved on a cluster of workstations, therefore, only very rough estimations of SUN SPARCstation 10 computation time can be given. For instance, the hard instance *ts225* took 5087 branch and cut nodes and about one year of SUN SPARCstation 10 computation time, the currently second largest is *fnl4461* (2092

Table 7
Computation of optimal solutions

Problem	BC	Cuts	Mrow	Mcol	Nlp	% Heu	% LP	Time
lin105	0	50	137	301	10	89.4	8.5	11
pr107	0	111	148	452	19	73.8	23.8	10
pr124	2	421	199	588	74	74.1	16.7	77
pr136	10	786	217	311	102	70.8	14.8	101
pr144	2	273	238	1043	52	71.3	25.3	43
pr152	20	1946	287	2402	371	37.9	44.5	303
u159	0	139	210	395	23	82.2	15.1	17
rat195	16	1730	318	483	217	49.7	26.2	463
d198	2	563	311	1355	66	79.3	13.7	129
pr226	0	296	344	3184	31	72.4	25.9	87
gil262	4	950	409	668	90	74.6	16.1	197
pr264	0	70	305	1246	17	82.5	15.8	47
pr299	18	3387	554	800	281	1.0	89.5	2394
lin318	4	1124	497	875	100	56.8	20.1	344
rd400	54	8474	633	1118	852	36.5	45.1	2511
pr439	92	10427	741	1538	1150	26.5	55.0	3278
pcb442	50	2240	608	895	486	52.4	31.0	530
d493	70	20291	845	1199	1105	13.9	27.8	7578
u574	2	2424	910	1588	140	42.0	30.9	1134
rat575	110	24185	851	1455	1652	21.2	40.1	7666
p654	2	969	870	2833	55	59.7	35.1	449
d657	220	67224	1056	2154	3789	2.1	41.4	37642
u724	40	14146	1112	1962	766	4.5	32.8	9912
rat783	6	2239	1097	1953	126	64.8	25.6	1039
pr1002	20	14713	1605	2781	572	4.0	43.5	18766
pcb1173	324	165276	1686	3362	5953	6.7	61.7	91422
rl1304	46	38772	2101	5305	1377	2.0	84.5	160098
nrw1379	614	226518	1942	3643	7739	7.4	39.0	155221
u1432	2	4996	2044	2956	96	33.6	53.5	1982
pr2392	2	11301	3553	6266	145	23.6	57.9	7056

branch and cut nodes, 1.9 years), and the currently largest is *pla7397* (2247 branch and cut nodes, 4 years). The proofs of optimality are available from the authors.

We do not know of any algorithmic approach other than the polyhedral branch and cut method which is able to solve even moderately sized instances from TSPLIB to optimality.

From the results presented above one may get the erroneous impression that todays algorithmic knowledge is sufficient to solve instances with up to a few thousand cities to optimality. Unfortunately, there are small instances that cannot be solved to optimality in a reasonable amount of time. See, for example, some non Euclidean instances described below. This is not surprising at all since the TSP is an NP-hard combinatorial optimization problem. Still the impression might remain that Euclidean instances of size up to, say, 1000 nodes can be solved routinely to optimality. Also this impression is wrong.

Some difficult Euclidean instances

Already from a quick look of Table 7 it is clear that, unlike in the case of the computation of heuristic solutions, there is a weak correlation between the computational effort and the instance size. Two small Euclidean instances from TSPLIB are not listed in Table 7, namely *pr76* and *ts225*. With the same implementation as used for the results of Table 7, solving *pr76* takes about 405 seconds and 92 nodes of the tree. As far as we know, no algorithm has found a certified optimal solution to *ts225* yet. We report on the computation of a quality guaranteed solution for this problem in Section 7.2. Clochard & Naddef [1993] observe that both these problems have the same special structure that might be the reason for the poor performance of branch and cut algorithms. They propose a possible explanation for why these problems are difficult and describe a generator that produces random Euclidean instances with the same structural property. Applying new separation heuristics for path inequalities combined with an elaborate branching strategy they obtained very encouraging results for the hard instance *pr76*.

Some difficult non Euclidean instances

It is actually not very difficult to create artificially hard instances for a branch and cut algorithm. As an example, take as the objective function of an instance a facet defining inequality for the TSP polytope that is not included in the list of inequalities that the separation procedure can produce. To give numerical examples, we considered the crown inequality described in Section 5. Table 8 shows the computational results for a few instances of this type. The names of these instances have the prefix *cro*.

Another kind of instances that are expected to be difficult are those that arise from testing if a graph is Hamiltonian. To provide difficult numerical examples, we considered some hypohamiltonian graphs that generalize the Petersen graph. A graph is *hypohamiltonian* if it is not Hamiltonian but the removal of any node makes it Hamiltonian. We applied the transformation described in Section 2 to

Table 8
Optimal solutions of non Euclidean instances

Problem	BC	Cuts	Nlp	% Heu	% LP	Time
cro12	38	57	83	9.7	41.0	0
cro16	204	277	390	6.0	40.9	5
cro20	1078	1657	1838	4.3	39.1	34
cro24	4064	10323	8739	3.5	32.8	369
cro28	19996	182028	68010	4.8	21.1	2659
NH58	40	287	276	10.0	55.5	8
NH82	58	489	505	6.3	62.5	20
NH196	294	2800	2817	1.1	69.0	650
H58	0	0	1	0.0	100.0	0
H82	0	0	1	0.0	100.0	1
H196	0	0	1	0.0	100.0	7

Table 9
Optimal solutions of 10,000 city random instances

BC	Cuts	Nlp	% Heu	% LP	% Pricing	Time
2	48	35	21.5	51.0	7.5	9080
22	88	64	16.0	58.8	4.2	9205
10	73	47	10.6	41.9	32.0	11817
0	43	31	8.9	62.7	5.0	7670
46	129	107	16.0	60.6	6.3	11825
52	132	115	4.2	30.4	56.2	22360
20	115	74	8.1	34.8	42.3	16318

make the tests. The results are also listed in Table 8. The instance names have the prefix *NH*. Finally, we added one edge to each graph considered before that makes it Hamiltonian and we ran the test once more. In this case the computation was very fast as can be seen in Table 8, where the modified instances appear with the prefix *H*.

Randomly generated instances

It is common in the literature that the performance of algorithms is evaluated on randomly generated problem instances. This is often due to the fact that real world instances are not available to the algorithm designers. For some combinatorial optimization problems, randomly generated instances are generally hard, for other problems such instances are easy. The symmetric traveling salesman problem seems to fall into the latter category. This is the case when, for example, the distances are drawn from a uniform distribution. To support this claim experimentally, we generated ten 10,000-city instances whose edge weights were taken from a uniform distribution of integers in the range [0, 50000]. We always stopped the computation after 5 hours. Within this time window, seven of them were solved to optimality. Table 9 contains the statistics of the successful runs. Since the computation of reduced costs of nonactive edges took a significant amount of time in some cases, we list the percentage of time spent for this in an extra column called '% Pricing'. The unaccounted percentage of the time is essentially spent in the initialization process. In all cases separation took negligible time.

However, in the Euclidean case, we could not observe a significant difference in difficulty between real-world and randomly created instances, whose coordinates are uniformly distributed on a square.

Instances arising from transformations

Recently Balas, Ceria & Cornuéjols [1993] reported on the solution of a difficult 43-city asymmetric TSP instance, which arises from a scheduling problem of a chemical plant. They solved the problem in a few minutes of a SUN SPARCstation 330 with a general purpose branch and cut algorithm that does no substantial exploitation of the structural properties of the asymmetric TSP. They also tried to solve the problem with a special purpose branch and bound algorithm for the asymmetric TSP, based on a additive bounding procedure described in Fischetti & Toth [1992], with an implementation of the authors. This algorithm could not find an optimal solution within a day of computation on the same computer. We transformed the asymmetric TSP instance to a symmetric one having 86 nodes, using the transformation described in Section 2 and solved it in less than a minute using only subtour elimination inequalities.

In a paper about a polyhedral approach to the rural postman problem, Corberán & Sanchis [1991] describe two problem instances which are based on the street map of the city of Albaida (Valencia). The two instances are obtained by declaring two different randomly chosen sets of edges as required. The underlying graph has 171 edges and 113 nodes, which represent all streets and intersections of the streets, respectively. The first instance has 99 required edges giving rise to 10 connected components of required edges. The second has 83 required edges in 11 connected components. We applied the transformation described in Section 2, thus producing TSP instances of 198 and 176 nodes, respectively. The solution time was 89 seconds for the first and 22 seconds for the second instance.

Combinatorial optimization problems arising in the context of the control of plotting and drilling machines are described in Grötschel, Jünger & Reinelt [1991]. While the drilling problems lead directly to TSP instances, the plotting problem is modeled as a sequence of Hamiltonian path and 'rural postman path' problems. One of the problem instances is shown in Figure 20.

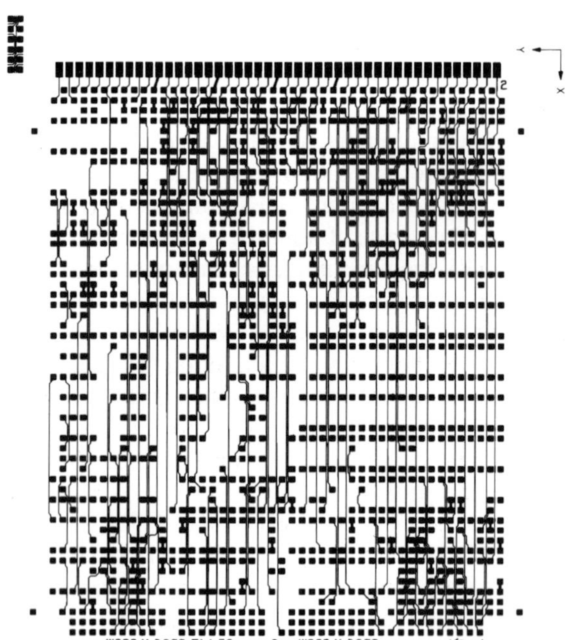

Fig. 20. Mask for a printed circuit board.

Table 10
Optimal solutions for mask plotting rural postman instances

Nre	BC	Time
10	0	0
87	2	28
258	224	9458

We use this mask to demonstrate the optimal solution of the three rural postman instances contained in it. The biggest of them has 258 required edges which correspond to the thin electrical connections between squares, the smallest of 10 required edges to the thick connections. The third instance has 87 required edges and corresponds to drawing the letters and digits at the bottom of the mask. Since the movements of the light source (see Section 3) are carried out by two independent motors in horizontal and vertical directions, we choose the Maximum metric (L_∞) for distances between points. (The mask gives also rise to two TSP instances of 45 and 1432 nodes, the Euclidean version of the latter is contained in the TSPLIB under the name *u1432*.) We solve the three rural postman instances independently, starting and ending each time at an origin outside the mask, so in addition to the required edges we have one required node in each case. Table 10 gives the statistics, with a column labeled 'Nre' for the number of required edges. All nodes except the origin have exactly one incident required edge in all three instances, so that the number of nodes in the TSP instance produced by the transformation is always 2Nre + 1.

7.2. *Provably good solutions*

A branch and cut algorithm as outlined in the previous section produces a sequence of increasing lower bounds as well as a sequence of Hamiltonian cycles of decreasing lengths. Therefore, at any point during the computation we have a solution along with a quality guarantee. Looking more closely at the optimization process we observe that quality guarantees of, say, 5% are obtained quickly whereas it takes a very long time to close the last 1%.

A typical example of this behavior is shown in Figure 21 for the problem *pcb442*. The jumps in the lower bounds are due to the fact that the validity of the LP-value as a global lower bound for the length of a shortest Hamiltonian cycle is only guaranteed after a pricing step in which all nonactive variables price out correctly. The lower bound obtained after about 17 seconds is slightly increasing over time, although this is not visible in the picture. After about 10 seconds, a solution is found which can be guaranteed to deviate at most 5.220% from the optimum. At the end of the root branch and cut node, the quality guarantee is 0.302%. (For this and the following experiments we have disabled the enumerative part of the algorithm. The implementation used here is the one by Jünger, Reinelt & Thienel [1994] using the CPLEX LP-software.)

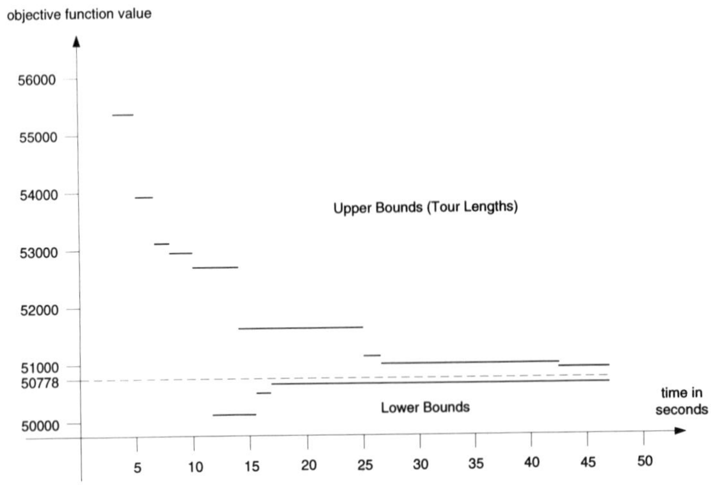

Fig. 21. Gap versus time plot for *pcb442*.

The phenomenon depicted in Figure 21 is indeed typical as the computational results in Table 11 show. Here for all the problems in our list we show the number of LPs solved (before the enumerative part would have been entered), the computation time in seconds, the guaranteed quality in percent, and the actual quality (in terms of the deviation of the known optimal solution) in percent.

Our approach for computing solutions of certified good quality fails miserably on the artificial Euclidean instance *ts225*. Table 12 shows the number of branch and cut nodes (BC) and the lower bounds (LB) after 200, 400, 600, 800 and 1000 minutes of computation. A Hamiltonian cycle of length 126643 (which we believe is optimal) is found after 72 minutes. No essential progress is made as the computation proceeds.

On the other hand, large real world instances can be treated successfully in this framework. As an example, we consider the Euclidean TSPLIB instance *d18512* whose nodes correspond to cities and villages in Germany. This instance was presented by Bachem & Wottawa [1991] along with a Hamiltonian cycle of length 672,721 and a lower bound on the value of an optimal solution of 597,832. Considering only subtour elimination and simple comb inequalities, we ran our standard implementation to the end of the root node computation, and obtained a Hamiltonian cycle of length 648,093 and a lower bound of 644,448 in 1295 minutes which results in a quality guarantee of about 0.57%. This Hamiltonian cycle is shown in Figure 22. Even using only subtour elimination inequalities, we obtained a lower bound of 642,082, i.e., a quality guarantee of less than 1%. In both cases we solved the first LP by the barrier method which was recently added to the CPLEX software.

When the size of the instance gets even larger, memory and time consumption prohibit the application of our method. For very large Euclidean instances, Johnson [1992] reports tours found by his implementation of a variant of the

Table 11
Computational results without branching

Problem	Nlp	Time	Guarantee	Quality
lin105	9	1	0.000	0.000
pr107	12	1	0.000	0.000
pr124	18	4	1.269	0.078
pr136	14	4	0.698	0.150
pr144	17	5	0.396	0.360
pr152	71	13	0.411	0.000
u159	21	7	0.202	0.000
rat195	73	60	0.430	0.130
d198	44	34	0.297	0.051
pr226	24	11	0.029	0.000
gil262	47	30	0.439	0.170
pr264	30	14	0.026	0.000
pr299	99	81	0.876	0.280
lin318	80	105	0.471	0.380
rd400	49	65	0.406	0.100
pr439	74	156	0.948	0.200
pcb442	32	39	0.302	0.185
d493	61	123	0.216	0.069
u574	86	173	0.182	0.073
rat575	60	128	0.444	0.207
p654	55	121	0.169	0.104
d657	80	248	0.779	0.033
u724	66	171	0.448	0.227
rat783	61	190	0.174	0.057
pr1002	110	485	0.249	0.024
pcb1173	92	520	0.361	0.030
rl1304	144	1239	1.025	0.421
nrw1379	92	736	0.386	0.290
u1432	132	1302	0.981	0.883
pr2392	148	3199	1.011	0.790

Table 12
Lower bounds for *ts225*

Time	200 min	400 min	600 min	800 min	1000 min
BC	2300	4660	6460	7220	8172
LB	123437	123576	123629	123642	123656

Lin–Kernighan heuristic, together with lower bounds obtained with a variant of the 1-tree relaxation method described above, which constitute excellent quality guarantees. Among the instances he considered are the TSPLIB instances *pla33810* and *pla85900*. For *pla33810*, he reports a solution of length 66,138,592 and a lower bound of 65,667,327. We applied a simple strategy for this instance. Trying to exploit the clusters in the problem data, we preselected a set of subtour elimination inequalities, solved the resulting linear program

Table 13
Lower bounds for *pla33810*

# Subtours	0	4	466	1114
Lower bound	65,354,778	65,400,649	65,579,139	65,582,859
Time	51,485	36,433	47,238	104,161

Fig. 22. A 0.0057-guaranteed solution of *d18512*.

containing them plus the degree equations on the Delaunay graph, priced out the nonactive edges and resolved until global optimality on the relaxation was established. As LP-solver, we used the program LOQO of Vanderbei [1992], because we found the implemented interior point algorithm superior to the simplex method. Table 13 shows the results for different sets of subtours. The implementation is rather primitive, the running time can be improved significantly.

7.3. Conclusions

In the recent years many new algorithmic approaches to the TSP (and other combinatorial optimization problems) have been extensively discussed in the literature. Many of them produce solutions of surprisingly good quality. However, the quality could only be assessed because optimal solutions or good lower bounds were known.

When optimization problems arise in practice we want to have confidence in the quality of the solutions. Quality guarantees become possible by reasonably efficient calculations of lower bounds. The branch and cut approach meets the goals of simultaneously producing good solutions as well as reasonable quality guarantees.

We believe that practical problem solving does not consist only of producing 'probably good' but *provably good* solutions.

Acknowledgements

We are grateful to Martin Grötschel, Volker Kaibel, Denis Naddef, George Nemhauser, Peter Störmer, Laurence Wolsey, and an anonymous referee who took the time to read an earlier version of the manuscript and made many valuable suggestions. Thanks are due to Sebastian Leipert, who implemented the transformation of the rural postman to the traveling salesman problem. We are particularly thankful to Stefan Thienel who generously helped us with our computational experiments, and heavily influenced the contents of Sections 6 and 7. This work was partially supported by EEC Contract SC1-CT91-0620.

References

Aarts, E.H.L., and J. Korst (1989). *Simulated Annealing and Boltzmann Machines,* John Wiley & Sons, Chichester.

Ablay, P. (1987). Optimieren mit Evolutionsstrategien. *Spektrum der Wissenschaft* 7, 104–115.

Althöfer, I., and K.-U. Koschnick (1991). On the convergence of "threshold accepting". *Appl. Math. and Opt.* 24, 183–195.

Applegate, D., R.E. Bixby, V. Chvátal and W. Cook (1994). Finding cuts in the TSP. *Preprint*, August 19, 1994.

Applegate, D., V. Chvátal and W. Cook (1990). Data Structures for the Lin–Kernighan Heuristic. Talk presented at the TSP-Workshop 1990, CRPC, Rice University.

Arthur, J.L., and J.O. Frendeway (1985). A computational study of tour construction procedures for the traveling salesman problem. Research report, Oregon State University, Corvallis.

Bachem, A., and M. Wottawa (1991). Ein 18512-Städte (Deutschland) traveling salesman problem. Report 91.97, Mathematisches Institut, Universität zu Köln.

Balas, E., and P. Toth (1985). Branch and bound methods. in: E.L. Lawler, J.K. Lenstra, A.H.G. Rinnooy Kan and D.B. Shmoys (eds.), *The Traveling Salesman Problem*, John Wiley & Sons, Chichester, pp. 361–401.

Balas, E., S. Ceria and G. Cornuéjols (1993). A lift-and-project cutting plane algorithm for mixed 0–1 programs. *Math. Program.* 58, 295–324.

Bartholdi, J.J., and L.K. Platzman (1982). An $O(n \log n)$ Planar Travelling Salesman Heuristic Based on Spacefilling Curves. *Oper. Res. Lett.* 4, 121–125.

Bayer, R. (1972). Symmetric binary b-trees: Data structure and maintenance algorithms. *Acta Informatica* 1, 290–306.

Beardwood, J., J.H. Halton and J.M. Hammersley (1959). The shortest path through many points. Proc. Cambridge Philos. Soc. 55, 299–327.

Bentley, J.L. (1992). Fast Algorithms for Geometric Traveling Salesman Problems. *ORSA J. Comput.* 4, 387–411.

Bland, R.E., and D.F. Shallcross (1989). Large traveling salesman problem arising from experiments in X-ray crystallography: a preliminary report on computation. *Oper. Res. Lett.* 8, 125–128.

Boyd, S.C., and W.H. Cunningham (1991). Small travelling salesman polytopes. *Math. Oper. Res.* 16, 259–271.

Boyd, S.C., W.H. Cunningham, M. Queyranne and Y. Wang (1993). Ladders for travelling salesman. Preprint, Department of Combinatorics and Optimization, University of Waterloo, Waterloo, Ontario, Canada, to appear in *SIAM J. Optimization.*

Boyd, S.C., and W.R. Pulleyblank (1990). Optimizing over the subtour polytope of the traveling salesman problem. *Math. Program.* 49, 163–187.

Burkard, R.E. (1990). Special cases of travelling salesman problems and heuristics. *Acta Math. Appl. Sin.* 6, 273–288.

Carpaneto, G., S. Martello and P. Toth (1984). An algorithm for the bottleneck traveling salesman problem. *Oper. Res.* 32, 380–389.

Carpaneto, G., M. Fischetti and P. Toth (1989). New lower bounds for the symmetric travelling salesman problem. *Math. Program.* 45, 233–254.

Cerny, V. (1985). A Thermodynamical Approach to the Travelling Salesman Problem: An Efficient Simulation Algorithm. *J. Optimization Theory Appl.* 45, 41–51.

Christof, T., M. Jünger and G. Reinelt (1991). A complete description of the traveling salesman polytope on 8 nodes. *Oper. Res. Lett.* 10, 497–500.

Christofides, N. (1976). Worst case analysis of a new heuristic for the travelling salesman problem. Report 388, Graduate School of Industrial Administration, Carnegie-Mellon University, Pittsburgh.

Christofides, N. (1979). The Travelling Salesman Problem. in: N. Christofides, A. Mingozzi, P. Toth and C. Sandi (eds.), *Combinatorial Optimization*, John Wiley & Sons, Chichester, pp. 131–149.

Chvátal, V. (1973). Edmonds polytopes and weakly Hamiltonian graphs. *Math. Program.* 5, 29–40.

Clarke, G., and J.W. Wright (1964). Scheduling of vehicles from a central depot to a number of delivery points. *Oper. Res.* 12, 568–581.

Clochard, J.M., and D. Naddef (1993). Using path inequalities in a branch and cut code for the symmetric traveling salesman problem. in: G. Rinaldi and L. Wolsey (eds.), *Integer Programming and Combinatorial Optimization 3*, Centro Ettore Majorana, Erice, pp. 291–311.

Collins, N.E., R.W. Eglese and B.L. Golden (1988). Simulated Annealing: An Annotated Bibliography. *Am. J. Math. Manage. Sci.* 8, 205–307.

Corberán, A., and J.M. Sanchis (1991). A polyhedral approach to the rural postman problem. Working paper, Facultad de Matemáticas, Universidad de Valencia.

Cormen, T.H., Ch.E. Leiserson and R.L. Rivest (1989). *Introduction to Algorithms,* MIT Press, Cambridge.

Cornuéjols, G., J. Fonlupt and D. Naddef (1985). The traveling salesman problem on a graph and some related polyhedra. *Math. Program.* 33, 1–27.

Cornuéjols, G., and G.L. Nemhauser (1978). Tight bounds for Christofides' TSP heuristic. *Math. Program.* 14, 116–121.

CPLEX (1993). *Using the CPLEX callable library and CPLEX mixed integer library,* CPLEX Optimization, Inc.

Cronin, T.M. (1990). The Voronoi diagram for the Euclidean Traveling Salesman Problem is Piecemeal Hyperbolic. CECOM Center for Signals Warfare, Warrenton.

Crowder, H., and M.W. Padberg (1980). Solving large-scale symmetric traveling salesman problems to optimality. *Manage. Sci.* 26, 495–509.

Cunningham, W., and A.B. Marsh III (1978). A Primal Algorithm for Optimum Matching. *Math. Program. Study* 8, 50–72.

Dantzig, G.B., D.R. Fulkerson and S.M. Johnson (1954). Solution of a large scale traveling-salesman problem. *Oper. Res* 2, 393–410.

Delaunay, B. (1934). Sur la sphère vide. *Izv. Akad. Nauk. SSSr, VII Ser., Otd. Mat. Estestv. Nauk 7* 6, 793–800.

Dillencourt, M.B. (1987a). Traveling Salesman Cycles are not Always Subgraphs of Delaunay Triangulations or of Minimum Weight Triangulations. *Inf. Process. Lett.* 24, 339–342.

Dillencourt, M.B. (1987b). A Non-Hamiltonian, Nondegenerate Delaunay Triangulation. *Inf. Process. Lett.* 25, 149–151.

Dreissig, W., and W. Uebach (1990). Personal communication.

Dueck, G. (1993). New Optimization Heuristics. The Great Deluge Algorithm and the Record-to-Record-Travel. *J. Comput. Phys.* 104, 86–92.

Dueck, G., and T. Scheuer (1990). Threshold Accepting: A General Purpose Algorithm Appearing Superior to Simulated Annealing. *J. Comput. Phys.* 90, 161–175.

Durbin, R., and D. Willshaw (1987). An analogue approach to the travelling salesman problem using an elastic net method. *Nature* 326, 689–691.

Edmonds, J. (1965). Maximum matching and a polyhedron with 0,1-vertices. *J. Res. Nat. Bur. Stand. B* 69, 125–130.

Edmonds, J., and E.L. Johnson (1970). Matching: a Well-Solved Class of Integer Linear Programs. in Proceedings of the Calgary International Conference on Combinatorial Structures and Their Applications, R.K. Guy et al. (eds.), Gordon and Breach, pp. 89–92.

Edmonds, J., and E.L. Johnson (1973). Matching, Euler tours and the Chinese postman. *Math. Program.* 5, 88–124.

Fischetti, M., and P. Toth (1992). An additive bounding procedure for the asymmetric travelling salesman problem. *Math. Program.* 53, 173–197.

Fleischmann, B. (1987). Cutting planes for the symmetric traveling salesman problem. Research Report, Universität Hamburg.

Fleischmann, B. (1988). A New Class of Cutting Planes for the Symmetric Travelling Salesman Problem. *Math. Program.* 40, 225–246.

Fritzke, B., and P. Wilke (1991). FLEXMAP – A neural network for the traveling salesman problem with linear time and space complexity. Research Report, Universität Erlangen-Nürnberg.

Garfinkel, R.S. (1985). Motivation and Modeling. in: E.L. Lawler, J.K. Lenstra, A.H.G. Rinnooy Kan and D.B. Shmoys (eds.), *The Traveling Salesman Problem* John Wiley & Sons, Chichester, pp. 307–360.

Gendreau, M., A. Hertz and G. Laporte (1992). New Insertion and Postoptimization Procedures for the Traveling Salesman Problem. *Oper. Res.* 40, 1086–1094.

Gilmore, P.C., E.L. Lawler and D.B. Shmoys (1985). Well-solved special cases. in: E.L. Lawler, J.K. Lenstra, A.H.G. Rinnooy Kan and D.B. Shmoys (eds.), *The Traveling Salesman Problem*, John Wiley & Sons, Chichester, pp. 87–143.

Glover, F. (1990). Tabu Search. *ORSA J. Comput.* 1, 190–206 (Part I), 2, 4–32 (Part II).

Goemans, M.X. (1993). Worst-case Comparison of Valid Inequalities for the TSP. Preprint, Department of Mathematics, Massachusetts Institute of Technology, Cambridge, to appear in *Math. Program.*.

Goldberg, D.E. (1989). *Genetic algorithms in search, optimization and machine learning,* Addison-Wesley.

Goldberg, A.V., and R.E. Tarjan (1988). A new approach to the maximum flow problem. *J. ACM* 35, 921–940.

Golden, B.L., and W.R. Stewart (1985). Empirical analysis of heuristics. in: E.L. Lawler, J.K. Lenstra, A.H.G. Rinnooy Kan and D.B. Shmoys (eds.), *The Traveling Salesman Problem* John Wiley & Sons, Chichester, pp. 207–249.

Gomory, R.E. (1958). Outline of an algorithm for integer solutions to linear programs. l Bull. Am. Math. Soc. 64, 275–278.

Gomory, R.E. (1960). Solving linear programming problems in integers. *Proc. Symp. Appl. Math.* 10, 211–215.

Gomory, R.E. (1963). An algorithm for integer solutions to linear programs. in: R.L. Graves and P. Wolfe (eds.), *Recent Advances in Math. Program.*, McGraw Hill, New York, pp. 269–302.

Gomory, R.E., and T.C. Hu (1961). Multi-terminal network flows. *SIAM J. Appl. Math.* 9, 551-570.

Graham, R.L. (1972). An efficient algorithm for determining the convex hull fo a finite planar set. *Inf. Process. Lett.* 1, 132–133.

Grötschel, M. (1977). *Polyedrische Charakterisierungen kombinatorischer Optimierungsprobleme,* Hain, Meisenheim am Glan.

Grötschel, M. (1980). On the symmetric traveling salesman problem: solution of a 120-city problem. *Math. Program. Studies* 12, 61–77.

Grötschel, M., and O. Holland (1987). A cutting plane algorithm for minimum perfect 2-matching. *Computing* 39, 327–344.

Grötschel, M., and O. Holland (1991). Solution of Large-scale Symmetric Traveling Salesman Problems. *Math. Program.* 51, 141–202.

Grötschel, M., M. Jünger and G. Reinelt (1984). A Cutting Plane Algorithm for the Linear Ordering Problem. *Oper. Res.* 32, 1195–1220.

Grötschel, M., M. Jünger and G. Reinelt (1991). Optimal Control of Plotting and Drilling Machines: A Case Study. *Z. Oper. Res. – Methods Models Oper. Res.* 35, 61–84.

Grötschel, M., L. Lovász and A. Schrijver (1981). The ellipsoid methods and its consequences in combinatorial optimization. *Combinatorica* 1, 169–197.

Grötschel, M., L. Lovász and A. Schrijver (1988). *Geometric Algorithms and Combinatorial Optimization,* Springer-Verlag, Berlin-Heidelberg.

Grötschel, M., and M.W. Padberg (1974). Zur Oberflächenstruktur des Traveling Salesman Polytopen. in H.J. Zimmermann et al. (eds.), Proc. Operations Research 4, Physica, Würzburg, pp. 207–211.

Grötschel, M., and M.W. Padberg (1977). Lineare Charakterisierungen von Traveling Salesman Problemen. *Z. Oper. Res.* 21, 33–64.

Grötschel, M., and M.W. Padberg (1978). On the symmetric traveling salesman problem: theory and computation. in R. Henn et al. (eds.), *Optimization and Operations Research*, Lecture Notes in Economics and Mathematical Systems 157, Springer, Berlin, pp. 105–115.

Grötschel, M., and M.W. Padberg (1979a). On the symmetric traveling salesman problem I: inequalities. *Math. Program.* 16, 265–280.

Grötschel, M., and M.W. Padberg (1979b). On the symmetric traveling salesman problem II: lifting theorems and facets. *Math. Program.* 16, 281–302.

Grötschel, M., and M.W. Padberg (1985). Polyhedral theory. in: E.L. Lawler, J.K. Lenstra, A.H.G. Rinnooy Kan and D.B. Shmoys (eds.), *The Traveling Salesman Problem*, John Wiley & Sons, Chichester, pp. 251–305.

Grötschel, M., and W.R. Pulleyblank (1986). Clique tree inequalities and the symmetric traveling salesman problem. *Math. Oper. Res.* 11, 537–569.

Guibas, L.J., and R. Sedgewick (1978). A dicromatic framework for balanced trees. in: *Proc. 19th Annu. Symp. on Foundations of Computer Science*, IEEE Computer Society, pp. 8–21.

Gusfield, D. (1987). Very simple algorithms and programs for all pairs network flow analysis. Preprint, Computer Science Division, University of California, Davis.

Hajek, B. (1985). A Tutorial Survey of Theory and Applications of Simulated Annealing. Proc. 24th IEEE Conf. on Decision and Control, pp. 755–760.

Hao, J., and J.B. Orlin (1992). A Faster Algorithm for Finding the Minimum Cut in a Graph. Proc. 3rd ACM–SIAM Symp. on Discrete Algorithms, Orlando, Florida, pp. 165–174.

Held, M., and R.M. Karp (1970). The Traveling Salesman Problem and Minimum Spanning Trees. *Oper. Res.* 18, 1138–1162.

Held, M., and R.M. Karp (1971). The Traveling Salesman Problem and Minimum Spanning Trees: Part II. *Math. Program.* 1, 6–25.

Hoffman, A.J., and P. Wolfe (1985). History. in: E.L. Lawler, J.K. Lenstra, A.H.G. Rinnooy Kan and D.B. Shmoys (eds.), *The Traveling Salesman Problem*, John Wiley & Sons, Chichester, pp. 1–15.

Hopfield, J.J., and D.W. Tank (1985). 'Neural' computation of decisions in optimization problems. *Biol. Cybern.* 52, 141–152.

Hu, T.C. (1965). Decomposition in Traveling Salesman Problems. Proc. IFORS Theory of Graphs, A34–A44.

Hurkens, C.A.J. (1991). Nasty TSP instances for classical insertion heuristics. University of Technology, Eindhoven.

Johnson, D.S. (1990). Local Optimization and the Traveling Salesman Problem. Proc. 17th Colloquium on Automata, Languages and Programming, Springer Verlag, 446–461.

Johnson, D.S. (1992). Personal communication.

Johnson, D.S., C.R. Aragon, L.A. McGeoch and C. Schevon (1991). Optimization by simulated annealing: An experimental evaluation. *Oper. Res.* 37, 865–892 (Part I), 39, 378–406 (Part II).

Johnson, D.S., J.L. Bentley, L.A. McGeoch and E.E. Rothberg (1994). Near-optimal solutions to very large traveling salesman problems. unpublished manuscript.

Johnson, D.S., and C.H. Papadimitriou (1985). Computational Complexity. in: E.L. Lawler, J.K. Lenstra, A.H.G. Rinnooy Kan and D.B. Shmoys, eds., *The Traveling Salesman Problem*, John Wiley & Sons, Chichester, pp. 37–85.

Johnson, D.S., C.H. Papadimitriou and M. Yannakakis (1988). How easy is local search?. *J. Comp. Syst. Sci.* 37, 79–100.

Jünger, M., G. Reinelt and S. Thienel (1994). Provably good solutions for the traveling salesman problem. *ZOR – Math. Meth. Oper. Res.* 40, 183–217.

Jünger, M., G. Reinelt and D. Zepf (1991). Computing Correct Delaunay Triangulations. *Computing* 47, 43–49.

Kaibel, V. (1993). Numerisch stabile Berechnung von Voronoi-Diagrammen. Diplomarbeit, Universität zu Köln.

Karger, D.R. (1993). Global min-cuts in $\mathcal{RNC}$, and other ramifications of a simple min-cut algorithm. Proc. 4th ACM–SIAM Symp. on Discrete Algorithms, pp. 21–30.

Karger, D.V., and C. Stein (1993). An $\tilde{O}(n^2)$ algorithm for minimum cuts. Proc. 25th ACM Symp. on the Theory of Computing, San Diego, CA, pp. 757–765.

Karp, R. (1977). Probabilistic analysis of partitioning algorithms for the traveling-salesman in the plane. *Math. Oper. Res.* 2, 209–224.

Karp, R.M., and J.M. Steele (1985). Probabilistic Analysis of Heuristics. in: E.L. Lawler, J.K. Lenstra, A.H.G. Rinnooy Kan and D.B. Shmoys, eds., *The Traveling Salesman Problem*, John Wiley & Sons, Chichester, pp. 181–205.

Kemke, C. (1988). Der Neuere Konnektionismus; Ein Überblick. *Inf. Spektrum* 11, 143–162.

Kirkpatrick, S. (1984). Optimization by simulated annealing: quantitative studies. *J. Statist. Phys.* 34, 975–986.

Kirkpatrick, S., C.D. Gelatt Jr. and M.P. Vecchi (1983). Optimization by simulated annealing. *Science* 222, 671–680.

Kiwiel, K.C. (1989). A Survey of Bundle Methods for Nondifferentiable Optimization. in: M. Iri & K. Tanabe (eds.) *Mathematical Programming. Recent Developments and Applications*, Kluwer Academic Publishers, Dordrecht, 263–282.

Knox, J., and F. Glover (1989). Comparative Testing of Traveling Salesman Heuristics Derived from Tabu Search, Genetic Algorithms and Simulated Annealing. Center for Applied Artificial Intelligence, Univ. of Colorado.

Knuth, D.E. (1973). *The art of computer programming, Volume 3: Sorting and searching*. Addison-Wesley, Reading, MA.

Kruskal, J.B. (1956). On the Shortest Spanning Subtree of a Graph and the Traveling Salesman Problem. *Proc. Am. Math. Soc.* 7, 48–50.

Lawler, E.L., J.K. Lenstra, A.H.G. Rinnooy Kan and D.B. Shmoys, eds. (1985). *The Traveling Salesman Problem,* John Wiley & Sons, Chichester.

Lenstra, J.K., and A.H.G. Rinnooy Kan (1974). Some Simple Applications of the Travelling Salesman Problem. BW 38/74, Stichting Mathematisch Centrum, Amsterdam.

Lin, S., and B.W. Kernighan (1973). An Effective Heuristic Algorithm for the Traveling-Salesman Problem. *Oper. Res.* 21, 498–516.

Litke, J.D. (1984). An improved solution to the traveling salesman problem with thousands of nodes. *Commun. ACM* 27, 1227–1236.

Mak, K.-T., and A.J. Morton (1993). A Modified Lin–Kernighan Traveling Salesman Heuristic. *Oper. Res. Lett.* 13, 127–132.

Malek, M., M. Guruswamy, H. Owens and M. Pandya (1989). Serial and Parallel Search Techniques for the Traveling Salesman Problem. *Annals of OR: Linkages with Artificial Intelligence*.

Malek, M., M. Heap, R. Kapur and A. Mourad (1989). A fault tolerant implementation of the traveling salesman problem. Research Report, Dept. of Electrical and Computer Engineering, Univ. of Texas at Austin.

Margot, F. (1992). Quick updates for p-OPT TSP heuristics. *Oper. Res. Lett.* 11.

Marsten, R. (1981). The design of the XMP linear programming library. *ACM Trans. Math. Software* 7, 481–497.

Martin, O., S.W. Otto and E.W. Felten (1992). Large-step Markov chains for the TSP incorporating local search heuristics. *Oper. Res. Lett.* 11, 219–224.

Maurras, J.F. (1975). Some results on the convex hull of Hamiltonian cycles of symmetric complete graphs. in: B. Roy (ed.), *Combinatorial Programming: Methods and Applications*, Reidel, Dordrecht, pp. 179–190.

Metropolis, N., A. Rosenbluth, M. Rosenbluth, A. Teller and E. Teller (1953). Equation of state calculation by fast computing machines. J. Chem. Phys. 21, 1087–1092.

Miller, D.L., J.F. Pekny and G.L. Thompson (1991). An exact branch and bound algorithm for the symmetric TSP using a symmetry relaxed two-matching relaxation. Talk presented at International Symposium on Mathematical Programming, Amsterdam.

Mühlenbein, H., M. Gorges-Schleuter and O. Krämer (1988). Evolution algorithms in combinatorial optimization. *Parallel Comput.* 7, 65–85.

Naddef, D. (1990). Handles and teeth in the symmetric traveling salesman polytope. in W. Cook and P.D. Seymour (eds.), *Polyhedral Combinatorics* DIMACS Series in Discrete Mathematics and Theoretical Computer Science 1, A.M.S., pp. 61–74.

Naddef, D. (1992). The binested inequalities for the symmetric traveling salesman polytope. *Math. Oper. Res.* 17, 882–900.

Naddef, D., and G. Rinaldi (1988). The symmetric traveling salesman polytope: New facets from the graphical relaxation. Report R. 248, IASI-CNR Rome.

Naddef, D., and G. Rinaldi (1991). The symmetric traveling salesman polytope and its graphical relaxation: Composition of valid inequalities. *Math. Program.* 51, 359–400.

Naddef, D., and G. Rinaldi (1992). The crown inequalities for the symmetric traveling salesman polytope. *Math. Oper. Res.* 17, 308–326.

Naddef, D., and G. Rinaldi (1993). The graphical relaxation: a new framework for the symmetric traveling salesman polytope. *Math. Program.* 58, 53–88.

Nagamochi, H., and T. Ibaraki (1992a). A linear-time algorithm for finding a sparse k-connected spanning subgraph of a k-connected graph. *Algorithmica* 7, 583–596.

H. Nagamochi and T. Ibaraki (1992b). Computing edge-connectivity in multigraphs and capacitated graphs. *SIAM J. Discrete Math.* 5, 54–66.

Nemhauser, G.L., and L.A. Wolsey (1988). *Integer and Combinatorial Optimization*, John Wiley & Sons, Chichester.

Norman, R.Z. (1955). On the convex polyhedra of the symmetric traveling salesman problem (abstract). *Bull. AMS* 61, 559.

Ohya, T., M. Iri and K. Murota (1984). Improvements of the Incremental Method for the Voronoi Diagram with Computational Comparison of Various Algorithms. *J. Oper. Res. Soc. Jap.* 27, 306–337.

Or, I. (1976). Traveling Salesman-Type Combinatorial Problems and Their Relation to the Logistics of Regional Blood Banking. Northwestern University, Evanston, Il.

Padberg, M.W., and M. Grötschel (1985). Polyhedral computations. in: E.L. Lawler, J.K. Lenstra, A.H.G. Rinnooy Kan and D.B. Shmoys, eds., *The Traveling Salesman Problem*, John Wiley & Sons, Chichester, pp. 307–360.

Padberg, M.W., and S. Hong (1980). On the symmetric traveling salesman problem: a computational study. *Math. Program. Studies* 12, 78–107.

Padberg, M.W., and M.R. Rao (1981). The Russian method for linear inequalities III: Boundend integer programming. GBA Working paper 81–39 New York University, New York, NY.

Padberg, M.W., and M.R. Rao (1982). Odd minimum cut sets and b-matchings. *Math. Oper. Res.* 7, 67–80.

Padberg, M.W., and G. Rinaldi (1987). Optimization of a 532 City Symmetric Traveling Salesman Problem by Branch and Cut. *Oper. Res. Lett.* 6, 1–7.

Padberg, M.W., and G. Rinaldi (1990a). An Efficient Algorithm for the Minimum Capacity Cut Problem. *Math. Program.* 47, 19–36.

Padberg, M.W., and G. Rinaldi (1990b). Facet Identification for the Symmetric Traveling Salesman Polytope. *Math. Program.* 47, 219–257.

Padberg, M.W., and G. Rinaldi (1991). A Branch and Cut Algorithm for the Resolution of Large-scale Symmetric Traveling Salesman Problems. *SIAM Rev.* 33, 60–100.

Padberg, M., and T.-Y. Sung (1988). A polynomial-time solution to Papadimitriou and Steiglitz's 'traps'. *Oper. Res. Lett.* 7, 117–125.

Papadimitriou, C.H. (1990). The Complexity of the Lin–Kernighan Heuristic for the Traveling Salesman Problem. University of California, San Diego, CA.

Plante, R.D., T.J. Lowe and R. Chandrasekaran (1987). The Product Matrix Traveling Salesman Problem: An Application and Solution Heuristics. *Oper. Res.* 35, 772–783.

Polyak, B.T. (1978). Subgradient Methods: A Survey of Soviet Research. in: C. Lemarèchal & R. Mifflin (eds.), *Nonsmooth Optimization*, Pergamon Press, Oxford, pp. 5–29.

Potvin, J.-Y., and J.-M. Rousseau (1990). Enhancements to the Clarke and Wright Algorithm for the Traveling Salesman Problem. Research report, University of Montreal.

Prim, R.C. (1957). Shortest Connection Networks and Some Generalizations. *The Bell System Tech. J.* 36, 1389–1401.

Pulleyblank, W.R. (1983). Polyhedral Combinatorics. in *Mathematical Programming The State of the Art*, A. Bachem et al., eds., Springer-Verlag, pp. 312–345.

Queyranne, M., and Y. Wang (1990). Facet tree composition for symmetric travelling salesman polytopes. Working paper 90–MSC–001, Faculty of Commerce, University of British Columbia, Vancouver, B.C..

Queyranne, M., and Y. Wang (1993). Hamiltonian path and symmetric travelling salesman polytopes. *Math. Program.* 58, 89–110.

Ratliff, H.D., and A.S. Rosenthal (1983). Order-Picking in a Rectangular Warehouse: A Solvable Case for the Travelling Salesman Problem. *Oper. Res.* 31, 507–521.

Rechenberg, I. (1973). *Evolutionsstrategie: Optimierung technischer Systeme nach Prinzipien der biologischen Evolution,* Frommann-Holzboog, Stuttgart.

Reinelt, G. (1991a). TSPLIB – A Traveling Salesman Problem Library. *ORSA J. Comput.* 3, 376–384.

Reinelt, G. (1991b). TSPLIB – Version 1.2. Report No. 330, Schwerpunktprogramm der Deutschen Forschungsgemeinschaft, Universität Augsburg.

Reinelt, G. (1992). Fast Heuristics for Large Geometric Traveling Salesman Problems. *ORSA J. Comput.* 2, 206–217.

Reinelt, G. (1994). *The Traveling Salesman – Computational Solutions,* Lecture Notes in Computer Science 840, Springer.

Rosenkrantz, D.J., R.E. Stearns and P.M. Lewis (1977). An analysis of several heuristics for the traveling salesman problem. *SIAM J. Comput.* 6, 563–581.

Ruján, P. (1988). Searching for optimal configurations by simulated tunneling. *Z. Phys. B – Condensed Matter* 73, 391–416.

Ruján, P., C. Evertsz and J.W. Lyklema (1988). A Laplacian Walk for the Travelling Salesman. *Europhys. Lett.* 7, 191–195.

Rumelhart, D.E., G.E. Hinton and J.L. McClelland (1986). *The PDP Research Group: Parallel Distributed Processing: Explorations in the Microstructure of Cognition,* MIT Press.

Sahni, S., and T. Gonzales (1976). P-complete approximation problems. *J. Assoc. Comput. Mach.* 23, 555-565.

Schramm, H. (1989). *Eine Kombination von Bundle- und Trust-Region-Verfahren zur Lösung nichtdifferenzierbarer Optimierungsprobleme,* Bayreuther Mathematische Schriften, Heft 30.

Segal, A., R. Zhang and J. Tsai (1991). A New Heuristic Method for Traveling Salesman Problem. University of Illinois, Chicago.

Shamos, M.I., and D. Hoey (1975). Closest point problems. Proc. 16th IEEE Annu. Symp. Found. Comput. Sci., pp. 151–162.

Shmoys, D.B., and D.P. Williamson (1990). Analyzing the Held–Karp TSP bound: A monotonicity property with application. *Inf. Process. Lett.* 35, 281–285.

Tarjan, R.E. (1983). *Data Structures and Network Algorithms,* Society for Industrial and Applied Mathematics, Philadelphia.

Ulder, N.L.J., E. Pesch, P.J.M. van Laarhoven, H.-J. Bandelt and E.H.L. Aarts (1990). Improving TSP Exchange Heuristics by Population Genetics. Preprint, Erasmus Universiteit Rotterdam.

van Dal, R. (1992). *Special Cases of the Traveling Salesman Problem,* Wolters-Noordhoff, Groningen.

van der Veen, J.A.A. (1992). Solvable Cases of the Traveling Salesman Problem with Various Objective Functions. Doctoral Thesis, Rijksuniversiteit Groningen, Groningen.

van Laarhoven, P.J.M. (1988). Theoretical and Computational Aspects of Simulated Annealing. PhD Thesis, Erasmus Universiteit, Rotterdam.

Vanderbei, R.J. (1992). LOQO User's Manual. Preprint, Statistics and Operations Research, Princeton University.

Volgenant, T., and R. Jonker (1982). A branch and bound algorithm for the symmetric traveling salesman problem based on the 1-tree relaxation. *Eur. J. Oper. Res.* 9, 83–89.

Voronoi, G. (1908). Nouvelles applications des paramètres continus à la théorie des formes quadratiques. Deuxième memoire: Recherche sur les parallélloèdres primitifs. *J. Reine Angew. Math.* 3, 198–287.

Warren, R.H. (1993). Special cases of the traveling salesman problem. Preprint, Advanced Concepts Center, Martin Marietta Corporation, King of Prussia, PA.

Wolsey, L.A. (1980). Heuristic analysis, linear programming and branch and bound. *Math. Program. Study* 13, 121–134.

M.O. Ball et al., Eds., *Handbooks in OR & MS, Vol. 7*

Chapter 5

Parallel Computing in Network Optimization

Dimitri Bertsekas
Laboratory for Information and Decision Systems, Massachusetts Institute of Technology, Cambridge, MA, U.S.A.

David Castañon
Department of Electrical, Computer and Systems Engineering, Boston University, Boston, MA, U.S.A.

Jonathan Eckstein
Mathematical Sciences Research Group, Thinking Machines Corporation, Cambridge, MA, U.S.A.

Stavros Zenios
Decision Sciences Department, Wharton School, University of Pennsylvania, Philadelphia, PA, U.S.A.

1. Introduction

Parallel and vector supercomputers are today considered basic research tools for several scientific and engineering disciplines. The novel architectural features of these computers — which differ significantly from the von Neumann model — are influencing the design and implementation of algorithms for numerical computation. Recent developments in large scale optimization take into account the architecture of the computer where the optimization algorithms are likely to be implemented. In the case of network optimization, in particular, we have seen significant progress in the design, analysis, and implementation of algorithms that are particularly well suited for parallel and vector architectures. As a result of these research activities, problems with several millions of variables can be solved routinely on parallel supercomputers. In this chapter, we discuss algorithms for parallel computing in large scale network optimization.

We have chosen to focus on a sub-class of network optimization problems for which parallel algorithms have been designed. In particular, for the most part, we address only *pure* networks (i.e., without arc multipliers). We also avoid discussion on large-scale problems with embedded network structures, like the *multicommodity* network flow problem, or the *stochastic* network problem. Nevertheless, we discuss parallel algorithms for both linear and nonlinear problems, and special attention is given to the assignment problem as well as other problems with bipartite structures (i.e., transportation problems). The problems we have chosen to discuss usually provide the building blocks for the development of parallel algorithms for the more complex problem structures that we are not addressing. Readers who are interested in a broader view of parallel optimization research — both for network structured problems and mathematical programming in general — should refer to several journal issues focused on parallel optimization which have been published recently on this topic [Mangasarian & Meyer, 1988, 1991; Meyer & Zenios, 1988; Rosen, 1990] or the textbook of Bertsekas & Tsitsiklis [1989].

1.1. Organization of this chapter

The introductory section discusses parallel architectures and broad issues that relate to the implementation and performance evaluation of parallel algorithms. It also defines the network optimization problems that will be discussed in subsequent sections. Section 2 develops the topic of parallel computing for linear network optimization problems and Section 3 deals with nonlinear networks. Concluding remarks, a brief overview of additional work for multicommodity network flows and stochastic network programs, as well as open issues, are addressed in Section 5.

Each of Sections 2–4 is organized in three subsections along the following thread. First, we present general methodological ideas for the design of specific algorithms for each problem class. Here, we present selectively those algorithms that have some potential for parallelism. The methodological development is followed by a subsection of parallelization ideas, i.e., specific ways in which each algorithm can be implemented on a parallel computer. Finally, computational results with the parallel implementation of some of the algorithms that have appeared in the literature are summarized and discussed.

1.2. Parallel architectures

Parallelism in computer systems is not a recent concept. ENIAC — the first large-scale, general-purpose, electronic digital computer built at the University of Pennsylvania — was designed with multiple functional units for adding, multiplying, and so forth. The primary motivation behind this design was to deliver the computing power infeasible with the sequential electronic technology of that time. The shift from diode valves to transistors, integrated circuits, and very large scale integrated circuits (VLSI) rendered parallel designs obsolete and uniprocessor systems were predominant through the late sixties.

The first milestone in the evolution of parallel computers was the Illiac IV project at the University of Illinois in the 1970's. A brief historical note on this project can be found in Desrochers [1987]. The array architecture of the Illiac prompted studies on the design of suitable algorithms for scientific computing. Interestingly, a study of this sort was carried out for linear programming [Pfefferkorn & Tomlin, 1976] — one of the first studies in parallel optimization. The Illiac never went past the stage of the research project, however, and only one machine was ever built.

The second milestone was the introduction of the CRAY 1 in 1976. The term *supercomputer* was coined at that time, and is meant to indicate the fastest available computer. The vector architecture of the CRAY introduced the notion of *vectorization* of scientific computing. Designing or restructuring of numerical algorithms to exploit the computer architecture — in this case vector registers and vector functional units — became once more a critical issue. Vectorization of an application can range from simple modifications of the implementation with the use of computational kernels that are streamlined for the machine architecture, to more substantive changes in data structure and the design of algorithms that are rich in vector operations.

Since the mid-seventies, supercomputers and parallel computers have been evolving rapidly in the level of performance they can deliver, the size of memory available, and the increasing number of parallel processors that can be applied to a single task. The Connection Machine CM–2, for example, can be configured with up to 65,536 very simple processing elements.

Several alternative parallel architectures have been developed. Today there is no single widely accepted model for parallel computation. A classification of computer architectures was proposed by Flynn [1972] and is used to distinguish between alternative parallel architectures. Flynn proposed the following four classes, based on the interaction among instruction and data streams of the processor(s):

1. SISD — *Single Instruction stream Single Data stream*. Systems in this class execute a single instruction on a single piece of data before moving on to the next piece of data and the next instruction. Traditional uniprocessor, scalar (von Neumann) computers fall under this category.

2. SIMD — *Single Instruction stream Multiple Data stream*. A single instruction can be executed simultaneously on multiple data. This of course implies that the operations of an algorithm are identical over a set of data and that data can be arranged for concurrent execution. An example of SIMD systems is the Connection Machine of Hillis [1985].

3. MISD — *Multiple Instruction stream Single Data stream*. Multiple instructions can be executed concurrently on a single piece of data. This form of parallelism has not received, to our knowledge, extensive attention from researchers. It appears in Flynn's taxonomy for the sake of completeness.

4. MIMD — *Multiple Instruction stream Multiple Data stream*. Multiple instructions can be executed concurrently on multiple pieces of data. The majority of parallel computer systems fall in this category. Multiple instructions indicate the presence of independent code modules that may be executing independently from each other. Each module may be operating either on a subset of the data of the problem, have copies of all the problem data, or access all the data of the problem together with the other modules in a way that avoids read/write conflicts.

Whenever multiple data streams are used (i.e., in the MIMD and SIMD systems) another level of classification is needed for the memory organization: In *shared memory* systems, the multiple data streams are accessible by all processors.

Typically, a common memory bank is available. In *distributed memory* systems, each processor has access only to its own local memory. Data from the memories of other processors need to be communicated by passing messages across some communication network.

Multiprocessor systems are also characterized by the number of available processors. 'Small-scale' parallel systems have up to 16 processors, 'medium-scale' systems up to 128, and 'large-scale' systems up to 1024. Systems with 1024 or more processors are considered 'massively' parallel. Finally, multiprocessors are also characterized as 'coarse-grain' versus 'fine-grain'. In the former case each processor is very powerful, typically on the workstation level, with at least several megabytes of memory. Fine-grain systems typically use very simple processing elements with a few kilobytes of local memory each. For example, the NCUBE system with 1024 processors is considered a coarse-grain massively parallel machine. The Connection Machine CM–2 with up to 64K processing elements is a fine-grain, massively parallel machine. Of course these distinctions are qualitative in nature, and are likely to change as technology evolves.

A mode of computing that deserves special classification is that of *vector computers*. While vector computers are a special case of SIMD machines, they constitute a class of their own. This is due to the frequent appearance of vector capabilities in many parallel systems. Also the development of algorithms or software for a vector computer — like, for example, the CRAY — poses different problems than the design of algorithms for a system with multiple processors that operate synchronously on multiple data — like, for example, the Connection Machine CM–2. The processing elements of a vector computer are equipped with functional units that can operate efficiently on long vectors. This is usually achieved by segmenting functional units so that arrays of data can be processed in a pipeline fashion. Furthermore, multiple functional units may be available both for scalar and vector operations. These functional units may operate concurrently or in a *chained* manner, with the results of one unit being fed directly into another without need for memory access. Using these machines efficiently is a problem of structuring the underlying algorithm with (long) homogeneous vectors and arranging the operations to maximize chaining or overlap of the multiple units.

1.2.1. Performance evaluation

There has been considerable debate on how to evaluate the performance of parallel implementations of algorithms. Since different algorithms may be suitable for different architectures, a valid way to evaluate the performance of a parallel algorithm is to implement it on a suitable parallel computer and compare its performance against the 'best' serial code executing on a von Neumann system for a common set of test problems (of course, the parallel and von Neumann computers should be of comparable prices). Furthermore, it is not usually clear what is the 'best' serial code for a given problem, and the task of comparing different codes on different computer platforms is tedious and time-consuming. Hence, algorithm designers have developed several measures to evaluate the performance of a parallel algorithm that are easier to observe. The most commonly used are (1) *speedup*, (2) *efficiency*, (3) *scalability* and (4) *sustained FLOPS rates*.

Speedup: This is the ratio of solution time of the algorithm executing on a single processor, to the solution time of the same algorithm when executing on multiple processors. (This is also known as *relative speedup*.) It is understood that the sequential algorithm is executed on one of the processors of the parallel system (although this may not be possible for SIMD architectures). *Linear* speedup is observed when a parallel algorithm on p processors runs p times faster than on a single processor. *Sub-linear* speedup is achieved when the improvement in performance is less than p. *Super-linear* speedup (i.e., improvements larger than p) usually indicates that the parallel algorithm takes a different — and more efficient — solution path than the sequential algorithm. It is often possible in such situations to improve the performance of the sequential algorithm based on insights gained from the parallel algorithms.

Amdahl [1967] developed a law that gives an upper bound on the relative speedup that can be expected from a parallel implementation of an algorithm. If k is the fraction of the code that executes serially, while $1-k$ will execute on p processors, then the best speedup that can be observed is:

$$S_p = \frac{1}{k + (1-k)/p}$$

Relative speedup indicates how well a given algorithm is implemented on a parallel machine. It provides little information on the efficiency of the algorithm in solving the underlying problem. An alternative measure of speedup is the ratio of solution time of the 'best' serial code on a single processor to the solution time of the parallel code when executing on multiple processors.

Efficiency: This is the ratio of speedup to the number of processors. It provides a way to measure the performance of an algorithm independently from the level of parallelism of the computer architecture. Linear speedup corresponds to 100% (or 1.00) efficiency. Factors less than 1.00 indicate sublinear speedup and superlinear speedup is indicated by factors greater than 1.00.

Scalability: This is the ability of an algorithm to solve a problem n times as large on np processors, as it would take to solve the original problem using p processors. Some authors [DeWitt & Gray, 1992] define *scaleup* as a measure of the scalability of a computer/code as follows:

$$\text{Scaleup}\ (p, n) = \frac{\text{Time to solve problem of size } m \text{ on } p \text{ processors}}{\text{Time to solve problem of size } nm \text{ on } np \text{ processors}}$$

FLOPS: This acronym stands for *Floating-point Operations per Second*. This measure indicates how well a specific implementation exploits the architecture of a computer. For example, an algorithm that executes at 190 MFLOPS (i.e., 190×10^6 FLOPS) on a CRAY X-MP that has a peak rate of 210 MFLOPS can be considered a successfully vectorized algorithm. Hence, little further improvements can be expected for this algorithm on this particular architecture. This measure does not necessarily indicate whether this is an efficient algorithm for solving problems. It is conceivable that an alternative algorithm can solve the same problem faster, even if it executes at a lower FLOPS rate. As of the writing of this chapter most commericially available parallel machines are able to deliver peak performance in the GFLOPS (i.e., 10^9 FLOPS) range. Dense linear algebra codes typically run at several GFLOPS, and similar performance has been achieved for dense transportation problems [McKenna & Zenios, [1990]. The current goal of high-performance computing is to design and manufacture machines that can deliver teraflops. Presently available systems can, in principle, achieve such computing rates.

For further discussion on performance measures see the feature article by Barr and Hickman [1993] and the commentaries that followed.

1.3. Synchronous versus asynchronous algorithms

In order to develop a parallel algorithm, one needs to specify a *partitioning sequence* and a *synchronization sequence*. The partitioning sequence determines those components of the algorithm that are independent from each other, and hence can be executed in parallel. These components are called *local algorithms*. On a multiprocessor system they are distributed to multiple processors for concurrent execution. The synchronization sequence specifies an order of execution that guarantees correct results. In particular, it specifies the data dependencies between the local algorithms. In a *synchronous* implementation, each local algorithm waits at predetermined points in time for a predetermined set of input data before it can proceed with its local calculations. Synchronous algorithms can often be inefficient, as processors may have to spend excessive amounts of time waiting for data from each other.

Several of the network optimization algorithms in this chapter have *asynchronous* versions. The main characteristic of an asynchronous algorithm is that the local algorithms do not have to wait at intermediate synchronization points. We thus allow some processors to compute faster and execute more iterations than others, some processors to communicate more frequently than others, and communication delays to be unpredictable. Also, messages might be delivered in a different order than the one in which they were generated. *Totally asynchronous* algorithms tolerate arbitrarily large communication and computation delays, whereas *partially asynchronous* algorithms are not guaranteed to work unless an upper bound is imposed on the delays. In asynchronous algorithms, substantial variations in performance can be observed between runs, due to the non-deterministic nature of the asynchronous computations. Asynchronous algorithms are most relevant to MIMD architectures, both shared memory and distributed memory.

Models for totally and partially asynchronous algorithms have been developed in Bertsekas and Tsitsiklis [1989]. The same reference develops some general convergence results to assist in the analysis of asynchronous algorithms and establishes convergence of both partially and totally asynchronous algorithms for several network optimization problems.

Asynchronous algorithms have, potentially, two advantages over their synchronous counterparts. First, by reducing the synchronization penalty, they can achieve a speed advantage. Second, they offer greater implementation flexibility and tolerance to changes of problem data. Experiences with asynchronous network flow algorithms is somewhat limited. A direct comparison between synchronous and asynchronous algorithms for nonlinear network optimization problems [Chajakis & Zenios, 1991; El Baz, 1989] has shown that asynchronous implementations are substantially more efficient than synchronous ones. Further work on asynchronous algorithms for the assignment and min-cost flow problem [Bertsekas & Castañon, 1989, 1990a, b] supports these conclusions. A drawback of asynchronous algorithms, however, is that termination detection can be difficult. Even if an asynchronous algorithm is guaranteed to be correct at the limit, it may be difficult to identify when some approximate termination conditions have been satisfied. Bertsekas and Tsitsiklis [Bertsekas & Tsitsiklis, 1989, chapter 8] address the problem of termination detection once termination has occurred. The question of ensuring that global termination of an asynchronous algorithm will occur through the use of approximate local termination conditions is surprisingly intricate, and has been addressed in Bertsekas & Tsitsiklis [1991] and Savari & Bertsekas [1994]. In spite of the difficulties in implementing and testing termination of asynchronous algorithms, the studies cited above have shown that these difficulties can be addressed successfully. Asynchronous algorithms for several network optimization problems have been shown to be more efficient than their synchronous counterparts when implemented on suitable parallel architectures.

1.4. Network optimization problems

We introduce here our notation and define the types of network optimization problems that will be used in later sections. The most general formulation we will be working with is the following *nonlinear network program* (NLNW):

$$\min \quad F(x) \tag{1}$$

$$\text{s.t.} \quad Ax = b \tag{2}$$

$$l \leq x \leq u, \tag{3}$$

where $F : \Re^m \longmapsto \Re$ is convex and twice continuously differentiable, A is an $n \times m$ node–arc incidence matrix, $b \in \Re^n$, l and $u \in \Re^m$ are given vectors and $x \in \Re^m$ is the vector of decision variables. The node–arc incidence matrix A specifies conservation of flow constraints (3) on some network $G = (\mathcal{N}, \mathcal{A})$ with $|\mathcal{N}| = n$ and $|\mathcal{A}| = m$. It can be used to represent pure networks, in which case each column has two non–zero entries : a '+1' and a '–1'. Generalized networks are also represented by matrices with two non–zero entries in each column : a '+1' and a real number that represents the arc multiplier. An arc (i, j) is viewed as an ordered pair, and is to be distinguished from the pair (j, i). We define the vector x as the lexicographic order of the elements of the set $\{x_{ij} \mid (i, j) \in \mathcal{A}\}$. x is the flow vector in the network G, and x_{ij} is the flow of the arc (i, j). For a given x_{ij}, i is the row of the corresponding column of the constraint matrix A with with entry '+1', while j denotes the row with negative entry '–1' for pure networks, or the arc's multiplier $-m_{ij}$ for generalized networks. Similarly, components of the vectors $l, u,$ and x are indexed by (i, j) to indicate the from- and to-node of the corresponding network edge.

As a special case we assume that the function $F(x)$ is separable. Hence, model (NLNW) can be written in the equivalent form:

$$\min \quad \sum_{(i,j)\in\mathcal{A}} f_{ij}(x_{ij}) \tag{4}$$

$$\text{s.t.} \quad Ax = b \tag{5}$$

$$l \leq x \leq u. \tag{6}$$

If the functions $f_{ij}(x_{ij})$ are linear we obtain the *min-cost network flow* problem. It can be expressed in algebraic form (MCF):

$$\min \quad \sum_{(i,j)\in\mathcal{A}} c_{ij}x_{ij} \tag{7}$$

$$\text{s.t.} \quad \sum_{j:(i,j)\in\mathcal{A}} x_{ij} - \sum_{j:(j,i)\in\mathcal{A}} m_{ji}x_{ji} = b_i \quad \forall\, i \in \mathcal{N} \tag{8}$$

$$l_{ij} \leq x_{ij} \leq u_{ij} \qquad \forall\, (i, j) \in \mathcal{A}. \tag{9}$$

This problem is a *generalized* network since the real-valued multipliers m_{ij} appear in the flow conservation equations. If all multipliers are equal to '1', then (MCF) is the *pure* min-cost network flow problem.

Some special forms of both the nonlinear and linear network problems are of interest. In particular, we will consider problems where G is bipartite, $G = (\mathcal{N}_O \cup \mathcal{N}_D, \mathcal{A})$. $\mathcal{N}_O$ is the set of origin nodes, with supply vector s; $\mathcal{N}_D$ is the set of destination nodes, with demand vector d. The nonlinear transportation problem can be written as (NLTR)

$$\min \quad \sum_{(i,j)\in\mathcal{A}} f_{ij}(x_{ij}) \tag{10}$$

$$\text{s.t.} \quad \sum_{j:(i,j)\in\mathcal{A}} x_{ij} = s_i \quad \forall\, i \in \mathcal{N}_O \tag{11}$$

$$\sum_{i:(i,j)\in\mathcal{A}} x_{ij} = d_j \quad \forall\, j \in \mathcal{N}_D \tag{12}$$

$$l_{ij} \leq x_{ij} \leq u_{ij} \quad \forall\, (i, j) \in \mathcal{A}. \tag{13}$$

Of course, if the functions $f_{ij}(x_{ij})$ are linear, we have a *linear transportation problem*. The special case of a transportation problem when the supply and demand vectors are everywhere equal to one (i.e., $s_i = 1$ for all $i \in N_O$ and $d_j = 1$ for all $j \in N_D$) is known as the *assignment problem*. (Assignment problems are more usually stated as maximization than as minimization problems.)

We will also need some additional terminology when describing the algorithms: A *path* P in a directed graph is a sequence of nodes $(n_1, n_2, \ldots, n_k)$ with $k \geq 2$ and a corresponding sequence of $k-1$ arcs such that the ith arc in the sequence is either (n_i, n_{i+1}) (in which case it is called a *forward* path of the arc) or (n_{i+1}, n_i) (in which case it is called a *backward* arc of the path). We denote by P^+ and P^- the set of forward and backward arcs of P, respectively. The arcs in P^+ and P^- are said to *belong* to P. Nodes n_1 and n_k are called the *start node* (or *origin*) and *end node* (or *destination*) of P, respectively. In the context of the min-cost flow problems, the *cost of a path* P is defined as $\sum_{(i,j)\in P^+} c_{ij} - \sum_{(i,j)\in P^-} c_{ij}$.

A *cycle* is a path whose start and end nodes are the same. A path is said to be *simple* if it contains no repeated arcs, and no repeated nodes, except possibly for the start and end nodes (in which case the path is called a *simple cycle*). A path is said to be *forward* (or *backward*) if all of its arcs are forward (respectively, backward) arcs. We refer similarly to a forward and a backward cycle. A directed graph that contains no cycles is said to be *acyclic*. A directed graph is said to be *connected* if for each pair of nodes i and j, there is a path starting at i and ending at j. A *tree* is a connected acyclic graph. A *spanning tree* of a directed graph Γ is a subgraph of Γ that is a tree and includes all the nodes of Γ. A *simple path flow* is a flow vector x of the form

$$x = \begin{cases} a & \text{if } (i,j) \in P^+ \\ -a & \text{if } (i,j) \in P^- \\ 0 & \text{otherwise,} \end{cases}$$

where a is some positive number and P is a simple path. Finally, some additional notation: $\nabla^2 F(x^k)$ and $\nabla F(x^k)$ denotes the Hessian matrix and gradient vector of the function $F(x)$ evaluated at x^k. The transpose of a matrix A is denoted by A^T. $A_{\cdot t}$ and $A_{t\cdot}$ denote the tth column and row respectively of A. e will be used to denote a conformable vector of all '1'.

2. Linear network optimization

2.1. Basic algorithmic ideas

We will discuss three main ideas underlying min-cost flow algorithms. These are:

(a) Primal cost improvement; here we try to iteratively improve the cost to its optimal value by constructing a corresponding sequence of feasible flows.

(b) Dual cost improvement; here we define a problem related to the minimum cost flow problem, called the *dual*, whose variables are called *prices*. We then try to iteratively improve the dual objective function to its optimal value by constructing a corresponding sequence of prices. Dual cost improvement algorithms also iterate on flows, which are related to the prices through a property called *complementary slackness*.

(c) Approximate dual coordinate ascent; here one tries to iteratively improve the dual cost in an approximate sense along price coordinate directions. In addition to prices, algorithms of this type also iterate on flows, which are related to prices through a property called *ϵ-complementary slackness*, an approximate form of the complementary slackness property.

2.1.1 Primal cost improvement — the simplex method

A first important algorithmic idea for the min-cost flow problem is to start from an initial feasible flow vector, and generate a sequence of feasible flow vectors, each having a better cost value than the preceding one. For several interesting algorithms, including the simplex method, two successive flow vectors differ by a simple cycle flow.

Let us define a path P to be *unblocked with respect to x* if $x_{ij} < u_{ij}$ for all forward arcs $(i, j) \in P^+$ and $l_{ij} < x_{ij}$ for all backward arcs $(i, j) \in P^-$. The following is a basic result shown in several sources, e.g. Papadimitriou & Steiglitz [1982], Rockafellar [1984]; Bertsekas [1991]:

Proposition 1. *Consider the min-cost flow problem and let x be a feasible flow vector which is not optimal. Then there exists a simple cycle flow that when added to x, produces a feasible flow vector with smaller cost than x; the corresponding cycle is unblocked with respect to x and has negative cost.*

The major primal cost improvement algorithm for the min-cost flow problem, the simplex method, uses a particularly simple way to generate unblocked cycles with negative cost. It maintains a spanning tree T of the problem graph, and a partition of the set of arcs not in T in two disjoint subsets, L and U. Each triplet (T, L, U), called a basis, defines a unique flow vector x satisfying the conservation of flow constraints (8), and such that $x_{ij} = l_{ij}$ for all $(i, j) \in L$ and $x_{ij} = u_{ij}$ for all $(i, j) \in U$. The flow of the arcs (i, j) of T is specified as follows:

$$x_{ij} = \sum_{v \in T_i} b_v - \sum_{\substack{(v,w)\in L \\ v\in T_i \\ w\in T_j}} l_{vw} - \sum_{\substack{(v,w)\in U \\ v\in T_i \\ w\in T_j}} u_{vw} + \sum_{\substack{(v,w)\in L \\ v\in T_j \\ w\in T_i}} l_{vw} + \sum_{\substack{(v,w)\in U \\ v\in T_j \\ w\in T_i}} u_{vw},$$

where T_i and T_j are the subtrees into which (i, j) separates T.

A basis will be called feasible if the corresponding flow vector is feasible, that is, satisfies $l_{ij} \le x_{ij} \le u_{ij}$ for all $(i, j) \in T$. We say that the feasible basis (T, L, U) is *strongly feasible* if all arcs $(i, j) \in T$ with $x_{ij} = l_{ij}$ are oriented away from the root (that is, the unique simple path of T starting at the root and ending at j passes through i) and if all arcs $(i, j) \in T$ with $x_{ij} = u_{ij}$ are oriented towards the root (that is, the unique simple path from the root to i passes through j). Strongly feasible trees are used to deal with the problem of degeneracy (possible cycling through a sequence of bases).

The simplex method starts with a strongly feasible basis and proceeds in iterations, generating another feasible basis and a corresponding basic flow vector at each iteration. Each basic flow vector has cost which is no worse than the cost of its predecessor. At each iteration (also called a *pivot*, following standard linear programming terminology), the method operates roughly as follows:

(a) It uses a convenient method to add one arc to the tree so as to generate a simple cycle with negative cost.

(b) It pushes along the cycle as much flow as possible without violating feasibility.

(c) It discards one arc of the cycle, thereby obtaining another strongly feasible basis to be used at the next iteration.

To detect negative cost cycles, the simplex method fixes a root node r and associates with r a scalar p_r, which can be chosen arbitrarily. A basis (T, L, U) specifies a vector p using the formula:

$$p_i = p_r - \sum_{(v,w)\in P_i^+} c_{vw} + \sum_{(v,w)\in P_i^-} c_{vw}, \qquad \forall\, i \in \mathcal{N},$$

where P_i is the unique simple path of T starting at the root node r and ending at i, and P_i^+ and P_i^- are the sets of forward and backward arcs of P_i, respectively. We call p the *price vector* associated with the basis. The price vector defines uniquely the reduced cost of each arc (i, j) by

$$r_{ij} = c_{ij} + p_j - p_i.$$

Given the strongly feasible basis (T, L, U) with a corresponding flow vector x and price vector p, an iteration of the simplex method produces another strongly feasible basis $(\overline{T}, \overline{L}, \overline{U})$ as follows:

Typical simplex iteration:

Find an in-arc $\overline{e} = (\overline{i}, \overline{j}) \notin T$ such that either

$$r_{\overline{i}\,\overline{j}} < 0 \qquad \text{if} \qquad \overline{e} \in L$$

or

$$r_{\overline{i}\,\overline{j}} > 0 \qquad \text{if} \qquad \overline{e} \in U.$$

(If no such arc can be found, x is primal optimal and p is dual optimal.) Let C be the cycle closed by T and $\overline{e}$. Define the forward direction of C to be the same as the one of $\overline{e}$ if $\overline{e} \in L$ and opposite to $\overline{e}$ if $\overline{e} \in U$ (that is, $\overline{e} \in C^+$ if $\overline{e} \in L$ and $\overline{e} \in C^-$ if $\overline{e} \in U$). Let also

$$\delta = \min\left\{ \min_{(i,j)\in C^-} \{x_{ij} - l_{ij}\},\ \min_{(i,j)\in C^+} \{u_{ij} - x_{ij}\} \right\},$$

and let $\hat{C}$ be the set of arcs where this minimum is obtained

$$\hat{C} = \{(i, j) \in C^- \mid x_{ij} - l_{ij} = \delta\} \cup \{(i, j) \in C^+ \mid u_{ij} - x_{ij} = \delta\}.$$

Define the *join of* C as the first node of C that lies on the unique simple path of T that starts from the root and ends at $\bar{i}$. Select as out-arc the arc e of $\hat{C}$ that is encountered first as C is traversed in the forward direction starting from the join node. The new tree $\overline{T}$ is obtained from T by adding $\bar{e}$ and deleting e. The corresponding flow vector $\bar{x}$ is obtained from x by

$$\bar{x}_{ij} = \begin{cases} x_{ij} & \text{if } (i, j) \notin C \\ x_{ij} + \delta & \text{if } (i, j) \in C^+ \\ x_{ij} - \delta & \text{if } (i, j) \in C^-. \end{cases}$$

The initial strongly feasible tree is usually obtained by introducing an extra node 0 and the artificial arcs $(i, 0)$, for all i with $b_i > 0$, and $(0, i)$, for all i with $b_i \le 0$. The corresponding basic flow vector x is given by $x_{ij} = l_{ij}$ for all $(i, j) \in \mathcal{A}$, $x_{i0} = b_i$, for all i with $b_i > 0$, and $x_{0i} = -b_i$, for all i with $b_i \le 0$. The cost of the artificial arcs is taken to be a scalar M, which is large enough to ensure that the artificial arcs do not affect the optimal solutions of the problem. This is known as the *Big M* initialization method. It can be shown that if M is large enough and the original problem (no artificial arcs) is feasible, the simplex method terminates with an optimal solution where all the artificial arcs carry zero flow. From this solution one can extract an optimal flow vector for the original problem.

2.1.2. Dual cost improvement

Duality theory for the min-cost flow problem can be developed by introducing a price vector $p = \{p_j \mid j \in \mathcal{N}\}$. We say that a flow-price vector pair (x, p) satisfies *complementary slackness (or CS for short)* if x satisfies $l_{ij} \le x_{ij} \le u_{ij}$ and

$$x_{ij} < u_{ij} \quad \Rightarrow \quad p_i - p_j \le c_{ij}, \quad \forall (i, j) \in \mathcal{A}, \tag{14}$$

$$l_{ij} < x_{ij} \quad \Rightarrow \quad p_i - p_j \ge c_{ij}, \quad \forall (i, j) \in \mathcal{A}. \tag{15}$$

Note that the above conditions imply that we must have $p_i - p_j = c_{ij}$ if $l_{ij} < x_{ij} < u_{ij}$.

The dual problem is obtained by a standard procedure in duality theory. We view p_i as a Lagrange multiplier associated with the conservation of flow constraint for node i and we form the corresponding Lagrangian function

$$\begin{aligned} L(x, p) &= \sum_{(i,j)\in\mathcal{A}} c_{ij} x_{ij} + \sum_{i\in\mathcal{N}} \left(b_i - \sum_{\{j|(i,j)\in\mathcal{A}\}} x_{ij} + \sum_{\{j|(j,i)\in\mathcal{A}\}} x_{ji} \right) p_i \\ &= \sum_{(i,j)\in\mathcal{A}} (c_{ij} + p_j - p_i) x_{ij} + \sum_{i\in\mathcal{N}} b_i p_i. \end{aligned} \tag{16}$$

Then, the dual function value $q(p)$ at a vector p is obtained by minimizing $L(x, p)$ over all capacity-feasible flows x:

$$q(p) = \min_x \left\{ L(x, p) \mid l_{ij} \le x_{ij} \le u_{ij}, (i, j) \in \mathcal{A} \right\}. \tag{17}$$

Because the Lagrangian function $L(x, p)$ is separable in the arc flows x_{ij}, its minimization decomposes into A separate minimizations, one for each arc (i, j). Each of these minimizations can be carried out in closed form, yielding

$$q(p) = \sum_{(i,j)\in\mathcal{A}} q_{ij}(p_i - p_j) + \sum_{i\in\mathcal{N}} b_i p_i, \tag{18}$$

where

$$\begin{aligned} q_{ij}(p_i - p_j) &= \min_{x_{ij}} \left\{ (c_{ij} + p_j - p_i) x_{ij} \mid l_{ij} \le x_{ij} \le u_{ij} \right\} \\ &= \begin{cases} (c_{ij} + p_j - p_i) l_{ij} & \text{if } p_i \le c_{ij} + p_j \\ (c_{ij} + p_j - p_i) u_{ij} & \text{if } p_i > c_{ij} + p_j. \end{cases} \end{aligned} \tag{19}$$

The dual problem is to maximize $q(p)$ subject to no constraint on p, with the dual functional q given by (18).

While there are several other duality results, the following proposition is sufficient for our purposes:

Proposition 2. *If a feasible flow vector x^* and a price vector p^* satisfy the complementary slackness conditions (1) and (2), then x^* is an optimal primal solution and p^* is an optimal dual solution. Furthermore, the optimal primal cost and the optimal dual objective value are equal.*

Dual cost improvement algorithms start with a price vector and try to successively obtain new price vectors with improved dual cost value. An important method of this type is known as the *sequential shortest path method*. It is mathematically equivalent to the classical primal–dual method of Ford and Fulkerson [1957]. Let us assume that the problem data are integer. The method starts with an integer pair (x, p) satisfying CS, and at each iteration, generates a new integer pair (x, p) satisfying CS such that the dual value of p is improved over the previous value.

To describe the typical iteration, for a pair (x, p) satisfying CS, define the *surplus* of a node i by

$$g_i = \sum_{\{j|(j,i)\in\mathcal{A}\}} x_{ji} - \sum_{\{j|(i,j)\in\mathcal{A}\}} x_{ij} + b_i.$$

An unblocked path is said to be an *augmenting path* if its start node has positive surplus and its end node has negative surplus. Consider the reduced costs of the arcs given by

$$r_{ij} = c_{ij} + p_j - p_i, \qquad \forall\, (i, j) \in \mathcal{A}. \tag{20}$$

We define the *length* of an unblocked path P by

$$L_P = \sum_{(i,j)\in P^+} r_{ij} - \sum_{(i,j)\in P^-} r_{ij}. \tag{21}$$

Note that since (x, p) satisfies CS, we have

$$r_{ij} \geq 0, \qquad \forall\, (i, j) \in P^+, \tag{22}$$

$$r_{ij} \leq 0, \qquad \forall\, (i, j) \in P^-. \tag{23}$$

Thus the length of P is nonnegative.

The sequential shortest path method starts each iteration with an integer pair (x, p) satisfying CS and with a set I of nodes i with $g_i > 0$, and proceeds as follows.

Sequential shortest path iteration:

Construct an augmenting path P with respect to x that has minimum length over all such paths that start at some node $i \in I$. Then, calculate

$$\delta = \min\left\{\min\{u_{ij} - x_{ij} \mid (i, j) \in P^+\}, \{x_{ij} - l_{ij} \mid (i, j) \in P^-\}\right\},$$

increase the flows of the arcs in P^+ by δ, and decrease the flows of the arcs in P^- by δ. Then, modify the node prices as follows: let $\overline{d}$ be the length of P and for each node $m \in \mathcal{N}$, let d_m be the minimum of the lengths of the unblocked paths with respect to x that start at some node in I and end at m ($d_m = \infty$ if no such path exists). The new price vector $\overline{p}$ is given by

$$\overline{p}_m = p_m + \max\{0, \overline{d} - d_m\}, \qquad \forall\, m \in \mathcal{N}. \tag{24}$$

The method terminates under the following circumstances:

(a) All nodes i have zero surplus; in this case it will be seen by Proposition 2 that the current pair (x, p) is primal and dual optimal.

(b) $g_i \leq 0$ for all i and $g_i < 0$ for at least one i; in this case the problem is infeasible, since $\sum_{i\in\mathcal{N}} b_i = \sum_{i\in\mathcal{N}} g_i < 0$.

(c) There is no augmenting path with respect to x that starts at some node in I; in this case it can be shown that the problem is infeasible.

We note that the shortest path computation can be executed using standard shortest path algorithms. The idea is to use r_{ij} as the length of each forward arc (i, j) of an unblocked path, and to reverse the direction of each backward arc (i, j) of an unblocked path and to use $-r_{ij}$ as its length [cf. the unblocked path length formula (21)]. Since by (22) and (23), the arc lengths of the residual graph are nonnegative, Dijkstra's method can be used for the shortest path computation. One can show the following result, which establishes the validity of the method [see e.g. Bertsekas, 1991].

Proposition 3. *Consider the min-cost flow problem and assume that a_{ij}, l_{ij}, u_{ij}, and b_i are all integer. Then, for the sequential shortest path method, the following hold:*

(a) Each iteration maintains the integrality and the CS property of the pair (x, p).

(b) If the problem is feasible, then after a finite number of iterations, the method terminates with an integer optimal flow vector x and an integer optimal price vector p.

(c) If the problem is infeasible, then after a finite number of iterations, the method terminates either because $g_i \leq 0$ for all i and $g_i < 0$ for at least one i, or because there is no augmenting path from any node of the set I to some node with negative surplus.

2.1.3. Approximate dual coordinate ascent

Our third type of algorithm represents a significant departure from the cost improvement idea; at any one iteration, it may worsen both the primal and the dual cost, although in the end it does find an optimal primal solution. It is based on an approximate version of complementary slackness, called *ϵ-complementary slackness*, and while it implicitly tries to solve a dual problem, it actually attains a dual solution that is not quite optimal.

The main idea of this class of methods was first introduced for the symmetric assignment problem, where we want to match n persons and n objects on a one-to-one basis so that the total benefit from the matching is maximized. We denote here by a_{ij} the benefit of assigning person i to object j. The set of objects to which person i can be assigned is a nonempty set denoted $A(i)$. An *assignment* S is a (possibly empty) set of person–object pairs (i, j) such that $j \in A(i)$ for all $(i, j) \in S$; for each person i there can be at most one pair $(i, j) \in S$; and for every object j there can be at most one pair $(i, j) \in S$. Given an assignment S, we say that person i is *assigned* if there exists a pair $(i, j) \in S$; otherwise we say that i is *unassigned*. We use similar terminology for objects. An assignment is said to be *feasible* if it contains n pairs, so that every person and every object is assigned; otherwise the assignment is called *partial*. We want to find an assignment $\{(1, j_1), \ldots, (n, j_n)\}$ with total benefit $\sum_{i=1}^{n} a_{ij_i}$, which is maximal.

It is well-known that the assignment problem is equivalent to the linear program

$$\max \sum_{i=1}^{n} \sum_{j \in A(i)} a_{ij}(x_{ij}) \tag{25}$$

$$\text{s.t.} \quad \sum_{j=1}^{n} x_{ij} = 1; \qquad \forall i = 1, \ldots, n \tag{26}$$

$$\sum_{i=1}^{n} x_{ij} = 1 \qquad \forall j = 1, \ldots, n \tag{27}$$

$$0 \leq x_{ij} \leq 1 \qquad \forall i = 1, \ldots, n, \; j \in A(i). \tag{28}$$

This linear program in turn can be converted into a min-cost flow problem of the form (7)–(9) involving n nodes $i = 1, \ldots, n$ corresponding to the n persons, another n nodes $j = 1, \ldots, n$ corresponding to the n objects, the graph with the set of arcs

$$\mathcal{A} = \{(i, j) | i = 1, \ldots, n, \; j \in A(i)\}$$

with corresponding costs $c_{ij} = -a_{ij}$ for all $(i, j) \in \mathcal{A}$, and upper and lower bounds $l_{ij} = 0, u_{ij} = 1$ for all $(i, j) \in \mathcal{A}$. The supplies at the nodes $i = 1, \ldots, n$ and $j = 1, \ldots, n$ are set to $b_i = 1$ and $b_j = -1$ respectively.

The *auction algorithm* for the symmetric assignment problem proceeds iteratively and terminates when a feasible assignment is obtained. At the start of the generic iteration we have a partial assignment S and a price vector $p = (p_1, \ldots, p_n)$ satisfying *ϵ-complementary slackness* (or ϵ-CS for short). This is the condition

$$a_{ij} - p_j \geq \max_{k \in A(i)} \{a_{ik} - p_k\} - \epsilon, \qquad \forall\ (i, j) \in S. \tag{29}$$

As an initial choice, one can use an arbitrary set of prices together with the empty assignment, which trivially satisfies ϵ-CS. The iteration consists of two phases: the *bidding phase* and the *assignment phase* described in the following.

Bidding phase: Let I be a nonempty subset of persons i that are unassigned under the assignment S. For each person $i \in I$:

1. Find a 'best' object j_i having maximum value, that is,

$$j_i = \arg\max_{j \in A(i)} \{a_{ij} - p_j\},$$

and the corresponding value

$$v_i = \max_{j \in A(i)} \{a_{ij} - p_j\}, \tag{30}$$

and find the best value offered by objects other than j_i

$$w_i = \max_{j \in A(i),\, j \neq j_i} \{a_{ij} - p_j\}. \tag{31}$$

[If j_i is the only object in $A(i)$, we define w_i to be $-\infty$ or, for computational purposes, a number that is much smaller than v_i.]

2. Compute the 'bid' of person i given by

$$l_{ij_i} = p_{j_i} + v_i - w_i + \epsilon = a_{ij_i} - w_i + \epsilon. \tag{32}$$

[We characterize this situation by saying that person i bid for object j_i, and that object j_i received a bid from person i. The algorithm works if the bid has any value between $p_{j_i} + \epsilon$ and $p_{j_i} + v_i - w_i + \epsilon$, but it tends to work fastest for the maximal choice of (32).]

Assignment phase: For each object j: Let $P(j)$ be the set of persons from which j received a bid in the bidding phase of the iteration. If $P(j)$ is nonempty, increase p_j to the highest bid:

$$p_j := \max_{i \in P(j)} l_{ij},$$

remove from the assignment S any pair (i, j) (if j was assigned to some i under S), and add to S the pair (i_j, j), where i_j is a person in $P(j)$ attaining the maximum above.

Note that there is some freedom in choosing the subset of persons I that bid during an iteration. One possibility is to let I consist of a single unassigned person. This version, known as the *Gauss–Seidel version* in view of its similarity with Gauss–Seidel methods for solving systems of nonlinear equations, usually works best in a serial computing environment. The version where I consists of all unassigned persons is the most well suited for parallel computation, and is known as the *Jacobi version*, in view of its similarity with Jacobi methods for solving systems of nonlinear equations.

The choice of bidding increment $v_i - w_i + \epsilon$ for a person i [cf. (32)] is such that ϵ-CS is preserved, as stated in the following proposition [see Bertsekas, 1979, 1988, 1991; Bertsekas & Tsitsiklis, 1989].

Proposition 4. *The auction algorithm preserves ϵ-CS throughout its execution, that is, if the assignment and price vector available at the start of an iteration satisfy ϵ-CS, the same is true for the assignment and price vector obtained at the end of the iteration.*

Furthermore, the algorithm is valid in the sense stated below.

Proposition 5. *If at least one feasible assignment exists, the auction algorithm terminates in a finite number of iterations with a feasible assignment that is within $n\epsilon$ of being optimal (and is optimal if the problem data is integer and $\epsilon < 1/n$).*

The auction algorithm can be shown to have an $O(A(n + nC/\epsilon))$ worst-case running time, where A is the number of arcs of the assignment graph, and

$$C = \max_{(i,j)\in\mathcal{A}} |a_{ij}|$$

is the maximum absolute object value; see Bertsekas [1979, 1988], Bertsekas & Tsitsiklis [1989]. Thus, the amount of work to solve the problem can depend strongly on the value of ϵ as well as C. In practice, the dependence of the running time on ϵ and C is often significant, particularly for sparse problems.

To obtain polynomial complexity, one can use *ϵ-scaling*, which consists of applying the algorithm several times, starting with a large value of ϵ and successively reducing ϵ up to an ultimate value that is less than $1/n$. Each application of the algorithm, called a *scaling phase*, provides good initial prices for the next application. In practice, scaling is typically beneficial, particularly for sparse assignment problems, that is, problems where the set of feasible assignment pairs is severely restricted. ϵ-scaling was first proposed in Bertsekas [1979] in connection with the auction algorithm. Its first analysis was given in Goldberg [1987] in the context of the ϵ-relaxation method, which is a related method and will be discussed shortly.

The ϵ-CS condition (29) can be generalized for the min-cost flow problem. For a flow vector x satisfying $l_{ij} \leq x_{ij} \leq u_{ij}$ for all arcs (i, j), and a price vector p it takes the form

$$x_{ij} < u_{ij} \quad \Rightarrow \quad p_i - p_j \leq a_{ij} + \epsilon \quad \forall\, (i, j) \in \mathcal{A}, \tag{33}$$

$$l_{ij} < x_{ij} \quad \Rightarrow \quad p_i - p_j \geq a_{ij} + \epsilon \quad \forall\, (i, j) \in \mathcal{A}. \tag{34}$$

It can be shown that if x is feasible and satisfies the ϵ-CS conditions (33) and (34) together with some p, then the cost corresponding to x is within $N\epsilon$ of being optimal, where N is the number of nodes; x is optimal if it is integer, if $\epsilon < 1/n$, and if the problem data is integer.

We now define some terminology and computational operations that can be used as building blocks in various auction-like algorithms.

Definition 1. *An arc (i, j) is said to be ϵ^+-*unblocked *if*

$$p_i = p_j + a_{ij} + \epsilon \qquad \textit{and} \qquad x_{ij} < u_{ij}.$$

*An arc (j, i) is said to be ϵ^--*unblocked *if*

$$p_i = p_j - a_{ji} + \epsilon \qquad \textit{and} \qquad l_{ji} < x_{ji}.$$

The push list *of a node i is the (possibly empty) set of arcs (i, j) that are ϵ^+-unblocked, and arcs (j, i) that are ϵ^--unblocked.*

Definition 2. *For an arc (i, j) [or arc (j, i)] of the push list of node i, let δ be a scalar such that $0 < \delta \leq u_{ij} - x_{ij}$ ($0 < \delta \leq x_{ji} - l_{ji}$, respectively). A δ-*push at node i on arc (i, j) *[(j, i), respectively] consists of increasing the flow x_{ij} by δ (decreasing the flow x_{ji} by δ, respectively), while leaving all other flows, as well as the price vector unchanged.*

In the context of the auction algorithm for the assignment problem, a δ-push (with $\delta = 1$) corresponds to assigning an unassigned person to an object; this results in an increase of the flow on the corresponding arc from 0 to 1. The next operation consists of raising the prices of a subset of nodes by the maximum common increment γ that will not violate ϵ-CS.

Definition 3. *A* price rise *of a nonempty, strict subset of nodes I (i.e., $I \neq \emptyset$, $I \neq \mathcal{N}$), consists of leaving unchanged the flow vector x and the prices of nodes not belonging to I, and of increasing the prices of the nodes in I by the amount γ given by*

$$\gamma = \begin{cases} \min\{\gamma^+, \gamma^-\}, & \textit{if } S^+ \cup S^- \neq \emptyset, \\ 0, & \textit{if } S^+ \cup S^- = \emptyset, \end{cases}$$

where S^+ and S^- are the sets of scalars given by

$$S^{+} = \{p_j + a_{ij} + \epsilon - p_i \mid (i, j) \in \mathcal{A} \text{ such that } i \in I,\ j \notin I,\ x_{ij} < u_{ij}\},$$

$$S^{-} = \{p_j - a_{ji} + \epsilon - p_i \mid (j, i) \in A \text{ such that } i \in I,\ j \notin I,\ l_{ji} < x_{ji}\},$$

and

$$\gamma^{+} = \min_{s \in S^{+}} s, \qquad \gamma^{-} = \min_{s \in S^{-}} s.$$

The ϵ-relaxation method, first proposed in Bertsekas [1986a, b], may be viewed as the extension of the auction algorithm to the min-cost flow problem. The ϵ-relaxation method uses a fixed positive value of ϵ, and starts with a pair (x, p) satisfying ϵ-CS. Furthermore, the starting arc flows are integer, and it will be seen that the integrality of the arc flows is preserved thanks to the integrality of the node supplies and the arc flow bounds. (Implementations that have good worst-case complexity also require that all initial arc flows be either at their upper or their lower bound; see e.g. Bertsekas & Tsitsiklis [1989]. In practice, this can be easily enforced, although it does not seem to be very important.) At the start of a typical iteration of the method we have a flow-price vector pair (x, p) satisfying ϵ-CS and we select a node i with $g_i > 0$; if no such node can be found, the algorithm terminates. During the iteration we perform several operations of the type described earlier involving node i.

Typical iteration of the ϵ-relaxation method:

Step 1: If the push list of node i is empty go to Step 3; else select an arc a from the push list of i and go to Step 2.

Step 2: Let j be the node of arc a which is opposite to i. Let

$$\delta = \begin{cases} \min\{g_i, u_{ij} - x_{ij}\} & \text{if } a = (i, j), \\ \min\{g_i, x_{ji} - l_{ji}\} & \text{if } a = (j, i). \end{cases}$$

Perform a δ-push of i on arc a. If as a result of this operation we obtain $g_i = 0$, go to Step 3; else go to Step 1.

Step 3: Perform a price rise of node i. If $g_i = 0$, go to the next iteration; else go to Step 1.

It can be shown that for a feasible problem, the method terminates finitely with a feasible flow vector, which is optimal if $\epsilon < 1/n$.

The ϵ-scaling technique discussed for the auction algorithm is also important in the context of the ϵ-relaxation method, and improves its practical performance. A complexity analysis of ϵ-scaling (first given in Goldberg [1987]; see also Bertsekas & Eckstein [1987, 1988], Bertsekas & Tsitsiklis [1989], Goldberg & Tarjan [1990]) shows that the ϵ-relaxation method, coupled with scaling, has excellent worst-case complexity.

It is possible to define a symmetric form of the ϵ-relaxation iteration that starts from a node with negative surplus and *decreases* (rather than increases) the price of that node. Furthermore, one can mix positive surplus and negative surplus iterations in the same algorithm. However, if the two types of iterations are mixed arbitrarily, the algorithm is not guaranteed to terminate finitely even for a feasible problem; for an example, see Bertsekas & Tsitsiklis [1989, p. 373]. For this reason, some care must be exercised in mixing the two types of iterations so as to guarantee that the algorithm eventually makes progress. With the use of negative surplus iterations, one can increase the parallelism potential of the method.

2.2. *Parallelization ideas*

2.2.1. *Primal cost improvement*

The most computation-intensive part of the primal simplex method is the selection of a new arc (i, j) to bring into the basis. Ideally, one would like to select an arc (i, j) which violates the optimality condition the most, so that it has the largest $|r_{ij}|$. However, such an algorithm would require computation of all the reduced costs r_{ij} of all arcs at each iteration, and would be very time-consuming. An alternative is to select the first nonbasic arc $(i, j) \notin T$ that has negative reduced cost and its flow at its lower bound, or has positive reduced cost and its flow is at its upper bound. Such an implementation may quickly find a candidate arc, but the progress towards optimality may be slow.

Successful network simplex algorithms [J.M. Mulvey, 1978b; Kennington & Helgason, 1980; Gibby, Glover, Klingman & Mead, 1983; Grigoriadis, 1986] often adopt an intermediate strategy, where a candidate list of arcs is generated, and the algorithm searches within this list for a candidate arc to bring into the basis. In this approach, selection of arcs becomes a two-phase procedure: First, a candidate list of arcs is constructed (typically, the maximum size of this list is a preset parameter). Second, the candidate list of arcs is scanned in order to find the arc in the candidate list which violates the optimality condition the most. While performing the second search at each iteration, arcs in the candidate list which no longer violate the optimality conditions are removed; eventually, the candidate list of arcs is empty. At this point, a new candidate list would be generated, and the iterations would be continued.

The main idea for accelerating the network simplex computations is the parallel computation of the candidate list of arcs. Using multiple processors, a larger list of candidate arcs can be generated and searched efficiently in a distributed manner to find a desirable pivot arc. Many approaches are possible, depending on how many arcs are considered between pivot steps and how the different candidate lists of arcs are generated. We will briefly overview three approaches: the work of Miller, Pekny & Thompson [1989] for transportation problems and the work of Peters [1990] and Barr & Hickman [1990] for min-cost network flow problems.

In Miller, Pekny & Thompson [1989], a parallel simplex algorithm is presented for transportation problems. This algorithm is structured around the concept of selecting a candidate list of pivot arcs in parallel, and then performing the individual pivots in that candidate list in a sequential manner. Each processor evaluates a fixed number of arcs (for transportation problems, this corresponds to a fixed number of rows in the transportation matrix) in order to find the best pivot arc among its assigned arcs. The union of the sets of arcs evaluated by the parallel processors may only be a subset of the network. The set of pivot arcs found by all the processors becomes the candidate arc list. If no admissible pivot arcs were found by any of the processors, a different subset of arcs is searched in parallel for admissible pivot arcs. For transportation problems, this is coordinated by selecting which rows will be searched in parallel. Once a candidate list of pivot arcs is generated (with length less than or equal to the number of processors), the pivots in this candidate list are executed and reevaluated in sequential fashion.

The algorithm of Miller, Pekny & Thompson [1989] is essentially a synchronous algorithm, with processor synchronization after each processor has searched its assigned arcs. Using this synchronization guarantees that all of the processors have completed their search before pivoting begins. Synchronization also facilitates recognition of algorithm convergence; at the synchronization point, if no candidate pivot arcs are found, the algorithm has converged to an optimal solution.

In Peters [1990], a different approach is pursued. Multiple processors are used to search in parallel the entire set of arcs in order to find the most desirable pivot arc in each iteration. At first glance, this appears inefficient. However, the parallel network simplex algorithm of Peters [1990] performs this search asynchronously, while the pivot operations are taking place. The processors are organized into a single pivot processor and a number of search processors. While the pivot processor is performing a pivoting operation, the search processors are using node prices (possibly outdated, since the pivoting operation may change them) to compute reduced costs and identify potential future pivot arcs. Each search processor is responsible for a subset of the arcs. Once a pivot operation is complete, the pivot processor obtains the combined results of the searches (which may not be complete), selects a new pivot arc, and performs a new pivot operation.

In contrast with Miller, Pekny & Thompson [1989], the algorithm of Peters [1990] has no inherent sequential computation time, since the search processors continue their search for pivot arcs while pivot operations are in progress. In order to guarantee algorithm convergence, the pivot processor must verify that the results of the search processors have indeed produced an acceptable pivot. In essence, one can view the operation of the search processors as continuously generating a small candidate list of pivot arcs; this list is then searched by the pivot processor to find the best pivot arc in the list. If this list is empty, the pivot processor will search the entire network to either locate a pivot arc or verify that an optimal solution has been reached. Thus, convergence of the asynchronous algorithm is established because a single processor (the pivot processor) selects valid pivot arcs, performs pivots and checks whether an optimal solution has been reached.

The algorithm of Barr & Hickman [1990] is a refinement of the asynchronous approach of Peters [1990]. Instead of having processors dedicated to search tasks or pivot tasks, Barr and Hickman create a shared queue of tasks which are allocated to available processors using a monitor [Hoare, 1974]. This approach offers the advantage that the algorithm can operate correctly with an arbitrary number of processors, since each processor is capable of performing each task. Barr and Hickman divide the search operation into multiple search tasks involving groups of outgoing arcs from disjoint subsets of nodes; they also divide the pivot operation into two steps: basis and flow update, and dual price update. Thus, the shared queue of tasks includes three types of tasks: searching a group of arcs to identify potential future pivot arcs, selecting a candidate pivot arc and performing a basis and flow update, and updating the prices on a group of nodes. Among these tasks, the task of selecting a candidate pivot arc and performing a basis and flow update is assigned the highest priority.

In Barr and Hickman's algorithm, an idle processor will first confirm that there is no current pivot operation in progress, and check the list of candidate arcs to determine whether any eligible pivot arcs have been found. If this is the case, the processor will begin a pivot operation and, after completing the basis and flow update, the same processor will perform the dual price update. In the interim, other processors will continue the search operations, generating candidates while using potentially-outdated prices. However, if a pivot is currently in progress, then the idle processor will select a search task and search a subset of arcs to generate pivot candidates. An additional refinement introduced by Barr and Hickman is that, when the number of dual price updates is sufficiently high, the task of updating dual prices is split into two tasks, and scheduled for two different processors. Convergence is achieved when all search tasks have been completed after the dual prices have been updated, and no candidate pivot arcs are identified.

As the discussion indicates, all of the above parallel network simplex algorithms perform pivots one at a time. An alternative approach was proposed in Peters [1990], where multiple pivot operations would be performed simultaneously in parallel. As Peters [1990] points out, this is possible provided the different pivots affect non-overlapping parts of the network; otherwise, significant synchronization overhead is incurred. Computational results using this approach [Peters, 1990] indicate that, for general network problems, little speedup is possible from parallel computation. In contrast, the approaches discussed above obtain significant reductions in computation time using parallel computation, as described in the computation experience section.

The above concepts for parallel network algorithms have been extended to the case of linear generalized network problems by Clark, Engquist, Finkel & Meyer [1987], Clark & Meyer [1987, 1989], Clark, Kennington, Meyer & Ramamurti [1987], Chen & Meyer [1988]. In their work, they develop variations of the parallel network simplex algorithms discussed above. In addition, due to the special structure of the basis for generalized network problems (discussed more extensively in the next chapter), it is often possible to perform parallel pivots which do not overlap. Clark and Meyer and their colleagues show that, for generalized networks, algorithms which perform multiple pivots in parallel often outperform the generalizations of the network simplex algorithms discussed above.

2.2.2. Dual cost improvement

In contrast with the network simplex algorithms of the previous section, there are two successful classes of parallel sequential shortest path algorithms for min-cost network flow problems: we denote these as *single node* and *multinode* parallel algorithms. By a single node parallel sequential shortest path algorithm, we mean an algorithm where a single augmenting path is constructed efficiently using parallel computation. In contrast, a multinode algorithm constructs several augmenting paths in parallel (from multiple sources), and potentially executes multiple augmentations in parallel.

The key step in single node parallel shortest path algorithms is the computation of a shortest path tree from a single origin. Most sequential implementations use a variation of Dijkstra's algorithm [Dijkstra, 1959], thereby exploiting the fact that only the distances to nodes which are closer than the length of the shortest path are needed. In a parallel context, alternative shortest path algorithms (e.g. Bellman–Ford [Bellman, 1958] or the recently-proposed parallel label-correcting algorithm of Bertsekas, Guerriero & Musmano [1994] or the auction algorithm of Polymenakos & Bertsekas [1993]) can be introduced which would be more amenable for parallel implementation. Nevertheless, the increased efficiency may be offset by the additional computation of shortest paths to all nodes.

Parallelization of Dijkstra's algorithm can take place at two levels: parallel scanning of multiple nodes, and parallel scanning of the arcs connected to a single node. The effectiveness of the first approach is limited by the nature of Dijkstra's algorithm, which uses parallel scanning only when the shortest distance to two nodes is the same. Thus, most parallel implementations of Dijkstra's algorithm have been limited to parallel scanning of the arcs connected to a single node [e.g. Kennington & Wang, 1988; Zaki, 1990]. The effectiveness of this approach to parallelization is limited by the density of the network; the work of Kennington & Wang [1988] and Zaki [1990] focuses on fully dense assignment problems with large numbers of persons, so that significant speedups can be obtained. For sparse problems, the effective speedup is very limited.

The alternative multinode approach is to compute multiple augmenting paths (starting from different sources with positive surplus), and to combine the outcome of the computations to perform multiple augmentations in parallel and to obtain a new set of prices which satisfy CS with the resulting flows. Balas, Miller, Pekny, and Toth [1991] introduced such an algorithm for the assignment problem. They developed a coordination protocol for combining the computations from multiple augmenting paths in order to increase the number of assignments and the node prices while preserving CS. In Bertsekas & Castañon [1990a], these results were extended to allow for different types of coordination protocols, including asynchronous ones. Subsequently, the results of Balas, Miller, Pekny & Toth [1991] and Bertsekas & Castañon [1990a] were extended to obtain multinode parallel algorithms for the general network flow problem [Bertsekas & Castañon [1990b]. We discuss these results below.

The following parallel algorithm [Bertsekas & Castañon, 1990b] is a direct generalization of the assignment algorithm of Balas, Miller, Pekny & Toth [1991] to min-cost network flow problems. It starts with a pair (x, p) satisfying CS and generates another pair $(\bar{x}, \bar{p})$ as follows:

Parallel synchronous sequential shortest path iteration:

Choose a subset $I = \{i_1, \ldots, i_K\}$ of nodes with positive surplus. (If all nodes have nonpositive surplus, the algorithm terminates.) For each i_k, $k = 1, \ldots, K$, let $\bar{p}(k)$ and $\bar{P}(k)$ be the price vector and augmenting path obtained by executing a sequential shortest path iteration starting at i_k, and using the pair (x, p). Then generate sequentially the pairs $(x(k), p(k))$, $k = 1, \ldots, K$, as follows, starting with $(x(0), p(0)) = (x, p)$:

For $k = 0, \ldots, K - 1$, if $\bar{P}(k + 1)$ is an augmenting path with respect to $x(k)$, obtain $x(k + 1)$ by augmenting $x(n)$ along $\bar{P}(k + 1)$, and set

$$p_j(k + 1) = \max\{p_j(k), \bar{p}_j(k)\}, \qquad \forall\ j \in \mathcal{N}.$$

Otherwise set

$$x(k + 1) = x(k), \qquad p(k + 1) = p(k).$$

The pair $(\bar{x}, \bar{p})$ generated by the iteration is

$$\bar{x} = x(K), \qquad \bar{p} = p(K).$$

In Bertsekas & Castañon [1990b], the above algorithm is shown to converge to an optimal solution of problem (MCF). Note that the protocol used for combining information from multiple parallel sequential shortest path computations consists of two conditions: First, multiple augmenting paths can be used provided that the paths do not have any nodes in common. Second, whenever multiple augmenting paths are used, the prices of nodes are raised to the maximum level associated with any of the paths used for augmenting the flow in the network. The first condition guarantees that the arc flows stay within bounds, while the second condition guarantees that CS is preserved.

The preceding algorithm can be parallelized by using multiple processors to compute the augmenting paths and the associated prices of an iteration in parallel. On the other hand the algorithm is synchronous in that iterations have clear 'boundaries'. In particular, all augmenting paths generated in the same iteration are computed on the basis of the same pair (x, p). Thus, it is necessary to synchronize the parallel processors at the beginning of each iteration, with an attendant synchronization penalty.

In Bertsekas & Castañon [1990b], an asynchronous version of the above algorithm is introduced. Let us denote the flow-price pair at the times $k = 1, 2, 3, \ldots$ by $(x(k), p(k))$. (In a practical setting, the times k represent 'event times', that is, times at which an attempt is made to modify the pair (x, p) through an iteration.) We require that the initial pair $(x(1), p(1))$ satisfies CS. The algorithm terminates when during an iteration, either a feasible flow is obtained or else infeasibility is detected.

kth Asynchronous primal–dual iteration:

At time k, the results of a primal–dual iteration performed on a pair $(x(\tau_k), p(\tau_k))$ are available, where τ_k is a positive integer with $\tau_k \leq k$; let $\bar{P}_k$ denote the augmenting path and $\bar{p}_k$ the resulting desired prices. The iteration (and the path $\bar{P}_k$) is said to be *incompatible* if $\bar{P}_k$ is not an augmenting path with respect to $x(k)$; in this case we discard the results of the iteration, that is, we set

$$x(k+1) = x(k), \qquad p(k+1) = p(k).$$

Otherwise, we say that the iteration (and the path $\bar{P}_k$) is *compatible*, we obtain $x(k+1)$ from $x(k)$ by augmenting $x(k)$ along $\bar{P}_k$, and we set

$$p_j(k+1) = \max\{p_j(k), \bar{p}_j(k)\}, \qquad \forall\, j \in \mathcal{N}.$$

Parallel implementation of this asynchronous algorithm is quite straightforward. The main idea is to maintain a 'master' copy of the current flow-price pair; this is the pair $(x(k), p(k))$ in the preceding mathematical description. To execute an iteration, a processor copies the current master flow-price pair; during this copy operation the master pair is locked, so no other processor can modify it. The processor performs a primal–dual iteration using the copy obtained, and then attempts to modify the master pair (which may be different from the start of the iteration); the results of the iteration are checked for compatibility with the current flows $x(k)$, and if compatible, they are used to modify the master flow-price pair. The times when the master pair is copied and modified correspond to the indices τ_k and k of the asynchronous algorithm, respectively, as illustrated in Figure 1.

In Bertsekas & Castañon [1990b], the asynchronous algorithm is shown to converge under the condition

$$\lim_{k\to\infty} \tau_k = \infty.$$

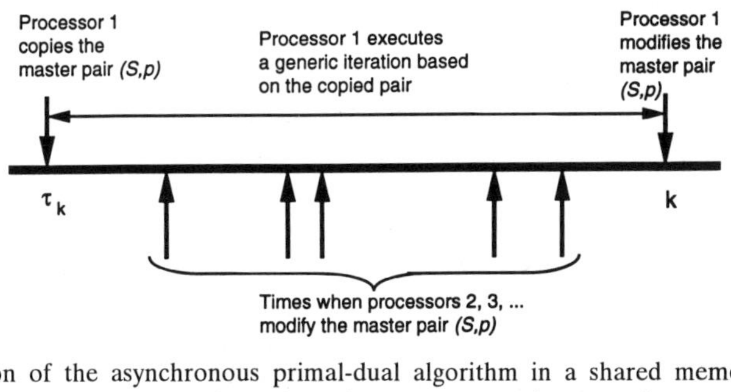

Fig. 1. Operation of the asynchronous primal-dual algorithm in a shared memory machine. A processor copies the master flow-price pair at time τ_k, executes between times τ_k and k a generic iteration using the copy, and modifies accordingly the master flow-price pair at time k. Other processors may have modified unpredictably the master pair between times τ_k and k.

This is a natural and essential condition, stating that the algorithm iterates with increasingly more recent information. A simpler sufficient condition guaranteeing convergence is that the maximum computation time plus communication delay in each iteration is bounded by an arbitrarily large constant.

2.2.3. Approximate dual coordinate ascent

The auction and ϵ-relaxation algorithms of Section 2.1.3 allow for a variety of parallel implementations. Most of the experimental work [(e.g. Phillips & Zenios, 1989; Bertsekas & Castañon, 1989; Zaki, 1990; Wein & Zenios, 1990, 1991] has focused on auction algorithms for assignment problems, for which the sequential performance of the auction algorithm is among the fastest. Thus, we restrict our parallelization discussion to the auction algorithm for assignment problems. Extensions of the concepts for parallel auction algorithms to parallel ϵ-relaxation algorithms for min-cost network flow problems have been developed recently by Li and Zenios [1991, 1992] for implementation on the Connection Machine CM-2, and by Narendran, DeLeone & Tiwari [1993] for implementation on the CM-5.

The auction algorithm for the assignment problem was designed to allow an arbitrary nonempty subset I of unassigned persons to submit a bid at each iteration. This gives rise to a variety of possible implementations, named after their analogs in relaxation and coordinate descent methods for solving systems of equations or unconstrained optimization problems [see e.g. Ortega & Rheinboldt, 1970; Bertsekas & Tsitsiklis, 1989]:

Jacobi where I is the set of all unassigned persons at the beginning of the iteration.

Gauss–Seidel where I consists of a single person, who is unassigned at the beginning of the iteration.

Block Gauss–Seidel where I is a subset of the set of all unassigned persons at the beginning of the iteration. The method for choosing the persons in the subset I may vary from one iteration to the next. This implementation contains the preceding two as special cases.

Similar to the dual improvement methods of the previous section, there are two basic approaches for developing parallel auction algorithms: in the first approach ('multinode'), the bids of several unassigned persons are carried out in parallel, with a single processor assigned to each bid; this approach is suitable for both the *Jacobi* and *block Gauss–Seidel* implementations. In the second approach ('single-node'), there is only one bid carried out at a time, but the calculation of each bid is done in parallel by several processors; this approach is suitable for the *Gauss–Seidel* implementation.

The above two approaches can be combined in a *hybrid* approach, whereby multiple bids are carried out in parallel, and the calculation of each bid is shared by several processors. This third approach, with proper choice of the number of processors used for each parallel task, has the maximum speedup potential.

An important characteristic of the auction algorithm is that it will converge to an optimal solution under a totally asynchronous implementation. The following result from Bertsekas & Castañon [1989] describes this asynchronous implementation:

We denote

$p_j(t)$ = price of object j at time t;
$r_j(t)$ = person assigned to object j at time t [$r_j(t) = 0$ if object j is unassigned];
$U(t)$ = set of unassigned persons at time t [$i \in U(t)$ if $r_j(t) \neq i$ for all objects j].

We assume that $U(t)$, $p_j(t)$, and $r_j(t)$ can change only at *integer* times t (t may be viewed as the index of a sequence of physical times at which events of interest occur.)

In addition to $U(t)$, $p_j(t)$, and $r_j(t)$, the algorithm maintains at each time t, a subset $R(t) \subset U(t)$ of unassigned persons that may be viewed as having a 'ready bid' at time t. We assume that by time t, a person $i \in R(t)$ has used prices $p_j(\tau_{ij}(t))$ and $p_j(\overline{\tau}_{ij}(t))$ from some earlier times $\tau_{ij}(t)$ and $\overline{\tau}_{ij}(t)$ with $\tau_{ij}(t) \leq \overline{\tau}_{ij}(t) \leq t$ to compute the best value

$$v_i(t) = \max_{j|(i,j)\in A} \left\{a_{ij} - p_j(\tau_{ij}(t))\right\}, \tag{35}$$

a best object $j_i(t)$ attaining the above maximum,

$$j_i(t) = \arg \max_{j|(i,j)\in A} \left\{a_{ij} - p_j(\tau_{ij}(t))\right\}, \tag{36}$$

the second best value

$$w_i(t) = \max_{j|(i,j)\in A, j\neq j_i(t)} \left\{a_{ij} - p_j(\overline{\tau}_{ij}(t))\right\}, \tag{37}$$

and has determined a bid

$$\beta_i(t) = a_{ij_i(t)} - w_i(t) + \epsilon. \tag{38}$$

(Note that ordinarily the best and second best values should be computed simultaneously, which implies that $\tau_{ij}(t) = \overline{\tau}_{ij}(t)$. In some cases, however, it may be more natural or advantageous to compute the second best value after the best value, with more up-to-date price information, which corresponds to the case $\tau_{ij}(t) \leq \overline{\tau}_{ij}(t)$ for some pairs (i, j).)

Assumption 1.

$$U(t) : \text{nonempty} \quad \Rightarrow \quad R(t') : \text{nonempty for some } t' \geq t.$$

Assumption 2. For all i, j, and t,

$$\lim_{t\to\infty} \tau_{ij}(t) = \infty.$$

The above assumptions guarantee that unassigned persons do not stop submitting bids and old information is eventually discarded. At each time t, if all persons are assigned ($U(t)$ is empty), the algorithm terminates. Otherwise, if $R(t)$ is empty nothing happens. If $R(t)$ is nonempty the following occur:

(a) A nonempty subset $I(t) \subset R(t)$ of persons that have a bid ready is selected.

(b) Each object j for which the corresponding bidder set

$$B_j(t) = \{i \in I(t) \mid j = j_i(t)\}$$

is nonempty, determines the highest bid

$$b_j(t) = \max_{i \in B_j(t)} \beta_i(t)$$

and a person $i_j(t)$ for which the above maximum is attained

$$i_j(t) = \arg\max_{i \in B_j(t)} \beta_i(t).$$

Then, the pair $(p_j(t), r_j(t))$ is changed according to

$$(p_j(t+1), r_j(t+1)) = \begin{cases} (b_j(t), i_j(t)) & \text{if } b_j(t) \geq p_j(t) + \epsilon \\ (p_j(t), r_j(t)) & \text{otherwise.} \end{cases}$$

The following proposition [Bertsekas & Castañon, 1989] establishes the validity of the above asynchronous algorithm.

Proposition 6. *Let Assumptions 1 and 2 hold and assume that there exists at least one complete assignment. Then for all t and all j for which $r_j(t) \neq 0$, the pair $(p_j(t), r_j(t))$ satisfies the ϵ-CS condition*

$$\max_{k \mid (i,k) \in A} \{a_{ik} - p_k(t)\} - \epsilon \leq a_{ij} - p_j(t), \qquad \text{if } i = r_j(t).$$

Furthermore, there is a finite time at which the algorithm terminates. The complete assignment obtained upon termination is within $n\epsilon$ of being optimal, and is optimal if $\epsilon < 1/n$ and the benefits a_{ij} are integer.

Notice that if $\tau_{ij}(t) = t$ and $U(t) = R(t)$ for all t, then the asynchronous algorithm is equivalent to the auction algorithm given in Section 2.1.3. The asynchronous model becomes relevant in a parallel computation context where some processors compute bids for some unassigned persons, while other processors simultaneously update some of the object prices and corresponding assigned persons. Suppose that a single processor calculates a bid of person i by using the values $a_{ij} - p_j(\tau_{ij}(t))$ prevailing at times $\tau_{ij}(t)$ and then calculates the maximum value at time t; see Figure 2. Then, if the price of an object $j \in A(i)$ is updated

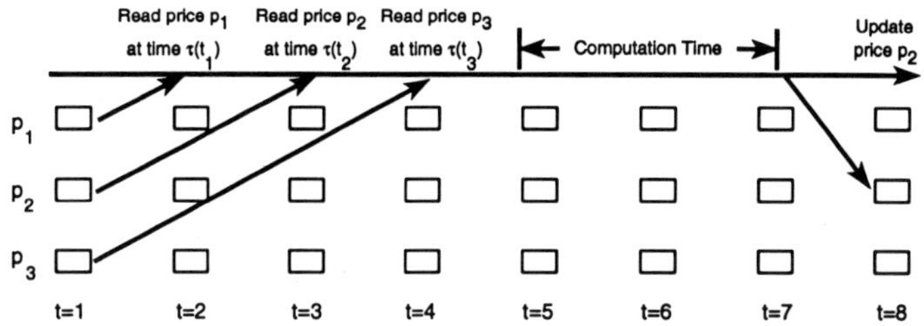

Fig. 2. Illustration of asynchronous calculation of a bid by a single processor, which reads from memory the values p_j at different times $\tau_{ij}(t)$ and calculates at time t the best object $j_i(t)$ and the maximum and second maximum values. The values of p_j may be out-of-date because they may have been updated by another processor between the read time $\tau_{ij}(t)$ and the bid calculation time t.

between times $\tau_{ij}(t)$ and t by some other processor, the maximum value will be based on out-of-date information. The asynchronous algorithm models this possibility by allowing $\tau_{ij}(t) < t$. A similar situation arises when the bid of person i is calculated cooperatively by several processors rather than by a single processor.

It is interesting to contrast the asynchronous auction algorithm described above with the asynchronous sequential shortest path algorithm of Bertsekas & Castañon [1990a]. In the sequential shortest path algorithm, computation of each shortest path must be based on a complete set of flows and prices (x, p) which may be outdated, but which represent a single previous state of the computation. In contrast, the asynchronous auction algorithm can use prices and assignments from many different times; thus, this algorithm requires much less coordination, and allows different processors to modify assignments and prices simultaneously.

Similar to the auction algorithms, the ϵ-relaxation algorithm of Section 2.1.3 also admits Gauss–Seidel, Jacobi and hybrid parallel implementations, as well as asynchronous implementations. For a convergence analysis of several synchronous and asynchronous implementations of ϵ-relaxation, the reader should consult Bertsekas & Tsitsiklis [1989].

2.3. *Computation experience*

2.3.1. *Primal cost improvement*

In this section, we overview the results of parallel computation experiments reported in Miller, Pekny & Thompson [1989], Peters [1990], and Barr & Hickman [1990] using parallel simplex algorithms for transportation and min-cost network flow problems.

As discussed previously, Miller, Pekny, and Thompson [1989] implemented a synchronous version of the transportation simplex algorithm for dense transportation problems on the BBN Butterfly Plus [Rettberg & Thomas, 1986] computer. The Butterfly Plus is a distributed-memory computer; each processor (Motorola 68020/68881) has 4 Megabytes of local memory. Processors may access information in other processor's memory, but such accesses are slower than accesses to local memory. The implementation of Miller, Pekny & Thompson [1989] divides the rows of the transportation matrix among the processors; each processor only stores the elements for which it is responsible. In addition, each processor keeps a complete local copy of the basis tree and dual variables.

The algorithm of Miller, Pekny & Thompson [1989] works as follows: At the beginning of each iteration, each processor searches a subset of its rows to locate suitable pivot arcs; the best arc found by each processor is stored in a *pivot queue*. A synchronization point is introduced to make sure that all processors have completed the search before going to the next part of the iteration. After synchronization, the processors check the pivot queue; if it is empty, a new search operation takes place using different rows. If the pivot queue is not empty, every processor uses the identical logic to select pivot elements and update its local copy of the basis tree and dual variables. Due to the distributed memory nature of the Butterfly Plus, it is more efficient to replicate these computations than to communicate the results among processors. At the end of an iteration, the processors start the search process for the next iteration.

The algorithm of Miller, Pekny & Thompson [1989] was tested on dense assignment and transportation problems with equal numbers of sources and sinks, using between 500 and 3000 sources and costs generated uniformly over a variable range. In order to evaluate the speedup, two sets of experiments were conducted: First, a single processor was used (although the data was stored over all 14 processors due to memory limitations); second, all 14 processors were used. Unfortunately, distributing the problem data across all 14 processors slows down the execution of the single-processor algorithm significantly, as most of the reduced cost computations require referencing to memory which is not collocated with the processor. In contrast, the parallel algorithm does not require use of any non-local memory references; this results in increased effectiveness of the parallel algorithm. In the computation experiments with arc costs in [0,1000] and [0,10000], the parallel algorithm ran between 3 and 7.5 times faster than the single-processor algorithm [Miller, Pekny & Thompson, 1989]; notably, this speedup increased with problem size. This is because the number of arcs to be searched increases quadratically with problem size, so that the search time becomes a larger fraction of the overall computation time.

Similar results were observed by Peters [1990] in test experiments using the parallel network simplex code PARNET. The PARNET code was implemented on a shared-memory multiprocessor (Sequent Symmetry S-81 [Sequent Computer Systems, 1987]). The PARNET implementation requires a minimum of 3 processors (1 pivot processor and 2 search processors), so that a direct comparison with a single-processor version was not possible. Instead, Peters [1990] compares the PARNET performance with the performance of NETFLO [Kennington & Helgason, 1980] on a variety of NETGEN [Klingman, Napier & Stutz, 1974] problems. The results indicate that PARNET is often 10–20 times faster using only three processors. However, this speedup is more indicative of the different search strategies used by NETFLO and PARNET than the advantage of using parallel processing. In essence, the use of multiple processors to search the en-

Table 1
Computation time in seconds on Sequent Symmetry for PARNET and NETFLO

NETGEN Problem	NETFLO (1)	PARNET (3)	PARNET (7)
110	430.50	28.19	12.42
122	802.60	72.28	21.91
134	195.40	23.94	6.26
150	802.60	71.62	15.13

tire set of arcs for candidate pivots reduces the total number of pivot iterations required for convergence. An even greater difference is observed when PARNET is implemented using 7 processors; although the number of search processors is tripled, the computation times compared to PARNET with three processors is often reduced by factors greater than 3! This superlinear speedup is again due to the use of different pivot strategies; by using more processors, more arcs can be searched between pivots. Table 1 summarizes some of the results reported in Peters [1990].

It should be noted that the above times represent a single run of the algorithm for each problem. Given the asynchronous nature of the PARNET algorithm, some variability in computation time should be expected, as small variations in the relative timing of search and pivot iterations may result in very different sequences of iterations.

Peters also examined the question of whether parallelization effectiveness increases with problem size. Two sets of experiments were conducted. In the first set, a set of network problems with constant number of arcs per node were generated (roughly 50 arcs per node, with problems ranging from 20,000 to 50,000 nodes). Interestingly, the results in Peters [1990] indicate that, once the number of processors is around 7, little additional speedup is obtained from using more processors. In essence, using 6 search processors provides enough processing to search the entire set of arcs while the pivot processor performs a pivot operation. In the second set of experiments, Peters generated problems with a constant number of nodes (1000), but an increasing number of arcs (from 25,000 to 500,000). In these experiments, Peters found that the number of processors which could be used efficiently increased from 6 in the sparsest problems to 12 in the denser problems. For additional details and results, the interested reader is referred to Peters [1990].

Barr and Hickman [1990] developed their parallel network simplex code, PPNET, as an extension of the sequential NETSTAR code written by Barr based on the ARC-II code of Barr, Glover & Klingman [1979]. Their experimental results on a comparable Symmetry S81 showed that the PPNET algorithm is roughly twice as fast as the PARNET algorithm of Peters [1990]. This is due in part to the superiority of the task scheduling approach, and to the division of the lengthier price update tasks among two different processors. Table 2 summarizes some of the results in Barr & Hickman [1990], and compares the performance of

Table 2
Computation time in seconds on Sequent Symmetry for PARNET and PPNET

NETGEN Problem	PPNET (1)	PARNET (3)	PPNET (3)
110	39.98	28.19	13.99
122	79.91	72.28	32.84
134	64.94	23.94	15.42
150	84.93	71.62	34.49

the PPNET and PARNET algorithms using 3 processors for the same NETGEN problems.

It should be noted that the PARNET times in Table 2 include input, initialization, computation, and output times. In contrast, the PPNET times exclude input and output times. Furthermore, all times reported are the results of single runs of the algorithms; the results of Barr & Hickman [1990] and Peters [1990] do not provide any indications concerning the run-to-run variability of the computation times.

Barr and Hickman [1990] conducted an exhaustive test of PPNET for NETGEN problems with 1,000 to10,000 nodes and 12,500 to 75,000 arcs, using from 1 to 10 processors. Their results show that, for most problems in this range, maximum speedups are obtained using 5 or 6 processors; adding additional processors can reduce the performance of the algorithm due to additional synchronization overhead associated with the monitor operations. With 6 processors, an average speedup of 4.68 was obtained. As in Peters [1990], the critical limitation in speedup is the time required to perform a pivot operation; once enough search processors are available to perform a complete search of the arcs during a pivot operation, adding additional processors merely creates extra overhead. In PPNET, this limit is reached sooner because the average duration of pivot operations is reduced as compared to PARNET due to the use of parallel dual price updates. Indeed, Barr and Hickman's experiments confirm that when 3 or more processors are used, a pivot is in progress over 85% of the time. For additional results and details, the reader is referred to Barr & Hickman [1990].

2.3.2. *Dual cost improvement*

Most of the work on parallel dual cost improvement algorithms has focused on implementations of the algorithm (JV) of Jonker & Volgenant [1987] for dense assignment problems. The JV algorithm is a hybrid algorithm consisting of two phases: an initialization phase (which uses an auction-like procedure for finding a partial assignment satisfying CS) and a sequential shortest path phase, which uses Dijkstra's algorithm. The first parallel implementation of the JV algorithm was performed by Kennington & Wang [1988] on an 8-processor Sequent Symmetry S81. In their implementation, both the initialization and sequential shortest path phases are executed in parallel. For the sequential shortest path phase, multiple processors are used to update node distances and to find which node to scan

Table 3
Computation time of the parallel JV algorithm on the Encore Multimax

1000 × 1000 assignment, cost range [0–1000]					
No. Processors	1	2	4	6	8
Shortest path time (100% dense)	112.7	77.5	60.0	54.0	52.0
Total time (100% dense)	132.5	87.9	65.7	59.0	56.2
Shortest path time (10% dense)	14.6	10.4	8.6	8.6	×
Total time (10% dense)	17.4	12.1	9.7	9.7	×

next in parallel. Experimental results in Kennington & Wang [1988] using dense assignment problems with 800, 1000 and 1200 persons indicate speedups in the range of 3–4 as the number of processors is increased from 1 to 8. However, this is the combined speedup of the complete algorithm; no information was provided concerning how much speedup was accomplished during the shortest path phase of the algorithm. Furthermore, the speedups increase with problem size, since the number of arcs which must be examined in parallel increases with problem size (for dense problems).

In order to shed some insight on the effectiveness of single-node parallelization for sequential shortest path methods, we developed a parallel implementation of the JV algorithm on the Encore Multimax, a 20-processor shared-memory parallel computer. In these experiments, we used 1000 person assignment problems of different densities, and varied the number of processors. Table 3 summarizes the results of the experiments using fully-dense and 10% dense assignment problems. Note that the speedup of the sequential shortest path phase is significantly smaller for the sequential shortest path phase than for the overall algorithm. Furthermore, the overall speedups are in the order of 2.5 for dense problems, and decrease to under 2 for 10% dense problems.

The experimental results discussed above have used shared-memory parallel architectures for implementing single-node parallelization. These architectures permit the use of sophisticated data structures to take advantage of problem sparsity in the algorithms; however, the above results illustrate that the effectiveness of single-node parallel algorithms is limited by the density of the network. As an alternative, these parallel algorithms can be implemented using massively parallel single-instruction, multiple-data stream (SIMD) processors such as the Connection Machine. Castañon, Smith & Wilson [1989] have developed parallel implementations of the JV algorithm on the DAP 510 (1024 single-bit processors) and the Connection Machine CM-2 (using only 1024 of its 65,536 processors). Both of these machines are array processors attached to a sequential processor. In order to minimize communications, the cost matrix was stored as a dense array, spread across the processors so that each processor has one row of the cost matrix. For the same problems discussed in Table 3, the computation times on the DAP 510 were 1 and 1.6 seconds for the dense and sparse problem, respectively. On the CM-2, the computation times were 18.7 and 29.1 seconds respectively.

Note the effectiveness of the massively parallel DAP architecture for solving both dense and sparse assignment problems (compared with the Encore Multimax results of Table 2.) The difference between the DAP and CM-2 times is due to the CM architecture, which is designed to work with 65,536 processors; our implementation required the use of only 1000 processors. In contrast, the DAP 510 architecture is optimized for 1024 processors. The above results highlight the advantage of SIMD architectures for implementing single-node parallel algorithms.

The theory of multinode parallel dual cost improvement algorithms has been worked out recently [Balas, Miller, Pekny & Toth, 1991; Bertsekas & Castañon, 1990a,b]; thus, computational experience with these methods is limited. For fully-dense assignment problems, Balas, Miller, Pekny & Toth [1991] developed a synchronous parallel sequential shortest path algorithms and implemented it on the 14-processor Butterfly Plus computer discussed previously. Subsequently, Bertsekas and Castañon [1990a] conducted further experiments on the Encore Multimax with other synchronous algorithms as well as an asynchronous variation of the sequential shortest path algorithm (corresponding closely to the theoretical algorithm of Section 2.1.3), in order to identify relative advantages and disadvantages of the synchronous and asynchronous algorithms. We discuss these experimental results below.

In Bertsekas & Castañon [1990a], three different parallel variations of the successive shortest path algorithm were evaluated:

(1) *Single path synchronous (SS)*: At each iteration, every processor finds a single shortest augmenting path from an unassigned person to an unassigned object.

(2) *Self-scheduled synchronous (SSS)*: At each iteration, every processor finds a variable number of shortest augmenting paths sequentially until the total number of augmenting paths equals some threshold number, which depends on the number of unassigned persons and the number of processors. This variation closely resembles the implementation of Balas, Miller, Pekny & Toth [1991].

(3) *Single path asynchronous (AS)*: At each iteration, each processor finds a single shortest augmenting path from an unassigned person to an unassigned object, but the processors execute the iterations asynchronously.

The purpose of the self-scheduled synchronous algorithm is to reduce the coordination overhead among processors, while improving computation load balance among the processors. Processors which find augmenting paths quickly will be assigned more work. Furthermore, synchronization is used less often than in single-path synchronous augmentation, because a fixed number of augmenting paths must be computed.

The experimental results of Bertsekas & Castañon [1990a] illustrate many of the limitations in reducing computation time using multinode parallelization in dual improvement algorithms. Figure 3 summarizes the computation time in the sequential shortest paths phase (averaged across three runs) of the three algorithms for a 1000 person, 30% dense assignment problem, cost range [1,1000]. When a single processor is used, the self-scheduled synchronous algorithm is the slowest because it must find additional shortest paths (as a result of incompatibility prob-

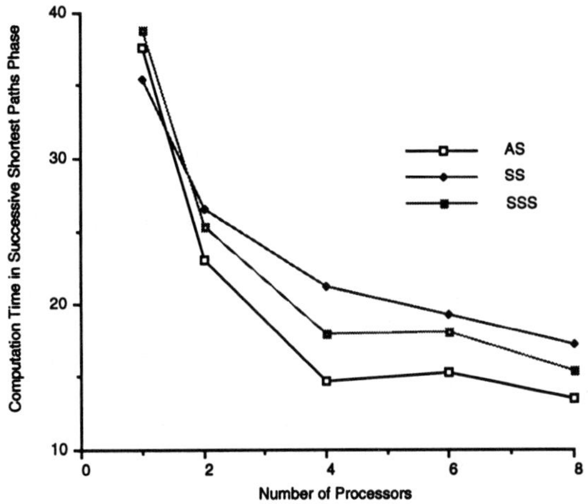

Fig. 3. Computation time of parallel multinode Hungarian algorithms for 1000 person, 30% dense assignment problem, with cost range [1,1000], as a function of the number of processors used on the Encore Multimax.

lems). As the number of processors increases, the reduced coordination required by the self-scheduled synchronous algorithm makes it faster than the single-path synchronous algorithm. For these experiments, the single path asynchronous algorithm is fastest.

In Bertsekas & Castañon [1990a], a measure of coordination overhead is proposed. This measure, called the wait time, is the time that a processor which has work to do spends waiting for other processors. For the synchronous algorithms, the majority of the wait time occurs at the synchronization point in each iteration, when processors which have already completed their augmenting path computations wait for other processors. For the asynchronous algorithm, it is the time waiting to get access to the master copy (while other processors are modifying it). Figure 4 shows the average wait time per processor for the results in Figure 3.

The above results indicate the existence of some fundamental limits in the speedups which can be achieved from parallel processing of dual improvement algorithms. As discussed in Bertsekas & Castañon [1990a], the principal limitation is the decreasing amount of parallel work in each iteration. Initially, there are many augmentations which must be found, so that many processors can be used effectively. However, as the algorithm progresses, there are fewer unassigned persons, so the number of parallel augmentations decreases significantly. Furthermore, the later augmentations are usually harder to compute, as they involve reversing many previous assignments. This limits the net speedup which can be achieved across all iterations. Other factors such as the synchronization overhead further limit the achievable speedup.

In terms of synchronous versus asynchronous algorithms, additional results in Bertsekas & Castañon [1990a] indicate that, as the problems become sparser, the

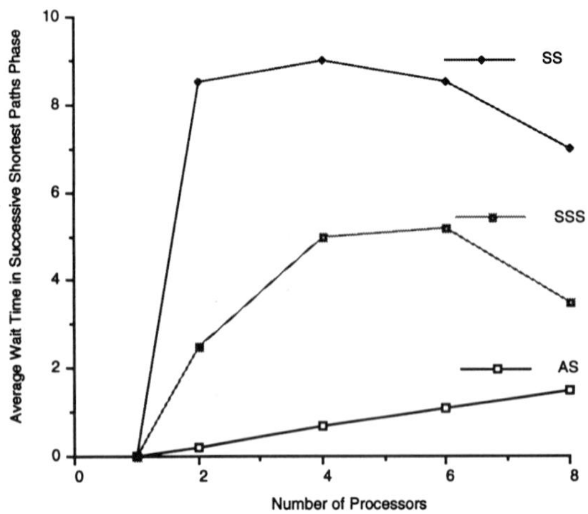

Fig. 4. Average wait time per processor for 1000 person, 30% dense assignment problem, with cost range [1,1000], as a function of the number of processors used on the Encore Multimax.

coordination overhead of the asynchronous algorithms may become larger than the overhead of the synchronous algorithms! This is because the time required to compute an augmenting path is much shorter for the sparse problems; this reduces the coordination overhead for synchronous algorithms because there is less variability in computation time across processors. However, it also increases the coordination overhead for asynchronous algorithms because more of the time is spent on modifying the master copy (rather than computation), so that the probability of access conflicts is increased. For a greater discussion of these topics, the reader is referred to Bertsekas & Castañon [1990a].

Although the above discussion focused on assignment problems, Bertsekas & Castañon [1990b] have extended the single-path synchronous algorithm to min-cost network flow problems, and evaluated its performance using NETGEN uncapacitated transshipment problems. Extensions of the self-scheduled synchronous algorithm and the asynchronous algorithm to these classes of problems are straightforward, based on the results of Section 2.2.3.

2.3.3. Approximate dual coordinate ascent

The auction algorithm for assignment problems has been tested widely across a variety of parallel implementations on different parallel computers. The simplicity of the auction algorithm makes it a natural candidate for parallel implementation. In addition, many variations exist which are amenable for either Gauss–Seidel, Jacobi or hybrid parallelization.

One of the earliest parallel implementations of the auction algorithm was reported by Phillips and Zenios [1989] on the Connection Machine CM-2 for dense assignment problems (also subsequently by Wein and Zenios [1990]). In their implementation, they used the large number of processors in the CM-2 to simultaneously compute bids for many different persons and to compute in parallel the bids of each person (thus using the hybrid approach discussed earlier). Their work was the first to illustrate the utility of massively parallel architectures for assignment problems.

A similar hybrid implementation was reported by Kempa, Kennington & Zaki [1989] on the Alliant FX/8 parallel computer. In their work, they experimented with various synchronous implementations of the auction algorithm for dense assignment problems. In their hybrid implementation, the vector processing capability of the Alliant's processors was used to compute in parallel the bid of each person, while the multiprocessor capability was used for computing in parallel multiple bids. For 1000-person dense assignment problems, with cost range [1,1000], Kempa and coworkers obtained total speedups of 8.6 for their hybrid auction algorithm using 8 vector processors. Subsequent work by Zaki [1990] on the Alliant FX/8 produced similar results.

In Castañon [1989], several synchronous and asynchronous implementations of the auction algorithm were developed and tested for dense and sparse assignment problems on different parallel computers (Encore Multimax, Alliant FX/8, DAP 510 and CM-2). Similar to the dual improvement algorithms, the SIMD implementations of the auction algorithm on the DAP 510 were extremely efficient, solving 1000-person dense assignment problems in under 3 seconds! Other important results in Castañon [1989] are evaluations of the relative computation advantage of asynchronous auction algorithms versus their synchronous counterparts.

In Bertsekas & Castañon [1989], where a number of variations of the auction algorithm were implemented and evaluated on the Encore Multimax. The auction algorithm variations tested were:

1. *Synchronous Gauss–Seidel Auction* : Parallelization using only one bidder at a time.
2. *Synchronous Jacobi Auction* : A block Gauss–Seidel parallelization where each processor generates a bid from a different person.
3. *Synchronous Hybrid Auction* : A hybrid parallel algorithm, where processors are hierarchically organized into groups; each group computes a single bid in parallel, while multiple groups generate different bids.
4. *Asynchronous Jacobi Auction* : A block Gauss–Seidel parallel algorithm, where each processor generates a bid from a different person.
5. *Asynchronous Hybrid Auction* : A hybrid parallel algorithm where processors are divided into search processors (for computing a single bid) and bid processors (for computing multiple bids).

The most interesting results in Bertsekas & Castañon [1989] are the comparisons between the synchronous and asynchronous algorithms. For example, for 1000 person, 20% dense assignment problems, the synchronous Jacobi auction algorithm achieves a maximum speedup of 4, whereas the asynchronous Jacobi auction algorithm achieves speedups of nearly 6 due to its lower synchronization overhead (which allows for efficient utilization of larger numbers of processors). This asynchronous advantage is the consequence of the strong asynchronous convergence results for the auction algorithm, which allow processors to perform computations with little need to maintain data consistency or integrity.

The asynchronous hybrid auction (AHA) algorithm of Bertsekas & Castañon [1989] is similar in structure to the parallel network simplex algorithms discussed previously [Peters, 1990; Barr & Hickman, 1990]. In each iteration of the AHA algorithm, some processors are designated as search processors, which evaluate arcs in the network and produce information for generation of bids. Other processors are bid processors, which process the information generated by the search processors and actually conduct the auction process. The search processors and the bid processors operate concurrently; thus, bids are often generated based on 'old' information. In contrast with the network simplex algorithms, the AHA algorithm uses multiple bid processors, and does not fix *a priori* whether a processor is a search or a bid processor. Rather, there is a task queue containing a set of search and bid tasks which must be performed; these tasks are generated as part of the auction process. Whenever a processor is available, it proceeds to the queue and selects the next task.

Unlike the network simplex algorithms, the AHA algorithm does not need to recompute the information provided by the search processors in order to 'validate' it (validation in the auction context would be very expensive, unlike in the network simplex context); instead, the AHA algorithm is based on the asynchronous convergence theory which guarantees that, if the information is outdated, the coordination mechanism in the auction algorithm will reject it. When tested across a set of 1000 person assignment problems with varying density, cost range [1,1000], the AHA algorithm was nearly twice as fast as the corresponding hybrid synchronous algorithm for every problem tested, highlighting the advantages of the asynchronous algorithm. For additional results on parallel auction algorithms, the reader should consult Bertsekas & Castañon [1989].

Unlike the auction algorithm for assignment problems, sequential implementations of the ϵ-relaxation method have been considerably slower than state-of-the-art min-cost network flow codes [Bertsekas & Eckstein, 1988]. Thus, there have been fewer parallel implementations of the algorithm. A notable exception is the work of Li and Zenios [1991, 1992] on the Connection Machine CM-2. They developed a modification of the ϵ-relaxation algorithm which assigns flows in fractional quantities, and implemented a parallel hybrid algorithm on the CM-2. Their results indicate that, for some problems derived from military transportation, the CM-2 implementation of ϵ-relaxation is substantially faster than their network simplex implementation on the Cray Y-MP; for other classes of problems, the network simplex code was faster [Li & Zenios, 1991, 1992].

2.4. *Summary*

In this section we have discussed results on parallel algorithms for min-cost network flow problems. We have roughly characterized parallelization approaches into two types: single-node and multinode, depending on the level of the function which is done in parallel. The experimental results indicate that both approaches are limited in their ability to use parallelization to reduce computation requirements.

The key limit in single-node approaches is imposed by the density of the network; for sparse networks, the amount of work which can be performed in parallel is a smaller part of the overall computation time. For network simplex algorithms, sparse problems have fewer arcs which must be inspected to find a desirable pivot, so that adding search processors beyond a point in these algorithms actually slows down performance because of increased synchronization requirements. Similarly, in sequential shortest path algorithms and auction algorithms, sparsity limits the number of arcs which must be examined per node, so that again the ratio of parallel work to total work is reduced.

In contrast, the key limit in multinode approaches is imposed by the time-varying load across iterations. For parallel dual cost improvement algorithms and auction algorithms, our convergence theories [Bertsekas & Castañon, 1989, 1990a, b] provide the basis for medium scale parallelization. The effectiveness of these parallel algorithms is not limited by sparsity, but rather by the fact that the number of nodes with positive surplus decreases with each iteration, so that the total amount of parallel work per iteration decreases. On average, the overall speedup using this approach for sparse problems is limited to factors of 3–5.

The computation experiments also highlight several critical issues such as the impact of parallel processor architectures and the choice of synchronous versus asynchronous algorithms. For dense network problems, the best processors are the massively parallel SIMD processors; however, these are limited in their ability to effectively use sparse data structures. When the problem density is below 1%, the best processors are shared-memory processors, which are best-suited for using multinode parallelism and sparse data structures. The multinode parallelism is limited to using at most 10–12 processors in parallel; with these few processors, there is little contention for shared memory and communications resources.

In terms of synchronous versus asynchronous algorithms, it is interesting to find that, in many problems, asynchronous algorithms offer computational advantages. This is particularly true of auction algorithms; indeed, the asynchronous convergence theory allows for efficient combination of single-node and multinode parallelism within the same algorithm. In contrast, the asynchronous theory for dual cost improvement algorithms requires excessive data integrity, and results in inefficient algorithms for sparse problems. Development of a less restrictive asynchronous convergence theory remains a problem for future research.

A class of parallel algorithms for linear network flow problems, mentioned here for completeness and discussed in greater detail in Section 3.2.2 and 3.3.2, combines nonlinear perturbations of the linear objective function and then uses parallel algorithms for the resulting nonlinear program. The solution of the min-cost network flow problem, combining the PMD algorithm of Censor & Zenios [1991] with the row-action algorithms of Censor & Lent [1981] and Zenios & Censor [1991] was suggested by Censor and Zenios [1992], and extensive computational studies were conducted by Nielsen and Zenios [1993b, 1994a].

3. Nonlinear network optimization

3.1. Basic algorithmic ideas

In this section, we discuss three approaches to parallel solution of convex nonlinear network optimization problems: primal truncated Newton methods, dual coordinate ascent methods, and alternating direction methods. To handle problem for which the objective function is convex, but not *strictly* convex, the dual coordinate ascent methods need to be embedded in some sort of convexification scheme, such as the proximal minimization algorithm. Therefore, we also include some discussion of proximal minimization and related methods. Other approaches to parallel nonlinear network optimization are certainly worthy of study; for brevity, we restrict ourselves to methods which have already been substantially investigated in the specific context of networks.

3.1.1. Primal methods
One of the most efficient primal algorithms for solving the nonlinear network optimization program (NLNW) is the primal truncated Newton (PTN) algorithm [Dembo & Steihaug, 1983a], implemented within the active set framework [Murtagh & Saunders, 1978]. The combination of both techniques for pure network problems is given in Dembo [1987] and for generalized networks in D. P. Ahlfeld, R. S. Dembo, J. M. Mulvey & S. A. Zenios [1987]. PTN has received considerable attention both in solving large scale problems, and in parallel computation. We describe the algorithm in two steps: First we give a model Newton's algorithm for unconstrained optimization problems. Second, we discuss the active set method which reduces a constrained optimization problem into a sequence of (locally) unconstrained problems in lower dimensions. Our general reference for this section is Gill, Murray & Wright [1981].

A truncated Newton algorithm for unconstrained optimization
Consider the unconstrained problem

$$\min_{x \in \Re^m} F(x), \tag{39}$$

where $F(x)$ is convex and twice continuously differentiable. The primal truncated Newton (PTN) algorithm starts from an arbitrary feasible point $x^0 \in \Re^n$ and generates a sequence $\{x^k\}$, $k = 1, 2, 3, \ldots$ such that

$$\lim_{k \to \infty} x_k = x^*,$$

where x^* belongs to the set of optimal solutions to (39) (i.e., $x^* \in X^* = \{x | F(x) \leq F(y), \forall\, y \in \Re^n\}$). The iterative step of the algorithm is the following:

$$x^{k+1} = x^k + \alpha^k d^k. \tag{40}$$

$\{d^k\}$ is a sequence of descent directions computed by solving the system of (Newton's) equations :

$$\nabla^2 F(x^k) d^k = -\nabla F(x^k). \tag{41}$$

This system is solved inexactly (hence the term *truncated*), i.e., a solution d^k is obtained that satisfies

$$\|\nabla^2 F(x^k) d^k - \nabla F(x^k)\|_\infty \leq \eta^k. \tag{42}$$

A scale independent measure of the residual error is :

$$r^k = \frac{\|\nabla^2 F(x^{k-1}) p^k + \nabla F(x^{k-1})\|_2}{\|\nabla F(x^{k-1})\|_2}. \tag{43}$$

The step direction is computed from (41) such that the condition $r^k \leq \eta^k$ is satisfied and the sequence $\{\eta^k\} \to 0$ as $k \to \infty$. $\{\alpha^k\}$ is a sequence of step sizes computed by solving

$$\alpha^k = \arg\min_{\alpha>0} \{F(x^k + \alpha d^k)\}, \tag{44}$$

(i.e., at iteration k the scalar α^k is the step size that minimizes the function $F(x)$ along the direction p^k starting from point x^k). Computing α^k from equation (44) corresponds to an exact minimization calculation that may be expensive for large scale problems. It is also possible to use an inexact linesearch. The global convergence of the algorithm is preserved if the step length computed by inexact solution of (44) produces a sufficient descent of $F(x)$, satisfying Goldstein–Armijo type conditions [see e.g. Bertsekas, 1982].

An active set algorithm for constrained optimization

Consider now the transformation of (NLNW) into a locally unconstrained problem. Following Murtagh & Saunders [1978] we partition the matrix A into the form:

$$A = [B\ S\ N]. \tag{45}$$

B is a non–singular matrix of dimension $n \times n$ whose columns form a basis. For the case of network problems a basis can be constructed using a greedy heuristic, as given in Dembo & Klincewicz [1985]. First-order estimates of the Lagrange multipliers that correspond to this basis are obtained by solving for p the system $p^T B = -\nabla F(x^k)$. S is a matrix of dimension $n \times r$. It corresponds to the superbasic variables, i.e., variables at their lower bound with negative reduced gradient, variables at their upper bound with positive reduced gradient, or free variables. For the set of superbasic variables it is possible to take a non-zero step that will reduce the objective function value. N is a matrix of dimension $n \times (m - n - r)$. It corresponds to the non-basic variables, i.e., variables that have positive reduced gradient and are at their lower bound, or have a negative reduced gradient and are at their upper bound.

We use $\mathcal{B}$, $\mathcal{S}$ and $\mathcal{N}$ to denote the sets of basic, superbasic and non–basic variables respectively. The vector x^k is partitioned into

$$x^k = [x_B^k \, x_S^k \, x_N^k]. \tag{46}$$

$x_B^k \in \Re^m$ are the basic variables, $x_S^k \in \Re^r$ are the superbasic variables and $x_N^k \in \Re^{n-m-r}$ denote non–basic variables. Non–basic variables — for a given partitioning (45)–(46) — are kept fixed to one of their bounds. If we now partition the step direction d as

$$d^k = [d_B^k \, d_S^k \, d_N^k] \tag{47}$$

we require $d_N^k \equiv 0$ (i.e., non–basic variables remain fixed) and furthermore d^k should belong to the nullspace of A (i.e., $Ad^k = 0$), so that d^k is a feasible direction. Hence d^k must satisfy

$$Bd_B^k + Sd_S^k = 0 \text{ or } d_B^k = -(B^{-1}S)d_S^k. \tag{48}$$

If the superbasic variables are strictly between their bounds and the basis B is maximal as defined in Dembo & Klincewicz [1985] (i.e., a non–zero step in the basic variables x_B is possible for any choice of direction $(d_B^k \, d_S^k \, 0)$), then the problem is locally unconstrained with respect to the superbasic variables. Hence a descent direction for d_S^k can be obtained by solving the (projected) Newton's equations:

$$(Z^T \nabla^2 F(x^k) Z) d_S^k = -Z^T \nabla F(x^k) + \eta^k e, \tag{49}$$

where Z is a basis for the nullspace of A defined as

$$Z = \begin{bmatrix} -B^{-1}S \\ I \\ 0 \end{bmatrix} \tag{50}$$

The primary computational requirement of the algorithm is in solving the system of equations (49) of dimension $r \times r$. This system is solved using a conjugate gradient method with a preconditioner matrix equal to the inverse of the diagonal of the reduced Hessian matrix $Z^T \nabla^2 F(x) Z$. Calculation of p_B^k from (48) involves only a matrix–vector product and is in general an easy computation. The partitioning of the variables and the matrix A into basic, superbasic and non–basic elements is also in general very fast. For some of the bigger problems reported in the literature the solution of system (49) takes as much as 99% of the overall solution time. Efforts in parallelizing the PTN algorithm have concentrated in the solution of projected Newton equations.

3.1.2. Dual coordinate ascent methods

Now consider a separable, nonlinear network optimization problem of the form (NLNW). where the cost component functions $f_{ij} : \Re \to \Re$ are for the moment assumed to be *strictly* convex and lower semicontinuous. For any flow $x \in \Re^m$, we let $g_i(x)$ denote the surplus at node i under the flow x (we make the dependence on x explicit because some nonlinear network algorithms manipulate several flow vectors simultaneously). Assume that (NLNW) is feasible and has an optimal solution. Defining for each $(i, j) \in \mathcal{A}$,

$$\bar{f}_{ij}(x_{ij}) = \begin{cases} f_{ij}(x_{ij}), & l_{ij} \le x_{ij} \le u_{ij} \\ +\infty & \text{otherwise} \end{cases}, \tag{51}$$

the problem may be written

$$\begin{aligned} &\text{minimize} \quad \bar{f}(x) \triangleq \sum_{(i,j)\in\mathcal{A}} \bar{f}_{ij}(x_{ij}) \\ &\text{such that} \quad Ax = b. \end{aligned} \tag{52}$$

Attaching a vector of Lagrange multipliers $p \in \Re^n$ to the equality constraints, one obtains the Lagrangian

$$\begin{aligned} L(x, p) &= \sum_{(i,j)\in\mathcal{A}} \bar{f}_{ij}(x_{ij}) - p^T(Ax - b) \\ &= \sum_{(i,j)\in\mathcal{A}} \left(\bar{f}_{ij}(x_{ij}) - (p_i - p_j)x_{ij}\right) - p^T b. \end{aligned}$$

Therefore, the dual functional of the problem is

$$q(p) = \inf_{x\in\Re^{|\mathcal{A}|}} \{L(x, p)\} = \sum_{(i,j)\in\mathcal{A}} q_{ij}(p_i - p_j) - p^T b, \tag{53}$$

where

$$q_{ij}(t_{ij}) = \inf_{x_{ij}\in\Re} \left\{\bar{f}_{ij}(x_{ij}) - t_{ij}x_{ij}\right\}. \tag{54}$$

Thus, we may define a problem dual to (NLNW) and (52) to be

$$\begin{aligned} &\text{maximize} \quad q(p) \\ &\text{such that} \quad p \in \Re^n \end{aligned}. \tag{55}$$

Now, each q_{ij} is the pointwise infimum of a set of affine (hence concave) functions in t_{ij}, and is necessarily concave. Thus, q is concave. Basic nonlinear programming duality theory implies that the optimal values of (NLNW) and (55) are equal. Readers familiar with aspects of convex analysis [Rockafellar, 1970, 1984] will recognize that q_{ij} is the negative of the convex conjugate of $\bar{f}_{ij}$. In simple terms, the following equivalences therefore hold:

$$\bar{f}_{ij} \text{ has a subgradient of slope } t_{ij} \text{ at } x_{ij} \tag{56}$$

$$\Longleftrightarrow q_{ij} \text{ has a supergradient of slope } -x_{ij} \text{ at } t_{ij} \tag{57}$$

$$\Longleftrightarrow x_{ij} \text{ attains the infimum in (54).} \tag{58}$$

The strict convexity of the f_{ij} implies strict convexity of the $\bar{f}_{ij}$. This means that at any two distinct points $x_{ij}^1, x_{ij}^2 \in [l_{ij}, u_{ij}]$, $\bar{f}_{ij}$ cannot have subgradients with the same slope. It follows that for each argument t_{ij}, q_{ij} has exactly one supergradient, that is, q_{ij} is differentiable. If follows that (55) is a differentiable, unconstrained minimization problem. The key idea in parallel dual methods for network optimization is to exploit this extremely advantageous dual structure.

The differentiability and convexity of q imply that, from any non-maximizing p, it is possible to increase q by changing only a single component p_i of p. This property suggests the following iteration for maximizing $q(p)$:

(i) Given a price vector $p(t)$, choose any node $i \in \mathcal{N}$.

(ii) Compute $p(t+1)$ such that $p_j(t+1) = p_j(t)$ for all $j \neq i$, while $p_i(t+1)$ maximizes $q(p)$ with respect to the i-coordinate, all other coordinates being held fixed at $p_j(t)$.

It can be shown that, so long as each node $i \in \mathcal{N}$ is selected an infinite number of times, that the sequence of price vectors $p(t)$ generated by this iteration converges to an optimal solution of the dual problem (55)[Bertsekas & Tsitsiklis, 1989, section 5.5]. Algorithms of this general sort have been discussed in such diverse sources as Bertsekas, Hossein & Tseng [1987], Censor & Lent [1981], Cottle, Duvall & Zikan [1986], Cryer [1971], and Hildreth [1957]. They are usually called *relaxation* methods, because maximizing the dual cost with respect to the coordinate p_i is equivalent to adjusting the primal solution x so that the flow constraint for node i is satisfied while constraints for other nodes are relaxed. To understand this, it is necessary to exploit the duality relationship between (NLNW)/(52) and (55) encapsulated by (56)–(58).

The equivalent conditions (56)–(58) are known as *complementary slackness*. Dual methods enforce these conditions at all times, where we define the *tension* t_{ij} of arc (i, j) to be $p_i - p_j$. We denote by $x(p)$ the unique flow vector satisfying complementary slackness with the price vector p on all arcs, whence

$$x_{ij}(p) = -\nabla q_{ij}(p_i - p_j).$$

Let c_{ij}^- be the derivative of f_{ij} at l_{ij}, and c_{ij}^+ be the derivative of f_{ij} at u_{ij}. For values of $t_{ij} = p_i - p_j$ below c_{ij}^-, the infimum in (54) is attained for $x_{ij} = l_{ij}$, so $x_{ij}(p)$ must be at its lower bound of l_{ij}. If $p_i - p_j \geq c_{ij}^+$, then the infimum is attained at $x_{ij} = u_{ij}$, and so $x_{ij}(p)$ is set to its upper bound. For $p_i - p_j \in [c_{ij}^-, c_{ij}^+]$, $x_{ij}(p) = -\nabla q_{ij}(p_i - p_j)$, by the concavity of q_{ij}, is a nondecreasing function of $p_i - p_j$. Thus, ∇q_{ij} is constant at $-l_{ij}$ for arguments below c_{ij}^-, constant at $-u_{ij}$ for arguments above c_{ij}^+, and nonincreasing in between, as shown in Figure 5. Accordingly, q_{ij} itself is linear with slope $-l_{ij}$ for arguments below c_{ij}^-, linear with slope $-u_{ij}$ for arguments above c_{ij}^+, and concave in between, as depicted in Figure 6. Now let us consider the partial derivative of the entire dual functional q with respect to the variable p_i:

$$\begin{aligned}
\frac{\partial}{\partial p_i} q(p) &= \frac{\partial}{\partial p_i}\left[\sum_{(k,l)\in\mathcal{A}} q_{kl}(p_k - p_l)\right] + \frac{\partial}{\partial p_i}\left[p^T b\right] \\
&= \sum_{j:(i,j)\in\mathcal{A}} \nabla q_{ij}(p_i - p_j) - \sum_{j:(j,i)\in\mathcal{A}} \nabla q_{ji}(p_j - p_i) + b_i \\
&= -\sum_{j:(i,j)\in\mathcal{A}} x_{ij}(p) + \sum_{j:(j,i)\in\mathcal{A}} x_{ji}(p) + b_i,
\end{aligned}$$

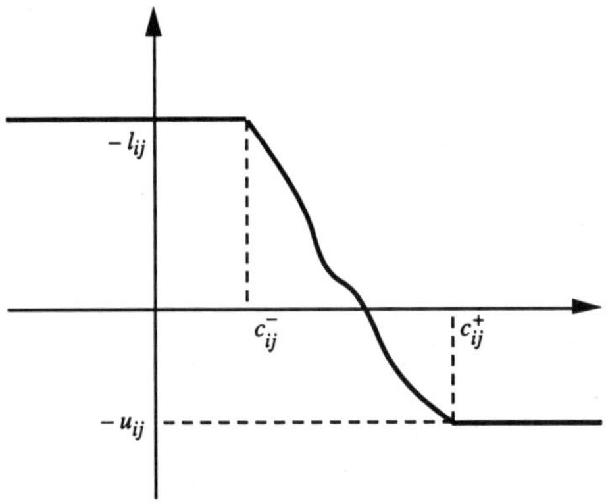

Fig. 5. Example of the function ∇q_{ij}.

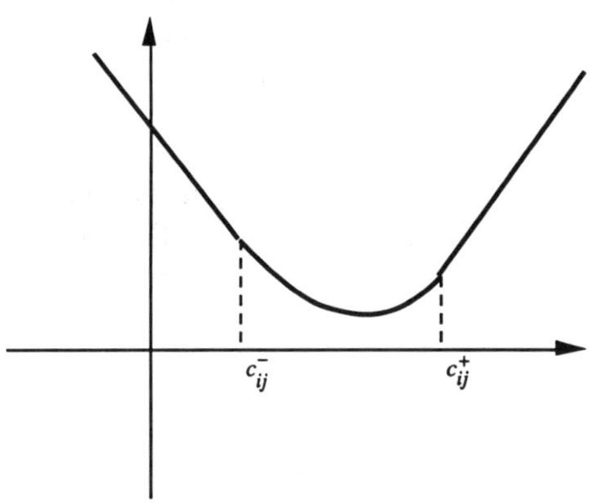

Fig. 6. Example of the function q_{ij}.

which is just the *surplus* $g_i(x(p))$ at node i under the flow $x(p)$. Thus, the coordinate ascent algorithm may be restated

(i′) Given $p(t)$, choose any node $i \in \mathcal{N}$.

(ii′) Compute $p(t+1)$, equal to $p(t)$ in all but the ith coordinate, such that $x(p(t+1)$ meets the flow balance constraint $g_i(x(p(t+1))) = 0$ at node i.

In the parlance of solving systems of nonlinear equations, we are 'relaxing' flow balance equation i at each iteration (although 'enforcing' might be a more appropriate word). Below, we will use the phrase 'relaxing node i' to denote applying the operation (ii′) at node i. Note that if $g_i(x(p))$ is positive, one must raise p_i in order to 'drive away' flow from i; raising p_i increases flow on the outgoing arcs of i, and reduces flow on i's incoming arcs. Conversely, if $g_i(x(p)) < 0$, then p_i must be *reduced* in order to attract flow to i.

A number of variations on the basic iteration are possible. For instance, instead of exactly maximizing q with respect to p_i, one may elect only to reduce the magnitude of its gradient by some factor $\delta \in [0, 1)$, as follows; see Bertsekas & Tsitsiklis [1989, section 5] and Bertsekas, Hossein & Tseng [1987]:

(i″) Given $p(t)$, choose any node $i \in \mathcal{N}$.

(ii″) If $g_i(x(p(t))) = 0$, do nothing. Otherwise, compute $p(t+1)$, equal to $p(t)$ in all but the ith coordinate, such that

$$0 \le g_i(x(p(t+1))) \le \delta g_i(x(p(t))) \text{ if } g_i(x(p(t))) > 0$$

$$\delta g_i(x(p(t))) \le g_i(x(p(t+1))) \le 0 \text{ if } g_i(x(p(t))) < 0.$$

Another possibility is to compute a maximizing value $\tilde{p}_i$ for p_i, but then set

$$p_i(t+1) = (1-\gamma)p_i(t) + \gamma \tilde{p}_i,$$

where $\gamma \in (0, 1)$ is some stepsize.

Algorithms in the dual relaxation family are also known as 'row-action' methods [Censor, 1981], in that, at each iteration, they select a single row of the constraint system $Ax = b$, and satisfy it at the expense of the other rows. In the case that the f_{ij} are differentiable, the flow $x(p(t+1))$ obtained by maximizing the dual cost along the p_i coordinate is equivalent to the projection of $x(p(t))$ onto the hyperplane given by the ith flow balance constraint with respect to a special distance measure, or 'D-function' derived from the f_{ij}. When the f_{ij} have certain special properties, and flow bound constraints are absent, it is actually possible to prove convergence of the method without any appeal to the dual problem; see Bregman [1967].

However, the main contribution of the row-action literature, from our perspective, is not in the handling of the equality constraints $Ax = b$, but in treatment of the *inequality* constraints $x \ge l$ and $x \le u$, for which some use of duality is required [Bregman, 1967; Censor & Lent, 1981]. Specifically, one may dualize the interval constraints $l_{ij} \le x_{ij} \le u_{ij}$ by attaching multipliers $v_{ij} \le 0$ and $w_{ij} \ge 0$, respectively. This yields a new dual functional

$$\hat{q}(p, v, w) = \\ \inf_{x \in \Re^m} \left\{ \sum_{(i,j) \in \mathcal{A}} f_{ij}(x_{ij}) - (p_i - p_j)x_{ij} - v_{ij}(x_{ij} - l_{ij}) - w_{ij}(u_{ij} - x_{ij}) \right\}$$

to be maximized with respect to $v \le 0$, $w \ge 0$, and $p \in \Re^n$. It is possible to construct an algorithm that successively maximizes the functional $\hat{q}(p, v, w)$ along not only the individual p_i coordinates, but also the individual v_{ij} and w_{ij} coordinates. Furthermore, for any optimal solution, the Karush–Kuhn–Tucker conditions guarantee that at most one of v_{ij} and w_{ij} will be nonzero for any (i, j), so the two variables may be condensed into a single quantity z_{ij} via $v_{ij} = (z_{ij})^-$ and $w_{ij} = (z_{ij})^+$. Censor and Lent [1981] give a procedure for handling such combined dual variables. When $z_{ij} \ge 0$, this procedure is equivalent to a dual-objective-maximizing step along the w_{ij} coordinate, followed by a maximizing step along the v_{ij} coordinate. When $z_{ij} < 0$, the roles of v_{ij} and w_{ij} are reversed.

The row action literature also suggests, for certain specific forms of the objective function, a special 'MART' step, which is an easily computed secant approximation to the true relaxation step [Censor, Pierro, Elfving, Herman & Iusem, 1990].

3.1.3. Proximal point methods

We now consider the problem (NLNW) with the assumption that the f_{ij} are convex, but not necessarily *strictly* convex. The dual coordinate methods described above rely on the differentiability of the dual problem (55), and hence on the strict convexity of the primal objective. When applied to problems whose objective functions are not strictly convex, including linear programs, such dual coordinate methods may 'jam' at a primal-infeasible (dual-suboptimal) point, due to the absence of dual smoothness. For purely linear problems, one may attempt to overcome this difficulty by using the special techniques of Section 2.1.3; however, it may be difficult to maintain much parallelism in the later stages of such methods. To address these deficiencies, and also to handle general (non-strictly) convex objectives, renewed attention has turned to the solution of non-strictly-convex problems via a *sequence* of strictly convex ones. So far, attention has centered on the *proximal minimization algorithm* (PMA) [Martinet, 1970; Rockafellar, 1976b] and variants thereof [Censor & Zenios, 1991]. There are several ways to motivate the PMA, but the simplest, perhaps, is to consider the following problem equivalent to (NLNW) [Bertsekas & Tsitsiklis, 1989, pp. 232–243]:

$$\begin{aligned} \text{minimize} \quad & \left[\sum_{(i,j)\in\mathcal{A}} f_{ij}(x_{ij})\right] + \frac{1}{2\lambda}\|x-y\|^2 \\ \text{such that} \quad & Ax = b \\ & l \le x \le u \\ & x, y \in \Re^m. \end{aligned} \tag{59}$$

Here, λ is a positive scalar. This problem is equivalent to (NLNW), the optimum being attained for $x = y = x^*$, where x^* solves (NLNW). One can imagine solving (59) by a block Gauss–Seidel method [Bertsekas & Tsitsiklis, 1989, section 3.3.5] by which one repeatedly minimizes the objective with respect to x, holding y fixed, and then minimizes with respect to y, holding x fixed. The latter operation just sets $y = x$, so one obtains the algorithm

$$\begin{aligned} x(t+1) &= \arg\min_{\substack{Ax=b \\ l\le x\le u}} \left\{ \sum_{(i,j)\in\mathcal{A}} f_{ij}(x_{ij}) + \frac{1}{2\lambda}\left(x_{ij} - y_{ij}(t)\right)^2 \right\} \\ y(t+1) &= x(t+1), \end{aligned}$$

or simply

$$x(t+1) = \arg\min_{\substack{Ax=b \\ l\le x\le u}} \left\{ \sum_{(i,j)\in\mathcal{A}} f_{ij}(x_{ij}) + \frac{1}{2\lambda}\left(x_{ij} - x_{ij}(t)\right)^2 \right\}. \tag{60}$$

Each of the successive subproblems in the method (60) is strictly convex due to the strong convexity of the terms $\frac{1}{2\lambda}\left(x_{ij} - x_{ij}(t)\right)^2$. Therefore, $x(t+1)$ is unique.

The convergence of the PMA can be proven even if λ is replaced by some $\lambda(t)$ that varies with t, so long as $\inf_{t\geq 0}\{\lambda(t)\} > 0$. This can be shown from first principles [Bertsekas & Tsitsiklis, 1989, pp. 232–243], or by appeal to the general theory of the proximal point algorithm [Rockafellar, 1976a, b]. The latter body of theory establishes that *approximate* calculation of each iterate $x(t+1)$ is permissible, an important practical consideration. Heuristically, one may think of (60) as an 'implicit' gradient method for (NLNW) in which the step $x(t+1) - x(t)$ is codirectional with the negative gradient of the objective not at $x(t)$, as is the case in most gradient methods, but at $x(t+1)$. Similar results using even weaker assumptions on $\{\lambda(t)\}$ are possible [Brézis & Lions, 1978].

Without the background of established proximal point theory [Rockafellar, 1976a, b], the choice of $\frac{1}{2\lambda}\|x - y\|^2$ as the strictly convexifying /term in (59) seems somewhat arbitrary. For example, $\frac{1}{2\lambda}\|x - y\|^3$ might in principle have served just as well. Recent theoretical advances [Censor & Zenios, 1991] indicate that any 'D-function' $D(x, y)$ of the form described in Censor & Lent [1981] may take the place of $\frac{1}{2}\|x - y\|^2$ in the analysis. Among the properties of such D-functions are

$$\begin{aligned}
&D(x, y) \geq 0 \quad \forall x, y\\
&D(x, y) = 0 \quad \Longleftrightarrow \quad x = y\\
&D(x, y) \text{ strictly convex in } x.
\end{aligned}$$

Of particular interest is the choice of $D(x, y)$ to be

$$\sum_{(i,j)\in\mathcal{A}} \left[x_{ij} \log\left(\frac{x_{ij}}{y_{ij}}\right) - (x_{ij} - y_{ij}) \right],$$

sometimes referred to as the *Kullback–Leibler divergence* of x and y. For additional information on the theory of such methods, see Teboulle [1992], Eckstein [1993] and Tseng & Bertsekas [1993].

3.1.4. *Alternating direction methods*

Alternating direction methods are another class of parallelizable algorithms that do not require strict convexity of the objective. Consider a general optimization problem of the form

$$\begin{aligned}
\text{minimize} \quad & h_1(x) + h_2(z)\\
\text{such that} \quad & z = Mx,
\end{aligned}$$

where $h_1 : \Re^r \to (-\infty, \infty]$ and $h_2 : \Re^s \to (-\infty, \infty]$ are convex, and M is a $r \times s$ matrix. A standard augmented Lagrangian approach to this problem, the *method of multipliers* (see Bertsekas [1982] for a comprehensive survey), is, for some scalar $\lambda > 0$,

$$\big(x(t+1), z(t+1)\big) = \operatorname*{arg\,min}_{(x,z)} \left\{ h_1(x) + h_2(z) + \langle \pi(t), Mx - z \rangle + \frac{\lambda}{2}\|Mx - z\|^2 \right\} \tag{61}$$

$$\pi(t+1) = \pi(t) + \frac{\lambda}{2}\left(Mx(t+1) - z(t+1)\right). \tag{62}$$

Here, $\{\pi(t)\} \subseteq \Re^q$ is a sequence of Lagrange multiplier estimates for the constraint system $Mx - z = 0$. The minimization in (61) is complicated by the presence of the nonseparable $z^T Mx$ term in $\|Mx - z\|^2$. However, one conceivable way to solve for $x(t+1)$ and $z(t+1)$ in (61) might be to minimize the augmented Lagrangian alternately with respect to x, with z held fixed, and then with respect to z with x held constant, repeating the process until both x and z converge to limiting values. Interestingly, it turns out to be possible to proceed directly to the multiplier update (62) after a single cycle through this procedure, without truly minimizing the augmented Lagrangian. The resulting method may be written

$$\begin{aligned}
x(t+1) &= \arg\min_x \left\{ h_1(x) + \langle \pi(t), Mx \rangle + \frac{\lambda}{2}\|Mx - z(t)\|^2 \right\} \\
z(t+1) &= \arg\min_z \left\{ h_2(z) - \langle \pi(t), z \rangle + \frac{\lambda}{2}\|Mx(t+1) - z\|^2 \right\} \\
\pi(t+1) &= \pi(t) + \frac{\lambda}{2}\left(Mx(t+1) - z(t+1)\right),
\end{aligned}$$

and is called the *alternating direction method of multipliers*. It was introduced in Glowinski & Marroco [1975], Gabay & Mercier [1976], Fortin & Glowinski [1983]; see also Bertsekas & Tsitsiklis [1989, pp. 253–261]. Note that the problem of nonseparability of x and z in (61) has been removed, and also that h_1 and h_2 do not appear in the same minimization. Gabay [1983] made a connection between the alternating direction method of multipliers and a generalization of an alternating direction method for solving discretized differential equations [Lions & Mercier, 1979]; see Eckstein [1989] and Eckstein & Bertsekas [1992] for comprehensive treatments.

One way to apply this method to a separable, convex-cost network problem of the form (NLNW) (without any assumption of *strict* convexity) is to let

$$\begin{aligned}
r &= m \\
s &= 2m \\
z &= (\eta, \xi) \in \Re^m \times \Re^m \\
M &= \begin{bmatrix} I \\ I \end{bmatrix} \\
h_1(x) &= \sum_{(i,j)\in\mathcal{A}} \bar{f}_{ij}(x_{ij}) \qquad \text{[as defined in (51)]} \\
h_2(\eta, \xi) &= \begin{cases} 0 & \sum_{j:(i,j)\in\mathcal{A}} \eta_{ij} - \sum_{j:(j,i)\in\mathcal{A}} \xi_{ji} = b_i \quad \forall i \in \mathcal{N} \\ +\infty & \text{otherwise.} \end{cases}
\end{aligned}$$

The idea here is to let $z = (\eta, \xi) \in \Re^m \times \Re^m$, where η_{ij} is the flow on (i, j) as perceived by node i, while ξ_{ij} is the flow on (i, j) as perceived by node j. The objective function term $h_2(\eta, \xi)$ essentially enforces the constraint that the perceived flows be in balance at each node, while the constraint $z = Mx$ requires that each η_{ij} and ξ_{ij} take a common value x_{ij}, that is, that flow be conserved along arcs. The function h_1 plays the role of the original objective function of (NLNW), and also enforces the flow bound constraints. Applying the alternating direction method of multipliers to this setup reduces, after considerable algebra [Eckstein, 1994, 1989], to the algorithm

$$\begin{aligned}
\hat{x}_{ij}(t) &= x_{ij}(t) + \frac{g_i(x(t))}{d(i)} - \frac{g_j(x(t))}{d(j)} \\
x_{ij}(t+1) &= \operatorname*{arg\,min}_{l_{ij} \le x_{ij} \le u_{ij}} \left\{ f_{ij}(x_{ij}) - (p_i - p_j)x_{ij} + \frac{\lambda}{2}\left(x_{ij} - \hat{x}_{ij}(t)\right)^2 \right\} \\
p_i(t+1) &= p_i(t) + \frac{\lambda}{d(i)} g_i(x(t+1)),
\end{aligned}$$

where $d(i)$ denotes the degree of node i in the network. This iteration is basic form of the *alternating step method*; see also Bertsekas & Tsitsiklis, [1989, p. 254]. The initial flow vector $x(0)$ and node prices $p(0)$ are arbitrary, and need fulfill neither feasibility nor complementary slackness conditions. For linear or quadratic f_{ij}, the one-dimensional minimization required to compute the $x_{ij}(t+1)$ can be done analytically. An 'overrelaxed' version of the method is also possible; letting $\{\rho(t)\}_{t=1}^{\infty}$ be a sequence of scalars such that

$$0 < \inf_{t \ge 1} \{\rho(t)\} \le \sup_{t \ge 1} \{\rho(t)\} < 2,$$

one can derive the more general alternating step method [Eckstein, 1994]

$$\hat{x}_{ij}(t) = y_{ij}(t) + \frac{g_i(y(t))}{d(i)} - \frac{g_j(y(t))}{d(j)} \tag{63}$$

$$x_{ij}(t+1) = \operatorname*{arg\,min}_{l_{ij} \le x_{ij} \le u_{ij}} \left\{ f_{ij}(x_{ij}) - (p_i - p_j)x_{ij} + \frac{\lambda}{2}\left(x_{ij} - \hat{x}_{ij}(t)\right)^2 \right\} \tag{64}$$

$$p_i(t+1) = p_i(t) + \frac{\lambda\rho(t)}{d(i)} g_i(x(t+1)) \tag{65}$$

$$y_{ij}(t+1) = (1-\rho(t))y_{ij}(t) + \rho(t)x_{ij}(t+1). \tag{66}$$

Here, the initial flow vector $y(0)$ and node prices $p(0)$ are again arbitrary. Approximate calculation of the $x_{ij}(t+1)$ is possible [Eckstein & Bertsekas, 1992; Eckstein, 1994].

For convex-cost transportation problems, a different application of the alternating direction method of multipliers is given in Bertsekas & Tsitsiklis [1989, exercise 5.3.10], Eckstein [1989, section 7.3], and Eckstein & Fukushima [1994]. This approach decomposes the problem network somewhat less aggressively than (63)–(66).

3.2. Parallelization ideas

3.2.1. Primal methods

Parallelization of the primal truncated Newton algorithm has concentrated on the linear algebra calculations required in solving Newton's equations (49).

The first approach, proposed in Zenios & Mulvey [1988b], views the matrix $Z^T \nabla^2 F(x^k) Z$ as a general, sparse matrix without any special structure. The conjugate gradient method is used to solve this system. In order to implement this method we need to form the products Zv, Hv and $Z^T v$, where v is a vector of the conjugate directions. On shared memory systems these products can be parallelized very efficiently by distributing the Hessian matrix row-wise across multiple processors, while distributing the Z matrix column-wise.

Another important feature of this algorithm is that it can be implemented efficiently on systems with vector features. To this end we need data structures that allow the algorithm to form compact, dense vectors from the sparse matrix representations of Z and H. Details on implementation designs for PTN on a vector architecture are rather technical. See Zenios & Mulvey [1986] for a description of the data structures and a complete discussion of implementations.

Similar parallelization ideas for PTN have been proposed in Lescrenier & Toint [1988] for cases where the objective function is *partially separable*, i.e., it is of the form $F(x) = \sum f_i(x)$, where each element function $f_i(x)$ has a Hessian matrix of lower rank than the original problem. These proposals have been very efficient in solving some large unconstrained optimization problems. However, they do not offer any real advantage over the methods discussed above for the network problems that are, usually, separable.

An alternative procedure for parallelizing the primal truncated Newton algorithm exploits the sparsity structure of the network basis and partitions Newton's equations in independent blocks. The blocks can then be distributed among multiple processors for solution using again the conjugate gradient solver. The *block-partitioned* truncated Newton method was developed in Zenios & Pinar [1992]. To describe this algorithm, we need to impose an ordering of the arcs (i, j). We will use $t \sim (i, j)$ to denote the lexicographic order of arc (i, j), and hence x_t is the tth variable that corresponds to flow x_{ij} on arc (i, j).

Block–partitioning of Newton's equations

We return now to equation (49), and try to identify a partitioning of the matrix $(Z^T \nabla^2 F(x^k) Z)$ into a block-diagonal form. Recall that

$$Z = \begin{bmatrix} -B^{-1}S \\ I \\ 0 \end{bmatrix} \tag{67}$$

and that the function $F(x) = \sum_{t=1}^{n} F_t(x_t)$ is separable, so that the Hessian matrix is diagonal. If we ignore momentarily the dense submatrix $(B^{-1}S)$ and assume that

$$Z = \hat{Z} = \begin{bmatrix} I \\ 0 \end{bmatrix} \tag{68}$$

(the identity I and null matrix 0 chosen such that Z is conformable to $\nabla^2 F(x^k)$) then the product

$$\hat{Z}^T \nabla^2 F(x^k) \hat{Z}$$

is a matrix of the form diag$[H_I\ 0]$, where H_I is a diagonal matrix with the tth diagonal element given by $\partial^2 F_t(x_t^k)/\partial x_t^2$. Hence, the complication in partitioning (49) is the presence of the submatrix $(B^{-1}S)$. The structure of this submatrix is examined next.

The structure of $(B^{-1}S)$

The matrix B is a basis for the network flows of problem (NLNW). It is well-known — see, e.g., Dantzig [1963], Kennington and Helgason [1980] or Glover, Klingman & Stutz [1973, 1978] — that the basis of a pure network problem is a lower triangular matrix. The graph associated with this basis matrix is a rooted tree. The basis of a generalized network is characterized by the following theorem [see, e.g., Dantzig, 1963, p. 421].

Theorem 1. *Any basis B of a generalized network problem can be put in the form*

$$B = \begin{bmatrix} B^1 & & & & & \\ & B^2 & & & & \\ & & \ddots & & & \\ & & & B^\ell & & \\ & & & & \ddots & \\ & & & & & B^L \end{bmatrix}$$

where each square submatrix B^ℓ is lower triangular with at most one element above the diagonal.

The graph associated with each submatrix B^ℓ is a tree with one additional arc, making it either a rooted tree or a tree with exactly one cycle, and is called a quasi–tree (abbreviated: q–tree). The graph associated with a generalized network basis is a forest of q–trees.

To describe the structure of $(B^{-1}S)$ we first define the *basic–equivalent–paths* (BEP) for a superbasic variable x_t with incident nodes (i, j). For pure network problems it is the set of arcs on the basis tree that lead from node j to node i. The arcs on BEP together with arc $t \sim (i, j)$ create a loop. In the case of a generalized network, it is the set of arcs that lead from nodes i and j to either the root of the tree or the cycle of a q-tree; the BEP includes all arcs on the cycle. The tth column of $(B^{-1}S)$ has non–zero entries corresponding to the BEP of the tth superbasic variable. The numerical values of $(B^{-1}S)$ are ± 1 for pure network problems and arbitrary real numbers for generalized networks; the numerical values are of no consequence to our development.

To illustrate the preceding discussion we show in Figure 7 the basis of a pure network problem together with the BEP for a superbasic arc and the corresponding column of $(B^{-1}S)$. Figure 8 illustrates the same definitions for generalized network problems.

The matrix $(B^{-1}S)$ can be partitioned into submatrices without overlapping

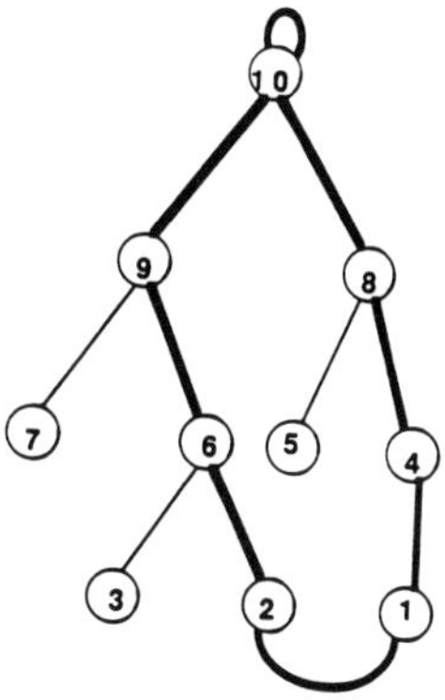

	1	2	3	4	5	6	7	8	9	10
1	*									
2		*								
3			*							
4	*			*						
5					*					
6		*	*			*				
7							*			
8				*	*			*		
9						*	*		*	
10								*	*	*

Basic-Equivalent-Path (BEP) for arc with incident nodes (1,2):

{(2,6),(6,9),(9,10),(10,10),(10,8),(8,4),(4,1)}

Sparsity pattern of $(B^{-1}S)$ corresponding to superbasic (1,2):

Row No.		Corresponding basic arc
1	*	(1,4)
2	*	(2,6)
3	0	
4	*	(4,8)
5	0	
6	*	(6,9)
7	0	
8	*	(8,10)
9	*	(9,10)
10	*	(10,10)

Fig. 7. Pure network basis: matrix and graph representation, and an example of a basic equivalent path.

rows if the columns of each submatrix have BEP with no basic arcs in common with the columns of any other submatrix.

Partitioning of $(B^{-1}S)$ for pure networks

Let β_t denote an ordered set of binary indices such that $(\beta_t)_l = 1$ if the lth arc $(i, j) \in \mathcal{B}$ is in the BEP of the tth arc $(i', j') \in \mathcal{S}$, and $(\beta_t)_l = 0$ otherwise. We seek a partitioning of the set $\mathcal{S}$ into K disjoint independent subsets, say $\mathcal{S}^k$, $k \in \mathcal{K} = \{1, 2, \ldots, K\}$ such that

$$\mathcal{S} = \bigcup_{k=1}^{K} \mathcal{S}^k \quad \text{and} \tag{69}$$

$$t \in \mathcal{S}^{k_1} \text{ and } \quad u \in \mathcal{S}^{k_2} \text{ iff } \beta_t \wedge \beta_u = 0 \,\forall\, k_1 \neq k_2 \in \mathcal{K} \tag{70}$$

(i.e., the sets β_t and β_u have no overlapping non–zeroes, and hence there is no common basic arc in the BEP of tth and uth superbasic arcs).

Escudero [1986b] was the first one to propose the partitioning of $\mathcal{S}$ into independent superbasic subsets $\mathcal{S}^k$ according to equation (69)–(70), for *replicated*

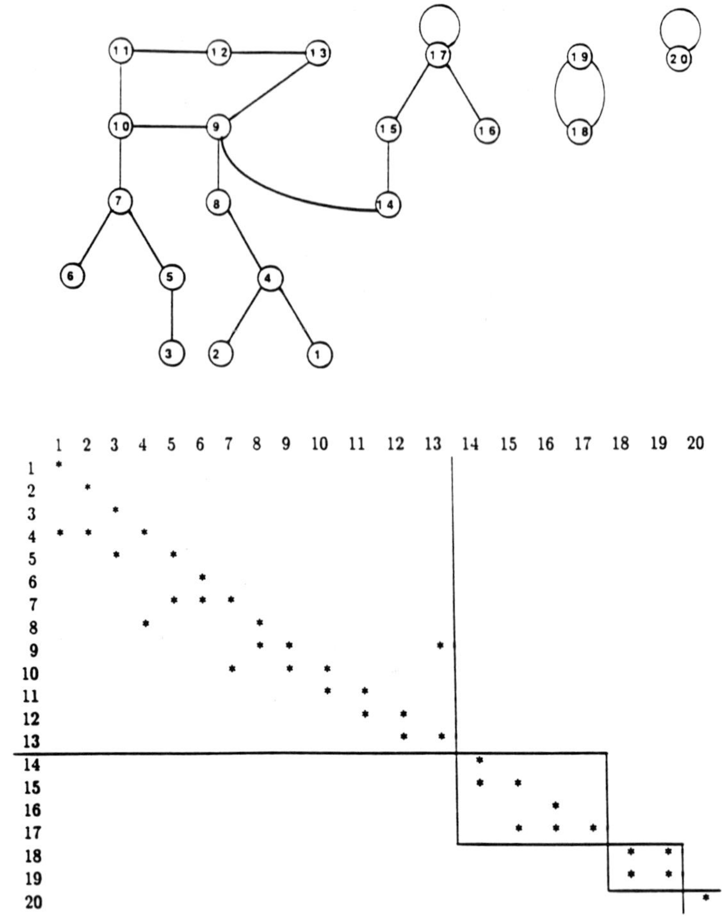

Basic-Equivalent-Path (BEP) for arc (9,14):

{(9,10),(10,11),(11,12),(12,13),(13,9),(14,15),(15,17),(17,17)}

Sparsity pattern of $(B^{-1}S)$ corresponding to superbasic (9,14):

Row No.	1	2	3	4	5	6	7	8	9	10	11	12	13	14	15	16	17	18	19	20
	0	0	0	0	0	0	0	0	*	*	*	*	*	*	*	0	*	0	0	0

Fig. 8. Generalized network basis: matrix and graph representation, and an example of a basic equivalent graph.

pure networks. Replicated networks consist of subnetworks with identical structure and are connected by linking arcs. (These linking arcs represent inventory flow for his problems that are multiperiod networks.) In the same reference Escudero gives a procedure for identifying the independent superbasic sets $\mathcal{S}^k$.

The problem of identifying independent superbasic sets can be formulated as a problems from graph theory (i.e., finding connected components or articulation points of the adjacency graph of the matrix $Z^T Z$). It can be solved efficiently using algorithms developed in Tarjan [1972]. For additional details see Zenios & Pinar [1992].

Partitioning of $(B^{-1}S)$ for generalized networks

The graph partitioning schemes discussed above for pure network problems can also be applied in the case of generalized networks. It is, however, possible to develop alternative — and much simpler — techniques to partition the superbasic set of generalized network problems that take advantage of the block structure of the generalized network basis.

Previously, we observed that the graph associated with the basis of a generalized network problem is a collection of quasi–trees. Suppose the basis matrix B consists of submatrices B^ℓ, $\ell = 1, \ldots, L$. We denote the graph (quasi–tree) associated with B^ℓ by $G_\ell = (N_\ell, E_\ell)$. The superbasic set $\mathcal{S}$ can be partitioned in subsets $\mathcal{S}^k$ defined by

$$\mathcal{S}^\ell = \{(i, j) \in \mathcal{S} | i, j \in N_\ell\} \quad \forall \ell = 1, 2, \ldots, L \tag{71}$$

with $\cup_{\ell=1}^{L} \mathcal{S}^\ell \subseteq \mathcal{S}$. This partitioning scheme will ignore any superbasic variables that connect basis submatrices. A partitioning scheme that includes additional superbasic variables is the following: Given indices k, $p(k) \leq L$ and $q(k) \leq L$, $p(k) \neq q(k)$ choose $B^{p(k)}$ and $B^{q(k)}$ and define

$$\mathcal{S}^{p_k q_k} = \{(i, j) \in \mathcal{S} | i \in N_{p(k)}, j \in N_{q(k)}\} \tag{72}$$

$$\mathcal{S}^{p_k} = \{(i, j) \in \mathcal{S} | i, j \in N_{p(k)}\} \tag{73}$$

$$\mathcal{S}^{q_k} = \{(i, j) \in \mathcal{S} | i, j \in N_{q(k)}\} \tag{73}$$

and finally $\mathcal{S}^k = \mathcal{S}^{p_k q_k} \cup \mathcal{S}^{p_k} \cup \mathcal{S}^{q_k}$. To ensure that two set $\mathcal{S}^{k_1}$, $\mathcal{S}^{k_2}$ are independent we require

$$B^{p(k_1)} \neq B^{p(k_2)} \neq B^{q(k_2)}$$
$$B^{q(k_1)} \neq B^{p(k_2)} \neq B^{q(k_2)}.$$

A procedure, that was found to work well in practice, to identify these independent subsets $\mathcal{S}^k$ of superbasic arcs is described in Phillips & Zenios [1989]. Finally, we point out that the partitioning techniques described in this section can be extended to handle non-separable problems as explained in the same reference.

3.2.2. Dual and proximal methods

Dual coordinate methods are amenable to both synchronous and asynchronous parallel implementation. In general, the basic idea is to perform the relaxation iterations for many nodes concurrently. On coarse-grain multiprocessors, each processor may be assigned to multiple nodes, whereas on extremely fine-grain machines, a cluster of processors might handle each node of the problem network.

The simplest synchronous approach involves the idea of a *coloring scheme* [Bertsekas & Tsitsiklis, 1989, pp. 21–27; see also Zenios & Mulvey, 1988a]. In a serial environment, one of the most natural implementations of dual relaxation is the *Gauss–Seidel* method, in which one lists the nodes of $\mathcal{N}$ in some fixed order, and cycles repeatedly through this list, relaxing each node in sequence. A *coloring* of the network graph $(\mathcal{N}, \mathcal{A})$ is some partition of the node set $\mathcal{N}$ into subsets (colors) such that no two nodes of the same color are adjacent in the network. Suppose now that one adopts a Gauss–Seidel node ordering in which all nodes of a given color are consecutive in the list.

From the form of (53), the maximizing value of p_i in each relaxation iteration depends only on the prices of the adjacent nodes p_j (for which there exists $(i, j) \in \mathcal{A}$ or $(j, i) \in \mathcal{A}$). and hence does not directly depend on the prices of any nodes having the same color as i. It follows that the price updates for all nodes of a given color may be performed simultaneously without altering the course of the algorithm. Such procedures work particularly well on transportation problems, for which only two colors suffice, one for the origin nodes, and one for the destination nodes. On more general networks, coloring is not guaranteed to be effective, but may still manage to employ a fairly large number of processors efficiently. For example, Zenios and Mulvey [1988a] present simulator-based results for up to 200 processors, using a greedy heuristic to color arbitrary problem networks. We also note that any network can be made bipartite by introducing an artificial node into the middle of each arc. Two colors will always suffice for the resulting expanded network.

The theory of synchronous dual methods other than those based on coloring schemes is subsumed in that of asynchronous methods, which we now summarize. We consider first a totally asynchronous environment in which there are no bounds on computational latency or communication delays. Each processor is associated with a single node i, and at time t stores the current price of i, $p_i(t)$, and also prices $p_j(i, t) = p_j(\tau_{ij}(t))$ for each neighbor j of i. These neighbor prices may be out of date, that is, $\tau_{ij}(t) \leq t$. As the algorithm progresses, the neighboring processors send messages carrying their prices to i, causing the $p_j(i, t)$ to be updated. As for the timing of the algorithm, one assumes only that for each time T and node i,

1. Node i is selected for relaxation at an infinite number of times after T.

2. All neighbors of i receive an infinite number of messages communicating the price $p_i(t)$ for some time $t > T$.

3. There exists some time $T' > T$ such that all messages carrying prices from before time T are no longer in transit at time T'.

Here, assumptions 1 and 2 intuitively say that processors never stop computing, nor do they ever cease successfully communicating their results to their neighbors. Assumption 3 says that outdated information is eventually purged from the communication system. The last two assumptions together imply $\tau_{ij}(t) \to \infty$ as

$t \to \infty$. To obtain convergence results in this general framework, one must take the (non-restrictive) step of fixing one node price p_i in each connected component of $(\mathcal{N}, \mathcal{A})$. In the following, we assume that the problem network is connected, and set $p_1 = 0$. Under these conditions, it still does not follow that $\{p(t)\}$ converges to an optimal solution. Instead, one can assert only that every limit point p_i^∞ of each coordinate sequence $\{p_i(t)\}$ is such that there exists an optimal solution p^* of (55) with $p_1^* = 0$ and $p_i^\infty = p_i^*$. To obtain true convergence, one must assume:

1. the set $P^* = \arg\max\{q(p) \mid p_1 = 0\}$ is bounded above in all coordinates;
2. when node i is relaxed, $p_i(t+1)$ is set to the *largest* value maximizing $q(p)$ with respect to the ith coordinate;
3. $p(0) \geq p^*$ (componentwise) for all $p^* \in P^*$.

There is an analogous set of assumptions that gives convergence if P^* is bounded *below* in all coordinates, and $p_i(t+1)$ is set as *small* as possible when relaxing node i. For a complete analysis, see Bertsekas & El Baz [1987] or Bertsekas & Tsitsiklis [1989, section 6.6].

We now consider *partially asynchronous* implementations, in which we assume a certain maximum time interval B between updates of each p_i, and that the prices of neighboring nodes used in each update are no more than B time units out of date, that is, $\tau_{ij}(t) \geq t - B$ for all i, j, and t. In this case, it is not necessary to fix the price of any node. Instead, convergence may be proven [Bertsekas & Tsitsiklis, 1989, section 7.2] under the assumption that, when relaxing node i,

$$p_i(t+1) = (1-\gamma)p_i(t) + \gamma\tilde{p}_i \quad ,$$

where $\gamma \in (0, 1)$ is fixed throughout the algorithm, and $\tilde{p}_i$ is, among all values maximizing $q(p)$ along the ith coordinate, the *farthest* from $p_i(t)$. This result can be applied, for instance, to a synchronous Jacobi implementation in which all nodes simultaneously perform an update based on $p(t)$, and then exchange price information with their neighbors.

The main difficulty with all synchronous parallel dual methods is the complexity of the line search needed to maximize $q(p)$ along a given coordinate. Even when the f_{ij} have a simple functional form, $q(p)$ tends to have a large number of 'breakpoints' along each coordinate, as the various arc flows $x_{ij}(p) = -\nabla q_{ij}(p_i - p_j)$ and $x_{ji}(p) = -\nabla q_{ji}(p_j - p_i)$ attain or leave their upper or lower bounds. If p_i must be moved across many such breakpoints, relaxing node i may be very time-consuming, possibly causing processors responsible for simultaneously relaxing other nodes be kept idle while waiting for the computation at i to be completed. To address this difficulty, Tseng [1990] has proposed a line search that is itself parallelizable, although it may result in small steps; see also Bertsekas & Tsitsiklis [1989, pp. 413–414]. Several other stepsize procedures are proposed in Zenios & Nielsen [1992].

For row-action methods in which inequality constraints are explicitly dualized, practical computational research has concentrated on the use of coloring schemes: once the nodes have been partitioned into colors, one can add one more color to handle the interval constraints [Zenios & Censor, 1991]. More general 'block iterative' parallel implementations have also been proposed; see Censor & Segman [1987] and the comprehensive survey in Censor [1988].

There are no parallelization ideas specific to proximal minimization methods; however, if dual coordinate methods are used to solve the sequence of strictly convex subproblems (60) generated by proximal minimization, any of the above parallelization approaches may be applicable.

3.2.3. Alternating direction methods

Alternating direction optimization methods are designed with massive, synchronous parallelism in mind. In the alternating step method, the primal update (63)–(64) for each arc (i, j) is completely independent of that for all other arcs; therefore, all such updates may be performed concurrently. Likewise, (66) may also be processed simultaneously for all arcs. The dual update (65) can be done concurrently for all nodes i. The simplicity of both the primal and dual calculations implies that they can be performed with little or no synchronization penalty. In fact, the most challenging part of (63)–(66) to efficiently parallelize is the computation of the surpluses $g_i(y(t))$ and $g_i(x(t+1))$ It turns out that, of these, only the $g_i(x(t+1))$ requires much effort, as the $g_i(y(t))$ can be found quickly and in parallel via the identity

$$g_i(y(t+1)) = (1 - \rho(t))g_i(y(t)) + \rho(t)g_i(x(t+1)).$$

This identity follows from (66) because $g_i(x)$ is an affine function of x [Eckstein, 1994, 1989].

Implementating the more 'aggregated' alternating direction method of Eckstein & Fukushima [1994] for nonlinear transportation problems is more complicated. For every origin node i, each iteration requires an (approximate) optimization over a simplex of dimension $d(i)$. All these optimizations are independent, and can be performed concurrently. Every destination node j requires a $d(j)$-element averaging operation at each iteration. Again, all these calculations can be performed at the same time.

3.3. Computational experiences

3.3.1. Primal methods

There has been substantial experience with vector and parallel computing using the truncated Newton algorithm and its parallel block variant. Experiments with the vectorization of truncated Newton on a CRAY X-MP/48 are reported in Zenios & Mulvey [1986], which used test problems derived from several applications: water distribution systems, matrix balancing and stick percolation. The performance of the vectorized algorithm was on average a factor of 5 faster than the scalar implementation. It is worth pointing out that a version of the program that was vectorized automatically by the compiler was only 15% faster than the scalar code. Substantial improvements in performance were achieved when appropriate data structures and a re-design of the implementation were developed.

A subsequent paper [Zenios & Mulvey, 1988b] reports on the performance of the parallel truncated Newton implementation on the CRAY X-MP/48 system. The algorithm achieved speedups of approximately 2.5 when executing on three processors. The observed speedup was very close to the upper bound provided by Amdhal's law, given that a fraction of the PTN algorithm was not parallelized.

The block-truncated Newton method was also tested empirically in Zenios & Pinar [1992]. The block-partitioning techniques have been tested on both pure and generalized network problems. For two sets of pure network test problems (i.e., water distribution and stick percolation models) it was observed that the superbasic sets did not yield good partitionings. While the partitioning methods used are very efficient, poor partitioning resulted in insignificant improvements in performance. For the generalized network test problems, however, very good partitionings were obtained. In this case even a serial implementation of the block-partitioned algorithm was superior to the non-partitioned code. Improvements varied from a factor of 1.5 to 5.3 with average improvement across all problems of 2.6. As expected, better performance was observed for the larger test problems.

Parallel implementations were carried out on a 4 processor Alliant FX/4 and the CRAY X-MP/48. In both cases some modest speedups in performance were observed, in the range of 2 to 3. This is far from the linear speedup of 4. The discrepancy is due to a load balancing effect: the blocks of independent superbasic sets are not of equal size. Hence some processors need more time to complete their calculations than others. Overall, however, the parallel block-partitioned algorithm was shown to be much faster than the serial non-partitioned algorithm. The larger test problems have 15K nodes and 37K arcs, and the projected Newton equations in the neighborhood of the optimal solution are of dimension 22K $\times$ 22K. This problem required approximately 1 hour on a 4-processor Alliant FX/4 and 15 min. on the CRAY X-MP/48 using the block-partitioned algorithm. The (non-partitioned) algorithm executing on a single processor of the Alliant required 7 hours.

Figure 9 illustrates the performance of the PTN algorithm when implemented without the block-partitioning ideas on a VAX 8700 mainframe, then implemented with block-partitioning on the same (serial) architecture, and finally implemented with parallelization of the block-partitioned algorithm on the Alliant FX/4.

3.3.2. Dual and proximal methods

Computational experience with dual methods has been fairly extensive. Results for serial workstations have been reported by Tseng [see, for example, Zenios & Censor, 1991]. A group of quadratic-cost test problems, known as TSENG1 through TSENG16, and especially TSENG1 through TSENG8, have become *de facto* standard test problems. These problems were generated using a version of NETGEN [Klingman, Napier & Stutz, 1974] altered to supply quadratic cost functions. TSENG1–TSENG8 are moderately ill-conditioned, positive definite, separable transportation problems ranging in size from 500 $\times$ 500 to 1250 $\times$ 1250. In TSENG1–TSENG4, the average node degree is about 10, whereas in TSENG5–TSENG8, the average node degree is about 20.

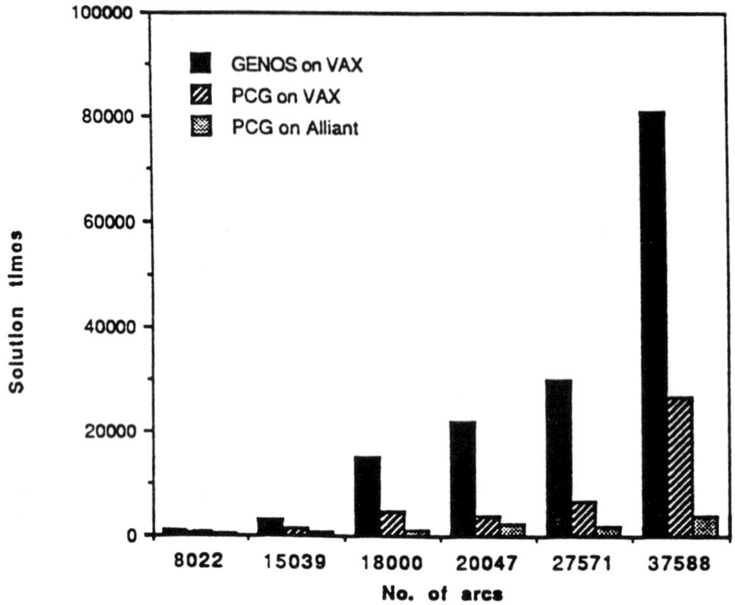

Fig. 9. Performance of Primal Truncated Newton and block- partitioned Truncated Newton on serial and parallel architectures.

On parallel machines, early computational testing has focused primarily on the Alliant FX/8, a high-performance 8-processor, shared-memory system, and the CM-2 (Zenios & Mulvey [1988a] also gives some early, simulator-based results).

For the Alliant FX/8, Chajakis & Zenios [1991] give a lengthy account of the details of implementing dual relaxation, both synchronously using a coloring scheme, and with various degrees of asynchronism. The algorithm seems to work faster the more asynchronism is allowed; the fully asynchronous, 8-processor version requiring between 3 and 15 seconds to solve each of the problems TSENG1–TSENG8. The speedups over a one-processor implementation averaged around 6 (75% efficiency), but were as low as 3 for TSENG2, which appears to be a hard problem for dual relaxation. These runs were done at an accuracy of 10^{-3}, meaning that the algorithm was terminated when the absolute value of the flow imbalance at all nodes was less than

$$\frac{10^{-3}}{n}\left(\sum_{i\in\mathcal{N}}|b_i|\right).$$

In other words,

$$\|r(x(p(t)))\|_\infty < (10^{-3})\frac{\|b\|_1}{n}$$

Runs for the same Alliant FX/8 implementation at an accuracy of 10^{-6} given in Zenios, Qi & Chajakis [1990] are longer by about an order of magnitude (and considerably more on the troublesome TSENG2). This phenomenon illustrates one drawback of dual coordinate methods — a 'tailing' effect by which final convergence near the optimum may be slow. At an accuracy of 10^{-8}, which is more standard in mathematical programming, tailing effects would be even more pronounced.

Another architecture on which dual methods have been extensively tested is the Connection Machine, starting with Zenios & Lasken [1988] on the CM-1, and continuing with McKenna & Zenios [1990], Zenios & Censor [1991] and Zenios & Nielsen [1992] on the CM-2. The work in Zenios & Lasken [1988] introduced the fundamental 'segmented scan' representation of sparse networks for the Connection Machine architecture, but studied only specialized cost functions conducive to efficient line search, with encouraging preliminary results. The Connection Machine, having a SIMD architecture, cannot be truly asynchronous in a hardware sense, but Zenios & Lasken [1988] established that, under such an architecture, it seems best to iterate on all nodes simultaneously, using price information possibly outdated by one time unit, rather than to use a coloring scheme. To prove convergence of such a method, one must appeal to the partially asynchronous convergence analysis of Bertsekas & Tsitsiklis [1989, section 7.2]. Still another architecture in which dual methods have been tested is a network of transputers [El Baz, 1989].

A Connection Machine dual relaxation implementation for general separable positive definite quadratic cost functions is discussed in Zenios & Nielsen [1992], which examines four different line searches: an iterative procedure based on the row-action literature, an exhaustive exact search method based on Helgason, Kennington & Lall [1980], a small-step procedure like that of Tseng [1990], and an original, Newton-like step. Of these, the small-step and Newton-like procedures seemed to give the most consistently good results. Solution times for the 16K-processor CM-2 appeared to be comparable to those for the Alliant FX/8 (e.g., using the Tseng-type line search, 12.4 seconds for TSENG8, versus 11.4 seconds for asynchronous relaxation of the FX/8). With twice as many processors, CM-2 solution times decreased by about 30%.

Row-action dual methods designed specifically for transportation problems with quadratic or entropy ($f_{ij}(x_{ij}) = x_{ij}[\log(x_{ij}/a_{ij}) - 1]$) cost structures are described in Zenios & Censor [1991]. The constraints are partitioned into three sets, the origin node flow balance equations, the destination node flow balance equations, and the arc flow bounds; a coloring scheme is used to alternate between the three sets. A simple grid data structure is used in place of the more general segmented scan representation of Zenios & Lasken [1988] and Zenios & Nielsen [1992]. At the fairly low accuracy level of 10^{-3}, a 32K-processor CM-2 gave very low run times, between 0.1 and 4.2 seconds, on TSENG1–TSENG8. Similar run times are given for test problems with an entropy cost structure, but at higher (10^{-6}) accuracy. For the quadratic case, McKenna & Zenios [1990] give an even more efficient, microcoded implementation of the same algorithm.

A direct comparison of dual methods on the FX/8 and CM-2 appears in Zenios, Qi & Chajakis [1990]. Here, the asynchronous FX/8 implementation of Chajakis &

Table 4
16K-processor CM-2 computational results for TSENG1–TSENG8

Problem	Run time (s)		
	Dual relaxation [133]	Row action [126]	Alternating direction [52]
TSENG1	6.77	1.55	2.11
TSENG2	n.a.	2.45	4.32
TSENG3	9.75	2.03	3.50
TSENG4	13.92	2.06	4.68
TSENG5	3.34	1.63	2.95
TSENG6	5.56	3.43	3.91
TSENG7	15.90	3.45	5.17
TSENG8	9.68	2.85	4.64

Accuracy is 10^{-3}; double precision arithmetic; C/PARIS implementation. The dual relaxation method uses the Newton-like linesearch, which appeared to be the most efficient.

Zenios [1991] is pitted against the row-action CM-2 method of Zenios & Censor [1991] at 10^{-6} accuracy on the quadratic TSENG1–TSENG8 test set. On average, a 32K-processor Connection Machine was about three times as fast as the (much cheaper) FX/8, but was actually significantly slower on two of the eight problems.

Comprehensive quantitative comparisons between the results of Chajakis & Zenios [1991], McKenna & Zenios [1990], Zenios & Censor [1991], Zenios, Qi & Chajakis [1990], Zenios & Lasken [1988], and Zenios & Nielsen [1992] are difficult because of the varying precision levels, programming environments, hardware configurations, and test problems. Table 4 gives a partial summary for TSENG1–TSENG8. It is taken from Eckstein [1993], and therefore also includes data for an alternating direction method.

3.3.3. *Alternating direction methods*

Alternating direction methods have been tested less exhaustively than dual relaxation approaches, the main results appearing in Eckstein [1993]. This paper tests the alternating step method on quadratic and linear cost networks. The algorithm is implemented on the CM-2 using two data structures, one resembling that of Zenios & Lasken [1988], and one similar to the simple transportation grid approach of Zenios & Censor [1991]. In agreement with preliminary results in Eckstein [1989, chapter 7], performance on purely linear-cost problems proved to be very disappointing. On the other hand, the results compare favorably with the dual relaxation implementations of Zenios & Nielsen [1992] on the purely quadratic TSENG1–TSENG8 problems, at similar (low) levels of accuracy. Performance was not quite as good as the row-action approach of Zenios & Censor [1991], which was specialized to transportation problems. Furthermore, the alternating step method was able to handle mixed linear-quadratic problems with as many as 20% linear-cost arcs without major degradation in performance, confirming the theoretical result that strict convexity is not necessary for its convergence.

Eckstein [1993] also derives and tests a version of (63)–(66) for networks with gains, finding performance similar to that obtained for pure networks.

Alternating direction methods seem to exhibit a similar 'tailing' behaviour to that of dual methods, in that convergence near the solution can sometimes be extremely slow. However, the phenomenon does not appear to operate identically in the two cases. For instance, the alternating step method has little difficulty with TSENG2, which produces a prolonged tailing effect under dual relaxation, but converges slowly near the optimum of TSENG1, which dual relaxation solves easily.

Recent results from Eckstein & Fukushima [1994] show that the amount of tailing in alternating direction methods depends strongly on the way the problem has been split. On transportation problems similar to TSENG1–TSENG8, an alternating direction method slightly different from that Eckstein [1993] converged without discernable tailing effects to considerably higher accuracy (10^{-6}).

The combination of PMD (proximal minimization with D-functions) with row-action algorithms for the parallel solution of min-cost network flow problems has been tested by Nielsen and Zenios [1993b]. They use both quadratic and entropic nonlinear perturbations, and report numerical results for a set of test problems with up to 16 million arcs on the Connection Machine CM-2. In general they find that for the smaller test problems — up to 20,000 arcs — the GENOS [Mulvey & Zenios, 1987] implementation of network simplex on a CRAY Y-MP is faster than the parallel code on a 16K CM-2. As problems get larger the difference in performance between the two codes is reduced. GENOS could not solve the extremely large test problems, with 0.5 to 16 million arcs. The parallel code solved these problems in times that range from 20 minutes to 1.5 hours. The same reference provides details on implementation of the PMD algorithm on the massively parallel machine, with particular discussion of termination criteria and the use of internal tactics that improve the performance of the algorithm. It also compares the quadratic with the entropic proximal algorithms. Experiences with the massively parallel implementation of PMD algorithms for problems with embedded network structures, i.e. two-stage and multi-stage stochastic network programs, are reported in Nielsen & Zenios [1994a, b]

4. Conclusions

Research activities in parallel optimization can be traced back to the early days of linear programming: Dantzig–Wolfe decomposition can be viewed as a parallel optimization algorithm. However, it was not until the mid-eighties that parallel computer architectures became practical. This prompted the current research in parallel optimization. Several new algorithms have been developed for the parallel solution of network optimization problems: linear and nonlinear problems, assignment problems, transportation problems and problems with embedded network structures like the multicommodity network flow problem and the stochastic programming problem with network recourse.

On the theoretical front this research has produced a broad body of theory on asynchronous algorithms. Furthermore, the insights obtained from looking into the parallel decompositions of network problems has occasionally produced algorithms that are very efficient even without the advantage of parallelism. In the domain of computational investigation, we have seen the design of general procedures and data structures that facilitate the parallel implementation of mathematical algorithms on a broad range of parallel architectures: from coarse-grain parallel vector computers, like the CRAY series of supercomputers, to massively parallel systems with thousands of processing elements, like the Connection Machines. The results is that problems with millions of variables can now be solved very efficiently.

The rapid progress towards the parallel solution of 'building-block' algorithms for network problems has motivated research in more complex problem structures that arise in several areas of application. For example, multicommodity network flow problems that arise in military logistics applications are now solvable within minutes of computer time on a parallel supercomputer. A few years ago these applications required multi-day runs with state-of-the-art interior point algorithms [Schultz & Meyer, 1991; Pinar & Zenios, 1992]. A new area of investigation, that has been prompted by development of parallel algorithms, deals with planning under uncertainty. The field of stochastic programming again goes back to the early days of linear programming [Dantzig, 1988]. Recently we have seen a renewed interest in the field of *robust optimization* as an alternative way for dealing with uncertain and noisy data [Mulvey, Vanderbei & Zenios, 1991] However, the size of these optimization problems grows exponentially with the number of scenarios and time-periods. Decomposition algorithms involving network subproblems, and implemented on suitable parallel architectures, are now being used to solve problems with thousands of scenarios and millions of variables [Mulvey & Vladimirou, 1989b; Nielsen & Zenios, 1993a; Jessup, Yang & Zenios, 1994].

References

Ahlfeld, D.P., R.S. Dembo, J.M. Mulvey and S.A. Zenios (1987). Nonlinear programming on generalized networks. *ACM Trans. Math. Software* 13, 350–368.

Amdahl, G. (1967). The validity of single processor approach to achieving large scale computing capabilities, in: *AFIPS Proc.* Vol. 30, 483–485.

Balas, E., D. Miller, J. Pekny and P. Toth (1991). A parallel shortest path algorithm for the assignment problem. *J. Assoc. Comput. Mach.* 38, 985–1004.

Barr, R.S., F. Glover and D. Klingman (1979). Enhancements of spanning tree labeling procedures for network optimization, *INFOR* 17, 16–34.

Barr, R.S., and B.L. Hickman (1990). A new parallel network simplex algorithm and implementation for large time-critical problems, Technical report, Department of Computer Science and Engineering, Southern Methodist University, Dallas, Texas.

Barr, R.S., and B.L. Hickman (1993). Reporting computational experiments with parallel algorithms: Issues, measures and experts' opinion. *ORSA J. Comput.* 5(1), 2–18.

Bellman, R. (1958). On a routing problem. *Q. Applied Math.* 16.

Bertsekas, D.P. (1979). A distributed algorithm for the assignment problem, Paper, Laboratory for

Information and Decision Systems, MIT, Cambridge, Mass.
Bertsekas, D.P. (1982). *Constrained Optimization and Lagrange Multiplier Methods*, Academic Press, New York, N.Y.
Bertsekas, D.P. (1986a). Distributed asynchronous relaxation methods for linear network flow problems, Technical report LIDS-P-1606, Laboratory for Information and Decision Systems, MIT, Cambridge, Mass.
Bertsekas, D.P. (1986b). Distributed relaxation methods for linear network flow problems, in: *Proc. 25th IEEE Conf. on Decision and Control*, Athens, Greece, pp. 2101–2106.
Bertsekas, D.P. (1988). The auction algorithm: A distributed relaxation method for the assignment problem. *Ann. Oper. Res.* 14, 105–123.
Bertsekas, D.P. (1991). *Linear Network Optimization: Algorithms and Codes*, MIT Press, Cambridge, Mass.
Bertsekas, D.P., and D.A. Castañon (1991). Parallel synchronous and asynchronous implementations of the auction algorithm. *Parallel Comput.* 17, 707–732.
Bertsekas, D.P., and D.A. Castañon (1993a). Parallel asynchronous hungarian methods for the assignment problem, *ORSA J. Comput.* 5.
Bertsekas, D.P., and D.A. Castañon (1993b). Parallel primal–dual methods for the minimum cost network flow problem, *Comput. Optimization Appl.* 2, 319–338.
Bertsekas, D.P., and J. Eckstein (1987). Distributed asynchronous relaxation methods for linear network flow problems, in: *International Federation of Automatic Control Congress*, Munich, Germany.
Bertsekas, D.P., and J. Eckstein (1988). Dual coordinate step methods for linear network flow problems. *Math. Program.* 42, 203–243.
Bertsekas, D.P., and D. El Baz (1987). Distributed asynchronous relaxation methods for the convex network flow problem. *SIAM J. Control Optimization* 25, 74–85.
Bertsekas, D.P., F. Guerriero and R. Musmano (1994). Parallel label correcting algorithms for shortest paths, Technical report, LIDS, M.I.T., Cambridge, Mass.
Bertsekas, D.P., P. Hossein and P. Tseng (1987). Relaxation methods for network flow problems with convex arc costs. *SIAM J. Control Optimization* 25, 1219–1243.
Bertsekas, D.P., and J.N. Tsitsiklis (1989). *Parallel and Distributed Computation: Numerical Methods*, Prentice Hall, Englewood Cliffs, N.J.
Bertsekas, D.P., and J.N. Tsitsiklis (1991). Some aspects of parallel and distributed iterative algorithms –a survey. *Automatica* 27, 3–21.
Bregman, L.M. (1967). The relaxation method of finding the common point of convex sets and its application to the solution of problems in convex programming. *USSR Comput. Math. Math. Phys.* 7, 200–217.
Brézis, H., and P.-L. Lions (1978). Produits infinis de resolvantes. *Isr. J. Math.* 29, 329–345.
Castañon, D.A. (1989). Development of advanced WTA algorithms for parallel processing, Technical report ONR N00014-88-C-0718, ALPHATECH, Inc., Burlington, Mass.
Castañon, D.A., B. Smith and A. Wilson (1989). Performance of parallel assignment algorithms on different multiprocessor architectures, Technical report TP-1245, ALPHATECH, Inc., Burlington, Mass.
Censor, Y. (1981). Row-action methods for huge and sparse systems and their applications. *SIAM Rev.* 23, 444–464.
Censor, Y. (1988). Parallel application of block-iterative methods in medical imaging and radiation therapy. *Math. Program.* 42, 307–325.
Censor, Y., and A. Lent (1981). An iterative row-action method for interval convex programming. *J. Optimization Theor. Appl.* 34, 321–353.
Censor, Y., A.R.D. Pierro, T. Elfving, G. Herman and A. Iusem (1990). On iterative methods for linearly constrained entropy maximization, in: A. Wakulicz (ed.), *Numerical Analysis and Mathematical Modelling*, Vol. 24, Banach Center Publications, PWN — Polish Scientific Publisher, Warsaw, pp. 145–163.
Censor, Y., and J. Segman (1987). On block-iterative entropy maximization. *J. Inf. Optimization*

Sci. 8, 275–291.

Censor, Y., and S.A. Zenios (1992). On the use of *D*-functions in primal–dual methods and the proximal minimization algorithm, in: A. Ioffe, M. Marcus and S. Reich (eds.), *Optimization and Nonlinear Analysis, Pitman Research Notes in Mathematics,* Ser. 244, Longman, PWN — Polish Scientific Publisher, Warsaw, pp. 76–97.

Censor, Y., and S.A. Zenios (1992). The proximal minimization algorithm with *d*-functions. *J. Optimization Theor. Appl.*, 73(3), 455–468.

Chajakis, E., and S. Zenios (1991). Synchronous and asynchronous implementations of relaxation algorithms for nonlinear network optimization. *Parallel Comput.* 17, 873–894.

Chang, M.D., M. Engquist, R. Finkel and R.R. Meyer (1987). A parallel algorithm for generalized networks, Technical report no. 642, Computer Science Department, University of Wisconsin, Madison, Wis.

Chen, R., and R. Meyer (1988). Parallel optimization for traffic assignment. *Math. Program.* 42, 327–345.

Clark, R., and R. Meyer (1987). Multiprocessor algorithms for generalized network flows, Technical report # 739, Computer Science Department, The University of Wisconsin–Madison, Madison, Wis.

Clark, R., and R. Meyer (1989). Parallel arc allocation algorithms optimizing generalized networks, Technical report # 862, Computer Science Department, The University of Wisconsin–Madison, Madison, Wis.

Clark, R.H., J.L. Kennington, R.R. Meyer and M. Ramamurti (1992). Generalized networks: Parallel algorithms and an empirical analysis. *ORSA J. Comput.* 4, 132–145.

Cottle, R.W., S.G. Duvall and K. Zikan (1986). A Lagrangean relaxation algorithm for the constrained matrix problem. *Nav. Res. Logist. Q.* 33, 55–76.

Cryer, C.W. (1971). The solution of a quadratic programming using systematic overrelaxation. *SIAM J. Control* 9, 385–392.

Dantzig, G. (1988). Planning under uncertainty using parallel computing. *Ann. Oper. Res.* 14, 1–16.

Dantzig, G.B. (1963). *Linear Programming and Extensions*, Princeton University Press, Princeton, N.J.

Dembo, R.S. (1987). A primal truncated newton algorithm for large-scale nonlinear network optimization. *Mathematical Programming Study 31*, pp. 43–72.

Dembo, R.S., and J.G. Klincewicz (1985). Dealing with degeneracy in reduced gradient algorithms. *Math. Program.* 31(3), 357–363.

Dembo, R.S., and T. Steihaug (1983). Truncated–newton algorithms for large-scale unconstrained optimization. *Math. Program.* 26, 190–212.

Desrochers, G.R. (1987). *Principles of Parallel and Multi-Processing*, McGraw Hill, New York, N.Y.

DeWitt, D., and J. Gray (1992). Parallel database systems: the future of high performance database systems. *Commun. Assoc. Comput. Mach.* 35(6), 85–98.

Dijkstra, E. (1959). A note on two problems in connection with graphs. *Numerical Math.* 1.

Eckstein, J. (1989). Splitting methods for monotone operators, with applications to parallel optimization, Ph.D. Thesis, Massachusetts Institute of Techonology, Cambridge, Mass, Report LIDS-TH-1877, Laboratory for Information and Decision Systems, M.I.T.

Eckstein, J. (1993). The alternating step method for monotropic programming on the Connection Machine CM-2. *ORSA J. Comput.* 5(1), 84–96.

Eckstein, J. (1993). Nonlinear proximal point algorithms using Bregman functions, with applications to convex programming, *Math. Oper. Res.* 18(1), 202–226.

Eckstein, J. (1994). Parallel alternating direction multiplier decomposition of convex programs. *J. Optimization Theor. Appl.* 80(1), 39–62.

Eckstein, J., and D.P. Bertsekas (1992). On the Douglas–Rachford splitting method and the proximal point algorithm for maximal monotone operators. *Math. Program.* 55(3), 293–312.

Eckstein, J., and M. Fukushima (1994). Some reformulations and applications of the alternating direction method of multipliers, in: W.W. Hager, D.W. Hearn and P.M. Pardalos (eds.), *Large-Scale Optimization: State of the Art*, Kluwer Scientific, pp. 119–138.

Edmonds, J., and R.M. Karp (1972). Theoretical improvements in algorithmic efficiency for network flow problems. *J. Assoc. Comput. Mach.* 19, 248–264.

El Baz, D. (1989). A computational experience with distributed asynchronous iterative methods for convex network flow problems, in: *Proc. 28th IEEE Conference on Decision and Control*, Tampa, Florida, Dec.

Escudero, L.F. (1986). Performance evaluation of independent superbasic sets on nonlinear replicated networks. *Eur. J. Oper. Res.* 23, 343–355.

Flynn, M. (1972). Some computer organizations and their effectiveness. *IEEE Trans. Comput.* C–21(9), 948–960.

Ford, L.R., and D.R. Fulkerson (1957). A primal–dual algorithm for the capacitated hitchcock problem. *Nav. Res. Logist. Q.* 4, 47–54.

Fortin, M., and R. Glowinski (1983). On decomposition-coordination methods using an augmented Lagrangian, in: M. Fortin and R. Glowinski (eds.), *Augmented Lagrangian methods: Applications to the Solution of Boundary Value Problems*, North-Holland, Amsterdam, pp. 97–146.

Gabay, D. (1983). Applications of the method of multipliers to variational inequalities, in: M. Fortin and R. Glowinski (eds.), *Augmented Lagrangian methods: Applications to the Solution of Boundary Value Problems*, North-Holland, Amsterdam, pp. 299–331

Gabay, D., and B. Mercier (1976). A dual algorithm for the solution of nonlinear variational inequalities via finite element approximations. *Comput. Math. Appl.* 2, 17–48.

Gibby, D., D. Glover, D. Klingman and M. Mead (1983). A comparison of pivot selection rules for primal simplex based network codes. *Oper. Res. Lett.* 2.

Gill, P.E., W. Murray and M.H. Wright (1981). *Practical Optimization*, Academic Press, London.

Glover, F., J. Hultz, D. Klingman and J. Stutz (1978). Generalized networks: A fundamental computer-based planning tool. *Manage. Sci.* 24(12), 1209–1220.

Glover, F., D. Klingman and J. Stutz (1973). Extensions of the augmented predecessor index method to generalized network problems. *Transport. Sci.* 7, 377–384.

Glowinski, R., and A. Marroco (1975). Sur l'approximation, par éléments d'ordre un, et la résolution, par pénalisation–dualité d'une classe de problèmes de Dirichlet non linéares. *Rev. Fr. Autom., Inf. Rech. Oper.* 9(R-2), 41–76.

Goldberg, A.V., (1987). Efficient graph algorithms for sequential and parallel computers, Report TR-374, Laboratory for Computer Science, MIT, Cambridge, Mass.

Goldberg, A.V., and R.E. Tarjan (1990). Solving minimum cost flow problems by successive approximation. *Math. Oper. Res.* 15, 430–466.

Grigoriadis, M.D. (1986). An efficient implementation of the network simplex method. *Math. Program. Study* 26.

Helgason, R., J. Kennington and H. Lall (1980). A polynomially bounded algorithm for singly constrained quadratic programs, *Math. Program.* 18, 338–343.

Hildreth, C. (1957). A quadratic programming procedure. *Nav. Res. Logist. Q.* 4, 79–85; Erratum, p. 361

Hillis, W.D. (1985). *The Connection Machine*, The MIT Press, Cambridge, Mass.

Hoare, C.A.R. (1974). Monitors: An operating systems scheduling concept. *Commun. Assoc. Comput. Mach.*, 17, 549–557.

Jessup, E., D. Yang and S. Zenios (1994). Parallel factorization of structured matrices arising in stochastic programming, to appear.

Jonker, R., and A. Volgenant (1987). A shortest augmenting path algorithm for dense and sparse linear assignment problems. *Computing* 38, 325–340.

Kempa, D., J. Kennington and H. Zaki (1989). Performance characteristics of the Jacobi and Gauus–Seidel versions of the auction algorithm on the Alliant FX/8, Technical report OR-89-008, Department of Mechanical and Industrial Engineering, University of Illinois, Champaign, Urbana, Ill.

Kennington, J., and Z. Wang (1988). Solving dense assignment problems on a shared memory multiprocessor, Report 88-OR-16, Dept. of Operations Research and Applied Science, Southern Methodist University, Dallas, Tex.

Kennington, J.L., and R.V. Helgason (1980). *Algorithms for Network Programming*, John Wiley, New York, N.Y.. Klingman, D., A. Napier and J. Stutz (1974). NETGEN — a program for generating large-scale (un)capacitated assignment, transportation, and minimum cost flow network problems. *Manage. Sci.* 20, 814–822.

Lescrenier, M., and P.L. Toint (1988). Large scale nonlinear optimization on the FPS164 and CRAY X-MP vector processors. *Int. J. Supercomput. Appl.*, 2, 66–81.

Li, X., and S. Zenios (1992). A massively parallel ϵ-relaxation algorithm for linear transportation problems, in: P. Pardalos (ed.), *Advances in Optimization and Parallel Computing*. Elsevier Science Publishers, Amsterdam, pp. 164–176,

Li, X., and S.A. Zenios (1994). Data-level parallel solution of min-cost network flow problems using ϵ-relaxations, to appear.

Lions, P.-L., and B. Mercier (1979). Splitting methods for the sum of two nonlinear operators. *SIAM J. Numerical Anal.* 16, 964–979.

Mangasarian, O.L., and R.R. Meyer, eds. (1988). Parallel Methods in Mathematical Programming, *Math. Program.* 42(2), 203–470.

Mangasarian, O.L., and R.R. Meyer, eds. (1991). *Parallel Optimization II. SIAM J. Optimization* 1(4), 425–674.

Martinet, B. (1970). Régularisation d'inequations variationelles par approximations successives. *Rev. Fr. Inf. Rech. Oper.* 4, 154–159.

McKenna, M., and S. Zenios (1990). An optimal parallel implementation of a quadratic transportation algorithm, in: *4th SIAM Conf on Parallel Processing for Scientific Computing*, SIAM, Philadelphia, PA, pp. 357–363.

Meyer, R.R., and S.A. Zenios, eds. (1988). *Parallel Optimization on Novel Computer Architectures. Ann. Oper. Res.*, 14, A.C. Baltzer Scientific Publishing Co., Switzerland.

Miller, D., J. Pekny and G.L. Thompson (1990). Solution of large dense transportation problems using a parallel primal algorithm. *Oper. Res. Lett.* 9(5), 319–324.

Mulvey, J., R. Vanderbei and S. Zenios (1994). Robust optimization of large scale systems, to appear.

Mulvey, J., and H. Vladimirou (1989). Evaluation of a parallel hedging algorithm for stochastic network programming, in: R. Sharda, B. Golden, E. Wasil, O. Balci and W. Stewart (eds.), *Impact of Recent Computer Advances on Operations Research*.

Mulvey, J., and S. Zenios (1987). GENOS 1.0: A generalized network optimization system. User's Guide, Report 87-12-03, Decision Sciences Department, the Wharton School, University of Pennsylvania, Philadelphia, Pa.

Mulvey, J.M. (1978). Pivot strategies for primal-simplex network codes. *J. Assoc. Comput. Mach.* 25(2).

Murtagh, B., and M. Saunders (1978). Large-scale linearly constrained optimization. *Math. Program.* 14, 41–72.

Narendran, B., R. DeLeone and P. Tiwari (1993). An implementation of the ϵ-relaxation algorithm on the CM-5, in: *Proc. Symp. on Parallel Algorithms and Architectures*.

Nielsen, S., and S. Zenios (1993a). A massively parallel algorithm for nonlinear stochastic network problems. *Oper. Res.* 41(2), 319–337.

Nielsen, S., and S. Zenios (1993b). Proximal minimizations with D-functions and the massively parallel solution of linear network programs. *Comput. Optimization Appl.* 1(4), 375–398.

Nielsen, S., and S. Zenios (1994a). Proximal minimizations with D-functions and the massively parallel solution of linear stochastic network programs, to appear.

Nielsen, S., and S. Zenios (1994b). Solving multistage stochastic network programs, to appear.

Ortega, J.M., and W.C. Rheinboldt (1970). *Iterative Solution of Nonlinear Equations in Several Variables*, Academic Press, New York, N.Y.

Papadimitriou, C.H., and K. Steiglitz (1982). *Combinatorial Optimization, Algorithms and Complexity*, Prentice Hall, Englewood Cliffs, N.J.

Peters, J. (1990). The network simplex method on a multiprocessor. *Networks* 20, 845–859.

Pfefferkorn, C., and J. Tomlin (1976). Design of a linear programming system for the Illiac IV,

Technical report, Department of Operations Research, Stanford University, Stanford, Calif.
Phillips, C., and S. Zenios (1989). Experiences with large scale network optimization on the Connection Machine, in: *The Impact of Recent Computing Advances on Operations Research* Elsevier Science Publishing, 9, 169–180.
Pinar, M., and S. Zenios (1992). Parallel decomposition of multicommodity network flows using linear-quadratic penalty functions. *ORSA J. Comput.* 4(3), 235–249.
Polymenakos, L.C., and D.P. Bertsekas (1993). Parallel shortest path auction algorithms, Report LIDS-P-2151, LIDS, MIT, Cambridge, Mass.
Rettberg, R., and R. Thomas (1986). Contention is no obstacle to shared-memory multiprocessing. *Commun. Assoc. Comput. Mach.* 29.
Rockafellar, R.T. (1970). *Convex Analysis*, Princeton University Press, Princeton, N.J.
Rockafellar, R.T. (1976a). Monotone operators and the proximal point algorithm. *SIAM J. Control Optimization* 14, 877–898.
Rockafellar, R.T. (1976b). Augmented Lagrangians and applications of the proximal point algorithm in convex programming. *Math. Oper. Res.* 1, 97–116.
Rockafellar, R.T. (1984). *Network Flows and Monotropic Optimization*, John Wiley, New York, N.Y.
Rosen, J., ed. (1990). *Supercomputing in Large Scale Optimization*, Ann. Oper. Res. A.C. Baltzer Scientific Publishing Co., Switzerland.
Savari, S.A., and D.P. Bertsekas (1994). Finite termination of asynchronous iterative algorithms, LIDS Report, MIT, Cambridge, Mass.
Schultz, G., and R. Meyer (1991). A structured interior point method. *SIAM J. Optimization* 1(4), 583–602.
Sequent Computer Systems (1987). *Symmetry™ Technical Summary*.
Tarjan, R. (1972). Depth-first search and linear graph algorithms. *SIAM J. Comput.* 1, 146–160.
Teboulle, M. (1992). Entropic proximal mappings with applications to nonlinear programming, *Math. Oper. Res.* 17, 670–690.
Tseng, P. (1990). Dual ascent methods for problems with strictly convex costs and linear constraints: a unified approach. *SIAM J. Control Optimization* 28, 214–242.
Tseng, P., and D.P. Bertsekas (1993). On the convergence of the exponential multiplier method for convex programming. *Math. Program.* 60, 1–19.
Wein, J., and S. Zenios (1990). Massively parallel auction algorithms for the assignment problem, in: *3rd Symp. on the Frontiers of Massively Parallel Computations*, pp. 90–99.
Wein, J., and S. Zenios (1991). On the massively parallel solution of the assignment problem. *J. Parallel Distrib. Comput.* 13, 221–236.
Zaki, H. (1990). A comparison of two algorithms for the assignment problem, Technical report ORL 90-002, Dept. Mech. Ind. Eng., University of Illinois, Champaign–Urbana, Ill.
Zenios, S., and Y. Censor (1991). Massively parallel row-action algorithms for some nonlinear transportation problems. *SIAM J. Optimization* 1, 373–400.
Zenios, S., and M. Pinar (1992). Parallel block-partitioning of truncated newton for nonlinear network optimization. *SIAM J. Sci. Stat. Comput.*, 13(5), 1173–1193.
Zenios, S., R. Qi and E. Chajakis (1990). A comparative study of parallel dual coordinate ascent implementations for nonlinear network optimization, in: T. Coleman and Y. Yi (eds.), *Large Scale Numerical Optimization* SIAM, pp. 238–255.
Zenios, S.A., and R.A. Lasken (1988). Nonlinear network optimization on a massively parallel Connection Machine, *Ann. Oper. Res.* 14, 147–163.
Zenios, S.A., and J.M. Mulvey (1986). Nonlinear network programming on vector supercomputers: a study on the CRAY X-MP. *Oper. Res.* 34, 667–682.
Zenios, S.A., and J.M. Mulvey (1988a). A distributed algorithm for convex network optimization problems. *Parallel Comput.* 6, 43–56.
Zenios, S.A., and J.M. Mulvey (1988b). Vectorization and multitasking of nonlinear network programming algorithms. *Math. Program.* 42, 449–470.
Zenios, S.A., and S. Nielsen (1992). Massively parallel algorithms for singly constrained nonlinear programs. *ORSA J. Comput.* 4, 166–181.

M.O. Ball et al., Eds., *Handbooks in OR & MS, Vol. 7*

Chapter 6

Probabilistic Networks and Network Algorithms

Timothy Law Snyder
Department of Computer Science, Georgetown University, Washington, DC 20057, U.S.A.

J. Michael Steele
Department of Statistics, The Wharton School, University of Pennsylvania, Philadelphia, PA 19104, U.S.A.

1. Introduction

The uses of probability in the theory of networks are extensive, and new applications emerge at an increasing rate. Still, when compared with the purely deterministic aspects of network theory, the part that calls upon probability theory is in its infancy. Certainly there are areas where the uses of probability have developed into a reasonably complete theory, but in many instances the results that have been obtained have to be regarded as fragmented and incomplete. This situation presents considerable opportunity for researchers, and the purpose of this chapter is to highlight aspects of the current state of the theory with an eye toward the developments and the tools that seem most likely to be of value in further investigations.

Probability enters into the theory of networks and network algorithms in several different ways. The most direct way is through probabilistic modeling of some aspect of the network. For example, in some freight management models the cost of transportation along the arcs of the network are modeled by random variables. In models such as these probability helps us grasp a little better a world that comes with its own physical randomness.

A second important way probability enters is through more stylized stochastic models where the aim is to provide deeper insight into our technical understanding of the methods of operations research. Here there is considerably less emphasis on building detailed models that hope to capture aspects of randomness that live in a specific application context; rather, the aim is to provide mathematically tractable models of reasonable generality that can be used to explore a variety of different computational or estimation methods. Among the types of issues that have been studied in such models are the efficacies of deterministic algorithms and of deterministic heuristic methods. Many of the 'average case' analyses of algorithms would fit into this second role for probability.

The third path by which probability enters into network theory is through randomized algorithms. This is the newest of the roles for probability, but it is a role that is of increasing importance. To make certain of the distinction that makes an algorithm 'randomized,' consider a version of depth-first search where one chooses the next vertex to be explored by selecting it at random from a set of candidates. Here one does not call on any modeling of the network, which may in fact be specified in a way that is completely deterministic. The use of probability here is purely technical in the sense that it is employed to serve our computations, not to model some external physical randomness, or even to capture the notion of an 'average case.'

In the material that follows, one does well to keep these differing uses of probability in clear sight. Still, the distinctions may not always be pristine, mostly because two or more roles for probability can be present in the same problem. As an example, consider the computation or estimation of the reliability polynomial $R(p)$ of a network. Here one begins with a simple, physically motivated stochastic model. Given a specific graph intended to represent a communication network, one models the possibility of degraded communication in the network by allowing edges to 'fail' with probability p. The key problem is the determination of the probability $R(p)$ that for each pair of vertices a and b in the graph there exists a path from a to b that consists only of edges that have not failed. As the problem sits, it offers a simple but useful stochastic model, and one can go about the calculation or estimation of $R(p)$ by whatever tools are at one's disposal. The multiplicity of roles for probability enters exactly when one starts to notice that there are randomized *algorithms* for the estimation of $R(p)$. This is just one example where there are several roles for probability in the context of a single problem.

There are even dicier instances where the role of probability in the design and analysis of algorithms starts to offer some ambiguity. For example, close cousins of the randomized algorithms are the algorithms that (a) assume that the input follows some stochastic model and (b) exploit that assumption in the computational choices that it makes. A natural example of this design is Karp's algorithm for the Euclidean traveling salesman problem which we take up in Section 4. Such algorithms are fairly called *probabilistic algorithms*, but in the absence of internally generated random choices the best practice is to preserve the distinction made above and to avoid calling them randomized algorithms; though, admittedly, there is no reason to press for a rigid nomenclature.

The central aim of this chapter is to engage at least some aspect of each of the major roles for probability in the theory of network algorithms. When choices must be made, an emphasis is placed on those ideas one can expect to continue to be used and developed in the future. In Section 2 we engage the probability theory of network characteristics, where one mainly sees probability in either of the first two roles described above, as elements of either a physical or an idealized stochastic model. The section first develops the background for several inequalities that have evolved in the area of percolation theory. The FKG inequality is the best known and most widely used of these; but, as applications in percolation theory have shown, the much newer BK inequality is also an instrument that belongs in every tool kit. The second part of Section 2 then looks at the computational problems of associated with more physical models of networks. In Section 3 we engage randomized algorithms in the context of several problems of concern to the basic themes of network theory. The first paradigm discussed there is one initiated in Karp & Luby [1985], which remains essential to the current technology of randomized algorithms. In Section 4 we focus on problems of geometric network theory. This is the area of network theory that seems to have progressed most extensively from the viewpoint of probability theory, but it also offers practical algorithmic insights on issues that have been of interest and concern even before there evolved an extensive theory of algorithms. The classic problems here include the behavior of traveling salesman tours, minimum spanning and Steiner minimum trees, and matchings.

2. Probability theory of network characteristics

There are three substantial probabilistic theories with lives of their own, yet which are intimately intertwined with the probability theory of networks. The most immediate of these is *network reliability*. This subject provides extensive investigation of the problem of calculating and bounding the probability of the existence of (s, t)-paths. Because network reliability is dealt with in a separate chapter of this volume and because the book of Colbourn [1987] provides an extensive treatment, we do not give many details of the subject. Still, in many probabilistic investigations of networks, one needs to keep in mind the existence and highlights of the large body of results provided by reliability theory, and several of the results reviewed here owe their motivation to the concerns of network reliability theory. A second theory that is closely connected to the theory of random networks is the theory of random graphs, which deals extensively with questions like the existence of long paths, connectedness, the existence of cycles, and many other issues that are of importance to the theory of networks. Since Bollobás [1985] provides an extensive treatment of the theory of random graphs, we do not go deeply into that subject here. A third closely-related field is percolation theory, and in many ways this subject has a claim on being the deepest of the three related fields. It certainly has been pursued extensively by a large number of mathematicians and physicists over a number of years.

2.1. *Tools from percolation theory*

In this section, we first recall what the aims of percolation theory have been over the years of its development. We then suggest some ways in which the theory may help researchers who are concerned with questions that are more at the heart of network theory. We will develop three elementary but central tools of percolation theory: the FKG inequality, the BK inequality, and Russo's formula. These powerful tools are the workhorses of percolation theory, yet they seem not to be well known to researchers in the more general areas of stochastic network theory.

Percolation theory evolved from questions in physics that are themselves of many different flavors, including the magnetization of materials, the formation of crystals, the transport of electrons in special materials, sustenance of chemical reactions, and the flow of fluids. The latter offers perhaps the least compelling physics, but it provides the easiest metaphor and is often called upon for illustration.

We consider the classical d-dimensional rectangular lattice, Z^d, and for each vertex v of the lattice we join v by an edge to each of its $2d$ nearest neighbors (in the sense of the usual Euclidean metric). If we use the traditional language of percolation theory, these edges are called 'bonds'. Bonds are viewed as being either 'open' or 'closed', and it is here where the probability modeling appears. To each bond e is associated an independent Bernoulli random variable X_e such that $P(X_e = 1) = p$ for some fixed $0 \leq p \leq 1$. The bonds for which $X_e = 1$ are regarded as being *open*, and the fundamental questions of the theory concern the components of lattice vertices connected by open edges. Among the main features that distinguish percolation theory from the theory of random graphs are the attention that is focused on subgraphs of the lattice and the interest that is focused on graphs with infinitely many edges.

A central quantity of interest in percolation theory is the *percolation probability* $\theta(p)$, defined as the probability that the origin is contained in an infinitely large connected component. One reason that $\theta(p)$ receives considerable attention is that it exhibits interesting critical phenomena that have close analogies with physical phenomena like the freezing of fluids. In particular, one can prove, for each dimension d, that there is a *critical constant* $p_c = p_c(d)$ depending on the dimension d such that $\theta(p) > 0$ if $p > p_c$ but $\theta(p) = 0$ if $p < p_c$. The work of Kesten [1980] culminated the efforts of a great many investigations and established the long conjectured result that $p_c(2) = 1/2$. This deep result required the development of techniques that would seem to offer useful insights for researchers in the theory of networks, and a well motivated exposition of Kesten's theorem can be found in Chapter 9 of Grimmett [1989].

The FKG inequality

The first tool we consider is named the FKG inequality, in respect of the work of Fortuin, Kasteleyn & Ginibre [1971]. Even though we will not call on the full generality of their result, it is worth noting that the FKG inequality has a beautiful generalization due to Ahlswede & Daykin [1978], and the full-fledged FKG inequality has already found elegant applications to problems of interest in network theory. In particular, one should note the articles by Shepp [1982] and Graham [1983].

The version of the inequality that we develop is actually a precursor of the FKG inequality due to Harris [1960], but Harris's inequality has the benefit of having very low overhead while still being able to convey the qualitative essence of its more sophisticated relatives. To provide a framework for Harris's inequality, we suppose that G is any graph and $\{X_e\}$ are identically distributed Bernoulli random variables associated with the edges of G. We think of the variables X_e as labels marking the edges of G that would be regarded in percolation theory as open edges. The random variables of interest in Harris's inequality are those that can be obtained as monotone non-decreasing functions of the variables $\{X_e\}$. In detail, if in a realization of the $\{X_e\}$ we change some of the $\{X_e\}$ that have value zero to have a value of one, then we require that the value of the function does not decrease. The classic example of such a variable is the indicator of an (s,t)-path of edges marked with ones. Harris's inequality confirms the intuitive fact that any pair X and Y of such monotone variables are positively correlated. Specifically, if X and Y are any non-decreasing random variables defined as functions of the edge variables X_e of G, then one has

$$E(XY) \geq E(X)E(Y).$$

This inequality is most often applied in the case of indicator functions. Since we will refer to this case later, we note that we can write Harris's inequality as

$$P(A \cap B) \geq P(A)P(B)$$

for all events A and B that are non-decreasing functions of the edge variables.

One can prove Harris's inequality rather easily by induction. If we write $X = f(\eta_1, \eta_2, \ldots, \eta_n)$ and $Y = g(\eta_1, \eta_2, \ldots, \eta_n)$, where f and g are monotonic and the $\{\eta_i\}$ are independent Bernoulli random variables, then by conditioning on η_n, we see that it suffices to prove Harris's inequality just in the case of $n = 1$. In this case we see that, for $q = 1 - p$,

$$\begin{aligned} &EXY - EX \cdot EY = \\ &\quad f(1)g(1)p + f(0)g(0)q - (f(1)p + f(0)q)(g(1)p + g(0)q) \end{aligned}$$

and since this factors as

$$pq\{f(1) - f(0)\}\{g(1) - g(0)\} \geq 0,$$

we obtain Harris's inequality.

One of the nice consequences of Harris's inequality is the fact that if m non-decreasing events $A_1, A_2, \ldots, A_m$ with equal probability have a *union* with large probability, then, *all* the events A_i must have fairly large probability. This so-called 'square root trick' noted in Cox & Durrett [1988] formally says that for each $1 \leq i \leq m$, we have

$$P(A_i) \geq 1 - \{1 - P(\textstyle\bigcup_{j=1}^m A_j)\}^{1/m}.$$

The proof of this inequality requires just one line where Harris's inequality provides the central step:

$$1 - P(\textstyle\bigcup_{j=1}^m A_j) = P(\textstyle\bigcap_{j=1}^m A_j^c) \geq \displaystyle\prod_{j=1}^m P(A_j^c) = (1 - P(A_1))^m.$$

To appreciate the value of this inequality, one should note that without the assumption that the $\{A_i\}$ are monotone, one could take the $\{A_i\}$ to be a partition of the entire sample space, making the left side equal to $1/m$, while the right side equals one. We see therefore that the FKG inequality helps us extract an important feature of monotone events. As a point of comparison with a more combinatorial result, one should note that the square root trick and the local LYM inequality of Bollobás and Thomason [cf. Bollobás, 1986] both address the way in which probabilities of non-decreasing sets (and their ideals) can evolve. For further results that call on the FKG and Harris inequalities one should consult Graham [1983] and Spencer [1993].

The BK inequality

The insights provided by the FKG and its sibling inequalities are valuable, but they are limited. The inequalities often just provide rigorous confirmation of intuitive results that one can justify by several means. A much deeper problem arises when one needs an inequality that goes in a direction opposite that of the FKG inequality. For this problem, the progress is much more recent and less well known.

As one can show by considering any dependent, non-decreasing events A and B, there is no hope of simply reversing the FKG inequality. In fact, the same examples can show that additional assumptions on A and B that fall short of independence are of no help, so some sort of additional structure, or some modification is needed for $A \cap B$. Van den Berg & Kesten [1985] discovered that the key to a useful reversal of the FKG inequality rests on a strengthening of the notion of $A \cap B$. The essence of their idea is that the event of A and B both occurring needs to be replaced with that of 'A and B both occurring, but for different reasons' or, as we will shortly define, A and B *occurring disjointly*.

The technical definition of disjoint occurrence takes some work, but it is guided by a canonical example. If A corresponds to the existence of an (s, t)-path and B corresponds to the existence of an (s', t')-path, then $A \cap B$ needs to be replaced by the event corresponding to the existence of (s, t)- and (s', t')-paths that *have no edge in common*. To make this precise in a generally applicable way, we have to be explicit about the underlying probability space. To keep ourselves from straying too far from network applications, we let Ω denote the set of $(0, 1)$-vectors $(x_1, x_2, \ldots, x_m)$, where m is the number of elements in a set S of edges that are sufficient to determine the occurrence of A. In many problems m cannot be bounded by anything sharper than the number of edges of G, but the bound can be useful even in such cases.

We define a measure on Ω via the Bernoulli edge variables X_e taken in some fixed order, so Ω taken with our probability measure P give us a product measure space $\{\Omega, P\}$. We now define the set $A \circ B$, the *disjoint occurrence of non-decreasing events A and B*, as follows:

$$A \circ B = \{\omega : \text{there exists } \omega_a \in A \text{ and } \omega_b \in B \\ \text{such that } \omega_a \cdot \omega_b = 0, \text{ and } \omega \geq \omega_a \text{ and } \omega \geq \omega_b \}.$$

Here, we use $\omega_a \cdot \omega_b$ to denote the usual inner product between vectors, so the combinatorial meaning of the last condition is that ω_a and ω_b share no 1's in their representation. In other words, for non-decreasing events A and B, ω_a and ω_b are able to bear respective witness that A and B occur, but they can base their testimony on disjoint sets of edges.

The BK Inequality. *If A and B are non-decreasing events in* $\{\Omega, P\}$, *then*

$$P(A \circ B) \leq P(A)P(B).$$

The systematic use of the BK inequality is just now becoming widespread even in percolation theory proper. In Grimmett [1989] one finds many proofs of older results of percolation theory that are rendered much simpler via the BK inequality.

Russo's formula

The last of the percolation theory tools that we will review is a formula due to Russo [1981] that tells how the probability of a non-decreasing event changes as one changes the probability of the events $\{X_e = 1\}$. To state the formula, we suppose as before that we have a graph G with edges that are 'open' with probability p in such a way that the indicator variables X_e are independent and identically distributed. In this context we will require that G is finite, and, to emphasize the use of p as a parameter, we will denote the governing probability measure by P_p.

Now, if A is any non-decreasing event, we introduce a new random variable N_A that we call 'the number of edges that are pivotal for A'. Formally, we define $N_A(\omega)$ as follows: (a) If $\omega \notin A$, then $N_A(\omega)$ is zero, and (b) if $\omega \in A$, then $N_A(\omega)$ equals the number of edges e such that, in the representation of ω as a $(0, 1)$-vector of edge indicators $\omega = (x_1, x_2, \ldots, x_m)$, we have $x_e = 1$, but, if we change x_e to 0 to get a new vector ω', then $\omega' \notin A$. In the latter case, we say that e is *pivotal* for A.

Russo's formula. *If A is any non-decreasing event defined on the Bernoulli process associated with a finite graph G, and if* N_A *denotes the number of edges that are pivotal for A, then*

$$\frac{\mathrm{d}}{\mathrm{d}p} P_p(A) = E_p(N_A).$$

This beautiful and intuitive formula can be used in many ways, but it is often applied to show that $P_p(A)$ cannot increase too rapidly as p increases. To see how one such bound can be obtained in a crude but general context, we first note that the differential equation of Russo's formula can be rewritten in integrated form for $0 \leq p_1 < p_2 \leq 1$ as

$$P_{p_2}(A) = P_{p_1}(A) \exp\Big(\int_{p_1}^{p_2} \frac{1}{p} E_p(N_A \mid A)\, \mathrm{d}p \Big).$$

If there is a set $S = \{e_1, e_2, \ldots, e_m\}$ of m edges such that the occurrence of A can always be determined by knowledge of S, then the integral representation and the trivial bound

$$P_p(e \text{ is pivotal for } A \mid A) \leq 1$$

provide a general inequality that bounds the rate of growth of $P_p(A)$ as a function of p:

$$P_{p_2}(A) \leq \left(\frac{p_2}{p_1}\right)^m P_{p_1}(A).$$

2.2. *Distributional problems of random networks*

In percolation theory the random variables associated with the edges are invariably Bernoulli, but in the network models that aim to model physical systems the network ingredients often are modeled by random variables with more general distributions, and the central questions in such models concern the distributions of larger network characteristics. Sadly, many of these distributional questions are analytically infeasible. Moreover, in many cases of practical interest these same questions are computationally intractable as well. We will illustrate some of the technology that has been developed for such problems by considering the problem of determining the distribution of the minimum-weight path from source to sink in a network with random edge weights.

Calculation of the distribution of the shortest paths

Formally, we let $G = (V, E)$ be an acyclic network with source vertex s and sink t, where edge weights are represented by independent random variables W_e for all $e \in E$. The stochastic quantity of interest is the distribution of the random variable $L(G)$, denoting the length of a shortest (s, t)-path in G. Valiant [1979] showed that the problem of determining the distribution of $L(G)$ is in general NP-hard, so at a minimum one must look to approximation methods. One natural approach to the distribution problem is to try to exploit the independence of the edge weights through the use of cut sets. This idea forms the basis of the simulation method of Sigal, Pritsker & Solberg [1979, 1980]. To describe their method for building a simulation estimate for $P(L(G) \geq t)$, we first let $C = e_1, e_2, \ldots, e_k$ be an *exact cut* in G, that is, we take C to be a set of edges such that every (s, t)-path in G shares exactly one edge with C. Such a cut always exists, and it offers us a natural method for exploiting the independence of the X_e. The key observation is that the edges of C induce a natural partition of the (s, t)-paths of G.

For each $1 \leq i \leq k$ and each $e_i \in C$ we let p_i be the set of all (s, t)-paths that contain e_i. Now, for any $t \in \mathbb{R}$, we consider the random variable defined by the conditional probability

$$R = P(L(G) \geq t \mid W_e,\ e \in E - C).$$

Since R satisfies $ER = P(L(G) \geq t)$, if we let r be the sample value of R based on a realization $\{w_e\}$ of $\{W_e : e \in E - C\}$, then by independence we have

$$\begin{aligned} r &= P\big(L(G) \geq t \mid w_e,\ e \in E - C\big) \\ &= P\left(\sum_{\substack{e \in p \\ e \neq e_i}} w_e + W_{e_i} \geq t \text{ for all } p \in p_i \text{ and for all } e_i \in C\right) \\ &= \prod_{i=1}^{k} P\left(W_{e_i} \geq t - \min_{p \in p_i}\left(\sum_{\substack{e \in p \\ e \neq e_i}} w_e\right)\right). \end{aligned}$$

Since the right hand side can be computed from the known distribution of the W_{e_i}, an estimate of $P(L(G) \geq t)$ is given by $n^{-1}\sum_{1 \leq i \leq n} r_i$, where the r_i are given by independent simulations of r. For some simple networks this process gives an effective method for the simulation estimation of the distribution of $L(G)$, and, at a minimum, the use of r is substantially more efficient than the completely naive estimation of $P(L(G) \leq t)$ via independent simulations of $L(G)$.

This method is reasonably crude, but with proper implementation it can provide answers in some situations of importance. Certainly, the computation of $\min_{p \in p_i}(\sum_{e \in p, e \neq e_i} w_e)$ for $1 \leq i \leq k$ should not be conducted naively since there can be an exponential number of (s, t)-paths in G. There are even moderately sized networks for which an exhaustive evaluation of the required sums is computationally prohibitive. One does much better to note that once the edges in $E - C$ have been sampled, the lengths of the shortest paths from s to e_i and from e_i to t can be computed for each i by using an appropriate deterministic single-source shortest path algorithm, such as that of Dijkstra [1959], or more recent refinements. Dijkstra's algorithm is easy to implement, has low overhead, and takes $O(|V|^2)$ steps in the worst case to compute all the required path lengths. Having the lengths of the shortest paths from s to the e_i and from the e_i to t allows the computations of the minima in the representation for r to be obtained in at most $O(|C|^2)$ additional steps.

The problem of estimating the distribution of $L(G)$ offers a typical instance of the trade-off one often meets in simulation estimations. First, there is a desire to have an efficient estimate of ER for which we would like a cut that provides for a low variance of R. This is a kind of efficiency that helps us minimize the number of independent realizations one must take in the simulation. Second, we would like to have efficiency in the computation of the estimate in the sense of computing the shortest (s, e_i)- and (e_i, t)-paths. The trade-off that faces us is that as $|C|$ increases, the variance of R decreases, but as the cut size increases so does the of cost of computing shortest paths and minima. There are often many different exact cuts on which one can base the simulation estimation of $P(L(G) \geq t)$ and the proper choice of the cut is an important design consideration.

Other distribution problems of random networks

Other studies have undertaken the difficult task of determining distributions of flows in networks with random capacities. Among these is the paper of Grimmett & Welsh [1982], which considered maximum flows in networks with independent and identically distributed capacities. Grimmett and Welsh found limit theorems for the cases where the networks are either complete graphs or branching trees. In subsequent work, Frieze & Grimmett [1985] looked at the shortest path problem under general independent models, and Kulkarni [1986] studied shortest paths in networks with exponentially distributed edge lengths. One point that emerges from these works is that the probability theory of network characteristics offers many individual problems of considerable challenge. So far there seems to have not been any attempt at providing the framework for the general theory of such characteristics. With the insights of several special problems in hand, perhaps it is time that work on a more general investigation was begun.

3. Probabilistic network algorithms

In this section we first provide an introduction to a general approach of Karp & Luby [1985] for the design of randomized algorithms. We then illustrate their method by showing how one can put the problem estimation of multiterminal network reliability into their framework. We then review briefly two recent randomized algorithms for maximum network flow. Finally, we review some of the work on randomized algorithms for perfect matching and maximal matching in graphs.

3.1. Karp–Luby structures for randomized algorithms

Karp & Luby [1985] provided a framework for randomized algorithms that is useful in a broad range of applications and which specifically offers an effective approach to some problems of network reliability. Their approach begins abstractly with a set S and a weight function $a : S \to \mathbb{R}^+$ which we then use to provide a weight for any $A \subset S$ by taking $a(A) = \sum_{x \in A} a(x)$. Clearly there are many important problems that can be framed in terms of the calculation of $a(A)$ for appropriate choices of a, S, and A; but, as one must suspect, we will have to impose some additional structures before this framework can show its value. We call (S, R, a) a *Karp–Luby Monte Carlo structure* if $R \subset S$ and we have the following three properties: (1) the 'total weight' $a(S)$ is known, (2) there is a 'sampling algorithm' that selects an item x at random from S according to the probability $a(x)/a(S)$ with independent selections at each invocation of the sampling algorithm, and (3) there is a 'recognition algorithm' that can test if a given element x of S is also an element of R.

For any such structure (S, R, a) one can estimate the weight $a(R)$ in the most straightforward way imaginable. One just selects n independent random elements $X_1, X_2, \ldots X_n$ of S by the sampling algorithm. Letting Y_i be 1 or 0 accordingly as $X_i \in R$ or not, one then takes as an estimator of $a(R)$ the value $\bar{Y} = a(S)(Y_1 + Y_2 + \cdots Y_n)/n$. As a consequence of the traditional Bernstein tail estimates of the binomial distribution, for any $\epsilon > 0$ and $\delta > 0$ we have

$$P(\frac{|\bar{Y} - a(R)|}{a(R)} > \epsilon) \leq \delta$$

provided that

$$n > \frac{9}{5}\epsilon^{-2} \log\left(\frac{2}{\delta}\right) \frac{a(S)}{a(R)}.$$

The punch line here is that once we are able to frame a problem in terms of a Karp–Luby structure, we can determine a δ-ϵ approximation in the sense of the preceding probability bound. Moreover, we can bound the expected computational cost of the algorithm by a polynomial in the parameters ϵ^{-1}, $\log(1/\delta)$, and the sensitivity ratio $a(S)/a(R)$.

3.2. Karp–Luby structures for network reliability

The multiterminal network reliability problem is a stylized model for communication reliability that has been studied from many perspectives, and it offers a good example of how one fits a natural problem into the framework of Karp–Luby structures. Given a connected graph $G = (V, E)$ and a special set of 'terminal vertices' $T = \{t_1, t_2, \ldots, t_k\} \subset V$, the motivating issue of multiterminal network reliability is to model the possibility of communication between the elements of T under random degradation of the network links. The probability modeling calls for a function $p : E \to [0, 1]$ that is viewed as giving for each $e \in E$ the probability $p(e)$ that the edge e is 'bad.' Under the assumption that the edges are made good or bad according to independent choices governed by p, the key problem is to determine the probability that for all pairs of elements of the set of terminals there is a path between them that consists only of good edges.

More formally, we consider the set of all mappings $s : E \to \{0, 1\}$ as the elements of our probability space, and we take the interpretation of this function as an assignment of a label of 0 on the good edges and 1 on the bad edges. The probability of a specific state s thus is given by $P(s) = \prod_{e \in E} p(e)^{s(e)}(1 - p(e))^{1-s(e)}$. The computational challenge is to calculate the probability that there is some pair of terminal vertices for which there does not exist a path between them in the graph consisting of the vertex set V and the set of all edges of G which are labeled 'good'. We call a state for which this event occurs a *failing state*, and we let F denote the set of all states s which have failure.

To provide a Karp–Luby structure so that we can use the strategy discussed in the preceding section, we first need the notion of a *canonical cut*. Let $s \in F$ be any failing state, and let $G(s) = (V, E(s))$ where $E(s)$ is the set of good edges for the state s. For any $1 \leq i \leq k$ we then let $C_i(s)$ denote the connected component of $G(s)$ that contains the terminal t_i. Since $s \in F$ there is some i for which $C_i(s)$ is not all of G and, further, because of the assumption that the full graph

$G = (V, E)$ is connected, there is at least one such $C_i(s)$ for which the graph induced by the removal of all the vertices of $C_i(s)$ from $G = (V, E)$ is connected. We let $i^*(s)$ denote the least such index i, and finally we let $g(s)$ denote the set of edges that have exactly one endpoint in $C_{i^*}(s)$. The set $g(s)$ is a T-cut in that it separates two terminals of T in the graph $G(s) = (V, E(s))$, and we call $g(s)$ the *canonical cut* for the state s.

We now have the machinery to specify the Karp–Luby structure for the multiterminal reliability problem. Let S be the set of all pairs (c, s) where $s \in F$ is a failing state and c is a T-cut for which each edge of c fails in state s. The weight function associated with a pair $(c, s) \in S$ is taken to be the probability of the state s, so $a((c, s)) = P(s)$. Although this weight function ignores the first component of (c, s), the presence of the first component turns out to be an essential in providing an effective sampling process. This choice of a and S permits us to write down a simple formula for $a(S)$. Since $a(S)$ is equal to the sum of all the probabilities of the states s where s fails for the cut c, we have

$$a(S) = \sum_{(c,s)\in S} a(c, s) = \sum_{c} \prod_{e\in c} p(e),$$

where the last sum is over all T-separating cut sets of $G = (V, E)$. The target set R is given by the set of all pairs $(g(s), s)$ where $s \in F$, and (S, R, a) will serve as our candidate for a Karp–Luby structure for the multiterminal network problem. To see the interest in this triple we first note that

$$a(R) = \sum_{s\in F} a(g(s), s) = \sum_{s\in F} P(s),$$

so the weight $a(R)$ corresponds precisely to the probability of interest.

For the effective use of (S, R, a), it would be handiest if we had at our disposal a list L of all the T-separating cut sets of $G = (V, E)$. When the list is not too large, the formula given above provides a way to calculate $a(S)$. Similarly, we also have at hand an easy way to test if $s \in F$ by examining the failure of each of the cuts. To complete our check that the (S, R, a) leads to a Karp–Luby structure in this nice case, it only remains to check that sampling from S is not difficult.

To choose an element $(c, s) \in S$ according to the required distribution, we first choose at random a $c \in L$ according to the probability distribution $\prod_{e\in c} p(e)/a(S)$. We then select a state function s such that $s(e) = 1$ for all $e \in c$ and by letting $s(e) = 1$ or $s(e) = 0$ with probability $p(e)$ or $1 - p(e)$, respectively.

We have completed the verification that (S, R, a) satisfies the constraints required of a Karp–Luby structure, but for it to serve as the basis for an effective randomized algorithm we also need to have a bound on the sensitivity ratio $a(S)/a(R)$. In many multiterminal reliability problems a sufficiently powerful bound is provided by the following inequality of Karp & Luby [1985]:

$$\frac{a(S)}{a(R)} \leq \prod_{e\in E} (1 + p(e)).$$

Thus far we have given a reasonably detailed view of the Karp–Luby structure and how it can be applied to a problem of computational interest in network theory. The development recalled here so far has the shortfalling that it seems to require an explicit list of the T-cuts of the network, and that list must be reasonably short. Karp & Luby [1985] go further and show that there are cases where this requirement can be avoided; in particular they show that if G is a planar graph, then the program still succeeds even without explicitly listing the cut sets.

3.3. Randomized max-flow algorithms

The theory of network flows is to many people what the theory of networks is all about, and there are two recent contribution of randomized algorithms to this important topic that have to be mentioned here, even though this survey cannot dig deeply enough into them to do real justice. The first of these is the algorithm of Cheriyan & Hagerup [1989] for maximum flow in the context where all the arc capacities are deterministic. The Cheriyan–Hagerup algorithm produces a maximum flow for any (non-random) single-source, single-sink input network. The algorithm takes $O(|V||E|\log|V|)$ time in the worst case, although this happens with probability no more than $|V|^{-\alpha|V|^2}$, where α is any constant. Most important is that the Cheriyan–Hagerup algorithm takes $O(|V||E|+|V|^2(\log|V|)^3)$ expected time, which, being $O(|V||E|)$ for all but relatively sparse networks, compares favorably with all known strongly polynomial algorithms. The algorithm is also strongly polynomial in the sense that the running time bound does not depend on the edge-capacity data.

The Cheriyan–Hagerup algorithm builds on some of the best deterministic algorithms and takes a step forward by introducing randomization at key stages. The algorithm calls on scaling techniques in the spirit of Gabow [1985], Goldberg & Tarjan [1988], and Ahuja & Orlin [1987] and also employs *pre-push labeling*, another device of the Goldberg and Tarjan max-flow algorithm [cf. Ahuja, Magnanti & Orlin, 1991]. The randomness of the Cheriyan and Hagerup algorithm arises in how the network is represented at a given moment during the course of the algorithm. The model used for network representation is the adjacency list model, in which the neighbors of each $v \in V$ are maintained by a list associated with v. One of the key ideas of the Cheriyan–Hagerup algorithm is to randomly permute each adjacency list at the outset of the algorithm, then randomly permute the adjacency list of vertex v whenever the label of v is updated. The net effect of the permutation is to lower the expected number of relabeling events that the algorithm must carry out during each phase, lowering the expected running time. One further interesting aspect of the Cheriyan–Hagerup algorithm is that Alon [1990] has provided a device that derandomizes the algorithm in a way that is effective for a large class of graphs.

A more recent contribution of a randomized algorithm for max-flow has been provided in Karp, Motwani & Nisan [1993]. Given a realization of an undirected network with independent identically distributed random capacities, the algorithm finds a network flow that is equal in value to the optimum flow value with high probability. The algorithm runs in linear time, which is significantly faster than the best known algorithms that are guaranteed to find an optimal flow.

The algorithm of Karp, Motwani, and Nisan is not simple, but at least some flavor for the design can be appreciated independently of the details. In the first stage of the algorithm, the max-flow problem on G is transformed to an instance of a probabilistic version of the transportation problem. The instance of the transportation problem is constructed so that its solution flow is forced to yield (1) a maximum flow that can be immediately transformed to a max-flow in G and (2) a flow that saturates the $(S, V - S)$ cut in G, where S is the set of sources. The second stage of the max-flow algorithm is a routine that attempts to solve the transportation problem. Here Karp, Motwani & Nisan [1993] introduce their so-called *mimicking method* which they outline in four steps: (1) before considering the realization of the random graph, consider instead the graph formed by replacing each random variable X_i with EX_i; (2) solve the resulting deterministic problem; (3) consider now the realization of the random graph, and attempt to solve the problem by 'mimicking' the solution from (2); and (4) fine-tune the result to get the optimum solution. Even though these steps have engaging and evocative descriptions, there is devil in the details which in the end leads to delicate analyses for which we must refer to the original.

3.4. Matching algorithms of several flavors

Information about matchings has a useful role in many aspects of the theory of networks. Moreover, some of most effective randomized algorithms are those for matching, so this survey owes the reader at least a brief look at randomized matchings for algorithms and related ideas.

The key observation of Lovász [1979] was that one can use randomization to test effectively for the positivity of a determinant, and this test can be used in turn to test for the existence of a perfect matching in a graph. To sketch the components of the method we first recall that with the graph $G = (V, E)$ we can associate an adjacency matrix D by taking $d_{ij} = 1$ if $(i, j) \in E$ and zero otherwise. From the adjacency matrix we can construct the Tutte matrix T for G by replacing the above-diagonal elements d_{ij} by the indeterminants x_{ij} and the below-diagonal elements d_{ij} by the indeterminants $-x_{ij}$. The construction of T is completed by putting zeros along the diagonal. The theorem of Tutte, for which he introduced this matrix, is that G has a perfect matching if and only if $\det T \neq 0$.

The core of the idea for testing if G has a perfect matching is then quite simple. One chooses random numbers for the values of the x_{ij} and then computes the determinant numerically, a process that is not more computationally difficult than matrix inversion. The savings come here from the fact that the determinant in the indeterminant variables x_{ij} can have exponentially many terms, but to test that the polynomial is not identically zero we only have to see that it is non-zero at a point. Naturally, to be true to the values of honest computational complexity theory one cannot rely on computation with real numbers, but by working over a finite field one comes quickly to the conclusion that there is merit to the idea.

Lovász [1979] generalized Tutte's theorem and went on to provide an algorithm that takes advantage of the idea just outlined in order to find maximal matchings in a general graph. Rabin and Vazirani [1989] pressed this idea further and provided an algorithm that is faster than that of Lovász. A computational virtue of both the Lovász and Rabin-Vazirani algorithms is that they are readily implemented as parallel algorithms.

Another development from the theory of matching that has had wide-ranging impact on the theory of combinatorial algorithms is the introduction of the method of rapidly mixing Markov chains. The development evolving from Jerrum & Sinclair [1986, 1989] calls on the idea that if one runs a Markov chain for a long time, then its location in the state space is well approximated by the stationary distribution of the Markov chain. This idea can be used to estimate the number of elements in a complicated set, say, the set of all matchings on a graph, if one can find a chain on a set of states that includes the set of matchings and for which a Markov chain can be constructed that converges rapidly to stationarity. This idea has undergone an extensive development over the last few years. For a survey of this field we defer to the recent volume of Sinclair [1993].

The final observation about matching in random graphs that deserves space in the awareness of researchers in network theory is that algorithms based on augmenting paths are likely to perform much better than their worst-case measures of performance would indicate. These algorithms, which exhibit the fastest worst-case running times, are also fast in expectation, sometimes out performing even the best heuristic algorithms. Many of the algorithms, including the algorithms of Even & Kariv [1975] and Micali & Vazirani [1980], run in linear expected time if the input graph is chosen uniformly from the set of all graphs. The reason behind this observation seems to be the expander properties of random graphs and the fact that in expander graphs one has a short path connecting any two typical points [cf. Motwani, 1989].

The proofs of these results come from an analysis of the lengths of augmenting paths. It is shown that, with high probability, every non-perfect matching in a random graph has augmenting paths that are relatively short. Since the bulk of augmenting path algorithms is spent carrying out augmentations, bounds on the lengths of augmenting paths translate to fast running times.

4. Geometric networks

One of the first studied and most developed parts of the theory of networks concerns networks that are embedded in Euclidean space. A *geometric network* is defined by the finite point set $S \subset \mathbb{R}^d$, with $d \geq 2$, and an associated graph, which is usually assumed to be the complete graph on S. The costs associated with the edges of the graph are typically the natural Euclidean lengths, though sometimes it is useful to consider functions of such lengths, for example, to take the cost of an edge to equal the square of its length. The central questions of the theory of geometric networks focus on the lengths of subgraphs; so for example, in the traveling salesman problem, we are concerned with the length of the shortest tour through the points of S. Also of central interest in this development is the theory of minimum spanning trees, Steiner trees, and several types of matchings. The key result in initiating the probabilistic aspects of this developments is the classic Beardwood, Halton, and Hammersley theorem.

Theorem [Beardwood, Halton & Hammersley, 1959]. *If X_i, $1 \leq i < \infty$ are independently and identically distributed random variables with bounded support in $\mathbb{R}^d$, then the length L_n under the usual Euclidean metric of the shortest path through the points $\{X_1, X_2, \ldots, X_n\}$ satisfies*

$$\frac{L_n}{n^{(d-1)/d}} \to \beta_{\mathrm{TSP},d} \int_{\mathbb{R}^d} f(x)^{(d-1)/d} \, \mathrm{d}x \text{ almost surely.}$$

Here, $f(x)$ is the density of the absolutely continuous part of the distribution of the X_i.

In addition to leading to algorithmic applications, the Beardwood, Halton, and Hammersley (BHH) theorem has led to effective generalizations, as well as new analytical tools. In this section, we review these tools, including the theory of subadditive Euclidean functionals, bounds on tail probabilities, related results in the theory of worst-case growth rates, and bounds on limit constants such as $\beta_{\mathrm{TSP},d}$.

One elementary point that may help avoid confusion in the limit theory offered by the Beardwood, Halton, Hammersley theorem is the observation that it is of a much deeper character than $\Theta(n^{(d-1)/d})$ results for L_n, which only require that there exist positive constants a and b such that $an^{(d-1)/d} \leq L_n \leq bn^{(d-1)/d}$. The latter results are sometimes useful, but they are almost trivial in comparison, unless one presses for very good values for a and b. The stronger asymptotic result that $L_n/n^{(d-1)/d}$ converges to a constant requires entirely different techniques and typically leads to much different applications.

A second comment concerns uses to which one can put results such as the Beardwood, Halton, Hammersley theorem and its relatives. The use of the BHH theorem in the polynomial-time probabilistic TSP algorithm of Karp [1976, 1977] is one of the primary reasons results like the BHH theorem are studied today. Part of the charm of TSP is that it is NP-hard, and it has been studied from many heuristic and approximation perspectives. Karp's algorithm has a special place in the theory of algorithms because given any $\epsilon > 0$, its expected running time is almost linear and with probability one it produces a tour of length no more than $(1+\epsilon)$ times the optimal tour length. Karp's algorithm played an important role in launching the field of probabilistic algorithms, and it certainly stimulated interest in the development of theorems that extend that of Beardwood, Halton, and Hammersley. Since then, theorems like the BHH theorem have been proved for other quantities, like the length of the minimum spanning and Steiner minimum trees, greedy and semi-matchings, and others. For further information on some

The restrictions that this theorem imposes on a Euclidean functional are as few as one can reasonably expect to yield a generally useful limit theorem, and because of this generality the restriction to uniformly distributed random variables is palatable. Moreover, since many of the probabilistic models studied in operations research and computer science also focus on the uniformly distributed case, the theorem has immediate applications. Still, one cannot be long content with a theory confined to uniformly distributed random variables. Fortunately, with the addition of just a couple of additional constraints, the limit theory of subadditive Euclidean functionals can be extended to quite generally distributed variables.

4.2. *Tail probabilities for the TSP and other functionals*

The theory just outlined has a number of extensions and refinements. The first of these that we consider is the work of Rhee & Talagrand [1989] on the behavior of the tail probabilities of the TSP and related functionals under the model of independent uniformly distributed random variables in the unit d-cube. In Steele [1981b], it was observed that Var L_n for $d = 2$ is bounded independently of n. This somewhat surprising result motivated the study of more detailed understanding of the tail probabilities $P(L_n \geq t)$, particularly the issue of determining if these probabilities decayed at the Gaussian rate $\exp(-cx^2/2)$. After introducing new methods from martingale theory and interpolation theory which led to interesting intermediate results, Rhee & Talagrand [1989] provided a remarkable proof that in $d = 2$, the TSP and many related functionals indeed have Gaussian tail bounds. The formal result can be stated as follows.

Theorem [Rhee & Talagrand, 1989]. *Let f be a Borel measurable function that assigns to each finite subset $F \subset [0, 1]^2$ a real value $f(F)$ such that*

$$f(F) \leq f(F \cup x) \leq f(F) + \min(d(x, y) : y \in F).$$

If X_i are independent and uniformly distributed in $[0, 1]^2$, then the random variable defined by $U_n = f(\{X_1, X_2, ..., X_n\})$ is such that there exists a constant K for which, for all $t \geq 0$,

$$P(|U_n - E(U_n)| > t) \leq \exp\left(\frac{-t^2}{K}\right).$$

4.3. *Worst-case asymptotics*

The probabilistic rates of growth just surveyed are replicated in worst-case settings. In this section, we survey some of the work that has been done on worst-case growth rates and draw parallels with the probabilistic rates.

Let $l(e)$ be the usual Euclidean length $|e|$ of the edge e. As a primary example of a worst-case growth rate, consider the *worst-case length* of an optimal traveling salesman tour in the unit d-cube:

$$\rho_{\mathrm{TSP}}(n) = \max_{\substack{S\subset[0,1]^d \\ |S|=n}} \min_T \left\{ \sum_{e\in T} l(e) : T \text{ is a tour of } S \right\}. \tag{4.5}$$

In words, $\rho_{\mathrm{TSP}}(n)$ is the maximum length, over all point sets in $[0, 1]^d$, that an optimal traveling salesman tour can attain. The minimized quantity in (4.5) is just the length of an optimal traveling salesman tour of the point set S, and the maximum is taken over all point sets S of size n. We note that there is no probability theory here, for the point sets and tours are deterministic. Steele & Snyder [1989] showed that, despite this, one obtains a rate of growth for ρ_{TSP} that is identical to the probabilistic growth rate in Theorem 1.

Theorem 3 [Steele & Snyder, 1989]. *As $n \to \infty$,*

$$\rho_{\mathrm{TSP}}(n) \sim \alpha_{\mathrm{TSP},d}\, n^{(d-1)/d},$$

where $\alpha_{\mathrm{TSP},d} > 0$ is a constant depending only on the dimension d.

4.4. Progress on the constants

Estimation of the limiting constants has a long history in both the worst-case and stochastic contexts. Few [1955] improved some very early work to provide the bound $\alpha_{\mathrm{TSP},2} \le \sqrt{2}$ and gave a dimension-d bound of $\alpha_{\mathrm{TSP},d} \le d\{2(d-1)\}^{(1-d)/2d}$, where $d \ge 2$. After other intermediate work, Karloff [1989] broke the $\sqrt{2}$ barrier in dimension two by showing that $\alpha_{\mathrm{TSP},2} \le 0.984\sqrt{2}$. The best bounds currently available in higher dimensions are those of Goddyn [1990].

Bounds on the worst-case constants are also available for other Euclidean network problems. Of particular note is the bound on $\alpha_{RST,d}$, the constant associated with the worst-case length of a rectilinear Steiner minimum tree in the unit d-cube. Chung & Graham [1981] proved that $\alpha_{RST,2} = 1$, which is significant in that it is the only non-trivial worst-case constant for which we have an exact expression. The problem of determining $\alpha_{RST,d}$ in higher dimensions is still open, with the current best-known bounds being $\max\{1,\ d/(4e)\} \le \alpha_{RST,d} \le d4^{(1-d)/d}$, for $d \ge 1$ [Snyder, 1991, 1992; Salowe, 1992].

In the case of the probabilistic models, there is recent progress due to Bertsimas and van Ryzin [1990], where asymptotic expressions as d gets large were obtained for the probabilistic minimum spanning tree and matching constants $\beta_{\mathrm{MST},d}$ and $\beta_{M,d}$. Specifically, they showed that $\beta_{\mathrm{MST},d} \sim \sqrt{d/2\pi e}$ and $\beta_{M,d} \sim (1/2)\sqrt{d/2\pi e}$ as $d \to \infty$. Still, the most striking progress on probabilistic constants was the determination of an exact expression for $\beta_{\mathrm{MST},d}$ for all $d \ge 2$ by Avram & Bertsimas [1992]. Their expression for $\beta_{\mathrm{MST},d}$ comes in the form of series expansion in which each term requires a rather difficult integration. The representation is still an effective one, and the first few terms of the series in dimension two have been computed to yield a numerical lower bound of $\beta_{\mathrm{MST},2} \ge 0.599$, which agrees well with experimental data. The proof of the series representation for $\beta_{\mathrm{MST},d}$ relies strongly on the fact that a greedy construction is guaranteed to yield an MST, and unfortunately these constructions are not possible for many objects of interest, including the TSP.

5. Concluding remarks

The theory of probabilistic networks and associated algorithms is rapidly evolving, but it is not yet a well consolidated field of inquiry. In surveying the literature, one finds relevant contributions growing in many separate areas, including the theory of random graphs, subadditive Euclidean functionals, stochastic networks, reliability, percolation, and computational geometry. Many of these areas make systematic use of tools and methodologies that remain almost unknown to the other areas, despite compelling relevance. The aim here has been to provide a view of part of the cross-fertilization that seems possible, but of necessity our focus has been on topics that allow for reasonably brief or self-contained description. Surely one can — and should — go much further. For more general information concerning probability theory applied to algorithms, one can consult the surveys of Karp [1977, 1991], Rinnooy Kan [1987], Hofri [1987], and Stougie [1990], as well as the bibliography of Karp, Lenstra, McDiarmid, and Rinnooy Kan [1985]. For more on percolation theory, Grimmett [1989] is a definitive reference.

Acknowledgements

This research was supported in part by Georgetown University 1991 Summer Research Award, by Georgetown College John R. Kennedy, Jr. Faculty Research Fund, and by the following grants: NSF DMS92-11634, NSA MDA904-91-H-2034, AFOSR-91-0259, and DAAL03-89-G-0092.

References

Ahlswede, R., and D.E. Daykin (1978). An inequality for the weights of two families of sets, their unions and intersections. *Z. Wahrscheinlichkeitstheor. Verw. Geb.* 43, 183–185.

Ahuja, R.K., T.L. Magnanti and J.B. Orlin (1991). Some recent advances in network flows. *SIAM Rev.* 33, 175–219.

Ahuja, R.K., and J.B. Orlin (1987). A fast and simple algorithm for the maximum flow problem. *Oper. Res.* 37, 748–759.

Alon, N. (1990). Generating pseudo-random permutations and maximum flow algorithms. *Inf. Process. Lett.* 35, 201–204.

Avram, F., and D. Bertsimas (1992). The minimum spanning tree constant in geometric probability and under the independent model: a unified approach. *Ann. Appl. Probab.* 2, 118–130.

Beardwood, J., J.H. Halton and J. Hammersley (1959). The shortest path through many points. *Proc. Camb. Philos. Soc.* 55, 299–327.

Bertsimas, D., and G. Van Ryzin (1990). An asymptotic determination of the minimal spanning tree and minimal matching constants in geometric probability. *Oper. Res. Lett.* 9, 223–231.

Bollobás, B. (1985). *Random Graphs*, Academic Press, New York, N.Y.

Bollobás, B. (1986). *Combinatorics*, Cambridge University Press, New York, N.Y.

Cheriyan, J., and T. Hagerup (1989). A randomized max-flow algorithm, *Proc. 30th IEEE Foundations of Computer Science*, IEEE, pp. 118–123.

Chung, F.R.K., and R.L. Graham (1981). On Steiner trees for bounded point sets. *Geom. Dedicata* 11, 353–361.

Colbourn, C.J. (1987). *The Combinatorics of Network Reliability*, Oxford University Press, New York, N.Y.

Cox, J.T., and R. Durrett (1988). Limit theorems for the spread of epidemics and forest fires. *Stochastic Process. Appl.* 30, 2, 171–191.

Dijkstra, E.W. (1959). A note on two problems in connexion with graphs. *Numer. Math.* 1, 269–271.

Even, S., and O. Kariv (1975). An $O(n^{2.5})$ algorithm for maximum matching in general graphs, *Proc. 16th IEEE Symp. Foundations of Computer Science*, IEEE, pp. 100–112.

Few, L. (1955). The shortest path and the shortest road through n points in a region. *Mathematika* 2, 141–144.

Fortuin, C.M., P.N. Kasteleyn and J. Ginibre (1971). Correlation inequalities on some partially ordered sets. *Commun. Math. Phys.* 22, 89–103.

Frieze, A.M., and G.R. Grimmett (1985). The shortest-path problem for graphs with random arc-lengths. *Discrete Appl. Math.* 10, 57–77.

Gabow, H.N. (1985). Scaling algorithms for network problems. *J. Comput. Systems Sci.* 31, 148–168.

Goddyn, L. (1990). Quantizers and the worst-case Euclidean traveling salesman problem. *J. Comb. Theor., Ser. B* 50, 65–81.

Goldberg, A.V., and R.E. Tarjan (1988). A new approach to the maximum-flow problem. *J. ACM* 35, 921–940.

Graham, R.L. (1983). Applications of the FKG inequality and its relatives, in: A. Bachem, M. Grötschel, and B. Korte (eds). *Mathematical Programming: The State of the Art, Bonn 1982*, Springer-Verlag, New York, NY, pp. 115–131.

Grimmett, G.R. (1989). *Percolation*, Springer-Verlag, New York, N.Y.

Grimmett, G.R., and D.J.A. Welsh (1982). Flow in networks with random capacities. *Stochastics* 7, 205–229.

Halton, J.H., and R. Terada (1982). A fast algorithm for the Euclidean traveling salesman problem, optimal with probability one. *SIAM J. Comput.* 11, 28–46.

Harris, T.E. (1960). A lower bound for the critical probability in a certain percolation process, *Proc. Camb. Philos. Soc.* 56, 13–20.

Hofri, M. (1987). *Probabilistic Analysis of Algorithms: On Computing Methodologies for Computer Algorithms Performance Evaluation*, Springer-Verlag, New York, N.Y.

Jerrum, M., and A. Sinclair (1986). The approximation of the permanent, *Proc. 18th Symp. on Theory of Computing* Association for Computing Machinery, pp. 235–243.

Jerrum, M., and A. Sinclair (1989). The approximation of the permanent. *SIAM J. Comput.* 18, 1149–1178.

Karloff, H.J. (1989). How long can a Euclidean traveling salesman tour be? *SIAM J. Disc. Math.* 2, 91–99.

Karp, R.M. (1976). The probabilistic analysis of some combinatorial search algorithms, in: J.F. Traub (ed.), *Algorithms and Complexity: New Directions and Recent Results*, Academic Press, New York, N.Y., pp. 1–19.

Karp, R.M. (1977). Probabilistic analysis of partitioning algorithms for the traveling salesman problem in the plane. *Math. Oper. Res.* 2, 209–224.

Karp, R.M. (1991). An introduction to randomized algorithms. *Discrete Appl. Math.* 34, 165–201.

Karp, R.M., J.K. Lenstra, C.J.H. McDiarmid and A.H.G. Rinnooy Kan (1985). Probabilistic analysis, in: M. O'hEigeartaigh, J.K. Lenstra and A.H.G. Rinnooy Kan (eds.), *Combinatorial Optimization: Annotated Bibliographies*, John Wiley and Sons, Chichester, pp. 52–88.

Karp, R.M., and M. Luby (1985). Monte Carlo algorithms for planar multiterminal network reliability, *J. Complexity* 1, 45–64.

Karp, R.M., R., Motwani and N. Nisan (1993). Probabilistic analysis of network flow algorithms. *Math. Oper. Res.* 18, 71–97.

Karp, R.M., and J.M. Steele (1985). Probabilistic analysis of heuristics, in: E.L. Lawler, J.K. Lenstra, A.H.G. Rinnooy Kan and D.B. Shmoys (eds.), *The Traveling Salesman Problem: A Guided Tour of Combinatorial Optimization*, John Wiley and Sons, New York, N.Y., pp. 181–206.

Kesten, H. (1980). The critical probability of bond percolation on the square lattice equals $\frac{1}{2}$. *Commun. Math. Phys.* 74, 41–59.

Kulkarni, V.G. (1986). Shortest paths in networks with exponentially distributed arc lengths. *Networks* 16, 255–274.

Lovász, L. (1979). On determinants, matchings, and random algorithms, in: L. Budach (ed.), *Fundamentals of Computing Theory*, Akademia-Verlag, Berlin.

Micali, S., and V.V. Vazirani (1980). An $O(|V|^{0.5}|E|)$ algorithm for finding maximum matchings in general graphs, *Proc. 21st IEEE Symp. on Foundations of Computer Science*, IEEE, pp. 17–27.

Motwani, R. (1989). Expanding graphs and the average-case analysis of algorithms for matchings and related problems, *Proc. 21st Symp. on Theory of Computing* Ass. Comput. Mach., pp. 550–561.

Rabin, M.O., and V.V. Vazirani (1989). Maximum matching in general graphs through randomization. *J. Algorithms* 10, 557–567.

Rhee, W.T., and M. Talagrand (1989). A sharp deviation inequality for the stochastic traveling salesman problem. *Ann. Probab.* 17, 1–8.

Rinnooy Kan, A.H.G. (1987). Probabilistic analysis of algorithms. *Ann. Discrete Math.* 31, 365–384.

Russo, L. (1981). On the critical percolation probabilities. *Z. Wahrscheinlichkeitstheor. Verw. Geb.* 56, 129–139.

Salowe, J.S. (1992). A note on lower bounds for rectilinear Steiner trees. *Inf. Proc. Lett.* 42, 151–152.

Shepp, L.A. (1982). The XYZ conjecture and the FKG inequality. *Ann. Probab.* 10, 824–827.

Sigal, E.C., A.A.B. Pritsker and J.J. Solberg (1979). The use of cutsets in Monte Carlo analysis of stochastic networks. *Math. Comput. Simulat.* 21, 376–384.

Sigal, E.C., A.A.B. Pritsker and J.J. Solberg (1980). The stochastic shortest route problem. *Oper. Res.* 28, 1122–1130.

Sinclair, A. (1993). *Algorithms for Random Generation and Counting: A Markov Chain Approach*, Birkhäuser Publishers, Boston, Mass.

Snyder, T.L. (1992). Minimal rectilinear Steiner trees in all dimensions. *Discr. Comp. Geometry* 8, 73–92.

Snyder, T.L. (1991). Lower bounds for rectilinear Steiner trees in bounded space, *Inf. Process. Lett.* 37, 71–74.

Spencer, J. (1993). The Janson Inequality, in: D. Miklos, V.T. Sos and T. Szonyi (eds.), *Combinatorics, Paul Erdös is Eighty*, Vol. 1, Bolyai Mathematical Studies, Keszthely (Hungary), pp. 421–432.

Steele, J.M. (1981a). Subadditive Euclidean functionals and non-linear growth in geometric probability. *Ann. Probab.* 9, 365–376.

Steele, J.M. (1981b). Complete convergence of short paths and Karp's algorithm for the TSP. *Math. Oper. Res.* 6, 374–378.

Steele, J.M. (1990a). Probabilistic and worst-case analyses of classical problems of combinatorial optimization in Euclidean space. *Math. Oper. Res.* 15, 749–770.

Steele, J.M. (1990b). Seedlings in the theory of shortest paths, in: J. Grimmett and D. Welsh (eds.), *Disorder in Physical Systems: A Volume in Honor of J.M. Hammersley*, Cambridge University Press, London, pp. 277–306.

Steele, J.M., and T.L. Snyder (1989). Worst-case growth rates of some classical problems of combinatorial optimization. *SIAM J. Comput.* 18, 278–287.

Stougie, L. (1990). Applications of probability theory in combinatorial optimization, Class Notes, University of Amsterdam.

Valiant, L.G. (1979). The complexity of enumeration and reliability problems. *SIAM J. Comput.* 12, 777–788.

Van den Berg, J., and H. Kesten (1985). Inequalities with applications to percolation and reliability. *J. Appl. Prob.* 22, 556–569.

M.O. Ball et al., Eds., *Handbooks in OR & MS, Vol. 7*

Chapter 7

A Survey of Computational Geometry

Joseph S.B. Mitchell
Applied Math, SUNY, Stony Brook, NY 11794-3600, U.S.A.

Subhash Suri
Bellcore, 445 South Street, Morristown, NJ 07960, U.S.A.

1. Introduction

Computational geometry takes an algorithmic approach to the study of geometrical problems. The principal motivation in this study is a quest for 'good' algorithms for solving geometrical problems. Of course, several practical and aesthetic factors determine what one means by a good algorithm, but the general trend has been to associate 'goodness' with the asymptotic efficiency of an algorithm in terms of the time and space complexity. Lately, however, the implementational ease and robustness also are becoming increasingly important considerations in the algorithm design.

Although many geometrical problems and algorithms were known before, computational geometry evolved into a cohesive discipline only in the mid to late seventies. An important event in this development was the publication of the Ph.D. thesis of M. Shamos [1978] in 1978. During its first decade, the field of computational geometry grew enormously as its fundamental structures were applied to a vast variety of problems in diverse disciplines, and many new tools and techniques were developed. In the process, new insights were gained into inter-relationships among some of these fundamental structures, which also led to a unification and consolidation of several disparate sets of ideas. In the last five or so years, the field has matured significantly, both in mathematical depth as well as in algorithmic ideas.

Computational geometry has had strong interactions with other fields, in mathematics as well as in applied computer science. A particularly fruitful interplay has taken place between computational geometry and combinatorial geometry. The latter is a branch of mathematics concerned primarily with the 'counting' of certain geometric structures. Examples include counting the maximum possible number of incidences between a set of lines and a set of points, and counting the number of lines that bisect a set of points. Both fields seem to have benefited from each other: combinatorial bounds for certain structures have been obtained by analyzing an algorithm that enumerates them and, conversely, the analysis of algorithms often depends crucially on the combinatorial bound on some geometric objects.

The field of computational geometry has also benefited from its interactions with other disciplines within computer science such as VLSI, database theory, robotics, computer vision, computer graphics, pattern recognition and learning theory. These areas offer a rich variety of problems that are inherently geometrical.

Due to its interconnections with many applications areas, the variety of problems studied in computational geometry is truly enormous. Our goal in this paper is quite modest: we survey the state-of-the-art in some selected areas of computational geometry, with a strong bias towards problems with an optimization component. In the process, we also hope to acquaint the reader with some of the fundamental techniques and structures in computational geometry.

Our paper has seven main sections. The survey proper begins in Section 3, while Section 2 introduces some foundational material. In particular, we briefly describe five key concepts and fundamental structures that permeate much of computational geometry, and therefore are somewhat essential to a proper understanding of the material in later sections. The structures covered are convex hulls, arrangements, geometric duality, Voronoi diagram, and point location data structures. The main body of our survey begins with Section 3, where we describe four popular geometric graphs: minimum and maximum spanning trees, relative neighborhood graphs, and Gabriel graphs. Section 4 is devoted to algorithms in path planning. The topic of path planning is a vast one, with problems ranging from finding shortest paths in a discrete graph to deciding the feasible motion of a complex robot in an environment full of complex obstacles. We briefly mention most of the major developments in path planning research over the last two decades, but to a large extent limit ourselves to issues related to shortest paths in a planar domain. In Section 5, we discuss the matching and the traveling salesman type problems in computational geometry. Section 6 describes results on a variety of problems related to shape analysis and pattern recognition. We close with some concluding remarks in Section 7. In each section, we also pose what in our opinion are the most important and interesting open problems on the topic. There are altogether twenty open problems in this survey.

2. Fundamental structures

2.1. Convex hulls

The *convex hull* of a finite set of points S is the smallest convex set containing S. In two dimensions, for instance, the convex hull is the smallest convex polygon containing all the points of S; see Figure 1 for an example. In higher dimensions, the convex hull is a polytope. Before we discuss the algorithms for computing a convex hull, we must address the question of representing it. There are several representations of a convex hull, depending upon how many features of the corresponding polytope are described. In the simplest representation, we may only

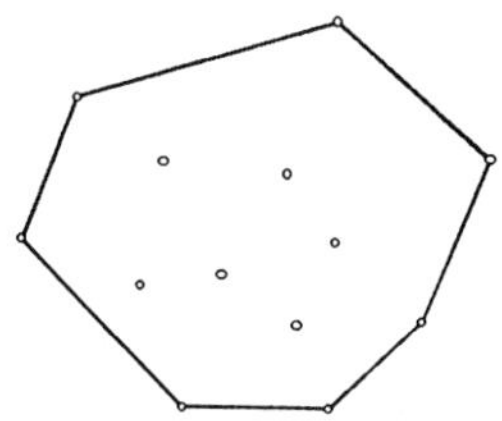

Fig. 1. A planar convex hull.

store the vertices of the convex hull. The other extreme of the representation is the *face lattice*, which stores all the faces of the convex hull as well as the incidence relationships among the faces. The intermediate forms of representation may store faces of only certain dimensions, such as the $(d-1)$-dimensional faces, also called the *facets*. The differences among these representations become significant only in dimensions $d \geq 4$, where the full lattice may have size $\Theta(n^{\lfloor d/2 \rfloor})$ while the number of vertices is obviously at most n. (Grünbaum's book [1967] is an excellent source for information on polytopes.)

In two dimensions, several $O(n \log n)$ time algorithms are known. Almost every algorithmic paradigm in computational geometry has been applied successfully to the planar convex hull algorithm: for instance, divide-and-conquer, incremental construction, planar sweep, and randomization have all been utilized to obtain $O(n \log n)$ time algorithms for planar convex hulls; see the textbook by Preparata & Shamos [1985]. The best theoretical bound for the planar convex hull problem is achieved by an algorithm of Kirkpatrick & Seidel [1986], which runs in time $O(n \log h)$, where h is the number of convex hull vertices.

In three dimensions, Preparata & Hong [1977] proposed an $O(n \log n)$ time algorithm based on the divide-and-conquer paradigm. Theirs was the only known optimal algorithm in three dimensions, until Clarkson & Shor [1989] developed a randomized incremental algorithm that achieved an expected running time $O(n \log n)$. A variant of the Clarkson-Shor algorithm was proposed by Guibas, Knuth & Sharir [1992], which admits a simpler implementation as well as an easier analysis. (These randomized algorithms are quite practical and considerably easier to implement than the divide-and-conquer algorithm.) Very recently, Chazelle & Matoušek [1992] have settled a long-standing open problem by announcing a deterministic $O(n \log h)$ time algorithm for the three-dimensional convex hull problem.

In higher dimensions, Chazelle [1991] recently proposed an algorithm whose worst-case running time matches the worst-case bound on the facet complexity of the convex hull in any dimension $d \geq 4$. Chazelle's algorithm builds on earlier ideas of Kallay [1984] and Seidel [1981], and runs in worst-case time $O(n \log n + n^{\lfloor d/2 \rfloor})$. The algorithm in Chazelle [1991] achieves the best worst-case performance, but it does not depend on the actual size of the face lattice. An algorithm by Seidel [1986] runs in time proportional to the size of the face lattice. In particular, the algorithm in Seidel [1986] takes $O(n^2 + F \log n)$ time for facet enumeration and $O(n^2 + L \log n)$ time for producing the face lattice, where F

is the number of facets in the convex hull and L is the size of the face lattice; Seidel's algorithm uses a method called 'gift-wrapping' and builds upon the earlier work by Chand & Kapur [1970] and Swart [1985].

There is a vast literature on convex hulls, and the presentation above has hardly scratched the surface. We have left out whole lines of investigation on the convex hull problem, such as the expected case analysis of algorithms and the average size of the convex hull; we refer the reader to Dwyer [1988], Devroye & Toussaint [1981], and Golin & Sedgewick [1988].

2.2. Arrangements

Arrangements refer to space partitions induced by lines, hyperplanes, or other algebraic varieties. A finite set of lines δ partitions the plane into convex regions, called 'cells,' which are bounded by straight line edges and vertices. The *arrangement of lines* $\mathcal{A}(\delta)$ refers to this collection of cells, along with their incidence relations. Similarly, a set of hyperplanes (or other surfaces, such as spheres) induce arrangements in higher dimensions.

An arrangement encodes the sign pattern for its generating set. In other words, an arrangement is a manifestation of equivalence classes induced by a set of lines or hyperplanes — all points of a particular cell have the same 'sign vector' with respect to all the hyperplanes in the set. This property is a key to solving many geometric problems in computational geometry. It often turns out that solving a geometric problem requires computing a particular cell or a family of cells in an arrangement of hyperplanes. We will say more about this in the section on Voronoi diagrams.

Combinatorial aspects of arrangements have been investigated for a long time; the arrangements in two and three dimensions were studied by Steiner [1967]. The interested reader may consult Grünbaum's book [1967] for a detailed discussion on arrangements. We will focus mainly on the computational aspects of arrangements. An arrangement of n hyperplanes in d-space can be computed in time $O(n^d)$ by an algorithm due to Edelsbrunner, O'Rourke & Seidel [1986]. This bound is asymptotically tight since a simple arrangement (where no more than d hyperplanes meet in a point) has $\Theta(n^d)$ complexity.

Although a single cell in an arrangement of hyperplanes can have complexity $\Theta(n^{\lfloor d/2 \rfloor})$, not many cells can be large: after all, the combined complexity of $\Theta(n^d)$ cells is only $O(n^d)$. There has been a considerable amount of work on bounding the complexity of a family of cells in an arrangement. For instance, in two dimensions, the total complexity of any m cells in an arrangement of n lines is roughly $O(m^{2/3}n^{2/3} + m + n)$, up to some logarithmic factors [Edelsbrunner, Guibas & Sharir, 1990; Aronov, Edelsbrunner, Guibas & Sharir, 1989]. Extensions to higher dimensions and related results can be found in Aronov, Matoušek & Sharir [1991], Edelsbrunner & Welzl [1986], Edelsbrunner, Guibas & Sharir [1990], and Pellegrini [1991]. Arrangements of bounded objects, such as line segments, triangles and tetrahedra, have also been studied; see Chazelle & Edelsbrunner [1992], and Matoušek, Miller, Pach, Sharir, Sifrony & Welzl [1991].

2.3. Geometric duality

Duality plays an important role in geometric algorithms, and it often provides a tool for transforming an unfamiliar problem into a familiar setting. In this section, we will give a brief description of two most frequently used transformations in computational geometry.

The first transform, $\mathcal{D}$, maps a point p to a hyperplane $\mathcal{D}(p)$ and vice versa:

$$p : (p_1, p_2, \ldots, p_d) \iff \mathcal{D}(p) : x_d = 2p_1x_1 + 2p_2x_2 + \cdots + 2p_{d-1}x_{d-1} - p_d \quad (1)$$

Thus, in the plane the point (a, b) maps to the line $y = 2ax - b$, and the line $y = mx + c$ maps to the point $(m/2, -c)$. This transformation preserves incidence and order: (1) a point p is incident to hyperplane h if and only if the dual point $\mathcal{D}(h)$ is incident to dual hyperplane $\mathcal{D}(p)$, and (2) a point p lies below hyperplane h if and only if the dual point $\mathcal{D}(h)$ lies below the dual hyperplane $\mathcal{D}(p)$.

The second transform, also called the 'lifting transform,' maps a points in R^d to a point in R^{d+1}. It maps a point $p = (p_1, p_2, \ldots, p_d)$ in R^d to the point $p^+ = (p_1, p_2, \ldots, p_d, p_1^2 + p_2^2 + \cdots + p_d^2)$. If we treat R^d as the hyperplane $x_{d+1} = 0$ embedded in R^{d+1}, then the lifting transform maps a point $p \in R^d$ onto its vertical projection on the unit paraboloid $U : x_{d+1} = x_1^2 + x_2^2 + \cdots + x_d^2$. The combination of the lifting transform and the duality map $\mathcal{D}$ maps a point $p \in R^d$ to the hyperplane

$$\mathcal{D}(p^+) : x_{d+1} = 2p_1x_1 + 2p_2x_2 + \cdots + 2p_dx_d - (p_1^2 + p_2^2 + \cdots + p_d^2) \quad (2)$$

The hyperplane $\mathcal{D}(p^+)$ is tangent to the paraboloid U at the point p^+. It turns out that this mapping is especially useful for computing Voronoi diagrams, the topic of our next section.

2.4. Voronoi diagram and Delaunay triangulation

The Voronoi diagram is perhaps the most versatile data structure in all of computational geometry. This diagram, along with its graph-theoretical dual, the *Delaunay Triangulation*, finds applications in problems ranging from associative file searching and motion planning to crystallography and clustering. In this section, we give a brief survey of some of the key ideas and results on these structures; for further details, consult Edelsbrunner's book [1987] or the survey by Aurenhammer [1991].

A Voronoi diagram encodes the 'nearest-neighbor' information for a set of 'sites.' We begin by explaining the concept in two dimensions. Given a set of n 'sites' or points $S = \{s_1, s_2, \ldots, s_n\}$ in the two-dimensional Euclidean plane, the Voronoi diagram of S partitions the plane into n convex polygons $V_1, V_2, \ldots, V_n$ such that any point in V_i is closer to s_i than to any other site:

$$V_i = \{x \mid d(x, s_i) \leq d(x, s_j), \text{ for all } j \neq i\},$$

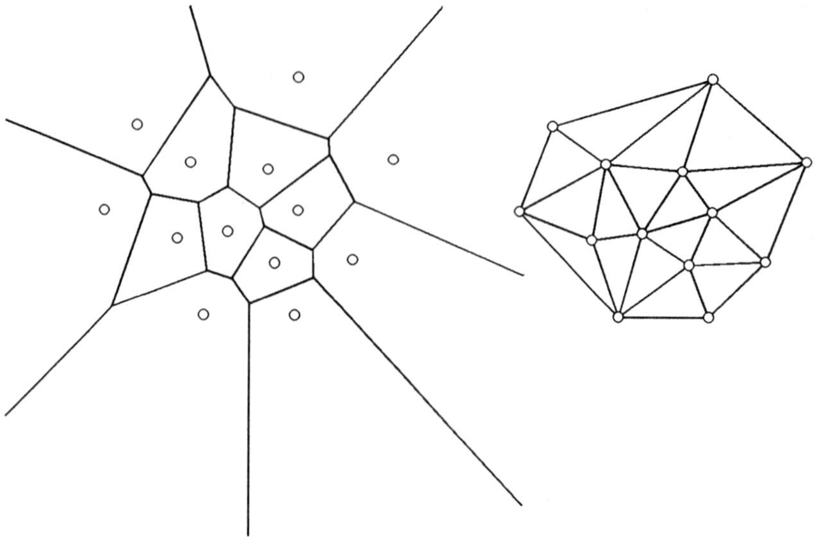

Fig. 2. The Voronoi diagram (left) and the Delaunay triangulation (right) of a set of points in the plane.

where $d(x, y)$ is the Euclidean distance between the points x and y. An interesting fact about Voronoi diagrams in the plane is their linear complexity: $O(n)$ vertices and edges.

The Delaunay triangulation of S is the graph-theoretic dual of its Voronoi diagram: two sites s_i and s_j are joined by an edge if the Voronoi polygons V_i and V_j share a common edge. Under a non-degeneracy assumption that no four points of S are co-circular, the dual graph is always a triangulation of S. Figure 2 shows an example of a Voronoi diagram and the corresponding Delaunay triangulation.

Just like convex hulls, algorithms based on several different paradigms are known for the construction of planar Voronoi diagrams and Delaunay triangulations, such as divide-and-conquer [Dwyer, 1987; Guibas & Stolfi, 1985], plane sweep [Fortune, 1987], and randomized incremental methods [Clarkson & Shor, 1989; Guibas, Knuth & Sharir, 1992]. They all run in $O(n \log n)$ time (worst-case for deterministic, and expected for randomized).

The concepts of Voronoi diagram and Delaunay triangulation extend naturally to higher dimensions, as well as to other metrics. In d dimensions, the Voronoi diagram of a set of points S is a tessellation of E^d by convex polyhedra. The polyhedral cell V_i consists of all those points that are closer to s_i than to any other site in S. The Delaunay triangulation of S is the geometric dual of the Voronoi diagram: there is a k-face for every $(d-k)$-face of the Voronoi diagram. In particular, there is an edge between s_i and s_j if the Voronoi polyhedra V_i and V_j share a common $(d-1)$-dimensional face. An equivalent way of defining the Delaunay triangulation is via the *empty-sphere* test: a $(d+1)$-tuple $(s^1, s^2, \ldots, s^{d+1})$ is a simplex (triangle) of the Delaunay triangulation of S if and only if the sphere determined by $(s^1, s^2, \ldots, s^{d+1})$ does not contain any other point of S. The Voronoi diagram of n points in d dimensions, $d \geq 3$, can have super-linear size: $\Theta(n^{\lceil d/2 \rceil})$ [Edelsbrunner, 1987].

It turns out that Voronoi diagrams and Delaunay triangulations are intimately related to convex hulls and arrangements via duality transforms. This relationship was first discovered by Brown [1980], who showed using an inversion map that the Voronoi diagram of a set $S \in R^d$ corresponds to the convex hull of a transformed set in R^{d+1}. Later, Edelsbrunner & Seidel [1986] extended and simplified this idea, using the paraboloid transforms mentioned in the previous section. We now sketch their idea.

Let $S = \{s_1, s_2, \ldots, s_n\}$ be a set of n points in R^d. We map S to a set of hyperplanes $\mathcal{D}(S^+)$ in R^{d+1}, using the combination of lifting and duality maps mentioned in Section 2.3. In particular, a point $s = (a_1, a_2, \ldots, a_d)$ maps to the hyperplane $\mathcal{D}(s^+)$ whose equation is

$$x_{d+1} = 2a_1x_1 + 2a_2x_2 + \cdots + 2a_dx_d - (a_1^2 + a_2^2 + \cdots + a_d^2).$$

Let $\mathcal{P}$ be the polyhedron defined by the intersection of the 'upper' halfspaces bounded by these hyperplanes. Then, the vertical projection of $\mathcal{P}$ onto the hyperplane $x_{d+1} = 0$ gives precisely the Voronoi diagram of S in R^d.

A similar (and perhaps easier to visualize) relationship exists between convex hulls and Delaunay triangulations, using only the lifting transform. We map the points $S = \{s_1, s_2, \ldots, s_n\}$ to their 'lifted' counterpart $S^+ = \{s_1^+, s_2^+, \ldots, s_n^+\}$. Now compute the convex hull of S^+. The triangles in the Delaunay triangulation of S correspond precisely to the facets of $CH(S^+)$ with downward normal.

Thus, both the Voronoi diagram and the Delaunay triangulation of a set of points in R^d may be computed using a convex hull algorithm in R^{d+1}. This relationship also explains why the worst-case size of both a Voronoi diagram in R^d and a convex hull in R^{d+1} is $\Theta(n^{\lfloor (d+1)/2 \rfloor})$.

2.5. Point location

Many problems in computational geometry often require solving a so-called point location problem. Given a partition of space into polyhedral cells and a query point q, the problem is to identify the cell containing q. For instance, if Voronoi diagrams are used for answering 'nearest neighbor' queries, one needs to locate the Voronoi polyhedron containing the query point. Typically, a large number of queries are asked with respect to the same cell complex; thus, it makes sense to preprocess the cell complex in order to speed up queries.

The problem has been studied intensely in two dimensions, where several optimal algorithms and data structures are now known. These algorithms can preprocess a planar map on n vertices into a linear space data structure and answer a point location query in $O(\log n)$ time [see Lipton & Tarjan, 1980; Kirkpatrick, 1983; Edelsbrunner, Guibas & Stolfi, 1986; Goodrich & Tamassia, 1991].

In higher dimensions, the point location problem is less well-understood, and no algorithm simultaneously achieves optimal bounds for preprocessing time, storage space, and query time. We give a brief summary of results and give pointers to relevant literature. We denote the performance of an algorithm by the triple $\{P, S, Q\}$, which refer to preprocessing time, storage space, and query time.

Preparata & Tamassia [1990] give an $\{O(n \log^2 n), O(n \log^2 n), O(\log^2 n)\}$ algorithm for point location in a convex cell complex of n facets in three dimensions. Using randomization, Clarkson [1987] gives an $\{O(n^{d+\varepsilon}), O(n^{d+\varepsilon}), O(\log n)\}$ algorithm for point location in an arrangement of n hyperplanes in d dimensions; the space and query bounds are worst-case, but the preprocessing time is expected. Chazelle & Friedman [1990] were later able to make Clarkson's algorithm deterministic, albeit at an increased preprocessing cost, resulting in an algorithm with resource bounds $\{O(n^{d(d+3)/2+2}), O(n^d), O(\log n)\}$.

3. Geometric graphs

3.1. Minimum spanning trees

The minimum spanning tree (MST) problem is one of the best-known problems of combinatorial optimization, and it has received a considerable attention in computational geometry as well. Given a graph $G = (V, E)$, with non-negative real-valued weights on edges, a minimum spanning tree of G is an acyclic subgraph of G that spans all the nodes in V and has minimum total edge weight. An MST has obvious applications in the design of computer and communication networks, transportation systems, and other kinds of networks. But applications of the minimum spanning tree extend far beyond network design problems. They are used in problems as diverse as network reliability, computer vision, automatic speech recognition, clustering and classification, matching and traveling salesman problems, and surface homogeneity tests.

Efficient algorithms for computing an MST have been known for a long time; a survey by Graham & Hell [1985] traces the history of MST and cites algorithms dating back to the beginning of the century. Although the algorithms of Kruskal [1956] and Prim [1957] are among the best known, an algorithm by Borøuvka preceded them by almost thirty years [Graham & Hell, 1985]. Using suitable data structures, these algorithms can be implemented in $O(|E| \log |V|)$ or $O(|V|^2)$ time. In the last two decades, several new implementations and variants of these basic algorithms have been proposed, and the fastest ones run in almost linear time in the size of the graph [Fredman & Tarjan, 1987; Gabow, Galil, Spencer & Tarjan, 1986].

The interest of computational geometry researchers in MST stems from the observation that in many applications the underlying graph is Euclidean: we want to compute a minimum spanning tree for a set of n points in R^d, for $d \geq 1$. The set of n points in this case completely specifies the graph, without an explicit enumeration of the edges. Since the edge-weights in this geometric graph are not entirely arbitrary, a natural question is if one can compute an MST in (asymptotically) less than n^2 steps, that is, without inspecting every edge.

Surprisingly, it turns out that for a set of n points in the plane, an MST can be computed in $O(n \log n)$ time. A key observation is the following lemma, which states that the edges of an MST are contained in the Delaunay triangulation graph; we omit the proof, but an interested reader may consult the book of Preparata-Shamos [1985].

Lemma 1. *Let S be a set of n points in the plane, and let* $\mathrm{DT}(S)$ *denote the Delaunay triangulation of S. Then,* $\mathrm{MST}(S) \subseteq \mathrm{DT}(S)$.

We recall from Section 2.4 that the Delaunay triangulation in two dimensions is a planar graph. Thus, by running an efficient graph MST algorithm on $\mathrm{DT}(S)$, we can find a minimum spanning tree of S in $O(n \log n)$ time. In fact, given the Delaunay triangulation, a minimum spanning tree of S can be extracted in linear time, by using an algorithm of Cheriton & Tarjan [1976], which computes an MST in linear time for planar graphs.

Lemma 1 holds in any dimension; however, it no longer serves a useful purpose for computing minimum spanning trees in higher dimensions since the Delaunay graph can have size $\Omega(n^2)$ in dimensions $d \geq 3$ [Preparata & Shamos, 1985]. Nevertheless, the underlying geometry can be exploited to compute an MST in subquadratic worst-case time. Yao [1982] has proposed a general method for computing geometric minimum spanning trees in time $O(n^{2-\alpha_d}(\log n)^{1-\alpha_d})$, where α_d is a dimension-dependent constant. Yao's algorithm is based on the following idea: if we partition the space around a point p into polyhedral cones of sufficiently small angular diameter, then there is at most one MST edge incident to p per cone, and this edge joins p to its nearest neighbor in that cone. In order to find these nearest neighbors efficiently, Yao utilizes a data structure that, after a polynomial-time preprocessing, can determine a nearest neighbor in logarithmic time. In the original paper of Yao [1982], the constant α_d had value of $2^{-(d+1)}$, thus, making his algorithm only slightly subquadratic; however, the interesting conclusion is that an MST can be computed without checking all the edges.

The exponent in the general algorithm of Yao has steadily improved, as better data structures have been developed for solving the nearest-neighbor problem. Recently, Agarwal, Edelsbrunner, Schwarzkopf & Welzl [1991] have shown also that computationally the twin problems of computing a minimum spanning tree and computing a bi-chromatic nearest neighbor are roughly equivalent. The constant α_d in the running time their algorithm is roughly $2/(\lceil d/2 \rceil + 1)$ [Agarwal, Edelsbrunner, Schwarzkopf & Welzl, 1991]. In three dimensions, the algorithm of Agarwal, Edelsbrunner, Schwarzkopf & Welzl [1991] computes an MST in $O(n^{4/3} \log^{4/3} n)$ time. An alternative and somewhat simpler, albeit randomized, algorithm of the same complexity is given by Agarwal, Matoušek & Suri [1992].

There also has been work on computing an approximation of the MST. Vaidya [1988] constructs in $O(\varepsilon^{-d} n \log n)$ time a spanning tree with length at most $1 + \varepsilon$ times the length of an MST. If the n points are independently and uniformly distributed in the unit d-cube, then the expected time complexity of Vaidya's algorithm is $O(n\alpha(cn, n))$, where α is the inverse Ackermann function.

The best lower bound known for the MST problem is $\Omega(n \log n)$, in any fixed dimension $d \geq 1$. (The lower bound holds in the algebraic tree model of computation for any input consisting of an unordered set of n points; $o(n \log n)$ time algorithms are possible for special configurations of points, such as the vertices of a convex polygon if the points are given in order along the boundary of the polygon.) It is an outstanding open problem in computational geometry to settle the asymptotic time complexity of computing a geometric minimum spanning tree in d-space.

Open Problem 1. *Given a set S of n unordered points in E^d, compute its Euclidean minimum spanning tree in $O(c_d n \log n)$ time, where c_d is a constant depending only on the dimension d. Alternatively, prove a lower bound that is better than $\Omega(n \log n)$.*

There is an obvious connection between MST and nearest neighbors: the MST neighbors of a points s include a nearest neighbor of s. Thus, the *all-nearest-neighbors* problems, which asks for a nearest neighbor for each of the points of S, is no harder than computing the MST(S). A few years ago, Vaidya [1989] gave an $O(c_d n \log n)$ time algorithm for the all-nearest-neighbors problems for any fixed dimension; the constant c_d in Vaidya's algorithm is of the order of 2^d.

Unfortunately, no reduction in the converse direction (given all nearest neighbors, compute MST) is known. However, the result of Agarwal, Edelsbrunner, Schwarzkopf & Welzl [1991] points out an equivalence between the MST and the *bi-chromatic closest pair* problem. The bi-chromatic closest pair problem is defined for two d-dimensional sets of points R and B, and it asks for a pair $r \in R$ and $b \in B$ that minimizes the distance over all such pairs. It is shown in Agarwal, Edelsbrunner, Schwarzkopf & Welzl [1991] that the asymptotic time complexities of the two problems are the same if they have the form $\Theta(n^{1+\varepsilon})$, for any $\varepsilon > 0$, otherwise, they are within a polylogarithmic factor of each other. This leads to the following open problem.

Open Problem 2. *Given two unordered sets of points $B, R \subset E^d$, compute a bi-chromatic closest pair of B and R in time $O(c_d n \log n)$, where $n = |B| + |R|$ and c_d is a constant depending only on the dimension d. Alternatively, prove a lower bound better than $\Omega(n \log n)$.*

3.2. Maximum spanning trees

A maximum spanning tree is the other extreme of the minimum spanning tree: it *maximizes* the total edge weight. In graphs, a maximum spanning tree can be computed using any minimum spanning tree algorithm, by simply negating all the edge weights. But what about a *geometric* maximum spanning tree? Is it possible to compute the maximum spanning tree of a set of points in less then quadratic time?

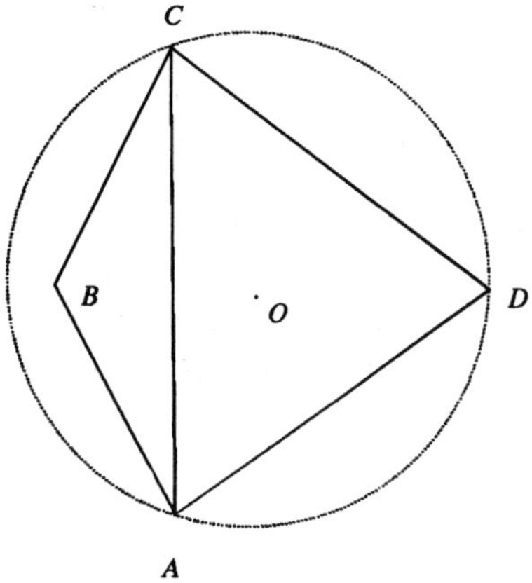

Fig. 3. MXST is not a subset of farthest-point Voronoi diagram.

As a first attempt, we could try to generalize Lemma 1. Instead of a Delaunay triangulation, we would consider the so-called *furthest-point* Delaunay triangulation, which is the graph-theoretic dual of the *furthest-point Voronoi diagram*. (In a furthest-point Voronoi diagram of a set of points S, the region associated with a site $s_i \in S$ consists of all the points x that satisfy $d(x, s_i) \geq d(x, s_j)$, for all $s_i \neq s_j$; see Preparata-Shamos [1985] for details.) Unfortunately, the maximum spanning tree edges do not necessarily lie in the furthest-point Delaunay triangulation. One of the reasons why this relationship does not hold is trivial: the furthest-point Delaunay triangulation only triangulates the convex hull vertices of S; the interior points of S have an empty furthest-point Voronoi polygon. The trouble in fact goes deeper: even if all points of S were to lie on its convex hull, the maximum spanning tree does not always lie in the Delaunay triangulation. Consider the example in Figure 3, which is due to Bhattacharya & Toussaint [1985]. In this figure, $\triangle ACD$ is an equilateral triangle and B lies on the line joining D with the center O of the circle ACD such that $2d(D, O) > d(D, B) > d(D, A)$. It is easy to check that the furthest-point Delaunay triangulation consists of triangles $\triangle ABC$ and $\triangle ACD$, and does not include the diagonal BD. On the other hand, the maximum spanning tree of $\{A, B, C, D\}$ necessarily contains the edge BD; the other two edges can be any two of the three edges of the equilateral triangle $\triangle ACD$.

An optimal $O(n \log n)$ time algorithm for computing a maximum spanning tree of n points in the plane was proposed a few years ago by Monma, Paterson, Suri & Yao [1990]. Their algorithm starts by computing the *furthest neighbor* graph: connect each point to its furthest neighbor. This results in a forest, whose components are called *clusters* in Monma, Paterson, Suri & Yao [1990]. Monma and coworkers show that these clusters can be cyclically ordered around their convex hull, and that the final tree can be computed by merging *adjacent* clusters, where merging two clusters means adding a longest edge between them. The algorithm in Monma, Paterson, Suri & Yao [1990] runs in $O(n)$ time if all the points lie on their convex hull.

Subquadratic algorithms for higher dimensional maximum spanning trees were obtained a little later by Agarwal, Matoušek & Suri [1992], who proposed randomized algorithms of expected time complexity $O(n^{4/3} \log^{7/3} n)$ for dimension $d = 3$, and $O(n^{2-\alpha_d})$ for dimension $d \geq 4$, where α_d is roughly $2/(\lceil d/2 \rceil + 1)$. Agarwal, Matoušek, Suri [1992] also present a simple approximation algorithm that computes in $O(\varepsilon^{(1-d)/2} n \log^2 n)$ time a spanning tree with total length at least $(1 - \varepsilon)$ times the optimal.

3.3. Applications of minimum and maximum spanning trees

We said earlier that minimum spanning trees have several applications; some are obvious, such as network design problems, and some are less obvious, such as pattern recognition, traveling salesman, and clustering problems. In this section, we mention some constrained clustering problems that can be solved efficiently using minimum and maximum spanning trees.

Given a set of points S in the plane, define a k-partition of S as a decomposition of S into k disjoint subsets $\{C_1, C_2, \ldots, C_k\}$. We want to find a k-partition that *maximizes* the minimum intercluster distance: $\min_{i,j} \min\{d(s,t) \mid s \in C_i, t \in C_j\}$. Asano, Bhattacharya, Keil & Yao [1988] show that an optimal k-partition is found by deleting the $(k-1)$ *longest* edges from the minimum spanning tree of S.

Next, consider the problem of partitioning a point set S into two clusters subject to the condition that the larger of the two diameters is minimized; recall that the *diameter* of a finite set of points is that maximum distance between any two points in the set. An $O(n \log n)$ time solution of this problem was proposed by Asano, Bhattacharya, Keil & Yao [1988], and also by Monma & Suri [1991], based on the maximum spanning tree. The method of Monma and Suri is particularly simple: compute a maximum spanning tree of S and 2-color its nodes (points). The partition induced by the 2-coloring is an optimal minimum diameter 2-partition.

A related bi-partition problem is to minimize the *sum* of measures of the two subsets. Monma & Suri [1991] gave an $O(n^2)$ time algorithm for computing a bi-partition of n points minimizing the sum of diameters. This result was subsequently improved to $O(n \log^2 n)$ time by Hershberger [1991]. An interesting problem in this class is to find a sub-quadratic algorithm for the two-disk covering of a point set with a minimum radius. The relevant results on this problem appear in Hershberger & Suri [1991] and Agarwal & Sharir [1991]; the former gives an $O(n^2 \log n)$ time algorithm to check the feasibility of a covering by two disks of given radii, and the latter gives an $O(n^2 \log^3 n)$ time algorithm for finding the minimum radius. It is an open problem whether a minimum radius two-disk covering of n points can be computed in sub-quadratic time.

Open Problem 3. *Given n points in the plane, give an* $o(n^2)$ *time algorithm for computing the minimum radius r such that all the points can be covered with two disks of radius r; also, find the corresponding disks.*

3.4. Gabriel and relative neighborhood graphs

Nearest neighbor relationships play an important role in pattern recognition problems. One of the simplest graphs encoding these relationships is the *nearest neighbor* graph, which has a (directed) edge from point a to point b if b is a nearest neighbor of a. The minimum spanning tree is a next step, which repeatedly applies the nearest neighbor rule until we obtain a connected graph. Building on this theme, several other classes of graphs have been introduced. We discuss two such graphs in this section: the Gabriel graph and the relative neighborhood graph. The Gabriel graph was introduced by Gabriel & Sokal [1969] in the context of geographical variation analysis, while the relative neighborhood graph was introduced by Toussaint [1990] in a graph-theoretical context. Matula & Sokal [1980] studied several properties of the Gabriel graphs, with applications to zoology and geography. A recent survey paper by Jaromczyk & Toussaint [1992] is a good source for additional information on these graphs. Let us first describe the Gabriel graph.

Let $S = \{s_1, s_2, \ldots, s_n\}$ be a set of points in the plane, and define the *circle of influence* of a pair $s_i, s_j \in S$ as

$$C(s_i, s_j) = \{x \in R^2 \mid d^2(x, s_i) + d^2(x, s_j) = d^2(s_i, s_j)\}.$$

We observe that $C(s_i, s_j)$ is the circle with diameter (s_i, s_j). The *Gabriel* graph $\mathrm{GG}(S)$ has the set of points S as its vertex set, and two vertices s_i and s_j have an edge between them if the circle $C(s_i, s_j)$ does not include any other point of S. In other words, (s_i, s_j) is an edge of $\mathrm{GG}(S)$ if and only if $d^2(s_i, s_k) + d^2(s_j, s_k) \geq d^2(s_i, s_j)$, for all s_k. This definition immediately implies that the Gabriel graph of S is a subgraph of the Delaunay triangulation $\mathrm{DT}(S)$; recall the empty circle definition of Delaunay triangulations (cf. Section 2.4).

Matula & Sokal [1980] give an alternative definition of the Gabriel graph: an edge (s_i, s_j) of $\mathrm{DT}(S)$ is in $\mathrm{GG}(S)$ if and only if (s_i, s_j) intersects its dual edge in the Voronoi diagram of S. This latter characterization leads immediately to an $O(n \log n)$ time algorithm for computing the Gabriel graph: first compute the Delaunay triangulation $\mathrm{DT}(S)$ and then delete all those edges that do not intersect their dual edges in the corresponding Voronoi diagram.

In dimensions $d \geq 3$, the complexity of the Gabriel graph depends on whether or not many points are co-spherical. (Note that this is not the case for Delaunay triangulation.) If no more than a constant number of points lie on a common $(d-1)$-sphere, then GG has only a linear number of edges. Computing this graph in less than quadratic time is still quite non-trivial. Slightly sub-quadratic algorithms are presented in Agarwal & Matoušek [1992]. Without the non-degeneracy assumption, the Gabriel graph can have $\Omega(n^2)$ edges even in three dimensions. The following example gives a lower bound construction for $d = 3$. Take two orthogonal, interlocking circles of radius 2, each passing through the center of the other. In particular, let the first circle lie in the xy-plane with $(0, -1, 0)$ as the center, while the second circle lies in the yz-plane with $(0, 1, 0)$ as the center. Place $n/2$ points on the first circle very close to the point $(0, 1, 0)$, and $n/2$ points on the second circle close to the point $(0, -1, 0)$. Then, it is easy to see that the Gabriel graph of these n points contains the complete bipartite graph between the two sets of $n/2$ points.

Open Problem 4. *Given n points in E^d such that only $O(d)$ points lie on a common sphere, give an $O(c_d n \log n)$ time algorithm to construct their Gabriel graph, where c_d is dimension-dependent constant. Alternatively, prove a lower bound better than $\Omega(n \log n)$.*

The basic construct in the definition of a relative neighborhood graph is the 'lune of influence'. Given two points $s_i, s_j \in S$, their *lune of influence* $L(s_i, s_j)$ is defined as follows:

$$L(s_i, s_j) = \{x \in R^2 \mid \max\{d(x, s_i),\ d(x, s_j)\} \leq d(s_i, s_j)\}.$$

Thus, the lune $L(s_i, s_j)$ is the common intersection of two disks of radius $d(s_i, s_j)$ centered on s_i and s_j. The *relative neighborhood* graph RNG(S) has an edge between s_i and s_j if and only if the lune $L(s_i, s_j)$ does not contain any other point of S. Again, it easily follows that RNG(S) $\subseteq$ DT(S); in fact, the relative neighborhood graph is also a subgraph of the Gabriel graph, since the circle of influence is a subset of the lune of influence. Thus, we have the following ordered relations among the four graphs we have discussed in this section:

$$\text{MST} \subseteq \text{RNG} \subseteq \text{GG} \subseteq \text{DT}.$$

Characterization of the DT edges not in RNG, however, is not so easy as it was for the Gabriel graph. In two dimensions, Supowit [1983] presents an $O(n \log n)$ time algorithm for extracting the RNG from the Delaunay triangulation. If the points form the vertices of a convex polygon, then the minimum spanning tree, relative neighbor graph, Gabriel graph, and Delaunay triangulation can each be computed in linear time. The bound for MST, GG, and DT is implied by a linear time algorithm for computing the Voronoi diagram of a convex polygon [Aggarwal, Guibas, Saxe & Shor, 1989], and the result on RNG is due to Supowit [1983].

In dimensions $d \geq 3$, the size of the relative neighborhood graphs depends on whether or not the points are co-spherical. If only a constant number of points lie on a common $(d-1)$-sphere, then the RNG has $O(n)$ edges, but without this restriction, it is easy to construct examples where RNG has $\Omega(n^2)$ edges in any dimension $d \geq 4$. In R^3, the best upper bound on the size of the relative neighborhood graph currently known is $O(n^{4/3})$ [Agarwal & Matoušek, 1992].

Open Problem 5. *Given a set S of n points in E^3 such that only a constant number of points lie on a common sphere, show that the relative neighborhood graph of S has only $O(n)$ edges. Alternatively, prove a super-linear lower bound on the size of the relative neighborhood graph.*

The size of the relative neighborhood graph is related to the number of bi-chromatic closest pairs [Agarwal & Matoušek, 1992].

Open Problem 6. *Given two unordered sets of points $B, R \subset E^3$, what is the maximum number of pairs (b, r) such that $r \in R$ is a closest neighbor of $b \in B$?*

4. Path planning

4.1. Introduction

The shortest path problem is a familiar problem in algorithmic graph theory. Given a graph $G = (V, E)$, whose edges have non-negative, real-valued weights associated with them, a shortest path between two nodes s and t is a path in G from s to t having the minimum possible total edge weight. The shortest path problem is to find such a path. Generalizations of this basic shortest path problem include the *single source* and the *all-pairs* shortest paths problems; the former asks for shortest paths to all the vertices of G from a specified *source* vertex s, and the latter asks for shortest paths between all pairs of vertices. The best-known algorithm for computing shortest paths is due to Dijkstra [1959]. If properly implemented, his algorithm can find a shortest path between two vertices in time $O(\min(n^2, m \log n))$; here n and m denote the number of vertices and edges of G. A considerable amount of research has been invested in improving this time complexity for sparse graphs, that is, graphs with $m \ll n^2$. Only a few years ago, Fredman & Tarjan [1987] succeeded in devising an implementation of Dijkstra's algorithm, using their Fibonnaci heap data structure, that achieved a worst-case running time of $O(m + n \log n)$; this time bound is optimal in a comparison-based model of computation.

The shortest path problem acquires a new richness when transported to a geometric domain. Unlike a graph, an instance of the geometric shortest path problem is typically specified through the description of some geometric objects that implicitly encode the graph. This raises the following rather interesting question: is it possible to compute a shortest path without explicitly constructing the entire graph? There are some intriguing possibilities associated with this question. For instance, a set of geometric objects can encode some very large, super-polynomial or even exponential, size graphs, implying that an efficient shortest path algorithm must necessarily avoid building the entire graph. Even if the graph is polynomial-size, considerable efficiency gain is possible if the shortest path problem can be solved by constructing only a small, relevant subset of the edges. We will address these issues in more detail later, but for now let us just say that there is a diverse variety of shortest path problems, depending upon the type of geometric objects considered, the metric used, and the dimension of the underlying geometric space. We start with a discussion of some common basic concepts.

4.2. Basic concepts

The most commonly studied shortest path problems in computational geometry typically involve a set of polyhedral objects, called *obstacles*, in Euclidean d-space, $d \geq 2$, and the goal is to find an obstacle-avoiding path of minimum length between two points. Much of our discussion will be limited to shortest paths in the plane ($d = 2$). A connected subset of the plane whose boundary consists of a union of a finite number of straight line segments will be called a *polygonal domain*. The boundary segments are called *edges*; their endpoints are called *vertices*. A polygonal domain P is called a *simple polygon* if it is simply-connected, that is, it is homeomorphic to a disk. A multiply-connected polygonal domain P is also called a *polygon with holes*.

4.2.1. Triangulation

A triangulation of a polygonal domain P is a decomposition of P into triangles with disjoint interiors, with each triangle having its corners among the vertices of P. (If we allow triangles whose corners are *not* among the vertices of P, the triangulation is called a *Steiner* triangulation; we do not use Steiner triangulations in this section.) It is a well-known fact that a polygonal domain can always be triangulated (without using Steiner points). Since a triangulation is a planar graph, the number of triangles is linearly related to the number of vertices of P.

A polygonal domain with n vertices can be triangulated in $O(n \log n)$ time [Preparata & Shamos, 1985]. This time complexity is worst-case optimal in the algebraic tree model of computation. The lower bound, however, does not apply if P is a simple polygon, raising the possibility than a better algorithm might be possible for triangulating a simple polygon. Indeed, the problem of triangulating a simple polygon became one of the most notorious problems in computational geometry in the eighties. Despite the discovery of numerous algorithms, the $O(n \log n)$ time bound remained unbeaten in the worst-case performance. Then in 1988, a breakthrough result by Tarjan and van Wyk [1988] produced an $O(n \log \log n)$ time triangulation algorithm. Finally, Chazelle [1991] recently managed to devise a linear-time algorithm for triangulating a simple polygon, thus settling the theoretical complexity of the problem.

For a polygon with holes, it is possible to perform a triangulation in running time dependent on the number of holes or the number of reflex, i.e., non-convex, vertices. In particular, a polygonal domain P can be triangulated in $O(n \log r)$ time, where r is the number of reflex vertices of P, or in time $O(n + h \log^{1+\varepsilon} n)$, where h is the number of holes in P and ε is an arbitrarily small positive constant [Bar-Yehuda & Chazelle, 1992].

4.2.2. Visibility

Visibility is a key concept in geometric shortest paths. We say that points s and t are (mutually) *visible* if the line segment joining them lies within the polygonal domain P. The relevance to shortest path planning is clear: If points s and t are

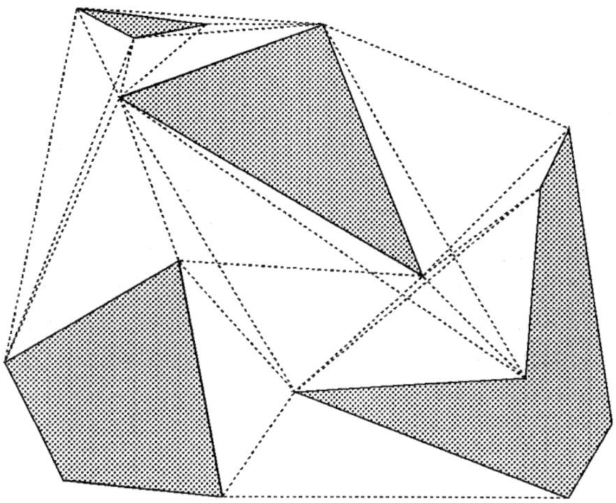

Fig. 4. A visibility graph.

visible to one another, then the shortest obstacle-avoiding path between them is simply the line segment joining them.

The *visibility polygon*, $V(s)$, with respect to a point $s \in P$ is defined to be the set of points that are visible to s. A visibility polygon can be found in time $O(n \log n)$ by applying the *sweep-line* paradigm of computational geometry: We simulate the sweeping of a ray $r(\theta)$ angularly about s, keeping track of the ordered *crossing list* of edges of P intersecting $r(\theta)$. When the ray $r(\theta)$ encounters a vertex of P, we insert and/or delete an edge from the crossing list, and we make any necessary updates to the visibility profile; the cost per update is $O(\log n)$. We can always know which vertex is encountered next by the sweeping ray if we sort the vertices of P angularly about s, in $O(n \log n)$ time. If P has h holes, then a recent result of Heffernan & Mitchell [1991] shows that one can compute a visibility polygon in optimal time, $O(n + h \log h)$.

The *visibility graph* (VG) of P is defined to be the graph whose nodes are the set of vertices of P and whose edges join pairs of vertices that are mutually visible. Refer to Figure 4. We let E_{VG} denote the number of edges in VG; note that $E_{\text{VG}} \leq \binom{n}{2}$ for an n-vertex domain P. Visibility graphs were first introduced in the work of Nilsson [1969], who used them for computing shortest paths for a mobile robot.

The most naive algorithm for computing the VG runs in time $O(n^3)$, checking each pair of vertices (u, v) for visibility by testing against all edges of P. A substantially improved algorithm is possible based on the idea of computing visibility polygon of each vertex. The visibility graph edges incident to a vertex v can be found in $O(n \log n)$ time by first constructing the visibility polygon of v, and hence the entire visibility graph can be computed in $O(n^2 \log n)$, using only $O(n)$ working space.

The state-of-the-art in VG construction remained at the $O(n^2 \log n)$ level until 1985, when Welzl [1985] (and, independently, Asano, Asano, Guibas, Hershberger & Imai [1986]) obtained algorithms whose worst-case running times were $O(n^2)$. These new algorithms rely on the trick of mapping the vertices of P to their *dual* lines, building the *arrangement* of these lines (in time $O(n^2)$ [Edelsbrunner, O'Rourke & Seidel, 1986], and then using the information present in the arrangement to read off the sorted order of vertices about each vertex v in total time $O(n^2)$. Thus, the $O(n)$ angular sorts are not independent of each other, as they can be done collectively in total time $O(n^2)$. Once the angular order is known for vertices about every other vertex, a further trick is necessary to produce the VG without the logarithmic overhead per pair — for example, Welzl [1985] uses a topological sort (available from the arrangement) to guide the construction of the visibility profiles about every vertex. Edelsbrunner & Guibas [1989] have shown how to use a method of 'topological sweep' to compute the VG in time $O(n^2)$ using only $O(n)$ working storage (i.e., avoiding the need to keep the entire line arrangement in memory during VG construction).

In the worst case, we know that it takes quadratic time ($O(n^2)$) to compute a visibility graph, since visibility graphs exist with this size. In some cases, however, the visibility graph is very sparse (linear in size). Thus, ideally, we would like an algorithm whose running time is *output-sensitive*, taking time proportional to the size (E_{VG}) of the output.

Ghosh & Mount [1991] have developed such an *output-sensitive* algorithm, achieving a time bound of $O(n \log n + E_{VG})$, with a working storage requirement of $O(E_{VG})$. Their algorithm begins with a triangulation of P and constructs VG edges by a careful analysis of the properties of 'funnel sequences'. Independently, Kapoor & Maheshwari [1988] obtained a similar bound and also show how one can compute the *subgraph of* VG *relevant for shortest path planning* in time $O(n \log n + E_{SP})$ and space $O(E_{SP})$, where E_{SP} is the size of the resulting subgraph. (In other words, only those edges of the VG that appear along some nontrivial shortest path are actually discovered and output.)

Overmars & Welzl [1988] give two very simple (easily implementable) algorithms for computing the visibility graph that use only $O(n)$ space. The first algorithm runs in time $O(n^2)$ and is based on 'rotation trees'; the second is output-sensitive, requiring time $O(E_{VG} \log n)$. See also Alt & Welzl [1988]. The main open problem in visibility graph construction is summarized below:

Open Problem 7. *Given a polygonal domain with n vertices and h holes, compute the visibility graph in time $O(h \log h + E_{VG})$, where E_{VG} is the number of edges of the resulting graph. Ideally, do this computation using only $O(n)$ working storage.*

Mitchell & Welzl [1990] have developed an on-line algorithm to construct a VG, by showing that one can update a VG when a new obstacle is inserted in time $O(n + k)$, where k is the number of VG edges that must be removed when the new obstacle is inserted. (Note that k may be as large as $\Omega(n^2)$.) Vegter [1990, 1991] shows that a VG can be maintained under both insertions and deletions, in time $O(\log^2 n + K \log n)$ per update, where K is the size of the change in the VG. We are left with an interesting open question:

Open Problem 8. *Devise a dynamic algorithm for maintaining a visibility graph in $O(\log n + K)$ time per insertion or deletion of an obstacle, where K denotes the number of changes in the visibility graph.*

4.3. Shortest obstacle-avoiding paths

The most basic kind of geometric shortest path problem is that of finding a shortest path from s to t for a point robot that is confined to the interior of a polygonal domain P. We assume that P has h holes (which can be thought of as the *obstacles*) and a total of n vertices. In this subsection, we measure path length according to the (usual) Euclidean metric; in the following subsections, we discuss variants on the objective function.

4.3.1. Paths in a simple polygon

Assume that there are no holes in P (i.e., $h = 0$). Then, there is a unique *homotopy* class of any path from s to t, and the shortest path from s to t will be the unique 'taut string' path. If we triangulate polygon P, then there is a unique path in the triangulation dual graph (which is a tree in this case), from the triangle containing s to the triangle containing t. This gives us a *sleeve* within P that is known to contain the shortest s-t path. Chazelle [1982] and Lee & Preparata [1984] show that, in time linear in the number of triangles defining the sleeve, one can 'pull taut' the sleeve, producing the unique shortest s-t path.

The algorithm proceeds incrementally, considering the effect of adding the triangles in order along the sleeve. At a general step of the algorithm, when we are about to add triangle Δabc, we know the shortest path from s to a vertex r (of the sleeve), and the (concave) shortest subpaths from r to a and from r to b, which define a region called the *funnel* with base ab and apex r. Refer to Figure 5. In order to add Δabc, we must 'split' the funnel according to the taut-string path from r to c, which will, in general, include a segment, uc, joining c to some vertex of tangency u along one of the concave chains of the funnel. We need to keep only one of the two funnels (based on ac and ab), since only one can lead through the sleeve to t, which allows us to charge off the work of searching for u to vertices that can be discarded.

Since a simple polygon can be triangulated in linear time [Chazelle, 1991], the result of Chazelle [1982] and Lee & Preparata [1984] establishes that shortest paths in a simple polygon can be found in $O(n)$ time, which is worst-case optimal. This result has been generalized in several directions:

– Guibas, Hershberger, Leven, Sharir & Tarjan [1987] show that one can construct the shortest path tree (and its extension into a 'shortest path map') rooted at a point s in $O(n)$ time, after which the length of a shortest path to any query point t can be reported in $O(\log n)$ time (and the shortest path can be output in time proportional to its size). Their result relies on using 'finger

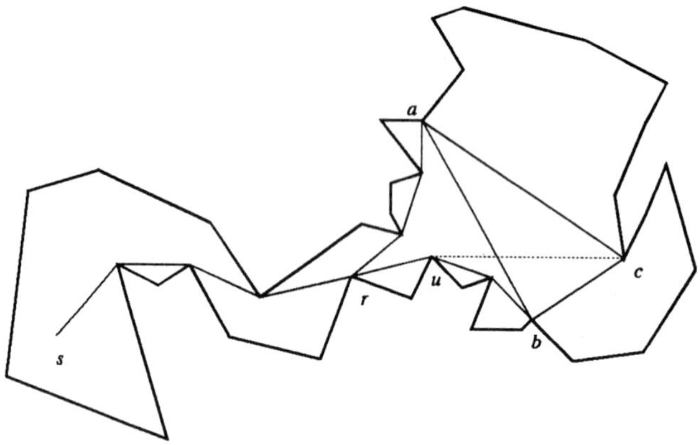

Fig. 5. Splitting a funnel.

search trees' to do funnel splitting, which now must be done *without* discarding either of the two new funnels. Hershberger & Snoeyink [1991] have given a considerably simpler algorithm to compute shortest path trees without any special data structures.

– Guibas & Hershberger [1989] show that a simple polygon can be preprocessed in time $O(n)$, into a data structure of size $O(n)$, such that one can answer shortest path length queries between a pair of points in $O(\log n)$ time. In fact, within the $O(\log n)$ query time, one can construct an implicit representation of the shortest path, so that one can output the path explicitly in time proportional to its length (number of vertices).

– ElGindy & Goodrich [1988] give a parallel algorithm to compute shortest paths in time $O(\log n)$, using $O(n)$ processors (in the CREW PRAM model). Goodrich, Shauck & Guha [1990] show how, with $O(n/\log n)$ processors and $O(\log n)$ time, one can compute a data structure that supports $O(\log n)$ (sequential) time shortest path queries between pairs of points in a simple polygon. They also give an $O(\log n)$ time algorithm using $O(n)$ processors to compute a shortest path tree. Hershberger [1992] builds on the results of Goodrich, Shauck & Guha [1990] and gives an algorithm for shortest path trees requiring only $O(\log n)$ time and $O(n/\log n)$ processors (CREW); he also obtains optimal parallel algorithms for related visibility and geodesic problems.

– Many other problems have been studied with respect to shortest path (*geodesic*) distances within a simple polygon. Aronov [1989] shows how to compute, in time $O(n \log^2 n)$, the Voronoi diagram of a set of point sites in a simple polygon if the metric is the geodesic distance. The *geodesic diameter* is the length of the longest shortest path between a pair of vertices; it can be computed in time $O(n \log n)$ [Suri, 1989; Guibas & Hershberger, 1989]. The *geodesic center* is the point within P that minimizes the maximum of the shortest path lengths to any other point in P; Pollack, Sharir & Rote [1989] give an $O(n \log^2 n)$ algorithm. Suri [1989] studies problems of computing geodesic furthest neighbors. The furthest-site Voronoi diagram for geodesic distance is computed in time $O(n \log n)$ by Aronov, Fortune & Wilfong [1988].

All of the above linear-time algorithms rely on a triangulation of a simple polygon. It's an interesting open problem whether a shortest path inside a simple polygon can be computed optimally without a triangulation.

Open Problem 9. *Given a simple polygon with n vertices, devise an $O(n)$ time algorithm for computing the shortest path between two points without triangulating the polygon.*

4.3.2. Paths in general polygonal spaces

In the general case in which P contains holes (*obstacles*), shortest paths can be computed using the visibility graph, as justified in the following straightforward lemma (proved in Lee [1978] and Mitchell [1986]).

Lemma 2. *Any shortest path from $s \in P$ to $t \in P$ in a polygonal domain P must lie on the visibility graph, VG, of P (where VG includes s and t, in addition to vertices of P, as nodes).*

This lemma implies that, after constructing the VG, we can search for shortest paths in time $O(E_{\mathrm{VG}} + n \log n)$, using Dijkstra's algorithm with appropriate data structures (e.g., Fibonacci heaps [Fredman & Tarjan, 1987] or relaxed heaps [Driscoll, Gabow, Shrairaman & Tarjan, 1988]). The result of Dijkstra's algorithm is a shortest path tree, SPT(s). In practice, it may be faster to apply the A* heuristic search algorithm [e.g., see Pearl, 1984], using the straight-line Euclidean distance as heuristic function, $h(\cdot)$ (which is a lower bound, so it implies an 'admissible' algorithm).

Since VG can be computed in time $O(E_{\mathrm{VG}} + n \log n)$ [Ghosh & Mount, 1991; Kapoor & Maheshwari, 1988], we conclude that Euclidean shortest paths among obstacles in the plane can be computed in time $O(E_{\mathrm{VG}} + n \log n) = O(n^2)$.

Special cases of these results are possible when the obstacles are *convex*, in which case the quadratic term can be written in terms of h (the number of obstacles) rather than n (the number of vertices); see Mitchell [1986] and Rohnert [1986a, b].

Another special case of relevance to VLSI routing problems [see Cole & Siegel, 1984; Leiserson & Maley, 1985; Gao, Jerrum, Kaufmann, Mehlhorn, Rülling & Storb, 1988] is to compute shortest paths among obstacles *of a given homotopy type*. Hershberger & Snoeyink [1991] generalize the shortest path algorithm for simple polygons to show that one can compute a shortest path among obstacles of a particular 'threading' in time proportional to the 'size' of the description of the homotopy type.

Shortest path maps. A *shortest path map*, SPM(s), is an implicit representation of the set of shortest paths from s to *all* points of P. The utility of SPM(s) is that it is a planar subdivision (of size $O(n)$) such that once we perform an $O(\log n)$ time point location query for t, the map tells us the length of a shortest s-t path and allows a path to be reported in time proportional to its size (number of bends). The general concept of shortest path maps applies to all metrics; here, we mention some facts relevant to Euclidean shortest paths among polygonal obstacles in the plane.

If our final goal is to compute a shortest path map, SPM(s), then we can obtain it in $O(n \log n)$ time, *given* the shortest path tree obtained by searching VG with Dijkstra's algorithm [1991].

An alternative approach is to build the (linear-size) SPM(s) directly, and avoid altogether the construction of the (quadratic-size) VG. Lee & Preparata [1984] use this approach to construct a shortest path map in optimal $O(n \log n)$ time for the case of obstacles that are parallel line segments (implying monotonicity of shortest paths with respect to the direction perpendicular to the segments). This approach also leads Reif & Storer [1985] to an $O(hn + n \log n)$ time, $O(n)$ space, algorithm for general polygonal obstacles based on adding the obstacles one-at-a-time, updating the SPM(s) at each step using a shortest path algorithm for simple polygons (without holes).

Mitchell [1991] shows how the Euclidean SPM(s) can be built in $O(kn \log^2 n)$ time, $O(n)$ space, where k is a quantity called the 'illumination depth' (and is bounded above by the number of obstacles touched by a shortest path). This algorithm is based on a general technique for solving geometric shortest path problems, called the *continuous Dijkstra* paradigm [see Mitchell, 1986, 1989, 1990b, 1991, 1992; Mitchell, Mount & Papadimitriou, 1987; Mitchell & Papadimitriou, 1991]. The main idea is to simulate the 'wavefront propagation' that occurs when running Dijkstra's algorithm in the continuum. The continuous Dijkstra paradigm has led to efficient algorithms for a variety of shortest path problems, as we mention later, including shortest paths on polyhedral surfaces [Mitchell, Mount & Papadimitriou, 1987], shortest paths through 'weighted regions' [Mitchell & Papadimitriou, 1991], maximum 'flows' in the continuum [Mitchell, 1990b], and rectilinear paths among obstacles in the plane [Mitchell, 1989, 1992].

A major open question in planar computational geometry is to devise a subquadratic-time algorithm for Euclidean shortest obstacle-avoiding paths. The only known lower bound is the trivial ($\Omega(n + h \log h)$) one.

Open Problem 10. *Given a polygonal domain with n vertices, compute a Euclidean shortest path between two points in* $O(n \log n)$ *time.*

4.4. Other notions of 'short'

Instead of measuring the length of a path as its Euclidean length, several other objective functions are possible, as we describe below.

4.4.1. Rectilinear metric

If we measure path length by the L_1 (or L_∞) metric ($d_1(p, q) = |p_x - q_x| + |p_y - q_y|$ or $d_\infty(p, q) = \max\{|p_x - q_x|, |p_y - q_y|\}$), or require that paths be *rectilinear* (with edges parallel to the coordinate axes), then subquadratic-time algorithms for shortest paths in the plane are known.

For the general case of a polygonal domain P, Mitchell [1989, 1992] shows how to apply the continuous Dijkstra paradigm to build the L_1 (or L_∞) shortest path map in time $O(n \log n)$ (and space $O(n)$).

Clarkson, Kapoor & Vaidya [1987] develop a method based on principles similar the use of visibility graphs in searching for L_2 optimal paths: They construct a sparse graph (with $O(n \log n)$ nodes and edges) that is *path preserving*, meaning that it suffices for searching for shortest paths. This allows them to apply Dijkstra's algorithm, obtaining an $O(n \log^2 n)$ time ($O(n \log n)$ space) algorithm for L_1 shortest paths. Alternatively, this approach yields an $O(n \log^{1.5} n)$ time ($O(n \log^{1.5} n)$ space) algorithm [Clarkson, Kapoor & Vaidya, 1987; Widmayer, 1989].

Fixed orientations and approximations. Methods for finding L_1 shortest paths generalize immediately to the case of *fixed orientation metrics* in which distances are measured in terms of the length of the shortest polygonal path whose links are restricted to a set of k fixed orientations [see Widmayer, Wu & Wong, 1987]. (The L_1 and L_∞ metrics are special cases in which there are four fixed orientations, equally spaced by 90 degrees.) The result is an algorithm for finding shortest obstacle-avoiding paths in time $O(kn \log n)$ [Mitchell, 1989, 1992].

We can apply the above result to get an approximation algorithm for Euclidean shortest paths by noting that the Euclidean metric is approximated to within accuracy $O(1/k^2)$ by the fixed orientation metric with k equally spaced orientations. The result is an algorithm that runs in time $O((n/\sqrt{\epsilon}) \log n)$ to produce a path guaranteed to have length within factor $(1+\epsilon)$ of the Euclidean shortest path length [Mitchell, 1989]. Clarkson [1987] also gives an approximation algorithm, using a related method, that computes an ϵ-optimal path in time $O(n/\epsilon + n \log n)$, after spending $O((n/\epsilon) \log n)$ time to build a data structure of size $O(n/\epsilon)$.

4.4.2. Minimum link paths

In some applications, the number of edges in a path, and not its length, is a more appropriate measure of the path complexity. In real life, for instance, while traveling in an unfamiliar territory, we tend to prefer directions with fewer turns, even at the expense of a slight increase in travel time. The technical motivation for minimizing the number of edges in a path arises from applications in robot motion planning, graph layouts, and telecommunication networks, where straight-line routing is often cheaper and preferable while 'turning' is an expensive operation [Niedringhaus, 1979; Reif & Storer, 1987; Suri, 1986, 1990]. One also encounters minimum link paths in solid modeling, where they are used for curve-compression and the approximation of univariate functions [Natarajan, 1991; Imai & Iri, 1988; Melkman & O'Rourke, 1988; Mitchell & Suri, 1992; Guibas, Hershberger, Mitchell & Snoeyink, 1991].

With this background, let us now formally define the notion of a minimum link path. We concentrate on two dimensions, but extensions to higher dimensions will be obvious. Given a polygonal domain P, a *minimum link path* between two points s and t is a polygonal path with fewest possible number of edges that connects s to t while staying inside P. The *link distance* between s and t, denoted $d_L(s, t)$, is the number of edges in a minimum link path from s to t. (It is possible that there is no path from s to t avoiding all the obstacles, in which case the link distance is defined to be infinite.)

Most of the results on minimum link paths are for the case of a simple polygon, and so we discuss that first.

Minimum link paths in a simple polygon. Like other shortest path problems, a considerable effort has been spent on the simple polygon. In this case, the obstacle space consists of the boundary of a simple polygon P and the free-space consists of the interior of the polygon. Evidently, the notion of link distance is closely related to the notion of visibility. After all, the visibility polygon $V(s)$ consists of precisely the set of points whose link distance to s is one. Building upon this idea, Suri [1986, 1990] introduced the concept of a *window partition* of a polygon. The window partition of P with respect to s is a partition of the interior of P into cells over which the link distance to s is constant. Clearly, $V(s)$ is the cell with link distance 1. The cells with link distance 2 are the regions of $P - V(s)$ that are visible from the *windows* of $V(s)$; a window of $V(s)$ is an edge that forms a boundary between $V(s)$ and $P - V(s)$. The cells with larger link distance are obtained by iterating this procedure. Figure 6 shows an example of a window partition.

Window partitions turn out to be a powerful tool for solving a number of minimum link path problems, both optimally as well as approximately. Suri [1986] presents a linear time algorithm for computing the window partition of a triangu-

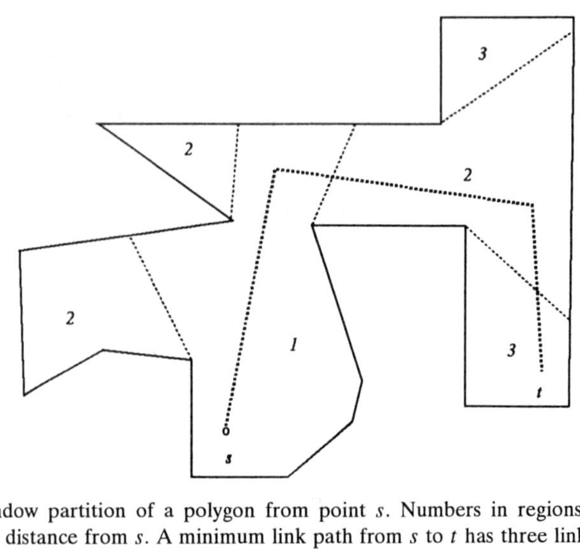

Fig. 6. The window partition of a polygon from point s. Numbers in regions denote their link distance from s. A minimum link path from s to t has three links.

lated simple polygon. Based on this construction, he derives the following results:

(i) The link distance from a fixed point s to all the vertices of P can be computed easily once the window partition from s has been computed: the link distance of a vertex v is k if the cell containing v has label k.

(ii) The window partition is a planar subdivision, which can be preprocessed in linear additional time to allow point location queries in logarithmic time (cf. Section 2.5). With this preprocessing, the link distance from s to a query point t can be determined in $O(\log n)$ time.

(iii) The graph-theoretic dual of the window partition is a tree, called the *window tree*. Suri [1990] observes that distances in the window tree nicely approximate the link distance between points. In particular, he shows how to calculate the link diameter of the polygon within ± 2 links in linear time; the *link diameter* is the maximum link distance between any two points of the polygon. More generally, the link-farthest neighbor of each vertex of P can also be computed within ± 2 links in linear time.

In all the cases above, a minimum link path can always be extracted in time proportional to the link distance. Suri [1987] and Ke [1989] propose $O(n \log n)$ time algorithms for computing the link diameter exactly. Another link-distance related concept that may have applications in shape analysis is *link center*: it is the set of points from which the maximum link distance to any point of P is minimized. Lenhart, Pollack, Sack, Seidel, Sharir, Suri, Toussaint, Whitesides & Yap [1988] proposed an $O(n^2)$ time algorithm, based on window partitions, for computing the link center. This was subsequently improved to $O(n \log n)$, independently by Ke [1989] and Djidjev, Lingas & Sack [1992].

Recently, Arkin, Mitchell & Suri [1992] have developed an $O(n^3)$ space data structure for answering link-distance queries in a simple polygon when *both* s and t are query points. Their data structure stores $O(n^2)$ window partitions. The query algorithm exploits information about the geodesic path between s and t. If it detects that the geodesic path has an *inflection edge* (i.e., the predecessor and the successor edges lie on opposite sides of the edge) or a *rotationally pinned edge* (i.e., the polygon touches the edges from both sides), then the link distance $d_L(s, t)$ is computed by searching the window partitions of the two polygon vertices that are associated with the inflection or pinned edge. If the path has neither an inflection nor a pinned edge, then it must be a *spiral* path, and this is the most complicated case. The query algorithm in this case uses *projection functions*, which are fractional linear forms, to track the other endpoint of a constant-turning path as its first endpoint moves linearly along an edge of P. The query algorithm of Arkin, Mitchell & Suri [1992] works even if s and t are convex polygons instead of just points, however, the query time becomes $O(\log k \log n)$ if the two polygons have a total of k edges. In particular, if the polygons have a fixed number of edges, the query time is asymptotically optimal.

Open Problem 11. *Devise a data structure to answer 2-point link distance queries in a simple polygon. The data structure should use no more than $O(n^2)$ time and space for its construction and answer queries in $O(\log n)$ time.*

Minimum link paths among obstacles. With multiple obstacles, determining the 'homotopy class' of an optimal path becomes a critical problem. Of course, the basic idea behind the window partition still holds: repeatedly compute visibility polygons until the destination point t is reached. However, unlike the simple polygon, where t is always separated from s by a unique window, there are multiple homotopically distinct paths in the general case. It requires a careful pruning technique to overcome a combinatorial explosion. There is essentially one result on link distance among general polygonal obstacles. Mitchell, Rote & Wöginger [1992] present an $O(E_{\mathrm{VG}} \log^2 n)$ time algorithm for finding a minimum link path between two fixed points among a set of polygonal obstacles with a total of n edges, where $E_{\mathrm{VG}} = O(n^2)$ is the size of the visibility graph. The result of Mitchell and coworkers is only a first step; the problem of computing link distances among obstacles is far from solved. The only lower bound known is $\Omega(n \log n)$ [Mitchell, Rote & Woeginger, 1992].

Open Problem 12. *Given a polygonal domain having n vertices, compute a minimum-link path between two given points in time $O(n \log n)$ (or any sub-quadratic bound).*

The assumption of orthogonal obstacles and rectilinear paths results in significant simplifications. De Berg [1991] shows how to preprocess a rectilinear simple polygon in $O(n \log n)$ time and space to support $O(\log n)$ time *rectilinear* link distance between two arbitrary query points. De Berg, van Kreveld, Nilsson & Overmars [1990] develop an $O(n \log n)$ space data structure for answering fixed-source link distance queries among orthogonal obstacles, with a total of n edges. The data structure requires $O(n^2)$ preprocessing time, and can answer a link distance query in $O(\log n)$ time. In fact, their data structure allows for the minimization of a *combined metric*, based on a fixed linear combination of the L_1 length and the link length: the cost of a rectilinear path is its L_1 length plus C times the number of turns, for some pre-specified constant $C \geq 0$. Subsequently, De Berg, Kreveld, and Nilsson [1991] generalized the result of De Berg, van Kreveld, Nilsson & Overmars [1990] to arbitrary dimensions. In d dimensions, their data structure requires $O((n \log n)^{d-1})$ space, $O(n^d \log n)$ preprocessing time, and supports fixed-source link distance queries in $O(\log^{d-1} n)$ time [De Berg, van Kreveld & Nilsson, 1991].

For the general link distance problem in higher dimensions, the only results known are approximations: Mitchell & Piatko [1992] show that one can get within a constant factor (2) of the link distance in polynomial time (for any fixed d).

4.4.3. Weighted regions

A natural generalization of the standard shortest obstacle-avoiding path problem is to consider varied terrain in which each region of the plane is assigned a weight that represents the cost per unit distance of traveling in that region. Clearly, the standard problem fits within this framework if we let obstacles have weight ∞ while free space has weight 1.

We can think of the 'weighted plane' as being a network with an (uncountably) infinite number of nodes, one per point of the plane. We join every pair of points with an edge, assigning a weight equal to the line integral of the weight function along the straight line segment joining the two points.

More formally, we consider the problem in which a planar polygonal subdivision $\mathcal{S}$ is given, with a weight $\alpha \in \{0, 1, \ldots, W, +\infty\}$ assigned to each face of the subdivision. We let n denote the total number of vertices describing the subdivision. Our objective is to find a path π from s to t that has minimum weighted length over all paths from s to t. (The weighted length of a path is given by the path integral of the weight function — it equals the weighted sum of its Euclidean lengths within each region.) This problem of finding an optimal path within a varied terrain is called the *Weighted Region Problem* (WRP), and was introduced by Mitchell & Papadimitriou [1986, 1991].

There are many potential applications of the WRP. The original motivation was to solve the minimum-time path problem for a point robot (without dynamic constraints) moving in a terrain of varied types: grassland, brushland, blacktop, marshland, bodies of water (obstacles to overland travel), and other types of terrain can each be assigned a weight according to the maximum speed at which a mobile robot can traverse the region. In this sense, the weights α denote a 'traversability index,' or the reciprocal of maximum speed.

Mitchell & Papadimitriou [1991] present a polynomial-time solution to the WRP, based on the continuous Dijkstra paradigm, that finds a path guaranteed to be within a factor of $(1 + \epsilon)$ of the optimal weighted length, where $\epsilon > 0$ is any user-specified degree of precision. The time complexity of the algorithm is $O(E \cdot S)$, where E is the number of 'events' in the simulation of Dijkstra's algorithm, and S is the complexity of performing a numerical search to solve the following subproblem: Find a $(1 + \epsilon)$-shortest path from s to t that goes through a given sequence of k edges of $\mathcal{S}$. It is known that $E = O(n^4)$ and that there are examples where E can actually achieve this upper bound (so that no better bound is possible) [Mitchell & Papadimitriou, 1991]. Mitchell and Papadimitriou also show that the numerical search can be done with a form of binary search that exploits the local optimality condition that an optimal path bends according to Snell's Law of Refraction when crossing a region boundary. This leads to a bound of $S = O(k^2 \log(nNW/\epsilon))$ on the time needed to perform a search on a k-edge sequence, where N is the largest integer coordinate of any vertex of the subdivision $\mathcal{S}$. Since one can show that $k = O(n^2)$, this yields an overall time bound of $O(n^8 L)$, where $L = \log(nNW/\epsilon)$ can be thought of as the bit complexity of the problem instance. Although the exponent looks particularly bad, we note that these are truly *worst-case* bounds; in the average case, we might expect that E behaves like n or n^2, and that k is effectively constant.

Many other papers are written on the WRP problem and its special cases; e.g., see Gewali, Meng, Mitchell & Ntafos [1990], Smith, Peng & Gahinet [1988], Alexander [1989] and Alexander & Rowe [1989, 1990]. A recent pair of papers by Kindl, Shing & Rowe [1991a, b] reports practical experience with a simulated annealing approach to the WRP. Papadakis & Perakis [1989, 1990] have generalized the WRP to the case of time-varying maps, where both the weights and the region boundaries may change over time; they obtain generalized local optimality conditions for this case and propose a search algorithm to find good paths.

4.5. Bicriteria shortest paths

The shortest path problem asks for paths that minimize some *one* objective function that measures 'length' or 'cost'. Frequently, however, our application actually wants us to find paths to minimize *two or more* different costs. For example, in mobile robotics applications, we may wish to find a path that simultaneously is short in (Euclidean) length and has few turns.

Multi-criteria optimization problems tend to be hard. Even the bicriteria path problem in a graph is NP-hard [Garey & Johnson, 1979]: Does there exist a path from s to t whose length is less than L and whose weight is less than W? Pseudo-polynomial time algorithms are known, and many heuristics have been devised [e.g., see Handler & Zang, 1980; Henig, 1985].

Several geometric versions of bicriteria shortest path problems have recently been investigated. Various optimality criteria are of interest, including any pair from the following list: Euclidean (L_2) length, rectilinear (L_1) length, other L_p metrics, the number of turns in a path (its link length), the total amount of integrated turning done by a path, etc.

For example, applications in robot motion planning problems may require us to find a shortest (L_2) path constrained to have at most k links. To date, no exact method is known for this problem. Part of the difficulty is that a minimum-link path will not, in general, lie on the visibility graph (or any simple discrete graph).

Arkin, Mitchell & Suri [1992] show that, in a simple polygon, one can always find an s-t path whose link length is within a factor of 2 of the link distance from s to t, while *simultaneously* having Euclidean length within a factor of $\sqrt{2}$ of the Euclidean shortest path length. (A corresponding result is not possible for polygons with holes.)

Mitchell, Piatko & Arkin [1992] study the problem of finding shortest k-link paths in a simple polygon, P. They exploit the local optimality condition on the turning angles at consecutive bends of a shortest k-link path in order to devise a binary search scheme, tracing paths according to this local optimality criterion, in order to find the turning angle at the first bend point. The results of these searches are then combined via dynamic programming recursions to yield an algorithm that produces a path whose length is guaranteed to be within a factor $(1+\epsilon)$ of the length of shortest k-link path, for any user-specified tolerance ϵ. The algorithm runs in time polynomial in n, k and logarithmic in $1/\epsilon$ and the largest integer coordinate of any vertex of P.

For polygons with holes, we pose an interesting open question:

Open Problem 13. *Given a polygonal domain (with holes), what is the complexity of computing a shortest k-link path between two given points? Is it* NP-*complete to decide if there exists a path with at most k links and Euclidean length at most L?*

Several recent papers have addressed the bicriteria path problem for a combination of *rectilinear* link distance and L_1 length, in an environment of rectilinear obstacles. In De Berg, van Kreveld, Nilsson & Overmars [1990] and De Berg, van Kreveld, Nilsson & Overmars [1992], efficient algorithms are given in two and higher dimensions for computing optimal paths according to a 'combined metric,' which takes a linear combination of rectilinear distance and L_1 path length. (Note that this is not the same as solving the problem of computing Pareto-optimal solutions.) Yang, Lee & Wong [1991, 1992] give an $O(n \log^2 n)$ algorithm for computing a shortest k-bend path, a minimum-bend shortest path, or any combined objective that uses a monotonic function of rectilinear link length and L_1 length in a planar rectilinear environment. In all of these rectilinear problems, there is an underlying grid graph which can serve as a 'path preserving graph'. This immediately implies the existence of polynomial-time solutions to the various problems studied by De Berg, van Kreveld, Nilsson & Overmars [1990], De Berg, van Kreveld & Nilsson [1991], and Yang, Lee & Wong [1991, 1992]; the contributions of these papers lie in their clever methods to solve the problems very efficiently.

Some lower bounds on bicriteria path problems have been established by Arkin, Mitchell & Piatko [1991]. In particular, they show that the following geometric versions are NP-hard: (1) Given a polygonal domain, find a path whose L_2 length is at most L, and whose 'total turn' is at most T; (2) Given a polygonal domain, find a path whose L_p length is at most λ_p and whose L_q length is at most λ_q $(p \neq q)$; and (3) Given a subdivision of the plane into red and blue polygonal regions, find a path whose travel through blue (resp. red) is bound by B (resp. R).

4.6. Higher dimensions

While the shortest obstacle-avoiding path problem is solved efficiently in the plane, Canny & Reif [1987; Canny, 1987] show that the problem of finding shortest obstacle-avoiding paths according to any L_p $(1 \leq p \leq \infty)$ metric in three dimensions is NP-hard, even when all of the obstacles are convex polytopes.

The difficulty lies in the structure of shortest paths in three dimensions: They do not (necessarily) lie on any kind of discrete visibility graph. In general, shortest paths in a three-dimensional polyhedral domain P will be polygonal, with bend points that lie *interior* to edges of obstacles. The manner in which a shortest path bends at an edge is well constrained: It must enter and leave at the same angle to the edge. This implies that any locally optimal subpath joining two consecutive obstacle vertices can be 'unfolded' at each obstacle edge that it touches, in such a way that the subpath becomes a straight segment.

The unfolding property of optimal paths can be exploited to yield polynomial-time algorithms in the special case in which the path must stay on a polyhedral *surface*. For the case of a *convex* surface, Sharir & Schor [1986] give an $O(n^3 \log n)$ time algorithm for computing shortest paths. Their algorithm has been improved by Mount [1985], who gives an $O(n^2 \log n)$ time algorithm for the same problem and shows how to use only $O(n \log n)$ space. For the case of shortest paths on a *nonconvex* polyhedral surface, O'Rourke, Suri & Booth [1985] give an $O(n^5)$ time algorithm. Mitchell, Mount & Papadimitriou [1987] improved the time bound to $O(n^2 \log n)$, giving an algorithm based on the continuous Dijkstra paradigm to construct a shortest path map for any given source point on an arbitrary polyhedral surface having n facets. Chen & Han [1990] improve the algorithm of Mitchell, Mount & Papadimitriou [1987], obtaining an $O(n^2)$ time (and $O(n)$ space) bound. (See Aronov & O'Rourke [1991] for the proof of the nonoverlap of the 'star unfolding,' required by Chen & Han [1990].)

For the case when the domain P has only a few convex obstacles, Sharir [1987] has given an $n^{O(k)}$ algorithm for shortest paths, based on a careful analysis of the structure of shortest paths, and a bound of $O(n^7)$ on the number of distinct edge sequences that correspond to shortest paths on the surface of a convex polytope. Mount [1990] has improved the bound on edge sequences to $O(n^4)$, which he shows to be tight. Schevon & O'Rourke [1989] show a tight bound of $\Theta(n^3)$ on the number of *maximal* edge sequences for shortest paths. Agarwal, Aronov, O'Rourke & Schevon [1990] give an $O(n^7 \log n)$ algorithm for computing all $O(n^4)$ edge sequences that correspond to shortest paths on a convex polytope.

For general three-dimensional polyhedral domains P, the best algorithmic results known are approximation algorithms. Papadimitriou [1985] gives a fully polynomial approximation scheme that produces a path guaranteed to be no longer than $(1+\epsilon)$ times the length of a shortest path. His algorithm requires time $O(n^3(L + \log(n/\epsilon))^2/\epsilon)$, where L is the number of bits in an integer coordinate of vertices of P. Clarkson [1987] also gives a fully polynomial approximation scheme, which improves upon that of Papadimitriou [1985] in the case that $n\epsilon^3$ is large.

While three-dimensional shortest path problems are known already to be hard, the proof [Canny & Reif, 1987] is based upon a construction in which the size of the SPM is exponential. This leaves open an interesting algorithmic question of a potentially practical nature, since we may hope that 'in practice' such huge SPM's will not arise:

Open Problem 14. *Given a polyhedral domain in 3 dimensions, compute a shortest path map in output-sensitive time.*

4.7. Kinetics and other constraints

Minimum time paths. Any real mobile robot has a bounded acceleration vector and a maximum speed. If we include these constraints in our model for path planning, then an appropriate objective is to minimize the *time* necessary for a (point) robot to travel from one point of free space to another, with the velocity vector known at the start and possibly constrained at the destination. In general, this kinodynamic planning problem is a very difficult optimal control problem. We are no longer in the nice situation of having optimal paths that are 'taut string' paths, lying on a visibility graph. Instead, the paths will be complicated curves in free space, and the complexity of finding such optimal paths remains open.

In a first step towards understanding the algorithmic complexity of computing time-optimal trajectories under dynamic constraints, Canny, Donald, Reif & Xavier [1988] have produced a polynomial-time procedure for finding a *provably good* approximating time-optimal trajectory that is within a factor of $(1+\epsilon)$ of being a minimum-time trajectory. Their method is fairly straightforward — they discretize the four-dimensional phase space that represents position and velocity. Special care is needed, however, to ensure that the size of the grid is bounded by a polynomial in $1/\epsilon$ and n and the analysis to prove the effectiveness of the resulting paths is quite tedious.

Canny, Rege & Reif [1991] give an *exact* algorithm for computing an optimal path when there is an upper bound on the L_∞ norm of the velocity and acceleration vectors. Their algorithm is based on characterizing a set of 'canonical solutions' (related to 'bang-bang' controls in one dimension) that are guaranteed to include an optimal solution path. Then, by writing an appropriate expression in the first-order theory of the reals, they obtain an exponential time, but polynomial space, algorithm. It remains an open question whether or not a polynomial-time algorithm exists.

Bounded turning radius. Related to the general problem of handling dynamic constraints, is the important problem of finding shortest paths subject to a bound on their *curvature*. Placing a lower bound on the curvature can be thought of as a means of handling an upper bound on the acceleration vector of a point robot whose speed is constant, or can be thought of as the realistic constraint imposed by the fact that many mobile robots have a bounded steering angle.

Fortune & Wilfong [1988] gave an exponential-time decision procedure to determine whether or not it is possible for a robot to move from a start to a goal among a set of given obstacles, while obeying a lower bound on the curvature of its path (and not allowing reversals). If the point following the path *is* allowed to reverse direction, then Laumond [1986] has shown that it is always possible to obtain a bounded curvature path if a feasible path exists.

Since the general problem seems to be extremely difficult, a restricted version has been studied: Wilfong [1988a, b] considers the case in which the robot is to follow a given *network* of lanes, with the robot allowed to turn from one segment to another along a (bounded curvature) circular arc if the two lanes intersect. In Wilfong [1988a], a polynomial-time algorithm is given for producing *some* feasible path; in Wilfong [1988b], the problem of finding a shortest feasible path is shown to be NP-complete, while a polynomial-time method is given for deforming a given feasible path into a shortest equivalent feasible path. (The time bound is $O(k^3n^2)$, where n is the number of vertices describing the obstacles, and k is the number of turns in the path.)

4.8. Optimal robot motion

Most of our discussion is focused on the case of point robots. When the robot is not a point, the problem usually becomes much harder. An exception is the case of a circular robot (which is often a very good assumption anyhow) or a non-rotating convex robot. In the case of a circular robot, the problem of finding a shortest path among obstacles is solved almost as in the point robot case — we simply must 'grow' the obstacles by the radius of the robot and 'shrink' the robot to a point. This is the standard 'configuration space' approach in motion planning, and leads to shortest path algorithms with time bounds comparable to the point robot case [Chew, 1985; Hershberger & Guibas, 1988; Mitchell, 1986].

Optimal motion of rotating non-circular robots is a very hard problem. Consider the simplest case of moving a line segment ('ladder') in the plane. The motion planning problem, which ignores any measure of 'cost' of motion, is solvable in time $O(n^2 \log n)$ [Yap, 1987]. A natural definition of cost of motion for a ladder is to consider the *work* necessary to move the ladder from one place to another, assuming that there is a uniform coefficient of kinetic friction. Optimal motion of a ladder is an open problem at this point: Papadimitriou & Silverberg [1987] and O'Rourke [1987] give solutions for restricted cases of moving a ladder among obstacles, and Icking, Rote, Welzl & Yap [1989] have characterized the solution for the general case *without* obstacles.

Open Problem 15. *Given a polygonal domain, compute an optimal motion of a ladder from one position to another.*

4.9. On-line algorithms and navigation without maps

In all of the path planning problems we have discussed so far, we have assumed that we know in advance the exact layout of the environment in which the robot moves — i.e., we assume we are given a perfect *map*. In most real problems, we cannot make this assumption. Indeed, if we are given a map or floorplan of where walls and obstacles are located, the map will invariably contain inaccuracies, and we may be interested also in being able to navigate around obstacles that may not be in the map. For example, for a robot moving in an office building, while the floorplan and desk layouts may be considered accurate and fixed, the location of a chair or a trashcan is something that we usually cannot assume to be known in advance.

When planning paths in the absence of perfect map information, we must have some model of *sensory inputs* that enable the robot to sense the local structure of its environment. Many different assumptions are possible here: visual sensors, range sensors (perhaps from sonar or computed from stereo imagery), touch sensors, etc. While numerous heuristic methods have been devised for sensor-based autonomous vehicle navigation [see Iyengar & Elfes, 1991], only recently has there been interest in these questions from the theory of algorithms community.

Necessarily the theoretical results require stringent assumptions before anything can be claimed and proven. One of the first papers was by Lumelsky & Stepanov [1987], who show that if a point robot is endowed only with a contact ('tactile') sensor, which can determine when it is in contact with an obstacle, then there is a strategy for 'feeling' one's way from a start to goal such that the resulting path length is at most 1.5 times the total perimeter length of the set of obstacles. (The strategy, called 'BUG2,' is closely related to the strategy of keeping one's hand on the wall when going through a maze.) No assumptions have to be made about the shapes of the obstacles. Lumelsky and Stepanov show that this ratio is (essentially) best possible for this model; see Datta & Krithivasan [1988] for some further work on an extension of the Lumelsky–Stepanov model.

An obvious complaint with the model of Lumelsky & Stepanov [1987] is that it does not bound the *competitive ratio* — the worst-case ratio of the length of the actual path to that of an optimal path. Among the first results that bound the competitive ratio is that of Papadimitriou & Yannakakis [1989], who show that if the obstacles are assumed to be squares, one can achieve a competitive ratio of $(\sqrt{26})/3$, and no strategy can achieve a ratio better than $3/2$. Further, by an adversary argument, they show that, for arbitrary (e.g., 'thin') aligned rectangular obstacles and a robot that has perfect line-of-sight vision, there is no strategy with a bounded competitive ratio. See also Eades, Lin & Wormald [1989].

Blum, Raghavan & Schieber [1991] show that if the obstacles are aligned (disjoint) rectangles in a square, n-by-n room, then there is a strategy using a tactile sensor that achieves competitive ratio $n2^{O(\sqrt{\ln n})}$. Bar-Eli, Berman, Fiat & Yan [1992] give a strategy that achieves competitive ratio $O(n \ln n)$, and show that no deterministic algorithm can yield an asymptotically better ratio (even if the robot is endowed with perfect vision).

Klein [1991] has shown that if one is navigating in a simple polygon of a special structure (called a 'street,' in which it is possible for two guards to traverse the boundary of the polygon, while staying mutually visible and never backing up), then there is a strategy for a robot with perfect visibility sensing to achieve competitive ratio $1 + (3/2)\pi$.

For the problem of finding a short path from s to t among arbitrary unknown obstacles, Mitchell [1990a] has given a method of computing the best possible *local strategy*, assuming that the robot has perfect vision and can remember everything that has been seen so far, and assuming that one assigns a cost per unit distance of some fixed constant, α, for travel in terrain that has not yet been seen.

If, instead of simply asking for a path from s to t, our objective is to traverse a path that allows the entire space to be mapped out, then Deng, Kameda & Papadimitriou [1991] have shown that no competitive strategy exists, in general. If the number of obstacles is bounded, then they give a competitive strategy.

4.10. Motion planning

There is a vast literature on the *motion planning problem* of finding any *feasible* path for a 'robot' moving in a geometrically constrained environment; see for instance Latombe [1991] and Hopcroft, Schwartz & Sharir [1987], and two survey articles Yap [1987] and Schwartz & Sharir [1990]. A general paradigm in this field is to think of the motion of a d-degree-of-freedom robot as described by the motion of a single point in a d-dimensional *configuration space*, C, in which the set of points representing feasible configurations of the system constitute 'free space,' $FP \subset C$.

A simple example of this concept is given by the planar problem of planning the motion of a circular robot among a set of polygonal obstacles: We think of 'shrinking' the robot to a point, while expanding the obstacles by the radius of the robot. The complement of the resulting 'fattened' obstacles represents the free space for the disk. One can use the Voronoi diagram of the set of polygonal obstacles (treating the polygons as the 'sources') to define a graph of size $O(n)$ (computable in time $O(n \log n)$ [Yap, 1987] that can be searched for a feasible path for a disk of any given size. This method, known as the 'retraction' method of motion planning [Yap, 1987], solves this particular instance of the problem in time $O(n \log n)$.

Abstractly, the motion planning problem is that of computing a path between two points in the topological space *FP*. In the first two of five seminal papers on the 'Piano Movers' Problem' (see Schwartz & Sharir [1983a–c, 1984] and Sharir & Ariel-Sheffi [1984], collected in Hopcroft, Schwartz & Sharir [1987]), Schwartz and Sharir show that the boundary of *FP* is a semi-algebraic set (assuming the original constraints of the problem are semi-algebraic). This then allows the motion planning problem to be written as a decision question in the theory of real closed fields [see Tarski, 1951], which can be solved by adding appropriate adjacency information to the *cylindrical decomposition* that is (symbolically) computed by the algorithm of Collins [1975]. For any fixed d and fixed degree of the polynomials describing the constraints, the complexity of the resulting motion planning algorithm is polynomial in n, the combinatorial size of the problem description.

Instead of computing a cell decomposition of *FP*, an alternative paradigm in motion planning is to compute a lower dimensional subspace, $FP' \subseteq FP$, and to define a 'retraction function' that maps *FP* onto FP'. Ó'Dùnlaing & Yap [1985] and Ó'Dùnlaing, Sharir & Yap [1983, 1986, 1987] have computed such retractions on the basis of Voronoi diagrams, obtaining efficient solutions to several low-dimensional motion planning problems.

Most recently, Canny [1987] has described a method of reducing the motion planning problem to a (one-dimensional) graph search problem, by means of a 'roadmap'; this is the currently best known method for general motion planning problems. The bottom line is that the motion planning problem can be solved in polynomial time (polynomial, that is, in the combinatorial complexity of the set of obstacles), for any *fixed* number of degrees of freedom of the robot.

Many lower bounds have also been established on motion planning problems. The first such results were by Reif [1987], who showed that the generalized movers' problem (with many independently movable objects) is PSPACE-hard. Hopcroft, Joseph & Whitesides [1984] give PSPACE-hardness and NP-hardness results for several planar motion planning problems. See also the recent lower bounds paper by Canny and Reif [1987].

5. Matching, traveling salesman, and watchman routes

Matching and traveling salesman are among the best known problems in combinatorial optimization. In this section, we survey some results on these problems where the underlying graph is induced by a geometric input.

5.1. Matching

5.1.1. Graph matching

By a classical result of Edmonds, an optimal weighted matching in a general graph can be computed in polynomial time. Specifically, if $G = (V, E)$ is a graph with real-valued edge weights, then a minimum-weight maximum-cardinality matching in G can be found in polynomial time. Edmonds' is a primal-dual algorithm, which works by growing and shrinking the so-called 'blossoms.' Exactly how these blossoms are maintained and manipulated critically determines the running time of the algorithm. The original algorithm proposed by Edmonds could be implemented to run in worst-case time $O(n^4)$, where $n = |V|$ [Edmonds, 1965]; this was later improved to $O(n^3)$ by Lawler [1976]. The last two decades have witnessed a flurry of research on further improving this time complexity, in particular, for sparse graphs. The latest result on this problem is due to H. Gabow, who presents an algorithm with the worst-case time complexity $O(n(m + n \log n))$, where the graph has n nodes and m edges [Gabow, 1990].

5.1.2. Matching in the Euclidean plane

A natural question from our point of view is this: can the $O(n^3)$ time bound for matching be improved if the graph is induced by a set of points in the plane? In other words, let S be a set of $2n$ points in the plane, and let G be the complete graph on the vertex set S, with the weight of an edge (u, v) being equal to the Euclidean distance between u and v. Does the geometry of the plane constrain an optimal matching sufficiently to admit a faster algorithm?

In the late seventies and early eighties, several conjectures were made regarding the relationship of minimum weight matching and other familiar geometric graphs, such as the Delaunay triangulation or minimum spanning tree [Shamos [1978]. In particular, it was conjectured that a minimum weight perfect matching of a set of points is a subset of the Delaunay triangulation of the points. Since triangulations are planar graphs, the validity of these conjectures would have immediately led to an order-of-magnitude improvement in the running time of the matching algorithm for the geometric case. Unfortunately, these conjectures all turned out to be false. Akl [1983] showed that none of the following graphs is guaranteed to contain a minimum-weight perfect matching: Delaunay triangulation, minimum-weight triangulation, greedy triangulation, minimum-weight spanning tree.

Nevertheless, it turns out that a faster algorithm is possible for the matching of points in the plane. Vaidya [1989] was able to improve the running time of Edmonds' algorithm from $O(n^3)$ to $O(n^{2.5} \log^4 n)$, using geometric data structures and a more careful choice of slack variables. He also gave improvements for bipartite matching and other metrics [Vaidya, 1989]. Vaidya's method depends on efficient solution to a bi-chromatic closest pair problem, where points may be deleted from one set and added to the other. Any improvements to the latter's solution would also improve the matching algorithm's running time.

Marcotte & Suri [1991] considered a special case where all the points are in a convex position, i.e., they form the vertices of a convex polygon. The problem retains much of its complexity even for this restricted class of input, as it can be shown that all the counterexamples of Akl [1983] still hold. But, surprisingly, Marcotte and Suri were able to devise a much simpler and significantly faster algorithm for matching. Their algorithm is based on divide-and-conquer and runs in time $O(n \log n)$. There are two key ideas in their algorithm: an extensibility lemma and vertex weights. The extensibility lemma establishes a geometric condition under which a certain subset of the edges can be immediately added to the optimal matching. The vertex weights are real numbers carefully chosen in such a way that we can invoke the extensibility lemma on the weighted nearest-neighbor graph. The algorithm in Marcotte & Suri [1991] also solves the assignment problem in the same time bound, and it also extends to the case where the points lie on the boundary of a simple polygon and the weight of an edge (u, v) is the length of the shortest path from u to v inside the polygon. X. He [1991] gives a parallel version of the Marcotte–Suri algorithm that runs in $O(\log^2 n)$ time with $O(n)$ processors on a PRAM.

There also are numerous approximation algorithms for matching. For uniform distributions of points, Bartholdi & Platzman [1983] and Dyer & Frieze [1984] describe fast heuristics that give matchings with total weight close to optimal as $n \to \infty$. Vaidya [1989] describes an approximation algorithm that has a guaranteed performance for any input and works for any fixed dimension. His algorithm produces a matching with weight at most $(1 + \varepsilon)$ times the weight of a minimum weight matching, and runs in time roughly $O(n^{1.5} \log^{2.5} n)$; the constant of proportionality is of the order of $(d/\varepsilon)^{1.5d}$, where d is the dimension of input space.

Despite the failure of earlier conjectures relating an optimal matching to other well-known geometric graphs, such as the Delaunay triangulation, it remains a reasonable question whether one can define certain easily constructed, sparse graphs that are guaranteed to contain an optimal geometric matching. The ultimate question, of course, is to determine the extent to which the geometry of the plane can be exploited in the matching problem.

Open Problem 16. *Give an $o(n^2)$ time algorithm for computing a minimum-weight complete matching for a set of $2n$ points in the plane.*

Interestingly, a result of Marcotte & Suri [1991] shows that, for the vertices of a convex polygon, finding a maximum-weight matching is substantially easier than finding a minimum-weight matching. A natural question then is: does the same hold for a general set of points.

Open Problem 17. *Give an $o(n^2)$ time algorithm for computing a maximum-weight complete matching for a set of $2n$ points in the plane.*

5.1.3. Non-crossing matching

There is a celebrated Putnam Competition problem on non-crossing matching [see Larson, 1983]]. Given two sets of points R (red) and B (blue) in the plane, with n points each, find a matching of R and B using straight line segments so that no two segments cross; clearly, we must assume that the points are in general position. There are several proofs of the fact that a non-crossing matching always exists. We give just one: pick a matching that minimizes the sum of all line segment lengths in the matching; by the triangle inequality, no two segments in this matching can cross. Akiyama & Alon [1989] extend this result to arbitrary dimensions: given d sets of points in d-space, each set containing n points, we can always find n pairwise disjoint simplices, each with one vertex from each set.

The algorithmic problem of finding such a matching was first considered by Atallah [1985], who gave an $O(n \log^2 n)$ time algorithm for the two-dimensional problem. Later, Hershberger & Suri [1992] were able to obtain an $O(n \log n)$ time algorithm for the same problem; this time bound is also optimal in the algebraic tree model of computation. Finding a non-intersecting simplex matching in d-dimensions, for $d \geq 3$, remains an open problem.

A minimum-weight matching in the plane is always non-crossing. On the other hand, a maximum-weight matching generally has many crossing edges. An interesting question is to compute a maximum-weight matching with *no crossings*. To the best of our knowledge, no polynomial time algorithm is known for this problem.

Open Problem 18. *Given $2n$ points in general position the plane, find a non-crossing maximum-weight matching.*

A very recent result of Alon, Rajagopalan & Suri [1992] gives a simple and efficient approximation algorithm for the above problem. Their algorithm produces a non-crossing matching of length at least $2/\pi$ times the longest matching, and takes $o(n^{5/2} \log n)$ time. Alternatively, they can find a non-crossing matching of length at least $(2/\pi)(1 - \varepsilon)$ times the optimal in $O(n \log n/\sqrt{\varepsilon})$, for any $\varepsilon > 0$. Somewhat surprisingly, Alon, Rajagopalan & Suri [1992] show that their approximate matching is within $2/\pi$ factor of even the longest *crossing* matching. Similar results are also obtained for the non-crossing *Hamiltonian path* problem and the non-crossing *spanning tree* problem.

5.2. Traveling salesman and watchman routes

It is well-known that the traveling salesman problem remains NP-complete even when restricted to the Euclidean plane [Papadimitriou, 1977]. The best heuristics known for approximating a Euclidean TSP are the same that work for graphs whose edge weights obey the triangle inequality. In particular, a performance ratio of 2 is achieved by double-traversing the MST, and a ratio of 1.5 is achieved by the heuristic of Christofides [1976]. It remains an outstanding open problem whether the ratio of 1.5 can be improved for the geometric problem.

Open Problem 19. *Give a polynomial time algorithm for approximating the Euclidean TSP of n points with a performance ratio strictly less than 1.5.*

Within computational geometry, the TSP has not received much consideration. A slightly related problem that elicited some interest was the question: Does the Delaunay triangulation of a set of points contain its traveling salesman tour? This, not too surprisingly, was answered negatively, first by Kantabutra [1983] for degenerate set of points, and later by Dillencourt [1987] for general-position points. The question of the 'Hamiltonicity' of Delaunay triangulations also arose in the context of pattern recognition and shape representation, in a paper by O'Rourke, Booth & Washington [1987]. Dillencourt [1987] shows that Delaunay triangulation graphs are 1-tough [1], partly explaining why in many practical cases the triangulations turned out to be Hamiltonian.

A problem that ties together the traveling salesman type issues with visibility issues is the *watchman route* problem. Given a polygonal region of the plane (possibly with holes), the problem is to compute a shortest cycle such that every point on the boundary of the region is visible from *some* point on the cycle. If the region is the interior of a simple polygon (without holes), and we are given a starting point through which the route must pass, then Chin & Ntafos [1991] give an $O(n^4)$ algorithm to compute an optimal route; for orthogonal polygons, the time bound improves to $O(n)$. Tan, Hirata & Inagaki [1991] have recently given an $O(n^3)$ algorithm for finding an optimal watchman route through a given point in a simple polygon. However, the problem becomes NP-complete for a polygon with holes or for a simple three-dimensional polyhedron [Chin & Ntafos, 1988]. Other results on watchman route type problems can be found in Ntafos [1990], Kranakis, Krizanc & Meertens [1990], Mitchell & Wynters [1991], and Gewali, Meng, Mitchell & Ntafos [1990].

6. Shape analysis, computer vision, and pattern matching

Applications in computer-aided design, machine vision, and pattern matching all have need to describe, represent, and reason about 'shapes'. Computational geometry has addressed many questions regarding shapes, such as: How can we compare two shapes? How can we detect when a shape is present within a digital image? How can a shape be represented efficiently in order to expedite basic geometric queries (such as intersection detection)?

Here, we think of a *shape* as being the image ('orbit') of some collection of points (countable or uncountable) under the action of some group of transfor-

[1] A connected graph G is called *1-tough* if deletion of any k nodes splits G in to at most k connected components.

mations, $\mathcal{T}$ (e.g., translation, rotation, rigid motions, etc.). Thus, a shape may be represented by a finite collection of points in d-space, or by a polygon, etc. A shape may be given to us in any of a number of forms, including a binary array (of 'pixels' that comprise the shape), a discrete set of points, a boundary description of a solid, a CSG (Constructive Solid Geometry) representation in terms of Boolean set operations on primitive solids (halfspaces), a Binary Space Partition tree, etc. When we speak of the 'complexity' of a shape, we mean the combinatorial size of its representation (e.g., the number of vertices defining the boundary of a polygon).

The fields of computer vision and pattern matching have motivated the study of many shape analysis problems in computational geometry over the last decade. Early work on shape analysis focussed on the use of planar convex hulls [Bhattacharya, 1980; Toussaint, 1980], decompositions of simple polygons into convex pieces [Chazelle, 1987], etc.

In the last five years, effort has concentrated on several problems in shape comparison based on precisely defined notions of distance functions and geometric matching. The goal has been to define a meaningful notion of shape resemblance that is efficiently computable.

6.1. *Shape comparison*

A very natural and general definition of shape distance can be based on the *Hausdorff metric*, which we now define precisely. Let A and B denote two given shapes, and let $\tau \in \mathcal{T}$ denote a transformation (in group $\mathcal{T}$), such as translation and/or rotation, under which we consider shapes to be equivalent. Then, the Hausdorff distance between shapes A and B is defined to be

$$d^{(H)}(A, B) = \min_{\tau \in \mathcal{T}} d_H(A, \tau(B)),$$

where d_H denotes standard Hausdorff distance

$$d_H(A, B) = \max\{\sup_{a \in A} \inf_{b \in B} \delta(a, b), \sup_{b \in B} \inf_{a \in A} \delta(a, b)\},$$

for some underlying distance function δ defined on pairs of points.

The problem of computing the Hausdorff distance between sets of points or between polygons, under various allowed transformations, has been addressed in several recent papers [Agarwal, Sharir & Toledo, 1992; Alt, Behrends & Blömer, 1991; Huttenlocher & Kedem, 1990; Huttenlocher, Kedem & Sharir, 1991; Huttenlocher, Kedem & Kleinberg, 1992; Rote, 1992].

Rote [1992] shows that the Hausdorff distance between two sets of points on the real line can be computed in time $O(n \log n)$, and this is best possible. Huttenlocher & Kedem [1990] show how to compute the Hausdorff distance between two sets of points (of sizes m and n) in the plane under translations τ in time $O((mn)^2 \alpha(mn))$, where $\alpha(\cdot)$ denotes the inverse Ackermann function. Huttenlocher, Kedem & Sharir [1991] improve the time bound to $O(mn(m + n)\alpha(mn) \log mn)$. They also show how to compute the Hausdorff distance between sets of (disjoint) line segments (under translation) in time $O((mn)^2 \log mn)$, assuming the underlying metric δ is L_1 or L_∞. Chew & Kedem [1992] have recently shown that the Hausdorff distance between two point sets can be computed in time $O(n^2 \log^2 n)$, assuming the underlying metric δ is L_1 or L_∞.

Alt, Behrends & Blömer [1991] study the problem of computing Hausdorff distance between simple polygons in the plane, under a variety of possible transformations τ with underlying metric $\delta = L_2$. They give an $O(n \log n)$ algorithm for computing Hausdorff distance between two simple polygons (without transformation), an $O((mn)^3(m+n)\log(m+n))$ algorithm for Hausdorff distance under translation, several algorithms with high-degree polynomial time bounds for various types of transformations, and *approximate* algorithms for these cases that require time $O(nm \log^2 nm)$.

Agarwal, Sharir & Toledo [1992] show how parametric search can be used to improve the complexity to $O((mn)^2 \log^3(mn))$ for the case of comparing two simple polygons under translation (and $\delta = L_2$).

Most recently, Huttenlocher, Kedem & Kleinberg [1992] examine the case of rigid body motions (translation and rotation) of point sets in the plane, and obtain an algorithm with time complexity of $O((m+n)^6 \log(mn))$.

One drawback of the Hausdorff metric for shape comparison is that it measures only the 'outliers' — points that are worst-case.

The polygon metric defined by Arkin, Chew, Huttenlocher, Kedem & Mitchell [1991] avoids some of the problems associated with the Hausdorff metric. Basically, Arkin, Chew, Huttenlocher, Kedem & Mitchell [1991] give an efficient ($O(n^2 \log n)$) means of computing the (L_2) distance between the 'turning functions' of two simple polygons (scaled to have the same perimeter), under all possible shifts of the origins of the parameterizations. (The *turning function* of a polygon measures the accumulated angle of the counterclockwise tangent as a function of the arc-length, starting from some reference point on the boundary.) Rote [1992] has suggested the use of the bounded Lipschitz norm for comparing two single-variable functions (e.g., turning functions), and gives an $O(n \log n)$ time method to compute it. The metrics given by Arkin, Chew, Huttenlocher, Kedem & Mitchell [1991] and Rote [1992] have the disadvantage of not applying as generally as does the Hausdorff; for example, neither metric extends readily to the case of polygons with holes or to higher dimensions.

Alt & Godau [1992] study the so-called *Fréchet-Metric* between curves, and give an $O(mn)$ algorithm to decide if the Fréchet distance between two fixed polygonal chains (of sizes m and n) is less than a given $\epsilon > 0$, and, using this with parametric search, they compute the Fréchet distance between two fixed chains in time $O(mn \log^2 mn)$. They do not optimize over a transformation group; it would be interesting to devise an efficient method to do so.

6.2. *Point pattern matching*

A special case of the shape comparison problem is that of matching two discrete sets of points: Find a transformation of a set of points B that makes it 'match' most closely a set of points A. Matching problems are present throughout the computer vision literature, since their solution forms a fundamental component in many object recognition systems [e.g., see Huttenlocher, 1988].

More precisely, the *point matching* problem in computer vision can be stated as follows: Given a set of n *image* points $A = \{a_1, \ldots, a_n\} \subset R^d$ and a set of m *model points* $B = \{b_1, \ldots, b_m\} \subset R^{d'}$, determine a *matching* μ (i.e., a list of pairs (a_i, b_j) such that no two pairs share the same first element or the same second element) and a transformation $\tau : R^d \to R^{d'}$, within an allowable class of transformations $\mathcal{T}$, such that the application of τ to point a_i brings it into 'correspondence' to point b_j, for each pair $(a_i, b_j) \in \mu$. The 'value' of a matching can be taken to be the number of pairs (a_i, b_j), or possibly a sum of weights.

The term 'correspondence' can take on several different meanings. In the *exact point matching problem* (also known as the 'image registration problem'), we require that $\tau(a_i) = b_j$ for every pair $(a_i, b_j) \in \mu$ of the matching. In the *inexact point matching problem* (also known as the 'approximate congruence problem'), we only require that $\tau(a_i)$ be *close* to b_j, for each $(a_i, b_j) \in \mu$. A natural definition of closeness is to define for each model point b_j, a 'noise region' B_j, and to say that $\tau(a_i)$ is 'close' to b_j if $\tau(a_i) \in B_j$. We let $\mathcal{R} = \{B_1, \ldots, B_m\}$ denote the set of noise regions. Refer to Figure 7.

The exact point matching problem has been solved in time $O(n^{d-2} \log n)$ for $d = d'$ and $\mathcal{T}$ the set of congruences (translations and rotations) [see Alt, Mehlhorn, Wagener & Welzl, 1988].

Baird [1984] formalizes the inexact point matching problem and provides algorithms for the case of similarity transformations and convex polygonal noise regions; his algorithms are worst-case exponential, and he leaves open the question of solving the problem in polynomial time. This open question is resolved in the work of Alt, Mehlhorn, Wagener & Welzl [1988] and the work of Arkin, Mitchell & Zikan [1989], where it is shown that many versions of the inexact matching problem can be solved in polynomial time, for various assumptions about the allowed transformations and the shapes of the noise regions. Arkin and coworkers also give lower bounds on the number of possible matches and generalize the problem to allow arbitrary piecewise-linear cost functions for the matching. Arkin,

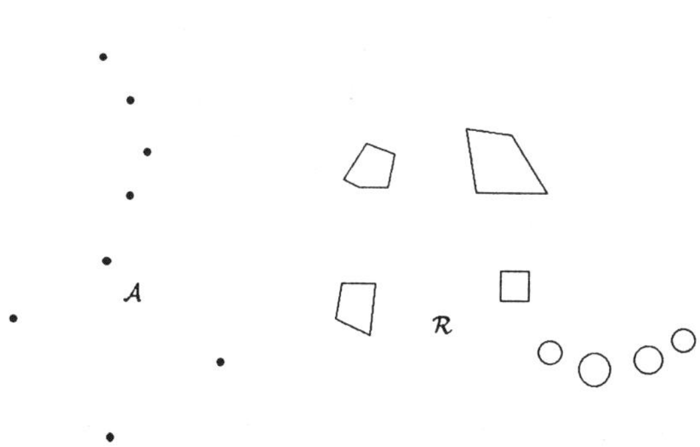

Fig. 7. A point matching problem.

Kedem, Mitchell, Sprinzak & Werman [1992] give improved algorithms and combinatorial bounds on the number of matches for several special cases of the inexact point matching problem in which the noise regions are assumed to be disjoint.

A major obstacle to making the existing methods of point matching practical is the very high degree polynomial time bounds. For example, even for the case of point matching under translation, the algorithm of Alt, Mehlhorn, Wagener & Welzl [1988] requires time $O(n^6)$. One possible direction for an improvement has been suggested by Heffernan & Schirra [1992], who show one can get low-degree polynomials (in n) if one allows an *approximate* decision procedure, which is allowed to give an 'I don't know' answer in response to situations in which the data is particularly 'bad'.

Zikan [1991] and Aurenhammer, Hoffmann & Aronov [1992] have studied problems in which the objective function is based on least-squares.

6.3. Shape approximation

A requirement for any system that analyzes physical models is the representation of geometric data, such as points, polygons, polyhedra, and general solids. One would like to have as compact a representation as possible, while still capturing the degree of precision required by the problem at hand. In particular, this issue is important for real-time systems whose algorithms have running times that depend on the size of the data according to some high degree polynomial (e.g., vision systems, motion planning systems, etc.).

For example, cartographers are interested in the question of approximating general planar objects, such as polygons with holes, sets of polygons, or general planar maps. A geographic terrain map may have millions of polygonal cells, some of which are large and open, others of which are tiny or quite contorted. Such would be the case if we were to look at an agricultural use map of the United States or at a segmentation of a digitized image. But, if we were to put on a pair of 'ϵ-blurring eyeglasses', what we would see in such a map is a subdivision with a few 'large' (in comparison with ϵ) cells, and blurred 'gray masses' where the cell structure is quite fine (in comparison with ϵ). Refer to Figure 8. We would like to replace the original subdivision with a new one of lower resolution (or perhaps a hierarchy of many different resolutions).

A standard approach to the map simplification problem is to take each polygonal curve that defines a boundary in the map and replace it by a simpler one, subject to the new curve being 'close' to the original curve. Cartographers have been interested in this 'line simplification problem' for some time [Douglas & Peuker, 1973; McMaster, 1987]. Computational geometers have defined and solved several instances of the problem; see Guibas, Hershberger, Mitchell & Snoeyink [1991], Hershberger & Snoeyink [1991], Imai & Iri [1986a, b, 1988], Melkman & O'Rourke [1988]. The general method has been to model the problem as an 'ordered stabbing' question, in which one wants to pass a polygonal curve through an ordered set of 'fattened' boundary elements (e.g., disks centered on vertices) from the original curve.

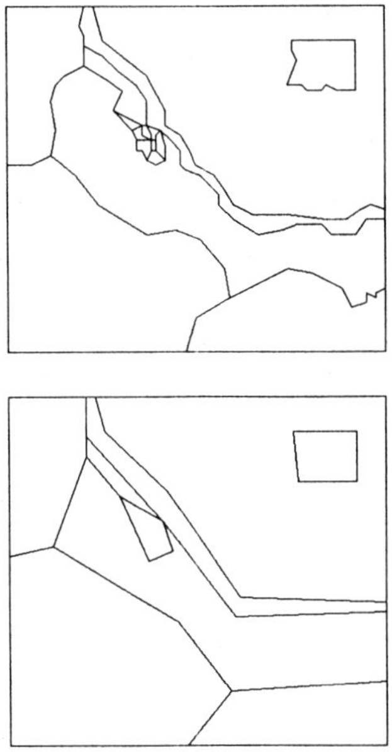

Fig. 8. The original map (top) and its simplification (bottom).

Guibas, Hershberger, Mitchell & Snoeyink [1991] have noted that simplifying each boundary curve of a map individually can cause all kinds of topological inconsistencies, such as islands becoming inland, intersections among boundaries that were previously disjoint, etc. Even the special case of the cartographer's problem in which one wants to approximate a *single* simple polygon, P, suffers from the potential problem that the approximating curve is not simple.

In particular, consider an 'ϵ-fattening' of the boundary of P to be the set of all ('gray') points within distance ϵ of the boundary of P. The boundary of the gray region can be computed in time $O(n \log n)$ by finding the Voronoi diagram of P. If the fattened gray region is an annulus, then we are lucky: The minimum-link cycle algorithms of Aggarwal, Booth, O'Rourke, Suri & Yap [1989], Wang and Chan [1986], or Ghosh & Maheshwari [1990] can be applied to give an exact answer to the problem in $O(n \log n)$ or $O(n)$ time. For larger values of ϵ, however, the fattening may create more holes, in which case, one wants a minimum-vertex *simple* polygon surrounding all the holes of the fattened region. Guibas, Hershberger, Mitchell & Snoeyink [1991] give an $O(n \log n)$ time algorithm to compute a simple polygon with at most $O(h)$ vertices more than optimal, where h is the number of holes in the fattening; they conjecture that the exact solution of the problem is NP-hard.

Mitchell & Suri [1992] have studied a related problem of finding a minimum-link subdivision that separates a given set of polygons. They give an $O(n \log n)$ time algorithm, based on computing minimum-link paths in the 'moats' between polygons, that produces an approximating subdivision (or a *separating family*) that is guaranteed to be within a constant factor of optimality. The exact solution of the problem has been shown to be NP-hard by Das & Joseph [1990b].

Polyhedral separation/approximation

The generalization of the boundary approximation problem to three dimensions is of primary importance for any real CAD applications. If we are given a polyhedral surface, how can we approximate it with a significantly simpler polyhedral surface?

One approach is to 'ϵ-fatten' the original surface and then look at simplifying surfaces that lie within the fattened region. Thus, we ask the following polyhedral separation question: Given two polyhedral surfaces, P and Q, find a polyhedral surface Σ of minimum facet complexity that separates P from Q.

Das & Joseph [1990, 1992; Das, 1990] have shown that this problem is NP-hard, even for convex surfaces P and Q. Mitchell & Suri [1992] have shown that if P and Q are convex, one can, in time $O(n^3)$, compute a separating surface whose facet complexity is guaranteed to be within a small, logarithmic, factor of the size of an optimal separator. While the preliminary results of Mitchell & Suri [1992] are interesting as a first step, many questions remain to be addressed, particularly with respect to nonconvex surfaces.

Open Problem 20. *Given two nonconvex polyhedra P and Q, with a total of n faces, find a polyhedral surface of $f(n)$ faces that separates P from Q such that $f(n)$ is within a small factor of the optimal.*

7. Conclusion

In this survey, we touched upon some of the major problem areas and techniques of computational geometry. Our emphasis was on optimization problems that should be of most interest to the Operations Research community. We certainly have not done justice to the field of computational geometry as a whole, and have left out entire subareas of intense research. But we hope to have supplied sufficient pointers to the literature that an interested reader can track down more detailed information on any particular subtopic.

Computational geometry is a very young discipline, and while it has matured extremely rapidly in the last ten years, we expect a steady stream of new results to continue. Particularly, as the interaction between more applied fields and computational geometry grows, entire new lines of investigation are expected to evolve.

Acknowledgements

We thank Joseph O'Rourke and Godfried Toussaint for several helpful comments that have improved the presentation of this survey.

Research is partially supported by grants from Boeing Computer Services, Hughes Research Laboratories, Air Force Office of Scientific Research contract AFOSR-91-0328, and by NSF Grants ECSE-8857642 and CCR-9204585.

References

Agarwal, P.K., H. Edelsbrunner, O. Schwarzkopf and E. Welzl (1991). Euclidean minimum spanning trees and bichromatic closest pairs, *Discrete Computational Geometry* 6, 407–422.

Agarwal, P.K., and J. Matoušek (1992). Relative neighborhood graphs in three dimensions. *Comput. Geom. Theory Appl.* 2(1), 51–14.

Agarwal, P.K., J. Matoušek and S. Suri (1992). Farthest neighbors, maximum spanning trees and related problems in higher dimensions, *Computational Geometry: Theory Appl.* 1, 189–201.

Agarwal, P.K., and M. Sharir (1994). Planar geometric location problems and maintaining the width of a planar set, *Algorithmica*, 11, 185–195.

Agarwal, P.K., M. Sharir and S. Toledo (1992). Applications of parametric searching in geometric optimization. *Proc. 3rd ACM–SIAM Symp. on Discrete Algorithms*, pp. 72–82.

Aggarwal, A., H. Booth, J. O'Rourke, S. Suri and C.K. Yap (1989). Finding minimal convex nested polygons, *Inf. Comput.* 83(1), 98–110, October

Aggarwal, A., L.J. Guibas, J. Saxe and P.W. Shor (1989). A linear-time algorithm for computing the Voronoi diagram of a convex polygon, *Discrete Comput. Geom.* 4, 591–604.

Akiyama, J., and N. Alon (1989). Disjoint simplices and geometric hypergraphs, in: G.S. Blum, R.L. Graham, and J. Malkevitch (eds.), *Combinatorial Mathematics; Proc. Third Int. Conf.*, New York, 1985, *Ann. NY Acad. Sci.* 555, 1–3.

Akl, S. (1983). A note on Euclidean matchings, triangulations and spanning trees. *J. Comb. Inf. Systems Sci.* 8(3), 169–174.

Alexander, R. (1989). Construction of optimal-path maps for homogeneous-cost-region path-planning problems. Ph.D. Thesis, Computer Science, U.S. Naval Postgraduate School, Monterey, CA.

Alexander, R., and N. Rowe (1989). Geometrical principles for path planning by optimal-path-map construction for linear and polygonal homogeneous-region terrain, Technical Report, Computer Science, U.S. Naval Postgraduate School, Monterey, CA.

Alexander, R.S., and N.C. Rowe (1990). Path planning by optimal-path-map construction for homogeneous-cost two-dimensional regions, in: *Proc. IEEE Int. Conf. on Robotics and Automation*, Cincinnati, OH, May 1990, pp. 1924–1929.

Alon, N., S. Rajagopalan and S. Suri (1993). Long non-crossing configurations in the plane, in: *Proc. 9th Annual ACM Sympos. Comput. Geom.*, pp. 257–263.

Alt, H., B. Behrends and J. Blömer (1991). Approximate matching of polygonal shapes, *Proc. 7th Annual ACM Symp. on Computational Geometry*, pp. 186–193.

Alt, H., and M. Godau (1992). Measuring the resemblance of polygonal curves, *Proc. 8th Annual ACM Symp. on Computational Geometry*, pp. 102–109.

Alt, H., K. Mehlhorn, H. Wagener and E. Welzl (1988). Congruence, similarity and symmetries of geometric objects, *Discrete Computat. Geom.* 3, 237–256.

Alt, H., and E. Welzl (1988). Visibility graphs and obstacle-avoiding shortest paths. *Z. Oper. Res.* 32, 145–164.

Arkin, E.M., L.P. Chew, D.P. Huttenlocher, K. Kedem and J.S.B. Mitchell (1991). An efficiently computable metric for comparing polygonal shapes. *IEEE Trans. Pattern Anal. Mach. Intell.* 13(3), 138–148.

Arkin, E.M., K. Kedem, J.S.B. Mitchell, J. Sprinzak and M. Werman (1992). Matching points into pairwise-disjoint noise regions: combinatorial bounds and algorithms. *ORSA J. Comput.* 4(4), 375–386.

Arkin, E.M., J.S.B. Mitchell and C.D. Piatko (1991). Bicriteria shortest path problems in the plane,

Proc. 3rd Can. Conf. on Computational Geometry, pp. 153–156.
Arkin, E.M., J.S.B. Mitchell and S. Suri (1992). Optimal link path queries in a simple polygon, *Proc. 3rd ACM–SIAM Symp. on Discrete Algorithms*, pp. 269–279. To appear: *Internat. J. Comput. Geom. Appl.*
Arkin, E.M., J.S.B. Mitchell and K. Zikan (1989). Algorithms for point matching problems. Manuscript, School Oper. Res. Indust. Engrg., Cornell Univ., Ithaca, NY.
Aronov, B. (1989). On the geodesic Voronoi diagram of point sites in a simple polygon. *Algorithmica* 4, 109–140.
Aronov, B., H. Edelsbrunner, L. Guibas and M. Sharir (1992). The number of edges of many faces in a line segment arrangement. *Combinatorica* 12(3), 261–274.
Aronov, B., S.J. Fortune and G. Wilfong (1993). Furthest-site geodesic Voronoi diagram, *Discrete Comput. Geom.* 9, 217–255.
Aronov, B., J. Matoušek and M. Sharir (1994). On the sum of squares of cell complexities in hyperplane arrangements, *J. Combin. Theory Ser. A* 65, 311–321.
Aronov, B., and J. O'Rourke (1992). Nonoverlap of the star unfolding, *Discrete Comput. Geom.* 8, 219–250.
Asano, Ta., Te. Asano, L.J. Guibas, J. Hershberger and H. Imai (1986). Visibility of disjoint polygons. *Algorithmica* 1, 49–63.
Asano, Te., B. Bhattacharya, J.M. Keil and F. Yao (1988). Clustering algorithms based on minimum and maximum spanning trees, *Proc. 4th Annual ACM Symp. on Computational Geometry*, pp. 252–257.
Atallah, M. (1985). A matching problem in the plane. *J. Comput. Systems Sci.* 31, 63–70.
Aurenhammer, F. (1991). Voronoi diagrams: A survey of a fundamental geometric data structure. *ACM Comput. Surv.* 23, 345–405.
Aurenhammer, F., F. Hoffmann and B. Aronov (1992). Minkowski-type theorems and least-squares partitioning, *Proc. 8th Annual ACM Symp. on Computational Geometry*, pp. 350–357.
Baird, H.S. (1984). *Model-Based Image Matching Using Location*. Distinguished Dissertation Series. MIT Press.
Bar-Eli, E., P. Berman, A. Fiat and P. Yan (1992). On-line navigation in a room, *Proc. 3rd ACM-SIAM Symp. on Discrete Algorithms*, Orlando, FL, pp. 237–249.
Bartholdi, J.J., III and L.K. Platzman (1983). A fast heuristic based on spacefilling curves for minimum-weight matching in the plane. *Inf. Process. Lett.* 17, 177–180.
Bar-Yehuda, R., and B. Chazelle (1992). Triangulating a set of non-intersecting and simple polygonal chains. Manuscript, Computer Science, Tel-Aviv University.
Ben-Or, M. (1983). Lower bounds for algebraic computation trees, *Proc. 15th Annual ACM Symp. on Theory of Computing*, pp. 80–86.
Bhattacharya, B.K. (1980). Applications of computational geometry to pattern recognition problems. Ph.D. Thesis, School Comput. Sci., McGill Univ., Montreal, PQ.
Bhattacharya, B.K., and G.T. Toussaint (1985). On geometric algorithms that use the furthest-point Voronoi diagram, in: G.T. Toussaint (ed.), *Computational Geometry*, North-Holland, Amsterdam, pp. 43–61.
Blum, A., P. Raghavan and B. Schieber (1991). Navigating in unfamiliar geometric terrain, *Proc. 23rd Annual ACM Symp. on Theory of Computing*, pp. 494–503.
Brown, K.Q. (1980). Geometric transforms for fast geometric algorithms, Ph.D. Thesis and Report CMU-CS-80-101, Dept. Comput. Sci., Carnegie-Mellon Univ., Pittsburgh, PA.
Canny, J. (1987). The complexity of robot motion planning. Ph.D. Thesis, Electrical Engineering and Computer Science, Massachusetts Institute of Technology.
Canny, J., B.R. Donald, J. Reif and P. Xavier (1988). On the complexity of kinodynamic planning, *Proc. 29th Annual IEEE Symp. on Found. Comput. Sci.*, pp. 306–316.
Canny, J., A. Rege and J. Reif (1991). An exact algorithm for kinodynamic planning in the plane. *Discrete Comput. Geom.* 6, 461–484.
Canny, J., and J.H. Reif (1987). New lower bound techinques for robot motion planning problems, *Proc. 28th Annual IEEE Symp. on Found. Comput. Sci.*, pp. 49–60.

Chand, D.R., and S.S. Kapur (1970). An algorithm for convex polytopes. *J. ACM* 17, 78–86.

Chazelle, B. (1982). A theorem on polygon cutting with applications, *Proc. 23rd Annual IEEE Symp. on Found. Comput. Sci.*, pp. 339–349.

Chazelle, B. (1987). Approximation and decomposition of shapes, in: J.T. Schwartz and C.-K. Yap (eds.), *Advances in Robotics, 1: Algorithmic and Geometric Aspects of Robotics*, Lawrence, Erlbaum Associates, Hillsdale, NJ, pp. 145–185.

Chazelle, B. (1991). An optimal convex hull algorithm and new results on cuttings, *Proc. 32nd Annual IEEE Symp. on Found. Comput. Sci.*, pp. 29–38.

Chazelle, B. (1991). Triangulating a simple polygon in linear time. *Discrete Comput. Geom.* 6, 485–524.

Chazelle, B., and H. Edelsbrunner (1992). An optimal algorithm for intersecting line segments in the plane. *J. ACM* 39, 1–54.

Chazelle, B., and J. Friedman (1990). A deterministic view of random sampling and its use in geometry. *Combinatorica* 10, 229–249.

Chazelle, B., and J. Matoušek (1992). Derandomizing an output-sensitive convex hull algorithm in three dimensions, Technical Report, Dept. Comput. Sci., Princeton Univ.

Chen, J., and Y. Han (1990). Shortest paths on a polyhedron, *Proc. 6th Annual ACM Symp. on Computational Geometry*, pp. 360–369.

Cheriton, D., and R.E. Tarjan (1976). Finding minimum spanning trees. *SIAM J. Comp.* 5, 724–742.

Chew, L.P. (1985). Planning the shortest path for a disc in $O(n^2 \log n)$ time, *Proc. 1st Annual ACM Symp. on Computational Geometry*, pp. 214–220.

Chew, L.P., and K. Kedem (1992). Improvements on geometric pattern matching problems, in: *Proc. 3rd Scand. Workshop Algorithm Theory*, Lecture Notes in Computer Science, Vol. 621, Springer-Verlag, pp. 318–325.

Chin, W., and S. Ntafos (1988). Optimum watchman routes. *Inf. Process. Lett.* 28, 39–44.

Chin, W.-P., and S. Ntafos (1991). Watchman routes in simple polygons. *Discrete Comput. Geom.*, 6(1), 9–31.

Christofides, N. (1976.) Worst-case analysis of a new heuristic for the traveling salesman problem, in: J.F. Traub (ed.), *Proc. Symp. on New Directions and Recent Results in Algorithms and Complexity*, Academic Press, New York, NY, pp. 441.

Clarkson, K.L. (1987). Approximation algorithms for shortest path motion planning, *Proc. Annual ACM Symp. on Theory of Computing*, pp. 55–65.

Clarkson, K.L. (1987). New applications of random sampling in computational geometry, *Discrete Comput. Geom.* 2, 195–222.

Clarkson, K.L., S. Kapoor and P.M. Vaidya (1987). Rectilinear shortest paths through polygonal obstacles in $O(n(\log n)^2)$ time, *Proc. 3rd Annual ACM Symp. on Computational Geometry*, pp. 251–257.

Clarkson, K.L., and P.W. Shor (1989). Applications of random sampling in computational geometry, II. *Discrete Comput. Geom.* 4, 387–421.

Cole, R., and A. Siegel (1984). River routing every which way but loose, *Proc. 25th Annual IEEE Symp. on Found. Comput. Sci.*, pp. 65–73.

Collins, G.E. (1975.) Quantifier elimination for real closed fields by cylindric algebraic decomposition, in: *Proc. 2nd GI Conf. on Automata Theory and Formal Languages*, Lecture Notes in Computer Science, Vol. 33, Springer-Verlag, Berlin, pp. 134–183.

Das, G. (1990). Approximation schemes in computational geometry. Ph.D. Thesis, Computer Science University of Wisconsin.

Das, G., and D. Joseph (1990). The complexity of minimum convex nested polyhedra, *Proc. 2nd Can. Conf. on Computational Geometry*, pp. 296–301.

Das, G., and D. Joseph (1992). Minimum vertex hulls for polyhedral domains, *Theoretical Comp. Sci.* 103, 107–135.

Datta, A., and K. Krithivasan (1988). Path planning with local information, in: *Proc. Conf. Found. Softw. Tech. Theoret. Comput. Sci.*, New Delhi, India, December 1988, Lecture Notes in Computer Science, Vol. 338, Springer-Verlag, Berlin, pp. 108–121.

de Berg, M. (1991). On rectilinear link distance. *Comput. Geom. Theory Appl.* 1, 13–34.

de Berg, M., M. van Kreveld, B.J. Nilsson and M.H. Overmars (1990). Finding shortest paths in the presence of orthogonal obstacles using a combined L^1 and link metric, in: *Proc. 2nd Scand. Workshop Algorithm Theory*, Lecture Notes in Computer Science, Vol. 447, Springer-Verlag, Berlin, pp. 213–224.

de Berg, M., M. van Kreveld, B.J. Nilsson and M.H. Overmars (1992). Shortest path queries in rectilinear worlds. *Int. J. Comput. Geom. Appl.* 2(3), 287–309.

Deng, X., T. Kameda and C. Papadimitriou (1991). How to learn an unknown environment, in: *Proc. 32nd Annual IEEE Symp. on Found. Comput. Sci.*, pp. 298–303.

Devroye, L., and G.T. Toussaint (1981). A note on linear expected time algorithm for finding convex hulls. *Computing* 26, pp. 361–366.

Dijkstra, E.W. (1959). A note on two problems in connexion with graphs. *Numer. Math.* 1, 269–271.

Dillencourt, M.B. (1987). A non-Hamiltonian, nondegenerate Delaunay triangulation, *Inf. Process. Lett.* 25, 149–151.

Djidjev, H.N., A. Lingas and J. Sack (1992). An $O(n \log n)$ algorithm for computing the link center of a simple polygon, *Discrete Comput. Geom.* 8(2), 131–152.

Douglas, D.H., and T.K. Peuker (1973). Algorithms for the reduction of the number of points required to represent a line or its caricature. *Can. Cartograph.* 10(2), 112–122.

Driscoll, J.R., H.N. Gabow, R. Shrairaman and R.E. Tarjan (1988). Relaxed heaps: An alternative to Fibonacci heaps with applications to parallel computation. *Commun. ACM* 31, 1343–1354.

Dwyer, R.A. (1987). A faster divide-and-conquer algorithm for constructing Delaunay triangulations. *Algorithmica* 2, 137–151.

Dwyer, R.A. (1988). Average-case analysis of algorithms for convex hulls and Voronoi diagrams. Ph.D. Thesis, Carnegie-Mellon University.

Dyer, M.E., and A.M. Frieze (1984). A partitioning algorithm for minimum weighted Euclidean matching. *Inf. Process. Lett.* 18, 59–62.

Eades, P., X. Lin and N.C. Wormald (1989). Performance guarantees for motion planning with temporal uncertainty, Technical Report, Dept. of Computer Science, Univ. of Queensland, St. Lucia, Queensland.

Edelsbrunner, H. (1987). *Algorithms in Combinatorial Geometry*. Springer-Verlag, Heidelberg.

Edelsbrunner, H., L. Guibas and M. Sharir (1990). The complexity of many cells in arrangements of planes and related problems. *Discrete Comput. Geom.* 5, 197–216.

Edelsbrunner, H., and L.J. Guibas (1989). Topologically sweeping an arrangement. *J. Comput. Syst. Sci* 38, 165–194 [Corrigendum in (1991), 42, 249–251].

Edelsbrunner, H., L.J. Guibas and M. Sharir (1990). The complexity and construction of many faces in arrangements of lines and of segments. *Discrete Comput. Geom.* 5, 161–196.

Edelsbrunner, H., L.J. Guibas and J. Stolfi (1986). Optimal point location in a monotone subdivision. *SIAM J. Comput.* 15, 317–340.

Edelsbrunner, H., J. O'Rourke and R. Seidel (1986). Constructing arrangements of lines and hyperplanes with applications. *SIAM J. Comput.* 15, 341–363.

Edelsbrunner, H., and R. Seidel (1986). Voronoi diagrams and arrangements. *Discrete Comput. Geom.* 1, 25–44.

Edelsbrunner, H., and E. Welzl (1986). On the maximal number of edges of many faces in an arrangement. *J. Comb. Theory Ser. A* 41, 159–166.

Edmonds, J. (1965). Maximum matching and a polyhedron with $0, 1$ vertices. *J. Res. NBS* 69B, 125–130.

ElGindy, H., and M.T. Goodrich (1988). Parallel algorithms for shortest path problems in polygons, *Visual Comput.* 3, 371–378.

Fortune, S., and G. Wilfong (1988). Planning constrained motion, *Proc. 20th Annual ACM Symp. on Theory of Computing,* pp. 445–459.

Fortune, S.J. (1987). A sweepline algorithm for Voronoi diagrams. *Algorithmica* 2, 153–174.

Fredman, M., and R.E. Tarjan (1987). Fibonacci heaps and their uses in improved network optimization problems. *JACM* 34, 596–615.

Gabow, H. (1990). Data structures for weighted matching and nearest common ancestors with linking, in: *Proc. 1st ACM–SIAM Symposium on Discrete Algorithms*, pp. 434–443.

Gabow, H., Z. Galil, T. Spencer and R.E. Tarjan (1986). Efficient algorithms for finding minimum spanning trees in undirected and directed graphs. *Combinatorica* 6, 109–122.

Gabriel, K.R., and R.R. Sokal (1969). A new statistical approach to geographic variation analysis, *Systematic Zoology* 18, 259–278.

Gao, S., M. Jerrum, M. Kaufmann, K. Mehlhorn, W. Rülling and C. Storb (1988). On continuous homotopic one layer routing, in: *Computational Geometry and its Applications*, Lecture Notes in Computer Science, Vol. 333, Springer-Verlag, Berlin, pp. 55–70.

Garey, M.R., and D.S. Johnson (1979). *Computers and Intractability: A Guide to the Theory of NP-Completeness*. W.H. Freeman, New York, NY.

Gewali, L., A. Meng, J.S.B. Mitchell and S. Ntafos (1990). Path planning in $0/1/\infty$ weighted regions with applications. *ORSA J. Comput.* 2(3), 253–272.

Ghosh, S.K., and A. Maheshwari (1990). An optimal algorithm for computing a minimum nested nonconvex polygon. *Inform. Process. Lett.* 36, 277–280.

Ghosh, S.K., and D.M. Mount (1991). An output-sensitive algorithm for computing visibility graphs, *SIAM J. Comput.* 20, 888–910.

Golin, M., and R. Sedgewick (1988). Analysis of a simple yet efficient convex hull algorithm, in: *Proc. 4th Annual Symp. on Computational Geometry*, pp. 153–163.

Goodrich, M., and R. Tamassia (1991). Dynamic trees and dynamic point location, in: *Proc. 23rd Annual ACM Symp. on Theory of Computing*, pp. 523–533.

Goodrich, M.T., S.B. Shauck and S. Guha (1993). Addendum to "parallel methods for visibility and shortest path problems in simple polygons", *Algorithmica* 9, 515–516.

Graham, R.L. (1972). An efficient algorithm for determining the convex hull of a finite planar set. *Inf. Process. Lett.* 1, 132–133.

Graham, R.L., and P. Hell (1985). On the history of minimum spanning tree problem. *Ann. Hist. Comput.* 7, 43–57.

Grünbaum, B. (1967). *Convex Polytopes*. Wiley, New York, NY.

Guibas, L.J., and J. Hershberger (1989). Optimal shortest path queries in a simple polygon. *J. Comput. Systems Sci.* 39, 126–152.

Guibas, L.J., J. Hershberger, D. Leven, M. Sharir and R.E. Tarjan (1987). Linear-time algorithms for visibility and shortest path problems inside triangulated simple polygons. *Algorithmica* 2, 209–233.

Guibas, L.J., J.E. Hershberger, J.S.B. Mitchell and J.S. Snoeyink (1993). Approximating polygons and subdivisions with minimum link paths, *Internat. Comp. Geom. Appl.* 3(4), 383–415.

Guibas, L.J., D.E. Knuth and M. Sharir (1992). Randomized incremental construction of Delaunay and Voronoi diagrams. *Algorithmica* 7, 381–413.

Guibas, L.J., and J. Stolfi (1985). Primitives for the manipulation of general subdivisions and the computation of Voronoi diagrams. *ACM Trans. Graph.* 4, 74–123.

Handler, G.Y., and I. Zang (1980). A dual algorithm for the constrained shortest path problem, *Networks* 10, 293–310.

He, Xin (1991). An efficient parallel algorithm for finding minimum weight matching for points on a convex polygon. *Inf. Process. Lett.* 37(2), 111–116.

Heffernan, P.J., and J.S.B. Mitchell (1991). An optimal algorithm for computing visibility in the plane, in: *Proc. 2nd Workshop Algorithms Data Structure*, Lecture Notes in Computer Science, Vol 519, Springer-Verlag, Berlin, pp. 437–448. To appear: *SIAM J. Comput.*

Heffernan, P.J., and S. Schirra (1992). Approximate decision algorithms for point set congruence, in: *Proc. 8th Annual ACM Symp. on Computational Geometry*, pp. 93–101.

Henig, M.I. (1985). The shortest path problem with two objective functions. *Eur. J. Oper. Res.* 25, 281–291.

Hershberger, J. (1992). Minimizing the sum of diameters efficiently, *Comp. Geom. Theory Appl.* 2(2), 111–118.

Hershberger, J. (1992). Optimal parallel algorithms for triangulated simple polygons, in: *Proc. 8th*

Annual ACM Symp. on Computational Geometry, pp. 33–42.

Hershberger, J., and L.J. Guibas (1988). An $O(n^2)$ shortest path algorithm for a non-rotating convex body. *J. Algorithms* 9, 18–46.

Hershberger, J., and J. Snoeyink (1991). Computing minimum length paths of a given homotopy class, in: *Proc. 2nd Workshop Algorithms Data Struct.* Lecture Notes in Computer Science, Vol. 519, Springer-Verlag, Berlin, pp. 331–342.

Hershberger, J., and J. Snoeyink (1992). Speeding up the Douglas-Peucker line simplification algorithm, *Proc. 5th Intl. Symp. Spatial Data Handling*, IGU Commission on GIS, pp. 134–143.

Hershberger, J., and S. Suri (1992). Applications of a semi-dynamic convex hull algorithm. *BIT*, 32, pp. 249–267.

Hershberger, J., and S. Suri (1991). Finding tailored partitions. *J. Algorithms* 12, 431–463.

Hopcroft, J.E., D.A. Joseph and S.H. Whitesides (1984). Movement problems for 2-dimensional linkages. *SIAM J. Comput.* 13, 610–629.

Hopcroft, J.E., J.T. Schwartz and M. Sharir (1987). *Planning, Geometry, and Complexity of Robot Motion*. Ablex Publishing, Norwood, NJ.

Huttenlocher, D.P. (1988). Three-dimensional recognition of solid objects from a two-dimensional image. Ph.D. Thesis and Report TR-1045, Electrical Engineering and Computer Science, Massachusetts Institute of Technology.

Huttenlocher, D.P., and K. Kedem (1990). Computing the minimum Hausdorff distance for point sets under translation, in: *Proc. 6th Annual ACM Symp. on Computational Geometry*, pp. 340–349.

Huttenlocher, D.P., K. Kedem and J.M. Kleinberg (1992). On dynamic Voronoi diagrams and the minimum Hausdorff distance for point sets under Euclidean motion in the plane, in: *Proc. 8th Annual ACM Sympos. Comput. Geom.*, pp. 110–120.

Huttenlocher, D.P., K. Kedem and M. Sharir (1993). The upper envelope of Voronoi surfaces and its applications, *Discrete Comput. Geom.*, pp. 9, 267–291.

Hwang, Y.-H., R.-C. Chang and H.-Y. Tu (1989). Finding all shortest path edge sequences on a convex polyhedron, in: *Proc. 1st Workshop Algorithms Data Structure*, Lecture Notes in Computer Science, Vol. 382, Springer-Verlag, Berlin, pp. 251–266.

Icking, C., G. Rote, E. Welzl and C. Yap (1993). Shortest paths for line segments. *Algorithmica*, 10, 182–200.

Imai, H., and M. Iri (1986a). Computational-geometric methods for polygonal approximations of a curve. *Comput. Vision, Graphics Image Process.* 36, 31–41.

Imai, H., and M. Iri (1986b). An optimal algorithm for approximating a piecewise linear function. *J. Inf. Process.* 9(3), 159–162.

Imai, H., and M. Iri (1988). Polygonal approximations of a curve-formulations and algorithms, in: G.T. Toussaint (ed.), *Computational Morphology*, North-Holland, Amsterdam, pp. 71–86.

Iyengar, S.S., and A. Elfes, eds. (1991). *Autonomous Mobile Robots: Perception, Mapping, and Navigation*. IEEE Computer Society Press, Los Alamitos, CA.

Jaromczyk, J., and G.T. Toussaint (1992). Relative neighborhood graphs and their relatives. *Proc. IEEE* 80(9), 1502–1517.

Jarvis, R.A. (1973). On the identification of the convex hull of a finite set of points in the plane. *Inf. Process. Lett.* 2, 18–21.

Kallay, M. (1984). The complexity of incremental convex hull algorithms in R^d. *Inf. Process. Lett.* 19, 197.

Kantabutra, V. (1983). Traveling salesman cycles are not always subgraphs of Voronoi duals. *Inf. Process. Lett.* 16, 11–12.

Kapoor, S., and S.N. Maheshwari (1988). Efficient algorithms for Euclidean shortest path and visibility problems with polygonal obstacles, in: *Proc. 4th Annual ACM Symp. on Computational Geometry*, pp. 172–182.

Ke, Y. (1989). An efficient algorithm for link-distance problems, in: *Proc. 5th Annual ACM Symp. on Computational Geometry*, pp. 69–78.

Kindl, M., M. Shing and N. Rowe (1991a). A stochastic approach to the weighted-region problem:

I. The design of the path annealing algorithm, Technical Report, Computer Science, U.S. Naval Postgraduate School, Monterey, CA.
Kindl, M., M. Shing and N. Rowe (1991b). A stochastic approach to the weighted-region problem: II. Performance enhancement techniques and experimental results, Technical Report, Computer Science, U.S. Naval Postgraduate School, Monterey, CA.
Kirkpatrick, D.G. (1983). Optimal search in planar subdivisions. *SIAM J. Comput.* 12, 28–35.
Kirkpatrick, D.G., and R. Seidel (1986). The ultimate planar convex hull algorithm? *SIAM J. Comput.* 15, 287–299.
Klein, R. (1992). Walking an unknown street with bounded detour, *Comput. Geom. Theory Appl.* 1, 325–351.
Kranakis, E., D. Krizanc and L. Meertens (1990). Link length of rectilinear watchman tours in grids, in: *Proc. 2nd Can. Conf. on Computational Geometry*, pp. 328–331.
Kruskal, J.B. (1956). On the shortest spanning tree of a graph and the traveling salesman problem. *Proc. Am. Math. Soc.* 7, 48–50.
Larson, L.C. (1983). *Problem-Solving Through Problems*. Springer Verlag, New York.
Latombe, J.C. (1991). *Robot Motion Planning*. Kluwer Academic Publishers.
Laumond, J.P. (1986). Feasible trajectories for mobile robots with kinematic and environment constraints, in: L.O. Herzberger and F.C.A. Groen (eds.), *Conf. on Intelligent Autonomous Systems* Amsterdam, December 8–11, 1986, Elsevier, Amsterdam, 346–354.
Lawler, E.L. (1976). *Combinatorial Optimization: Networks and Matroids*. Holt, Rinehart and Winston, New York, NY.
Lee, D.T. (1978). Proximity and reachability in the plane. Report R-831, Dept. Elect. Engrg., Univ. Illinois, Urbana, IL.
Lee, D.T., and F.P. Preparata (1984). Euclidean shortest paths in the presence of rectilinear barriers. *Networks* 14, 393–410.
Leiserson, C.E., and F.M. Maley (1985). Algorithms for routing and testing routability of planar VLSI layouts, in: *Proc. 17th Annual ACM Symp. on Theory of Computing,* pp. 69–78.
Lenhart, W., R. Pollack, J.-R. Sack, R. Seidel, M. Sharir, S. Suri, G.T. Toussaint, S. Whitesides and C.K. Yap (1988). Computing the link center of a simple polygon. *Discrete Comput. Geom.* 3, 281–293.
Lipton, R.J., and R.E. Tarjan (1980). Applications of a planar separator theorem. *SIAM J. Comput.* 9, 615–627.
Lumelsky, V.J., and A.A. Stepanov (1987). Path-planning strategies for a point mobile automaton moving amidst unknown obstacles of arbitrary shape. *Algorithmica* 2, 403–430.
Marcotte, O., and S. Suri (1991). Fast matching algorithms for points on a polygon. *SIAM J. Comput.* 20, 405–422.
Matoušek, J., N. Miller, J. Pach, M. Sharir, S. Sifrony and E. Welzl (1991). Fat triangles determine linearly many holes, in: *Proc. 32nd Annual IEEE Symp. on Foundations of Computer Science*, pp. 49–58.
Matula, D.W., and R.R. Sokal (1980). Properties of Gabriel graphs relevant to geographic variation research and clustering of points in the plane. *Geogr. Anal.* 12, 205–222.
McMaster, R.B. (1987). Automated line generation. *Cartographica* 24(2), 74–111.
Melkman, A., and J. O'Rourke (1988). On polygonal chain approximation, in: G.T. Toussaint (ed.), *Computational Morphology*, North-Holland, Amsterdam, pp. 87–95.
Mitchell, J., and E. Welzl (1990). Dynamically maintaining a visibility graph under insertions of new obstacles, Manuscript, School Oper. Res. Indust. Engrg., Cornell Univ., Ithaca, NY.
Mitchell, J.S.B. (1986). Planning shortest paths, Ph.D. Thesis, Stanford Univ., Stanford, CA.
Mitchell, J.S.B. (1991). An algorithmic approach to some problems in terrain navigation. S. Sitharama, Iyengar and A. Elfes (eds.), *Autonomous Mobile Robots: Perception, Mapping, and Navigation*, IEEE Computer Society Press, Los Alamitos, CA, pp. 408–427.
Mitchell, J.S.B. (1990b). On maximum flows in polyhedral domains. *J. Comput. Systems Sci.* 40, 88–123.

Mitchell, J.S.B. (1991). A new algorithm for shortest paths among obstacles in the plane. *Ann. Math. Artif. Intell.* 3, 83–106.

Mitchell, J.S.B. (1992). L_1 shortest paths among polygonal obstacles in the plane. *Algorithmica* 8, 55–88.

Mitchell, J.S.B., D.M. Mount and C.H. Papadimitriou (1987). The discrete geodesic problem. *SIAM J. Comput.* 16, 647–668.

Mitchell, J.S.B., and C.H. Papadimitriou (1991). The weighted region problem: finding shortest paths through a weighted planar subdivision. *J. ACM* 38, 18–73.

Mitchell, J.S.B., G. Rote and G. Woeginger (1992). Minimum-link paths among obstacles in the plane. *Algorithmica*, 8, 431–459.

Mitchell, J.S.B., and S. Suri (1992). Separation and approximation of polyhedral surfaces, in: *Proc. 3rd ACM-SIAM Symp. on Discrete Algorithms*, pp. 296–306. To appear: *Comput. Geom. Theory Appl.*

Mitchell, J.S.B., and E.L. Wynters (1991). Watchman routes for multiple guards, in: *Proc. 3rd Can. Conf. on Computational Geometry*, pp. 126–129.

Mitchell, J.S.B. (1989). An optimal algorithm for shortest rectilinear paths among obstacles, Manuscript, School Oper. Res. Indust. Engrg., Cornell Univ., Ithaca, NY.

Mitchell, J.S.B., and C. Piatko (1992). Approximation methods for link distances in higher dimensions, Manuscript, School Oper. Res. Indust. Engrg., Cornell Univ., Ithaca, NY.

Mitchell, J.S.B., C.D. Piatko and E.M. Arkin (1992). Computing a shortest k-link path in a simple polygon, in: *Proc. 33rd Annual IEEE Symp. on Foundations of Computer Science*, pp. 573–582.

Monma, C., M. Paterson, S. Suri and F. Yao (1990). Computing Euclidean maximum spanning trees. *Algorithmica* 5, 407–419.

Monma, C., and S. Suri (1991.) Partitioning points and graphs to minimize the maximum or the sum of diameters, in: *Graph Theory, Combinatorics and Applications, Proc. 6th Int. Conf. on Theory an Application of Graphs*, Vol. 2, Wiley, New York, NY, pp. 899–912.

Mount, D. (1985). On finding shortest paths on convex polyhedra. Technical Report 1495, Department of Computer Science, University of Maryland.

Mount, D.M. (1990). The number of shortest paths on the surface of a polyhedron, *SIAM J. Comput.* 19, 593–611.

Natarajan, B.K. (1991). On comparing and compressing piece-wise linear curves, Technical report, Hewlett Packard.

Niedringhaus, W.P. (1979). Scheduling with queueing, the space factory problem, Technical report, Princeton University.

Nilsson, N. (1969). A mobile automaton, An application of artificial intelligence techniques, in: *Proc. IJCAI*, pp. 509–520.

Ntafos, S. (1992). Watchman routes under limited visibility, *Comp. Geom. Theory Appl.* 1(3), 149–170.

Ó'Dúnlaing, C., M. Sharir and C.K. Yap (1986). Generalized Voronoi diagrams for moving a ladder: I. topological analysis. *Commun. Pure Appl. Math.* 39, 423–483.

Ó'Dúnlaing, C., M. Sharir and C.K. Yap (1987). Generalized Voronoi diagrams for moving a ladder: II. efficient construction of the diagram. *Algorithmica* 2, 27–59.

Ó'Dúnlaing, C., and C.K. Yap (1985). A 'retraction' method for planning the motion of a disk, *J. Algorithms* 6, 104–111.

Ó'Dúnlaing, C., and M. Sharir C.K. Yap (1983). Retraction: a new approach to motion-planning, in: *Proc. 15th Annual ACM Symp. on Theory of Computing,* pp. 207–220.

O'Rourke, J. (1987). Finding a shortest ladder path: a special case, IMA Preprint Series 353, Inst. Math. Appl., Univ. Minnesota, Minneapolis, MN.

O'Rourke, J., H. Booth and R. Washington (1987). Connect-the-dots: a new heuristic. *Comput. Vision, Graphics Image Process.* 39, pp. 258–266.

O'Rourke, J., and C. Schevon (1989). Computing the geodesic diameter of a 3-polytope, in: *Proc. 5th Annual ACM Symp. on Computational Geometry*, pp. 370–379.

O'Rourke, J., S. Suri and H. Booth (1985). Shortest paths on polyhedral surfaces, in: *Proc. 2nd*

Symp. on Theoretical Aspects of Computing Science, Lecture Notes in Computer Science, Vol. 182, Springer-Verlag, Berlin, pp. 243–254.

Overmars, M.H., and E. Welzl (1988). New methods for computing visibility graphs, in: *Proc. 4th Annual ACM Symp. on Computational Geometry*, pp. 164–171.

Papadakis, N., and A. Perakis (1989). Minimal time vessel routing in a time-dependent environment, *Transp. Sci.* 23(4), 266–276.

Papadakis, N., and A. Perakis (1990). Deterministic minimal time vessel routing. *Oper. Res.* 38(3), 426–438.

Papadimitriou, C.H. (1977). The Euclidean traveling salesman problem is NP-complete, *J. Theor. Comput. Sci.* pp. 237–244.

Papadimitriou, C.H. (1985). An algorithm for shortest-path motion in three dimensions, *Inf. Process. Lett.* 20, 259–263.

Papadimitriou, C.H., and E.B. Silverberg (1987). Optimal piecewise linear motion of an object among obstacles, *Algorithmica* 2, 523–539.

Papadimitriou, C.H., and M. Yannakakis (1989). Shortest paths without a map, in: *Proc. 16th Internat. Colloq. Automata Lang. Program.* Lecture Notes in Computer Science, Vol. 372, Springer-Verlag, Berlin, pp. 610–620.

Pearl, J. (1984). *Heuristics: Intelligent Search Strategies for Computer Problem Solving*. Addison-Wesley, Reading, MA.

Pellegrini, M. (1991). On the zone of a co-dimension p surface in a hyperplane arrangement, in: *Proc. 3rd Can. Conf. on Computational Geometry*, pp. 233–238.

Pollack, R., M. Sharir and G. Rote (1989). Computing of the geodesic center of a simple polygon. *Discrete Comput. Geom.* 4, 611–626.

Preparata, F.P., and S.J. Hong (1977). Convex hulls of finite sets of points in two and three dimensions. *Commun. ACM* 20, 87–93.

Preparata, F.P., and M.I. Shamos (1985). *Computational Geometry: an Introduction*. Springer-Verlag, New York, NY.

Prim, R.C. (1957). Shortest connection networks and some generalizations. *Bell Systems Tech. J.* 36, 1389–1401.

Reif, J.H. (1987). Complexity of the generalized movers problem, in: J. Hopcroft, J. Schwartz and M. Sharir (eds.), *Planning, Geometry and Complexity of Robot Motion*, Ablex Pub. Corp., Norwood, NJ, pp. 267–281.

Reif, J.H., and J.A. Storer (1985). Shortest paths in Euclidean spaces with polyhedral obstacles, Report CS-85-121, Dept. Comput. Sci., Brandeis Univ., Waltham, MA.

Reif, J.H., and J.A. Storer (1987). Minimizing turns for discrete movement in the interior of a polygon. *IEEE J. on Robotics and Automation*, pp. 182–193.

Rohnert, H. (1986a). A new algorithm for shortest paths avoiding convex polygonal obstacles, Report A86/02, Fachber. Inf., Univ. Saarlandes, Saarbrücken.

Rohnert, H. (1986b). Shortest paths in the plane with convex polygonal obstacles, *Inf. Process. Lett.* 23, 71–76.

Rote, G. (1991). Computing the minimum Hausdorff distance between two point sets on a line under translation. *Inform. Process. Lett.* 38, 123–127.

Rote, G. (1992). A new metric between polygons, and how to compute it, in: *Proc. 19th Internat. Colloq. Automata Lang. Program.* Lecture Notes in Computer Science, Vol. 623, pp. 404–415.

Schwartz, J.T., and M. Sharir (1983a). On the 'piano movers' problem I: the case of a two-dimensional rigid polygonal body moving amidst polygonal barriers. *Commun. Pure Appl. Math.* 36, 345–398.

Schwartz, J.T., and M. Sharir (1983b). On the 'piano movers' problem II: general techniques for computing topological properties of real algebraic manifolds, *Adv. Appl. Math.* 4, 298–351.

Schwartz, J.T., and M. Sharir (1983c). On the 'piano movers' problem III: coordinating the motion of several independent bodies: the special case of circular bodies moving amidst polygonal barriers. *Int. J. Rob. Res.* 2(3), 46–75.

Schwartz, J.T., and M. Sharir (1984). On the 'piano movers' problem V: the case of a rod moving in three-dimensional space amidst polyhedral obstacles. *Commun. Pure Appl. Math.* 37, 815–848.

Schwartz, J.T., and M. Sharir (1990). Algorithmic motion planning in robotics, in: J. van Leeuwen (ed.), *Algorithms and Complexity*, Handbook of Theoretical Computer Science, Vol. A, Elsevier, Amsterdam, pp. 391–430.

Seidel, R. (1981). A convex hull algorithm optimal for point sets in even dimensions, Report 81/14, Dept. Comput. Sci., Univ. British Columbia, Vancouver, BC.

Seidel, R. (1986). Constructing higher-dimensional convex hulls at logarithmic cost per face, in: *Proc. 18th Annual ACM Symp. on Theory Comput.* pp. 404–413.

Shamos, M.I. (1978). Computational geometry. Ph.D. Thesis, Dept. of Computer Science, Yale University.

Sharir, M. (1987). On shortest paths amidst convex polyhedra, *SIAM J. Comput.* 16, 561–572.

Sharir, M., and E. Ariel-Sheffi (1984). On the 'piano movers' problem IV: various decomposable two-dimensional motion planning problems. *Commun. Pure Appl. Math.* 37, 479–493.

Sharir, M., and A. Schorr (1986). On shortest paths in polyhedral spaces. *SIAM J. Comput.* 15, 193–215.

Smith, T., G. Peng and P. Gahinet (1988). A family of local, asynchronous, iterative, and parallel procedures for solving the weighted region least cost path problem, Technical Report, Department of Computer Science, Univ. of California, Santa Barbara, CA.

Supowit, K.J. (1983). The relative neighborhood graph with an application to minimum spanning trees. *J. ACM* 30, 428–448.

Suri, S. (1986). A linear time algorithm for minimum link paths inside a simple polygon. *Comput. Vision Graphics Image Process.* 35, 99–110.

Suri, S. (1987). Minimum link paths in polygons and related problems. Ph.D. Thesis, Dept. Comput. Sci., Johns Hopkins Univ., Baltimore, MD.

Suri, S. (1989). Computing geodesic furthest neighbors in simple polygons. *J. Comput. Systems Sci.* 39, 220–235.

Suri, S. (1990). On some link distance problems in a simple polygon. *IEEE Trans. Robotics Autom.* 6, 108–113.

Swart, G.F. (1985). Finding the convex hull facet by facet. *J. Algorithms* 6, 17–48.

Tamassia, R., and F.P. Preparata (1990). Dynamic maintenance of planar digraphs, with applications, *Algorithmica* 5, 509–527.

Tan, X.H., T. Hirata and Y. Inagaki (1991). An incremental algorithm for constructing shortest watchman routes, in: *Proc. 2nd Annual SIGAL Int. Symp. on Algorithms*, Lecture Notes in Computer Science, Vol. 557, Springer-Verlag, Berlin, pp. 163–175.

Tarjan, R.E., and C.J. Van Wyk (1988). An $O(n \log\log n)$-time algorithm for triangulating a simple polygon. *SIAM J. Comput.* 17, 143–178 [Erratum in (1988), 17, 106].

Tarski, A. (1951). *A decision method for elementary algebra and geometry*. Univ. of California Press, Berkeley, CA.

Toussaint, G.T. (1980). Pattern recognition and geometrical complexity, in: *Proc. 5th Int. Conf. on Pattern Recognition*, pp. 1324–1347.

Toussaint, G.T. (1980). The relative neighborhood graph of a finite planar set. *Pattern Recognition* 12, 261–268.

Toussaint, G. (1990). Computing geodesic properties inside a simple polygon, Technical Report, School of Computer Science, McGill University.

Vaidya, P.M. (1988). Minimum spanning trees in k-dimensional space. *SIAM J. Comput.* 17, 572–582.

Vaidya, P.M. (1989). Approximate minimum weight matching on points in k-dimensional space. *Algorithmica* 4, 569–583.

Vaidya, P.M. (1989). Geometry helps in matching. *SIAM J. Comput.* 18, 1201–1225.

Vaidya, P.M. (1989). An $O(n \log n)$ algorithm for the all-nearest-neighbors problem. *Discrete Comput. Geom.* 4, 101–115.

Vegter, G. (1990). The visibility diagram: A data structure for visibility problems and motion planning, in: *Proc. 2nd Scand. Workshop on Algorithm Theory*, Lecture Notes in Computer

Science, Vol. 447, Springer-Verlag, Berlin, pp. 97–110.

Vegter, G. (1991). Dynamically maintaining the visibility graph, in: *Proc. 2nd Workshop on Algorithms Data Structure*, Lecture Notes in Computer Science, Vol. 519, Springer-Verlag, Berlin, pp. 425–436.

Wang, C.A., and E.P.F. Chan (1986). Finding the minimum visible vertex distance between two nonintersecting simple polygons, in: *Proc. 2nd Annual ACM Symp. on Computational Geometry*, pp. 34–42.

Welzl, E. (1985). Constructing the visibility graph for n line segments in $O(n^2)$ time. *Inf. Process. Lett.* 20, 167–171.

Widmayer, P. (1989). Network design issues in VLSI, Technical Report, Institut für Informatik, University Freiburg, Rheinstrasse 10–12, 7800, Freiburg, West Germany.

Widmayer, P., Y.F. Wu and C.K. Wong (1987). On some distance problems in fixed orientations. *SIAM J. Comput.* 16, 728–746.

Wilfong, G. (1988). Motion planning for an autonomous vehicle, in: *IEEE Int. Conf. on Robotics and Automation*, pp. 529–533.

Wilfong, G. (1988). Shortest paths for autonomous vehicles. Technical Report, AT& T Bell Labs.

Yang, C., D. Lee and C. Wong (1992). Rectilinear paths among rectilinear obstacles revisited, Technical Report, Dept. of EE & CS, Northwestern Univ. To appear: *SIAM J. Comput.*

Yang, C.D., D.T. Lee and C.K. Wong (1991). On bends and lengths of rectilinear paths: a graph-theoretic approach, in: *Proc. 2nd Workshop on Algorithms Data Structure*, Lecture Notes in Computer Science, Vol. 519, Springer-Verlag, Berlin, pp. 320–330.

Yao, A. (1982). On constructing minimum spanning trees in k-dimensional spaces and related problems. *SIAM J. Computing* 11, 721–736.

Yao, A.C. (1981). A lower bound to finding convex hulls. *J. ACM* 28, 780–787.

Yap, C.K. (1987). Algorithmic motion planning, in: J.T. Schwartz and C.-K. Yap (eds.) *Advances in Robotics, 1: Algorithmic and Geometric Aspects of Robotics*, Lawrence Erlbaum Associates, Hillsdale, NJ, pp. 95–143.

Yap, C.K. (1987). An $O(n \log n)$ algorithm for the Voronoi diagram of a set of simple curve segments. *Discrete Comput. Geom.* 2, 365–393.

Zikan, K. (1991). Least-squares image registration. *ORSA J. Comput.* 3, 169–172.

M.O. Ball et al., Eds., *Handbooks in OR & MS, Vol. 7*

Chapter 8

Algorithmic Implications of the Graph Minor Theorem

Daniel Bienstock
Department of Civil Engineering, Columbia University, New York, NY 10027, U.S.A.

Michael A. Langston
Department of Computer Science, University of Tennessee, Knoxville, TN 37996, U.S.A.

1. Introduction

In the course of roughly the last ten years, Neil Robertson and Paul Seymour have led the way in developing a vast body of work in graph theory. One of their most celebrated results is a proof of an old and intractable conjecture in graph theory, previously known as Wagner's Conjecture, and now known as the Graph Minor Theorem. The purpose of this chapter is to describe some of the algorithmic ramifications of this powerful theorem and its consequences.

Significantly, many of the tools used in the proof of the Graph Minor Theorem can be applied to a very broad class of algorithmic problems. For example, Robertson and Seymour have obtained a relatively simple polynomial-time algorithm for the disjoint paths problem (described in detail later), a task that had eluded researchers for many years. Other applications include combinatorial problems from several domains, including network routing, utilization and design. Indeed, it is a critical measure of the value of the Graph Minor Theorem that so many applications are already known for it. Only the tip of the iceberg seems to have surfaced thus far. Many more important applications are being reported even as we write this.

The entire graph minors project is immense, containing almost 20 papers whose total length may exceed 600 pages. Thus we focus here primarily on some of the main algorithmic ideas, although a brief sketch of related issues is necessary. We assume the reader is familiar with basic concepts in graph theory [Bondy & Murty, 1976]. Except where noted otherwise, all graphs we consider are finite, simple and undirected.

2. A brief outline of the graph minors project

Three of the key notions employed are *minors*, *obstructions* and *well-quasi-orders*, and we examine them in that order.

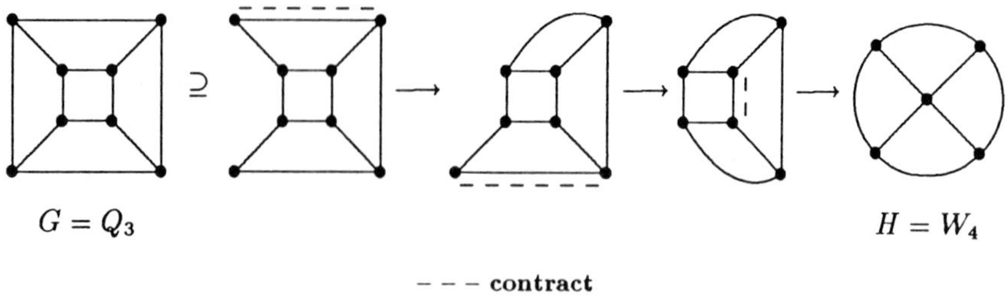

Fig. 1.

Minors. Given graphs H and G, we say that H is a minor of G (or that G contains H as a minor) if a graph isomorphic to H can be obtained by removing from G some vertices and edges and then contracting some edges in the resulting subgraph. Thus every graph is a minor of itself, and the single vertex graph is a minor of every nonempty graph. For a slightly less trivial example, see Figure 1, which illustrates that the wheel with four spokes (W_4) is a minor of the binary three-cube (Q_3).

A concept related to minor containment is *topological* containment. We say that a graph G is a *subdivision* of a graph H if G may be obtained by subdividing edges of H (an edge $\{u, v\}$ is subdivided by replacing $\{u, v\}$ with a path with ends u and v and whose internal vertices are new). We say that G topologically contains H if G contains a subgraph that is a subdivision of H. Thus topological containment is a special case of minor containment (we can only contract edges at least one of whose endpoints have degree two). Observe that W_4 is not topologically contained in Q_3.

Topological containment has been heavily studied by graph theorists. Perhaps the most famous theorem in this regard is Kuratowski's [1930]: a graph is planar if and only if it does not topologically contain K_5 or $K_{3,3}$. We note here that these two graphs are *minimally* nonplanar, that is, every graph topologically (and properly) contained in either of them is planar.

For the sake of exposition, let us view this theorem in terms of minors. Clearly, every minor of a planar graph is also planar. That is, the class of planar graphs is *closed* in the minor order. Consequently, no planar graph contains a K_5 or $K_{3,3}$ minor. Moreover, every proper minor of either of these two graphs is planar, and neither one contains the other as a minor. But can there be other minimal excluded minors? The answer is negative, for if G were such a purported graph, then G would be nonplanar, and thus it would contain, topologically (and therefore as a minor), either K_5 or $K_{3,3}$. In summary, a graph is planar if and only if it does not contain a K_5 or $K_{3,3}$ minor.

We note in passing two other points of interest concerning planarity. One is that planarity can be tested in polynomial time (in fact in linear time [Hopcroft & Tarjan, 1974]). The other is that a problem of natural interest is to try to extend Kuratowski's theorem to higher surfaces. (A surface is obtained from the sphere by 'gluing' onto it a finite number of 'handles' and/or 'crosscaps' [Massey, 1967].) A graph can be embedded on a given surface if it can be drawn on that surface without crossings. Given a surface S, we can characterize those graphs embeddable in S by a finite list of excluded graphs in the topological order? In the 1930s, Erdös conjectured that the answer is yes. No results were obtained on this conjecture until much later, first with a proof for the case when S is the projective plane [Archdeacon, 1980], and then for the case when S is non-orientable [Glover, Huneke & Wang, 1979].

Obstructions. Kuratowski's theorem may be regarded as a characterization of planarity by means of excluded graphs, henceforth termed obstructions. Characterizations of this nature abound in combinatorial mathematics and optimization. Some familiar examples include the max-flow min-cut theorem, Seymour's description of the clutters with the max-flow min-cut property and Farkas' lemma. In all these, the presence of a desired feature is characterized by the absence of some obstruction. Besides being aesthetically pleasing, theorems of this type are desirable because such characterizations provide evidence of 'tractability' of the problem at hand, giving hope for a polynomial-time test for the desired feature.

The graph minors project contains several such theorems, many of which turn out to be at the heart of both proofs and applications. As expected, there are algorithmic aspects to these theorems. As a very introductory example, one can test in polynomial time if any graph can be embedded on the torus.

Well-quasi-orders. A class Q, equipped with a transitive and reflexive relation $\leq$, is called a *quasi-order*. For example, the class of all graphs is a quasi-order, where $\leq$ is the minor containment relation. There has been some confusion as to the difference between quasi-orders and partial-orders; it suffices to look at graph minors to understand this difference. It is convenient to regard isomorphic copies of a given graph as different entities and so, for distinct graphs G and H, we can simultaneously have $G \leq H$ and $H \leq G$. Thus $\leq$ is not a partial-order, because the minor relation is not anti-symmetric.

A quasi-order with class Q and relation $\leq$ is a *well-quasi-order* if (1) for every infinite sequence $a_1, a_2, \ldots$ of elements of Q, there exist integers $1 \leq i < j$ such that $a_i \leq a_j$, and (2) there exists no infinite descending chain $b_1 > b_2 > \ldots$ of distinct elements of Q.

Example 2.0.1. Let Q be the set of all closed intervals of the real line, with non-negative integer endpoints; i.e., $Q = \{[a, b] : 0 \leq a \leq b$ and a, b integer$\}$. For $I = [a, b]$, $J = [c, d]$, we write $I \leq J$ if either J contains I and $a = c$ or if I and J are disjoint with $b < c$. Clearly $(Q, \leq)$ is a quasi-order. To see that it is a well-quasi-order, note first that (2) is satisfied. To prove that (1) holds, consider any sequence $S = I_1, I_2, \ldots$ with the property that there do not exist integers $1 \leq i < j$ such that $I_i \leq I_1 = [a, b]$. Clearly S contains no members of the form $[c, d]$ with $b < c$. It is also clear that S contains finitely many members of the form $[c, d]$ with $d < b$. Finally, for each integer x with $a \leq x \leq b$, S contains finitely many members of the form $[x, y]$. For if $I_{j(x)} = [x, y^*]$ is the first such member, then all remaining members of the form $[x, y]$ satisfy $y < y^*$. We conclude that S is finite, as desired.

We use this example to illustrate that, given a well-quasi-order, we in general do not have an absolute bound on the size of an antichain (set of pairwise incomparable elements), we merely know it must be finite. In particular, 'finite' does not necessarily imply 'small.' A result of relevance is Kruskal's proof [1960] of a conjecture of Vázsonyi, namely, that trees form a well-quasi-order under topological containment.

2.1. The Graph Minor Theorem and some of its consequences

We can now state the Graph Minor Theorem and some of the most important consequences arising from its proof. Very little progress has been made on this result, formerly a conjecture attributed to K. Wagner, until the work of Robertson and Seymour.

The Graph Minor Theorem. *The class of all graphs is a well-quasi-order under the minor relation.*

Corollary 2.1.1. Let C be a class of graphs closed under minors. Then C can be characterized by a finite list of minor obstructions.

To see that this follows from the theorem, let S be the set of minor-minimal graphs not in C. Then S is an antichain, and thus it is finite. Hence $G \in C$ if and only if G does not contain as a minor any graph isomorphic to a member of S.

Let v denote a vertex in a graph G. Let the edges incident on v be $\{v, u_i\}$, $1 \leq i \leq p$, and $\{v, w_j\}$, $1 \leq j \leq q$, where $2 \leq p, q$. Let H be the graph obtained by replacing v with two new vertices, u and w, replacing the edges incident on v with $\{u, u_i\}$, $1 \leq i \leq p$, and $\{w, w_j\}$, $1 \leq j \leq q$, and adding a new edge $\{u, w\}$. We say H is obtained from G by *splitting* v.

Corollary 2.1.2 [Robertson & Seymour, 1990]. Let C be a class of graphs closed under minors. Then C can be characterized by a finite number of topological obstructions. Moreover, each such obstruction can be obtained from a minor obstruction with vertex splittings.

Corollary 2.1.3 [Robertson & Seymour, 1990]. Let S be any surface. Then the class of graphs embeddable in S can be characterized by a finite number of topological obstructions.

This follows from the last corollary since embeddability in S is closed under minors.

We remark here that the proof of Erdös' Conjecture (Corollary 2.1.3) does not require a solution to Wagner's Conjecture. Rather, the tools used to settle the latter are a superset of those used for the former. Also, the number of topological obstructions can be quite large (indeed, for the projective plane there are 103, and for higher surfaces there are many more).

Theorem 2.1.4 [Robertson & Seymour, 1994]. *Let H be a fixed graph. Then there is a polynomial-time algorithm for testing, for any input graph G, whether G contains H as a minor.*

The running time of the algorithm is $O(n^3)$, where $n = |V(G)|$. The constant hidden in the big Oh is a very rapidly growing function of the size of H.

Corollary 2.1.5. Let C be a class of graphs closed under minors. Then membership in C can be tested in polynomial time.

With regards to Corollary 2.1.5, the testing algorithm would make use of Theorem 2.1.4, by testing minor containment of all obstructions. Thus the proof of this corollary is intrinsically *nonconstructive*. We know the desired polynomial-time algorithm exists, but we cannot implement it unless we have the list of obstructions, a task towards which the graph minors project provides no clues. Moreover, even if the obstructions were available, the resulting algorithm would be very impractical. Means of avoiding these problems in many general cases have been developed by Fellows and Langston, and are discussed in Section 6.

Another way to interpret Corollary 2.1.5 is to regard minor closure as a certificate of tractability. Once a given graph property is found to be closed, then an effort can be launched to find an efficient, direct algorithm for testing that particular property.

Corollary 2.1.6. Let S be a given surface. Then graph embeddability in S can be tested in polynomial time.

Of special interest are the consequences towards the disjoint paths problem: given vertices s_i and t_i $(1 \le i \le k)$, not necessarily distinct, in a graph G, find pairwise vertex-disjoint paths between s_i and t_i $(1 \le i \le k)$. This problem is NP-hard for general k [Karp, 1975]. For $k = 1$, the problem is trivial. For $k = 2$, a complex solution has been known for some time [Seymour, 1980; Shiloach, 1980]. But the techniques used in Robertson & Seymour [1994] to obtain Theorem 2.1.4 yield the following.

Theorem 2.1.7. *The disjoint paths problem can be solved in polynomial time for every fixed k.*

The algorithms of Theorems 2.1.4 and 2.1.7 are constructive. The disjoint paths problem is addressed in more detail in Section 5.

This concludes our outline of some of the major results that stem from the graph minors project. In the sequel, we provide more information on topics as promised above, and discuss the important graph parameters treewidth and pathwidth.

3. Treewidth

Treewidth plays a critical role in the graph minors project. It may be said that it measures the complexity of a graph, in the sense that a graph of small treewidth can be recursively decomposed, by removing a few vertices, into two graphs of roughly equal size. A consequence of this is that many NP-hard problems can be efficiently solved in graphs of small treewidth with dynamic programming.

A *tree decomposition* of the graph G consists of a pair (T, X), where T is a tree (which is not part of G, but merely another graph) and $X = \{X_t : t \in V(T)\}$ is a family of subsets of $V(G)$, one for each vertex of T, satisfying the following properties:

(1) for every edge $\{u, v\}$ of G, there exists $t \in V(T)$ with $u, v \in X_t$, and

(2) for every pair y, z of vertices of T, if w is any vertex in the path between y and z in T, then $X_y \cap X_z \subseteq X_w$.

The *width* of (T, X) is $\max\{|X_t| - 1 : t \in V(T)\}$. The *treewidth* of G is the minimum integer w such that there is a tree decomposition of G of width w. Graphs of treewidth at most k have also been called *partial k-trees*.

These definitions may be restated in a way that is more familiar to some readers. Condition (2) states that, for every vertex v of G, the set of vertices $t \in V(T)$ such that $v \in X_t$ forms a subtree of T, say T_v. Condition (1) then states that for every edge $\{u, v\}$ of G, the subtrees T_u and T_v must intersect. Now consider the graph H with vertex set $V(G)$, and such that $\{a, b\}$ is an edge of H whenever T_a and T_b intersect. Then H is chordal, and every chordal graph can be obtained as such an intersection graph of subtrees of a tree [Gavril, 1974]. Furthermore, the clique number of H is precisely the width of (T, X) plus one. Consequently, we arrive at an equivalent definition of treewidth: it is the minimum, over all chordal supergraphs H of G, of the clique number of H minus one.

It is NP-hard to compute treewidth [Arnborg & Proskurowski, 1987]. But the family of graphs with treewidth at most k is minor-closed for every fixed k, and hence polynomial-time recognizable.

Example 3.0.1. *Series-parallel graphs*. These may be iteratively defined as follows: the graph consisting of a single edge is series-parallel, and given a series-parallel graph, we obtain a new one by adding an edge in parallel to an existing edge, or by subdividing an existing edge. Series–parallel graphs have treewidth at most 2, which may be shown inductively. For example, let G be series-parallel, and let (T, X) be a tree decomposition of G of width at most 2. Let $\{u, v\}$ be an edge of G. Suppose we subdivide $\{u, v\}$ by introducing a new vertex w, to obtain a new graph G'. Let r be a vertex of T such that $u, v \in X_r$. Now define a new tree T' to consist of T, together with the edge $\{r, q\}$, where q is a new vertex. Set $Y_t = X_t$, for $t \in V(T)$, and $Y_q = \{u, v, w\}$. It is seen that (T', Y) is a tree decomposition of G', of width at most 2.

Example 3.0.2. *Grids*. Let $m > 1$ be an integer. The m-grid is the graph with vertex set $\{(i, j) : 1 \leq i \leq m, 1 \leq j \leq m\}$ and edge set $\{\{(i, j), (i+1, j)\} : 1 \leq i \leq$

$m-1, 1 \leq j \leq m\} \cup \{\{(i, j), (i, j+1)\} : 1 \leq i \leq m, 1 \leq j \leq m-1\}$. It can be shown that the m-grid has treewidth m. To see that the treewidth is at most m, consider the tree decomposition (T, X), where T is the path with vertices $1, 2, \ldots, m^2 - m$, and, for $1 \leq i \leq m-1, 1 \leq j \leq m$, $X_{m(i-1)+j} = \{(i, k) : j \leq k \leq m\} \cup \{(i+1, k) : 1 \leq k \leq j\}$. It is easily verified that this is indeed a tree decomposition of width m. The lower bound (treewidth at least m) is more difficult to establish [Robertson & Seymour, 1991].

Grids play an important role in the graph minors project, for it can be shown that for every $n \geq 1$, any planar graph with n vertices is a minor of the m-grid, with $m = O(n^2)$. Moreover, grids are useful in proving the following results.

Theorem 3.0.3 [Robertson & Seymour, 1986]. *Let H be a planar graph. Then there exists a number $w(H)$, such that any graph with no minor isomorphic to H has treewidth at most $w(H)$.*

Corollary 3.0.4 [Robertson & Seymour, 1990]. Given an infinite list $G_1, G_2, \ldots$ of graphs, with G_1 planar, there exist $i < j$ such that G_i is a minor of G_j.

Of course this corollary follows from the Graph Minor Theorem even without G_1 being planar. But using Theorem 3.0.3 its proof can proceed as follows: assuming that G_1 is not a minor of G_i, $i > 1$, it follows that G_1 has bounded treewidth. Thus, G_i has 'simple' structure (think of it as a 'thickened' tree) and now a generalization of the proof technique in Kruskal's theorem finishes the job. The main point here is, once more, that small treewidth implies simple structure. We underscore the planarity requirement in Theorem 3.0.3. Is there some 'structure theorem' if this requirement is dropped? This question is studied in Robertson & Seymour [1990], the result can be informally described as follows.

Let us first restate the above concerning the question of planar graphs. Let H be a fixed planar graph. If G does not contain a minor isomorphic to H, then G can be obtained by pasting together small graphs in a tree-like structure. Now suppose H is not planar. We obtain a corresponding structure theorem for graphs with no H-minor by replacing the phrase 'small graphs' with the phrase 'graphs with simple structure.' This simple structure is in fact rather complex to state precisely. It is convenient to view it as parameterized by H itself; since H is fixed, this is loosely termed 'simple.' We obtain a graph of simple structure by starting with a graph of 'small' genus, up to a 'small' number of troublesome vertices (that is, a graph than can be embedded on a surface of low genus after removing only a few vertices). We attach to this graph a 'small' number of necklace-like graphs. In these necklaces, the beads are 'small' graphs, and the beads are attached to one another in a ring-like fashion.

We have obviously used the word 'small' rather liberally. In the case of genus, small means smaller than the genus of H. In the other cases (the number of troublesome vertices, the number of necklaces and the sizes of the beads in the necklaces), it is again helpful to think of these numbers as parameters of H. Thus, for a fixed H, we would be able to write done explicit upper bounds for all parameters instead of the word 'small'. The proof techniques involved in obtaining these results are quite complex and interesting in their own right.

It is worth pointing out one further structure theorem. In Alon, Seymour & Thomas [1994] it is shown that for any fixed graph H, if G has no minor isomorphic to H, then G satisfies a 'separator theorem' that generalizes Lipton & Tarjan's [1979] planar separator theorem.

3.1. Bounded treewidth and efficient algorithms

Suppose we are given a tree decomposition (T, X) of a graph G, whose width is bounded by some small constant w. It is often the case that we can exploit the structure of (T, X) to derive polynomial-time, dynamic-programming algorithms for solving problems that are NP-hard on general graphs.

The generic approach would be as follows: first direct all edges of T away from a given root r. Next, for any vertex t of T, let T_t denote the subtree of T rooted at t. The key point is to notice that $\bigcup\{X_v : v \in V(T_t)\}$ can 'interact' with the rest of G only through X_t, which is 'small'. Thus one seeks to generate a dynamic-programming strategy that stores all relevant information about the subgraph of G induced by $\bigcup\{X_v : v \in V(T_t)\} = Y_t$, by listing all possible 'states' of X_t. We may assume that, without loss of generality, T is binary (degree at most 3), and thus it is easy for the recursion to move up the three T. This approach, which frequently yields polynomial-time algorithms, has been taken by many authors, and it is essentially impossible to list all papers on this topic.

Let us discuss some folklore examples. The algorithms we sketch are not the best possible. We simply want to illustrate how polynomiality is achieved. Using notation as before, for any vertex t of T, we denote by G_t the subgraph of G induced by Y_t

Example 3.1.1. The *vertex cover problem*. Given a graph G we seek a subset of vertices C, with minimum cardinality, such that every edge of G has at least one end in C. Now suppose we are given a width-w tree decomposition (T, X) of G. Let t be a vertex of T and, for each subset Z of X_t, let $f(Z, t)$ denote the smallest cardinality of a vertex cover of G_t, with the added restriction that we use Z in the cover. Thus a table with 2^{w+1} entries can be used to record this information for each t. It is not difficult to see how to update the table as we move up the tree T. Furthermore, $\min\{f(Z, r) : Z \subseteq X_r\}$ solves the vertex cover problem. The run time of the algorithm is exponential in w, but linear in $|E(G)|$.

Example 3.1.2. The *travelling salesman problem* in graphs. Let G be a graph with weights on the edges, where we seek a Hamiltonian cycle of minimum length. One possible recursion is based on the following states. Let Z be a subset of X_t, and consider (for each $0 \le m \le |X_t|$) $2m$ distinct vertices o_i, d_i, $1 \le i \le m$, contained in $X_t - Z$. Let P be the set of pairs $\{o_i, d_i\}$. Then we denote by $f(t, Z, P)$ the minimum total length of a family of vertex-disjoint paths R_i, $1 \le i \le m$, in Y_t, such that R_i has ends o_i and d_i, and $\bigcup\{V(R_i) : 1 \le i \le m\} = Y_t - Z$.

Example 3.1.3. The *vertex coloring problem*. Given a graph G, we seek to partition the vertices of G into a minimum number of independent sets (the chromatic number of G). Here the recursion works as follows: for each partition π of X_t, we denote by $s(t, \pi)$ the minimum cardinality of a partition of $Y(t)$ into independent sets, such that the intersection of X_t with the color classes yields the partition π.

Many graph problems are not amenable to this generic dynamic programming approach. For example, any problem that is NP-hard on trees cannot be solved this way (trees have treewidth 1). A partial characterization of the solvable problems, together with a prototype algorithm, can be found in Arnborg, Lagergren & Seese [1991].

The fact that the algorithms described above exist is not, of course, a consequence of Robertson and Seymour's work; nevertheless, it is clearly related to the concepts of treewidth and tree decomposition, and so it seems appropriate to touch on this topic here.

3.2. Branchwidth, tangles and graph searching

An interesting graph parameter that is related to treewidth is *branchwidth*. Branchwidth may be computationally more tractable than treewidth (at least in terms of approximation). Further, concepts related to branchwidth play a crucial role in the development of the graph minors project, and so it is useful to review them here.

Given a graph G, a *graph decomposition* of G consists of a pair (T, f), where T is a binary tree, and f is an injective map from $E(G)$ to the leaves of T. Notice that if e is an edge of T, then there is a partition of $E(G)$ into two classes (A, B) that corresponds to e — this is defined by the partition of the leaves of T into the two subtrees of $T - e$. The *order* of e is defined as the number of vertices of G that have incident edges both in A and in B. The width of (T, f) is the maximum order of any edge in T. See Figure 2.

The branchwidth of G is the minimum width of any branch decomposition of G. Computing the branchwidth of a graph is NP-hard. It is not very difficult to show the following.

Theorem 3.2.1 [Robertson & Seymour, 1991]. *The branchwidth and treewidth of any graph differ by at most a factor of 3/2.*

Does this theorem help in approximating treewidth? Historically, it first appeared that the answer to this question should be positive, since in Robertson & Seymour [1991] a min–max characterization of branchwidth was given, while no such characterization of treewidth was yet known. However, the complete picture is more complex.

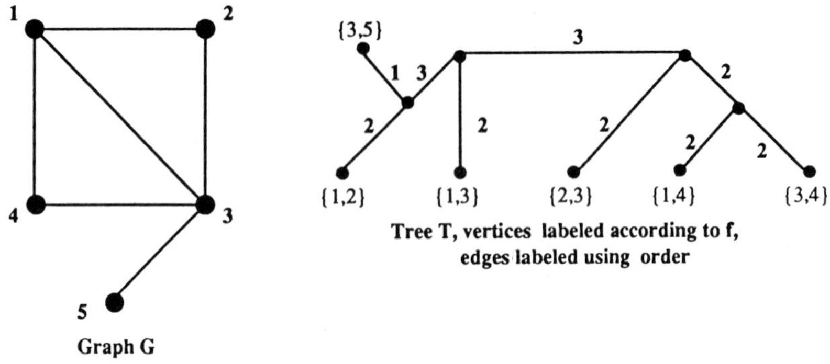

Fig. 2.

Theorem 3.2.2 [Seymour & Thomas, 1994]. *The branchwidth of a planar graph can be computed in polynomial time.*

In any case, the tools developed in Robertson & Seymour [1991] were extremely useful towards completing the proof of the Graph Minor Theorem. Moreover, the ideas behind the min-max characterization of branchwidth were later adapted to yield a similar characterization of treewidth (and also of pathwidth, a concept we cover later). These min–max formulae were then used to improve some of the original theorems in the graph minors project.

It is enlightening to describe the relationship of treewidth to a class of graph searching games. Regard a graph as a system of roads. A fugitive resides in the vertices and can travel along edges. We wish to capture the fugitive (whose position is always known) using a fixed number of guards, who always occupy vertices and travel using helicopters. In one time unit, some of the guards can move to a different subset of vertices. During the move the fugitive can scurry, infinitely fast, to a new vertex, travelling along any path that is not blocked by an unmoving guard. Our objective is to corner the fugitive in such a way that no escape is possible.

The minimum number of guards needed for this purpose is called the *search number* of the graph. Thus, for example, a tree has search number 2 (place a guard at any vertex and observe which subtree is occupied by the fugitive, then corner the fugitive into smaller and smaller subtrees). The graph in Figure 3 has search number 3. To see this, notice that with at most 2 guards the fugitive can always

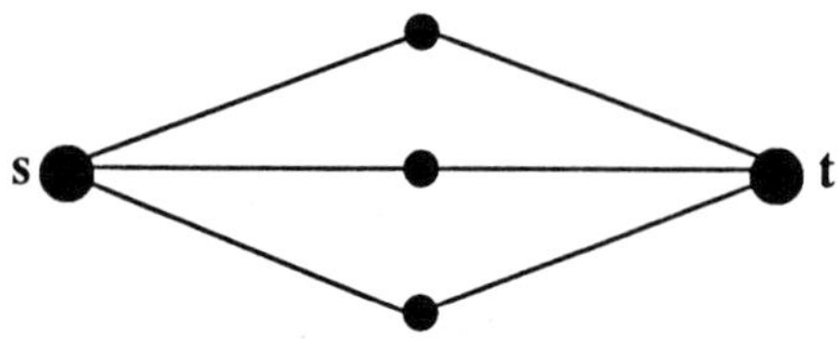

Fig. 3.

occupy a vertex in $V(G) - \{s, t\}$. But using three guards, with two guards we can first isolate the fugitive in one of the paths with ends s, t, and with the third guard we then capture the fugitive.

Theorem 3.2.3 [Seymour & Thomas, 1994]. *For any graph G, the treewidth of G equals the search number of G minus 1.*

Intuitively, a tree decomposition (T, X) of G yields a search strategy of G involving a number of guards equal to the width of (T, X) plus one: we corner the fugitive into smaller and smaller subgraphs of the form G_t (recall the notation of Section 2.1) by placing guards in the subset X_t. The definition of tree decomposition shows that this strategy indeed works and requires the desired number of guards. This argument yields one of the two bounds needed to prove Theorem 3.2.3.

The proof of the other bound also yields a min-max formula for treewidth, as follows. Let k be an integer. A *haven of order* k is a function g that assigns to each subset Y of k or fewer vertices, the vertex set of one of the components of $G - Y$, denoted $g(Y)$. The function g, in addition, satisfies that whenever X and Y are subsets of vertices with $|Y| \leq k$ and $X \subseteq Y$, then $g(Y) \subseteq g(X)$. The *tangle number* of G is the maximum order of a tangle. As examples, the tangle number of a tree is 1, and the tangle number of the complete graph K_n is $n - 1$.

Theorem 3.2.4. *For any graph G, the tangle number of G equals the treewidth of G.*

Intuitively, think of a tangle of order k as the escape plan of the fugitive in case k guards are being used: if the guards occupy Y, the fugitive hides in $g(Y)$. The proof of Theorem 3.2.4 and the formalization of the preceding argument are quite complex, and are not given here. These tools are related to some of the concepts in Robertson & Seymour [1991].

Given that it is NP-hard to compute treewidth, Theorem 3.2.4 is at first glance surprising. But observe that (for arbitrary k) the complete description of a tangle requires exponential time. Nevertheless, this theorem and its connection to graph searching may turn out to be useful (from a computational point of view) towards estimating treewidth.

Graph searching games similar to the one given above have previously been considered by researchers in the computer science community. They are of interest in that they provide a worst-case scenario for the process of immunizing a network against a computer virus. We return to this topic later.

3.3. Treewidth and signal routing problems

A problem that frequently arises in communications networks is the following: suppose we are given nodes $1, 2, \ldots, n$, and an $n \times n$ traffic matrix M (where m_{ij} is the traffic rate between i and j). We wish to design a binary tree T, with leaves precisely $1, 2, \ldots, n$, to carry the traffic. Notice that given such a tree T, each edge e of T will carry a certain total amount of traffic or *congestion* (namely, the traffic between nodes separated in T by e). The tree T is to be chosen so that the maximum such congestion is minimized. The resulting optimal congestion level is called the *carvingwidth* of M [Seymour & Thomas, 1994]. We remark that the use of trees as communications networks is widespread and natural in many applications because of their simplicity and ease of fabrication. We are essentially designing a tree (a very simple structure) to realize a more complex pattern (the traffic requirements).

Not surprisingly, it is NP-hard to compute carvingwidth. However, if M is planar (that is, if the graph G with vertices $\{1, \ldots, n\}$ and an edge $\{i, j\}$ whenever $m_{ij} > 0$ is planar) then carvingwidth can be solved in polynomial time. In particular, there is a nice min–max characterization of carvingwidth in this special case [Seymour & Thomas, 1994]. Further, the tools and proof techniques are essentially the same as those used to prove Theorem 3.2.2.

The above results suggest that there is a deep connection between treewidth and carvingwidth. Let us consider the case where M is a $\{0, 1\}$-matrix; i.e., M is the adjacency matrix of G. Then computing carvingwidth corresponds to finding good graph embeddings [Hong, Mehlhorn & Rosenberg, 1983]. Let *congestion* (G) denote the carvingwidth of M.

Theorem 3.3.1 [Bienstock, 1990]. *If G has treewidth k and maximum degree d, then* $\Omega(\max\{k, d\} \le \text{congestion}(G) \le O(kd)$.

Thus, for graphs of bounded degree, treewidth and congestion are of the same order of magnitude.

There is another parameter that arises in routing problems and is related to treewidth. Consider a binary tree T with leaves labeled $\{1, 2, \ldots, n\}$ as above. If $m_{ij} > 0$, then it is desirable that the path in T between i and j be short. The dilation of T is the maximum length of any such path, and *dilation* (G) is the minimum dilation over all binary trees T.

Theorem 3.3.2 [Bienstock, 1990]. *If G has treewidth k and maximum degree d, then* $\Omega(\log k + \log d) \le \text{dilation}(G) \le O(\log k + \log^* n \log d)$.

Thus, approximating treewidth is tantamount to approximating dilation within a very small additive error ($\log^* n$ is an extremely slowly growing function of n).

3.4. On computing treewidth

Given that the concepts associated with treewidth are extremely useful in a wide range of applications, and that computing treewidth (and branchwidth, carvingwidth, etc.) is NP-hard, what can one say about computability of treewidth? There are at least three ways of approaching this problem: (1) approximation algorithms, (2) testing for small treewidth, and (3) experimental results.

A polynomial-time approximation algorithm that does not depend on fixing the treewidth has very recently been devised by Robertson, Seymour and Thomas.

Theorem 3.4.1 [Thomas, 1991]. *There is a polynomial-time algorithm that, given a graph G and an integer k, either proves that G has treewidth at least k or provides a tree decomposition of G of width at most* $k^2 + 2k - 1$.

The proof of this result relies on a minimax characterization from Seymour & Thomas [1994] and the decomposition method of Robertson & Seymour [1994]. We stress that in this theorem the parameter k is not fixed, but is instead part of the input. The algorithm is fairly reasonable (its run time does not involve excessive constants or very high degrees) and its main application lies in testing whether the treewidth of a given graph is small. This is important since many of the above applications require bounded treewidth.

Are there sharper approximation algorithms? Ideally, one seeks a polynomial-time algorithm that approximates treewidth up to a constant factor. Until branch-width was shown to be NP-hard, a natural approach was to seek an exact algorithm for this parameter instead. In any case, approximating treewidth remains a crucial open problem. Moreover, it seems likely that the tools required for this task would also be of use towards other NP-hard problems involving cuts (such as graph bisection) for which no constant-factor approximation algorithms are known.

Next we turn to the problem of testing for small treewidth. Recall that the property (for any given k) of having treewidth at most k is closed under minors. Thus, according to Corollary 2.1.5, there is a polynomial-time algorithm to compute (exactly) the treewidth of a graph known to have small treewidth. But this approach is nonconstructive and perhaps not useful. Another approach would be to use a dynamic-programming scheme as described in Section 3.1. However, such an algorithm may be quite unreasonable. A third approach is the recent result of Reed.

Theorem 3.4.2 [Reed, 1990] *For each fixed k, we can test whether G has treewidth at most k in* $O(n \log n)$ *time.*[1]

In terms of experimental results concerning the computation of treewidth, no major results are available. An intriguing possibility is the use of integer programming to compute tangles. It is easy to see that the existence of a tangle (of a given order) can be described by a system of equations in 0-1 variables (but an exponential numbers of those, unfortunately). A possible research problem of interest would be to describe the polyhedral structure of the convex hull of tangles.

4. Pathwidth and cutwidth

In the development of the graph minors project, treewidth was preceded by another graph parameter, *pathwidth*. The pathwidth of a graph can be much larger

[1] This can now be done in $O(n)$ time.

than its treewidth. Several important applications of pathwidth arose well before the graph minors project.

The definition of pathwidth is similar to that of treewidth, except that we restrict ourselves to tree decompositions (T, X) where T is a *path* (such tree decompositions are called *path decompositions*). Thus if (T, X) is a path decomposition of G, then every vertex v of G is mapped into a subpath P_v of T (i.e., each vertex essentially mapped into an interval), so that whenever $\{u, v\}$ is an edge of G, then P_u and P_v intersect. The width of the path decomposition is the maximum number of subpaths P_v that are incident with any vertex of T minus one. There is a connection similar to that between treewidth and chordal graphs: pathwidth equals the smallest clique number over all *interval* supergraphs of G minus one [Golumbic, 1980]. For example, paths have pathwidth 1, and a complete binary tree on $m > 1$ levels has path with $\lceil m/2 \rceil$.

In terms of graphs minors, the most important theorem involving pathwidth is an analogue to Theorem 3.0.3.

Theorem 4.0.1 [Robertson & Seymour, 1983]. *For every forest F there is a number $p(F)$, such that if a graph G does not have a minor isomorphic to F, then G has pathwidth less than $p(F)$.*

The original proof of Theorem 4.0.1 employed a function p that was very rapidly growing in $|V(F)|$. This result has been improved [Bienstock, Robertson, Seymour & Thomas, 1991] to show that $p(F) = |V(F)| - 1$, which is best possible.

Recall that treewidth is related to graph searching and embedding problems. The same is true for pathwidth, and again these connections chronologically preceded those for treewidth.

First we consider graph searching. There are two versions of this game that have been known in the literature for some time. Here the main difference is that the guards do not know where the fugitive is. In one version, called *edge searching*, the portion of the graph 'secured' by the guards can be extended by sliding a guard along an edge leading out of this portion. In the other, called *node searching*, an edge is cleared by placing a guard at each end, simultaneously. For either kind of game one can define the search number of a graph, as previously, to be the minimum number of guards needed to catch the fugitive. It is shown in Kirousis & Papadimitriou [1986] that the edge-search number and the node-search number never differ by more than 1, and that the node-search number always equals pathwidth plus 1. A different version of the game, called *mixed searching*, is considered in Bienstock & Seymour [1991]. In mixed searching, moves from both edge and node searching are allowed. This enables one to obtain short proofs for the monotonicity of both edge and node searching (monotonicity here means that no search strategy of a graph need ever repeat the same step).

With regards to graph embedding problems, the connection here is via the NP-hard *cutwidth problem*, defined as follows. Given a graph G on n vertices, suppose we label the vertices with the integers $1, 2, \ldots, n$. The width of this labeling is the maximum, over $1 \leq h \leq n-1$, of the number of edges $\{i, j\}$ with $i \leq h$ and $h < j$.

The objective is to find a labeling with minimum width (defined as the cutwidth of G). This problem originally arose in the design of linear arrays, an early form of printed circuit. In Makedon & Sudborough [1989] it is shown that if G has pathwidth p and maximum degree d, then the cutwidth of G is $\Omega(\max\{p, d\})$ and $O(pd)$, a result similar to Theorem 3.3.1. Thus, for graphs of bounded pathwidth, there is a polynomial-time algorithm that approximates cutwidth up to a constant factor.

It also turns out that pathwidth is linear-time equivalent to the gate matrix layout problem [Deo, Krishnamoorthy & Langston, 1987], another problem with application to printed circuits. This problem can be stated as follows. Suppose we are given a $\{0, 1\}$ matrix M. Let $M(\pi)$ result from permuting the columns of M according to some permutation π, and suppose we replace the 0 entries of $M(\pi)$ in each row, between the first and last occurrences of 1 in that row, with 1's. The maximum number of 1's in any column of the resulting matrix is called the *width* of π. Then we seek a permutation π of minimum width (this corresponds to laying out devices in a chip so as to minimize the number of wire tracks required to effect desired connections). Call this number the layout width of M. To see the connection with pathwidth, let G denote the clique graph of the transpose of M; i.e., the graph with vertices of rows of M, and a clique arising from the 1's in each column. Then it is easy to verify that the layout width of M is exactly the pathwidth of G (refer to the interval graph interpretation of pathwidth above).

As with treewidth, it is NP-hard to compute the pathwidth of a graph, and approximation algorithms are known only for very special cases [Yan, 1989]. Again, there is a min-max formula for pathwidth with corresponding obstructions. These obstructions (an appropriate name might be 'linear tangles') are described in detail in Bienstock, Robertson, Seymour & Thomas [1991], and it suffices here to say that they are closely related to the tangles of Section 3.2.

Much is known about the nature of obstructions for pathwidth k. For $k = 0$ there is one; for $k = 1$ there are two; for $k = 2$ there are 110 [Kinnersley & Langston, 1991]; and for $k = 3$ there are at least 122 million! Moreover, all tree obstructions are known.

The approximate computation of pathwidth for general graphs is an interesting open problem, and once more we point out the possible use of integer programming techniques in this context. Notice that the existence of a path decomposition of given width corresponds directly to the solvability of a system of linear equations in $\{0, 1\}$ variables (as opposed to the treewidth case, where it is easiest to describe the obstructions in this manner).

5. Disjoint paths

Recall the definition of the disjoint paths problem. We are given, in a graph G, vertices s_i and t_i ($1 \le i \le k$), not necessarily distinct. We seek pairwise vertex-disjoint paths between s_i and t_i ($1 \le i \le k$). In this section we outline how graph minors theory yields an algorithm with complexity $O(n^3)$ for this problem, for each fixed value of k.

It is worthwhile first to compare this problem to that of H-minor containment: given G, test whether it has a minor isomorphic to H. For each fixed H, this problem can be reduced to the disjoint paths problem. The resulting algorithm will, however, have high complexity (the degree depends on $|V(H)|$, still polynomial for fixed H, but perhaps not very appealing).

Similarly, the disjoint paths problem is somewhat reminiscent of the H-minor containment problem, where H consists of k independent edges. In any case, Robertson and Seymour reduced both problems to a more general one, called the Folio problem [Robertson & Seymour, 1994].

We next briefly outline one of the main ideas in the disjoint paths algorithm. Our intent is not to try to present an accurate description of the algorithm, but rather to illustrate the deep connection between the disjoint paths problem and issues related to graph minors. The argument is most persuasive when restricted to planar graphs. Thus we assume a planar input graph, G.

If G has 'not very large' treewidth (a condition that can be tested in polynomial time), then the problem is fairly simple: one can apply a dynamic programming approach as in Section 2.1. Suppose on the other hand that G has very large treewidth. Then, by Theorem 3.0.3, G contains an enormous square grid minor H; i.e., a minor isomorphic to the m-grid where m is very large. For simplicity, assume H is actually a subgraph of G (the exact situation is not very different). Since there are at most $2k$ vertices s_i, t_i, we may even assume that H is 'far away' from all the s_i and t_i. (For example, none of the internal faces of H, as embedded in G, topologically contain any of the s_i and t_i. See Figure 4.)

Now let v be a vertex of H located near the middle of H. Then removing v from G should not alter the situation; that is, G contains the desired family of disjoint paths if and only if $G - v$ does. To see this, assume we are given the desired family of disjoint paths, where one of these paths, p_1, contains v. Suppose we perturb slightly p_1 around v. This perturbation will then cause a ripple effect:

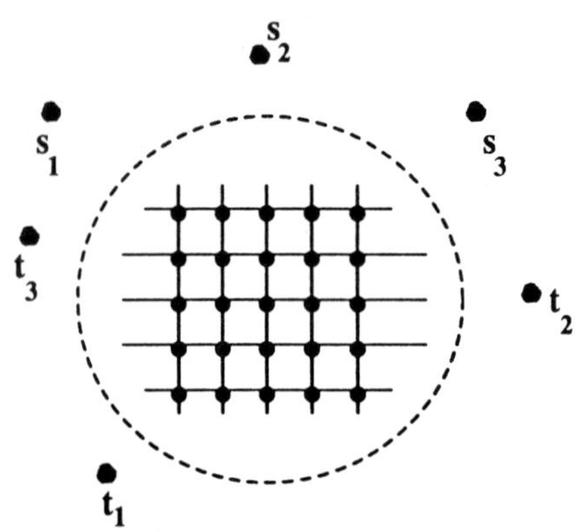

Fig. 4.

we will have to move other paths in order to preserve disjointness. But the fact that H is a very large square grid, and far away from all the s_i and t_i, ensures that a global way of shifting the paths does exist, and we can indeed remove v from G without changing the problem, as desired. Consequently, we have now reduced the problem to an equivalent one in a smaller graph. Now we can go back and repeat the treewidth test, and so on until the problem is solved after at most a linear number of vertex removal steps. There remains the algorithmic problem of constructing the square grid minors when needed — but here the fact that G has very high treewidth makes the task easy.

How do we bypass the planarity assumption? The argument that yields H is just as above. But if G is not planar all vertices near the middle of H may be crucial; i.e., always needed in the disjoint paths. Moreover, the 'far away' requirement for H may not work out. But in any case, it turns out that one can always find an 'irrelevant' vertex. With such a vertex at hand we continue as above. The proof of all this uses several deep structure theorems that are far too complex to describe here. See Robertson & Seymour [1990] for an excellent detailed survey of the disjoint paths algorithm.

5.1. Some new developments concerning disjoint paths

There are some interesting variants of the disjoint paths problem on planar graphs (in fact, on graphs embedded on surfaces) that have recently been studied. The algorithms and theorems involved do not follow from graph minors theory, but we describe them here for completeness.

Some problems have been solved by Schrijver [1991]. The problems were initially motivated by certain issues in circuit routing, as follows. Suppose we are given a chip that contains some devices (think of these as right-angle polygons). The chip also contains a system of tracks, forming a grid, for the purpose of routing wires. Our problem is to route wires on this grid so as to realize connections between given pairs of terminals on these devices. These wires cannot touch one another or a device (other than at their ends) and, moreover, we are even given a sketch of how each wire must look; i.e., how the wire must thread its way among the devices. The algorithmic question is: can we find a wire routing that meets all these requirements? A polynomial-time algorithm for this problem was given by Cole and Siegel [1984] and Leiserson & Maley [1985].

The problem can be substantially generalized as follows. We are given a planar graph G, a collection of faces $F_1, F_2, \ldots, F_m$ of G, a collection of vertices s_i, t_i $(1 \leq i \leq k)$ of G, each located in the boundary of some F_j, and a collection of paths q_i $(1 \leq i \leq k)$ between s_i and t_i. Do there exist vertex-disjoint paths p_i $(1 \leq i \leq k)$ between s_i and t_i, such that p_i is homotopic to q_i in $\Re^2 - F_1 \cup F_2 \ldots \cup F_1$?

Schrijver has presented an $O(n^2 \log n)$ algorithm for this problem. We stress here that, unlike the version of the disjoint paths problem discussed before, the parameter k is *not* assumed to be fixed. At first glance this seems surprising, since the (standard) disjoint paths problem is NP-hard for planar graphs. But notice that in this new version we are told how each path must 'look like'. Reed has improved the algorithm so as to achieve linear run time. The algorithm can also be partially extended to handle disjoint trees (rather than paths) that join specified vertex sets, and also to higher surfaces.

Another area of interest concerns the disjoint paths problem on directed graphs. Ding, Schrijver & Seymour [1994] have considered the following case: we are given a planar digraph D, verticess_i, t_i $(1 \leq i \leq k)$ all located on the boundary of one face F, and subsets of edges A_i $(1 \leq i \leq k)$. We seek vertex-disjoint $s_i - t_i$ paths p_i, all of whose edges are contained in A_i $(1 \leq i \leq k)$. They presented a necessary and sufficient condition for the existence of such paths (which extends one given in Robertson & Seymour [1986]), together with a polynomial-time algorithm for the problem.

6. Challenges to practicality

We close this chapter with a discussion of several unusual aspects of algorithms provided by the Graph Minor Theorem. Recall that if F is a minor-closed family of graphs, then we know from the developments already sketched that F can be recognized in polynomial time. Letting n denote the number of vertices in G, the general bound is $O(n^3)$. If F excludes a planar graph, then the bound is reduced to $O(n^2)$. Interestingly, such algorithms suffer from novel shortcomings:

- the algorithms require immense constants of proportionality,
- only the complexity of decision problems is established, and
- there is no general means for finding (or even recognizing) correct algorithms.

We tackle each of these issues in turn, illustrating algorithmic techniques with simple examples. We make no pretence that these examples reflect the state of the art. The interested reader is referred to Fellows & Langston [1988, 1989] for more complex methods.

6.1. Constants of proportionality

The theory developed by Robertson and Seymour proceeds in distinct structural stages. The theorems that employ this structural information introduce stunningly enormous constants of proportionality into polynomial-time decision algorithms. These huge structural constants can sometimes be eliminated by proving problem-specific structural bounds.

Example 6.1.1. Consider the gate matrix layout problem mentioned in the last section. It is known that, for any fixed value of k, there is a surjective map from Boolean matrices to graphs such that all matrices mapped to the same graph have the same layout cost, that the 'yes' family of graphs in the image of the map is minor-closed, and that planar obstructions exist. Thus gate matrix layout is decidable for any fixed k in $O(n^2)$ time, but with a gigantic structural constant c_k bounding the treewidth of any graph in F_k entering into the constant of proportionality of the algorithm. (This constant is computed by a nine step procedure that involves several compositions of towers of 2's functions [Robertson & Seymour, 1986]. As we have previously noted, however, the family of matrices with gate matrix layout cost k turn out to correspond to the family of graphs with pathwidth $k-1$, which is a proper subset of the family of graphs with treewidth $k-1$. Thus a direct consideration of the needed structural bound allows the constant c_k to be replaced by $k-1$.

A more general approach is to prove structural bounds specific to a particular class of obstructions. These bounds then apply to any family with an obstruction in that class.

Theorem 6.1.2 [Fellows & Langston, 1989]. *Any minor-closed family that excludes a cycle of length l has treewidth at most $l-2$ and can be recognized in $O(n)$ time.*

Example 6.1.3. Reconsider the vertex cover problem, where we seek to determine whether all edges in an input graph G can be covered by at most k vertices, for some fixed k. As discussed in Example 3.1.1, this problem could be solved by finding a tree decomposition and then applying dynamic programming. Both of these steps could require $O(n^2)$ time without special tools. Moreover, the tree decomposition width is the enormous c_k. But the family of 'yes' instances is minor-closed and excludes C_{2k+1}, the cycle of length $2k+1$. By applying the technique used in the proof of Theorem 6.1.2, only a (linear time) depth-first search is needed to obtain a tree decomposition of width at most $2k-1$, followed by a finite number of obstruction tests, each taking linear time. Thus both the structural constant and the time complexity are reduced.

6.2. Decision problems versus search problems

Algorithms based on finite obstruction sets only solve decision problems. In practice, one is usually more concerned with *search* problems, where the goal is to search for evidence that an input is a 'yes' instance. For example, a 'yes' or 'no' response is sufficient to answer the decision version of vertex cover. For the search version, however, we want a satisfying cover (set of k or fewer vertices) when any exist. Fortunately, decision algorithms can be converted into search algorithms for the vast majority of problems amenable to the work of the graph minors project. The general idea is often termed *self-reduction*, whereby the decision algorithm is used as a subprogram by the search algorithm.

Example 6.2.1. In the decision version of the longest path problem, we seek to know whether an input graph contains a simple path of length k or more. The problem is NP-complete in general, but solvable in $O(n)$ time for any fixed k, because the 'no' family is minor-closed and excludes a cycle of length $k+1$. When solving this problem in a practical setting, of course, we are concerned with finding a sufficiently long path when any exist, that is, solving the search version of the problem. To accomplish this, we need only self-reduce as follows. First, accept the input and pose the decision version of the problem. If the response is 'no,' then halt - no satisfying evidence exists. If the response is 'yes', then perform the following series of operations for each edge in the graph:

1. temporarily remove the edge and pose the decision problem again;
2. if the new graph is a 'no' instance, replace the edge (it is needed in any sufficiently long path);
3. if the new graph is a 'yes' instance, permanently remove the edge (some sufficiently long path remains).

Thus, $O(n^2)$ calls to an $O(n)$ decision algorithm suffice, yielding an $O(n^3)$ time search algorithm.

6.3. Nonconstructivity

As mentioned in Section 2, a guarantee of polynomial-time decidability provided by minor-closure is nonconstructive. But need this be the case? To consider such a question, we must decide on a finite representation for an arbitrary minor-closed family. (After all, it would of course be impossible to construct algorithms if the representation were not finite!) A reasonable choice is the Turing machine, the standard model of complexity theory. Unfortunately, a reduction from the halting problem affirms that nonconstructivity cannot be eliminated in a general sense.

Theorem 6.3.1 [Fellows & Langston, 1989]. *There is no algorithm to compute, from a finite description of a minor-closed family represented by a Turing machine that accepts precisely the graphs in the family, the set of obstructions for that family.*

So we must settle for something less. In the following, the term *known* refers to an algorithm that can, at least in principle, be coded up and run.

Theorem 6.3.2 [Fellows & Langston, 1989]. *Let P_D denote a decision problem whose 'yes' instances are minor-closed. Let P_S denote the corresponding search problem. If algorithms are known to self-reduce P_S to P_D and to check whether a candidate solution satisfies P_S, then an algorithm is known that solves both P_D and P_S.*

The proof of this has an interesting wrinkle, in that the resultant (known) algorithms generate and make use of incomplete obstruction sets, yet they cannot be used to generate complete sets or even to check the completeness of proffered sets!

Example 6.3.3. Consider the NP-complete modified cutwidth problem, in which we are given a graph G and a positive integer k, and are asked whether G can be laid out with its vertices along a straight line so that no plane that cuts the line *on* an arbitrary vertex can cut more than k edges. Until recently, the fastest known algorithm for both the decision and the search versions of this problem had time complexity $O(n^k)$. Thus modified min-cut is technically in P for any fixed value of k. This can be improved on, but nonconstructively, because the family of line graphs of 'yes' instances is minor-closed. But modified min-cut is easy to self-reduce and easy to check. Thus the decision and search versions of modified min-cut can be solved in $O(n^3)$ time constructively (with known algorithms).

Acknowledgments

We wish to express our appreciation to Jean Blair, Heather Booth, Rajeev Govindan, Eric Kirsch, Scott McCaughrin and Siddharthan Ramachandramurthi for carefully reviewing an early draft of this chapter. We also wish to thank an anonymous reviewer for many helpful comments.

Postscript

Progress on the topics we have discussed continues apace. By the time this chapter reaches print, we are confident that many more relevant results will have been announced. We apologize in advance to those authors whose recent work has thus been unfortunately omitted from this treatment.

References

Archdeacon, D. (1980). A Kuratowski Theorem for the Projective Plane, Ph. D. Thesis, Ohio State University.

Arnborg, S., J. Lagergren and D. Seese (1991). Easy problems for tree decomposable graphs. *J. Algorithms* 12, 308–340.

Arnborg, S., and A. Proskurowski (1987). Complexity of finding embeddings in a k-tree. *SIAM J. Alg. Disc. Meth.* 8 277–284.

Alon, N., P.D. Seymour and R. Thomas (1994). A separator theorem for non-planar graphs, to appear.

Bienstock, D. (1990). On embedding graphs in trees. *J. Comb. Theory Ser. B* 49, 103–136.

Bondy, J.A. and U.S.R. Murty (1976). *Graph Theory with Applications*, London, Macmillan.

Bienstock, D., N. Robertson, P.D. Seymour and R. Thomas (1991). Quickly excluding a forest. *J. Comb. Theory Ser. B* 52 , 274–283.

Bienstock, D. and P.D. Seymour (1991). Monotonicity in graph searching. *J. Algorithms* 12, 239–245.

Cole, R., and A. Siegel (1984). River routing every which way, but loose, *Proc. 25th Annu. Symp. on Foundations of Computer Science*, pp. 65–73.

Deo, N., M.S. Krishnamoorthy and M.A. Langston (1987). Exact and approximate solutions for the gate matrix layout problem. *IEEE Trans. Comput-Aided Design Integrated Circuits Syst.* 6, 79–84.

Ding, G., A. Schrijver and P.D. Seymour (1994) Disjoint paths in a planar graph — a general theorem, to appear.

Fellows, M.R. and M.A. Langston (1988). Nonconstructive tools for proving polynomial-time decidability. *J. ACM* 35, 727–739.

Fellows, M.R. and M.A. Langston (1989). On search, decision and the efficiency of polynomial-time algorithms. *Proc. 21st Annu. ACM Symp. on Theory of Computing*, pp. 501–512.

Gavril, F. (1974). The intersection graphs of subtrees in trees are exactly the chordal graphs. *J. Comb. Theory Ser. B* 16, 47–56.

Glover, H., P. Huneke and C.S. Wang (1979). 103 Graphs that are irreducible for the projective plane. *J. Comb. Theory Ser. B* 27, 332–370.

Golumbic, M.O. (1980). *Algorithmic Graph Theory and Perfect Graphs*, Academic Press.

Hong, J., K. Mehlhorn and A. Rosenberg (1983). Cost trade-offs in graph embeddings, with applications. *J. ACM* 30, 709–728.

Hopcroft, J.E., and R.E. Tarjan (1974). Efficient planarity testing. *J. ACM* 21, 549–568.

Karp, R.M. (1975). On the complexity of combinatorial problems. *Networks* 5, 45–68.

Kinnersley, N.G., and M.A. Langston (1991). Obstruction set isolation for the gate matrix layout problem, Technical Report CS-91-126, Department of Computer Science, University of Tennessee.

Kirousis, L.M., and C.H. Papadimitriou (1986). Searching and pebbling. *J. Theor. Comput. Sci.* 47, 205–218.

Kruskal, J. (1960). Well-quasi-ordering, the tree theorem, and Vázsonyi's conjecture. *Trans. Am. Math. Soc.* 95, 210–225.

Kuratowski, C. (1930). Sur le problème des courbes gauches en topologie, *Fund. Math.* 15, 271–283.

Leiserson, C.E., and F.M. Maley (1985). Algorithms for routing and testing routability of planar VLSI-layouts. *Proc. 17th Annu. ACM Symp. on Theory of Computing*, pp. 69–78.

Lipton, R.J., and R.E. Tarjan (1979). A separator theorem for planar graphs, *SIAM J. Appl. Math.* 36, 177–189.

Massey, W.S. (1967). *Algebraic Topology: An Introduction*, Springer, New York, N.Y.

Makedon, F., and I.H. Sudborough (1989). On minimizing width in linear layouts. *Discr. Appl. Math.* 23, 243–265.

Reed, B. (1990). Personal communication.

Robertson, N., and P.D. Seymour (1990). An outline of a disjoint paths algorithm, in: B. Korte, L. Lovász, H.-J. Prömel and A. Schrijver (eds.), *Algorithms and Combinatorics*, Springer-Verlag, pp. 267–292.

Robertson, N., and P.D. Seymour (1983). Graph Minors. I. Excluding a Forest, *J. Comb. Theory Ser. B* 35, 39–61.

Robertson, N., and P.D. Seymour (1990). Graph Minors. IV. Treewidth and Well-Quasi-Ordering. *J. Comb. Theory Ser. B* 48, 227–254.

Robertson, N., and P.D. Seymour (1986). Graph Minors. V. Excluding a Planar Graph. *J. Comb. Theory Ser. B* 41, 92–114.

Robertson, N., and P.D. Seymour (1986). Graph Minors. VI. Disjoint Paths Across a Disk. *J. Comb. Theory Ser. B* 41, 115–138.

Robertson, N., and P.D. Seymour (1990). Graph Minors. VIII. A Kuratowski theorem for general surfaces. *J. Comb. Theory Ser. B* 48, 255–288.

Robertson, N., and P.D. Seymour (1991). Graph Minors. X. Obstructions to tree decomposition. *J. Comb. Theory Ser. B* 52, 152–190.

Robertson, N., and P.D. Seymour (1994). Graph Minors. XIII. The disjoint paths problem, to appear.

Thomas, R. (1991). Personal communication.

Schrijver, A. (1991). Decomposition of graphs on surfaces and a homotopic circulation theorem. *J. Comb. Theory Ser. B* 51, 161–210.

Seymour, P.D. (1980). Disjoint paths in graphs. *Discr. Math.* 29, 239–309.

Shiloach, Y. (1980). A polynomial solution to the undirected two paths problem. *J. ACM* 27, 455–456.

Seymour, P.D., and R. Thomas (1994). Graph searching and a minimax theorem for treewidth, to appear.

Seymour, P.D., and R. Thomas (1994). Call routing and the rat-catcher, to appear.

Yan, X. (1989). Approximating the pathwidth of outerplanar graphs, M.S. Thesis, Washington State University.

M.O. Ball et al., Eds., *Handbooks in OR & MS, Vol. 7*

Chapter 9

Optimal Trees

Thomas L. Magnanti
Sloan School of Management and Operations Research Center, MIT, Cambridge, MA 02139, U.S.A.

Laurence A. Wolsey
C.O.R.E., Université Catholique de Louvain, 1348 Louvain-la-Neuve, Belgium

1. Introduction

Trees are particular types of graphs that on the surface appear to be quite specialized, so much so that they might not seem to merit in-depth investigation. Perhaps, surprisingly, just the opposite is true. As we will see in this chapter, tree optimization problems arise in many applications, pose significant modeling and algorithmic challenges, are building blocks for constructing many complex models, and provide a concrete setting for illustrating many key ideas from the field of combinatorial optimization.

A tree[1] is a connected graph containing no cycles. A tree (or subtree) of a general undirected graph $G = (V, E)$ with a node (or vertex) set V and edge set E is a connected subgraph $T = (V', E')$ containing no cycles. We say that the tree spans the nodes V'. For convenience, we sometimes refer to a tree by its set of edges with the understanding that the tree also contains the nodes incident to these edges. We say that T is a spanning tree (of G) if T spans all the nodes V of G, that is, $V' = V$. Recall that adding an edge $\{i, j\}$ joining two nodes in a tree T creates a unique cycle with the edges already in the tree. Moreover, a graph with n nodes is a spanning tree if and only if it is connected and contains $n - 1$ edges.

Trees are important for several reasons:

(i) Trees are the minimal graphs that connect any set of nodes, thereby permitting all the nodes to communicate with each other without any redundancies (that is, no extra arcs are needed to ensure connectivity). As a result, if the arcs of a network have positive costs, the minimum cost subgraph connecting all the

[1] Throughout this chapter, we assume familiarity with the basic definitions of graphs including such concepts as paths and cycles, cuts, edges incident to a node, node degrees, and connected graphs. We also assume familiarity with the max-flow min-cut theorem of network flows and with the elements of linear programming. The final few sections require some basic concepts from integer programming.

nodes is a tree that spans all of the nodes, that is, it is a spanning tree of the network.

(ii) Many tree optimization problems are quite easy to solve; for example, efficient types of greedy, or single pass, algorithms are able to find the least cost spanning tree of a network (we define and analyze this problem in Section 2). In this setting, we are given a general network and wish to find an optimal tree within this network. In another class of models, we wish to solve an optimization problem defined on a tree, for example, find an optimal set of facility locations on a tree. In this setting, dynamic programming algorithms typically are efficient methods for finding optimal solutions.

(iii) Tree optimization problems arise in a surprisingly large number of applications in such fields as computer networking, energy distribution, facility location, manufacturing, and telecommunications.

(iv) Trees provide optimal solutions to many network optimization problems.

Indeed, any network flow problem with a concave objective function always has an optimal tree solution (in a sense that we will define later). In particular, because (spanning) tree solutions correspond to basic solutions of linear programs, linear programming network problems always have (spanning) tree solutions.

(v) A tree is a core combinatorial object that embodies key structural properties that other, more general, combinatorial models share. For example, spanning trees are the maximal independent sets of one of the simplest types of matroids, and so the study of trees provides considerable insight about both the structure and solution methods for matroids (for example, the greedy algorithm for solving these problems, or linear programming representations of the problems). Because trees are the simplest type of network design model, the study of trees also provides valuable lessons concerning the analysis of more general network design problems.

(vi) Many optimization models, such as the ubiquitous traveling salesman problem, have embedded tree structure; algorithms for solving these models can often exploit the embedded tree structure.

Coverage

This paper has two broad objectives. First, it describes a number of core results concerning tree optimization problems. These results show that even though trees are rather simple combinatorial objects, their analysis raises a number of fascinating issues that require fairly deep insight to resolve. Second, because the analysis of optimal trees poses many of the same issues that arise in more general settings of combinatorial optimization and integer programming, the study of optimal trees provides an accessible and yet fertile arena for introducing many key ideas from the branch of combinatorial optimization known as polyhedral combinatorics (the study of integer polyhedra).

In addressing these issues, we will consider the following questions:

- Can we devise computationally efficient algorithms for solving tree optimization problems?
- What is the relationship between various (integer programming) formulations of tree optimization problems?

• Can we describe the underlying mathematical structure of these models, particularly the structure of the polyhedra that are defined by relaxing the integrality restrictions in their integer programming formulations?

• How can we use the study of optimal tree problems to learn about key ideas from the field of combinatorial optimization such as the design and analysis of combinatorial algorithms, the use of bounding procedures (particularly, Lagrangian relaxation) as an analytic tool, and basic approaches and proof methods from the field of polyhedral combinatorics?

We begin in Section 2 with a taxonomy of tree optimization problems together with illustrations of optimal tree applications in such fields as telecommunications, electric power distribution, vehicle routing, computer chip design, and production planning. In Section 3, we study the renowned minimum spanning tree problem. We introduce and analyze a greedy solution procedure and examine the polyhedral structure of the convex hull of incidence vectors of spanning trees. In the context of this discussion, we examine the relationship between eight different formulations of the minimum spanning tree problem that are variants of certain packing, cut, and network flow models.

In Section 4, we examine another basic tree optimization problem, finding an optimal rooted tree within a tree. After showing how to solve this problem efficiently using dynamic programming, we then use three different arguments (a network flow argument, a dynamic programming argument, and a general 'optimal' inequality argument from the field of polyhedral combinatorics) to show that a particular linear programming formulation defines the convex hull of incidence vectors of rooted trees. Because the basic result in this section is fairly easy to establish, this problem provides an attractive setting for introducing these important proof techniques.

In Section 5, we consider two other tree models that can be solved efficiently by combinatorial algorithms — a degree constrained minimum spanning tree problem (with a degree constraint imposed upon a single node) and a directed version of the minimum spanning tree problem. For both problems, we describe an efficient algorithmic procedure and fully describe the underlying integer polyhedron.

In Sections 6-9 we consider more general models that are, from the perspective of computational complexity theory, difficult to solve. For each of these problems, we provide a partial description of the underlying integer polyhedron and describe one or more solution approaches.

We begin in Section 6 by studying a network version of the well-known Steiner tree problem. Actually, we consider a more general problem known as the node weighted Steiner tree problem. Generalizing our discussion of the spanning tree problem in Section 3, we examine the relationship between the polyhedron defined by five different formulations of the problem. For one model, we show that the objective value for a linear programming relaxation of the Steiner tree problem has an optimal objective value no more than twice the cost of an optimal Steiner tree. Using this result, we are able to show that a particular spanning tree heuristic always produces a solution whose cost is no more than twice the cost of an optimal

Steiner tree. In this discussion, we also comment briefly on solution methods for solving the Steiner tree problem.

In Section 7, we study the problem of packing rooted trees in a given tree. This model arises in certain applications in production planning (the economic lot-sizing problem) and in facility location on a tree (for example, in locating message handling facilities in a telecommunications network). We show how to solve uncapacitated versions of this problem by dynamic programming and, in this case, we completely describe the structure of the underlying integer polyhedron. For more complex constrained problems, we show how to 'paste' together the convex hull of certain subproblems to obtain the convex hull of the overall problem (this is one of the few results of this type in the field of combinatorial optimization). We also describe three different solution approaches for solving the problem — a cutting plane procedure, a column generation procedure, and a Lagrangian relaxation procedure.

In Section 8, we consider the more general problem of packing subtrees in a general graph. This problem arises in such varied problem settings as multi-item production planning, clustering, computer networking, and vehicle routing. This class of models permits constraints that limit the number of subtrees or that limit the size (number of nodes) of any subtree. Our discussion focuses on extending the algorithms we have considered previously in Section 7 when we considered optimal subtrees of a tree.

In Section 9, we briefly introduce one final set of models, hierarchical tree problems that contain two types of edges — those with high reliability versus those with low reliability (or high capacity versus low capacity). In these instances, we need to connect certain 'primary' nodes with the highly reliable (or high capacity) edges. We describe an integer programming formulation of this problem that combines formulations of the minimum spanning tree and Steiner tree problems as well as a heuristic algorithm; we also give a bound on how far both the heuristic solution and the optimal objective value of the linear programming relaxation can be from optimality.

Section 10 is a brief summary of the chapter and Section 11 contains notes and references for each section.

Notation

Frequently in our discussion, we want to consider a subset of the edges in a graph $G = (V, E)$. We use the following notation. If S and T are any two subsets of nodes, not necessarily distinct, we let $E(S, T) = \{e = \{i, j\} \in E : i \in S \text{ and } j \in T\}$ denote the set of edges with one end node in S and the other end node in T. We let $E(S) \equiv E(S, S)$ denote the set of edges whose end nodes are both in S. $\bar{S} = V \setminus S$ denotes the complement of S and $\delta(S)$ denotes the cutset determined by S, that is, $\delta(S) = E(S, \bar{S}) = \{e = \{i, j\} \in E : i \in S \text{ and } j \in \bar{S}\}$. For any graph G, we let $V(G)$ denote its set of nodes and for any set of edges $\hat{E}$ of any graph, we let $V(\hat{E})$ denote the set of nodes that are incident to one of the edges in $\hat{E}$.

At times, we consider directed graphs, or digraphs, $D = (V, A)$ containing a set A of directed arcs. In these situations, we let $\delta^+(S) = \{e = (i, j) \in A :$

$i \in S$ and $j \in \bar{S}\}$ denote the cutset directed *out of* the node set S and let $\delta^-(S) = \{e = (i, j) \in A : i \in \bar{S}$ and $j \in S\}$ denote the cutset directed *into* the node set S. We also let $A(S) = \{e = (i, j) \in A : i \in S$ and $j \in S\}$ and define $V(D)$ and $V(\bar{A})$ for any set $\bar{A}$ of arcs, respectively, as the nodes in the digraph D and the nodes in the digraph that are incident to one of the arcs in $\bar{A}$.

As shorthand notation, for any node v, we let $\delta(v) = \delta(\{v\})$, $\delta^+(v) = \delta^+(\{v\})$, and $\delta^-(v) = \delta^-(\{v\})$.

We also let $\mathbf{1}$ denote a vector of ones, whose dimension will be clear from context, let R^m denote the space of m-dimensional real numbers, Z^m denote the space of m-dimensional integer vectors, and $\{0, 1\}^m = \{x \in Z^m : 0 \leq x \leq \mathbf{1}\}$.

The set notation $A \subset B$ denotes $A \subseteq B$ and $A \neq B$.

For any set S, we let $\text{conv}(S)$ denote the convex hull of S, that is, the set of points $x = \{\sum_{j=1}^{k} \lambda_j s^j : \sum_{j=1}^{k} \lambda_j = 1$ and $\lambda_j \geq 0$ for $j = 1, \ldots, k$ and some points $s^j \in S\}$ obtained as weighted combinations of points in S.

Recall that a polyhedron in R^n is the set of solutions of a finite number of linear inequalities (and equalities). If a polyhedron is bounded, then it also is the convex hull of its extreme points. If each extreme point is an integer vector, we say that the polyhedron is an *integer polyhedron*.

Let A and B be two given matrices and b be a column vector, all with the same number of rows. Frequently, we will consider systems of inequalities $Ax + Dy \leq b$ defined by two sets x and y of variables. We refer to the set of points $\{x : Ax + Dy \leq b$ for some vector $y\}$ as the set of x-feasible solutions to this system. Note that $Q = \{x : Ax + Dy \leq b$ for some vector $y\}$ is the projection of the polyhedron $P = \{(x, y) : Ax + Dy \leq b\}$ onto the space of x-variables. As is well known, Q itself is a polyhedron, that is, can be expressed as the set of solutions of a finite number of inequalities involving only the x-variables.

2. Tree optimization problems

Tree optimization problems arise in many forms. In this chapter, we consider two generic problem types:

(a) *Optimal trees*. Given a graph $G = (V, E)$ with node set V and edge set E and with a weight w_e defined on each edge $e \in E$, find a tree T in G that optimizes (maximizes or minimizes) the total weight of the edges in T. The tree might also have a designated root node r and have various constraints imposed on the root node or on the subtrees created if we eliminate the root node and its incident edges.

(b) *Optimal subtrees of a tree (packing subtrees in a tree)*. Given a tree T, suppose we wish to find a set of node disjoint subtrees of T, each with a designated root node. Each node v has a weight w_v^r that depends upon the root node r of the subtree that contains it, and we wish to optimize (maximize or minimize) the total weight of the nodes in the subtrees. Note that this model permits edge weights as well as node weights since once we have selected a root node for each subtree, each edge of the tree has a uniquely associated node, namely the first node on the

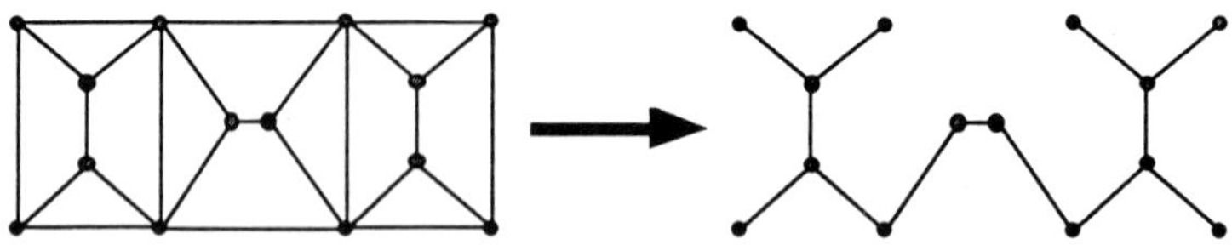

Fig. 1. Optimal tree in a graph.

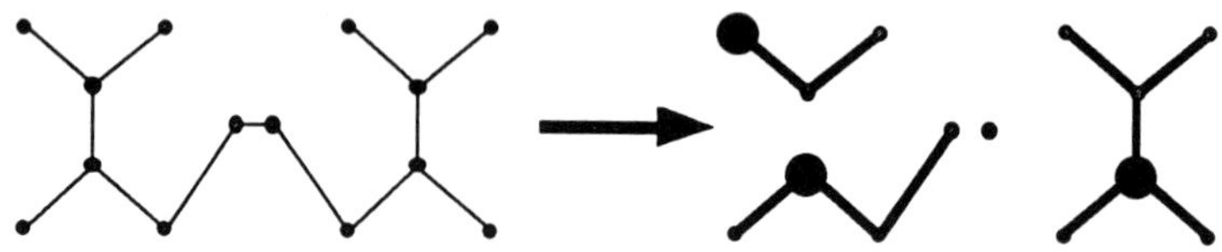

Fig. 2. Optimal rooted subtrees of a tree.

path connecting that edge to the root. Therefore, by associating the edge weight with this node, we can formulate the tree packing problem with edge weights as an equivalent model with weights defined only on the nodes.

Figures 1 and 2 give a schematic representation of both of these problem types. We might also consider another problem type: packing subtrees in a general network. In principle, we might view this problem as a composite of the other two: first, we find a tree in the network and then we pack subtrees in this tree.

Both the optimal tree problem and optimal subtrees in a tree problem arise in several different forms, depending upon the restrictions we impose upon the set of (feasible) trees/subtrees we wish to consider. The following constraints arise in several practical problem settings (the root node in these constraints is either a designated single root in the optimal tree problem, or the root node of any subtree in the packing subtree problem):

- A size constraint imposed upon the number of nodes in any (sub)tree.
- Topological constraints imposed on any (sub)tree [e.g., maximum or minimum node degrees, restrictions that certain specific nodes must be included in a (sub)tree]. In particular, we might impose a degree constraint on the root node.
- A size constraint imposed upon the subtrees formed by removing the root node of any (sub)tree. More generally, each node might have an associated weight, and we might wish to restrict the total node weight of any subtree.
- Bounds (maximum and/or minimum) imposed upon the number of subtrees in the packing subtrees of a tree problem.
- A flow requirement imposed upon the (sub)trees, together with capacity constraints imposed upon the total throughput of any edge or node; and/or
- The availability of multiple types of edges and root nodes, with restrictions imposed upon the types of facilities (edges and nodes) used. For example, each type of edge or node might have an associated cost or capacity and we might seek a minimum cost solution that loads the network with certain required capacity. Or, we might impose multi-layered requirements, for example, certain primary nodes

be connected by high capacity (or reliability) edges and secondary nodes by any type of facility (high or low capacity).

These constraints arise for a variety of reasons. In several applications, root nodes represent service facilities, for example, plants or warehouses in a production system, hospitals in an urban network, centralized computers in a computer network, or multiplexers in a communication network. A size constraint on each subtree might model capacity limitations on the service facility and a cardinality constraint on the number of subtrees might model limited availability of service facilities or a managerial or engineering decision to limit the number of facilities. In some applications, the edges adjacent to the root node represent a physical entity such as a loading dock in a warehouse or a communication port in a centralized computer. In these settings, a degree constraint on the root node might represent a physical limit on the availability of these facilities. In addition, nodes in the subtrees formed by removing the edges adjacent to the root node might represent the customers served by each entity (port) at the root node. For reliability reasons, or to model capacities, we might wish to limit the size (number of nodes) in each subtree.

We could, of course, add even further wrinkles on these various problem types. For example, rather than viewing just the root nodes as 'special' and imposing various restrictions on them or on the subtrees formed by removing them from the solution, we could consider layered problems with root nodes, first-level nodes (those adjacent to the root nodes), second-level nodes, and so forth, and impose restrictions on the nodes at each level. Models with constraints imposed upon the root node appear to capture many of the issues encountered in practice and studied by researchers in the past, so we focus on these versions of the problems.

Certain special versions of these problems are either standard topics in the literature or arise frequently as subproblems in more general combinatorial optimization applications.

- *Minimum spanning tree problem*. In this basic 'tree in a network model', we wish to find a tree that spans (contains) all the nodes of a graph G and that minimizes the overall weight of the edges in the tree. In this in this case, we impose no topological or capacity restrictions on the tree.
- *Rooted subtree problem*. Given a tree and a root node r, we wish to find a subtree rooted at (containing) this node and that minimizes the overall weight of the nodes (and/or arcs) in the subtree. This problem is a core model for the class of subtrees of a tree problems, much like the minimum spanning tree is a core model for the class of optimal tree problems.
- *Steiner tree problem*. Let $G = (V, E)$ be a given graph with a weight (cost) defined on each edge $e \in E$. Given a set T of terminal nodes that need to be connected to each other, we wish to find a tree of G that contains these nodes and whose total edge weight is as small as possible. Note that the optimal tree might contain nodes, called Steiner nodes, other than the terminal nodes T.
- *K-median problem*. Find K or fewer node disjoint subtrees of a network, each with a root, that minimizes the total edge weight of the edges in the subtrees. The K-median problem on a tree is the version of this problem defined on a tree.

• *C-capacitated tree problem*. In this version of the rooted subtrees of a network problem, each subtree is limited to containing at most C nodes. The C-capacitated problem on a tree is the version of this problem defined on a tree.

In order to make this problem taxonomy more concrete, we next briefly consider a few important application contexts.

Designing telecommunications and electric power networks

Suppose that we wish to design a network that connects customers in a telecommunications or electrical power network. The links are very expensive (in part, because we might need to dig trenches to house them) and the routing costs are negligible: once we have installed the line facilities (edges), the routing cost is very small. Therefore, we want to connect the customers using the least expensive tree. If all the nodes of the network are customers, the problem is the classical minimum spanning tree problem. If we need to connect only some of the nodes, and can use other nodes as intermediate nodes, the problems become the classical Steiner tree problem. In a multi-layered version of this class of problems, we wish to connect certain 'key' users using highly reliable communication lines or high voltage transmission lines. We can use less reliable lines or low voltage lines at a lower cost to connect the other users of the system.

Similar types of applications arise in other settings as well. For example, in constructing a highway infrastructure in a developing country, our first objective might be to solve a minimum spanning tree problem or Steiner tree problem to connect all the major cities as cheaply as possible. Or, we might wish to solve a multi-layered problem, ensuring that we connect all the major cities by highways and all the cities, whether major or not, through the use of any combination of highways or secondary access roads.

Facility location

In a distribution system of geographically dispersed customers on a network, we wish to establish warehouses (distribution centers) to fulfill customer orders. Suppose that for administrative reasons (simplicity of paperwork or ease in monitoring and controlling the system), we wish to service each customer from a single warehouse. Moreover, suppose we wish to satisfy a contiguity property: if a warehouse at location r services customer i along a path that passes though the location of customer j, then warehouse i must also service customer j. So, each feasible solution is set of subtrees, each with a root node which is the location of the warehouse serving the customers in that subtree.

This basic facility location problem arises in many alternate forms and in many guises. For example, we might impose capacities (for example, a limit on the number of customers served) on each service facility or we might restrict the total number of service facilities. These versions of the problem would be C-capacity subtree and K-median problems.

Figure 3 illustrates another application context that arises in telecommunications. Most current telecommunications systems use a tree (typically of copper cable) to connect subscribers to physical devices called local switches that route

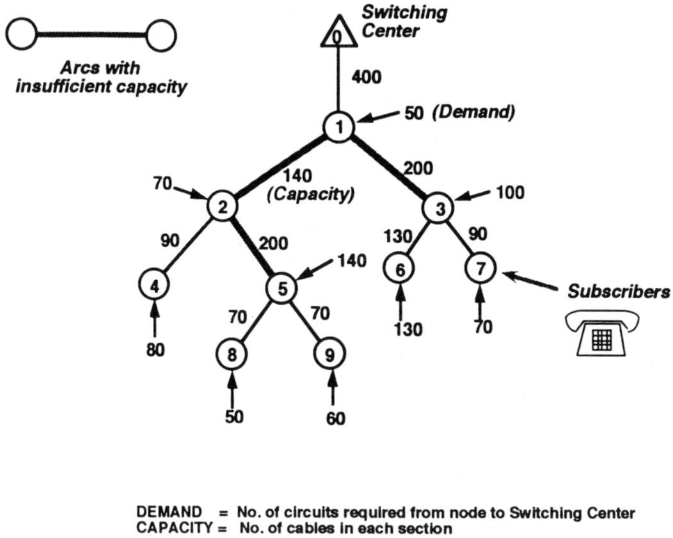

Fig. 3. Local access telecommunication network.

calls between the subscribers. Each subscriber is connected to a switching center in the 'local access' tree by a dedicated circuit (telephone line). Each edge of the tree has a capacity (the number of physical telephone lines installed on that edge) and as the demand for services increases, the network might have insufficient capacity. In this example the nodes 3, 6, and 7 require a total of 300 circuits and the edge $\{1,3\}$ between these nodes and the switch has a capacity of only 200 circuits. Two other edges in the tree have insufficient capacity: (a) nodes 5, 8, and 9 require 250 circuits, and the edge $\{2, 5\}$ has a capacity of only 200 circuits; and (b) nodes 2, 4, 5, 8, and 9 require 400 circuits, and the edge $\{1, 2\}$ has a capacity of only 140 circuits

One way to meet the excess demand is to add additional capacity (copper cable) on edges with insufficient capacity. Another option is to add sophisticated equipment known as concentrators (or alternative equipment known as multiplexers, or remote switches) that compress messages so that they require less circuit capacity (that is, so that calls can share lines). Figure 4 shows one possible solution for fulfilling the demand of all the nodes in Figure 3. In this case, we have added 100 extra lines on edge $\{1,3\}$ and added a concentrator at node 5 that serves the subscribers of nodes 2, 4, 5, 8, and 9. This concentrator uses a compression ratio of 10 to 1 so the 400 circuits assigned to it use only 40 circuits on the downstream path 5–2–1–0 connecting node 5 to the switching center. Consequently, this path has sufficient capacity for all the subscribers that use it.

Note two properties of the solution shown in Figure 4. First, the solution assigns all of the demand at each node either directly to the switching center (the nodes 1, 3, 6, and 7) or to the concentrator at node 5 (the nodes 2, 4, 5, 8, and 9). In

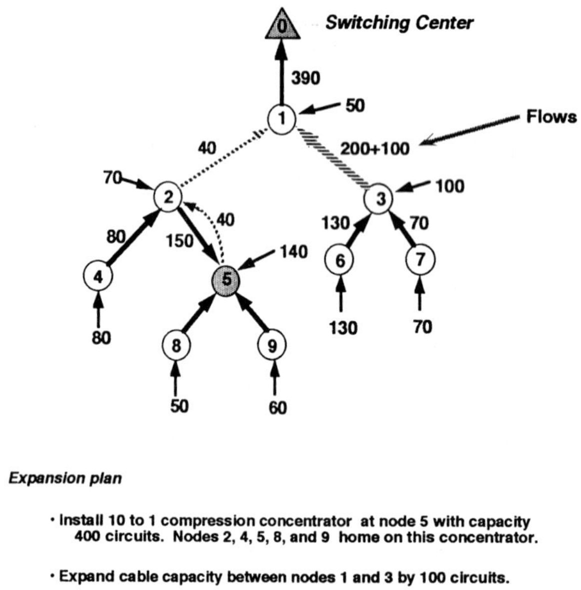

Fig. 4. Local access expansion strategy as subtree packing.

addition, the solution satisfies a contiguity property: if the solution assigns node u to the switching center (or to the concentrator) and node v lies on the path connecting node u to the switching center (concentrator), then it also assigns node v to the switching center (concentrator). These two assumptions imply that the solution decomposes the tree into a set of rooted subtrees; one contains the switching center and all others contain a single concentrator. Therefore, the problem is an application of packing rooted subtrees in a tree.

Routing problems

Figure 5 shows a solution to an illustrative vehicle routing problem. In this application context, we are given a fleet of vehicles domiciled at a depot, node 0, and wish to find a set of tours (cycles that are node disjoint except at the depot) that contain all the customer nodes $1, 2, \ldots, n$. We incur a cost for traversing any edge and wish to find the minimum cost routing plan. Note that if we are given any set of tours and eliminate all of the edges incident to the depot, the resulting graphical structure is a set of node disjoint paths that contain all the nodes $1, 2, \ldots, n$. Therefore, the solution is a very special type of tree packing problem, one in which each tree must be a path. Since we wish to include every node in the solution, we might refer to this problem as a path partitioning problem since we are partitioning the non-depot nodes into paths.

In the simplest version of this problem, the customers are identical and each vehicle has sufficient capacity to serve all of the customers, so any path partitioning of the nodes $1, 2, \ldots, n$ will be feasible. If we impose additional restrictions on the

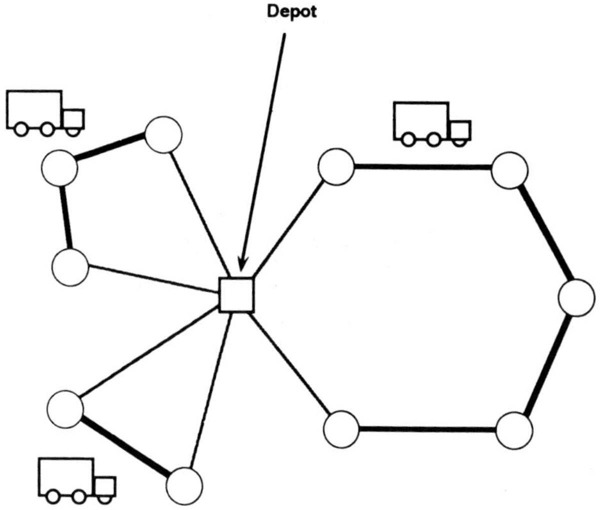

Fig. 5. Vehicle routing as packing paths in a network.

solution, then the problem becomes a special version (because the trees must be paths) of one of the alternative tree problems we have introduced previously.

For example, if we have K available vehicles, the problem becomes a K-median version of the path partitioning problem since we wish to use at most K paths to cover the nodes $1, 2, \ldots, n$. In particular, if $K = 1$, the vehicle routing problem becomes the renowned traveling salesman problem; in this case, any feasible solution to the associated path partitioning problem is a Hamiltonian path (that is, a single path containing all the nodes).

If the customers are identical, that is, have the same demands, which by scaling we can assume are all one unit, and each vehicle has a capacity of C units, then each tour, and so each path in the path partitioning problem, can contain at most C customer points. Therefore, the problem becomes a C-capacitated subtree version of the path partitioning problem.

In practice, applications often have other important problem features; for example, (a) the demands typically vary across the customers, or (b) each edge might have an associated travel time and we might have a limit on the overall travel time of every route (and so each path in the associated path partitioning problem) or we might have specified time windows on the delivery (pick-up) time of each customer. In these instances, the Hamiltonian paths in any feasible solution will have other constraints imposed upon them, for example, restrictions on the total length (travel time) of each path.

We might note that the 'vehicle routing' problem, and so its associated tree problems, arise in other application contexts — for example, machine scheduling. In this setting, we associate each vehicle with a machine and the customers are jobs that we wish to perform on the machines. Each 'vehicle tour' becomes a sequence of jobs that each machine will process. That is, the machine 'visits' the jobs. When we have K identical machines, the problem becomes a K-median version of the problem. When we impose capacities on the machines and processing times of the jobs (which correspond to demands), we obtain other versions of the vehicle routing problem and, therefore, of tree packing problems.

Clustering

In cluster analysis, we wish to partition (cluster) a set of data so that data in the same partition (cluster) are 'similar' to each other. Suppose that we can represent the data in an n-dimensional space (the dimensions might, for example, represent different symptoms in a medical diagnosis). Suppose we view two points as close to each other if they are close to each other in Euclidean distance, and measure similarity of a set of points as the length of the (Euclidean) minimum spanning tree that connects these points. Then if we want to find the best k clusters of the points, we need to find the best set of k disjoint trees that contain all the points; we connect the points in any one of these k sets by a minimum spanning tree defined on these points. In practice, we might solve this problem for all values of k and then use some secondary criteria (e.g., human judgment) to choose the best value of k.

VLSI design

The designers of very large scale integrated (VLSI) chips often wish to connect a set of modules on the surface of a chip using the least total length of wire. The physical layout of the chip usually requires that wires can be routed only along 'channels' that are aligned along north/south or east/west directions of the surface. Thus the wiring distance metric between a pair of modules is the rectilinear or manhattan metric. If we represent each module location as a point (although physically the modules occupy a nonzero area), this problem is a Steiner tree problem with the set of module location points as the terminal nodes T; in theory, the Steiner nodes could be anywhere on the chip's surface (see Figure 6a). Since we are measuring distances between nodes according to the rectilinear norm, researchers often refer to this type of problem as the rectilinear Steiner tree problem.

One popular model for this application models the surface of the chip as a grid graph (see Figure 6b) with each node chosen as either the location of a module or the intersection point of some north/south and east/west line that passes through one of the module locations. Wires can run only along the edges of the grid graph. In Figure 6.b, nodes 1, 2, 5, 6 and 9 are the terminal nodes and nodes 3, 4, 7 and 8 are Steiner nodes arising from lines passing through these nodes. This derived grid graph model is a special case of the classical Steiner tree problem.

In practice, the design of a chip usually involves many different sets of modules, each set needing to be connected together. When multiple sets of modules use any channel in their Steiner tree solution, multiple wires will use the same edge of the underlying grid graph. In this application setting, each channel on the chip surface can accommodate only a limited number of wires; so this more general problem is a variant of the Steiner tree problem with several rectilinear Steiner tree problems defined on the same grid graph, but with a limit on the number

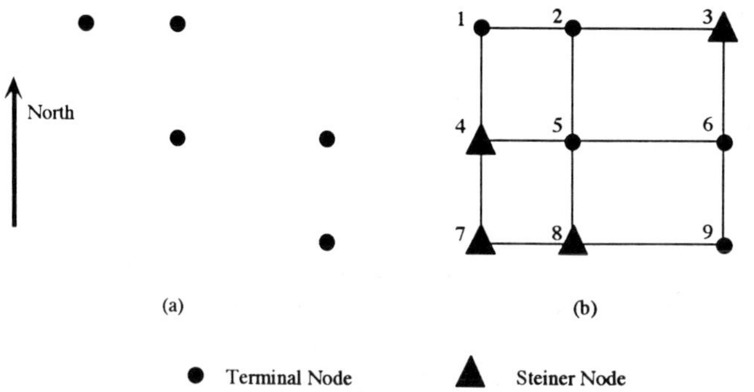

Fig. 6. (a) Set of points representing a rectilinear Steiner tree problem; (b) grid graph representation of rectilinear Steiner tree example.

of Steiner trees that use any edge of the graph. That is, the problem becomes a problem of 'packing' Steiner trees into the capacitated edges.

In this discussion, we have considered a rectilinear Steiner tree model for connecting modules of a computer chip; the same type of model arises in the design of printed circuit boards.

Production planning

Suppose we wish to find a production and inventory plan that will meet the demand $d_t > 0$ of product over each of T time periods $t = 1, 2, \dots, T$. If we produce x_t units of the product in period t, we incur a fixed (set-up) plus variable cost: that is, the cost is $f_t + c_t x_t$. Moreover, if we carry s_t units of inventory (stock) from period t to period $t + 1$, we incur an inventory cost of $h_t s_t$. We wish to find the production and inventory plan that minimizes the total production and inventory costs. We refer to this problem as the single item uncapacitated economic lot-sizing problem. We can view this problem as defined on the network shown in Figure 7. This network contains one node for each demand period and one node that is the source for all production.

On the surface, this problem might not appear to be a tree optimization model. As shown by the following result, however, the problem has a directed spanning tree solution, that is, it always has at least one optimal production plan whose set of flow carrying arcs (that is, those corresponding to $x_t > 0$ and $s_t > 0$) is a spanning tree with exactly one arc directed into each demand node.

Theorem 2.1. *The single item uncapacitated economic lot-sizing problem always has a directed spanning tree solution.*

Proof. First note that since the demand d_t in each period is positive, at least one of x_t and s_{t-1} is positive in any feasible solution. Consider any given feasible solution to the problem and suppose that it is not a directed spanning tree solution. We will show we can construct a directed spanning tree solution with a cost as small

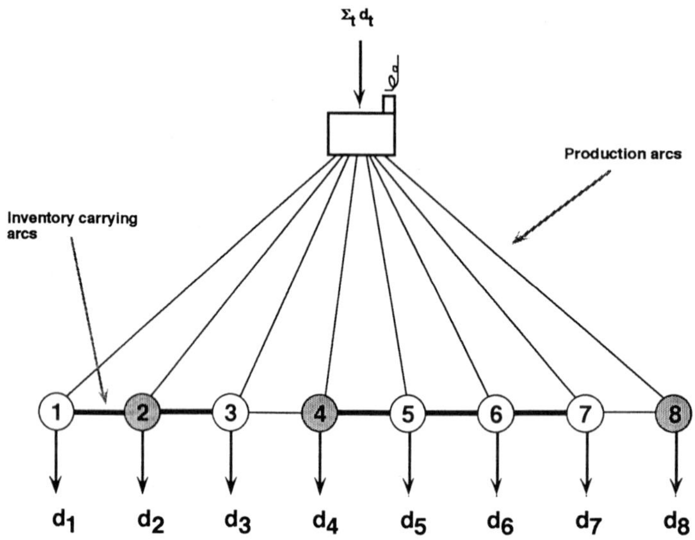

Fig. 7. Production lot-sizing as packing (Rooted) subtrees in trees.

as the cost of the given solution. Suppose $x_t > 0$, $s_{t-1} > 0$ and x_τ is the last period prior to period t with $x_\tau > 0$. Let $\epsilon = min\{x_\tau, s_{t-1}\}$. Note that if $x_\tau < s_{t-1}$, then $s_{\tau-1} > 0$.

Consider the two costs c_t and $c_{\tau t} = c_\tau + h_\tau + h_{\tau+1} + \ldots + h_{t-1}$. If $c_t \leq c_{\tau t}$, we set $x_t \leftarrow x_t + \epsilon$, $x_\tau \leftarrow x_\tau - \epsilon$ and $s_j \leftarrow s_j - \epsilon$ for all $j = \tau, \ldots, t-1$; if $c_t > c_{\tau t}$, we set $x_t \leftarrow 0$, $x_\tau \leftarrow x_\tau + x_t$ and $s_j \leftarrow s_j + x_t$ for all $j = \tau, \ldots, t-1$. In both cases, we obtain a solution with at least as small a cost as the given solution and with one less period with $x_q > 0$ and $s_{q-1} > 0$. (If $\epsilon = x_\tau \leq s_{t-1}$ and $c_t \leq c_{\tau t}$, then $q = \tau$; otherwise, $q = t$.) By repeating this process as many times as necessary, we eventually obtain a directed spanning tree solution with a cost as small as that of the given solution. □

Note that for a directed spanning tree production plan, we never simultaneously produce in any period and carry inventory into that period. Therefore, whenever we produce, we must produce for an integral number of periods (from the current period until just before the next production period). This property permits us to use a simple dynamic programming algorithm to solve the problem very efficiently. Let v_t be the value of a minimum cost solution of the problem for periods $1, 2, \ldots, t$ assuming $s_t = 0$. Set $v_0 = 0$. Then

$$v_t = \min_{1 \leq \tau \leq t} \left\{ v_{\tau-1} + f_\tau + \sum_{\tau \leq j \leq t} c_{\tau j} d_j \right\}$$

since in any optimal directed spanning tree solution, we must produce for the last time in some period τ, carry zero inventory into period τ, and incur inventory carrying costs for all periods $\tau, \tau+1, \ldots, t-1$.

Finally, we might note that we can view this problem as a subtree optimization problem on the line graph containing only the demand nodes $1, 2, \ldots, T$ (see Figure 7). Since whenever we produce, we always produce for an integral number of consecutive periods, the problem always has an optimal solution that decomposes the line graph into a collection of interval subgraphs each containing a consecutive set of nodes: if a subgraph has nodes $t, t+1, \ldots, q$, its root is node t and the weight of any node j in the subgraph is given by: $w_t^t = f_t + c_t d_t$, and $w_j^t = c_{tj} d_t$, for $j \neq t$. In this model, we choose the root node of any interval as the leftmost node of the interval since we do not permit backlogging (that is each $s_t \geq 0$). If we permit backlogging, then we could choose any node τ in the interval as its root (that is, as the production point) for that interval and the weight w_j^τ for any node j to the left of the root (that is, $j < \tau$) would be the cost of supplying the demand d_j by backlogged production from time t.

2.1. Trees and network flow problems

As illustrated by the production planning example we have just considered, trees can arise as solutions to optimization problems that on the surface are unrelated to trees; that is, the problem is not defined on a tree nor does it explicitly seek a tree solution. This example is a special case of a more general result in the field of linear programming. Consider any optimization problem of the form $\min\{cx : Nx = b, 0 \leq x \leq u\}$ and suppose that each column of the n by m matrix N has at most two nonzero entries, which have the values ± 1; moreover, if a column has two nonzero entries, one is a $+1$ and the other a -1. Define a directed graph G with $n+1$ nodes, numbered 0 to n, and with m directed arcs: G contains one node for each row of N, plus one additional node (node 0), and one arc for each column of N. We define the graph as follows: if a column of N has a $+1$ entry in row i and a minus one entry in row j, the graph G contains arc (i, j). If a column has a single nonzero entry and it is $+1$ in row i, the graph contains the arc $(i, 0)$, and if its single nonzero entry is -1 in row j, the graph has the arc $(0, j)$. We can then interpret the variables x as flows on the arcs of this graph; the jth constraint of $Nx = b$ is a mass balance constraint stating that the total flow out of node j minus the total flow into that node equals the supply b_j at that node. We wish to find a flow that meets these mass balance restrictions, the bounding constraints $0 \leq x \leq u$, and that has the smallest possible flow cost cx.

From the theory of linear programming, we know that this linear program always has an optimal solution corresponding to a basis B of N. That is, the solution has the property that we can set each variable x_e not in B to either value 0 or u_e, and then solve for the basic variables from the system $Nx = b$. But now we observe that any basis corresponds to a spanning tree of the graph G if we ignore the orientation of the arcs. Recall that each column of B corresponds to an arc in G. Let $A(B)$ denote the set of arcs corresponding to the columns of B. If the graph defined by the edges $A(B)$ contains an undirected cycle C, then as we traverse this cycle we encounter any arc $e = (i, j)$ either first at node i or first at node j. Let $y_e = +1$ in the former case, $y_e = -1$ in the latter case, and

$y_e = 0$ if arc e does not belong to the cycle C. Then $By = 0$ and so B is not a basis matrix of N. Therefore, the subgraph T of G corresponding to any basis matrix cannot contain any cycles. Consequently, it either is a tree or a collection of disjoint subtrees.

The fact that network flow problems always have tree solutions, in the sense we have just discussed, has far reaching implications. It permits us to solve these problems very effectively by searching efficiently among tree solutions (the simplex method has this interpretation) using the underlying graph to implement the algorithms using graphical methods in place of more complex matrix operations. Although we will not discuss these methods in this chapter, we note that the fact that network flow problems have spanning tree solutions, and are solvable efficiently using tree manipulation methods, is one of the primary reasons why tree optimization problems are so important in both theory and practice.

The optimal tree property of network flows also has polyhedral implications. It implies that the extreme points of the system $\{x : Nx = b, 0 \leq x \leq u\}$ are integer whenever the vectors b and u are integer. In this case, it is easy to show that the incidence vector of any basic solution to the linear program is integer since (i) the flows on all arcs e not corresponding to the basis are set to value 0 or u_e, which are integer; (ii) setting the values x_e of any nonbasic arc $e = (i, j)$ to value $\overline{x}_e = 0$ or $\overline{x}_e = u_e$ has the effect of updating the vector b by adding the integer $\overline{x}_e$ to b_j and subtracting $\overline{x}_e$ from b_i, so the updated value $\overline{b}$ of the vector b, once we have made this assignment of variables, remains integer; and (iii) solving for the values of the arcs in a tree for any integer vector $\overline{b}$ gives integer values for the following reason. Note that the tree always has at least one node v (actually at least two nodes) with a single arc e in the tree incident to it (that is, a degree one node in the tree). Therefore, the value of x_e is $\pm\overline{b}_v$. If we set x_e to this value, we update the b vector by adding and subtracting the value $\pm\overline{b}_v$ from each of the two components of b corresponding to the nodes v and q that are incident to arc e. If we now eliminate node v and arc e from the tree, we obtain a new tree with one less node. The new tree will again have at least one degree one node so we can find an integer value for one other component of the vector x. By repeating this process, we determine integer values for all the basic (tree) variables.

This discussion shows that every basic solution is integer valued, and the theory of linear programming implies that every extreme point to the system $\{x : Nx = b, 0 \leq x \leq u\}$ is integer valued.

3. Minimum spanning trees

In this section we study the minimum spanning tree problem. We begin by describing a fundamental solution method, known as the greedy algorithm, for solving this problem. We establish the validity of this algorithm in two ways: (i) using a direct combinatorial argument, and (ii) using a mathematical programming lower bounding argument based upon relaxing part of the problem constraints. Both of these arguments are representative of methods used frequently in the field of combinatorial optimization. In order to highlight the interaction between algorithms and theory, we then use this algorithm to give a constructive proof of a polyhedral representation of the spanning tree polytope; namely, we show that the extreme points of a basic 'packing' or 'subtour breaking' linear programming formulation of the problem are the incidence vectors of spanning trees.

We then introduce and study variants of two other 'natural' formulations of the minimum spanning tree problem: a cutset model and a flow model. We show how to improve the formulation of both of these models, using the notion of multicuts in the cutset formulation and using multicommodity and directed versions of the flow formulation. These modeling variations produce a hierarchy of models for the minimum spanning tree problem. When formulated as integer or mixed integer programs, all these models are equivalent; some of them give better (more accurate) linear programming relaxations of the problem and in our discussion, we show the relationship between these relaxations and the linear programming relaxation of the basic packing formulation.

3.1. The greedy algorithm

The greedy algorithm is a simple one-pass procedure for solving the minimum spanning tree problem: the algorithm orders the edges in a given graph $G = (V, E)$ from smallest to largest weight (breaking ties arbitrarily) and considers the edges in this order one at a time, at each stage either accepting or rejecting an edge as a member of the tree it is constructing. The decision rule for each edge is very simple: if the edge forms a cycle with those already chosen, the method discards it from further consideration; otherwise, the method adds it to the tree it is forming. To illustrate this algorithm, consider the example shown in Figure 8a, with the edges ordered from smallest to largest weight as $a, b, c, d, e, f, i, j, g, h$. The method accepts the edges a, b, c, d, e, and f since they do not form any cycles. Edge i forms a cycle with edges a, b and e and edge j forms a cycle with edges c, d, and f, so the algorithm rejects those arcs. It then accepts edge g and rejects edge h since it forms a cycle with the edges e, f, and g. Figure 8b shows the tree that the algorithm produces.

Does the greedy algorithm solve the minimum spanning tree problem? If so, how might we prove this result? We will answer these questions by considering two different proof techniques, one combinatorial and one based upon a mathematical programming bounding argument. In the process, we show that the greedy algorithm actually solves a more general 'componentwise' optimization tree problem and show the relationship between the greedy algorithm and the proof that a polyhedron is an integer polyhedron.

Combinatorial argument

In our example, the greedy algorithm chooses the edges of a greedy tree T_{greedy} in the order a, b, c, d, e, f, g. Suppose that T is any spanning tree. We will show that we can find a sequence of trees $T = T_0, T_1, T_2, \ldots, T_k = T_{\text{greedy}}$ satisfying the property that for $j = 1, 2, \ldots, k-1$, each tree T_{j+1} has a weight at least as small

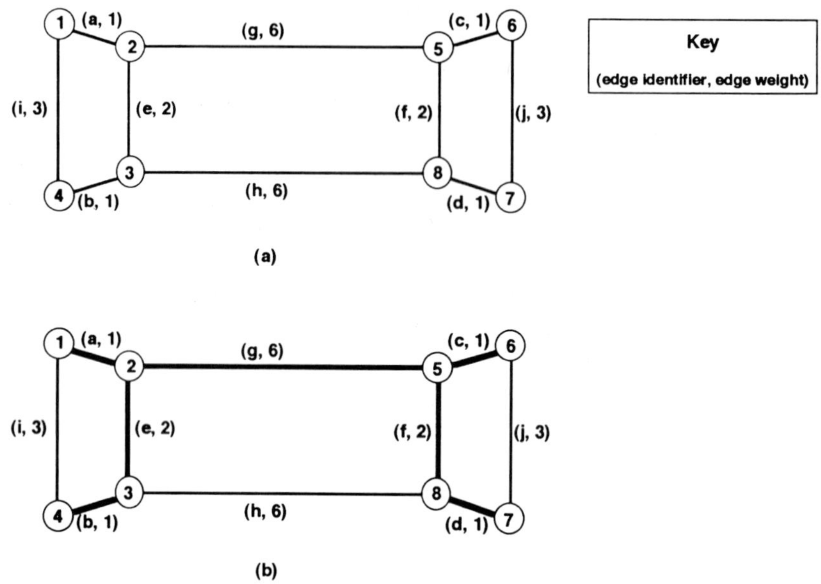

Fig. 8. Minimum spanning tree example. (a) data; (b) an optimal tree (bold edges).

as its predecessor T_j. Therefore, the weight of T_{greedy} is as small as the weight of T and since T is an arbitrary spanning tree, T_{greedy} is a minimum spanning tree.

We first note that if the tree T does not contain the edge a, adding a to the tree T creates a unique cycle and removing any edge from this cycle creates another tree T_1. Since the greedy algorithm chooses edge a as a minimum weight edge in the graph, the weight of the tree T_1 is at least as small as the weight of the tree T. So for any tree T, we can find a tree T_1 whose weight is at least as small that of T and that contains the edge a. Now suppose that after several steps of adding edges from T_{greedy}, one at a time, to the trees $T_0, T_1, \ldots$ we have obtained a tree T_4, say, whose weight is no more than T and that contains the first five edges a, b, c, d, e of the greedy tree. Adding edge f to this tree creates a cycle. The steps of the greedy algorithm imply that this cycle must contain at least one edge q other than the edges a, b, c, d, e. Moreover, the greedy algorithm also implies that the weight of the edge f is as small as the weight of the edge q (otherwise the algorithm would have added q before adding the edge f). Therefore, we can replace edge q in T_4 by the edge f, creating a new tree T_5 with the edges a, b, c, d, e, f whose weight is as small as T_4.

Continuing in this way, we eventually add each of the edges of the greedy tree and the resulting tree $T_k = T_{\text{greedy}}$ has a weight as small as T. Therefore, we have proved that the weight of the greedy tree is as small as the weight of any other tree T, so T_{greedy} is a minimum spanning tree.

We might note that in this argument we have actually proved a stronger result. Let T_1 and T_2 be any two trees of a graph G and suppose that we order the weights of the edges in each tree from smallest to largest. That is, $w_a \leq w_b \leq \ldots \leq w_g$ are the weights of the edges a, b, c, d, e, f, g in T_1 and $w_\alpha \leq w_\beta \leq \ldots \leq w_\eta$ are the weights of the edges $\alpha, \beta, \gamma, \delta, \epsilon, \nu, \eta$ in T_2. We say that T_1 is componentwise as small as T_2 if $w_a \leq w_\alpha, w_b \leq w_\beta, \ldots, w_g \leq w_\eta$. Note that if T_1 is componentwise as small as T_2, then the total weight of T_1 is at least as small as the total weight of T_2. Also note the transitivity property: if T_1 is componentwise as small as T_2 and T_2 is componentwise as small as T_3, then T_1 is componentwise as small as T_3. We refer to a tree as componentwise minimum if it is componentwise as small as any other tree. We might note that there is no a priori guarantee that a componentwise minimum spanning tree exists. Indeed, most classes of combinatorial optimization problems do not contain componentwise optimal solutions.

Consider two subsequent trees T_q and T_{q+1} in the argument we have just given for showing that the greedy algorithm produces a minimum spanning tree. We obtained T_{q+1} by replacing one edge of T_q by an edge with a weight at least as small. Therefore, T_{q+1} is componentwise as small as T_q. But if each tree T_{q+1} in the sequence is componentwise smaller than its predecessor T_q, then the final tree T_{greedy} is componentwise as small as the tree T. Since T was an arbitrary tree, we have established the following property.

Theorem 3.1. *The greedy algorithm produces a componentwise minimum spanning tree.*

We next give an alternative proof that the greedy algorithm generates a minimum spanning tree, and in doing so illustrate ideas from the field of mathematical programming that have been proven to have wide applicability in combinatorial optimization.

Lower bounding argument

Consider the following integer programming formulation of the minimum spanning tree problem:

$$\min \sum_{e \in E} w_e x_e \tag{3.1}$$

subject to

$$\sum_{e \in E} x_e = n - 1 \tag{3.2}$$

$$\sum_{e \in E(S)} x_e \leq |S| - 1 \quad \text{for any nonempty set } S \subset V \text{ of nodes} \tag{3.3}$$

$$x_e \geq 0 \text{ and integer for all edges } e. \tag{3.4}$$

In this formulation, the 0–1 variable x_e indicates whether we select edge e as part of the chosen spanning tree (note that the second set of constraints with $|S| = 2$ implies that each $x_e \leq 1$). The constraint (3.2) is a cardinality constraint implying that we choose exactly $n - 1$ edges, and the 'packing' constraints (3.3) imply that the set of chosen edges contain no cycles (if the chosen solution contained a cycle, and S were the set of nodes on this cycle, then the solution would violate this constraint). Note that as a function of the number of nodes in the network, this model contains an exponential number of constraints. Nevertheless, as we will show, we can solve it very efficiently by applying the greedy algorithm.

To develop a bound on the weight of a minimum spanning tree, suppose that we associate a 'Lagrange multiplier' μ_V with constraint (3.2) and nonnegative Lagrange multipliers μ_S with constraints (3.3), and add the weighted combination of these constraints to the objective function, creating the following optimization problem:

$$\min \sum_{e\in E} w_e x_e + \mu_V\left[\sum_{e\in E} x_e - (n-1)\right] + \\ + \sum_{\phi\subset S\subset V} \mu_S\left[\sum_{e\in E(S)} x_e - (|S|-1)\right] \tag{3.5}$$

subject to

$$\sum_{e\in E} x_e = n-1 \tag{3.6}$$

$$\sum_{e\in E(S)} x_e \le (|S|-1) \quad \text{for any nonempty set } S\subset V \text{ of nodes} \tag{3.7}$$

$$x_e \ge 0 \text{ and integer for all edges } e. \tag{3.8}$$

Note that for any feasible solution x to the problem and any value of the multiplier μ_V, the term $\mu_V[\sum_{e\in E} x_e - (n-1)]$ is zero. Moreover, for any feasible solution to the problem and any nonnegative choice of the multipliers μ_S for $S\subset V$, the last term in (3.5) is nonpositive. Therefore, the optimal value of this modified problem is a lower bound on the weight of any minimum spanning tree. Moreover, if we remove the constraints from this problem except for the 0–1 bounds on the variables, the problem's optimal objective value cannot become any larger, so the optimal objective value of the problem

$$\min \sum_{e\in E} w_e x_e + \mu_V\left[\sum_{e\in E} x_e - (n-1)\right] + \\ + \sum_{\phi\subset S\subset V} \mu_S\left[\sum_{e\in E(S)} x_e - (|S|-1)\right] \tag{3.9}$$

subject to

$$0\le x_e\le 1 \text{ for all edges } e \tag{3.10}$$

is also a lower bound on the weight of any minimum spanning tree.

Let us record this result formally as the following property.

Lower bounding property. *If μ_V is any scalar (positive, negative, or zero) and μ_S for each nonempty node set $S\subset V$ is any nonnegative scalar, then the optimal objective value of the problem (3.9)–(3.10) is a lower bound on the weight of any minimum spanning tree.*

In order to use this bounding property, let us collect together the terms in the objective function (3.9) by defining a 'reduced weight' for any edge e as follows:

$$w_e^\mu = w_e + \sum_{E(S) \text{ contains edge } e} \mu_S.$$

The last term in this expression contains the multiplier μ_V associated with the constraint $\sum_{e\in E} x_e = n-1$ (corresponding to $S = V$). Using the reduced weight notation, we can write the lower bounding problem (3.9)–(3.10) as follows:

$$\min \sum_{e\in E} w_e^\mu x_e - \mu_V(n-1) - \sum_{\phi\subset S\subset V} \mu_S(|S|-1) \tag{3.11}$$

subject to

$$0 \le x_e \le 1 \quad \text{for all edges } e. \tag{3.12}$$

Observe that this problem is easy to solve. If $w_e^\mu < 0$, set $x_e = 1$; if $w_e^\mu > 0$, set $x_e = 0$; and if $w_e^\mu = 0$, set x_e to any value between 0 and 1. Since the problems (3.9)–(3.10) and (3.11)–(3.12), which are equivalent, provide us with a lower bound on the weight of *any* minimum spanning tree, if we can find a spanning tree whose weight equals the value of the lower bound, we can be guaranteed that the tree is a minimum spanning tree. To show that the greedy algorithm generates a minimum spanning tree, we use the greedy tree, together with a set of multipliers μ_S, to provide a certificate of optimality; that is, we use the tree and the multiplier data to ensure that the tree is a minimum spanning tree without the need to make any further computations (in particular, we need not explicitly consider the exponential number of other spanning trees).

Certificate of optimality: *Suppose that the incidence vector y of a spanning tree T and the set of multipliers, μ_V unconstrained and $\mu_S \ge 0$ for all nonempty sets $S \subset V$, satisfy the following 'complementary slackness' properties:*

$$(a)\ \mu_S\left[\sum_{e\in E(S)} y_e - (|S|-1)\right] = 0 \quad \text{for all nonempty } S \subset V \tag{3.13}$$

$$(b)\ w_e^\mu = 0 \quad \text{if } y_e > 0 \tag{3.14}$$

$$(c)\ w_e^\mu \ge 0 \quad \text{if } y_e = 0. \tag{3.15}$$

Then T is a minimum spanning tree.

Proof. Since y is a feasible spanning tree, $\sum_{e\in E} y_e = (n-1)$; when combined with condition (a), this result implies that the objective function (3.9), or equivalently (3.11), equals the weight $\sum_{e\in E} w_e y_e$ of the tree y. Therefore, if we can show that y solves the lower bounding problem (3.9)–(3.10), then we know that its weight is as small as the weight of any spanning tree and so it is a minimum spanning tree. But since the only constraints in problem (3.11)–(3.12) are the bounding conditions $0 \le x_e \le 1$ for all edges e, conditions (b) and (c) imply that y solves this problem. □

Table 1
Appropriate Lagrange multipliers

Set S	μ_S
{1,2}	1
{3,4}	1
{5,6}	1
{7,8}	1
{1,2,3,4}	4
{5,6,7,8}	4
{1,2,3,4,5,6,7,8}	−6

Table 2
Edge reduced weights

Edge e	Reduced weight $w_e^{\mu} = w_e + \sum_{E(S) \text{ contains edge } e} \mu_S$
a	$1+1+4-6=0$
b	$1+1+4-6=0$
c	$1+1+4-6=0$
d	$1+1+4-6=0$
e	$2+4-6=0$
f	$2+4-6=0$
g	$6-6=0$
h	$6-6=0$
i	$3+4-6=1$
j	$3+4-6=1$

Example 3.1. As an illustration of this result, consider the tree generated by the greedy algorithm for our example. Suppose that we define the multipliers μ_S as shown in Table 1.

We set the multipliers of all other node sets S to value zero. With this choice of multipliers, the edges have the reduced weights shown in Table 2.

Let us make a few observations about the greedy solution y and the multipliers we have chosen. First, each step in the greedy algorithm introduces an edge joining two nodes i and j, and therefore forms a connected component $S(i, j)$ of nodes. For example, when the algorithm added edge f, it formed a connected component containing the nodes 5, 6, 7, and 8. The number of edges in this component is $|S(i, j)| - 1 = 4 - 1 = 3$, so the set $S = S(i, j)$ of nodes satisfies the constraint $\sum_{e \in E(S)} y_e = |S| - 1$. Consequently, the multipliers and greedy solution y satisfy the first optimality condition (a). The only multipliers that we have set to nonzero values correspond to these sets (and to the overall set V). Consequently, since the greedy algorithm adds exactly $n - 1$ edges to the spanning tree, at most $n - 1$ of the multipliers are nonzero.

Note that in this case, since $y_a = y_b = y_c = y_d = y_e = y_f = y_g = 1$, and $y_h = y_i = y_j = 0$, the reduced weights satisfy the optimality conditions (b) and (c). Since the greedy solution y and the multipliers also satisfy condition (a), we have in hand a certificate showing that the greedy solution is optimal.

In this case, the reduced weight for any edge not in the greedy tree is the difference in weight between that edge and the largest weight of any other edge in the cycle formed by adding that edge to the tree. For example, if we add edge i to the greedy tree, it forms a cycle with the edges a, b, and e; edge e has the largest weight in this cycle and so the reduced weight for the edge i is $3 - 2 = 1$. The reason for this is that the edge i is contained in exactly the same sets $E(S)$ whose defining nodes S have positive multipliers as the edge e and so the difference in their reduced weights $w_e^\mu = w_e + \sum_{E(S) \text{ contains edge } e} \mu_S$ is the difference in their original weights.

To conclude this discussion, we might indicate how we chose values for the multipliers μ_S so that every edge in the greedy tree has a zero reduced weight. We set $\mu_V = -w_g$, the negative of the weight of the last edge added to the tree. When the greedy algorithm adds an edge $\alpha = \{i, j\}$ at any step, it combines the previous components of nodes containing nodes i and j, forming (in our earlier notation) a new connected component $S(i, j)$. To determine the multiplier μ_S, we consider what the algorithm does at a later step. At some subsequent step, it adds another edge $\beta = \{p, q\}$ that enlarges the component $S(i, j)$ by connecting it to some other component. We set $\mu_{S(i,j)} = w_\beta - w_\alpha \geq 0$ (the provisions of the greedy algorithm ensure that edge β weighs at least as much as edge α). After the greedy algorithm has added edge α, at later steps it adds other edges $\beta, \gamma, \ldots, \phi, \nu$ that form increasingly larger sized components containing edge α. By our choice of the multipliers, these are the only node sets S, with $\alpha \in E(S)$, that receive nonzero multipliers. But our choice of the multipliers implies that $\sum_{E(S) \text{ contains edge } \alpha} \mu_S = (w_\beta - w_\alpha) + (w_\gamma - w_\beta) + \ldots + (w_\phi - w_\nu) + w_\nu = -w_\alpha$. Therefore, the reduced weight of edge α is zero. Since α is an arbitrary edge in the greedy tree, this choice of multipliers ensures that every edge in the greedy tree receives a zero reduced weight.

This argument applies in general to the greedy tree produced by any application of the greedy algorithm for any network and, therefore, provides an alternative proof that the greedy tree is a minimum spanning tree.

3.2. Polyhedral representations

In the previous section, we showed how to use lower bounding information about the integer programming model (3.1)–(3.4) to demonstrate that the greedy algorithm generates a minimum spanning tree. In this section, we study the linear programming relaxation of this problem obtained by removing the integrality restrictions imposed on the variables. We also introduce several other formulations of the minimum spanning tree problem and study the polyhedra defined by their linear programming relaxations.

The study of integer programming models like (3.1)–(3.4) has become a fruitful topic within the field of combinatorial optimization; indeed, we will see its use in many of the following sections of this chapter as we consider more complex tree optimization problems. In general, because integer programming problems are hard to solve, the optimization community frequently solves some type of more easily solved relaxation of the problem. In the last subsection we considered one such type of relaxation, known as Lagrangian relaxation.

Perhaps the most popular type of relaxation is the linear programming relaxation obtained by eliminating the restriction that the decision variables x in the model (3.1)–(3.4) need to be integer. In general, since we have eliminated the integrality restriction from the model, the linear programming relaxation will have a lower optimal objective value than does the integer program. As we show next, for the minimum spanning tree problem, this is not the case. In this setting, the integer program and linear program have the same optimal objective value since any solution to the integer program (in particular, the greedy solution) solves the linear programming relaxation.

Although we have not noted this important result before, we have actually already established it. For suppose that we start with the linear programming relaxation of the minimum spanning tree formulations (3.1)–(3.4), obtained by replacing the constraints '$x_e \geq 0$ and integer for all edges e' by the relaxed constraints '$x_e \geq 0$ for all edges e', and apply the same lower bounding arguments that we used for proving that the greedy solution solves the integer programming model. Then we also see that the greedy solution solves the linear programming relaxation. In fact, we might interpret our lower bounding argument as follows: let z^{ST} denote the optimal objective value of the spanning tree integer program and let z^{lp} denote the optimal objective function value of its linear programming relaxation. Suppose that we form the following linear programming dual of the linear programming relaxation.

$$\max \; -\mu_V(n-1) - \sum_{S \neq V} \mu_S(|S|-1) \tag{3.16}$$

subject to

$$-\sum_{EA(S) \text{ contains edge } (i,j)} \mu_S \leq w_e \quad \text{for all edges } e \tag{3.17}$$

$$\mu_S \geq 0 \quad \text{for all } S \neq V. \tag{3.18}$$

As before, the expression to the lefthand side of the inequality (3.17) contains the term μ_V. Since the linear program is a relaxation of the integer programming model $z^{\text{lp}} \leq z^{ST}$ and by linear programming duality theory the value of the linear program equals the value of its dual linear program (3.16)–(3.18). The multiplier values that we defined in the lower bounding argument satisfy the constraints (3.17) and (3.18) of the linear programming dual problem. Moreover, note that by the way we have defined the multipliers, if w_α is the weight of any edge added to the greedy tree, then w_α contributes to the objective function (3.16) in one or more terms: (i) it contributes $-w_\alpha$ to μ_S, corresponding to the set S of nodes in a single component that is formed when we add edge α to the tree; and (ii) it contributes w_α to the two the multipliers μ_Q and μ_P of the sets Q and P of nodes that define the components that are combined when we add edge α to the tree. But since $(|S|-1) = (|Q|-1) + (|P|-1) + 1$, the overall effect is to add w_α to the objective function. Since this is true for every edge α that we add to the greedy tree, the objective function of the linear program has the same value $\sum_{e \in T_{\text{greedy}}} w_e x_e$ as the greedy tree. Therefore, $z^{\text{lp}} = z^{ST}$ and so the greedy tree, which is feasible in the linear program, solves this problem.

Note that this development has not only shown that the optimal value of the integer programming problem (3.1)–(3.4) and its linear programming relaxation are the same, but has also shown that the linear programming relaxation always has an integer optimal solution (which is an incidence vector of a spanning tree) for any choice of the objective function weights. Moreover, we can always choose the weights so that any particular spanning tree is a solution to the linear program. A standard result in linear programming shows that the incidence vectors of spanning trees must then be the extreme points of the linear programming relaxation. Therefore, we have established the following important result.

Theorem 3.2. *The extreme points of the polyhedron defined by the linear programming relaxation of the spanning tree model (3.2)–(3.4) are the 0–1 incidence vectors of spanning trees.*

Alternative formulations

Generally, it is possible to devise many valid formulations of integer programming problems, using alternate sets of variables and/or constraints.

One formulation might be preferable to another for some purposes, for example, because it is easier to study theoretically or because it is solvable by a particular solution procedure. The model (3.1)–(3.4) that we have introduced in this section is a 'natural packing' formulation. It is a 'natural' formulation because it uses the natural 0–1 decision variables, indicating whether or not we include the underlying edges in the spanning tree; we refer to it as a packing formulation because the constraints $\sum_{e \in E(S)} x_e \leq |S| - 1$ restrict the number of edges that we can pack within any set S of nodes. In this discussion we examine two alternative approaches for formulating the minimum spanning tree problem, one involving cuts and another involving flows.

Cutset formulations. Let $\mathcal{S}$ denote the set of incidence vectors of spanning trees of a given graph $G = (V, E)$. In formulating the spanning tree problem as an integer program, we used the property that a spanning tree on an n node graph is any subgraph containing $n - 1$ edges and no cycles. Therefore, at most $|S| - 1$ edges in any tree can connect any set S of nodes, and so $\mathcal{S} = \{x \in Z^{|E|} : 0 \leq x \leq \mathbf{1}, \sum_{e \in E} x_e = n - 1$, and $\sum_{e \in E(S)} x_e \leq |S| - 1$ for any nonempty $S \subset V\}$. As an alternative, we could use another equivalent definition of a spanning tree: it is a connected subgraph containing $n - 1$ edges. This definition leads to a cutset formulation: $\mathcal{S} = \{x \in Z^{|E|} : 0 \leq x \leq \mathbf{1}, \sum_{e \in E} x_e = n - 1$, and $\sum_{e \in \delta(S)} x_e \geq 1$ for all nonempty node sets $S \subset V\}$. As we saw in Theorem 3.2, if

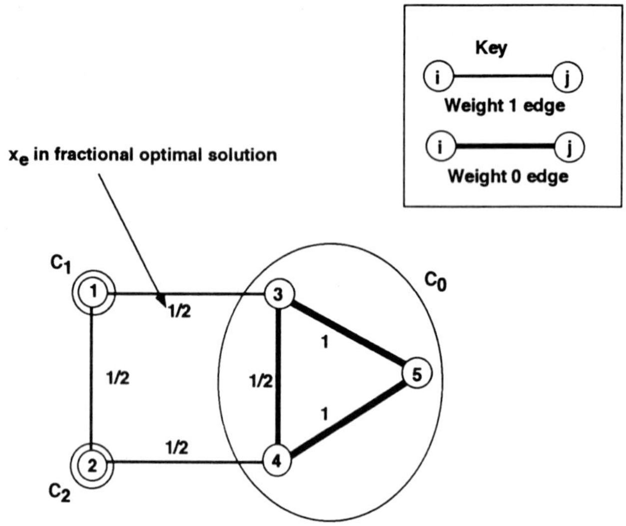

Fig. 9. Fractional optimal solution to cut relaxation.

we relaxed the integrality restrictions in the packing (subtour) formulation, then the resulting polyhedron P_{sub} equals conv($\mathcal{S}$). Let P_{cut} denote the polyhedron formed if we remove the integrality restrictions from the cutset formulation, that is, $P_{\text{cut}} = \{x \in R^{|E|} : 0 \leq x \leq \mathbf{1}, \sum_{e \in E} x_e = n - 1$, and $\sum_{e \in \delta(S)} x_e \geq 1$ for all nonempty nodes sets $S \subset V\}$. As we show next, the polyhedron P_{cut} can be larger than the convex hull of S.

Proposition 3.3. $P_{\text{cut}} \supseteq P_{\text{sub}}$. *In general,* P_{cut} *has fractional extreme points and so is larger than* P_{sub}.

Proof. Note that for any set S of nodes, $E = E(S) \cup \delta(S) \cup E(\overline{S})$. If $x \in P_{\text{sub}}$, then $\sum_{e \in E(S)} x_e \leq |S| - 1$ and $\sum_{e \in E(\overline{S})} x_e \leq |\overline{S}| - 1$. Therefore, since $\sum_{e \in E} x_e = n - 1$, $\sum_{e \in \delta(S)} x_e \geq 1$. Consequently, $x \in P_{\text{cut}}$ and so $P_{\text{cut}} \supseteq P_{\text{sub}}$.

To establish the second part of the proposition, consider Figure 9. Recall that a polyhedron has integer extreme points if and only if a linear program defined over it has an integer optimal objective value for all choices of integer objective coefficients (whenever the optimal value is finite). In Figure 9, if we define a linear program $\min\{\gamma x : x \in P_{\text{cut}}\}$ by setting the objective coefficients of edges $\{1, 2\}$, $\{1, 3\}$ and $\{2, 4\}$ to value 1 and the coefficients of edges $\{3, 4\}$, $\{4, 5\}$, and $\{3, 5\}$ to value 0, then the optimal solution is the fractional solution x shown in the figure with an objective value of 3/2. Note that x belongs to P_{cut} but not to P_{sub} since the edges $E(S)$ in the 3-node set $S = \{3, 4, 5\}$ have a weight 5/2. □

This result shows that the linear programming relaxation of a 'simple' cutset formulation does not define the convex hull of incidence vectors of spanning trees.

We next show that if we replace simple cutsets by 'multicuts,' then the linear programming relaxation will define conv(S).

Given a $k+1$ node partition of V, that is disjoint nonempty sets $C_0, C_1, \ldots, C_k$ of nodes whose union is V, we call the set of edges, denoted as $\delta(C_0, C_1, \ldots, C_k)$, with one end point in one of these node sets and the other end point in another, a *multicut*. Note that a (usual) cut is just a multicut with $k = 1$. Since any spanning tree T contains at least k edges joining the sets $C_0, C_1, \ldots, C_k$, $|\delta(C_0, C_1, \ldots, C_k) \cap T| \geq k$. Therefore, any spanning tree lies in the multicut polyhedron $P_{\text{mcut}} = \{x \in R^{|E|} : 0 \leq x \leq \mathbf{1}, \sum_{e \in E} x_e = n-1$, and $\sum_{e \in \delta(C_0, C_1, \ldots, C_k)} x_e \geq k$ for all node partitions $C_0, C_1, \ldots, C_k$ of $V\}$. The following result gives a strengthening of Proposition 3.3.

Theorem 3.4. $P_{\text{mcut}} = \text{conv}(\mathcal{S})$.

Proof. Consider any set S of $n-k$ nodes and node partition with $C_0 = S$, and C_j for $j = 1, \ldots, k$ as singletons each containing one of the nodes of $\overline{S}$.

Since $E = E(S) + \delta(C_0, C_1, \ldots, C_k)$, if $0 \leq x \leq 1$ and $\sum_{e \in E} x_e = n-1$, then $\sum_{e \in E(S)} x_e \leq |S| - 1$ if and only if $\sum_{e \in \delta(C_0, C_1, \ldots, C_k)} x_e \geq n - 1 - (|S| - 1) = k$. Consequently, $x \in P_{\text{sub}}$ if and only if $x \in P_{\text{mcut}}$. Therefore, $P_{\text{mcut}} = P_{\text{sub}}$ and, by Theorem 3.1, $P_{\text{mcut}} = \text{conv}(\mathcal{S})$. □

This proof uses our prior results concerning the relationship between the greedy algorithm and the polyhedron P_{sub}. The following alternative proof, based upon a lower bounding or duality argument, is also instructive; it shows a direct relationship between the greedy algorithm and the polyhedron P_{mcut}.

Alternative Proof. Consider the following dual to the linear program $\min\{wx : x \in P_{\text{mcut}}\}$:

$$\max \sum k\mu_{C_0,\ldots,C_k}$$

subject to

$$\sum_{e \in \delta(C_0,\ldots,C_k)} \mu_{C_0,\ldots,C_k} \leq w_e \qquad \text{for all } e \in E$$

$$\mu_{C_0,\ldots,C_k} \geq 0 \qquad \text{whenever } k \leq n-2.$$

In this formulation, $\mu_{C_0,\ldots,C_k}$ is the dual variable corresponding to the constraint $\sum_{e \in \delta(C_0,\ldots,C_k)} x_e \geq k$, $\mu_{\{1\},\ldots,\{n\}}$ is the dual variable corresponding to the constraint $\sum_{e \in E} x_e = n-1$, and the sum in the objective function of this problem is taken over all multicuts $\{C_0, \ldots, C_k\}$.

Suppose the greedy algorithm chooses a set S_1 of edges of weight w_1, S_2 of weight w_2, and so forth and that $w_1 \leq w_2 \leq \ldots \leq w_r$. Let $S_0 = \phi$. Suppose further that $(C_0^i, \ldots, C_{k_i}^i)$ are the connected components of the graph spanned by the edges $\cup_{j=1}^i S_j$, and that we set $\mu_{\{1\},\ldots,\{n\}} = w_1$, $\mu_{C_0^i,\ldots,C_{k_i}^i} = w_{i+1} - w_i$, for $i = 1, \ldots, r-1$, and $\mu_{C_0,\ldots,C_k} = 0$ for any other multicut $\{C_0, \ldots, C_k\}$. It is easy to see that the values for $\mu_{C_0,\ldots,C_k}$ provide an optimal dual solution of the same value as the greedy solution, and therefore, as we have argued before, all the extreme points of the polyhedron P_{mcut} are integral[2]. □

Example 3.2. Consider again the example in Figure 8. In this case:

• $S_0 = \phi$, $(C_0^0, \ldots, C_{k_0}^0) = \{\{1\}, \{2\}, \ldots, \{8\}\}$, and $k_0 = 7$. Since $w_1 = 1$, $\mu_0 \equiv \mu_{\{1\},\ldots,\{n\}} = 1$.

• $S_1 = \{\{1, 2\}, \{3, 4\}, \{5, 6\}, \{7, 8\}\}$, $(C_0^1, \ldots, C_{k_1}^1) = \{\{1, 2\}, \{3, 4\}, \{5, 6\}, \{7, 8\}\}$, and $k_1 = 3$. Since $w_2 = 1$, $\mu_1 = \mu_{\{1,2\},\{3,4\},\{5,6\},\{7,8\}} = 2 - 1 = 1$.

• $S_2 = \{\{2, 3\}, \{5, 6\}\}$, $(C_0^2, \ldots, C_{k_2}^2) = \{\{1, 2, 3, 4\}, \{5, 6, 7, 8\}\}$ and $k_2 = 1$.

Since $w_3 = 6$, $\mu_2 \equiv \mu_{\{1,2,3,4\},\{5,6,7,8\}} = 6 - 2 = 4$.

Observe that $\sum k_i \mu_i = 14$ equals the objective value of the greedy solution.

Note that in the first proof of Theorem 3.4, we have actually proved a slightly stronger version of the theorem. In selecting the multicuts in the definition of P_{mcut}, we can choose all but one of them to be a singleton. Therefore, the multicuts have one large node set S; the remaining node sets are singletons. To contrast the results in Proposition 3.3 and Theorem 3.4, once again consider the example in Figure 9. Let $C_0 = \{3, 4, 5\}$, $C_1 = \{1\}$, and $C_2 = \{2\}$. Note that $\sum_{e\in\delta(C_0,C_1,C_2)} x_e = 3/2$ so this solution does not satisfy the multicut constraint $\sum_{e\in\delta(C_0,C_1,C_2)} x_e \geq 2$.

Flow formulations. Another way to conceive of the minimum spanning tree problem is as a special version of a network design problem: in this setting, we wish to send flow between the nodes of the network and view the edge variable x_e as indicating whether not we install the edge e to be available to carry any flow. We consider three such flow models: a single commodity model, a multicommodity model, and an extended multicommodity model. In each of these models, although the edges are undirected, the flow variables will be directed. That is, for each edge $e = \{i, j\}$, we will have flow in both the directions i to j and j to i.

In the *single commodity model*, one of the nodes, say node 1, serves as a source node; it must send one unit of flow to every other node. Let f_{ij} denote the flow on edge $e = \{i, j\}$ in the direction i to j. The model is:

$$\min\ wx \tag{3.19}$$

subject to

$$\sum_{e\in\delta^+(1)} f_e - \sum_{e\in\delta^-(1)} f_e = n - 1 \tag{3.20}$$

[2] This proof also shows that if we eliminate the cardinality constraint $\sum_{e\in E} x_e = n - 1$ from the polyhedron (the polyhedron still has the constraint $\sum_{e\in E} x_e \geq n - 1$), the resulting polyhedron, which we denote as P^+_{mcut} also has integer extreme points (corresponding to spanning trees). In this case, if any weight $w_e < 0$, then we let x_e approach $+\infty$, showing that the linear program $\min \sum_{e\in E}\{w_e x_e : e \in P^+_{\mathrm{mcut}}\}$ has no optimal solution. If each $w_e \geq 0$, then the argument in the proof shows that this linear program has an optimal solution with x as the incidence vector of a spanning tree.

$$\sum_{e\in\delta^-(v)} f_e - \sum_{e\in\delta^+(v)} f_e = 1 \qquad \text{for all } v \neq 1, v \in V \tag{3.21}$$

$$f_{ij} \leq (n-1)x_e \qquad \text{for every edge } e = \{i, j\} \tag{3.22}$$

$$f_{ji} \leq (n-1)x_e \qquad \text{for every edge } e = \{i, j\} \tag{3.23}$$

$$\sum_{e\in E} x_e = n - 1 \tag{3.24}$$

$$f \geq 0, \text{ and } 0 \leq x_e \leq 1 \qquad \text{for all edges } e \tag{3.25}$$

$$x_e \text{ integer} \qquad \text{for all edges } e \in E. \tag{3.26}$$

In this model, equations (3.20) and (3.21) model flow balances at the nodes; the forcing constraints (3.22) and (3.23) are redundant if $x_e = 1$ and state that the flow on edge e, in both directions, is zero if $x_e = 0$. Note that this model imposes no costs on the flows. The mass balance equations imply that the network defined by any solution x (that is, those edges with $x_e = 1$) must be connected. Since the constraint (3.24) states that the network defined by any solution contains $n - 1$ edges, every feasible solution must be a spanning tree. Therefore, when projected into the space of the x variables, this formulation correctly models the minimum spanning tree problem.

If we ignore the integrality restrictions on the x variables, the constraints of this model determine a polyhedron. Let P_{flo} denote the set of feasible solutions to this linear program, that is to the system (3.20)–(3.25), in the x-space. We will use Figure 10 to illustrate the fact that this formulation of the minimum spanning tree problem is quite weak in the sense that the linear program $min\{\gamma x : x \in P_{\text{flo}}\}$ can be a poor representation of the weight of a minimum spanning tree as determined by the integer program min $\{\gamma x : x$ is an incidence vector of a spanning tree$\}$.

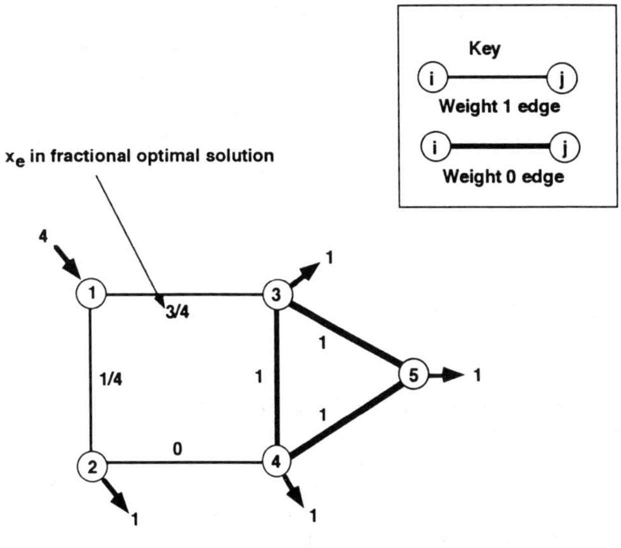

Fig. 10. Fractional optimal solution to the flow relaxation.

This is the same example used in Figure 9; the edge weights γ_e all have values zero or one as shown. The optimal weight of a minimum spanning tree is 2 and, as we saw before, the optimal weight of the linear programming relaxation of the cut formulation is 3/2. In this case, suppose we choose node 1 as the root node. It has a supply of 4 units. The optimal solution sends the three units of demand destined for nodes 3, 4, and 5 on edge $\{1, 3\}$ and the one unit of demand destined for node 2 on edge $\{1, 2\}$. To do so, it requires a capacity $(n - 1)x_e = 3$ on edge $\{1, 3\}$ and a capacity of $(n - 1)x_e = 1$ on edge $\{1, 2\}$, giving a weight $x_e = 3/4$ on edge $\{1, 3\}$ and a weight of 1/4 on edge $\{1, 2\}$. So in this case, the linear programming relaxation of the flow formulation has value 1, which is even less than value of the linear programming relaxation of the cut formulation.

As we will see, even though the differences between the minimum spanning tree problem, the cut formulation, and the flow formulation apply in general, we can considerably strengthen the cut and flow formulations. Indeed, we will derive tight cut and flow models, that is, formulations whose underlying polyhedrons equal the convex hull of the incidence vectors of the spanning trees. We begin by introducing a directed version of P_{sub}.

Note that if we select any node r as a root node for any spanning tree, then we can direct the edges of the tree so that the path from the root node to any other node is directed from the root to that node. To develop a model for this directed version of the problem, we consider a digraph $D = (V, A)$ formed by replacing each edge $\{i, j\}$ in E by arcs (i, j) and (j, i) in A. Let $y_{ij} = 1$ if the tree contains arc (i, j) when we root it at node r.

The *directed model* is:

$$\sum_{e\in A(S)} y_e \le |S| - 1 \quad \text{for any nonempty set } S \subset V \text{ of nodes} \tag{3.27}$$

$$\sum_{e\in\delta^-(v)} y_e = 1 \quad \text{for all } v \in V \setminus \{r\} \tag{3.28}$$

$$\sum_{e\in A} y_e = n - 1 \tag{3.29}$$

$$y_e \ge 0 \quad \text{for all arcs } e \in A \tag{3.30}$$

$$x_e = y_{ij} + y_{ji} \quad \text{for all edges } e \in E. \tag{3.31}$$

Note that the constraints (3.28) in this model imply that $\sum_{k\neq r}\sum_{e\in\delta(k)} y_e = n-1$ which, combined with equation (3.29), implies that $\sum_{e\in\delta(r)} y_e = 0$. Therefore, $y_e = 0$ for every arc e directed into the root node.

Let P_{dsub} denote the feasible set for this model in x-space. Since every tree can be directed, $\mathcal{S} \subseteq P_{\text{dsub}}$ and so $\text{conv}(\mathcal{S}) = P_{\text{sub}} \subseteq P_{\text{dsub}}$. Note that if $x \in P_{\text{dsub}}$, then since y_{ij} and y_{ji} occur together in the constraints (3.27) and (3.29), $x \ge 0$ satisfies constraints (3.2) and (3.3) and so $x \in P_{\text{sub}}$ implying that $P_{\text{dsub}} \subseteq P_{\text{sub}}$. We have therefore established the following result[3].

[3] In this proof, we have used the integrality property $\text{conv}(\mathcal{S}) = P_{\text{sub}}$. In establishing Theorem 6.4, we give a proof of a more general result without using the integrality property.

Proposition 3.5. $P_{\text{sub}} = P_{\text{dsub}}$.

Now consider a *directed cut model* with some arbitrary node r chosen as a 'root node':

$$\sum_{e \in \delta^+(C)} y_e \geq 1 \qquad \text{for all } C \subset V \text{ with } r \in C \tag{3.32}$$

$$\sum_{e \in A} y_e = n - 1 \tag{3.33}$$

$$y_e \geq 0 \qquad \text{for all arcs } e \in A \tag{3.34}$$

$$x_e = y_{ij} + y_{ji} \qquad \text{for all edges } e \in E. \tag{3.35}$$

Let P_{dcut} denote the set feasible solutions of this model in x-space. The constraint (3.32) in this model states that every directed cut $\delta(C)$ that separates the root node r from any set of nodes $\overline{C}$ must contain at least one arc.

Before studying the directed cut model, let us make one observation about its formulation. If we set $C = V \setminus \{k\}$ in constraint (3.32), then for each node $k \neq r$, this model contains the constraint $\sum_{e \in \delta^-(k)} y_e \geq 1$. In contrast, the directed formulation states these constraints as equalities (3.28). Note that if we add the inequalities $\sum_{e \in \delta^-(k)} y_e \geq 1$ over all nodes $k \neq r$, we obtain the inequality $\sum_{e \in A(V \setminus \{r\})} y_e + \sum_{e \in \delta^+(r)} y_e \geq n - 1$. If any vector y were to satisfy any inequality $\sum_{e \in \delta^-(k)} y_e \geq 1$ in (3.32) as a strict inequality, it would also satisfy the last inequality as a strict inequality as well, contradicting equation (3.33). Therefore, every feasible solution y in the directed model satisfies equation (3.28), that is, $\sum_{e \in \delta^-(k)} y_e = 1$ for all $k \neq r$. (As before, these equalities imply that the weight $y_e = 0$ for every arc e directed into the root node.) These conclusions tell us something about the relationship between the directed model and the directed cut model. As the next result shows, we can say even more.

Proposition 3.6. $P_{\text{dsub}} = P_{\text{dcut}}$.

Proof. Let S be any set of nodes containing node r. Since $A = A(\overline{S}) \cup \delta^+(S) \cup_{k \in S} \delta^-(k)$, we can rewrite equation (3.29) as

$$\sum_{e \in A(\overline{S})} y_e + \sum_{e \in \delta^+(S)} y_e + \sum_{k \in S} \sum_{e \in \delta^-(k)} y_e = n - 1. \tag{3.36}$$

The constraints (3.28), which as we have just seen are valid in both the directed model and the directed cut model, and the condition that $y_e = 0$ for every edge directed into the root node r, imply that the last term on the left-hand side of this equation equals $|S| - 1$ and so equation (3.36) becomes

$$\sum_{e \in A(\overline{S})} y_e + \sum_{e \in \delta^+(S)} y_e = n - 1 - (|S| - 1) = |\overline{S}|. \tag{3.37}$$

But this inequality implies that (3.27) is valid for $\overline{S}$ if and only if (3.32) is valid for S. Therefore, the constraints (3.32) are equivalent to the constraints (3.27) applied to all sets that do not contain the root node r.

If r is not contained in S, then the last term in equation (3.36) equals $|S|$ and so the right-hand side of equation (3.37) becomes $|\bar{S}| - 1$. But then, since the second term in this equation is nonnegative, the equation implies that $\sum_{e \in A(\bar{S})} \leq |S| - 1$ and so the inequality (3.27) is valid whenever r does not belong to S as well.

But these two conclusions imply that a vector y is feasible in the directed model if and only if it is feasible in the directed cut model and, therefore, $P_{\text{dsub}} = P_{\text{dcut}}$. □

The max-flow min-cut theorem immediately provides us with a reformulation, called the *directed multicommodity flow model*, of the directed cut formulation. In this model every node $k \neq r$ defines a commodity: one unit of commodity k originates at the root node r and must be delivered to node k. Letting f_{ij}^k be the flow of commodity k in arc (i, j), we formulate this model as follows:

$$\sum_{e \in \delta^-(r)} f_e^k - \sum_{e \in \delta^+(r)} f_e^k = -1 \quad \text{for all } k \neq r \tag{3.38}$$

$$\sum_{e \in \delta^-(v)} f_e^k - \sum_{e \in \delta^+(v)} f_e^k = 0 \quad \text{for all } v \neq r, v \neq k, \text{and all } k \tag{3.39}$$

$$\sum_{e \in \delta^-(k)} f_e^k - \sum_{e \in \delta^+(k)} f_e^k = 1 \quad \text{for all } k \neq r \in V \tag{3.40}$$

$$f_{ij}^k \leq y_{ij} \quad \text{for every arc } (i, j) \text{ and all } k \neq r \tag{3.41}$$

$$\sum_{e \in A} y_e = n - 1 \tag{3.42}$$

$$y_{ij} + y_{ji} = x_e \quad \text{for every edge } e \in E \tag{3.43}$$

$$f_e^k \geq 0, \text{ and } y_e \geq 0 \quad \text{for all arcs } e \in A \text{ and all } R \neq r \tag{3.44}$$

$$y_e \text{ integer} \quad \text{for all arcs } e \in A. \tag{3.45}$$

In this model, the variable y_{ij} defines a capacity for the flow of each commodity k in arc (i, j). The *forcing constraint* (3.41) implies that we can send flow of each commodity on arc (i, j) only if that arc is a member of the directed spanning tree defined by the variables y. We let P_{dflo} denote the set of feasible solutions of the linear programming relaxation of this model in the x-space, that is, of the constraints (3.38)–(3.44).

Proposition 3.7. $P_{\text{dflo}} = P_{\text{dcut}}$.

Proof. From the max-flow min-cut theorem, $\sum_{e \in \delta^+(C)} y_e \geq 1$ for all C with $r \in C$ and $k \notin C$ if and only if the digraph has a feasible flow of 1 unit from the root to node k with arc capacities y_{ij}, that is, if and only if the system (3.38)–(3.41) has a feasible solution with $f^k \geq 0$. This observation implies the proposition. □

We obtain a closely related formulation by eliminating the y_{ij} variables. The resulting formulation is (3.38)–(3.40), plus $f \geq 0$, plus

$$\sum_{e \in E} x_e = n - 1 \tag{3.46}$$

$$f_{ij}^k + f_{ji}^{k'} \leq x_e \quad \text{for all } k, k' \text{ and all } e \in E. \tag{3.47}$$

We refer to this model as the *extended multicommodity flow formulation* and let $P_{\mathrm{mc'flo}}$ denote its set of feasible solutions in x-space. Observe that since we have eliminated the variables y_e in constructing the extended multicommodity flow formulation, this model is defined on the undirected graph $G = (V, E)$, even though for each commodity k, we permit flows f_{ij}^k and f_{ji}^k in both directions on edge $e = \{i, j\}$. The *bidirectional flow inequalities* (3.47) in this formulation link the flow of different commodities flowing in different directions on the edge $\{i, j\}$. These constraints model the following fact: in any feasible spanning tree, if we eliminate edge $\{i, j\}$, we divide the nodes into two components; any commodity whose associated node lies in the same component as the root node does not flow on edge $\{i, j\}$; any two commodities whose associated nodes both lie in the component without the root both flow on edge $\{i, j\}$ in the same direction. So, whenever two commodities k and k' both flow on edge $\{i, j\}$, they both flow in the same direction and so one of f_{ij}^k and $f_{ji}^{k'}$ equals zero.

Note that equalities (3.42) and (3.43) imply (3.46) and the inequalities (3.41) and equalities (3.43) imply (3.47). Therefore, $P_{\mathrm{dflo}} \subseteq P_{\mathrm{mc'flo}}$. Conversely, suppose the vectors x and f are feasible in the extended multicommodity flow problem; for each edge $e = \{i, j\}$, choose an arc direction (i, j) arbitrarily, and define $y_{ij} = \max_{k \neq r} f_{ij}^k$ and $y_{ji} = x_e - y_{ij}$. Then y and f are feasible in the directed flow formulation. Therefore, we have established the following result.

Proposition 3.8. $P_{\mathrm{dflo}} = P_{\mathrm{mc'flo}}$.

We obtain one final formulation, which we refer to as the *undirected multicommodity flow model*, by replacing the inequalities (3.47) by the weaker constraints

$$f_{ij}^k \leq x_e \text{ for all } k \neq r \text{ and } e \in E.$$

Let P_{mcflow} denote the set of feasible solutions of this model in the x-space. As in the proof of Proposition 3.6, the max-flow min-cut theorem implies that this model is equivalent to a cut formulation in the following sense.

Proposition 3.9. $P_{\mathrm{mcflow}} = P_{\mathrm{cut}}$.

Let us pause to reflect upon the results we have just established. In this section we have examined the polyhedra defined by the linear programming relaxation of nine different formulations of the minimum spanning tree problem. Figure 11 shows the relationship between these polyhedra.

Six of the polyhedra — the packing, multicut, extended multicommodity flow, directed spanning tree, directed cut, and directed flow polyhedra — are the same (the latter three when transformed into the space of x variables); each has integer extreme points. The cut and multicommodity flow are the same; like the weaker flow formulation, they can have fractional extreme points and so they do not define the spanning tree polyhedron.

There are several lessons to be learned from this development. First, for undirected formulations, multicuts improve upon models formulated with cuts

$$\left\{\begin{array}{l} P_{\text{sub}} \\ P_{\text{mcut}} \\ P_{\text{mc'flo}} \\ P_{\text{dsub}} \\ P_{\text{dcut}} \\ P_{\text{dflo}} \end{array}\right\} \subseteq \left\{\begin{array}{l} P_{\text{cut}} \\ P_{\text{mcflo}} \end{array}\right\} \subseteq P_{\text{flo}}$$

Fig. 11. Relationship between underlying polyhedra.

and the bidirectional flow inequalities (3.47) improve upon the multicommodity flow formulation (at the expense of adding considerably more constraints to the formulation). Second, (extended) multicommodity flow formulations, which have the disadvantage of introducing many additional flow variables, improve upon single commodity flow formulations. Third, even though the polyhedra P_{sub} and $P_{\text{mc'flo}}$ are the same, we obtain $P_{\text{mc'flo}}$ by projecting out the flow variables from the multicommodity flow formulation; this formulation, and indeed each of the flow formulations we have examined, are 'compact' in the sense that the number of variables and constraints in these models is polynomial in the size of the underlying graph. The subtour formulation and cut formulations contain fewer variables, but are 'exponential' in the sense that the number of constraints in these models grows exponentially in the graph's size. We obtain the compact formulations by introducing 'auxiliary' flow variables beyond the natural 0–1 edge variables. Finally, we have shown that a directed flow model gives a better representation of the spanning tree polyhedron than does the undirected flow model (unless we add the bidirectional constraints (3.47) to the undirected model).

These observations provide powerful modeling lessons that extend well beyond the realm of spanning trees. For example, they apply to several network design and vehicle routing problems. Later in this chapter, we will illustrate the use of these ideas again as we study other versions of optimal tree problems (especially variations of the Steiner tree problem).

Linear programs, cutting planes, and separation

To conclude this section, we consider an important algorithmic implication of the formulations we have considered. As we have noted, both the basic packing formulation (3.1)–(3.4) and the directed multicommodity flow formulation (3.38)–(3.44) are large optimization models; the number of constraints in the packing formulation grows exponentially in the number $|V|$ of nodes in the underlying graph; the directed cutset formulation has $|V|^3$ flow variables and forcing constraints (3.41) — 1 million for a problem with 100 nodes. Fortunately, we need not explicitly solve either model since we can use the greedy algorithm to solve the minimum spanning tree problem very efficiently.

What happens, however, when the spanning tree problem is a subproblem in a more general model that cannot be solved using a combinatorial algorithm like the greedy algorithm? Examples are spanning tree problems with additional constraints imposed upon the network topology (see Section 8) or the traveling salesman problem, with its embedded spanning tree structure. In solving these problems, many algorithms attempt to solve the linear programming relaxation of the problem (and then often use an enumeration procedure such as branch and bound). If we model spanning tree solutions using the packing formulation, any feasible point in the linear programming relaxation of these more general problems must both lie in the set P_{sub} and satisfy any additional constraints imposed upon the problem variables. Since we cannot possibly even list all the constraints of the packing formulation for any reasonably sized problem, linear programming algorithms often adopt a 'cutting plane' or 'constraint generation' procedure that works as follows.

We first formulate and solve a linear programming model ignoring all but a few of the packing constraints (3.3). If the solution $\overline{x}$ of this model happens to satisfy all the packing constraints that we have ignored, then $\overline{x} \in P_{\text{sub}}$. To discover if this is the case, we would like to determine if the solution $\overline{x}$ violates any packing constraint: that is, if possible, we would like to find a set of nodes S for which $\sum_{e\in E(S)} \overline{x}_e > |S| - 1$. If we found such a set, we would add the constraint $\sum_{e\in E(S)} x_e \leq |S| - 1$ to the linear programming model. In polyhedral combinatorics, the problem of finding such a violated constraint, or cut as it is known, is called the *separation problem* since we are finding a violated inequality that separates the vector $\overline{x}$ from the polyhedron P_{sub}.

Solving the separation problem

How can we solve the separation problem? The directed cutset formulation and the development in this section have implicitly provided us with an answer to this question. As we have seen, $P_{\text{sub}} = P_{\text{dflo}}$. Stated in another way, $\overline{x} \in P_{\text{sub}}$ if and only if $\overline{y}$, with $\overline{x}_e = \overline{y}_{ij} + \overline{y}_{ji}$ for all edges $e = \{i, j\}$, together with some flow vector $\overline{f}$ is feasible in the directed multicommodity flow formulation (3.38)–(3.44). Suppose we set the capacity of each arc (i, j) in this model to $\overline{x}_e$; then $\overline{x} \in P_{\text{sub}}$ if and only if the capacitated network has a flow of one unit from the root node r to every other node k (since we can always find the maximum flow between two nodes by sending flow only in one of the arcs (i, j) and (j, i)). Any maximum flow algorithm that finds a minimum cut as well as a maximum flow will give us this answer: for suppose for each node k, we find a maximum flow from the root node to node k. If the maximum flow has value $\theta < 1$, then by the max-flow min-cut theorem, some directed cut $\delta^+(S)$ has a capacity less than 1, that is, $\sum_{e\in\delta(S)} \overline{x}_e = \sum_{e\in\delta^+(S)} \overline{x}_e < 1$. But since $\sum_{e\in E} \overline{x}_e = |V| - 1$, either $\sum_{e\in S} \overline{x}_e > |S| - 1$ or $\sum_{e\in\overline{S}} \overline{x}_e > |\overline{S}| - 1$ and so the minimum cut provides us with a violated packing inequality. We could then add this inequality to the linear program, solve it again, and repeat the separation procedure.

Most violated constraint [4]

The method we have described is guaranteed to find a violated packing inequality, or show that none exists, by solving $|V| - 1$ maximum flow problems.

[4] Readers can skip the next two paragraphs without any loss of continuity.

It is not, however, guaranteed to find a most violated packing inequality, that is, one that maximizes $\sum_{e\in E(S)} \overline{x}_e - (|S|-1)$ over all node sets S. We next show that by cleverly defining an ancillary maximum flow network, we can find a most violated constraint using the same number of maximum flow computations. To find a most violated inequality we wish to maximize $\sum_{e\in E(S)} \overline{x}_e - (|S|-1)$, or, equivalently, minimize $|S| - \sum_{e\in E(S)} \overline{x}_e$ over all nonempty node sets $S \subset V$. Since $\sum_{e\in E} \overline{x}_e = |V| - 1$, a constant, this minimization is equivalent to minimizing $|S| + \sum_{e\notin E(S)} \overline{x}_e$. We will show how to minimize this quantity. If its minimum has a value of at least $|V|$, then the solution $\overline{x}$ satisfies all the packing constraints.

As before, we solve $|V| - 1$ maximum flow (minimum cut) problems, each defined on a directed version $G^* = (V, A)$ of the network $G = (V, E)$. In the kth such problem, we find a cut that separates the root node r from node k. For every edge $e = \{i, j\}$ in E, G^* contains the arcs (i, j) and (j, i); we set the capacity of any directed arc (p, q) obtained from edge e to $(1/2)\overline{x}_e + (1/2)\sum_{e\in\delta(q)} \overline{x}_e$. For each node $v \in V$, we also include an arc (v, k), which might be a parallel arc, with a capacity of one unit. Note that if S is any node set with $r \in S$ and $k \in \overline{S}$, then the unit capacity arcs we just added contribute a capacity of $|S|$ to the cut $\delta^+(S)$. Moreover, the other arcs in $\delta(S)$ have a capacity $\sum_{e\notin E(S)} \overline{x}_e$ since our definition of capacities double counts $\overline{x}_e$ for each edge $e = \{p, q\}$ with both of its endpoints in $\overline{S}$ — this accounts for the factor of 1/2 in the definition of the arc capacities. Therefore, the minimum capacity cut in the kth maximum flow problem minimizes $|S| + \sum_{e\notin E(S)} \overline{x}_e$ over all cuts $\delta(S)$ separating the root node r from node k. Consequently, by solving $|V| - 1$ maximum flow problems, one for each node $k \neq r$, we solve the linear programming separation problem.

3.3. *A linear programming bound*

In Section 3.2, we saw that $P_{\mathrm{mcut}} = P_{\mathrm{sub}}$ (as well as several other equivalent polyhedra) are the convex hull of incidence vectors of spanning trees. We also saw, through an example, that P_{cut} might be a larger polyhedron. Therefore, for any vector $w = (w_e)$ of objective coefficients, the optimal value of the linear program $z^{\mathrm{LP}}_{\mathrm{cut}} = \min_{x\in P_{\mathrm{cut}}} wx$ will generally be less than the optimal value of the linear program $z^{\mathrm{LP}}_{\mathrm{mcut}} = \min_{x\in P_{\mathrm{mcut}}} wx$. That is, $z^{\mathrm{LP}}_{\mathrm{mcut}}/z^{\mathrm{LP}}_{\mathrm{cut}} = r \geq 1$. How large can the ratio r become? In our example in Figure 9, $r = 2/1.5 = 4/3$.

Recall that $P_{\mathrm{cut}} = \{x \in R^{|E|} : \sum_{e\in E} x_e = n - 1$, and $\sum_{e\in\delta(S)} x_e \geq 1$ for all nonempty node sets $S \subset V\}$ and that $P_{\mathrm{mcut}} = \{x \in R^{|E|} : 0 \leq x \leq \mathbf{1}, \sum_{e\in E} x_e = n - 1$, and $\sum_{e\in\delta(C_0,C_1,\ldots,C_k)} x_e \geq k$ for all node partitions $C_0, C_1, \ldots, C_k$ of $V\}$. Throughout this analysis, we assume $w \geq 0$.

To obtain a bound on $z^{\mathrm{LP}}_{\mathrm{mcut}}/z^{\mathrm{LP}}_{\mathrm{cut}}$, let us first consider the polyhedra P_{mcut} and P_{cut} without the cardinality constraint $\sum_{e\in E} x_e = n - 1$ and the upper bound constraints $x_e \leq 1$ imposed upon the edges e. Let P^+_{mcut} and P^+_{cut} denote these polyhedra. Note that any integer point in either of these polyhedra corresponds to a 'supertree', that is, a graph that contains a spanning tree (i.e., a spanning tree plus additional edges which can, if any $x_e > 1$, duplicate some edges). Let

$z^{\text{LP}}_{\text{mcut}+} = \min_{x \in P^+_{\text{mcut}}} wx$ and $z^{\text{LP}}_{\text{cut}+} = \min_{x \in P^+_{\text{cut}}} wx$ be the optimal objective values of the linear programs with objective function coefficients w over these polyhedra. We wish to establish a bound on the value of these linear programs.

Let $\bar{x}$ be any point in P^+_{cut} and let $\delta(C_0, C_1, \ldots, C_k)$ be any multicut. Suppose we add the constraints $\sum_{e \in \delta(C_q, V \setminus C_q)} x_e \geq 1$ for all $q = 0, 1, \ldots, k$. In doing so, we include each edge in $\delta(C_0, C_1, \ldots, C_k)$ twice, and so the resulting inequality is

$$2 \sum_{e \in \delta(C_0, C_1, \ldots, C_k)} x_e \geq k + 1.$$

or, equivalently, $[2k/(k+1)] \sum_{e \in \delta(C_0, C_1, \ldots, C_k)} x_e \geq k$.

Note that since $k+1 \leq |V|$, $1 - 1/|V| \geq 1 - 1/(k+1) = k/(k+1)$. Consequently, the point $\bar{x}$ also satisfies the following inequality

$$2\left(1 - \frac{1}{|V|}\right) \sum_{e \in \delta(C_0, C_1, \ldots, C_k)} x_e \geq k.$$

Therefore, the point $\bar{z} = 2(1 - 1/|V|)\bar{x}$ belongs to P^+_{mcut}. Note that this point has an objective value of $w[2(1 - 1/|V|)]\bar{x} = w\bar{z}$. Thus, for *any* point $x \in P^+_{\text{cut}}$, the point $z = 2(1 - 1/|V|)x$ with objective value wz belongs to P^+_{mcut}, that is, $[2(1 - 1/|V|)]wx = wz \geq z^{\text{LP}}_{\text{mcut}+}$. Since this inequality is valid for all points $x \in P^+_{\text{cut}}$, including any optimal point, it implies the following bound:

Proposition 3.10. $z^{\text{LP}}_{\text{mcut}+}/z^{\text{LP}}_{\text{cut}+} \leq 2(1 - 1/|V|)$.

This result shows that, in general, the objective value for any linear program defined over P^+_{mcut} is no more than twice the objective value of a linear program with the same objective function defined over the polyhedron P^+_{cut}. If the underlying graph is a cycle with $|V|$ edges, and if all the edge costs $w_e = +1$, then the optimal solution to $\min\{wx : e \in P^+_{\text{cut}+}\}$ sets $x_e = 1$ for all but one edge and the optimal solution to $\min\{wx : e \in P^+_{\text{cut}+}\}$ sets $x_e = 1/2$ for all edges. Therefore, $z^{\text{LP}}_{\text{mcut}+}/z^{\text{LP}}_{\text{cut}+} = (|V| - 1)/(|V|/2) = 2(1 - 1/|V|)$ achieves the bound in this proposition.

Now consider the problems $z^{\text{LP}}_{\text{mcut}} = \min_{x \in P_{\text{mcut}}} wx$ and $z^{\text{LP}}_{\text{cut}} = \min_{x \in P_{\text{cut}}} wx$. We first make the following observations:

(1) In the multicut polyhedron P_{mcut}, for any edge $\bar{e}$, the upper bound constraints $x_{\bar{e}} \leq 1$ are redundant since the constraint $\sum_{e \in E} x_e = n - 1$ and the multicut constraint $\sum_{e \neq \bar{e}} x_e \geq n - 2$ (here C_0 contains just the end nodes of edge $\bar{e}$ and all the other C_k are singletons) imply that $x_{\bar{e}} \leq 1$.

(2) If $w_e \geq 0$, then the cardinality constraint $\sum_{e \in E} x_e = n - 1$ is redundant in the linear program $z^{\text{LP}}_{\text{mcut}} = \min_{x \in P_{\text{mcut}}} wx$ in the sense that the problem without the cardinality constraint always has a solution satisfying this constraint. This result follows from the second proof of Theorem 3.4 which shows that since $w_e \geq 0$ for all $e \in E$, the dual linear program has an optimal solution in which the dual variable on the constraint $\sum_{e \in E} x_e = \sum_{\delta(\{1\}, \ldots, \{n\})} x_e = n - 1$ is nonnegative.

We can now establish the following result.

Proposition 3.11. $z^{\text{LP}}_{\text{mcut}}/z^{\text{LP}}_{\text{cut}} \le 2(1-1/|V|)$.

Proof. Since the polyhedron P_{cut} contains one more constraint than the polyhedron P^{+}_{cut}, $z^{\text{LP}}_{\text{cut}} \ge z^{\text{LP}}_{\text{cut+}}$ and as we have just seen, since $w \ge 0$, $z^{\text{LP}}_{\text{mcut}} = z^{\text{LP}}_{\text{mcut+}}$. Therefore, Proposition 3.10 shows that

$$\left[2\left(1-\frac{1}{|V|}\right)\right] z^{\text{LP}}_{\text{cut}} \ge \left[2\left(1-\frac{1}{|V|}\right)\right] z^{\text{LP}}_{\text{cut+}} \ge z^{\text{LP}}_{\text{mcut+}} = z^{\text{LP}}_{\text{mcut}}.$$

□

Observe that since the polyhedron P_{mcut} is integral, $z^{\text{LP}}_{\text{mcut}} = Z$, the optimal value of the spanning tree problem. Therefore, Proposition 3.11 bounds the ratio of the optimal integer programming value to the optimal objective value of the linear programming relaxation of the cut formulation. In Section 6.3, we consider a generalization of this bound. Using a deeper result than we have used in this analysis (known as the parsimonious property), we are able to show that the bound of $2(1-1/|V|)$ applies to Steiner trees as well as spanning trees.

4. Rooted subtrees of a tree

In Section 2, we considered the core tree problem defined on a general network. We now consider the core problem encountered in packing trees within a tree: the rooted subtree problem. Given a tree T with a root node r and a weight w_v on each node v of T, we wish to find a subtree T^* of T that contains the root and has the largest possible total weight $\sum_{j\in T^*} w_v$. We permit T^* to be the empty tree (with zero weight). In Section 4.2 we consider an extension of this model by introducing capacity restrictions on the tree T^*.

4.1. Basic model

We begin by setting some notation. Let $p(v)$, the predecessor of v, be the first node $u \ne v$ on the unique path in T connecting node v and the root r, and let $S(v)$ be the immediate successors of node v; that is, all nodes u with $p(u) = v$. For any node v of T, let $T(v)$ denote the subtree of T rooted at node v; that is, $T(v)$ is the tree formed if we cut T by removing the edge {p(v),v} just above node v.

Dynamic programming solution

The solution to this problem illustrates the type of dynamic programming procedure that solves many problems defined on a tree. For any node v of T, let $H(v)$ denote the optimal solution of the rooted subtree problem defined on the tree $T(v)$ with node v as the root. If v is a leaf node of T, $H(v) = \max\{0, w_v\}$

since the only two rooted subtrees of $T(v)$ are the single node $\{v\}$ and the empty tree. The dynamic programming algorithm moves 'up the tree' from the leaf nodes to the root. Suppose that we have computed $H(u)$ for all successors of node v; then we can determine $H(v)$ using the following recursion:

$$H(v) = \max\left\{0, w_v + \sum_{u \in S(v)} H(u)\right\}. \tag{4.1}$$

This recursion accounts for two cases: the optimal subtree of $T(v)$ rooted at node v is either (a) empty, or (b) contains node v. In the latter case, the tree also contains (the possibly empty) optimal rooted subtree of each node u in $S(v)$. Note that since each node u, except the root, is contained in exactly one subset $S(v)$, this recursion is very efficient: it requires one addition and one comparison for each node of T. After moving up the tree to its root and finding $H(r)$, the optimal value of subtree problem defined over the entire tree T, we can determine an optimal rooted subtree T^* by deleting from T all subtrees $T(u)$ with $H(u) = 0$.

Example 4.1. For the example problem shown in Figure 12 with root $r = 1$, we start by computing $H(4) = 4$, $H(6) = 0$, $H(7) = 2$, $H(8) = 4$, $H(10) = 0$, and $H(11) = 3$ for the leaf nodes of the tree. We then find that $H(9) = \max\{0, -5 + 0 + 3\} = 0$, $H(5) = \max\{0, -1 + 4 + 0\} = 3$, $H(2) = \max\{0, -5 + 4 + 3\} = 2$, $H(3) = \max\{0, -1 + 0 + 2\} = 1$, and finally $H(1) = \max\{0, 2 + 2 + 1\} = 5$. Since $H(9) = H(6) = 0$, as shown in Figure 12b, the optimal rooted tree does not contain the subtrees rooted at these nodes.

Variants and enhancements of this recursion apply to many other problems defined on trees. We will examine one such example in Section 4.2 where we consider the addition of capacity constraints to the subtree of a tree problem. In later sections, we consider several other similar dynamic programming algorithms.

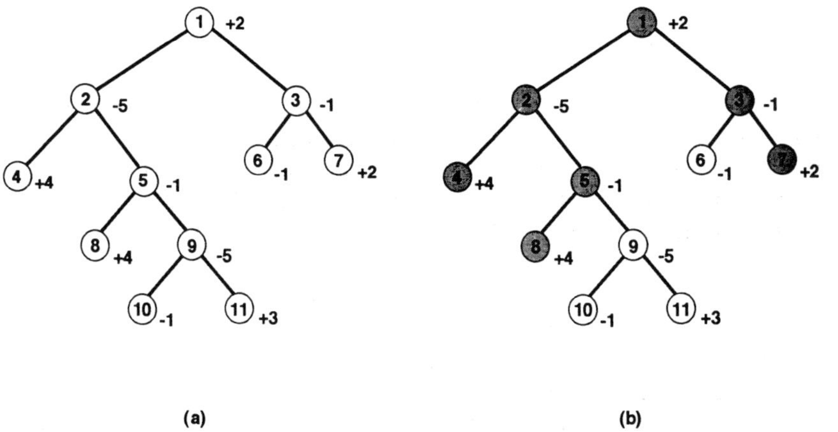

Fig. 12. (a) Given tree with node weights w_e; (b) optimal rooted subtree (shaded nodes).

Polyhedral description

Let x_v be a zero-one variable indicating whether ($x_v = 1$) or not ($x_v = 0$) we include node v in a rooted subtree of T, and let X denote the set of incidence vectors $x = (x_v)$ of subtrees rooted at node r (more precisely, X is the incidence vectors of nodes in a subtree rooted at node r — for convenience we will refer to this set as the set of subtrees rooted at node r). Note that points in X satisfy the following inequalities

$$0 \leq x_r \leq 1, 0 \leq x_v \leq x_{p(v)} \text{ for all nodes } v \neq r \text{ of } T. \tag{4.2}$$

Since every point in X satisfies these inequalities, so does the convex hull, conv(X), of X, that is, $\{x : 0 \leq x_r \leq 1, 0 \leq x_v \leq x_{p(v)}$ for all nodes $v \neq r$ of $T\}$ $\supseteq$ conv(X). We will show that the inequalities in (4.2) actually completely describe r-rooted subtrees of T in the sense that the convex hull of X equals the set of solutions to (4.2). That is, we will establish the following result:

Theorem 4.1. *The set of solutions to the linear inequality system (4.2) is the convex hull of the 0–1 incidence vectors of subtrees of T rooted at node r.*

This theorem shows that the extreme points of the polyhedron $P = \{x \in R^{|V|} : 0 \leq x_r \leq 1, 0 \leq x_v \leq x_{p(v)}$ for all nodes $v \neq r$ of $T\}$ are the 0–1 incidence vectors of subtrees of T rooted at node r. Notice that $X = P \cap Z^{|V|}$, that is, every integer point in P is the incidence vector of a subtree rooted at node r and so we will establish the theorem if we show that every extreme point of the polyhedron P is integer valued.

We will prove Theorem 4.1 using three different arguments that underlie many results in the fields of combinatorial optimization and polyhedral combinatorics: a network flow argument, a dynamic programming argument, and an argument based upon the nature of optimal solutions to the optimization problem $\min\{wx : x \in X\}$ as we vary the weights w. All three of these arguments rely on basic results from linear programming.

Approach 1 (Network flow argument). Consider the (primal) linear programming problem $\max\{wx : 0 \leq x_r \leq 1, 0 \leq x_v \leq x_{p(v)}$ for all nodes $v \neq r$ of $T\}$ for any choice w of integer node weights. Since this problem has one inequality (in addition to the nonnegativity restriction) for each node v of T, its linear programming dual problem $\min\{y_r : y_v - \sum_{u \in S(v)} y_u \geq w_v$ and $y_v \geq 0$ for all nodes v of $T\}$ has one dual variable y_v for each node of v. Note that the dual problem is a network flow problem since each variable y_q for $q \neq r$ appears in exactly two constraints: in constraint $v = q$ with a coefficient of $+1$ and in the constraint $v = p(q)$ with a coefficient of -1. Since node r has no predecessor, it appears in only the constraint $v = r$ with a coefficient of $+1$. The theory of network flows shows that whenever the cost data of any network flow problem are integral, its dual linear program always has an integer optimal solution (see the discussion at the end of Section 2). In this case, since the network flow problem $\min\{y_r : y_v - \sum_{u \in S(v)} y_u \geq w_v$ and $y_v \geq 0\}$ has integer objective function coefficients (zeros and the single +1 coefficient for y_r), for *any* choice of integer weights w, the dual problem $\max\{wx : 0 \leq x_r \leq 1, 0 \leq x_v \leq x_{p(v)}$ for all nodes $v \neq r$ of $T\} = \max\{wx : x \in P\}$ has an integer optimal solution. But then the theory of linear programming theory shows that every extreme point of P is integer valued, and so by our previous observation, the extreme points are the incidence vectors of subtrees rooted at node r.

Approach 2 (Dynamic programming argument). This argument is similar to the linear programming duality argument that we gave in the last section when we studied the minimum spanning tree problem, though it uses information provided by the dynamic programming solution procedure instead of the greedy procedure to set values for the linear programming dual variables. Consider the (primal) linear programming problem $\max\{wx : 0 \leq x_r \leq 1, 0 \leq x_v \leq x_{p(v)}$ for all nodes $v \neq r$ of $T\}$. As we noted in the last subsection, its linear programming dual problem is $\min\{y_r : y_v - \sum_{u \in S(v)} y_u \geq w_v$ and $y_v \geq 0$ for all nodes v of $T\}$.

Let T^* be the optimal rooted subtree of T determined by the dynamic programming recursion (4.1) and let $x_v = 1$ if node v belongs to T^* and let $x_v = 0$ otherwise. This solution is feasible in the primal linear program and its objective value is $H(r)$, the optimal objective value determined by the dynamic programming recursion. Since the values of $H(v)$, for v in T, satisfy the recursion (4.1), $H(v) \geq w_v + \sum_{u \in S(v)} H(u)$. Therefore, the choice $y_v = H(v)$, for all $v \in V$, of the variables in the dual linear program is feasible. Moreover, since the dual objective value is $y_r = H(r)$, we have shown that for every choice of objective function coefficients w, the primal problem has an integer optimal solution (since its objective value equals the objective value of some dual feasible solution). But then linear programming theory implies that the extreme points of the polyhedron $P = \{x : 0 \leq x_r \leq 1, 0 \leq x_v \leq x_{p(v)}$ for all nodes $v \neq r$ of $T\}$ are the 0–1 incidence vectors of subtrees of T rooted at node r.

Approach 3 (Optimal inequality argument). This proof uses an argument from the field of polyhedral combinatorics. Let w be any n-vector of weight coefficients and consider the optimization problem max $\{wy : y \in Y\}$ for any finite set Y. Let $Q = \{a_j x \leq b_j$ for $j = 1, 2, \ldots, m\}$ be any bounded polyhedron that contains Y. Suppose that we can show that for any choice of $w \neq 0$, the set of optimal solutions to the problem max $\{wy : y \in Y\}$ all lie on the same inequality $a_j y \leq b_j$ of the polyhedron Q. We refer to any such inequality (for a given choice of w) as an *optimal inequality*. Note that for a particular choice of w, the polyhedron P might contain more than one optimal inequality and as we vary w, we will find different optimal inequalities. We might distinguish between two types of inequalities: an inequality $a_j y \leq b_j$ is *binding* if it is satisfied as an equality by all points in Y and is *nonbinding* otherwise.

Nonbinding optimal inequalities are useful for the following reason: suppose w defines a *facet* of the convex hull conv(Y) of Y in the sense that for some constant w_0, $Y \subseteq \{y : wy \leq w_0\}$ and the set of solutions of the system $\{y : y \in Y$ and $wy = w_0\}$ has dimension one less than the dimension of conv(Y). Then the set of optimal solutions to the optimization problem $\max\{wy : y \in Y\}$ are just the points of Y on the facet. In this case, any optimal inequality $a_j y \leq b_j$ contains all the points on the facet. This result implies that if the polyhedron Q contains a nonbinding optimal inequality for every choice of the weight vector $w \neq 0$, then the polyhedron is the convex hull of Y.

To utilize this proof technique, we set $Y = X$ and $Q = P$, with X and P as defined previously. Note that in this case every inequality $x_v \leq x_{p(v)}$ in the definition of P is nonbinding since we can choose a rooted feasible tree with $x_v = 0$ and $x_{p(v)} = 1$. The inequality $x_r \leq 1$ is nonbinding because the zero vector is a feasible rooted tree and the inequalities $0 \leq x_v$ are nonbinding since the tree with each $x_v = 1$ is feasible. Since each defining inequality of P is nonbinding, to use the optimal inequality argument, we wish to show that for every choice of $w \neq 0$, the polyhedron P contains an optimal inequality.

Let $z = \max\{wx : x \in X\}$ with $w \neq 0$. Since the zero vector is feasible, $z \geq 0$. If $z > 0$, then all optimal solutions satisfy the condition $x_r = 1$ and so $x_r \leq 1$ is an optimal inequality. So suppose that $z = 0$. Note that since we are assuming that the weight vector $w \neq 0$, some component of w must be negative. Otherwise, setting $x_v = 1$ for all v gives a solution with $z > 0$. So we suppose $w_v < 0$ for some node v of T.

If $x_v = 0$ in every optimal solution to the problem, then $x_v \geq 0$ is an optimal inequality. So suppose that T^* is an optimal subtree solution containing node v. If the weight of the subtree $T(v) \cap T^*$ rooted at node v is negative, then by eliminating $T(v) \cap T^*$ from T^*, we would obtain a rooted tree with a weight that exceeds 0, contrary to our assumption that $z = 0$ (this situation corresponds to the fact that $H(v) \geq 0$ in the dynamic programming solution to the problem). Therefore, since $w_v < 0$, at least one successor u of node v must satisfy the property that the total weight of the subtree $T(u) \cap T^*$ rooted at node u is positive (that is $H(u) > 0$ in the dynamic program). We claim that $x_u \leq x_v$ is an optimal inequality. If $x_v = 0$ in any optimal solution, then $x_u = x_v = 0$. So suppose we have a optimal tree with $x_v = 1$. If $x_u = 0$, we could add $T(u) \cap T^*$ to this optimal tree and obtain a feasible solution with $z > 0$, again contrary to our assumption that $z = 0$. Therefore, $x_u = 1$ and so $x_u \leq x_v$ is an optimal inequality.

These arguments show that for any choice of $w \neq 0$, one of the inequalities in the system (4.2) is a nonbinding optimal inequality to the problem max $\{wx : x \in X\}$, which implies the conclusion of the theorem.

4.2. Constrained subtrees

In some application contexts, we are not free to choose every rooted subtree as a possible solution to the rooted subtree of a tree problem. For example, the root node might correspond to a concentrator in a telecommunication system that is serving demands d_v at the nodes v of the tree. If the concentrator has a limited throughput capacity C, then the node incidence vector $x = (x_v)$ of any feasible rooted tree must satisfy the following capacity constraint:

$$\sum_{v \in V(T)} d_v x_v \leq C. \tag{4.3}$$

Recall that $V(T)$ denotes the set of nodes in the tree T.

We assume that each demand d_v is a positive integer that does not exceed the capacity C. We refer to this problem as the capacitated (rooted) subtree of a tree problem.

As an important special case, each node v has a demand $d_v = 1$ and so the capacity constraint becomes a cardinality constraint

$$\sum_{v \in V(T)} x_v \leq K \tag{4.4}$$

stating that the chosen tree can contain at most $K = C$ nodes. We refer to this version of the problem as the cardinality constrained rooted subtree of a tree problem.

A solution procedure

To find an optimal capacitated subtree of a tree, we can once again use a dynamic programming algorithm. Let $H(u, q)$ denote the optimal objective value for the constrained subtree of a tree problem defined by the tree $T(u)$ with node u as the root and with the integer $q \leq C$ as the capacity. $H(r, C)$ is the optimal objective value for the original problem.

We can solve the original problem by again working from the leaf nodes toward the root using a dynamic programming recursion:

$$H(v, q) = \max\{0, w_v + \max_{\{q_u : \sum_{u \in S(v)} q_u \leq q - d_v\}} \sum_{u \in S(v)} H(u, q_u)\}. \tag{4.5}$$

We initiate the recursion by setting $H(u, q) = \max\{w_u, 0\}$ for any leaf node with $d_u \leq q$ and $H(u, q) = 0$ for any leaf node with $d_u > q$.

This recursion says that the optimal solution on the tree $T(v)$ either does not use node v and so has value zero or uses node v, consuming d_v units of the capacity and leaving $q - d_v$ units of capacity to be shared by the subtrees rooted on the successor nodes $S(v)$ of node v.

Note that for the cardinality version of the problem, this recursion is particularly simple since its says that if we include node v in the optimal subtree of $T(v)$, then at most $q - 1$ nodes are available for distribution to the subtrees on node v's successor nodes.

If we order the successors of each node, it is easy to implement the basic dynamic programming recursion using $O(nC^2)$ computations (that is, the number of computations is a polynomial with a constant times nC^2 as the leading term). For the special cardinality version of the problem, this implementation requires $O(n^3)$ computations. Let $u_1, \ldots u_r$ be the successors of node v. We will find the optimal value of $H(v, q)$ by building up the solution from the optimal solution over the subtree $T(u_1)$, then $T(u_1)$ and $T(u_2)$, then $T(u_1), T(u_2)$, and $T(u_3)$, and so forth until we consider all the successors. To do so, we let $G_j(v, t)$ denote the objective value of an optimal forest containing rooted subtrees from the trees $T(u_1), T(u_2), \ldots, T(u_j)$, given that the nodes in the solution contain a total demand of at most t. For each index $j = 1, 2, \ldots, r$, let $G_j(v, 0) = 0$. Then for all $t = 1, 2, \ldots, C$, $G_1(v, t) = H(u_1, t)$ and for each $j = 2, 3, \ldots, r$, $G_j(v, t) = \max_{0 \le s \le t}\{G_{j-1}(v, s) + H(u_j, t - s)\}$. Finally, we find $H(v, q) = \max\{0, w_v + G_r(v, q - d_u)\}$.

Polyhedral considerations

Let X^C and X^K denote the set of feasible solutions to the capacitated rooted subtree of a tree problem, that is, the rooted subtree of a tree problem with the additional constraints (4.3) or (4.4). Let $P^C \supseteq X^C$ denote the polyhedra defined by the constraints (4.2) and (4.3) and let $P^K \supseteq X^K$ denote the polyhedra defined by the constraints (4.2) and (4.4).

As shown by the following example, unlike the (uncapacitated) subtree of a tree problem, in general the polyhedra P^C and P^K are not integral.

Example 4.2. Consider the cardinality constrained problem with $K = 4$ for the example shown in Figure 12. In this case the optimal tree T^* contains nodes $1, 3$, and 7 and has a weight 3. The fractional solution $\hat{x}_1 = 1$, $\hat{x}_2 = \hat{x}_3 = \hat{x}_4 = \hat{x}_5 = \hat{x}_7 = \hat{x}_8 = 1/2$, and $\hat{x}_6 = \hat{x}_9 = \hat{x}_{10} = \hat{x}_{11} = 0$ satisfies all the constraints of the system (4.2) as well as the inequality (4.4) and so lies in P^K. The point $\hat{x}$ is the unique solution to the linear program $\max\{\sum_{v \in V(T)} w_v x_v : x = (x_v) \in P^K\}$ and so by linear programming theory it is a fractional extreme point of the polyhedron P^K.

In order to obtain a polyhedral description of $\mathrm{conv}(X^C)$ or of $\mathrm{conv}(X^K)$, we need to add additional valid inequalities to P^C and P^K. Finding enough valid inequalities to completely describe these convex hulls appears to be quite difficult. To illustrate how complicated the convex hulls are, we will illustrate two sets of valid inequalities.

Let us call a subtree T' of T rooted at node r a *tree cover* if $\sum_{v \in V(T')} d_v$ exceeds C. Note that for the cardinality constrained problem, a tree cover is just a subtree rooted at node r with at least $K + 1$ nodes. The cover condition implies that we cannot include all the nodes of T' in any feasible rooted subtree of T.

To eliminate this possibility, we can use the following *tree cover inequality*:

$$\sum_{v \in T'} (x_{p(v)} - x_v) \ge x_r.$$

Proposition 4.2. *Feasible points in X^C satisfy every tree cover inequality.*

Proof. Consider any tree cover inequality. The null tree $\overline{x} = 0$ clearly satisfies it. For any other feasible solution $\overline{x}$, $\overline{x}_r = 1$. Note that the constraints $x_v \le x_{p(v)}$ in the system (4,2) imply that every term on the left-hand side of the cover inequality is nonnegative. Therefore, if the solution $\overline{x}$ to the problem violates this inequality, then $\overline{x}_{p(v)} - \overline{x}_v = 0$ for all $v \in V(T')$. But then since $\overline{x}_r = 1$, $\overline{x}_v = 1$ for all the nodes $v \in V(T')$, contradicting the fact that T' is a cover. $\square$

To illustrate the tree cover inequalities, consider the cardinality constrained tree problem of Example 4.2 once again (with $K = 4$). In this case, let T' contain the five nodes 1, 2, 4, 5, and 8. The tree cover inequality for T' is:

$$(x_5 - x_8) + (x_2 - x_4) + (x_2 - x_5) + (x_1 - x_2) \geq x_1$$

or

$$x_2 \geq x_4 + x_8.$$

Since $\hat{x}_2 = \hat{x}_4 = \hat{x}_8 = 1/2$, the fractional solution $\hat{x}$ from Example 4.2 does not satisfy this inequality. Therefore, by adding the tree cover inequalities to the polyhedron P^K, we obtain a better representation of the cardinality constrained polyhedron $\text{conv}(P^K)$.

We obtain the second set of inequalities by considering another special class of subtrees rooted at node r. Given a subtree T' containing the root node, let $L(T')$ denote its leaf nodes, and for any node set $S \subset V(T)$, let $Cl(S)$, the *closure* of S, be the tree determined by the union of the paths from each node $v \in S$ to the root.

Proposition 4.3. *Let T' be a subtree of T rooted at node r. If for some positive integer q, the closure $Cl(S)$ is a tree cover of the tree T for all node sets $S \subset L(T')$ with $|S| = q$, then any feasible solution of X^C satisfies the following* leaf cover *inequality:*

$$\sum_{u \in L(T')} x_u \leq (q-1)x_r.$$

Proof. This inequality is clearly valid for $\overline{x} = 0$. If $\sum_{u \in L(T')} \overline{x}_u \geq q\overline{x}_r$ and $\overline{x}_r = 1$, then the subtree of T' defined by the nodes u with $\overline{x}_u = 1$ would contain a tree cover and so be infeasible. So every feasible solution satisfies every leaf cover inequality. □

Example 4.3. Consider the cardinality constrained subtree problem with $K = 5$ on the tree T shown in Figure 13. In this case, we can obtain four tree covers by deleting any single leaf node from T. The tree cover inequalities for these subtrees (once we have collected terms) are:

$$x_1 + x_2 \geq x_4 + x_5 + x_6$$
$$x_1 + x_2 \geq x_4 + x_5 + x_7$$
$$x_1 + x_3 \geq x_4 + x_6 + x_7$$
$$x_1 + x_3 \geq x_5 + x_6 + x_7.$$

Let $T' = T$ and $q = 3$. Then the tree T' satisfies the conditions of the leaf cover inequality and so the following leaf cover inequality is valid:

$$x_4 + x_5 + x_6 + x_7 \leq 2x_1.$$

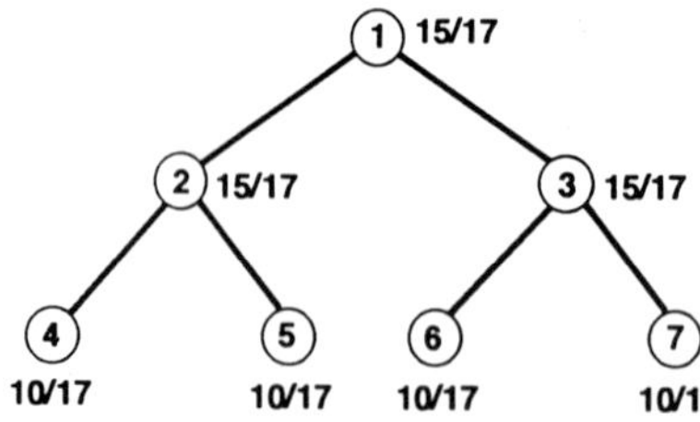

Fig. 13. Fractional values x_v violating a leaf inequality.

The fractional solution shown in Figure 13 satisfies all four tree cover inequalities as well as the inequalities in the system (4.2). However, it violates this leaf cover inequality.

Let P^K_{TC} and $P^K_{TC,LC}$ denote the polyhedra defined by adding the tree cover inequalities and both the tree cover and leaf cover inequalities to P^K. This example and Example 4.2 show that in general,

$$\mathrm{conv}(X^K) \subset P^K_{TC,LC} \subset P^K_{TC} \subset P^K.$$

That is, in general, as in this case, by adding the tree cover inequalities and then the leaf cover inequalities to P^K, we obtain better polyhedral approximations to $\mathrm{conv}(X^K)$.

As this discussion has shown, the addition of a single inequality, even a very simple cardinality constraint, can add considerably to the complexity of an integer polyhedron.

5. Polynomially solvable extensions/variations

In the last two sections, we considered two core tree optimization problems — the minimum spanning tree problem and the rooted subtree of a tree problem. For each of these problems, we were able to develop an algorithm (the greedy algorithm and a dynamic programming algorithm) that solves the problem using a number of computations that is polynomial in the problem's size (as measured by the number of nodes and edges in the underlying graph). In this section, we consider two other polynomially-solvable versions of tree problems; both are variations of the minimum spanning tree problem. In one problem, the degree-constrained spanning tree problem, we impose a degree constraint on one of the nodes and in the other problem, defined over a directed graph, we seek a directed version of a minimum spanning tree, a so-called optimal branching or optimal arborescence. We not only describe polynomial algorithms for these problems, but also show that we can describe the convex hull of incidence vectors of feasible solutions of these problems by using minor variants of the formulations we have already studied for the minimum spanning tree problem.

5.1. Degree-constrained spanning trees

Suppose we wish to find a minimum spanning tree in a graph G, but require that the tree satisfies the property that the degree of a particular node r, which we will call the root node, is fixed at value k. That is, the root node must have k incident edges. We call any such tree a *degree-constrained minimum spanning tree*. In this section, we show how to solve this problem efficiently for all values of k and, for any fixed k, we give a polyhedral description of the incidence vectors of degree-constrained spanning trees.

To solve the degree-constrained minimum spanning tree, we consider a parametric minimum spanning tree problem with edge weights chosen as follows: the weight of every edge $e = \{r, j\}$ incident to the root node is $w_e + \theta$ for some scalar parameter θ, and the weight of any edge e not incident to the root node is the w_e. Note that if θ is sufficiently large, the solution to the unconstrained minimum spanning tree problem will be a spanning tree T_0 in G containing the fewest possible number $k_{\min}$ of edges incident to node r. (Why?) As we decrease θ, the edges incident to the root node become more attractive. For each edge $e = \{r, j\}$ not in T^0, let P_j denote the path in T^0 that connects node j to the root node. As we decrease θ, for the first time at some point $\theta = \theta_j^-$ (that is, θ_j minus any positive amount), the edge $e = \{r, j\}$ has a weight smaller than one of the edges on the path P_j. Among all choices of node j, we choose one with a maximum value of θ_j and add edge $e = \{r, j\}$ to T_0, dropping the edge from P_j whose weight is larger than $w_e + \theta_j^-$. Note that since we add the same amount θ to the cost of every edge incident to the root node, the edge we drop will not be incident to the root. Therefore, this operation gives us a new tree T_1 with $k_{\min} + 1$ edges incident to the root node. We then repeat this step using T_1 in place of T_0 (and so with new paths P_j). By continuing in this way, for each $i = 1, 2, \ldots$, we obtain a spanning tree T_i with $k_{\min} + i$ edges until the tree T_i contains every edge $\{r, j\}$ in G emanating from the root node. In Theorem 5, we show that each intermediate tree T_i is a minimum spanning tree for the value of θ at which we converted the tree T_{i-1} into the tree T_i.

Suppose we maintain and update two labels for each node: (i) a precedence label that indicates for each node j the next node q on the unique path in the tree T_i connecting node j to the root node, and (ii) a second label l that denotes the maximum weight of the edges on P_j, not including the edge incident to the root. With these labels, determining the next edge $e = \{r, j\}$ to add to the tree at each step and finding the edge on the path P_j to drop from the tree requires at most n operations. Since updating the node labels when we add edge e to T_i requires at most n operations as well (the node labels change only for the nodes on P_j), this algorithm requires $O(n^2)$ operations plus the amount of work needed to find the initial spanning tree T_0.

We note that as a byproduct, the algorithm that we have just described provides a polyhedral characterization of degree-constrained minimal spanning trees. Let $Q = \{x \geq 0 : \sum_{e \in E} x_e = n - 1,\ \text{and } \sum_{e \in E(S)} x_e \leq |S| - 1 \text{ for all } S \neq V\}$ be the polyhedron whose extreme points are incidence vectors of spanning trees (see Section 3).

Theorem 5.1. *Suppose that node r in a given graph has k or more incident edges. Then the incidence vectors of spanning trees with degree k at the root node r are the extreme points of the polyhedron $P = Q \cap \{x : \sum_{e\in\delta(r)} x_e = k\}$.*

Proof. From the theory of linear programming, this result is true if the optimization problem $\min\{wx : x \in P\}$ has an integer solution for every choice of the weight vector w. As we have argued in Section 3, for any value of θ, the optimal objective value of the problem $\min\{wx + \theta(\sum_{e\in\delta(r)} x_e - k) : x \in Q\}$ is a lower bound on the optimal objective value of this problem . At some point in the course of the algorithm we have just described at a value $\theta = \theta^*$, the solution x^* to the problem $\min\{wx + \theta^*(\sum_{e\in\delta(r)} x_e - k) : x \in Q\}$ is a spanning tree with k edges incident to node r. But then since the lower bound $wx^* + \theta^*(\sum_{e\in\delta(r)} x_e^* - k)$ equals the cost wx^* of the vector x^* and x^* is feasible to the problem $\min\{wx : x \in P\}$, x^* solves this problem. Therefore, this problem has an integer solution for every choice of the cost vector w and so P has integer extreme points. □

Note that for every value of k the polyhedron $P = Q \cap \{x : \sum_{e\in\delta(r)} x_e = k\}$ is a slice through the polyhedron Q along the hyperplane $\{x : \sum_{e\in\delta(r)} x_e = k\}$. The last theorem shows that every such slice has integer extreme points.

5.2. *Optimal branchings*

Let $D = (V, A)$ be a directed graph with a designated root node r. A *branching* is a directed subgraph (V, B) that satisfies two properties: (i) it is a tree if we ignore arc orientations, and (ii) it contains a directed path from the root to each node $v \neq r$. It is easy to see that we obtain an equivalent definition if we replace (ii) by the following condition: (ii′) the network has exactly one arc directed into each node $v \neq r$. Given a weight w_e on each arc e of D, we would like to solve the optimal branching problem of finding a branching with the smallest possible total arc weight.

For convenience, we usually refer to a branching by its arc set B.

5.2.1. *Branching models*

In Section 3.2, we introduced the following integer programming packing model of the optimal branching problem. For $e \in A$, we let $y_e = 1$ if e is in the branching, and $y_e = 0$ otherwise.

$$\min \sum_{e\in A} w_e y_e$$

subject to

$$\sum_{e\in A(S)} y_e \leq |S| - 1 \quad \text{for all nonempty sets } S \subset V$$

$$\sum_{e\in\delta^-(v)} y_e = 1 \qquad \text{for all } v \in V \setminus \{r\}$$

$$\sum_{e\in\delta^-(r)} y_e = 0$$

$$y \geq 0 \qquad \text{and integer.}$$

Notice that the equality constraints in this formulation remain the same if we replace the constraint $\sum_{e\in\delta^-(r)} y_e = 0$ by $\sum_{e\in A} y_e = n - 1$. For simplicity in the following discussion, we assume that we have eliminated all the arcs directed into the root node r and so we can delete the constraint $\sum_{e\in\delta^-(r)} y_e = 0$.

Let P be the polyhedron defined by this system if we ignore the integrality restrictions on the variables y. The results in Section 3 imply that if the digraph is symmetric in the sense that $(j, i) \in A$ whenever $(i, j) \in A$ and $w_{ij} = w_{ji}$, then the linear programming relaxation of this problem always has an integer optimal solution. We also showed that the greedy algorithm for the (undirected) minimum spanning tree problem solves this special case. In this section, we establish a more general polyhedral result: the extreme points of P are the incidence vectors of branchings and so the linear programming relaxation of the problem always has an integer optimal solution, even for the asymmetric case. We also develop an algorithm for finding an optimal branching.

Rather than work with the polyhedron P directly, we will consider an equivalent cutset formulation. Let $Q = \{y \in R^{|A|} : y \geq 0, \sum_{e\in\delta^-(v)} y_e = 1$ for all $v \in V \setminus \{r\}$, and $\sum_{e\in\delta^+(S)} y_e \geq 1$ for all nonempty sets S with $r \in S \subset V\}$. Our discussion of directed models of the minimum spanning tree problem in Section 3 shows that, as formalized by the following proposition, we can formulate the branching model in an equivalent cutset form.

Proposition 5.2. $P = Q$.

In the following discussion, we consider a related class of subgraphs (V, B') of D, called *superbranchings*, that contain one or more paths directed from the root node to every other node (and so contain one or more arcs directed into each node $v \neq r$). Since any superbranching B' contains a directed path to any node $v \neq r$, it contains a directed arc across every cutset, that is, the superbranching and every set $S \subset V$ of nodes containing the root node r satisfy the cutset condition $|\delta^+(S) \cap B'| \geq 1$.

Note that if B' is a superbranching and $\overline{B} \subseteq B'$ is a branching on a subset $\overline{V}$ of the nodes V, then if $\overline{V} \subset V$ we can extend $\overline{B}$ to a branching on a larger subset of nodes by adding any arc from $\delta^+(\overline{V}) \cap B'$ to $\overline{B}$. This observation implies that every superbranching contains a branching.

5.2.2. Finding an optimal branching

As we have already noted, any branching rooted at node r satisfies three properties: (i) node r has no incoming arc, (ii) every node $v \neq r$ has exactly one incoming arc, and (iii) every directed cutset $\delta^-(S)$, with $r \notin S$, contains at least one arc (or, equivalently, since $P = Q$, if we ignore the orientation of arcs, the branching contains no cycles). As the first step of an algorithm for solving the branching problem, we will ignore the last condition. The resulting problem is very easy to solve: for each node $v \neq r$, we simply choose a minimum weight arc that is directed into that node. We refer to this solution as a *node greedy solution*. If the node greedy solution NG contains no cycles, then it solves the optimal branching problem; otherwise, this solution contains one or more cycles C with $r \notin C$. Note that, since every node has exactly one incoming arc, the cycle will be directed. Moreover, the set $S = V(C)$ violates condition (iii)[5].

Our next result tells us even though the node greedy solution (typically) does not solve the branching problem, it contains considerable information about an optimal solution — enough to permit us to devise an efficient algorithm for finding an optimal branching.

To simplify our notation, we first transform the weights so that the minimum weight arc e directed into each node $v \neq r$ has weight zero. Suppose we subtract any constant q_v from the weight of all the arcs directed into any node v. Since any branching has exactly one arc directed into each node, the transformed and original problems have exactly the same optimal solutions, since *any* feasible solution to the transformed problem costs q_v less than that of the same solution in the original problem. If we choose q_v as the minimum weight of the arcs in $\delta^-(v)$, then the minimum weight arc directed into node v in the transformed digraph has weight zero.

Proposition 5.3. *Suppose that a node greedy solution NG contains a directed cycle C not containing the root node r. Then the optimal branching problem always has an optimal solution with exactly $|C| - 1$ arcs from the cycle C (and so exactly one arc in the directed cutset $\delta^-(V(C))$).*

Proof. Note that by our transformation of arc weights, every arc in C has weight zero.

Let B be any optimal branching. Since B is a branching, the arc set $B^-(C) \equiv \delta^-(V(C)) \cap B$ satisfies the cutset condition $|B^-(C)| \geq 1$. If $|B^-(C)| = 1$, B satisfies the conclusion of the proposition; so, assume $|B^-(C)| > 1$. We will use an induction argument on the size of $|B^-(C)|$. Suppose $|B^-(C)| = k$ and the induction hypothesis (that is, the conclusion of the proposition) is true for all cycles C satisfying the condition $|B^-(C)| \leq k - 1$.

Since $|B^-(C)| > 1$, B contains at least two arcs (i, j) and (p, q) directed into the cycle C, that is, with $i, p \notin V(C)$ and $j, q \in V(C)$. By definition, the graph B contains a (unique) directed path from the root node to every other node; therefore, it must contain a path from the root node to node j or node q, say node j, that does not pass through the other node. Let P_j denote this path. Let $B' = C \cup (B \setminus (p, q))$. Note that B' contains a path from the root node r to node q: the path P_j plus the arcs on the path from node j to node q in the cycle C.

[5] Recall that $V(C)$ denotes the set of nodes in the cycle C.

Therefore, B' contains a path to every node $v \neq r$ and so it is a superbranching. Consequently, it contains a branching $\overline{B}$. Since the arcs in C that we added to the branching B in this construction have zero weight, the weight $w(\overline{B})$ of $\overline{B}$ is no more than the weight $w(B)$ of B and so $\overline{B}$ is also an optimal branching. But since in constructing B', we eliminated the arc (p, q) from B and added the arcs in C, $|\overline{B}^-(C)| \equiv |\delta^-(V(C)) \cap \overline{B}| \leq k-1$. Therefore, the induction hypothesis implies that the problem has an optimal branching B^* containing exactly $|C|-1$ arcs from $|C|$. □

Figure 14 gives an example of a branching problem. In this case, the node greedy solution contains a cycle C_1 on the nodes 4, 5, 6 and 7, and so Proposition 5.3 implies that we can find an optimal branching B_1 containing three arcs in this cycle C_1 and exactly one arc directed into this cycle. Moreover, since we have transformed the costs so that each arc in this cycle has zero weight, we are indifferent as to the set of arcs that we choose from the cycle.

The node greedy solution in Figure 14 also contains a cycle C_2 on the nodes 8 and 9. Since C_2 contains two nodes, Proposition 5.3 implies that the problem has an optimal branching B_2 containing exactly one arc in the cycle C_2 and exactly one arc directed into this cycle as well.

Note that if we use the construction in the proof of Proposition 5.3 as applied to the cycle C_2 and the optimal branching B_1, which contains $|C_1|-1$ arcs from C_1, we produce an optimal branching B_2 by adding $|C_2|-1$ arcs from C_2 to B_1 and deleting some arcs from the set $\delta^-(V(C_2))$. But since the cycles C_1 and C_2 are node disjoint, $\delta^-(V(C_2)) \cap C_1 = \phi$ and so the branching B_2 continues to contain $|C_1|-1$ arcs from the cycle C_1. Therefore, we obtain an optimal branching containing $|C_1|-1$ arcs from the cycle C_1 and $|C_2|-1$ arcs from the cycle

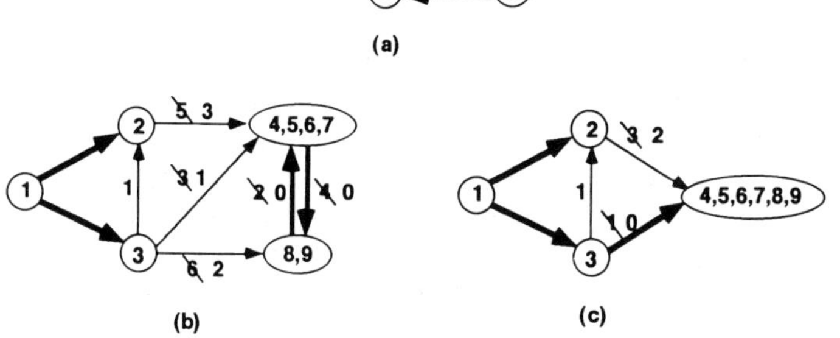

Fig. 14. Determining an optimal branching.

C_2. Therefore, we can simultaneously obtain an optimal branching that satisfies the conclusion of Proposition 5.3 for both the cycles C_1 and C_2. This argument applies to any set of disjoint cycles and so permits us to establish the following strengthening of Proposition 5.3.

Proposition 5.4. *Suppose that a node greedy solution NG contains node disjoint directed cycles $C_1, C_2, \ldots, C_J$, none containing the root node r. Then the optimal branching problem always has an optimal branching containing exactly $|C_j| - 1$ arcs from each cycle C_j, for $j = 1, 2, \ldots J$ [and so exactly one arc in each directed cutset $\delta^-(V(C_j))$].*

These observations imply that if we contract all the nodes of each cycle from a node greedy solution into a *pseudo node*, then in the resulting reduced digraph we once again need to find an optimal branching. Any optimal branching in the reduced digraph has exactly one arc (i, j) directed into any cycle C in the node greedy solution. Let $(k, j) \in C$ be the arc directed into node j in the node greedy solution. The proofs of Propositions 5.3 and 5.4 imply that we obtain an optimal solution of the original branching problem by 'expanding' every pseudo node; that is, adding the arcs $C \setminus \{(k, j)\}$ into the optimal branching of the reduced problem.

The discussion shows that we can solve the optimal branching problem using the following algorithm:

(i) *Transform costs.* By subtracting a node-dependent constant γ_v from the weight of the arcs directed into every node $v \neq r$, transform the weights so that the minimum weight arc directed into each node has weight zero.

(ii) *Solve a relaxed problem.* Find a node greedy solution *NG* on the transformed digraph. If the node greedy solution contains no cycles, stop. The node greedy solution is optimal and the optimal branching has a weight $\sum_{v \in V} \gamma_v$. If the node greedy solution contains any cycle, then contract the digraph D into a reduced digraph D_1 by replacing every such cycle C by a single (pseudo) node. Any arc incident into or out of a node in any cycle C becomes an arc, with the same weight, incident into or out of the pseudo node corresponding to that arc. (Note that the reduced digraph might contain parallel arcs; we can eliminate all but the least weight arc of any parallel arcs).

(iii) *Solve the reduced problem.* Find an optimal branching B^R on the reduced digraph and use it to create an optimal branching in the original digraph by expanding every pseudo node.

This algorithm reduces the optimal branching problem to solving an optimal branching problem on the smaller digraph. To solve the reduced problem, we would once again apply the algorithm starting with step (i). Eventually, the node greedy solution will contain no cycles (in the limit, it would have only a single edge). At that point, the algorithm will terminate

Proposition 5.4 and a simple induction argument show that the algorithm finds an optimal branching.

Example 5.1. To illustrate the algorithm, consider the example in Figure 14. As we have noted before, the bold-faced arcs in Figure 14a are the arcs in the node greedy solution. This solution contains two cycles; contracting them gives the digraph D_1 shown in Figure 14b. Note that we have eliminated the parallel arcs created by the arcs (3, 7), (5, 9), and (8, 6).

We next reapply the algorithm on the reduced digraph. First, we subtract $\gamma_{4,5,6,7} = 2$ from the weights of arcs directed into the pseudo node $\{4, 5, 6, 7\}$ and $\gamma_{8,9} = 4$ from the arcs directed into the pseudo node $\{8, 9\}$. The bold arcs in Figure 14b define the node greedy solution on the reduced digraph. Since this solution contains a cycle (on the two pseudo nodes), we contract it, giving the reduced digraph in Figure 14c. We decrease the weights of the arcs directed into the pseudo node $\{4, 5, 6, 7, 8, 9\}$ by $\gamma_{4,5,6,7,8,9} = 1$. Since the node greedy solution (the bold arcs) on the digraph D_2 shown in Figure 14c contains no cycle, we stop.

To recover the solution to the original problem, we need to retrace our steps by expanding pseudo nodes, beginning with the optimal branching on the final reduced digraph D_2 (see Figure 15). Since the arc $(3, \{4, 5, 6, 7, 8, 9\})$ directed into the pseudo node $\{4, 5, 6, 7, 8, 9\}$ corresponds to arc $(3, \{4, 5, 6, 7\})$ in the digraph D_1 (see Figure 14b), we delete arc $(\{8, 9\}, \{4, 5, 6, 7\})$ from the cycle in the node greedy solution shown in Figure 14b. Figure 15b gives the resulting solution.

We next expand the two pseudo nodes $\{4, 5, 6, 7\})$ and $\{8, 9\}$. As we expand these pseudo nodes, we eliminate the arcs (7, 4) and (8, 9) from the cycles that defined them in Figure 14a. Figure 15c shows the resulting solution. Note that its weight equals $\gamma_{4,5,6,7} + \gamma_{8,9} + \gamma_{4,5,6,7,8,9} = 2 + 4 + 1$.

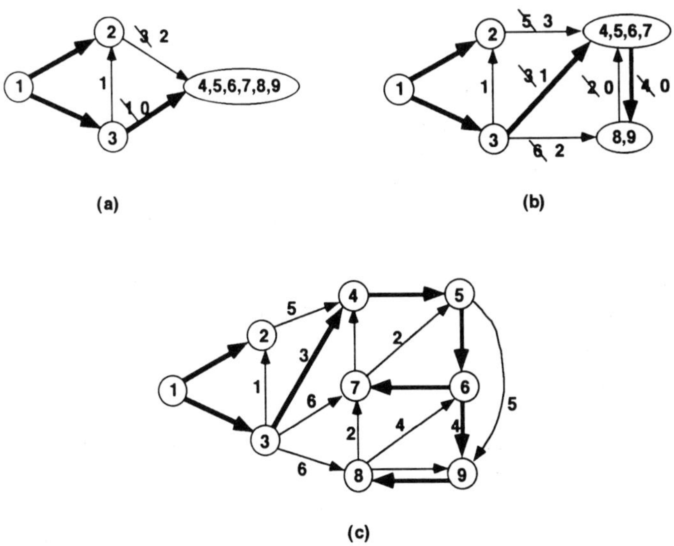

Fig. 15. Expanding pseudo nodes.

5.2.3. Polyhedral implications

Using a combinatorial argument (the development in the proofs of Propositions 5.3 and 5.4), we have shown that the branching algorithm produces an optimal branching. We now give an alternative linear programming-based proof.

It is easy to see by inspection that the branching shown in Figure 15c is optimal. Any branching must contain at least one arc directed into the node set {4, 5, 6, 7} and at least one arc directed into the node set {8, 9}. Since the minimum weight arc in $\delta^-(\{4,5,6,7\})$ is arc (8, 7) with weight 2 and in $\delta^-(\{8,9\})$ is arc (6, 9) with weight 4, and the sets $\delta^-(\{4,5,6,7\})$ and $\delta^-(\{8,9\})$ are disjoint, *any* feasible branching must have a weight of at least 6. Note that we achieve a weight of 6 only if we choose both the arc (8, 7) from $\delta^-(\{4,5,6,7\})$ and (6, 9) from $\delta^-(\{8,9\})$. Otherwise, the weight of the chosen arcs is at least 7. But if we choose the arcs (8, 7) and (6, 9), then the cutset $\delta^-(\{4,5,6,7,8,9\})$ must contain at least one other arc and so the total weight of the branching will exceed 7. Therefore, in every case, the weight of any branching must be at least 7. But since the solution in Figure 15c achieves this lower bound, it must be optimal.

This argument is reminiscent of the lower bounding argument we have used in Section 3 for proving that the greedy algorithm solves the (undirected) minimum spanning tree problem. We will use a version of this argument to not only provide another proof that the branching algorithm produces an optimal branching, but also to give a constructive proof of the following integrality theorem.

Theorem 5.5. *The polyhedron $P = Q$ is the convex hull of incidence vectors of branchings.*

Proof. Let us first set some notation. Let $\mathcal{S}$ denote the set of nonempty subsets of $V \setminus \{r\}$ and let $\mathcal{S}_p \subseteq \mathcal{S}$ denote the set of all node sets that define some pseudo node. That is, if $S \in \mathcal{S}_p$ then S is the set of nodes from the original graph in one of the pseudo nodes.

We will establish the result using the cutset representation $Q = \{y \in R^{|A|} : y \geq 0, \sum_{e\in\delta^-(v)} y_e = 1$ for all $v \in V \setminus \{r\}$, and $\sum_{e\in\delta^-(S)} y_e \geq 1$ for all $S \in \mathcal{S}\}$.

Consider the linear program:

$$\min \left\{ \sum_e w_e y_e : y \in Q \right\}, \tag{5.1}$$

and its linear programming dual:

$$\max \sum_{S\in\mathcal{S}} \alpha_S \tag{5.2}$$

subject to

$$\sum_{S\in\mathcal{S}} \sum_{\delta^-(S)\ni e} \alpha_S \geq w_e \qquad \text{for all arcs } e \in A \tag{5.3}$$

$$\alpha_S \geq 0 \qquad \text{for all } S \in \mathcal{S} \text{ with } |S| \geq 2. \tag{5.4}$$

When $S = \{v\}$ and $v \neq r$, the dual variable α_S corresponds to the constraint $\sum_{e\in\delta^-(v)} y_e = 1$ in Q. For any set of nodes $S \in \mathcal{S}$ with $|S| \geq 2$, α_S is the dual variable corresponding to the constraint $\sum_{e\in\delta^-(S)} y_e \geq 1$.

Let γ_v for $v \in V \setminus \{r\}$ and γ_S for $S \in \mathcal{S}_p$ be the node-dependent constants used in the weight transformation steps in the branching algorithm. These constants correspond to the nodes v of the original digraph D at the first step of the algorithm and to pseudo nodes (whose node sets are S) at later steps. For any node $v \neq r$, suppose that we set $\alpha_{\{v\}} = \gamma_v$ and for any set $S \in \mathcal{S}_p$ corresponding to a pseudo node, suppose that we set $\alpha_S = \gamma_S \geq 0$. Define $\alpha_S = 0$ otherwise.

The steps of the branching algorithm imply that final weight $\overline{w}_e$ for each arc e is nonnegative. The branching algorithm also implies that if edge e is directed into node $v^+(e)$, then $\overline{w}_e = w_e - \sum_{\{S\in\mathcal{S}_p \text{ with } \delta^-(S)\ni e\}} \gamma_S - \gamma_{v^+}(e)$. The last term includes the node weight γ_{v^+} of the node that the edge e is directed into. Therefore, the variables γ_v and γ_S are feasible in the linear programming dual. As we have already seen, the branching found by the algorithm has a cost $\sum_{v\neq r} \gamma_v + \sum_{S\in\mathcal{S}_p} \gamma_S$. Therefore, the algorithm constructs an integer feasible solution to the linear program (5.1) and its weight equals the weight of a feasible dual solution. As a result, linear programming theory implies that the branching solution solves the linear program (5.1); therefore, the linear program has an integer solution (the incidence vector of a branching) for every choice of objective function, and, consequently, the incidence vector of branchings are the extreme points of the underlying polyhedron $Q = P$. □

6. The Steiner tree problem

As we noted in Section 2, the (undirected) Steiner tree (ST) problem is defined by a graph $G = (V, E)$, with edge weights w_e on the edges $e \in E$, and with a subset of the nodes T called *terminal nodes*. The objective is to find a minimum weight subtree of G that spans all the nodes in T. The subtree might or might not include some of the the other (optional) nodes $S = V \setminus T$, which we refer to as *Steiner nodes*.

Two special versions of the Steiner tree problem are easy to solve. When $|T| = 2$ and $w_e \geq 0$ for $e \in E$, the problem reduces to the shortest path problem, and when $T = V$, it is the spanning tree problem that we examined in Section 3. In general, the problem is difficult to solve (in the parlance of computational complexity theory, it is NP-complete), and so we will not be able to solve it using combinatorial or dynamic programming algorithms of the type we have considered in the earlier sections. For more complicated models like this, an important step is often to find a 'good' linear programming representation of the problem. Starting with an initial integer or mixed integer programming formulation, the goal is to add new valid constraints or new variables to the model so that the resulting linear program provides a tight bound on the value of the optimal solution — in some cases even an integral and, hence, optimal solution.

This section is divided into three parts. We begin by examining different integer programming formulations of a slight generalization of (*ST*), called the node weighted Steiner tree problem (*NWST*), and showing that the values of the linear programming relaxations of all these formulations are the same. This discussion generalizes our development in Section 3 of alternate formulations of the minimum spanning tree problem. We then briefly discuss computational studies based on these formulations. Finally, we consider the strength of a linear programming relaxation for the Steiner problem and present results on the worst case behavior of simple heuristics for (*ST*) and (*NWST*).

6.1. The node weighted Steiner tree problem (NWST)

Given a graph $G = (V, E)$ with edge weights w_e for $e \in E$, we designate one node r as a root node. Node r will be the only terminal node (that is, $T = \{r\}$). For all the other nodes $j \in V \setminus \{r\}$, we incur a profit d_j if the Steiner tree contains node j.

To formulate the node weighted Steiner tree problem (*NWST*), we let $z_j = 1$ if node j is in the Steiner tree, and $z_j = 0$ otherwise; $x_e = 1$ if edge e is in the tree and $x_e = 0$ otherwise. The first formulation we present, called *the subtour packing formulation*, is a natural generalization of the subtour (packing) formulation of the spanning tree problem, namely:

$$\min \sum_{e\in E} w_e x_e - \sum_{i\in V} d_i z_i \tag{6.1}$$

subject to

$$\sum_{e\in E(U)} x_e \leq \sum_{i\in U\setminus\{k\}} z_i \quad \text{for all } U \subset V \text{ and for all } k \in U \tag{6.2}$$

$$\sum_{e\in E} x_e = \sum_{i\in V\setminus\{r\}} z_i \tag{6.3}$$

$$z_r = 1 \tag{6.4}$$

$$0 \leq x_e \leq 1, 0 \leq z_i \leq 1 \tag{6.5}$$

$$x, z \text{ integer.} \tag{6.6}$$

Note that the first set of constraints (6.2), called *generalized subtour elimination constraints*, imply that the solution contains no cycles on the subgraph of G determined by the selected nodes (those with $z_i = 1$). The second set of constraints (6.3) ensures that the solution spans the set of selected nodes, and so defines a Steiner tree. We let P_{sub} denote the set of feasible solutions to (6.2)–(6.5) in the (x, z)-space.

Before proceeding, let us make one observation about this formulation. Note that if $r \in U$ and $k \neq r \in U$, then

$$\sum_{i\in U\setminus\{r\}} z_i = \sum_{i\in U\setminus\{k\}} z_i + z_k - z_r \leq \sum_{i\in U\setminus\{k\}} z_i$$

since $z_r = 1$ and $z_k \leq 1$. Therefore, when $r \in U$, all the constraints (6.2) with $k \neq r \in U$ are redundant.

It is instructive to view the node weighted Steiner tree problem in another way: as a spanning tree problem with additional constraints. To develop this interpretation, suppose that we let $\bar{z}_r = z_r$ and complement the node variables z_i for $i \neq r$ in the previous model; that is, let $\bar{z}_i = 1 - z_i$ and replace z_i with $1 - \bar{z}_i$, creating the following alternative formulation:

$$\min \sum_{e\in E} w_e x_e + \sum_{i\in V} d_i \bar{z}_i - \sum_{i\in V} d_i \tag{6.7}$$

subject to

$$\sum_{e\in E(U)} x_e + \sum_{i\in U\setminus\{k\}} \bar{z}_i \leq |U| - 1 \quad \text{for all } U \subset V \text{ and for all } k \in U \tag{6.8}$$

$$\sum_{e\in E} x_e + \sum_{i\in V\setminus\{r\}} \bar{z}_i = |V| - 1 \tag{6.9}$$

$$\bar{z}_r = 1 \tag{6.10}$$

$$0 \leq x_e \leq 1, \ 0 \leq \bar{z}_i \leq 1 \tag{6.11}$$

$$x, \bar{z} \text{ integer}. \tag{6.12}$$

To interpret this problem as a variant of the minimum spanning tree problem, suppose that we add a supernode 0 to the underlying graph G with an edge $\{0, i\}$ between this node and every node i in G. Let $\bar{G}$ denote the resulting graph. Then the zero-one variable $\bar{z}_i$ for $r \neq r$ indicates whether or not we include edge $\{0, i\}$ in the minimum spanning tree on $\bar{G}$ (observe that we *include* edge $\{0, i\}$ in the tree, that is, $\bar{z}_i = 1$, when we *exclude* node i in the previous formulation, that is, $z_i = 0$). We always include edge $\{0, r\}$.

The constraint (6.8) with $U = \{i, j\}$ for any two nodes $i \neq r$ and $j \neq r$ implies that $x_e + \bar{z}_i \leq 1$ for all $e \in \delta(i)$. That is, if the chosen tree contains edge $\{0, i\}$, then it contains no edge from G incident to that node. In the formulation (6.1)–(6.8), this statement says that if we exclude node i from the tree, then the tree cannot contain any edge incident to that node. The inequalities $x_e + \bar{z}_i \leq 1$ and the fact that any solution to the model (6.7)–(6.12) contains edge $\{0, r\}$ implies that any spanning tree solution to the formulation (6.7)–(6.12) has the following form: if we remove node 0 and its incident edges, the resulting forest is a subtree containing the node r as well as a set of isolated nodes j (those with $\bar{z}_j \leq 1$).

Next consider the subtour constraint (6.8) in this model for any set U that does not contain the root node r. In the spanning tree formulation, we would have written this constraint without the $\bar{z}_i$ variables as $\sum_{e\in E(U)} x_e \leq |U| - 1$. The inequality (6.8) is a strengthening of that inequality: if $\bar{z}_i = 0$ for all nodes $i \in U$, then the set $E(U)$ can contain as many as $|U| - 1$ edges; every time we add an edge $\{0, i\}$ to the tree for any node $i \in U$, node i become isolated, so the effective size of $|U|$ decreases by one and the tree can contain one fewer edge from $E(U)$. If the set U contains node r, then as we saw previously, the constraints (6.8) with $k \neq r$ are redundant, so the only effective inequality in the complimented model (6.7)–(6.12) is $\sum_{e\in E(U)} x_e + \sum_{i\in U\setminus\{r\}} \bar{z}_i \leq |U| - 1$. Note that these constraint are exactly the usual subtour breaking constraints on the node set $U \cup \{0\}$, given that the solution contains the edge $\{0, r\}$.

Our next model, the so-called *multicut formulation*, is a generalization of the multicut formulation of the minimum spanning tree problem. Let P_{mcut} be the set of solutions in the (x, z) variables to the constraints (6.3)–(6.5) as well as the additional inequalities

$$\sum_{e \in \delta(C_0, \ldots, C_s)} x_e \geq \sum_{j=1}^{s} z_{i_j} \tag{6.13}$$

over all node partitions $(C_0, C_1, \ldots, C_s)$ of V with $r \in C_0$. In this expression, as in our earlier discussion, $\delta(C_0, C_1, \ldots, C_s)$ is the multicut defined by the node partition (i.e., the set of edges having endpoints in different subsets C_i, C_j), and $i_j \in C_j$ for $j = 1, \ldots, s$. These constraints say that if k of the sets $(C_1, \ldots, C_s)$ contain nodes of the Steiner tree, and C_0 contains the root node, then the multicut must contain at least k edges of the Steiner tree.

Proposition 6.1. $P_{\text{sub}} = P_{\text{mcut}}$.

Proof. Summing the inequalities (6.2) for the sets $U = C_0, C_1, \ldots, C_s$ with $k = i_j$ gives

$$\sum_{t=0}^{s} \sum_{e \in E(C_t)} x_e \leq \sum_{t=0}^{s} \sum_{i \in (C_t \setminus \{i_t\})} z_i .$$

Taking $i_0 = r$ in this expression and subtracting from (6.3) gives (6.13). Thus, $P_{\text{sub}} \subseteq P_{\text{mcut}}$.

Conversely, suppose first that $r \in U$. Consider the inequality (6.13) with $C_0 = U$, and with $C_1, \ldots, C_k$ as singletons whose union is $V \setminus U$. Subtracting from (6.3) gives $\sum_{e \in E(U)} x_e \leq \sum_{i \in U \setminus \{r\}} z_i$. As we have noted previously, this inequality implies (6.3) for all nodes $k \in U$.

If $r \notin U$, take $C_0 = \{r\}$, $C_1 = U$, and $C_2, \ldots, C_k$ as singletons whose union is $V \setminus (U \cup \{r\})$. Subtracting (6.13) from (6.3) gives $\sum_{e \in E(U)} x_e \leq \sum_{i \in U \setminus \{i_1\}} z_i$ for $i_1 \in U$. Thus, $P_{\text{mcut}} \subseteq P_{\text{sub}}$. $\square$

The next formulation again uses the observation that we can orient any Steiner tree to obtain a directed Steiner tree with node r as its root. In this formulation, the vector y represents an arc incidence vector of this directed Steiner tree. The *directed subtour formulation* is given by the constraints (6.2)–(6.5) and the *the dicut inequality*

$$\sum_{(i,j) \in \delta^-(j)} y_{ij} = z_j \quad \text{for } j \in V \setminus \{r\} \tag{6.14}$$

$$y_{ij} + y_{ji} = x_e \quad \text{for } e = \{i, j\} \in E \tag{6.15}$$

$$y_{ij},\ y_{ji} \geq 0 \quad \text{for } e = \{i, j\} \in E. \tag{6.16}$$

Constraints (6.14) say that if node j is in the Steiner tree, one arc of the directed Steiner tree enters this node. We let P_{dsub} denote the set of feasible solutions in the (x, z) space.

Similarly *the directed cut formulation* is given by the constraints (6.3)–(6.5), (6.15), (6.16) and

$$\sum_{(i,j)\in\delta^+(C)} y_{ij} \geq z_k \text{ for all } k \in V, \text{ and all } C \text{ with } r \in C \subseteq V \setminus \{k\}. \quad (6.17)$$

These constraints say that if k is in the Steiner tree, any directed cut separating nodes r and k must contain at least one arc of the directed Steiner tree. We let P_{dcut} denote the set of feasible solutions in the (x, z) space.

Proposition 6.2. $P_{\text{dsub}} = P_{\text{dcut}}$.

Proof. The proof mimics the proof of Proposition 3.6. In this case, when $r \in \overline{S}$ the right-hand side of equation (3.36) becomes $\sum_{k\in V\setminus\{r\}} z_k$ instead of $n - 1$ and by equation (6.14) the last term on the left-hand side of equation (3.36) equals $\sum_{k\in\overline{S}} z_k$. Therefore, the right-hand side of equation (3.37) becomes $\sum_{k\in V\setminus\{r\}} z_k - \sum_{k\in\overline{S}} z_k = \sum_{i\in S} z_i$. So this equation becomes

$$\sum_{e\in A(S)} y_e + \sum_{e\in\delta^+(\overline{S})} y_e = \sum_{i\in S} z_i.$$

This equality implies that y satisfies inequality (6.2) for S if and only if it satisfies the inequality (6.17) for $\overline{S}$.

If $r \in S$, then the arguments used in the proof of Proposition 3.6 show that the right-hand side of the last displayed equation is $\sum_{i\in S\setminus\{r\}} z_i$, and since the last term on the left-hand side of this equation is nonnegative, the equation implies (6.2) for $k = r$, which as we have seen implies (6.2) for all $k \in S$.

These arguments show that $y \in P_{\text{dsub}}$ if and only if $y \in P_{\text{dcut}}$. □

The four formulations we have described so far all have an exponential number of constraints. The next formulation we consider has only a polynomial number of constraints, but an additional $0(n^3)$ variables. We obtain this *directed flow formulation* by remodelling the cut inequalities (6.17). More specifically, viewing the values of the variables y_{ij} as capacities, (6.16) and (6.17) say that every cut separating nodes r and k has capacity at least z_k. By the max-flow min-cut theorem, these inequalities are valid if and only if the network has a feasible flow of z_k units from node r to node k. Letting f_{ij}^k denote the flow destined for node k in arc (i, j), we obtain P_{dflo} which is the set of feasible solutions in the (x, z)-space of the constraints (6.3)–(6.5), (6.15), (6.16) and the equations describing such a flow:

$$\sum_{i\in\delta^+(j)} f_{ji}^k - \sum_{i\in\delta^-(j)} f_{ij}^k = 0 \qquad \text{for all } j \neq r, j \neq k, k \neq r \qquad (6.18)$$

$$\sum_{i\in\delta^+(k)} f_{ki}^k - \sum_{i\in\delta^-(k)} f_{ik}^k = -z_k \qquad \text{for all } k \neq r \qquad (6.19)$$

$$0 \leq f_{ij}^k \leq y_{ij} \qquad \text{for all } (i, j) \in A, k \neq r. \qquad (6.20)$$

As was the case for the minimum spanning tree problem, the directed flow and directed cut formulations are equivalent in the following sense.

Proposition 6.3. $P_{\text{dcut}} = P_{\text{dflo}}$.

The directed formulation is as strong as the undirected formulation (e.g., $P_{\text{dsub}} \subseteq P_{\text{sub}}$) since it contains additional constraints. Are the directed formulations P_{dsub} (respectively P_{dcut}, P_{dflo}) stronger than the undirected formulations P_{sub} (respectively P_{mcut})? As for the case of trees, it turns out that these polyhedra are identical even though, in general, they are no longer integral.

Theorem 6.4. $P_{\text{sub}} = P_{\text{dsub}}$.

Proof. As we have just noted, $P_{\text{dsub}} \subseteq P_{\text{sub}}$. To show the converse, we need to show that for every $(x', z') \in P_{\text{sub}}$, there exists a y with $(x', z', y) \in P_{\text{dsub}}$. In other words, we need to verify that the system

$$\sum_{i \in \delta^-(r)} y_{ir} = 0$$

$$\sum_{i \in \delta^-(j)} y_{ij} = z'_j \qquad \text{for } j \in V \setminus \{r\}$$

$$y_{ij} + y_{ji} = x'_e \qquad \text{for } e = \{i, j\} \in E$$

$$y_{ij},\ y_{ji} \geq 0 \qquad \text{for } e = \{i, j\} \in E$$

has a feasible solution. This is a feasibility transportation problem with a supply 0 for node $j = r$, a supply z'_j for each node $j \in V \setminus \{r\}$, and demand x'_e for each edge $e \in E$. Equation (6.3) implies that the sum of supplies equals the sum of the demands. To determine if this system has a feasible solution, we can formulate a maximum flow problem with a node for each node $v \in V$, a node for each edge $e \in E$, and a directed arc (v, e) of infinite capacity whenever v is incident to edge e. We also introduce a source node s and terminal node t and directed arcs (s, v) with capacity z'_v (with $z'_r = 0$) for each $v \in V$ and directed arcs (e, t) with capacity x'_e for each $e \in E$. The transportation problem has a feasible flow if and only if the maximum flow problem has a flow of value $\sum_{e \in E} x'_e$ from node s to node t, which by the max-flow min-cut theorem is true if and only if the capacity of every cutset is at least $\sum_{e \in E} x'_e$.

Let $S \subseteq V$ and $F \subseteq E$ be the supplies and demands on the t side of a cut. The cut has infinite capacity if it has an edge between $V \setminus S$ and F. Thus, the cut capacity is finite only if $F \subseteq E(S)$. For given sets S and F, the capacity is $\sum_{i \in S \setminus \{r\}} z'_i + \sum_{e \in E \setminus F} x'_e$ which is minimized when $F = E(S)$ giving $\sum_{i \in S \setminus \{r\}} z'_i + \sum_{e \in E \setminus E(S)} x'_e$. This sum is at least $\sum_{e \in E} x'_e$ if and only if $\sum_{e \in E(S)} x'_e \leq \sum_{i \in S \setminus \{r\}} z'_i$. But (6.2) implies that this inequality is valid for all (x', z'), establishing the claim. □

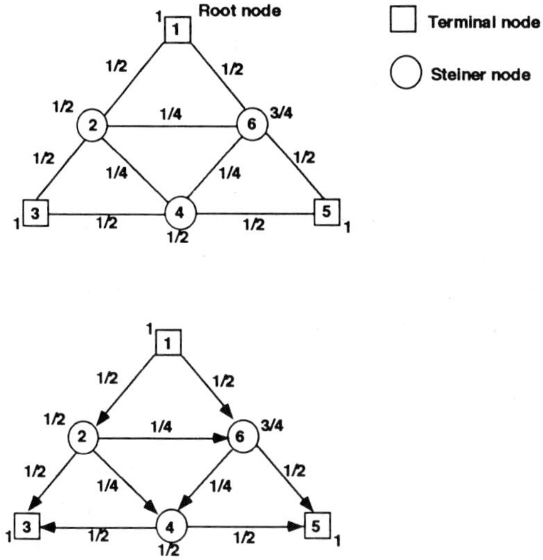

Fig. 16. Fractional solution that violates the subtour, multicut, and dicut formulations. (a) Undirected version; (b) directed version.

Example 6.1. Figure 16 shows a fractional solution in the (x, z) variables satisfying constraints (6.3)–(6.5), and a compatible directed (x, z, y) solution satisfying constraints (6.14)–(6.16) as well.

Observe that the fractional solution violates the subtour inequality with $C = \{2, 3, 4\}$ and $k = 3$, the multicut inequality with $C_0 = \{1\}, C_1 = \{2, 3, 4\}, C_2 = \{5\}, C_3 = \{6\}$ and $i_1 = 3$, and the dicut inequality with $C = \{1, 5, 6\}$ and $k = 3$, each by $1/4$ unit.

The final *directed branchings* formulation is also of polynomial size, but based on the separation problem for the generalized subtour inequalities (6.2). It is described by the constraints (6.3)–(6.5) and the constraints:

$$\sum_{i \in \delta^-(j)} y_{ij}^k \leq z_j \qquad \text{for } j \in V \setminus \{k\}, \text{ and all } k \tag{6.21}$$

$$\sum_{i \in \delta^-(k)} y_{ik}^k \leq 0 \qquad \text{for all } k \tag{6.22}$$

$$y_{ij}^k + y_{ji}^k = x_e \qquad \text{for } e = (i, j) \in E, \text{ and all } k \tag{6.23}$$

$$y_{ij}^k,\ y_{ji}^k \geq 0 \qquad \text{for } e = (i, j) \in E, \text{ and all } k \tag{6.24}$$

Let P_{dbran} denote the set of (x, z) solutions to this model.

Note that once the Steiner tree is fixed, then for each node k in the tree, we can orient the edges to construct a directed Steiner tree rooted at node k. Suppose we define $y_{ij}^k = 1$ if the arc (i, j) belongs to this branching and $y_{ij}^k = 0$ otherwise. If k is not in the tree, we can choose r as the root. Consequently, this formulation is valid.

Proposition 6.5. $P_{\text{dbran}} = P_{\text{sub}}$.

Proof. Summing the inequalities (6.21) over $C \setminus \{k\}$ and adding (6.22) gives

$$\sum_{e \in A(C)} y_e^k + \sum_{(i,j) \in \delta^-(C)} y_{ij}^k \leq \sum_{i \in C \setminus \{k\}} z_i.$$

Thus $P_{\text{dbran}} \subseteq P_{\text{sub}}$.

Conversely, note that (x, z) satisfies the generalized subtour inequality (6.2) for all node sets C containing node k if and only if

$$\eta^k = \max_{\nu \in \{0,1\}^{|V|}} \left\{ \sum_{e=(i,j) \in E} x_e \nu_i \nu_j - \sum_{j \neq k} z_j \nu_j : \nu_k = 1 \right\} = 0.$$

In this expression, $\nu_i = 1$ if $i \in C$ and $\nu_i = 0$ otherwise. This problem can be solved as a linear program, namely:

$$\eta^k = \max \sum_{e \in E} x_e \alpha_e - \sum_{j \neq k} z_j \nu_j$$

subject to

$$\begin{aligned} \alpha_e - \nu_i &\leq 0 \quad \text{for all } i \in V \text{ and all } e \in \delta^+(i) \\ \alpha_e - \nu_j &\leq 0 \quad \text{for all } j \in V \text{ and all } e \in \delta^-(j) \\ \nu_i &\geq 0 \quad \text{for all } j \in V. \end{aligned}$$

Since the constraints in this system are homogeneous, ν^k equals either 0 or $+\infty$. Because the constraint matrix for this model is totally unimodular, its extreme rays have 0–1 coefficients. Thus the solution (x, z) satisfies the subtour inequality if and only if $\eta^k = 0$, which by linear programming theory is true if and only if the following dual problem is feasible

$$\begin{aligned} y_{ij}^k + y_{ji}^k &= x_e \\ -\sum_{i \in \delta^-(j)} y_{ij}^k &\geq -z_j \quad \text{for all } j \in V \setminus \{k\} \\ -\sum_{i \in \delta^-(k)} y_{ik}^k &\geq 0 \\ y_{ij}^k &\geq 0. \end{aligned}$$

The conclusion establishes the claim. □

We close this discussion with three observations.

(1) As we noted before, the formulation P_{dflo} contains $0(n^3)$ variables ($n = |V|$), which leads to impractically large linear programs. However, this formulation indicates how to carry out separation for the formulations P_{dcut} and P_{sub}. Just as the directed flow formulation provided us with a separation procedure for the subtour inequalities of the spanning tree polyhedron, P_{dflo} provides a separation procedure, via a series of maximum flow computations, for the inequalities (6.17) or (6.2).

(2) To obtain a formulation of the minimum spanning tree problem from each of the formulations we have considered in this section, we would add the constraints $z_v = 1$ for all nodes $v \neq r$ to each of these models. Since, as we have shown, the node weighted Steiner tree polyhedra for each of these formulations are the same, their intersection with the constraints $z_v = 1$ will be the same as well. Therefore, the results in this section generalize those in Section 3. Moreover, the formulation P_{dbran} with the additional stipulation that $z_v = 1$ for all $v \neq r$ provides us with yet another integer formulation of the spanning tree polyhedron that is equivalent to the six integer formulations we examined in Section 3.

(3) In addition to the models we have considered in this discussion, we could formulate straightforward extensions of the single commodity flow and cut formulations of the spanning tree problem for the node-weighted Steiner tree problem. These formulations will, once again, be weak, in the sense that their linear programming relaxations will provide poor approximations for the underlying integer polyhedra. We could also state a multicommodity flow formulation with bidirectional forcing constraints as in (3.47). The arguments given in Section 3 show that this formulation is equivalent to the directed formulation.

6.2. The Steiner problem

What happens when we specialize the node weighted formulations for the Steiner problem by taking $z_i = 1$ for all $i \in T$ and setting $d_j = 0$ for all $j \in V \setminus T$? The first obvious alternative is to work with the six extended formulations we have just examined.

A second approach is to find a formulation without the node variables z. Note that formulation P_{dcut} easily provides one such formulation. We simply eliminate the cardinality constraint (6.3) and the dicut constraints (6.17) whenever $k \notin T$. The resulting formulation is (6.15), (6.16) and

$$\sum_{(i,j)\in\delta^+(C)} y_{ij} \geq 1 \text{ for all } C \text{ with } r \in C \subseteq V \text{ and } (V \setminus C) \cap T \neq \phi \qquad (6.25)$$

The integer solutions of this formulation are Steiner trees and their supersets. Eliminating constraints in a similar fashion for the formulation P_{mcut}, gives

$$\sum_{e\in\delta(C_0,\ldots,C_s)} x_e \geq s \qquad (6.26)$$

over all node partitions $(C_0, C_1, \ldots, C_s)$ of V with $r \in C_0$ and $C_i \cap T \neq \phi$ for $i = 1, \ldots, s$. Note, however, that the resulting directed cut and multicut formulations no longer are equivalent. The fractional solution shown in Figure 16 satisfies all the multicut inequalities (6.26), but is infeasible for the dicut polyhedron (6.15), (6.16) and (6.25).

For the other four formulations, there is no obvious way to eliminate constraints to obtain a formulation of this type.

A third approach would be to find an explicit description of $Q_{\text{sub}} = \text{proj}_x(P_{\text{sub}})$ and, ideally, of $Q_{ST} = \text{conv}(Q_{\text{sub}} \cap \{x : x \text{ integer}\})$.

Researchers have obtained various classes of facet-defining inequalities for Q_{ST}. (Facet-defining inequalities are inequalities that are necessary for describing a region defined by linear inequalities. All the others are implied by the facet-defining inequalities and are thus theoretically less important. However, in designing practical algorithms, facets might be hard to find or identify, and so we often need to use inequalities that are not facet-defining).

Proposition 6.6 (Steiner partition inequalities). *Let $C_1, \ldots, C_s$ be a partition of V with $T \cap C_i \neq \phi$ for $i = 1, \ldots, s$, then the multicut inequality*

$$\sum_{e \in \delta(C_1,\ldots,C_s)} x_e \geq s - 1$$

defines a facet of Q_{ST} if

(i) the graph defined by shrinking each node set C_i into a single node is two-connected, and

(ii) for $i = 1, \ldots, s$, the subgraph induced by each C_i is connected.

Another class of facet-defining inequalities are the 'odd holes'. Consider a graph $G^t = (V, E)$ on $2t$ nodes, with t odd, V composed of terminal nodes $T = \{u_1, \ldots, u_t\}$ and Steiner nodes $V \setminus T = \{v_1, \ldots, v_t\}$, and $E \supseteq E_t = \{(u_i, v_i)_{i=1}^t, (v_i, v_{i+1})_{i=1}^t, (v_i, u_{i+1})_{i=1}^t\}$. In this expression, $v_{t+1} = v_1$ and $u_{t+1} = u_1$.

Proposition 6.7. *The inequality*

$$\sum_{e \in E_t} x_e + 2 \sum_{e \in E \setminus E_t} x_e \geq 2(t - 1)$$

is a facet defining inequality for G^t.

In the odd hole (V, E_t), each terminal node u_i is at least two edges away from any other terminal node u_j, so using only the edges in E_t to connect the terminal nodes requires at least $2(t - 1)$ edges. Every edge in $E \setminus E_t$ that we add to the Steiner tree can replace two such edges. This fact accounts for the factor of 2 on the edges in $E \setminus E_t$.

Example 6.1. The graph shown in Figure 16 is itself an odd hole with $t = 3$. Note that the fractional edges values in this figure belong to Q_{sub} and satisfy all the multicut inequalities. This solution does, however, violate the following odd hole inequality:

$$x_{12} + x_{16} + x_{26} + x_{24} + x_{23} + x_{34} + x_{46} + x_{45} + x_{56} \geq 2(3 - 1) = 4.$$

Another known extensive class are the so called combinatorial design facets. All three classes are valid for Q_{sub} and thus would be generated implicitly by any vector that satisfies all violated generalized subtour inequalities for P_{sub}. However, surprisingly, the separation problem for the Steiner partition inequalities is NP-complete.

Now consider the algorithmic question of how to solve the Steiner problem. The latest and perhaps most successful work has been based on the formulations we have examined. Three recent computational studies with branch and cut algorithms have used the formulations P_{sub} and P_{dcut} with the edge variables x_e eliminated by substitution. Two other recent approaches have been dual. One involves using dual ascent heuristics to approximately solve the P_{dcut} formulation. Another has been to use Lagrangian relaxation by dualizing the constraints $x_e + \bar{z}_i$ for all edges $e \in \delta(i)$ in the model (6.7)–(6.12). If we further relax (drop) the variables $\bar{z}_i$ from the constraints (6.8), the resulting subproblem is a minimum spanning tree problem.

6.3. Linear programming and heuristic bounds for the Steiner problem

Considerable practical experience over the last two decades has shown that a formulation of an integer program is effective computationally only when the optimal objective value of the linear programming relaxation is close to the optimal value of the integer program (within a few percent). Moreover solution methods often rely on good bounds on the optimal solution value. These considerations partly explain our efforts to understand different formulations.

Just how good a lower bound does the linear programming relaxation P_{sub} provide for the optimal value of the Steiner problem? Unfortunately, nothing appears to be known about this question. However a bound is available for the weaker cut formulation introduced at the end of Section 3, and which extends naturally to (SP), namely

$$Z = \min \sum_{e \in E} w_e x_e$$

subject to

$$\sum_{e \in \delta(S)} x_e \geq 1 \qquad \text{for } S \subset V \text{with } S \cap T \neq \phi,\ T \setminus S \neq \phi$$

$$x_e \in \{0, 1\} \quad \text{for } e \in E.$$

Note that this formulation is a special case of the survivable network problem formulation:

$$Z = \min \sum_{e/inE} w_e x_e$$

subject to

$$\sum_{e \in \delta(U)} x_e \geq r_U \quad \text{for } U \subset V$$

$$x_e \geq 0,\ x_e \text{ integral for } e \in E$$

treated in greater detail in Chapter 10. Here we just note that we obtain the Steiner problem by taking $r_U = 1$ whenever $\phi \subset U \subset V, U \cap T \neq \phi, T \setminus U \neq \phi$, and $r_U = 0$ otherwise.

Consider a restriction of the problem obtained by replacing the inequalities by equalities for the singleton sets $U = \{i\}$ for $i \in D \subseteq V$. The resulting problem is:

$$Z(D) = \min \sum_e w_e x_e$$

subject to

$$\sum_{e \in \delta(U)} x_e \geq r_U \quad \text{for } U \subset V$$

$$\sum_{e \in \delta(\{i\})} x_e = r_{\{i\}} \quad \text{for } i \in D$$

$$x_e \geq 0, \; x_e \text{ integral for all } e \in E.$$

Thus $Z = Z(\phi)$. We let $Z^{\mathrm{LP}}(D)$ denote the value of the corresponding linear programming relaxation and let $Z^{\mathrm{LP}} = Z^{\mathrm{LP}}(\phi)$ be the value of the basic linear programming relaxation. The following surprising result concerning this linear program is known as the 'parsimonious property'.

Theorem 6.8. *If the distances w_e satisfy the triangle inequality, then $Z^{\mathrm{LP}} = Z^{\mathrm{LP}}(D)$ for all $D \subseteq V$.*

We obtain a particularly interesting case by choosing D to be the set of Steiner nodes, i.e., $D = V \setminus T$. Since $\sum_{e \in \delta(\{i\})} x_e = 0$ for $i \notin T$, the problem reduces to:

$$Z(V \setminus T) = \min \sum_e w_e x_e$$

subject to

$$\sum_{e \in \delta(U)} x_e \geq 1 \quad \text{for } \phi \subset U \subset T$$

$$x_e \geq 0, \; x_e \text{ integral for } e \in E(T),$$

namely, the spanning tree problem on the graph induced by the terminal nodes T. Applying Theorem 6.8 to the graph $G' = (T, E(T))$ shows that

$$Z^{\mathrm{LP}}(V \setminus T) = \min \sum_e w_e x_e$$

subject to

$$\sum_{e \in \delta(U)} x_e \geq 1 \quad \text{for } U \subset T$$

$$\sum_{e \in \delta(\{i\})} x_e = 1 \quad \text{for } i \in T$$

$$x_e \geq 0 \quad \text{for } e \in E(T).$$

If we multiply the right hand side of the constraints by two, the resulting model is a well-known formulation for the traveling salesman problem (two edges, one 'entering' and one 'leaving', are incident to each node and every cutset contains at least two edges) on the graph $G' = (T, E(T))$. The corresponding linear programming relaxation is known as the Held and Karp relaxation; we let $Z^{\mathrm{HK}}(T)$ denote its value. We have established the following result.

Proposition 6.9. *If the distances w_e satisfy the triangle inequality, then* $Z^{\mathrm{LP}} = (1/2)Z^{\mathrm{HK}}(T)$.

This result permits us to obtain worst case bounds both for the value of the linear programming relaxation Z^{LP} and of a simple heuristic for the Steiner tree problem.

To apply this result, we either need to assume that the distances satisfy the triangle inequality or we can first triangularize the problem using the following procedure: replace the weight w_e on each edge $e = (i, j)$ by the shortest path distance d_e between nodes i and j. (To use this result, we need to assume that the shortest path distances exist: that is, the graph contains no negative cost cycles. We will, in fact, assume that $w_e \geq 0$ for all edges e).

Proposition 6.10. *The Steiner tree problem with the distances $w_e \geq 0$ and the Steiner tree problem with the shortest path distances d_e have the same solutions and same optimal objective values.*

Proof. Since $d_e \leq w_e$ for every edge e, the optimal value of the triangularized problem cannot exceed the optimal value of the problem with the distances w_e. Let ST be an optimal Steiner tree for the problem with distances w_e. If $d_e = w_e$ for every edge $e \in ST$, then ST also solves the triangularized Steiner problem and we will have completed the proof. Suppose $d_{\bar{e}} < w_{\bar{e}}$ for some edge $\bar{e} \in ST$. Then we delete edge $\bar{e}$ from ST and add the edges not already in ST that lie on the shortest path joining the end nodes of edge $\bar{e}$. The resulting graph G^* might not be a tree, but we can eliminate edges (which have costs $w_e \geq 0$) until the graph G^* becomes a tree. The new Steiner tree has a cost less than the cost of ST; this contradiction shows that $d_e = w_e$ for every edge $e \in ST$ and thus completes the proof.[6] □

The tree heuristic for the Steiner tree problem

If the distances w_e satisfy the triangle inequality, construct a minimum cost spanning tree on the graph induced by the terminal nodes T. If the distances w_e do not satisfy the triangle inequality,

Step 1. Compute the shortest path lengths $\{d_e\}$.

Step 2. Compute the minimum spanning tree with lengths $\{d_e\}$ on the complete graph induced by T.

Step 3. Construct a subnetwork of G by replacing each edge in the tree by a corresponding shortest path.

Step 4. Determine a minimum spanning tree in this subnetwork.

Step 5. Delete all Steiner nodes of degree 1 from this tree.

[6] This same argument applies, without removing the edges at the end, to any network survivability problem, even when some of the costs are negative (as long as the graph does not contain any negative cost cycles, so that shortest paths exist). If we permit the solution to the Steiner tree problem to be any graph that contains a Steiner tree, that is, a Steiner tree plus additional edges, then the result applies to this problem as well when some of the costs are negative.

Let z^H be the cost of the heuristic solution and let $Z_\Delta^{\mathrm{HK}}(T)$ and Z_Δ^{LP} denote the Held and Karp value for the graph $G' = (T, E(T))$ and the optimal linear programming value when we use the distances d_e in place of w_e.

Theorem 6.11. *If $w_e \geq 0$ for all edges e, then $Z \leq z^H \leq (2 - 2/|T|)Z^{\mathrm{LP}}$.*

Proof. Since by Proposition 6.10, the optimal value of the Steiner tree problem is the same with the costs w_e and d_e, and z^H is the value of a feasible solution for the triangularized problem, $Z \leq z^H$.

Let x^* be an optimal solution of the Held and Karp relaxation with the shortest path distances d_e. It is easy to show that x^* also satisfies the conditions $\sum_{e\in E(S)} x_e^* \leq |S| - 1$ for all $S \subset T$, and $\sum_{e\in E(T)} x_e^* = |T|$. Now $\bar{x} = (1 - 1/|T|)x^*$ satisfies $\sum_{e\in E(S)} x_e \leq |S| - 1$ for all $S \subset T$, $\sum_{e\in E(T)} x_e = |T| - 1$, and $x \geq 0$, so $\bar{x}$ lies in the spanning tree polytope (see Section 3) and thus $w\bar{x} \geq z^H$. However Proposition 6.9 shows that $Z_\Delta^{\mathrm{HK}}(T) = wx^* = 2Z_\Delta^{\mathrm{LP}}$. Thus $z^H \leq w\bar{x} = w(1 - 1/|T|)x^* = 2(1 - 1/|T|)Z_\Delta^{\mathrm{LP}}$. But since $w \geq d$, $Z_\Delta^{\mathrm{LP}} \leq Z^{\mathrm{LP}}$, implying the conclusion of the theorem □

In Theorem 6.11, we have shown that $Z \leq z^H \leq (2 - 2/|T|)Z^{\mathrm{HK}} = (2 - 2/|T|)Z^{\mathrm{LP}}$. We could obtain the conclusion of the theorem, without the intermediate use of Z^{HK}, by noting that the linear programming value $Z^{\mathrm{LP}}(V \setminus T)$ is the same as the value of the linear programming relaxation of the cutset formulation, without the cardinality constraint $\sum_{e\in E} x_e = n - 1$, of the minimum spanning tree problem on the graph $G' = (T, E(T))$. As we have shown in Section 3.3, the optimal objective value of this problem, which equals, Z^H is no more than $(2 - 2/|T|)Z^{\mathrm{LP}}(V \setminus T)$, the value of the linear program on the graph $G' = (T, E(T))$. Therefore, $Z \leq z^H \leq (2 - 2/|T|)Z^{\mathrm{LP}}(V \setminus T)$. But by the parsimonious property, $Z^{\mathrm{LP}} = Z^{\mathrm{LP}}(V \setminus T)$ and so $Z \leq z^H \leq (2 - 2/|T|)Z^{\mathrm{LP}}$.

A linear programming/tree heuristic for the node weighted Steiner tree problem

Various heuristics based upon minimum spanning tree computations also apply to the node weighted Steiner tree problem. We consider the model (6.1)–(6.6) with the objective function

$$z^{NWST} = \min \sum_{e\in E} w_e x_e + \sum_{i\in V} \pi_i(1 - z_i).$$

We assume that $w_e \geq 0$ for all $e \in E$. In this model, $\pi_i \geq 0$ is a penalty incurred if node i is not in the Steiner tree T. Note that we can impose $i \in T$ (or $z_i = 1$) by setting π_i to be very large. As before, we assume that r is a node that must necessarily be in any feasible solution — that is, $z_r = 1$.

Step 1. Solve the linear programming relaxation of the cut formulation (the analog of P_{cut} in Section 3).

$$\min \sum_{e\in E} w_e x_e + \sum_{i\in V} \pi_i(1 - z_i)$$

subject to

$$\sum_{e\in\delta(S)} x_e \geq z_i \quad \text{for all } i \text{ and } S \text{ with } r \notin S, i \in S$$
$$0 \leq z_i \leq 1 \quad \text{for all } i \in V$$
$$x_e \geq 0 \quad \text{for all } e \in E.$$

Let (x^*, z^*) be an optimal solution to this problem.

Step 2. Let $U = \{i : z_i^* \geq 2/3\}$.

Step 3. Apply the tree heuristic for the Steiner tree problem with the terminal nodes U, producing a heuristic solution $(\tilde{x}, \tilde{z})$ with value z^H.

Theorem 6.12 *Let Z^{NWST} denote the optimal objective value for the node weighted Steiner tree problem and let z^H denote the objective value of the linear programming/tree heuristic solution when applied to the problem. Then $z^H/z^{NWST} \leq 3$.*

Proof. First observe that if $Q = \{i : \tilde{z}_i = 1\}$ is the set of nodes in the heuristic solution, then $Q \supseteq U$ and so

$$\begin{aligned}\sum_{i\in V} \pi_i(1-\tilde{z}_i) &= \sum_{i\in V\setminus Q} \pi_i \\ &\leq \sum_{i\in V\setminus U} \pi_i \\ &\leq 3\sum_{i\in V\setminus U} \pi_i(1-z_i^*) \\ &\leq 3\sum_{i\in V} \pi_i(1-z_i^*).\end{aligned}$$

The second inequality uses the definition of U, and the third inequality the nonnegativity of π.

Now let $\hat{x} = (3/2)x^*$. Observe that if $i \in U \setminus \{r\}$ and $S \ni i$, then

$$\sum_{e\in\delta(S)} \hat{x}_e = \tfrac{3}{2}\sum_{e\in\delta(S)} x_e^* \geq \tfrac{3}{2} z_i^* \geq 1,$$

so $\hat{x}$ is a feasible solution of the cut formulation

$$z^{\mathrm{LP}}(U) = \min \sum_{e\in E} w_e x_e$$

subject to

$$\sum_{e\in\delta(S)} x_e \geq 1 \quad \text{for all } i \text{ and } S \text{ with } r \notin S, i \in U \cap S$$
$$x_e \geq 0$$

and thus $z^{\mathrm{LP}}(U) \leq w\hat{x}$. Also, by Theorem 6.11 $w\tilde{x} \leq 2z^{\mathrm{LP}}(U)$ and so $w\tilde{x} \leq 2w\hat{x} = 3wx^*$. Thus, $z^H = w\tilde{x} + \sum_{i\in V} \pi_i(1-\tilde{z}_i) = w\tilde{x} + \sum_{i\in V\setminus Q} \pi_i \leq 3wx^* + 3\sum_{i\in V} \pi(1-z_i^*) \leq 3z^{NWST}$. □

7. Packing subtrees of a tree

In Section 2 we saw how to view several problems — lot-sizing, facility location, and vehicle routing — as packing subtrees of a graph. The special case when the underlying graph is itself a tree is of particular interest; we treat this problem in this section. Our discussion builds upon our investigation in Section 4 of the subproblem of finding a best (single) optimal subtree of a tree.

Our principal results will be threefold: (i) to show how to use the type of dynamic programming algorithm we developed in Section 4 to solve the more general problem of packing subtrees of a tree, (ii) to discuss several algorithms for solving problems of packing trees in a tree, and (iii) to understand if and when we can obtain integral polyhedra, or tight linear programming formulations. In the next section we develop extensions of the algorithms introduced in this section to tackle harder problems on general graphs.

7.1. Simple optimal subtree packing problem

We start by examining the simplest problem. Given a tree $G = (V, E)$, suppose we are given a finite set $F^1, \ldots, F^q$ of subtrees, with each $F^j \subseteq V$ for $j = 1, 2, \ldots, q$. Each subtree F^j has an associated value c_j. The *simple optimal subtree packing problem* (SOSP) is to find a set of node disjoint subtrees of maximum value. Suppose that A is the node-subtree incidence matrix, i.e., $a_{ij} = 1$ if node i is in subtree F^j, and $a_{ij} = 0$ otherwise. Figure 17 shows an example of such a matrix. Letting $\lambda_j = 1$ if we choose subtree F^j, and $\lambda_j = 0$ otherwise, and letting λ be the vector (λ_j), we can formulate problem (SOSP) as the following optimization model:

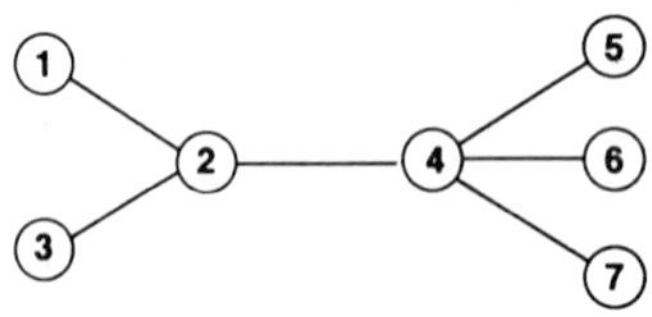

A tree graph G=(V,E)

Nodes ↓ \ Subtrees →	1	2	3	4	5	6	7	8
1	1	0	0	0	0	0	1	1
2	1	1	1	0	0	1	1	1
3	1	0	0	0	0	1	1	0
4	1	1	1	1	0	1	0	0
5	0	1	0	1	0	0	0	0
6	0	0	1	0	0	0	0	0
7	0	1	0	1	1	0	0	0
c values →	4	5	3	3	1	2	3	2

A node-subtree incidence matrix

Fig. 17. Optimal subtree packing problem.

$$\max\left\{\sum_j c_j\lambda_j : A\lambda \leq \mathbf{1},\ \lambda \in \{0,1\}^q\right\}.$$

To describe a dynamic programming algorithm, we first introduce some notation. Given the tree G, we arbitrarily choose a root node r and thereby obtain a partial ordering $(V, \preceq)$ on the nodes by defining $u \preceq v$ if and only if node u lies on the path (r, v). We define the predecessor $p(u)$ of node u as the last node before u on the path (r, u), $S(u) = \{w : p(w) = u\}$ as the set of successors of node u, and $S(F^j)$ as the set of successors of subtree F^j, i.e., $S(F^j) = \{w : p(w) \in F^j, w \notin F^j\}$. We also define the root $r(F^j)$ of F^j to be the unique node in F^j satisfying the condition that $r(F^j) \preceq u$ for all $u \in F^j$.

For the example shown in Figure 17, with node 1 as the root, $p(1) = \phi$, $p(2) = 1$, $p(3) = 2$,, and so forth; the set of successors of node 2 is $S(2) = \{3, 4\}$. The set of successors of the subtree F^2 on the nodes 2, 4, 5, 7, is $S(F^2) = \{3, 6\}$ and the root of this tree is $r(F^2) = 2$.

Algorithm for the SOSP problem

The algorithm is based on the following recursion:

$$H(u) = \max\left\{\sum_{w\in S(u)} H(w),\ \max_{\{j:r(F^j)=u\}}\left[c_j + \sum_{w\in S(F^j)} H(w)\right]\right\}.$$

In this expression, $H(u)$ is the value of an optimal packing of the subtree induced by the node set $V^u = \{v : u \preceq v\}$ and the set of subtrees $\{F^j\}$ with $F^j \subseteq V^u$. The recursion follows from the observation that in an optimal solution of value $H(u)$,

(i) if node u is not in any subtree, the solution is composed of optimal solutions for each subgraph induced by V^w with $w \in S(u)$. Thus $H(u) = \sum_{w\in S(u)} H(w)$; or

(ii) if node u is in one of the subtrees $F^j \subseteq V^u$, then necessarily $r(F^j) = u$, and the optimal solution must be composed of F^j and optimal solutions for each subtree induced by V^w with $w \in S(F^j)$. Thus $H(u) = \max_{\{j:r(F^j)=u\}}[c_j + \sum_{w\in S(F^j)} H(w)]$

Starting from the leaves and working in towards the root, the dynamic programming algorithm recursively calculates the values $H(v)$ for all $v \in V$. $H(r)$ is the value of the optimal solution. To obtain the subtrees in an optimal solution, we iterate back from the root r to see how the algorithm obtained the value $H(r)$.

Example 7.1. We consider the (SOSP) problem instance shown in Figure 17 with node 1 chosen as the root. Working in towards the root, the algorithm gives:

$$H(7) = \max\{0, c_5\} = 1$$
$$H(6) = 0$$
$$H(5) = 0$$
$$H(3) = 0$$
$$H(4) = \max\{H(5) + H(6) + H(7), c_4 + H(6)\} = 3$$

$$H(2) = \max\{H(3) + H(4), c_2 + H(3) + H(6),\\ c_3 + H(3) + H(5) + H(7), c_6 + H(5) + H(6) + H(7)\} = 5$$
$$H(1) = \max\{H(2), c_1 + H(5) + H(6) + H(7), c_7 + H(4),\\ c_8 + H(3) + H(4)\} = 6.$$

Thus the optimal solution value is 6. To obtain the corresponding solution, we observe that $H(1) = c_7 + H(4)$, $H(4) = c_4 + H(6)$, $H(6) = 0$, so subtrees 7 and 4 give an optimal packing of value 6.

The linear program $\max\{\sum_j c_j\lambda_j : A\lambda \le \mathbf{1}, \lambda \ge 0\}$ has a primal solution with $\lambda_4 = \lambda_7 = 1$, and $\lambda_j = 0$ otherwise. Observe that if we calculate the values $\pi_u = H(u) - \sum_{w\in S(u)} H(w)$ for $u \in V$, i.e., $\pi_1 = H(1) - H(2) = 6 - 5 = 1$, etc., $\pi = (1, 2, 0, 2, 0, 0, 1)$ is a dual feasible solution to this linear program and its objective value $\sum_{j\in V} \pi_j$ equals $H(r) = 6$.

It is easy to see that this observation concerning the dual variables $\pi_u = H(u) - \sum_{w\in S(u)} H(w)$ holds in general. The recursion for $H(u)$ implies that $\pi_u \ge 0$. For a tree F^j with $r(F^j) = u$, $\sum_{v\in F^j} \pi_v = \sum_{v\in F^j}(H(v) - \sum_{w\in S(v)} H(w)) = H(u) - \sum_{w\in S(F^j)} H(w)$, and the recursion implies that the last term is greater than or equal to c_j. This observation permits us to verify that the algorithm is correct, and the primal-dual argument used in Sections 3 and 4 immediately leads to a proof of an integrality result.

Theorem 7.1. *Given a family of subtrees of a tree, if A is the corresponding node-subtree incidence matrix, then the polyhedron $\{x : Ax \le \mathbf{1}, x \ge 0\}$ is integral.*

7.2. More general models

In Section 2, we described three problems of packing subtrees of a tree, namely the lot-sizing problem, the facility location problem with the contiguity property, and the telecommunications application shown in Figure 2.3. In each of these application contexts, we are not given the subtrees and their values explicitly; instead, typically the problem is defined by an exponential family of subtrees, each whose value or cost we can calculate, and a function prescribing the value or cost of each subtree. We now consider the issues of how to model and solve such problems.

The optimal subtree packing problem

The optimal subtree packing problem (OSP) is defined by a tree $G = (V, E)$, families $\mathcal{F}^k$ of subtrees associated with each node $k \in V$, and a value function $c^k(F)$ for $F \in \mathcal{F}^k$. Each nonempty tree of $\mathcal{F}^k$ contains node k. We wish to choose a set of node disjoint subtrees of maximum value, selecting at most one tree from each family.

OSP differs from SOSP in that neither the subtrees in each family $\mathcal{F}^k$ nor their costs $c^k(F)$ need be given explicitly. We now show how to model the three problems mentioned previously as OSP problems. For the first two problems, the objective is a linear function, that is, a function of the form $c^k(F) = \sum_{j\in F} c^k_j$.

Uncapacitated lot-sizing (ULS)

Given a finite number of periods, demands $\{d_t\}_{t=1}^T$, production costs $\{p_t\}_{t=1}^T$, storage costs $\{h_t\}_{t=1}^n$ (which can be transformed by substitution to be zero without any loss of generality), and set-up (or fixed) costs $\{f_t\}_{t=1}^T$ (if production is positive in the period), find a minimum cost production plan that satisfies the demands. From Theorem 2.1 (see Figure 7), we know that this problem always has a directed spanning tree solution. Taking the tree graph to be a path from node 1 to node $T = |V|$, the family of subpaths $\mathcal{F}^k$ associated with node k are of the form $(k, k+1, \ldots, t)$ for $t = k, \ldots, T$ corresponding to a decision to produce the demand for periods k up to t in period k. The costs are $c_k^k = f_k + p_k d_k$, $c_j^k = p_k d_j$ for $j > k$ and $c_j^k = \infty$ for $j < k$.

Facility location on a tree

Given a tree graph, edge weights α_e for $e \in E$, a distance function $d_{ij} = \sum_{e \in Path(i,j)} \alpha_e$ and weights f_j for $j \in V$, locate a set of depots $U \subseteq V$ and assign each node to the nearest depot to minimize the sum of travel distances and node weights: $\min_{U \subseteq V}\{\sum_{j \in U} f_j + \sum_{i \in V}(\min_{j \in U} d_{ij})\}$. Here we take $c_k^k = f_k$ and $c_j^k = d_{kj}$ for $j \neq k$. In this model, the constraints $\sum_{k \in V} x_j^k \leq 1$ become equalities.

Each of the previous two applications is a special case of an OSP problem with linear costs that we can formulate as the following integer program:

$$\max\left\{\sum_{k \in V}\sum_{j \in V} c_j^k x_j^k : \sum_{k \in V} x_j^k \leq 1 \text{ for } j \in V,\ x^k \in X^k \text{ for } k \in V\right\}.$$

In this model, X^k is the set of node incidence vectors of all subtrees rooted at node k plus the vector 0 corresponding to the empty tree. We can interpret the coefficient c_j^k in this model as the cost of assigning node j to a subtree rooted at node k. In practice, frequently node k will contain a 'service' facility: c_k^k will be the fixed cost of establishing a service facility at node k and c_j^k will be the cost of servicing node j from node k.

In Section 4 we showed that if $p(j,k)$ denotes the predecessor of node j on the path from node j to node k, then $\text{conv}(X^k) = \{x^k \in R_+^{|V|} : x_k^k \leq 1, x_j^k \leq x_{p(j,k)}^k$ for $j \neq k\}$. Thus, we obtain the formulation

$$\max \sum_{k \in V} c^k x^k \tag{7.1}$$

subject to

$$\sum_{k \in V} x_j^k \leq 1 \qquad \text{for } j \in V \tag{7.2}$$

$$x_j^k \leq x_{p(j,k)}^k \qquad \text{for } j \neq k, k \in V \tag{7.3}$$

$$x_j^k \geq 0 \qquad \text{for } j, k \in V \tag{7.4}$$

$$x_j^k \text{ integer} \qquad \text{for } j, k \in V. \tag{7.5}$$

Later in this section, we consider the effectiveness of this formulation, particularly the strength of its linear programming relaxation.

A more complex model

The telecommunications example requires a more intricate modeling approach since the objective function $c^k(F)$ is nonlinear with respect to the nodes in the subtree F. In addition as we have seen in Section 4.2, in many models it is natural to consider situations in which some of the subtrees rooted at node k are infeasible. We now describe a formulation that allows us to handle both these generalizations.

This formulation contains a set X^k of incidence vectors of the subtrees in $\mathcal{F}^k$. We let x^k be a an incidence vector of a particular subtree rooted at node k. The formulation also contains a set of auxiliary variables w^k that model other aspects of the problem. For example, for the telecommunications problem, the variables w^k correspond to the flow and capacity expansion variables associated with a subtree $F \in \mathcal{F}^k$ defined by x^k. In Section 8 we consider other applications: for example, in a multi-item production planning problem, the variables x^k indicate when we produce product k and the variables w^k model the amount of product k that we produce and hold in inventory to meet customer demand. In this more general problem setting, we are given a set W^k that models the feasible combinations of the x^k and w^k variables. We obtain the underlying trees from W^k by projecting out the w^k variables, that is, $\mathrm{proj}_{x^k}(W^k) = X^k$

For any particular choice x^k of the x variables, let $c^k(x^k) = \max\{e^k x^k + f^k w^k : (x^k, w^k) \in W^k\}$ denote the optimal value of the tree defined by x^k obtained by optimizing over the auxiliary variables w. Once again, we assume that $0 \in X^k$ and $c^k(0) = 0$. We are interested in solving the following *optimal constrained subtree packing problem (OCSP)*:

$$\max \sum_k e^k x^k + \sum_k f^k w^k \tag{7.6}$$

subject to

$$\sum_{k \in V} x_j^k \le 1 \qquad \text{for } j \in V \tag{7.7}$$

$$(x^k, w^k) \in W^k \qquad \text{for } k \in V. \tag{7.8}$$

This modeling framework permits us to consider applications without auxiliary variables as well. In this case, $W^k = X^k$ can model constraints imposed upon the x variables and so only a subset of the subtrees rooted at k will be feasible. In Section 4.2 we considered one such application, a capacitated model with knapsack constraints of the form: $\sum_{j \in V} d_j x_j^k \le C$.

7.3. Algorithmic approaches

We briefly outline four algorithmic strategies for solving OCSP.

Algorithm A. Primal column generation algorithm using subtree optimization

We know from Theorem 7.1 that if we could explicitly write out all feasible subtrees and calculate their values for each $k \in V$, the resulting linear program

$$z^A = \max\{\sum_k c^k\lambda^k : \sum_k A^k\lambda^k \leq \mathbf{1},\ \lambda^k \geq 0 \text{ for } k \in V\}, \tag{7.9}$$

called the *Master Problem*, or its corresponding integer program with λ^k integer, would solve OCSP. In this model, A^k is the node-tree incidence vector for all feasible subtrees rooted at node k and $\lambda^k = (\lambda_j^k)$ is vector that, when integer, tells which tree we choose (i.e., if $\lambda_j^k = 1$ we choose the tree with incidence vector A_j^k). $c^k = (c_j^k)$ is a vector of tree costs; that is, c_j^k is the cost of the tree with the incidence vector A_j^k. When the problem has auxiliary variables w, $c_j^k = c^k(A_j^k) = \max\{e^k A_j^k + f^k w^k : (A_j^k, w^k) \in W^k\}$ is the optimal cost of the tree rooted at node k with the incidence vector A_j^k.

Typically, this linear program is impractical to formulate explicitly because of the enormous number of subtrees and/or the difficulty of evaluating their values. Therefore, we might attempt to solve it using the idea of a column generation algorithm; that is, work with just a subset of the columns (subtrees), and generate missing ones if and when we need them.

At iteration t, we have a *Restricted Master Problem*:

$$\max \sum_{k \in V} c^{k,t}\lambda^{k,t}$$

subject to

$$\sum_{k \in V} A^{k,t}\lambda^{k,t} \leq \mathbf{1}$$

$$\lambda^{k,t} \geq 0.$$

In this model each $A^{k,t}$ is the incidence matrix of some of the feasible subtrees rooted at node k with an associated cost $c^{k,t}$, and so $A^{k,t}$ is a submatrix of A^k. $\lambda^{k,t}$ denotes the corresponding subvector of the vector λ^k. Let π^t be optimal dual variables for Restricted Master linear program. Since the Restricted Master Problem contains only a subset of the feasible trees, we would like to know whether we need to consider other subtrees or not.

Let x^k denote the incidence vector of a generic column of A^k corresponding to a feasible subtree rooted at node k. The subproblem for each k is:

$$\mu_k^t = \max\{e^k x^k + f^k w^k - \pi^t x^k, (x^k, w^k) \in W^k\}.$$

If $\mu_k^t \leq 0$ for all $k \in V$, linear programming theory tells us that we have an optimal solution of the Master Problem.

However, if $\mu_k^t > 0$ for some k, then we add one or more subtrees to the Restricted Master Problem, update the matrices giving $A^{k,t+1}$ and $c^{k,t+1}$ and we pass to iteration $t + 1$. □

Because of Theorem 7.1, we observe that

(i) the Restricted Master Problem produces an integer feasible solution at each iteration (that is, each $\lambda^{k,t}$ is a 0–1 vector).

(ii) the Restricted Master Problem is an SOSP problem, so we can solve it using the dynamic programming algorithm presented at the beginning of this section, rather than using the Simplex algorithm.

Since, as we have already noted, Theorem 7.1 implies that the Master Problem (as a linear program) is equivalent to OCSP (an integer program), we have the following result:

(iii) the algorithm terminates with an optimal solution of OCSP.

Algorithm B. Lagrangian relaxation

As we have seen in Section 3, Lagrangian relaxation is an algorithmic approach for finding a good upper bound (a lower bound in that discussion since we were minimizing) for a maximization problem by introducing some of the problem constraints into the objective function with associated prices (also called dual variables or Lagrange multipliers). Specifically, we start with the formulation (7.6)–(7.8) of OCSP. Dualizing the packing constraints (7.7) leads to the so-called Lagrangian subproblem:

$$L(\pi) = \max \sum_{k \in V} (e^k x^k + f^k w^k - \pi x^k) + \sum_{j \in V} \pi_j$$

subject to

$$(x^k, w^k) \in W^k \text{ for all } k.$$

Observe that the Lagrangian subproblem separates into independent subproblems for each k, namely,

$$\mu_k = \max\{e^k x^k + f^k w^k - \pi x^k, (x^k, w^k) \in W^k\}.$$

Thus,

$$L(\pi) = \sum_{j \in V} \pi_j + \sum_k \mu_k.$$

To obtain a 'best' upper bound, we solve the Lagrangian dual problem:

$$z^B = \min_{\pi \geq 0} L(\pi).$$

We can find an optimal value for π by using standard algorithms from the theory of Lagrangian relaxation (e.g., subgradient optimization) that generate values π^t for π at each iteration. □

What can be said about this algorithm? To resolve this question, we first observe that we can rewrite the Master Problem (7.9) as

$$z^A = \max \left\{ \sum_k c^k \lambda^k : \sum_k A^k \lambda^k \leq \mathbf{1}, \right.$$
$$\left. \mathbf{1}\lambda^k = 1 \text{ for } k \in V,\ \lambda^k \geq 0 \text{ for } k \in V \right\}.$$

This model contains additional constraints $\mathbf{1}\lambda^k = 1$ for $k \in V$. Since $0 \in X^k$ with cost 0, and since every nonempty tree $x^k \in X^k$ contains node k, the kth constraint of $\sum_k A^k\lambda^k \leq \mathbf{1}$ implies that $\mathbf{1}\lambda^k = 1$ for $k \in V$ is redundant. More precisely, the row of the node-subtree incidence matrix A^k corresponding to node k has $+1$ for each subtree in X^k and so the row corresponding to this constraint implies that $\sum_j \lambda_j^k \leq 1$. Note that $1 - \sum_j \lambda_j^k$ is the weight placed on the null tree in X^k.

Linear programming theory shows that the optimal value of the linear programming relaxation of this modified Master Problem, equals the value of its linear program dual, that is,

$$\begin{aligned} z^A &= \min\left\{\sum_i \pi_i + \sum_k \mu_k : \pi A^k + \mu_k \mathbf{1} \geq c^k \ \text{ for all } k, \pi \geq 0\right\} \\ &= \min_{\pi \geq 0, \mu}\left\{\sum_i \pi_i + \sum_k \mu_k : \mu_k \geq c(x^k) - \pi x^k \ \text{ for all } x^k \in X^k\right\} \\ &= \min_{\pi \geq 0} L(\pi). \end{aligned}$$

The final equality is due to the fact that for a given value of π, the optimal choice of each μ_k is given by $\mu_k = \max_{x^k \in X^k}\{c(x^k) - \pi x^k\} = \max\{e^k x^k + f^k w^k - \pi x^k : (x^k, w^k) \in W^k\}$ and so the objective function $\sum_{i \in V} \pi_i + \sum_k \mu_k$ is equal to $L(\pi)$.

This discussion shows that $z^A = z^B$, and so

(i) the optimal value of the Lagrangian dual gives the optimal value of the OCSP problem, and

(ii) the optimal solution π of the Lagrangian dual is an optimal dual solution for the Master Problem (7.9).

Algorithm C. Dual cutting plane algorithm using subtree separation

The goal in this approach is to implicitly solve the linear program

$$z^C = \max \sum_k e^k x^k + \sum_k f^k w^k \tag{7.10}$$

subject to

$$\sum_k x^k \quad \leq \mathbf{1} \tag{7.11}$$

$$(x^k, w^k) \in \text{conv}(W^k) \quad \text{for all } k \tag{7.12}$$

by a cutting plane algorithm.

Let $\{(x^k, w^k) : G^k x^k + H^k w^k \leq b^k, 0 \leq x_j^k \leq 1, w^k \geq 0 \text{ for all } k, j \in V\}$ be a representation of $\text{conv}(W^k)$ by a system of linear inequalities. Since it is impractical to write out all the constraints explicitly, we work with a subset of them, and generate missing ones if and when needed.

At iteration t, we therefore have a relaxed linear program:

$$\max \sum_{k\in V} (e^k x^k + f^k w^k)$$

subject to

$$\sum_{k\in V} x^k \leq \mathbf{1}$$

$$G^{k,t} x^k + H^{k,t} w^k \leq b^{k,t} \quad \text{for } k \in V$$

$$0 \leq x_j^k \leq 1, \; w^k \geq 0 \quad \text{for } k, j \in V.$$

After finding a solution $(x^{k,t}, w^{k,t})$ to the linear programming relaxation, we then solve the separation problem for each k, i.e., we check to see if $(x^{k,t}, w^{k,t}) \in \text{conv}(W^k)$ or not. If $(x^{k,t}, w^{k,t}) \in \text{conv}(W^k)$ for all k, the algorithm terminates. If not, we obtain one or several violated inequalities that we add to the formulation. We then update the matrices giving $G^{k,t+1}$, $H^{k,t+1}$ and $b^{k,t+1}$ and we pass to iteration $t+1$. □

Note that if we always generate a violated facet-defining inequality of $\text{conv}(W^k)$, the cutting plane algorithm will terminate in a finite number of iterations having satisfied all the constraints in $G^k x^k + H^k w^k \leq b^k$, even though we have not written all the constraints explicitly.

How does the optimal objective function for this approach compare with the values we obtain from the column generation and Lagrangian relaxation procedures? Linear programming duality (results on 'partial duality') implies that

$$z^C = \min_{\pi \geq 0} \left\{ \sum_i \pi_i + \right.$$
$$\left. + \sum_k \max_{x^k, w^k} \{ e^k x^k + f^k w^k - \pi x^k : (x^k, w^k) \in \text{conv}(W^k) \} \right\}$$

But since optimal solutions of linear programs occur at extreme points, this optimization problem has the same value as:

$$\min_{\pi \geq 0} \left\{ \sum_i \pi_i + \sum_k \max_{x^k, w^k} \{ e^k x^k + f^k w^k - \pi x^k : (x^k, w^k) \in W^k \} \right\}$$

which is just the Lagrangian dual problem.

Thus, we have shown that $z^A = z^B = z^C$. Stated differently, we have shown that for any objective function coefficients $(e^1, f^1, \ldots, e^n, f^n)$, the linear programming value z^C of (7.10)–(7.12) equals the optimal value of the corresponding OCSP integer program (7.6)–(7.8). This implies that

(i) the polyhedron (7.11)–(7.12) is integral (i.e. in all its extreme points $(x^k, w^k) \in W^k$), and thus

(ii) Algorithm C terminates with such an extreme point, and thus it gives an optimal solution to OCSP.

In summary, this discussion shows that if we (i) solve the linear programming relaxation of the column generation model, (ii) solve the Lagrangian dual problem, or (iii) use the cutting algorithm until we have a point lying in $\text{conv}(W^k)$ for all k, then we obtain the same optimal objective function values: $z^A = z^B = z^C$. Algorithms A and C terminate with an optimal solution to OCSP since they work in the space of x and w variables. Lagrangian relaxation provides us with an optimal set of dual prices π, which we can then use to find an optimal solution x and w by solving a linear programming feasibility problem (we will not provide the details.)

Example 7.2. We solve an instance of the OCSP problem for the graph shown in Figure 18. The initial profit matrix is

$$\begin{pmatrix} 6 & 2 & 3 & 4 & 1 \\ 9 & 2 & 7 & 6 & 0 \\ 8 & 2 & 1 & 4 & 9 \\ 2 & 1 & 3 & 4 & 4 \\ 1 & 6 & 2 & 9 & 1 \end{pmatrix}.$$

The entry c_j^k in row k and column j is the value if node j is in a subtree rooted at node k. All subgraphs are feasible.

We use the column generation algorithm starting with an initial set of five single node subtrees. Thus, the initial Master Problem is of the form:

$$\max\ 6\lambda_1^1 + 2\lambda_1^2 + 1\lambda_1^3 + 4\lambda_1^4 + 1\lambda_1^5$$

subject to

$$\begin{aligned}
&1\lambda_1^1 + 0\lambda_1^2 + 0\lambda_1^3 + 0\lambda_1^4 + 0\lambda_1^5 \le 1\\
&0\lambda_1^1 + 1\lambda_1^2 + 0\lambda_1^3 + 0\lambda_1^4 + 0\lambda_1^5 \le 1\\
&0\lambda_1^1 + 0\lambda_1^2 + 1\lambda_1^3 + 0\lambda_1^4 + 0\lambda_1^5 \le 1\\
&0\lambda_1^1 + 0\lambda_1^2 + 0\lambda_1^3 + 1\lambda_1^4 + 0\lambda_1^5 \le 1\\
&0\lambda_1^1 + 0\lambda_1^2 + 0\lambda_1^3 + 0\lambda_1^4 + 1\lambda_1^5 \le 1\\
&\lambda \ge 0.
\end{aligned}$$

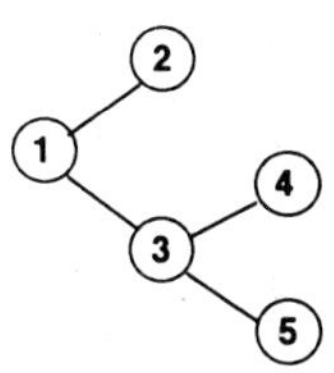

Fig. 18. Tree for the OCSP problem instance.

Solving the Restricted Master by linear programming (or the dynamic programming algorithm for SOSP, we obtain

$$H(5) = 1,\ H(4) = 4,\ H(3) = 6,\ H(2) = 2,\ H(1) = 14,$$

and dual variables $\pi^1 = (6, 2, 1, 4, 1)$.

The subproblem profit matrix $\bar{c}^1$, with $c^1_{kj} = (c^k_j - \pi^1_j)$ is now

$$\begin{pmatrix} 0 & 0 & 2 & 0 & 0 \\ 3 & 0 & 6 & 2 & -1 \\ 2 & 0 & 0 & 0 & 8 \\ -4 & -1 & 2 & 0 & 3 \\ -5 & 4 & 1 & 5 & 0 \end{pmatrix}.$$

Thus, we have modified the original values c^k_j by a cost π_j if node j appears in the subtree. For instance $\bar{c}^1_{43} = c_{43} - \pi_3 = 3 - 1 = 2$. Solving the 5 subproblems, we obtain $\mu^1_1 = 2$, $\mu^1_2 = 11$, $\mu^1_3 = 10$, $\mu^1_4 = 5$, and $\mu^1_5 = 6$. Introducing the column with the greatest reduced price of 11, namely $F^2 = \{1, 2, 3, 4\}$ rooted at 2 of value 24, we update the Restricted Master problem, obtaining

$$\max\ 6\lambda^1_1 + 2\lambda^2_1 + 1\lambda^3_1 + 4\lambda^4_1 + 1\lambda^5_1 + 24\lambda^2_2$$

subject to

$$\begin{aligned} &1\lambda^1_1 + 0\lambda^2_1 + 0\lambda^3_1 + 0\lambda^4_1 + 0\lambda^5_1 + 1\lambda^2_2 \le 1 \\ &0\lambda^1_1 + 1\lambda^2_1 + 0\lambda^3_1 + 0\lambda^4_1 + 0\lambda^5_1 + 1\lambda^2_2 \le 1 \\ &0\lambda^1_1 + 0\lambda^2_1 + 1\lambda^3_1 + 0\lambda^4_1 + 0\lambda^5_1 + 1\lambda^2_2 \le 1 \\ &0\lambda^1_1 + 0\lambda^2_1 + 0\lambda^3_1 + 1\lambda^4_1 + 0\lambda^5_1 + 1\lambda^2_2 \le 1 \\ &0\lambda^1_1 + 0\lambda^2_1 + 0\lambda^3_1 + 0\lambda^4_1 + 1\lambda^5_1 + 0\lambda^2_2 \le 1 \\ &\lambda \ge 0. \end{aligned}$$

For this problem,

$$H = (25, 2, 6, 4, 1) \text{ and } \pi^2 = (17, 2, 1, 4, 1).$$

Returning to the subproblems, $\mu^2_5 = 12$ is the maximum violation, so we next introduce the subtree $F^5 = \{3, 4, 5\}$ with value 12.

On the third iteration, the Restricted Master gives

$$H = (25, 2, 12, 4, 1) \text{ and } \pi^3 = (11, 2, 7, 4, 1).$$

The subproblem cost matrix $\bar{c}^3 = (c^k_j - \pi^3_j)$ is now

$$\begin{pmatrix} -5 & 0 & -9 & 0 & 0 \\ -2 & 0 & -5 & 2 & -1 \\ -3 & 0 & -11 & 0 & 8 \\ -9 & -1 & -9 & 0 & 3 \\ -10 & 4 & -10 & 5 & 0 \end{pmatrix}.$$

All the subproblems have an optimal value of 0, so the corresponding primal solution $\{1, 2, 3, 4\}$ rooted at 2 and $\{5\}$ rooted at 5 with value $H(1) = 25$ is optimal.

Algorithm D. Dynamic programming

Dynamic programming provides another algorithmic approach for solving OSP with linear costs (that is, the model (7.1)–(7.3)). The algorithm is a more complicated version of the dynamic program for solving SOSP in that it implicitly treats an exponential number of subtrees. To develop one such algorithm, let u be any node of the given tree T and let $T(u)$ be the subtree of T rooted at node u. Any feasible solution to OSP contains (i) subtrees that are disjoint from $T(u)$, (ii) subtrees contained in $T(u)$, and (iii) at most one subtree T^k, rooted at some node k, that has nodes both in and out of $T(u)$. Note that in case (iii), $T(u) \cap T^k$ is a subtree of $T(u)$ that contains node u. Let $C(u)$ denote the assignment values of the nodes in the subtree $T(u)$, that is, $C(u) = \sum_{j \in T(u)} c_j^{k(j)}$. In this expression, $k(j)$ denotes the node to which the solution assigns node j. If the solution does not assign node j to any node, we set $k(j) = 0$. In this case, $c_j^0 = 0$. Among all feasible solutions to the problem, let $H(u, k)$ denote the maximum value of $C(u)$, given that the solution assigns node u to node k (note that node k might or might not belong to $T(u)$).

Let $H(u)$ denote the value of the maximum packing of the tree $T(u)$, that is, the value of OSP restricted to the subtree $T(u)$. Let r be a root node of the given tree T. We wish to find $H(r)$.

As before, $S(u)$ denotes the successors of node u in T. We can find $H(u, k)$ for all nodes u and k by working towards the root from the leaves of T using the following recursion:

$$H(u,k) = c_u^k + \sum_{w \in S(u)} \max\{H(w,k), H(w)\} \text{ if } k = u \text{ or } k \notin T(u)$$

$$H(u,k) = c_u^k + H(\bar{w},k) + \sum_{w \in S(u),\ w \neq \bar{w}} \max\{H(w,k), H(w)\}$$
$$\text{if } k \in T(\bar{w}) \text{ and } \bar{w} \in S(u)$$

$$H(u,k) = \sum_{w \in S(u)} H(w) \text{ if } k = 0 \text{ (that is, node } u \text{ belongs to no subtree)}$$

$$H(u) \quad = \min_{k \in T(u)} \{H(u,k)\}.$$

Note that this recursion uses the fact that if node w is any successor of node u and a solution assigns node u to itself ($k = u$), to no node, or to a node k not in $T(w)$, then the solution must assign node w to node k, to no node, or to a node in $T(w)$. $H(w, k)$ gives the optimal cost of the first case and $H(w)$ gives the optimal cost of the last two cases. If the solution assigns node u to a node k in $T(\bar{w})$ for one of is successor nodes $\bar{w}$, then it must assign node $\bar{w}$ to node k as well.

Note that since this recursion compares $H(w, k)$ to $H(w)$ for any two nodes w and k of T once, for a tree with n nodes, it requires $O(n^2)$ computations.

When each node v has an associated demand $d(v)$ and each subtree is restricted to contain nodes with a total demand of at most C for some positive integer capacity C, we can use an extension of the dynamic programming argument that we introduced in Section 4 for the rooted subtree of a tree problem. In this case, we let $H(u, k, t)$ be the maximum value $C(u)$ of a packing on the subtree $T(u)$, given a capacity of t for the packing, and given that the solution assigns node u to node k. If we use the dynamic programming approach suggested in Section 4, the overall algorithm will require $O(n^2C^2)$ computations. In the special case in which each demand $d(v)$ is one, and so C is a restriction on the number of nodes in any subtree, the algorithm requires $O(n^4)$ computations. The well-known p-median problem is a variant of this problem; recall that in this case, the solution can contain at most p subtrees. The dynamic programming algorithm for this case is similar to those for these other cases, using, for example, a recursion with the definition of $H(u, k, q)$ as the maximum value $C(u)$ of the subtree $T(u)$, given that the solution assigns node u to node k and that the packing of $T(u)$ contains (or intersects) at most q subtrees. The resulting algorithm requires $O(n^2p^2)$ computations.

7.4. Polyhedral results

Our discussion of algorithms in the previous subsection has shown that when the costs are linear, the linear programming relaxation of this integer program solves OSP with linear costs.

Theorem 7.2. *If X^k is the set of all subtrees rooted at node k, and $X = \{(x^1, \ldots, x^n) : \sum_k x^k \leq 1, x^k \in X^k$ for all $k \in V\}$, then* conv(X) *is described by the constraints (7.2)–(7.4).*

This is an apparently surprising result because it is very unusual that, when a set of integral polyhedra (the conv(X^k)) are combined with a set of additional constraints (in this case, the packing constraints $\sum_{k \in V} x_j^k \leq 1$), the resulting polyhedron is integral. As we have seen in our discussion of the column generation algorithm, even though we are dealing with an exponential number of subtrees, the essential reason is again Theorem 7.1.

Our discussion has also established the following more general result.

Theorem 7.3. *If*

$$W = \Big\{(x^1, w^1, \ldots, x^n, w^n) : (x^k, w^k) \in W^k \text{ for all } k \in V, \sum_{k \in V} x_j^k \leq \mathbf{1} \text{ for all } j \in V\Big\},$$

then

$$\mathrm{conv}(W) = \Big\{(x^1, w^1, \ldots, x^n, w^n) : (x^k, w^k) \in \mathrm{conv}(W^k) \text{ for all } k \in V, \sum_{k \in V} x_j^k \leq 1 \text{ for all } j \in V\Big\}$$

This generalization of Theorem 7.2 tells us that except for the packing inequalities, the polyhedron describing the convex hull of solutions to OCSP has no facet-defining inequalities that link the subtrees associated with different nodes k. Thus, it suffices to study each set W^k independently.

Finally, we obtain an important Corollary to Theorem 7.3.

Corollary. *There is a polynomial algorithm to optimize over W if and only if there is a polynomial algorithm to optimize over W^k for each k.*

In this section we have shown how the structure of the problem of packing subtrees of a tree leads not only to surprising integrality results, but also allows us to successfully adapt three of the most popular algorithmic strategies from integer programming for solving OCSP. Theorem 7.1 implies that we need not make any distinction between the linear programming and integer programming formulations of the tree packing problem when the underlying graph is itself a tree, since the feasible region of the linear program is an integer polyhedron. In Section 8 we examine the more complicated problem of packing subtrees or subgraphs into a general graph. In this more general setting, the linear programming relaxation of the analogous integer programming model does not have integer extreme points, and so we need to develop extensions of the algorithms presented in this section.

8. Packing subtrees of a general graph

In this section we show how to model several of the problems mentioned in Section 2 as problems of packing (or partitioning) subtrees, forests, or subgraphs of an arbitrary graph. Our purpose in this discussion is not to be comprehensive. Rather, we wish to (i) show that packing problems on general graphs arise in a number of important problem settings, (ii) show how to use the models and solution approaches we have considered in the previous sections to begin to address these problems, and (iii) suggest some of the results researchers have obtained for these problems. One lesson will become apparent: the polyhedral representations of these problems can be quite complex.

We again use the basic optimal capacitated subtree formulation

$$z^{\mathrm{PSG}} = \max\left\{\sum_k e^k x^k + \sum_k f^k w^k : \sum_k x_j^k \leq 1,\, (x^k, w^k) \in W^k \text{ for all } k\right\}.$$

We refer to this formulation as the *Packing Subtrees in a Graph* (PSG) model since the underlying graph need not be a tree. In this model, x^k again represents the incidence vector of nodes of the kth subgraph (usually a tree or a forest), and w^k typically represents edge or flow variables associated with this subgraph (k no longer necessarily represents a node v). As we show next, several generic application contexts are special cases of this model.

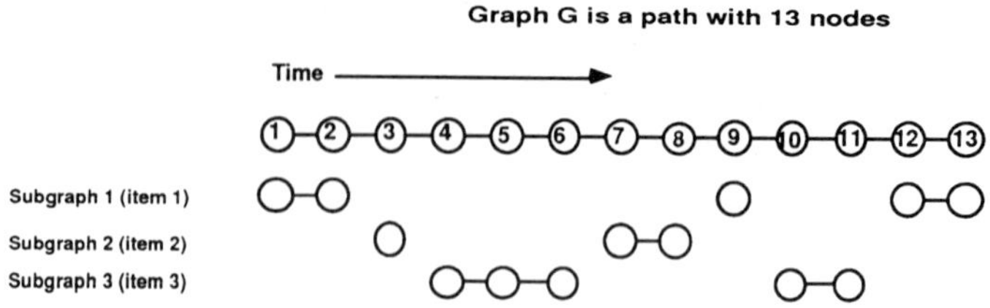

Fig. 19. Multi-item lot-sizing as packing forests (paths).

8.1. Applications

(1) Multi-item lot-sizing. Suppose we are given demands d_t^k for items $k = 1, \ldots, K$ over a time horizon $t = 1, \ldots, T$. All items must be produced on a single machine, and the machine can produce only one item in each period. Given production, storage and set-up costs for each item in each period, we wish to find a minimum cost production plan. The graph G we consider for this application is a path with K nodes, one for each time period. Each item defines a set of subgraphs on G; each is a set of paths from G. Figure 19 shows the graph and a possible set of subgraphs for a 13 period, 3 item problem (we could choose item K as a dummy item: when the machine is 'processing' this item, it is idle). In this figure, the production plan corresponding to this solution produces item 1 in periods 1, 2, 9, 12, and 13, item 2 in periods 3, 7, and 8, and item 3 in periods 4–6, 10, and 11. This solution indicates, for example, that in period 2 we produce the demand of item 1 for periods 2 through 8 and carry forward inventory of this item in periods 2–7.

(2) Clustering. Given a graph $G = (V, E)$, edge costs c_e for $e \in E$, node weights d_i for $i \in V$, positive integers K and C, we wish to find K clusters satisfying the property that the sum of the node weights in each cluster does not exceed C, in a way that minimizes the sum of the weights of edges between clusters (maximizes the sum of weights of edges within clusters). Figure 20 shows a feasible solution with 3 clusters and a capacity of 10.

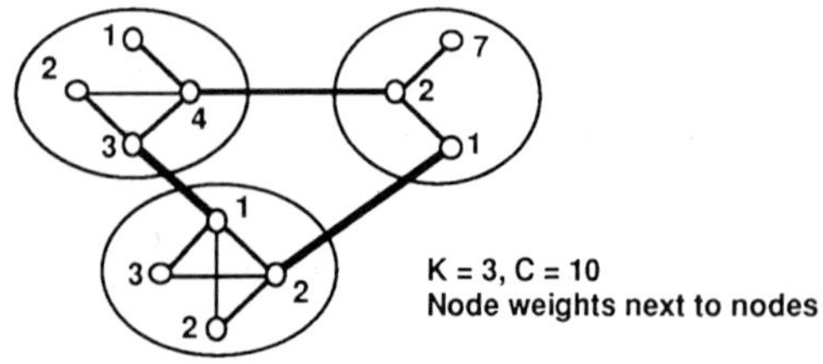

Fig. 20. Clustering solution with three clusters.

(3) The C-capacitated tree problem. Given a graph $G = (V, E)$, a root node 0, edge costs c_e for $e \in E$, find a minimum cost spanning tree in which each subtree on the nodes $V \setminus \{0\}$ contains at most C nodes. That is, if we delete the root node and all its incident edges, the spanning tree decomposes into a forest on the nodes $V \setminus \{0\}$. Each tree in this forest can contain at most C nodes.

(4) Capacitated trees. Given a graph $G = (V, E)$, a root node 0, edge costs c_e for every $e \in E$, positive demands d_i for $i \in V \setminus \{0\}$ and a positive integer C, we wish to find a minimum cost spanning tree satisfying the property that the total demand in each subtree on the nodes $V \setminus \{0\}$ does not exceed C.

(5) Capacitated vehicle routing. Given a graph $G = (V, E)$, a depot node 0, edge costs c_e for each $e \in E$, K vehicles of capacity C and client orders d_i for $i \in V \setminus \{0\}$, we wish to find a set of tours (cycles) for each vehicle that (i) each contain the depot, (ii) collectively contain all the nodes, (iii) are disjoint on the node set $V \setminus \{0\}$, and (iv) satisfy the property that the total demand on each cycle (that is, total amount delivered by each vehicle) does not exceed C.

(6) VLSI design. The global routing problem in VLSI design can be viewed as a problem of packing Steiner trees (or nets) with packing constraints imposed upon the edges. The problem, defined by a graph $G = (V, E)$, a collection of n terminal sets $T_k \subseteq V$ for $k = 1, 2, \ldots, n$, and edge (channel) capacities u_e, is to find a set of Steiner trees $\{S_k\}_{k=1}^n$ so that (i) S_k contains the terminal nodes T_k, and (ii) the number of Steiner trees containing edge e does not exceed u_e.

Models (1), (3) and (4) fit into the framework of the general model (PSG), and model (2) fits if we allow general subgraphs. If we remove the depot, model (5) requires the packing of trees (paths) on $G \setminus \{0\}$. Model (6) is somewhat different in that the packing is over the edges rather than the nodes.

8.2. Algorithmic strategies

How have researchers tackled each of these problems? Before attempting to answer this question, we first briefly discuss three general solution methods for solving PSG, each extending the algorithms we considered in the last section.

Algorithm A. Column generation algorithm

To correctly model PSG, we write the Master Problem with the additional 'convexity' constraints

$$z^A = \max\left\{\sum_k c^k\lambda^k : \sum_k A^k\lambda^k \leq \mathbf{1}, \mathbf{1}\lambda^k = 1,\ \lambda^k \geq 0 \text{ for } k \in V\right\}.$$

If we associate dual variables (π, σ) with the packing and convexity constraints, the kth subproblem becomes

$$\omega(\pi, \sigma) = \max\{e^k x^k + f^k w^k - \pi^t x^k - \sigma_k, (x^k, w^k) \in W^k\}. \tag{8.1}$$

In this model, A^k is the node-subtree incidence matrix of all feasible edge incidence vectors x^k satisfying the conditions $(x^k, w^k) \in W^k$ for some w^k, and $c^k = c^k(x^k) = \max_{w^k}\{e^k x^k + f^k w^k : (x^k, w^k) \in W^k\}$. When the column generation algorithm terminates, the vector $(\lambda^1, \ldots, \lambda^n)$ might not be integer, and so we might need to use an additional branching phase to solve the problem.

To implement a branching phase, we might wish to form two new problems (branch) by setting some fractional variable λ_i^k to 0 in one problem and to 1 in the other. Typically, however, in doing so we encounter a difficulty: when we set $\lambda_i^k = 0$ and then solve the subproblem (8.1) at some subsequent iteration, we might once again generate the subgraph $S_i \in W^k$ associated with the variable λ_i^k. To avoid this difficulty, we can add a constraint to W^k when solving the subproblem, chosen so that the subtree S_i will not be feasible and so the subproblem will no longer be able generate it. The following inequality will suffice:

$$\sum_{j \in S_i} x_j - \sum_{j \notin S_i} x_j \leq |S_i| - 1$$

since the solution with $x_j = 1$ for all nodes $j \in S_i$ and $x_j = 0$ for all nodes $j \notin S_i$ does not satisfy the inequality.

Unfortunately, this scheme leads to a highly unbalanced enumeration tree (setting the λ_i^k to zero eliminates very few feasible solutions). A better strategy is to choose two subgraphs S_i and S_j whose associated variables λ_i^k and $\lambda_j^{k'}$ are fractional. Consider any pair of nodes u, v satisfying the conditions $u, v \in S_i, u \in S_j$, but $v \notin S_j$. In an optimal solution either (i) u and v lie in the same subgraph, or (ii) they do not. In the first case, for each subproblem k we can impose the condition $x_u = x_v$; this condition implies that any subgraph S contains either both or neither of u and v. Since $v \notin S_j$, this constraint will eliminate the variable $\lambda_j^{k'}$ corresponding to the set S_j from the formulation. In the second case (ii), all subgraphs satisfy the condition $x_u + x_v \leq 1$; since S_i contains both nodes u and v, this constraint will eliminate the variable λ_i^k corresponding to the set S_i from the formulation. So, imposing the constraints $x_u = x_v$ or $x_u + x_v \leq 1$ on each subproblem permits us to branch as shown in Figure 21.

A third, related approach is to branch directly on the original problem variables, i.e., node or edge variables x^k or w^k.

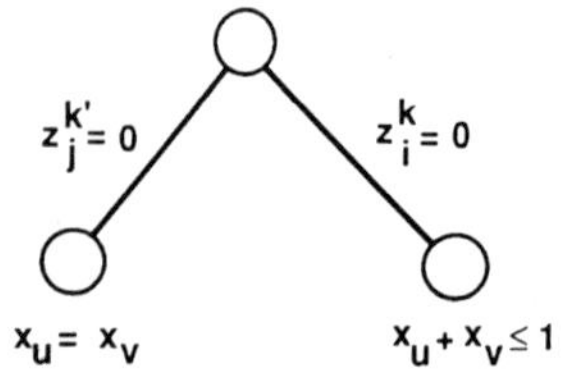

Fig. 21. A branching scheme.

Algorithm B. Lagrangian relaxation of the packing constraints

As we saw in Section 7, if we attach a Lagrange multiplier π_j to each packing constraint $\sum_k x_j^k \leq 1$, and bring these constraints into the objective function, we obtain a Lagrangian subproblem that decomposes into a separate problem for each k (since the packing constraint was the only constraint coupling the sets W^k). The resulting Lagrangian subproblem becomes $L(\pi) = \sum_k \mu^k(\pi) + \sum_i \pi_i$, and $\mu^k(\pi) = \{\max(e^k - \pi)x^k + f^k w^k : (x^k, w^k) \in W^k\}$.

As before, for each value of the Lagrange multipliers π, the optimal objective value $L(\pi)$ of the Lagrangian subproblem is an upper bound an the optimal objective value of PSG. To find the multiplier value providing the sharpest upper bound on the optimal objective value, we would solve the Lagrangian dual problem stated earlier:

$$z^B = \min_{\pi \geq 0} L(\pi).$$

To implement this approach we need an exact optimization algorithm for solving a linear optimization problem over W^k. We would use standard procedures to solve the Lagrangian dual problem and to continue from its solution using branch and bound.

Algorithm C. A cutting plane algorithm plus branch and bound

One way to approach this problem could be to apply the cutting plane algorithm from the previous section, that is, start with a partial polyhedral representation of each polyhedral set W^k and solve the linear programming relaxation of the problem. We then check to see if the solution to this problem satisfies all of the constraints defining each set W^k. If not, we determine a violated constraint for some W^k (i.e., solve the separation problem) and add a new constraint to the partial polyhedral representation of W^k.

Assuming the availability of an exact separation algorithm for W^k, the cutting plane algorithm C described in the last section will terminate with value:

$$z^C = \max \left\{ \sum_k e^k x^k + \right. \\ \left. + \sum_k f^k w^k : \sum_{k \in V} x^k \leq \mathbf{1}, (x^k, w^k) \in \mathrm{conv}(W^k) \text{ for all } k \right\}$$

However, in contrast to the earlier case, the final solution (x^k, w^k) might not be integer, and a further branch and bound, or branch and cut, phase might be necessary. To implement this approach, we could branch on the variables (x^k, w^k), and add other global cuts in standard fashion. (We describe several such cuts later in this section.)

As we have shown in Section 7, each of these three algorithms provides the same upper bound at the initial node of a branch and bound tree.

Theorem 8.1 *For problem PSG, the bounds satisfy* $z^{PSG} \leq z^A = z^B = z^C$.

In practice, the effectiveness of these algorithms depends in part on how good an approximation

$$\left\{(x, w) : \sum_{k \in V} x^k \leq \mathbf{1}, (x^k, w^k) \in \text{conv}(W^k) \text{ for all } k\right\}$$

provides to conv(W). Put somewhat differently, if z^A is a tight upper bound on z^{PSG}, the branch and bound tree might permit us to rapidly find an optimal solution and prove its optimality. To tackle the more difficult problems, it is usually necessary to find 'strong' valid inequalities (e.g., facets) for conv(W), linking together the different sets W^k. We would also need to integrate these inequalities into the Lagrangian or cutting plane approaches, thereby also improving the branching phase of these algorithms.

We now discuss some approaches that researchers have used for solving the six example problems we introduced at the outset of this section, and comment on the usefulness of the model PSG.

(1) Multi-item lot-sizing. In this context, the graph G is a path from $1, \ldots, n$. For each item k, we need to find a set of intervals (or subpaths) in which item k is produced. The sets W^k assume the form:

$$W^k = \{(x^k, s^k, v^k) : s^k_{t-1} + v^k_t = d^k_t + s^k_t \text{ for all } t, \;\; v^k_t \leq M x^k_t \text{ for all } t, \\ s^k, v^k \geq 0, x^k_j \in \{0, 1\} \text{ for all } j \text{ and } k\}$$

with $x^k_t = 1$ if item k is produced in period t; $w^k = (s^k, v^k)$ are the unit stock and production variables, and

$$c^k(x^k) = \min_{s,v}\left\{\sum_t (p^k_t v^k_t + h^k_t s^k_t + f^k_t x^k_t) : (s^k, v^k, x^k) \in W^k\right\}.$$

For this problem, both optimization and separation over W^k are well understood and can be implemented rapidly. In particular, in Theorem 7.2 we showed that formulation (7.2)–(7.4) provides an integer polyhedron for the uncapacitated lot-sizing problem (ULS) based upon packing subpaths of a path. Researchers have successfully used both cutting plane and Lagrangian relaxation based algorithms, as well as some heuristic algorithms based on column generation, in addressing these problems. Little or nothing is known about facet-defining inequalities linking the items (and so the sets W^k).

(2) Clustering. For this problem class, each subgraph in the partition is totally unstructured. The set W^k is of the form:

$$W^k = \left\{(x^k, y^k) : \sum_{i \in V} d_i x^k_i \leq C, \;\; y^k_e \leq x^k_i \text{ for } e = (i, j) \in E, \right. \\ \left. y^k_e \leq x^k_j \text{ for } e = (i, j) \in E, \;\; x^k_j, y^k_e \in \{0, 1\} \text{ for all } j, e, \text{ and } k.\right\}$$

The variables x_i^k and y_e^k indicate whether node i or edge e belongs to the kth cluster. The constraints impose a common capacity C on each cluster, and state that any edge e can belong to a cluster only if both its endpoints do. One approach for solving this problem has been to use column generation (Algorithm A), using branch and cut with valid inequalities for W^k to solve the subproblems to optimality. We will describe one family of valid inequalities that can be used to improve the linear programming approximation of W^k in the subproblem. It is important to note that for the model we are considering (with an unlimited number of clusters), it is necessary to solve only one subproblem (they are all the same).

Proposition 8.2. *Let $T = (V', E')$ be a subtree of G whose node set V' forms a minimal cover, i.e., $\sum_{i \in V'} d_i > C$ and no subset of V' satisfies this property. Let $deg(i)$ be the degree of node i in T. Then the inequality*

$$\sum_{e \in E'} y_e^k \leq \sum_{i \in V'} (deg(i) - 1) x_i^k$$

is valid for W^k.

Because V' forms a cover, some node $r \in V'$ must satisfy the condition $x_r^k = 0$. Suppose we root the tree arbitrarily at node r, and sum the inequalities $y_e^k \leq x_i^k$ over all edges $e \in E'$ with i as the endpoint of edge e closest to node r: we obtain $\sum_{e \in E'} y_e^k \leq \sum_{i \in V'} (deg(i) - 1) x_i^k + x_r^k$. (Note that if we collect terms, the coefficient of node r is $deg(r)$.) Since $x_r^k = 0$, the inequality is valid. But since the inequality is independent of the node r satisfying the condition $x_r^k = 0$, all feasible solutions satisfy it.

Limited computational experience has shown that the addition of these inequalities to the linear programming relaxation of the subproblem can be effective in practice. One study, modeling a problem in compiler design, found that on 7 out of 10 problems, the final Restricted Master linear program in the column generation approach gave an integer solution; for the other three problems, the column generation approach found a good feasible solution with a small amount of branching.

Some models use a specialized objective function: they wish to minimize a weighted sum of the edges between the clusters. For these problems, we might use an alternative approach, using the node variables x^k and edge variables w_e ($= 1 - \sum_k y_e^k$). That is, we no longer keep track of which cluster each edge belongs to, but simply keep track of the edges in the cutsets between the clusters. One advantage of this approach is that it would permit us to create branch and cut algorithms by drawing upon the extensive work conducted on facets of cut polytopes.

The next three problems are all closely related. The C-capacitated tree problem is the special case of the general capacitated tree problem with $d_i = 1$ for all $i \in V \setminus \{0\}$. The vehicle routing problem is the restriction of the general capacitated tree problem in which each subtree must essentially be a path (whose endpoints are joined to the depot node 0 to form a tour).

The most successful approaches to date for these problems have all avoided an explicit representation using distinct vehicle variables x^k. Researchers have used column generation and dynamic programming to treat vehicle routing problems with tightly constrained schedules (see Desrosiers, Dumas, Solomon & Soumis [1995]). have worked entirely in the space of the edge variables y_e for $e \in E$. This approach implicitly treats the weights and capacities using 'generalized subtour inequalities' of the form:

$$\sum_{e \in E(S)} y_e \leq |S| - f(S).$$

In this expression, $f(S)$ is the minimum number of trees, or vehicles, needed to satisfy the demand of all nodes in S. Thus $f(S) = 1$ for the tree problem of Section 3, and $f(S) = \lceil \sum_{i \in S} d_i / C \rceil$ for the capacitated tree problems and vehicle routing problems since any feasible solution must allocate at least $f(S)$ vehicles (subtrees for the capacitated tree problem) to the nodes in S, implying that the edges in $E(S)$ must contain at least $f(S)$ components. Researchers have obtained even sharper inequalities (that is, with larger values for $f(S)$) by solving the NP-hard bin-packing problem of finding the minimum number of bins of capacity C needed to contain the set of weights $\{d_i\}_{i \in S}$.

(3) The C-capacitated tree problem. The starting formulation is:

$$\min \sum_{e \in E} c_e y_e$$

subject to

$$\sum_{e \in E(S)} y_e \leq |S| - \lceil |S|/C \rceil \quad \text{for all } S \subset V \setminus \{0\}$$

$$\sum_{e \in E(S)} y_e \leq |S| - 1 \quad \text{for all } S \subset V \text{ with } 0 \in S$$

$$\sum_{e \in E} y_e = n - 1$$

$$y_e \in \{0, 1\} \quad \text{for all } e \in E.$$

We refer to the convex hull of the (integer solutions) of this problem as the capacitated tree polyhedron.

Researchers have tackled this problem using a set of facet-defining inequalities as cutting planes. The inequalities are numerous and fairly complex, so we simply illustrate a few of them with examples. Each of the inequalities is determined by relationships between edges in particular subgraphs (so called supporting subgraphs) of the underlying network. This is an approach we have used before; for example, in studying the minimum spanning tree problem, we considered subtour inequalities. In this case, the subgraphs consist of all edges $E(S)$ with both endpoints in any node set S and the inequality stated that no feasible solution could use more than $|S| - 1$ of these edges.

Figure 22 shows an example of the C-capacitated tree problem with 7 nodes and with $C = 3$. The supporting *multistar* graph divides the nodes into two classes:

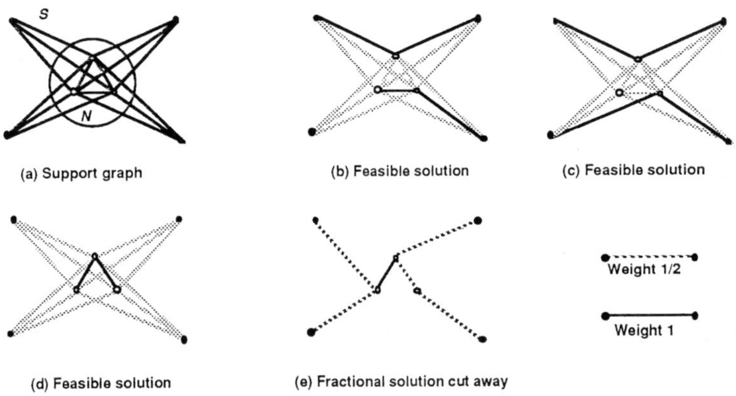

Fig. 22. Multistar with $C = 3$.

a set of nucleus nodes N all connected to each other and a set of satellite nodes S, each connected to every node in the nucleus. If $0 \notin N$ and $0 \notin S$, the only feasible solutions are those shown in Figure 22b–d as well as subsets of these solutions. Note that if we let $E(N, S)$ denote the set of edges with one endpoint in N and the other in S, then every feasible solution satisfies the *multistar inequality*

$$3 \sum_{e \in E(N)} y_e + \sum_{e \in E(N,S)} y_e \leq 6.$$

The general version of this inequality is

$$K \sum_{e \in E(N)} y_e + \sum_{e \in E(N,S)} y_e \leq (C-1)|N|.$$

The constant K in this inequality is required to be less than C and its value depends upon specific problem data (for our example, $K = C$). The fractional solution shown in Figure 22(e) satisfies all the constraints of the starting formulation, but not the multistar inequality.

Figure 23 shows a second example: in this case $C = 5$ and the supporting *clique cluster* graph has three cliques (complete graphs) C_1, C_2, and C_3, all sharing exactly one common node and none containing node 0. The figure shows feasible solutions that satisfy and a fractional solution that does not satisfy the valid *clique cluster inequality*

$$\sum_{e \in C_1} y_e + \sum_{e \in C_2} y_e + \sum_{e \in C_3} y_e \leq 6.$$

The general inequality for t cliques is

$$\sum_{1 \leq j \leq t} \sum_{e \in C_j} y_e \leq \text{constant}.$$

The constant is determined by the structure of the clique cluster and the value of C.

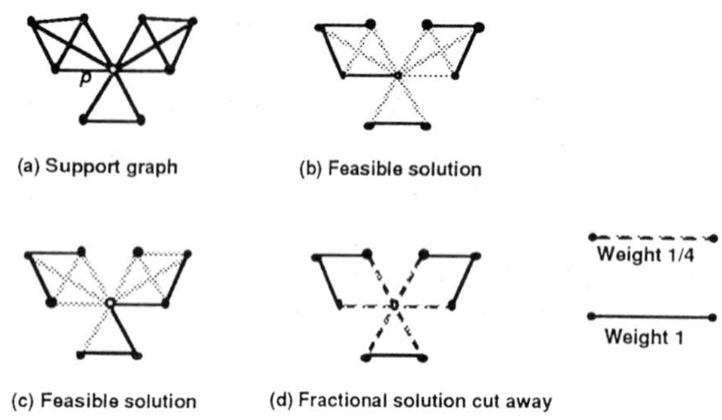

Fig. 23. Clique cluster with $C = 5$.

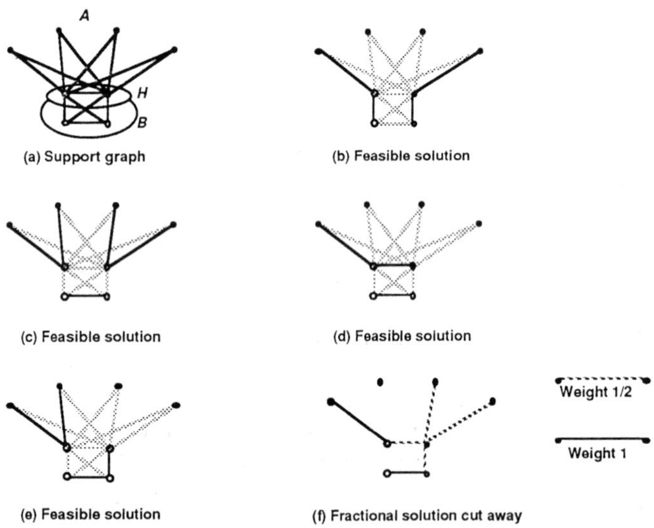

Fig. 24. Ladybug with $C = 3$.

Figure 24 shows a third example: in this case $C = 3$ and the supporting *ladybug* graph has a set B of body nodes all connected to each other, a set H of head nodes and a set A of antenna nodes. The head and antenna nodes form a multistar and each body node is connected to each head node. We assume that the ladybug graph does not contain node 0. The figure shows several feasible solutions and a fractional solution that satisfies all the other constraints we have considered but not the valid *ladybug* inequality which is

$$2 \sum_{e \in E(B)} y_e + 2 \sum_{e \in E(B,H)} y_e + 3 \sum_{e \in E(H)} y_e + \sum_{e \in E(H,A)} y_e \leq 6.$$

The ladybug inequality in general is similar: the edges in $E(H)$ have a weight of C, the edges in $E(H, A)$ have a weight of 1. The edges $E(B)$ and $E(B, H)$ have the same weight d. The value of d and the right-hand side of the inequality all depend upon the values of C, $|B|$, and $|H|$.

All three classes of these inequalities define facets of the C-capacitated tree polyhedron, assuming mild restrictions on C and the sizes of the support graphs. Because these conditions are complicated, we will not present them in this discussion.

Computational experience has shown that multistar and certain partial multistar extensions of them are useful in a cutting plane approach for solving the C-capacitated tree problem.

(4) The capacitated tree problem. The starting formulation is:

$$\min \sum_{e \in E} c_e y_e$$

subject to

$$\sum_{e \in E(S)} y_e \leq |S| - f(S) \qquad \text{for all } S \subset V \setminus \{0\}$$

$$\sum_{e \in E(S)} y_e \leq |S| - 1 \qquad \text{for all } S \subset V \text{ with } 0 \in S$$

$$\sum_{e \in \delta(i)} y_e \geq 1$$

$$y_e \in \{0, 1\} \qquad \text{for all } e \in E.$$

The function $f(S)$ can be any of the functions we introduced before, for example, $f(S) = \lceil \sum_{i \in S} d_i / C \rceil$. In this model $\delta(i)$ denotes the edges with one endpoint at node i. We can view one approach for solving this problem as a combination of a cutting plane algorithm and Lagrangian relaxation. The first difficulty is the exponential number of constraints. Unlike the tree problem, no polynomial separation algorithm is known for the generalized subtour inequalities, so researchers have used heuristics to find violated inequalities. Suppose we have added constraints corresponding to the set $S^1, \ldots, S^t$ to the formulation. One solution approach would be to remove these constraints by incorporating them into the objective function with Lagrange multipliers giving, as a Lagrangian subproblem, a branching problem:

$$\min \sum_{e} c_e y_e - \sum_{t=1}^{r} \left[\sum_{e \in E(S^r)} y_e - (|S^r| - f(S^r)) \right] \lambda_t$$

subject to

$$\sum_{e \in \delta(i)} y_e \geq 1 \qquad \text{for } i \in V \setminus \{0\}$$

$$\sum_{e \in E} y_e = n - 1$$

$$y_e \in \{0, 1\} \qquad \text{for } e \in E$$

which we could then solve using the algorithm we presented in Section 5.

(5) Capacitated vehicle routing. The starting formulation is very similar to that of the previous model, and that of the travelling salesman problem. Let 0 be the depot and $V_0 = V \setminus \{0\}$. With a fixed number K of vehicles, we have:

$$\min \sum_{e \in E} c_e y_e \tag{8.2}$$

subject to

$$\sum_{e \in \delta(0)} x_e = 2K \tag{8.3}$$

$$\sum_{e \in \delta(i)} x_e = 2 \qquad \text{for } i \in V_0 \tag{8.4}$$

$$\sum_{e \in \delta(S)} x_e \geq 2\alpha(S) \qquad \text{for } S \subseteq V_0,\ S \neq \phi \tag{8.5}$$

$$x_e \in \{0, 1\} \qquad \text{for } e \in E(V_0),\ x_e \in \{0, 1, 2\} \text{ for } e \in \delta(0). \tag{8.6}$$

In this model $\alpha(S) \in \{\lceil \sum_{i \in S} d_i / C \rceil, r(S), R(S)\}$ with $\lceil \sum_{i \in S} d_i / C \rceil \leq r(S) \leq R(S)$. The first term in this expression for $\alpha(S)$ is the basic capacity bound; $r(S)$ is the bin packing bound, i.e., the minimum number of bins needed to pack the set of demands $\{d_i\}_{i \in S}$; and $R(S)$ is the same bound taking into account the requirement that we must pack all the demands into K bins (this bound accounts for the demand $\{d_i\}_{i \notin S}$). Note that for the travelling salesman problem, $C = \infty$, and then (8.5) becomes $\sum_{e \in \delta(S)} x_e \geq 2$, the basic cut (or, equivalently, subtour elimination) constraint.

A generalization of the so-called *comb inequalities* from the TSP problem applies to this problem. The support graph for these inequalities contains a *handle* H and a set $\{T_j\}_{j=1}^{s}$ of *teeth* satisfying the conditions:

$$T_i \cap T_j = \phi$$
$$T_j \cap H \neq \phi$$
$$T_j, H \subseteq V_0$$
$$\sum_{i \in T_j} d_i \leq C.$$

See Figure 25.

Let $2P(H)$ be the minimum number of times any set of subpaths must intersect H in order for any solution to satisfy all the demands in H given that the demands in each tooth T_j are consecutive on one of the subpaths; see Figure 25.

Proposition 8.3. *The generalized comb inequality*

$$\sum_{e \in \delta(H)} x_e \geq 2P(H) - \sum_{j=1}^{s} \left(\sum_{e \in \delta(T_j)} x_e - 2 \right)$$

is valid for CVRP.

Note that if a vehicle visits the clients in a tooth one after the other, $\sum_{e \in \delta(T_j)} x_e = 2$ for each j, and the inequality is valid by definition of $P(H)$. If one or more

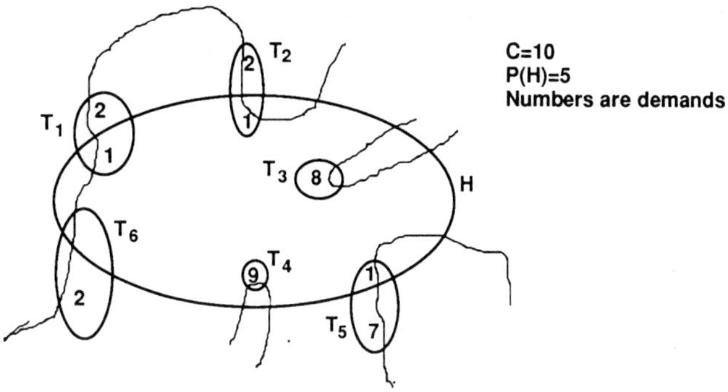

Fig. 25. A feasible solution and a comb.

vehicles visits the tooth T_j more than once, then $\sum_{e\in\delta(T_j)} x_e = 2 + k_j$ for some integer $k_j \geq 1$. It is then necessary to verify that this solution never reduces the intersections with the boundary of H by more than $\sum_{j=1}^{s} 2k_j$.

Observe that for the TSP, when $T_j \setminus H \neq \phi$ for $j = 1, \ldots, s$ and s is odd, then $2P(H) = s + 1$, and so this inequality becomes the usual comb inequality.

(6) Packing Steiner trees. Packing Steiner trees is important in VLSI design, and recently researchers have solved some previously unsolved problems to optimality using a polyhedral approach. Suppose that the edge capacities c_e all equal 1, so we are seeking a minimum cost packing of edge disjoint Steiner trees. We present two results concerning valid inequalities that have proven to be useful in recent computational successes.

Proposition 8.4. *Every nontrivial facet-defining inequality of the Steiner tree polyhedron yields a facet-defining inequality of the Steiner tree packing polyhedron.*

This result implies in particular that the Steiner partition inequalities (Proposition 6.3) provide facets for this problem.

The next class of inequalities involve more than a single Steiner tree. Consider two terminal disjoint sets T_1 and T_2. We refer to a cycle $F \subseteq E$ as an *alternating cycle* with respect to T_1 and T_2 if $F \subseteq E(T_1, T_2)$. We refer to an edge (u, v) as a *diagonal edge* if $u, v \in V(F)$, but $(u, v) \notin F$. We let $y_e^k = 1$ if edge e is in Steiner tree k.

Proposition 8.5. *Let F be an alternating cycle with respect to T_1 and T_2, and $F_1 \subseteq E(T_2)$, $F_2 \subseteq E(T_1)$ be two sets of diagonal edges. Then*

$$\sum_{e\in E\setminus(F\cup F_1)} y_e^1 + \sum_{e\in E\setminus(F\cup F_2)} y_e^2 \geq \frac{|V(F)|}{2} - 1$$

is a valid inequality for the Steiner tree packing polyhedron.

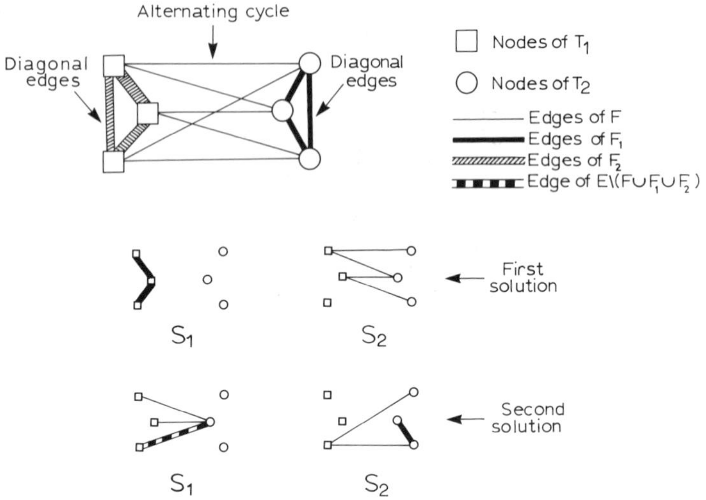

Fig. 26. Alternating cycle for packing Steiner trees.

Figure 26 shows an alternating cycle F of length 6, as well as two tight feasible solutions in which S_1 and S_2 are the Steiner trees spanning T_1 and T_2.

9. Trees-on-trees

Each of the models we have examined so far uses a single type of facility (edge) to construct a tree, or a packing of trees, in a given graph. Some applications need to distinguish between different types of edges. For example, electrical power systems often must connect major users with high voltage (or highly reliable) transmission lines, but can use cheaper, low voltage (or less reliable) lines to connect other users. Roadway systems often need to use highways to connect major cities, but can use secondary roads to connect smaller cities.

These applications give rise to a set of hierarchical models that are generalizations of the models we have considered earlier in this chapter. To illustrate the analysis of this class of models, we will consider a particular type of hierarchical problem.

9.1. Tree-on-tree model

Suppose we are given an undirected graph $G = (V, E)$ with two types of nodes, primary P and secondary S: $P \cup S = V$ and $P \cap S = \phi$. We wish to find a minimum cost spanning tree in G. The problem differs from the usual spanning tree problem, however, because we need to designate any edge in the spanning tree either as a primary (high capacity, more reliable) edge or as a secondary (low capacity, less reliable) edge. Designating edge $\{i, j\}$ as a primary edge costs

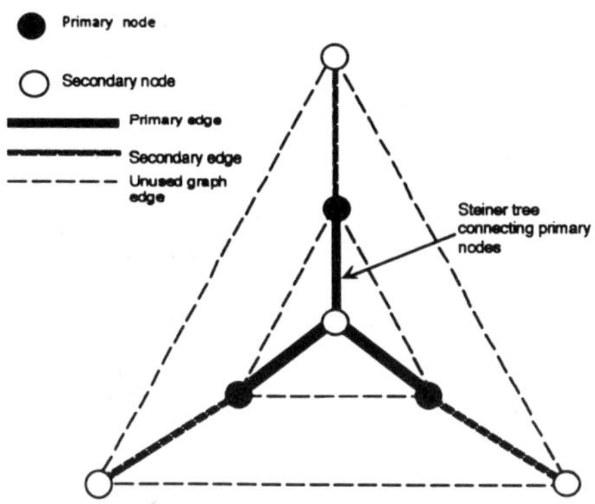

Fig. 27. Steiner tree on a spanning tree.

$a_{ij} \geq 0$ and as a secondary edge costs $b_{ij} \geq 0$; we assume $b_{ij} \leq a_{ij}$. The spanning tree we choose must satisfy the property that the unique path joining every pair of primary nodes contains only primary edges.

As shown in Figure 27, we can interpret the solution to this 'tree-on-tree' problem as a Steiner tree with primary edges superimposed on top of a spanning tree. The Steiner tree must contain all the primary nodes (as well, perhaps, as some secondary nodes).

Note that if the costs of the secondary edges are zero, then the problem essentially reduces to a Steiner tree problem with edge costs a_{ij} (the optimal solution to the tree-on-tree problem will be a Steiner tree connected to the other nodes of the network with zero-cost secondary edges). Therefore, the tree-on-tree problem is at least as hard as the Steiner tree problem and so we can expect that solving it will be difficult (at least from a complexity perspective) and that its polyhedral structure will be complicated. If the costs a and b are the same, the problem reduces to a minimum spanning tree problem. Therefore, the tree-on-tree problem encompasses, as special cases, two of the problems we have considered previously in this chapter.

In this section, we develop and analyze a heuristic procedure for solving this tree-on-tree problem; we also analyze a linear programming representation of the problem. In the context of this development, we show how to use some of the results developed earlier in this chapter to analyze more complex models.

To model the tree-on-tree problem, we let x_{ij} and y_{ij} be 0–1 variables indicating whether or not we designate edge $\{i, j\}$ as a primary or secondary edge in our chosen spanning tree; both these variables will be zero if the spanning tree does not include edge $\{i, j\}$. Let S denote the set of incidence vectors of spanning trees on the given graph and let ST denote the set of incidence vectors of feasible Steiner trees on the graph (with primary nodes as terminal nodes and secondary nodes as Steiner nodes). Let $x = (x_{ij})$ and $y = (y_{ij})$ denote the vectors of decision variables. In addition, let $c_{ij} = a_{ij} - b_{ij}$ denote the incremental cost of upgrading a secondary edge to a primary edge. With this notation, we can formulate the tree-on-tree problem as the following integer program:

$$z^{\mathrm{ip}} = \min\ cx + by$$

subject to

$$x \leq y$$
$$x \in ST$$
$$y \in S.$$

The forcing constraint $x_{ij} \leq y_{ij}$ states that if the chosen Steiner tree contains edge $\{i, j\}$, that is, $x_{ij} = 1$, then the chosen spanning tree must also contain this edge. Thus, if edge $\{i, j\}$ belongs to the Steiner tree, then $x_{ij} = y_{ij} = 1$ and so, as required, edge $\{i, j\}$ contributes $a_{ij} = c_{ij} + b_{ij}$ to the solution's cost.

9.2. Heuristic Analysis

In an attempt to solve the tree-on-tree problem, we consider a heuristic that chooses the better solution obtained from two other heuristics:

Spanning tree heuristic: Solve a minimum spanning tree problem with respect to the costs a_{ij} and designate each edge in the minimum spanning tree as a primary edge.

Steiner tree completion heuristic: Solve a Steiner tree problem (optimally or approximately) with the primary nodes P as the terminal (required) nodes and designate every edge on this tree T as a primary edge; using the costs b_{ij}, find a least cost spanning tree T' that contains T; designate each edge in $T \setminus T'$ as a secondary edge.

Composite heuristic: Choose the smaller cost solution found by the spanning tree and Steiner tree completion heuristics.

A few comments are in order concerning these heuristics. After using the spanning tree heuristic, we might try to improve upon the solution by downgrading some primary edges into secondary edges: we can downgrade every edge not on a path joining any two primary nodes. Even though we would use this procedure in practice, examples show that this improvement does not affect the worst-case analysis we will be considering, and so we use a simpler procedure without this downgrading phase. Second, to complete the Steiner tree into a minimum spanning tree in the Steiner tree completion heuristic, we can use a version of the spanning tree greedy algorithm: starting with the edges in the Steiner tree as members of the spanning tree, add cheapest cost secondary edges one at a time that do not create cycles with the edges already chosen.

We might consider two versions of the problem, an *unrelated cost model* in which the costs a and b are arbitrary, and a *proportional cost model* in which the costs of primary to secondary edge are proportional to each other; that is, for some constant r, $a_{ij} = rb_{ij}$ for all edges $\{i, j\}$. First, let us consider the proportional cost model.

Let z^S, z^{ST}, z^{CH} be the objective values of the solutions produced by the spanning tree, the Steiner tree completion, and the composite heuristics. To streamline our notation, assume by scaling that $z^S = r$. Since $z^S = r$ is the cost of a minimum spanning tree containing primary edges, the cost of a minimum spanning tree with secondary edges is 1. We also let s, an unknown, denote the cost of an optimal Steiner tree connecting the primary nodes with *secondary edges*.

In terms of this notation, if we solve the Steiner tree to optimality, we have,

$$z^S = r$$
$$z^{ST} \leq rs + 1.$$

The specified upper bound on z^{ST} is valid because the incremental cost of completing an optimal Steiner tree T, at least cost, using secondary edges can be no more than the cost of a minimum spanning tree with secondary edges. (To establish this result, note that if we set the cost of every edge $\{i, j\}$ in T from $b_{ij} \geq 0$ to zero, the greedy algorithm on the entire graph could generate the tree produced by the completion procedure as a minimum spanning tree. The assertion is true since reducing some edge costs from b_{ij} to 0 cannot increase the length of a minimum spanning tree.)

If we eliminate the forcing constraints $x \leq y$ from the integer programming formulation of the tree-on-tree problem, the problem decomposes into a minimum spanning tree problem with secondary costs b and a Steiner tree problem with respect to incremental costs c. Since we are considering the proportional cost model, $c = a - b = rb - b = b(r - 1)$. Therefore, the cost of an optimal Steiner tree with respect the incremental costs c is $(r - 1)s$. Since removing constraints cannot increase the optimal costs, we obtain a lower bound on z^{ip}:

$$z^{\text{ip}} \geq 1 + (r - 1)s.$$

We next use the previous upper and lowerbounds to analyze the composite heuristic.

Theorem 9.1. *For the tree-on-tree problem with proportional costs, if we solve the Steiner tree problem in the Steiner tree completion heuristic to optimality, then*

$$\frac{z^{CH}}{z^{\text{ip}}} \leq \frac{4}{3}.$$

Proof. Combining the upper bounds on z^S and z^{ST} and the lower bound on z^{ip} shows that

$$\frac{z^{CH}}{z^{\text{ip}}} \leq \frac{\min\{r, rs + 1\}}{1 + (r - 1)s}.$$

For a given value of r, the first term on the right-hand side of this expression decreases with s and the second term increases with s. Therefore, we maximize the right-hand side of this expression by setting $r = rs + 1$ or $s = (r - 1)/r$. With this choice of s, the bound on z^{CH}/z^{ip} becomes

$$\frac{z^{CH}}{z^{\text{ip}}} \leq \frac{r}{1 + (r - 1)s} = \frac{r^2}{r + (r - 1)^2}.$$

To maximize the right-hand side over r, we set the derivative of the right-hand side to zero, giving $r = 2$ and so $z^{CH}/z^{\text{ip}} \leq 4/3$. □

Note that when $|P| = 2$, the Steiner tree problem becomes a shortest path problem and so we can solve it to optimality. We can also solve the Steiner tree problem to optimality for specialized classes of network, in particular so-called series-parallel networks. Therefore, the 4/3 bound applies to these situations.

In general, we will not be able to solve the Steiner tree problem to optimality, but will instead use an approximation procedure to solve the problem. Let us suppose that for the problem class that we wish to investigate, we can obtain a heuristic solution to the Steiner tree problem with a guaranteed performance bound of ρ; that is, the cost of the solution we generate is never more than $\rho \geq 1$ times the cost of an optimal solution. For example, as we have seen in Section 6, for problems satisfying the triangle inequality, we can use an heuristic with a performance guarantee of $\rho = 2$. For Euclidean graphs $\rho = 2$ and, as we have just noted, for series-parallel graphs, $\rho = 1$.

In this case, we obtain the following upper bound on the cost z^{ST} of the Steiner tree completion heuristic:

$$z^{ST} \leq \rho r s + 1.$$

An analysis similar to the one we used to analyze the situation when we could solve the Steiner tree problem optimally permits us to establish the following result.

Theorem 9.2. *For the tree-on-tree problem with proportional costs, if we solve the Steiner tree problem in the Steiner tree completion heuristic using a heuristic with a performance guarantee of ρ, then*

$$\frac{z^{CH}}{z^{\text{ip}}} \leq \frac{4}{4-\rho} \text{ if } \rho \leq 2$$

and

$$\frac{z^{CH}}{z^{\text{ip}}} \leq \rho \text{ if } \rho > 2$$

For the unrelated cost model, a similar analysis permits us to obtain the following result.

Theorem 9.3. *For the tree-on-tree problem with unrelated costs, if we solve the Steiner tree problem in the Steiner tree completion heuristic using a heuristic with a performance guarantee of ρ, then*

$$\frac{z^{CH}}{z^{\text{ip}}} \leq \rho + 1.$$

Although we will not establish this fact in this discussion, examples show that the bound in Theorems 9.1, 9.2, and 9.3 are tight — that is, some examples achieve the worst-case bounds.

9.3. Linear programming bounds

Let P_{ST} and P_S be any polyhedral approximations to the spanning tree and Steiner tree polytopes in the sense that $ST \subseteq P_{ST}$ and $S \subseteq P_S$. For example, these polyhedra can be any of the possibilities we have considered in Sections 3 and 6. With respect to these polyhedra, we can consider the following linear programming relaxation of the tree-on-tree problem:

$$z^{\text{lp}} = \min\ cx + by$$

subject to

$$x \leq y$$
$$x \in P_{ST}$$
$$y \in P_S.$$

Note that the polyhedra P_{ST} and P_S contain the constraints $0 \leq x_{ij} \leq 1$, and $0 \leq y_{ij} \leq 1$ for all edges $\{i, j\}$.

To see how well this linear program represents the tree-on-tree problem, we would like to bound $z^{\text{ip}}/z^{\text{lp}}$. The bound we will obtain depends upon how well the polyhedra P_{ST} and P_S represent ST and S. As we have seen in Section 3, we can choose several equivalent polyhedra so that P_S is the convex hull of S. Suppose we choose one of these polyhedra. Moreover, suppose that our choice P_{ST} permits us to obtain a performance guarantee of $\theta \geq 1$ whenever we optimize any linear function over P_{ST}, that is for all choices γ of objective functions coefficients,

$$\min\{\gamma x : x \in ST\} \leq \theta \min\{\gamma x : x \in P_{ST}\}.$$

Note that if we eliminate the forcing constraints $x \leq y$ from the linear programming relaxation, then the problem separates into two independent linear programming subproblems, one defined over P_S with cost coefficients b and one defined over P_{ST} with cost coefficients c. Since P_S equals the convex hull of spanning tree solutions, the first linear program has an optimal objective value equal to 1, the value of a minimum spanning tree using secondary edges. Our performance guarantee implies that the optimal objective value equal of the second linear program is no less than $(r-1)s/\theta$. (Recall that $(r-1)s$ is the cost of an optimal Steiner tree connecting the primary nodes using incremental costs c.)

As before, eliminating the forcing constraints cannot increase the optimal objective value, so we obtain the following lower bound on the objective value of the linear programming relaxation:

$$z^{\text{lp}} \geq 1 + \left[\frac{(r-1)s}{\theta}\right].$$

As we have noted before, if we solve the Steiner tree problem to optimality, $z^{\text{ip}} \leq z^{ST} \leq rs + 1$; moreover, $z^{\text{ip}} \leq r$, the cost of a minimum spanning tree with primary edges. Combining these results gives us the bound

$$\frac{z^{\text{ip}}}{z^{\text{lp}}} \leq \frac{\theta \min\{r, rs+1\}}{\theta + (r-1)s}.$$

Using an analysis similar to that used in the development of Theorems 9.1 and 9.2 permits us to establish the following result.

Theorem 9.4. *Suppose that for any vector γ, the cost of an optimal Steiner tree with edge costs γ is at most θ times the optimal value of the linear program* $\min\{\gamma y : y \in P_{ST}\}$ *defined over the polyhedron P_{ST}. Then for the tree-on-tree problem with proportional costs,*

$$\frac{z^{\mathrm{ip}}}{z^{\mathrm{lp}}} \le \frac{4}{4-\theta} \quad \text{if } \theta \le 2$$

and

$$\frac{z^{\mathrm{ip}}}{z^{\mathrm{lp}}} \le \theta \quad \text{if } \theta > 2.$$

Similarly, we can obtain the following result for the unrelated cost model.

Theorem 9.5. *Suppose that for any vector γ, the cost of an optimal Steiner tree with edge costs γ is at most θ times the optimal value of the linear program* $\min\{\gamma y : y \in P_{ST}\}$ *defined over the polyhedron P_{ST}. Then for the tree-on-tree problem with unrelated costs,*

$$\frac{z^{\mathrm{ip}}}{z^{\mathrm{lp}}} \le \theta + 1.$$

In Section 6, we analyzed one formulation of the Steiner tree problem with $\theta = 2$. Theorems 9.4 and 9.5 show that by using this same formulation for the tree-on-tree problem, we obtain the same worst-case bound $4/(4-2) = 2$ for the proportional cost tree-on-tree problem and a bound of 3 for the unrelated cost model.

Using flow models to represent P_{ST} and P_S and a dual ascent procedure to approximately solve the resulting linear programming relaxation of the tree-on-tree problem combined with an associated linear programming-based heuristic, researchers have been able to solve large-scale problems (with up to 500 nodes and 5000 arcs) to near optimality (guaranteed within 2% of optimality). This computational experience is comparable (in problem size, performance guarantee, and algorithm execution time) to the computational experience for solving the Steiner tree subproblem itself.

10. Summary

In this chapter, we have considered a variety of tree optimization problems. Motivated by applications in telecommunications, production planning, routing, and VLSI design introduced in Section 2, we set out to examine a variety of issues in modeling and algorithm design. Rather than attempting to summarize all the results we have presented, in these concluding remarks, we will focus on a few lessons to be learned from our discussion — about trees, about modeling, or about algorithm design.

The algorithms we have considered either directly exploit a problem's underlying combinatorial structure and/or build upon insights derived from the problem's representation(s) as mathematical programs. For example, typically, when a problem is defined on a tree, a dynamic program will provide an efficient solution procedure. Algorithms in this category include dynamic programs for the optimal rooted subtree problem (Section 4.1) and for packing subtrees in a tree (Section 7). Greedy algorithms traditionally exploit a problem's underlying combinatorial structure: for example, the fact that whenever we add an edge to a spanning tree, we create a new tree by deleting any edge in the cycle that this edge produces. The basic greedy algorithm for the minimum spanning tree problem (Section 3) exploits this property and the modified greedy algorithm (with node shrinking) for the optimal branching problem (Section 5.2) exploits the combinatorial property that if the node greedy solution contains a cycle C, then the problem has an optimal solution containing $|C| - 1$ arcs from C.

Often in combinatorial optimization, whenever we can solve a problem efficiently (using a number of computations that are polynomial in the problems size — e.g., the number of nodes and edges of the associated graph), we are able to completely describe the underlying integer polyhedron, for example, by showing that the algorithm that generates an integer solution solves a linear programming formulation of the problem. This is the case for each of the problems mentioned in the last paragraph.

For other problems, such as the capacitated version of the optimal rooted subtree of a tree problem (Section 4.2), dynamic programming algorithms might require more than a polynomial number of computations. The number of computations required for the dynamic programming algorithm for the capacitated rooted subtree of a tree problem we have given grows as a polynomial in the capacity C, which is exponential in $log(C)$, the number of bits necessary to store the capacity (and therefore, in the sense of computational complexity theory, the dynamic program is not a polynomial algorithm). As a general rule, in these situations, the underlying integer polyhedra will be quite complex; our discussion of several valid inequalities for this capacitated rooted subtree problem illustrates this point.

Our analysis of the core minimum spanning tree problem in Section 3 provides useful lessons concerning the use of a mathematical programming model to analyze an algorithm even when the mathematical program itself is not required for creating or even stating an algorithm. In particular, the use of linear programming lower bounds (assuming a minimization form of the problem) and linear programming dual variables has permitted us to show that the algorithm finds an optimal solution. The same type of linear programming bounding argument applies to many other problem situations, as illustrated, for example, by our discussion of the rooted subtree of a tree problem.

For more complex models such as those we considered in Sections 6, 8, and 9, the underlying integer polyhedra are generally quite complicated. The problems are also generally difficult to solve, at least in the theoretical sense of computational complexity worst-case analysis. One major stream of research for addressing such situations has been to develop 'good' linear programming representations of these problems, typically by adding new valid inequalities to 'natural' starting formulations. As we have noted, often by developing good linear programming representations, empirically we are able to obtain optimal or near optimal solutions fairly efficiently.

For two classes of models, Steiner tree models (with costs satisfying the triangle inequality) and tree-on-tree models, we have been able to establish worst-case bounds on the ratio between the objective values of the underlying integer program and its linear programming relaxation and between the optimal objective value of the problem and the optimal objective value of certain heuristic solution methods. In each case, we were able to do so by using a common technique in combinatorial optimization: relating the optimal objective value of the problem to some convenient and more easily analyzed (linear programming or Lagrangian) relaxation of it.

Our development has frequently introduced and compared alternate modeling and algorithmic approaches. In the context of packing subtrees on a tree, we showed the equivalence between three popular general modeling approaches — column generation, Lagrangian relaxation, and cutting planes. As we have seen, these three approaches all are capable of solving this class of problems: they all find the optimal objective value. When applied to the more general problem of packing subtrees on general graphs, the three solution methods also provide identical initial bounds on the optimal objective function value of problem. In this broader context, we typically need to embed the starting solution into an enumeration procedure or in a branch and cut method that adds valid inequalities to improve the initial bounds. Lagrangian relaxation and column generation are attractive solution methods whenever we can identify an easily solvable subproblem (for example, if the subproblems are any of the polynomially solvable problems we have mentioned above). Column generation, like cutting plane methods, has the advantage of working directly in the space of decision variables, whereas Lagrangian relaxation works in a dual space of Lagrange multipliers. Column generation has the further advantage of possibly providing a feasible solution to the problem before optimality is attained. Lagrangian relaxation, on the other hand, has the advantage of not having to solve a complex master problem, but rather typically uses simple multiplier adjustment methods (subgradient optimization, simple heuristic methods) to find the optimal Lagrange multipliers. Cutting planes require the solution of comparatively expensive linear programs at each stage (reoptimization from stage to stage is, however, much easier than solving the linear program form scratch); cutting planes have the advantage of being rather universal, however, (not requiring easily solvable subproblems) as long as we can solve the separation problem of finding a violated inequality at each iteration, either optimally or heuristically.

As this discussion shows, having different solution procedures at our disposal offers us the flexibility of exploiting various characteristics of the problem we are solving.

Alternate models can be valuable for several reasons. First, as we have just seen, some models (those that better identify special underlying substructure) are better suited for use with different algorithms. Second, alternate models can offer different theoretical or applied insight; for example, the fact that the number of variables and constraints in the multicommodity flow formulation of the minimum spanning tree is polynomial in the number of nodes and edges of the associated graph immediately implies, from the theory of linear programming, that the problem is solvable in polynomial time without any insight into the problem's combinatorial structure. In addition, alternate, but equivalent models, often can work in concert with each other. For example, as we saw in our discussion of the minimum spanning tree problem, the multicommodity flow formulation permits us to efficiently solve the separation problem that arises when we apply a cutting plane algorithm to the subtour formulation.

Our development has also highlighted one other fact; we have seen that tree optimization problems provide a concrete problem setting for introducing a number of important methodologies and proof (and solution) techniques from combinatorial optimization that are applicable more widely. We have seen, for example, how to use dynamic programming or combinatorial methods (the greedy algorithm) to define dual variables for underlying linear programming representations of integer programs. We have introduced the optimal inequality argument from the field of polyhedral combinatorics. We have used linear programming and Lagrangian relaxation lower bounds to show that certain polyhedra are integer. In our study of the Steiner problem, we have seen how to develop worst-case bounds by relaxing some of the constraints of a linear programming model and we have seen how to use the 'parsimonious property' to establish worst-case bounds. For the tree-on-tree problem, we have seen how to combine two heuristics to develop a composite heuristic with attractive worst-case error bounds.

Tree optimization problems and their variants are conceptually simple. As we have seen, simple yet elegant solution methods are able to solve some versions of this general problem class and very good mathematical (linear programming) representations are available for many of these problems. In this sense, many tree optimization problems are well solved. And yet, new insights about tree optimization problems continue to surface; moreover, this deceptively simple problem setting also poses significant algorithmic and modeling challenges that have the potential, as in the past, to not only draw upon a wealth of knowledge from the general field of combinatorial optimization, but to also stimulate new results and new methods of analysis.

11. Notes and references

Section 1. No previous source has dealt with the range of topics we have considered in this paper. Many books in the fields of network flows, graph theory, and combinatorial optimization, as well as several survey articles that we cite below, treat a number of particular topics though.

Section 2. For general background on the application domains we have considered in this discussion, see the following sources: clustering [Hartigan, 1975], computer and communications networks [Bertsekas & Gallager, 1992; Schwartz, 1977; Tanenbaum, 1985], facility location (Francis, McGinnis & White, 1992; Mirchandani & Francis, 1990], production planning [Graves, Rinnooy Kan & Zipkin, 1993], routing [Bodin, Golden, Assad & Ball, 1983; Lawler, Lenstra, Rinnooy Kan & Shmoys, 1985], VLSI design [Leighton, 1983; Hu & Kuh, 1985; Lengauer, 1990]. Wagner & Whitin [1958] proposed the dynamic programming recursion for the production planning problem. For a recent account of network flows, see Ahuja, Magnanti & Orlin [1993].

Section 3. The greedy algorithm and the combinatorial proof for the spanning tree problem are due to Kruskal [1956). This result can also be found earlier in the Russian literature [Boruvka, 1926] and Jarnićk [1930]. Prim [1957) and Sollin [see Berge & Ghouila-Houri, 1962] have developed other efficient algorithms for the spanning tree problem (see Ahuja, Magnanti & Orlin [1993] for a discussion of these methods). Edmonds [1971) has described the matroid polytope of which the tree polytope is a special case, and used the primal-dual proof to prove integrality of the tree polyhedron. His idea of a certificate of optimality has been of crucial importance in combinatorial optimization that predates the development of NP-completeness [Cook, 1971] and Karp [1972].

Wong [1980] first spurred interest in alternative formulations for the tree polyhedron. In the context of the travelling salesman problem, he showed the equivalence of the subtour model and the multicommodity flow model. In Section 6 in our investigation of formulations for the Steiner tree problem, we consider generalizations of this work. The study of alternative formulations has become an important topic in combinatorial optimization; see Martin [1987] and Nemhauser & Wolsey [1988].

The minimum spanning tree is just one of many problems defined on trees; for example, we could choose a 'bottleneck' objective function of maximizing the minimum edge weight in the chosen tree. Camerini, Galbiati & Maffioli [1984] have provided a survey of the computational status of many such alternative tree optimization problems

Section 4. In our discussion, we have introduced the core tree problem as a prototype for a nonserial (non shortest path) dynamic program. Groeflin & Liebling [1981] have presented a more general model and used the network flow argument to prove integrality. The dynamic programming argument is again of the primal-dual type due to Edmonds that we cited above. Lovász [1979] first used the so-called optimal inequality argument which other researchers have recently rediscovered and used extensively.

Section 5. The degree constrained spanning tree problem and the optimal branching problem are special cases of the matroid intersection problem and are thus covered in Edmonds [1969]. The algorithm for the degree constrained spanning tree problem is due to Volgenant [1989]. Edmonds [1967] first treated optimal branchings. The branching algorithm, though not our analysis of it, is based on a presentation in Lawler [1976]. The proof of integrality we have given uses ideas of an optimal inequality proof due to Goemans [1992].

Section 6. The Steiner problem has received considerable attention in recent years; Maculan [1987] and Winter [1987] have presented surveys on exact and heuristic algorithms, respectively. See also, Hwang & Richards [1992] and Hwang, Richards & Winter [1992]. Very recently several researchers — Goemans [1994b], Myung & Goemans [1993], Lucena & Beasley [1992] and Margot, Prodon & Liebling [1994] — have modeled and analyzed the node weighted Steiner tree problem. The first two of these papers show the equivalence of a large number of formulations.

The polyhedral structure of the Steiner problem is treated in Chopra & Rao [1994a,b] who developed various families of facet-defining inequalities, including both Steiner partition and odd hole inequalities. Goemans [1994b] introduced combinatorial design facets.

As was suggested in Section 4, most optimization problems on trees turn out to be easy, so it is natural to ask whether there is a larger class of graphs on which problems that are NP-hard on general graphs remain polynomially solvable. Series-parallel graphs (also known as two-trees) and more generally k-trees often have this property: see, for example, Takamizawa, Nishizeki & Saato [1982], Arnborg, Lagergren & Seese [1991]. The critical point in analyzing these problems is the fact that by eliminating a k-clique in a k-tree it is possible to decompose the graph, and thereby derive a recursive algorithm. Given that there are efficient algorithms on such graphs, we might also expect polyhedral results. For instance, Goemans [1994b] and Margot, Prodon & Liebling [1994] show that the formulation P_{sub} is integral for the node-weighted Steiner problem on series parallel graphs. Prodon, Liebling & Groeflin [1985], Goemans [1994a], and Schaffers [1991] contain related polyhedral results on such graphs.

Lucena & Beasley [1992] and Chopra & Gorres [1990] have conducted computational studies based on the formulation P_{sub}, Chopra, Gorres & Rao [1992] have used formulation P_{dcut} and Balakrishnan, Magnanti & Mirchandani [1994b] have used a directed flow formulation. Wong [1984] earlier developed a dual ascent algorithm for the directed Steiner problem using P_{dcut}, and Beasley [1989] developed a Lagrangian approach with spanning tree subproblems. Beasley has solved randomly generated sparse problems with up to 2500 nodes and 62500 edges, Chopra et al. have handled graphs with up to 300 nodes and average degrees of 2,5 and 10, as well as complete Euclidean graphs with between 100 and 500 nodes, and Balakrishnan et al. have solved problems with up to 500 nodes and 5000 edges.

Much work has been done on heuristics for the Steiner problem. One motivation comes from VLSI design and the routing of nets, see Korte, Promel & Steger [1990]. The worst case bound of 2 for the tree heuristic has been known for years. The parsimonious property and our proof of the bound is due to Goemans & Bertsimas [1990]. The Held and Karp relaxation for the TSP first appeared in Held & Karp [1971]. Recently, researchers have developed improved worst case heuristics for the Steiner problem. Zelikovsky [1993] derives a bound of 11/6, and Berman & Ramaiyer [1992] show how to reduce it to about 1.75.

Theorem 6.12 is due to Bienstock, Goemans, Simchi-Levi & Willianson [1993]. Geomans & Willianson [1992] have developed a heuristic with a worst-case bound of 2. When we must pay to include a node in the Steiner tree, i.e., the objective function is of the form $\sum_{e \in E} w_e + \sum_{i \in V} \pi_i z_i$ with each $\pi \geq 0$, there is little hope of finding a heuristic with a constant worst-case bound. Klein & Ravi [1993] have presented a heuristic for which $z^H / z^{NWST} \leq 2 \log |T|$.

Section 7. The dynamic programming algorithm for the OSP and the polyhedral proofs of Theorems 1 and 2 appear in Bárány, Edmonds & Wolsey [1986]. The relationship between subtree of tree incidence matrices and clique matrices of chordal graphs appears in Golumbic [1980]. Since chordal graphs are perfect, this analysis provides an alternative proof of Theorem 1. For the extensive literature on facility location on trees, see Mirchandani & Francis [1990]. Results for the Constrained Subtree Packing problem OCSP are based on Aghezzaf, Magnanti & Wolsey [1995]. Balakrishnan, Magnanti & Wong [1995] report on computational experience, using Lagrangian relaxation, with the telecommunications model cited in Section 2.

Section 8. The equality of the objective values of the three algorithms can be derived by consulting standard texts. The important question of how to continue when the column generation approach gives a fractional solution has received some attention in Vance, Barnhart, Johnson & Nemhauser [1992] and in Hansen, Jaumard and De Aragao [1992], see also Desrosiers, Dumas, Solomon & Soumis [1995].

During the last decade, researchers have intensively studied the multi-item lotsizing problem. Work on the polyhedral structure of the single-item problem can be found in Bárány, Van Roy & Wolsey [1984], Van Hoesel, Wagelmans & Wolsey [1994], and Leung, Magnanti & Vachani [1989]. Thizy and Van Wassenhove [1986] have used the Lagrangian relaxation approach for a closely related multi-item problem. Cattryse, Maes and Van Wassenhove [1990] have examined heuristics based on a column generation approach, and Pochet & Wolsey [1991] have reported computational results with the cutting plane approach.

Wagner & Whiten [1958) proposed a dynamic programming algorithm for solving the single-item dynamic lotsizing problem. Aggarwal & Park [1993), Federgrun & Tsur [1991), Wagelmans, Van Hoesel & Kolen [1992) have proposed very efficient algorithms for this problem. Their analysis shows how to exploit special structure to develop algorithms that are more efficient than those that apply to general trees.

The valid inequalities and the column generation approach described for the clustering problem can be found in Johnson, Mehrotra and Nemhauser [1993]. The literature on the polyhedral structure of cut polytopes is extensive. Work closely related to the clustering problem includes Chopra & Rao [1993], Groetschel & Wakayabashi [1989], and De Souza, Ferreira, Martin, Weismantel & Wolsey [1994].

Araque, Hall & Magnanti [1990] have studied the polyhedral structure of the capacitated tree problem and the unit demand vehicle routing problem including the inequalities we have presented. Araque [1989] has studied the use of these inequalities for solving unit demand capacitated vehicle routing problems, but little or nothing has been reported for non-unit demands. However, several papers contain results on the polyhedral structure of the VRP polyhedron with constant capacity but arbitrary demands, including Cornuéjols & Harche [1993]. Our presentation of the comb inequalities is taken from Pochet [1992]. Gavish [1983, 1984] has designed several algorithms for the capacitated tree problem including the approach we described.

Work on the Steiner tree packing polyhedron is due to Groetschel, Martin & Weismantel [1992a, b]. These authors have developing a branch and cut approach based on valid inequalities and separation heuristics including both Steiner tree inequalities and the cycle inequalities we have presented; they solve seven unsolved problems from the literature all to within 0.7% of optimality (four to optimality).

Section 9. The results in this section are drawn from Balakrishnan, Magnanti and P. Mirchandani [1994a, b, 1995] who treat not only the tree-on-tree problem, but also more general problem of 'overlaying' solutions to one problem on top of another other. Previously, Iwainsky [1985] and Current, Revelle & Cohon [1986] had introduced and treated the tree-on tree problem and a more specialized path on tree problem. Duin & Volgenant [1989] have shown how to convert the tree-on-tree problem into an equivalent Steiner tree problem. Therefore, in principle, any Steiner tree algorithm is capable of solving these problems. The computational experience we have cited in Section 9 [see Balakrishnan, Magnanti & Mirchandani, 1993] uses a specialized dual-ascent approach directly on the tree-on-tree formulation.

Acknowledgment

We are grateful to Michel Goemans, Leslie Hall, Prakash Mirchandani, S. Raghavan, and D. Shaw for constructive feedback on an earlier version of this paper.

The preparation of this paper was supported, in part, by NATO Collaborative Research Grant CRG 900281.

References

Aggarwal, A., and J. Park (1993). Improved algorithms for economic lot-size problems. *Oper. Res.* 41, 549–571.

Aghezzaf, E.H., T.L. Magnanti and L.A. Wolsey (1995). Optimizing constrained subtrees of trees. *Math. Program.* 69, (to appear).

Ahuja, R., T. Magnanti and J. Orlin (1993). *Network Flows: Theory, Algorithms, and Applications*, Prentice-Hall, Englewood Cliffs, NJ.

Araque, J.R. (1989). Solution of a 48-city routing problem by branch-and-cut, Unpublished Manuscript, Department of Applied Mathematics and Statistics, SUNY at Stony Brook, Stony Brook, N.Y.

Araque, J.R., L.A. Hall and T.L. Magnanti (1990), Capacitated trees, capacitated routing and associated polyhedra. Core DP 9061, Louvain-la-Neuve, Belgium

Arnborg, S., J. Lagergren and D. Seese (1991). Easy problems for tree decomposable graphs. *J. Algorithms* 12, 308–340

Balakrishnan, A., T. Magnanti and P. Mirchandani (1994a), Modeling and heuristic worst-case performance analysis of the two-level network design problem, *Manage. Sci.* 40, 846–867.

Balakrishnan, A., T. Magnanti and P. Mirchandani (1994b). A dual-based algorithm for multi-level network design, *Manage. Sci.* 40, 567–581.

Balakrishnan, A., T. Magnanti and P. Mirchandani (1995), Heuristics, LPs, and network design analyses of trees on trees, *Opers. Res.* 43, to appear.

Balakrishnan, A., T.L. Magnanti and R.T. Wong (1995). A decomposition algorithm for expanding local access telecommunications networks, *Oper. Res.* 43, in press.

Bárány, I., J. Edmonds and L.A. Wolsey (1986), Packing and covering a tree by subtrees. *Combinatorica* 6, 245–257.

Bárány, I., T.J. Van Roy and L.A. Wolsey (1984), Uncapacitated lot-sizing: The convex hull of solutions. *Math. Program.*, Study 22, 32–43.

Beasley, J.E. (1989). An SST-based algorithm for the Steiner problem on graphs. *Networks* 19, 1–16

Berman, P., and V. Ramaiyer (1992). Improved approximation of the Steiner tree problem. *Proc. 3rd Annu. ACM–SIAM Symp. on Discrete Algorithms*, pp. 325–334.

Berge, C., and A. Ghouila-Houri (1962). *Programming, Games, and Transportation Networks*, John Wiley, New York.

Bertsekas, D., and R. Gallager (1992). *Data Networks*, 2nd edition, Prentice-Hall, Englewood Cliffs, NJ.

Bienstock, D., M. Goemans, D. Simchi-Levi and D. Willianson (1993). A note on the prize-collecting travelling salesman problem. *Math. Program.* 59, 413–420.

Bodin, L., B. Golden, A. Assad and M. Ball, Routing and Scheduling of Vehicles and Crews: The State of the Art. *Comput. Oper. Res.* 10, 69–211.

Boruvka, O. (1926). Příspekěv křešení otázky ekonomické stavby elektrovodních sítí. *Elektrotech. Obzor* 15, 153–154.

Camerini, P.M., G. Galbiati and F. Maffioli (1984). The complexity of weighted multi-constrained spanning tree problems. *Colloq. Math. Soc. János Boltai, Theory of Algorithms Pécs*, 44.

Cattryse, D., J. Maes and L.N. Van Wassenhove (1990). Set partitioning and column generation heuristics for capacitated dynamic lot-sizing. *Eur. J. Oper. Res.* 46, 38–47.

Chopra, S., and E. Gorres (1990). On the node weighted Steiner tree problem. Department of Managerial Economics and Decision Sciences, J. Kellogg School of Management, Northwestern University.

Chopra, S., and M.R. Rao (1994a). The Steiner tree problem I: Formulations, compositions and extensions of facets. New York University, *Math. Program.* 64, 209–230.

Chopra, S., and M.R. Rao (1994b). The Steiner tree problem II: Properties and classes of facets. New York University, *Math. Program.* 64, 231–246.

Chopra, S., and M.R. Rao (1993). The Partition Problem. *Math. Program.* 59, 87–116.

Chopra, S., E.R. Gorres and M.R. Rao (1992). Solving the Steiner Tree problem on a graph using branch and cut. *ORSA J. Comput.* 4, 320–335.

Cook, S.A. (1971). The complexity of theorem-proving procedures. *Proc. 3rd Annu. ACM Symp. on Theory of Computing*, pp. 151–158.

Cornuéjols, G., and F. Harche (1993). Polyhedral study of the capacitated vehicle routing problem. *Math. Program.* 60, 21–52.

Current, J.R., C.S. Revelle and J.L. Cohon (1986). The hierarchical network design problem. *Eur. J. Oper. Res.* 27, 57–66.

Desrosiers, J., Y Dumas, M. Solomon and F. Soumis (1995). Time Constrained Routing and Scheduling, in: M. Ball, T. Magnanti, C. Monma and G. Nemhauser (eds.), *Network Routing*, Handbooks in Operations Research and Management Science, Vol. 8, North-Holland, Amsterdam, Chapter 2.

de Souza, C., C. Ferreira, A. Martin, R. Weismantel and L.A. Wolsey (1993). Formulations and valid inequalities for node capacitated graph partitioning problem, *Core Discussion*, Paper DP9437, Université Catholique de Louvain, Louvain.

Duin, C., and A. Volgenant (1989). Reducing the hierarchical network design problem. *Eur. J. Oper. Res.* 39, 332–344.

Edmonds, J. (1967). Optimum branchings. *J. Res. Nat. Bur. Stand.* 71B, 233–240

Edmonds, J. (1970). Submodular functions, matroids and certain polyhedra, in: R. Guy et al. (eds.), *Combinatorial Structures and their Applications*, Gordon and Breach, New York, pp. 69–87.

Edmonds, J. (1971). Matroids and the greedy algorithm. *Math. Program.* 1, 127–136.

Eppen, G.D., and R.K. Martin (1987). Solving multi-item lot-sizing problems using variable definition. *Oper. Res.* 35, 832–848.

Federgrun, A., and M. Tsur (1991). A simple forward algorithm to solve general dynamic lot-size models with n periods in $o(n \log n)$ or $O(n)$ time. *Manage. Sci.* 37, 909–925.

Francis, R.L., L.F. McGinnis and J.A. White (1992). *Facility Layout and Location: An Analytic Approach*, Prentice-Hall, Englewood Cliffs, NJ.

Gavish, B. (1983). Formulations and algorithms for the capacitated minimal directed tree problem. *J. ACM* 30, 118–132

Gavish, B. (1984). Augmented Lagrangean based algorithms for centralized network design. *IEEE Trans. Commun.* 33, 1247–1275

Goemans, M.X. (1994a). Arborescence polytopes for series-parallel graphs. *Discrete Appl. Math.*.

Goemans, M.X. (1994b). The Steiner polytope and related polyhedra. *Math. Program.* 63, 157–182.

Goemans, M.X. (1992). Personal communication.

Goemans, M.X., and D. Bertsimas (1993). Survivable networks, linear programming relaxations and the parsimonious property. *Math. Program.* 60, 145–166.

Goemans, M., and D. Willianson (1992). A general approximation technique for constrained forest problems, *Proc. 3rd ACM–SIAM Symp. on Discrete Algorithms*, to appear.

Goemans, M.X., and Y-S. Myung (1993). A catalog of Steiner tree formulations. *Networks* 23, 19–28.

Golumbic, M.C. (1980). *Algorithmic Graph Theory and Perfect Graphs*, Academic Press, New York.

Graves, S., A.H.G. Rinnooy Kan and P. Zipkin, eds. (1993). *Logistics of Production and Inventory*, Handbooks of Operations Research and Management Science, Vol. 4, North-Holland, Amsterdam.

Groeflin, H., and T.M. Liebling (1981). Connected and alternating vectors: Polyhedra and algorithms. *Math. Program.* 20, 233–244.

Groetschel, M., A. Martin and R. Weismantel (1992a). Packing Steiner trees: Polyhedral investigations, Preprint SC 92-8 Konrad-Zuse Zentrum, Berlin.

Groetschel, M., A. Martin and R. Weismantel (1992b). Packing Steiner trees: a cutting plane algorithm and computational results, Preprint SC 92-9 Konrad-Zuse Zentrum, Berlin.

Groetschel, M., C. Monma and M. Stoer (1995). Design of survivable networks, in: M. Ball, T. Magnanti, C. Monma and G. Nemhauser (eds.), *Network Models*, Handbooks in Operations Research and Management Science, Vol.7, North-Holland, Amsterdam, Chapter 10.

Groetschel, M., and Y. Wakayabashi (1989). A cutting plane algorithm for a clustering problem. *Math. Program.* 45, 59–96.

Hansen, P., B. Jaumard and M.P. de Aragao (1992). Mixed integer column generation algorithms and the probabilistic maximum satisfiability problem, in: E. Balas, G. Cornuéjols and R. Kannan (eds.) *Proc. 2nd IPCO Conf.*, Carnegie Mellon University, pp. 165–180.

Hartigan, J.A. (1975). *Clustering Algorithms*, John Wiley, New York.

Held, M., and R.M. Karp (1971). The travelling salesman problem and minimum spanning trees: Part II. *Math. Program.* 1, 6–25.

Hu, T.C., and E.S. Kuh, eds. (1985). *VLSI Layout: Theory and Design*, IEEE Press, New York, NY.

Hwang, F.W., and D.S. Richards (1992). Steiner tree problems. *Networks* 22, 55–90.

Hwang, F.W., D.S. Richards and P. Winter (1992). *The Steiner tree problem*. Annals of Discrete Mathematics 53, North-Holland, Amsterdam.

Iwainsky, A. (1985). Optimal Trees–A Short Overview on Problem Formulations, in Optimization of Connection Structures in Graphs, Central Institute of Cybernetics and Information Processes, Berlin.

Jarnick, V. (1930). Ojistém problému minimálním. *Acta Soc. Nat. Moravicae* 6, 57–63.

Johnson, E.L., A. Mehrotra and G.L. Nemhauser (1993), Min-cut clustering. *Math. Program.* 62, 133–151

Karp, R.M. (1972). Reducibility among Combinatorial Problems, in: R.E. Miller and J.W. Thatcher (eds.), *Complexity of Computer Computations,* Plenum Press, New York, pp. 85–103.

Klein, P., and R. Ravi (1993). A nearly best possible approximation for node weighted Steiner trees, in: G. Rinaldi and L.A. Wolsey (eds.), *Proc. 3rd IPCO Conf.*, pp. 323–331.

Korte, B., H.J. Promel and A. Steger (1990). Steiner Trees in VLSI Layout, in: B. Korte, L. Lovász, H.J. Promel and A. Schijver (eds.), *Paths, Flows and VLSI Layout*, Springer, Berlin pp. 185–214.

Kruskal, J.B. (1956). On the shortest spanning tree of a graph and the travelling salesman problem. *Proc. Am. Math. Soc.* 7, 48–50.

Lawler, E.L., J.K. Lenstra, A.H.G. Rinnooy Kan and D.B. Shmoys, eds. (1985). *The Traveling Salesman Problem: A Guided Tour of Combinatorial Optimization*, John Wiley, New York.

Lawler, E.L. (1976). *Combinatorial Optimization: Networks and Matroids*, Holt, Rinehart and Winston, New York.

Leighton, T.E. (1983). *Complexity Issues in VSLI*, M.I.T. Press, Cambridge, MA.

Lengauer, T. (1990). *Combinatorial Algorithms for Integrated Circuit Layout*, John Wiley, New York.

Leung, J., T. Magnanti and R. Vachani (1989). Facets and algorithms for capacitated lot sizing. *Math. Program.* 45, 331–360

Lovász, L. (1979). Graph theory and integer programming. *Ann. Discrete Math.* 4, 141–158.

Lucena, A., and J.E. Beasley (1992). A branch and cut algorithm for the Steiner problem in graphs, Report, The Management School, Imperial College, London.

Maculan, N. (1987). The Steiner problem in graphs. *Ann. Discrete Math.* 31, 185–212.

Magnanti, T., and R. Vachani (1990). A strong cutting plane algorithm for production scheduling with changeover costs. *Oper. Res.* 38, 456–473.

Margot, F., A. Prodon and Th.M. Liebling (1994). Tree polytope on 2-trees. *Math. Program.*, 63, 183–192.

Martin, R.K. (1987). Generating alternative mixed-integer programming models using variable redefinition. *Oper. Res.* 35, 331–359.

Martin, R.K. (1991). Using separation algorithms to generate mixed integer model reformulations. *Oper. Res. Lett.* 10, 119–128

Mirchandani, P.B., and R.L. Francis (1990). *Discrete Location Theory*, John Wiley, New York.

Nemhauser, G.L., and L.A. Wolsey (1988). *Integer and Combinatorial Optimization*, John Wiley, New York.

Pochet, Y. (1992). A common derivation of TSP and VRP inequalities, Talk presented at the 3rd Cycle FNRS day on Combinatorial Optimization, Core, Louvain-la-Neuve, December 11th.

Pochet, Y., and L.A. Wolsey (1991). Solving multi-item lot-sizing problems with strong cutting planes. *Manage. Sci.* 37, 53–67.

Prim, R.C. (1957). Shortest connection networks and some generalizations. *Bell System Tech. J.* 36, 1389–1401.

Prodon, A., T.M. Liebling and H. Groeflin (1985). Steiner's problem in two trees, RO 850315, Department of Mathematics, Ecole Polytechnique Fédérale de Lausanne.

Schwartz, M. (1977). *Computer and Communication Network Design and Analysis*, Prentice-Hall, Englewood Cliffs, NJ.

Takamizawa, K., T. Nishizeki and N. Saato (1982). Linear time computability of combinatorial problems on series-parallel graphs. *J. ACM* 29, 623–641.

Tanenbaum, A.S. (1989). *Computer Networks*, 2nd edition, Prentice-Hall, Englewood Cliffs, NJ.

Thizy, J.M., and L.N. Van Wassenhove (1986). A subgradient algorithm for the multi-item capacitated lot-sizing problem. *IEE Trans.* 18, 114–123.

Vance, P.H., C. Barnhart, E.L. Johnson and G.L. Nemhauser (1992). Solving binary cutting stock

problems by column generation and branch-and-bound, Computational Optimization Center COC-92-09, Georgia Institute of Technology.
Van Hoesel, C.P.M., A.P.M. Wagelmans and L.A. Wolsey (1991). Polyhedral characterization of the economic lot-sizing problem with start-up costs, *SIAM J. Discrete Math.* 7, 141–151.
Volgenant, A. (1989). A Lagrangian approach to the degree-constrained minimum spanning tree problem. *Eur. J. Oper. Res.* 39, 325–331.
Wagner, H.M., and T.M. Whitin (1958). A dynamic version of the economic lot size model. *Manage. Sci.* 5, 89–96.
Wagelmans, A.P.M., C.P.M. van Hoesel and A.W.J. Kolen (1992). Economic lot-sizing: an $O(n \log n)$ algorithm that runs in linear time in the Wagner-whitin case. *Oper. Res.* 40, Supplement 1, 145–156
Winter, P. (1987). Steiner problem in networks: a survey. *Networks* 17, 129–167.
Wong, R.T. (1980). Integer programming formulations of the travelling salesman problem. *Proc. IEEE Conf. on Circuits and Computers*, pp. 149–152.
Wong, R.T. (1984). A dual ascent approach for Steiner tree problems on a directed graph. *Math. Program.* 28, 271–287
Zelikovsky, A.Z. (1993). An 11/6 approximation algorithm for the network Steiner problem. *Algorithmica* 9, 463–470.

M.O. Ball et al., Eds., *Handbooks in OR & MS, Vol. 7*

Chapter 10

Design of Survivable Networks

M. Grötschel
Konrad-Zuse-Zentrum für Informationstechnik Berlin, Heilbronner Str. 10, D-10711 Berlin, Germany

C.L. Monma
Bell Communications Research, 445 South Street, Morristown, NJ 07960, U.S.A

M. Stoer
Telenor Research, P.O. Box 83, N-2007 Kjeller, Norway

1. Overview

This chapter focuses on the important practical and theoretical problem of designing survivable communication networks, i.e., communication networks that are still functional after the failure of certain network components. We motivate this topic in Section 2 by using the example of fiber optic communication network design for telephone companies. A very general model (for undirected networks) is presented in Section 3 which includes practical, as well as theoretical, problems, including the well-studied minimum spanning tree, Steiner tree, and minimum cost k-connected network design problems.

The development of this area starts with outlining structural properties in Section 4 which are useful for the design and analysis of algorithms for designing survivable networks. These lead to worst-case upper and lower bounds. Heuristics that work well in practice are also described in Section 4. Polynomially-solvable special cases of the general survivable network design problem are summarized in Section 5.

Section 6 contains polyhedral results from the study of these problems as integer programming models. We present large classes of valid and often facet-defining inequalities. We also summarize the complexity of the separation problem for these inequalities. Finally we provide complete and nonredundant descriptions of a number of polytopes related to network survivability problems of small dimensions.

Section 7 contains computational results using cutting plane approaches based on the polyhedral results of Section 6 and the heuristics described in Section 4. The results show that these methods are efficient and effective in producing optimal or near-optimal solutions to real-world problems.

A brief review of the work on survivability models of directed networks is given in Section 8. We also show here how directed models can help to solve undirected cases.

2. Motivation

In this section we set the stage for the topic of this chapter by considering an application to designing communication networks for telephone companies based on fiber optic technology. We will use this to introduce the concept of survivability in network design and to motivate the general optimization models described in the next section. It will become clear later that our models capture many other situations that arise in practice and in theory as well.

Fiber optic technology is rapidly being deployed in communication networks throughout the world because of its nearly-unlimited capacity, its reliability and cost-effectiveness. The high capacity of new technology fiber transmission systems has resulted in the capability to carry many thousands of telephone conversations and high-speed data on a few strands of fiber. These advantages offer the prospect of ushering in many new information networking services which were previously either technically impossible or economically infeasible.

The economics of fiber systems differ significantly from the economics of traditional copper-based technologies. Copper-based technologies tend to be bandwidth limited. This results in a mesh-like network topology, which necessarily has many diverse paths between any two locations with each link carrying only a very small amount of traffic. In contrast, the high-capacity fiber technologies tend to suggest the design of sparse 'tree-like' network topologies. These topologies have only a few diverse paths between locations (often just a single path) and each link has a very high traffic volume. This raises the possibility of significant service disruptions due to the failure of a single link or single node in the network. The special report 'Keeping the phone lines open' by Zorpette [1989] describes the many man-made and natural causes that can disrupt communication networks, including fires, tornados, floods, earthquakes, construction or terrorist activities. Such failures occur surprisingly frequently and with devastating results as described in this report and in the popular press [e.g., see Newark Star Ledger, 1987, 1988a, 1988b; New York Times, 1988, 1989; Wall Street Journal, 1988].

Hence, it is vital to take into account such failure scenarios and their potential negative consequence when designing fiber communication networks. Recall that one of the major functions of a communication network is to provide connectivity between users in order to provide a desired service. We use the term 'survivability' to mean the ability to restore network service in the event of a catastrophic failure, such as the complete loss of a transmission link or a facility switching node. Service could be restored by means of routing traffic around the damage through other existing facilities and switches, if this contingency is provided for in the network architecture. This requires additional connectivity in the network topology and a means to automatically reroute traffic after the detection of a failure.

A network topology could provide protection against a single link failure if it remains connected after the failure of any single link. Such a network is called 'two-edge connected' since at least two edges have to be removed in order to disconnect the network. However, if there is a node in the network whose removal does disconnect the network, such a network would not protect against a single node failure. Protection against a single node failure can be provided in an analogous manner by 'two-node connected' networks.

In the case of fiber communication networks for telephone companies, two-connected topologies provide an adequate level of survivability since most failures usually can be repaired relatively quickly and, as statistical studies have revealed, it is unlikely that a second failure will occur in their duration. However, for other applications it may be necessary to provide higher levels of connectivity.

One simple and cost effective means of using a two-connected topology to achieve an automatic recovery from a single failure is called diverse protection routing. Most fiber transmission systems employ a protection system to back up the working fiber systems. An automatic protection switch detects failure of a working system and switches the working service automatically to the protection system. By routing the protection system on a physically diverse route from the working system, one provides automatic recovery from a single failure at the small cost of a slightly longer path for the protection system than if it were routed on the same physical path as the working system. One would suspect, and studies have proven, that the additional cost of materials and installation of diverse fiber routes would be acceptable for the benefits gained [see Wu, Kolar & Cardwell, 1988; Wu & Cardwell, 1988; Kolar & Wu, 1988; and Cardwell, Wu & Woodall, 1988].

We consider the problem of designing minimum cost networks satisfying certain connectivity requirements that model network survivability. A formal model will be described in the next section. This model allows for different levels of survivability that arise in practical and theoretical models including the minimum spanning tree, Steiner tree, and minimum cost k-connected network design problems.

The application described here requires three distinct types of nodes: special offices, which must be protected from single edge or node failures, ordinary offices, which need only be connected by a single path, and optional offices, which may be included or excluded from the network design depending only upon cost considerations. The designation of office type is performed by a planner based on a cost/benefit analysis. Normally, the special offices are highly important and/or high-revenue-producing offices, with perhaps a high proportion of priority services. It may not be economically possible to ensure the service of ordinary or optional offices in the face of potential failures. In fact, some offices only have one outlet to the network and so it would be technologically impossible to provide recovery if this path were blocked by a network failure.

We note that two-connected network topologies are cost effective. For example, a typical real-world problem which we solved has a survivable network with cost of only 6% above the minimum spanning tree; however, a single failure in the tree network could result in the loss of 33% of the total traffic while the survivable network could lose at most 3% from any single failure. We also note that the heuristic methods described in Section 4 and the polyhedral cutting plane methods described in Section 6 are efficient and very effective in generating optimal and near-optimal network topologies as we describe in Section 7.

We conclude this section by pointing out that the topology design problem considered here is just the first step in the overall design of fiber communications networks. There are other issues that need to be addressed once the topology is in place. For instance, demands for services are usually in units of circuits, called DS0 rate, or bundles of 24 circuits, called DS1 rate. Fiber optic transmission rates come in units of 28 DS1s or 672 DS0s, called DS3 rate. Hence, it is necessary to place multiplexing equipment to convert between the three rates, and to route these demands through the network. Another issue is that the fiber signals need to be amplified using repeater equipment if the signals travel beyond a given threshold distance.

Furthermore, the network is generally organized into a facility hierarchy. That is, offices are grouped together into clusters, with each cluster having one hub office to handle traffic between clusters. This grouping considers such factors as community-of-interest and geographic area. Groups of clusters can be grouped into sectors, with each sector having one gateway office, which is a hub building designated to handle inter-sector traffic. This hierarchy allows traffic to be concentrated into high capacity routes to central locations where the demands are sorted according to destination. This concept of hub routing has proven to give near-optimal results, see Wu & Cardwell [1988]. A two-connected network allows for dual homing, i.e., for the possibility of splitting the demand at an office between a home hub and foreign hub to protect against a hub failure.

Given the complexity of the overall design problem, these issues are generally handled in a sequential fashion with the topology question being decided first. This process is often iterated until an acceptable network design is obtained. In this chapter, we only deal with the network topology design aspect and not the multiplexing or bundling aspects. For an overview of this combined process and a description of a computer-based planning tool [Bellcore, 1988] incorporating all of these features, see Cardwell, Monma & Wu [1989].

3. Integer programming models of survivability

In this section, we formalize the undirected survivable network design problems that are being considered in this chapter. Variants that are based on directed networks will be briefly treated in Section 8.

3.1. The general model for undirected networks

A set V of *nodes* is given representing the locations of the offices that must be interconnected into a network in order to provide the desired services. A collection E of *edges* is also specified that represent the possible pairs of nodes between which a direct transmission link can be placed. We let $G = (V, E)$ be the (undirected) graph of possible direct link connections. Each edge $e \in E$ has a nonnegative *fixed cost* c_e of establishing the direct link connection. For our range of applications, loops are irrelevant. Thus we assume in the following that graphs have no loops. Parallel transmission links occur in practice; therefore, our graphs may have parallel edges. For technical reasons to be explained later, we will however restrict ourselves in this paper to simple graphs (i.e., loopless graphs without parallel edges) when we consider node survivability problems and node connectivity.

The cost of establishing a network consisting of a subset $F \subseteq E$ of edges is the sum of the costs of the individual links contained in F. The goal is to build a minimum-cost network so that the required survivability conditions, which we describe below, are satisfied. We note that the cost here represents setting up the topology for the communication network and includes placing conduits in which to lay the fiber cable, placing the cables into service, and other related costs. We do not consider costs that depend on how the network is implemented such as routing, multiplexing, and repeater costs. Although these costs are also important, it is (as mentioned in Section 2) usually the case that a topology is designed first and then these other costs are considered in a second stage of optimization.

If $G = (V, E)$ is a graph, $W \subseteq V$ and $F \subseteq E$, then we denote by $G - W$ and $G - F$ the graph that is obtained from G by deleting the node set W and the edge set F, respectively. For notational convenience we write $G - v$ and $G - e$ instead of $G - \{v\}$ and $G - \{e\}$, respectively. The difference of two sets M and N is denoted by $M \setminus N$.

For any pair of distinct nodes $s, t \in V$, an $[s, t]$-*path* P is a sequence of nodes and edges $(v_0, e_1, v_1, e_2, \ldots, v_{l-1}, e_l, v_l)$, where each edge e_i is incident with the nodes v_{i-1} and $v_i (i = 1, \ldots, l)$, where $v_0 = s$ and $v_l = t$, and where no node or edge appears more than once in P. A collection $P_1, P_2, \ldots, P_k$ of $[s, t]$-paths is called *edge-disjoint* if no edge appears in more than one path, and is called *node-disjoint* if no node (except for s and t) appears in more than one path. In standard graph theory two parallel edges are not considered as node-disjoint paths. For our applications it is sensible to do so. However, this modification would lead to considering nonstandard variations of node connectivity, reformulations of Menger's theorem etc. In order not to trouble the reader with these technicalities we have decided to restrict ourselves to simple graphs when node-disjoint paths are treated. The results presented in this paper carry, appropriately stated, over to the case where parallel edges are considered as node-disjoint paths. This theory is developed in Stoer [1992].

Two different nodes s, t of a graph are called *k-edge* (resp. *k-node*) *connected* if there are k edge-disjoint (resp. node-disjoint) paths between s and t. A graph with at least two nodes is called *k-edge* or *k-node connected* if all pairs of distinct nodes of G are k-edge or k-node connected, respectively. For our purposes, the graph K_1 consisting of just one node is k-edge and k-node connected for every natural number k. For a graph $G \neq K_1$, the largest integer k such that G is k-edge connected (resp. k-node connected) is denoted by $\lambda(G)$ (resp. $\kappa(G)$) and is called the *edge connectivity* (resp. *node connectivity*) of G. An *articulation set* of a connected graph G is a set of nodes whose removal disconnects G, and an *articulation node* is a single node disconnecting G.

The survivability conditions require that the network satisfy certain edge and node connectivity requirements. To specify these, three nonnegative integers r_{st}, k_{st} and d_{st} are given for each pair of distinct nodes $s, t \in V$. The numbers r_{st} represent the *edge survivability requirements*, and the numbers k_{st} and d_{st} represent the *node survivability requirements*; this means that the network $N = (V, F)$ to be designed has to have the property that, for each pair $s, t \in V$ of distinct nodes, N must contain at least r_{st} edge-disjoint $[s, t]$-paths, and that the removal of at most k_{st} nodes (different from s and t) from N must leave at least d_{st} edge-disjoint $[s, t]$-paths. (Clearly, we may assume that $k_{st} \leq |V|-2$ for all $s, t \in V$, and we will do this throughout this chapter). These conditions ensure that some communication path between s and t will survive a prespecified level of combined failures of both nodes and links. The levels of survivability specified depend on the relative importance placed on maintaining connectivity between different pairs of offices.

Given $G = (V, E)$ and $r, k, d \in \mathbf{Z}_+^{E_V}$, extend the functions r and d to functions operating on sets by setting

$$\mathrm{con}(W) := \max\{r_{st} \mid s \in W,\ t \in V \setminus W\} \tag{1}$$

and

$$d(Z, W) := \max\{d_{st} \mid s \in W \setminus Z,\ t \in V \setminus (Z \cup W)\} \text{ for } Z, W \subseteq V. \tag{2}$$

We call a pair (Z, W) with $Z, W \subseteq V$ *eligible* (with respect to k) if $Z \cap W = \emptyset$ and $|Z| = k_{st}$ for at least one pair of nodes s, t with $s \in W$ and $t \in V \setminus (Z \cup W)$.

Let us now introduce a variable x_e for each edge $e \in E$, and consider the vector space $\mathbf{R}^E$. Every subset $F \subseteq E$ induces an *incidence vector* $\chi^F = (\chi_e^F)_{e \in E} \in \mathbf{R}^E$ by setting $\chi_e^F := 1$ if $e \in F$, $\chi_e^F := 0$ otherwise; and vice versa, each 0/1-vector $x \in \mathbf{R}^E$ induces a subset $F^x := \{e \in E \mid x_e = 1\}$ of the edge set E of G. If we speak of the incidence vector of a path in the sequel we mean the incidence vector of the edges of the path. We can now formulate the network design problem introduced above as an integer linear program with the following constraints.

$$\begin{array}{lll}
\text{(i)} \displaystyle\sum_{i \in W} \sum_{j \in V \setminus W} x_{ij} & \geq \mathrm{con}(W) & \text{for all } W \subseteq V, \emptyset \neq W \neq V, \\
\text{(ii)} \displaystyle\sum_{i \in W} \sum_{j \in V \setminus (Z \cup W)} x_{ij} & \geq d(Z, W) & \text{for all eligible } (Z, W) \text{ of subsets of } V, \\
\text{(iii)} \ 0 \leq x_{ij} \leq 1 & & \text{for all } ij \in E, \\
\text{(iv)} \ x_{ij} \text{ integral} & & \text{for all } ij \in E.
\end{array} \tag{3}$$

Note that if $N - Z$ contains at least d_{st} edge-disjoint $[s, t]$-paths for each pair s, t of distinct nodes in V and for each set $Z \subseteq V \setminus \{s, t\}$ with $|Z| = k_{st}$, and if $r_{st} = k_{st} + d_{st}$, then all node survivability requirements are satisfied, i.e., inequalities of type (3ii) need not be considered for node sets $Z \subseteq V \setminus \{s, t\}$ with $|Z| < k_{st}$. It follows from Menger's theorem (see [Frank, 1995]) that, for every feasible solution x of (3), the subgraph $N = (V, F^x)$ of G defines a network that satisfies the given edge and node survivability requirements.

This model, introduced in Grötschel & Monma [1990], generalizes and unifies a number of problems that have been investigated in the literature either from a practical or theoretical point of view. We mention here some of these cases.

The classical *network synthesis problem* for multiterminal flows (see Chapter 13) is obtained from (3) by dropping the constraints (3ii) and (3iv). In the standard formulation of the network synthesis problem, the upper bounds $x_e \leq 1$ are not present. But our model allows parallel edges in the underlying direct-link graph, or equivalently, allows the upper bound in constraints (3iii) to take on any nonnegative values for each edge. This linear programming problem has a number of constraints that is exponential in the number of nodes of G. For the case $c_{ij} = c$ for all $ij \in E$, where c is a constant, Gomory & Hu [1961] found a simple algorithm for its solution. Bland, Goldfarb & Todd [1981] pointed out that the separation problem for the class of inequalities (3i) can be solved in polynomial time by computing a minimum capacity cut; thus, it follows by the ellipsoid method that the classical network synthesis problem can be solved in polynomial time. (See Grötschel, Lovász & Schrijver [1988] for details on the ellipsoid algorithm and its applications to combinatorial optimization.) The *integer network synthesis problem*, i.e., the problem obtained from (3) by dropping constraints (3ii) and the upper bounds $x_e \leq 1$, was solved by Chou & Frank [1970], for the case $c_{ij} = c$ for all $ij \in V \times V$, and $r_{ij} \geq 2$ for all ij.

The *minimum spanning tree problem* can be phrased as the task to find a minimum-cost connected subset $F \subseteq E$ of edges spanning V (see Chapter 12). This problem can be viewed as a special case of (3) as follows. We drop the constraints (3ii) and set $r_{st} := 1$ for all distinct $s, t \in V$ in constraints (3i).

Similarly, the closely related *Steiner tree problem* is to find a minimum-cost connected subset $F \subseteq E$ of edges that span a specified subset $S \subseteq V$ of nodes. This problem is a special case of (3) where we drop constraints (3ii) and set $r_{st} := 1$ in constraints (3i), for all $s, t \in S$, and $r_{st} := 0$ otherwise. (See also Chapter 12.) Let us remark at this point that the Steiner tree problem is well-known to belong to the class of NP-hard problems. As it is a special case of (3), our general problem of designing survivable networks is NP-hard as well. (See Garey & Johnson [1979] for details on the theory of NP-completeness.)

The problem of finding a *minimum-cost k-edge connected network* in a given graph is a special case of (3) where all inequalities (3ii) are dropped and where $r_{st} = k$ for all distinct $s, t \in V$. The problem of finding an *optimal k-node connected network*, for $k \leq |V| - 1$, is a special case of (3) where we drop the constraints (3i) and set $k_{st} := k - 1$ and $d_{st} := 1$ for all distinct $s, t \in V$.

3.2. A brief discussion of the model

The rest of this chapter is mainly devoted to studying various aspects of model (3) and some of its special cases. Among other subjects, we describe heuristics to compute upper and lower bounds for the optimum value of (3). To compute a lower bound, one is naturally led to dropping the integrality constraints (3iv) and solving the LP-relaxation (3i), (3ii) and (3iii) of the survivable network design problem.

Two questions immediately arise. Can one solve this LP despite the fact that it has exponentially many constraints? Can one find a better (or a series of increasingly better) LP-relaxations that are solvable in polynomial (or practically reasonable) time?

We will address the first question in Section 7 and the second in Section 6. But here we would like to give a glimpse at the approach that leads to answering these questions. The method involved is known as polyhedral combinatorics. It is a vehicle to provide (in some sense) the best possible LP-relaxation. We want to demonstrate now how the second question leads to the investigation of certain integral polyhedra in a very natural way. See Grötschel & Padberg [1985] and Pulleyblank [1989] for a survey on polyhedral combinatorics.

To obtain a better LP-relaxation of (3) than the one arising from dropping the integrality constraints (3iv), we define the following polytope. Let $G = (V, E)$ be a graph, let $E_V := \{st \mid s, t \in V, s \neq t\}$, and let $r, k, d \in Z_+^{E_V}$ be given. Then

$$\mathrm{CON}(G; r, k, d) := \mathrm{conv}\{x \in \mathbf{R}^E \mid x \text{ satisfies (3i)–(3iv)}\} \tag{4}$$

is the polytope associated with the network design problem given by the graph G and the edge and node survivability requirements r, k, and d. (Above 'conv' denotes the convex hull operator.) In the sequel, we will study $\mathrm{CON}(G; r, k, d)$ for various special choices of r, k and d. Let us mention here a few general properties of $\mathrm{CON}(G; r, k, d)$ that are easy to derive.

Let $G = (V, E)$ be a graph and $r, k, d \in \mathbf{Z}_+^{E_V}$ be given as above. We say that $e \in E$ is *essential with respect to* $(G; r, k, d)$ (short: $(G; r, k, d)$-*essential*) if $\mathrm{CON}(G - e; r, k, d) = \emptyset$. In other words, e is essential with respect to $(G; r, k, d)$ if its deletion from G results in a graph such that at least one of the survivability requirements cannot be satisfied. We denote the set of edges in E that are essential with respect to $(G; r, k, d)$ by $\mathrm{ES}(G; r, k, d)$. Clearly, for all subsets $F \subseteq E \setminus \mathrm{ES}(G; r, k, d)$, $\mathrm{ES}(G; r, k, d) \subseteq \mathrm{ES}(G - F; r, k, d)$ holds. Let $\dim(S)$ denote the *dimension* of a set $S \subseteq \mathbf{R}^n$, i.e., the maximum number of affinely independent elements in S minus 1. Then one can easily prove the following two results [see Grötschel & Monma, 1990].

Theorem 1. *Let* $G = (V, E)$ *be a graph and* $r, k, d \in \mathbf{Z}_+^{E_V}$ *such that* $\mathrm{CON}(G; r, k, d) \neq \emptyset$. *Then*

$$\mathrm{CON}(G; r, k, d) \subseteq \{x \in \mathbf{R}^E \mid x_e = 1 \textit{ for all } e \in \mathrm{ES}(G; r, k, d)\}, \quad \textit{and}$$
$$\dim(\mathrm{CON}(G; r, k, d)) = |E| - |\mathrm{ES}(G; r, k, d)|.$$

An inequality $a^T x \leq \alpha$ is *valid* with respect to a polyhedron P, if $P \subseteq \{x \mid a^T x \leq \alpha\}$; the set $F_a := \{x \in P \mid a^T x = \alpha\}$ is called the *face* of P defined by $a^T x \leq \alpha$. If $\dim(F_a) = \dim(P) - 1$ and $F_a \neq \emptyset$ then F_a is a *facet* of P, and $a^T x \leq \alpha$ is called *facet-defining* or *facet-inducing*.

Theorem 2. *Let* $G = (V, E)$ *be a graph and* $r, k, d \in \mathbf{Z}_+^{E\nu}$ *such that* $\mathrm{CON}(G; r, k, d) \neq \emptyset$. *Then*

(a) $x_e \leq 1$ *defines a facet of* $\mathrm{CON}(G; r, k, d)$ *if and only if* $e \in E \setminus \mathrm{ES}(G; r, k, d)$;

(b) $x_e \geq 0$ *defines a facet of* $\mathrm{CON}(G; r, k, d)$ *if and only if* $e \in E \setminus \mathrm{ES}(G; r, k, d)$ *and* $\mathrm{ES}(G; r, k, d) = \mathrm{ES}(G - e; r, k, d)$.

Theorems 1 and 2 solve the dimension problem and characterize the trivial facets. But these characterizations are (in a certain sense) algorithmically intractable as the next observation shows, which follows from results of Ling & Kameda [1987].

Remark 1. *The following three problems are* NP-*hard.*

Instance: A graph $G = (V, E)$ *and vectors* $r, k, d \in \mathbf{Z}_+^E$.

Question 1: Is $\mathrm{CON}(G; r, k, d)$ *nonempty?*

Question 2: Is $e \in E$ $(G; r, k, d)$*-essential?*

Question 3: What is the dimension of $\mathrm{CON}(G; r, k, d)$*?*

However, for most cases of practical interest in the design of survivable networks, the sets $\mathrm{ES}(G; r, k, d)$ of essential edges can be determined easily, and thus the trivial LP-relaxation of (3) can be set up without difficulties by removing the redundant inequalities identified by Theorem 2.

3.3. A model used in practice

The model discussed so far mixes node and edge connectivity, and provides, for each pair of nodes, the possibility to specify particular connectivity requirements. It is, as mentioned before, a quite general framework that models many practical situations simultaneously. This generality demands a considerable amount of data. In our real-world application, it turned out that the network designers were either interested in node connectivity or in edge connectivity requirements but not in both simultaneously. Also, the data for implementing the general model were not available in practice. A specialized version, to be described below, proved to be acceptable from the point of view of data acquisition and was still considered a reasonable model of reality by practitioners.

To model these (slightly more restrictive) survivability conditions, we introduce the concept of node types. For each node $s \in V$ a nonnegative integer r_s, called the *type* of s, is specified. For any $W \subseteq V$, the integer $r(W) := \max\{r_v \mid V \in W\}$ is called the *type* of W. We say that the network $N = (V, F)$ to be designed satisfies the *node survivability conditions* if, for each pair $s, t \in V$ of distinct nodes, N contains at least

$$r_{st} := \min\{r_s, r_t\} \tag{5}$$

node-disjoint $[s, t]$-paths. Similarly, we say that $N = (V, F)$ satisfies the *edge survivability conditions* if, for each pair $s, t \in V$ of distinct nodes, N contains r_{st} edge-disjoint $[s, t]$-paths. These conditions ensure that some communication path between s and t will survive a prespecified level of node or link failures. We will discuss these special cases in more detail later. To make the (somewhat clumsy) general notation easier, we introduce further symbols and conventions to denote these node- or edge-survivability models. Given a graph $G = (V, E)$ and a vector of node types $r = (r_s)_{s \in V}$ we assume — without loss of generality — that there are at least two nodes of the largest type. If we say that we consider the kNCON problem (for G and r) then we mean that we are looking for a minimum-cost network that satisfies the node survivability conditions and where $k = \max\{r_s \mid s \in V\}$. Similarly, we speak of the kECON problem (for G and r), when we consider the edge survivability conditions. When we want to leave it open whether the survivability problem is specified by node types or by a $V \times V$ matrix r (k and d), we speak of ECON (resp., NCON) problems.

Let $G = (V, E)$ be a graph. For $Z \subseteq V$, let $\delta_G(Z)$ denote the set of edges with one end node in Z and the other in $V \setminus Z$. It is customary to call $\delta_G(Z)$ a *cut*. If it is clear with respect to which graph a cut $\delta_G(Z)$ is considered, we simply drop the index and write $\delta(Z)$. We also write $\delta(v)$ for $\delta(\{v\})$. If X, Y are subsets of V with $X \cap Y = \emptyset$, we set $[X : Y] := \{ij \in E \mid i \in X,\ j \in Y\}$, thus $\delta(X) = [X : V \setminus X]$. For any subset of edges $F \subseteq E$, we let $x(F)$ stand for the sum $\sum_{e \in F} x_e$. Consider the following integer linear program for a graph $G = (V, E)$ with edge costs c_e for all $e \in E$ and node types r_s for all $s \in V$ [using (5) in the definition of con(W) in (1)]:

$$\min \quad c^T x$$

subject to

$$\begin{array}{lll}
\text{(i)} & x(\delta(W)) \geq \text{con}(W) & \text{for all } W \subseteq V, \emptyset \neq W \neq V; \\
\text{(ii)} & x(\delta_{G-Z}(W)) \geq \text{con}(W) - |Z| & \\
& & \text{for all pairs } s, t \in V,\ s \neq t, \text{ and} \\
& & \text{for all } \emptyset \neq Z \subseteq V \setminus \{s, t\} \text{ with } |Z| \leq r_{st} - 1, \\
& & \text{and for all } W \subseteq V \setminus Z \text{ with } s \in W,\ t \notin W; \\
\text{(iii)} & 0 \leq x_{ij} \leq 1 & \text{for all } ij \in E; \\
\text{(iv)} & x_{ij} \text{ integral} & \text{for all } ij \in E.
\end{array} \tag{6}$$

It follows from Menger's theorem that the feasible solutions of (6) are the incidence vectors of edge sets F such that $N = (V, F)$ satisfies all node survivability conditions; i.e., (6) is an integer programming formulation of the kNCON problem. Deleting inequalities (6ii) we obtain, again from Menger's theorem, an integer programming formulation for the kECON problem. The inequalities of type (6i) will be called *cut inequalities* and those of type (6ii) will be called *node cut inequalities.*

The polyhedral approach to the solution of the kNCON (and similarly the kECON) problem consists of studying the polyhedron obtained by taking the convex hull of the feasible solutions of (6). We set

$$k\text{NCON}(G; r) := \text{conv}\{x \in \mathbf{R}^E \mid x \text{ satisfies (6i)–(6iv)}\},$$
$$k\text{ECON}(G; r) := \text{conv}\{x \in \mathbf{R}^E \mid x \text{ satisfies (6i), (6iii) and (6iv)}\}.$$

To tie this notation with the previously introduced more general concept, note that

$$k\text{ECON}(G;r) = \text{CON}(G;r',0,0),$$

where $r' \in \mathbf{R}^{V\times V}$ with $r'_{st} := \min\{r_s, r_t\}$ for all $s, t \in V$. Also, if there are no parallel edges then

$$k\text{NCON}(G;r) = \text{CON}(G;r',k',d'),$$

where $k'_{st} := \max\{0, r'_{st} - 1\}$ for all $s, t \in V$ and $d' := r' - k'$.

This survivability model, the polyhedra, and the integer and linear programming problems associated with the kECON and kNCON problems, will be studied in more detail in the sequel.

4. Structural properties and heuristics

In this section we describe some heuristic approaches for the solution of ECON and NCON problems. There are standard methods like greedy and interchange heuristics and heuristics that are motivated by techniques for the approximate solution of other NP-hard problems, like the (much investigated) traveling salesman problem. Some heuristics are more special-purpose since they make use of structural properties of k-connected graphs. A few structural results and their uses are reviewed in Sections 4.1 and 4.2. Section 4.2 concentrates on the design of practically-effective heuristics for ECON and NCON problems, while Section 4.3 discusses lower bounds and heuristics with worst-case performance guarantees.

4.1. Lifting and the structure of optimum solutions

Connectivity is a very rich and active topic of graph theory and it is conceivable that the knowledge that has been accumulated in this area can be exploited further for the design of effective approximation algorithms. We do not attempt to cover the work on connectivity here in detail and refer to Frank [1995] for a comprehensive survey of connectivity results. We will just mention a few structural properties of connected graphs that have been employed for the design of practically-useful heuristics. We begin by describing a 'local' construction technique. Lovász [1976] & Mader [1978] [see also Frank, 1992b] have proved so-called 'lifting theorems' that show that simple manipulations of a graph can be made without destroying certain edge-connectivity properties. These manipulations are useful for construction heuristics for the ECON problem (where parallel edges are allowed) for general connectivity requirements r_{vw}.

In order to state these results we have to introduce a few definitions. Let $G = (V, E)$ be a graph, and let x be a node of G that we call *special*. We assume that x is adjacent to distinct nodes y and z. The graph G' obtained from G by deleting the edges xy and xz and adding the edge yz is called *lifting* of G at x.

If the edge-connectivity of G' between any two nodes of $V' - x$ is not smaller than that of G, then the lifting is called *admissible*.

Theorem 3 [Mader, 1978]. *Let $G = (V, E)$ be a graph with special node x.*

(a) If the degree of x is at least 4 and x is not an articulation node then G has an admissible lifting at x.

(b) If x is an articulation node and no single edge incident to x is a cut then G has an admissible lifting at x.

Theorem 4 [Lovász, 1976]. *Let $G = (V, E)$ be a Eulerian graph, let x be a (special) node of even degree, and let y be any node adjacent to x in G. Then there is another neighbor node z of x, such that the lifting at x involving y and z is admissible.*

Consider the kECON and kNCON problems for the complete graph K_n where $r_x = k$ for all nodes x, and where the nonnegative costs satisfy the triangle inequality, i.e., $c_{xz} \leq c_{xy} + c_{yz}$ for all nodes x, y and z. Using the Lifting Theorems 3 and 4, Monma, Munson & Pulleyblank [1990] showed the following result for kECON for K_n with $k = 2$.

Theorem 5. *Given costs satisfying the triangle inequality, there is an optimal k-edge connected spanning network satisfying the following conditions:*

(a) all nodes are of degree k or $k + 1$; and

(b) removing any set of at most k edges does not leave all of the resulting connected components k-edge connected.

It is easy to see that if the cost function satisfies the triangle inequalities, there is an optimal 2-node connected solution with cost equal to an optimal 2-edge connected solution. In fact, Monma, Munson & Pulleyblank [1990] show that the term 'k-edge' can be replaced by 'k-node' throughout Theorem 5 to obtain a similar result for kNCON for K_n with $k = 2$. Furthermore, they show that these 'characterize' the optimal 2-connected networks in the following sense: given any 2-connected graph $G = (V, E)$ satisfying conditions (a) and (b) of Theorem 5 for $k = 2$, then there exist costs satisfying the triangle inequality such that G is the unique optimal solution.

Bienstock, Brickell & Monma [1990] showed that Theorem 5 holds for arbitrary k. They also showed that 'k-edge' can be replaced by 'k-node' throughout Theorem 5 with the technical restriction that $|V| \geq 2k$ in order for (a) to hold. This technical restriction is necessary since without it there is an infinite family of examples where condition (a) fails.

We note that the cost of an optimal k-edge connected solution may be strictly less than the cost of an optimal k-node connected solution for any $k \geq 3$, and that the conditions in Theorem 5 do not characterize the optimal solutions as they do in the case $k = 2$. The proof of Theorem 5 for the k-edge connected case uses the Lifting Theorem 3. The proof of Theorem 5 in the k-node connected case is much more difficult and requires the use of pairs of liftings as well as a number of further technical results.

4.2. Construction and improvement heuristics

We now describe some heuristics for constructing feasible networks and heuristics for improving the cost of a feasible solution for the kECON and kNCON problems. This is, to a large extent, a summary of the work of Monma & Shallcross [1989] for the 'low-connected' survivable network design problem, that is, problems with node types in $\{0, 1, 2\}$. The performance on real-world problems is described in Section 7. These heuristics were inspired by the wide variety of heuristics for the traveling salesman problem and other combinatorial optimization problems and were designed to work on sparse underlying graphs. The heuristics are used in a local search approach to obtain low-cost network designs; see Papadimitriou & Steiglitz [1982] for a general discussion of local search procedures for combinatorial optimization problems. It is obvious how these heuristics can be generalized and applied to the general survivable network design problem, and so we just briefly mention one such extension.

One useful structural fact is that any edge-minimal two-connected graph $G = (V, E)$ can be constructed by an 'ear decomposition' procedure; see Lovász & Plummer [1986]. That is, first find a cycle C in G. Then repeatedly find a path P, called an ear, that starts at a node v in the solution, passes through nodes not yet in the solution, and ends up at a node w in the solution. All edge-minimal two-edge connected graphs can be constructed in this manner. If the nodes v and w are required to be distinct, then every edge-minimal two-node connected graph can be constructed in this manner.

The ear decomposition approach can be employed to construct a feasible 2-connected subgraph of a graph $G = (V, E)$ using costs to add ears in a greedy fashion. We call this the *greedy ears construction heuristic.* The first step is to construct a partial solution consisting of a cycle C spanning the set of nodes of type 2, called the 'special nodes'. This is done by randomly selecting a special node v, and then selecting a special node w whose shortest path P to v is the longest among all special nodes. Let node u be the node next to w on the path P. We will construct a short cycle through the edge uw by finding a shortest path from u to w not using the edge uw. (There must be such a path; if not, there would not two node-disjoint paths between the special nodes v and w and so the problem would be infeasible). The next step is to repeatedly add 'short' ears to the current partial solution until all special nodes are on this two-connected network. This is done by first selecting a special node z, not yet in the solution, whose shortest path P to the partial solution is longest among all special nodes not yet included. We will find another shortest path Q from z to the partial solution that does not use any edges of P and that terminates on the partial solution at a node w other than v. (Again, such a path must exist for the problem to be feasible). The combination of paths P and Q must contain an ear, which is added to the partial solution.

A second construction heuristic uses the ear decomposition approach to construct a feasible 2-connected subgraph of a graph $G = (V, E)$ in a random fashion. We call this the *random sparse construction heuristic*. The first step is to construct an initial random cycle C spanning a subset of the special nodes. This is done by randomly choosing a special node v, and constructing a depth-first-search tree T rooted at v. Form a cycle by randomly choosing an edge of the form vz that is not in T. (There must be such an edge or else v is not on any cycle and so the problem is infeasible.) Next, random ears are repeatedly added until all special nodes are on the two-connected network. This is done by first constructing a depth-first-search forest F rooted at the nodes that are in the partial solution. A node v is said to be *allowed* if v is not yet in the solution, but it has an edge vw in E but not F, where w is in the solution. Randomly choose a node v from among the allowed nodes. (Again, there must be such a node or the problem is infeasible.) Let T be the tree in the forest F containing v, and let z be the root of T. The random ear chosen is the path from v to z in T, together with the edge vw. Since this method does not use cost information, generally it does not produce a low cost solution. However, this method is useful for generating starting random initial solutions on which to apply the improvement methods.

Improvement heuristics apply local transformations to any feasible network in order to reduce the cost while preserving feasibility. These transformations are applied until a locally optimal network is obtained; that is, no further improvements of these types are possible. Six local transformations are described in the sequel. These transformations are general enough to cover a wide range of feasible topologies, yet fast enough to be performed in real time, even on a personal computer.

Every two-connected network contains at least one cycle, and often is made up of many interconnected cycles. Furthermore, replacing the edges in a cycle C by edges forming a cycle C' on the same nodes preserves the feasibility of a solution. So it is natural to draw upon the extensive research on the traveling salesman problem [see Lawler, Lenstra, Rinnooy Kan & Shmoys, 1985, or Reinelt, 1994, or Chapter 3 of this handbook] for finding a near-optimal cycle. This is the basis of the two-optimal cycle and three-optimal cycle improvement heuristics. The pretzel, quetzel and degree improvement heuristics alter the structure of the two-connected part of the solution in less obvious ways. The one-optimal improvement heuristic alters the structure of the entire solution. These heuristics are described below.

The *two-optimal interchange heuristic* attempts to replace two edges of a cycle C by two edges not in the cycle to form a new cycle C' of lower cost. Similarly, the *three-optimal interchange heuristic* attempts to replace three edges of a cycle C by three edges not in the cycle to form a new cycle C' of lower cost.

These improvement heuristics replace one cycle C by another cycle C' on the same nodes, and so do not change the fundamental structure underlying the solution. The *pretzel transformation* replaces an edge uv of a cycle C by two crossing edges ux and vy to form a 'pretzel' P, where the nodes u, v, x and y appear in order on the cycle.

The *quetzel transformation* is the reverse of the pretzel transformation; that is, a pretzel P is replaced by a cycle C by removing two crossing edges ux and vy and adding an edge uv.

As mentioned before, if the costs satisfy the triangle inequality, then there is an optimal two-connected solution where all nodes are of degree two or three. The proof of this result employs the Lifting Theorems 3 and 4. We will describe now the algorithmic use of these theorems in the form of *degree improvement transformations*.

Let node u be of degree four or more, and let nodes a, b, c, and d be four of its neighbors. There are three cases to consider. In Case a, there are node-disjoint paths P_{ab} and P_{bc} from a to b, and from b to c, respectively, which miss node u. In this case, edge ub is a chord and can be deleted. So we may assume that no three of the nodes a, b, c and d have such paths. Therefore, the paths P_{ab} and P_{bc} must intersect in node v and the paths P_{bc} and P_{cd} must intersect in node w. If nodes v and w are different, we are in Case b, and edges bu and uc are removed, and edge bc is added; this preserves the connectivity requirements and does not increase the cost if the triangle inequality holds. If nodes v and w are the same node, we are in Case c. Let a', b', c' and d' be the neighbors of $v(= w)$. Edges $au, ub, b'v$ and vc' are removed, and edges ab and $b'c'$ are added; this preserves the connectivity requirements and does not increase the cost if the triangle inequality holds. In all cases, the degree of node u is decreased, and no other degrees are increased; so repeated application of these transformations guarantees that an optimal solution where all degrees are two or three will be obtained so long as the triangle inequality holds.

All of the previous improvement heuristics operated only on the two-connected part of the solution. The *one-optimal interchange heuristic* considers the entire solution. This heuristic attempts to remove an edge uv from the current feasible solution and replace it with another edge of the form ux not currently in the solution. Such an interchange is made only if the resultant network is feasible and of lower cost.

These heuristics were tested on randomly generated problems and on real-world problems of telephone network design; see Monma & Shallcross [1989] for details. They could decrease the cost of manually constructed solutions by about 10% in a test of these methods on a real-world application. Comparison with the optimal solutions computed by the cutting plane algorithm (to be described in Section 7) on the same examples show that the gap was usually very small, about 1% of the optimal value. The highest running time for a sparse 116-node problem was 156 seconds on an IBM-PC/AT.

Ko & Monma [1989] modified the low-connectivity heuristics of Monma and Shallcross to the design of k-edge or k-node connected networks. A first feasible solution is constructed either by deleting edges successively from the whole graph while maintaining feasibility, or by adding successively k edge-disjoint $[1, j]$-paths of minimum overall length, for all nodes $j \neq 1$. The output is necessarily k-edge connected. The local transformations for the low-connectivity case carry over to the high-connectivity case, except that the feasibility checks have to be implemented differently.

These heuristics could only be tested on random examples, as real-world examples were not yet available. When the cost of the best heuristic solution was compared with the optimal solution produced by our cutting plane algorithm, the gap was approximately 6%. Running times on a VAX 8650 ranged between 13 s for dense graphs of 20 nodes and $k = 3$ and 120 s for dense graphs of 40 nodes and $k = 5$.

Let us remark that to our knowledge the first heuristics for the design of minimum-cost survivable networks under general connectivity requirements date back to Steiglitz, Weiner & Kleitman [1969]. Their heuristic consists of a randomized starting routine and an optimization routine where local transformations are applied to a feasible solution as follows. Given a random ordering of the nodes, the *starting routine* adds edges between the first node with the highest connectivity requirement and the first node with the next highest connectivity requirement. In each step the connectivity requirements are updated. If the solution is feasible, the *optimizing routine* tries to improve this solution by successively replacing one pair of edges with another pair of edges to obtain another feasible solution of lower cost until no more improvements can be made this way.

They applied their heuristics to two real-world problems with 10 nodes and 58 nodes, and connectivity requirements in $\{3, 4, 5, 6\}$ and $\{6\}$, respectively. (Unfortunately, the data are not available any more.) The 58-node problem took about 12 minutes (on a UNIVAC 1108) per local optimum. Since no lower bounds on the optimal value for these problems are given, we cannot say how well these heuristics work.

4.3. *Heuristics with performance guarantees*

The last remark leads us to an important issue: quality guarantees, i.e., worst-case performance and lower bounds. The polyhedral approach, to be described later, can be viewed as a technique to obtain very good lower bounds for the optimum value of a kECON or kNCON problem. But sometimes nice performance guarantees for heuristics or estimates for optimum values can be given with less elaborate techniques.

Let us first relate the 2-edge connectivity problems to the traveling salesman problem. Since every Hamiltonian cycle is 2-node (and thus 2-edge) connected, the optimum TSP-value, CTSP say, is not smaller than the optimum value COPT of the 2ECON problem with node types $r_v = 2$ for all v. On the other hand, using Theorem 1, Monma, Munson & Pulleyblank [1990] were able to show that if the triangle inequality holds, COPT can be bounded by CTSP from below by a constant factor, more precisely

$$\tfrac{3}{4}\text{CTSP} \leq \text{COPT} \leq \text{CTSP}.$$

To solve the 2ECON problem approximately, Frederickson & JáJá [1982] modified the Christofides heuristic for the traveling salesman problem and proved that, when the triangle inequality holds, the solution attains a cost CCHR with the same worst-case bound as in the traveling salesman problem, namely

$$\text{CCHR} \leq \tfrac{3}{2}\ \text{COPT}.$$

Another type of lower bound for the k-edge connected network design problem can be derived from the subtour elimination polytope, which is a natural linear programming relaxation of the traveling salesman polytope. Let CSUB denote the value of an optimal solution to the subtour elimination linear program; see Chapter 3. Goemans & Bertsimas [1993] showed that

$$\frac{k}{2}\text{CSUB} \leq \text{COPT}.$$

For $k = 2$, this was previously shown by Cunningham, see Monma, Munson & Pulleyblank [1990]. These results make use of the Lifting Theorem 4.

For edge connectivity problems with varying edge connectivity requirements the following heuristics are known. Goemans & Bertsimas [1993] developed a tree heuristic with worst-case guarantee for a version of kECON problem with general node types $r \in \mathbf{Z}_+^V$, where edges may be used more than once. This means that a feasible solution to this problem is a vector $x \in \mathbf{Z}_+^E$ of nonnegative integers that satisfies all cut inequalities of $x(\delta(W)) \geq r(W)$, but not necessarily the upper bounds $x_e \leq 1$. A component $x_e \geq 2$ may be interpreted as 'edge e used x_e times'.

Let $\rho_1 < \rho_2 < \ldots < \rho_p$ be the ordering of the distinct node types in r, let $\rho_0 := 0$, and let V_k be the set of nodes of type at least ρ_k, $k = 1, \ldots, p$.

Tree heuristic

Compute, for all pairs of nodes u, v, the shortest path length c'_{uv} with respect to the given costs c.

Set $x_e := 0$ for all $e \in V \times V$.

For $k = 1$ to p do

- *compute $T_k = (V_k, E_k)$ as the minimum spanning tree of the complete graph induced by V_k with respect to costs c'_e;*
- *set $x_e := x_e + (\rho_k - \rho_{k-1})$ for all $e \in E_k$.*

For each edge $e = (u, v)$ with $c'_e < c_e$ and $x_e > 0$, decrease x_e to 0 and increase by x_e the weights on the edges of a shortest $[u, v]$-path.

Output x_e for $e \in E$.

Note that $x \in \mathbf{Z}_+^E$ satisfies all cut inequalities $x(\delta(W)) \geq r(W)$. Let y be the solution to the LP-relaxation of the ECON problem consisting of cut inequalities (2.4i) and nonnegativity constraints $x_e \geq 0$. Goemans & Bertsimas show that

$$\frac{x^T c}{y^T c} \leq \left(2 - \frac{2}{|V_1|}\right)\left(\sum_{k=1}^{p} \frac{\rho_k - \rho_{k-1}}{\rho_k}\right).$$

This bound is tight, if we consider for instance a 2ECON problem on a cycle of costs 1. Goemans & Bertsimas also describe a more refined tree heuristic with a better worst-case guarantee.

Agrawal, Klein & Ravi [1991] state a heuristic for an ECON problem, where edge-connectivity requirements are given by $r \in \mathbf{Z}_+^{V \times V}$ and the use of multiple parallel edges is allowed in the solution. They prove that their algorithm outputs a solution whose cost is approximately within $2 \log R$ of the optimal, where R is the highest requirement value.

The worst-case guarantees for the heuristics of Goemans & Bertsimas [1993] and of Agrawal, Klein & Ravi [1991] were found by reduction to an ECON problem with costs satisfying the triangle inequality (see Step 1 and 4 of the tree heuristic). This reduction does not work, if the use of parallel edges is forbidden in the solution. For the case that the use of parallel edges is forbidden, the edge connectivity requirements are given by a node type vector r, and the cost function is arbitrary (but nonnegative), Goemans, Mihail, Vazirani & Williamson [1992] proposed a heuristic based on a primal-dual approach (using the cut inequalities). This heuristic has an approximation factor of

$$2 \sum_{i=1}^{p} \mathcal{H}(\rho_i - \rho_{i-1}),$$

where $\mathcal{H}$ is the harmonic function $\mathcal{H}(k) = 1 + \frac{1}{2} + \frac{1}{3} + \cdots + \frac{1}{k}$, and where ρ_i $(i = 1, \ldots, p)$ is defined as above. In particular, for k-edge connectivity problems, the approximation factor is $2 \ln k$, and for $k = 2$ it is 3.

For k-edge connectivity problems one can do even better. Khuller & Vishkin [1994] gave a simple heuristic for finding a minimum-cost k-edge connected subgraph of a graph (where parallel edges do not appear in the solution). This heuristic has a worst-case guarantee of 2, even when the costs do not satisfy the triangle inequality. Worst-case guarantees for heuristics for the NCON problem are not known.

5. Polynomially solvable special cases

We have already remarked that the general problem of designing survivable networks is NP-hard; in fact, quite a number of special cases are known to be NP-hard. The intention of this section is to give an overview of those special cases where polynomial time solution procedures are known.

There are basically three ways to restrict the general model. One either considers special choices of node types, special cost functions or special classes of graphs. It turns out that some combinations of these restrictions lead to easy (but still interesting) problems.

5.1. Node type restrictions

Let us start by considering restrictions on the node types. $G = (V, E)$ may be any graph with costs $c_e \in \mathbf{R}$ for all $e \in E$.

If $r_v = 1$ for all $v \in V$ and $c_e > 0$ for all $e \in E$, the 1NCON and 1ECON problems for G and r are equivalent to finding a minimum cost spanning tree. This problem is well known to be solvable in polynomial time; see Chapter 12. If the costs are not necessarily positive, we look for a minimum cost connected subgraph. This can be found by first choosing all edges of nonpositive cost, shrinking all resulting components to single nodes, and computing a minimum spanning tree in the resulting graph.

If two nodes have type 1, say nodes u and v, all other nodes have type 0, and if all costs are positive, then the 1ECON (and 1NCON) problem for G and r is nothing but the problem of computing a shortest $[u, v]$-path; see Chapter 1 for polynomial time methods for that problem. If costs are general, we seek an edge set of minimum cost that contains a path from u to v. This can be solved in polynomial time by shrinking away the components induced by the nonpositive edges and then computing a shortest $[u, v]$-path in the resulting graph.

The 'slightly' more general case, when $r_v \in \{0, 1\}$ for all $v \in V$, is the Steiner tree problem, which is NP-hard even if $c_e = 1$ for all $e \in E$. If there is a fixed number of nodes of type 0 or a fixed number of nodes of type 1, then the Steiner tree problem is polynomially solvable, see Lawler [1976].

The shortest path problem has an extension that is solvable in polynomial time. Namely, if two nodes have type k, say $r_u = r_v = k$, all others type 0, and if all costs are positive, then the kNCON problem asks for a collection of k node-disjoint $[u, v]$-paths that are of minimum total cost and, similarly, the kECON problem requires finding a minimum collection of k edge-disjoint $[u, v]$-paths. Both problems can be solved with min-cost flow algorithms in polynomial time; see Chapter 1. As above, it is trivial to extend this to the case where costs are arbitrary.

We do not know of any other (nontrivial) case where special choices of node types lead to polynomial time solvability.

5.2. Cost restrictions

When edge costs are restricted to be in $\{1\}$ or $\{0, 1\}$, and the underlying graph is complete, two well-known problems of graph theory are obtained, namely, first, the problem of constructing graphs of certain connectivity properties having a minimum number of edges, and, second, the 'augmentation problem' of extending a given graph with as few as possible edges until it satisfies certain connectivity properties. We are going to survey some results for these problems.

Consider the following problem:

Problem 1. *Given a set V of nodes, and a requirement $r_{st} \geq 0$, for each pair $s, t \in V$, find a graph $G = (V, E)$ such that each pair s, t is at least r_{st}-edge connected in G and such that $|E|$ is as small as possible.*

Chou & Frank [1970] gave a polynomial-time algorithm to solve this problem when G may contain parallel edges, and when the edge-connectivity requirements are all at least 2. Frank & Chou [1970] solved a similar problem when no parallel edges but additional nodes are allowed in the construction. If neither parallel edges nor further nodes are allowed, it is not known whether one can solve Problem 1 in polynomial time.

The node connectivity problem analogous to Problem 1 is open. We are only aware of a result of Harary [1962] who proved that, given n and k, the minimum number of edges in a k-connected graph on n nodes (without parallel edges) is $\lceil kn/2 \rceil$. Such a graph can be constructed easily.

To our knowledge these are the only solved cases with uniform edge costs.

The following problem often runs under the name *augmentation problem* in the graph theory literature.

Problem 2. *Given a graph, augment it by a minimum number of edges so that the new graph meets certain connectivity requirements.*

This type of problem was solved by Eswaran & Tarjan [1976] for 2-edge connected graphs. The k-edge connected graph augmentation problem was studied by Watanabe & Nakamura [1987], Ueno, Kajitani & Wada [1988], Cai & Sun [1989], and Naor, Gusfield & Martel [1990]. Frank [1992a] solved the general edge connectivity case. All these edge-connectivity augmentation algorithms allow the use of parallel edges. The solutions are algorithmic and can be found in polynomial time.

Frank [1992a] proved a nice min-max result for the minimum number of edges needed to augment a given graph G to satisfy given edge connectivity requirements r_{ij}. Let us define the *deficit* def(A) of a node set A as

$$\operatorname{def}(A) := \max_{u \in A, v \notin A} r_{uv} - |\delta_G(A)|.$$

The deficit of A is the smallest number of edges that have to be added to $\delta_G(A)$ in order to connect A sufficiently to all other nodes. Clearly, if several disjoint sets A_i $(i = 1, \ldots, t)$ have deficit def(A_i), then a lower bound on the number of edges to be added is

$$\tfrac{1}{2} \sum_{i=1}^{t} \operatorname{def}(A_i).$$

Frank [1992a] shows that, under certain assumptions, the best such lower bound is exactly the minimum number of edges needed in an augmentation.

Theorem 6. *Given a graph $G = (V, E)$ and edge connectivity requirements r_{st} for all pairs of nodes $s, t \in V$. Then the following holds:*

If some component of G with node set A has deficit at most 1 and all proper subsets of A have nonpositive deficit, then the minimum number of edges to be added to G is def(A) plus the minimum number of edges to be added to $G - A$.

If no such component exists in G then the minimum number of edges to be added to G to satisfy the edge connectivity requirements, is

$$\max_{A_1, \ldots, A_t \text{ disjoint}} \tfrac{1}{2} \sum_{i=1}^{t} \operatorname{def}(A_i).$$

Frank's proof is constructive and results in a polynomial-time algorithm to create a minimum cost augmentation.

Augmentation to k-node connected graphs has been solved for $k = 2$ by Eswaran & Tarjan [1976], and for $k = 3$ by Hsu & Ramachandran [1991]. For $k = 4$, Hsu [1992] solved the problem, when the given graph is already 3-node connected. For general node connectivity requirements it is not even known whether the augmentation problem is NP-complete or not.

When directed graphs are considered, augmentation to k-edge connected digraphs is polynomially solvable, see Frank (1992a) (with parallel edges allowed), and so is augmentation to k-node connected digraphs, see Frank & Jordán [1993]. For general edge- or node-connectivity requirements, the augmentation problem in directed graphs was shown to be NP-complete by Frank [1992a].

5.3. Special classes of graphs

It is often the case that NP-hard problems become easy when restricted to graphs with special structural properties. In the case of the ECON and NCON problems, there are only very few (and relatively simple) classes of graphs known where some versions can be solved in polynomial time. Most notable is the class of series-parallel graphs.

Series-parallel graphs are created from a single edge by iterative application of the following two operations:

- addition of parallel edges, and
- subdivision of edges.

(For our purposes we note that all series-parallel graphs, except K_2, are 2-connected.) *Outerplanar graphs* are a subclass of series-parallel graphs, namely those graphs that can be drawn in the plane as one cycle with noncrossing chords. *Halin graphs* are planar graphs that can be drawn in the plane as a tree without nodes of degree 2 plus one cycle connecting all leaves of the tree.

The Steiner tree problem can be solved in linear time on series-parallel graphs by a recursive algorithm. This was mentioned by Takamizawa, Nishizeki & Saito [1982], and stated explicitly by Wald & Colbourn [1983]. By a modification of this recursive algorithm, the 2NCON problem can also be solved, where node types 0, 1, and 2 are allowed.

Winter has developed linear-time algorithms for 2ECON and 2NCON problems with node types 0 and 2 on outerplanar, series-parallel, and Halin graphs, see Winter [1985a, b, 1986]. In his survey article, Winter [1987] mentioned that he also found linear-time algorithms for Halin graphs that solve the 3ECON and 3NCON problems with node types 0 and 3.

If there exist polynomial-time algorithms to solve edge- or node-connectivity problems on special classes of graphs then it is, in principle, possible to find a complete characterization by linear inequalities of the associated polytopes. Such characterizations are known for series-parallel graphs, and are listed in the following. Complete descriptions of Steiner tree polytopes and related polyhedra, using auxiliary variables, can be found in Prodon, Liebling & Gröflin

[1985], Goemans [1994a, b], Goemans & Myung [1993], and Margot, Prodon & Liebling [1994]. Using projection techniques one can obtain a complete characterization for the 1ECON polyhedron on series-parallel graphs without auxiliary variables, see also Goemans [1994b]. The projection technique and the inequalities generated by it are described in Section 8. A nonredundant description for the 1ECON polytope on series-parallel graphs is, however, yet unknown. Cornuéjols, Fonlupt & Naddef [1985] found a complete description of the dominant of the 2-edge connected subgraph polytope of series-parallel graphs. A complete description of the 2-edge-connected Steiner subgraph polytope on series-parallel subgraphs is given in Baïou & Mahjoub [1993]. For odd k, Chopra [1994] investigated the k-edge connected subgraphs of a given outerplanar graph, and found a complete description of the dominant of the associated polyhedron by the so-called *lifted outerplanar partition inequalities*. Chopra's result is as follows:

Theorem 7. *For outerplanar graphs $G = (V, E)$ and uniform node types $r \in \{k\}^V$, k odd, the dominant of the kECON$(G;r)$ polytope (that is, kECON$(G;r) + \mathbf{R}_+^E$) is completely characterized by the inequalities:*

$$\tfrac{1}{2}\sum_{i=1}^{p} x(\delta(W_i)) \geq p \cdot \lceil k/2 \rceil - 1 \quad \text{for all partitions } \{W_1, \ldots, W_p\} \text{ of } V_j,$$
$$x_e \geq 0 \quad \text{for all } e \text{ in } E.$$

This inequality can be lifted to an inequality valid and nonredundant for the dominant of kECON(K_n, r) by computing the coefficients of the missing edges as the shortest-path value between their endpoints, and using as 'lengths' the coefficients on E).

6. Polyhedral results

Except for the results of Grötschel & Monma [1990] mentioned in Section 3, there is not much known about the polytope CON$(G; r, k, d)$ for general edge and node survivability requirements r, k and d. We will thus concentrate on the kNCON and kECON problems that have been investigated in more depth and survey some of the known results. Particular attention has been paid to the low-connectivity case, that is, where $r \in \{0, 1, 2\}^E$. See Grötschel & Padberg [1985] and Pulleyblank [1989] for a general survey of polyhedral combinatorics and the basics of polyhedral theory.

Let us mention again the idea behind this approach and its goal. We consider an integer programming problem like (3) or (6). We want to turn such an integer program into a linear program and solve it using the (quite advanced) techniques of this area. To do this, we define a polytope associated with the problem by taking the convex hull of the feasible (integral) solutions of a program like (3) or (6). Let P be such a convex hull. We know from linear programming theory that, for any objective function c, the linear program $\min\{c^T x \mid x \in P\}$ has an optimum vertex solution (if it has a solution). This vertex solution is, by definition, a feasible solution of the initial integer program and thus, by construction, an optimum solution of this program.

The difficulty with this approach is that $\max c^T x, x \in P$ is a linear program only 'in principle'. To provide an instance to an LP-solver, we have to find a different description of P. The polytope P is defined as the convex hull of (usually many) points in $\mathbf{R}^E$, but we need a complete (linear) descriptions of P by means of linear equations or inequalities. The Weyl–Minkowski theorem tells us that both descriptions are in a sense equivalent, in fact, there are constructive procedures that compute one description of P from the other. However, these procedures are inherently exponential and nobody knows how to make effective use of them, in particular, for NP-hard problem classes. Moreover, there are results in complexity theory, see Papadimitriou & Yannakakis [1982], that indicate that it might be much harder to find a complete linear description of such a polytope P than to solve $\min c^T x, x \in P$.

At present, no effective general techniques are known for finding complete or 'good partial' descriptions of such a polytope or large classes of facets. There are a few basic techniques like the derivation of so-called Chvátal cuts (see Chvátal [1973]). But most of the work is a kind of 'art'. Valid inequalities are derived from structural insights and the proofs that many of these inequalities define facets use technically complicated, ad-hoc arguments.

If large classes of facet-defining inequalities are found, one has to think about their algorithmic use. The standard technique is to employ such inequalities in the framework of a cutting plane algorithm. We will explain this in Section 7. It has turned out in the recent years that such efforts seem worthwhile. If one wants to find true optimum solutions or extremely good lower bounds, the methods of polyhedral combinatorics are the route to take.

6.1. Classes of valid inequalities

We will now give an overview of some of the results of Grötschel, Monma & Stoer [1992a–c] and Stoer [1992] concerning classes of valid inequalities for the kECON and kNCON problems. We will motivate these inequalities and mention how they arise.

As before, we consider a loopless graph $G = (V, E)$, and in the kECON case possibly with multiple edges. We assume that for each node $v \in V$ a nonnegative integer r_v, its node type, is given, that $k = \max\{r_v \mid v \in V\}$ and that at least two nodes are of type k. Recall that $r(W) = \max\{r_v \mid v \in W\}$ is called the node type of W.

We start out by repeating those classes we have already introduced in Section 3. Clearly, the *trivial inequalities*

$$0 \le x_e \le 1 \qquad \text{for all } e \in E \tag{7}$$

are valid for kECON$(G;r)$ and kNCON$(G;r)$ since problem (6) is a 0/1-

optimization problem. The *cut inequalities*

$$x(\delta(W)) \geq \mathrm{con}(W) \qquad \text{for all } W \subseteq V,\ \emptyset \neq W \neq V, \tag{8}$$

where $\mathrm{con}(W)$ is given by (1), or equivalently by $\min\{r(W), r(V \setminus W)\}$, are valid for $k\mathrm{ECON}(G;r)$ and $k\mathrm{NCON}(G;r)$, since the survivable network to be designed has to contain at least $\mathrm{con}(W)$ edge-disjoint paths that connect nodes in W to nodes in $V \setminus W$. (Recall that $r_{st} = \min\{r_s, r_t\}$, $s, t \in V$.) In the node connectivity case we require that upon deletion of any set Z of nodes there has to be, for all pairs $s, t \in V \setminus Z$, at least $r_{st} - |Z|$ more paths connecting s and t in the remaining graph. This requirement leads to the *node cut inequalities*

$$x(\delta_{G-Z}(W)) \geq \mathrm{con}(W) - |Z| \tag{9}$$

$$\begin{aligned}&\text{for all pairs } s, t \in V, s \neq t \text{ and}\\ &\text{for all } \emptyset \neq Z \subseteq V \setminus \{s,t\} \text{ with } |Z| \leq r_{st} - 1\\ &\text{and for all } W \subseteq V \setminus Z \text{ with } s \in W,\ t \notin W.\end{aligned}$$

These inequalities are valid for $k\mathrm{NCON}(G;r)$ but — of course — not for $k\mathrm{ECON}(G;r)$.

How does one find further classes of valid inequalities? One approach is to infer inequalities from structural investigations. For instance, the cut inequalities ensure that every cut separating two nodes contains at least r_{st} edges. These correspond to partitioning the node set into two parts and guaranteeing that there are enough edges linking them. We can generalize this idea as follows. Let us call a system $W_1, \ldots, W_p$ of nonempty subsets of V with $W_i \cap W_j = \emptyset$ for $1 \leq i < j \leq p$, and $W_1 \cup \ldots \cup W_p = V$ a *partition* of V and let us call

$$\begin{aligned}&\delta(W_1, \ldots, W_p) :=\\ &\quad := \{uv \in E \mid \exists i, j,\ 1 \leq i, j \leq p,\ i \neq j \text{ with } u \in W_i,\ v \in W_j\}\end{aligned}$$

a *multicut* or *p-cut* (if we want to specify the number p of *shores* $W_1, \ldots, W_p$ of the multicut). Depending on the numbers $\mathrm{con}(W_1), \ldots, \mathrm{con}(W_p)$, any survivable network (V, F) will have to contain at least a certain number of edges of the multicut $\delta(W_1, \ldots, W_p)$. For every partition it is possible to compute a lower bound of this number, and thus to derive a valid inequality for every node partition (resp. multicut). This goes as follows.

Suppose $W_1, \ldots, W_p$ is a partition of V such that $\mathrm{con}(W_i) \geq 1$ for $i = 1, \ldots, p$. Let $I_1 := \{i \in \{1, \ldots p\} \mid \mathrm{con}(W_i) = 1\}$, and $I_2 := \{i \in \{1, \ldots, p\} \mid \mathrm{con}(W_i) \geq 2\}$. Then the *partition inequality* (or multicut inequality) induced by $W_1, \ldots W_p$ is defined as

$$\begin{aligned}&x(\delta(W_1, \ldots, W_p)) =\\ &\quad = \tfrac{1}{2}\sum_{i=1}^{p} x(\delta(W_i)) \geq \begin{cases} \left\lceil \tfrac{1}{2}\sum_{i \in I_2} \mathrm{con}(W_i) \right\rceil + |I_1| & \text{if } I_2 \neq \emptyset,\\ p - 1 & \text{if } I_2 = \emptyset.\end{cases}\end{aligned} \tag{10}$$

Every partition inequality is valid for $k\mathrm{ECON}(G;r)$ and thus for $k\mathrm{NCON}(G;r)$.

Just as the cut inequalities (8) can be generalized as outlined above to partition inequalities (10), the node cut inequalities (9) can be generalized to a class of inequalities that we will call node partition inequalities, as follows.

Let $Z \subseteq V$ be some node set with $|Z| \geq 1$. If we delete Z from G then the resulting graph must contain an $[s, t]$-path for every pair of nodes s, t of type larger than $|Z|$. In other words, if $W_1, \ldots, W_p$ is a partition of $V \setminus Z$ into node sets with $r(W_i) \geq |Z| + 1$ then the graph G' obtained by deleting Z and contracting $W_1, W_2, \ldots, W_p$ must be connected. This observation gives the following class of *node partition inequalities* valid for kNCON$(G; r)$, but not for kECON$(G; r)$:

$$\tfrac{1}{2}\sum_{i=1}^{p} x(\delta_{G-Z}(W_i)) \geq p - 1 \qquad (11)$$

for every node set $Z \subseteq V$, $|Z| \geq 1$ and every partition $W_1, \ldots, W_p$ of $V \setminus Z$ such that $r(W_i) \geq |Z| + 1$, $i = 1, \ldots, p$.

If $r(W_i) \geq |Z| + 2$ for at least two node sets in the partition, then the right-hand side of the node partition inequality can be increased. This leads to further generalizations of the classes (10) and (11), but their description is quite technical and complicated, see Stoer [1992]. So we do not discuss them here.

We now mention another approach to finding new classes of valid inequalities. The idea here is to relax the problem in question by finding a (hopefully easier) further combinatorial optimization problem such that every solution of the given problem is feasible for the new problem. One can then study the polytope associated with the new combinatorial optimization problem. If the relaxation is carefully chosen — and one is lucky — some valid inequalities for the relaxed polytope turn out to be facet-defining for the polytope one wants to consider. These inequalities are trivially valid.

In our case, a relaxation that is self-suggesting is the so-called r-cover problem. This ties the survivability problem to matching theory and, in fact, one can make good use of the results of this theory for the survivability problem.

The survivability requirements imply that if $v \in V$ is a node of type r_v, then v has degree at least r_v for any feasible solution of the kECON problem. Thus, if we can find an edge set of minimum cost such that each node has degree at least r_v (we call such a set an *r-cover*), we obtain a lower bound for the optimum value of the kECON problem. Clearly, such an edge set can be found by solving the integer linear program

$$\begin{array}{llll} \min & c^T x & & \\ \text{(i)} & x(\delta(v)) & \geq r_v & \text{for all } v \in V, \\ \text{(ii)} & 0 \leq x_e & \leq 1 & \text{for all } e \in E, \text{ and} \\ \text{(iii)} & x_e \text{ integer} & & \text{for all } e \in E, \end{array} \qquad (12)$$

which is obtained from (3i), (3iii) and (3iv) by considering only sets of cardinality one in (3i). The inequalities (12i) are called *degree constraints*. This integer program can be turned into a linear program, i.e., the integrality constraints (12iii) are replaced by a system of linear inequalities, using Edmonds' polyhedral results on b-matching, see Edmonds [1965]. Edmonds proved that, for any vector $b \in Z_+^V$, the vertices of the polyhedron defined by

$$\begin{array}{llll} \text{(i)} & y(\delta(v)) & \leq b_v & \text{for all } v \in V, \\ \text{(ii)} & y(E(H)) + y(\bar{T}) & \leq \left\lfloor \frac{1}{2} \sum_{v \in H} (b_v + |\bar{T}|) \right\rfloor & \text{for all } W \subseteq V \\ & & & \text{and all } \bar{T} \subseteq \delta(H), \text{ and} \\ \text{(iii)} & 0 \leq y_e \leq 1 & & \text{for all } e \in E \end{array} \tag{13}$$

are precisely the incidence vectors of all (1-capacitated) *b-matchings* of G, i.e., of edge sets M such that no node $v \in V$ is contained in more than b_v edges of M. For the case $b_v := |\delta(v)| - r_v$, the b-matchings M are nothing but the complements $M = E \setminus F$ of r-covers F of G. Using the transformation $x := 1 - y$ and $T := \delta(H) \setminus \bar{T}$ we obtain the system

$$\begin{array}{llll} \text{(i)} & x(\delta(v)) & \geq r_v & \text{for all } v \in V, \\ \text{(ii)} & x(E(H)) + x(\delta(H) \setminus T) & \geq \left\lfloor \frac{1}{2} \sum_{v \in H} (r_v - |T|) \right\rfloor & \text{for all } H \subseteq V \\ & & & \text{and all } T \subseteq \delta(H), \text{ and} \\ \text{(iii)} & 0 \leq x_e \leq 1 & & \text{for all } e \in E. \end{array} \tag{14}$$

(14) gives a complete description of the convex hull of the incidence vectors of all r-covers of G. We call the inequalities (14ii) *r-cover inequalities*. Since every solution of the kECON problem for G and r is an r-cover, all inequalities (14ii) are valid for $k\text{ECON}(G; r)$. It is a trivial matter to observe that those inequalities (14ii) where $\sum_{v \in H} r_v - |T|$ is even are redundant. For the case $r_v = 2$ for all $v \in V$, Mahjoub [1994] described the class of r-cover inequalities, which he calls odd wheel inequalities.

Based on these observations one can extend inequalities (14ii) to more general classes of inequalities valid for $k\text{ECON}(G; r)$ (but possibly not valid for the r-cover polytope). We present here one such generalization.

Let H be a subset of V called the *handle*, and $T \subseteq \delta(H)$ with $|T|$ odd and $|T| \geq 3$. For each $e \in T$, let T_e denote the set of the two end nodes of e. The sets $T_e, e \in T$, are called *teeth*. Let $H_1, \ldots, H_p$ be a partition of H into nonempty pairwise disjoint subsets such that $r(H_i) \geq 1$ for $i = 1, \ldots, p$, and $|H_i \cap T_e| \leq r(H_i) - 1$ for all $i \in \{1, \ldots, p\}$ and all $e \in T$. Let $I_1 := \{i \in \{1, \ldots, p\} \mid r(H_i) = 1\}$ and $I_2 = \{i \in \{1, \ldots, p\} \mid r(H_i) \geq 2\}$. We call

$$x(E(H)) - \sum_{i=1}^{p} x(E(H_i)) + x(\delta(H) \setminus T) \geq \left\lceil \frac{1}{2} \sum_{i \in I_2} (r(H_i) - |T|) \right\rceil + |I_1| \tag{15}$$

the *lifted r-cover inequality* (induced by $H_1, \ldots, H_p, T$). All inequalities of type (15) are valid for $k\text{ECON}(G; r)$.

The names 'handle' and 'teeth' used above derive from the observation that there is some relationship of these types of inequalities with the 2-matching, comb and clique tree inequalities for the symmetric traveling salesman polytope; see Chapter 3. In fact, comb inequalities for the traveling salesman problem can be transformed in various ways to facet-defining inequalities for 2ECON and kNCON polyhedra, as mentioned in Grötschel, Monma & Stoer [1992a], Boyd & Hao [1993], and Stoer [1992].

Another technique for finding further classes of valid and facet-defining inequalities will be mentioned in Section 6.3.

To develop successful cutting plane algorithms, it is not enough to know some inequalities valid for the polytope over which one wants to optimize. The classes of inequalities should contain large numbers of facets of the polytope. Ideally, one would like to use classes of facet-defining inequalities only.

In our case, it turned out to be extremely complicated to give (checkable) necessary and sufficient conditions for an inequality in one of the classes described above to define a facet of $k\text{NCON}(G;r)$ or $k\text{ECON}(G;r)$. Lots of technicalities creep in, when general graphs G, as opposed to complete graphs, are considered. Nevertheless, it could be shown that large subsets of these classes are facet-defining also for the relatively sparse graphs that come from the applications, see Figures 4 and 7 for examples. These results provide a theoretical justification for the use of these inequalities in a cutting plane algorithm. Details about facet results for the inequalities described above can be found in Grötschel & Monma [1990], Grötschel, Monma & Stoer [1992a–c], Stoer [1992].

6.2. Separation

Note that — except for the trivial inequalities — all classes of valid inequalities for the kECON and kNCON problem described in Section 6.1 contain a number of inequalities that is exponential in the number of nodes of the given graph. So it is impossible to input these inequalities into an LP-solver. But there is an alternative approach. Instead of solving an LP with all inequalities, we solve one with a few 'carefully selected' inequalities and we generate new inequalities as we need them. This approach is called a cutting plane algorithm and works as follows.

We start with an initial linear program. In our case, it consists of the linear program (12) without the integrality constraints (12iii). We solve this LP. If the optimum solution y is feasible for the kECON or kNCON problem, then we are done. Otherwise we have to find some inequalities that are valid for $k\text{ECON}(G;r)$ or $k\text{NCON}(G;r)$ but are violated by y. We add these inequalities to the current LP and repeat.

The main difficulty of this approach is in efficiently generating violated inequalities. We state this task formally.

Separation Problem (for a class C of inequalities). *Given a vector y decide whether y satisfies all inequalities in C and, if not, output an inequality violated by y.*

A trivial way to solve Problem 3 is to substitute y into each of the inequalities in C and check whether one of the inequalities is violated. But in our case this is too time consuming since C is of size exponential in $|V|$. Note that all the classes C

described before have an implicit description by means of a formula with which all inequalities can be generated. It thus may happen that algorithms can be designed that check violation much more efficiently than the trivial substitution process. We call an algorithm that solves Problem 3 an (exact) *separation algorithm* for C, and we say that it runs in polynomial time if its running time is bounded by a polynomial in $|V|$ and the encoding length of y.

A deep result of the theory of linear programming, see Grötschel, Lovász & Schrijver [1988], states (roughly) that a linear program over a class C of inequalities can be solved in polynomial time if and only if the separation problem for C can be solved in polynomial time. Being able to solve the separation problem thus has considerable theoretical consequences.

This result makes use of the ellipsoid method and does not imply the existence of a 'practically efficient' algorithm. However, the combination of separation algorithms with other LP solvers (like the simplex algorithms) can result in quite successful cutting plane algorithms; see Section 7.

Our task now is to find out whether reasonable separation algorithms can be designed for any of the classes (8), (9), (10), (11), (14ii), and (15).

There is some good and some bad news. The good news is that for the cut inequalities (8), the node cut inequalities (9) and the r-cover inequalities (14ii), exact separation algorithms are known that run in polynomial time; see Grötschel, Monma & Stoer [1992c].

When C is the class of cut inequalities and y is a nonnegative vector, separation can be solved by any algorithm determining a cut $\delta(W)$ of minimum capacity $y(\delta(W))$ in a graph. Fast min-cut algorithms are described in Hao & Orlin [1992] and Nagamochi & Ibaraki [1992]. Both algorithms do not need more than $O(|V|^3)$ time. The so-called Gomory–Hu tree storing one minimum (s,t)-cut for each pair of nodes s, t in a tree structure can be computed in $O(|V|^4)$ time, see Gomory & Hu [1961].

When C is the class of cut and node cut inequalities (8) and (9), the separation problem can be reduced to a sequence of minimum (s,t)-cut computations in a directed graph. This polynomial-time method is described in Grötschel, Monma & Stoer [1992c].

The polynomial-time exact separation algorithm for the r-cover inequalities is based on the Padberg–Rao procedure for solving the separation problem for the capacitated b-matching inequalities, see Padberg & Rao [1982]. The 'trick' is to reverse the transformation from the b-matching to the r-cover problem described in (13) and (14) and call the Padberg–Rao algorithm. It is easy to see that y satisfies all r-cover inequalities (14ii) if and only if its transformation satisfies all b-matching inequalities (13ii). The Padberg–Rao procedure is quite complicated to describe, so we do not discuss it here.

The bad news is that it was shown in Grötschel, Monma & Stoer [1992b] that the separation problems for partition inequalities (10), node partition inequalities (11) and lifted r-cover inequalities (15) are NP-hard. (A certain generalization of partition inequalities for kECON problems with $k \leq 2$ is, however, polynomial-time separable, see Section 8).

Thus, in these cases we have to revert to *separation heuristics*, i.e., fast procedures that check whether they can find an inequality in the given class that is violated by y, but which are not guaranteed to find one even if one exists. We discuss separation heuristics in more detail in Section 7.

6.3. Complete descriptions of small cases

For the investigation of combinatorially defined polyhedra, it is often useful to study small cases first. A detailed examination of such examples provides insight into the relationship between such polyhedra and gives rise to conjectures about general properties of these polyhedra. Quite frequently a certain inequality is found, by numerical computation, to define a facet of a kECON or kNCON polytope of small dimension. Afterwards it is often possible to come up with a class (or several classes) of inequalities that generalize the given one and to prove that many inequalities of these classes define facets of combinatorial polytopes of any dimension.

By means of a computer program, we have computed complete descriptions of all kECON and kNCON polytopes of small dimensions. To give a glimpse of these numerically obtained results, we report here the complete descriptions of all 2ECON and all 2NCON polytopes of the complete graphs on five vertices K_5. More information about small kECON polytopes can be found in Stoer [1992].

6.3.1. The 2ECON polytope for K_5

Let us begin with the polytopes 2ECON$(K_5; r)$ where $r = (r_1, \ldots, r_5)$ is the vector of node types. The node types r_i have value 0, 1 or 2, and by assumption, at least two nodes are of highest type 2. Clearly, we can suppose that $r_i \geq r_{i+1}$ for $i = 1, \ldots, 4$. These assumptions result in ten node type vectors to be considered. It is obvious that, if a node type vector r componentwise dominates a vector r' (i.e., $r_i \geq r'_i$ for all i), then 2ECON$(K_n; r)$ is contained in 2ECON$(K_n; r')$.

Figure 1 provides a comprehensive summary of our findings. In this figure, a polytope 2ECON$(K_5; r)$ is depicted by its node type vector $r = (r_1, \ldots, r_5)$. A line linking two such vectors indicates that the polytope at the lower end of the line directly contains the polytope at the upper end of the line and that no other 2ECON polytope is 'in between'. For example, the polytope 2ECON$(K_5; (2, 2, 2, 1, 0))$ is directly contained in 2ECON$(K_5; (2, 2, 1, 1, 0))$ and 2ECON$(K_5; (2, 2, 2, 0, 0))$, and it contains directly 2ECON$(K_5; (2, 2, 2, 1, 1))$ and 2ECON$(K_5; (2, 2, 2, 2, 0))$. Next to the right or left of a node type vector r, a box indicates which inequalities are needed to define the corresponding 2ECON $(K_5; r)$ polytope completely and nonredundantly. The notation is as follows:

- The type of inequality appears in the first column.
  - 'p' stands for 'partition inequality', see (10),
  - 'cut' stands for 'cut constraint', see (8),
  - 'rc' stands for 'lifted r-cover inequality', see (15),
  - 'd' stands for 'degree constraint', see (12i),
  - 'rc+1', 'I1', and 'I2', stand for new types of inequalities explained later.

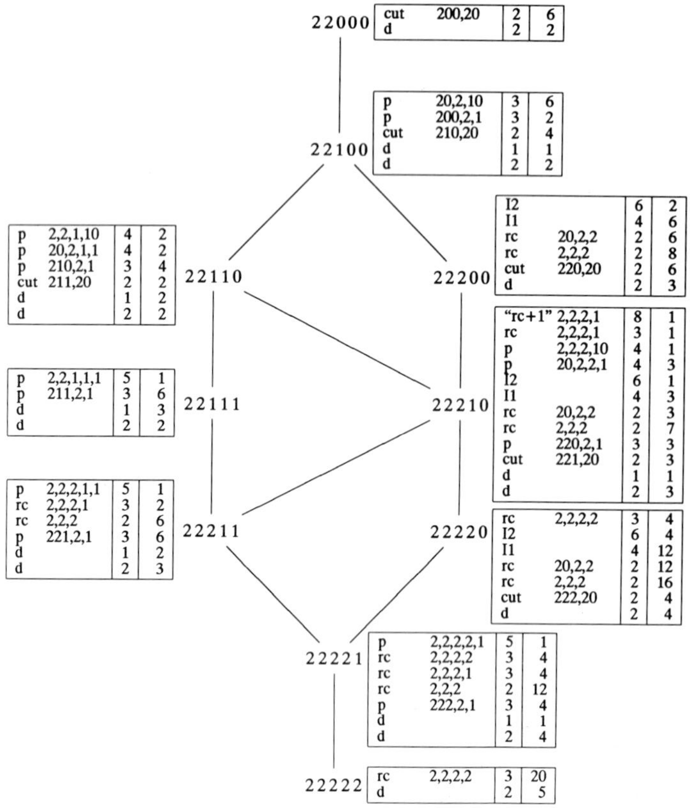

Fig. 1. 2ECON($K_5; r$) polyhedra.

- The next column lists, for each partition of the handle ('rc') or the whole node set ('p'), the node types in each node set. The different sets are separated by commas.
  - For instance, 'p 200,2,1' stands for a partition inequality induced by a partition of V, whose first node set contains one node of type 2 and two nodes of type 0, whose second set contains exactly one node of type 2, and whose last set contains exactly one node of type 1.
  - 'rc 20,2,2' stands for a lifted r-cover inequality induced by a handle that is partitioned into three node sets, the first one containing two nodes, a node of type 2 and a node of type 0, the second node set containing exactly one node of type 2, and the third node set containing exactly one node of type 2; the number of teeth can be computed with the help of the right-hand side.
- The next column gives the right-hand side.
- The last column contains the number of inequalities of the given type.

We do not list the trivial inequalities $0 \leq x_e \leq 1$, since they always define facets of the considered polytopes, see Theorem 2.

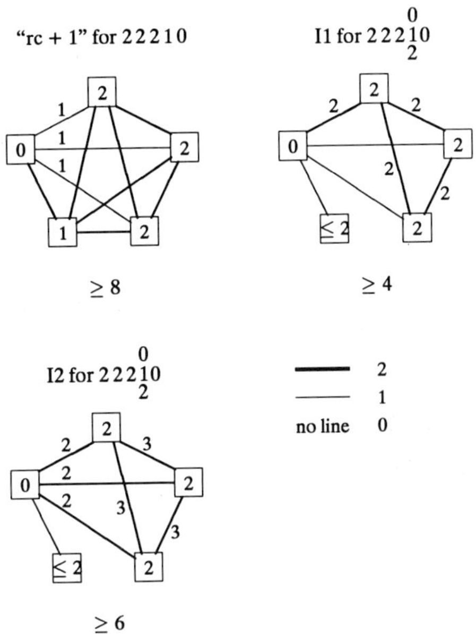

Fig. 2. Inequalities for 2ECON($K_5; r$).

The inequalities denoted by 'rc+1', 'I1', and 'I2' in 2ECON(K_5; (2, 2, 2, 0, 0)), and 2ECON(K_5; (2, 2, 2, 1, 0)), are depicted in Figure 2. All except I2 have coefficients in {0, 1, 2}. The coefficients of inequality I2 in Figure 2 take values in {0, 1, 2, 3}. Edges of coefficient 0 are not drawn, edges of coefficient 1 are drawn by thin lines, and edges of coefficient 2 are drawn by bold lines. To make this distinction somewhat clearer, we additionally display the coefficients of all thin or of all bold lines.

This numerical study of 2ECON polytopes of K_5 reveals that degree, cut, partition, and lifted r-cover inequalities play a major role in describing the polytopes completely and nonredundantly. But at the same time we found three new classes of (complicated) inequalities.

6.3.2. The 2NCON polytope for K_5

We now turn our attention to the 2NCON polytopes of the complete graph K_5. It turned out that only two further classes of inequalities are needed to describe all polytopes 2NCON ($K_5; r$) completely and nonredundantly. These are the classes of node cut and node partition inequalities (9) and (11). Figure 3 displays the complete descriptions of the 2NCON polytopes for K_5 in the same way as Figure 1 does for the 2ECON polytopes. The (new) entries for the node cut and node partition inequalities read as follows.

'ncut 20,20' denotes a node cut inequality $x(\delta_{G-z}(W))$, where both W and contains a node of type 2 and a node of type 0, and $V \backslash (W \cup \{z\})$ contains a node of type 2 and a node of type 0; the '.' in 'ncut 2.,2.' represents a node of any type;

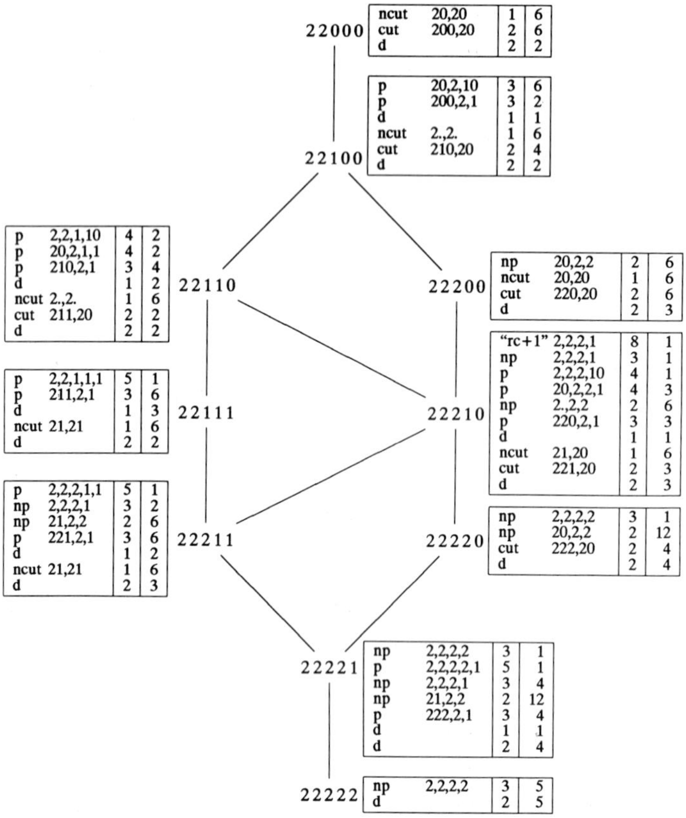

Fig. 3. 2NCON($K_5;r$) polyhedra.

and 'np 20,2,2' denotes a node partition inequality induced by a partition of $V\setminus\{z\}$, where z is some node in V, and the first shore of the partition consists of a node of type 2 and a node of type 0, the second shore consists of a node of type 2, and the third shore consists of a node of type 2.

This concludes our examination of polyhedra for small instances of 2NCON and 2ECON problems. For further such results, see Stoer [1992].

7. Computational results

For applied mathematicians, the ultimate test of the power of a theory is its success in helping solve the practical problems for which it was developed. In our case, this means that we have to determine whether the polyhedral theory for the survivable network design problem can be used in the framework of a cutting plane algorithm to solve the design problems of the sizes arising in practice. The results reported in Grötschel, Monma & Stoer [1992b] show that the design problems for Local Access Transport Area (LATA) networks arising at Bell Communications Research (Bellcore) can be solved to optimality quite easily. There is good reason to hope that other network design problems of this type can also be attacked successfully with this approach. Moreover, the known heuristics also seem to work quite well, at least for the low connectivity case.

7.1. Outline of the cutting plane algorithm

We have already mentioned in Section 6.2 how a cutting plane approach for our problem works. Let us repeat this process here a little more formally.

We assume that a graph $G = (V, E)$ with edge cost $c_e \in \mathbf{R}$ for all $e \in E$ and node types $r_v \in \mathbf{Z}_+$ for all $v \in V$ is given. Let $k := \max\{ r_v \mid v \in V \}$. We want to solve either the kNCON or the kECON problem for G and r and the given cost function, i.e., we want to solve

$$\min_{x \in k\text{NCON}(G;r)} c^T x \quad \text{or} \quad \min_{x \in k\text{ECON}(G;r)} c^T x.$$

We do this by solving a sequence of linear programming relaxations that are based on the results we have described in Section 6. The initial LP (in both cases) consists of the degree constraints and trivial inequalities, i.e.,

$$\begin{aligned} &\min \quad c^T x \\ &x(\delta(v)) \geq r_v \quad \text{for all } v \in V; \\ &0 \leq x_e \leq 1 \quad\quad \text{for all } e \in E. \end{aligned} \tag{16}$$

If a current LP-relaxation has been solved and y is a basic optimum solution, we check whether y is in $k\text{NCON}(G;r)$ or $k\text{ECON}(G;r)$. If it is, the problem is already solved to optimality. Otherwise we call our separation routines and try to find inequalities in one of the classes described in Section 6 that are violated by y. If one of the separation algorithms is successful, we add the inequalities found to the current LP and repeat the procedure. If no violated inequality can be identified, there are two options. We may simply stop and report a lower bound or we may resort to an enumerative procedure like branch and bound.

The best option to select in the case that the cutting plane phase has not produced an optimum solution of the kECON or kNCON problem depends on the demands of practice. Before starting the cutting plane algorithm, one usually runs some heuristics to produce (hopefully) good feasible solutions, see Section 4 for a description of such heuristics. By comparing the lower bound L from the cutting plane algorithm with the upper bound U from the heuristics, one can easily get an idea about the quality of the solutions. If the percentage deviation of these values, usually taken as $100 * (U - L)/L$, is less than some threshold, say 5%, and if the cost data were somewhat fuzzy anyway, the best heuristic solution might simply be considered as an appropriate good solution of the practical problem. If the data are precise and an optimum solution is needed, branch and bound has a good chance to terminate in a 'reasonable' amount of time.

If, however, the deviation of U and L is large, no simple advice can be given. Either the heuristic or the cutting plane algorithm or both may have produced poor bounds. Further research is usually necessary to determine the reasons for failure, to detect special structures of the problem in question that may have resulted in traps for the heuristics or poor performance of the cutting plane algorithms. If such structural insight can be obtained, one has a good chance to improve the result, at least for the current problem instance.

The last case is definitely unsatisfactory, but since we are dealing with hard problems, we have to expect such behavior every now and then. For our real world applications we can report that, for the LATA network design problems, the cutting plane algorithm always found an optimum integral solution, except in three cases that were easily solved by branch and bound or manual interaction. Random problems with low and high connectivity requirements and random cost structure were solved to optimality extremely quickly. But there is one large scale practical problem with 494 nodes, and 1096 edges and highly structured topology and costs, where we ran into considerable difficulties. Using special purpose separation heuristics etc., we were finally able to solve two versions of the problem to optimality, with quite some effort however.

7.2. *Implementation details*

We will not discuss implementation details of the heuristics outlined in Section 4; see Monma & Shallcross [1989] for details. This is quite straightforward, although of course, the use of good data structures and search strategies may result in considerable running time improvements.

We will focus here on implementation issues of the cutting plane algorithm, a number of which are vital for obtaining satisfactory running time performances. Before starting the cutting plane algorithm we try to reduce the size of the problem instance by decomposing it. In fact, the practical problems we solved have rather sparse graphs of possible direct links, and the survivability requirements often force certain edges to be present in every feasible solution. Such edges can be fixed and removed from the problem by appropriately changing certain node types. This removal may break the original problem into several smaller ones that can be solved independently.

There are further ways of decomposing a problem into independent subproblems like decomposing on articulation nodes, on cut sets of size two, and on articulation sets of size two. In each of these cases one can perform the decomposition or determine that no such decomposition is possible, using polynomial time methods like depth-first search or connectivity algorithms. All of this is quite easy, though a precise description of the necessary transformations would require considerable space. Details can be found in Grötschel, Monma & Stoer [1992b] and Stoer [1992].

The main purpose of this decomposition step is to speed up the computation by getting rid of some trivial special cases that the cutting plane algorithm does not need to check any more, and by reducing the sizes of the problems to be solved.

At the end of this preprocessing phase we have decomposed the original problem into a list of subproblems for which we call the cutting plane algorithm. The optimal solution of the original problem can then be composed from the optimal solutions of the subproblems in a straightforward manner.

An issue of particular importance is the implementation of the separation algorithms. Good heuristics have to be found that 'solve' the separation problems for those classes that are not known to be separable in polynomial time. And even if polynomial exact separation routines are known, separation heuristics may help considerably to speed up the overall program. Further problems are to determine the order in which the heuristic and exact separation routines are to be called, when to stop the separation process in case many violated inequalities have been found, which cutting planes to add and which to eliminate from the current LP. These issues cannot be decided by 'theoretical insight' only. Practical experience on many examples is necessary to come up with recipes that result in satisfactory overall performance of such an algorithm. We outline some of the techniques used in the sequel.

Let G be a graph with node types $r \in \mathbf{Z}_+^V$, and let y be some point in $\mathbf{R}^E$ with $0 \leq y_e \leq 1$. Our aim is to find a partition (resp. node partition, lifted r-cover) inequality violated by this point. By the NP-completeness results mentioned in Section 6.2 it seems hopeless to find an efficient exact algorithm for the separation of these inequalities; therefore we have to use heuristics. Nevertheless, it is possible to solve the separation problem for a certain subclass of these inequalities, namely cut constraints (resp., node cut and r-cover constraints) in polynomial time. So in our heuristics we often use 'almost violated' inequalities of these subclasses and transform them into violated inequalities of the larger class. Here an almost violated inequality is an inequality $a^T x \geq b$ with $a^T y \leq b + \alpha$ for some 'small' parameter α (we used $\alpha = 0.5$). $a^T x \geq b$ is a violated inequality, if $a^T y < b$.

The heuristic that we applied has the following general form:

Heuristic for finding violated partition inequalities

(1) Shrink all or some edges $e \in E$ with the property that any violated partition inequality using this edge can be transformed into some at-least-as-violated partition inequality not using this edge. ('Using e' means; e has coefficient 1 in the partition inequality.)

(2) Find some violated or almost violated cut constraints in the resulting graph.

(3) Attempt to modify these cut constraints into violated partition inequalities.

Exactly the same approach is used for separating node partition (11) and lifted r-cover inequalities (15), except that we have to use other shrinking criteria and, in Step 2, plug in the appropriate subroutine for separating node cut (8), resp., r-cover constraints (14ii).

Shrinking is important for reducing graph sizes before applying a min-cut algorithm and for increasing sizes of shores. If we are looking for related partition inequalities, we test whether edge $e = uv$ satisfies one of the following shrinking criteria.

Shrinking criteria

(1) $y_e \geq q := \max\{r_w : w \in V\}$.

(2) $y_e \geq r_v$ and $y_e \geq y(\delta(v)) - y_e$.

(3) $y_e \geq \max\{y(\delta(v)) - y_e, y(\delta(u)) - y_e\}$ and there is a node $w \notin \{u, v\}$ with $r_w \geq \max\{r_u, r_v\}$.

If these criteria are satisfied for edge e, we shrink it by identifying u and v, giving type con($\{u, v\}$) to the new node, and identifying parallel edges by adding their y-values.

It can be shown, that if cases (1) or (2) apply, then any violated partition inequality using e can be transformed into some at-least-as-violated partition inequality not using e. In case (3), edge e has the same property with respect to cut inequalities. Similar shrinking criteria can be found for node partition inequalities and lifted r-cover inequalities.

In the reduced graph G' we now find violated or almost violated cut constraints (resp., node partition and r-cover constraints) using the Gomory–Hu algorithm (or, for r-cover constraints the Padberg–Rao algorithm). These inequalities, defined for G', are transformed back into the original graph G. For instance, a cut inequality in G', $x(\delta_{G'}(W')) \geq r(W')$ is first transformed into a cut inequality $x(\delta_G(W)) \geq r(W)$ in G by blowing up all shrunk nodes in W'. This provides the enlarged node set W. Secondly, this cut inequality is transformed into a (hopefully) violated partition inequality by splitting W or $V \setminus W$ into smaller disjoint node sets $W_1, \ldots, W_p$.

We also check whether the given cut inequality satisfies some simple necessary criteria for defining a facet of kECON$(G;r)$ (or kNCON$(G;r)$). If this is not so, it can usually be transformed into a partition inequality that defines a higher-dimensional face of the respective polyhedron. A similar approach is taken for node partition and lifted r-cover inequalities. More details can be found in Grötschel, Monma & Stoer [1992b] and in Stoer [1992]. Typically, in the first few iterations of the cutting plane algorithm, the fractional solution y consists of several disconnected components. So, y violates many cut and partition inequalities but usually no lifted r-cover inequalities. We start to separate lifted r-cover inequalities only after the number of violated partition inequalities found drops below a certain threshold. Node partition inequalities are used only after all other separation algorithms failed in finding more than a certain number of inequalities.

To keep the number of LP constraints small, all inequalities in the current LP with non-zero slack are eliminated. But since all inequalities ever found by the separation algorithms are stored in some external pool they can be added again if violated at some later point.

7.3. Computational results for low-connectivity problems

In this section we describe the computational results based on the practical heuristics described in Section 4 and the cutting plane approach described earlier

Table 1
Data for LATA problems

	Original graphs					Reduced graphs				
Problem	0	1	2	Nodes	Edges	0	1	2	Nodes	Edges
LATADMA	0	12	24	36	65/0	0	6	15	21	46/4
LATA1	8	65	14	77	112/0	0	10	14	24	48/2
LATA5S	0	31	8	39	71/0	0	15	8	23	50/0
LATA5L	0	36	10	46	98/0	0	20	9	29	77/1
LATADSF	0	108	8	116	173/40	0	28	11	39	86/26
LATADS	0	108	8	116	173/0	0	28	11	39	86/3
LATADL	0	84	32	116	173/0	0	11	28	39	86/6

in this section. We consider the low-connectivity case with node types in $\{0, 1, 2\}$ here and the high connectivity case in the next section. All running times reported are for a SUN 4/50 IPX workstation (a 28.5 MIPS machine). The LP-solver used is a research version of the CPLEX-code provided to us by Bixby [1992]. This is a very fast implementation of the simplex algorithm.

To test our code, network designers at Bellcore provided the data (nodes, possible direct links, costs for establishing a link) of seven real LATA networks that were considered typical for this type of application. The sizes ranged from 36 nodes and 65 edges to 116 nodes and 173 edges; see Table 1. The problem instances LATADL, LATADS, and LATADSF are defined on the same graph. The edges have the same costs in each case, but the node types vary. Moreover, in LATADSF, 40 edges were required to be in the solution. (The purpose was to check how much the cost would increase if these edges had to be used, a typical situation in practice, where alternative solutions are investigated by requiring the use of certain direct links.)

Table 1 provides information about the problems. Column 1 contains the problem names. For the original graphs, columns 2, 3, and 4 contain the numbers of nodes of type 0, 1, and 2, respectively; column 5 lists the total number of nodes, column 6 the number of edges and the number of edges required to be in any solution (the forced edges). All graphs were analysed by our preprocessing procedures described in Section 7.2. Preprocessing was very successful. In fact, in every case, the decomposition and fixing techniques ended up with a single, much smaller graph obtained from the original graph by splitting off side branches consisting of nodes of type 1, replacing paths where all interior nodes are of degree 2, by a single edge, etc. The data of the resulting reduced graphs are listed in columns $6, \ldots, 10$ of Table 1.

To give a visual impression of the problem topologies and the reductions achieved, we show in Figure 4 a picture of the original graph of the LATADL problem (with 32 nodes of type 2 and 84 nodes of type 1) and in Figure 5 a picture of the reduced graph (with 39 nodes and 86 edges) after preprocessing. The nodes of type 2 are displayed by squares, and the nodes of type 1 are displayed by circles. The 6 forced edges that have to be in any feasible solution are drawn bold.

Fig. 4. Original graph of LATADL-problem.

LATA1 is a 2ECON problem, while the other six instances are 2NCON problems. All optimum solutions of the 2ECON versions turned out to satisfy all node-survivability constraints and thus were optimum solutions of the original 2NCON problems — with one exception. In LATA5L one node is especially attractive because many edges with low cost lead to it. This node is an articulation node of the optimum 2ECON solution. In the following, LATA5LE is the 2ECON version of problem LATA5L.

Table 2 contains some data about the performance of our code on the eight test instances.

We think that it is worth noting that each of these real problems, typical in size and structure, can be solved on a 28-MIPS machine in less than thirty seconds including all input and output routines, drawing the solution graph, branch and cut, etc.

A detailed analysis of the running times of the cutting plane phase is given in Table 3. All times reported are in percent of the total running time (without the branch & cut phase). The last column TT\RED shows the running times of the cutting plane phase of our algorithm applied to the full instances on the original graphs (without reduction by preprocessing). By comparing the last two columns, one can clearly see that substantial running time reductions can be achieved by our preprocessing algorithms on the larger problems.

A structural analysis of the optimum solutions produced by our code revealed that — except for LATADSF, LATA5LE, and LATA1 — the optimum survivable networks consist of a long cycle (spanning all nodes of type 2 and some nodes

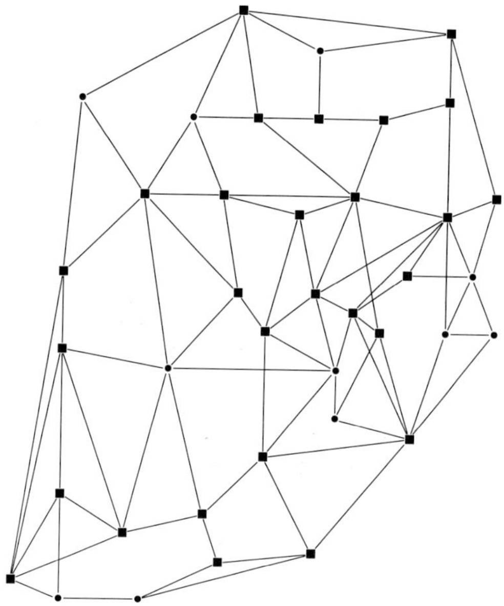

Fig. 5. Reduced graph of LATADL-problem.

Table 2
Performance of branch & cut on LATA problems

Problem	IT	P	NP	RC	C	COPT	GAP	T	BN	BD	BT
LATADMA	12	65	3	7	1489	1489	0	1			
LATA1	4	73	0	1	4296	4296	0	1			
LATA5S	4	76	0	0	4739	4739	0	1			
LATA5LE	7	120	0	0	4574	4574	0	1			
LATA5L	19	155	12	0	4679	4726	0.99	2	4	2	4
LATADSF	7	43	0	0	7647	7647	0	1			
LATADS	17	250	0	4	7303.60	7320	0.22	4	28	9	17
LATADL	14	182	0	28	7385.25	7400	0.20	3	32	10	21

IT = number of iterations (= calls to the LP-solver); P = number of partition inequalities (6.4) used in the cutting plane phase; NP = number of node partition inequalities (6.5) used in the cutting plane phase; RC = number of lifted r-cover inequalities (6.9) used in the cutting plane phase; C = value of the optimum solution after termination of the cutting plane phase; COPT = optimum value; GAP = $100 \times (\text{COPT} - \text{C})/\text{COPT}$ (= percent relative error at the end of the cutting plane phase); T = total running time including input, output, preprocessing, etc., of the cutting plane phase (not including branch & cut), in rounded seconds; BN = number of branch & cut nodes generated; BD = maximum depth of the branch & cut tree; BT = total running time of the branch & cut algorithm including the cutting plane phase in seconds.

of type 1) and several branches connecting the remaining nodes of type 1 to the cycle. The optimum solution of the LATADL instance is shown in Figure 6, with the 2-connected part (the long cycle) drawn bold.

Table 3
Relative running times of cutting plane algorithm on LATA problems

Problem	PT (%)	LPT (%)	CT (%)	MT (%)	TT (s)	TT\RED (s)
LATADMA	2.0	39.2	41.2	17.6	1	1
LATA1	3.8	34.6	34.6	26.9	1	4
LATA5S	3.8	34.6	34.6	26.9	1	1
LATA5LE	0.0	42.9	41.1	16.1	1	1
LATA5L	0.7	37.1	55.2	7.0	2	5
LATADSF	2.1	21.3	57.4	19.2	1	4
LATADS	0.0	44.7	49.0	6.4	4	17
LATADL	1.0	26.3	66.2	6.5	3	18

PT = time spent in the preprocessing phase; CT = time spent in the separation routines; LPT = time used by the LP-solver; MT = miscellaneous time for input, output, drawing, etc.; TT = total time; TT\RED = total time of the algorithm when applied to the original instance without prior reduction by preprocessing.

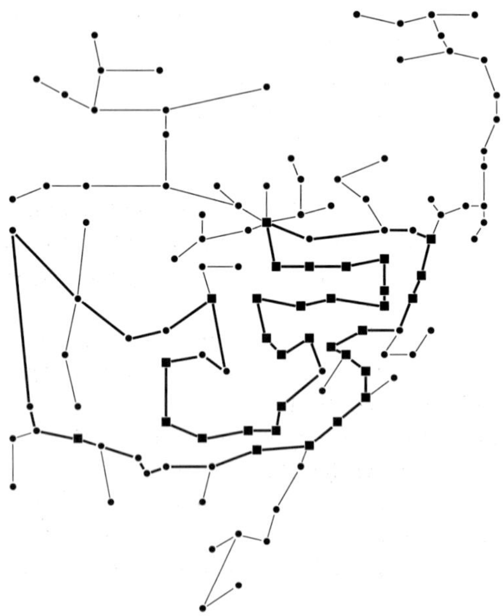

Fig. 6. Solution of LATADL-problem.

From the view of a telephone network designer, a long cycle connecting all nodes of type 2 is not a desirable feature in a communication network because routing paths are very long, resulting in delays, and because each link has to carry a high traffic load, resulting in high costs for terminal electronics, multiplexers, etc. But since the network installation costs form part of the whole network cost, the lowest network installation cost (as found by our algorithm) provides a lower bound for

Table 4
Comparison of heuristic values with optimal values

Problem	COPT	CHEUR	GAP
LATADMA	1489	1494	0.34
LATA1	4296	4296	0
LATA5S	4739	4739	0
LATA5LE	4574	4574	0
LATA5L	4726	4794	1.44
LATADSF	7647	7727	1.05
LATADS	7320	7361	0.56
LATADL	7400	7460	0.81

the whole network cost, and the subgraph minimizing the installation cost could be modified, e.g., by adding some more links of low cost, to produce a network with shorter routing paths between each pair of nodes. This is the design approach taken in the software package distributed by Bellcore [1988]; first a survivable network topology of low cost in computed by heuristics, then this topology is modified to account also for costs associated with the expected traffic in the network.

We ran a few tests on randomly generated problems of higher density and 50–100 nodes. Here our code did perform reasonably well but not as well as on sparse problems. (That is not of great importance, since our goal was to solve real-world problems and not random problems.) More serious is a dramatic increase in running time when many nodes of type 0 are added, as is the case in the ship problem treated in the next section. But the problems that we address here mainly and that come up in the design of fiber optic telephone networks have very few nodes of type 0, if any.

Another motivation for our work was to find out how well the heuristics of Monma & Shallcross [1989] described in Section 4 perform. It turned out that they do very well. Table 4 compares the values CHEUR of the solutions produced by the heuristics with the optimum values COPT computed by our code. The percent relative error GAP ($= 100 \times$ (CHEUR $-$ COPT)/COPT) is always below 1.5%. In three cases the heuristics found an optimum solution. This result definitely justifies the present use of these heuristics in practice. We note that these heuristics are very fast, typically taking only a few seconds on a IBM PC/AT.

7.4. Computational results for high-connectivity problems

At present, we have a first preliminary version of a code for solving survivability problems with higher connectivity requirements. In order to test our code for general kNCON problems, we first used a set of random problems. Later, we also obtained test data for a real-world 3NCON problem, which arose in the design of a communication network on a ship. Both types of test problems have their 'drawbacks', however. The random problems turned out to be too easy (most of them were already solved in the first iteration), and the ship problem confronted us with so many new difficulties (with respect to space, running time, and quality of solutions) that we have to redesign our separation strategies completely to solve variants of the ship problem to optimality.

7.4.1. Random problems

We first report about our computational results on random kECON problems. We used the same set of random data as Ko & Monma [1989] used for their high-connectivity heuristics. So we will be able to compare results later.

The test set of Ko & Monma consists of five complete graphs of 40 nodes and five complete graphs of 20 nodes, whose edge costs are independently drawn from a uniform distribution of real numbers between 0 and 20. For each of these 10 graphs, a minimum-cost k-edge connected subgraph for $k = 3, 4, 5$ is to be found. The next table reports the number of iterations (minimum and maximum) and the average time taken by our code to solve these problems for $k = 3, 4$, and 5, respectively. Only the time for the cutting plane phase is given.

# Nodes	# Iterations			Average time (s)		
	$k = 3$	4	5	3	4	5
20 nodes:	1–2	1–5	1–4	0.43	0.51	0.58
40 nodes:	1–2	1–2	1–4	1.54	1.95	2.36

All problems except one 3ECON instance on 20 nodes were solved in the cutting plane phase. In fact 20 of the 30 problems were already solved in the first iteration with the initial LP (16). For the instances not solved in the first iteration, at most four lifted r-cover inequalities (15) had to be added to obtain the optimal solution. Except for one 3ECON instance, no partition inequalities were added. So, the average solution time is mainly the solution time for the first LP.

All optimal solutions for the kECON problems were at the same time feasible and hence optimal for the corresponding kNCON problems, except the one 3ECON problem which could not be solved in the cutting plane phase. There the optimal solution (obtained by branch & cut) is 3-edge connected, but not 3-node connected.

These excellent results were surprising, because we always thought high-connectivity problems to be harder than low-connectivity problems. But this does not seem to be true for random costs.

The high-connectivity heuristics of Ko & Monma did not perform quite as well as the low-connectivity heuristics, but still reasonably well. The relative gap between the heuristic (h) and the optimal solution value (o), namely $100 \times (h - o)/o$, computed for the above set of random problems, ranged between 0.8 and 12.8 with an average of 11% error (taken over all problems).

7.4.2. Ship problems

One real-world application of survivable network design, where connectivities higher than two are needed, is the design of a fiber communication network that

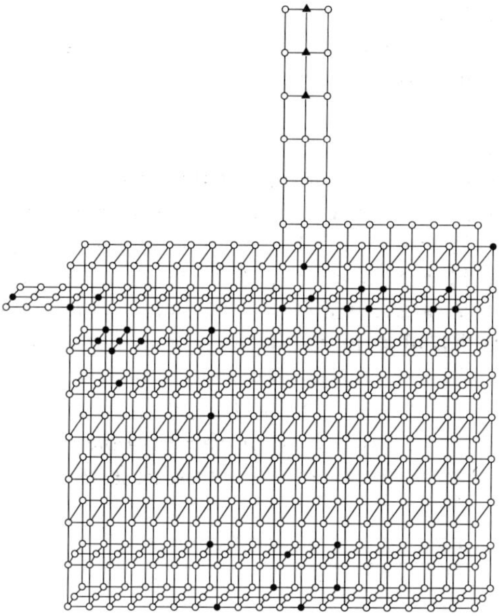

Fig. 7. Grid graph of the ship problem.

connects locations on a military ship containing various communication systems. The reason for demanding high survivability of this network is obvious.

The problem of finding a high-connected network topology minimizing the cable installation cost can be formulated as a 3NCON problem. We will describe the characteristics of this problem in the following.

We obtained the graph and edge cost data of a generic ship model. It has the following features. The graph of possible link installations has the form of a three-dimensional grid with 15 layers, 494 nodes, and 1096 edges, which is depicted in Figure 7. The problem to be solved in this graph is a 3NCON problem with the following node types and costs.

Of the grid's 494 nodes, only 33 are of nonzero type, called 'special nodes'. They are drawn by filled circles or triangles. The 33 special nodes symbolize the various communication systems to be interconnected by the network. To evaluate the dependence of network topology cost on the required survivability, the ship problem appears in three different versions depending on the node types of the 33 special nodes. The three nodes depicted by triangles in the tower of the ship always have type 3, the other 29 special nodes are all given either type 1, type 2, or type 3. We call the three resulting versions of the ship problem 'ship13', 'ship23', and 'ship33', respectively. The remaining 461 nodes are nodes of type 0. They represent possible fiber junction boxes where the fiber cable may be routed.

The cost structure is highly regular. The costs are proportional to the distances between nodes, with the feature that horizontal distances are much higher than vertical distances. (The grid shown in Figure 7 has been scaled. Also, contrary to the graphical representation, the horizontal layers do not always have the same distance from each other.) With this cost structure, it is much cheaper to route vertically than horizontally. Since there exist many shortest paths between any two nodes, there will also exist many optimum solutions to the survivable network problem. So the problem is highly degenerate. Degeneracy together with the size of the ship problem caused us to run into difficulties. In fact, when we first applied our code to the 'ship13' problem, with the initial LP consisting only of the degree constraints for the special nodes, the fractional solutions did not get connected for a long time.

Our first idea was to heuristically reduce the size of the problem in some way. Unfortunately, none of the decomposition techniques described earlier applied, except at the tower of the ship, where nodes of type 3 are separated by a cut of size 3. We cut out some of the 'unnecessary' nodes of type 0 in the lower left and right hand corner of the grid, and also deleted some of the horizontal layers of the grid containing only nodes of type 0.

It is not obvious at all that corners of a grid may be cut out and layers may be deleted without affecting the optimum objective function value of the problem. We could prove such a result only for Steiner tree problems (1NCON problems), not for 2NCON or 3NCON problems. But nevertheless, we used these reductions heuristically to cut down problem sizes in the hope that some optimal solution of the original graph is still contained in the reduced graph. For the 'ship23' problem, the optimal solution of the reduced problem turned out to be optimal for the non-reduced problem, too.

Figure 8 shows the reduced graph of the 'ship13' problem. The result of the reductions can be seen from Table 5, whose columns list, from left to right, the problem names, and, for the original ship graph and the reduced ship graphs, the number of nodes of type 0, 1, 2, and 3, the total number of nodes and the total number of edges/number of forced edges. The forced edges are those edges contained in some cut of size 3 separating two nodes of type 3, which must be contained in any feasible solution.

An optimal solution for the reduced 'ship23' problem is shown in Figure 9.

Table 5 shows that the reductions are enormous, yet there are still many more nodes of type 0 than nodes of nonzero type in each problem.

When we applied our code to the reduced graphs, the fractional solutions still looked frequently like paths beginning at some special node and ending in some

Table 5
Sizes of ship problems

	Original graph					Reduced graph				
Problem	0	1	2	3	Edges	0	1	2	3	Edges
ship13	461	30	0	3	1096/0	128	28	0	3	325/3
ship23	461	0	30	3	1096/0	249	0	30	3	607/3
ship33	461	0	0	33	1096/0	300	0	0	33	719/9

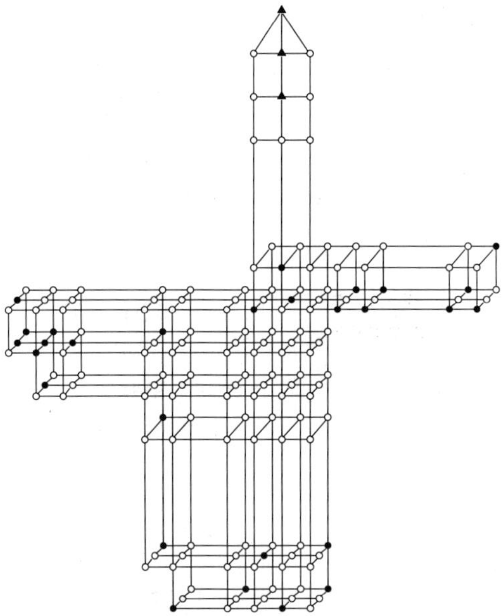

Fig. 8. Reduced grid graph of the 'ship13' problem.

node of type 0. To cure this problem, we made use of the following type of inequalities.

$$x(\delta(v)\backslash\{e\}) \geq x_e$$

for all nodes v of type 0 and all $e \in \delta(v)$. These inequalities (we call them *con0 inequalities)* describe algebraically that nodes of type 0 do not have degree 1 in an edge-minimal solution. This is not true for all survivable networks, but it is true for the optimum solution if all costs are positive. So, although these inequalities are not valid for the kNCON polytope, we used them to force the fractional solutions into the creation of longer paths. Another trick to obtain better starting solutions was to use cuts of a certain structure in the initial LP.

Table 6 gives some preliminary computational results of our cutting plane algorithm on the three reduced and not reduced versions of the ship problem.

Although Table 6 shows that the code is still rather slow, it could at least solve two of the ship problems. In order to obtain better results and running times, some more research most be done, especially on finding better starting solutions, devising faster separations heuristics that exploit the problem structure, and, maybe, inventing new classes of inequalities for high-connectivity problems. The table also shows that the speedup on the (heuristically) reduced problems was significant.

Table 7 shows the percentage of time spent in the different routines.

We do not understand yet why our code solves the ship23 problem rather easily and why there is still a gap after substantial running time of our cutting plane algorithms for the ship33 problem. Probably, the 'small' changes of a

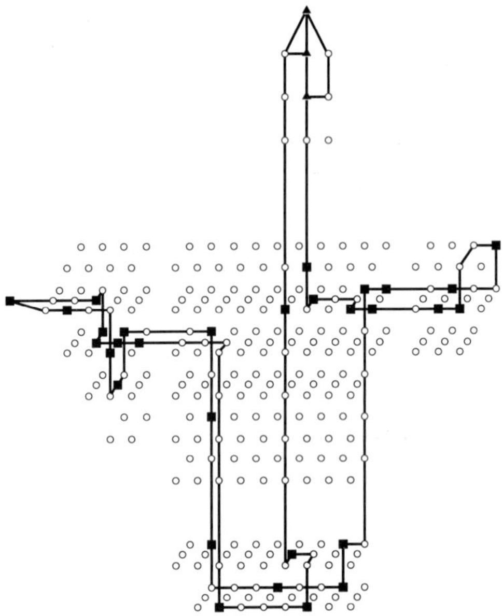

Fig. 9. Optimum solution of reduced 'ship23' problem.

Table 6
Performance of cutting plane algorithm on ship problems

Problem	VAR	IT	PART	RCOV	LB	UB	GAP (%)	Time (min:s)
ship13	1088	3252	777261	0	211957.1	217428	2.58	10122:35
ship23	1088	15	4090	0	286274	286274	0	27:20
ship33	1082	42	10718	1	461590.6	483052	4.64	55:26
ship13red	322	775	200570	0	217428	217428	0	426:47
ship23red	604	12	2372	0	286274	286274	0	1:54
ship33red	710	40	9817	0	462099.3	483052	4.53	34:52

Problem = problem name, where 'red' means reduced; VAR = number of edges minus number of forced edges; IT = number of LPs solved; PART = number of partition inequalities added; RCOV = number of r-cover inequalities added; LB = lower bound (= optimal LP value); UB = upper bound (= heuristic value); GAP = (UB − LB)/LB.

few survivability requirements result in more dramatic structural changes of the polyhedra and thus of the inequalities that should be used. It is conceivable that our code has to be tuned according to different survivability requirements settings. We should mention that we did not attempt to solve ship13 and ship33 by entering the branching phase of our code. The gaps are not small enough yet for the enumerative stage to have a decent perspective. Further details of our attempts to solve network design problems with higher connectivity requirements can be found in Grötschel, Monma & Stoer [1992c].

Table 7
Relative running times on ship problems

Problem	PT (%)	LPT (%)	CT (%)	MT (%)	Time (min:s)
ship13	0.0	75.6	23.9	0.5	10122:35
ship23	0.0	13.1	86.4	0.4	27:20
ship33	0.0	31.2	68.2	0.6	55:26
ship13red	0.0	68.5	30.1	1.4	426:47
ship23red	0.1	39.2	58.6	1.9	1:54
ship33red	0.0	41.1	58.4	0.5	34:52

Problem = problem name where 'red' means reduced; PT = time spent for reduction of problem; LPT = time spent for LP solving; CT = time spent for separation; MT = time on miscellaneous items, input, output, etc.

Summarizing our computational results, we can say that for survivability problems with many nodes of type 0 and highly regular cost structure (such as the ship problems) much still remains to be done to speed up our code and enhance the quality of solutions. But for applications in the area of telephone network design, where problem instances typically are of moderate size and contain not too many nodes of type 0, our approach produces very good lower bounds and even optimum solutions in a few minutes. This work is a good basis for the design of a production code for the 2ECON and 2NCON problems coming up in fiber optic network design and a start towards problems with higher and more varying survivability requirements and larger underlying graphs.

8. Directed variants of the general model

There are many possible variants of the general model described in Section 3 for the design of networks with connectivity constraints. A natural variant is to consider networks with directed links. As we will see below, there are practical and theoretical reasons for considering survivability in directed graphs.

8.1. Survivability models for directed networks

In order to model directed links, we let $D = (N, A)$ denote a *directed graph* consisting of a set V of *nodes* (just as in the undirected case) and a set A of *directed arcs*. Each *arc* $a = (u, v) \in A$ represents a link directed from node u to node v. For example, this could model certain communications facilities that allow only the one-way transfer of information. Of course, there may be arcs directed each way between any given pair of nodes. Each arc $a \in A$ has a nonnegative *fixed cost* c_a of establishing the link connection. The directed graph may have parallel arcs (in each direction). As before, the cost of establishing a network consisting of a subset $B \subseteq A$ of arcs is the sum of costs of the individual links contained in B.

The goal is to build a minimum-cost network so that the required survivability conditions are satisfied.

The survivability requirements demand that the network satisfy the same types of edge and node connectivity requirements as in the undirected case. We simply replace the notion of an undirected path by a directed one. The previous definitions and model formulations are essentially unchanged.

The problem of designing a survivable directed network has not received as much attention in the literature as the undirected case. We briefly summarize some recent efforts along these lines.

Dahl [1991] has given various formulations for the directed survivable network design problem with arc connectivity requirements. He mainly studies the bi-Steiner problem, which is the problem of finding a minimum-cost directed subgraph that contains two arc-disjoint paths from a given root to each node of a set of terminal nodes. This problem has applications in the design of hierarchical subscriber networks, see Lorentzen & Moseby [1989].

Chopra [1992] modeled a directed version of the 2ECON problem, which becomes the 'undirected' 2ECON problem after 'projection' into a lower-dimensional space. He showed that all partition inequalities and further inequalities can be generated by the projection of certain directed cut inequalities. Chopra's model can be generalized to higher edge connectivity requirements, as shown below.

8.2. Projection

The last remarks show that directed versions of the kECON and kNCON problems are not only interesting in their own right, but they are sometimes also useful in solving their undirected counterparts. We will illustrate this now by pointing out the value of projections. For many combinatorial problems, good polyhedral descriptions can be obtained by transferring the original problem into higher dimensions, that is, by formulating it with additional (auxiliary) variables, which may later be projected away. This was done successfully for the 2-terminal Steiner tree problem in directed graphs, see Ball, Liu & Pulleyblank [1987]. There the formulation with auxiliary variables contains a polynomial number of simple constraints, which by projection are turned into an exponential number of 'weird' constraints. The general idea of projection was described by Balas & Pulleyblank [1983].

For the 2ECON problem, Chopra [1992] has found a formulation in directed graphs using $2|E|$ integer variables and directed cut constraints, which he called the DECON problem, see (17) below. The directed cut constraints, see (17i), used in the formulation of the DECON problem have the advantage that they can be separated in polynomial time, whereas the separation of the inequalities appearing in our undirected 2ECON problem is NP-hard.

Projection of the directed cut constraints and nonnegativity constraints of the DECON problem gives a new class of inequalities for the 2ECON problem (we call these Prodon inequalities) which contain as a subclass the partition inequalities (10). For the Steiner tree problem, where $r_v \in \{0, 1\}$, these new inequalities have been found by Prodon [1985]. In the following we show how the Prodon inequalities are derived from the DECON model by projection.

In order to do this, we must first introduce some terminology. Let a graph $G = (V, E)$ and node types $r_v \in \{0, 1, 2\}$ be given, where at least two nodes are of highest (positive) node type. This may either be a 2ECON or a 1ECON problem. From G we construct a directed graph $D = (V, A)$ by replacing each undirected edge ij with two directed edges (i, j) and (j, i). Furthermore, we pick some node $w \in V$ of highest node type. Let $\delta^-(W)$ be the set of arcs directed into node set W. If (x, y) is a solution to the following system of inequalities (where $x \in \mathbf{Z}^E$ and $y \in \mathbf{Z}^A$),

$$
\begin{array}{lll}
\text{(i)} & y(\delta^-(W)) & \geq 1 \text{ for all } W \subseteq V, \emptyset \neq W \neq V, \text{with} \\
 & & \quad \text{con}(W) = 2 \text{ (or } r(W) = 1 \text{ and } w \notin W); \\
\text{(ii)} & y_{(i,j)} & \geq 0 \text{ for all } (i, j) \in A; \\
\text{(iii)} & y_{(i,j)} \text{ integral} & \quad \text{for all } (i, j) \in A; \\
\text{(iv)} & -y_{(i,j)} - y_{(j,i)} + x_{ij} & = 0 \text{ for all } ij \in E; \\
\text{(v)} & x_{ij} & \leq 1 \text{ for all } ij \in E.
\end{array}
\qquad (17)
$$

then the integer vector x is feasible for the 2ECON problem, and vice versa: if some integer vector x is feasible for the 2ECON problem, then an integer vector y can be found so that (x, y) satisfies (17i)–(17v). So the projection of system (17) onto x-variables gives a formulation of the 2ECON problem. (Originally, Chopra considered this system without the upper bound constraints.) If no node is of type 2, a feasible vector y is just the incidence vector of a subgraph of D containing a Steiner tree rooted at w. If all nodes are of type 2, then y is the incidence vector of a strongly connected directed subgraph of D ('strongly connected' means that between each distinct pair s, t of nodes there exists a directed (s, t)-path and a directed (t, s)-path).

Without the integrality constraints (17iii) and upper bound constraints (17v), we obtain a relaxation, which, after projection onto x-variables, gives a relaxation of the 2ECON problem. The projection works as follows. Let us define

(1) $\mathcal{F}$ as the set of those $W \subseteq V$ that appear in the formulation of inequalities (17i),
(2) $b_W \geq 0$ as the variables assigned to each inequality (17i) for $W \in \mathcal{F}$,
(3) $a_{ij} \in \mathbf{R}$ as the variables assigned to each equation (17iv) for $ij \in E$,
(4) $s(\mathcal{F}; b; i; j)$ as the sum of b_W over all $W \in \mathcal{F}$ with $i \in W$ and $j \notin W$, and
(5) C as the cone of variables $a \in \mathbf{R}^E$ and $b := (b_W)_{W \in \mathcal{F}}$ satisfying

$$
\begin{aligned}
a_{ij} &\geq s(\mathcal{F}; b; i; j), \quad \text{for all } ij \in E, \\
a_{ij} &\geq s(\mathcal{F}; b; j; i), \quad \text{for all } ij \in E, \\
b &\geq 0.
\end{aligned}
$$

If $(a, b) \in C$, and if all inequalities of type (17i) and all inequalities of type (17iv) are added with coefficients b_W and a_{ij} respectively, then we obtain an inequality

$$\sum_{(i,j)\in A} u_{(i,j)} y_{(i,j)} + \sum_{ij\in E} a_{ij}x_{ij} \geq \sum_{W\in\mathcal{F}} b_W,$$

where the $u_{(i,j)}$ are non-positive coefficients of the variables $y_{(i,j)}$. In fact, C was defined exactly in such a way, that the $u_{(i,j)}$ are non-positive. The above inequality is valid for the system given by all inequalities (17i), (17ii), and (17iv). Since $y \geq 0$,

$$\sum_{ij\in E} a_{ij}x_{ij} \geq \sum_{W\in\mathcal{F}} b_W,$$

is valid for 2ECON$(G;r)$. It can also be proved with the general projection technique of Balas & Pulleyblank [1983] that

$$\begin{array}{ll} \sum_{ij\in E} a_{ij}x_{ij} \geq \sum_{W\in\mathcal{F}} b_W & \text{for all } (a,b)\in C, \\ x \geq 0 & \end{array} \tag{18}$$

is exactly the projection of system (17i), (17ii) and (17iv) onto the x-variables. Not all $(a,b) \in C$ are needed in the formulation of (18). The following system is clearly sufficient to describe the projection of (17i), (17ii) and (17iv) onto x-variables:

$$\begin{array}{lll} \text{(i)} & \sum_{ij\in E} a_{ij}x_{ij} \geq \sum_{W\in\mathcal{F}} b_W & \text{for all } b \geq 0 \text{ and} \\ & & a_{ij} := \max\{s(\mathcal{F};b;i;j), s(\mathcal{F};b;j;i)\}, \\ \text{(ii)} & x \geq 0. & \end{array} \tag{19}$$

We call inequalities (19i) *Prodon inequalities* (induced by b), because this class of inequalities was discovered by Prodon [1985] for 1ECON$(G;r)$.

The class of Prodon inequalities properly contains the class of partition inequalities (10). Namely, a partition inequality

$$x[W_1 : \ldots : W_p] \geq \begin{cases} p & \text{if at least two } W_i \text{ contain nodes of type 2} \\ p-1 & \text{otherwise} \end{cases}$$

(where $W_1, \ldots, W_p$ is a partition of V into p node sets with $r(W_i) \geq 1$) can also be written as a Prodon inequality, if b_W is set to 1 for all W_i that are in $\mathcal{F}$ and $b_W := 0$ for all other sets in $\mathcal{F}$. By definition of $\mathcal{F}$, if at least two sets W_i contain nodes of type 2, then $W_i \in \mathcal{F}$ for all W_i, and if only one set, say W_p, contains nodes of type 2 (and therefore the 'root' w), then $W_1, \ldots, W_{p-1}$ are in $\mathcal{F}$, but W_p is not. This explains the differing right-hand sides in both cases.

But not every facet-defining Prodon inequality is also a partition inequality. For instance, the inequality depicted in Figure 10 is not a partition inequality, but can be written as a Prodon inequality induced by $b_W := 1$ for the sets $\{1\}, \{2\}, \{5\}, \{7\}, \{3,5,6\}, \{4,6,7\}$, and $b_W := 0$ for all other sets W in $\mathcal{F}$. So the coefficients on all depicted edges are 1, and the right-hand side is 6. Here, nodes 1 and 2 are nodes of type 2; nodes 5 and 7 are nodes of type 1; all others are of type 0. The Prodon inequality of Figure 10 can be proved to be facet-defining for 2NCON$(G;r)$, where G consists exactly of the depicted nodes and edges.

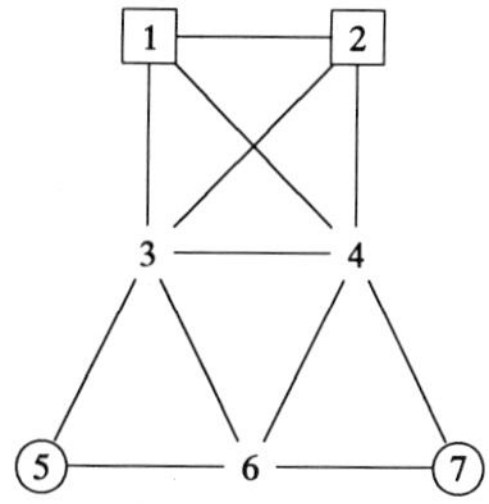

Fig. 10. Prodon inequality.

We show in the following remark that no Prodon inequality except the cut inequalities are facet-defining if there are no nodes of type 1.

Remark 2. *If (G, r) is an instance of the 2ECON problem, where node types r_v only take values 0 and 2 for all $v \in V$, then no Prodon inequalities except the cut constraints define facets of 2ECON$(G; r)$.*

Proof. Let $\sum_{ij} a_{ij}x_{ij} \geq \sum_{W \in \mathcal{F}} b_W$ be a Prodon inequality. By definition,

$$a_{ij} \geq \tfrac{1}{2}s(\mathcal{F}; b; i; j) + \tfrac{1}{2}s(\mathcal{F}; b; j; i),$$

which is the same as 1/2 times the sum of all b_W over $W \in \mathcal{F}$ with $ij \in \delta(W)$. Therefore,

$$a^T x \geq \tfrac{1}{2} \sum_{W \in \mathcal{F}} b_W \, x(\delta(W)).$$

Since $x(\delta(W)) \geq \text{con}(W) = 2$ for all $W \in \mathcal{F}$, this expression is at least $\sum_{W \in \mathcal{F}} b_W$ for all $x \in$ 2ECON$(G; r)$. So our Prodon inequality is implied by the sum of some cut inequalities, and must itself be a cut inequality, if it is to be facet-defining. □

The projection technique applied to the Steiner tree polytope is not new. Goemans & Myung [1993] list various formulations of the Steiner tree problem, all of which use auxiliary variables, among them system (17) [without (v)]. They show that upon projection to variables $x \in \mathbf{R}^E$ all these models generate the same inequalities. Goemans [1994b] investigates, again for the Steiner tree polytope, facet properties of a subclass of inequalities obtained by such a projection, which are, in fact, a subclass of the class of Prodon inequalities.

8.2.1. Higher connectivity requirements

The DECON model can be generalized to higher connectivity requirements, when node types are in $\{0, 1, 2, 4, 6, \ldots\}$. The directed model in this case requires $(1/2)\min\{r_u, r_v\}$ directed arc-disjoint (u, v)-paths between each pair of nodes u, v whose node types are at least 2, and one directed path from a specified root of highest node type to each node of type 1. This does appropriately model the undirected kECON problem, because of a theorem of Nash-Williams [1960], which says that undirected graphs containing r_{uv} edge-disjoint paths between each pair of nodes u and v can be oriented in such a way, that the resulting directed graph contains, between each u and v, $\lfloor r_{uv}/2 \rfloor$ arc-disjoint paths. The inequalities resulting from the projection of directed cut inequalities in this model do *not* generalize all partition inequalities of type (10) when $k \geq 4$.

8.2.2. Separation of Prodon inequalities

We close this section by observing that the separation problem for Prodon inequalities can be performed in polynomial time. The separation algorithm also makes use of projection and works in the same way as one iteration of Benders' decomposition method, see Benders [1962]. This observation is meant to show that projection is not only of theoretical value but also of computational interest.

Suppose a point x^* with $0 \leq x^* \leq 1$ is given, for which it has to be decided whether there is a Prodon inequality violated by this point or not. This can be decided by solving the following LP derived from (17):

$$\min \quad z$$

subject to

$$\begin{array}{llll} \text{(i)} & y(\delta^-(W)) + z & \geq 1 & \text{for all } W \in \mathcal{F}; \\ \text{(ii)} & y_{(i,j)} & \geq 0 & \text{for all } (i,j) \in A; \\ \text{(iii)} & -y_{(i,j)} - y_{(j,i)} - zx^*_{ij} & = -x^*_{ij} & \text{for all } ij \in E; \\ \text{(iv)} & z & \geq 0. & \end{array} \tag{20}$$

This LP has the feasible solution $y = 0$ and $z = 1$. If its optimal value is 0, and y^* is an optimal solution, then (x^*, y^*) is feasible for the system (17), hence x^* satisfies all Prodon inequalities (by the projection result). If the optimal value is non-zero, then the optimal dual variables b_W for inequalities (20i) and a_{ij} for equations (20iii) define a Prodon inequality violated by x^*. More explicitly, the optimal dual variables b_W $(W \in \mathcal{F})$ and $a \in \mathbf{R}^E$ satisfy

$$\begin{array}{lll} -a_{ij} + \displaystyle\sum_{W \in \mathcal{F}: i \in W, j \notin W} b_W \leq 0 & \text{for all } (i,j) \in A, \\ -a_{ij} + \displaystyle\sum_{W \in \mathcal{F}: j \in W, i \notin W} b_W \leq 0 & \text{for all } (i,j) \in A, \\ -a^T x^* + \displaystyle\sum_{W \in \mathcal{F}} b_W > 0. & \end{array} \tag{21}$$

The first two inequalities imply that a_{ij} is at least the maximum of $s(\mathcal{F}; b; i; j)$ and $s(\mathcal{F}; b; j; i)$ for each $ij \in E$. This implies that a and b induce the Prodon inequality

$$\sum_{ij \in E} a_{ij} x_{ij} \geq \sum_{W \in \mathcal{F}} b_W.$$

From the last inequality in (21) it follows that x^* violates this Prodon inequality.

The LP (20) can be solved in polynomial time, since there exist polynomial separation algorithms for the directed cut inequalities (20i). Therefore, the Prodon inequalities can also be separated in polynomial time. We have, however, not yet made use of these inequalities.

References

Agrawal, A., Ph. Klein and R. Ravi (1991). When trees collide: An approximation algorithm for the generalized Steiner tree problem on networks. *Proc. 23rd Annu. Symp. on Theory of Computing*, pp. 134–144, May 1991.

Baïou, M., and A.R. Mahjoub (1993). The 2-edge connected Steiner subgraph polytope of a series-parallel graph, Département d'Informatique, Université de Bretagne Occidentale, France, October 1993.

Balas, E., and W.R. Pulleyblank (1983). The perfectly matchable subgraph polytope of a bipartite graph. *Networks* 13, 495–516.

Ball, M.O., W.G. Liu and W.R. Pulleyblank (1987). Two-terminal Steiner tree polyhedra, Technical Report 87466-OR, University of Bonn.

Bellcore (1988). FIBER OPTIONS: Software for designing survivable optical fiber networks, Software Package, Bell Communications Research.

Benders, J.F. (1962). Partitioning procedures for solving mixed-variable programming problems. *Numer. Math.* 4, 238–252.

Bienstock, D., E.F. Brickell and C.L. Monma (1990). On the structure of minimum-weight k-connected spanning networks, *SIAM J. Discrete Math.* 3, 320–329.

Bixby, R.E. (1992). Implementing the simplex method: The initial basis, *ORSA J. Comput.* 4, 267–284.

Bland, R.G., D. Goldfarb and M.J. Todd (1981). The ellipsoid method: a survey. *Oper. Res.* 29, 1039–1091.

Boyd, S.C., and T. Hao (1993). An integer polytope related to the design of survivable communication networks, *SIAM J. Discrete Math.* 6(4), 612–630.

Cai, G.-R., and Y.-G. Sun (1989). The minimum augmentation of any graph to a k-edge connected graph. *Networks* 19, 151–172.

Cardwell, R.H., C.L. Monma and T.H. Wu (1989). Computer-aided design procedures for survivable fiber optic networks, *IEEE J. Selected Areas Commun.* 7, 1188–1197.

Cardwell, R.H., T.H. Wu and W.E. Woodall (1988). Decreasing survivable network cost using optical switches, in: *Proc. GLOBECOM '88*, pp. 93–97.

Chopra, S. (1992). Polyhedra of the equivalent subgraph problem and some edge connectivity problems, *SIAM J. Discrete Math.* 5(3), 321–337.

Chopra, S. (1994). The k-edge connected spanning subgraph polyhedron, *SIAM J. Discrete Math.* 7(2), 245–259.

Chou, W., and H. Frank (1970). Survivable communication networks and the terminal capacity matrix, *IEEE Trans. Circuit Theor.* CT-17(2), 192–197.

Chvátal, V. (1973). Edmonds polytopes and a hierarchy of combinatorial problems, *Discrete Math.* 4, 305–337.

Cornuéjols, G., J. Fonlupt and D. Naddef (1985). The traveling salesman problem on a graph and some related integer polyhedra. *Math. Program.* 33, 1–27.

Dahl, G. (1991). Contributions to the design of survivable directed networks, Ph.D. Thesis, University of Oslo. Technical Report TF R 48/91, Norwegian Telecom, Research Dept., Kjeller, Norway.

Edmonds, J. (1965). Maximum matching and a polyhedron with 0,1-vertices, *J. Res. Nat. Bur. Stand. Ser. B* 69, 125–130.

Eswaran, K.P., and R.E. Tarjan (1976). Augmentation problems. *SIAM J. Comput.* 5(4), 653–665.

Frank, A. (1992a). Augmenting graphs to meet edge-connectivity requirements. *SIAM J. Discrete Math.* 5(1), 25–53.

Frank, A. (1992b) On a theorem of Mader. *Ann. Discrete Math.* 101, 49–57.

Frank, A. (1995). *Connectivity and network flows*, in: R. Graham, M. Grötschel and L. Lovász (eds.), *Handbook of Combinatorics*, North Holland, Amsterdam, Chapter 2, to appear.

Frank, A., and T. Jordán (1993). Minimal Edge-Coverings of Pairs of Sets, Research Institute for Discrete Mathematics, University of Bonn, Germany, June 1993.

Frank, H., and W. Chou (1970). Connectivity considerations in the design of survivable networks. *IEEE Trans. Circuit Theor.* CT-17(4), 486–490.

Frederickson, G.N., and J. JáJá (1982). On the relationship between the biconnectivity augmentation and traveling salesman problem. *Theor. Comput. Sci.* 19, 189–201.

Garey, M.R., and D.S. Johnson (1979). *Computers and Intractability: A Guide to the Theory of NP-Completeness*, Freeman, San Francisco, Calif.

Goemans, M.X. (1994a). Arborescence polytopes for series-parallel graphs, *Discrete Appl. Math.* 51(3), 277–289.

Goemans, M.X. (1994b). The Steiner tree polytope and related polyhedra, *Math. Program.* A63(2), 157–182.

Goemans, M.X., and D.J. Bertsimas (1993). Survivable networks, linear programming relaxations and the parsimonious property. *Math. Program.* 60(2), 145–166.

Goemans, M.X., M. Mihail, V. Vazirani, D. Williamson (1992). An approximation algorithm for general graph connectivity problems, preliminary version. *Proc. 25th ACM Symp. on the Theory of Computing*, San Diego, CA, 1993, pp. 708–717.

Goemans, M.X., and Y.-S. Myung (1993). A catalog of Steiner tree formulations, *Networks* 23(1), 19–28.

Gomory, R.E., and T.C. Hu (1961). Multi-terminal network flows. *J. Soc. Ind. Appl. Math.* 9, 551–570.

Grötschel, M., L. Lovász and A. Schrijver (1988). Geometric algorithms and combinatorial optimization, Springer, Berlin.

Grötschel, M., and C.L. Monma (1990). Integer polyhedra associated with certain network design problems with connectivity constraints. *SIAM J. Discrete Math.* 3, 502–523.

Grötschel, M., C.L. Monma and M. Stoer (1992a). Facets for polyhedra arising in the design of communication networks with low-connectivity constraints, *SIAM J. Optimization* 2, 474–504.

Grötschel, M., C.L. Monma and M. Stoer (1992b). Computational results with a cutting plane algorithm for designing communication networks with low-connectivity constraints, *Oper. Res.* 40, 309–330.

Grötschel, M., C.L. Monma and M. Stoer (1992c). Polyhedral and computational investigations for designing communication networks with high survivability requirements, ZIB-Preprint SC 92-24, Konrad-Zuse-Zentrum für Informationstechnik Berlin.

Grötschel, M. and M.W. Padberg (1985). Polyhedral theory, in: E.L. Lawler, J.K. Lenstra, A.H.G. Rinnooy Kan and D. Shmoys (eds.), *The traveling salesman problem*, Wiley, Chichester, pp. 251–305.

Hao, J., and J.B. Orlin (1992). A faster algorithm for finding the minimum cut in a graph. *Proc. 3rd Annu ACM–SIAM-Symp on Discrete Algorithms*, Orlando, Florida, pp. 165–174.

Harary, F. (1962). The maximum connectivity of a graph. *Proc Nat. Acad. Sci., USA* 48, 1142–1146.

Hsu, T.S. (1992). On four-connecting a triconnected graph (extended abstract). *Proc. 33rd Annu. IEEE Symp. on the Foundations of Computer Science*, pp. 70–79.

Hsu, T.S., and V. Ramachandran (1991). A linear-time algorithm for triconnectivity augmentation (extended abstract). *Proc. 32rd Annu. Symp. on the Foundations of Computer Science*, pp. 548–559.

Khuller, S., and U. Vishkin (1994). Biconnectivity approximations and graph carvings. *J. ACM* 41(2), 214–235.

Ko, C.-W., and C.L. Monma (1989). Heuristic methods for designing highly survivable communication networks, Technical report, Bell Communications Research.

Kolar, D.J., and T.H. Wu (1988). A study of survivability versus cost for several fiber network architectures. *Proc. ICC '88*, pp. 61–66.

Lawler, E. (1976) Combinatorial Optimization: Networks and Matroids. Holt, Reinhart and Winston, New York.

Lawler, E.L., J.K. Lenstra, A.H.G. Rinnooy Kan and D. Shmoys (1985). *The traveling salesman problem*, Wiley, Chichester.

Ling, F., and T. Kameda (1987). Complexity of graph connectivity functions, Technical Report, School of Computing Science, Simon Fraser University, Burnaby, British Columbia.

Lorentzen, R., and H. Moseby (1989). Mathematical models and algorithms used in the subscriber network planning tool ABONETT, Norwegian Telecommunications Research Dept., TF-report 66/89.

Lovász, L. (1976). On some connectivity properties of Eulerian graphs. *Acta Math. Acad. Sci. Hung.* 28, 129–138.

Lovász, L., and M.D. Plummer (1986). *Matching theory*. Annals of Discrete Mathematics, Vol. 29, North-Holland, Amsterdam.

Mader, W. (1978). A reduction method for edge-connectivity in graphs. *Ann. Discrete Math.* 3, 145–164.

Mahjoub, A.R. (1994). Two edge connected spanning subgraphs and polyhedra, *Math. Program.* A64(2), 199–208.

Margot, F., A. Prodon and Th.M. Liebling (1994). Tree polyhedron on 2-trees, *Math. Program.* A63(2), 183–191.

Monma, C.L., B.S. Munson and W.R. Pulleyblank (1990). Minimum-weight two-connected spanning networks. *Math. Program.* 46, 153–171.

Monma, C.L., and D.F. Shallcross (1989). Methods for designing communication networks with certain two-connected survivability constraints. *Oper. Res.* 37, 531–541.

Nagamochi, H., and T. Ibaraki (1992). Computing edge-connectivies in multigraphs and capacitated graphs. *SIAM J. Discrete Math.* 5(1), 54–66.

Naor, D., D. Gusfield and C. Martel (1990). A fast algorithm for optimally increasing the edge-connectivity. *Proc. 31st Annu. Symp. on the Foundation of Computer Science*, pp. 698–707.

Nash-Williams, C.St.J.A. (1960). On orientations, connectivity, and odd vertex pairings in finite graphs. *Can. J. Math.* 12, 555–567.

Newark Star Ledger (1987). Damage to fiber cable hinders phone service, September 22, 1987.

Newark Star Ledger (1988a). Cable snaps, snags area phone calls, February 26, 1988.

Newark Star Ledger (1988b). Phone snafu isolates New Jersey; long-distance cable snaps, November 19, 1988.

New York Times (1988). Phone system feared vulnerable to wider disruptions of service, May 26, 1988.

New York Times (1989). Experts say phone system is vulnerable to terrorists, February 8, 1989.

Padberg, M.W., and M.R. Rao (1982). Odd minimum cut sets and *b*-matchings. *Math. Oper. Res.* 7, 67–80.

Papadimitriou, C.H., and K. Steiglitz (1982). *Combinatorial optimization: algorithms and complexity*, Prentice-Hall, Englewood Cliffs, N.J.

Papadimitriou, C.H., and M. Yannakakis (1982). The complexity of facets and some facets of complexity *J. Assoc. Comput. Mach.* 29, 285–309.

Prodon, A., Th.M. Liebling and H. Gröflin (1985). Steiner's problem on two-trees, Technical Report RO-830315, École Polytechnique Fédérale de Lausanne, Switzerland.

Prodon, A. (1985). A polyhedron for Steiner trees in series-parallel graphs, Technical Report, École Polytechnique Fédérale de Lausanne, Switzerland.

Pulleyblank, W.R. (1989). Polyhedral combinatorics, in: G.L. Nemhauser, A.H.G. Rinnooy Kan and M.J. Todd (eds.), *Optimization*, Handbook in Operations Research and Management Science, Vol. 1, North-Holland, Amsterdam, 371–446.

Steiglitz, K., P. Weiner and D.J. Kleitman (1969). The design of minimum cost survivable networks. *IEEE Trans. Circuit Theor.* 16, 455–460.

Stoer, M. (1992). Design of survivable networks, Ph.D. thesis, University of Augsburg. *Lecture Notes in Mathematics*, Springer, Heidelberg, Vol. 1531.

Takamizawa, K., T. Nishizeki and N. Saito (1982). Linear-time computability of combinatorial problems on series-parallel graphs. *J. Ass. Comput. Mach.* 29(3), 623–641.

Ueno, S., Y. Kajitani and H. Wada (1988). Minimum augmentation of a tree to a k-edge-connected graph, *Networks* 18, 19–25.

Wald, J.A., and C.J. Colbourn (1983). Steiner trees, partial 2-trees and minimum IFI networks. *Networks* 13, 159–167.

Wall Street Journal (1988). Fire in fiber gateway sparks flight delays, problems at brokerages, May 11, 1988.

Watanabe, T., and A. Nakamura (1987). Edge-connectivity augmentation problems. *Comput. System Sci.* 35(1), 96–144.

Winter, P. (1985a). Generalized Steiner tree problem in Halin networks. *Proc. 12th Int. Symp. on Mathematical Programming*, MIT.

Winter, P. (1985b). Generalized Steiner problem in outerplanar networks, *BIT* 25, 485–496.

Winter, P. (1986). Generalized Steiner problem in series-parallel networks, *J. Algorithms* 7, 549–566.

Winter, P. (1987). Steiner problem in networks: A survey. *Networks* 17(2), 129–167.

Wu, T.H., and R.H. Cardwell (1988). Optimum routing in fiber network design: models and applications. *Proc. ICC '88*, pp. 251–257.

Wu, T.H., D.J. Kolar and R.H. Cardwell (1988). Survivable network architectures for broadband fiber optic networks: model and performance comparison. *IEEE J. Lightwave Technol.* 6, 1698–1709.

Zorpette, G. (1989). Keeping the phone lines open. *IEEE Spectrum*, June 1989, pp. 32–36.

M.O. Ball et al., Eds., *Handbooks in OR & MS, Vol. 7*

Chapter 11

Network Reliability

Michael O. Ball
College of Business and Management and Institute for Systems Research, University of Maryland, College Park, MD 20742–1815, U.S.A.

Charles J. Colbourn
Department of Combinatorics and Optimization, University of Waterloo, Waterloo, Ont. N2L 3G1, Canada

J. Scott Provan
Department of Operations Research, University of North Carolina, Chapel Hill, NC 27599–3180, U.S.A.

1. Motivation

Network reliability encompasses a range of issues related to the design and analysis of networks which are subject to the random failure of their components. Relatively simple, and yet quite general, network models can represent a variety of applied problem environments. Network classes for which the models we cover are particularly appropriate include data communications networks, voice communications networks, transportation networks, computer architectures, electrical power networks and command and control systems.

The advent of the digital computer led to significant reliability modeling efforts [Moore & Shannon, 1956]. Early computer memories were made up of large numbers of individual components such as relays or vacuum tubes. Computer systems which failed whenever a single component failed were extremely unreliable, since the probability of at least one component out of thousands failing is quite high, even if the component failure probability is low.

Much initial work in highly reliable systems concentrated on systems whose failure could cause massive damage or loss of human life. Examples include aircraft and spacecraft systems, nuclear reactor control systems and defense command and control systems. More recently, it has been recognized that very high reliability systems make economic sense in a wide range of industries. Examples include telecommunications networks, banking systems, credit verification systems and order entry systems.

The ultimate objective of research in the area of network reliability is to give design engineers procedures to enhance their ability to design networks for which reliability is an important consideration. Ideally, one would like to generate network design models and algorithms which take as input the characteristics of network components as well as network design criteria, and produce as output an 'optimal' network design. Since explicit expressions for the reliability of a network are very complex, typical design models use *surrogates* in place of explicit reliability expressions. For example, in Chapter 10 of this volume, Grötschel and Monma address network design problems where the surrogate used is network connectivity. In this chapter we treat the network reliability analysis problem, which is the problem of evaluating a measure of the reliability of a network. Analysis models are typically used in conjunction with network design procedures. For example, once a network design is produced using the techniques described in Chapter 10, models we describe might be used to determine the value of the network's reliability. If the reliability value were not satisfactory then the design model might be resolved with different design criteria. Alternatively, a designer might manually adjust the design. After a modified design was generated by one of the aforementioned techniques, the value of the network's reliability would be recomputed to determine if it is satisfactory. This process might iterate several times.

For other integrated treatments of network reliability we refer the reader to the book by Colbourn [1987], which gives a comprehensive treatment of the mathematics of network reliability, the book by Shier [1991], which treats the subject from an algebraic perspective, the two collections of papers edited by Rai & Agrawal [1990a, b], and the recent issue of IEEE Communications Magazine [Bose, Taka & Hamilton, 1993], which discusses issues of telecommunications network reliability.

1.1. Application areas

Interest in network reliability, particularly telecommunications network reliability, has increased substantially in recent years [Daneshmand & Savolaine, 1993]. Rapid advancement in telecommunications technology has led to an environment in which telecommunications services are a vital component of business, national security and public services. These technological advances have both provided customers with a broader range of services and made certain basic services more economical. On the other hand, much of the new technology involves capacity concentration, e.g. fiber optic communications links, high capacity digital switches. Such concentration widens the impact of the failure of a single network element. It is the combination of increased dependence on networks and increased network vulnerability to individual failures that has brought network reliability to the forefront of public interest.

We now describe some specific application settings from telecommunications as well as other areas.

1.1.1. Backbone level of packet switched networks

Packet switched networks were first developed in the 1960's to allow sharing of high speed communications circuits among many data communications users [Frank & Frisch, 1971; Frank, Kahn & Kleinrock, 1972]. Since the traffic associated with individual users tended to be bursty in nature, traffic on individual circuits could be dynamically allocated over time to a variety of users. ARPANET was the first major packet switched network. Much of the research on network reliability in the early 1970s and beyond was motivated by ARPANET. Most of the reliability measures used for ARPANET are 'connectivity' measures. That is, they define the network as operating as long as the network is connected or, in the case of specific user communities, as long as a specified subset of nodes is connected. Such measures are justified since ARPANET employed dynamic routing so that traffic could be rerouted around failed links as long as the network remained connected. However, even though traffic could be rerouted, congestion could occur and delays could increase due to the decrease in overall network capacity.

When one compares ARPANET with the backbone networks of commercial packet switched networks in use in the 1980s, such as Telenet and Tymnet, it is clear that these networks are much denser than ARPANET. As a result the probability of network disconnection is much lower. However, the increased link density is primarily motivated by larger traffic loads. The implication is that capacity and congestion issues must be taken more explicitly into account in defining reliability measures. To address this concern, some recent research has involved the definition and calculation of so-called performability measures [see for example Li & Silvester, 1984; Sanso & Soumis, 1991; Yang & Kubat, 1990]. Rather than defining the network as operating as long as it is connected, performability measures define the network as operating as long as its performance, possibly measured in terms of average delay, satisfies certain criteria.

1.1.2. Backbone level of circuit switched networks

By far the largest telecommunications networks in existence today are the circuit switched networks that make up the world's public telephone systems. In circuit switched networks, a communications channel is dedicated to a pair of users for the length of their call. As overall network capacity is reduced due to component failures the number of communications channels that the network can support is reduced. Thus, users are adversely affected in that it becomes more likely that when a call is attempted no circuit is available. This phenomenon is known as call blocking. This is to be contrasted with packet switched networks where the effect of failures is increased transmission delay. Of course, in either case, if the network becomes disconnected then it becomes impossible for certain pairs of users to communicate. Some of the earliest work in network reliability involved modeling of circuit switched networks [Lee, 1955] where network links are defined to be failed if they are blocked. Connectivity based measures were then used in conjunction with this failure definition. More recently, network performability measures have been defined [Sanso, Soumis & Gendreau, 1990]. In this case network performance is defined in terms of blocking rather than delay.

1.1.3. Interconnection networks

A special case of circuit-switched networks arises in the design of networks connecting parallel processors and memories in parallel computer architectures. Parallel computer systems have multiple components of the same type for the purpose of increasing overall throughput. However, parallel architectures also naturally have superior reliability characteristics. Typically, these fault tolerant and parallel computer systems are modeled as networks for the purpose of reliability analysis. Whereas much of the work in network reliability analysis motivated by telecommunications networks has concentrated on algorithms for analyzing general network topologies, most network reliability work motivated by computer architectures has concentrated on designing and analyzing the highly structured networks associated with particular computer architectures. Connectivity-based models are used both for failures due to congestion and to component wearout. Lee's pioneering work in telephone switching [Lee, 1955] anticipated the extensive use of connectivity-based measures for general interconnection networks [Agrawal, 1983; Hwang & Chang, 1982]. These measures have been particularly important in designing redundancy into interconnection networks [Blake & Trivedi, 1989a, b; Botting, Rai & Agrawal, 1989; Kini, Kumar & Agrawal, 1991; Varma & Raghavendra, 1989]; the surrogate for overall system performance here is the average connectivity of an input to an output [Colbourn, Devitt, Harms & Kraetzl, 1994].

1.1.4. Metropolitan area fiber networks

A recently developed technology that is transforming the world's telecommunications networks is fiber optics (see Flanagan [1990] for example). Fiber optic communications channels transmit communications signals via light waves traveling over glass fibers. The principal advantage of this communications medium over traditional cables is a significant increase in transmission capacity. In addition there are certain performance advantages in terms of signal quality, particularly relative to terrestrial microwave radio systems. Because of these very significant advantages most public telephone systems are rapidly replacing their existing transmission networks with networks based on fiber optics. However, it has quickly become apparent that there are major reliability concerns that must be addressed. In particular, due to the extremely high capacity of fiber optic circuits, the resultant fiber optic networks tend to be much sparser than traditional networks. The net effect is that previously reliability could be ignored in designing large scale networks since the networks tended to be very dense and, consequently, naturally had acceptable levels of reliability. Now, if reliability is not explicitly considered in network design, networks can result for which single link failures can cause major disruptions. It is this phenomenon that has motivated much of the work described in Chapter 10.

Fiber optic circuits have redundant channels and rerouting capability built in. In addition, as has been mentioned, they are very sparse. As a result it is felt that connectivity based measures are appropriate for quantifying their reliability.

1.1.5. Other applications

The richness of network models have led to their use in modeling several other reliability applications. In Colbourn, Nel, Boffey & Yates [1994] network reliability model is used to model random spread of fire. In this context, once a fire has established itself in a room or building there is a possibility that it spreads through a barrier (wall) to an adjacent room or building. A network model is employed in which the link failure probability is interpreted as the probability that the fire spreads from a compartment through a wall to an adjacent compartment.

Sanso & Soumis [1991] discuss network reliability models in several application settings. A major theme is to stress the importance of routing in all of the application settings. In particular, in all cases analyzed, the network supports a diverse set of users and each user's traffic follows one or more routes through the network. The implication is that reliability can only be accurately evaluated if routing considerations are incorporated into the reliability measure. To accomplish this, it is necessary to consider performability measures. One of the more interesting applications areas discussed is urban transportation networks. In this context, incidents, such as highway accidents, cause the failure of network nodes and links. Although it is rare that urban transportation networks become disconnected, it is quite common for node and link failures to cause major congestion.

Several innovative applications have been developed based on bipartite network models (see for example Colbourn & Elmallah [1993], Colbourn, Provan & Vertigan [1994], Harms & Colbourn [1990], Ball & Lin [1993]). The underlying networks include a set of resource nodes and a set of user nodes. A resource node and a user node are adjacent if the resource node is capable of providing services to the user node. Reliability models have been formulated to study the effects of resource node failures. Applications have been studied in which the resource nodes are processors, personnel and emergency services vehicles and the users are tasks, jobs and emergency calls respectively. Many of the reliability tools developed for connectivity-based measures of network performance generalize to this setting.

1.1.6. Causes of failures

In most classical reliability analysis, failure mechanisms and the causes of failure are relatively well understood. For example, in electronic systems long term wear would result from continual exposure to heat. Such wear randomly causes failure over the range of exposed components. Reliability analysis typically involves the study of these random processes and characterization of associated failure distributions. Although some failure mechanisms associated with network reliability applications have these characteristics, many of the most important do not. For example, many well-publicized failures associated with fiber optic networks have been caused by natural disasters such as fires or human error such as the severing of a communications line by a back-hoe operator. As a result it is difficult to model failure mechanisms in order to come up with failure rates. Typically, component failure rates are estimated based on historical data.

1.2. Basic definitions

Due both to the inability to model failure mechanisms and the inherent difficulty of computing network reliability, time independent, discrete probability models are typically employed in network reliability analysis. In the most commonly studied model to which we devote most of our attention, network components (nodes and edges) can take on one of two states: operative or failed. The state of a component is a random event that is independent of the states of other components. Similarly, in the simplest models, the network itself is in one of two states, operative or failed. The reliability analysis problem is: given the probabilities that each component is operative, compute a measure of network reliability. We treat some generalizations of this model. In particular, we look at models in which components can take on one of several state values or models in which a quantity is associated with the operative state. The state values typically either correspond to distances or capacities. The simple two-state model is sufficient for the consideration of connectivity measures, but when more complex performability measures are considered, more complex component states must be considered.

In the two-state model, the component's probability of operation or, simply, reliability, could have one of several possible interpretations. The most common interpretations are (1) the component's availability, and (2) the component's reliability. Generally, throughout this chapter, we use the term reliability to mean the probability that a component or system operates. Here we discuss a more specific definition. Availability is used in the context of repairable systems. In these settings, components alternate between being in the operative state and being failed and under repair. The component's (steady-state) *availability* is defined as the limit as t approaches infinity of the probability that the component is operating at time t. If a component's on/off behavior obeys the assumptions of an alternating renewal process [see Barlow & Proschan, 1981] then the availability is equal to

$$\frac{\text{mean time to failure}}{\text{mean time to failure} + \text{mean time to repair}}.$$

That is, the availability can be estimated by estimating both the mean time to failure and the mean time to repair.

The definition of component reliability does not involve considerations of repair. Rather, a length of time t is specified and the *reliability* of a component is defined to be the probability that the component does not fail within time t.

Other interpretations of a component's probability of operation are possible. For example, in Lee [1955] the probability that a circuit is not blocked is used as the probability that the corresponding edge operates. The preceding discussion carries over to multi-state components as well. For example, suppose that one were using an availability model in a context where edges could take on one of three capacity levels. Then the probability associated with a particular capacity level, *cap*, would be the limit as t approaches infinity of the probability that the component had capacity level *cap* at time t.

Of course, the interpretation of the component level reliabilities in turn determine the appropriate interpretation of the network reliability measures calculated. In the remainder of this paper we simply refer to the probability of operation or reliability and are not specific about the interpretation.

The input to all network reliability analysis problems includes a network $G = (V, E)$, where V is a set of nodes and E is a set of undirected edges or a set of directed arcs. For connectivity measures, for each $e \in E$, p_e, the probability that e operates is input. For the multi-state systems we discuss, a length, capacity or duration distribution function is defined for each edge. In most cases we use finite discrete distributions.

It is sometimes convenient to consider the general system reliability context. Here, a system which is made up of a set of components and a random variable, X_e, associated with each component, e. The value of X_e indicates the 'health' of e; the health of the system is a function of the X_e values. In the network reliability context, the system is the network and the components are arcs or edges. A function Φ maps the states of the components into system states. Thus, $\Phi(\boldsymbol{X})$ is a random variable which provides information on the overall health of the system. A variety of system reliability measures may be defined in terms of Φ. Of course, several options exist for defining Φ itself.

A simple, but very general, model is the *stochastic binary system* (SBS). Each component in the component set, $T = \{1, 2, \ldots, m\}$, can take on either of two states: operative or failed. X_e has value 1 if e operates and 0 if e fails. Φ maps a binary component state vector $\boldsymbol{x} = (x_1, x_2, \ldots, x_m)$ into the system state by

$$\Phi(\boldsymbol{x}) = \begin{cases} 1 \text{ if } \boldsymbol{x} \text{ is an operating system state,} \\ 0 \text{ if } \boldsymbol{x} \text{ is a failed system state.} \end{cases}$$

An SBS is *coherent* if $\Phi(\boldsymbol{1}) = 1$, $\Phi(\boldsymbol{0}) = 0$ and $\boldsymbol{x}^1 \geq \boldsymbol{x}^2$ implies $\Phi(\boldsymbol{x}^1) \geq \Phi(\boldsymbol{x}^2)$. The third property implies that the failure of any component can only have a detrimental effect on the operation of the system. The computational problem of interest is to compute:

$$Rel(\text{SBS}, \boldsymbol{p}) = \Pr[\Phi(\boldsymbol{X}) = 1]$$

given some representation of $\Phi()$. At times we consider reliability problems where $p_e = p$ for all e in which case we replace $\boldsymbol{p}$ by p in the above notation. For any stochastic coherent binary system (SCBS), define a *pathset* as a set of components whose operation implies system operation, and a *minpath* as a minimal pathset; similarly, define a *cutset* to be a set of components whose failure implies system failure, and a *min-cut* to be a minimal cutset.

1.3. Network reliability measures

Network reliability measures that we study are either the probability of certain random events or the expected value of certain random variables. The majority of the research in network reliability as well as the majority of this paper is devoted connectivity measures, specifically to the *k-terminal measure*. A set of nodes K and a node $s \in K (k = |K|)$ are given. Given a network G and an edge reliability vector $\boldsymbol{p}$, the k-terminal reliability measure is defined as

$$Rel(G, s, K, \boldsymbol{p}) = \\ = \Pr[\text{there exist operating paths from } s \text{ to each node in } K].$$

Two important special cases of the measure are the *two-terminal measure* for which $|K| = 2$ and the *all-terminal measure* for which $K = V$. The two-terminal and all-terminal measures are denoted by $Rel_2(G, s, t, \boldsymbol{p})$ and $Rel_A(G, s, \boldsymbol{p})$ respectively. We call the node s the source node and the nodes in $K \setminus \{s\}$ the terminals. When the appropriate values are obvious from the context, we may leave one or more of the arguments out of the $Rel()$ notation. Other connectivity measures have been analyzed [see for example Ball, 1980; Colbourn, 1987]. The details are omitted here, not because they are unimportant, but because their coverage would not provide substantial additional insight.

In addition to connectivity measures we discuss measures that apply to more general multi-state problems. In such cases Φ and/or the X_e can take on values other than 0 and 1. Included in this category are stochastic flow, shortest path and PERT measures. An important subclass consists of performability measures. Performability measures provide an evaluation of a network's reliability relative to some performance criterion. Several performance criteria have been considered. For example, for packet switched networks a commonly used criterion is average message or packet delay. Viewed in terms of our general model, Φ gives the value of average message delay as a function of $\boldsymbol{X}$ where X_e is the capacity of edge e. In this case there is another key input, namely, the traffic load on the network. There does not appear to be a generally accepted, precise set of criterion that distinguish performability measures from other multi-state measures. However, we feel that one key element is that the measure should evaluate the ability of the network to carry out a certain 'assigned task', e.g. to handle a traffic load. In general, if Φ is the performance criterion then two classes of performability measures are commonly considered:

- $\Pr[\Phi \geq \alpha]$ or $\Pr[\Phi \leq \alpha]$, the probability that a threshold is met; and
- $\mathrm{Ex}[\Phi]$, the expected value of the criterion random variable.

We discuss general techniques that apply to a wide range of performability measures. In addition, we analyze in detail three multi-state problems: shortest path, maximum flow and PERT. For these problems, together with G, we are given a source node, s and a terminal node, t. For the stochastic shortest path problem, X_e is the length of arc e and Φ_{PATH} is the length of a shortest s,t-path. For the stochastic max flow problem, X_e is the capacity of arc e and Φ_{FLOW} is the value of a max s,t-flow. For the stochastic PERT problem, X_e is the duration of arc e and Φ_{PERT} is the value of a max-duration s,t-path. The reliability measures of interest are $\mathrm{Ex}[\Phi]$ in all cases and $\Pr[\Phi_{\mathrm{PATH}} \leq \alpha]$, $\Pr[\Phi_{\mathrm{FLOW}} \geq \alpha]$ and $\Pr[\Phi_{\mathrm{PERT}} \leq \alpha]$ where α is defined appropriately in each case. We also discuss work which produces complete distributions of Φ.

2. Computational complexity and relationships among problems

We start by discussing the differences between directed and undirected problems and the impact of node failures in Section 2.1 and 2.2, respectively. We then address issues of computational complexity in the remaining sections.

2.1. Directed vs. undirected networks

The general technique of replacing an undirected edge $\{i, j\}$ with the two corresponding anti-symmetric directed arcs (i, j) and (j, i), applies quite generally to network reliability problems. Specifically,

Undirected to directed transformation: Suppose that the directed graph G' is obtained from the undirected graph G be replacing each undirected edge by the corresponding pair of anti-symmetric directed arcs. As part of this transformation each directed arc inherits the appropriate stochastic properties of the undirected edge, e.g. failure probability, capacity distribution, etc. Then the system reliability of G and G' are equal for each of the following measures $Rel(G, s, K, p)$ and $\Pr[\Phi \geq t]$; $\Pr[\Phi \leq t]$ and $\text{Ex}[\Phi]$ for Φ equal to Φ_{FLOW} or Φ_{PATH}.

This transformation is similar to transformations used in network flows. It is interesting and somewhat surprising that it applies in this context since effectively, this transformation allows us to treat the states of the anti-symmetric pair of arcs as independent random variables when in fact they are not independent. For the proof of this result in the case of connectivity, see Nakazawa [1979]. and Ball [1980]; in the case of shortest paths see Hagstrom [1983] and in the case of flows see Hagstrom [1984]. This result does not necessarily hold in the context of more complex performability measures.

2.2. Node failures

In many applications, nodes as well as edges can fail. Consequently, one is led to consider models that can handle both node and edge failures. Fortunately, in the case of directed networks, a node i can be replaced by two nodes, i_1 and i_2, and the directed arc (i_1, i_2), where all arcs previously directed into i are directed into i_1 and all arcs previously directed out of i are directed out of i_2. Using this transformation a problem with unreliable nodes and arcs can be transformed into a problem with only unreliable arcs and perfectly reliable nodes. The transformation applies to all the measures to which the previous transformation applied where in each case arc (i_1, i_2) inherits the characteristics of node i. When carrying out the transformation for a terminal i, only the replacement node i_2 should be a terminal. Similarly, when carrying out this transformation for a source node i, only the replacement node i_1 should be a source See Ball [1980] or Colbourn [1987] for a general discussion of this transformation.

The transformations given in this section and the previous one indicate that, from a practical standpoint, one would prefer codes for directed network reliability analysis over codes for undirected network reliability analysis. By properly preparing input data, directed network codes can be used to analyze directed and undirected problems and problems with and without node failures.

2.3. An introduction to the complexity of reliability analysis

The computational problems most often studied by computer scientists and others interested in algorithms are recognition problems, such as determining if a graph contains a Hamiltonian cycle, and optimization problems, such as finding a minimum cost traveling salesman tour. Reliability analysis problems are fundamentally different. They compute a value that depends on the structure of a network as well as related data. Consequently, the analysis of their complexity involves concepts related to, but different from, the machinery used to analyze recognition and optimization problems: the classes P, NP and NP-complete.

In order to most easily relate reliability analysis problems to more familiar combinatorial problems we consider the special case of the reliability analysis problem that arises when all individual component reliabilities are equal, i.e. $p_e = p$ for all components e. In this case, $Rel(\text{SBS}, p)$ can written as a polynomial in p with the following form:

$$Rel(\text{SBS}, p) = \sum_{i=0}^{m} F_i p^{m-i}(1-p)^i .$$

This polynomial is the *reliability polynomial.* The associated computational problem, which we call the *functional reliability analysis problem*, takes as input a representation of an SBS and produces as output the vector $\{F_i\}$.

The general term in the reliability polynomial, $F_i p^{m-i}(1-p)^i$, is the probability that exactly $m-i$ components operate and the system operates. Thus, we can interpret F_i as the number of operating system states having i failed components or more precisely:

$$F_i = \left|\left\{x : \sum_k x_k = m - i \text{ and } \Phi(x) = 1\right\}\right|.$$

We can see that the problem of determining each of the coefficients F_i is a counting problem. Whereas the output of the Hamiltonian cycle recognition problem is 'yes' if the input graph contains a Hamiltonian cycle, and 'no' if the graph does not, the output of the Hamiltonian cycle counting problem is the number of distinct Hamiltonian cycles contained in the graph. NP and NP-complete are classes of recognition problems. The corresponding classes of counting problems are #P and #P-complete. It is clear that any counting problem is at least as hard as the corresponding recognition problem. For example, if one knows the number of Hamiltonian cycles in a graph then one can immediately answer the question: 'Is the number of Hamiltonian cycles greater than zero?'. Thus, the counting versions of any NP-complete problems are trivially NP-hard. In fact, it seems to be a general rule that the counting problems associated with NP-complete problems are #P-complete. However, such a relationship has not been formally established.

On the other hand there are certain recognition problems solvable in polynomial time whose corresponding counting problems are #P-complete. For example, the problem of determining whether a bipartite graph contains a perfect matching is polynomially solvable but the problem of determining the number of perfect matchings in a bipartite graph is #P-complete [Valiant, 1979]. To make the presentation simpler, we do not delve further into detailed complexity issues but rather simply indicate whether problems are NP-hard or polynomial.

Many practical applications require the use of models with unequal component reliabilities. For the case of unequal component reliabilities, where all probabilities are rational numbers, we define the *rational reliability analysis problem* as follows. The input consists of a representation of an SBS and, for each component i a pair of integers a_i, b_i. The output is a pair of integers a, b where $a/b = Rel(\text{SBS}, \{a_i/b_i\})$.

We start by establishing the following:

Functional to rational reducibility: For any rational reliability analysis problem, r-Rel, and its corresponding functional reliability analysis problem, f-Rel, f-Rel can be reduced in polynomial time to r-Rel.

To see this, we proceed as follows. An instance of f-Rel consists of a representation of an SBS. The required output is the set of coefficients $\{F_i\}$ of the reliability polynomial. To transform f-Rel to r-Rel we select $m + 1$ rational probabilities $0 < p_0 < p_1 < \cdots < p_m < 1$. For $j = 0, 1, \cdots, m$, we denote by $r_j = Rel(\text{SBS}, p_j)$, the solution to the corresponding rational reliability analysis where all component reliabilities are set equal to p_j. We now can set up the following system of equations:

$$\sum_{i=0}^{m} F_i p_j^{m-i}(1 - p_j)^i = r_j \quad \text{for } j = 0, 1, \cdots, m.$$

Having solved $m + 1$ rational reliability analysis problems, the p_j's and the r_j's are known. We have a system of $m + 1$ linear equations in $m + 1$ unknowns, the F_i's. The coefficient matrix has the Vandemonde property and consequently is non-singular so that the F_i's can be efficiently determined.

We now investigate more carefully the structure of the reliability polynomial for SCBSs. Given an SCBS, we define:

m = number of components in the system,
c = cardinality of a minimum cardinality cutset,
C_c = number of minimum cardinality cutsets,
ℓ = cardinality of a minimum cardinality pathset,
N_ℓ = number of minimum cardinality pathsets.

It can immediately be seen that the coefficients of the reliability polynomial have the following properties:

$$0 \le F_i \le \binom{m}{i} \quad \text{for } i = 0, 1, \cdots, m,$$

$$F_i = \binom{m}{i} \quad \text{for } i < c,$$
$$F_i = \binom{m}{i} - C_c \quad \text{for } i = c,$$
$$F_i = N_\ell \quad \text{for } i = m - \ell,$$
$$F_i = 0 \quad \text{for } i > m - \ell.$$

These properties imply that by computing the reliability polynomial we immediately determine important properties of the SCBS. For example, by examining the reliability polynomial we can determine the size of a minimum cardinality cutset. Thus, if the minimum cardinality cutset recognition problem is NP-hard, then computing the reliability polynomial is NP-hard. This line of reasoning leads to the following result.

Complexity of reliability of analysis: For any SCBS if any one of the following five conditions hold then the functional and rational reliability analysis problems are NP-hard.

1. The minimum cardinality pathset recognition problem is NP-hard.
2. The minimum cardinality pathset counting problem is NP-hard.
3. The minimum cardinality cutset recognition problem is NP-hard.
4. The minimum cardinality cutset counting problem is NP-hard.
5. The problem of determining a general coefficient of the reliability polynomial is NP-hard.

We now use the framework just established to investigate network reliability problems.

2.4. The complexity of network reliability analysis

We present results concerning the complexity of network reliability analysis problems for the following problem classes: k-terminal, 2-terminal and all-terminal.

2.4.1. k-Terminal

A minimum cardinality pathset for the k-terminal measure is a minimum cardinality Steiner tree. It is well known [Karp, 1972] that the associated recognition problem is NP-hard for both directed and undirected networks so the associated functional and rational reliability analysis problems are NP-hard. Valiant [1979] gives an alternate proof of this result by showing that computing $SN(K) = \sum F_i = |\{S : S \text{ is a subgraph that contains a path to each node in } K\}|$ is NP-hard. Here K is the set of terminals.

2.4.2. Two-terminal

The minimum cardinality pathset and cutset recognition problems associated with the 2-terminal measure are the shortest path and minimum cut problems respectively. Polynomial algorithms are known for both of these problems [Moore, 1959; Ford & Fulkerson [1962]]. Valiant [1979] first showed that the 2-terminal reliability analysis problems were NP-hard. His reduction, which we now describe, is a good illustration of the proof techniques used in this area. The proof given below reduces the problem of computing $SN(K)$ to the 2-terminal rational reliability analysis problem:

Proof. Given a graph $G = (N, A)$, a source node s and a set of terminal nodes K, construct G' by adding a node t and edges (u, t) for each $u \in K$. We assign a failure probability of $1 - p$ to each (u, t) and a failure probability of $\frac{1}{2}$ to all edges in the original network. Note that if all edges in A have failure probability equal to $\frac{1}{2}$ then all random states of the edges in A have probability $\frac{1}{2}^{|A|}$. If we define A_i as the number of subgraphs in G in which s is connected to exactly i members of K we now have: $Rel(G', s, t) = \sum_i \sum_{S \subseteq K, |S|=i}$ Pr[there exist operating paths from s to S, but to no other nodes in $K - S$ and (u, t) operates for at least one $u \in S] = 2^{-|A|} \sum A_i p^i$. By evaluating this reliability for $|K|$ different values of p we can set up a system of $|K|$ equations in $|K|$ unknowns, the A_i. The A_i can then be determined. The reduction is now complete since $A_{|K|} = SN(K)$. This reduction shows that the rational problem is NP-hard. A slight extension also shows that the functional problem is NP-hard. □

Provan & Ball [1983] give an alternate proof of this result by showing the problem of determining the number of minimum cardinality s, t-cuts is NP-hard.

2.4.3. All-terminal

For the directed all-terminal measure (*reachability*), the minimum cardinal pathset and cutset problems are the minimum cardinality spanning arborescence and minimum cardinality s-directed cut problems respectively. Both of these are polynomially solvable [Edmonds, 1967]. Provan and Ball [1983] showed that the problem of counting minimum cardinality s-directed cuts is NP-hard, which in turn implies that the associated reliability analysis problems are NP-hard. For the undirected case, the minimum cardinality pathset and cutset recognition and counting problems are all polynomially solvable. However, Provan and Ball [1983] showed that the problem of computing a general term in the reliability polynomial is NP-hard, implying that the undirected reliability analysis problems are NP-hard.

Table 1 summarizes the known complexity results for the five counting and recognition problem classes listed in the previous section for the k-terminal, all-terminal and 2-terminal problems.

In light of these negative results, much research has been aimed at the analysis of structured networks. The widest class of networks known to be solvable in polynomial time involve series-parallel graphs and certain generalizations. Section 3 treats the solvable cases in more detail. Recent research has addressed the complexity of reliability analysis over structured networks, specifically directed acyclic networks and planar networks. Provan [1986] shows that the undirected two-terminal reliability problem remains NP-hard over planar networks having node

Table 1

Complexity results related to k-terminal, two-terminal and all-terminal reliability analysis

k-Terminal	2-Terminal		All-terminal	
			directed	undirected
min. card. pathset rec.	! Karp [1972]	∗ Moore [1959]	∗	∗
min. card. cutset rec.	∗ Even & Tarjan [1975], Ford & Fulkerson [1962]	∗ Even & Tarjan [1975], Ford & Fulkerson [1962]	∗ Even & Tarjan [1975], Ford & Fulkerson [1962]	∗ Even & Tarjan [1975], Ford & Fulkerson [1962]
min. card. pathset count.	! Karp [1972]	∗ Ball & Provan [1983]	∗ Kirchoff [1958]	∗ Kirchoff [1958]
min. card. cutset count.	! Provan and Ball [1983]	! Provan and Ball [1983]	! Ball & Provan [1983]	∗ Ball & Provan [1983], Bixby [1975]
gen. term. count.	! Karp [1972]	! Valiant [1979]	! Provan and Ball [1983]	! Provan and Ball [1983]

(∗ = polynomial, ! = NP-hard)

degrees bounded by 3 and the directed 2-terminal reliability analysis problems remain NP-hard over acyclic planar networks having node degrees bounded by 3. Vertigan [1994a, b] has recently shown that directed and undirected all-terminal reliability analysis problems are NP-hard when restricted to planar networks. There is a simple formula for the directed all-terminal reliability analysis problem of acyclic networks [Ball & Provan, 1983].

The results of this section indicate that polynomial algorithms are only likely to exist for network reliability problems restricted to small classes of networks. Due to this fact a large amount of research has been devoted to the study of network reliability bounds and Monte Carlo approaches, the subjects of Sections 4 and 5, respectively.

3. Exact computation of reliability

In this section, we examine exact algorithms for computing reliability measures. We have seen that for general networks, the computation of all of the reliability measures of interest here is NP-hard. For this reason, we explore two main directions: exponential time exact algorithms for general networks, and polynomial time exact algorithms for restricted classes of networks.

Both directions rely on a simple but important observation: there exist graph transformations that leave the values of various reliability measures unchanged, and these can often be used to simplify the network used in the exact computation of reliability. Our first topic is such simplifying transformations.

3.1. Transformations and reductions

An edge or arc that appears in no minpath is *irrelevant*: the operation or failure of the network is not affected by the operation or failure of such an irrelevant edge. The easiest simplifying transformation is the *deletion of irrelevant edges*. By definition, the transformation is reliability-preserving. Now for the transformation to be of practical use, we must be able to apply it efficiently (in polynomial time in the size of the network). For all-, k-, and two-terminal reliability, loops are always irrelevant. For k- and two-terminal reliability, so also is any edge having an endpoint in a 2-connected component containing no terminal; such edges can be found easily and deleted. For the directed reliability problems, the identification of irrelevant arcs is by no means an easy problem. Provan & Kulkarni [1989] have shown that determining whether an arc is irrelevant for s, t-connectedness is NP-hard, although the general undirected problem admits an efficient solution.

We focus on the undirected problems here. An edge or arc that appears in *every* minpath is *mandatory*. After irrelevant edges have been deleted, any bridge (edge cutset of size one) that remains is mandatory. Let $G = (V, E)$ with terminal set $K \subseteq V$, and bridge $e \in E$ with operation probability p_e. The *contraction* $G \cdot e$ of an edge $e = \{x, y\}$ in G is obtained by removing e, identifying x and y and making the resulting node a terminal whenever $K \cap \{x, y\} \neq \emptyset$. The *deletion* $G - e$ of an edge e is the graph obtained from G by simply removing the edge e. The reliability of G, $Rel(G)$, satisfies $Rel(G) = p_e Rel(G \cdot e)$ when e is a mandatory edge. Thus the mapping from G to $G \cdot e$ is a reliability-preserving transformation *with multiplicative factor* p_e.

Two edges e, f having the same endpoints are in *parallel*. Since any minpath contains at most one of the two, and interchanging e and f is a natural bijection between the minimal pathsets containing one and the minpaths containing the other, the replacement of e and f by a single edge g having $p_g = 1 - (1 - p_e)(1 - p_f)$ is reliability-preserving. This is a *parallel reduction*. The notion of parallel reductions can be generalized when e and f are 'substitutes'; see Hagstrom [1990].

Two edges $e = \{x, y\}$ and $f = \{y, z\}$ are in *series* when y is a node of degree 2. In this case, any min-cut contains at most one of e or f, and interchanging e and f is a natural bijection between the min-cuts containing e and those containing f. Thus a reliability-preserving transformation is obtained by removing the node y and the edges e, f, and adding the edge $g = \{x, z\}$ with $p_g = p_e p_f$ *provided* that y is not a terminal node. This is a *series reduction.* More generally, when two edges are 'complements', similar reductions can be applied; see Hagstrom [1990].

When a degree two *terminal* node is present, one cannot apply a series reduction. However, when x, y and z are all terminals, the same structural replacement is reliability-preserving with factor $1 - (1 - p_e)(1 - p_f)$ if the new edge g is given probability $p_g = p_e p_f / (1 - (1 - p_e)(1 - p_f))$. This is a *degree-2 reduction*. There remain cases when y is a terminal, but at least one of x or z is not. Generalizations of the series and degree-2 reductions, the *polygon-to-chain reductions*, are available [Resende, 1986; Wood, 1985].

In essence, each of the simplifications thus far can be viewed as the replacement of some subnetwork by a subnetwork that has equivalent reliability characteristics, or characteristics that scale the original reliability measure by a fixed amount. With this in mind, consider a network $G = (V, E)$ with terminal set $K \subseteq V$. An induced subnetwork $H = (W, F)$ of G is *s-attached* if there is a set $A \subseteq W$ with $|A| \leq s$, for which every edge of G with one endpoint in $V \setminus W$ and one endpoint in W has an endpoint in A; in other words, only the nodes in A attach the subnetwork to the remainder of the network. A general class of simplifications arises by examining s-attached subgraphs for small s, and replacing each by a simpler s-attached subgraph.

A 1-attached subnetwork is connected to the rest of the network at a single cutnode. If the 1-attached subnetwork contains no terminal, all edges in it are irrelevant. On the other hand, if both the subnetwork and the remainder of the network contain terminals, the cutnode itself may be treated as a terminal since it is connected to the terminals in any minpath. So add the cutnode as a terminal. Then the network can be split into two subnetworks H and $G \setminus (W - A)$, and the reliability measure is the product of the measures for these two. This generalizes the notion of transformation to one that partitions the network into two or more subnetworks.

For 2-attached subnetworks, we view the replacement of the subnetwork as the determination of an equivalent edge. If H is a subnetwork attached at $\{x, y\}$, and H contains no terminals, we can determine the two-terminal reliability of H from x to y, and replace H by an edge $\{x, y\}$ whose operation probability is the two-terminal reliability found. When H contains terminals, the situation is more complicated, as it no longer suffices to know whether x can reach y; one must also know whether all of the internal terminals can reach x or y or both. Nevertheless, by permitting an edge to carry a number of probability values, rather than just an operation probability, transformations have been developed [Wald & Colbourn, 1983a, b]. The number of values that must be maintained here is independent of the size of the network, but grows exponentially with the number of attachment nodes; see Section 3.2. A number of specific methods for employing the reduction of 2-attached and 3-attached subnetworks have been examined [Agrawal & Satyanarayana, 1985; Arnborg & Proskurowski, 1989; Fu & Yau, 1962; Hagstrom, 1983, 1984; Shogan, 1978]. Rosenthal [Rosenthal, 1977; Rosenthal & Frisque, 1977] was apparently the first to develop a general framework for these transformations, and for generalizations to k-attached subnetworks.

3.2. *Efficient algorithms for restricted classes*

Our goal first and foremost is to obtain polynomial time algorithms for calculating reliability measures whenever possible. In view of the complexity results, we cannot hope at the present time to obtain efficient methods for networks in general. However, we can expect to treat restricted classes of networks efficiently.

From Section 3.1, we have a large collection of reliability-preserving reductions that, at least in their simplest forms, can be applied in polynomial time. Any set of reductions succeeds in reducing some (typically small) class of networks to a single node, and thus when the reductions can be applied efficiently, to an efficient algorithm for this class. For example, the elimination of irrelevant edges and the contraction of mandatory edges together give an algorithm for k-terminal reliability of trees. A better example is obtained by using also series and parallel reductions. Then the two-terminal reliability of series-parallel networks can be calculated. This result dates back at least to Lee [1955]. When instead one adds degree-2 and parallel reductions, an all-terminal reliability algorithm for series-parallel graphs is immediate using characterization theorems for series-parallel graphs [Duffin, 1965; Wald & Colbourn, 1983a].

For k-terminal reliability of series-parallel graphs, two linear time algorithms exist. Satyanarayana & Wood [1985] employed series, parallel and degree-2 reductions, along with a type of 2-attached subnetwork reductions called the polygon-to-chain reductions, to reduce an arbitrary series-parallel network with terminals to a single node. Agrawal & Satyanarayana [1984, 1985] extended this to the directed reliability measures. By employing in addition the reduction of certain 3-attached networks, Politof & Satyanarayana [1983, 1986] extended these linear time algorithms to larger subclasses of the planar networks.

Wald & Colbourn [1983a, b] obtained a linear time algorithm for k-terminal reliability by a different method. They generalized the notion of a transformation to permit each edge to carry a fixed finite number of reliability values, and developed a scheme for replacing an arbitrary 2-attached subnetwork by an edge having six associated values. When two subnetworks are attached at s nodes, the reliability of their union can be determined completely from the values on each of the subnetworks. This can in fact be accomplished by a dynamic programming algorithm whose running time is linear in the number of nodes, but exponential in the number of attachment nodes. Wald and Colbourn's method exploits the fact that series-parallel graphs can be recursively decomposed at node cutsets of size two; that is, series-parallel graphs have tree-width two. The extension of the linear time algorithm to any fixed tree-width is almost immediate [Arnborg & Proskurowski, 1989; Mata-Montero, 1990], *except* for the difficulty of recognizing networks of a given tree-width. El-Mallah & Colbourn [1985] establish an algorithm for certain networks of tree-width three, and observe that the algorithms of Politof & Satyanarayana [1983, 1986] can be seen as algorithms on networks of small tree-width.

Obtaining efficient algorithms by restricting the tree-width accounts for the majority of efficient algorithms in the literature. Planar graphs do not have tree-width independent of the number of nodes, although tree-width is bounded by $O(\sqrt{n})$ for an n-node planar network. This underlies an algorithm due to Bienstock [1986] for planar networks that improves on general exact algorithms, but remains exponential time.

Beyond graphs with fixed tree-width, little is known for undirected reliability measures. Gilbert [1959] developed an elegant recursive method for computing all-terminal reliability of complete networks when all edges operate with the same probability. The method requires linear time. It generalizes in a natural way to k-terminal reliability [Colbourn, 1987], and to complete bipartite [Colbourn, 1987] and related networks [Brown, Colbourn & Devitt, 1993]. The method depends essentially on the observation that the number of nonisomorphic induced subgraphs is bounded by a polynomial in the number of nodes. A dynamic programming method then need only examine a polynomial number of subnetworks.

Turning to directed reliability measures, one important algorithm stands out. Ball & Provan [1983] develop a linear time algorithm for computing reachability of acyclic directed networks. The algorithm is based on the observation that such a network is operational (for reachability) if and only if every non-root node has at least one of its incoming arcs operational.

There has recently been an extensive investigation of efficiently solvable classes of *nodal* reliability problems. Nodal two-terminal reliability (without edge failures) admits efficient algorithms for permutation graphs and interval graphs [AboElFotoh & Colbourn, 1990]. From this one can obtain efficient algorithms for two-terminal reliability with edge failures when the network has a line graph that is an interval or permutation graph, using a transformation from edge failure problems to node failure problems in AboElFotoh & Colbourn [1989b].

The classes for which efficient exact algorithms are known are quite sparse, and do not appear to be those in which we expect to find most practical problems. Nevertheless, the presence of such exact algorithms even for sparse classes can be used to accelerate exact algorithms for larger classes that simplify the network in some way; and they have applications in computing bounds.

3.3. *State-based methods*

When reliability-preserving transformations fail to reduce the network into a restricted class for which an efficient exact method is known, we are forced to resort to potentially exponential time methods. The first main class of these exact methods examines the possible states of the network.

A *state* of a network $G = (V, E)$ is a subset $S \subseteq E$ of operational edges. The conceptually simplest exact algorithm is *complete state enumeration*. Let $\mathcal{O}$ be the set of all operational states (pathsets). Then

$$Rel(G) = \sum_{S \in \mathcal{O}} \prod_{e \in S} p_e \prod_{e \notin S} (1 - p_e).$$

By generating all states, and determining which is operational, the reliability is 'easily' (but not efficiently) computed.

Of course, large groups of the states are easily seen to be operational without listing them out one by one. Hence an immediate improvement is obtained by generating the states in a more clever way. A basic ingredient in this is the Factoring Theorem:

$$Rel(G) = p_e Rel(G \cdot e) + (1 - p_e) Rel(G - e)$$

for any edge e of G. *Factoring*, also called *pivotal decomposition*, was explicitly introduced in Moskowitz [1958] and Mine [1959]. Factoring carried out until the networks produced have no edges is just complete state enumeration. However, some simple observations result in improvements. When $G - e$ is failed, any sequence of contractions and deletions results in a failed network, and hence there is no need to factor $G - e$. Moreover, although we may be unable to simplify G with a reliability-preserving transformation, we may well be able to simplify $G \cdot e$ or $G - e$.

Factoring with elimination of irrelevant edges, contraction of mandatory edges, and series, parallel and degree-2 reductions forms the basis of many exact algorithms in the literature [Resende, 1988, 1986; Satyanarayana and M.K. Chang, 1983; Wood, 1982, 1989]. Satyanarayana & Chang [1983] analyzed the number of times a factoring step is required in an optimal factoring strategy [see also Wood, 1985; Huseby, 1990]. For complete graphs, complete state enumeration examines $2^{\binom{n}{2}}$ states, while a factoring algorithm using series and parallel reductions examines only $(n - 1)!$.

We state the factoring method more explicitly here:

procedure **factor** (graph G);
  apply reliability-preserving transformations to G that
    delete irrelevant edges
    contract mandatory edges
    apply series reductions
    apply parallel reductions
    apply degree-2 reductions
    apply other reductions such as polygon-to-chain
  maintaining G with each edge having the probability resulting from the sequence of reductions, and also maintaining a multiplicative factor **mult** that results from the reductions
  if G has only one terminal remaining, **return(mult)**
  else
    select an edge e of G
    return(mult*(p_e***factor** (G·e)+(1 − p_e)***factor**(G–e)))
  end

Further improvements are possible by partitioning the graph into its biconnected or triconnected components at each step [Wood, 1989].

3.4. Path- and cut-based methods

Once basic reductions are done, complete state enumeration could be applied to finish the computation. Unless the reductions have succeeded in dramatically reducing the size of the graph, however, this remains impractical. It may nevertheless be possible to generate all minpaths of the network, and hence a method employing just the minpaths is in order. Suppose then that the minpaths $P_1, \ldots, P_h$ of G have been listed. Let E_i be the event that all edges in minpath P_i are operational. Then the reliability is just the probability that one (or more) of the events $\{E_i\}$ occurs. Unfortunately, the $\{E_i\}$ are not disjoint events, and hence we cannot simply sum their probabilities of occurrence. To be specific, $\Pr[E_1 \text{ or } E_2]$ is $\Pr[E_1] + \Pr[E_2] - \Pr[E_1 \text{ and } E_2]$. Now $Rel(G) = \Pr[E_1 \text{ or } E_2 \text{ or } \ldots \text{ or } E_h]$, and hence

$$Rel(G) = \sum_{j=1}^{h} (-1)^{j+1} \sum_{I \subseteq \{1,\ldots,h\}, |I|=j} \Pr[E_I]. \tag{1}$$

where E_I is the event that all paths P_i with $i \in I$ are operational. This is a standard inclusion–exclusion expansion.

The algorithmic consequences of this formulation are immediate. Having a list of minpaths, one computes the probability of each subset of the minpaths occurring. To compute the reliability, one need only evaluate the above sum. In doing so, observe that an odd number of minpaths contributes positively to the sum, while an even number contributes negatively. This essentially reduces our problem to the production of the set of all minpaths. This algorithm has been suggested by a number of authors; see, for example, Fu & Yau [1962], Kim, Case & Ghare [1972], Lee [1955], Lin, Leon & Huang [1976], Misra & Rao [1970]. A naive implementation of this approach is, in fact, worse than complete state enumeration. The number of pathsets, h, may be exponential in n, and hence just the minpath generation requires exponential time. However, the more serious defect is that generating all subsets of minpaths in the naive manner takes 2^h time, which leaves us with a *doubly* exponential time algorithm for the reliability.

With a little care, this doubly exponential behavior can be avoided. Every subset of the minpaths corresponds to a subgraph whose edge set is the union of the edge sets of the minpaths. With this in mind, an *i-formation* of a subgraph is a set of i minpaths whose union is the subgraph. A formation is *odd* when i is odd, *even* when i is even. A graph having a formation is a *K-subgraph.* Every odd formation of the subgraph contributes positively to the reliability, and every even formation contributes negatively. The *signed domination* of G with terminal set K of nodes, $sdom(G, K)$, is the number of odd formations of G minus the number of even formations of G. The *domination* $dom(G, K)$ is the absolute value of the signed domination. We usually write $sdom(G)$ and $dom(G)$ with the terminal set K understood. With these definitions, Satyanarayana & Prabhakar [1978] simplified the expression for the reliability substantially:

$$Rel(G) = \sum_{H \subseteq G} sdom(H)\Pr[H],$$

where H varies over all states of G. This simplification is substantial, as it entails only the generation of all states rather than the generation of all subsets of the minpaths. However, some effort is still required if we are to improve on complete state enumeration. In particular, we now require the signed domination of each state.

In each of the directed reliability problems, Satyanarayana and his colleagues [Satyanarayana, 1982; Satyanarayana & Hagstrom, 1981a, b; Satyanarayana & Prabhakar, 1978; Willie, 1980] completely determined the signed domination of each state. We outline the derivation in the reachability case here [Satyanarayana & Hagstrom, 1981a, b].

The first goal is to determine which states have signed domination zero, and can therefore be ignored in the reliability expression. With this in mind, a state (subgraph) is *relevant* whenever it contains no irrelevant arcs. A subgraph containing irrelevant arcs has no formations whatsoever, and hence has signed domination zero. Thus we restrict our attention to relevant subgraphs. Among the relevant subgraphs, many have signed domination zero: precisely the cyclic subgraphs (subgraphs with some directed cycle) [Satyanarayana & Prabhakar, 1978]. Moreover, an acyclic relevant digraph with m arcs and n nodes has signed domination $sdom(G) = (-1)^{m-n+1}$.

This study of domination in directed reliability problems is a remarkably clever application of combinatorial arguments. A method which naively requires doubly exponential time has been reduced to requiring the generation of the acyclic relevant graphs, and a trivial calculation for each. In practice, this affords a substantial improvement on complete state enumeration. Nevertheless, the number of acyclic subdigraphs is potentially exponential in n. Hence, despite a very significant reduction in computational effort, a very large computational task remains.

The use of signed domination in undirected problems arises in quite a different way. In undirected problems, cyclic relevant graphs have nonzero domination. Thus inclusion–exclusion algorithms using minpaths would require algorithms to compute the signed domination. However, the current algorithm to compute the signed domination of a single graph is the *same* as the algorithm which computes the reliability recursively using factoring. In fact, optimal factoring strategies using factoring with series and parallel reductions employ a number of factoring steps *equal* to the domination [Satyanarayana and M.K. Chang, 1983].

Let us once again suppose that we have an enumeration $P_1, \ldots, P_h$ of the minpaths, and let E_i be the event that all edges/arcs in minpath P_i are operational. As we have remarked, the events $\{E_i\}$ are not disjoint. We examine the strategy of forming a set of disjoint events. Let $\bar{E}_i$ denote the complement of event E_i. Now define the event $D_1 = E_1$, and in general, $D_i = \bar{E}_1 \cap \bar{E}_2 \cap \ldots \cap \bar{E}_i - 1 \cap E_i$. The events D_i are disjoint, and hence are often called 'disjoint product' events. Moreover, $Rel(G) = \sum_{i=1}^{h} \Pr[D_i]$. In employing this approach, one must obtain a formula for $\Pr[D_i]$ in terms of the states of the edges/arcs. Each event E_i can be written as a Boolean expression which is the product of the states of the edges/arcs in minpath P_i. Hence D_i can also be written as a Boolean expression. For this reason, algorithms using disjoint products are sometimes called 'Boolean algebra' methods.

There is a wide variety of Boolean algebra methods. The pioneering paper here is by Fratta & Montanari [1973]. Subsequent improvements have employed two basic ideas [Abraham, 1979; Aggarwal, Misra & Gupta, 1975; Aggarwal & Rai, 1981; Ball & Nemhauser, 1979; Ball & Provan, 1985; Dotson & Gobien, 1979; Locks, 1980, 1982; Parker & McCluskey, 1975; Rai & Aggarwal, 1978; Tsuboi & Aruba, 1975]. Firstly, observe that the expression for event D_i is a complex Boolean expression, involving complements of events E_i and not just edge states and complements of edge states. Evaluation of D_i requires simplification of the Boolean expression to one which involves only edge states and their complements. Most methods are primarily concerned with making this simplification efficient, and with producing resulting expressions which are as small as possible. Secondly, in order to make simplification easier, most methods employ some simple strategy for reordering the minpaths prior to defining the events $\{D_i\}$. The particular events defined depend heavily on the ordering of the minpaths chosen. A typical heuristic here is to sort the minpaths, placing minpaths with the fewest edges/arcs first. Despite these heuristic improvements, there is no known polynomial bound in general for the length of the simplified Boolean expression for $Rel(G)$ in terms of the number of minpaths. Provan [1991] develops a general theory for the size of the Boolean expression in special cases.

In the case of all-terminal reliability and reachability, however, such a polynomial bound is achievable using an algorithm of Ball & Nemhauser [1979; see also Ball & Provan, 1985, 1988] to produce a Boolean formula describing disjoint events based on minpaths in which the number of terms equals the number of minpaths.

Colbourn & Pulleyblank [1989] give an algorithm for ordering the minpaths in k-terminal reliability problems so that the number of terms does not exceed the number of spanning trees. However, this may exceed the number of minpaths by an exponential factor; see also Chari [1993]. Ball & Provan [1985, 1988] treat the optimal ordering of minpaths in a general setting.

Our goal to this point has been to compute directly the probability of obtaining a pathset. An indirect way to do this is to compute instead the probability of obtaining a cutset. The reliability is then one minus the cutset probability. Let us suppose that we have an enumeration of min-cuts. Let $C_1, \ldots, C_g$ be the min-cuts, and let E_i be the event that all edges in min-cut C_i fail. Once again, we can apply the strategy of inclusion–exclusion [Jensen & Bellmore, 1969; Nelson, Batts & Beadles, 1970] or the strategy of disjoint products [Ball, 1979; Ball & Van Slyke, 1977; Hänsler, McAuliffe & Wilkov, 1974; Rai, 1982]. The advantage of this approach computationally is that the number of min-cuts is often much smaller than the number of minpaths. In fact, many methods generate both minpaths and min-cuts, and proceed with the smaller collection of the two [see, for example, Dotson & Gobien, 1979]. Although one may typically prefer working with min-cuts because of the smaller number, no current theory analogous to domination accounts for which states are relevant. This is a serious drawback to approaches based on cutsets.

A recent algorithm due to Provan & Ball [1984] determines two-terminal reliability in time which is polynomial in the number of min-cuts. Their algorithm has much the same flavor as a dynamic programming strategy suggested by Buzacott [Buzacott, 1980, 1983, 1987; Buzacott & Chang, 1984].

Buzacott's algorithm applies more generally than the Provan–Ball strategy, but does not perform nearly as well when the number of min-cuts is small. Ball & Provan [1987] develop a common extension of both methods.

Every algorithm mentioned here requires exponential time in the worst case, whether it enumerates states, minpaths, or min-cuts. A complete graph on n nodes, for example, has $2^n - 1$ min-cuts, n^{n-2} spanning trees, and $2^{\binom{n}{2}}$ states. If exponential algorithms are the best one can hope for, it is reasonable to consider algorithms that explore a *relatively* small number of states. Among the many algorithms mentioned here, three are noteworthy. Methods based on domination ensure that only relevant states are examined, and thereby improve on almost all other inclusion–exclusion methods. Methods based on factoring (for the undirected case) also generate only relevant states. The Satyanarayana–Chang approach obtains the best possible computation time using series and parallel reductions. In fact, in the all-terminal case, since the number of spanning trees exceeds the domination, the algorithm improves on all methods based on minpaths. Finally, methods based on disjoint products, although difficult to analyze in general, give two important algorithms: the Ball–Nemhauser algorithm to compute all-terminal reliability in time polynomial in the number of minpaths, and the Provan–Ball algorithm which computes two-terminal reliability in time polynomial in the number of min-cuts.

This useful device of measuring complexity in terms of the number of minpaths, min-cuts, or relevant states enables one to see that the methods singled out are indeed improvements on the vast remainder of reliability algorithms.

4. Bounds on network reliability

Essentially all reliability problems of interest are NP-hard, and hence the fact that the exact algorithms described are quite inefficient comes as no surprise. Nevertheless, in assessing the reliability of a network, it is imperative that the assessment can be completed in a 'reasonable' amount of time. The conflicting desires for fast computation and for great accuracy have led to a varied collection of methods for estimating reliability measures.

Two main themes arise: the *estimation* of reliability by Monte Carlo sampling techniques, and the *bounding* of reliability. In the first, the goal is to obtain an accurate estimate of a reliability measure by examining a small fraction of the states, chosen randomly. This leads to a point estimate of the reliability measure, along with confidence intervals for the estimate. Bounding is different, both in technique and in result. Current techniques for bounding attempt to find combinatorial or algebraic structure in the reliability problem, permitting the deduction of structural information upon examination of a small fraction of the states. Unlike Monte Carlo methods, the states examined are not chosen randomly. The goal of bounding is to produce *absolute* upper and lower bounds on the reliability measure.

It is perhaps misleading to draw a line between Monte Carlo methods and bounding techniques, since a number of the Monte Carlo methods employ bounding as a vehicle to limit the size of the sample space. In this section, we explore bounding methods, leaving Monte Carlo techniques for Section 5. We first examine the case when all edges operate with the same known probability, independently. We then examine the case when edges operate independently with arbitrary (but still known) probabilities. It must be stressed that the equal-probability situation is more of combinatorial interest than of immediate practical application. Nevertheless, an understanding of the underlying combinatorial structure has proved fundamental in the development of more general bounds and approximation methods.

4.1. Equal edge failure probabilities

In this section, we treat bounds that are valid when every edge has the same operation probability p; in this case, as we have seen, reliability can be expressed as a polynomial in p. A subgraph with operational edges $E' \subseteq E$ now arises with probability $p^{|E'|}(1-p)^{|E-E'|}$. Consequently, the probability of obtaining a subgraph depends only on the number of edges it contains. Then let N_i denote the number of operational subgraphs with i edges. The probability of network operation, denoted $Rel(G, p)$ or simply $Rel(p)$ is then

$$Rel(p) = \sum_{i=0}^{m} N_i p^i (1-p)^{m-i}.$$

Thus the probability is a polynomial in p, which we saw before, called the *reliability polynomial.* This formulation is in terms of pathsets. Another formulation is obtained by examining cutsets. Letting C_i be the number of i-edge cutsets (leaving $m-i$ operational edges),

$$Rel(p) = 1 - \sum_{i=0}^{m} C_i (1-p)^i p^{m-i}.$$

Still another formulation, and probably the most common, is obtained by examining complements of pathsets. Let F_i denote the number of sets of i edges for which the $m-i$ remaining edges form a pathset. Then

$$Rel(p) = \sum_{i=0}^{m} F_i (1-p)^i p^{m-i}.$$

4.1.1. Basic observations

The first goal in introducing the reliability polynomial is to obtain a compact encoding of reliability information to compare candidate topologies. Moore & Shannon [1956] pioneered this approach in their study of electrical relays. Moskowitz [1958] and Mine [1959] employed a simple strategy for computing reliability polynomials in certain two-terminal problems on series-parallel networks. The application to computer networks, and the reliability polynomials introduced here, were studied by Kel'mans [1967]. Kel'mans was apparently the first to make a fundamental observation about comparing two networks via their reliability polynomial. He proved that for two graphs G and H, their reliability polynomials may 'cross'; that is, one may be more reliable for one value of p, while the other is more reliable for another value of p. Colbourn, Harms & Myrvold [1993] proved that they can, in fact, cross twice. Kel'mans [1979, 1981] and Myrvold, Cheung, Page & Perry [1991] proved that for a given number of nodes and edges, in certain cases there is no graph that is most reliable for all edge operation probabilities. Thus reliability is more than simple graph parameters; it is truly a function of link reliabilities.

4.1.2. Computing some coefficients exactly

In Section 2.3, we saw that being able to compute the size ℓ of a minimum cardinality pathset and the size c of a minimum cardinality cutset enables us to determine a number of coefficients exactly. When ℓ is efficiently computable, further information can often be obtained by computing N_ℓ. In the k-terminal problem, this is truly hopeless; one would have to count minimal Steiner trees, an NP-hard problem. In the other two cases, however, efficient algorithms exist.

In the all-terminal problem, N_ℓ is the number of spanning trees. Kirchoff [1958] in 1847 developed an elegant method for counting spanning trees; it is developed in a computationally useful form in Brooks, Smith, Stone & Tutte [1940]. In the two-terminal problem, N_ℓ is the number of shortest s, t-paths. Ball & Provan [1983] developed an efficient strategy for computing this number. Brecht & Colbourn [1988] establish that for any fixed $k \geq 0$, pathsets of size $\ell + k$ in the two-terminal problem can be counted efficiently.

An efficient algorithm to compute c enables us to determine a number of additional coefficients exactly as well. This problem is tractable in each of the three cases of interest, using the same method in each case. Menger's theorem [Menger, 1927] guarantees that the minimum s, t-cut has size c exactly when the maximum number of edge-disjoint s, t-paths is c. This problem is easily recast as a network flow problem, with all edge capacities equal to 1.

Once again, having computed c it would be valuable to compute C_c, the number of minimum cardinality cutsets. Provan and Ball have shown that in the two-terminal case, computing just this coefficient is #P-complete [Provan & Ball, 1983]; since the k-terminal problem includes the two-terminal problem, computing C_c in either of these problems is apparently intractable. However, in the all-terminal problem, Ball & Provan [1982] devised a method for computing C_c efficiently. Lomonosov & Polesskii [1972] and Bixby [1975] have shown that every n-node graph G has $C_c(G) \leq \binom{n}{2}$. In addition, observe that for any i and any edge e of G, $C_i(G) = C_i(G \cdot e) + C_{i-1}(G - e)$. These two facts were used by Ramanathan & Colbourn [1987] to develop an efficient method for counting cutsets of size $c + k$ for any fixed $k \geq 0$.

4.1.3. Simple bounds

The computation of many of the coefficients leaves us with many coefficients about which we have said nothing. Kel'mans [1965, 1967] observed that when p is close to zero, for all-terminal reliability we have

$$Rel_{\mathrm{A}}(p) \approx N_{n-1} p^{n-1} (1-p)^{m-n+1}$$

and when p is close to 1,

$$Rel_{\mathrm{A}}(p) \approx 1 - C_c p^{m-c} (1-p)^c.$$

In the setting of the reliability polynomials introduced, a precise statement of the Kel'mans approximations is valid for all p:

$$N_{n-1} p^{n-1} (1-p)^{m-n+1} \leq Rel_{\mathrm{A}}(p) \leq 1 - C_c p^{m-c} (1-p)^c.$$

Thus the Kel'mans approximations can be viewed as absolute bounds on the reliability polynomial. The essential observation here is that for extreme values of p, either the term involving N_{n-1} or the term involving C_c dominates the remaining terms. Another observation is inherent in the Kel'mans approach. We know that $N_i + C_{m-i} = \binom{m}{i}$, and hence we have $0 \leq N_i, C_i \leq \binom{m}{i}$.

This observation leads us to a simple set of bounds first formulated by Jacobs [1959] and improved to this current form by Van Slyke & Frank [1972]:

$$Rel_{\mathrm{A}}(p) \geq N_{n-1}p^{n-1}(1-p)^{m-n+1} + N_{m-c}p^{m-c}(1-p)^{c} + + \sum_{i=m-c+1}^{m} \binom{m}{i} p^{i}(1-p)^{m-i}.$$

$$Rel_{\mathrm{A}}(p) \leq N_{n-1}p^{n-1}(1-p)^{m-n+1} + \sum_{i=n}^{m} \binom{m}{i} p^{i}(1-p)^{m-i}.$$

In the lower bound, each 'unknown' N_i is approximated by zero; the known coefficients are for $i < n-1$ (zero), $i = n-1$ (the number of spanning trees), $i = m-c$ (the $m-c$-edge subgraphs whose complement is *not* a minimum cardinality cutset), and $i > m-c$ (all possible i-edge subgraphs). In the upper bound, the unknown N_i are approximated by $\binom{m}{i}$. The extension to two-terminal reliability is straightforward; simply substitute ℓ for $n-1$ throughout. The extension to k-terminal reliability is complicated by the difficulty of computing ℓ. A lower bound is nevertheless obtained by underestimating N_ℓ as 0; an upper bound is obtained merely by using an underestimate for ℓ itself.

These bounds are *extremely* weak, and provide useful information only when p is very near 0 or 1.

4.1.4. Coherence

Bounding the unknown N_i depends very heavily on knowledge of the combinatorial structure of the collection of operational subgraphs. Most reliability problems of interest to us have the property of *coherence.* With this in mind, consider the set $\mathcal{D} = \{D_1, D_2, D_3, \ldots, D_r\}$ in which each D_i is a set of edges for which $E - D_i$ is operational. The set $\mathcal{D}$ has a set for each operational subgraph, in which the edges *not* in the operational subgraph are listed. Defining $\mathcal{D}$ in this way, we produce a hereditary family of sets (or *complex*) called the *$\mathcal{F}$-complex* (that is, if $S \in \mathcal{D}$ and $S' \subseteq S$, then $S' \in \mathcal{D}$). The fact that the family of sets produced is hereditary is precisely the property of coherence. It is also no coincidence that F_i, defined earlier, is precisely the number of sets of cardinality i in $\mathcal{D}$. In fact, the reliability polynomial for a hereditary family $\mathcal{D}$ is completely prescribed by its *F-vector* $(F_0, F_1, \ldots, F_d)$ where $d = m - \ell$.

The property of coherence also suggests a particularly appropriate way of viewing the family $\mathcal{D}$, as a partial order whose relation is set inclusion. The key to using coherence in obtaining bounds is the following. Consider all of the i-sets in a hereditary family $\mathcal{D}$. Since the family is hereditary, there must be a number of $i-1$-sets contained in the collection of i-sets; the minimum number of such induced $i-1$-sets is a lower bound on F_{i-1}. The minimization of F_{i-1} as a function of F_i is a well-known problem in extremal set theory, apparently first studied by Sperner [1928]. He proved that $F_{i-1} \geq (i/(m-i+1))F_i$.

Birnbaum, Esary & Saunders [1961] also prove this result, and observe that it has obvious consequences to the coefficients in the reliability polynomial. Bauer,

Boesch, Suffel & Tindell [1982] observe that this has a very simple interpretation: the fraction of operational subgraphs with i edges over all subgraphs with i edges is nondecreasing as i increases. They also observe that Sperner's theorem can be used at little computational effort to improve the simple bounds. We assume that ℓ, c, $F_{m-\ell}$, and F_c are available. Then

$$Rel(p) \geq \sum_{i=0}^{c-1} \binom{m}{i} p^{m-i}(1-p)^i + F_c p^{m-c}(1-p)^c + \\ + \sum_{i=c+1}^{m-\ell} F_{m-\ell} \frac{\binom{m}{i}}{\binom{m}{m-\ell}} p^{m-i}(1-p)^i.$$

$$Rel(p) \leq \sum_{i=0}^{c-1} \binom{m}{i} p^{m-i}(1-p)^i + \sum_{i=c}^{m-\ell-1} F_c \frac{\binom{m}{i}}{\binom{m}{c}} p^{m-i}(1-p)^i + \\ + F_{m-\ell} p^{\ell}(1-p)^{m-\ell}.$$

This bounding technique applies to any coherent system, and affords significant improvements on the simple bounds [Colbourn, 1987].

One method to improve on these bounds is to improve on Sperner's theorem. The best possible result in this direction was obtained by Kruskal [1963] and independently by Katona [1966]. Simplified proofs of this key result have been given by Daykin [1974] and Frankl [1984]. The Kruskal–Katona theorem places a lower bound $F_i^{i-1/i}$ on F_{i-1} given F_i; alternatively, it places an upper bound $F_{i-1}^{i/i-1}$ on F_i given F_{i-1}. The form of the bound is of little importance here, except to note that $x^{j/i}$ can be efficiently calculated, and that whenever $x \geq y$, $x^{j/i} \geq y^{j/i}$. Van Slyke & Frank [1972] used the Kruskal–Katona theorem to bound individual coefficients in the reliability polynomial. Recall that F_c is $\binom{m}{c} - C_c$. For all-terminal reliability, we can therefore compute F_c exactly; in the remaining two cases, we cannot hope to compute F_c, but we can easily compute F_{c-1}. In general, let us suppose that we can compute a sequence of coefficients $F_0, F_1, \ldots, F_s$ efficiently. Then the Kruskal–Katona theorem gives us an upper bound on F_{s+1}. Then given an upper bound on F_{s+1}, we proceed in the same way to obtain upper bounds on F_{s+2}, F_{s+3} and so on.

Lower bounds are obtained symmetrically. We compute some sequence of coefficients $F_{m-\ell}, F_{m-\ell+1}, \ldots, F_m$ efficiently. For all-terminal and two-terminal reliability, $F_{m-\ell}$ is the number of spanning trees and shortest paths, respectively. In the k-terminal problem, we can take $\ell = k - 1$ (for example) in order to compute this sequence. In any event let $d = m - \ell$. Knowing F_d, the Kruskal–Katona theorem gives a lower bound on F_{d-1}, namely $F_d^{d-1/d}$. This application of the Kruskal–Katona theorem, first done by Van Slyke & Frank [1972], gives us the *Kruskal--Katona* bounds.

4.1.5. Shellability

The Kruskal–Katona theorem is best possible for hereditary families of sets. We therefore have no hope of improving on the Kruskal–Katona bounds without additional information. Such additional information could come in a number of ways. One would be efficient algorithms for computing (or even bounding more tightly) one or more of the unknown F_i. Another would be to observe that the particular hereditary family which arises has some special combinatorial structure. This latter approach is promising, because although complements of pathsets in coherent systems produce a hereditary family, not all hereditary families arise in this way.

In fact, the $\mathcal{F}$-complex in an all-terminal reliability problem is a matroid, the *cographic matroid* of the graph. For now, we restrict our attention to the all-terminal problem. No progress appears to have been made on characterizing F-vectors of cographic matroids, and so one might ask what the F-vector of a matroid can be in general. Even on this problem, no progress has been made directly. However, we can identify a class of hereditary systems that are intermediate between matroids and hereditary systems in general, and results *are* available here.

Provan & Billera [1980] prove a powerful result about the structure of matroids, which (together with later results) constrains their F-vectors; they observe that matroids are 'shellable' complexes. The importance of the Provan–Billera result in our reliability investigations is that they suggest the possibility of exploiting shellability to improve on the Kruskal–Katona bounds. Of course, this requires that we obtain structure theorems for shellable systems. An *interval* $[L, U]$ is a family of subsets $\{S : L \subseteq S \subseteq U\}$. An interval partition of a complex is a collection of disjoint intervals for which every set in the complex belongs to precisely one interval. A complex is *partitionable* if it has an interval partition $[L_i, U_i]$; $1 \leq i \leq J$ with U_i a base for all i. Shellable complexes are all partitionable.

Ball & Nemhauser [1979] developed the application of the partition property to reliability. Consider a shellable complex with b bases; let $\{[L_i, U_i]|1 \leq i \leq b\}$ be an interval partition for this complex. $[L_i,U_i]$ is a compact encoding of all sets in this interval; the probability that any one of these sets arises is then $p^{m-|U_i|}(1-p)^{|L_i|}$. In other words, $|L_i|$ edges must fail, and $m - |U_i|$ edges must operate; the state of the remaining edges is of no consequence. Every U_i is a base in the complex; hence the cardinality of each U_i is the same, the rank d of a base. However, the ranks of the L_i are not all identical; we therefore define $H_i = |\{L_j : 1 \leq j \leq b, |L_j| = i\}|$. This gives rise to an *H-vector* $(H_0, \ldots, H_d)$. The coefficient H_i counts intervals in the partition whose lower set has rank i.

This gives yet another form of the reliability polynomial:

$$Rel(p) = p^{\ell} \sum_{i=0}^{d} H_i (1-p)^i .$$

Here, ℓ is the cardinality of a minimum cardinality pathset (spanning tree), and $d = m - \ell$ is then the rank of a base. More concretely, in an n-node m-edge graph, $\ell = n - 1$ and $d = m - n + 1$.

Naturally, any information about the H-vector also provides information about the reliability polynomial. However, to place the H-vector in appropriate context, it is worthwhile examining the relation between the H-vector and the F-vector for a shellable complex. The H-vector for any complex can be defined directly in terms of the F-vector [see, for example, Stanley, 1978]. In the partitionable case, however, the correspondence is easily seen combinatorially.

Consider the sets of rank k in the complex. These are counted by F_k. Now any interval $[L_i, U_i]$ accounts for certain of these sets. Let r be the rank of L_i. If $r > k$, the interval accounts for 0 of the sets in F_k; however, if $r \leq k$, it accounts for $\binom{d-r}{k-r}$ of the sets. Hence, we find that $F_k = \sum_{r=0}^{k} H_r \binom{d-r}{k-r}$. Equating the F-vector and H-vector forms of the reliability polynomial gives an expression for H_i in terms of the F-vector, namely:

$$H_k = \sum_{r=0}^{k} F_r (-1)^{k-r} \binom{d-r}{k-r}.$$

This expression allows us to efficiently compute $H_0, \ldots, H_s$ from $F_0, \ldots, F_s$. Another obvious, but useful, fact is that $F_d = \sum_{i=0}^{d} H_i$.

Following pioneering research of Macaulay [1927], Stanley [Billera, 1977; Stanley, 1975, 1977, 1978, 1984] has studied H-vectors in an algebraic context, as 'Hilbert functions of graded algebras.' Stanley obtained a lower bound $H_i^{<i-1/i>}$ on H_{i-1} given H_i that is tight for shellable complexes in general; this in turn gives an upper bound $H_{i-1}^{<i/i-1>}$ on H_i given H_{i-1}. For our purposes, three things are important. First of all, for $k \geq j \geq i$, $x^{<k/i>} = x^{<j/i><k/j>}$. Secondly, given x, j and i we can compute $x^{<j/i>}$ efficiently. Thirdly, whenever $x \geq y$, $x^{<j/i>} \geq y^{<j/i>}$.

Stanley's theorem can be used to obtain efficiently computable bounds on the reliability polynomial. Given a prefix $(F_0, \ldots, F_s)$ of the F-vector, we can efficiently compute a prefix $(H_0, \ldots, H_s)$ of the H-vector.

Knowing this prefix, we obtain some straightforward bounds; these apply to shellable systems in general, but we present them here in the all-terminal case.

$$Rel(p) \geq p^{n-1} \sum_{i=0}^{s} H_i (1-p)^i.$$

$$Rel(p) \leq p^{n-1} \left[\sum_{i=0}^{s} H_i (1-p)^i + \sum_{i=s+1}^{d} H_s^{<i/s>} (1-p)^i \right].$$

This exploits information about the size of the minimum cardinality cutset and, where available, the number of such cutsets. This simple formulation ignores a substantial piece of information, the number of spanning trees. This is introduced by recalling that $F_d = \sum_{i=0}^{d} H_i$. Ball & Provan [1982, 1983] develop bounds that incorporate this additional information; they suggest a very useful pictorial tool for thinking about the problem. Associate with each H_i a 'bucket'. Now suppose we have F_d 'balls'. Our task is to place all of the balls into buckets, so that the number of balls in the ith bucket, n_i, satisfies $n_i \leq n_{i-1}^{<i/i-1>}$.

How do we distribute the balls so as to maximize or minimize the reliability polynomial? These distributions, when found, give an upper and a lower bound on the reliability polynomial. Consider carefully the sum in the reliability polynomial: $\sum_{i=0}^{d} H_i(1-p)^i$. Since $0 < p < 1$, the sum is larger when the lower order coefficients are larger. In fact, for two H-vectors $(H_0, \ldots, H_d)$ and $(J_0, \ldots, J_d)$, whenever $\sum_{j=0}^{i} H_j \geq \sum_{j=0}^{i} J_j$ for all i, the reliability polynomial for the H_i dominates the reliability polynomial for the J_i.

This last simple observation suggests the technique for obtaining bounds. In the pictorial model, an upper bound is obtained by placing balls in the leftmost possible buckets (with buckets $0, \ldots, d$ from left to right); symmetrically, a lower bound is obtained by placing balls in the rightmost possible buckets. We are not totally without constraints in making these placements, as we know in advance the contents of buckets $0, \ldots, s$.

With this picture in mind, we give a more precise description. We produce coefficients $\overline{H}_i$ for an upper bound polynomial, and $\underline{H}_i$ for a lower bound polynomial, using the prefix $(H_0, \ldots, H_s)$ and F_d. The steps are:

1. For $i = 0, \ldots, s$, set $\underline{H}_i = H_i = \overline{H}_i$.
2. For $i = s+1, s+2, \ldots, d$, set

$$\underline{H}_i = \min\left[r : \sum_{j=0}^{i-1} \underline{H}_j + \sum_{j=i}^{d} r^{<j/i>} \geq F_d\right].$$

$$\overline{H}_i = \max\left[r : r \leq \overline{H}_{i-1}^{<i/i-1>} \text{ and } \sum_{j=0}^{i-1} \overline{H}_j + r \leq F_d\right].$$

An explanation in plain text is in order. In each bound, we determine the number of balls in each bucket from 0 to d in turn; as we remarked, the contents of buckets $0, \ldots, s$ are known. For subsequent buckets, the upper bound is determined as follows. The number of balls which can go in the current bucket is bounded by Stanley's theorem, and is also bounded by the fact that there is a fixed number of balls remaining to be distributed. If there are more balls remaining than we can place in the current bucket, we place as many as we can. If all can be placed in the current bucket, we do so; in this case, all balls have been distributed and the remaining buckets are empty. The lower bound is determined by placing as few balls as possible.

The method leads to a very powerful set of bounds, the *Ball–Provan* bounds:

$$Rel(p) \geq p^{n-1} \sum_{i=0}^{d} \underline{H}_i (1-p)^i.$$

$$Rel(p) \leq p^{n-1} \sum_{i=0}^{d} \overline{H}_i (1-p)^i.$$

Unlike the Kruskal–Katona bounds, in the case of the Ball–Provan bounds it is not generally the case that $\underline{H}_i \leq H_i \leq \overline{H}_i$. Brown, Colbourn & Devitt

[1993] observe that a number of simple network transformations can be used to determine bounds $L_i \leq H_i \leq U_i$ efficiently. Incorporating these coefficient bounds on the H-vector into the Ball–Provan process can result in substantial improvements.

4.1.6. Polyhedral complexes and matroid ports

The Ball–Provan bounds as developed here apply to all-terminal reliability and to reachability. For reachability, Provan [1986] observes that the $\mathcal{F}$-complex is a 'polyhedral complex', and uses a theorem of Billera & Lee [1981] to obtain efficiently computable bounds on reliability that are tight for polyhedral complexes.

The matroid structure of the all-terminal problem and the polyhedral structure of reachability both lead to dramatic improvements over the Kruskal–Katona bounds for general coherent reliability problems. Building on a structure theorem of Colbourn & Pulleyblank [1989], Chari [1993] characterized two-terminal complexes in terms of 'matroid ports', and generalized these to develop some remarkable structure theorems about $\mathcal{F}$-complexes from k-terminal problems. For an n-node connected graph having k terminals, and edge set E, $|E| = m$, let $\mathcal{F}$ be its $\mathcal{F}$-complex. The blocking complex $\mathcal{F}^*$ of $\mathcal{F}$ is $\{E \setminus S : S \in 2^E \setminus \mathcal{F}\}$. The F-vector $(F_0, \ldots, F_m)$ of $\mathcal{F}$ and the F-vector $(F_0^*, \ldots, F_m^*)$ of $\mathcal{F}^*$ satisfy $F_i + F_{m-i}^* = \binom{m}{i}$ for $0 \leq i \leq m$. Chari [1993] shows that the subcomplex $\mathcal{F}^{(m-n+1)}$ obtained by removing all sets from $\mathcal{F}$ of cardinality exceeding $m - n + 1$, is a shellable complex. Hence, given bounds on the single coefficient F_{m-n+1}, the Ball–Provan strategy can be applied to this k-terminal problem in order to bound $(F_0, \ldots, F_{m-n+1})$. What about the remaining coefficients? Chari further proves that $\mathcal{F}^{*(n-2)}$, obtained from $\mathcal{F}^*$ by removing all sets of cardinality exceeding $n-2$, is also shellable. Hence the Ball–Provan bounds can be applied again to bound $(F_0^*, \ldots, F_{n-2}^*)$, or equivalently to bound $(F_m, \ldots, F_{m-n+2})$.

All of the approaches developed for equal edge failure probabilities to date examine extremal results for complexes that are more general than those actually arising in reliability problems on graphs. It remains a very active area of research to determine least or most reliable graphs, rather than complexes, given the values of some specified graph parameters. Even the characterization of least reliable graphs for specified numbers of nodes and edges remains unresolved, however [Boesch, Satyanarayana & Suffel, 1990].

4.1.7. The standard form

We consider yet another form of the reliability polynomial. Until this point, the underlying framework has been the notion of state enumeration, in which either operational or failed states are enumerated. Many reliability algorithms do not operate in this manner, but instead use path or cut enumeration.

We have seen earlier that one of the more useful exact algorithms employs the theory of domination. We return briefly to this theory, to develop an interpretation of coefficients in the reliability polynomial. Extending the notion of domination, the *i-parity* $P_i(G)$ is defined as follows. Let S_i be all i-edge subgraphs of G. Then

$P_i(G) = \sum_{H \in S_i} sdom(H, K)$. Satyanarayana & Khalil [1986] have established that $P_i(G) = P_{i-1}(G \cdot e) + P_i(G - e) - P_{i-1}(G - e)$.

In obtaining reliability via pathsets, every formation is considered exactly once. Letting $\{G_1, \ldots, G_t\}$ denote all K-subgraphs of G, $Rel_k(G) = \sum_{i=1}^{t} sdom(G_i, K)$ $\Pr[G_i]$. Hence when all edge probabilities are equal, we obtain another form of the reliability polynomial, the *standard* form:

$$Rel(p) = \sum_{i=0}^{m} P_i p^i.$$

In other words, the parities are precisely the coefficients of this reliability polynomial, the *P-vector*.

The characterization of P-vectors arising from reliability polynomials has not been widely studied. Two remarks are of interest here. First, Satyanarayana & Chang [1983] establish that the coefficients in the P-vector for all-terminal reliability alternate in sign. Second, Brown & Colbourn [1988] conjecture that the P-vector is log concave; that is, a P-vector for an all-terminal reliability polynomial satisfies $P_i^2 \geq P_{i-1}P_{i+1}$ for every i. They prove a partial result in support of this conjecture. Little else is currently known about the P-vector. One can easily see that $P_i = 0$ for $i < n - 1$, and that P_{n-1} is the number of spanning trees of the network. However, essentially nothing is known about the remaining coefficients in the P-vector (except, of course, those relations inherited by equivalence with the H-vector and F-vector).

4.2. Arbitrary edge failure probabilities

When edges fail with different probabilities, the $\mathcal{F}$-complex contains all of the information about reliability, given the operation probability of each edge. However, the F-vector and the reliability polynomial are no longer applicable. This has led to a number of techniques for using the network structure to obtain bounds. We explore the major techniques here, referring the interested reader to Colbourn [1987] for proofs and further discussion.

4.2.1. Edge-packing

Let $G = (V, E)$ be a graph (or digraph or multigraph). An *edge-partition* of G into k graphs $G_1, \ldots, G_k$ with $G_i = (V, E_i)$, where the edge set E is partitioned into k classes $E_1, \ldots, E_k$. An *edge-packing* of G by k graphs $G_1, \ldots, G_k$ is obtained by partitioning the edge-set E into $k + 1$ classes $E_1, \ldots, E_k, U$ and defining $G_i = (V, E_i)$. The main observation here is straightforward:

Edge packing lower bound: If G has an edge-packing by k graphs $G_1, \ldots, G_k$, and *Rel* is any coherent reliability measure,

$$Rel(G) \geq 1 - \prod_{i=1}^{k} (1 - Rel(G_i)). \tag{2}$$

Inequality (2) is in general not an equality because there are operational states of G in which no G_i is operational. Some notes are in order on the effective use these lower bounds. Consider an edge-packing of G by $G_1, \ldots, G_k$. If any G_i is non-operational, coherence ensures that $Rel(G_i) = 0$; in this event, the inclusion of G_i in the edge-packing does not affect the bound, and G_i can be omitted. Thus we need only be concerned with edge-packings by *operational* subgraphs.

Our goal is to obtain efficiently computable bounds; hence, it is necessary that we compute (or at least bound) $Rel(G_i)$ for each G_i. One solution to this, suggested by Polesskii [1971], is to edge-pack G with minpaths. The reliability of a minpath is easily computed. This suggests a solution in which we edge-pack G with as many minpaths as possible, and then apply the bound; this basic strategy has been explored extensively.

While subgraph counting bounds require that edges have the same operation probability, no such assumption is needed here; one need only compute the probability of a minpath as the product of the edge operation probabilities over edges in the minpath. With this in mind, one might modify our edge-packing problem to require packing by the most reliable minpaths rather than by the largest number of minpaths.

Any edge-packing by operational subgraphs $G_1, \ldots, G_k$ for which $Rel(G_i)$ is easily computed provides an efficiently computable lower bound. This leads to problems such as edge-packing by series-parallel graphs, or by partial k-trees for fixed k. This latter approach seems not to have been studied in the literature; hence, we concentrate on edge-packing by minpaths.

Polesskii [1971] pioneered the use of edge-packing lower bounds, in the all-terminal reliability problem. Here an edge-packing by minpaths is a set of edge-disjoint spanning trees. Using a theorem of Tutte [1961] and Nash-Williams [1961], Polesskii observed that a c-edge-connected n-node graph has at least $\lfloor \frac{c}{2} \rfloor$ edge-disjoint spanning trees; hence when all edge operation probabilities are the same value p, the all-terminal reliability of the graph is at least $1-(1-p^{n-1})^{\lfloor c/2 \rfloor}$. When edge probabilities are not all the same, Polesskii's bound extends in a natural way. Using Edmonds's matroid partition algorithm [1965, 1968], a maximum cardinality set of edge-disjoint spanning trees, or its minimum cost analogue [Clausen & Hansen, 1980], can be found in polynomial time. Applying inequality (2) then yields a lower bound on all-terminal reliability. Naturally, to obtain the best possible bound from (2), one wants not only a large number of edge-disjoint minpaths, but also minpaths that are themselves reliable. Edmonds's algorithm need not yield a set of spanning trees giving the best edge-packing bound using minpaths. In fact, the complexity of finding the set of spanning trees leading to the best edge-packing bound remains open.

Edge-packing as a general technique was pursued much later. Brecht & Colbourn [1988] and Litvak and Ushakov [Kaustov, Litvak & Ushakov, 1986; Litvak, 1983] independently developed edge-packing lower bounds for two-terminal reliability. For two-terminal reliability, minpaths are just s,t-paths. Menger's theorem [Dirac, 1966; Menger, 1927] asserts that the maximum number of edge-disjoint s,t-paths is the cardinality of a minimum s,t-cut. Thus using network flow techniques, a maximum edge-packing can be found [Ford & Fulkerson, 1962; Edmonds & Karp, 1972]. Here the problem of finding the best edge-packing, even when all edge operation probabilities are equal, is complicated by the fact that minpaths exhibit great variation in cardinality. In fact, Raman [1991] has shown that finding the best edge-packing by s, t-paths is NP-hard. For this reason, heuristics have been examined to find 'good' edge-packings. Brecht & Colbourn [1988] examine the use of minimum cost network flow routines [Fujishige, 1986; Tardós, 1985] using edge cost $-\ln p_i$ on an edge of probability p_i, and report improvements over (general) edge-packings of maximum cardinality.

Turning to k-terminal reliability, the situation is not as satisfactory. Here a minpath is a subtree in which each leaf is a terminal, i.e. a *Steiner tree*. Colbourn [1988] showed that determining the maximum number of Steiner trees in an edge-packing is NP-hard. No heuristics for finding 'good' edge-packings by Steiner trees appear to have been studied. For directed networks, edge-packing (or more properly, arc-packing) bounds can be obtained using directed s, t-paths found by network flow techniques (for s, t-connectedness), and by using arc-disjoint rooted spanning arborescences (directed rooted spanning trees) found by Edmonds's branchings algorithm [Edmonds, 1972; Fulkerson & Harding, 1976; Lovász, 1976] (for reachability). See Ramanathan & Colbourn [1987] for a discussion of the reachability bounds.

Until this point, we have examined lower bounds based on edge-packings by minpaths. Let us now turn to upper bounds. Not surprisingly, inequality (2) has a 'dual' form for upper bounds obtained by interchanging the role of pathsets and cutsets:

Edge packing upper bound: Let $G = (V, E)$ be a graph (or digraph or multigraph). Let *Rel* be a coherent reliability measure. Let $C_1, \ldots, C_s$ be an edge-packing of G by cutsets. Then

$$Rel(G) \leq \prod_{i=1}^{s} \left(1 - \prod_{e \in C_i} (1 - p_e)\right) \tag{3}$$

where p_e is the operation probability of edge e.

The inequality (3) is in general not an equality since the failure of any cut in the edge-packing causes G to fail, but the failure of G can occur even when no cutset in the packing is failed.

Brecht & Colbourn [1988] and Litvak & Ushakov [1983] first studied edge-packing upper bounds for the two-terminal reliability problem. A theorem of Robacker [1956; Fulkerson, 1968, 1971] gives the necessary dual to Menger's theorem: the maximum number of edge-disjoint s, t-cuts is the length of a shortest s, t-path. Finding a maximum set of edge-disjoint min-cuts is straightforward: simply label each node with its distance from s. If t gets label ℓ, form cutset C_i containing all edges between nodes labeled $i - 1$ and vertices labeled i, for $1 \leq i \leq \ell$. The result is ℓ edge-disjoint s, t-cuts. Finding a 'good' set of min-cuts for the edge-packing upper bound appears to be more difficult than for the lower bound. Recently, Wagner [1990] gave a polynomial time algorithm for finding a minimum cost set of edge-disjoint s, t-cutsets of maximum cardinality. Nel and Strayer [1993] report that, while using Wagner's mincost algorithm improves in general upon the bounds from edge-packings found by the labeling method above, it is often not competitive with a simple greedy algorithm that repeatedly takes the least reliable cut disjoint from those chosen thus far.

Turning to upper bounds on all- and k-terminal reliability using edge-packings by min-cuts, we encounter a major difficulty: even for all-terminal reliability, finding a maximum packing by min-cuts is NP-hard [Colbourn, 1988]. Thus it is particularly surprising that by directing the reliability problems, we *are* able to find a maximum arc-packing by cutsets for the reachability problem using an efficient algorithm of Fulkerson's [Fulkerson, 1974; Edmonds, 1967]. Thus an all-terminal reliability upper bound can be obtained by using the *arc*-packing bound for reachability.

Two potential methods to improve the edge-packing strategy stand out. The first is to consider packings by more reliable subgraphs; the second is to extend the sets of pathsets and cutsets being examined to permit some edge intersection (thereby losing the independence of the sets of edges in the packing). We treat the second extension, which has been explored more extensively, in the next subsection. For the first, little work appears to have been done. Using the efficient exact algorithm for reachability of acyclic rooted directed graphs, Ramanathan & Colbourn [1987] obtained improvements in reachability upper bounds, and also in all-terminal upper bounds. However, the use of edge-packings by general pathsets or cutsets has not proceeded far, in part because of the scarcity of exact algorithms for restricted classes, and in part because of the difficulty of finding suitable edge-packings.

4.2.2. Noncrossing and consecutive cuts

The use of edge-disjoint pathsets and cutsets until this point is motivated primarily by the necessity to compute the probability that one of the pathsets operates (as in the edge packing lower bound) or that one of the cutsets fail (as in the edge packing upper bound). Lomonosov & Polesskii [1971] devised a method that permits cutsets to share edges, while retaining an efficient method for computing the probability that one of the cutsets fails. For a graph $G = (V, E)$, a partition (A, B) of V forms a cutset, containing all edges having one end in A and the other in B. Two such cutsets (A, B) and $(\hat{A}, \hat{B})$ are *noncrossing* if at least one of $A \cap B$, $A \cap \hat{B}$, $\hat{A} \cap B$ and $\hat{A} \cap \hat{B}$ is empty. A collection of cuts is *noncrossing*, or *laminar*, if every two cutsets in the collection are noncrossing. In an n-node graph with k terminals, a set of noncrossing cutsets contains at most $n - 1 + k - 2 \leq 2n - 3$ noncrossing cuts [Colbourn, Nel, Boffey & Yates, 1994].

A *cut basis* of an n-node graph is a set of $n - 1$ cuts $C_1, \ldots, C_{n-1}$ for which every cut can be written as the modulo 2 sum of these $n - 1$ cuts. Gomory & Hu [1961] give an algorithm for finding a cut basis $C_1, \ldots, C_{n-1}$ in which $\sum_{i=1}^{n-1} |C_i|$ is minimum; moreover, their cut basis is a set of noncrossing cuts. Lomonosov

& Polesskii [1971] showed that for any cut basis $C_1, \ldots, C_{n-1}$, the all-terminal reliability satisfies

$$Rel(G) \leq \prod_{i=1}^{n-1} \left(1 - \prod_{e \in C_i} (1 - p_e) \right).$$

The use of cut bases for the k-terminal problem has been studied by Polesskii [1990b], generalizing the method outlined here. The restriction to a basis, however, limits the number of cuts that can be employed to one fewer than the number of terminals. A more general extension is obtained by permitting the use of sets of noncrossing cuts. Shanthikumar [1988] used consecutive cuts in obtaining a two-terminal upper bound. This has been extended to k-terminal reliability (actually to s, T-connectedness) in Colbourn, Nel, Boffey & Yates [1994]. The bound is obtained by establishing that the probability that none of the noncrossing cuts fail agrees with the k-terminal *nodal* reliability of a special type of graph, a *directed path graph*. A simple dynamic programming strategy then produces the bound in polynomial time.

Bounds using noncrossing cuts extend the edge-packing strategies essentially by considering a larger set of cuts, but still a polynomial number of them.

4.2.3. Transformation and graph approximation

We have thus far seen two methods for extending the edge-packing strategy: packing with non-minpaths or cutsets, and relaxing the edge-disjointness requirement. In this subsection, we examine a third extension that is perhaps less immediate than the previous two.

We have seen that transformations can be used to 'simplify' a network, in order to reduce the time required in exact algorithms. Such transformations *preserve* the value of the reliability measure. Other transformations on networks may have the property that they guarantee *not to increase* the reliability measure; these *D-transformations* preserve lower bounds on the reliability measure (that is, computing a lower bound *after* applying such a transformation gives a lower bound on the reliability of the network *before* the transformation). Similarly, *I-transformations* guarantee *not to decrease* the reliability measure, and hence preserve upper bounds.

A trivial D-transformation is deleting an edge or arc in a network; it follows from coherence and statistical independence that the reliability measure cannot increase upon such a deletion. Similarly, the operation of *splitting* a node x into two nodes x_1 and x_2, and replacing each edge $\{y, x\}$ by *either* $\{y, x_1\}$ *or* $\{y, x_2\}$, we cannot increase the reliability. These trivial transformations have remarkable consequences. AboElFotoh & Colbourn [1989a] observe that the edge-packing lower bound for two-terminal reliability can be obtained by using just edge deletion and node splitting (delete all edges not on any path in the packing, and split non-terminals as necessary that are on more than one path of the packing). The result of these transformations is a parallel combination of s, t-paths, a very simple series-parallel graph. The edge-packing upper bound for two-terminal reliability is similar, using the I-transformation that identifies two nodes [AboElFotoh & Colbourn, 1989a]. The use of transformations to obtain the two-terminal edge-packing bounds permits one to stop the transformation process 'early'. Once the network has been transformed into a series-parallel network, for example, the reliability can be calculated exactly in polynomial time and there is no need for further transformations. AboElFotoh & Colbourn [1989a] remark that the approach is very sensitive to the order and location in which the transformations are applied, and suggest some detailed heuristics for the transformations introduced so far.

Lomonosov [1974] simplified the presentation of the Lomonosov–Polesskii upper bound that uses cut bases. He introduced an I-transformation, which we call the *Lomonosov join*. Let x, y, z be three nodes and let $\{x, y\}$ be an edge. The Lomonosov join removes edge $\{x, y\}$ and adds the edges $\{\{x, z\}, \{y, z\}\}$ each with the same operation probability as the deleted edge. Lomonosov proved that when x, y, z are all terminals, this cannot decrease the reliability, and Colbourn [1992] showed that we only require that z is a terminal. This leads to upper bounds for all-terminal [Brown, Colbourn & Devitt, 1993] and k-terminal [Colbourn, 1992] reliability. The use of transformations also permits a better bound than the Lomonosov–Polesskii bound to be obtained, by applying transformations only until the network is series-parallel.

A further I-transformation was studied by Lomonosov & Polesskii [1972]. Given an arbitrary graph G, treat every nonadjacent pair of nodes as being connected by an edge of failure probability one; then G is essentially a complete graph. For any two nodes x, y, consider the adjacencies of x and y with the remaining nodes $\{v_1, \ldots, v_{n-2}\}$. Suppose that $\{x, v_i\}$ has failure probability q_i and that $\{y, v_i\}$ has failure probability q'_i. A transformation of the probabilities is carried out by setting the failure probabilities for both $\{x, v_i\}$ and $\{y, v_i\}$ to $\sqrt{q_i q'_i}$, for all $1 \leq i \leq n-2$. Lomonosov and Polesskii [1972] show that this is an I-transformation, and by repeated application that the most reliable graph $G = (V, E)$ with $\prod_{e \in E} q_e = \theta$ on n nodes is the complete graph with each edge having failure probability $\theta^{-\binom{n}{2}}$.

A number of other transformations have been studied for their applications in reliability bounds [Boesch, Satyanarayana & Suffel, 1990; Brown, Colbourn & Devitt, 1993; Colbourn & Litvak, 1991; Polesskii, 1990a]. One particular application of transformations is in the analysis of blocking probabilities of 'channel graphs' [Caccetta & Kraetzl, 1991]. Transformations for channel graphs apply to s, t-connectedness as well [Hwang & Odlyzko, 1977; Kraetzl & Colbourn, 1993]. Leggett [Leggett, 1968; Leggett & Bedrosian, 1969] was one of the first to use a transformation-based approach in developing a bound, but his bounds are in error [Harms & Colbourn, 1985].

One pair of transformations, the delta-wye and wye-delta transformations, merit special mention. A delta (or Δ) in a network is a set of three edges $\{\{x, y\}, \{x, z\}, \{y, z\}\}$ forming a triangle, and a wye (or Y) is a set of three edges $\{\{x, w\}, \{y, w\}, \{z, w\}\}$ for which w is incident only to these three edges. Wye-delta and delta-wye transformations are just the replacement of one configuration by the other. In 1962, Lehman [1963] provided two methods for determining probabilities on the edges of the wye (delta) given the edge probabilities on the delta (wye, respectively). His two methods are *not* exact in general, but rather he showed that one of the transformations is an I-transformation and the other is a D-transformation, provided the central node of the wye is not a terminal. Surprisingly, which of the two transformations is the I-transformation depends on the numerical probabilities. Thus the wye-delta and delta-wye transformations seem to differ from the earlier transformations mentioned, as there does not appear to be a simple combinatorial justification for the transformations.

Epifanov [1966] subsequently showed that every planar network can be reduced to a single edge by repeated applications of wye-delta, delta-wye, series, parallel and degree-1 transformations; see also Feo & Provan [1993] and Truemper [1989]. This leads to remarkably accurate bounds for two-terminal reliability for planar networks [Chari, Feo & Provan, 1990]. See Traldi [1994a], Litvak [1981a] and Politof [1983] for other delta-wye transformations.

Perhaps the most powerful aspect of developing bounds by composing simple transformations is the potential ease of extension to more complicated reliability measures. For example, whenever edge deletion and node splitting are I-transformations for a specified reliability or performance measure, we have efficiently computable edge-packing bounds; see Carey & Hendrickson, [1984], Litvak [1983] and Litvak & Ushakov [1984] for some examples. If in addition the measure can be calculated exactly for series-parallel networks in polynomial time, we have efficient series-parallel approximations. Colbourn and Litvak [1991] discuss more general measures of performance using this transformational approach. Using the Lomonosov join, the bounds for static measures of reliability discussed here can be extended in part to time-dependent reliability measures [Colbourn & Lomonosov, 1991].

4.2.4. Miscellaneous bounds

There are a number of further bounding techniques that have been explored which do not admit an easy classification as 'edge-packing' or transformation-based bounds. Among efficiently computable bounds, the most notable is the *k-cycle bound* introduced by Lomonosov and Polesskii [1972] and sharpened by Lomonosov [1974]. Using a random graph model introduced by Erdös & Rényi [1959, 1960], Lomonosov [1974] examined a graph evolution process. Suppose that at time 0 each edge is absent, but has an exponentially distributed time of arrival in the graph. What is the first time at which the graph becomes connected? Lomonosov established an equivalence between this graph evolution process and the static evaluation of all-terminal reliability, and by examining the expected time at which a transition is made from a network state with ℓ components to a state with $\ell - 1$ components (for $\ell = n, \ldots, 2$), he established a lower bound on all-terminal reliability. See Colbourn [1987] and Lomonosov [1974] for details.

Classical bounds due to Bonferroni (see Prékopa, Boros & Lih [1991]; Shier [1991]) can be obtained using the inclusion–exclusion formula (1) of section 3.4. By truncating the sum after $\ell < h$ terms, an upper bound is obtained when ℓ

is odd, and a lower bound is obtained when ℓ is even. The Bonferroni bounds require knowledge of all minpaths, an exponential quantity. Two-terminal bounds have been developed by Prékopa, Boros & Lih [1991] that use 'binomial moments' to improve upon the Bonferroni bounds.

Bounds have also been studied in the case that statistical dependence cannot be assumed. Hailperin 1965] develops a general linear programming formulation for reliability measures when the worst possible dependencies are permitted. Efficient implementations of Hailperin's method have been developed for two-terminal reliability by Zemel [1982] and Assous [1986], and for all-terminal reliability by Carrasco & Colbourn [1986]. Under worst case assumptions about statistical dependencies, however, the bounds appear to have little or no practical import unless the information about dependencies specified is substantial.

Finally, there is an extensive literature on bounds that require exponential time in the worst case to compute. We have focussed on efficient methods, so do not give a complete survey of exponential time methods here. Undoubtedly the most influential method is due to Esary & Proschan [1963]. They observed that if one examines *all* minpaths in the network, and computes the probability that at least one path fails *under the assumption that paths are independent*, one obtains an upper bound on the reliability. This is a remarkable contrast to the edge-packing strategy, where the same technique was applied to a subset of all paths, but a *lower* bound was obtained. Esary & Proschan [1963] also prove the dual statement to obtain a lower bound from all cuts. At the present time, no algorithm is known to compute the Esary–Proschan bounds, or to improve upon them, in polynomial time, except for upper bounds on s, t-connectedness [Colbourn, Devitt, Harms & Kraetzl, 1991]. This is not to say, however, that they are typically more accurate than the efficiently computable bounds; our experience suggests the contrary.

A further recent direction to obtain bounds is to examine a limited subset of all states, and to compute a bound based upon the states examined. By concentrating on most probable states, one expects that a small fraction of all states need to be examined in order to see most of the contribution to the reliability. Shier [1991] gives an excellent introduction to this subject; see also Lam & Li [1986], Shier & Whited [1987], Yang & Kubat [1989, 1990]. While accuracy/time tradeoffs are observed empirically here, there appears to be no guarantee that a prescribed accuracy can be achieved in polynomial time.

Along the same lines, Nel & Colbourn [1990] observe that one can apply factoring a limited number of times, and apply any or all of the bounding techniques discussed earlier. If the number of edges factored on is bounded by $\log n$, where n is the number of nodes, the process remains polynomial in time — but one expects improved accuracy. Of course, the notions of most probable states and efficiently computable bounds can be combined; see Nel & Colbourn [1990] for some steps in this direction.

4.2.5. Postoptimization on bounds

So far we have discussed basic strategies for obtaining bounds. Even the thumbnail description of each cannot fail to convince one that there is great variety in the available bounds. It is natural to combine the bounds to obtain better, or more general, bounds. The preferred way is to find a general theory in which a number of bounds are unified. Failing that, one wants at least to deduce whatever information about reliability is possible from the many different bounds provided. For example, if one knows the probability of reaching u from s, and independently the probability of reaching t from u, what can be said about the probability of reaching t from s? Using a remarkable theorem of Ahlswede & Daykin [1978], Brecht and Colbourn [1986, 1989] develop methods for improving lower bounds. They observe that if an network G is connected for terminal set K_1 with probability p_1, connected for terminal set K_2 with probability p_2, and $K_1 \cap K_2 \neq \emptyset$, then G is connected for terminal set $K_1 \cup K_2$ with probability at least $p_1 p_2$. This gives a multiplicative triangle inequality for two-terminal reliability, that Brecht & Colbourn [1989] found to be effective in improving upon two-terminal reliability bounds that were computed by other methods (edge-packing in particular). The key here is the *post*optimization, techniques to improve upon arbitrary bounds.

A somewhat analogous method for upper bounds, called *renormalization*, has been studied [Harms & Colbourn, 1993a]. For two-terminal reliability, the probability x_e that one terminal, s, cannot reach the other terminal, t, through a specified edge e is bounded above by the probability that edge e itself fails plus the probability that e operates times the computed probability x_f for every edge f incident to e. Two-terminal upper bounds can be used to determine initial estimates of the probability x_e for each edge e. Then each inequality may force the reduction of some x_e value. Renormalization obtains an upper bound by underestimating the effect of intersections of s, t-paths, and by examining s, t-walks rather than just s, t-paths. For s, t-connectedness of acyclic directed graphs, the lack of directed cycles ensures that all s, t-walks are s, t-paths; in this case, renormalization is a polynomial time method that guarantees an improvement on the Esary–Proschan upper bound [Colbourn, Devitt, Harms & Kraetzl, 1991]. Renormalization is essentially a postoptimization strategy, but can be used by itself commencing with initial overestimates on x_e of the edge failure probability of e.

One final postoptimization strategy appears to apply only when all edge operation probabilities are equal. Nevertheless, we outline the idea here. Colbourn & Harms [1988] observe that if one evaluates the polynomial $\sum_{i=0}^{d} F_i p^{m-i}(1-p)^i$ at a fixed value for p, one obtains a linear combination of $F_0, \ldots, F_d$. Hence, if one knows an upper or lower bound on the value of the reliability polynomial at a specified value of p, one obtains a linear constraint. Thus any bounding method whatsoever, when applied to a network with all edge operation probabilities equal, yields a linear constraint of this form. All of the linear inequalities so produced are met simultaneously, and hence one can combine bounds of all different sorts using linear programming. If all the basic bounds used are efficiently computable, one may use a polynomial time algorithm for linear programming [Fishman & Kulkarni, 1990; Khachiyan, 1979] to retain polynomial running time overall. Colbourn & Harms [1988] note that the linear programming bound so obtained occasionally improves upon all of the basic bounds used to supply constraints.

5. Monte Carlo methods

Due to the extreme intractability of exact computation of the various reliability measures covered in this paper, and to the present inability of polynomial-time bounding algorithms to provide very tight bounds on these measures, it is often necessary to turn to *simulation* techniques in order to obtain accurate estimates. This, of course, comes at a price: the estimates obtained have a certain degree of uncertainty. Nevertheless, this price is typically well justified by the superior results given by simulation methods over deterministic techniques. Due to the relatively simple structure of these problems it is natural to use the powerful and well-studied *Monte Carlo* method of simulating the stochastic behavior of the system. The first use of Monte Carlo methods in network reliability seems to have occurred in the context of percolation problems by Frisch, Hammersley & Welsh [1962], with early work in Dean [1963], Hammersley [1963], Hammersley & Handscomb [1964] and Levy & Moore [1967]. Most of the significant current techniques, however, were developed within the past decade or so.

5.1. Crude sampling

We first establish some notation to be used throughout the section. Let $(\Phi, \boldsymbol{p})$ be an instance of a particular reliability problem with structure function Φ, and with $\boldsymbol{p} = (p_1, \ldots, p_m)$ being the vector of component operating probabilities. Let $\boldsymbol{q} = (q_1, \ldots, q_m) = (1 - p_1, \ldots, 1 - p_m)$ be the vector of failure probabilities, and denote by $P(x)$ the probability that a particular state vector x appears, that is,

$$P(x) = \prod_{x_e=1} p_e \prod_{x_e=0} q_e.$$

We are interested in obtaining an estimate $\hat{R}$ for the true system reliability $R = \Pr[\Phi = 1]$.

The crude method of sampling is fairly straightforward. A sample of K vectors $x^k = (x_1^k, \ldots, x_m^k)$, $k = 1, \ldots, K$ is drawn from the distribution P, by drawing mK independent samples $\hat{U}_{kj}$, $k = 1, \ldots, K$, $j = 1, \ldots, m$ from a uniform random number generator and then setting

$$x_j^k = \begin{cases} 1 & \hat{U}_{kj} \leq p_j \\ 0 & \hat{U}_{kj} > p_j \end{cases} \qquad k = 1, \ldots, K, \ j = 1, \ldots, m.$$

Let $\hat{K}$ be the number of vectors x^k for which $\Phi(x^k) = 1$. Then an unbiased estimator for R is $\hat{R} = \hat{K}/K$, with variance $R(1-R)/K$. Reduction of this variance can be obtained by a number of standard Monte Carlo sampling techniques, such as *antithetic* and *control variates*, and *conditional, importance*, and *stratified sampling*. Since these techniques belong more in the area of probability theory than network theory, we refer the reader to a standard text such as Hammersley & Handscomb [1964] for their treatment.

We do wish to review some of the major techniques which have been applied to network reliability problems. An excellent treatment of the first four of these schemes is found in Fishman [1986a], and we refer the reader to that paper for further details.

5.2. Dagger sampling

Dagger sampling was developed by Kumamoto, Tanaka, Inoue & Henley [1980], and can be thought of as an 'm-dimensional' extension of antithetic sampling. The idea, common to several Monte Carlo techniques, is to 'spread out' the individual edge failures in such a way that repeats of sample states is minimized. The procedure is given below.

Dagger sampling method

1. Let $(N_e : e \in E)$ be a vector of integers chosen proportionally to the (rational) q_e's.
2. Choose sample size K^* so that for each edge e the sequence of K^* replications can be broken into exactly N_e *subblocks* of size K^*/N_e.
3. For each edge e, choose at random exactly one replication in each of these N_e subblocks for which that edge fails. This gives a failure pattern for the K^* replications in which the frequency of failures of each edge is exactly proportional to the average failure rate of that edge.
4. Make a final pass through the K^* replications, computing the proportion of replications corresponding to system operation. This proportion is an unbiased estimator of R.

5.3. Sequential construction/destruction

The Sequential Construction/Destruction Method of Easton & Wong [1980], and later improved by Fishman [1986a] and Elperin, Gertsbakh & Lomonosov [1991], is based on considering an *ordering* of the edges of the graph. The edges begin as all failed, and then edges are successively 'repaired', i.e. caused to operate, one by one in the specified ordering, until the system becomes operational. The reliability estimate is then a function of how long it takes for the system to become operational. This can result in better estimates than could be obtained by the crude method.

The sample space for the sequential construction method consists of a *pair* (x, π), where x is a state vector for the system, and $\pi = (\pi(1), \ldots, \pi(m))$ is a *permutation* of the edge indices of E such that for some index k we have

$$x_{\pi(1)} = \ldots = x_{\pi(k)} = 1, \quad x_{\pi(k+1)} \ldots x_{\pi(m)} = 0.$$

If the state vector x is chosen according to the prescribed state probabilities, and the permutation π is chosen independently and uniformly over all matching permutations, then the probability of a particular pair (x, π) occurring is

$$\rho(x, \pi) = \frac{P(x)}{k!(m-k)!} = \frac{1}{m!}\binom{m}{k}P(x),$$

where k is the number of operating elements in x. The sequential construction method samples a permutation $\hat{\pi}$, and considers simultaneously the collection $\mathcal{P}_{\hat{\pi}}$ of *all* possible state pairs (x, π) with $\pi = \hat{\pi}$ and x consistent with $\hat{\pi}$ according to the above criterion. The sample reliability value $\hat{R}$ for this set is then the *conditional* probability of system operation with respect to $\mathcal{P}_{\hat{\pi}}$, that is, the ratio of the sum probabilities of the pairs $(x, \pi) \in \mathcal{P}_{\hat{\pi}}$ for which $\Phi(x) = 1$ divided by the probability of $\mathcal{P}_{\hat{\pi}}$. The details are given below.

Sequential construction method

1. Choose a sample permutation $\hat{\pi} = (\hat{\pi}(1), \ldots, \hat{\pi}(m))$ over the set of permutations of $\{1, \ldots, m\}$. Define the vectors $x^{(k)}$, $k = 1, \ldots, m$ by

$$x^{(k)}_{\hat{\pi}(1)} = \ldots = x^{(k)}_{\hat{\pi}(k)} = 1, \qquad x^{(k)}_{\hat{\pi}(k+1)} = \ldots = x^{(k)}_{\hat{\pi}(m)} = 0.$$

2. Determine the first index $r = 0, \ldots, m$ for which $\Phi(x^{(r)}) = 1$.
3. The contribution $\hat{R}$ to the estimator of R is now

$$\hat{R} = \frac{\sum_{k=1}^{m} \Phi(x^{(k)})\rho(x^{(k)}, \hat{\pi})}{\sum_{k=1}^{m} \rho(x^{(k)}, \hat{\pi})} = \frac{\sum_{k=r}^{m} \binom{m}{k} P(x^{(k)})}{\sum_{k=1}^{m} \binom{m}{k} P(x^{(k)})}.$$

4. Accumulate the set of $\hat{R}$ values, and divide by the number of sample permutations chosen. This yields an unbiased estimator of R.

The estimator obtained for each sample permutation chosen has smaller variance than that obtained in one sample of the crude algorithm. The main computational effort occurs in Step 2, and depends critically on how fast one can update the value of $\Phi(x^{(k)})$, that is, how easily one can determine system operation as the edges are repaired one by one. Notice, however, that the edge repair needs to be performed only until the point at which the system operates for (assuming coherence of the system) further edge repairs do not change the operational state of the system. Thus the amount of work done may be considerably less than the order of m. In the case of connectivity reliability, moreover, Fishman [1986a] has shown that the determination of the index r can be done almost as easily as the determination of $\Phi(x)$ for one value of x, so that the sequential samples come at about the same cost as a single sample in the crude method. Finally, with equal edge failures we have the added advantage that the denominator in the expression in Step 3 above is always 1, and so an extra computational step can be saved.

One can develop a sequential *destruction* method analogous to the sequential construction method given above by starting with all components *operating* and sequentially 'destroying' edges until the system fails. This may be advantageous in the situation where the system tends to fail after relatively few edges fail, so that fewer destruction iterations are performed than construction iterations in the reverse process.

5.4. Sampling using bounds

This is a powerful hybrid of the classical importance sampling and control variate schemes in Monte Carlo. It was first used to solve network reliability problems by Van Slyke & Frank [1972], and expanded upon by Kumamoto, Tanaka & Inoue [1977] and later Fishman [1986a, b, 1989a]. It can in principle be applied to any reliability problem where the system function Φ has associated with it a *lower bounding* function Φ^L and an *upper bounding* function Φ^U having the properties

- $\Phi^L(x) \leq \Phi(x) \leq \Phi^U(x)$ for every state vector x;
- For $k = 0, \ldots, m$ and any assignment $\hat{x}^{(k)} = (\hat{x}_1, \ldots, \hat{x}_k)$ of values for the first k components of x, the values

$$R_k^L(x^{(k)}) \equiv \Pr[\Phi^L = 1 |\ X_1 = \hat{x}_1, \ldots, X_k = \hat{x}_k]$$

and

$$R_u^U(x^{(k)}) \equiv \Pr[\Phi^U = 1 |\ X_1 = \hat{x}_1, \ldots, X_k = \hat{x}_k]$$

can be computed in polynomial time.

The values $R_0^L = \Pr[\Phi^L = 1]$ and $R_0^U = \Pr[\Phi^U = 1]$ are the unconditional operating probabilities for the bounding functions Φ^U and Φ^L, and typically have the most straightforward reliability computation algorithms. For connectivity measures on undirected graphs, however, the values of R_k^L and R_k^U can be obtained by computing the R_0^L and R_0^U values on the graph obtained by *deleting* all edges e_k for which $\hat{x}_k = 0$, and *contracting* all edges e_k for which $\hat{x}_k = 1$. Thus computation of these values is usually no more difficult than computing the unconditional reliabilities.

The values R_0^L and $1 - R_0^U$ represent easily computable measures of events in which the structure function Φ has known values. What bounds-based sampling does is to draw samples from the *remaining* space

$$\mathcal{X} = \{x \in \{0, 1\}^E : \Phi^U(x) = 1,\ \Phi^L(x) = 0\}$$

in proportion to their probability in the original space. The known probability R_0^L of the unsampled space is then added to the estimate of probability within $\mathcal{X}$ to obtain an estimate for the required measure R. The estimate obtained has variance that is better by a factor of $(R_0^U - R_0^L)^2/4$ than that obtained by crude sampling for the same number of samples. The associated Monte Carlo scheme is given below.

Bounds-based sampling method

1. Take samples $\hat{x} = (\hat{x}_1, \ldots, \hat{x}_m)$ from the space $\mathcal{X}$ by successively drawing, for $k = 1, \ldots, m$, the component state $\hat{x}_k$ with operating probability

$$\begin{aligned}\hat{p}_{e_k} &= \Pr[x_k = 1 |\, x_1 = \hat{x}_1, \ldots, x_{k-1} = \hat{x}_{k-1} \text{ and } \Phi^U(x) = 1,\ \Phi^L(x) = 0] \\ &= \left[\frac{R_k^U(x^{(k-1)}, 1) - R_k^L(x^{(k-1)}, 1)}{R_{k-1}^U(x^{(k-1)}) - R_{k-1}^L(x^{(k-1)})}\right] p_{e_k}.\end{aligned}$$

2. Compute the proportion $\hat{R}$ of those samples for which $\Phi(x) = 1$. The number $R_0^L + \hat{R}(R_0^U - R_0^L)$ is now an unbiased estimator of R.

A simple example of this scheme, investigated in in Fishman [1986a] and Van Slyke & Frank [1972], uses the fact that for any state vector x, at least ρ elements must be operating in order for the system to operate, and at least γ elements must be failed in order for the system to fail, where ρ is the minimum cardinality of a pathset for the system and γ is a minimum cardinality cutset for the system. Then we define $\Phi^U(x)$ to be equal to equal to 1 when at least ρ elements of x are operating and 0 otherwise, and we define $\Phi^L(x)$ to be equal to equal to 0 when at least γ elements of x are failed, and 1 otherwise. The evaluations of R^L and R^U are then k-out-of-m reliability problems, which are known to have efficient probability computation algorithms, and the sample space $\mathcal{X}$ is just the space of state vectors x having at least ρ, but not more that $m - \gamma$, elements operating. In Fishman [1986a] the disjoint pathset and cutset bounds are exploited to give a stronger bounded Monte Carlo sampling scheme. Here if $C_1, \ldots, C_r$ are disjoint cuts and $P_1, \ldots, P_s$ are the disjoint paths, then

$$\Phi^L(x) = \begin{cases} 0 \text{ if } \textit{any} \text{ of the cutsets } C_1, \ldots, C_r \text{ fail} \\ 1 \text{ otherwise.} \end{cases}$$

$$\Phi^U(x) = \begin{cases} 1 \text{ if } \textit{any} \text{ of the pathsets } P_1, \ldots, P_s \text{ operate} \\ 0 \text{ otherwise} \end{cases}$$

The values R_0^L and R_0^U can be computed as described earlier, and this extends to the computation of $R_k^L(x^{(k)})$ and $R_k^U(x^{(k)})$ as well.

5.5. The coverage method

The accuracy of the Monte Carlo schemes given above is generally measured by the *variance* of the estimator $\hat{R}$. The variance estimate in each case turns out to be roughly of the form α/K, where K is the number of samples and α is some constant which depends upon R and the type of sampling being done. Analytic comparisons between these estimators are then made based on the relative values of α. With the rate of decrease of variance linear in K, however, the time to sample can become unacceptable when very small variances are required. To obtain computationally efficient estimates that also have guaranteed accuracy, it is necessary to use a different criteria for the accuracy of a Monte Carlo estimate.

The coverage method, developed by Karp & Luby [1985], is based on a more demanding criterion of effectiveness of a Monte Carlo scheme, and thus is able to obtain a correspondingly stronger statement concerning the behavior of the method. Specifically, let ϵ and δ be positive scalars. Suppose we are interested in computing some reliability measure value R and let $\hat{R}$ be the outcome from some Monte Carlo scheme for estimating R. Then the estimate $\hat{R}$ is an ϵ-δ *approximation* for R if

$$\Pr[\frac{|R - \hat{R}|}{R} > \epsilon] \leq \delta.$$

A Monte Carlo scheme is called a *fully polynomial randomized approximation scheme* (FPRAS) if in addition, the time to obtain the estimate $\hat{R}$ is of the order of ϵ^{-1}, $\log(\delta^{-1})$, and the size of the problem instance. In rough terms, a FPRAS is an algorithm which efficiently produces an estimate of R whose percentage error can be guaranteed to be sufficiently small with high probability.

The Karp–Luby Monte Carlo scheme is actually a variant of the importance and stratified sampling methods, and makes use of the *min-cuts* of the system to improve on the crude sampling scheme. To be consistent with the Karp–Luby paper we consider the computation of $R = \Pr[\Phi = 0]$, i.e. the probability of system failure, although one can develop an analogous scheme from the viewpoint of system operation as well. The idea is to embed the set F of failure events into a *universal weighted space* $(\mathcal{U}, w)$, where w is a nonnegative weight function on the elements of $\mathcal{U}$, which satisfies the following criteria:

- $w(F) = \Pr(F) = R$;
- $w(\mathcal{U})$ is efficiently (polynomial-time) computable; further, samples can be efficiently drawn from $\mathcal{U}$ with probability proportional to their weight;
- It can be efficiently recognized when an element in $\mathcal{U}$ is also in F;
- $w(\mathcal{U})/w(F)$ is bounded above by some value M for all instances in the problem class.

If any sample is drawn from $\mathcal{U}$, and the estimate $\hat{R}$ is produced by multiplying the proportion of this sample which is contained in F by $w(\mathcal{U})$, then $\hat{R}$ is an unbiased estimator of R. In Karp & Luby [1985] it is further established that for any positive scalars ϵ and δ, if the sample size is at least $M \ln(2/\delta) 4.5/\epsilon^2$, then the resulting estimator $\hat{R}$ is an ϵ-δ approximation for R. In other words, this sampling scheme is a FPRAS.

We now describe the coverage method as it applies to the s, t-connectedness reliability problem, although the same techniques can be applied in a wide range of situations. Let (G, s, t, p) be an instance of the s, t-connectedness reliability problem, and let $\mathcal{C}$ be the collection of minimal s, t-cuts for G. Define the universal weighted space $\mathcal{U}$ to consist of the pairs (x, C) with x a state vector, $C \in \mathcal{C}$, and $x_e = 0$ for all $e \in C$. The weight assigned to each pair (x, C) is simply $P(x)$. Now each failure state x of the system appears in the elements of $\mathcal{U}$ as many times as the number of min-cuts on which x fails; in order to embed F in $\mathcal{U}$, it is necessary to assign to each x a *unique* $C \in \mathcal{C}$. In the s, t-connectedness problem this is done by finding the set of elements which can be reached from s by a path of operating edges (with respect to x) and setting $C \equiv C(x)$ to be the set of edges from X to $V \setminus X$. The elements of F now appear in $\mathcal{U}$ as (x, C) such that $C = C(x)$, and an element of $\mathcal{U}$ can be determined in linear time to correspond to an element of F by checking the condition $C = C(x)$. The coverage method for the s, t-connectedness problem is given below.

Coverage method

1. Determine the collection $\mathcal{C}$ of s, t-cutsets of G. For each $C \in \mathcal{C}$ compute $w(C) = \prod_{e \in C} q_e$ = the total weight of all elements of $\mathcal{U}$ with second component equal to C, and then compute $w(\mathcal{U}) = \sum_{C \in \mathcal{C}} w(C)$.
2. Draw elements (x, C) from $\mathcal{U}$ in proportion to their weights by first drawing a C from $\mathcal{C}$ with probability $w(C)/w(\mathcal{U})$ and then drawing x by setting $x_e = 0$, $e \in C$, and sampling the states of the other components of x according to their original component probabilities.
3. Compute the proportion $\hat{K}$ of times that a sample (x, C) has $C = C(x)$. Then $\hat{R} = \hat{K} w(\mathcal{U})$ is an unbiased estimator for R.

The above scheme is *not* a FPRAS, for two reasons. First, it is necessary to enumerate the entire set $\mathcal{C}$ of min-cuts, and the cardinality of this set generally grows exponentially in the size of the problem (in fact Provan & Ball [1984] give a method of computing R exactly from this list). Second, the boundedness condition for $w(\mathcal{U})/w(F)$ is not satisfied for general instances of the problem. Karp and Luby, however, go on to modify the above procedure for the class if s, t-connectedness reliability problems where the graph G is planar and undirected, has its facial boundaries bounded in cardinality and sum probability, and satisfies the condition that $\prod_{e \in E}(1 + q_e)$ is bounded above by some fixed M. We do not go into the details here; the general idea is to expand $\mathcal{C}$ to include cuts which are 'almost minimal', in such a way that the associated space $\mathcal{U}$ defined above satisfies the required properties with respect to F. The planarity of G is then employed to allow elements of the expanded space $\mathcal{U}$ to be sampled efficiency and with the correct probabilities, so that the modified scheme becomes an FPRAS for s, t-connectedness reliability.

5.6. Estimating the coefficients of the reliability polynomial

One problem with the methods given thus far is that they only estimate the reliability for a single vector $\boldsymbol{p}$ of probabilities. Of greater interest, frequently, is some estimate of the *functional form* of the reliability polynomial. This makes the most sense in the case when the edge failure probabilities are all the same probability p, so that the system reliability can be written in one of two polynomial forms:

$$Rel(p) = \sum_{i=0}^{m} F_i \, p^{m-i}(1-p)^i$$

$$= p^{\ell} \sum_{i=0}^{m} H_i (1-p)^i$$

In this case a more useful Monte Carlo scheme would be one that estimated each of the coefficients F_i or H_i, for then one could use these estimates to derive an estimate of reliability for any desired value of p.

Two papers have dealt specifically with computing the coefficients of the reliability polynomial. The work of Van Slyke & Frank [1972] and Fishman [1987a] concerns the F_i-coefficients for k-terminal reliability. Van Slyke and Frank uses standard stratified sampling to estimate the F_i values, by sampling separately states having exactly i operating components. Fishman improves this by extending the sequential construction method. He actually estimates the values

$$\mu_i = \frac{F_i}{\binom{m}{i}}$$

$$= \Pr[\text{the system operates } \textit{given } i \text{ elements fail}]$$

by noting that an unbiased estimator for the differences $\mu_i - \mu_{i-1}$ is simply the proportion of times that the index r obtained in Step 2 of the Sequential Construction Method is equal to i. An unbiased estimator for μ_k, and hence F_k, can then obtained by summing the appropriate difference estimators.

Nel & Colbourn [1990] investigate the all-terminal reliability problem, and provide a scheme for estimating the H_i coefficients for this problem. Since the sum of the H_i coefficients is equal to the number of *spanning trees* in the graph G, as opposed to the number of *connected sets* of G, which is the sum of the F_i coefficients, then the number of states contributing to the estimators of the H_i coefficients is much smaller than those which need to be sampled to estimate the F_i coefficients. Let $\mathcal{U} = \{[L_i, U_i] | \ i = 1, \ldots, b\}$ be any shelling of the $\mathcal{F}$-complex of G. From the definition of H_i as the number of L_j's of cardinality i, it follows that for any uniform sampling of intervals $[L_j, U_j]$ in $\mathcal{U}$, the proportion of L_j's of cardinality i is an unbiased estimator of H_i. Nel and Colbourn go on to give a technique for sampling uniformly from the collection of intervals of a 'canonical' shelling of the $\mathcal{F}$-complex of G, based on a uniform sampling of spanning trees in G [Aldous, 1990; Broder, 1989; Colbourn, Day & Nel, 1989].

Describing the reliability function when general edge failures are present is problematic, since the polynomial form itself has an exponentially large number of terms. Fishman [1989a], however, develops a method for partially describing the reliability function by giving the system reliability as a function of a small number of component reliabilities. In particular, suppose that we are interested in knowing the reliability R as a function of the operating probabilities $p_1, \ldots, p_k$ of a chosen set of k edges $e_1, \ldots, e_k$, given *specific* operating probabilities $\tilde{p}_{k+1}, \ldots, \tilde{p}_m$ for the remaining edges $e_{k+1}, \ldots, e_m$. Then we compute the 'coefficients' of the partial description of the reliability by performing a variant of stratified sampling (or conditional sampling, depending on the viewpoint). The procedure is as follows: for *each state vector* $\hat{x}^{(k)} = (\hat{x}_1, \ldots, \hat{x}_k)$ on the edges $e_1, \ldots, e_k$, sample the *strata* of states where edge $x_i = \hat{x}_i$, $i = 1, \ldots, k$ and the remaining edges operate according to their given probabilities. We then compute an estimate $\hat{R}(\hat{x}^{(k)})$ for the associated reliability. When edges operate independently, the strata sampling is fairly straightforward, and can frequently make use of the other improvement schemes given earlier in this section. An estimate for the required functional form for R can now be written in the form

$$\hat{R} = \sum_{\hat{x}^{(k)} \in \{0,1\}^k} \prod_{i:\, \hat{x}_i = 1} p_i \prod_{i:\, \hat{x}_i = 0} q_i \; \hat{R}(\hat{x}^{(k)})$$

As well as its descriptive value, this functional form is useful in measuring the 'criticality' of the edges on which the function is defined, by testing the derivative effects on the function of changing a particular component reliability. Although criticality measures have drawn a significant amount of attention in reliability theory, their treatment is beyond the scope of this chapter.

It is apparent from our discussion here that Monte Carlo methods have been explored largely independently of the development of bounds; however, we emphasize that bounds and Monte Carlo approaches appear to operate most effectively when used in conjunction with each other, as was done in Section 5.4.

6. Performability analysis and multistate network systems

The previous four sections have been concerned with connectivity measures. In the context of communications networks, the underlying assumption of these measures is that as long as a path exists between a pair of nodes then satisfactory communication can take place between those nodes. In many practical problem settings this is not the case. Specifically, issues such as delay, path-length, capacity, and the like can be of vital importance: the network must not just be connected, but it must function at an acceptable performance level. This viewpoint has led to research on performability measures. To study measures of this type additional information, such as edge lengths and edge capacities, are associated with the network components. In addition, it is possible that information representing the load on the system must be specified, e.g. a set of origin-destination traffic requirements. In general, the assessment of such information changes the nature of the reliability problem from a binary 'operate-fail' type of probabilistic statement to one involving multiple system or component states. In many cases this simply results in a more complex variant of the binary-state problem, but it also includes problems involving average behavior and/or continuous state and system variables, which require substantially different solution techniques. We refer to this more general type of reliability problem as a multistate problem and intend it to include performability measures as well as other measures.

The general format for the multistate network problems considered in this paper is as follows: We are given a network $G = (V, E)$, together with a set of *random variables* $\{X_e : e \in E\}$ associated with the edges of the network.

The value assigned to an edge random variable represents a parameter such as length, capacity, or delay. We do not place any *a priori* restrictions on the type of distribution these random variables must have, although in most cases it is assumed that each edge random variable can take a finite number of states. The analogue to the 'operate-fail' model of connectivity measures is the 'two-state system,' where each random variable takes on a 'good' and a 'bad' state. Generally, in the 'good' state the edge operates with a specified capacity, length, etc. and in the 'bad' state the edge fails and has zero capacity, infinite length, etc. This turns out to provide a realistic model in many situations. Random variables may also be assigned to nodes of the network to represent demand, throughput, or delays at the node itself. We do not touch upon those models here, except to mention that they can often be modeled as problems with stochastic edge parameters only.

Corresponding to any vector $\boldsymbol{x} = (x_e : e \in E)$ of assignments for the edge parameter random variables the system itself is given a *system state* value $\Phi(\boldsymbol{x})$, which represents some measure of system performance. Thus, the system state value is also a random variable, whose distribution is some complex function of the distribution of the individual parameter values. The goal of a multistate system evaluation problem is to compute, or estimate, some characteristic of the random variable representing the system state. This could involve a complete description of the system state distribution, or the probability that a certain threshold of system performance has been attained, or the mean, variance, or selected moments of the system state distribution. For the two-state system described above, the threshold measure would be the most analogous system operation characteristic. In fact, the binary systems considered in the previous sections are a special case of this more general format, where the X_e are 0-1 variables with $Pr[X_e = 1] = p_e$, $Pr[X_e = 0] = 1 - p_e$, and

$$Rel(\text{SBS}, \boldsymbol{p}) = \text{Ex}[\Phi] = \text{Pr}[\Phi \geq 1].$$

We make more use of this connection later in this section.

Performability analysis the name given to reliability analysis in several applied areas, including computer and communications systems, where some of the most important practical multi-state measures are considered. Performability measures can involve sophisticated indicators of network performance such as lost call traffic for circuit switched networks and packet or message delay for packet switched networks. The evaluation of the performance measure function Φ itself is usually nontrivial, involving some variant of a multicommodity flow algorithm. Methods to compute expected performance or threshold probability for these measures need to be more general-purpose — and correspondingly tend to be less effective — than methods for the more elementary systems highlighted below.

The following measures are perhaps the most important and widely studied of the multistate network measures, and are treated extensively in this section.

Shortest path

Input: Graph $G = (V, E)$, nodes s and t.

Random parameter: d_e = length of edge e.

System value: Φ_{PATH} = length of shortest (s, t)-path from s to t.

Maximum flow

Input: Directed graph $G = (V, E)$ with nodes s and t.

Random parameter: c_e = the capacity of edge e.

System value: Φ_{FLOW} = the maximum s, t-flow in G.

PERT network performance

Input: Directed acyclic graph $G = (V, E)$ with source node s and sink node t.

Random parameter: t_e = time to complete task associated with edge e.

System value: Φ_{PERT} = minimum time to complete the project, where the project starts at point s, ends at point t, and no task can be started from node v until all tasks to node v are completed. Equivalently, Φ_{PERT} = length of *longest* (s, t) path in G with edge lengths t_e.

Although these measures are more simplistic than general performability measures, they capture many of the features important to the more sophisticated measures. As well as cataloguing the extensive research papers for these problems, we can also use them to illustrate how the extensive work on connectivity measures can be adapted to the multistate context. This relationship can serve as the basis for analysis of more complex multistate measures.

Investigations of stochastic path and flow problems began about the same time as those of binary-state reliability problems. The PERT problem was probably the first of these problems to draw significant attention, and has certainly been the most popular of the stochastic network problems. An excellent account of the current state of computational methods in PERT optimization can be found in Elmaghraby [1989a], and an extensive bibliography on the subject can be found in Adlakha & Kulkarni [1989]. The problem was first introduced in Malcolm, Roseboom, Clark & Fazar [1959] in the context of project evaluations; early work on stochastic PERT problems also appears in Charnes, Cooper & Thompson [1964], Fulkerson [1962] and Hartley & Wortham [1966]. The first analysis of stochastic shortest path problems was probably in Frank [1969], and early work concerning stochastic flow problems appears in Douilliez & Jamoulle [1972] and Frank & Hakimi [1965].

6.1. General purpose algorithms for performability analysis and multistate systems

In this section we give two important general-purpose algorithms for dealing with performability and multistate reliability measures.

6.1.1. The most probable states method

The most probable states method is the current method of choice in performability analysis, for it is one that can be applied to a very general class of multistate problems [Li & Silvester, 1984; Lam & Li, 1986; Yang & Kubat, 1989,

1990]. The only requirement is an efficient method for evaluating the related performance measure, Φ. We describe the application of the most probable states method to computing the performability measure Ex[Φ] where larger values of Φ are 'better than' smaller values of Φ. The application to $\Pr[\Phi \leq \alpha]$ follows in a similar manner. Suppose that the network states are ordered $\boldsymbol{x}^1, \ldots, \boldsymbol{x}^s$, such that $\Pr[\boldsymbol{x}^1] \geq \Pr[\boldsymbol{x}^2] \ldots$ The most probable states method is based on enumerating states in this order. Define $lp_\Phi(k)$ and $up_\Phi(k)$ to be any lower and upper bounds, respectively, on $\min_{j=k}^s \Phi(\boldsymbol{x}^j)$ and $\max_{j=k}^s \Phi(\boldsymbol{x}^j)$. The upper and lower bounds typically used here are easily computable and, in most cases, are trivial bounds that are independent of k.

For 2-state systems if p_e is the probability of the 'good' state for edge e and $1 - p_e$ the probability of the 'bad' state for edge e; typical assumptions are that $p_e \geq 1 - p_e$ and as a result $\Phi(\boldsymbol{x}^1) \geq \Phi(\boldsymbol{x}^i) \geq \Phi(\boldsymbol{x}^{2^n})$ so that we can set $lp_\Phi(k) = \Phi(\boldsymbol{x}^{2^n})$ and $up_\Phi(k) = \Phi(\boldsymbol{x}^1)$ for all k. The most probable states bounds are defined by

$$LP_\Phi = \sum_{k=1}^{\bar{k}} \Phi(\boldsymbol{x}^k)\Pr[\boldsymbol{x}^k] + (1 - \sum_{k=1}^{\bar{k}} \Pr[\boldsymbol{x}^k])lp_\Phi(\bar{k}+1)$$

$$UP_\Phi = \sum_{k=1}^{\bar{k}} \Phi(\boldsymbol{x}^k)\Pr[\boldsymbol{x}^k] + (1 - \sum_{k=1}^{\bar{k}} \Pr[\boldsymbol{x}^k])up_\Phi(\bar{k}+1)$$

Here, $\bar{k}$ can be defined dynamically based on some stopping criterion. The most typical criterion is to require that the difference between the upper and lower bounds be within some tolerance. Lower and upper bounds for the threshold value measures can be defined in a similar way.

Li & Silvester [1984] first explored the idea of generating most probable states, and Lam & Li [1986] developed algorithms for the effective generation of the next most probable state. Gaebler and Chen [1987] developed a more efficient technique, which has been refined by Shier and his colleagues [Bibelnieks, Jarvis, Lakin & Shier, 1990; Shier, Valvo & Jamison, 1992]. At present, their strategy appears to lead to the most efficient generation algorithms.

Yang & Kubat [1989] describe a method for enumerating states, i.e. $\boldsymbol{x}^i$, in order of decreasing probability for 2-state systems in $O(n)$ time per state. Specifically, they maintain a partial binary enumeration tree where each node represents the assignment of a 'good' or 'bad' state to each edge in some set S. The branching step is to choose an edge j, not in S, and create two new nodes, one with j assigned the 'good' state and one with j assigned the 'bad' state. At each iteration of the algorithm a new leaf node is created with corresponds to a (complete) network state $\boldsymbol{x}^i$. In order to generate these leaf nodes in the correct order two values are associated with each node in the enumeration tree: the probability of the highest probability leaf node, not yet enumerated, in the left sub-tree rooted at the node and the corresponding value for the right sub-tree. These values allow the algorithm to choose the appropriate leaf node to generate at each iteration in $O(n)$ time and can be updated in $O(n)$ time per iteration. Just as Monte Carlo algorithms employ state space reduction by using efficiently computable bounds, a variety of simple bounds can be incorporated in the Yang–Kubat method to reduce the number of states that must be generated to obtain desired accuracy in bounds [Harms & Colbourn, 1993b].

Sanso & Soumis [1991] suggest that rather than most probable states it is often more appropriate to enumerate the 'most important' states. The motivation is that in some situations certain lower probability states, which might not otherwise be enumerated, could significantly affect performance measures. Such states might correspond to situations in which the system exhibits extremely poor performance. Specifically, $\Pr[\boldsymbol{x}]$ might be relatively small but $\Phi(\boldsymbol{x})$ could be very large or very small. This is particularly relevant when computing bounds on $\mathrm{Ex}[\Phi]$. Jarvis & Shier [1993] also examine a most relevant states method.

6.1.2. State-space partitioning

One of the most effective heuristics to date for computing and bounding reliabilities for multistate systems is one introduced by Douilliez & Jamoulle [1972] and developed further by Shogan [1977a] and Alexopoulos [1993]. It can be applied to binary systems as well, although it gives particularly good results when used in the multistate context.

The major requirement on the system state function Φ is that it be *coherent*, that is, that Φ is a nondecreasing function of $\boldsymbol{x}$. (Note that this is a straightforward generalization of the condition required for a binary system to be coherent.) To describe the method, we first generalize the concept of interval introduced in Section 4.1.5. Let $\boldsymbol{a} = (a_1, \ldots, a_n)$ and $\boldsymbol{b} = (b_1, \ldots, b_n)$ be vectors of component values, with $a_j \le b_j$, $j = 1, \ldots, n$. An *interval* $[\boldsymbol{a}, \boldsymbol{b}]$ is the set of all states $\boldsymbol{x}$ having $a_j \le x_j \le b_j$, $j = 1, \ldots, n$. The entire state space for the problem can be represented by the interval $[\boldsymbol{a}^l, \boldsymbol{b}^u]$, where $\boldsymbol{a}^l$ and $\boldsymbol{b}^u$ are, respectively, the vectors of smallest and largest component values.

An interval $[\boldsymbol{a}, \boldsymbol{b}]$ is called *feasible* if *every* state $\boldsymbol{x} \in [\boldsymbol{a}, \boldsymbol{b}]$ has $\Phi(\boldsymbol{x}) \ge \alpha$, *infeasible* if *no* state $\boldsymbol{x} \in [\boldsymbol{a}, \boldsymbol{b}]$ has $\Phi(\boldsymbol{x}) \ge \alpha$, and *undetermined* otherwise. It is clear that the operating probabilities for feasible and infeasible intervals are easy to calculate, the former being simply the sum of probabilities of the events in the interval,

$$\prod_{j=1}^{n} \left(\sum_{a_j \le x_j \le b_j} \Pr[x_j] \right),$$

and the latter being 0.

Now suppose the $[\boldsymbol{a}, \boldsymbol{b}]$ is an undetermined interval. For any state $\boldsymbol{\xi} \in [\boldsymbol{a}, \boldsymbol{b}]$ which has $\Phi(\boldsymbol{\xi}) \ge \alpha$ we know that the subinterval $[\boldsymbol{a}, \boldsymbol{\xi}]$ must be feasible, since Φ is coherent. Further, the remaining set of outcomes $[\boldsymbol{a}, \boldsymbol{b}] \setminus [\boldsymbol{a}, \boldsymbol{\xi}]$ can in turn be subdivided into intervals, the number of which is equal to the number of coordinates of $\boldsymbol{\xi}$ which have $a_j < \xi_j < b_j$. Thus we can compute the value $\Pr[\Phi \le \alpha]$ by starting with the interval $[\boldsymbol{a}^l, \boldsymbol{b}^u]$, partitioning it into a feasible interval plus a number of other intervals, and then computing the reliability recursively for these intervals, using the same technique. Summing up the reliabilities of the intervals gives the system reliability.

This procedure is in effect a type of *branch-and-bound* procedure, where the nodes of the branching tree consist of the unprocessed intervals. At each stage an unprocessed interval is chosen and its type is established. A feasible or infeasible interval corresponds to a node that can be 'fathomed', while undetermined intervals are further decomposed, producing a 'branching' and adding additional nodes to the tree. Moreover, as in classic branch and bound methods *lower* and *upper bounds* can be maintained on the actual reliability, by keeping, respectively, the sum of the reliabilities of the *feasible* intervals and 1− the sum of the reliabilities of the *infeasible* intervals. Further, at any stage in the branching process there is always the option of computing or estimating the reliability values of the remaining undetermined intervals by most probable state, combinatorial, or Monte Carlo methods, and by adding these probabilities to those of the intervals already fathomed, thereby obtaining an even more accurate estimate of system reliability.

The effectiveness of the above procedure depends upon how quickly the type of an interval can be determined, how large the search tree becomes, and in the case of a partial search, how much of the probability resides in the determined intervals and how good the estimates of probability are for the undetermined intervals. For network problems the above questions can be dealt with by finding and manipulating the associated network objects such as shortest paths, min-cuts, and critical paths, and by using approximation techniques such as those outlined later in this section. The techniques given in Alexopoulos [1993] and Shogan [1977a] give exceptionally good estimates of reliability for PERT and shortest path problems, and presumably will give similar results for stochastic max flow and more general performability measures.

6.2. Elementary properties of multistate systems

The remainder of the section concentrates on the shortest path, max flow, and PERT systems described above, although the reliability computation techniques surveyed often have application to more general multistate systems. We first note that all three of these problems have special cases which correspond precisely to the 2-terminal reliability measure $Rel_2(G, s, t, \boldsymbol{p})$. Specifically, each edge parameter takes on values of 0 or 1 corresponding to the operational state of the edge in the binary problem. The 2-terminal reliability function then has the following interpretations:

Shortest path: If edge failure corresponds to $d_e = 1$ and edge operation corresponds to $d_e = 0$, then $Rel_2(G, s, t, \boldsymbol{p}) = \Pr[\Phi_{\mathrm{PATH}} = 0]$.

Maximum flow: If edge failure corresponds to $c_e = 0$ and edge operation corresponds to $c_e = 1$, then $Rel_2(G, s, t, \boldsymbol{p}) = \Pr[\Phi_{\mathrm{FLOW}} \geq 1]$.

PERT network performance: If edge failure corresponds to $t_e = 0$, edge operation corresponds to $t_e = 1$, *and* every s, t-path in G has the same length n, then $Rel_2(G, s, t, \boldsymbol{p}) = \Pr[\Phi_{\mathrm{PERT}} = n]$.

It follows that these problems are NP-hard for any class of graphs for which the associated 2-terminal reliability problem is NP-hard. Similar arguments show that the computation of Ex[Φ] is also NP-hard for these same classes of graphs.

The type of distribution allowed for the edge random variables is of critical concern in the computational efficiency of the techniques discussed here, and hence it is necessary to outline the computational efficiency of computing and manipulating distributions for network problems. The following two operations play a major role in most of the computational schemes for multistate network reliability, and the difficulty of computing the corresponding distributions is of primary importance.

• the *sum* $X_1 + X_2$ of two independent random variables (their distribution being the *convolution*);

• the *maximum* $\max(X_1, X_2)$ or *minimum* $\min(X_1, X_2)$ of two independent random variables X_1 and X_2.

A class of edge distributions for a multistate network problem must typically satisfy one or more of the following three criteria, depending on the type of analysis being performed:

1. The computation of a given cdf value of an element in the class must be able to be performed to a given number of digits accuracy in time polynomial in the number of digits and the size of the input describing the distribution.

2. Given a set of distributions in the class and a sequence of k successive min, max, and sum operations starting with these distributions, the distribution resulting from this sequence of operations must also be in the class, and further, it must be possible to find the description of the resulting distribution in time polynomial in the size of the input describing the original distributions.

3. The expected value, variance, or more generally any specified moment must be computable (in terms of digits of accuracy) in polynomial time.

Typically, it is assumed that the random variables take on *discrete* distributions, in particular, ones having a finite number of values. Although the computation of the distribution resulting from a single min, max, or sum operation is elementary, the computation of the distribution for a series of k of these operations is known to be NP-hard, even when each of the original variables has only two values. What is necessary to ensure efficient computation of the min, max, or sum distributions is that the random variables take on the consecutive values $\{1, 2, \ldots, x_q\}$ (or more generally consecutive multiples of some common denominator) on every edge of the graph, for some fixed q. Hagstrom [1990] has in fact shown that in many cases multistate edge distributions such as this can be efficiently reduced to two-state distributions with edge 'operation' probabilities all equal to $1/2$.

There are also two classes of infinite-valued distributions which are among the most general known to satisfy (1)–(3) above. The first is discrete-valued, and can be described as 'mixtures of negative binomial' distributions, having pdf's of the form

$$f(x) = \sum_{i=1}^{q} \sum_{j=1}^{r} a_{ij} p^{jx} (1 - p^j)^i \quad x = 0, 1, \ldots$$

for $0 < p < 1$ and appropriately chosen values of a_{ij}. There is also a class of continuous distributions which satisfy the required properties. These distributions can be described as 'mixtures of Erlang' distributions (also known as *Coxian* distributions [Cox, 1955; Sahner & Trivedi, 1987]). They are the continuous analogy to the 'mixture of negative binomial' class described above, and have cdf's of the form

$$F(t) = \sum_{i=1}^{q} \sum_{j=1}^{r} a_{ij} t^i e^{-jt} \quad 0 \leq t < \infty$$

for appropriately chosen values of a_{ij}.

We now give a sketch of the major evaluation and bounding techniques for multistate network problems. In most cases, it turns out that the same technique apply to two or more of the above problems. As well, most of the techniques have binary-state versions which have been presented in Sections 3 and 4. Thus we organize the discussion by technique rather than by subject, and follow, when possible, the format given for the binary version of the problem. For most of the discussion we concentrate on the evaluation of $\Pr[\Phi \geq \alpha]$ for the particular system value function Φ and specific system value α of interest.

6.3. *Transformations and reductions*

One of the 'reductions' popular in multistate problems that does not have an analogue in binary-state problems is to transform the multistate problem into one in which each edge has a 'failed' state, where it is essentially deleted from the network; and an 'operating' state, where it takes on a single length, capacity, or completion time. Although it generally does not provide any improvement in complexity of the associated reliability computation algorithm, it does allow conceptually easier application of factoring and other enumeration methods. The method for the PERT problem is given in Hagstrom [1988], and for the shortest path problem in Mirchandani [1976]. Specifically, if edge random variable X_e takes on values $x_1 \leq x_2 < \ldots < x_q$ with $\Pr[X_e = x_i] = p_i,\ i = 1, \ldots, q$, then the edge e can be replaced by

Shortest path: q parallel edges, the ith edge having length x_i and operating probability $p_i(1 - \sum_{j=1}^{i-1} p_j)^{-1}$.

Maximum flow: q series edges, the ith edge having capacity x_i and operating probability $p_i(1 - \sum_{j=1}^{i-1} p_j)^{-1}$.

PERT network performance: q parallel edges, the ith edge having completion time x_i and operating probability $p_i(1 - \sum_{j=i+1}^{q} p_j)^{-1}$.

The handling of irrelevant and mandatory edges (Section 3.1) also carries over to the problems above. Irrelevant edges can be deleted, since the state of an edge which does not lie on any s, t-path affects neither path length, flow, or project completion time. Mandatory edges can similarly be contracted in the maximum flow problem. In particular, if mandatory edge e is contracted for instance graph G when computing $\Pr[\Phi_{\text{FLOW}} \geq \alpha]$, then the multiplicative factor applied to the problem on the contracted graph $G \cdot e$ is $\Pr[c_e \geq \alpha]$. In the shortest path and PERT problems mandatory edges cannot be immediately contracted, since the value of the edge affects the length of the shortest or longest path in the contracted graph. They do, however, induce a partition of G into 1-attached subnetworks which can be evaluated separately (see below).

Series and parallel reductions have powerful analogues in path and flow problems (Martin [1965] for PERT). To summarize the use of these reductions, let e and f be two edges which are either in series or in parallel, and let g be the edge which replaces these two edges in the series or parallel reduction.

Shortest path: For a *series* reduction, d_g is the *convolution* of d_e and d_f. For a *parallel* reduction, d_g is the *minimum* of d_e and d_f.

Maximum flow: For a *series* reduction, c_g is the *minimum* of c_e and c_f. For a *parallel* reduction, c_g is the *convolution* of c_e and c_f.

PERT network performance: For a *series* reduction, t_g is the *convolution* of t_e and t_f. For a *parallel* reduction, t_g is the *maximum* of t_e and t_f.

More complicated subnetwork reductions have been considered for the PERT problem in Hartley & Wortham [1966], Ringer [1969] and Ringer [1971]. The 1- and 2-attached subnetworks also have multistate analogues [Elmaghraby, 1977; Shogan, 1982]. First, let H be a 1-attached subnetwork of G, with r the attachment point, and let Φ^H and $\Phi^{\bar{H}}$ be the system value functions for the appropriate problem when applied to the subnetworks H and $G \setminus H$ with terminals s, v and v, t, respectively. Then the system value function Φ^G satisfies

Shortest path and **PERT network performance:** Φ^G is the *convolution* of Φ^H and $\Phi^{\bar{H}}$.

Maximum flow: Φ^G is the *minimum* of Φ^H and $\Phi^{\bar{H}}$.

Second, let H be a 2-attached subnetwork of G, with attachment points x and y, and let Φ^{xy} and Φ^{yx} be the appropriate system value function for the subgraph H when oriented from x to y and from y to x, respectively. Then the system reliability function of G is the same as that obtained by replacing the subgraph H by the two edges (x, y) and (y, x) having edge random parameters distributed as Φ^{xy} and Φ^{yx}, respectively.

Another reduction, discussed for PERT and reliability problems in Elmaghraby, Kamburowski & Stallmann [1989b], but applicable as well to shortest path and maximum flow problems, is called the *node reduction* [Elmaghraby, Kamburowski & Stallmann, 1989b]. It is actually an edge contraction, the edge having the property that it is either the only out-edge of its tail or the only in-edge of its head. The essential feature of this contraction is that it does not introduce any spurious paths, as could occur when an arbitrary edge is contracted in a directed graph. Thus the associated problem can be reduced to k subproblems on networks with one less edge and node, where k is the number of states taken on by the contracted edge. This is covered in more detail next.

6.4. *Efficient algorithms for restricted classes*

The evaluation of Φ for series-parallel graphs can be accomplished in the same manner as it is done for the 2-terminal problem, with series and parallel reductions performed as indicated above. The complexity is $O(Rn)$, where R is the worst-case complexity of performing a max/min or convolution at any time in the algorithm. Thus the complexity of these algorithms depends critically upon the time to perform the series and parallel operations. For the three types of distributions given at the beginning of the section, R is linear in nq (in the finite case) or nqr (in the two infinite cases). It is generally believed that polynomial algorithms exist as well for graphs with bounded tree-width, although this has not been treated specifically.

A second interesting class of stochastic network reliability problems which have efficient solution algorithms were observed by Nádas for PERT problems [1979], who called them 'complete tracking' problems. They are characterized by edge random variables of the form $X_e = a_e Z + b_e$, where a_e and b_e are edge parameters and Z is a common random variable. For PERT problems it turns out the resulting system reliability $\Pr[\Phi \leq \alpha]$ is equal to the maximum of the values

$$w(P) = \frac{\sum_{e\in P} b_e - t}{\sum_{e\in P} a_e}$$

taken over all s,t-paths P in G. Computing this maximum can be done in polynomial time by solving a modification of the 'minimal cost-to-time ratio cycle problem' [see for example Lawler, 1976, pp. 94–97]. The associated problem for shortest paths involves minimizing $w(P)$ over all s,t-paths P, and that for the maximum flow problem involves minimizing $w(C)$ over all s,t-*cuts* C; and likewise can be solved in polynomial time.

A third efficient special-case algorithm has been proposed by Ball, Hagstrom, and Provan [1994] to solve the max flow and PERT problems on 'almost critical systems'. These are two-state threshold systems, where the components have specified 'operating' and 'failed' capacity or duration values and the system tries to maintain a given system state level α. The system is *1-critical* if it is 'minimal' with respect to withstanding any single component failure, that is, it can maintain the given operating level α whenever any single component fails, but every component is a member of a two-component set whose failure renders the system unable to maintain this operating level. In Ball, Hagstrom & Provan [1994] it is shown that the probability of failure of a 1-critical flow or planar PERT system is computable in polynomial time, although the problem becomes NP-hard if either 1-criticality or planarity (in the PERT case) is relaxed.

6.5. State-based methods

Enumerative methods for computing multistate system reliability are necessarily restricted to problems having a finite number of states for each edge. Specifically, let each edge e_j have associated random parameter X_j taking on values $1, 2, \ldots, q$, with probabilities $p_{ji} = \Pr[X_j = i]$, $i = 1, \ldots, q$, and let the system value function Φ take on values $1, \ldots, K$. Then the two classic stochastic measures $\Pr[\Phi \leq \alpha\ ]$, $\alpha \in \{1, \ldots, K\}$, and $\mathrm{Ex}[\Phi]$ can be written:

$$\Pr[\Phi \leq \alpha\ ] = \sum_{\substack{(i_1,\ldots,i_n)\in\{1,\ldots,q\}^n \\ \Phi(i_1,\ldots,i_n)\leq\alpha}} \prod_{j=1}^{n} p_{j,i_j}$$

$$\mathrm{Ex}[\Phi] = \sum_{(i_1,\ldots,i_n)\in\{1,\ldots,q\}^n} \Phi(i_1, \ldots, i_n) \prod_{j=1}^{n} p_{j,i_j}$$

the number of terms in the above two measures can be on the order of $q^{|E|}$, and hence these problems become intractable on a considerably smaller scale than even those of the binary state reliability measures. It is worth noting (and this was developed for the maximum flow problem in Somers [1982]) that when all but a small number of edges have deterministic edge values (or equivalently, have only a single length/capacity/duration) then the above enumeration can be performed in polynomial time.

The factoring method given in Section 3.3 has an analogous, if somewhat cumbersome, form here. Namely, if for 'pivotal edge' e and edge parameter value x, define $G_{e,x}$ to be the network having edge e fixed at value x. Then the Factoring Theorem for the above two measures becomes

$$\begin{aligned} \Pr[\Phi \leq \alpha] &= \sum_x \Pr[\Phi_{G_{e,x}} \leq \alpha] \\ \mathrm{Ex}[\Phi] &= \sum_x \mathrm{Ex}[\Phi_{G_{e,x}}] \end{aligned} \tag{4}$$

The Factoring Theorem was first applied to stochastic network problems in Douilliez & Jamoulle [1972] (in a somewhat disguised form), and has been applied effectively to PERT and shortest path problems in Elmaghraby, Kamburowski & Stallmann [1989b], Fisher, Saisi & Goldstein [1985], Hagstrom [1990] and Hagstrom & Kumar [1984], and to the maximum flow problem in Lee [1980]. It can be combined elegantly with series-parallel reductions by means of the node reduction method. The technique is given in Elmaghraby [1989a], and is based on the simple observation that in any acyclic graph without parallel edges, there is always at least one node reduction which can be performed, say by contracting edge e one of whose endpoints v has only e as an in (out) edge. Let $e_1, \ldots, e_k$ be the other edges incident with node v. Then the system $G_{e,x}$ given above is simply $G \cdot e$, with the variables associated with $e_1, \ldots, e_k$ modified as follows:

Shortest path: d_{e_i} is *shifted* by an amount x, that is, $d'_{e_i} = d_{e_i} + x$, $i = 1, \ldots, k$.

Maximum flow: c_{e_i} is *capped* at value x, that is, $c'_{e_i} = \max\{c_{e_i}, x\}$, $i = 1, \ldots, k$.

PERT network performance: t_{e_i} is likewise shifted by an amount x, $i = 1, \ldots, k$.

The distributions of these modified random variables are easily computed for finite-state distributions, for example by treating them as convolutions or maximums having one of the random variables one-valued. Thus (4) reduces the particular problem on network G to k subproblems, where k is the number of values taken by edge e, all on the same network $G \cdot e$ but with different distributions on the edges $e_1, \ldots, e_k$. The complexity of computing the cdf value or mean in this case depends critically on the *number* of such node reductions which must be performed, in tandem with performing available series and parallel reductions, in order to reduce the network to a single edge. Bein, Kamburowski & Stallmann [1992] have given an $O(n^3)$ algorithm for determining the *minimum* number of node reductions which must be performed in order to reduce a graph in the above manner. This number, therefore, in some sense also represents the 'complexity' of a network with respect to path and flow problems.

There have also been some papers which use path- and cut-enumeration-based techniques to solve multistate flow problems. Evans [1976] uses a lattice of cutsets to compute maximum flow, Mirchandani [1976] extends the disjoint products procedure of Section 3.4, and Hagstrom [1984] extends the inclusion–exclusion results of Satyanarayana & Prabhakar [1978] to stochastic path problems.

The enumeration techniques given above are of little use in solving problems involving infinite or continuous edge random variables. When the edge random variables have *exponential* distributions, Kulkarni and colleague [Kulkarni, 1986; Kulkarni & Adlakha, 1987, 1985] give an interesting procedure for solving shortest path problems, maximum flow problems on planar networks, and PERT problems. We illustrate for the shortest path problem, the flow and PERT problems having similar, though more involved, solution techniques. We have a set of 'runners', who begin at the source node s and each proceeds to traverse one of the edges going out of s. After an interval of time with known (Erlang) distribution, one of the runners reaches the end of his edge. At this point the runner who has finished his edge stops, and simultaneously runners begin running along edges pointing away from the newly-reached node. (Runners are not sent along edges which go to previously visited nodes.) Due to the 'memoryless' exponential distribution of running times, it follows that at the point at which the runners begin running from the newly-reached node one can assume that *all* of the runners have just started along their edge. In short, the process of these runners traversing the network is a *continuous time Markov chain*. A state of this Markov chain corresponds to a possible set of nodes which the runners have visited together with a corresponding set of edges on which the runners are currently running, and the absorbing states are those in which the sink node t has been labeled. The average time to absorption for this Markov chain is now precisely the expected length of a shortest path. The absorbing state probability, moreover, can be calculated easily by the appropriate ordering of the states of the Markov chain. Although the computation time is linear in the number of states of the Markov chain, this number grows exponentially in the size of the network. For networks of the order of 15–20 edges, however, this method is fairly effective. As well, it has been applied to other stochastic settings such as reliability [Bailey & Kulkarni, 1986], minimum spanning trees [Kulkarni, 1988], and min cost flow [Corea & Kulkarni, 1990] (for a unified framework, see Bailey [1991]).

6.6. Bounding techniques

Due to the particularly intractable nature of multistate network reliability problems, the dominant focus of research in this area is in developing techniques for bounding the various system measures of interest. These techniques often differ substantially from those used for binary-state problems.

The historically first technique used for approximating reliability in stochastic networks, and the one which has enjoyed the most attention, arises from the intuitively appealing notion that the expected value $\mathrm{Ex}[\Phi]$ of the shortest path length/max flow/project completion time should be able to be obtained by replacing each *random* edge length/capacity/completion time by a *deterministic* parameter whose value is the *expectation* of this value, and then solving the *deterministic* version of the problem. This was in fact the solution technique proposed in the original treatment of PERT in Malcolm, Roseboom, Clark & Fazar [1959]. It seems to be part of the folklore that the value obtained by this technique is an *upper bound* to the true value of $\mathrm{Ex}[\Phi]$ in the PERT problem and a *lower bound* in the shortest path and maximum flow problems. (A unified account of this can be found in Weiss [1986].)

Most of the succeeding research concentrated on approximating the cdf value $F(\alpha) = \Pr[\Phi \leq \alpha]$ for Φ. An early technique along the same lines as Malcolm and coworkers was suggested by Charnes, Cooper & Thompson [1964]. It applied originally to the PERT problem with continuous edge parameter values, but can be modified to apply to shortest path/maximum flow problems and discretely distributed edge random variables as well. Specifically, suppose that one wants to compute the number α for which $F_{\mathrm{PERT}}(\alpha) = \beta$ for some specified probability β. Replace each edge random variable T_e by the value t_e for which $\Pr[T_e \leq t_e] = \beta$, and solve for the deterministic shortest completion time. The resulting value is again an upper bound on the actual project completion time having the given cdf value β.

Improvements in the above schemes for the PERT problem compute, for each node v in the graph, distributions for the intermediate random variable

$$\Phi_v = \text{ the longest path to node } v.$$

In all cases the computations are done in *topological order* $v_1 = s, v_2, \ldots, v_n = t$, that is, all edges pointing into v_i come from nodes v_j with $j < i$. The first of these schemes was proposed by Fulkerson [1962]. It computes a lower bound E_i^L on $\mathrm{Ex}[\Phi_{v_i}]$ using the fairly straightforward recursive formula

$$\begin{aligned} E_1^L &= 0 \\ E_i^L &= \sum_{\beta} p(\beta) \min_{j<i}(E_j^L + \beta_{v_j,v_i}) \qquad j = 2, \ldots, n \end{aligned}$$

the sum being taken over all vectors $\beta = (\beta_{ji} : j < i)$ of parameter values which can be taken by the edges pointing into node v_i (terms where (v_j, v_i) does not exist are ignored). Improvements on this technique primarily involved estimating the cdf values $F_i(\alpha)$ for Φ_{v_i}, in particular, computing *upper* and *lower* bounds $F_i^U(\alpha)$ and $F_i^L(\alpha)$, respectively. These values can be used in turn to compute *lower* and *upper* bounds, respectively for the values $\mathrm{Ex}[\Phi_{v_i}]$, using the elementary formula

$$\mathrm{Ex}[\Phi] = \sum_{\alpha}(1 - F(\alpha)).$$

All of them use the same type of recursive formula, with $F_1^U(\alpha) = F_1^L(\alpha) = 1$ for all nonnegative α (and zero otherwise). The first of these was given by Kleindorfer [1971], namely,

$$F_i^U(\alpha) = \min_{j<i} \sum_{x_{ij}} \Pr[t_{v_j,v_i} = x_{ij}] F_j^U(\alpha - x_{ij}) \qquad i = 2, \ldots, n$$

and

$$F_i^L(\alpha) = \prod_{j<i} \sum_{x_{ij}} \Pr[t_{v_j,v_i} = x_{ij}] F_j^L(\alpha - x_{ij}) \qquad i = 2, \ldots, n.$$

Shogan [1977a] gives bounds of the form

$$F_i^U(\alpha) = \sum_{\beta} p(\beta) \min_{j<i} F_j^U(\alpha - \beta_{ji}) \qquad i = 2, \ldots, n$$

and

$$F_i^L(\alpha) = \sum_{\beta} p(\beta) \prod_{j<i} F_j^L(\alpha - \beta_{ji}) \qquad i = 2, \ldots, n,$$

again with the sum taken over all vectors β of parameter values of edges pointing into v_i. He shows that the bounds apply under distributions where the indexing edges in the summand have a certain degree of dependence (i.e. are 'associated'), and that the bounds are strictly better that those of Kleindorfer (the lower bounds being equal under complete independence of edge parameters). The corresponding bounds on $\mathrm{Ex}[\Phi_{v_i}]$ are therefore likewise ordered, and his lower bound is strictly better than that of Fulkerson, with Kleindorfer's and Fulkerson's lower bounds not uniformly comparable. The bounds of Fulkerson, Kleindorfer, and Shogan can also be modified to apply to the shortest path problem, as long as the underlying graph is acyclic, and to the case of continuous edge parameters. Kleindorfer's bound is *a priori* polynomial-time computable, and Clingen [1964] gives a polynomial-time computation of Fulkerson's bound, which can modified to compute Shogan's bound as well. Related research along these lines is found in Clingen [1964], Robillard and M. Trahan [1977], Dodin & Elmaghraby [1984], Dodin [1985a], and the work of Agnew as reported in Elmaghraby [1989a]. The node reduction technique given in Section 6.3 has been used effectively in very similar bounding schemes for PERT problems [Elmaghraby, Kamburowski & Stallmann, 1989b], and Dodin [1985b] uses essentially the *inverse* of the node reduction technique to obtain yet another similar bounding technique.

Bounds for the flow problem require a different approach. A lower bound on $\mathrm{Ex}[\Phi_{\mathrm{FLOW}}]$ was first given in Aneja & Nair [1980], with further refinements given in Carey & Hendrickson [1984] and precise conditions for tightness of the bound given in Nagamochi & Ibaraki [1991]. It uses the chain formulation of flow [see Ford & Fulkerson, 1962, p. 8]. Here we assume that the random capacity c_e is binary, with the 'operating' state representing normal capacity c_e with probability p_e, and the failed state representing capacity zero with probability $1 - p_e$. Let $\Gamma_1, \ldots, \Gamma_r$ be the set of all s,t-chains in G, and let $h_1, \ldots, h_r$ be an assignment of flow for each of the r s,t-chains. This flow is valid if for each edge e in G, the sum of the chain flows on chains passing through e is less than or equal to the capacity of e, and the value of the flow is $\sum_{k=1}^{r} h_k$. Consider any *fixed* chain flow $h_1, \ldots, h_r$ which is valid for the normal set of capacities (c_e: $e \in E$). In the random model, a particular chain Γ_k can therefore provide the requested portion h_k of flow if and only if all of its edges are operating, and provides zero flow otherwise. The (marginal) probability of this occurring is therefore $\gamma_k = \prod_{e\in\Gamma_k} p_e$. First, the value of $\mathrm{Ex}[\Phi_{\mathrm{FLOW}}]$ is clearly at least as great as the expected value of the random flow obtained by allowing the flow along each chain Γ_k to be at most h_k regardless of the operating condition of the other chains sharing edges with Γ_k — 'in the absence of rerouting', as phrased in Carey & Hendrickson [1984]. This expectation in turn equals the sum of the expected flow values on each of the chains $\Gamma_1, \ldots, \Gamma_r$ taken as *independent* random variables. Summarizing,

$$\mathrm{Ex}[\Phi_{\mathrm{FLOW}}] \geq \sum_{k=1}^{r} h_k \gamma_k.$$

Aneja & Nair [1980] and Carey & Hendrickson [1984] give heuristics for the problem of finding a chain flow $h_1, \ldots, h_r$ which maximizes the right-hand side of this expression, in order to provide the best lower bound of this type, and Nagamochi & Ibaraki [1991] give conditions under which this bound is tight. The latter paper also gives a polynomial algorithm for finding the maximizing chain flow under these conditions.

The methods of edge-packing and noncrossing cuts provide effective techniques for bounding in the multistate network problems as well [Spelde, 1977]. Specifically, let $P_1, \ldots, P_q$ and $\Gamma_1, \ldots, \Gamma_r$ be a collection of disjoint s,t-paths and s,t-cuts, respectively. These two collections provide natural *upper* and *lower bounding* functions Φ^L and Φ^U for the actual function Φ depending on the problem:

Shortest path:

$$\begin{aligned}\Phi^U_{\text{PATH}} &= \text{the length of the shortest of the paths } P_1, \ldots, P_r \\ &= \min_{i=1,\ldots,r} \sum_{e \in P_i} d_e \\ \Phi^L_{\text{PATH}} &= \text{the sum of the lengths of the shortest edges from} \\ &\quad\ \text{each of the cuts } \Gamma_1, \ldots, \Gamma_r \\ &= \sum_{i=1}^{r} \min_{e \in \Gamma_i} d_e\end{aligned}$$

Maximum flow:

$$\begin{aligned}\Phi^U_{\text{FLOW}} &= \text{the minimum capacity of the cuts } \Gamma_1, \ldots, \Gamma_r \\ &= \min_{i=1,\ldots,r} \sum_{e \in \Gamma_i} c_e \\ \Phi^L_{\text{FLOW}} &= \text{the maximum flow through the set of paths } P_1, \ldots, P_r \\ &= \sum_{i=1}^{r} \min_{e \in P_i} c_e\end{aligned}$$

PERT network performance:

$$\begin{aligned}\Phi^U_{\text{PERT}} &= \text{the sum of the completion times of longest edges from} \\ &\quad\ \text{each of the cuts } \Gamma_1, \ldots, \Gamma_r \\ &= \sum_{i=1}^{r} \max_{e \in \Gamma_i} t_e \\ \Phi^L_{\text{PERT}} &= \text{the completion time of the longest of the paths } P_1, \ldots, P_r \\ &= \max_{i=1,\ldots,r} \sum_{e \in P_i} t_e\end{aligned}$$

These functional bounds in turn provide natural bounds for both the cdf and the expectation of the actual function Φ, in particular,

$$\Pr[\Phi^U \leq \alpha] \leq \Pr[\Phi \leq \alpha] \leq \Pr[\Phi^L \leq \alpha]$$

and

$$\text{Ex}[\Phi^L] \leq \text{Ex}[\Phi] \leq \text{Ex}[\Phi^U].$$

Further, each of the functions described above consists of one min/max and one convolution, and so the cdf's and expectations for all of these functions can be computed in polynomial time.

Two of the miscellaneous bounding techniques have been studied in the context of multistate problems as well. Prékopa & Boros [1991] have extended the Bonferroni bounds to the maximum flow problem, and Nádas [1979] has applied the Hailperin technique to PERT problems. We do not go into the details.

Bounding techniques using shellability and polyhedral combinatorics have been difficult to extend to multistate problems, due to their strongly combinatorial nature. One limited avenue of extension was investigated by Provan [1986]. In that paper bounds were found for reliability in the context where the components are represented by variables $y_1, \ldots, y_n$ constrained by the system

$$\begin{aligned} Ay &= b \\ y &\geq 0 \end{aligned} \tag{5}$$

where A is an $m \times n$ matrix and b an m-vector. The 'failure' of component i corresponds to the variable y_i being removed from the system (or equivalently, set to zero) and the system operates when the remaining variables are sufficient to satisfy the system (5). The system value functions Φ_{PATH}, Φ_{FLOW}, and Φ_{PERT}, when in addition edge parameters have only two states, can all be represented in terms of a linear system in the form (5). Unfortunately, there is an additional restriction that the system (5) be *nondegenerate*, that is, all solutions to (5) have at least m nonzero components. The three problems given here do not have this property. Many variations of these problems do have representations corresponding to a nondegenerate linear system, such as requiring specified shortest path lengths or flow values from a source to *every* point, or more generally having flow satisfy a set of 'nondegenerate' supplies or demands at each node of the network. Furthermore by *perturbing* the linear system representing a degenerate problem, one arrives at a nondegenerate problem which provides a lower bound on the actual reliability, and hence the lower bound techniques of Section 4.1 apply.

6.7. Monte Carlo methods

Multistate problems, and particularly PERT problems, have always been prime candidates for Monte Carlo methods. As with the deterministic schemes, many of the Monte Carlo schemes given in Section 5 can be extended to multistate problems. In the interest of brevity, we only touch upon the major multistate Monte Carlo sampling schemes. For a general account of Monte Carlo schemes the reader is again referred to Hammersley & Handscomb [1964], and for applications to the PERT problem the accounts in Elmaghraby [1977] and the two surveys [Adlakha & Kulkarni, 1989; and Elmaghraby, 1989a] for details and additional information.

Monte Carlo schemes for multistate problems deal almost exclusively with the PERT problem. The earliest Monte Carlo treatment of PERT problems seems to be by Van Slyke [1963], who extends the crude sampling technique of Section 5.1 to multistate problems. Specifically, suppose that an estimate for, say, the sample mean $\mathrm{Ex}[\Phi]$ is desired. For each edge e let $F_e(\alpha) = \Pr[X_e \leq \alpha]$ be the cdf for the random variable X_e. The sample value $\hat{x}_e$ for this random variable is chosen by drawing a sample $\hat{U}_e$ from a uniform random number generator, and setting $\hat{x}_e = F_e^{-1}(\hat{U}_e)$ (or $\min\{x \mid F_e(x) \geq U_e\}$). After the entire sample state vector $\hat{x}$ is generated, $\Phi(\hat{x})$ is computed using the appropriate deterministic algorithm. The average over all sample system values is then an unbiased estimator for Ex[Φ].

Early work on variance reduction for the naive Monte Carlo method concentrated on applying classical variance-reduction techniques. Burt, Gaver & Perlas [1970], and later Burt & Garman [1971a] investigate the improvement to the PERT problem gained by using classical techniques such as antithetic variates, control variates, stratified sampling, regression, and conditional sampling. Again, as these are primarily probabilistic rather than network methods we refer the reader to Hammersley & Handscomb [1964] for details.

One of the most frequently used network techniques for solving multistate problems is based on the conditional sampling method. The idea is to determine a 'small' set of edges such that if the edge parameters on this set of edges is fixed, then the conditional system probability or expectation can be determined analytically. One then samples only from this small set, and the resulting conditional probabilities or expectations are averaged to produce the overall sample system measure. Burt & Garman [1971b] suggest conditioning on the *common edges* of the network, that is, edges which share two or more s, t-paths. In acyclic networks these can be found efficiently, and after fixing these lengths, the remaining network can be analyzed as if it were a collection of disjoint paths (see Section 5). Unfortunately, in a reasonably complex graph all, or nearly all, of the edges in the graph are common, and so this does not lead to a significant improvement in the sampling. Sigal, Pritsker, and Solberg [Sigal, 1977; Sigal, Pritsker, and Solberg, 1980a, b] suggest the use of *uniformly directed* s, t*-cuts* in a conditional sampling scheme. A uniformly directed s, t-cuts is an edge-set C having the property that each s, t-path in G intersects C in *exactly* one edge. (As a technical point, these should be called *exact* s, t-cuts, although in PERT networks the definition given by Sigal and coworkers is equivalent to that given here. See Provan & Kulkarni [1989] for the precise distinction between these two terms.) The importance of conditioning on the edges of a uniformly directed s, t-cut is that this allows the activity of the s, t-paths to be analyzed independently on each side of the cut. As well, every graph always contains at least one uniformly directed s, t-cut. Kulkarni & Provan [1985, 1989] show how the conditional system measure can be found efficiently after the non-cut edges have been sampled, and also how the maximum cardinality uniformly directed s, t-cut can be found to use in such a sampling scheme. Fishman has combined the use of uniformly directed s, t-cuts and *quasi-random sampling* to improve the method of Sigal and coworkers still further. Additional work along this line is found in Adlakha [1986, 1987].

Most of the Monte Carlo techniques given above also apply to the shortest path problem as well, and generally do not require that the underlying graph be acyclic as do many of the bounding techniques for the PERT problem. Fishman and colleagues [Alexopoulos & Fishman, 1991; Fishman, 1987b 1989b; Fishman & Shaw, 1989] seems to be the only group to explicitly address the maximum flow problem using Monte Carlo methods. The approach uses the multistate extension of the bounds-based sampling method of Section 5.4 to compute a cdf value

$\Pr[\Phi \le \alpha]$ for threshold value α. We give it in its general form, for the technique applies just as easily to both the PERT and shortest path problems. As above, for $e \in E$ let the edge parameter X_e have cdf $F_e(x)$. Suppose that one has functional bounds Φ^L and Φ^U for the multivariate measure Φ, satisfying

- $\Phi^L(x) \le \Phi(x) \le \Phi^U(x)$ for every state vector x.
- For $k = 0, \dots, m$, any assignment $\hat{x}^{(k)} = (\hat{x}_1, \dots, \hat{x}_k)$ of values for the first k components of x, the conditional cdf values

$$R_k^L(x^{(k)}) \equiv \Pr[\Phi^L \le \alpha |\, x_1 = \hat{x}_1, \dots, x_k = \hat{x}_k]$$

and

$$R_u^U(x^{(k)}) \equiv \Pr[\Phi^U \le \alpha |\, x_1 = \hat{x}_1, \dots, x_k = \hat{x}_k]$$

can be computed in polynomial time.

The space $\mathcal{X}$ of importance in the multivariate version of the problem is now

$$\mathcal{X} = \{x \in \{0,1\}^E : \Phi^L(x) \le \alpha,\ \Phi^U(x) > \alpha\}$$

and the modification to the Bounds-based Sampling Method of Section 5.4 becomes

1. Take samples $\hat{x} = (\hat{x}_1, \dots, \hat{x}_m)$ from the space $\mathcal{X}$ by successively drawing, for $k = 1, \dots, m$, the component state $\hat{x}_k$ from the cdf

$$\begin{aligned}\Pr[x_k \le \beta |\, x_1 &= \hat{x}_1, \dots, x_{k-1} = \hat{x}_{k-1} \text{ and } \Phi^L(x) \le \alpha,\ \Phi^U(x) > \alpha] \\ &= \left[\frac{R_k^L(x^{(k-1)}, \beta) - R_k^U(x^{(k-1)}, \beta)}{R_{k-1}^L(x^{(k-1)}) - R_{k-1}^U(x^{(k-1)})}\right] F_{e_k}(\beta).\end{aligned}$$

2. Compute the proportion $\hat{R}$ of those samples for which $\Phi(x) \le \alpha$. The number $R_0^U + \hat{R}(R_0^L - R_0^U)$ is now an unbiased estimator of R.

The edge-packing and noncrossing cuts functional bounds provide excellent bounding functions for this method in the multistate context as well, since the conditional cdf's for these functions can be computed easily as well. Although in the papers by Fishman *et al.* these bounds are used only for the maximum flow problems, they apply as well to PERT and shortest path problems, essentially as stated here.

7. Using computational techniques in practice

After reading the previous six sections it would be understandable if one were confused in deciding how to apply the myriad of reliability measures, algorithms, bounds, and so on to the solution of real problems. In this final section we provide some guidance on this issue.

7.1. *Where does reliability fit in?*

Although this entire chapter has been devoted to reliability it would be inaccurate to give the impression that reliability is the only criterion of interest, or even the most important criterion in the design of most networks. In fact, there are typically several competing interests, including cost, overall system capacity or throughput, and various performance criteria [Frank, Kahn & Kleinrock, 1972]. The most typical scenario encountered during a network design session is:

Minimize	cost
subject to:	throughput constraint(s) performance constraint(s) reliability constraint(s)

The throughput constraints typically state that the network must have capacity sufficient to support traffic requirements specified for a set of origin/destination pairs. The predominant performance measure for packet switched networks is some measure of message or packet delay [Gavish & Altinkemer, 1990] and the predominant performance measure for circuit switched networks is some measure of lost or blocked calls [Sanso, Soumis & Gendreau, 1990]. Ideally the performance and reliability constraints would restrict an accurate expression for performance and reliability, respectively, to be within certain limits. However, as stated in the introduction, since computing the values of performance and reliability measures are typically very difficult, surrogates are usually employed. A typical surrogate for delay is a path length restriction and a typical surrogate for reliability is a connectivity restriction. Even if exact measures of performance and reliability were included as constraints, the model given above would contain approximations since throughput, performance and reliability are treated as separate constraints. Ideally, one would like a design that satisfied certain throughput and performance criteria even in the presence of failures. All of these shortcomings lead to the use of detailed performance and reliability analysis algorithms. That is, once an initial design is obtained it is typically refined, either manually or automatically, based on the results of performance and reliability analysis.

The issue of whether specific performance or reliability constraints are required and, of whether performance and reliability analysis algorithms are required, depends on the specifics of the problem setting. For example, if networks designed based on only cost and performance criteria naturally were very reliable, then reliability analysis and reliability constraints might be unnecessary. This scenario might occur under any of the following circumstances:

1. the network components were themselves highly reliable;
2. occasional network failures were not particularly disruptive; and
3. networks designed based on other (non-reliability) criteria tended to be very dense and consequently had high reliability (this might occur if the link capacities were small relative to the overall throughput required).

An analogous set of statements could be made relative to performance and throughput criteria. For example, the models described in Chapter 10 employ reliability constraints but not delay or throughput constraints because the fiber optic links used have very high capacity (and speed) so that capacity and delay criteria are met without the need of specific model consideration. On the other hand, since the link capacities are so high network designs naturally tend to be very sparse so that explicit reliability constraints are necessary.

7.2. Choosing a measure

We have given detailed treatment to several different reliability measures and, in fact, there are still other measures which we have not mentioned and have given only cursory treatment to. One might be left with the dilemma of which measure to choose. The following considerations are fundamental to this decision:

1. significance and nature of performance criteria;
2. community of users served; and
3. philosophy of service: good on the average or good at the extremes.

A fundamental decision is whether to use a connectivity measure or a performability measure. Connectivity measures include all versions of the k-terminal measure as well as certain related measures such as network resilience, the expected fraction of node pairs communicating [Ball, 1980; Colbourn, 1987]. Performability measures explicitly model variations in performance caused by network failures. A connectivity measure would be useful whenever the network performance was considered to be satisfactory as long as the network was connected. This is considered to be the case in many fiber optic networks. Connectivity measures are also useful in measuring the probability that the network is able to provide some minimum level of service, i.e. the probability that the network can handle essential or emergency calls. The study of performability measures was motivated by network settings where it is possible that component failures could degrade system performance to unsatisfactory levels while, nonetheless, the network remained connected. In such cases the use of performability measures is necessary.

The choice among the two-terminal, k-terminal and all-terminal depends on the community of users of interest. The two-terminal measure measures the ability of the network to satisfy the communications needs of a specific pair of user terminals. Thus, the measure can be viewed as a user specific measure. On the other hand the all-terminal measure takes a system provider perspective. It measures the performance of the system relative to its ability to provide service all possible terminal pairs. In a certain sense, the all-terminal measure is extremely 'conservative'. Specifically, the all-terminal measures is smaller than the smallest two-terminal value and, generally, could be much less.

Another system-wide reliability measure is the minimum over all two-terminal values. This can be interpreted as the reliability level guaranteed by the network to all users. Another related value is the average over all two-terminal values, which equals the resilience. The choice between the minimum and average relates to the overall philosophy of the service provider; specifically, is the objective to provide satisfactory service 'on the average' or to guarantee a certain service level?

The general k-terminal measure addresses the interests of a community of users located at a subset of network nodes. Depending on the node subset chosen it could provide system wide information or user specific information.

Performability measures are always defined in terms of some performance measure, Φ. Thus, the starting point for any performability analysis is the choice of an appropriate performance measure. Certain important policy considerations are embodied in the choice of the performance measure itself. Suppose that call blocking probability was the 'general' measure of interest. Similar to the choice that existed for connectivity measures, we have a choice between using i.) the overall network call blocking probability and ii.) the minimum, taken over all node pairs, of the call blocking probability experienced for calls between node pair $\{i, j\}$. A similar tradeoff would exist for measures of network delay. Once Φ had been determined one must choose between using Ex[Φ] or Pr[$\Phi \leq$ or $\geq \alpha$]. The above two choices both reflect the philosophy of service consideration mentioned earlier. Let us compare the two extremes. On the one hand, there is the 'double average' case where we use definition i.) for Φ and Ex[Φ] as the performability measure. If the value this measure was 0.02, then one could interpret this value as, 'On the average 2% of the calls attempted in the network are blocked'. At the other extreme, we could use definition ii.) for Φ and Pr[$\Phi \leq 0.01$] as the performability measure. If the value of this measure was 0.98, then one could interpret this value as, '98% of the time, each user of the network will be able to complete at least 99% of the calls attempted'. The fact the Φ is defined as the minimum blocking probability over all node pairs implies that the value of the measure is a service guarantee for 'each user of the network'. As was mentioned before the choice between these two measures would depend on the network provider's philosophy of service. Alternatively the network designer might be interested in both measures. The additional computational burden to compute both measures rather than one would be relatively small, when using a most probable states approach, since the computational bottleneck is the evaluation of Φ for all states enumerated.

7.3. Choosing the right algorithm

In addition to presenting the reader with a variety of reliability measures, this chapter has also provided numerous choices for computing each of the measures mentioned, including exact algorithms, analytic bounds and Monte Carlo simulation. Thus, one might be left with a sense of confusion in this area as well. The choice among algorithms depends largely on the size and structure of the networks involved. Ideally one would like to employ an algorithm that gives the exact reliability value. However, efficient (polynomially bounded) algorithms are only available for certain structured classes of graphs as are described in Section 3. For general graphs enumerative algorithms can only solve problems of limited size. For other situations approximate algorithms must be employed. The choice between analytic bounds and Monte Carlo is a more subtle one. Analytic bounds depend on specific problem results. Consequently, as described in Section 4, such bounds only exist for certain problem classes. Furthermore, their quality and the associated algorithm running time may differ depending on the specific problem class in question. In certain cases, the bounds available are very good and they can be computed very quickly. However, in other cases this is not true. The advantages of Monte Carlo are that some Monte Carlo approaches can be constructed for virtually all measures of interest and a Monte Carlo algorithm can be run for a long or short period of time with a commensurate increase or decrease in the level of accuracy. Some of the more advanced Monte Carlo methods make use of problem structure and, consequently, do not have the first advantage.

One important issue to point out relative to analytic bounds is whether unequal failure probabilities are allowed. Some bounding methods bound the reliability polynomial and as a result only give information for the case where all failure probabilities are equal. On the other hand, it should be pointed out that the reliability polynomial provides information over the entire range of values for that one failure probability.

A related issue that should be considered is whether the directed or undirected graph model is used. As was pointed out in Section 2, the directed graph model is more general in that for a wide class of measures it allows for the easy modeling of both directed and undirected problems and of problems with and without node failures.

For the more complex performability measures, e.g. expected lost call traffic and expected message delay, the only viable approach is some variant of the most probable states method. Other techniques given in Section 6, apply to some of the simpler performability measures. The relative sparsity of techniques available for performability analysis indicate that this is a fertile domain for future work in reliability analysis. Furthermore, we feel that many of the powerful concepts developed in the study of connectivity measures should be of use in further research on performability analysis.

7.4. Design criteria

The design criterion most commonly used as a surrogate for a constraint on all-terminal reliability is a connectivity constraint, that is, a constraint that the network connectivity must be at least c. The generalization of this criterion to SCBSs (and to other classes of systems, including to k-terminal problems and certain performability measures) is a constraint that the SCBS contain no cutset of size $c-1$ or smaller. Probably the most common scenario is the case where $c=2$. With $c=2$, the design criterion states that the system should contain no single point of failure.

Let us examine the reliability level guaranteed by this design criterion. If there are no cutsets of size $c-1$ or smaller then a lower bound on the system reliability would be obtained by assuming that every set of elements of size c was a cutset. For an m-element SCBS the reliability of this system would be equal to the reliability of a K-out-of-N system with $\mathrm{K}=m-c+1$ and $\mathrm{N}=m$. Assuming equal component failure probabilities the system reliability would be:

$$\sum_{i=m-c+1}^{m} \binom{n}{i} p^i (1-p)^{m-i}$$

This value is the best lower bound possible on the system reliability level guaranteed by a c-connectivity constraint, assuming no additional system structure is known. For $c = 2$, this bound is achieved for the case of all-terminal reliability by a simple cycle.

It is usually very worthwhile to compute this bound and to take it into account during network design. It is often the case that the value of the bound is lower than expected. Then either the design criterion (the value of c) must be increased or a more detailed reliability analysis must be carried out together with possible design modifications. Once more complex measures of reliability are warranted in network design, the task of the network designer must not be simply to obtain a numerical measure of reliability, but rather to gain insight into the impact of the underlying network topology on the network's ability to perform its desired functions [Colbourn, 1991; Colbourn & Nel, 1990]. Ultimately, the large number of techniques developed here for producing numerical measures of reliability contribute primarily not by giving algorithms to produce numbers, but rather by providing tools for capturing in part how network structure affects network performance.

Acknowledgements

Research of the first author was supported by NSF grant No. CDR-8803012 and was carried out, in part, while the first author was visiting the Department of Operations Research at the University of North Carolina at Chapel Hill. Research of the second author was supported by NSERC Canada under grant A0579. Research of the third author was supported by NSF grant No. CCR-9200572. The authors would like to thank Christos Alexopoulos for providing many valuable comments.

References

AboElFotoh, H.M.F., and C.J. Colbourn (1989a). Series-parallel bounds for the two-terminal reliability problem. *ORSA J. Comp.* 1, 205–222.

AboElFotoh, H.M.F., and C.J. Colbourn (1989b). Computing the two-terminal reliability for radio broadcast networks. *IEEE Trans. Reliab.* R-38, 538–555.

AboElFotoh, H.M.F., and C.J. Colbourn (1990). Efficient algorithms for computing the reliability of permutation and interval graphs. *Networks* 20, 883–898.

Abraham, J.A. (1979). An improved algorithm for network reliability. *IEEE Trans. Reliab.* R-28, 58–61.

Adlakha, V.G. (1986). An improved conditional Monte Carlo technique for the stochastic shortest path problem. *Managenent Sci.* 32, 1360–1367.

Adlakha, V.G. (1987). A Monte Carlo technique with quasirandom points for the stochastic shortest path problem. *Am. J. Math. Managenent Sci.* 7, 325 358.

Adlakha, V.G., and V.G. Kulkarni (1989). A classified bibliography of research on PERT networks. *INFOR* 27, 272–296.

Aggarwal, K.K. (1985). Integration of reliability and capacity in performance measure of a telecommunication network. *IEEE Trans. Reliab.* R-34, 184–186.

Aggarwal, K.K., Y.C. Chopra and J.S. Bajwa (1982). Capacity consideration in reliability analysis of communication systems. *IEEE Trans. Reliab.* R-31, 177–181.

Aggarwal, K.K., K.B. Misra and J.S. Gupta (1975). A fast algorithm for reliability evaluation. *IEEE Trans. Reliab.* R-24, 83–85.

Aggarwal, K.K., and S. Rai (1981). Reliability evaluation in computer-communication networks. *IEEE Trans. Reliab.* R-30, 32–35.

Agrawal, A., and R.E. Barlow (1984). A survey of network reliability and domination theory. *Oper. Res.* 32, 478–492.

Agrawal, A., and A. Satyanarayana (1984). An $O(|E|)$, time algorithm for computing the reliability of a class of directed networks. *Oper. Res.* 32, 493–515.

Agrawal, A., and A. Satyanarayana (1985). Network reliability analysis using 2-connected digraph reductions. *Networks* 15, 239–256.

Agrawal, D.P. (1983). Graph theoretical analysis and design of multistage interconnection networks. *IEEE Trans. Comput.* C-32, 637–648.

Ahlswede, R., and D.E. Daykin (1978). An inequality for the weights of two families of sets, their unions, and their intersections. *Z. Wahrscheinl. Geb.* 43, 183–185.

Aho, A.V., J.E. Hopcroft and J.D. Ullman (1974). *The Design and Analysis of Computer Algorithms,* Addison-Wesley.

Aldous, D.J. (1990). The random walk construction of uniform spanning trees and uniform labeled trees. *SIAM J. Discr. Math.* 3, 450–465.

Alexopoulos, C. (1988). Maximum flows and critical cutsets in stochastic networks with discrete arc capacities, Ph.D. thesis, University of North Carolina at Chapel Hill, N.C.

Alexopoulos, C. (1993). State space partitioning methods for stochastic stortest path problems, Tech. Rep. J-93-01, School of Industrial and Systems Engineering, Georgia Institute of Technology.

Alexopoulos, C., and G.S. Fishman (1991). Characterizing stochastic flow networks using the Monte Carlo method. *Networks* 21, 775–798.

Alexopoulos, C., and G.S. Fishman (1993). Sensitivity analysis in stochastic flow networks using the Monte Carlo method. *Networks* 23, 605–621.

Andreatta, G. (1986). Shortest path models in stochastic networks. *Stochastics Comb. Optimization,* 178–186.

Aneja, Y.P., and K.P.K. Nair (1980). Maximum expected flow in a network subject to arc failures. *Networks* 10, 45–57.

Arnborg, S., and A. Proskurowski (1989). Linear time algorithms for NP-hard problems restricted to k-trees. *Discr. Appl. Math.* 23, 11–24.

Assous, J.Y. (1986). First and second order bounds for terminal reliability. *Networks* 16, 319–329.

Ayanoglu, E., and C.-L. I (1989). A method of computing the coefficients of the network reliability polynomial. *Proc. Globecom89,* pp. 331–337.

Bailey, M.P. (1991). Constant access systems: a general framework for greedy optimization on stochastic networks, Technical Report, Department of Operations Research, U.S. Naval Postgraduate School, Monerey, CA.

Bailey, M.P., and V.G. Kulkarni (1986). A recursive algorithm for computing exact reliability measures. *IEEE Trans. Reliab.* R-35, 36–40.

Ball, M.O. (1979). Computing network reliability. *Oper. Res.* 27, 823–836.

Ball, M.O. (1980). Complexity of network reliability computations. *Networks* 10, 153–165.

Ball, M.O., and G.L. Nemhauser (1979). Matroids and a reliability analysis problem, Math. *Oper. Res.* 4, 132–143.

Ball, M.O., and J.S. Provan (1982). Bounds on the reliability polynomial for shellable independence systems. *SIAM J. Algebraic Discr. Methods* 3, 166–181.

Ball, M.O., and J.S. Provan (1983). Calculating bounds on reachability and connectedness in stochastic networks. *Networks* 13, 253–278.

Ball, M.O., and J.S. Provan (1985). Properties of systems which lead to efficient computation of reliability. *Proc. IEEE Global Telecommunications Conference,* pp. 866–870.

Ball, M.O., and J.S. Provan (1987). Computing k-terminal reliability in time polynomial in the number of (s, K)-quasicuts. *Proc. 4th Army Conf. on Applied Mathematics and Computing,* pp. 901–907.

Ball, M.O., and J.S. Provan (1988). Disjoint products and efficient computation of reliability. *Oper. Res.* 36, 703–716.

Ball, M.O., J.N. Hagstrom and J.S. Provan (1994). Threshold reliability of networks with small failure sets, *Networks,* to appear.

Ball, M., and F. Lin (1993). A reliability model applied to emergency service vehicle location. *Oper. Res.* 41, 18–36.

Ball, M.O., J.S. Provan and D.R. Shier (1991). Reliability covering problems. *Networks* 21, 345–358.

Ball, M.O., and R.M. Van Slyke (1977). Backtracking algorithms for network reliability analysis. *Ann. Discr. Math.* 1, 49–64

Barlow, R.E. (1982). Set theoretic signed domination for coherent structures, Report ORC 82-1, Univ. California, Berkeley, Calif.

Barlow, R.E., and S. Iyer (1988). Computational complexity of coherent systems and the reliability polynomial. *Prob. Engrg. Inf. Sci.* 2, 461–469.

Barlow, R.E., and F. Proschan (1981). *Statistical Theory of Reliability and Life Testing,* To Begin With, Silver Spring, Md.

Bauer, D., F. Boesch, C. Suffel and R. Tindell (1982). Combinatorial optimization problems in the analysis and design of probabilistic networks, Stevens Research Report in Mathematics 8202, Stevens Institute of Technology, Hoboken, N.J.

Beichelt, F., and L. Spross (1989) Bounds on the reliability of binary coherent systems. *IEEE Trans. Reliab.* R-38, 425–427.

Beichelt, F., and P. Tittmann (1991). A generalized reduction method for the connectedness probability of stochastic networks. *IEEE Trans. Reliab.* R-40, 198–204.

Bein, W.W., J. Kamburowski and M.F.M. Stallmann (1992). Optimal reduction of directed acyclic graphs. *SIAM J. Comput.* 21, 1112–1129.

Belovich, S.G., and V.K. Konangi (1991). A linear-time approximation method for computing the reliability of a network. *Comput. Networks ISDN Syst.* 21, 121–127.

Ben-Dov, Y. (1981). Optimal testing procedures for special structures of coherent systems. *Managenent Sci.* 27, 1410–1420.

Bibelnieks, E., J.P. Jarvis, R.J. Lakin and D.R. Shier (1990). Algorithms for approximating the performance of multimode systems. *Proc. INFOCOM90,* pp. 741–748.

Bienstock, D. (1986). An algorithm for reliability analysis of planar graphs. *Networks* 16, 411–422.

Bienstock, D. (1988). Some lattice-theoretic tools for network reliability analysis, Math. *Oper. Res.* 13, 467–478.

Bienstock, D. (1988). Asymptotic analysis of some network reliability models. *SIAM J. Discr. Math.* 1, 14–21.

Biggs, N.L. (1974). *Algebraic Graph Theory,* Cambridge University Press.

Billera, L.J. (1977). Polyhedral theory and commutative algebra, in: A. Bachem, M. Grötschel and B. Korte (eds.), *Mathematical Programming: The State Of The Art,* Springer-Verlag, pp. 57–77.

Billera, L.J., and C.W. Lee (1981). The number of faces of polytope pairs and unbounded polyhedra. *Eur. J. Comb.* 2, 307–332.

Birnbaum, Z.W., and J.D. Esary (1965) Modules of coherent binary systems. *J. SIAM* 13, 444-462.

Birnbaum, Z.W., J.D. Esary and S.C. Saunders (1961). Multi-component systems and structures and their reliability. *Technometrics* 3, 55–77.

Bixby, R. (1975). The minimum number of edges and vertices in a graph with edge-connectivity N and MN-bonds. *Networks* 5, 259–298.

Blake, J.T., and K.S. Trivedi (1989a). Multistage interconnection network reliability. *IEEE Trans. Comput.* C-38, 1600–1603.

Blake, J.T., and K.S. Trivedi (1989b). Reliability analysis of interconnection networks using hierarchical composition. *IEEE Trans. Reliab.* R-38, 111–119.

Bodin, L. (1970) Approximation to system reliability using a modular decomposition. *Technometrics* 12, 335.

Boesch, F.T. (1986). *Synthesis of reliable networks — a survey*. *IEEE Trans. Reliab.* R-35, 240–246.

Boesch, F.T., A. Satyanarayana and C.L. Suffel (1990). Some alternate characterizations of reliability domination. *Prob. Engrg. Inf. Sci.* 4, 257–276.

Boesch, F.T., A. Satyanarayana and C.L. Suffel (1990). Least reliable networks and the reliability domination. *IEEE Trans. Commun.* 38, 2004–2009.

Bose, R., M. Taka and C. Hamilton, eds. (1993). Dependability of Network Services. *IEEE Commun. Mag.*, Special Issue, 31.

Botting, C., S. Rai and D.P. Agrawal (1989). Reliability computation of multistage interconnection networks. *IEEE Trans. Reliab.* R-38, 138–145.

Brecht, T.B., and C.J. Colbourn (1988). Lower bounds for two-terminal network reliability. *Discr. Appl. Math.* 21, 185-198.

Brecht, T.B., and C.J. Colbourn (1986). Improving reliability bounds in computer networks. *Networks* 16, 369–380.

Brecht, T.B., and C.J. Colbourn (1989). Multiplicative improvements in reliability bounds. *Networks* 19, 521–530.

Broder, A. (1989). Generating random spanning trees. *Proc. Symp. Foundations of Computer Science*, IEEE, pp. 442–447.

Brooks, R.L., C.A.B. Smith, A.H. Stone and W.T. Tutte (1940). The dissection of rectangles into squares. *Duke Math. J.* 7, 312–340.

Brown, J.I., and C.J. Colbourn (1988). A set system polynomial with reliability and coloring applications. *SIAM J. Discr. Math.* 1, 151–157.

Brown, J.I., and C.J. Colbourn (1994). Log concavity and the reliability polynomial, to appear.

Brown, J.I., and C.J. Colbourn (1992). Roots of the reliability polynomial. *SIAM J. Discr. Math.* 5, 571–585.

Brown, J.I., C.J. Colbourn and J.S. Devitt (1993). Network transformations and bounding network reliability. *Networks* 23, 1–17.

Brown, D.B. (1971). A computerized algorithm for determining the reliability of redundant configurations. *IEEE Trans. Reliab.* R-20, 121–124.

Bukowski, J.V. (1982). On the determination of large scale system reliability. *IEEE Trans. Syst. Man Cybern.* SMC-12, 538–548.

Burt, J.M., Jr., D.P. Gaver and M. Perlas (1970). Simple stochastic networks: some problems and procedures. *Nav. Res. Logist. Q.* 17, 439–460.

Burt, J.M., Jr. and M.B. Garman (1971). Monte Carlo techniques for stochastic PERT network analysis. *INFOR* 9, 248–262.

Burt, J.M., Jr. and M.B. Garman (1971). Conditional Monte Carlo: A simulation technique for stochastic network analysis. *Managenent Sci.* 18, 207–217.

Busacker, R.G., and P.J. Gowen (1961). A procedure for determining a family of minimal cost flow patterns, ORO Research Report 15, Johns Hopkins University.

Buzacott, J.A. (1980). A recursive algorithm for finding reliability measures related to the connection of nodes in a graph. *Networks* 10, 311–327.

Buzacott, J.A. (1983). The ordering of terms in cut-based recursive disjoint products. *IEEE Trans. Reliab.* R-32, 472–474.

Buzacott, J.A. (1987). Node partition formula for directed graph reliability. *Networks* 17, 227–240.

Buzacott, J.A., and J.S.K. Chang (1984). Cut-set intersections and node partitions. *IEEE Trans. Reliab.* R-33, 385–389.

Caccetta, L., and M. Kraetzl (1991). Blocking probabilities of certain classes of channel graphs, in: V.R. Kulli (ed.), *Advances in Graph Theory*, Vishwa International, Delhi, pp. 70–103.

Carey, M., and C. Hendrickson (1984). Bounds on expected performance with links subject to failure. *Networks* 14, 439–456.

Carrasco, E.H., and C.J. Colbourn (1986). Reliability bounds for networks with statistical dependence. *Proc. INFOCOM86*, Miami, pp. 290–292.

Chari, M.K. (1993). Steiner complexes, matroid ports, and K-connectedness reliability. *J. Comb. Theor. (B)* 59, 41–68.

Chari, M.K., T.A. Feo and J.S. Provan (1990). A computational study of wye-delta bounds for (s, t)-connectedness reliability, preprint, University of North Carolina.

Charnes, A., W.W. Cooper and G.L. Thompson (1964). Critical path analysis via chance constrained and stochastic programming. *Oper. Res.* 12, 460–470.

Chiou, S.-N., and V.O.K. Li (1986). Reliability analysis of a communication network with multimode components. *IEEE Trans. Sel. Areas Commun.* SAC-4, 1156–1161.

Clausen, J., and L.A. Hansen (1980). Finding k edge-disjoint spanning trees of minimum total weight in a network: an application of matroid theory. *Math. Program. Study* 13, 88–101.

Clements, G.F., and B. Lindström (1968). A generalization of a combinatorial theorem of Macaulay. *J. Comb. Theor.* 7, 230–238.

Clingen, C.T. (1964). A modification of Fulkerson's PERT algorithm. *Oper. Res.* 12, 629–632.

Colbourn, C.J. (1987). *The Combinatorics of Network Reliability*, Oxford University Press, Oxford, New York.

Colbourn, C.J. (1987). Network resilience. *SIAM J. Algebraic Discr. Methods* 8, 404–409.

Colbourn, C.J. (1988). Edge-packings of graphs and network reliability. *Discr. Math.* 72, 49-61.

Colbourn, C.J. (1991). Network reliability: numbers or insight?. *Ann. Oper. Res.* 33, 87–93.

Colbourn, C.J. (1992). A note on bounding k-terminal reliability. *Algorithmica* 7, 303–307.

Colbourn, C.J., R.P.J. Day and L.D. Nel (1989). Unranking and ranking spanning trees of a graph. *J. Algorithms* 10, 271–286.

Colbourn, C.J., J.S. Devitt, D.D. Harms and M. Kraetzl (1991). Renormalization for channel graphs. *Acta. XI Congr. de Metodologias en Ingenieria de Sistemas*, Santiago, Chile, pp. 171–174.

Colbourn, C.J., J.S. Devitt, D.D. Harms and M. Kraetzl (1994). Assessing reliability of multistage interconnection networks, to appear.

Colbourn, C.J., M. Elbert, E. Litvak and T. Weyant (1992). Performability analysis of large-scale packet-switching networks. *Proc. Int. Conf. on Communications (SUPERCOMM/ICC)*, Chicago, pp. 416–419.

Colbourn, C.J., and E.S. Elmallah (1993) Reliable assignments of processors to tasks and factoring on matroids. *Discr. Math.* 114, 115–129.

Colbourn, C.J., and D.D. Harms (1988). Bounds for all-terminal reliability in computer networks. *Networks* 18, 1–12.

Colbourn, C.J., D.D. Harms and W.J. Myrvold (1993). Reliability polynomials can cross twice. *J. Franklin Inst.* 330, 629–633.

Colbourn, C.J., and E.I. Litvak (1991). Bounding network reliability by graph transformations, in: F. Roberts, F. Hwarg and C. Monma (eds.), *Reliability of Computer and Communications Networks*, AMS/ACM, pp. 91–104.

Colbourn, C.J., and M.V. Lomonosov (1991). Renewal networks: connectivity and reliability on a time interval. *Probab. Engrg. Inf. Sci.* 5, 361–368.

Colbourn, C.J., and L.D. Nel (1990). Using and abusing bounds for network reliability. *Proc. IEEE Telecommunications Conference (Globecom90)*, IEEE Press, pp. 663–667.

Colbourn, C.J., L.D. Nel, T.B. Boffey and D.F. Yates (1994). Probabilistic estimation of damage from fire spread, to appear.

Colbourn, C.J., J.S. Provan and D.L. Vertigan (1994). The complexity of computing the Tutte polynomial on transversal matroids, to appear.

Colbourn, C.J., and W.R. Pulleyblank (1989). Matroid Steiner problems, the Tutte polynomial and network reliability. *J. Comb. Theor.* B41, 20–31.

Corea, G.A., and V.G. Kulkarni (1990). Minimum cost routing on stochastic networks. *Networks* 18, 527–536.

Cox, D.R. (1955). The use of complex probability in the theory of stochastic processes. *Proc. Camb. Philos. Soc.* 51, 313-319.

Crapo, H.H. (1967). A higher invariant for matroids. *J. Comb. Theor.* 2, 406–417.

Daneshmand, M., and C. Savolaine (1993). Measuring outages in telecommunications switched networks. *IEEE Commun. Mag.* 31, 34–38.

Daykin, D.E. (1974). A simple proof of the Kruskal–Katona theorem. *J. Comb. Theor.* A 17, 252–253.

Dean, P. (1963). A new Monte Carlo method for percolation probabilities on a lattice. *Proc. Camb. Philos. Soc.* 59, 397–410.

deMercado, J., N. Spyratos and B.A. Bowen (1976). A method for calculation of network reliability. *IEEE Trans. Reliab.* R-25, 71–76.

Devitt, J.S., and C.J. Colbourn (1992). On implementing an environment for investigating network reliability, in: *Computer Science and Operations Research*, Pergamon Press, pp. 159–173.

Dirac, G.A. (1966). Short proof of Menger's theorem. *Mathematika* 13, 42–44.

Dodin, B.M. (1985a). Approximating the distribution function in stochastic networks, Computers and *Oper. Res.* 12, 207–223.

Dodin, B.M. (1985b), Bounding the project completion time distribution in PERT networks. *Oper. Res.* 33, 862–881.

Dodin, B.M., and S.E. Elmaghraby (1984). Approximating the criticality indices of the activities in PERT networks. *Oper. Res.* 32, 493–515.

Dotson, W.P., and J.O. Gobien (1979). A new analysis technique for probabilistic graphs. *IEEE Trans. Circuits Syst.* CAS-26, 855–865.

Douilliez, P., and E. Jamoulle (1972). Transportation networks with random arc capacities. *R.A.I.R.O.* 3, 45–49.

Duffin, R.J. (1965). Topology of series-parallel networks. *J. Math. Anal. Appl.* 10, 303–318.

Easton, M.C., and C.K. Wong (1980). Sequential destruction method for Monte Carlo evaluation of system reliability. *IEEE Trans. Reliab.* R-29, 27–32.

Edmonds, J. (1965). Minimum partition of a matroid into independent subsets. *J. Res. Nat. Bur. Stand.* 69B, 67–72.

Edmonds, J. (1967). Optimum branchings. *J. Res. Nat. Bur. Stand.* 71B, 233–240.

Edmonds, J. (1968). Matroid partition, in: G.B. Dantzig and A.F. Viennott (eds.), *Mathematics of the Decision Sciences*, American Mathematical Society, pp. 335–345.

Edmonds, J. (1972). Edge-disjoint branchings, in: R. Rustin (ed.), *Combinatorial Algorithms*, Algorithmics Press, pp. 91–96.

Edmonds, J., and R.M. Karp (1972). Theoretical improvements in algorithmic efficiency for network flow problems. *J. ACM* 19, 248–264.

Elmaghraby, S.E. (1977). *Activity Networks: Project Planning and Control by Network Models*, J. Wiley & Sons.

Elmaghraby, S.E. (1967). On the expected duration of PERT-type networks. *Managenent Sci.* 13, 299–306.

Elmaghraby, S.E. (1989a). The estimation of some network parameters in the PERT model of activity networks: review and critique, in: R. Słowiński and J. Weglarz (eds.), *Advances in Project Scheduling*

Elmaghraby, S.E., J. Kamburowski and M.F.M. Stallmann (1989b). On the reduction of acyclic digraphs and its applications, OR Report 233, Oper. Res. Ser., North Carolina State University, Raleigh, N.C.

El Mallah, E.S., and C.J. Colbourn (1985). Reliability of Δ–Y networks. *Proc. 16th Southeastern Conf. on Combinatorics, Graph Theory, Computing*, pp. 49–54.

Elperin, T., I. Gertsbakh and M.V. Lomonosov (1991). Network reliability estimation using graph evolution models. *IEEE Trans. Reliab.* R-40, 572–581.

Epifanov, G.V. (1966). Reduction of a plane graph to an edge by a star–triangle transformation. *Sov. Math. Dokl.* 166, 13–17.

Erdös, P., and A. Renyi (1959). On random graphs I. *Publ. Math. Debrecen* 6, 290–297.

Erdös, P., and A. Renyi (1960). On the evolution of random graphs. *Magyar Tud. Akad. Mat. Kutató Int. Közl.* 5, 17–61.

Esary, J.D., and F. Proschan (1963). Coherent structures of non-identical components. *Technometrics* 5, 191–209.

Evans, J.R. (1976). Maximum flow in probabilistic graphs — the discrete case. *Networks* 6, 161–183.

Even, S., and R.E. Tarjan (1975). Network flow testing and graph connectivity. *SIAM J. Comput.* 4, 507–518.

Feo, T.A., and R. Johnson (1990) Partial factoring: an efficient algorithm for approximating two-terminal reliability on complete graphs. *IEEE Trans. Reliab.* R-39, 290–295.

Feo, T.A., and J.S. Provan (1993). Delta-wye transformations and the efficient reduction of two-terminal planar graphs. *Oper. Res.* 41, 572–582.

Fisher, D.D., D. Saisi and W.M. Goldstein (1985). Stochastic PERT networks: OP diagrams, critical paths and the project completion time. *Comput. Oper. Res.* 12, 471–482.

Fishman, G.S. (1985). Estimating network characteristics in stochastic activity networks. *Managenent Sci.* 31, 579–593.

Fishman, G.S. (1986a). A comparison of four Monte Carlo methods for estimating the probability of (s, t)-connectedness. *IEEE Trans. Reliab.* R-35, 145–155.

Fishman, G.S. (1986b). A Monte Carlo sampling plan for estimating network reliability. *Oper. Res.* 34, 581–594.

Fishman, G.S. (1987a). A Monte Carlo sampling plan for estimating reliability parameters and related functions. *Networks* 17, 169–186.

Fishman, G.S. (1987b). The distribution of maximum flow with applications to multistate reliability systems *Oper. Res.* 35, 607–618.

Fishman, G.S. (1989a). Estimating the s–t reliability function using importance and stratified sampling. *Oper. Res.* 37, 462–473.

Fishman, G.S. (1989b). Monte Carlo estimation of the maximum flow distribution in a network with discrete stochastic arc capacity levels. *Nav. Res. Logist. Q.* 36, 829–849.

Fishman, G.S. (1990). How errors in component reliability affect system reliability. *Oper. Res.* 38, 728–732.

Fishman, G.S., and V.G. Kulkarni (1990). Improving Monte Carlo efficiency by increasing variance, Report UNC/OR/TR 90-21, University of North Carolina, N.C.

Fishman, G.S., and V.G. Kulkarni (1991). Bounds on conditional reliability, Report UNC/OR/TR 90-22, University of North Carolina, N.C.

Fishman, G.S., and T.D. Shaw (1989). Evaluating reliability of stochastic flow networks. *Probab. Engrg. Inf. Sci.* 3, 493–509.

Flanagan, T. (1990). Fiber Network Survivability. *IEEE Commun. Mag.* 28, 46-53.

Fong, C.C., and J.S. Buzacott (1987). An algorithm for symbolic reliability calculation with pathsets or cutsets. *IEEE Trans. Reliab.* R-36, 34–37.

Feng, H., and S.-P. Chan (1990). A method of reliability evaluation for computer-communication networks. *Proc. 1990 IEEE Int. Symp. on Circuits and Systems*, pp. 2682–2684.

Ford, L.R., and D.R. Fulkerson (1962). *Flows in Networks,* Princeton University Press.

Frank, H. (1969). Shortest paths in probabilistic networks. *Oper. Res.* 17, 583–599.

Frank, H., and W. Chou (1974). Network properties of the ARPA computer network. *Networks* 4, 213–239.

Frank, H., and I.T. Frisch (1971). *Communication, Transmission and Transportation Networks,* Addison-Wesley.

Frank, H., and S.L. Hakimi (1965). Probabilistic flow through a communications network. *IEEE Trans. Circuit Theor* CT-12, 413–414.

Frank, H., R.E. Kahn and L. Kleinrock (1972). Computer Communication Network Design-Experience with Theory and Practice. *AFIPS Conf. Proc.* 40, 255–270.

Frank, O., and W. Gaul (1982). On reliability in stochastic graphs. *Networks* 12, 119–126.

Frankl, P. (1984). A new short proof for the Kruskal–Katona theorem. *Discr. Math.* 48, 327–329.

Fratta, L., and U.G. Montanari (1978). A recursive method based on case analysis for computing network terminal reliability. *IEEE Trans. Commun.* COM-26, 1166–1177.

Fratta, L., and U.G. Montanari (1973). A Boolean algebra method for computing the terminal reliability in a communication network. *IEEE Trans. Circuit Theor.* CT-20, 203–211.

Frisch, H.L., J.M. Hammersley and D.J.A. Welsh (1962). Monte Carlo estimates of percolation probabilities for various lattices. *Phys. Rev.* 126, 949–951.

Fu, Y., and S.S. Yau (1962). A note on the reliability of communication networks, J. SIAM 10, 469–474.

Fujishige, S. (1986). A capacity-rounding algorithm for the minimum cost circulation problem: a dual framework of the Tardos algorithm. *Math. Program.* 35, 298–308.

Fulkerson, D.R. (1962). Expected critical path lengths in PERT Networks. *Oper. Res.* 10, 808–817.

Fulkerson, D.R. (1968). Networks, frames, blocking systems, in: G.B. Dantzig and A.F. Viennott (eds.), *Mathematics of the Decision Sciences*, American Math. Society, pp. 303–334.

Fulkerson, D.R. (1971). Blocking and anti-blocking pairs of polyhedra. *Math. Program.* 1, 168–194.

Fulkerson, D.R. (1974). Packing rooted directed cuts in a weighted directed graph. *Math. Program.* 6, 1–13.

Fulkerson, D.R., and G.H. Harding (1976). On edge-disjoint branchings. *Networks* 6, 97–104.

Gadasin, V. (1973). Reliability of communication networks represented by semi-oriented graphs. *Engrg. Cybern.* 11, 66–76.

Gadasin, V. (1976). Reliability of networks with semi-oriented structure. *Engrg. Cybern.* 14, 98–104.

Gaebler, R., and R. Chen (1987). An efficient algorithm for enumerating states of a system with multimode unreliable components, Technical Report, US Sprint, Overland Park, Kans.

Garey, M.R., and D.S. Johnson (1979). *Computers and Intractability: A Guide to the Theory of NP-complete/ness*, Freeman, San Francisco, Calif.

Gavish, B., and K. Altinkemer (1990). Backbone Network Design Tools with Economic Tradeoffs. *ORSA J. Comput* 2, 236-252.

Gilbert, E.N. (1959). Random graphs. *Ann. Math. Stat.* 30, 1141–1144.

Gomory, R.E., and T.C. Hu (1961). Multiterminal network flows. *SIAM J. Applied Math.* 9, 551–570.

Gondran, M., and M. Minoux (1984). *Graphs and Algorithms,* Wiley-Interscience.

Greene, C., and D.J. Kleitman (1978) Proof techniques in the theory of finite sets, in: G.-C. Rota (ed.), *Studies in Combinatorics*, MAA, pp. 22–79.

Grimmett, G.R., and D.J.A. Welsh (1982), Flow in networks with random capacities. *Stochastics* 7, 205–229.

Grötschel, M., L. Lovász and A. Schrijver (1981). The ellipsoid method and its consequences in combinatorial optimization. *Combinatorica* 1, 169–197.

Hagstrom, J.N. (1983). Combinatorial Properties for Directed Network Reliability with a Path Length Criterion, Technical Report 82-14, College of Business Administration, University of Illinois, Chicago.

Hagstrom, J.N. (1983). Using the decomposition tree of a network in reliability computation. *IEEE Trans. Reliab.* R-32, 71–78.

Hagstrom, J.N. (1984). Using the decomposition tree for directed network reliability computations. *IEEE Trans. Reliab.* R-33, 390–395.

Hagstrom, J.N. (1984). Note on Independence of Arcs in Antiparallel for Network Flow Problems. *Networks* 14, 567-570.

Hagstrom, J.N. (1988). Computational complexity of PERT Problems. *Networks* 18, 139–147.

Hagstrom, J.N. (1990). Computing the probability distribution of project duration in a PERT network. *Networks* 20, 231–244.

Hagstrom, J.N. (1990). Directed network reliability: domination and computing coefficients of the success-marginal expansion. *Networks* 20, 65–78.

Hagstrom, J.N. (1990). Component state dependence and error in reliability computation, preprint, University of Illinois, Chicago.

Hagstrom, J.N. (1990). Redundancy, substitutes and complements in system reliability, preprint, University of Illinois, Chicago.

Hagstrom, J.N., and P. Kumar (1984). Reliability computation on a probabilistic network with Path Length Criterion, Technical Report, College of Business Administration, University of Illinois, Chicago.

Hailperin, T. (1965). Best possible inequalities for the probability of a logical function of events. *Am. Math. Monthly* 72, 343–359.

Halpern, J. (1977). The sequential covering problem under uncertainty. *INFOR* 15, 76–93.

Hammersley, J.M. (1963). A Monte Carlo solution of percolation in the cubic crystal. *Methods Comput. Phys.* 1, 281–298.

Hammersley, J.M., and D.C. Handscomb (1964). *Monte Carlo Methods*, Methuen & Co. Ltd., London.

Hänsler, E. (1972). A fast recursive algorithm to calculate the reliability of a communication network. *IEEE Trans. Commun.* COM-20, 637–640.

Hänsler, E., G.K. McAuliffe and R.S. Wilkov (1974). Exact calculation of computer network reliability. *Networks* 4, 95–112.

Hariri, S., and C.S. Raghavendra (1987). SYREL: A symbolic reliability algorithm based on path cutset methods. *IEEE Trans. Comput.* C-36, 1224–1232.

Harms, D.D., and C.J. Colbourn (1985). The Leggett bounds for network reliability. *IEEE Trans. Circuits Syst.* CAS-32, 609–611.

Harms, D.D., and C.J. Colbourn (1993a). Renormalization of two-terminal reliability. *Networks* 23, 289–298.

Harms, D.D., and C.J. Colbourn (1993b). Evaluating performability: most probable states and bounds. *Proc. Telecomm. Syst. Conf.*, Nashville, Tenn.

Harms, J.J., and C.J. Colbourn (1990). Probabilistic single processor scheduling. *Discr. Appl. Math.* 27, 101–112.

Hartley, H.O., and A.W. Wortham (1966). A statistical theory for PERT critical path analysis. *Managenent Sci.* 12, B469–B481.

Hayhurst, K.J., and D.R. Shier (1991). A factoring approach for the stochastic shortest path problem. *Oper. Res. Lett.* 10, 329–334.

Hoffman, A.J., and J.B. Kruskal (1956). Integral boundary points of convex polyhedra, in: H.W. Kuhn and A.W. Tucker (eds.), *Linear Inequalities and Related Systems*, Princeton University Press, pp. 233–246.

Höpfinger, E., and V. Steinhardt (1976). *On the Exact Evaluation of Finite Activity Networks with Stochastic Durations of Activities*, Lecture Notes in Economics and Mathematical Systems, Vol. 117, Optimization and Operations Research, Springer-Verlag, New York, N.Y.

Hu, T.C. (1969). *Integer Programming and Network Flows,* Addison-Wesley.

Huseby, A.B. (1984). A unified theory of domination and signed domination with application to exact reliability computations, Statistical Research Report 3, University of Oslo.

Huseby, A.B. (1989). Domination theory and the Crapo β-invariant. *Networks* 19, 135–149.

Huseby, A.B. (1990). On regularity, amenability and optimal factoring strategies for reliability computations, preprint, University of Oslo.

Hwang, F.K., and A.M. Odlyzko (1977). A probability inequality and its application to switching networks. *Bell Syst. Tech. J.* 56, 821–826.

Hwang, K., and T.P. Chang (1982). Combinatorial reliability analysis of multiprocessor computers. *IEEE Trans. Reliab.* R-31, 469–473.

Jacobs, I.M. (1959). Connectivity in probabilistic graphs, Technical Report 356, Electronics Research Laboratory, MIT.

Jaeger, F., D. Vertigan and D.J.A. Welsh (1990). On the computational complexity of the Jones and Tutte polynomials. *Math. Proc. Camb. Phil. Soc.* 108, 35–53.

Jain, S.P., and K. Gopal (1988) An efficient algorithm for computing global reliability of a network. *IEEE Trans. Reliab.* R-37, 488-492.

Jarvis, J.P., and D.R. Shier (1993). An algorithm for approximating the performance of telecommunications systems. *Proc. Telecomm. Syst. Conf.,* Nashville, Tenn.

Jensen, P.A., and M. Bellmore (1969). An algorithm to determine the reliability of a complex system. *IEEE Trans. Reliab.* R-18, 169–174.
Jerrum, M. (1981). On the complexity of evaluating multivariate polynomials, Ph.D. thesis, Department of Computer Science, University of Edinburgh. [Report CST-11-81, Computer Science, University of Edinburgh.]
Jewell, W.S. (1958). Optimal flow through networks, Interim Report 8, Massachusetts Institute of Technology.
Johnson, R. (1984). Network reliability and acyclic orientations. *Networks* 14, 489–506.
Kamburowski, J. (1985). Bounds in temporal analysis of stochastic networks. *Found. Control Eng.* 10, 177–189.
Kamburowski, J. (1986). On the computational complexity of the shortest route and maximum flow problems in stochastic networks. *Found. Control Eng.* 11, 167–175.
Kamburowski, J. (1987). An overview of the computational complexity of the PERT shortest route and maximum flow problems in stochastic networks. *World Sci. Publishing*, 187–196.
Kamburowski, J. (1989). PERT networks under incomplete probabilistic information, in: R. Slowinski and J. Weglarz (eds.), *Advances in Project Scheduling*, Elsevier Scientific Publishing Co., pp. 433–466.
Karmarkar, N. (1984). A new polynomial time algorithm for linear programming. *Combinatorica* 4, 373–396.
Karp, R.M. (1972) Reducibility among combinatorial problems, in: R.E. Miller and J.W. Thatcher (eds.), *Complexity of Computer Computations*, Plenum, pp. 85–103.
Karp, R.M., and M. Luby (1985). Monte Carlo algorithms for the planar multiterminal network reliability problem. *J. Complexity* 1, 45–64.
Katona, G. (1966). A theorem of finite sets, in: P. Erdös and G. Katona (eds.), *Theory of Graphs*, Akademia Kiadó, Budapest, pp. 187–207.
Kaustov, V.A., Ye.I. Litvak and I.A. Ushakov (1986). The computational effectiveness of reliability estimates by the method of nonedge-intersecting chains and cuts. *Sov J. Comput. Syst. Sci.* 24, 70–73.
Kel'mans, A.K. (1965). Some problems of network reliability analysis. *Autom. Remote Control* 26, 564–573.
Kel'mans, A.K. (1967). Connectivity of probabilistic networks. *Autom. Remote Control* 29, 444–460.
Kel'mans, A.K. (1970). On estimation of the probabilistic characteristics of random graphs. *Autom. Remote Control* 32, 1833–1839.
Kel'mans, A.K. (1972). Asymptotic formulas for the probability of k-connectedness of random graphs. *Theor. Prob. Appl.* 27, 243–254.
Kel'mans, A.K. (1979). The graph with the maximum probability of remaining connected depends on the edge-removal probability. *Graph Theor Newslett.* 9, 2–3.
Kel'mans, A.K. (1981). On graphs with randomly deleted edges. *Acta Math. Acad. Sci. Hung.* 37, 77–88.
Kel'mans, A.K. (1985). On the analysis and synthesis of probabilistic networks. *Selected Transl. Math. Stat. Prob. (AMS)* 16, 127–138.
Kesten, H. (1987). Surfaces with minimal random weights and maximal flows: a higher-dimensional version of first-passage percolation. *Ill. J. Math.* 31, 99–166.
Khachiyan, L.G. (1979). A polynomial algorithm in linear programming. *Sov. Math. Dokl.* 20, 191–194.
Kim, Y.H., K.E. Case and P.M. Ghare (1972). A method for computing complex system reliability. *IEEE Trans. Reliab.* R-21, 215–219.
Kini, N.M., A. Kumar and D.P. Agrawal (1991). Quantitative reliability analysis of redundant multistage interconnection networks. *Reliab. Comput. Commun. Networks*, AMS/ACM, pp. 153–170.
Kirchoff, G. (1958). Über die Auflösung der Gleichungen auf welche man bei der Untersuchung der Linearen Verteilung Galvanischer Ströme Geführt wird. *Poggendorg's Ann. Phys. Chem.*, 1847, 72, 497–508. [On the solution of equations obtained from the investigation of the linear

distribution of galvanic currents. *IRE Trans. Circuit Theor.* CT-5, 4–7.]

Klein-Hanevelid, K.W. (1985). Distributions with known marginals and duality of mathematical programming with application to PERT. *Proc. 7th Conf. on Probability Theory,* VNU Sci. Press, 221–236.

Klein-Hanevelid, K.W. (1986). Robustness against dependence in PERT: an application of duality and distribution with known marginals. *Math. Program. Studies*, 153–182.

Kleindorfer, G.B (1971). Bounding distributions for a stochastic acyclic network *Oper. Res.* 19, 1586–1601.

Kraetzl, M., and C.J. Colbourn (1993). Transformations on channel graphs. *IEEE Trans. Commun.* 41, 664–666.

Krishnamurthy, E.V., and G. Komissar (1972). Computer-aided reliability analysis of complicated networks. *IEEE Trans. Reliab.* R-21, 86–89.

Kruskal, J.B. (1963). The number of simplices in a complex, in: R. Bellman (ed.), *Mathematical Optimization Techniques*, University of California Press, pp. 251–278.

Kubat, P. (1986). Reliability Analysis for Integrated Networks with Application to Burst Switching. *IEEE Trans. Commun.* COM-34, 564–568.

Kubat, P. (1989). Estimation of Reliability for Communication/Computer Networks–Simulation/Analytic Approach. *IEEE Trans. Commun.* COM-37, 927–933.

Kubat, P., U. Sumita and Y. Masuda (1988). Dynamic Performance Evaluation of Communication/Computer Systems with Highly Reliable Components. *Probab. Engrg. Inf. Sci.* 2, 185–213.

Kulkarni, V.G. (1986). Shortest paths in networks with exponentially distributed arc lengths. *Networks* 16, 255–274.

Kulkarni, V.G. (1988). Minimum spanning trees in undirected networks with exponentially distributed arc weights. *Networks* 18, 111–124.

Kulkarni, V.G. (1990). Generating random combinatorial objects. *J. Algorithms* 11, 185–207.

Kulkarni, V.G., and V.G. Adlakha (1985). Maximum flow in networks with exponentially distributed arc capacities. *Stochastic Models* 1, 263–290.

Kulkarni, V.G., and V.G. Adlakha (1987). Markov and Markov-regenerative PERT networks. *Oper. Res.* 34, 769–781.

Kulkarni, V.G., and J.S. Provan (1985). An improved implementation of conditional Monte Carlo estimation of path lengths in stochastic networks. *Oper. Res.* 33, 1389–1393.

Kumamoto, H., K. Tanaka and K. Inoue (1977). Efficient evaluation of system reliability by Monte Carlo method. *IEEE Trans. Reliab.* R-26, 311–315.

Kumamoto, K., K. Tanaka, K. Inoue and E.J. Henley (1980). Dagger sampling Monte-Carlo for system unavailability evaluation. *IEEE Trans. Reliab* R-29, 122–125.

Lam, Y.F., and V.O.K. Li (1986). Reliability modeling and analysis of communications networks with dependent failures. *IEEE Trans. Commun.* COM-34, 82–84.

Lam, Y.F., and V.O.K. Li (1986). An improved algorithm for performance analysis of networks with unreliable components. *IEEE Trans. Commun.* COM-34, 496–497.

Lawler, E.L. (1976). *Combinatorial Optimization: Networks and Matroids*, Holt, Rinehart, and Winston.

Lee, C.Y. (1955). Analysis of switching networks. *Bell Syst. Tech. J.* 34, 1287–1315.

Lee, S.H. (1980). Reliability evaluation of a flow network. *IEEE Trans. Reliab.* R29, 24–26.

Leggett, J.D. (1968). Synthesis of reliable networks, Ph.D. thesis, Moore School of Engineering, University of Pennsylvania.

Leggett, J.D., and S.D. Bedrosian (1969). Synthesis of reliable networks. *IEEE Trans. Circuit Theor.* CT–16, 384–385.

Lehman, A.B. (1963). Wye-delta transformations in probabilistic networks. *SIAM J.* 11, 773-805.

Levy, L.I., and A.H. Moore (1967). A Monte Carlo technique for obtaining system reliability confidence limits from test data. *IEEE Trans. Reliab.* R-16, 69–72.

Li, V.O.K., and J. A. Silvester (1984). Performance Analysis of Networks with Unreliable Components. *IEEE Trans. Commun.* COM-32, 1105-1110.

Lin, P.M., B.J. Leon and T.C. Huang (1976). A new algorithm for symbolic system reliability analysis. *IEEE Trans. Reliab.* R-25, 2–15.

Litvak, Ye.I. (1975). The probability of connectedness of a graph *Engrg. Cybern.* 13, 121–125.

Litvak, Ye.I. (1981a). A generalized triangle–star transformation of properties of complex networks. *Engrg. Cybern.* 19, 158–162.

Litvak, Ye.I. (1981b). Two-sided estimates of the minimum cost of a flow in two-terminal networks. *Engrg. Cybern.* 19, 132–135.

Litvak, E.I. (1983). A generalized theorem on negative cycles and estimates of the quality of flow transport in a network. *Sov. Math. Dokl.* 27, 369–371.

Litvak, Ye.I., and I.A. Ushakov (1984). Estimation of the parameters of structurally complex systems. *Engr. Cybern.* 22, 35–49.

Liu, C.I., and Y. Chow (1983). Enumeration of connected spanning subgraphs of a planar graph. *Acta Math. Hung.* 41, 27–36.

Locks, M.O. (1980). Recursive disjoint products, inclusion–exclusion, and min–cut approximations. *IEEE Trans. Reliab.* R-29, 368–371.

Locks, M.O. (1982). Recursive disjoint products: a review of three algorithms. *IEEE Trans. Reliab.* R-31, 33–35.

Lomonosov, M.V. (1974). Bernoulli scheme with closure. *Probl. Inf. Transm.* 10, 73–81.

Lomonosov, M.V. (1989). Tender-spot of a reliable network, preprint, Ben Gurion University of the Negev.

Lomonosov, M.V., and V.P. Polesskii (1971). An upper bound for the reliability of information networks. *Probl. Inf. Transm.* 7, 337–339.

Lomonosov, M.V., and V.P. Polesskii (1972). Lower bound of network reliability. *Probl. Inf. Transm* 8, 118–123.

Lomonosov, M.V., and V.P. Polesskii (1972). Maximum of the connectivity probability. *Probl. Inf. Transm.* 8, 324–328.

Loui, R.P. (1983). Optimal paths in graphs in graphs with stochastic or multidimensional weights *Commun. ACM* 26, 670–676.

Lovász, L. (1976). On two minimax theorems in graph theory. *J. Comb. Theor.* B21, 96–103.

Macaulay, F.S. (1927). Some properties of enumeration in the theory of modular systems. *Proc. London Math. Soc.* 26, 531–555.

Malcolm, D.G., J.H. Roseboom, C.E. Clark and W. Fazar (1959). Application of a technique for research and development program evaluation. *Oper. Res.* 7, 646–669.

Martin, J.J. (1965). Distribution of the Time Through a Directed Acyclic Network. *Oper. Res.* 13, 46–66.

Mata-Montero, E. (1990). Reliability of partial *k*-tree networks, Ph.D. thesis, University of Oregon.

Menger, K. (1927). Zur allgemeine Kurventheorie. *Fund. Math.* 10, 96–115.

Meyer, J.F. (1992). Performability: a retrospective and some pointers to the future. *Performance Evaluation* 14, 139–156.

Mihail, M., and A.L. Buchsbaum (1994). Monte Carlo and Markov chain techniques for network reliability and sampling, to appear.

Mine, H. (1959). Reliability of physical systems. *IRE Trans. Circuit Theor.* CT-6, 138–151.

Mirchandani, P.B. (1976). Shortest distance and reliability of probabilistic networks. *Comp. Oper. Res.* 3, 347–355.

Mirchandani, P., and Hossein Soroush (1987). *Stochastics in Combinatorial Optimization*, World Scientific Publishing Co., pp. 128–177.

Misra, K.B. (1970). An algorithm for the reliability of redundant networks. *IEEE Trans. Reliab.* R-19, 146–151.

Misra, K.B., and T.S.M. Rao (1970). Reliability analysis of redundant networks using flow graphs. *IEEE Trans. Reliab.* R-19, 19–24.

Monhor, D. (1987). An approach to PERT: application of Dirichlet distribution. *Optimization* 18, 113–118.

Moore, E.F. (1959). The Shortest Path through a Maze. *Ann. Comput. Lab. Harvard Univ.* 30, 285-292.

Moore, E.F., and C.E. Shannon (1956). Reliable circuits using less reliable relays. *J. Franklin Inst.* 262, 191–208; 263, 281–297.

Moskowitz, F. (1958). The analysis of redundancy networks. *AIEE Trans. Commun. Electron.* 39, 627–632.

Murray, K., A. Kershenbaum and M.L. Shooman (1993). Communications network reliability analysis: approximations and bounds. *Proc. Reliability and Maintainability Symp.*, IEEE, pp. 268–275.

Myrvold, W.J., K.H. Cheung, L. Page and J. Perry (1991). Uniformly reliable graphs do not always exist. *Networks* 21, 417–419.

Nádas, A., (1979). Probabilistic PERT. *IBM J. Res. Dev.* 23, 339–347.

Nagamochi, H., and T. Ibaraki (1991). Maximum flows in probabilistic networks. *Networks* 21, 645–666.

Nakazawa, H. (1979). Equivalence of a nonoriented line and a pair of oriented lines in a network. *IEEE Trans. Reliab.* R-28, 364–367.

Nash-Williams, C.St.J.A. (1961). Edge-disjoint spanning trees of finite graphs. *J. London Math, Soc.* 36, 445–450.

Nash-Williams, C.St.J.A. (1964). Decomposition of finite graphs into forests. *J. London Math. Soc.* 39, 12.

Nel, L.D., and C.J. Colbourn (1990). Locating a facility in an unreliable network. *INFOR* 26, 363-379.

Nel, L.D., and C.J. Colbourn (1990). Combining Monte Carlo estimates and bounds for network reliability. *Networks* 20, 277–298.

Nel, L.D., and H.J. Strayer (1993). Two-terminal reliability bounds based on edge-packings by cutsets. *J. Comb. Math. Comb. Comput.* 13, 3–22.

Nelson, A.C., J.R. Batts and R.L. Beadles (1970). A computer program for approximating system reliability. *IEEE Trans. Reliab.* R-19, 61–65.

Neufeld, E.M., and C.J. Colbourn (1985). The most reliable series-parallel networks. *Networks* 15, 27–32.

von Neumann, J. (1956). Probabilistic logics and the synthesis of reliable organisms from unreliable components. *Automata Studies*, pp. 43–98.

Novikov, V.F., A.L. Raikin and Yu.B. Shufchuk (1988). The effect of reliability on communication delays in a network with maximal flows. *Autom. Remote Control* 49, 1613–1619.

Onaga, K. (1986). Bounds on the average terminal capacity of probabilistic nets. *IEEE Trans. Reliab.* R 35, 252–259.

Page, L.B., and J.E. Perry (1988). A practical implementation of the factoring theorem for network reliability. *IEEE Trans. Reliab.* R-37, 259–267.

Page, L.B., and J.E. Perry (1989). Reliability of directed networks using the factoring theorem. *IEEE Trans. Reliab.* R-38, 556–562.

Parker, K.P., and E.J. McCluskey (1975). Probabilistic treatment of general combinational networks. *IEEE Trans. Comput.* C-24, 668–670.

Pippenger, N. (1990). Developments in 'The synthesis of reliable organisms from unreliable components'. *Proc. AMS Symp. Pure Math.* 50, 311–324.

Polesskii, V.P. (1971). A lower boundary for the reliability of information networks. *Probl. Inf. Transm.* 7, 165–171.

Polesskii, V.P. (1990a). Bounds on probability of connectedness of a random graph. *Probl. Inf. Transm.* 26, 75–82.

Polesskii, V.P. (1990b). Bounds on probability of group connectedness of a random graph. *Probl. Inf. Transm.* 26, 161–169.

Polesskii, V.P. (1992). Lower bounds on the probability of connectedness of random graphs generated by 2-connected graphs with a given base spectrum. *Probl. Inf. Transm.* 28, 175–183.

Politof, T. (1983). A characterization and efficient reliability computation of Δ–Y reducible networks, Ph.D. thesis, University of California at Berkeley, Calif.

Politof, T., and A. Satyanarayana (1986). Efficient algorithms for reliability analysis of planar networks — a survey. *IEEE Trans. Reliab.* R-35, 252–259.

Politof, T., and A. Satyanarayana (1986). Network reliability and inner-four-cycle-free graphs. *Math. Oper. Res.* 7, 97–111.

Politof, T., and A. Satyanarayana (1990). A linear-time algorithm to compute the reliability of planar cube-free networks. *IEEE Trans. Reliab.* R-39, 557–563.

Prékopa, A., and E. Boros (1991). On the existence of a feasible flow in a stochastic transportation network. *Oper. Res.* 39, 119–129.

Prékopa, A., E. Boros and K.-W. Lih (1991). The use of binomial moments for bounding network reliability, in: F. Roberts, F. Hwang and C. Monma (eds.), *Reliability of Computer and Communications Networks*, AMS/ACM, pp. 197–212.

Provan, G.M. (1990). A logic-based analysis of Dempster-Shafer theory. *Int. J. Approx. Reason.* 4, 451–495.

Provan, J.S. (1986). Polyhedral combinatorics and network reliability. *Math. Oper. Res.* 11, 36–61.

Provan, J.S. (1986). Bounds on the reliability of networks. *IEEE Trans. Reliab.* R-35, 260–268.

Provan, J.S. (1986). The complexity of reliability computations in planar and acyclic graphs. *SIAM J. Comp.* 15, 694–702.

Provan, J.S. (1991). Boolean decomposition schemes and the complexity of reliability computations, in: F. Roberts, F. Hwang and C. Monma (eds.), *Reliability of Computer and Communications Networks*, AMS/ACM, pp. 213–228.

Provan, J.S., and M.O. Ball (1983). The complexity of counting cuts and of computing the probability that a graph is connected. *SIAM J. Comp.* 12, 777–788.

Provan, J.S., and M.O. Ball (1984). Computing network reliability in time polynomial in the number of cuts. *Oper. Res.* 32, 516–526.

Provan, J.S., and L.J. Billera (1980). Decompositions of simplicial complexes related to diameters of convex polyhedra. *Math. Oper. Res.* 5, 579–594.

Provan, J.S., and V.G. Kulkarni (1989). Exact cuts in networks. *Networks* 19, 281–289.

Rai, S. (1982). A cutset approach to reliability evaluation in communication networks. *IEEE Trans. Reliab.* R-31, 428–431.

Rai, S., and D. Agrawal, eds. (1990a). *Distributed Computing Network Reliability*, IEEE Computer Society Press, Los Alamitos, Calif.

Rai, S., and D. Agrawal, eds. (1990b). *Advances in Distributed System Reliability,* IEEE Computer Society Press, Los Alamitos, Calif.

Rai, S., and K.K. Aggarwal (1978). An efficient method for reliability evaluation of a general network. *IEEE Trans. Reliab.* R-27, 206–211.

Raman, V. (1991). Finding the best edge-packing for two-terminal reliability is NP-hard. *J. Comb. Math. Comb. Comput.* 9, 91–96.

Ramanathan, A., and C.J. Colbourn (1987). Bounds on all-terminal reliability via arc-packing. *Ars Combinatoria* 23A, 91–94.

Ramanathan, A., and C.J. Colbourn (1987). Counting almost minimum cutsets and reliability applications. *Math. Program.* 39, 253–261.

Ramesh, A., M.O. Ball and C.J. Colbourn (1987). Bounding all-terminal reliability in planar networks. *Ann. Discr. Math.* 33, 261–273.

Resende, L.I.P. (1988). Implementation of a factoring algorithm for reliability evaluation of undirected networks. *IEEE Trans. Reliab.* R-37, 462–468.

Resende, M.G.C. (1986). A program for reliability evaluation of undirected networks via polygon-to-chain reductions. *IEEE Trans. Reliab.* R-35, 24–29.

Ringer, L.J. (1969). Numerical operators for statistical PERT critical path analysis. *Management Sci.* 16, 136–143.

Ringer, L.J. (1971). A statistical theory for PERT in which completion times of activities are inter-dependent. *Management Sci.* 17, 717–723.

Robacker, J.T. (1956). Minmax theorems on shortest chains and disjoint cuts of a network, Memo RM-1660-PR, The Rand Corporation.

Robillard, P., and M. Trahan (1977). Expected Completion Time in PERT Networks. *Oper. Res.* 24, 177–182.

Rosenthal, A. (1975). A computer scientist looks at reliability computations, in: R.E. Barlow, J.B. Russell and N.B. Singparwalla (eds.), *Reliability and Fault Tree Analysis*, SIAM, pp. 133–152.

Rosenthal, A. (1977). Computing the reliability of complex networks. *SIAM J. Appl. Math.* 32, 384–393.

Rosenthal, A., and D. Frisque (1977). Transformations for simplifying network reliability calculations. *Networks* 7, 97–111.

Roskind, J., and R.E. Tarjan (1985). A note on finding minimum cost edge-disjoint spanning trees. *Math. Oper. Res.* 10, 701–708.

Sahner, R.A., and K.S. Trivedi (1987). Performance and reliability analysis using directed acyclic graphs. *IEEE Trans. Software Engrg.* 10, 1105–1114.

Sanso, B., M. Gendreau and F. Soumis (1992). An algorithm for network dimensioning under reliability consideration, *Ann. Oper. Res.* 36, 263–274.

Sanso, B., and F. Soumis (1991). Communication and Transportation Networks Reliability Using Routing Models. *IEEE Trans. Reliab.* R-40, 29–38.

Sanso, B., and F. Soumis (1990). A Centralized Optimization Routing Model and Applications in Telecommunication Networks, Centre de Recherche sur les Transports Research Report #CRT-657, Université de Montreal.

Sanso, B., F. Soumis and M. Gendreau (1991). On the Evaluation of Telecommunications Network Reliability Using Routing Models, *IEEE Trans. Commun.* 39, 1494–1501.

Satyanarayana, A. (1982). A unified formula for analysis of some network reliability problems. *IEEE Trans. Reliab.* R-31, 23–32.

Satyanarayana, A., and M.K. Chang (1983). Network reliability and the factoring theorem. *Networks* 13, 107–120.

Satyanarayana, A., and J.N. Hagstrom (1981a). A new algorithm for the reliability analysis of multi-terminal networks. *IEEE Trans. Reliab.* R-30, 325–334.

Satyanarayana, A., and J.N. Hagstrom (1981b). Combinatorial properties of directed graphs useful in computing network reliability. *Networks* 11, 357–366.

Satyanarayana, A., and Z. Khalil (1986). On an invariant of graphs and the reliability polynomial. *SIAM J. Alg. Disc. Methods* 7, 399–403.

Satyanarayana, A., and A. Prabhakar (1978). New topological formula and rapid algorithm for reliability analysis of complex networks. *IEEE Trans. Reliab.* R-27, 82–100.

Satyanarayana, A., L. Schoppmann and C.L. Suffel (1992) A reliability-improving graph transformation with applications to network reliability. *Networks* 22, 209–216.

Satyanarayana, A., and R.K. Wood (1985). A linear time algorithm for computing k-terminal reliability in series-parallel networks. *SIAM J. Comput.* 14, 818–832.

Shanthikumar, J.G. (1987). Reliability of systems with consecutive minimal cutsets. *IEEE Trans. Reliab.* 36, 546–550.

Shanthikumar, J.G. (1988). Bounding network reliability using consecutive minimal cutsets. *IEEE Trans. Reliab.* R-37, 45–49.

Shier, D.R. (1985). Iterative algorithms for calculating network reliability, in: Y. Alavi et al. (eds.), *Graph Theory with Applications to Algorithms and Computer Science,* Wiley, pp. 741–752.

Shier, D.R. (1988). A new algorithm for performance analysis of communication systems. *IEEE Trans. Commun.* COM-36, 516–519.

Shier, D.R. (1991). *Network Reliability and Algebraic Structures*, Oxford University Press, Oxford, New York.

Shier, D.R. (1991). Algebraic methods for bounding network reliability, in: F. Roberts, F. Hwang and C. Monma (eds.), *Reliability of Computer and Communications Networks*, AMS/ACM, pp. 245–259.

Shier, D.R., and J.D. Spragins (1985). Exact and approximate dependent failure models for telecommunications networks. *Proc. INFOCOM85*, pp. 200–205.

Shier, D.R., E. Valvo and R. Jamison (1992). Generating the states of a binary stochastic system. *Discr. Appl. Math.* 38, 489–500.

Shier, D.R., and D.E. Whited (1987). Algebraic methods applied to network reliability problems. *SIAM J. Algebraic Discr. Methods* 8, 251–262.

Shogan, A.W. (1976). Sequential bounding of the reliability of a stochastic network *Oper. Res.* 24, 1027–1044.

Shogan, A.W. (1977a). Bounding distributions for a Stochastic PERT network *Networks* 7, 359–381.

Shogan, A.W. (1977b). A recursive algorithm for bounding network reliability. *IEEE Trans. Reliab.* R-26, 322–327.

Shogan, A.W. (1978). A decomposition algorithm for network reliability analysis. *Networks* 8, 231–251.

Shogan, A.W. (1982). Modular decomposition and reliability computation in stochastic transportation networks having cutnodes. *Networks* 12, 255–275.

Shooman, A.M., and A. Kershenbaum (1992). Methods for communication-network reliability analysis: probabilistic graph reduction. *Proc. Symp. on Reliability and Maintainability*, pp. 441–448.

Shooman, M.L. (1968). *Probabilistic Reliability: An Engineering Approach,* McGraw-Hill, New York, N.Y.

Sigal, C.E. (1977) The stochastic shortest route problem, Ph.D. Dissertation, Purdue University.

Sigal, C.E., A.A.B. Pritsker and J.J. Solberg (1980a). The stochastic shortest route problem. *Oper. Res.* 28, 1122–1129.

Sigal, C.E., A.A.B. Pritsker and J.J. Solberg (1980b). The use of cutsets in Monte Carlo analysis of stochastic networks. *Math. Comput. Simulation* 21, 376–384.

Soh, S., and S. Rai (1991) CAREL: Computer aided reliability evaluator for distributed computing networks. *IEEE Trans. Parallel Distrib. Syst.* 2, 199–213.

Soi, I.M., and K.K. Aggarwal (1981). Reliability indices for topological design of computer communication networks. *IEEE Trans. Reliab.* R-30, 438–443.

Somers, J.E. (1982). Maximum flow in a network with a small number of random arc capacities. *Networks* 12, 242-253.

Spelde, H.G. (1977). Bounds for the distribution function of network variables. *Proc. 1st Symp. of Operations Research*, University of Heidelberg, 3, 113–123.

Sperner, E. (1928). Ein Satz über Untermengen einer endlichen Menge. *Math. Z.* 27, 544–548.

Sperner, E. (1930). Über einen kombinatorischen Satz von Macaulay und seine Anwendung auf die Theorie der Polynomideale. *Abh. Math. Sem. Univ. Hamburg* 7, 149–163.

Spragins, J.D. (1977). Dependent failures in data communication systems. *IEEE Trans. Commun.* COM-25, 1494–1499.

Spragins, J.D., J.C. Sinclair, Y.J. Kang and H. Jafari (1986). Current telecommunication network reliability models: a critical assessment. *IEEE Trans. Selected Areas Commun.*, SAC-4, 1168–1173.

Stanley, R.P. (1975). The upper bound conjecture and Cohen–Macaulay rings. *Studies Appl. Math.* 14, 135–142.

Stanley, R.P. (1977). Cohen–Macaulay complexes, in: M. Aigner (ed.), *Higher Combinatorics*, Reidel, pp. 51–64.

Stanley, R.P. (1978). Hilbert functions of graded algebras. *Adv. Math.* 28, 57–83.

Stanley, R.P. (1983). *Combinatorics and Commutative Algebra,* Birkhäuser Verlag.

Stanley, R.P. (1984). An introduction to combinatorial commutative algebra, in: D.M. Jackson and S.A. Vanstone (eds.), *Enumeration and Design*, Academic Press, pp. 3–18.

Stepanov, V.E. (1969). Combinatorial algebra and random graphs. *Theor. Probab. Appl.* 14, 373–399.

Strayer, H.J., and C.J. Colbourn (1992). Bounding network reliability via surface duality, preprint.

Strayer, H.J., and C.J. Colbourn (1994). Consecutive cuts and paths, and bounds on k-terminal reliability, to appear.

Tardós, E. (1985). A strongly polynomial minimum cost circulation algorithm. *Combinatorica* 5, 247–256.

Tarjan, R.E. (1974). A good algorithm for edge-disjoint branching. *Inf. Process. Lett.* 3, 51–53.

Theologou, O.R., and J.G. Cartier (1991). Factoring and reductions for networks with imperfect vertices. *IEEE Trans. Reliab.* R-40, 210–217.

Tiwari, R.K., and M. Verma (1980). An algebraic technique for reliability evaluation. *IEEE Trans. Reliab.* R-29, 311–313.

Traldi, L. (1994a). On the star–delta transformation in network reliability, preprint, Lafayette College.

Traldi, L. (1994b), Generalized activities and k-terminal reliability, preprint, Lafayette College.

Truemper, K. (1989). On the delta-wye reduction of planar graphs. *J. Graph Theory* 13, 141-148.

Tsuboi, T., and K. Aruba (1975). A new approach to computing terminal reliability in large complex networks. *Electr. Commun. Japan* 58A, 52–60.

Tsukiyama, S., I. Shirakawa, H. Ozaki and H. Ariyoshi (1980). An algorithm to enumerate all cutsets of a graph in linear time per cutset. *J. ACM* 27, 619–632.

Tutte, W.T. (1961). On the problem of decomposing a graph into n connected factors. *J. London Math. Soc.* 36, 221–230.

Ushakov, I., and Ye.I. Litvak (1977). An upper and lower estimate of the parameters of two-terminal networks. *Engrg. Cybern.* 15.

Valiant, L.G. (1979). The complexity of computing the permanent. *Theor. Comput. Sci.* 8, 189–201.

Valiant, L.G. (1979). The complexity of enumeration and reliability problems. *SIAM J. Comp.* 8, 410–421.

Valvo, E.J., D.R. Shier and R.E. Jamison (1987). Generating the most probable states of a communication system. *Proc. INFOCOM87*, 1128–1136.

Varma, A., and C.S. Raghavendra (1989). Reliability analysis of redundant-path interconnection networks. *IEEE Trans. Reliab.* R-38, 130–137.

Van Slyke, R.M. (1963). Monte Carlo methods and the PERT problem. *Oper. Res.* 11, 839–860.

Van Slyke, R.M., and H. Frank (1972). Network reliability analysis: part I. *Networks* 1, 279–290.

Veeraraghavan, M., and K.S. Trivedi (1991) An improved algorithm for symbolic reliability analysis. *IEEE Trans. Reliab.* R-40, 347–358.

Vertigan, D. (1994a) The computational complexity of Tutte invariants for planar graphs, preprint, Mathematical Institute, Oxford University.

Vertigan, D. (1994b) Bicycle dimension and special points of the Tutte polynomial, preprint, Mathematical Institute, Oxford University.

Vtorova-Karevskaya, B.Ya. and Ye.I. Litvak (1979). The reliability of systems admitting two kinds of failures. *Engrg. Cybern.* 17.

Wagner, D.K. (1990). Disjoint (s, t)-cuts in a network. *Networks* 20, 361–372.

Wald, J.A., and C.J. Colbourn (1983a). Steiner trees, partial 2-trees, and minimum IFI networks. *Networks* 13, 159–167.

Wald, J.A., and C.J. Colbourn (1983b). Steiner trees in probabilistic networks. *Microelectron. Reliab.* 23, 837–840.

Weiss, G. (1986). Stochastic bounds on distributions of optimal value functions with applications to PERT, network flows and reliability. *Oper. Res.* 34, 595–605.

Welsh, D.J.A. (1976). *Matroid Theory,* Academic Press.

Whipple, F. (1928). On a theorem due to F.S. Macaulay. *J. London Math. Soc.* 8, 431–437.

Whited, D.E., D.R. Shier and J.P. Jarvis (1990). Reliability computations for planar networks. *ORSA J. Comput.* 2, 46–60.

Whitney, H. (1932). A logical expansion in mathematics. *Bull. Am. Math. Soc.* 38, 572–579.

Willie, R.R. (1980). A theorem concerning cyclic directed graphs with applications to network reliability. *Networks* 10, 71–78.

Wing, O., and P. Demetriou (1964). Analysis of probabilistic networks. *IEEE Trans. Commun. Tech.* COM-12, 38–40.

Wong, R.T. (1984). A dual ascent approach for Steiner tree problems on a directed graph. *Math. Program.* 28, 271–287.
Wood, R.K. (1982). Polygon-to-chain reductions and extensions for reliability evaluation on undirected networks, Ph.D. thesis, University of California at Berkeley, Calif.
Wood, R.K. (1985). A factoring algorithm using polygon-to-chain reductions for computing k-terminal network reliability. *Networks* 15, 173–190.
Wood, R.K. (1989). Triconnected decomposition for computing K-terminal network reliability. *Networks* 19, 203–220.
Yang, C.-L., and P. Kubat (1989). Efficient computation of most probable states for communication networks with multimode components, IEEE *Trans. Commun.* COM-37, 535–538
Yang, C.-L., and P. Kubat (1990). An algorithm for network reliability bounds. *ORSA J. Comput.* 2, 336–345.
Yarlagadda, R., and J. Hershey (1991) Fast algorithm for computing the reliability of a communication network. *Int. J. Electron.* 70, 549–564.
Yoo, Y.B., and N. Deo (1988). A comparison of algorithms for terminal-pair reliability. *IEEE Trans. Reliab.* R-37, 210–215.
Zemel, E. (1982). Polynomial algorithms for estimating network reliability. *Networks* 12, 439–452.

Biographical Information

Ravindra K. AHUJA is an Associate Professor in the Department of Industrial nad Management Engineering at Indian Institute of Technology, Kanpur. He visited MIT Sloan School of Management from 1986 to 1988 to collaborate with Professor J.B. Orlin on the design of faster algorithms for several network flow problems. This collaboration stimulated the development of the book, *Network Flows: Theory, Algorithms and Applications*, which he co-authored with T.L. Magnanti and J.B. Orlin. This book won the Lanchester prize of 1993; this prize is given to the best English Language published contribution of the year in Operations Research. Dr. Ahuja's research interests include network flows, combinatorial optimization, genetic algorithms, and computational testing of algorithms. Chapter 1.

Michael O. BALL holds a joint appointment in the College of Business and Management and the Institute for Systems Research at the University of Maryland, College Park. He received his B.S.E. and M.S.E. degrees in engineering science from Johns Hopkins University and his Ph.D. degree in operations research from Cornell University. He was previously a member of the technical staff at Bell Laboratories and has held visiting positions at the University of Waterloo, the University of North Carolina and the Center for Operations Research and Econometrics at the University of Louvain. His research interests are in the areas of network optimization and network reliability analysis, particularly applied to the design of telecommunications networks and transportation systems. Dr. Ball is Area Editor for optimization for *Operations Research*. (Chapter 11.)

Dimitri P. BERTSEKAS received a combined B.S.E.E. and B.S.M.E. from the National Technical University of Athens, Greece in 1965, the M.S.E.E. degree from George Washington University in 1969 and the Ph.D. degree in system science from the Massachusetts Institute of Technology in 1971.

Dr. Bertsekas has held faculty positions with the Engineering-Economic Systems Department, Stanford University (1971–1974) and the Electrical Engineering Department of the University of Illinois, Urbana (1974–1979). He is currently Professor of Electrical Engineering and Computer Science at MIT. He consults regularly with private industry and has held editorial positions in several journals. He was elected Fellow of the IEEE in 1983.

Professor Bertsekas has done research in a broad variety of subjects in control theory, operations research, optimization, and systems analysis. He has worked in the areas of estimation and control of stochastic systems, linear, nonlinear and dynamic programming, data communication networks, and parallel and distributed computation, and neural networks and their applications. He has written numerous papers in each of these areas.

He has written several books. He is the author of *Dynamic Programming and Stochastic Control*, Academic Press, 1976, *Constrained Optimization and Lagrange Multiplier Methods*, Academic Press, 1982, *Dynamic Programming: Deterministic and Stochastic Models*, Prentice-Hall, 1987, *Linear Network Optimization: Algorithms and Codes*, MIT Press 1991, *Dynamic Programming and Optimal Control*, Athena Scientific, 1995; and co-author of *Stochastic Optimal Control: The Discrete-Time Case*, Academic Press, 1978, *Data Networks*, 1987, and *Parallel and Distributed Computation: Numerical Methods*, Prentice-Hall, 1989. (Chapter 5.)

Daniel BIENSTOCK received the Ph.D. in Operations Research from MIT in 1985. He is a Professor at the Department of Industrial Engineering and Operations Research at Columbia University, where he has been since 1989. Prior to that he was at the Combinatorics and Optimization group at Bell Communications Research (1986–1989) and the Graduate School of Industrial Administration, Carnegie Mellon University (1985–1986).

His current research centers on high performance computing issues in optimization, particularly integer programming and discrete optimization problems arising in telecommunications.

He is Associate Editor of the *SIAM Journal on Discrete Mathematics*, the *ORSA Journal on Computing*, and *Networks*. (Chapter 8.)

David A. CASTAÑON received his B.S. degree in Electrical Engineering from Tulane University in 1971, and his Ph.D. degree in Applied Mathematics from MIT in 1976. He was a research scientist from 1976–1981 at MIT's Laboratory for Information and Decision Systems, and a senior research scientist from 1982–1990 at Alphatech, Inc. in Burlington, MA. Since 1990, he has been Associate Professor in Electrical, Computer and Systems Engineering at Boston University, Boston, MA. His research interests include stochastic control and estimation, game theory, optimization, and parallel and distributed computing. (Chapter 5.)

Charles J. COLBOURN was born in Toronto, Canada in 1953. He completed university degrees from the University of Waterloo and the University of Toronto, earning a Ph.D. in Computer Science from Toronto in 1980. He is currently Professor and Chair in the Department of Combinatorics and Optimization, University of Waterloo, Canada. He is the author of numerous scientific papers, and of the book *The Combinatorics of Network Reliability*, Oxford, 1987. (Chapter 11.)

Jonathan ECKSTEIN has been a scientist at Thinking Machines Corporation since 1991, specializing in parallel numerical optimization. He received his Ph.D. in operations research from MIT in 1989, following a Masters in the same subject in 1986. He then worked for two years as an Assistant Professor at Harvard Business school. Prior to his graduate work, Eckstein received his undergraduate degree in Mathematics from Harvard University in 1980, and then worked as a software developer, mathematical analyst, and energy auditor for Xenergy, Inc., a Boston-area energy conservation consulting and software firm. Eckstein has published scholarly papers on set-valued operator theory, nonlinear programming, portfolio design, truck routing, and parallel linear and mixed integer programming. (Chapter 5.)

Bert GERARDS was born on October 2, 1954 in Heerlen, the Netherlands. He obtained a Master's Degree in Mathematics at the University of Technology in Eindhoven in 1981. After that he taught at the Henric van Veldeke College, a high school in Maastricht, for the period of one year. In 1982, he became Research Assistant at ZWO (the Dutch Organization for the Advancement of Pure Research) and as such he started his Ph.D.-research under the supervision of Alexander Schrijver. This research was carried out at the universities of Amsterdam (for a period of one year) and Tilburg. In 1984, Bert Gerards took a position as Assistant Professor at the University of Tilburg, which he held until 1989. There he obtained his Doctoral Degree in 1988. In the summer of 1988 he was Visiting Assistant Professor at the University of Waterloo (Ontario). Since September 1989 he is a Senior Researcher at CWI in Amsterdam.

Gerards' research interests are graph theory, combinatorial optimization and polyhedral combinatorics.

Martin GRÖTSCHEL holds a chair in Applied Mathematics at the Technische Universität Berlin and is Vice President of the Konrad-Zuse-Zentrum für Informationstechnik Berlin. His fields of research interests are Optimization, Discrete Mathematics and Operations Research. He is involved in many joint research projects with partners from industry, ranging from the design of telecommunication networks, transportation and logistics, energy optimization, VLSI design, to optimization of production and flexible manufacturing systems. He received several scientific awards for his work including the Beckurts, the Dantzig, the Fulkerson and the Leibniz prize. (Chapter 10.)

Richard V. HELGASON is an Associate Professor in the Department of Computer Science and Engineering at Southern Methodist University. He has had extensive industrial experience in the development of large-scale scientific computer programs. He has conducted research in mathematical optimization algorithms and their efficient implementation. His current interests lie in the areas of network flows and computational geometry. He is the co-author with Jeff Kennington of the book *Algorithms for Network Programming.* (Chapter 2).

Michael JÜNGER studied Computer Science and Operations Research at the University of Bonn and Staford University where he received an M.S. degree in 1980. In 1983, after doctoral studies in Bonn and Augsburg, he received a doctoral degree in Applied Mathematics from the University of Augsburg. From 1983 to 1990 he was a member of the research staff at the University of Augsburg. In 1990 he became Professor of Mathematical Methods of Operations Research at the University of Paderborn, and in 1991, he was appointed Professor and Director of the Institute of Computer Science at the University of Cologne. Dr. Jünger is Associate Editor of *Mathematical Programming, Operations Research Letters* and the *ORSA Journal Computing.* His research interests are in Mathematical Programming, in particular the design, analysis, and evaluation of algorithms for combinatorial optimization problems, with special emphasis on optimal, respectively quality guaranteed, solutions. His current research projects include algorithms for the traveling salesman problem, the maximum cut problem, the minimum cut problem, and some combinatorial optimization problems related to graph drawing. (Chapter 4.)

Jeffery L. KENNINGTON is a Professor in the Department of Computer Science and Engineering at Southern Methodist University. For over two decades he and his colleagues have been developing and empirically testing algorithms for various optimization models which possess an underlying network or graphical structure. Some of the models that he has investigated include the shortest path problem, the assignment problem, the pure network problem, the generalized network problem, the multicommodity network problem, and the network with side constraint model. The experimental software developed for these investigations has been used by many other research groups world-wide, and several of these codes have been incorporated into commercial application systems.

He is the co-author with Dick Helgason of the 1980 John Willey book, *Algorithms for Network Programming.* He was the Program Chair for the famous 1984 Dallas ORSA/TIMS Meeting where the interior point algorithm for linear programming was first formally presented to the operations research community. Since 1988 he has served as the Area Editor for Computing for *Operations Research*, and since 1992 he has served on the Editorial Board of *Computational Optimization and Applications*. (Chapter 2.)

Michael A. LANGSTON was born on April 21, 1950, in Glen Rose, Texas. He received the B.S. degree in Mathematics from Texas A&M University in 1972, the M.S. degree in Systems and Information Science from Syracuse University in 1975, and the Ph.D. degree in Computer Science from Texas A&M University in 1981. From 1981 to 1989 he served on the faculty at Washington State University. Since that time he has been on the faculty at the University of Tennessee, where he holds the rank of Professor.

Dr. Langston has authored or co-authored over eighty refereed papers, with publications appearing in *IEEE Transactions on Computers*, *Journal of the ACM*, *Operations Research*, *SIAM Journal on Computing* and elsewhere. He has delivered invited presentations at over forty universities, research laboratories and international meetings.

His work has been supported by the National Science Foundation, the Office of Naval Research and the Oak Ridge National Laboratory. His current research interests include the analysis of algorithms, concrete complexity theory, graph theory, parallel computing and VLSI design. (Chapter 8.)

Thomas L. MAGNANTI is George Eastman Professor of Management Science at the Massachusetts Institute of Technology (MIT) and directs two interdepartmental MIT Programs: the Operations Research Center (as co-director) and Decision Sciences Program. He has previously served as founding co-director of MIT's Leaders for Manufacturing Program, head of the MIT Sloan School's Management Sciences Area, and MIT Class of 1960 Faculty Fellow. Dr. Magnanti's research focuses on large scale optimization (networks, combinatorial optimization, nonlinear programming) in such varied fields as communications, logistics, manufacturing, and transportation. His numerous publications include co-authorship of the books, *Applied Mathematical Programming* and *Network Flows: Theory, Algorithms, and Applications*. As an educator, he has served on approximately 70 Ph.D. committees (25 as supervisor) and advocated and developed programs combining management and technology. He is a past President of the Operations Research Society of America (ORSA) and past Editor-in-Chief of the Society's flagship journal *Operations Research*. Currently, he is an Advisory Editor of several journals and book service series. Dr. Magnanti has received two citations for distinguished service: MIT's Gordon Y. Billard Award and ORSA's Kimball Medal. He has also received the 1993 Lanchester Prize for best publication in the field of operations research. Professor Magnanti is a member of the US National Academy of Engineering. (Chapters 1 and 9.)

Joe MITCHELL received a B.S. in Applied Mathematics and Physics from Carnegie-Mellon University in 1981. During 1981–1986, he worked with Hughes Research Laboratories, while studying at Stanford University. He obtained his Ph.D. in Operations Research from Stanford University in 1986, under the advisorship of Christos Papadimitriou. Joe joined the faculty of the School of Operations Research and Industrial Engineering at Cornell in 1986, where he remained until 1991, when he joined the faculty of the University at Stony Brook, where he is currently an Associate Professor. Joe's research interests include the study of algorithms for geometry and network optimization. He currently works on several projects in applied computational geometry, especially for route planning, collision detection, pattern recognition, and manufacturing. He is an Associate Editor of the *ORSA Journal on Computing* and the *International Journal on Computational Geometry and Applications*. (Chapter 7.)

Clyde MONMA is the Executive Director of the Network Design and Security Research Department at Bellcore. His current research interests include Mathematical Modeling, Optimization Methods and Software Implementations, especially directed towards Telecommunications Applications. He received the 1986 Leonard G. Abraham Prize Paper Award in the Field of Communications Systems from the IEEE Communications Society jointly with Diane Sheng for their work on the paper *Backbone Network Design and Performance Analysis: A Methodology for Packet Switching Networks*. He has published over 50 papers and has presented numerous invited lectures at national and international conferences. He is currently the Editor-in-Chief of the *SIAM Journal of Discrete Mathematics*. (Chapter 10.)

James B. ORLIN is a Professor of Operations Research at MIT. He currently serves as the head of the Management Sciences Area of the Sloan School, and previously served as the head of the Sloan School Ph.D. program. Dr. Orlin's research interests include network optimization, combinatorial optimization, and computational molecular biology. He has been particularly interested in developing faster polynomial-time algorithms for a variety of network optimization problems. He has published approximately 50 papers in the fields of operations research, mathematical programming and theoretical computer science. Together with R.K. Ahuja and T.L. Magnanti, he co-authored the book, *Network Flows: Theory, Algorithms, and Applications*. Dr. Orlin was a winner of a 1984 Fulbright Grant for research in the Netherlands in 1984. He was awarded a Presidential Young Investigator award from NSF in 1985. He is also co-winner of the Lanchester Prize for his co-authored book of network flows. (Chapter 1.)

J. Scott PROVAN is Paul Ziff Professor in the Operations Research Department at the University of North Carolina at Chapel Hill. Dr. Provan received his Ph.D. in Operations Research from

Cornell University in 1977. He was Assistant Professor in the Applied Mathematics and Statistics Department at the State University of New York at Stony Brook from 1977 through 1982, spending his final two years as a NRC Postdoctoral Associate at the National Bureau of Standards (currently the National Institute of Standards and Technology) in Gaithersburg, Maryland. He has been at the University of North Carolina since 1982, and visited the University of Waterloo in Ontario, Canada in 1988–89.

Dr. Provan's areas of research include: network and combinatorial reliability, Steiner tree and other network synthesis problems, polyhedral combinatorics, combinatorial listing and enumeration algorithms, and other network and combinatorial optimization problems. (Chapter 11.)

M.R. REDDY is a Research Associate in the Department of Industrial & Management Engineering at Indian Institute of Technology, Kanpur. He completed his B.S. in Mechanical Engineering from S.V. University, Tirupati, India, and M.S. in Industrial & Management Engineering from Indian Institute of Technology, Kanpur. His masters thesis consisted of identifying and compiling applications of network optimization problems which have been scattered in many journals belonging to a number of engineering disciplines. Mr. Reddy's research interests include combinatorial optimization and operations management. (Chapter 1.)

Gerhard REINELT studied Computer Science and Operations Research at the University of Bonn where he received a diploma in Computer Science in 1981. In 1984, after doctoral studies in Bonn and Augsburg, he received a doctoral degree in Applied Mathematics from the University of Augsburg. From 1984 to 1992 he was a member of the research staff at the University of Augsburg, where he completed his habilitation in 1991. In 1992, he became Professor of Computer Science at the University of Heidelberg. Dr. Reinelt is Associate Editor of the *SIAM Journal on Discrete Mathematics*. His research interests are in Mathematical Programming, in particular the design, analysis, and evaluation of algorithms for combinatorial optimization problems. His current research projects include algorithms for the traveling salesman problem, the maximum cut problem and variants of the linear ordering problem. In addition, he is currently working on the generation and algorithmic exploitation of complete facet descriptions of polyhedra associated with small instances of combinatorial optimization problems and on the implementation of parallel mixed-integer algorithms for solving industrial production scheduling problems. (Chapter 4.)

Giovanni RINALDI studied Operations Research and Computer Science at the University of Rome where he received a master in Electrical Engineering in 1976. From 1977 to 1981 he was Research Assistant at the Center for Studies in System Control and Computer Science in Rome. In 1979 he was co-founder of a software consulting company, in which he remained as co-president and consultant until 1982. In 1982 he became Researcher at the Institute of System Analysis and Computer Science (IASI) of the Italian National Research Council (CNR) in Rome. Since 1991 he is Research Director of the CNR at IASI. His research interests are in Mathematical Programming, in particular in polyhedral combinatorics and in the design of algorithms for the exact solution of large scale combinatorial optimization problems. His current research projects include algorithms for the vehicle routing problem, the maximum cut problem, the minimum cut problem, and some applications of logic programming for the design of efficient urban traffic control systems. (Chapter 4.)

Timothy Law SNYDER is Chair of Computer Science and Adjunct Associate Dean for Science Education at Georgetown University. He received his Ph.D. in Applied and Computational Mathematics from Princeton University in 1987. Snyder researches problems from combinatorial optimization and their properties in Euclidean space, with special attention to the TSP and Steiner tree problems. He is also active in digital signal processing research and computer music, and the design of new ways to engage the modern mind in the appreciation and enjoyment of modern science. (Chapter 6.)

J. Michael STEELE is C.F. Koo Professor of Statistics in the Wharton School of the University of Pennsylvania. His research focuses on the interface of probability theory and the theory of algorithms. He has served as chair of the National Academy of Sciences panel on Probability and Algorithms, and with David Aldous he co-edited the NAS report on the state of the field. He also served as a key organizer for the recent 'Special Year in the Emerging Applications of Probability' at the Institute for Mathematics and Its Applications. A considerable part of the 'Special Year' focused on probability as it relates to combinatorial optimization. (Chapter 6.)

Mechthild STOER is employed at Telenor Research in Norway. She has a doctoral degree in mathematics, with a dissertation on the design of survivable networks, with Martin Grötschel as advisor. Her interest is to apply combinatorial optimization inside the area of telecommunications. (Chapter 10.)

Subhash SURI received a B.S. in Electronics Engineering from the University of Roorkee, India, in 1981. During the next two years, he worked as a programmer analyst at Tata Engineering and Locomotive Co. In 1983, he entered the graduate program in the department of Computer Science at the Johns Hopkins University, where he received an M.S. and a Ph.D. in 1984 and 1987, respectively. Upon finishing his studies, Subhash joined Bell Communications Research, where he was a Member of the Technical Staff in the Applied Research Laboratory. In 1994, Subhash left Bellcore to become an Associate Professor in the Department of Computer Science at Washington University in St. Louis. His current research interests include computational geometry, data structures, robotics, computer graphics, and network design. (Chapter 7.)

Laurence A. WOLSEY is Professor of Applied Mathematics in the Engineering Faculty at the Université Catholique de Louvain in Louvain-la-Neuve, Belgium. He has a research appointment at CORE (Center for Operations Research and Econometrics), and has been President of Core since September 1992.

His field of research is mixed integer programming and discrete optimization, with special interest in cutting plane and decomposition methods for the solution of practical production scheduling and sequencing, and network design problems.

In 1988 with T.J. Van Roy, he was awarded the Orchard-Hays prize by the Mathematical Programming Society for their work on solving mixed integer programs. He is author with G.L. Nemhauser of *Integer and Combinatorial Optimization*, Wiley 1988, for which they were awarded the Lanchester Prize by the Operations Society of America in 1989. In 1994 he was awarded the EURO Gold Medal. He is at present co-Editor of *Mathematical Programming* and an Associate Editor of *Operations Research*. (Chapter 9.)

Stavros A. ZENIOS is Professor of Management Science at the University of Cyprus, where he serves as the Dean of the School of Economics and Management since 1994. He was an Associate Professor of Decision Sciences, and Principal Investigator with the HERMES Laboratory for Financial Modeling, at the Wharton School of the University of Pennsylvania (1986–1995). He also held academic appointments at MIT, Universities of Bergamo and Milano, and the University of Haifa. Dr Zenios' research focuses on large-scale optimization, and on the use of parallel computers for the solution of large-scale problems arising in operations research applications. He is also involved in the development of management science models in finance, and especially for portfolio management. He developed models for organizations such as the World Bank, Union Bank of Switzerland, Metropolitan Life and others. He (co)authored more than hundred refereed articles, and edited eight books, in the areas of his expertise, including *Financial Optimization* for Cambridge University Press. He authored (with Y. Censor) the book *Parallel Optimization: Theory, Algorithms and Applications*, Oxford University Press. He is associate editor for several journals, including *Journal of Economic Dynamics and Control*, *SIAM Journal on Optimization*, *ORSA Journal on Computing* and *Naval Research Logistics*. He holds B.Sc. degrees in Mathematics and in Electrical Engineering, and received his Ph.D. from Princeton University (1986). (Chapter 5.)

Subject Index

Handbooks in Operations Research and Management Science Contents of Previous Volumes

Volume 1. Optimization
Edited by G.L. Nemhauser, A.H.G. Rinnooy Kan and M.J. Todd
1989. xiv + 709 pp. ISBN 0-444-87284-1

Volume 2. Stochastic Models
Edited by D.P. Heyman and M.J. Sobel
1990. xv + 725 pp. ISBN 0-444-87473-9

Volume 3. Computing
Edited by E.G. Coffman, Jr., J.K. Lenstra and A.H.G. Rinnooy Kan
1992. x + 682 pp. ISBN 0-444-88097-6

Volume 4. Logistics of Production and Inventory
Edited by S.C. Graves, A.H.G. Rinnooy Kan and P.H. Zipkin
1993. xiii + 760 pp. ISBN 0-444-87472-0

Volume 5. Marketing
Edited by J. Eliashberg and G.L. Lilien
1993. xiv + 894 pp. ISBN 0-444-88957-4

Volume 6. Operations Research and the Public Sector
Edited by S.M. Pollock, M.H. Rothkopf and A. Barnett
1994. xv + 723 pp. ISBN 0-444-89204-4